Contents

Index of **TABLES**

Index of **FIGURES**

Index of **MAPS**

Editor's Introduction

The Military Balance 2021 is published at the end of a year defined by the coronavirus pandemic. While the pandemic affected nearly every country, tackling this common challenge did little to improve relations between states. Conflict and confrontation did not abate, and great-power competition continued to drive some countries' defence planning considerations and procurement decisions.

The unstable security environment was also manifest in continued strain in defence relations between states – even between allies when it came to NATO – and persistent challenges to the remaining elements of the post-Second World War rules-based order, particularly arms-control arrangements. The United States withdrew from the Open Skies Treaty in November 2020, just over a year after it formally pulled out of the Intermediate-range Nuclear Forces Treaty, in this case similarly accusing Russia of breaching the treaty. Furthermore, the failure of the Trump administration's plan to include China in a successor to the Russia–US New START agreement left the incoming Biden administration little time to extend the treaty.

Meanwhile, wars continued in Libya, Syria and Yemen. Turkey sent military support to Libya in 2020, backing the recognised government in Tripoli, while opposition forces received support from Egypt, Russia and the UAE. While regional states' contributions were modest they were nonetheless able to sustain their presence, indicating their developing military capabilities. In Syria's Idlib province, Turkey's army clashed with government troops for the first time in February and in late 2020 was mounting joint patrols with Russia to monitor a ceasefire. In Yemen, the Iranian-backed Houthis displayed growing military competence through the use of weapons including uninhabited aerial vehicles (UAVs), long-range rockets and cruise missiles. Conflict continued in Ukraine, and also in Africa, including the long-running war in the Sahel. Fighting in Ethiopia threatened stability in East Africa, not only because of cross-border incidents, but also because Addis Ababa had been a significant contributor to regional peacekeeping efforts. In Nagorno-Karabakh, an old conflict re-ignited, with Azerbaijan regaining a swathe of territory.

The coronavirus effect

Although primarily a public-health issue, the coronavirus pandemic had a significant effect on defence establishments. In many countries, troops were deployed to support civil authorities in tasks including planning and logistics support; armed forces also provided medical personnel and infrastructure. Some defence procurements were delayed. The pandemic had other implications, with fear of infection leading training and exercises to be scaled down or cancelled, while some deployments were postponed and others extended. Elsewhere, outbreaks challenged readiness, such as when sailors from ships of the US 7th Fleet fell ill. In short order, however, armed forces adapted. By June, the US Navy was able to send three carrier strike groups to sea in the Asia-Pacific for the first time since 2017.

Defence spending

The pandemic will also have an effect on defence spending, though not immediately. Although several countries in Asia and the Middle East quickly adjusted public spending, it will likely take until 2022–23 for the full effect of governments' financial responses to the pandemic to translate into defence budget cuts. Indeed, in the West, the immediate effect of the pandemic seems to have been that some governments, such as France and Germany, accelerated spending to support local suppliers to the defence sector. As of November, most of the countries that had announced their 2021 budgets had maintained their defence allocations.

Global defence spending increased in 2020 to reach US$1.83 trillion, growing by 3.9% in real terms. This growth was only slightly lower than in 2019, notwithstanding the pandemic and subsequent 4.4% contraction in global economic output. Strengthened or stable growth in Latin America, North America and sub-Saharan Africa offset regional slowdowns in Europe and, to a lesser extent, Asia. Real growth in China's defence budget slowed to 5.2% in 2020, down from 5.9% in 2019, while wider spending growth in Asia also slowed, from 3.8% to 3.6%, as countries funded pandemic-relief efforts. That said, China's 2020 increase, amounting to a nominal US$12 billion, was still greater than the combined defence budget increases in all other Asian states. Indeed, increases in the US and Chinese defence budgets accounted for almost two-thirds of the total increase in global defence spending in 2020.

Sharpened threat perceptions in Europe have helped boost European defence spending, after Russia's 2014 seizure of Crimea and involvement in the conflict in Ukraine's east. European NATO members have increased their defence expenditure as a proportion of GDP. This trend continued in 2020, with their spending reaching 1.64% of GDP, up from 1.25% of GDP in 2014. However, despite the coronavirus pandemic causing a 7% average economic contraction among members in 2020, only nine European NATO members met NATO's recommendation that they spend 2.0% of GDP on defence.

European defence

European and NATO cohesion has been tested by continued transatlantic disagreement on issues including defence spending, escalating tensions between NATO members Greece and Turkey, and the still-unfolding effects of the UK's decision to leave the European Union. Although the EU aspired to a more active geopolitical role, the outcome of discussions in July 2020 on its multi-year budget saw funding for the European Defence Fund, for military mobility projects and for the European Peace Facility lowered by almost 40%, 75% and just over 45% respectively when compared to the figures that were circulated at the start of the budget process.

As the EU's membership contracted with the UK's departure, NATO's expanded when North Macedonia joined in March. NATO maintained focus on the NATO Readiness Initiative, intended to improve the operational readiness of existing forces. Meanwhile, the US again exhorted NATO members to spend more, with then-President Donald Trump labelling Germany 'delinquent' on this issue. It is unlikely that Washington's aim will change with a new administration, even if there is a modification in tone and a desire in both Washington and European capitals to repair relations. After all, these calls began long before Trump took office. Looking further ahead, NATO's project to define a vision for 2030 had China as a clear focus. China's rise would, according to the NATO Secretary-General, 'fundamentally' change the global balance of power.

China and Russia

China's military modernisation continues to drive procurement and R&D efforts in the US and is also shaping defence policies in the Asia-Pacific. An example of this was Australia's mid-2020 Defence Strategic Update with its emphasis on conventional deterrence and more capable strike systems, even if the document made few direct references to China.

Beijing is apparently intent on achieving primacy in its littoral areas. The People's Liberation Army Navy (PLAN) maintained an 'over-the-horizon' presence, with China's maritime paramilitary forces taking the lead and using facilities on Chinese-occupied features in the Spratly Islands as forward operating bases in the South China Sea. Naval shipbuilding continued at pace, with the second of the new Type-075 amphibious ships launched, as well as the eighth Type-055 cruiser and 25th Type-052D destroyer. Meanwhile, the PLAN's second aircraft carrier began sea trials; a third, larger, carrier is under construction.

China's air force also continues to integrate more advanced systems. In October, a modified H-6 bomber was observed carrying what appeared to be a very large air-launched ballistic missile. China is likely developing a replacement bomber, assumed to have a low-observable design. At the same time, the air force is fielding more J-10C *Firebird*, J-16 and J-20 combat aircraft while increased numbers of Y-20 heavy transports mean that the air force has effectively doubled its heavy transport fleet in the last four years.

Russia is also integrating more modern systems into its inventory, though on a more modest scale. With continuing problems in fielding new-generation equipment, such as the *Armata* family of armoured vehicles and the Su-57 *Felon* combat aircraft, there is emphasis on modernising existing platforms and integrating new weapons. One example is the Kh-69 medium-range cruise missile. While it may be carried on the Su-34 *Fullback* initially, the weapon's configuration seems driven by the requirement for internal carriage, likely on the Su-57.

Further and faster

In 2020, Russia's navy continued tests of the hypersonic 3M22 *Tsirkon* missile. Service entry of this weapon would mark an important development in Russian naval strike capabilities. *Kinzhal* (AS-24 *Killjoy*) air-launched ballistic missiles have been observed on MiG-31 *Foxhounds*, while Russia has already fielded the *Avangard* (SS-19 mod 4 *Stiletto*) hypersonic glide vehicle (HGV). China also continues to develop its hypersonic systems, though it remains unclear whether its DF-17 HGV has reached initial operating capability. Similar developments continue in the US, including the Long-range Hypersonic Weapon intended to form part of the US Army's artillery-modernisation programme.

Hypersonic weapons are, along with greater numbers of cruise missiles and the threat from UAVs, prompting Western armed forces to take greater interest in air and missile defence. Like long-range artillery, this capability was comparatively neglected during counter-insurgency campaigns in relatively permissive operating environments. Finland and Sweden are looking to improve such capabilities, while Australia's Defence Strategic Update indicated increased investment in integrated air and missile defence and 'very high-speed and ballistic missile defence'. Meanwhile, Azerbaijan's use of loitering munitions and UAVs in the short Nagorno-Karabakh war highlighted, for armour, the importance of defence against top-attack and, for armed forces more broadly, the value of effective mobile, and layered, air and missile defence.

That said, it is increasingly clear that military capabilities like these form only part of the toolkit of the West's potential adversaries. Strategies and capabilities will also be required to effectively tackle the activities of paramilitary forces, such as China's Coast Guard and maritime militia, and also address the employment of state power through non-state actors, such as the use – ostensibly by Russia's Wagner Group – of MiG-29 *Fulcrum* and Su-24 *Fencer* combat aircraft. There is also the challenge from other, more insidious activities below the threshold of military force, such as information and influence operations conducted through cyberspace. Along with worries about more capable weapons, this is perhaps one reason why states as diverse as Australia and Finland are now asserting that there is reduced warning time of a crisis.

These threats pose problems that are best addressed collectively, or at least not by individual countries alone, and could bolster cooperation among 'like-minded' states. However, the coronavirus pandemic showed that a common challenge does not always lead to collective action. That said, the pandemic could, perhaps paradoxically, help states in developing the capabilities required to tackle 'grey zone' challenges, not least as it highlighted the need to boost societal resilience, and even to broaden conceptions of defence and security. This does not negate the requirement for modernised and new military capabilities, but it could mean that some armed forces will need to make clearer arguments for new equipment. However, effectively tackling the pandemic required the close integration of military, government and civil-sector capabilities, as well as sharper focus on resilience. If states were able to accomplish this, it could indicate progress towards more efficiently harnessing the levers of state power in a way that could be useful for future security and defence challenges at home and abroad, in the grey zone as well as during peace and war.

Domain trends

Defence Economics

- The coronavirus pandemic will have a significant effect on public and therefore defence spending, but not immediately. While several countries in Asia and in the Middle East moved quickly to adjust public spending, it will likely take until 2022–23 for the full economic effects of governments' financial responses to the pandemic to translate into defence budget cuts elsewhere. A similar lag was seen after the 2008 financial crisis. Perhaps paradoxically, the immediate effect of the pandemic seems to have been that some Western governments – for example France and Germany – accelerated spending to support local suppliers, while sharper threat perceptions also led to increased spending.
- Global defence spending increased in 2020, growing by 3.9% in real terms, notwithstanding the pandemic and subsequent global economic contraction. Regional slowdowns in Asia and Europe were offset by strengthening or stable growth in North America, Latin America and Sub-Saharan Africa. Most of the countries that have already announced their 2021 budgets have managed to maintain defence allocations for next year, despite wider fiscal pandemic-related concerns.
- US defence budget growth slowed in 2020, increasing by 6.3% in real terms to reach US$738 billion, which was slightly lower than the 6.6% increase in 2019. However, the slowdown in China's defence budget growth was more marked. Beijing's defence funding in 2020 increased by 5.2% in real terms compared to 5.9% in 2019. This is one of the slowest rates of growth in China's defence funding since the early 1990s, although China's 2020 increase – amounting to US$12bn – was still greater than the combined defence budget increases of all other countries in the Asia region.
- As a proportion of GDP, average spending among European NATO members has increased from 1.25% of GDP in 2014 to 1.52% in 2019 and will increase to 1.64% of GDP in 2020. Therefore, despite the coronavirus pandemic causing a 7.0% average economic contraction among members in 2020, European states are still unable, on average, to meet NATO's recommendation to spend 2% of GDP on defence.

Land

- Operations in Libya and Nagorno-Karabakh highlight the importance of providing effective mobile air and missile defence. The continuing proliferation of relatively cheap armed uninhabited aerial vehicles (UAVs) and loitering munitions has prompted a hasty re-investment in short-range air defence (SHORAD) capabilities by armed forces that had reduced this capability in previous years. The US Army placed an order for an interim capability (IM-SHORAD), based on the *Stryker* vehicle, in 2020.
- The relevance of heavy armour in modern combat operations has once again been called into question, following the US Marine Corps divestment of its main battle tank fleet, and suggestions that the United Kingdom might be considering following suit. However, supporters of this approach have struggled to make the case for its utility against opponents that might remain invested in heavy armour. Moreover, the marine-corps plan is predicated on the retention of main battle tanks by the US Army.
- The coronavirus pandemic has complicated and restricted training activity for many armed forces that is essential to maintaining combat readiness, such as opposing force and live-fire exercises. This has boosted efforts in many countries to further develop synthetic training environments, using virtual reality and digital programmes to simulate combat training.
- Land forces are further integrating intelligence, information, space and cyber units into traditional combat formations. The US Army's Multi-Domain Task Force (MDTF) concept brings together long-range artillery, space, cyber, intelligence, information operations and intelligence, surveillance and reconnaissance (ISR) UAV capabilities into a brigade-level formation. In February 2020, the Israel Defense Forces (IDF) inaugurated their own Multi-dimensional Unit as part of the IDF's ongoing *Tnufa* programme. Drawing on armour, infantry, engineering, intelligence and air-force personnel, the unit was first deployed to Israel's northern border in the summer of 2020.

Maritime

- There is continuing interest in developing aircraft carriers, and some states are considering different platforms to deliver carrier-based naval power. Japan undertook initial work to modify the *Izumo* to operate F-35B *Lightning* II combat aircraft while South Korea confirmed plans for a potentially F-35B-equipped light aircraft carrier and the US again raised the prospect of developing smaller aircraft carriers to complement its large nuclear-powered ships. At the same time, the UK is planning the first operational deployment of its new carrier capability in 2021. China's second aircraft carrier began its sea trials in 2020, though work on India's new carrier *Vikrant* was delayed due to the coronavirus pandemic.
- A number of surface-combatant projects are having to balance requirements for high-end capabilities and general-purpose maritime security. Examples include Germany's high-end MKS-180 frigates and the US with its *Constellation*-class (FFG(X)) frigate programme. In the UK, the high-capability Type-26 frigate is being complemented by the general-purpose Type-31. France and Italy are supplementing their larger FREMM frigates with FDI medium frigates and PPA patrol vessels respectively, while Japan is building a 5,500-tonne Future Multi-Mission frigate.
- The need to support extended blue-water missions means efforts are increasing to recapitalise logistic support

capabilities, including vessels with increased multi-purpose capabilities to help support limited maritime-security and peace-support missions. Australia, Canada, China, France, Italy, the Netherlands, New Zealand, the UK and the US, as well as a number of other Asian navies, are looking to bolster logistic support capabilities.

- Russian navy tests during 2020 of the *Tsirkon* hypersonic weapon have focused attention on the potential impact of such weapons in a naval context. They may enter service during this decade and countering or defending against them is a particular concern. China is also developing such capabilities. The US Navy's Conventional Prompt Strike programme may see a hypersonic weapon deployed towards the end of this decade.
- Ongoing maritime disputes in the eastern Mediterranean have been fuelled by growing competition over offshore energy resources. There are particular tensions between Greece and Turkey, as well as friction within Europe and among various Middle Eastern states over policy towards the conflict in Libya, while Russia's growing presence and maritime ambition in the Mediterranean are adding to tensions with NATO.

Aerospace

- In September 2020, the United States Air Force (USAF) revealed that a technology demonstrator related to its Next-Generation Air Dominance programme had already been flown. Analysts thought that Lockheed Martin was possibly the industrial lead. The project's key aim is to provide the USAF with the ability to operate and prevail in contested air space in a conflict with a peer or near-peer rival.
- Projects continued regarding future combat-aircraft requirements. France, Germany and Spain are working on the Next-Generation Fighter while the UK, with involvement from Italy and Sweden, is working on a project that has been dubbed *Tempest*. The UK-led group continued to explore additional 'partners', while the French-led team wanted to remain at three (with Germany and Spain) in the near term. Some also remain concerned over Europe's still-divided approach.
- Air forces are refocusing on the need to counter advanced surface-to-air missile threats. Flight testing of an extended-range variant of the US AGM-88E Advanced Anti-Radiation Guided Missile – the AARGM-ER – began in mid-2020. This missile development is intended to increase range, and likely average fly-out speed, to better counter the latest generation of long-range surface-to-air missile systems. The AARGM-ER will be integrated on the F/A-18E/F and EA-18G and the F-35.
- The current generation of combat aircraft are still being procured. Germany signed a contract for 38 Eurofighter Tranche 4 aircraft to replace earlier models. The plan is to buy 55 more Tranche 4/5 aircraft, and also F/A-18F *Super Hornet*s and EA-18G *Growler*s to replace the *Tornado* when it is withdrawn from service. Meanwhile, Spain is looking to replace 20 F/A-18As with additional Eurofighters. All four Eurofighter partners (Germany, Italy, Spain and the UK) will also introduce active electronically scanned array radar on the aircraft. In Russia, delays to the Su-57 *Felon* have led to moves to buy additional Su-35S *Flanker* Ms. Meanwhile, Sweden is planning to upgrade some of its JAS39C/D *Gripen*s to keep them in service alongside the JAS39E.
- Plans moved ahead to team crewed and uninhabited combat aircraft as Boeing's 'Loyal Wingman' started taxi trials in Australia in October 2020. The project is being developed for the Royal Australian Air Force. Most, if not all, of the crewed combat aircraft now in development include complementary UAV concepts, with the plan that these support crewed combat aircraft, and provide a broader range of combat options.

Cyber

- In some countries, such as Australia and Germany, signals-intelligence organisations retain the lead on offensive cyber operations, rather than military structures. In some cases, joint units are being formed that integrate intelligence organisations with military units under the lead of the intelligence agency, an example being the UK's National Cyber Force. Although cyber-focused units and capabilities continue to be established, problems remain in generating adequate personnel numbers, as well as shortfalls in defence-industrial capacity.
- The effective integration of defensive and offensive cyber operations with cyber intelligence, surveillance and reconnaissance, as well as kinetic strike capabilities, imposes a requirement for sophisticated battle management systems.
- Cyberspace is both a key integrator and an attack vector for the Multi-Domain Operations (MDO) concept that will underpin some military-modernisation efforts over the next decade. The US is working on a Joint Warfighting Concept for All-Domain Operations, based on MDO. The UK defence ministry's new Integrated Operating Concept emphasises the need for integration across all combat domains at the tactical level and there are also efforts to incorporate cyber capability under what it refers to as Multi-Domain Integration.
- Space assets are essential for military operations in cyberspace. Satellites rely on cyberspace for the flow of data linking command, control, communications, computers and intelligence, surveillance, reconnaissance (C4ISR) systems. The US Space Force is exploring artificial intelligence-enabled cyber defence solutions because space-based assets are vulnerable to distributed denial of service attacks or spoofing or capture via cyberspace. Space-based assets like satellites not only provide crucial cyber ISR but are also capable of delivering offensive cyber effects.
- A number of countries are experimenting with integrating cyber and electromagnetic warfare capability. The US Army has created a new Intelligence, Information Operations, Cyberspace, Electronic Warfare and Space Operations battalion as part of its Multi-Domain Task Force concept. The UK has said that cyber and electromagnetic activities are interdependent. Similar integration efforts will likely increase as MDO concepts are adopted by more Western armed forces.

Chapter One

The future of maritime competition

New dynamics of maritime competition are emerging. The contemporary maritime domain continues to become more congested, contested and complex, while the centre of gravity of maritime power appears to be shifting further away from Western states. This will change the way operations are conducted at sea and the types of naval capabilities that states seek to procure. Rapid technological change and the proliferation of advanced weaponry are only accelerating this transformation.

Sea power during the Cold War

The United States and its allies held a preponderance of sea power for most of the Cold War. However, the Soviet Navy under Admiral Sergei Gorshkov – its commander from 1956–85 – built a globally capable force combining missile-armed submarines, surface ships and long-range land-based aviation. This was a significant challenge to Western naval strategy and led to a more defensive posture in US and NATO maritime forces during the 1970s.

In the 1980s, President Ronald Reagan reasserted US naval dominance by expanding the fleet as part of the '600-ship navy' strategy and implementing a more assertive maritime strategy. This posture – intended to challenge the Soviet Navy globally, but particularly in its northern bastions – was enabled by capability enhancements such as the *Aegis* combat system, the long-range combination of the F-14 *Tomcat* carrier-based interceptor and its AIM-54 *Phoenix* missile, and towed array sonars that exploited the Western advantage in submarine acoustic signatures and the relative noisiness of Soviet submarine designs of the time.

While this maritime posture added to the strategic pressure on the Soviet Union in the culminating years of the Cold War, the US strategy was not tested in combat; indeed, the only significant combat experience at sea after the advent of the missile age was the Falklands conflict in 1982. This confrontation heightened concerns about the potential vulnerability of surface ships to anti-ship missiles, particularly sea-skimming missiles. However, some of the vulnerabilities exposed were chiefly due to specific capability gaps, such as the Royal Navy's lack of an organic airborne early-warning capacity.

No contest at sea

After the end of the Cold War, the maritime domain seemed an almost benign conduit for the projection of (essentially Western) power. Operational focus shifted from traditional blue-water operations to encompass the ability not only to project power from the sea but also to operate in the littoral zone, which was becoming the main contested maritime space. The military and diplomatic power of the carrier strike group seemed clear: six US aircraft-carrier groups took part in *Operation Desert Storm* against Iraq in 1991; and in 1996, Washington ostentatiously deployed two carriers to the waters off Taiwan in response to pressure from Beijing on Taipei. Though this may have had a deterrent effect at the time, it is now widely viewed as having spurred Beijing's subsequent push to develop an area-denial capability.

Now that they were not facing any major state-based threat, European navies switched investment from sea-control missions to power-projection and presence missions. While many retained a core of high-end platforms, such as the Franco-Italian *Horizon* and British Type-45 air-defence destroyers, fleet inventories dwindled and procurement shifted more towards amphibious and general maritime-security capabilities. Examples include the French *Mistral*-class amphibious assault ships and latterly the large German F125 *Baden-Württemberg*-class frigates aimed particularly at littoral and peace-support missions. At the same time, the US Navy developed the Littoral Combat Ship (LCS). Although the programme became mired in controversy over both capabilities and cost, the LCS was intended to be a relatively simple and inexpensive small combatant that would operate in the littoral zone but also potentially carry out maritime-security and presence missions.

In some cases, a cooperative approach developed to tackle the challenge of countering mainly non-state-based threats to maritime security. This was perhaps most evident in the various multinational groupings that coalesced around the threat of piracy off the coast of Somalia from around 2008. Not only Western states but also China, Russia and others, notably India, deployed naval forces to the waters off Somalia, illustrating that piracy was a shared concern. For China's People's Liberation Army Navy (PLAN), these missions brought the strategic benefit of gaining experience in long-range deployments, but neither China nor Russia was keen to participate in formal structures with Western navies. Indeed, the late 2000s saw navies following different courses. In the West – with defence establishments preoccupied by land-focused campaigns in Iraq and Afghanistan – the focus seemed to be more on seeking to deliver operational relevance in the maritime-security sphere than on capability development. But among a number of Indo-Pacific navies, not least the PLAN, this period saw not only incremental progress in overseas deployments but also significant investments to modernise forces and boost combat and support capabilities. Russian naval forces were also starting to emerge slowly from the years of reduced investment in the 1990s.

High-intensity warfare redux

For Western states, heightened focus on the importance of blue-water operations has revived the requirement for high-end capabilities and related training and skills, not least in the field of anti-submarine warfare. However, maritime-security requirements remain, and are now complicated by the proliferation of sea-denial capabilities among state and non-state actors (with the latter including Hizbullah in Lebanon and the Houthis in Yemen). This has sharpened the dilemmas facing naval forces over where to focus future investments, not only operationally, in terms of deployments, but also in terms of procurements, and the debate and challenge particularly in the West of balancing numbers versus levels of capability.

The increased premium on high-end capabilities has been exemplified by the US Navy's decision that its LCS programme was – in the context of a potential near-peer confrontation – insufficiently robust in terms of survivability or firepower and that it should be replaced by the more capable (and expensive) FFG(X) frigate programme. Another example has been the adoption by both the Australian and Canadian navies of the UK Type-26 design as the basis for their future surface combatants.

The diffusion of more capable weapons has added an extra dimension to the challenge. Much attention has focused on China's development of the 'carrier-killer' DF-21D (CH-SS-5 Mod 5) anti-ship ballistic missile as well as an anti-ship version of the longer-range DF-26 ballistic missile, but China's navy also fields a broader family of anti-ship missile capabilities delivered from various platforms. Likewise, Russia is distributing across its fleet the *Kalibr* family of anti-ship and land-attack cruise missiles, perhaps soon also to be joined by hypersonic weapons. Both are also improving maritime air defence. Other states are also introducing more modern systems, such as Vietnam, which has deployed new submarines and anti-ship cruise missiles in its own version of a sea-denial posture.

New rules of the game

The spread of faster, more precise, more manoeuvrable and longer-range anti-ship missiles, coupled with advances in submarine capabilities and emerging technological challenges such as swarming technologies and autonomous systems, is likely to have implications for most nations – even those whose naval ambitions are focused only locally or on maritime-security tasks.

These developments will make it increasingly difficult for maritime forces to determine the balance between effectively delivering on a task and operational risk. In anticipation of this problem, naval theoreticians are now considering whether the lethality and reach of the threats facing forward-deployed naval forces in high-intensity confrontations – of offensive versus defensive capabilities, including land-based systems intended to outrange and outgun those at sea – are now so altered that they require a different set of capabilities to deliver effect on, and from, the sea. At the same time, potential forms of confrontation are evolving. State-based maritime competition is not just about high-intensity confrontation, or even primarily so in certain scenarios of contesting control at sea and pressing rival visions for the maritime domain.

Increased economic activity offshore has increased pressure on the legal frameworks meant to facilitate peaceful use of the sea, and these frameworks have also become more contested, not least the United Nations Convention on the Law of the Sea. In this context, some states have been adopting asymmetric, sub-threshold strategies to pursue their aims. Moreover, recent encounters have underlined that navies are now no longer the only players. Coastguards, paramilitary organisations and maritime militias – and even, in the case of China, an apparently state-directed fishing fleet – add to the complexities of judging the appropriate forces and responses to devote to particular challenges.

At the same time, the seizure of Ukrainian Navy units by Russian Federal Security Service units in the Kerch Strait in November 2018, and the harassment of US naval vessels in international waters by craft from the Islamic Revolutionary Guards Corps Navy in April 2020, highlighted that the sea still provides an arena for 'encounter' events. Without any realistic prospect of new codes of conduct for the most contentious situations, the diffusion of increased capability increases the risk of escalation in these encounters and places a premium on better training. In a more contested and congested maritime domain, the maritime forces of potential protagonists are encountering each other more frequently.

Inter-state competition can make it more challenging to create cooperative maritime coalitions, as exemplified by the initially slow and somewhat disparate responses to the frictions and challenges to navigation in and around the Strait of Hormuz in the summer of 2019. And if it proved problematic to generate broad coalitions in a globally significant waterway such as the Gulf, it might also prove difficult to gain similar traction if comparable requirements emerge in other areas of the maritime environment, such as under the surface. Indeed, the recent focus on the potential vulnerability of the undersea cable network raises issues of how to monitor, identify and, if necessary, defend against attacks on these vital information arteries. If anything, this and the recent events in the Gulf point to an increased need for persistent surveillance so that hostile activities can be identified, attributed and tackled.

Concentrated and distributed responses

The renewed challenge of great-power competition at sea is generating new patterns of activity by conventional naval forces. US and NATO commanders have highlighted increased activity by the Russian Navy. The US Navy has reconstituted its 2nd Fleet, which is focused on operations in the North Atlantic, and is also deploying back into the Arctic and the 'High North'. In July 2020, as the US pushed back publicly on the perception that it had been incapacitated by COVID-19 infections (notably as a result of the high-profile sidelining of the carrier USS *Theodore Roosevelt*), two US carriers manoeuvred together in the South China Sea – a rare occurrence during the last two decades. This followed concerns on the part of military commanders that both China, with a carrier deployment close to Taiwan and Japan, and Russia, with multiple ship deployments close to UK waters and elsewhere, were testing whether the ripple effects of the pandemic were affecting the resilience of their competitors, not least at sea, and whether this elevated level of probing might prove to be enduring.

Increasing the concentration of combat power is one means by which navies, particularly those of the US and its allies and partners, are responding to the dynamics of renewed great-power competition and the area-denial challenges posed at sea, especially by China and Russia. In particular, the US Navy has been refocusing its training and deployments on complex multi-carrier operations. However, balancing this effort with ongoing rotational requirements, and maintenance schedules, will be a challenge.

There is also increased emphasis on more closely integrating allied capabilities, notably those of France and the UK but also Japan and Australia. This approach offers the prospect of a more coordinated approach by these players to pursuing global-presence ambitions, as well as adding critical mass for potential high-intensity confrontation. The formations made possible as a result – such as multi-carrier groupings – strengthen the strategic rationale for the relatively limited high-end, and particularly carrier, capabilities of medium powers such as France and the UK. There has also been an increase in the number of multilateral exercises in the Indo-Pacific – such as those involving India, Japan and the US – and cooperative agreements between various groupings of these states in response to the growth in PLAN capabilities.

While concentrating force is one part of the answer, it is not sufficient given the scale of the evolving technological challenge, so the US Navy and others are also pushing for a more disaggregated or distributed response. For the US, this means that both defensive systems and offensive missile capability are being deployed more widely within the fleet. It complements the concept of distributed maritime operations, intended to combine these capabilities with doctrinal and wider operational changes including enhanced cooperative-engagement systems. It can also be seen in the new vision for the US Marine Corps, which heralds more dispersed and agile capabilities, and reduced reliance on large amphibious platforms.

In this context, the US is leveraging not just amphibious but also land-based air capabilities in a more networked, cross-domain concept of operations. It is one that is also heavily dependent on cyber and space to deliver effect. Underlying all this is the fact that the US Navy is refocusing on a more competitive high-intensity sea-control mission. However, a distributed, dispersed approach with greater numbers of platforms is, for the US, also key to maintaining the presence role required to address the diverse, low-level, grey-zone challenges at sea. Other major maritime states, meanwhile, may also be moving towards cross-domain operations, but without the US Navy's ability to actually deliver on a distributed approach: put simply, they have less to go around.

The naval forces of potential adversaries have been pursuing similar approaches. In Russia's case, it has been in the distribution of land-attack cruise missiles – the '*Kalibr*-isation' of its fleet. For Iran, it has been in the development of fast attack craft with potential 'swarming' capabilities. In the case of China, it is not only the prodigious recent output of modern naval platforms that has posed a numbers-game challenge that even the US has struggled to counter (China has launched a third more in naval tonnage than the US since 2015). It is also the other organisations – its coastguard and maritime militia – it has been employing in pursuit of its ambitions, and the sheer number of high-capability offensive systems that it has developed and deployed.

Integrating uninhabited platforms

In December 2016, the US Navy raised the end-goal for its future fleet from 308 to 355 'battle-force' vessels. This was a response to the revival of great-power competition and the particular challenge being posed by the emerging capability of the PLAN. While the 355-ship target subsequently assumed somewhat totemic status, there has been uncertainty over the US ability both to afford it and deliver it – at least in terms of traditional crewed platforms. There have even been debates over whether it was an appropriate target in the first place, or sufficient on its own to satisfy the likely scale and breadth of future requirements.

This reflected the strategic and technological challenge the US Navy has been facing in alighting on a credible, durable and effective future force structure that would be capable of counterbalancing the PLAN's assumed development trajectory. This problem is shared by other major navies, albeit at different scales.

The debates in and around the Pentagon over the future size and shape of the fleet – on issues such as the number and types of aircraft carriers, the balance between large and small surface combatants, and the integration of optionally crewed, uninhabited and autonomous capabilities and platforms into maritime operations – have underscored that the US Navy is at an inflection point in terms of future force development.

These deliberations led in October 2020 to the unveiling of 'Battle Force 2045'. Perhaps in an echo of the Reagan 600-ship navy, it envisaged a fleet of some 500 vessels. There would be a different balance between capital ships, submarines and smaller combatants, but perhaps the chief technological facilitators this time would be the incorporation of significant numbers of uninhabited surface vessels (USVs) – which could also be optionally manned – and uninhabited underwater vehicles (UUVs). In addition, the increased adoption of uninhabited aerial vehicles (UAVs) would extend the reach of carrier air wings. In this vision, the 355-ship target for traditional battle-force vessels would be merely a waypoint (in 2035) in the pursuit of the higher goal.

Uninhabited and autonomous systems would seem to offer answers across the spectrum of maritime requirements. At the lower end, they promise more persistent presence and more comprehensive surveillance and maritime awareness. At the higher end, they offer the prospect of greater reach, capable of overcoming the threat from long-range offensive capabilities, and of changing the risk calculus of operating in a high-intensity area-denial environment.

Significant hurdles remain, not least from the US Congress over the ultimate cost. Questions have been raised about whether the navy has the necessary doctrine and concepts in order to fit USVs and UUVs effectively into the overall pattern of fleet operations; about their reliability; about the infrastructure potentially needed to support them; and even about the safety and security of potentially weaponised platforms operating autonomously at sea, not least on remote and isolated presence missions in sub-threshold scenarios.

Constrained budgets are a key driver for the US Navy's pursuit of uninhabited capabilities. However, even more limited resources in other navies may force them to limit their ambitions. The Royal Navy, for example, seems to be adopting a more incremental approach to the potential for USVs. It has, however, raised the prospect of using one of the navy's new aircraft carriers as a test bed for UAVs.

Questions remain over the maturity of uninhabited and autonomous systems, at least in terms of some missions and operational concepts. However, there is a growing sense among the traditional naval powers that to effectively overcome the emerging strategic and technological challenges in the maritime domain, the experimentation in this area needs now to be followed by decisions on mainstream procurement. Moreover, this quandary over balancing the trade-offs between different capability investments does not only affect the traditional naval powers.

The PLAN's push towards the 'world-class' naval status sought by Beijing may be the main driver for much of the decision-making in other navies on future-force plans. However, the PLAN could also be facing its own inflection point in terms of its ambition and ability to deploy more capable formations at range. It too may need to rebalance its aspirations between power projection based on carrier formations and more defensive sea control or sea denial closer to home. This will depend on how it perceives other navies' evolving plans and responses; on the lessons it has taken from its integration of new capabilities, ranging from uninhabited and increasingly networked systems to more lethal offensive capabilities; and on how it envisages employing its broader range of maritime capabilities – such as the coastguard and maritime militia – within the arena of sub-threshold or grey-zone operations.

Charting new courses

In some ways the new dynamics may resemble the traditional business of navies, including surface and subsurface warfare, power projection and even encounter warfare – but overlaid with new factors deriving from profound technological change and renewed geopolitical tensions.

Navies today are being asked to do more, with renewed high-end requirements adding to enduring lower-level demands that have evolved to include states and non-state actors seeking to exploit sub-threshold ambiguities. Here, too, the proliferation of new systems is raising the bar of capability requirements and complicating the search for a new equilibrium.

Within fleets, any likely new balance will be determined by the relative weight placed on the value and roles of traditional platforms versus new technologies, particularly uninhabited and autonomous systems, and the degree to which they match and can deliver national maritime priorities. Between fleets, it will involve the development of new relationships of partnership and interoperability or of competition. This will be the case whether navies are intending to cope with the cost implications and the challenges posed by the new range of area- or sea-denial capabilities, and still retain – with an acceptable level of risk – the ability to manoeuvre and deliver effect and sea control; or whether they are intending to be effective in the face of apparently novel concepts, tactics and tools in the contested space below the threshold of overt armed confrontation. They will have to make these choices at a time when some of the distinctions between higher-end and lower-end security threats are becoming increasingly blurred and capability requirements across the board are on the rise.

Battle management systems

Battle management systems (BMS) are becoming progressively more important to the conduct of warfare. Today, these effectively comprise digitised decision-support systems intended to improve command and control and information management, and manage the battle space across all domains. They are designed to boost situational awareness by providing an integrated common operating picture. By enabling faster assessments and engagement decisions, it is hoped that they will increase the tempo of operations to out-manoeuvre and to out-pace an enemy's decision-making cycle.

The most ambitious developments are seen among armed forces looking to conduct multi-domain operations. That said, they are also of broader relevance. Almost all armed forces that aspire to integrate advanced weapons and systems into their inventories, and to think and act faster than adversaries, will look to employ some form of battle management system. For many, these plans are accompanied or enabled by similar focus on developing and improving the combat management and command and control (C2) systems that underpin and supply essential data to BMS.

However, there are challenges to the effective use of these systems. More data from more sensors is leading to greater use of technologies including artificial intelligence (AI). In turn, these raise legal and ethical questions around autonomous data sifting and selection as well as decision-making. Meanwhile, the electromagnetic spectrum (EMS) on which modern BMS rely remains vulnerable to electronic attack, which drives related developments in offensive and defensive capability. It is also leading some armed forces to train for operations where their access to BMS, or the EMS more broadly, might be degraded or denied entirely. This places a premium on improved electromagnetic protection and more resilient architectures, for instance of satellites. It also means that personnel have to be better trained so they can operate when access to the EMS is contested, as well as better trained to use BMS in complex multi-domain operations and integrate them into the battle space, particularly in coalition operations. For Western defence organisations, an additional challenge is that these systems are increasingly being integrated into the armed forces of their principal strategic competitors, Russia and China.

Evolution

The growing complexity of military capabilities and operations after 1945 led to a need for automation in many areas of battle management. This coincided with improving information-technology capabilities to enable the automation of C2 processes and the increasingly rapid dissemination of decisions through an expanding communications infrastructure, as well as the introduction of more complex weapons systems. The US development of the Global Command and Control System (GCCS), which entered service in the late 1990s–early 2000s, was an important indication of the capabilities intended for a modern BMS. It comprises a suite of hardware and software used for the planning and execution of operations and operates across the US Department of Defense's classified and non-classified networks. Variants of the GCCS architecture are used by all US armed services.

Although nations have worked to develop and increasingly integrate sovereign C2 systems, including more advanced cooperative engagement capability systems, there have also been multilateral developments, exemplified by NATO's Air Command and Control System (ACCS). The ACCS initiative began in 1999 to produce a scalable hardware and software architecture to equip NATO's European members, also providing Combined Air Operations Centres for use in wartime. As well as fulfilling national requirements related to integrated air and missile defence, the architecture can federate national recognised air pictures developed by national ACCSs and produce what NATO calls an integrated and interoperable capability. There is the potential for similar developments of common scalable C2 and BMS systems that could be employed by NATO states' armed forces. Such efforts could reduce development costs through economies of scale and burden sharing while improving the coordination of planning and operations by multinational deployed forces. Indeed, interoperability requirements set by NATO are designed to ensure that even disparate member states' systems are able to effectively share information within alliance C2 networks. That said, as the US accelerates the sophistication and multi-domain integration of its BMS, there is the risk that its NATO allies could face an interoperability challenge in simply keeping up.

The BMS advantage

BMS are designed to collate and analyse disparate information and present this to commanders, their staff and subordinates, at varying levels within the command chain. Modern BMS are designed to link platforms, sensors and systems in order to fuse data including – at the most basic level – cartographic information on an area of operations, friendly- and enemy-force dispositions and the scheme of manoeuvre. As they have emerged, these systems have been scalable – able to be employed at the strategic, operational and tactical levels. For instance, at the tactical level a system might provide integrated secure communications, feeds from intelligence, surveillance and reconnaissance sensors, targeting information, orders and plans, and situational awareness regarding the location and movement of friendly and enemy forces. At the operational and strategic levels, the system would have to integrate and present, more rapidly, an exponentially greater amount of data from an expanding range of sources across domains. As operational demands increase, and as technological capacities grow, it will be increasingly important to seamlessly synchronise and integrate these systems in order to achieve the objective of providing comprehensive situational awareness and a common operating picture across the land, sea, air, space and cyber domains and to improve the speed of information exchange. There is a chance that automated BMS could accelerate this process, speeding up decision-making and perhaps also giving greater space for decisions, overall improving the speed at which combatants can navigate the OODA (Observe, Orient, Decide, Act) Loop, first proposed by former US Air Force Colonel John Boyd, in a bid to prevail over their adversaries.

However, there is also the risk that greater volumes of information could compress decision-making space unless adequate support is given by BMS or indeed if personnel are not trained to adequately use or to fully exploit these systems' capabilities. Moreover, over-reliance on advanced technology could also render operational and tactical decision-making vulnerable to attempts to disrupt C2. This could include conventional jamming to disrupt the EMS. Russia's continued investment in electronic-warfare (EW) capabilities, including deployable systems, reflects this trend. Moreover, BMS will face a growing risk from activities designed to spoof, as well as disrupt and deny, use of the EMS. These threats will grow as faster and more complex weapons, and increasingly cross-domain military operations, drive the adoption of AI for decision support in BMS architectures. While affording key benefits by reducing the workload for human operators, increasing technical capabilities risk increasing the potential attack vectors for adversary operations, such as cyber attacks. It was noteworthy, in the US Department of Defense's (DoD) 2020 report on 'Military and Security Developments Involving the People's Republic of China', that China's People's Liberation Army (PLA) was reported as actively exploring the potential of using space, cyber and EW to target adversary C2 networks.

Adoption beyond Western states

China places a high value on C2 automation. The PLA has, according to the Pentagon, already started to embrace 'big data analytics that fuse a variety of data to improve automation and to create a comprehensive, real-time picture' as part of its drive to 'fight and win informatized wars'. The report also noted that the PLA is looking at how AI can enhance autonomy in command and control. It states that future command, control, communications, computers, intelligence, surveillance and reconnaissance (C4ISR) systems 'will seek to use AI to collect, fuse, and transmit big data for more effective battlespace management and to generate optimal courses of action'. It is likely that China's investments in this area owe much to observation of the importance of these technological developments to the recent performance and likely future trajectory of the US armed forces. An example of this is the investment the PLA Navy (PLAN) has ploughed into developing weapons and sensors that appear to be analogous to the US Navy's *Aegis* combat management system. The range of sensors and weapons apparent on its Type-052D (*Luyang* III) destroyers has led analysts to consider it likely that the PLAN will be using software to synthesise these capabilities. Indeed, it is possible that the PLA seeks not just to match US and allied BMS technology, but that it is looking to surpass this with the application of advanced technologies like machine-learning to bring about what a senior executive of the Aviation Industry Corporation of China (AVIC) has called OODA 3.0, in order to improve 'information agility'. While these remarks were made in the context of combat air developments, they likely reflect the development trajectory within China's armed forces. Indeed, as was noted in a 2020 report by the US Congressional Research Service, 'the close parallels between U.S. and Chinese AI development have some DOD leaders concerned about the prospects for retaining conventional U.S. military superiority as envisioned in current defense innovation guidance'. China is, more broadly, pursuing a national AI development plan. At the same time, China has looked to improve joint combat capabilities by pursuing institutional reforms designed to streamline command and control, such as the establishment of five Theatre Commands in 2016, in place of the previous seven military regions.

Russia has followed a somewhat similar path, though while it began the process of streamlining and improving C2 before China, its overall progress is perhaps more vulnerable to issues relating to defence funding and industrial capacity. Nonetheless, Russia's post-2008 military reforms have seen the armed forces' C2 apparatus change significantly. Advanced technologies have been introduced, and there has also been significant organisational change designed to improve the speed and coordination of decision-making. Based on the lessons of the wars in Chechnya and Georgia, Russian defence policymakers decided to streamline the armed forces' command and control architecture and joint strategic commands (OSK) were created after 2010. OSK commanders controlled all forces within their area of responsibility, with the exception of strategic-level assets such as the Strategic Rocket Forces. That said, Moscow has also developed automated capabilities like the National Defence Management Centre (NDMC). Based in Moscow and established in 2014, this facility integrates the Russian General Staff and Ministry of Defence with other government departments. This may have been part of a bid to improve central coordination as Russia had placed control of combat assets under the OSK commanders. The NDMC will also eventually link the operational-strategic levels with the tactical levels by connecting to subordinate command centres and reportedly has a high degree of automation through its use of supercomputers. The creation of the NDMC reflects the general adoption in the Russian armed forces of combat capabilities enabled by improved digital networking.

Since 2000, Russia's defence industry has been working on a unified tactical command and control system (YeSU-TZ), though analysts of Russia's armed forces have judged that industry has had problems in meeting both the envisaged timelines and dealing with the practical problems in generating such a new system, as well as the cost of rolling it out in adequate numbers. In December 2017, defence minister Sergei Shoigu reported that YeSU-TZ had been successfully used in combat training and a contract was reportedly signed in December 2018. One challenge for Moscow is now in coordinating the various systems that have been developed. As well as YeSU-TZ, Russia's Ground Forces are developing the *Akatsiya*-M BMS for use at the operational and tactical levels, while the Airborne Forces use the *Andromeda*-D BMS for employment at similar levels. Russian contractors are now marketing such equipment internationally: Sozvezdie, the manufacturer of YeSU-TZ, lists online for apparent export what it terms 'automated army tactical command and control systems (ATCCS) designed to provide automated command and control for land forces from section-to-battalion levels'. Traditionally Russian equipment has been competitively priced when compared to its Western equivalents, and Moscow has also been less discriminating regarding the countries to which it supplies materiel, opening procurement possibilities for nations unable or unwilling to procure Western C2 technology. Other nations have moved forward with similar domestic efforts – the PAK-IBMS (Pakistan Integrated BMS) equips Pakistan's army and provides mission planning, blue-force tracking and messaging at tactical and operational levels. India too is pursuing efforts at improving and automating command and control tools. In both cases, success will depend not just on delivering a technical solution, but on integrating this across the armed services. Domestic C2 and BMS developments, as well as the potential proliferation of these, have implications for US and allied armed forces which may have to face adversaries in future conflicts with a high degree of C2 sophistication.

Future trends

While the automation of command and control systems has changed the conduct of operations since the end of the Second World War, there remain risks for future development. The digitisation on which BMS rely renders them potentially vulnerable to cyber attack. The use of digital C2 systems places a premium on sophisticated Communications/Transmission Security (COMSEC/TRANSEC) protocols in order to prevent malicious code penetrating the BMS, as well as robust cyber-security tools to detect and neutralise an attack. Cyber attacks could take the form of attempts to deny or disrupt the use of all or part of a BMS, or indeed more subtle actions such as spoofing. This range of sophisticated challenges increase the possibility that, as well as AI technologies being integrated in order to improve the operational capacity of BMS, AI-enabled cyber defences will be as much a part of any future BMS architecture. Moreover, with dominance over the electromagnetic spectrum no longer assured, cyber-security and COMSEC/TRANSEC measures will be vital, but so too will the ability for forces to be able to fight and prevail in environments where digital C2 is compromised or denied outright. Because of this, while BMS may be of growing importance to operations, they must not become so indispensable that military tasks cannot be performed without them. This perceived vulnerability will spur greater Western attention on EMS protection as well as capabilities designed to, in turn, deny adversary use of the EMS.

The advent of Multi-Domain Operations (MDO), together with the growing adoption of AI-enabled capabilities, is likely to influence the future trajectory of BMS technology. The US Army defines MDO as providing 'commanders numerous options for executing simultaneous and sequential operations [...] and the rapid continuous integration of capabilities across all domains'. The US DoD has developed this further with the concept of Joint All-Domain Command and Control (JADC2) which according to the US Congressional Research Service is intended to 'connect sensors from all of the military services – Air Force, Army, Marine Corps, Navy and Space Force – into a single network'. This new system would be developed to overcome insufficiencies in the current command and control structure when dealing with the modern conflict environment. Two large JADC2 exercises were carried out in late 2019 and in mid-2020. The USAF has offered its Advanced Battle Management System (ABMS) as a foundation for the JADC2. That said, there remain questions over the JADC2 concept, including over technical feasibility and reliability, access to bandwidth in contested environments and the implications of reducing human involvement in decision-making. In April 2020, meanwhile, the US Government Accountability Office highlighted development risks for ABMS, including that the air force needed to 'develop a plan to attain mature technologies when needed for each ABMS development area'. Similar risks will likely apply to advanced BMS developments elsewhere, because in some cases wholly new technologies and systems are being pursued.

Artificial intelligence is one such area. It is likely that all future BMS development, whether projects that hold joint cross-domain ambitions like JADC2 or smaller, single-service programmes, will be influenced by AI. These systems will be able to retrieve, process and present data to commanders in increasingly efficient ways, learning from previous operations, missions and tasks. However, as these decision-making aids improve in capability, and as data grows in volume, there will have to be attendant levels of trust in their technical capacities. In turn, these raise legal and ethical questions around autonomous data sifting and selection as well as decision-making. Factors like this make it more likely that AI will be integrated incrementally. Initially, the technology will probably be adopted for applications ranging from decision support on operations to decision support over inventory management matters. AI is also likely to be initially employed for repetitive, straightforward aspects of command and control which can be easily learned by the system.

Nonetheless, the pace of innovation in AI technologies makes it more likely that they will be applied to increasingly sophisticated aspects of command and control. This raises ethical questions regarding the extent to which, if any, C2 and BMS will be allowed to take decisions and initiate action without human intervention. In this regard, AI development and integration may – at least in Western states – face similar challenges to those seen in the debate over use of uninhabited aerial vehicles, particularly over issues relating to the level of human involvement, and degrees of automation and autonomy, not least for cases involving weapons release.

There have been significant technical advances as automated C2 systems have developed. However, while these might improve Western states' BMS capacities in an emerging age of multi-domain operations, a levelling technological playing field raises new challenges, as states like Russia and China look to develop their own advanced BMS tools and at the same time improve their capabilities to deny effective use of the EMS by others. An additional challenge, at least for Western states, is that these countries may not perceive similar constraints in their export policies. Another is that they may feel similarly less constrained when it comes to the legal and ethical debates that in the West will likely influence the sophistication of BMS as well as the speed with which, and the degree to which, these are able to be augmented with AI technologies.

China's civil-military integration in the 21st century

Trade and technology competition is accelerating between China and the United States. Moreover, both Beijing and Washington are increasingly focused on ascendancy in the research and development (R&D) of emerging technologies. Both are looking to effectively integrate these technologies into their armed forces and, in doing so, transforming them for high-technology-enabled aspects of future warfare. There is particular attention on the central role of the civilian sector in R&D and on civil–military integration (CMI), so that armed forces and defence establishments can benefit rapidly from technological developments in the civilian sector. Concerns in the US have arisen about whether China's formulation of military–civil fusion (MCF) is more suited to this new technological competition than the approach of the US Department of Defense, which includes partnering with the private sector, outreach to academia, supporting its own R&D centres and working with others that are federally funded. China's approach looks to seamlessly integrate the civilian and defence economies, and indeed restructure the latter, so that it is better able to leverage commercial innovation and resources to develop capabilities for the People's Liberation Army (PLA) as well as for export. However, although this approach has received high-level political attention in Beijing, it remains a work in progress and China has faced challenges in implementation.

Focus on high technology

Both the US and China recognise the importance of technology for their military modernisation. The 2018 US National Defense Strategy highlighted 'rapid technological advancements and the changing character of war'. China's 2019 defence white paper, meanwhile, said that military competition was changing because of rapid advances in new and advanced military technologies based on information technologies. Defence documents in both countries point to the military application of a host of potentially transformative technologies and enablers such as artificial intelligence (AI), quantum technologies, big data, autonomy, robotics, directed energy, hypersonic systems and biotechnology. Some of these technologies will be important to the PLA as it looks to develop capabilities suitable for 'systems confrontation'. Recent PLA studies and white papers indicate that China sees this as important for modern warfare, where an adversary's operational system would be destroyed or degraded such that it undermined operational effectiveness. The PLA must be prepared to confront opponents in multiple domains including space and cyberspace, an earlier white paper indicated in 2015. Targets could include elements of operational systems, including command-and-control nodes as well as conventional capabilities and networks. China likely envisages these operations being conducted through both kinetic and non-kinetic means. Its pursuit of integrated operations across services and domains is perhaps analogous to the US concept of multi-domain operations. As well as helping generate more effective military capabilities, advanced technologies are seen as key to integrating these across the air, land, sea, space and cyber domains.

While government agencies are important in generating some of these new developments, such as in hypersonic flight and directed-energy and quantum technologies, the private sector is increasingly important in developing dual-use technologies that are relevant to armed forces. Furthermore, private-sector firms are now seen as vital innovation centres, and defence organisations are increasingly cognisant of the need to not only harness technologies from commercial applications, but also benefit from the innovation and risk-taking culture that is seen in some private-sector enterprises. China's President Xi Jinping has made clear that the PLA needs to be open to innovation as it pursues the aim to generate 'world class' armed forces by 2049. In 2014, Xi emphasised the need to study global innovations; by 2017, amid increasing international competition, he was saying that 'only the innovator wins'. According to Xi, whoever leads in scientific and technological innovation will have the advantage in 'national defence construction'.

CMI in China: from spin-off to spin-on

China began concerted efforts to integrate civilian and military defence technology and industrial sectors after Deng Xiaoping's call for civil–military integration during the early 1980s. This was, however, less an effort to improve defence technology and more an effort to engage in commercial ventures. Aside from the March 1986 High-Technology Research and Development Plan (the 863 Plan), which was intended to advance China's civil–military capabilities in automation, biotechnology, energy, information technology, lasers, new materials and space technology, this era was largely characterised by spin-offs, where military technology and know-how was transferred to civilian applications in order to give China's developing defence industry additional financial resources. The 863 Plan helped to fund China's National University of Defense Technology *Tianhe*-1A and *Tianhe*-2 supercomputers in 2010 and 2013 respectively. In 2015 the 863 Plan was folded into a national key R&D development plan.

Military and security developments during the 1990s sharpened focus in Beijing on the US lead in defence-related technology. Beijing paid particular attention to the US use of advanced airpower and long-range precision-strike capabilities in *Operation Desert Storm*. In the mid-1990s, Beijing suffered a further blow when PLA activities were unable to sway Taiwanese voting behaviour in the 1996 national elections. Nor did its military drills prove a sufficient deterrent to the US to prevent it from sending two carrier battle groups to the waters off Taiwan. The bombing of the Chinese Embassy in Belgrade during the Kosovo campaign in May 1999 further alerted China's Central Military Commission that it needed to develop advanced technology systems. The New High-Technology Weapons Plan (the 995 Plan) was a direct response to the Belgrade bombing, and prioritised the development of new and asymmetric capabilities. The Central Military Commission established a new High-Technology and Engineering Leadership group to manage the development of these weapons systems. At the same time, China began to pursue dual-use technologies and 'spin-on' policies in a bid to leverage its growing economy and civilian sector in order to both maintain economic development and also modernise the PLA. Successive five-year defence plans, as well as the 2006–20 National Medium and Long Term Plan for Science and Technology Development, promote domestic innovation to develop new or improve existing technologies while also importing and absorbing foreign technology. With their reference to the integration of civilian dual-use technologies into the defence sector, these documents signalled the intent to pursue 'spin-on' projects.

Military–civil fusion under Xi

China looked primarily to the US for inspiration in its early civil–military integration efforts, and under its current leadership has looked to develop military–civil fusion. In March 2015, at the third session of the Twelfth National People's Congress, Xi said that the development of military–civil fusion should become a national strategy and that implementation should move from 'early' to 'deep fusion'. China's 2015 military strategy, issued in May, indicated that China would move beyond 'integration' towards a system that sought to create an 'all-element, multi-domain and cost-efficient pattern of CMI'. Xi elaborated further in a speech to the 19th Party Congress in 2017. China would, he said, 'deepen reform of defence-related science, technology, and industry, achieve greater military–civil integration, and build integrated national strategies and strategic capabilities'. The July 2016 Central Committee 'Opinion on the Integrated Development of Economic and Defense Construction' was the first document to publicly elaborate MCF. That year, the CMC Science and Technology Commission was established during reforms to the Central Military Commission (CMC). Several other high-level agencies have been established to promote MCF. In 2017, the Central Commission for Integrated Military Civilian Development was set up to oversee the strategy and its implementation. That same year, the 13th Five Year Plan for Science and Technology Military–Civil Fusion Development detailed an integrated system for advanced R&D in AI, biotechnology, advanced electronics, quantum technologies, advanced energy, advanced manufacturing, future networks [and] new materials so that China could 'capture commanding heights of international competition'.

This effort at 'fusion' has diverged from previous approaches in a number of ways. Firstly, it advocates not only that civilian companies should integrate directly into the defence sector and into the PLA's supply chain, but that the state-led defence sector should also reform. Indeed, under Xi, defence-related state-owned enterprises have been instructed to slim down, to improve their competitiveness on the global defence market, and also to increase their financial efficiency and independence.

Secondly, the fusion policy has risen in prominence due to China's pursuit of advanced technologies as a way of obtaining an advantage in its competition with the US. CMI had already been mentioned, in the 2015 Military Strategy, as a means by which China could strengthen progress in key technology areas. China's defence-technological and -industrial base continues its effort to build conventional capabilities similar to those of the US. However, this will take time, and the US will look to retain a capability advantage. This is one reason to pursue 'leapfrog' capabilities, including by using advanced technologies. It is likely that China's leadership will be looking to its civilian private sector to pursue technologies that could be developed for militarily relevant applications. Certainly, some Chinese firms are globally competitive, for instance in applications of artificial intelligence and the analysis of big data. There is little publicly available evidence of coordination between China's technology firms and the defence sector. However, analysts have said that one possible example may be the joint laboratory, set up in 2018, between technology firm Baidu and state-owned China Electronics Technology Group Corporation for the application of big data, AI and cloud computing to command and control. As the PLA looks to improve its technical modernisation, technologies like these – many developed by private-sector firms – will become increasingly important for military development.

Thirdly, MCF is also intended to leverage innovation in the private sector to reform and modernise other PLA capabilities, besides weaponry. Emerging dual-use technologies will play an increasingly important role in improving logistics and communications as well as discipline. In the civilian realm, big data and AI have been utilised for social management and as part of Xi's anti-corruption campaign. In the PLA, too, there have been reports of pilot social-credit programmes to monitor loyalty to the Chinese Communist Party. More broadly, military–civil fusion is now central to major national policies including Made in China 2025 and the Next Generation Artificial Intelligence Plan; this calls for common standards across the civilian and military sectors.

Is MCF unique, or just well timed?

Some analysts have indicated similarities between MCF and CMI efforts in the US. One is China's establishment of the Central Military Commission Science and Technology Commission and subordinate 'rapid response small groups'. The latter have been described by some US analysts as similar to the US Defense Innovation Unit and the former as analogous to the Defense Advanced Research Projects Agency (DARPA). Nonetheless, China's effort is distinct. While the US promotes partnerships between defence industries and the Pentagon, and private-sector firms retain a strong degree of independence, China's MCF is bolstered by political will and financial resources as well as a supportive regulatory framework. China's academic sector plays an important role not only in collaborating with the PLA on advanced technology R&D, and has been reported to engage with foreign academics to broaden access to cutting-edge innovation. That said, China's ability to send students overseas to gain experience and gather knowledge, at least in the US, was reduced after Washington imposed stricter visa regulations in 2018. China's MCF strategy may face obstacles in engaging private-sector firms, such as adapting military culture and institutions to exploit new technologies. However, the emphasis on party loyalty has, according to analysts, on occasion been able to drive technology transfer.

China's current effort at MCF also reflects national strategic ambition. In the US, some moves on this scale, intended to achieve security and military-capability goals, have been framed as 'offsets'. The third of these offset strategies, introduced in 2014, sought to 'sustain and advance America's military dominance' by exploiting emerging and innovative technologies for military purposes and integrating them with new operational concepts. An added challenge for the third offset was that, unlike the first two, it relied more on private-sector R&D than that developed in defence laboratories. However, the 'third offset' declined in prominence under the Trump administration. There was criticism of a 'disconnect' between the administration and both the Strategic Capabilities Office (SCO) and the Defence Innovation Unit, which had been set up to leverage new and innovative capabilities, and to connect the Pentagon with Silicon Valley firms respectively. (A plan for the SCO to move under DARPA was halted in 2019.) A larger issue is whether established defence industries are reluctant to move beyond legacy platforms to emerging technological capabilities and whether the complexity of the US defence acquisition process discourages companies from doing business with the Pentagon.

China's approach to MCF may give it an advantage in this respect. The PLA has been reformed since Xi took office, and Beijing has had the opportunity to reshape the armed forces' organisational structure, leadership bodies, and military capabilities and training systems, among other initiatives. This overhaul was driven by a long-term strategic vision to modernise the PLA such that it can compete with the US armed forces in the Asia-Pacific region by 2035, and have 'world class' armed forces by 2049. This military transformation drive may have given the PLA the organisational structure it requires to leverage emerging technological developments and the time to develop new concepts and doctrines. For instance, the disbandment of the Central Military Commission's four general departments in 2016, and the subsequent creation of 15 smaller functional departments, sought not only to improve the leadership's oversight of the CMC but also to break long-established silos of power and tackle specific areas of improvement. For example, the Science and Technology Commission now falls directly under CMC oversight, in order to improve the management of military innovation and promote civil–military integration in defence R&D.

Challenges to implementation

Nonetheless, while MCF might hold more promise than previous attempts at CMI, challenges remain to its full implementation and China's efforts at civil–military integration may experience some of the same problems as other states. For instance, although there has been media focus on some US technology firms' reluctance to participate in defence-related projects, some of China's technology companies are equally commercially driven and resistant, if perhaps less vocally, to open integration with the PLA. The competition between China and the US in technology and trade is making it harder for Chinese private-sector companies to access international markets amid concerns about China's military and political goals and the MCF that is designed to support these objectives.

According to retired PLA Rocket Force official Chen Lufan, China's MCF policy is 'troubled by a string of problems', such as poor communication, incomplete policies and regulations and the excessively high authority needed to approve integration. According to Chen, these obstacles are impeding the development of MCF and it has yet to meet the PLA's key military capability needs including in terms of defence concepts, novel energy sources, materials science, sensors, and guidance and command-and-control technologies.

Political will may give China the ability to bring together the civilian and military sectors, but political control and ideology in China may, perhaps paradoxically, prove to be stumbling blocks in the long term. At the meeting of the Academicians of the Chinese Academy of Sciences and that of the Chinese Academy of Engineering, both in June 2018, Xi spoke about the need to foster creative thinking and more open-minded, highly qualified and innovation-driven talent in order to establish a forward-thinking and innovation-focused military institutional culture. He also indicated awareness that doing this effectively is a long-term effort.

Similarly, at the joint discussion of the Association for Science and Technology and the Science and Technology Committee, during the First Session of the 12th National Committee of the Chinese People's Political Association for Science and Technology, Xi pointed out the need to break the risk-averse nature of China's science and technology system, stating that China must create an 'atmosphere … that encourages bold, daring and courageous innovation … that not only values success but also tolerates failure'. However, Xi's focus on ensuring loyalty to the Party within the PLA may complicate this aspiration. Not only has the Party's political work in the PLA been strengthened since his first term in office, but there have been attendant measures to improve loyalty to the Party, even fealty to Xi himself, in wider society. However, while MCF may primarily be an effort to drive national civil and military development, the importance of access to foreign skills and technology should not be underestimated. Increasing strategic competition between China and the US, as well as heightened scrutiny of Chinese academic collaboration and of trade with Western states, will undoubtedly restrict China's ability to harness foreign civil technology and skills for military purposes. While this could prompt a redoubling of attempts to generate technologies indigenously, it could also lead to increased attempts to gather information illicitly. Overall, it will increasingly limit one strand of China's MCF policy, and may affect China's ability to reach ascendancy in technology R&D and in the competition for defence innovation. Neither the US nor China has developed a CMI model that guarantees technological dominance. In light of the weak results of previous attempts at CMI in China, Beijing may hope that the latest attempt through MCF will be different. While long-standing institutional and financial blocks to efficient civil–military industrial collaboration have been tackled, forcing creativity in a command-led system is more difficult. Moreover, should the interests of China's private sector diverge from those of government, the MCF model may face similar challenges to CMI in the US. In that case, the future of MCF could depend on Beijing's ability to coerce the innovative sectors of China's economy. This will be important as China moves, in the 2020s, into the next phase in MCF development strategy, from a preparatory stage to 'deep implementation'.

Fractures in the arms-control environment

The Trump administration's belated effort to extend the bilateral New START nuclear-arms regime left only limited opportunity for the Biden administration to extend the treaty. New START is the last remaining bilateral nuclear-arms-control treaty limiting Russian and US strategic nuclear forces. Besides its sluggishness over the New START extension, Washington had in November 2020 also withdrawn from the Open Skies Treaty, alleging continued Russian violations, and accuses Russia of selective implementation and circumvention of the Vienna Document. Meanwhile, emerging technologies applicable to strategic systems are a further complication in relation to arms control, besides poor US–Russian ties.

Arms-control relations between the two had already been strained by the US withdrawal, on 2 August 2019, from the Intermediate-range Nuclear Forces (INF) Treaty. Washington cited Russia's 9M729 (SSC-8 *Screwdriver*) ground-launched land-attack cruise missile as the basis for pulling out. The US has similarly accused Moscow of selectively complying with the Vienna Document and of breaching Open Skies, both of which support transparency and confidence-building. Though the US sought to strengthen the Vienna Document with new proposals (tabled by 32 countries), its withdrawal from Open Skies raised questions over the latter's survival. The risk was that Russia may see little value in remaining, given its concern that the US might receive flight imagery from other NATO states.

When it decided to pursue New START extension, the Trump administration's initial gambit was that China should be included in the negotiations, insisting that Moscow persuade China to join talks. The US bid appeared bound to fail from the outset: China was opposed to a strategic arms treaty and Moscow said it respected this view. Beijing pointed to the relative disparity between Chinese and US strategic arsenals. However, Washington's concerns were broader. The Pentagon's annual report on China highlighted its large inventory of medium- and intermediate-range ballistic missiles. US withdrawal from the INF allows it to address this, should it wish, by fielding intermediate-range ground-launched ballistic and cruise missiles in Asia.

After this, US negotiators sought a one-year New START extension and a deal to freeze both sides' warhead totals, including non-strategic nuclear weapons and any nuclear warheads connected to systems not covered by the treaty. Russia's President Vladimir Putin also proposed a 12-month extension, without additional conditions. That position appears to have shifted, with Russia accepting the possibility of a freeze and declaration of total warheads, but balking at proposed verification measures. Regardless, both countries will also need to consider how to now deal with new offensive and defensive strategic systems. Moscow has long maintained that US ballistic-missile defence needs to be addressed in the context of strategic arms control, though this argument has so far gained no purchase in Washington.

There is also the question of new systems and technologies. Some new systems fall under existing precepts. Russia's *Sarmat* intercontinental ballistic missile and *Avangard* hypersonic glide vehicle are covered by New START. But wholly new systems, or revived older ones, also cause concern. US desires to constrain new Russian nuclear-weapons systems, such as the developmental *Burevestnik* (SSC-X-9 *Skyfall*) nuclear-powered, nuclear-armed cruise missile, might become unstuck given Moscow's view that Washington remains intransigent on defensive missile systems. Indeed, the *Burevestnik* nuclear-powered cruise missile is only one of several missile and technology developments that pose questions for arms control more generally. An additional complication is that several technologies, such as hypersonic glide vehicles and cruise missiles, are applicable equally to conventional- and nuclear-weapon delivery. Elsewhere, the introduction of counter-space systems also places at risk satellite-based ballistic-missile detection and other command-and-control constellations.

For strategic-arms-control advocates, the first step in at least halting any further deterioration will be a New START extension. The treaty could be extended for any amount of time up until 5 February 2026, instead of expiring on 5 February 2021. Doing this would at least provide room for the Biden administration to review its options, and for Moscow to consider to what extent it might show flexibility.

Chapter Two
Comparative defence statistics

Defence budgets: top 15 in 2020† US$bn

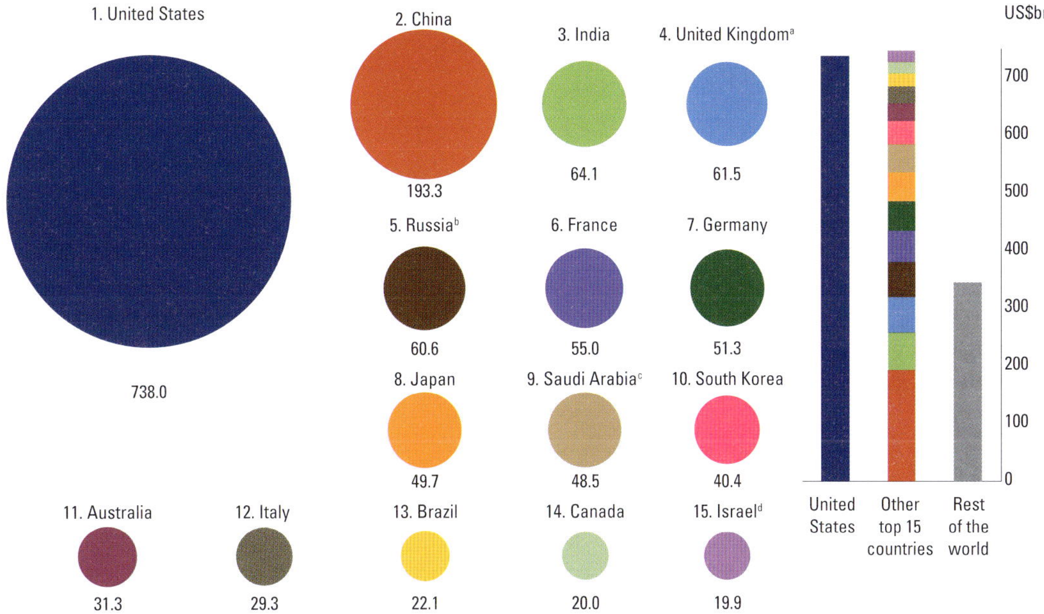

[a] Includes Armed Forces Pension Scheme; [b] Total defence expenditure including National Guard, Federal Border Service and military pensions; [c] Excludes security expenditure; [d] Includes US Foreign Military Assistance. Note: US dollar totals are calculated using average market exchange rates for 2020, derived using IMF data. The relative position of countries will vary not only as a result of actual adjustments in defence spending levels, but also due to exchange-rate fluctuations between domestic currencies and the US dollar. The use of average exchange rates reduces these fluctuations, but the effects of such movements can be significant in a number of cases. Security spending removed from Middle East estimates (including Saudi Arabia and Iraq). Armed Forces Pension Scheme and Special Appropriations (pensions) added to UK and Australia budgets respectively.

2020 top 15 defence budgets as a % of GDP*

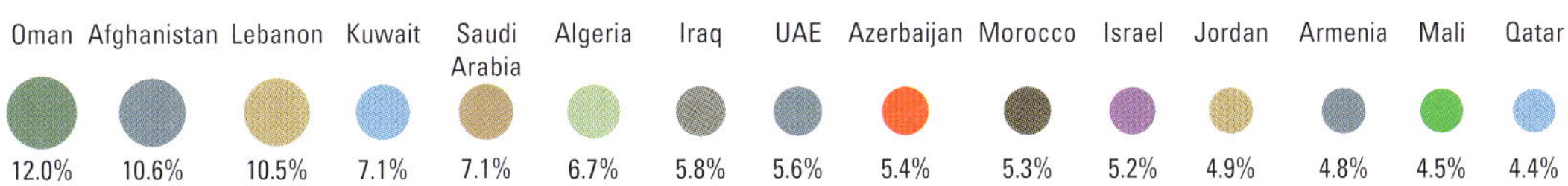

* Analysis only includes countries for which sufficient comparable data is available. Notable exceptions include Cuba, Eritrea, Libya, North Korea and Syria.

Planned global defence-spending by region 2020†

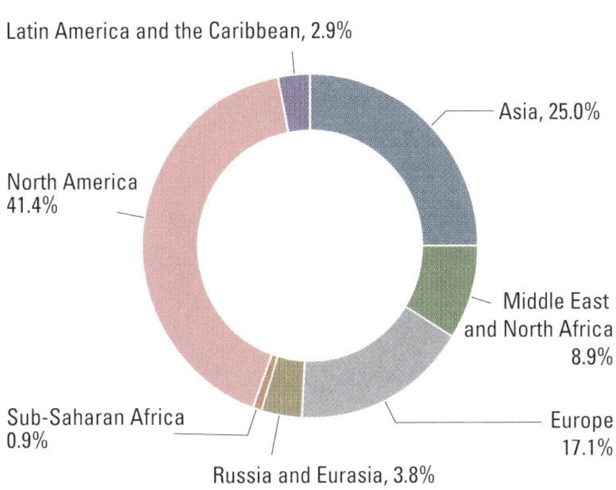

Planned defence-spending by country 2020†

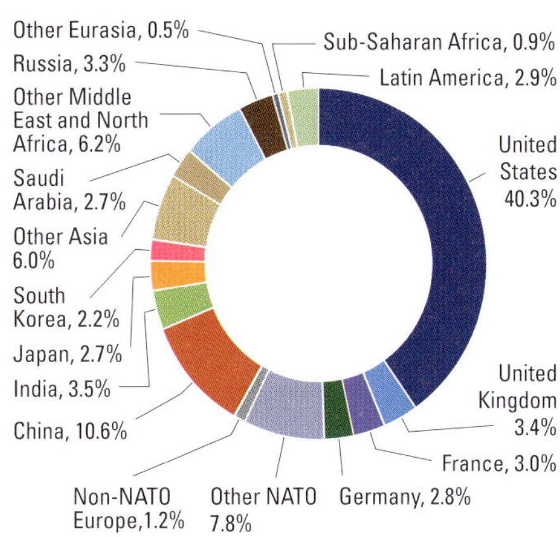

† At current prices and exchange rates

Real global defence-spending changes by region, 2018–20*

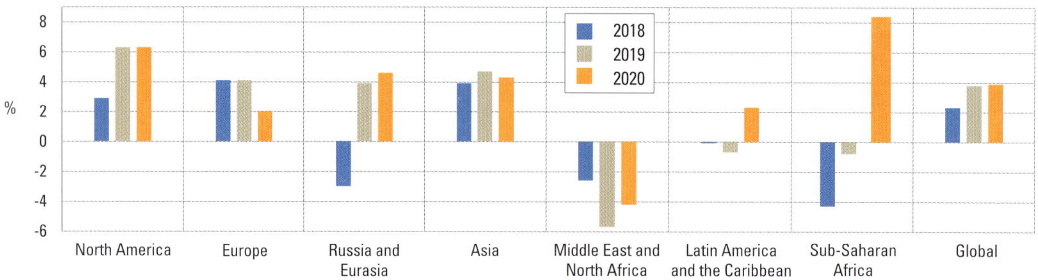

* Excludes states for which insufficient data is available

NATO members and selected regions, 2020 defence budget (% GDP) and investment (% defence budget)

NATO recommends members spend 2% of annual GDP on defence and allocate 20% of the defence budget to equipment or investment spending – i.e. the top right quadrant of the below chart.

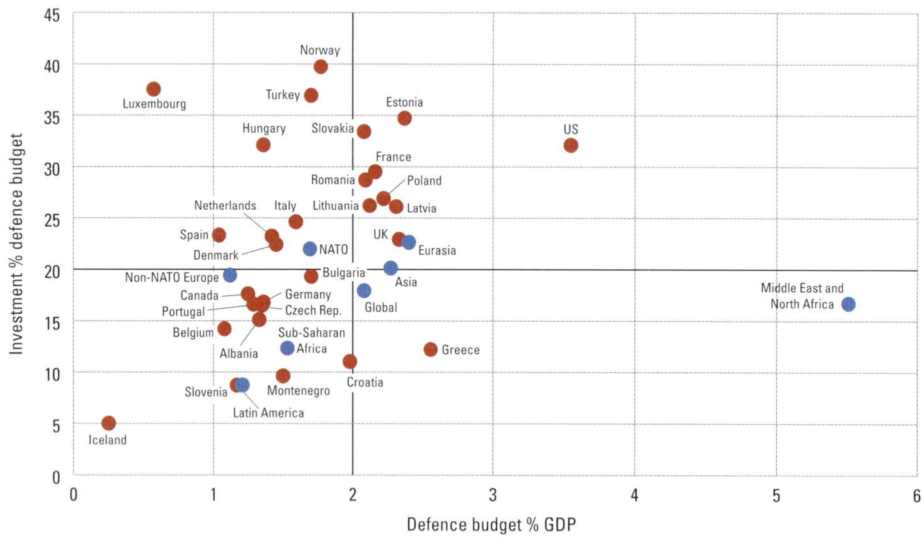

Composition of real defence-spending increases 2019–20‡

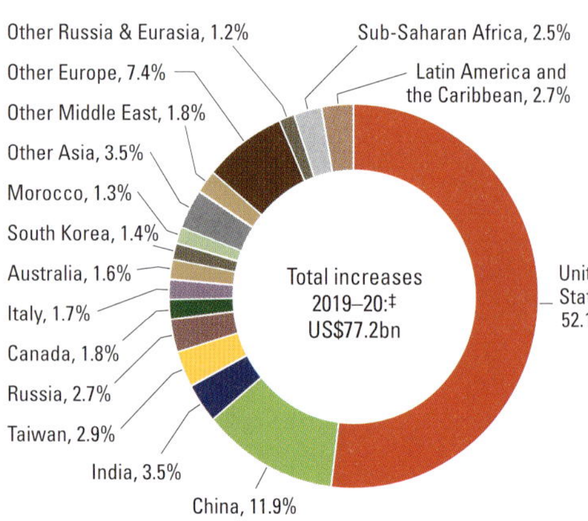

Composition of real defence-spending reductions 2019–20‡

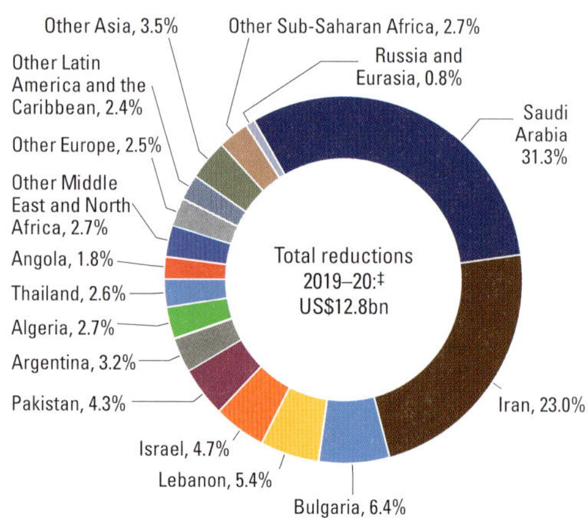

‡ At constant 2015 prices and exchange rates

The AESA advantage

The introduction of Active Electronically Scanned Array (AESA) radars has significantly increased airborne radar capabilities, with AESA systems offering greater reliability, performance and flexibility than their mechanical counterparts. Mechanically scanned airborne intercept or multi-mode radars physically move the antenna, sometimes rapidly, to track targets. In contrast, AESA radars do not need to be moved, instead using an array of hundreds of transmit-receive modules (TRM), each akin to a mini-radar, for target tracking. Even a fixed-array AESA radar offers a greater field of regard than an equivalent mechanically scanned radar, delivering increased situational awareness. In the air-to-air role, AESA systems are better able to manage multiple targets that may be flown in such a way as to try to break radar lock. Interest in the potential of AESA technology for multi-role fighters increased in the 1970s and developmental systems were tested in the 1980s. AESAs were introduced into service on new designs of multi-role fighter aircraft from the beginning of this century and have since formed a central element of upgrades to designs already in service.

AESAs: Fixed, repositioned or multiple

Three design approaches have been adopted for AESA arrays: a non-moving array; a single array that can be repositioned; and multiple arrays. The first is the least technically complex, and does not re-introduce some of the mechanics that risk greater failure rates; the second offers greater fields of regard, while accepting the complexity of moving the array; the third avoids the complexity of moving the array, but introduces the challenge of integrating multiple radar pictures from each array.

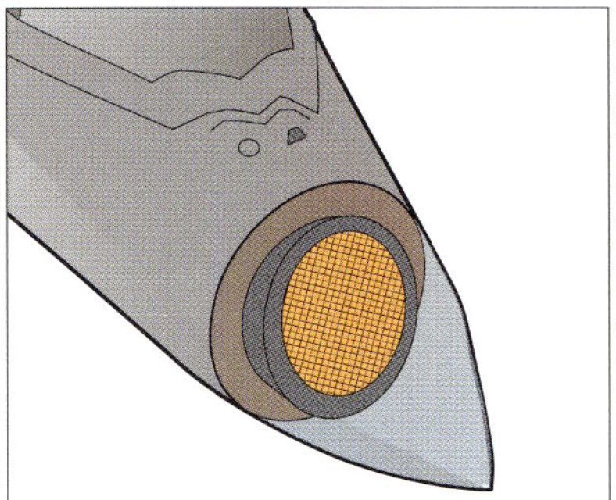

Fixed (F-35)

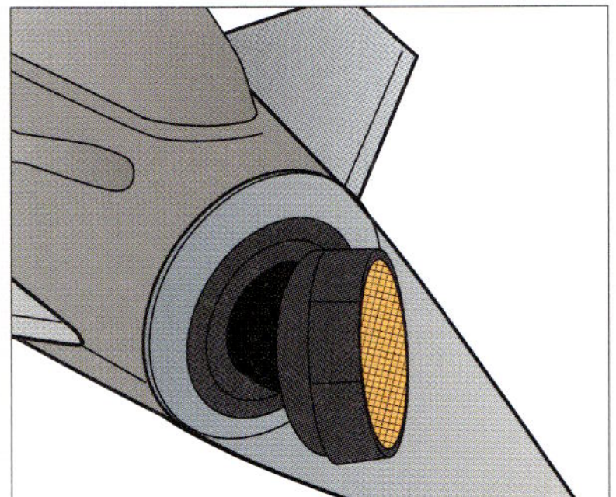

Repositioned (*Typhoon*)

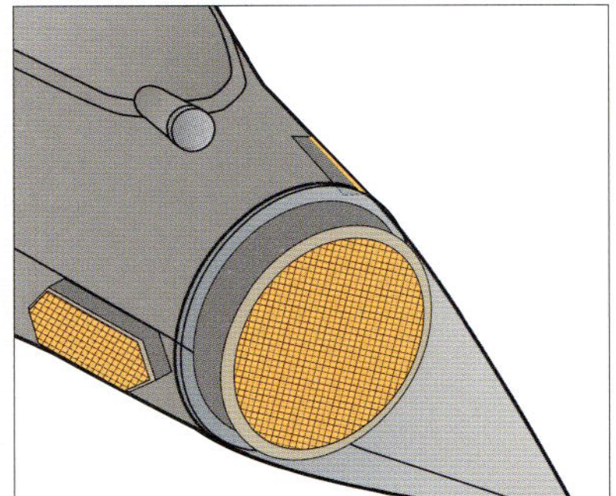

Multiple (Su-57)

Selected AESA-equipped combat aircraft

Designed with:

US: F-22 (APG-77), F-35 (APG-81)

Russia: Su-57 (N036)

China: J-20 (KLJ-5), J-16 (KLJ-X)

Japan: F-2 (APG-1)

Upgrade (in-service or planned):

US: F-15C (APG-63 V3), F-15E (APG-82 V1), F-16E (APG-80), F-16C (APG-83), F/A-18E/F (APG-79), F/A-18C/D (APG-79 V4)

China: J-10C, J-11B, JF-17 Block III (KLJ-7A)

Germany, Italy, Spain, UK: *Typhoon* (Captor-E variants)

France: *Rafale* (RBE2 AESA)

Sweden: *Gripen* E/F (ES-05), *Gripen* C/D (PS-05 AESA)

Situational awareness

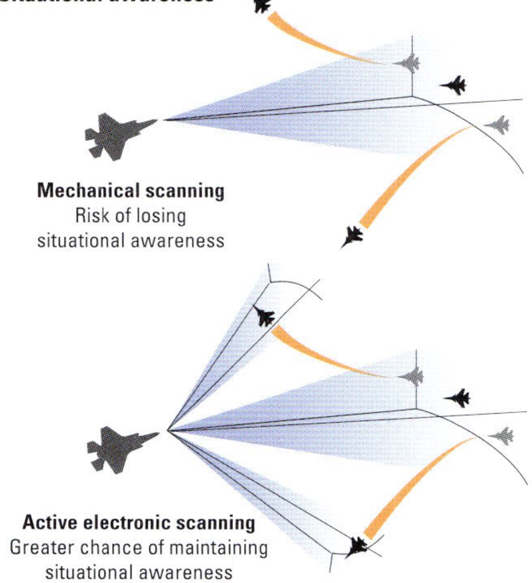

Fields of regard: fixed vs repositioning

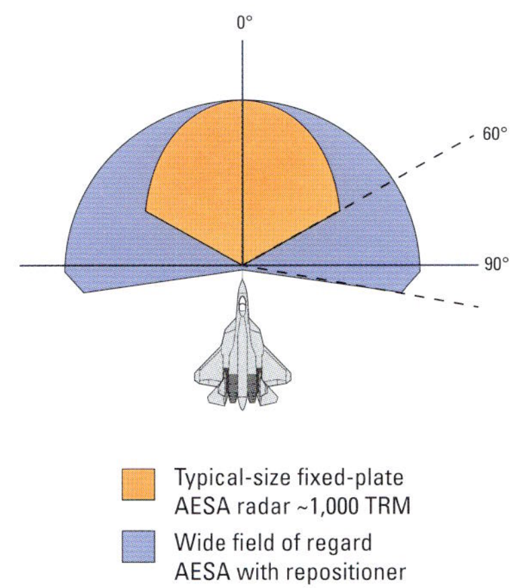

Selected developments in infantry fighting vehicle design

Infantry fighting vehicles (IFVs), alongside main battle tanks (MBTs), form the backbone of many armoured manoeuvre forces. IFVs are designed and equipped to transport an infantry squad into battle and are armed with, at a minimum, a 20mm cannon. This calibre of weaponry, together with their armour protection and the role that these allow the IFV to perform, distinguish these vehicles from armoured personnel carriers (APCs), according to IISS classifications. As well as possessing the firepower to support infantry, effective IFV designs are able to engage lighter armoured fighting vehicles such as APCs. They are also intended to keep pace with accompanying MBTs, having sufficiently large protected volume to transport an infantry squad into battle. However, it has proven challenging, over time, to maintain this balance while keeping the design within acceptable weight and cost limits; indeed, many current IFVs are Cold War designs approaching obsolescence. Together with the anticipated demands of future conflicts, including for better weaponry and protection, these factors have prompted a number of replacement programmes.

United States

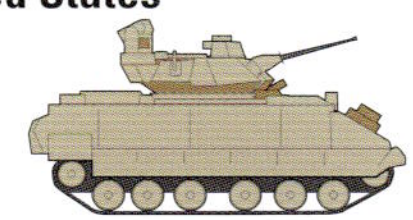

BAE Systems (originally United Defense) M2A3 *Bradley*
- 29 tonnes (32 with additional armour)
- Seven dismounts
- Entry into service: 1981 (basic M2 *Bradley*)
- ■ Stabilised 25mm M242 gun
- ■ Independent commander's thermal sight
- ■ Two ready-to-fire BGM-71 TOW anti-tank guided missiles (ATGMs)
- ■ Aluminium armour with appliqué passive and reactive armour
- ■ 600hp diesel engine

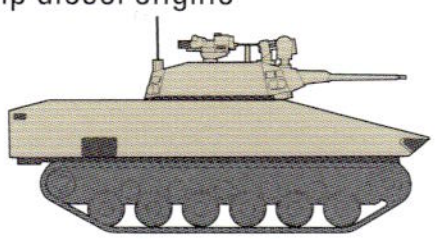

General Dynamics *Griffin* III
- Candidate for Optionally Manned Fighting Vehicle (OMFV) programme
- 36 tonnes
- At least six dismounts
- Intended OMFV entry into service: n.k. (was 2026)
- ■ Stabilised 50mm gun
- ■ 12.7mm remote weapons station
- ■ Independent commander's thermal sight
- ■ Missile
- ■ Steel armour, likely appliqué composite armour
- ■ *Iron Fist* hard-kill active protection system
- ■ 800hp diesel engine

United Kingdom

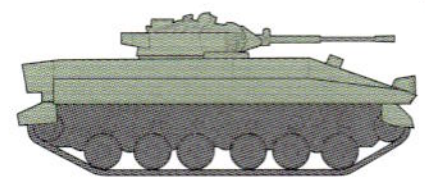

BAE Systems (originally GKN Defence) FV510 *Warrior*
- 24.5 tonnes
- Seven dismounts
- Entry into service: 1984
- ■ Unstabilised 30mm RARDEN gun
- ■ Aluminium armour with appliqué passive and reactive armour
- ■ 550hp diesel engine

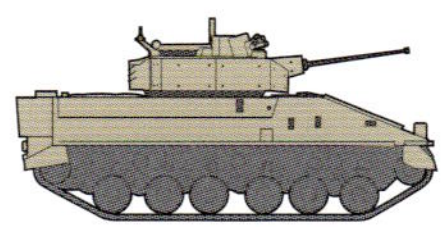

LMUK *Warrior* CSP
- Weight n.k. (est. 26–28 tonnes)
- Upgrade of FV510 *Warrior*
- Seven dismounts
- Intended entry into service: None set
- ■ Stabilised 40mm CTAS gun
- ■ Upgraded fire-control systems
- ■ Aluminium armour with appliqué passive and reactive armour
- ■ 550hp diesel engine

Germany

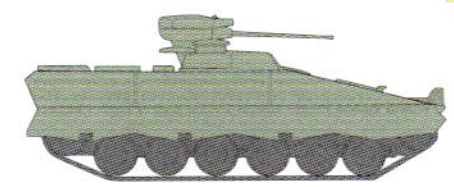

Rheinmetall *Marder* 1A5
- 37 tonnes
- Six dismounts
- Entry into service: 1971 (basic *Marder*)
- ■ Unstabilised 20mm RH202 gun
- ■ One ready-to-fire MILAN ATGM
- ■ Steel armour
- ■ 591hp diesel engine

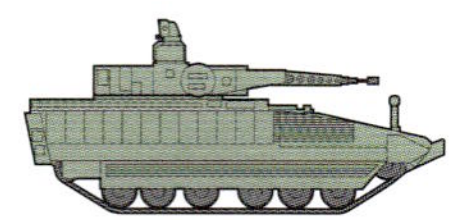

PSM *Puma*
- 31–43 tonnes, depending on configuration
- Six dismounts
- Entry into service: 2015
- ■ Stabilised 30mm MK30-2 gun with airburst capability
- ■ Independent commander's thermal sight
- ■ Two ready-to-fire *Spike* long-range ATGMs
- ■ Steel armour with appliqué composite and explosive reactive armour
- ■ Laser-warning receivers and MUSS soft-kill active protection system
- ■ 1,100hp diesel engine

Russia

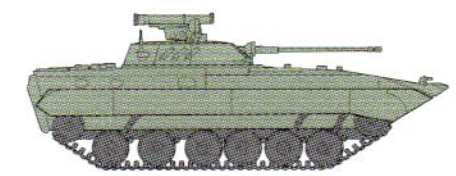

Kurganmashzavod BMP-2
- 14 tonnes
- Seven dismounts
- Entry into service: 1980
- ■ Stabilised 30mm 2A42 gun
- ■ One ready-to-fire 9M113 *Konkurs* ATGM
- ■ Steel armour
- ■ 300hp diesel engine
- ■ Amphibious

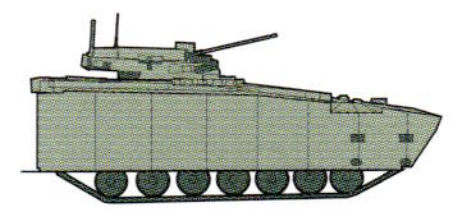

Kurganmashzavod B18/B-19 *Kurganets-25*
- 25 tonnes
- Eight dismounts
- Acceptance trials to be completed 2022
- ■ Stabilised LShO-57 57mm gun with airburst capability
- ■ Independent commander's thermal sight
- ■ Four ready-to-fire 9M113 *Kornet* ATGMs
- ■ Eight ready-to-fire *Bulat* guided missiles
- ■ Aluminium armour with appliqué *Relikt* explosive reactive armour
- ■ *Afghanit* hard-/soft-kill active protection system
- ■ 800hp diesel engine
- ■ Amphibious

■ Firepower ■ Protection ■ Mobility

Naval vertical launch missile systems

Vertical launch systems (VLSs) provide large numbers of ready-use weapons launched from cells in the deck. They are capable of rapid fire and are more reliable and easier to maintain than trainable missile launchers, which they have largely supplanted. By far the most ubiquitous VLS is the United States' Mk 41, which first become operational in 1986. A key operational advantage of systems like the Mk 41 is that they can host different types of missiles without modification, greatly increasing a platform's multi-mission flexibility.

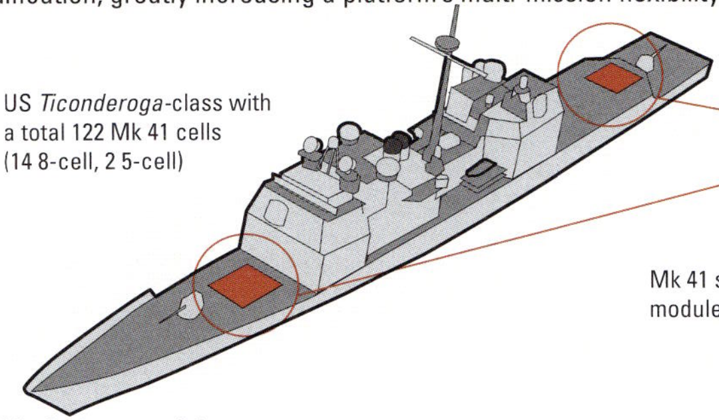

US *Ticonderoga*-class with a total 122 Mk 41 cells (14 8-cell, 2 5-cell)

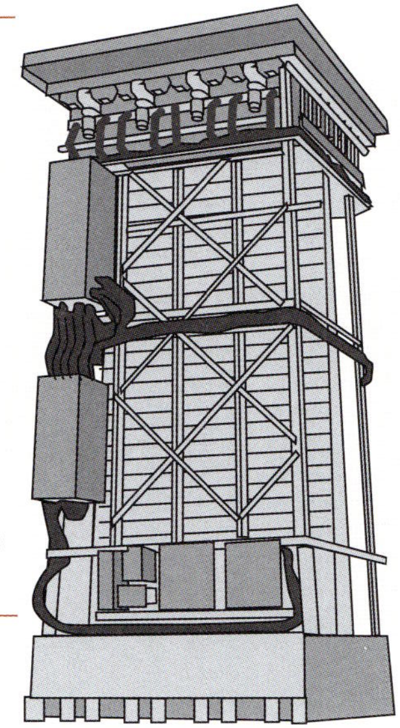

Mk 41 strike length module height 7.6 metres

Multi-mission potential

Today, a key operational advantage of VLSs like the Mk 41 is that they can host different types of missiles without modification, greatly increasing a platform's multi-mission flexibility.

Both the Mk 41 and the European *Sylver* VLS are available in modules with different cell lengths belowdecks. This enables the VLS to carry missiles capable of self-defence, area air defence and ballistic-missile defence, as well as land-attack cruise missiles (LACMs) and, in the Mk 41's case, anti-submarine rockets. Cells within these launchers can house shorter-range air-defence missiles, in quad-packs in order to increase even further versatility/missile numbers.

Potential missile payloads for Mk 41 tactical and strike-length variants

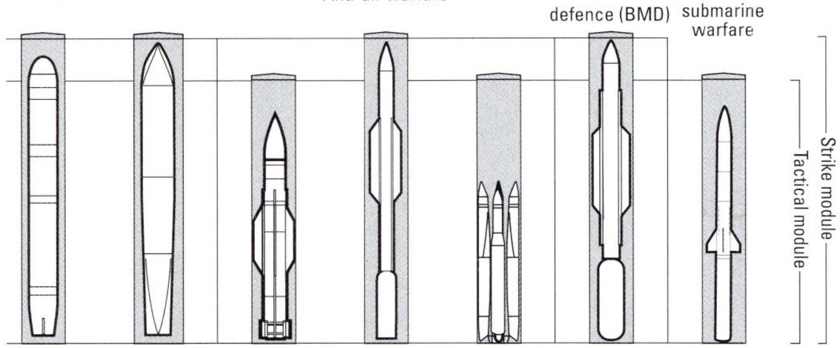

Other VLS systems

- Some VLS systems are designed specifically for only one missile type.
- VLSs for submarines allow a greater volume and variety of weapons to be deployed.
- US *Zumwalt*-class cruisers have Mk 57 peripheral VLSs distributed around hull edge for better battle-damage resistance.
- Early Mk 41 installations had reload cranes able to handle some weapons. Most VLS systems today cannot be reloaded at sea, limiting 'magazine depth' and the locations at which vessels have to reload.

Surface ship multi-mission VLS systems

The US retains an advantage in multi-mission VLS firepower, with its large surface combatants housing between 80 and 122 cells each. China has begun fielding its own multi-mission VLS system similar to the Mk 41 in its most modern surface combatants. Russia's 3S14 UKSK VLS chiefly houses anti-ship missiles and cruise missiles, LACMs, and anti-submarine weapons, with air-defence weapons utilising separate VLS launchers.

	2000		2010		2020	
	Ships	Cells	Ships	Cells	Ships	Cells
Mk 41 (all variants)						
Australia	2	16	12	96	11	208
Canada	4	32	3	24	0	0
Chile	0	0	0	0	2	16
Denmark	0	0	0	0	3	96
Germany	4	64	7	160	7	160
Japan	11	472	21	872	30	1,164
Netherlands	0	0	4	160	4	160
New Zealand	0	0	2	16	2	16
Norway	0	0	5	40	4	32
South Korea	0	0	8	352	9	432
Spain	0	0	4	192	5	240
Thailand	0	0	2	16	3	24
Turkey	0	0	2	16	8	96
United States	76	6,930	81	8,180	90	9,044
***Sylver* family (all variants)**						
Algeria	0	0	0	0	1	8
Egypt	0	0	0	0	1	16
France	1	32	3	128	9	320
Italy	0	0	3	128	11	256
Morocco	0	0	0	0	1	16
Saudi Arabia	0	0	3	48	3	48
Singapore	0	0	6	192	6	192
United Kingdom	0	0	2	96	6	288
K-VLS						
South Korea	0	0	6	96	11	260
3S14 UKSK						
India	0	0	2	16	6	48
Russia	0	0	0	0	17	160
Chinese VLS						
China	0	0	0	0	15	1,008

Key defence statistics

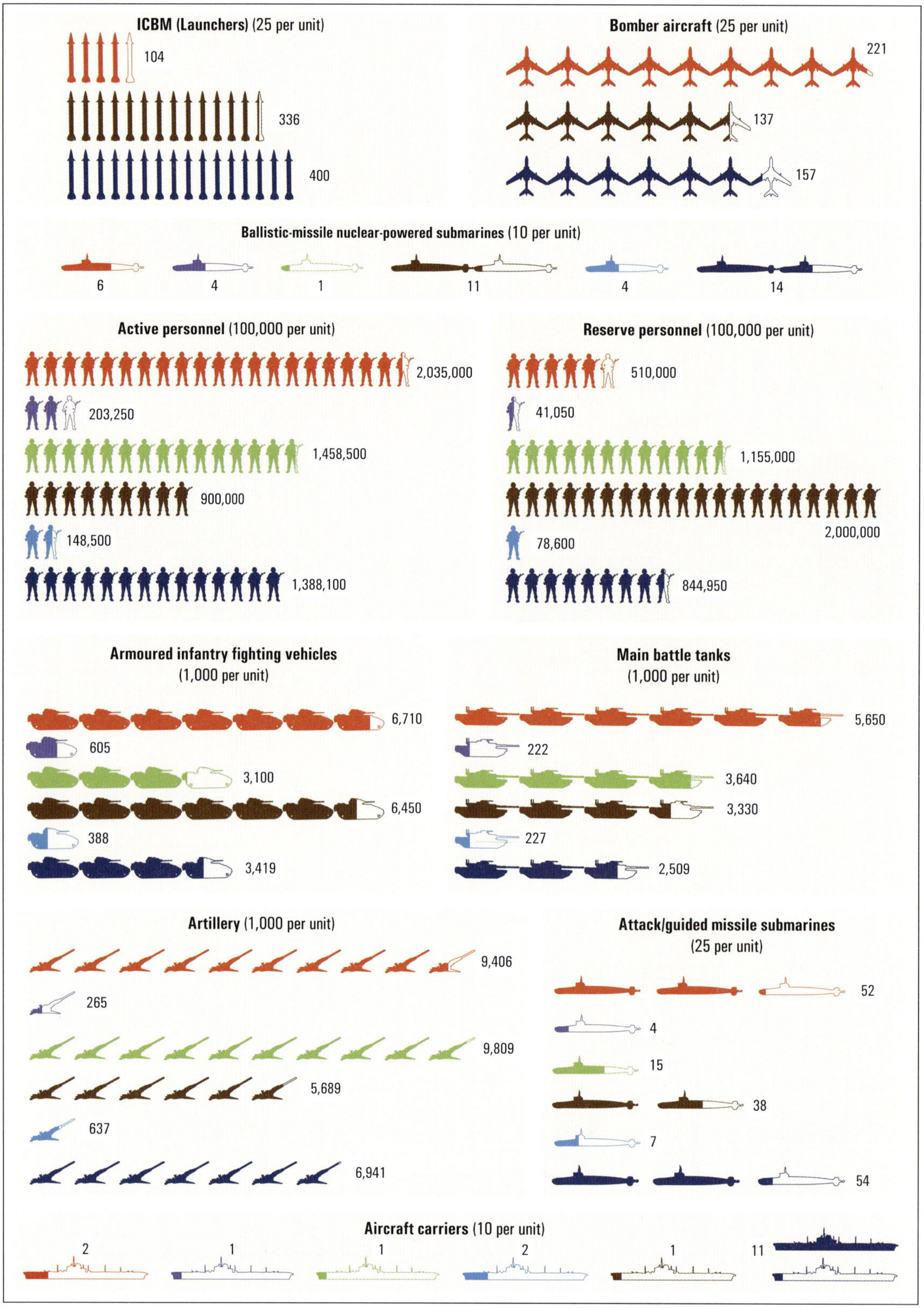

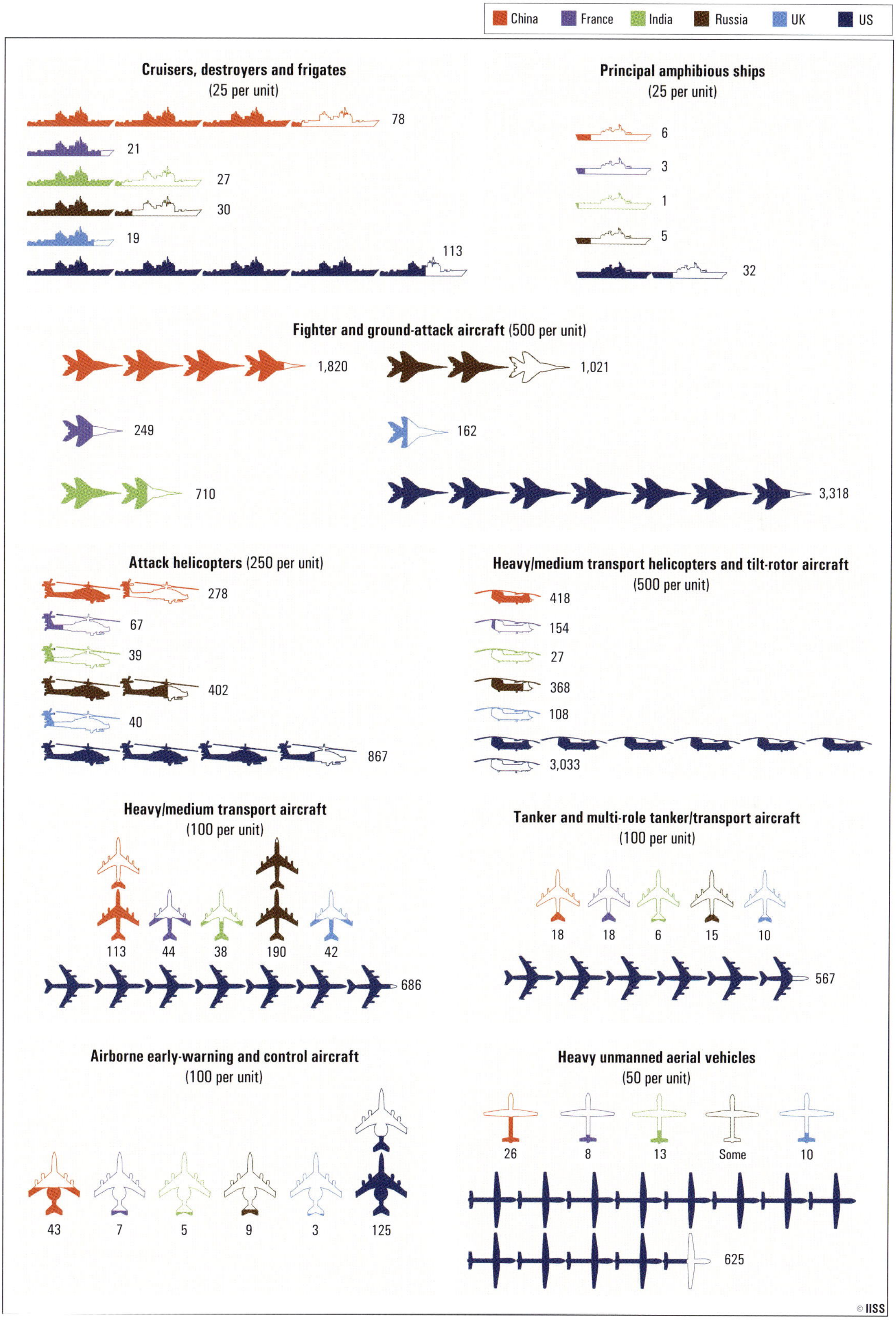
China
France
India
Russia
UK
US
Cruisers, destroyers and frigates
(25 per unit)
78
21
27
30
19
113
Principal amphibious ships
(25 per unit)
6
3
1
5
32
Fighter and ground-attack aircraft (500 per unit)
1,820
1,021
249
162
710
3,318
Attack helicopters (250 per unit)
278
67
39
402
40
867
Heavy/medium transport helicopters and tilt-rotor aircraft
(500 per unit)
418
154
27
368
108
3,033
Heavy/medium transport aircraft
(100 per unit)
113
44
38
190
42
686
Tanker and multi-role tanker/transport aircraft
(100 per unit)
18
18
6
15
10
567
Airborne early-warning and control aircraft
(100 per unit)
43
7
5
9
3
125
Heavy unmanned aerial vehicles
(50 per unit)
26
8
13
Some
10
625
© IISS

Chapter Three

North America

- In late 2020 the US Army announced the selection of variants of both the Standard Missile (SM)-6 and *Tomahawk* land-attack cruise missile as part of its Mid-Range Capability (MRC) concept. Intended to form an element of the army's Long Range Precision Fires portfolio, the MRC will provide the army with a middle option between the planned Precision Strike Missile (PrSM) and the planned Long-Range Hypersonic Weapon. All three of these programmes are expected to begin initial fielding in 2023.
- In March 2020, the US Marine Corps unveiled a new 'Force Design 2030' to address the priorities set out in the 2018 National Security Strategy. Notable changes include a 12,000 reduction in establishment strength, divestment of the Marines' main battle tank fleet and fewer infantry battalions, but increased numbers of rocket artillery and uninhabited aerial vehicle units.
- The first tranche of active-duty personnel officially transferred from the US Air Force to the Space Force in late 2020, coinciding with the service establishing its first field command – Space Operations Command. Additional Air Force personnel will continue to transfer in 2021, with Army and Navy personnel scheduled to follow in fiscal years 2022 and 2023. In March 2020, the Counter Communications System Block 10.2 – the first openly declared offensive counter-space capability for the US Space Force – reached initial operating capability.
- Following the January 2020 Iranian missile attacks on US bases in Iraq, the Pentagon deployed additional *Patriot* surface-to-air-missile batteries to Iraq. Force-protection demands in the Middle East continue to place heavy demands on the US Army's air-defence force. Of the 11 non-forward-deployed *Patriot* battalions in the US Army, more than a third are now deployed to the Middle East, straining reset timelines and leaving comparatively few units available for potential contingency operations elsewhere.
- By the mid-2020 deadline, three bids were submitted for Canada's Future Fighter Capability Project: Boeing's F/A-18E/F *Super Hornet*, Lockheed Martin's F-35A *Lightning* II and Saab's *Gripen* E. Intended to replace the air force's CF-18 multi-role combat aircraft, the project was launched in 2017 following the abandonment of a previous plan to purchase the F-35A. Deliveries are envisaged from 2025. In the interim, deliveries of ex-Australian F/A-18s continue, to help sustain the current CF-18 fleet.

US Marine Corps: Force Design 2030 planned restructuring

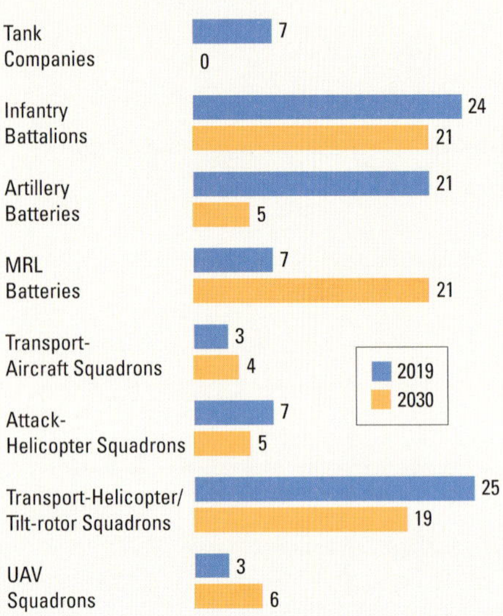

Active military personnel
(25,000 per unit)

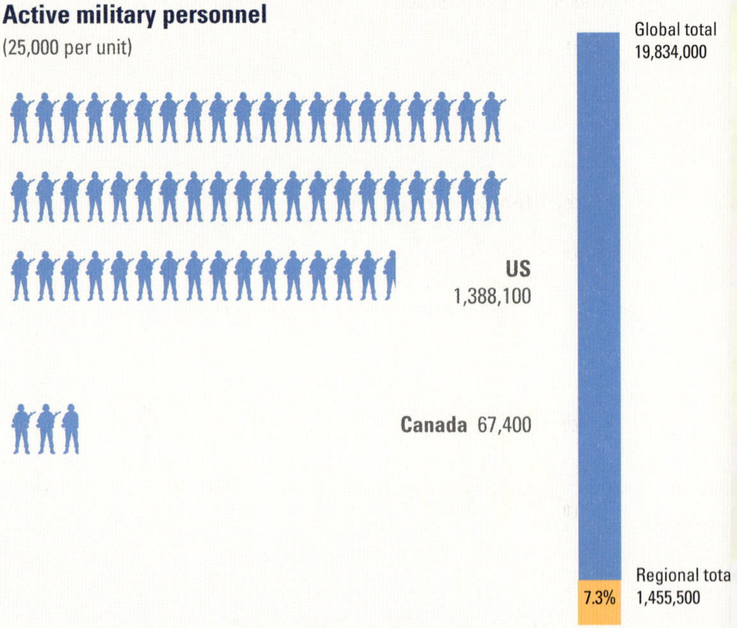

US artillery modernisation: ranges of selected systems

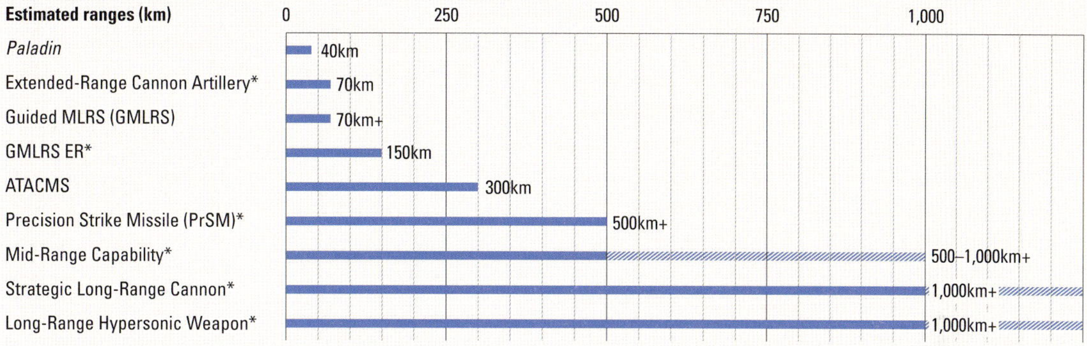

*In development

US Army missile defence: dispositions as of November 2020

	Patriot Battalions	THAAD Batteries
Permanently Forward Deployed	4	2
Rotationally Deployed	4	1
Continental United States	7	4
Total	**15**	**7**

US: Selected militarily-relevant satellites

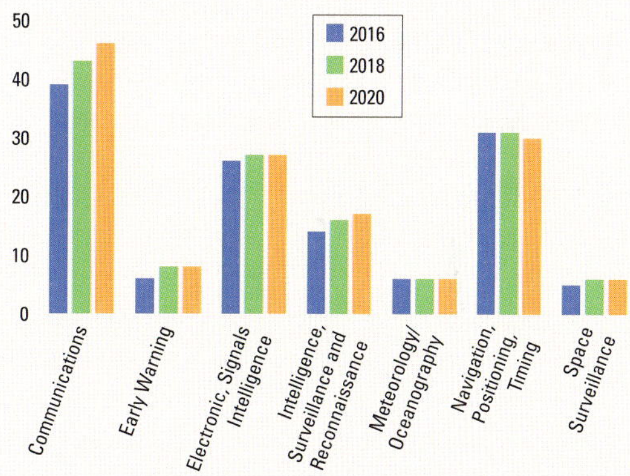

US real-terms defence budget trend, 2010–20

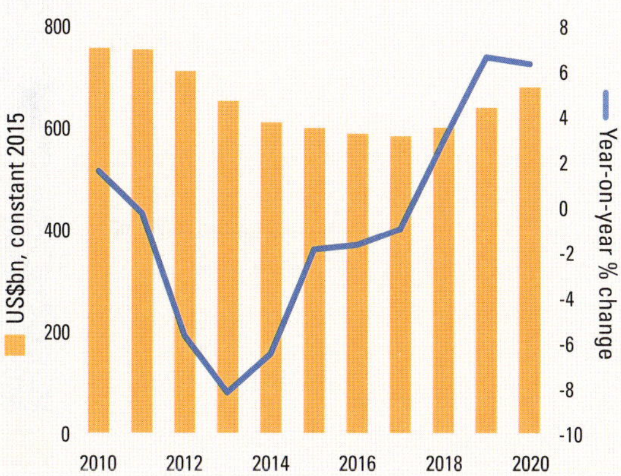

Canada real-terms defence budget trend, 2010–20

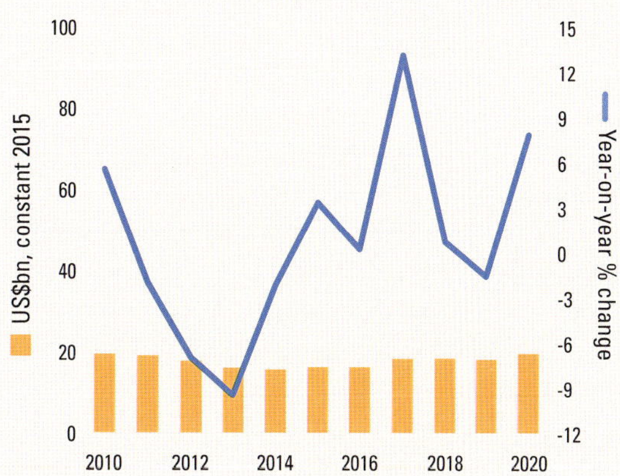

North America

UNITED STATES

Then United States Secretary of Defense Mark Esper stated in July 2020 that his top priority remained implementation of the January 2018 National Defense Strategy (NDS), though the Pentagon has also been occupied in 2020 with efforts to sustain military readiness during the coronavirus pandemic.

The key theme of the 2018 NDS was that great-power competition had returned and that there needed to be renewed emphasis on strengthening US and allied deterrence in relation to nuclear and conventional capabilities and advanced technologies. In that regard, the NDS follows logically from the Trump administration's 2017 National Security Strategy, but was designed as a broader whole-of-government document. The unclassified summary of the NDS said that the 'central challenge to US prosperity and security' was 'long-term, strategic competition by … revisionist powers'. It was increasingly clear, the document continued, that 'China and Russia want to shape a world consistent with their authoritarian model – gaining veto authority over other nations' economic, diplomatic, and security decisions'. This assessment of the global strategic environment informs the NDS's emphasis on lethality, resilience and innovation for high-end combat and deterrence.

Esper identified ten goals in order to address these concerns, including reallocating and redeploying forces in line with the NDS, modernising the armed forces (and investing in 'game-changing' technologies) and developing a 'modern joint warfighting concept, and ultimately, doctrine'. The latter is intended to help the armed forces move to multi- or all-domain operations, as well as 'Joint All Domain Command and Control', though as former US Air Force chief of staff David Goldfein acknowledged in July 2020, these are in many ways just variations on the theme of building a faster, more seamless and less vulnerable means of detecting, identifying and engaging targets across the armed services and, to the extent possible, across Washington's military alliances, or improving overall battle management.

Yet the objective of refocusing US defence postures and capabilities is somewhat inevitably influenced by the competing priorities of the United States' continuing global military deployments and the country's allies and security partners. Indeed, the NDS relates a familiar list of defence priorities that remind observers of the range of tasks the US still seeks to undertake, including 'maintaining favorable regional balances of power in the Indo-Pacific, Europe, the Middle East, and the Western Hemisphere; defending allies from military aggression and bolstering partners against coercion, and fairly sharing responsibilities for common defense'. If anything, this underscores that the pace, and process, of defence transformation remains measured. A 2019 report from the US Center for Strategic and Budgetary Assessments concluded that most major areas of defence spending saw shifts of only 1–2% after the NDS. Moreover, the military services have not changed significantly in size under the NDS, nor have signature weapons programmes been cancelled yet, while spending priorities for major existing programmes have followed largely predictable trajectories. The Pentagon's largest requests in the Fiscal Year (FY) 2021 budget were a familiar list of programmes, some of which pre-date the NDS. Programmes with a request of at least US$1 billion include the F-35 *Lightning* II combat aircraft, the *Columbia*-class submarine, the KC-46A *Pegasus* tanker, the B-21 *Raider* bomber, the next-generation infrared satellite and the next-generation air-dominance programme.

Department of Defense prioritisation

Esper issued two memos in early 2020 and used speeches through the year to continue laying out his vision for how the department should realise the ambitions of the NDS. He advocated 'ruthless prioritisation' of defence expenditure, including divesting of legacy and underperforming assets, and called for reforms to the 'fourth estate' (the non-service or intelligence elements of the Department of Defense, DoD) and a review of combatant commands in order to 'focus on strategic priorities, harvest opportunities to reduce costs, and realign forces/manpower in order to support NDS priorities and rebuild readiness'. Esper's efficiency agenda saw him, together with then air-force chief Goldfein, emphasise the selective retirement of aircraft that were proving expensive and difficult to maintain. However, though Esper has had a history in generating savings when army secretary through his 'night-court' reviews, the fact remains that no major programmes were ended during his tenure as defence secretary – even as flattening US defence-budget projections (including those made by the Trump administration before COVID-19 struck in force) suggest that prioritisation will be more important than ever.

Esper began a review of combatant commands in January 2020. This was due to report in late 2020, in time to inform the next presidential budget requests. There has been some change to the United States' global military footprint, but the US will in the near term nonetheless sustain an overseas military footprint comprising between 60,000 to 80,000 uniformed personnel in each of three main regions: the Middle East, Northeast Asia and Europe, though numbers in the broader Middle East may be slightly lower after personnel reductions in Afghanistan, Iraq and Syria.

While Esper's combatant-command review promised to evaluate force and personnel levels according to the priorities of the NDS – the review reportedly identified possible changes to US Africa Command's posture – there are also other influences. For instance, although Esper said that the Pentagon would reposition to other locations some 12,000 personnel stationed in Germany and that these changes were in line with NDS priorities, when asked the same day, President Donald Trump said that Germany was 'delinquent' in its defence spending. Meanwhile, the US demanded at least a fivefold increase in the roughly US$1bn per year that South Korea has been paying in host-nation support for US forces on its territory. The issue remained unresolved at the time of writing, with Seoul proposing a more modest funding increase. Similar disagreements continue with other NATO allies, even as President Trump claimed credit for inducing them to spend more on defence since 2016, notwithstanding that US exhortations for Europeans to spend more on defence pre-dated his presidency, and that defence spending began to increase before Trump assumed office as threat perceptions changed in Europe after Russia's annexation of Crimea in 2014.

Enduring outcomes?

By 2020, increased defence budgets and continued attention by defence leaders had contributed to improved military readiness. Though there is, in general, only limited data released on this issue, it was reported in budget documentation that many aircraft 'mission capable rates' were back up to the goal of about 80%, after often being in the 50–60% range. Despite reductions in large-scale exercises and a few specific lockdowns due to the coronavirus pandemic, such as with the USS *Theodore Roosevelt* aircraft carrier in spring 2020 and US Marine Corps units on Okinawa over the summer, readiness indicators appear to have remained generally acceptable.

Even if NDS implementation slows or is further modified over the coming years, some of its accomplishments will likely endure. Indeed, a measure of Esper's hopes in this regard came in a July 2020 speech when, after setting out his modernisation goals, he closed by saying that 'our collective efforts will ensure the irreversible implementation of the NDS'.

The need to reinvigorate US military capabilities for peer and near-peer warfare pre-dated Trump and Esper. It had its roots in the so-called 'rebalance' to the Asia-Pacific that began late in President Barack Obama's first term, the so-called 'Third Offset' (by which the US sought to assure military advantage by pursuing new technologies) and the development of the Pentagon's '4+1' threat framework (with China and Russia joining the previous priorities of Iran, North Korea and transnational extremism) in Obama's second term. These continuities permeate other aspects of Trump administration policy, such as sending two aircraft-carrier battle-groups at once to the South China Sea in summer 2020. As such, it is possible that strategies such as this could also persist under the incoming Biden administration. War plans for dealing with military contingencies involving China and Russia are now being more regularly reviewed, which will likely continue even if particular formulations from the Trump years, such as emphasis on succeeding in the 'contact' and 'blunt' phases of a future conflict, may be re-appraised.

The same is true for technological innovation and investment. The Pentagon planned a US$3.2bn aggregate budget for hypersonic weapons in the FY2021 budget and intends to start fielding these weapons by 2023. In addition, US$1.5bn was allocated for microelectronics and 5G networks; US$1.7bn for autonomous vehicles; and US$0.8bn for artificial intelligence (AI). The United States' Joint Artificial Intelligence Center is now operational, and the DoD released an AI strategy in 2019.

It was reported in early 2020 that a new low-yield nuclear warhead had been fielded. Later confirmed by then-Under Secretary of Defense for Policy John Rood, the development of the W76-2 low-yield submarine-launched ballistic missile (SLBM) can be seen in the context of a statement in the 2018 Nuclear Posture Review, intended to dissuade potential adversaries that 'believe that employment of low-yield nuclear weapons will give them an advantage' over the US. Meanwhile, two conventionally armed intermediate-range missiles have been tested in the wake of Washington's decision to withdraw from the Intermediate-range Nuclear Forces Treaty.

The Trump administration had at the time of writing avoided committing to a renewal of the New START (Strategic Arms Reduction Treaty), meaning that accord would expire in early February 2021. However, a joint announcement with Moscow in October 2020 suggested an extension for a year as negotiations proceeded on a possible new treaty. The administration also emphasised the potential of space-based missile-defence systems in its 2019 Missile Defense Review, returning to an idea that was widely seen as provocative and destabilising when first articulated in the 1980s and which, analysts considered, would likely provoke a response by China and/or Russia if such systems were deployed by the US.

In 2019, the DoD created Space Command and then Space Force, the first new US military service since 1947. Though it currently remains small in terms of personnel numbers, and will likely remain so in relation to the other services, its personnel strength will increase when specialist personnel transfer from the air force from late 2020 and other 'common career' personnel join from early 2021.

Though scrutiny was expected of the DoD's programmes and plans during Trump's term, particularly in 2020 the complex subject of civil–military relations raised challenges that were perhaps unforeseen. Then defence secretary Esper talked publicly about the need to dominate the country's urban 'battlespace' during civil unrest in spring 2020, while Chairman of the Joint Chiefs of Staff General Mark Milley wore combat uniform when he appeared with the president at St John's Episcopal Church in Washington DC. Both episodes drew criticism from Republicans as well as Democrats and civilians as well as retired military leaders. Later, both leaders opposed invoking the 1807 Insurrection Act to justify the use of active-duty troops to quell protests in spring 2020. General Milley later apologised, saying that his presence with Trump outside the church in Washington DC 'created a perception of the military involved in domestic politics', while Esper wrote in his July 2020 review of his first year in office that the armed forces 'remain apolitical' in carrying out their mission. Nevertheless, though US defence policy may be under duress, the structural and physical dimensions of the armed forces and their global operations did not change significantly.

How much change under Biden?

In November 2020, Christopher Miller took over as defence secretary. Miller's tenure was to be short, after Joe Biden's victory in the November 2020 presidential election. The precise direction of the new administration's defence policy remained unclear at the time of writing, but while policy reviews may begin in a number of areas, it is unlikely that there will be significant change to the broad objectives pursued in recent years of reshaping the armed forces for great-power competition, improving readiness and accelerating the generation of new capabilities.

US Army

The US Army is continuing the process of adaptation set out in the late 2018 US Army Strategy and echoing that year's National Military Strategy, which highlighted the re-emergence of great-power competition and the need to deter or defeat conventional attacks. As a result, the army is shifting its focus to preparing for high-intensity conventional combat operations against peer and near-peer opponents. However, it has to balance this requirement against the need to retain the counter-insurgency skill set developed in the recent wars in Afghanistan and Iraq.

Structural changes

The army is introducing new types of unit, such as the brigade-sized Multi-Domain Task Force (MDTF). The MDTFs will have the capability to apply traditional fires, such as artillery, as well as managing effects such as cyber operations. This is in keeping with the army's interest in multi-domain operations, including lethal and non-lethal elements. The army is framing the MDTFs, along with related capability developments such as long-range fires, under its AimPoint initiative. The director of Army Futures Command said in April 2020 that AimPoint was 'resource informed' and

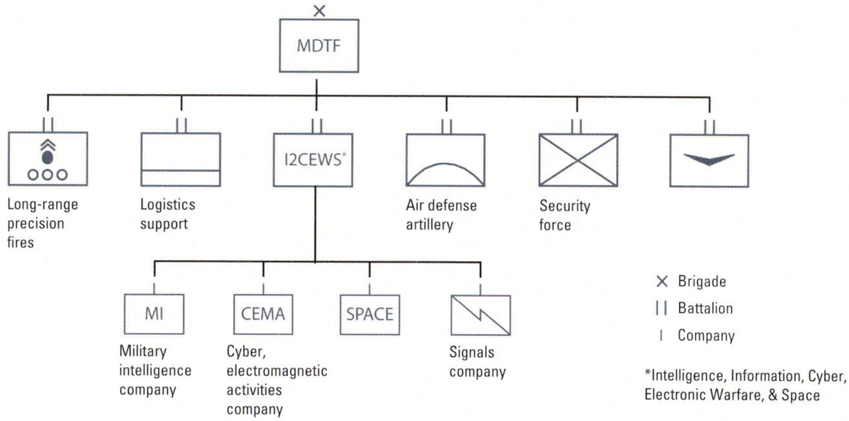

Source: US Army

© IISS

▲ Figure 1 **Multi-Domain Task Force (MDTF)**

was 'an architect's design for a force structure in the future'. Forces and capabilities may both be tailored to meet regional requirements. The concept is designed to enable the generation of higher-echelon capabilities and formations that had been 'mortgaged' during recent conflicts because the army's Brigade Combat Teams (BCTs) 'were so powerful relative to our opponent'. AimPoint informed the reactivation of V Corps headquarters; this was announced in February 2020. While the main body will remain in Fort Knox, Kentucky, a forward HQ will be set up in Poland. This is intended to provide command and control at the level required to coordinate US, allied and partner-nation tactical formations. Notwithstanding this refocus on high-intensity war fighting, the army now has six Security Force Assistance Brigades (SFABs), after the 5th SFAB was activated in May 2020 (there is one SFAB in the National Guard). Like special-forces units, the SFABs are aligned to particular global regions, where they specialise in working with foreign armed forces in order to improve local capabilities, including in irregular warfare.

The reactivation of V Corps was not the only significant structural change in 2020. The primary combat units in the regular force structure are ten division headquarters and 31 BCTs, but the refocus on high-intensity conventional combat operations led the army to convert two of its Infantry Brigade Combat Teams into armoured formations. More field-artillery units are also being added to the regular army structure, such as the 1/77 Field Artillery Regiment, which was activated in Germany in August 2020. This unit is the second Multiple-Launch Rocket System-equipped regiment of the Grafenwohr-based 41st Field Artillery Brigade. Under AimPoint, the army is also looking to generate Theatre Fires Commands that will use advanced technologies to integrate a range of relevant long- and extended-range artillery capabilities. In line with the Pentagon's review of US European Command, which the DoD said was accelerated after President Trump's decision to reduce the United States' presence in Germany, other unit movements included the relocation to the US, from Germany, of most of the 2nd Cavalry Regiment.

However, the regular force is only one element of the US Army's strength. Indeed, the reserve component, consisting of the federally controlled Army Reserve and the state-focused Army National Guard, outnumbers the regular force. The Army Reserve contains a considerable portion of the service's support capabilities, such as logistics, healthcare and military police. The Army National Guard – with 27 BCTs and eight divisional headquarters – focuses primarily on combat capabilities and has roughly 336,000 personnel spread among the 50 US states and various territories. Together, the army's reserve components total some 525,000 personnel and represent a significant segment of the US Army's overall capability.

Force modernisation

The army's modernisation strategy over the next decade is focused on six key areas.

- **Long Range Precision Fires** (LRPF) is a top priority, with the army recognising that its artillery capabilities have atrophied during recent wars. New systems such as the

Precision Strike Missile (PrSM) are intended to replace older weapons and provide greater range and effectiveness. The army is also exploring the possibility of using some of its new fires capability, including PrSM, to engage maritime targets. It was reported in late 2020 that a Mid-Range Capability concept, with variants of both the SM-6 and *Tomahawk* cruise missile, would be developed to sit between LRPF and hypersonic systems. (A similar capability that would have been made possible through an upgrade to the ATACMS surface-to-surface missile system was reportedly delayed.)

- **The Next Generation Combat Vehicle** (NGCV) programme includes several systems, such as the Optionally Manned Fighting Vehicle, robotic vehicles and, eventually, an *Abrams* replacement. While both the *Abrams* main battle tank (MBT) and *Bradley* armoured infantry fighting vehicle have benefited from numerous upgrades since they entered service, they are in essence 40-year-old designs.
- The **Future Vertical Lift** programme includes future attack and reconnaissance (FARA) and future long-range assault aircraft (FLRAA), as well as uninhabited versions. The army indicated in 2020 that the first aircraft under the FARA project could be fielded in 2028 and the first FLRAA in 2030.
- The army is also pursuing improved **networking** in the form of an integrated hardware and software system intended to enable high-tempo multi-domain operations and improve resilience against cyber and electronic-warfare threats.
- The **Air and Missile Defence** programme has a focus on protecting manoeuvre formations threatened by air platforms or missiles. Testing on short-range air-defence systems and command-and-control technologies was due to take place in mid-2020 but was reportedly delayed for a short time in order to improve COVID-19 protection for personnel.
- Meanwhile, **soldier lethality** will focus on improving the weapons, survivability and communications capability of small units and individual soldiers.

These modernisation priorities are gaining additional relevance as the army examines more closely capabilities that could contribute to deterrence in the western Pacific. While the plan to develop an anti-ship capability for some artillery formations is of relevance to this theatre, the army still faces considerable challenges in the Pacific, not least that of distance. Currently there are few units permanently based west of Hawaii or south of Korea and Japan. In the near term, the Pentagon is exploring new basing and prepositioning options, alongside the development of new weapons considered appropriate for the geographic scale of the western Pacific.

The structural change planned for the US Marine Corps is another consequence of the US armed forces' increased focus on the Pacific. This will have implications for the army. In early 2020, the USMC announced that it would divest some capabilities in order to refocus resources toward the types of operations envisioned taking place in the Pacific. This included eliminating all tank units. While the long-term implications of this change are still being determined, it could result in an increased requirement for the army and marines to plan and operate together. If, for example, future marine operations require the support of heavy armour, the army might be the only US source of MBTs.

Another evolving issue, and one that will certainly influence army budgets, is the effect of the coronavirus pandemic. There has already been some impact on programme development and military exercises, but the financial ramifications for defence spending have yet to become clear. Going into FY2021, the army has requested a budget of US$178bn within the DoD budget of US$740.5bn, roughly 24% of overall DoD spending.

US naval forces

For much of 2020, there was uncertainty over the trajectory of the United States' naval policy and force-structure development, as well as long-term shipbuilding plans. Indeed, it is possible that the navy is now at a critical juncture over both fleet composition and how it incorporates new technologies – particularly uninhabited and autonomous systems – in order to meet the challenges of great-power competition, particularly from China.

Future force structure

The much-anticipated Integrated Naval Force Structure Assessment (INFSA) was first scheduled for late 2019 but had still not been issued by October 2020. INFSA had been meant to generate a more comprehensive set of long-term goals than its predecessor assessments, underscoring greater inter-service integration, particularly between the US Navy and the USMC. The principal issue for many observers was whether and how it might modify the previous Force Structure Assessment goal, which was unveiled in December 2016 for a 355-ship naval force.

Amid signs of friction within the Pentagon over the direction and possible cost of INFSA, a Future Naval Force Study set up by defence secretary Esper increasingly became the focus of efforts in 2020. What Esper unveiled in October was a plan for 355 ships by 2035 and 500 – including significant numbers of uninhabited platforms fulfilling a range of tasks – by 2045.

In broad terms, the plan called for accelerated procurement of a larger number of nuclear-powered attack submarines; a force of 8–11 full-size aircraft carriers supplemented by up to six light carriers; more smaller surface combatants for greater flexibility in fulfilling different missions; more amphibious ships, but again including smaller ones; the incorporation of those uninhabited platforms; and more logistics ships to support a more diverse and dispersed fleet. The cost of it all will supposedly be met in large part by the navy finding savings elsewhere.

Some of the variables in this vision – not least over future aircraft-carrier numbers – mean questions and challenges in the US Congress are almost inevitable, not least on affordability. Budgetary constraints were already forcing new approaches to procurement. In January 2019, the navy announced a two-carrier block order for the third and fourth *Ford*-class carriers, an approach it said would reduce the overall bill for the ships by US$4bn, to a total of US$24bn. However, the post-delivery tests and trials of the lead ship, USS *Gerald R. Ford*, have been beset by technical problems, mainly on the ship's weapons elevators, and it is unlikely that the carrier will be ready for deployment before 2023. Meanwhile, after it emerged from maintenance in August 2020, the USS *Carl Vinson* was scheduled to be the first carrier due to take the F-35C variant of the *Lightning* II to sea on deployment in 2021.

At the same time, the navy has embarked on the expensive *Columbia* programme to replace the current *Ohio*-class of nuclear-powered ballistic-missile submarines. This programme is set to consume a significant proportion of the shipbuilding budget and complicates further the navy's force planning and particularly any ambition to accelerate construction of other submarines.

At the end of 2019, the navy placed an order for nine new *Virginia*-class nuclear-powered guided-missile submarines (SSGNs) with an option on a tenth. This order, a slightly lower number than some observers had anticipated, included eight Block V boats with the Virginia Payload Module containing additional launch tubes for up to 28 *Tomahawk* land-attack cruise missiles. This configuration would in part compensate for the anticipated retirement later this decade of the *Ohio*-class SSGNs.

Another milestone was the navy's selection at the end of April 2020 of Fincantieri and Marinette Marine to design and build its new-generation frigate, the FFG(X), which is based on a modification of the Italian variant of the Franco-Italian multi-mission FREMM frigate design. The assumption has been that a class of 20 FFG(X)s would be built, although the new force-structure proposals suggest that greater numbers might be procured of what was announced as the *Constellation*-class, or another design.

Meanwhile, US naval forces were also affected by the coronavirus pandemic. There was much international focus on the sidelining in late March of the carrier USS *Theodore Roosevelt* in Guam, due to an outbreak of COVID-19 aboard. The handling of the incident, including the removal of the commanding officer and the subsequent related resignation of the then acting secretary of the navy, led to much scrutiny. Nonetheless, the navy was able to maintain a significant presence by sustaining other ships and groups at sea, though the long-term effect of this on readiness remains to be seen. High-profile deployments into the South China Sea, the Barents Sea and the Black Sea in 2020 continued to emphasise the US armed forces' focus on great-power competition.

Amphibious capability

In July 2020, the second *America*-class large amphibious ship with enhanced aviation facilities, the USS *Tripoli*, was commissioned. However, the devastating fire aboard the USS *Bonhomme Richard* while in the later stages of a maintenance period will complicate the navy's deployment plans for these vessels. The *Bonhomme Richard* was being upgraded to support the operation of the F-35B short-take-off and vertical-landing version of the *Lightning* II.

As a sign of the anticipated evolution of requirements, the navy was also pressing ahead with plans for a new light amphibious warship to supplement its current fleet and in particular support the USMC's new concept of more distributed forward operations, which are being developed to face the challenge of China in the western Pacific. The corps also revealed in March 2020 in its Force Design 2030 proposals its further thinking on transforming to meet these aims. These included reducing the overall size of the force by 12,000 personnel, divesting itself of MBTs and potentially reducing the number of deployed F-35Bs in order to free up resources for new capability investments, not least more uninhabited systems and rocket artillery.

Coast Guard

The US Coast Guard's (USCG) plans to recapitalise its icebreaker capabilities came under the spotlight after the White House issued a statement in June 2020 calling for a review. This was a year after a contract was awarded for the first of a new class of heavy icebreakers or Polar Security Cutters (PSCs), part of a programme for a fleet of new heavy and medium icebreakers.

The document called for a review of the number and mix of future vessels. Options include nuclear propulsion and the possible fitting of defensive armaments, as well as a study of new base locations, both domestic and international, to support the new force.

The USCG operates just one heavy icebreaker, USCGC *Polar Star*, which is already more than 40 years old, and a medium icebreaker, USCGC *Healy*, which had to abandon an Arctic mission in August 2020 after a fire in its propulsion system. Currently, construction of the first new PSC is scheduled to start in 2021 for delivery in 2024.

US Air Force

The United States Air Force (USAF) continued in 2020 to try and balance immediate needs and demands with future requirements. The service is refocusing on meeting a near-peer threat and recapitalising its aircraft and weapons inventories. At the same time, it is managing ageing aircraft and an acquisition rate of replacement types that is lower than ideal. Compounding these challenges is a shortfall in combat aircrew.

Going faster

In August 2020, General Clarence Brown, the new USAF chief of staff, published a paper called 'Accelerate Change or Lose' that laid out the demands of the emerging security environment for the air domain. Brown flagged as 'historically anomalous' the nearly three decades of uncontested dominance the service has enjoyed since *Operation Desert Storm*. Brown cautioned that any air campaign against a peer or near-peer rival would likely be 'highly contested' and that there would be greater combat losses than in any recent experience.

The air force once aspired to meet such a challenge by fielding an all 'fifth generation' force of tactical fighter aircraft. However, it is now having to accept that the core of its inventory out to the late 2030s will consist of mixed fleets of fourth-generation types, such as the Boeing F-15 *Eagle*, and fifth-generation aircraft such as the Lockheed Martin F-22 *Raptor* and F-35 *Lightning* II. Moreover, it is possible that the crewed element of the Next-Generation Air Dominance (NGAD) programme could begin to enter service well within this time frame. In this eventuality, the air force would have to manage three 'generations' of aircraft with associated support, logistics and inter-operability demands. Indeed, in September 2020 the air force said it had flown a previously classified demonstrator aircraft as part of the NGAD programme.

Two months earlier, funding was allocated for the first F-15 acquisition since 2004. The F-15EX is being purchased to replace the F-15C in the homeland air-defence role in order to avoid giving the F-35A this task. At least part of the rationale for this investment is that an upgraded F-15 will provide enough capability to meet many operational needs, although low-observable platforms would be required for the demands of peer-on-peer combat. This new-build 'legacy' purchase is effectively being used in order to sustain combat mass. It is also possible that the air force's thinking on the F-15 was influenced by the opportunity to take advantage of upgrades funded by export customers combined with assessments of F-35 maintenance costs and acquisition rates.

Going further

The requirements of potential combat in the Indo-Pacific theatre and China's emergence as a peer competitor are guiding the USAF's approach to high-end air warfare. These factors are also steering the design and performance parameters for the

NGAD air vehicle, not least the goal of an extended operational radius and increased combat persistence compared to the F-35A. These requirements are also likely to drive platform size so that the maximum amount of fuel and weapons can be carried within the design's required performance parameters.

Having opted to supplement its combat-aircraft fleet and replace the F-15C with the F-15EX rather than with the F-35A, the air force is now beginning to consider replacement options for remaining early-model (Block 30 standard and below) F-16 *Fighting Falcons*. While it was previously assumed that the F-35A would be the sole replacement for these aircraft, a question mark was placed against this in April 2020 by General Mike Holmes, commander of Air Combat Command, who suggested that a final decision had yet to be taken.

A vital but troubled combat enabler for the Indo-Pacific is the Boeing KC-46A *Pegasus* air-to-air refuelling aircraft. As of September 2020, a total of 38 aircraft had been delivered to the air force at five air bases. So far, McConnell air base has the largest fleet with 21 KC-46s. However, the air force is still at least three years from introducing the type into front-line operational service. There have been quality-control issues with production aircraft. There have also been problems with the remote operator's station, though the air force and the manufacturer agreed in April 2020 a way ahead, with a redesigned 'remote vision system' to be ready by 2023. It is understood that the air force will not approve full-rate KC-46 production until it is satisfied with the performance of this system.

Although the air force may have revived the production of a fourth-generation fighter design in the form of the F-15EX, it is elsewhere looking to reduce the number of ageing aircraft in the inventory. In the case of the tanker fleet, its FY2021 funding submission proposed cutting 13 KC-135s and 16 KC-10s, irrespective of problems with the KC-46A, while also trimming 17 B-1Bs from the bomber fleet. However, the tanker proposal encountered trouble during the mark-up stage with the House Armed Services Committee. Although the KC-10 proposal was allowed through, the plan to cut KC-135 numbers was blocked, though the air force would be allowed to retire specific KC-135 airframes if damage meant they were beyond economical repair.

Although the prototype B-21 *Raider* bomber is now anticipated to be flown for the first time in 2022, the move to reduce the B-1B *Lancer* fleet is intended to allow the air force to focus funding on improving the readiness of a reduced pool of aircraft. At one point during 2019, just under 25% of the B-1B fleet was deemed to be fully mission-capable.

Meanwhile, the air force is also moving to reshape its intelligence, surveillance and reconnaissance inventory to better address the demands of highly contested airspace. Again, however, the FY2021 funding process did not proceed smoothly, and the House Armed Services Committee placed caveats on plans to divest two-thirds of the Northrop Grumman RQ-4 *Global Hawk* uninhabited aerial vehicle fleet. The air force also decided to stop buying the General Atomics MQ-9 *Reaper* and instead is looking to identify a successor more capable of operating in heavily defended air space.

DEFENCE ECONOMICS

FY2021 budget request: holding steady

The DoD FY 2021 Budget Proposal, submitted on 10 February 2020, requested US$740.5 billion for defence, effectively holding the budget in real terms level with the FY2020 allocation of US$738bn. The core DoD budget amounted to US$705.4bn out of the FY2021 total, with the remainder comprising 'defense-related' expenditure and discretionary spending for the Department of Energy's (DoE's) management of the national nuclear stockpile.

The US Senate passed its version of the FY2021 National Defense Authorization Act (NDAA) on 23 July in a bipartisan 86–14 vote, two days after the House of Representatives passed its version by a similarly overwhelming margin of 295–125. Minor

The Bipartisan Budget Act of 2019 and closing the OCO loophole

The parameters of the Fiscal Year (FY) 2020 and FY2021 budgets were set by the August 2019 Bipartisan Budget Act. The legislation raised discretionary-spending caps for the last two years of the Budget Control Act (BCA) from US$576.2 billion and US$590.2bn to US$666.5bn and US$671.5bn respectively. To accommodate these increases, the legislation also reduced Overseas Contingency Operations (OCO) funding from the requested amounts of US$165bn and US$155.8bn in FY2020 and FY2021 to US$71.5bn and US$69bn respectively. This move essentially eliminated the controversial practice of making conspicuous requests for funding or base-budget activities in the OCO line item to circumvent the previously lower BCA caps.

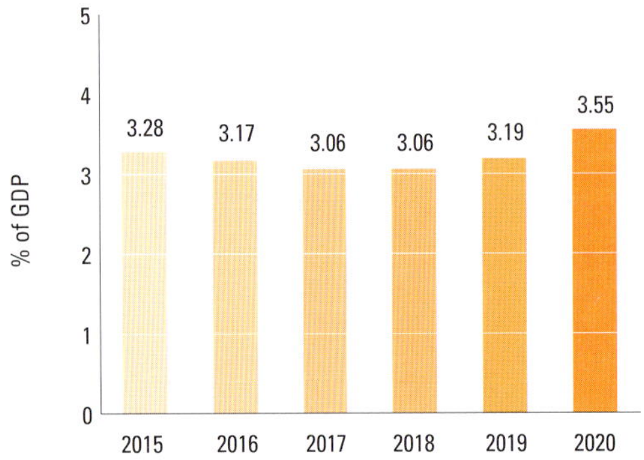

[1] Figures refer to the National Defense (050) Budget Function (Outlays) as a % of GDP

▲ Figure 2 **US defence budget** as % of GDP[1]

differences in the bills still need to be reconciled at the time of writing, but the House and Senate versions agree on the main points and are largely in line with the DoD FY2021 budget request.

Given the FY2020 core DoD budget amounted to US$704.6bn, the FY2021 allocation represents a negligible 0.1% nominal increase. The flat budget was cited as a driver for the DoD to make what it termed 'numerous hard choices' in order to fund the priorities outlined in the 2018 National Defense Strategy (NDS). With spending power effectively lower (notwithstanding that an exceptionally low inflation rate of 0.7% was expected in 2020), there was further justification for then Secretary of Defense Mark Esper's Defense-Wide Review, which generated almost US$5.7bn in savings in the FY2021 budget, and wider efficiency measures generating a further US$2.3bn in cost reductions. The money saved was directed towards the NDS priorities, including hypersonic weapons, artificial intelligence, fifth-generation communications technologies, nuclear modernisation, space, missile defence and improving combat readiness.

While the defence budget's flat top-line growth attracted headlines, especially after three years of growth and investment in force readiness, the main story of the FY2020 and, especially, FY2021 budgets is not about the top line. Rather, it is about how individually and collectively these budgets reflect the trade-offs and tensions facing the DoD as it seeks to meet near-term challenges while transforming its forces, organisation, processes and culture to better position itself for accelerating great-power competition.

The FY2020 budget was not fully enacted until 81 days into the fiscal year due to protracted negotiations between Congress and President Donald Trump over the use of DoD funds for construction of the border wall. Similarly, in September 2020, Representative Mac Thornberry, the senior Republican on the House Armed Services Committee, acknowledged that enactment of the 2021 NDAA will be delayed until after the 3 November 2020 general election due to objections from the Trump administration to provisions requiring the DoD to change the names of military bases currently named for Confederate generals.

FY2021 budget: 'ruthless prioritisation' and the National Defense Strategy

The main objective of the FY2020 and FY2021 budgets is to ensure the 'irreversible implementation' of the 2018 NDS, which seeks to shift the DoD's strategic and operational focus in the direction of great-power competition with Russia and especially China. There is emphasis on the development, diffusion and multi-domain deployment of the advanced technologies that are changing the character of war.

The document advocates investing in capabilities that will ensure sustained US competitive advantage in key capability areas. The NDS also stresses the need for improved talent recruitment, retention and training, and stronger alliances and partnerships.

In many ways the FY2021 budget advances these initiatives. For example, both the House and Senate versions of the budget include a 3% increase in pay and variable increases in housing allowances for service personnel. The FY2021 NDAA also funds the European Deterrence Initiative as part of Overseas Contingency Operations funding and establishes the Pacific Deterrence Initiative, both of which reinforce Washington's alliance networks, which are designed to deter, dissuade and contain its two principal strategic competitors.

Allocations for research, development, testing and evaluation (RDT&E), the budget category most associated with the development of new technologies and novel capabilities, were increased. Not only is the FY2021 proposed RDT&E budget of US$106.6bn the largest RDT&E budget ever, it is also US$5.8bn more than the amount forecast in the FY2020 Future Years Defense Program. Hypersonic capabilities, micro-electronics, autonomous systems and artificial intelligence are singled out as being particularly important to the Pentagon's efforts to establish

Table 1 **US National Defense Budget Function and other selected budgets, 2000, 2010–21**

US$ in billions, current-year dollars	National Defense Budget Function		Atomic Energy Defense Activities	Other Defense Activities	Total National Defense			Department of Homeland Security	Department of Veterans' Affairs	Total Federal Government Outlays	Total Federal Budget Surplus/ Deficit
						Discretionary					
FY	BA	Outlay	BA	BA	BA	BA	Outlay	BA	BA		
2000	290.3	281.00	12.4	1.3	304.0	300.8	294.4	13.8	45.5	1,789.0	236.2
2010	695.6	666.7	18.2	7.3	721.2	714.1	693.5	45.4	124.3	3,457.1	-1,294.4
2011	691.5	678.1	18.5	7.0	717.0	710.1	705.6	41.6	122.8	3,603.1	-1,299.6
2012	655.4	650.9	18.3	7.7	681.4	669.6	677.9	45.9	124.0	3,526.6	-1,076.6
2013	585.2	607.8	17.5	7.4	610.2	600.4	633.4	61.9	136.0	3,454.9	-679.8
2014	595.7	577.9	18.4	8.2	622.3	606.2	603.5	44.1	165.7	3,506.3	-484.8
2015	570.8	562.5	19.0	8.5	598.4	585.9	589.7	45.3	160.5	3,691.9	-442.0
2016	595.7	565.4	20.1	8.3	624.1	606.8	593.4	46.0	163.3	3,852.6	-584.7
2017	626.2	568.9	21.4	8.7	656.3	634.1	598.7	62.3	178.8	3,981.6	-665.4
2018	694.5	600.7	23.3	9.0	726.8	700.9	631.2	103.0	191.8	4,109.0	-779.1
2019	712.6	654.0	24.0	9.1	745.7	718.8	686.0	62.8	194.2	4,448.3	-984.2
2020*	721.5	689.6	25.9	9.7	757.2	746.0	724.5	66.0	214.0	4,789.7	-1,083.4
2021*	716.2	729.3	27.6	9.7	753.5	740.5	767.1	52.2	238.4	4,829.4	-966.1

Notes

FY = Fiscal Year (1 October–30 September)

* (request)

[1] The National Defense Budget Function subsumes funding for the DoD, the Department of Energy Atomic Energy Defense Activities and some smaller support agencies (including Federal Emergency Management and Selective Service System). It does not include funding for International Security Assistance (under International Affairs), the Veterans Administration, the US Coast Guard (Department of Homeland Security), nor for the National Aeronautics and Space Administration (NASA). Funding for civil projects administered by the DoD is excluded from the figures cited here.

[2] Early in each calendar year, the US government presents its defence budget to Congress for the next fiscal year, which begins on 1 October. The government also presents its Future Years Defense Program (FYDP), which covers the next fiscal year plus the following five. Until approved by Congress, the budget is called the Budget Request; after approval, it becomes the Budget Authority (BA).

and maintain military advantage. The FY2021 increases in Military Personnel and RDT&E budget allocations come largely at the expense of the Military Construction and Family Housing appropriation, which fell from US$20bn to US$9.5bn. Smaller reductions were also evident in Operations and Maintenance (O&M), the line item usually associated with readiness and legacy systems, and Procurement, which is not projected to return to FY2020 levels in nominal terms until 2025. Therefore, in real terms the procurement budget will decrease at a time when the DoD is in the middle of a reported modernisation 'bow-wave' caused by the concurrence of higher production rates in existing programmes (such as the F-35 combat aircraft) and the start of major new programmes (such as the long-range stand-off (LRSO) cruise missile).

Nevertheless, the FY2021 budget dedicates US$137bn to procurement, and procurement efforts by the armed services and DoD agencies indicate increasing investment in platforms, systems and capability areas thought to be particularly relevant to great-power competition. Missile defence constitutes US$20.3bn in investment, which is shared across the armed services. The navy receives the most funding and among its biggest procurement priorities, much like the air force, is the F-35. Overall, the FY2021 budget envisions the procurement of 79 F-35s of different variants.

The NDAA will also establish an independent budget of US$15.4bn for the Space Force, which was separated from other services in order to enhance the United States' competitiveness in this critical domain. US$15.3bn of the Space Force's budget has been transferred from other accounts, while US$10.3bn is dedicated to space RDT&E, reinforcing the DoD's shift from readiness to modernisation.

'Ruthless prioritisation' and the trade-offs of DoD transformation

The ambition to fully commit to NDS implementation comes with trade-offs that will have consequences for the armed forces in the short and long term. The National Defense Strategy Commission argued that implementation would require a year-on-year 3–5% real-terms increase in defence spending. However, planning assumptions for FY2022–FY2025 released in February 2020 with the FY2021 budget request included year-on-year constant growth that fluctuated between 2% and 2.4%. However, the cost of recovering from the coronavirus pandemic might render these numbers moot.

As a result, then-Secretary of Defense Esper pressed the DoD to implement 'ruthless and relentless prioritization' in terms of its expenditure, but also on which tasks the organisation directs its time and labour. According to Esper, the DoD had to be able to make tough choices, including 'divesting legacy systems, low priority tasks, and activities that deliver little value'.

The budget proposal reveals the degree to which the armed services began to address Esper's appeals by cancelling and delaying programmes and retiring some capabilities in the short term in order to introduce more advanced capabilities by the end of the decade. The air force has identified reductions in several aircraft types over the next five years, including B-1 bombers, fourth-generation fighter aircraft, *Global Hawk* uninhabited aerial vehicles and refuelling aircraft. The navy's FY2021 budget reduces surface-combatant procurements from 12 to eight, reduces the number of *Virginia*-class attack submarines to be procured in FY2021 from two to one and notes equipment retirements. The army, meanwhile, has similarly identified programmes in order to free up funds for investment in more advanced capabilities by the end of the decade.

Nuclear-weapons modernisation

Nuclear-weapons modernisation is one of the DoD's most pressing priorities in the FY2021 budget. The principal programmes are the development of *Columbia*-class nuclear-powered ballistic-missile submarines, the B-21 strategic bomber, the LRSO cruise missile and the Ground Based Strategic Deterrent. The FY2021 budget includes US$28.9bn for nuclear modernisation, including US$7bn on nuclear command, control and communications. In addition, the National Nuclear Security Administration's (NNSA's) budget is set to increase by nearly US$3bn year-on-year to US$19.8bn, US$15.6bn of which is dedicated to the weapons account. The increase in the NNSA's budget comes out of the DoD's base budget, adding an additional dimension to Esper's calls for prioritisation, at least for FY2021.

COVID-19 and the CARES Act

Even in this environment, there remain concerns about the effect that COVID-19 could have on defence modernisation, for instance on long-term and costly programmes such as the LRSO. The DoD was granted US$10.5bn in emergency supplemental funding as part of the US$2.1 trillion Coronavirus Aid, Relief, and Economic Security (CARES) Act. Most of this funding was allocated to defence-wide health programmes and related O&M priorities, though the Act also included US$1bn for Defense Production Act purchases, of which US$688 million formed a defence-industrial base (DIB) fund. The fund aimed to offset any financial distress in the DIB caused by COVID-19 by providing direct investment to sectors most severely affected to sustain essential capabilities and support employment.

The Pentagon also used the CARES Act to address vulnerabilities in defence production, allocating US$1.45bn for supply-chain protection. This was a central theme of the September 2018 report 'Assessing and Strengthening the Manufacturing and Defense Industrial Base and Supply Chain Resiliency of the United States', which highlighted the presence of Chinese companies in the US defence supply chain.

The pandemic injected urgency into DoD efforts to build supply-chain resilience and the CARES Act delivered a platform by which it began to address this issue. In June 2020, the Pentagon announced a partnership with the US International Development

▼ Figure 3 **FY2020 US DoD CARES Act – Defence-industrial base fund**

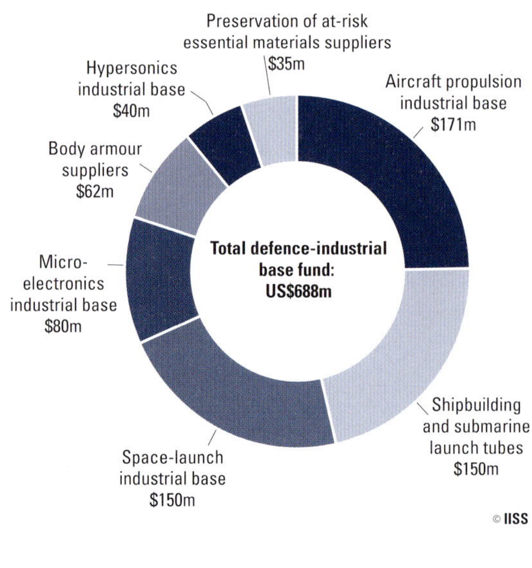

© IISS

Finance Corporation (DFC) to spend US$100m of CARES Act funding over two years to 'subsidize federal loans to create, maintain, protect, expand and restore domestic industrial-based capabilities to support the national COVID-19 response'. The chief executive officer (CEO) of the DFC said in a 14 May speech that following the executive order delegating authorities to the CEO under the Defense Production Act (DPA), the organisation would then 'have the tools under the DPA to re-shore critical industries in our country'.

CANADA

Following victory in Canada's October 2019 general election, Prime Minister Justin Trudeau's administration reaffirmed its commitment to its 2017 defence-policy review. Entitled 'Strong, Secure, Engaged', the review promised to transform Canada's armed forces, detailing financial commitments and a major equipment-procurement programme. It outlined continued support for alliance and coalition ties and a broad-based, if relatively modest, global deployment posture.

Canada's ambitious procurement programme is in part the legacy of delays to previous modernisation efforts, as well as policy changes. In some cases, similarly ambitious initiatives by previous governments had been affected by economic downturns. Concerns have been aired that a similar fate awaits the current programme when the economic cost of COVID-19 is finally calculated. However, there is, arguably, a different strategic outlook this time, with the re-emergence of great-power competition. At the same time, provisions to benefit the economy and industrial base have been embedded into the procurement programmes more deeply than before.

Procurement plans

In August 2020, General Dynamics Land Systems–Canada started work on the first new Armoured Combat Support Vehicle for the army, to be delivered by the end of the year, with a total of 360 altogether planned. Other planned land-systems improvements include ground-based air defence and surveillance, where some of the LAV *Coyote* fleet are to be replaced by improved surveillance systems integrated onto the LAV 6.0 chassis. However, naval and air capabilities are central to the procurement effort. In 2018, the government selected the Lockheed Martin bid, with a design based on the BAE Systems Type-26 frigate, for the Canadian Surface Combatant (CSC) project. The plan is to cut steel on the first vessel in 2023, with 15 delivered in total. However, concerns continue to be raised – including in parliament – over the eventual cost of the programme, prompting a further investigation by the parliamentary budget officer.

The CSC programme was first announced a decade ago as part of Canada's National Shipbuilding Strategy (then the National Shipbuilding Procurement Strategy). This plan also includes new multi-purpose vessels and icebreakers for the Canadian Coast Guard and for the navy the *Harry DeWolf*-class of new Arctic Offshore Patrol Ships (AOPS) and two Joint Support Ships (JSS) based on the German navy's *Berlin*-class. The navy's first AOPS, *Harry DeWolf*, was delivered in July 2020, though this was later than originally planned. *Harry DeWolf* is expected to commission, after trials, in mid-2021. Six are on order for the navy (plus two for the coast guard).

The keel-laying ceremony for the first JSS, the future HMCS *Protecteur*, took place in January 2020. Work on early blocks began in 2018 and the vessel is expected to be delivered in 2023. Although this programme is also running late, it will significantly enhance the navy's afloat support and its ability to deploy at range or for sustained periods. Currently, the converted MV *Asterix* is providing an interim capability.

In July 2020, three companies submitted bids for Canada's new combat aircraft: Boeing with the F/A-18 *Super Hornet* Block III, Lockheed Martin with the F-35A *Lightning* II and Saab with the *Gripen* E. Canada has a requirement for 88 aircraft and a final decision is scheduled for the end of 2022, with first deliveries commencing in 2025. The original plan had been to buy 65 F-35A *Lightning* IIs, though the government decided in 2017 to instead procure second-hand F/A-18s from Australia as an interim measure to supplement its ageing CF-18s (F/A-18s). However, to help ensure that Canada will continue to be able to fulfil its commitments under the North American Aerospace Defence Command arrangement, the United States cleared the possible sale of upgrades to part of the original CF-18 fleet, including an active electronically-scanned array radar and various weapons enhancements, which could help bridge the gap until the arrival of new aircraft.

Deployments

Canada continued in 2020 to pursue its broad global deployment posture in the face of the challenges presented by the onset of the coronavirus pandemic. Deployments to support NATO come under the banner of *Operation Reassurance*, a centrepiece of which is leading the NATO Enhanced Forward Battlegroup in Latvia. This was maintained despite some issues with rotations. Meanwhile, in September 2020 another detachment of six CF-18 *Hornet* combat aircraft arrived for a four-month deployment to the NATO enhanced air-policing mission in Romania.

Other missions were more affected by the fallout from the pandemic. *Operation Unifier*, to support and train Ukrainian security forces, was suspended in early April 2020 and its strength cut from 200 to 60 personnel. Nevertheless, in June, a further 90 personnel deployed to *Unifier* and the mission was resumed. Canada's various commitments in the Middle East have recently totaled up to 850 personnel. However, the Canadian-led NATO training and capacity-building mission in Iraq was suspended following the killing of Iran's Quds Force commander Qasem Soleimani. The spread of COVID-19 then led Iraq to suspend all training. In all, some 400 Canadian personnel left the region. Capacity was rebuilt, but Canada passed the lead-nation role to Denmark in November 2020 and was scaling back its contribution.

Canada has also been seeking to enhance its forward presence in the Indo-Pacific. Two frigates took part in the latest US-led *Rim of the Pacific* exercise. One of the ships subsequently remained in the region to conduct presence and maritime-security missions, including supporting United Nations sanctions on North Korea.

Canada CAN

Canadian Dollar $		2019	2020	2021
GDP	C$	2.30tr	2.16tr	
	US$	1.74tr	1.60tr	
per capita	US$	46,272	42,080	
Growth	%	1.7	-7.1	
Inflation	%	1.9	0.6	
Def exp [a]	C$	29.6bn	30.8bn	
	US$	22.3bn	22.9bn	
Def bdgt [b]	C$	24.9bn	27.0bn	
	US$	18.7bn	20.0bn	
US$1=C$		1.327	1.348	

[a] NATO definition

[b] Department of National Defence and Veterans Affairs

Real-terms defence budget trend (US$bn, constant 2015)

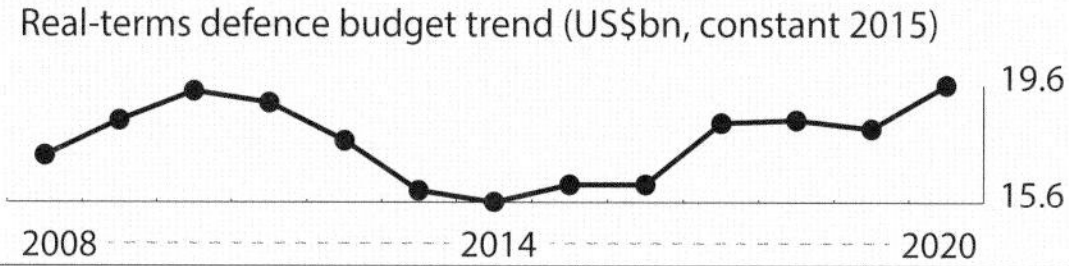

Population 37,694,085

Age	0–14	15–19	20–24	25–29	30–64	65 plus
Male	8.2%	2.7%	3.0%	3.5%	23.5%	8.7%
Female	7.8%	2.6%	2.8%	3.2%	23.7%	10.3%

Capabilities

Canada's armed forces are focused principally on territorial defence, as well as contributing capabilities to international missions, principally through NATO. The 2017 defence review reaffirmed commitments to NATO, but also to modernisation, including cyber power. The review promised to increase regular and reserve forces, with particular growth in the areas of cyber and intelligence. Canada's deployments, although relatively small scale, underscore a determination to maintain both international engagement and power-projection capability. Canada's leadership of a NATO battlegroup in Latvia highlights its capacity to deploy medium-sized land formations. It has also contributed to NATO's air-policing mission. Meanwhile, the deployments of frigates and submarines to the European theatre and to the Pacific demonstrated a continuing blue-water capability. The 2017 review pledged to deliver on a range of delayed procurements. It raised the target for a new-generation fighter to 88 aircraft with the choice between the F/A-18E/F *Super Hornet*, the F-35A *Lightning* II and the *Gripen* E. In the interim, Canada has been supplementing its existing fighter force with former Australian F/A-18 *Hornets*. In October 2018, the government selected the Lockheed Martin-led consortium and its BAE Systems Type-26 frigate design as the preferred bidder for Canada's future surface combatant and is pursuing a range of programmes to recapitalise its naval forces with new support ships and Arctic- and offshore-patrol vessels. Canada maintains a well-developed range of mainly small and medium-sized defence firms. The strongest sector is in combat vehicles and components, though the government is using its latest naval procurements to establish a long-term national shipbuilding strategy.

ACTIVE 67,400 (Army 23,800 Navy 8,300 Air Force 12,000 Other 23,300) Paramilitary 4,500

RESERVE 35,600 (Army 25,600 Navy 4,600 Air 2,100 Other 3,300)

ORGANISATIONS BY SERVICE

Space

EQUIPMENT BY TYPE

SATELLITES • SPACE SURVEILLANCE 1 *Sapphire*

Army 23,800

FORCES BY ROLE

MANOEUVRE

Mechanised

1 (1st) mech bde gp (1 armd regt, 2 mech inf bn, 1 lt inf bn, 1 arty regt, 1 cbt engr regt, 1 log bn)

2 (2nd & 5th) mech bde gp (1 armd recce regt, 2 mech inf bn, 1 lt inf bn, 1 arty regt, 1 cbt engr regt, 1 log bn)

COMBAT SUPPORT

1 engr regt

3 MP pl

AIR DEFENCE

1 SAM regt

EQUIPMENT BY TYPE

ARMOURED FIGHTING VEHICLES

MBT 82: 42 *Leopard* 2A4 (trg role); 20 *Leopard* 2A4M (upgraded); 20 *Leopard* 2A6M (52 *Leopard* 1C2 in store)

RECCE ε120 LAV-25 *Coyote*

IFV 550 LAV 6.0

APC 443

APC (T) 268: 235 M113; 33 M577 (CP)

APC (W) 175 LAV *Bison* (incl 10 EW, 32 amb, 32 repair, 64 recovery)

AUV 507: 7 *Cougar*; 500 TAPV

ENGINEERING & MAINTENANCE VEHICLES

AEV 23: 5 *Buffalo*; 18 *Wisent* 2

ARV 12 BPz-3 *Büffel*

ANTI-TANK/ANTI-INFRASTRUCTURE

MSL • MANPATS TOW-2

RCL 84mm *Carl Gustaf*

ARTILLERY 287

TOWED 163 **105mm** 126: 98 C3 (M101); 28 LG1 MkII; **155mm** 37 M777

MOR 124: **81mm** 100; **SP 81mm** 24 LAV *Bison*

UNINHABITED AERIAL VEHICLES • ISR • Light 5 RQ-21A *Blackjack*

Reserve Organisations 25,600

Canadian Rangers 5,250 Reservists

Provide a limited military presence in Canada's northern, coastal and isolated areas. Sovereignty, public-safety and surveillance roles

FORCES BY ROLE

MANOEUVRE

Other

5 (patrol) ranger gp (209 patrols)

Army Reserves 20,350 Reservists

Most units have only coy-sized establishments

FORCES BY ROLE

COMMAND

10 bde gp HQ

MANOEUVRE
Reconnaissance
18 recce regt (sqn)
Light
51 inf regt (coy)
COMBAT SUPPORT
16 fd arty regt (bty)
3 indep fd arty bty
10 cbt engr regt (coy)
1 EW regt (sqn)
4 int coy
10 sigs regt (coy)
COMBAT SERVICE SUPPORT
10 log bn (coy)
3 MP coy

Royal Canadian Navy 8,300

EQUIPMENT BY TYPE
SUBMARINES 4
SSK 4 *Victoria* (ex-UK *Upholder*) (of which 1 in long-term refit) with 6 single 533mm TT with Mk 48 HWT
PRINCIPAL SURFACE COMBATANTS • FRIGATES 12
FFGHM 12 *Halifax* with 2 quad lnchr with RGM-84L *Harpoon* Block II AShM, 2 8-cell Mk 48 mod 0 VLS with RIM-162C ESSM SAM, 2 twin 324mm SVTT Mk 32 mod 9 ASTT with Mk 46 LWT, 1 Mk 15 *Phalanx* Block 1B CIWS, 1 57mm gun (capacity 1 CH-148 *Cyclone* ASW hel)
MINE WARFARE • MINE COUNTERMEASURES 12
MCO 12 *Kingston* (also used in patrol role)
LOGISTICS AND SUPPORT 10
AORH 1 *Asterix* (*Resolve*) (capacity 2 CH-148 *Cyclone* ASW hel)
AX 9: **AXL** 8 *Orca;* **AXS** 1 *Oriole*

Reserves 4,600 reservists

24 units tasked with crewing 10 of the 12 MCOs, harbour defence & naval control of shipping

Royal Canadian Air Force (RCAF) 12,000

FORCES BY ROLE
FIGHTER/GROUND ATTACK
4 sqn with F/A-18A/B *Hornet* (CF-18AM/BM)
ANTI-SUBMARINE WARFARE
2 sqn with CH-148 *Cyclone*
MARITIME PATROL
2 sqn with P-3 *Orion* (CP-140M *Aurora*)
SEARCH & RESCUE/TRANSPORT
3 sqn with AW101 *Merlin* (CH-149 *Cormorant*); C-130H/H-30 (CC-130) *Hercules*
1 sqn with DHC-5 (CC-115) *Buffalo*
TANKER/TRANSPORT
1 sqn with A310/A310 MRTT (CC-150/CC-150T)
1 sqn with KC-130H
TRANSPORT
1 sqn with C-17A (CC-177) *Globemaster*
1 sqn with CL-600 (CC-144B)
1 sqn with C-130J-30 (CC-130) *Hercules*
1 (utl) sqn with DHC-6 (CC-138) *Twin Otter*
TRAINING
1 OCU sqn with F/A-18A/B *Hornet* (CF-18AM/BM)
1 OCU sqn with C-130H/H-30/J (CC-130) *Hercules*
1 OCU sqn with CH-148 *Cyclone*
1 OCU sqn with Bell 412 (CH-146 *Griffon*)
1 sqn with P-3 *Orion* (CP-140M *Aurora*)
TRANSPORT HELICOPTER
5 sqn with Bell 412 (CH-146 *Griffon*)
3 (cbt spt) sqn with Bell 412 (CH-146 *Griffon*)
1 (Spec Ops) sqn with Bell 412 (CH-146 *Griffon* – OPCON Canadian Special Operations Command)
1 sqn with CH-47F (CH-147F) *Chinook*

EQUIPMENT BY TYPE
AIRCRAFT 96 combat capable
FGA 82: 61 F/A-18A (CF-18AM) *Hornet*; 21 F/A-18B (CF-18BM) *Hornet*
ASW 14 P-3 *Orion* (CP-140M *Aurora*)
SAR 3 C295W (CC-295)
TKR/TPT 5: 2 A310 MRTT (CC-150T); 3 KC-130H
TPT 48: **Heavy** 5 C-17A (CC-177) *Globemaster* III; **Medium** 26: 7 C-130H (CC-130) *Hercules;* 2 C-130H-30 (CC-130) *Hercules;* 17 C-130J-30 (CC-130) *Hercules;* **Light** 10: 6 DHC-5 (CC-115) *Buffalo*; 4 DHC-6 (CC-138) *Twin Otter;* **PAX** 7: 3 A310 (CC-150 *Polaris*); 4 CL-600 (CC-144B/C)
TRG 4 DHC-8 (CT-142)
HELICOPTERS
ASW 17 CH-148 *Cyclone*
MRH 68 Bell 412 (CH-146 *Griffon*)
TPT 29: **Heavy** 15 CH-47F (CH-147F) *Chinook*; **Medium** 14 AW101 *Merlin* (CH-149 *Cormorant*)
RADAR 53
AD RADAR • NORTH WARNING SYSTEM 47: 11 AN/FPS-117 (range 200nm); 36 AN/FPS-124 (range 80nm)
STRATEGIC 6: 4 Coastal; 2 Transportable
AIR-LAUNCHED MISSILES
AAM • IR AIM-9L *Sidewinder*
ARH AIM-120C AMRAAM
BOMBS
Laser-guided: GBU-10/GBU-12/GBU-16 *Paveway* II; GBU-24 *Paveway* III
INS/GPS-guided: GBU-31 JDAM; GBU-38 JDAM; GBU-49 *Enhanced Paveway* II

NATO Flight Training Canada

EQUIPMENT BY TYPE
AIRCRAFT
TRG 45: 26 T-6A *Texan* II (CT-156 *Harvard* II); 19 *Hawk* 115 (CT-155) (advanced wpns/tactics trg)

Contracted Flying Services – Southport

EQUIPMENT BY TYPE
AIRCRAFT
TPT • Light 7 Beech C90B *King Air*
TRG 11 G-120A
HELICOPTERS
MRH 9 Bell 412 (CH-146)
TPT • Light 7 Bell 206 *Jet Ranger* (CH-139)

Canadian Special Operations Forces Command 1,500

FORCES BY ROLE

SPECIAL FORCES

1 SF regt (Canadian Special Operations Regiment)

1 SF unit (JTF 2)

COMBAT SERVICE SUPPORT

1 CBRN unit (Canadian Joint Incident Response Unit – CJIRU)

TRANSPORT HELICOPTER

1 (spec ops) sqn, with Bell 412 (CH-146 *Griffon* – from the RCAF)

EQUIPMENT BY TYPE

NBC VEHICLES 4 LAV *Bison* NBC

HELICOPTERS • MRH 10 Bell 412 (CH-146 *Griffon*)

Canadian Forces Joint Operational Support Group

FORCES BY ROLE

COMBAT SUPPORT

1 engr spt coy

1 (close protection) MP coy

1 (joint) sigs regt

COMBAT SERVICE SUPPORT

1 (spt) log unit

1 (movement) log unit

Paramilitary 4,500

Canadian Coast Guard 4,500

Incl Department of Fisheries and Oceans; all platforms are designated as non-combatant

EQUIPMENT BY TYPE

PATROL AND COASTAL COMBATANTS 74

PSOH 1 *Leonard J Cowley*

PSO 1 *Sir Wilfred Grenfell* (with hel landing platform)

PCO 13: 2 *Cape Roger*; 1 *Gordon Reid*; 9 *Hero*; 1 *Tanu*

PCC 1 *Harp*

PBF 1 Response Boat-Medium (RB-M)

PB 57: 1 *Post*; 1 *Quebecois*; 1 *Vakta*; 10 Type-300A; 36 Type-300B; 1 *S. Dudka*; 1 *Simmonds* (on loan from RCMP); 6 *Baie de Plaisance*

AMPHIBIOUS • LANDING CRAFT 4

UCAC 4 Type-400

LOGISTICS AND SUPPORT 32

ABU 7

AG 4

AGB 16

AGOS 5

HELICOPTERS • MRH 7 Bell 412EP • **TPT** 19: **Medium** 1 S-61; **Light** 18: 3 Bell 206L *Long Ranger*; 15 Bell 429

DEPLOYMENT

CYPRUS: UN • UNFICYP (*Operation Snowgoose*) 1

DEMOCRATIC REPUBLIC OF THE CONGO: UN • MONUSCO (*Operation Crocodile*) 7

EGYPT: MFO (*Operation Calumet*) 55; 1 MP team

IRAQ: *Operation Inherent Resolve* (*Impact*) 100; 1 SF trg gp; 1 med unit; **NATO** • NATO Mission Iraq 250; 1 hel flt with 3 Bell 412 (CH-146 *Griffon*) hel

KUWAIT: *Operation Inherent Resolve* (*Impact*) 2 C-130J-30 *Hercules* (CC-130J)

LATVIA: NATO • Enhanced Forward Presence (*Operation Reassurance*) 527; 1 mech inf bn HQ; 1 mech inf coy(+); 1 cbt spt coy; LAV 6.0; M777

MALI: UN • MINUSMA (*Operation Presence*) 4

MIDDLE EAST: UN • UNTSO (*Operation Jade*) 6

NORTH SEA: NATO • SNMG 1: 1 FFGHM

ROMANIA: NATO • Air Policing 135; 6 F/A-18A *Hornet* (CF-18)

SERBIA: NATO • KFOR • *Joint Enterprise* (*Operation Kobold*) 5

SOUTH SUDAN: UN • UNMISS (*Operation Soprano*) 9

UKRAINE: *Operation Unifier* 200

FOREIGN FORCES

United Kingdom BATUS 400; 1 trg unit; 1 hel flt with SA341 *Gazelle* AH1

United States 140

United States US

United States Dollar $		2019	2020	2021
GDP	US$	21.4tr	20.8tr	
per capita	US$	65,254	63,051	
Growth	%	2.2	-4.3	
Inflation	%	1.8	1.5	
Def exp [a]	US$	730bn	785bn	
Def bdgt [b]	US$	685bn	738bn	741bn

[a] NATO definition

[b] National Defense Budget Function (50) Outlays. Includes DoD funding, as well as funds for nuclear-weapons-related activities undertaken by the Department of Energy. Excludes some military retirement and healthcare costs.

Real-terms defence budget trend (US$bn, constant 2015)

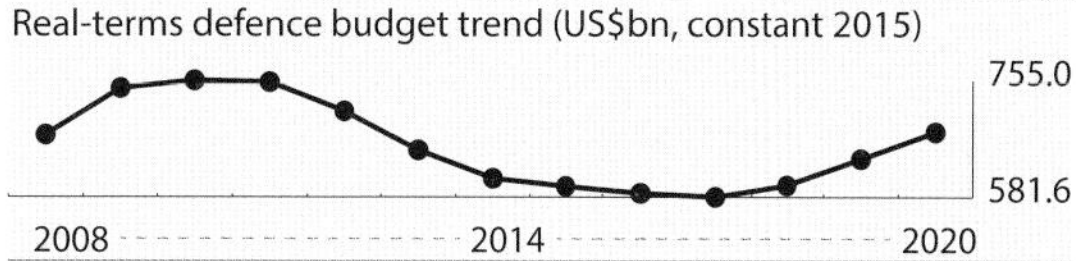

Population 332,639,102

Age	0–14	15–19	20–24	25–29	30–64	65 plus
Male	9.4%	3.2%	3.4%	3.6%	22.1%	7.5%
Female	9.0%	3.1%	3.2%	3.5%	22.6%	9.3%

Capabilities

The United States remains the world's most capable military power, with a unique ability to project power on a global basis. The Pentagon's 2018 National Defense Strategy refocused priorities on renewed 'great-power competition' and called for a reversal in reductions in the size of the joint force. A Nuclear Posture Review in 2018 backed the development of low-yield warheads and a nuclear-capable sea-launched cruise missile. A missile-defence review was published in January 2019 envisaging a number of new programmes and technologies, including space-based systems. In August 2019 the Pentagon established a new Space Command followed by the establishment of a Space Force in December 2019. In June 2020 it also published a Defense Space Strategy. The US is NATO's most capable member, though an announced drawdown of its presence in Germany disconcerted some European allies. This ostensibly formed part of a review of all US Combatant Commands. The US maintains an all-volunteer force, including significant reserves, with high levels of training throughout all command and services. The Pentagon is trying to improve readiness. Modernisation priorities include a renewal of strategic nuclear capabilities, including a new class of ballistic-missile submarine and a new long-range bomber, recapitalisation of air assets across the services and also improved naval capabilities that are likely to include both crewed and uninhabited platforms. In August 2019, the US withdrew from the Intermediate-range Nuclear Forces Treaty, and three weeks later conducted a ground-launched cruise-missile test. The US also continues to actively develop its defensive and offensive cyber capabilities. The country has the world's most capable defence industry, with a dominant position in the international defence market.

ACTIVE 1,388,100 (Army 485,400 Navy 346,500 Air Force 331,400 Space Force 2,400 US Marine Corps 180,950 US Coast Guard 41,450)

RESERVE 844,950 (Army 524,800 Navy 101,900 Air Force 176,450 Marine Corps Reserve 35,500 US Coast Guard 6,300)

ORGANISATIONS BY SERVICE

US Strategic Command

HQ at Offutt AFB (NE). Five missions: US nuclear deterrent; missile defence; global strike; info ops; ISR

US Navy

EQUIPMENT BY TYPE

SUBMARINES • STRATEGIC • SSBN 14 *Ohio* with up to 20 UGM-133A *Trident* D-5/D-5LE nuclear SLBM, 4 single 533mm TT with Mk 48 ADCAP mod 6/7 HWT

US Air Force • Global Strike Command

FORCES BY ROLE

MISSILE

9 sqn with LGM-30G *Minuteman* III

BOMBER

5 sqn with B-52H *Stratofortress*

2 sqn with B-2A *Spirit* (+1 ANG sqn personnel only)

EQUIPMENT BY TYPE

SURFACE-TO-SURFACE MISSILE LAUNCHERS

ICBM • Nuclear 400 LGM-30G *Minuteman* III (1 Mk12A or Mk21 re-entry veh per missile)

AIRCRAFT

BBR 66: 20 B-2A *Spirit*; 46 B-52H *Stratofortress*

AIR-LAUNCHED MISSILES

ALCM • Nuclear AGM-86B

Strategic Defenses – Early Warning

North American Aerospace Defense Command (NORAD) – a combined US–CAN organisation

EQUIPMENT BY TYPE

RADAR

NORTH WARNING SYSTEM 50: 14 AN/FPS-117; 36 AN/FPS-124

SOLID STATE PHASED ARRAY RADAR SYSTEM (SSPARS) 5: 2 AN/FPS-123 Early Warning Radar located at Cape Cod AFS (MA) and Clear AFS (AK); 3 AN/FPS-132 Upgraded Early Warning Radar located at Beale AFB (CA), Thule (GL) and Fylingdales Moor (UK)

SPACETRACK SYSTEM 10: 1 AN/FPS-85 Spacetrack Radar at Eglin AFB (FL); 6 contributing radars at Cavalier AFS (ND), Clear (AK), Thule (GL), Fylingdales Moor (UK), Beale AFB (CA) and Cape Cod (MA); 3 Spacetrack Optical Trackers located at Socorro (NM), Maui (HI), Diego Garcia (BIOT)

PERIMETER ACQUISITION RADAR ATTACK CHARACTERISATION SYSTEM (PARCS) 1 AN/FPQ-16 at Cavalier AFS (ND)

DETECTION AND TRACKING RADARS 5 located at Kwajalein Atoll, Ascension Island, Australia, Kaena Point (HI), MIT Lincoln Laboratory (MA)

GROUND BASED ELECTRO OPTICAL DEEP SPACE SURVEILLANCE SYSTEM (GEODSS) Socorro (NM), Maui (HI), Diego Garcia (BIOT)

STRATEGIC DEFENCES – MISSILE DEFENCES

SEA-BASED: *Aegis* engagement cruisers and destroyers

LAND-BASED: 40 ground-based interceptors at Fort Greely (AK); 4 ground-based interceptors at Vandenburg AFB (CA)

Space

EQUIPMENT BY TYPE

SATELLITES 141

COMMUNICATIONS 46: 6 AEHF; 6 DSCS-III; 2 *Milstar*-I; 3 *Milstar*-II; 5 MUOS; 1 PAN-1 (P360); 5 SDS-III; 2 SDS-IV; 6 UFO; 10 WGS SV2

NAVIGATION/POSITIONING/TIMING 31: 12 NAVSTAR Block IIF; 9 NAVSTAR Block IIR; 7 NAVSTAR Block IIRM; 3 NAVSTAR Block III

METEOROLOGY/OCEANOGRAPHY 6 DMSP-5

ISR 17: 5 FIA *Radar*; 5 *Evolved Enhanced/Improved Crystal* (visible and infrared imagery); 2 *Lacrosse* (*Onyx* radar imaging satellite); 1 NRO L-71; 1 NRO L-76; 1 ORS-1; 1 *TacSat*-4; 1 *TacSat*-6

ELINT/SIGINT 27: 2 *Mentor* (advanced *Orion*); 3 Advanced *Mentor*; 4 *Mercury*; 1 NRO L-67; 1 *Trumpet*; 4 Improved *Trumpet*; 12 SBWASS (Space Based Wide Area Surveillance System; Naval Ocean Surveillance System)

SPACE SURVEILLANCE 6: 4 GSSAP; 1 SBSS (Space Based Surveillance System); 1 ORS-5

EARLY WARNING 8: 4 DSP; 4 SBIRS *Geo*-1

COUNTERSPACE • EW Counter Communications System (CCS)

US Army 485,400

FORCES BY ROLE

Sqn are generally bn sized and tp are generally coy sized

COMMAND

4 (I, III, V & XVIII AB) corps HQ
1 (2nd) inf div HQ

SPECIAL FORCES

(see USSOCOM)

MANOEUVRE

Armoured

2 (1st Armd & 1st Cav) armd div (3 (1st–3rd ABCT) armd bde (1 armd recce sqn, 2 armd bn, 1 armd inf bn, 1 SP arty bn, 1 cbt engr bn, 1 CSS bn); 1 SP arty bde HQ; 1 log bde; 1 (hy cbt avn) hel bde)
1 (1st) inf div (2 (1st & 2nd ABCT) armd bde (1 armd recce sqn, 2 armd bn, 1 armd inf bn, 1 SP arty bn, 1 cbt engr bn, 1 CSS bn); 1 SP arty bde HQ; 1 log bde; 1 (cbt avn) hel bde)
1 (3rd) inf div (2 (1st & 2nd ABCT) armd bde (1 armd recce sqn, 2 armd bn, 1 armd inf bn, 1 SP arty bn, 1 cbt engr bn, 1 CSS bn); 1 lt inf bn; 1 SP arty bde HQ; 1 log bde; 1 (cbt avn) hel bde)

Mechanised

1 (4th) inf div (1 (3rd ABCT) armd bde (1 armd recce sqn, 2 armd bn, 1 armd inf bn, 1 SP arty bn, 1 cbt engr bn, 1 CSS bn); 2 (1st & 2nd SBCT) mech bde (1 armd recce sqn, 3 mech inf bn, 1 arty bn, 1 cbt engr bn, 1 CSS bn); 1 SP arty bde HQ; 1 log bde; 1 (hy cbt avn) hel bde)
1 (7th) inf div (2 (1st & 2nd SBCT, 2nd ID) mech bde (1 armd recce sqn, 3 mech inf bn, 1 arty bn, 1 cbt engr bn, 1 CSS bn))
1 (1st SBCT, 25th ID) mech bde (1 armd recce sqn, 3 mech inf bn, 1 arty bn, 1 cbt engr bn, 1 CSS bn)
2 (2nd & 3rd CR) mech bde (1 armd recce sqn, 3 mech sqn, 1 arty sqn, 1 cbt engr sqn, 1 CSS sqn)

Light

1 (10th Mtn) inf div (3 (1st–3rd IBCT) lt inf bde (1 recce sqn, 3 inf bn, 1 arty bn, 1 cbt engr bn, 1 CSS bn); 1 log bde; 1 (cbt avn) hel bde)
1 (25th) inf div (2 (2 & 3rd IBCT) inf bde (1 recce sqn, 2 inf bn, 1 arty bn, 1 cbt engr bn, 1 CSS bn); 1 log bde; 1 (cbt avn) hel bde)
5 (Sy Force Assist) inf bde(-)

Air Manoeuvre

1 (82nd) AB div (1 (1st AB BCT) AB bde (1 recce bn, 1 mech coy; 3 para bn, 1 arty bn, 1 cbt engr bn, 1 CSS bn); 2 (2nd & 3rd AB BCT) AB bde (1 recce bn, 3 para bn, 1 arty bn, 1 cbt engr bn, 1 CSS bn); 1 (cbt avn) hel bde; 1 log bde)
1 (101st) air aslt div (3 (1st–3rd AB BCT) AB bde (1 recce bn, 3 para bn, 1 arty bn, 1 cbt engr bn, 1 CSS bn); 1 (cbt avn) hel bde; 1 log bde)
1 (173rd AB BCT) AB bde (1 recce bn, 2 para bn, 1 arty bn, 1 cbt engr bn, 1 CSS bn)
1 (4th AB BCT, 25th ID) AB bde (1 recce bn, 2 para bn, 1 arty bn, 1 cbt engr bn, 1 CSS bn)

Other

1 (11th ACR) trg armd cav regt (OPFOR) (2 armd cav sqn, 1 CSS bn)

COMBAT SUPPORT

4 MRL bde (2 MRL bn)
1 MRL bde (5 MRL bn)
4 engr bde
2 EOD gp (2 EOD bn)
10 int bde
2 int gp
4 MP bde
1 NBC bde
3 (strat) sigs bde
4 (tac) sigs bde
1 (I2CEWS) cbt spt bn

COMBAT SERVICE SUPPORT

2 log bde
3 med bde
1 tpt bde

ISR

1 ISR avn bde

HELICOPTER

2 (cbt avn) hel bde
1 (cbt avn) hel bde HQ

AIR DEFENCE

5 SAM bde

Reserve Organisations

Army National Guard 336,100 reservists

Normally dual-funded by DoD and states. Civil-emergency responses can be mobilised by state governors. Federal government can mobilise ARNG for major domestic emergencies and for overseas operations

FORCES BY ROLE

COMMAND

8 div HQ

SPECIAL FORCES

(see USSOCOM)

MANOEUVRE

Reconnaissance

1 armd recce sqn

Armoured

5 (ABCT) armd bde (1 armd recce sqn, 2 armd bn, 1 armd inf bn, 1 SP arty bn, 1 cbt engr bn, 1 CSS bn)

Mechanised

2 (SBCT) mech bde (1 armd recce sqn, 3 mech inf bn, 1 arty bn, 1 cbt engr bn, 1 CSS bn)

Light

14 (IBCT) lt inf bde (1 recce sqn, 3 inf bn, 1 arty bn, 1 cbt engr bn, 1 CSS bn)
6 (IBCT) lt inf bde (1 recce sqn, 2 inf bn, 1 arty bn, 1 cbt engr bn, 1 CSS bn)
1 (Sy Force Assist) inf bde(-)
4 lt inf bn

Air Manoeuvre

1 AB bn

COMBAT SUPPORT

8 arty bde
1 SP arty bn
8 engr bde

1 EOD regt
3 int bde
3 MP bde
1 NBC bde
2 (tac) sigs bde
17 (Mnv Enh) cbt spt bde

COMBAT SERVICE SUPPORT
10 log bde
17 (regional) log spt gp

HELICOPTER
8 (cbt avn) hel bde
5 (theatre avn) hel bde

AIR DEFENCE
3 SAM bde

Army Reserve 188,700 reservists

Reserve under full command of US Army. Does not have state-emergency liability of Army National Guard

FORCES BY ROLE

SPECIAL FORCES
(see USSOCOM)

COMBAT SUPPORT
4 engr bde
4 MP bde
2 NBC bde
2 sigs bde
3 (Mnv Enh) cbt spt bde

COMBAT SERVICE SUPPORT
9 log bde
11 med bde

HELICOPTER
2 (exp cbt avn) hel bde

Army Stand-by Reserve 700 reservists

Trained individuals for mobilisation

EQUIPMENT BY TYPE

ARMOURED FIGHTING VEHICLES
MBT 2,509: 750 M1A1 SA *Abrams*; 1,605 M1A2 SEPv2 *Abrams*; 154 M1A2C *Abrams*; (ε3,700 more M1A1/A2 *Abrams* in store)
ASLT 134 M1128 *Stryker* MGS
RECCE 1,745: ε1,200 M3A2/A3 *Bradley*; 545 M1127 *Stryker* RV (ε800 more M3 *Bradley* in store)
IFV 2,931: ε14 LAV-25; ε2,500 M2A2/A3 *Bradley*; 334 M7A3/SA BFIST (OP); 83 M1296 *Stryker Dragoon* (ε2,000 more M2 *Bradley* in store)
APC 10,549
APC (T) 5,002: 2 AMPV (in test); ε5,000 M113A2/A3 (ε8,000 more in store)
APC (W) 2,613: 1,773 M1126 *Stryker* ICV; 348 M1130 *Stryker* CV (CP); 188 M1131 *Stryker* FSV (OP); 304 M1133 *Stryker* MEV (Amb)
PPV 2,934: 2,633 *MaxxPro Dash*; 301 *MaxxPro* LWB (Amb)
AUV 16,516+: 7,500+ JLTV; 2,900 M1117 ASV; 465 M1200 *Armored Knight* (OP); 5,651 M-ATV

ENGINEERING & MAINTENANCE VEHICLES
AEV 531: 113 M1 ABV; 250 M9 ACE; 168 M1132 *Stryker* ESV
ARV 1,195+: 360 M88A1; 835 M88A2 (ε1,000 more M88A1 in store); some M578
VLB 341: ε230 M60 AVLB; 51 M1074 Joint Assault Bridge (in test); 20 REBS; 40 *Wolverine* HAB
MW 3+: *Aardvark* JSFU Mk4; some *Husky* 2G; 3+ *Hydrema* 910 MCV-2; M58/M59 MICLIC; M139; *Rhino*

NBC VEHICLES 234 M1135 *Stryker* NBCRV

ANTI-TANK/ANTI-INFRASTRUCTURE
MSL
SP 1,133: 133 M1134 *Stryker* ATGM; ε1,000 M1167 HMMWV TOW
MANPATS FGM-148 *Javelin*
RCL 84mm *Carl Gustaf*

ARTILLERY 5,469
SP 155mm 1,023: 900 M109A6; 123 M109A7 (ε500 more M109A6 in store)
TOWED 1,339: **105mm** 821 M119A2/3; **155mm** 518 M777A2
MRL 227mm 600: 375 M142 HIMARS; 225 M270A1 MLRS
MOR 2,507: **81mm** 990 M252; **120mm** 1,076 M120/M1064A3; **SP 120mm** 441 M1129 *Stryker* MC

SURFACE-TO-SURFACE MISSILE LAUNCHERS
SRBM • Conventional MGM-140A/B ATACMS; MGM-168 ATACMS (All launched from M270A1 MLRS or M142 HIMARS MRLs)

AMPHIBIOUS
PRINCIPAL AMPHIBIOUS SHIPS 7
LSL 7 *Frank Besson* (capacity 24 *Abrams* MBT)
LANDING CRAFT 70
LCT 34 LCU 2000 (capacity 5 M1 *Abrams* MBT)
LCM 36 LCM 8 (capacity either 1 M1 *Abrams* MBT or 200 troops)

AIRCRAFT
ISR 47: 8 EMARSS-G; 4 EMARSS-V; 8 EMARSS-M; 19 RC-12X *Guardrail* (5 trg); 8 RO-6A ARL-E
SIGINT 2 CL-600 *Artemis*
ELINT 9: 4 EMARSS-S; 4 EO-5C ARL-M (COMINT/ELINT); 1 TO-5C (trg)
TPT 156: **Light** 152: 113 Beech A200 *King Air* (C-12 *Huron*); 28 Cessna 560 *Citation* (UC-35A/B); 11 SA-227 *Metro* (C-26E); **PAX** 4: 1 Gulfstream IV (C-20F); 2 Gulfstream V (C-37A); 1 Gulfstream G550 (C-37B)
TRG 4 T-6D *Texan* II

HELICOPTERS
ATK 714: 360 AH-64D *Apache*; 354 AH-64E *Apache*
SAR 294: 19 HH-60L *Black Hawk*; 275 HH-60M *Black Hawk* (medevac)
TPT 2,907: **Heavy** 450 CH-47F *Chinook*; **Medium** 1,935: 200 UH-60A *Black Hawk*; 943 UH-60L *Black Hawk*; 792 UH-60M *Black Hawk*; **Light** 522: 457 UH-72A *Lakota*; 65 UH-1H/V *Iroquois*
TRG up to 20 TH-67 *Creek*

UNINHABITED AERIAL VEHICLES 416
CISR • Heavy ε180 MQ-1C *Gray Eagle*
ISR • Medium 236 RQ-7B *Shadow*

AIR DEFENCE
SAM 1,183+
Long-range 480 MIM-104D/E/F *Patriot* PAC-2 GEM/PAC-2 GEM-T/PAC-3/PAC-3 MSE
Short-range *Iron Dome*; NASAMS
Point-defence 703+: FIM-92 *Stinger*; 703 M1097 *Avenger*
GUNS • Towed • 20mm *Phalanx* (LPWS)

MISSILE DEFENCE • Long-range 42 THAAD
AIR-LAUNCHED MISSILES
ASM AGM-114K/L/M/N/R *Hellfire* II; APKWS

US Navy 346,500

Comprises 2 Fleet Areas, Atlantic and Pacific. 6 Fleets: 2nd – Atlantic; 3rd – Pacific; 4th – Caribbean, Central and South America; 5th – Arabian Sea, Persian Gulf, Red Sea; 6th – Mediterranean; 7th – Indian Ocean, East Asia, W. Pacific; plus Military Sealift Command (MSC); Naval Reserve Force (NRF). For Naval Special Warfare Command, see US Special Operations Command

EQUIPMENT BY TYPE

SUBMARINES 68
STRATEGIC • SSBN 14 *Ohio* (opcon US STRATCOM) with up to 20 UGM-133A *Trident* D-5/D-5LE nuclear SLBM, 4 single 533mm TT with Mk 48 ADCAP mod 6/7 HWT
TACTICAL 54
SSGN 51:
4 *Ohio* (mod) with 22 7-cell MAC VLS with UGM-109C/E *Tomahawk* Block III/IV LACM , 4 single 533mm TT with Mk 48 ADCAP mod 6/7 HWT
6 *Los Angeles* Flight II with 1 12-cell VLS with UGM-109C/E *Tomahawk* Block III/IV LACM, 4 single 533mm TT with Mk 48 ADCAP mod 6/7 HWT
22 *Los Angeles* Flight III with 1 12-cell VLS with UGM-109C/E *Tomahawk* Block III/IV LACM, 4 single 533mm TT with Mk 48 ADCAP mod 6/7 HWT
10 *Virginia* Flight I/II with 1 12-cell VLS with UGM-109C/E *Tomahawk* Block III/IV LACM, 4 single 533mm TT with Mk 48 ADCAP mod 6/7 HWT
8 *Virginia* Flight III with 2 6-cell VPT VLS with UGM-109C/E *Tomahawk* Block III/IV LACM, 4 single 533mm TT with Mk 48 ADCAP mod 6/7 HWT
1 *Virginia* Flight IV with 2 6-cell VPT VLS with UGM-109C/E *Tomahawk* Block III/IV LACM, 4 single 533mm TT with Mk 48 ADCAP mod 6/7 HWT
SSN 3 *Seawolf* with 8 single 660mm TT with UGM-109C/E *Tomahawk* Block III/IV LACM/Mk 48 ADCAP mod 6/7 HWT

PRINCIPAL SURFACE COMBATANTS 124
AIRCRAFT CARRIERS • CVN 11:
1 *Gerald R. Ford* with 2 octuple Mk 29 mod 5 GMLS with RIM-162D ESSM SAM, 2 Mk 49 mod 3 GMLS with RIM-116C RAM Block 2 SAM, 3 Mk 15 *Phalanx* Block 1B CIWS (typical capacity 75+ F/A-18E/F *Super Hornet* FGA ac; F-35C *Lightning* II FGA ac; E-2D *Hawkeye* AEW&C ac; EA-18G *Growler* EW ac; MH-60R *Seahawk* ASW hel; MH-60S *Knight Hawk* MRH hel)
10 *Nimitz* with 2 8-cell Mk29 GMLS with RIM-162 ESSM SAM, 2 Mk 49 GMLS with RIM-116 RAM Block 2 SAM, 3 Mk 15 *Phalanx* Block 1B CIWS (typical capacity 55 F/A-18E/F *Super Hornet* FGA ac; F-35C *Lightning* II FGA ac; 4 EA-18G *Growler* EW ac; 4 E-2C/D *Hawkeye* AEW ac; 6 MH-60R/S *Seahawk/Knight Hawk* hel)

CRUISERS • CGHM 24:
22 *Ticonderoga* with *Aegis* Baseline 5/6/8/9 C2, 2 quad lnchr with RGM-84D *Harpoon* Block 1C AShM, 16 8-cell Mk 41 VLS (of which 2 only 5-cell and fitted with reload crane) with RGM-109C/E *Tomahawk* Block III/IV LACM/SM-2 Block III/IIIA/IIIB/IV SAM/SM-3 Block IA/B SAM/SM-6 Block I SAM, 2 triple 324mm SVTT Mk 32 ASTT with Mk 54 LWT, 2 Mk 15 *Phalanx* Block 1B CIWS, 2 127mm guns (capacity 2 MH-60R *Seahawk*/MH-60S *Knight Hawk* hels)
2 *Zumwalt* with 20 4-cell Mk 57 VLS with RGM-109C/E *Tomahawk* Block III/IV LACM/RIM-162 ESSM SAM/SM-2 Block IIIA SAM/ASROC A/S msl, 2 155mm guns (capacity 2 MH-60R *Seahawk* ASW hel or 1 MH-60R *Seahawk* ASW hel and 3 *Fire Scout* UAV)

DESTROYERS 68:
DDGHM 40:
5 *Arleigh Burke* Flight IIA with *Aegis* Baseline 5/9 C2, 12 8-cell Mk 41 VLS with RGM-109C/E *Tomahawk* Block III/IV LACM/SM-2 Block III/IIIA/IIIB/IV SAM/SM-3 Block IA/B SAM/SM-6 Block I SAM/ASROC A/S msl, 2 triple 324mm SVTT Mk 32 ASTT with Mk 54 LWT, 2 Mk 15 *Phalanx* Block 1B CIWS, 1 127mm gun (capacity 2 MH-60R *Seahawk*/MH-60S *Knight Hawk* hels)
35 *Arleigh Burke* Flight IIA with *Aegis* Baseline 6/7/9 C2, 12 8-cell Mk 41 VLS with RGM-109C/E *Tomahawk* Block III/IV LACM/SM-2 Block III/IIIA/IIIB/IV SAM/SM-3 Block IA/B SAM/SM-6 Block I SAM/ASROC A/S msl, 2 triple 324mm SVTT Mk 32 ASTT with Mk 54 LWT, 1 Mk 15 *Phalanx* Block 1B CIWS, 1 127mm gun (capacity 2 MH-60R *Seahawk*/MH-60S *Knight Hawk* hels) (of which 1 vessel also with 1 Mk 15 SeaRAM with RIM-116C RAM Block 2 and 1 vessel also with 1 *Optical Dazzling Interdictor, Navy* (ODIN) LWS)
DDGM 28 *Arleigh Burke* Flight I/II with *Aegis* Baseline 5/9 C2, 2 quad lnchr with RGM-84D *Harpoon* Block 1C AShM, 12 8-cell Mk 41 VLS (of which 2 only 5-cell and fitted with reload crane) with RGM-109C/E *Tomahawk* Block III/IV LACM/SM-2 Block III/IIIA/IIIB/IV SAM/SM-3 Block IA/B SAM/SM-6 Block I SAM/ASROC A/S msl, 2 triple 324mm SVTT Mk 32 ASTT with Mk 54 LWT, 2 Mk 15 *Phalanx* Block 1B CIWS (of which 3 vessels with 1 Mk 15 SeaRAM with RIM-116C RAM Block 2, 1 Mk 15 *Phalanx* Block 1B instead of 2 *Phalanx*), 1 127mm gun, 1 hel landing platform

FRIGATES 21:
FFGHM 1 *Independence* with 2 quad lnchr with NSM (RGM-184A) AShM, 1 11-cell SeaRAM lnchr with RIM-116C Block 2 SAM, 1 57mm gun (capacity 2 MH-60R/S *Seahawk/Knight Hawk* hel and 3 MQ-8 *Fire Scout* UAV)
FFHM 20:
10 *Freedom* with 1 21-cell Mk 49 lnchr with RIM-116C RAM Block 2 SAM, 1 57mm gun (capacity 2 MH-60R/S *Seahawk/Knight Hawk* hel or 1 MH-60 with 3 MQ-8 *Fire Scout* UAV)
10 *Independence* with 1 11-cell SeaRAM lnchr with RIM-116C Block 2 SAM, 1 57mm gun (capacity 2 MH-60R/S *Seahawk/Knight Hawk* hel and 3 MQ-8 *Fire Scout* UAV)

PATROL AND COASTAL COMBATANTS 83
PCFG 10 *Cyclone* with 1 quad Mk 208 lnchr with BGM-176B *Griffin B* SSM
PCF 3 *Cyclone*
PBF 64: 12 Mk VI; 25 Combatant Craft Assault; 2 Combatant Craft Heavy; 25 Combatant Craft Medium Mk 1
PBR 6 Riverine Command Boat
MINE WARFARE • MINE COUNTERMEASURES 8
MCO 8 *Avenger*
COMMAND SHIPS
LCC 2 *Blue Ridge* with 2 Mk 15 *Phalanx* Block 1B CIWS (capacity 3 LCPL; 2 LCVP; 700 troops; 1 med hel) (of which 1 vessel partially crewed by Military Sealift Command personnel)
AMPHIBIOUS
PRINCIPAL AMPHIBIOUS SHIPS 32:
LHA 2 *America* with 2 8-cell Mk 29 GMLS with RIM-162D ESSM SAM, 2 Mk 49 GMLS with RIM-116C RAM Block 2 SAM, 2 Mk 15 *Phalanx* Block 1B CIWS (capacity 5 F-35B *Lightning* II FGA ac; 12 MV-22B *Osprey* tpt ac; 4 AH-1Z *Viper* atk hel; 2 MH-60S *Knight Hawk* MRH; 4 CH-53E *Sea Stallion* tpt hel; 2 UH-1Y *Iroquois* tpt hel; up to 1,800 troops)
LHD 7 *Wasp* (1 other non-operational following serious fire in 2020) with 2 8-cell Mk 29 GMLS with RIM-7M/P *Sea Sparrow* SAM, 2 Mk 49 GMLS with RIM-116C RAM Block 2 SAM, 2 Mk 15 *Phalanx* Block 1B CIWS (capacity: 6 AV-8B *Harrier* II FGA or F-35B *Lightning* II FGA ac; 4 CH-53E *Sea Stallion* hel; 6 MV-22B *Osprey* tpt ac; 4 AH-1W/Z hel; 3 UH-1Y hel; 3 LCAC(L); 60 tanks; 1,687 troops)
LPD 11 *San Antonio* with 2 21-cell Mk 49 GMLS with RIM-116C RAM Block 2 SAM (1 vessel also fitted with 1 *Solid-State Laser Technology Maturation* (SSL-TM) LWS) (capacity 2 CH-53E *Sea Stallion* hel or 2 MV-22 *Osprey*; 2 LCAC(L); 14 AAAV; 720 troops)
LSD 12:
4 *Harpers Ferry* with 2 Mk 49 GMLS with RIM-116C RAM Block 2 SAM, 2 Mk 15 *Phalanx* Block 1B CIWS (capacity 2 CH-53E *Sea Stallion* hel; 2 LCAC(L); 40 tanks; 500 troops)
8 *Whidbey Island* with 2 Mk 49 GMLS with RIM-116C RAM Block 2 SAM, 2 Mk 15 *Phalanx* Block 1B CIWS (capacity 2 CH-53E *Sea Stallion* hel; 4 LCAC(L); 40 tanks; 500 troops)
LANDING CRAFT 141:
LCU 32 LCU 1610 (capacity either 1 M1 *Abrams* MBT or 350 troops)
LCM 8 LCM 8
LCP 33 Maritime Positioning Force Utility Boat (MPF-UB)
LCAC 68 LCAC(L) (MLU ongoing) (capacity either 1 MBT or 60 troops)
LOGISTICS AND SUPPORT 12
AFDL 1 *Dynamic*
AGOR 6 (all leased out): 2 *Ocean*; 3 *Thomas G. Thompson*; 1 *Kilo Moana*
ARD 2
AX 1 *Prevail*
ESB 2 *Lewis B. Puller* (capacity 4 MH-53/MH-60 hel)
UUV (1 *Cutthroat* for testing)

Naval Reserve Forces 101,900

Selected Reserve 59,150

Individual Ready Reserve 42,750

Naval Inactive Fleet

Notice for reactivation:
60–90 days minimum (still on naval-vessel register)

EQUIPMENT BY TYPE
AMPHIBIOUS • LHA 3 *Tarawa*
LOGISTICS AND SUPPORT 4
AOE 2 *Supply*
ARS 2 *Safeguard*

Military Sealift Command (MSC)

Fleet Oiler (PM1)

EQUIPMENT BY TYPE
LOGISTICS AND SUPPORT 15
AOR 15 *Henry J. Kaiser* with 1 hel landing platform

Special Mission (PM2)

EQUIPMENT BY TYPE
LOGISTICS AND SUPPORT 21
AGM 3: 1 *Howard O. Lorenzen*; 1 *Invincible* (commercial operator); 1 Sea-based X-band radar
AGOR 6 *Pathfinder*
AGOS 5: 1 *Impeccable* (commercial operator); 4 *Victorious*
AGS 1 *Waters*
ARC 1 *Zeus*
AS 5 (long-term chartered, of which 1 *Dominator*, 4 *Arrowhead*)

Prepositioning (PM3)

EQUIPMENT BY TYPE
LOGISTICS AND SUPPORT 20
AG 2: 1 *V Adm K.R. Wheeler*; 1 *Fast Tempo*
AKR 10: 2 *Bob Hope*; 1 *Stockham*; 2 *Watson*
AKRH 5 *2nd Lt John P. Bobo*
ESB 1 *Lewis B. Puller* (capacity 4 MH-53 hel/4 MV-22 tiltrotor; 250 troops)
ESD 2 *Montford Point*

Service Support (PM4)

EQUIPMENT BY TYPE
LOGISTICS AND SUPPORT 12
AH 2 *Mercy* with 1 hel landing platform
ARS 2 *Safeguard*
AS 4: 1 *Carolyn Chouest*; 2 *Emory S Land*; 1 *Malama* (long-term chartered)
ATF 4: 1 *Gary Chouest*; 3 *Powhatan*

Sealift (PM5)

(At a minimum of 4 days' readiness)

EQUIPMENT BY TYPE
LOGISTICS AND SUPPORT 25
AOT 1 *Maersk Peary* (long-term chartered)

AK 9: 3 *Sgt Matej Kocak;* 1 *1st Lt Harry L. Martin;* 1 *LCpl Roy M. Wheat;* 2 *LTC John U.D. Page;* 1 *Maj. Bernard F. Fisher;* 1 *CPT David I. Lyon*
AKR 15: 5 *Bob Hope;* 2 *Gordon;* 2 *Shughart;* 6 *Watson*

Fleet Ordnance and Dry Cargo (PM6)

EQUIPMENT BY TYPE
LOGISTICS AND SUPPORT 16
AOE 2 *Supply*
AKEH 14 *Lewis and Clark*

Expeditionary Fast Transport (PM8)

EQUIPMENT BY TYPE
LOGISTICS AND SUPPORT 14
AP 2 *Guam*
EPF 12 *Spearhead*

Dry Cargo and Tankers

EQUIPMENT BY TYPE
LOGISTICS AND SUPPORT 7
AK 3 (long-term chartered, of which 1 MV *Maersk Valencia;* 1 *Sea Eagle;* 1 SLNC *Corsica*)
AOT 4 (long-term chartered, of which 2 *Empire State;* 1 SLNC *Pax;* 1 SLNC *Goodwill*)

US Maritime Administration (MARAD)

National Defense Reserve Fleet

EQUIPMENT BY TYPE
LOGISTICS AND SUPPORT 27
AGOS 2 *General Rudder*
AGM 2: 1 *Pacific Collector;* 1 *Pacific Tracker*
AK 12: 2 *Cape Ann* (breakbulk); 1 *Cape Chalmers* (breakbulk); 2 *Cape Farewell;* 1 *Cape Fear;* 2 *Cape Jacob;* 1 *Cape May* (heavy lift); 1 *Cape Nome* (breakbulk); 1 *Del Monte* (breakbulk); 1 *Savannah*
AOT 5: 1 *Lawrence H. Gianella;* 3 *Paul Buck;* 1 *Petersburg*
AP 4: 1 *Empire State* VI; 1 *Golden Bear;* 1 *Kennedy;* 1 *State of Maine*
AX 2: 1 *Freedom Star;* 1 *Kings Pointer*

Ready Reserve Force

Ships at readiness up to a maximum of 30 days

EQUIPMENT BY TYPE
LOGISTICS AND SUPPORT 42
ACS 6: 2 *Flickertail State;* 1 *Gopher State;* 3 *Keystone State*
AK 2 *Wright* (breakbulk)
AKR 33: 1 *Adm W.M. Callaghan;* 4 *Algol;* 4 *Cape Capella;* 1 *Cape Decision;* 4 *Cape Ducato;* 1 *Cape Edmont;* 1 *Cape Henry;* 2 *Cape Hudson;* 2 *Cape Knox;* 4 *Cape Island;* 1 *Cape Orlando;* 1 *Cape Race;* 1 *Cape Trinity;* 2 *Cape Trinity;* 2 *Cape Victory;* 2 *Cape Washington*
AOT 1 *Petersburg*

Naval Aviation 98,600

10 air wg. Average air wing comprises 8 sqns: 4 with F/A-18; 1 with MH-60R; 1 with EA-18G; 1 with E-2C/D; 1 with MH-60S

FORCES BY ROLE
FIGHTER/GROUND ATTACK
22 sqn with F/A-18E *Super Hornet*
11 sqn with F/A-18F *Super Hornet*
1 sqn with F-35C *Lightning* II
ANTI-SUBMARINE WARFARE
12 sqn with MH-60R *Seahawk*
3 ASW/ISR sqn with MH-60R *Seahawk;* MQ-8B *Fire Scout*
ELINT
1 sqn with EP-3E *Aries* II
ELINT/ELECTRONIC WARFARE
13 sqn with EA-18G *Growler*
MARITIME PATROL
12 sqn with P-8A *Poseidon*
1 (special projects) sqn with P-8A *Poseidon*
AIRBORNE EARLY WARNING & CONTROL
4 sqn with E-2C *Hawkeye*
5 sqn with E-2D *Hawkeye*
COMMAND & CONTROL
2 sqn with E-6B *Mercury*
MINE COUNTERMEASURES
2 sqn with MH-53E *Sea Dragon*
TRANSPORT
2 sqn with C-2A *Greyhound*
TRAINING
1 (FRS) sqn with EA-18G *Growler*
1 (FRS) sqn with C-2A *Greyhound;* E-2C/D *Hawkeye;* TE-2C *Hawkeye*
1 sqn with E-6B *Mercury*
2 (FRS) sqn with F/A-18C/D *Hornet;* F/A-18E/F *Super Hornet*
1 (FRS) sqn with F-35C *Lightning* II
1 (FRS) sqn with MH-53 *Sea Dragon*
2 (FRS) sqn with MH-60S *Knight Hawk;* HH-60H *Seahawk*
2 (FRS) sqn with MH-60R *Seahawk*
1 (FRS) sqn with P-3C *Orion;* P-8A *Poseidon*
6 sqn with T-6A/B *Texan* II
2 sqn with T-44C *Pegasus*
5 sqn with T-45C *Goshawk*
3 hel sqn with TH-57B/C *Sea Ranger*
1 (FRS) UAV sqn with MQ-8B *Fire Scout;* MQ-8C *Fire Scout*
TRANSPORT HELICOPTER
14 sqn with MH-60S *Knight Hawk*
1 tpt hel/ISR sqn with MH-60S *Knight Hawk;* MQ-8B *Fire Scout*
ISR UAV
1 sqn with MQ-4C *Triton*

EQUIPMENT BY TYPE
AIRCRAFT 989 combat capable
FGA 720: 30 F-35C *Lightning* II; 10 F-16A *Fighting Falcon;* 4 F-16B *Fighting Falcon;* 5 F/A-18B *Hornet;* 60 F/A-18C *Hornet;* 25 F/A-18D *Hornet;* 313 F/A-18E *Super Hornet;* 273 F/A-18F *Super Hornet*
ASW 111: 11 P-3C *Orion;* 100 P-8A *Poseidon*
EW 158 EA-18G *Growler**
ELINT 9 EP-3E *Aries* II
AEW&C 94: 49 E-2C *Hawkeye;* 45 E-2D *Hawkeye*
C2 16 E-6B *Mercury*

North America

TKR 3: 1 KC-130R *Hercules*; 1 KC-130T *Hercules*; 1 KC-130J *Hercules*
TPT • Light 54: 4 Beech A200 *King Air* (C-12C *Huron*); 6 Beech A200 *King Air* (UC-12F *Huron*); 8 Beech A200 *King Air* (UC-12M *Huron*); 27 C-2A *Greyhound*; 2 DHC-2 *Beaver* (U-6A); 7 SA-227-BC *Metro* III (C-26D)
TRG 580: 44 T-6A *Texan* II; 231 T-6B *Texan* II; 7 T-38C *Talon*; 55 T-44C *Pegasus*; 241 T-45C *Goshawk*; 2 TE-2C *Hawkeye*

TILTROTOR • TPT 6 CMV-22B *Osprey*

HELICOPTERS
ASW 269 MH-60R *Seahawk*
MRH 260 MH-60S *Knight Hawk* (Multi Mission Support)
MCM 28 MH-53E *Sea Dragon*
ISR 3 OH-58C *Kiowa*
TPT 13: **Heavy** 2 CH-53E *Sea Stallion*; **Medium** 3 UH-60L *Black Hawk*; **Light** 8: 5 UH-72A *Lakota*; 2 UH-1N *Iroquois*; 1 UH-1Y *Venom*
TRG 119: 43 TH-57B *Sea Ranger*; 76 TH-57C *Sea Ranger*

UNINHABITED AERIAL VEHICLES • ISR 108
Heavy 58: 5 MQ-4C *Triton*; 20 MQ-8B *Fire Scout*; 29 MQ-8C *Fire Scout*; 4 RQ-4A *Global Hawk* (under evaluation and trials); **Medium** 35 RQ-2B *Pioneer*; **Light** 15 RQ-21A *Blackjack*

AIR-LAUNCHED MISSILES
AAM • IR AIM-9M *Sidewinder*; **IIR** AIM-9X *Sidewinder* II; **SARH** AIM-7 *Sparrow* (being withdrawn); **ARH** AIM-120C-5/C-7/D AMRAAM
ASM AGM-65F *Maverick*; AGM-114B/K/M *Hellfire*; APKWS
AShM AGM-84D *Harpoon*; AGM-119A *Penguin* 3; AGM-158C LRASM
ARM AGM-88B/C/E HARM/AARGM
ALCM • Conventional AGM-84E/H/K SLAM/SLAM-ER

BOMBS
Laser-guided: GBU-10/12/16 *Paveway* II; GBU-24 *Paveway* III
INS/GPS guided: GBU-31/32/38 JDAM; Enhanced *Paveway* II; GBU-54 Laser JDAM; AGM-154A/C/C-1 JSOW

Naval Aviation Reserve

FORCES BY ROLE

FIGHTER/GROUND ATTACK
1 sqn with F/A-18C/D *Hornet*

ANTI-SUBMARINE WARFARE
1 sqn with MH-60R *Seahawk*

ELECTRONIC WARFARE
1 sqn with EA-18G *Growler*

MARITIME PATROL
2 sqn with P-3C *Orion*

TRANSPORT
6 log spt sqn with B-737-700 (C-40A *Clipper*)
1 log spt sqn with Gulfstream IV (C-20G); Gulfstream V/G550 (C-37A/B)
4 sqn with C-130T *Hercules*
1 sqn with KC-130T *Hercules*

TRAINING
2 (aggressor) sqn with F-5F/N *Tiger* II
1 (aggressor) sqn with F/A-18C/D *Hornet*

TRANSPORT HELICOPTER
1 sqn with MH-60S *Knight Hawk*

EQUIPMENT BY TYPE

AIRCRAFT 83 combat capable
FTR 31: 2 F-5F *Tiger* II; 29 F-5N *Tiger* II
FGA 35: 30 F/A-18C *Hornet*; 5 F/A-18D *Hornet*
ASW 12 P-3C *Orion*
EW 5 EA-18G *Growler**
TKR 5 KC-130T *Hercules*
TPT 40: **Medium** 18 C-130T *Hercules*; **PAX** 22: 17 B-737-700 (C-40A *Clipper*); 1 Gulfstream IV (C-20G); 1 Gulfstream V (C-37A); 3 Gulfstream G550 (C-37B)

HELICOPTERS
ASW 7 MH-60R *Seahawk*
MRH 11 MH-60S *Knight Hawk*
MCM 7 MH-53E *Sea Dragon*

US Marine Corps 180,950

3 Marine Expeditionary Forces (MEF), 3 Marine Expeditionary Brigades (MEB), 7 Marine Expeditionary Units (MEU) drawn from 3 div. An MEU usually consists of a battalion landing team (1 SF coy, 1 lt armd recce coy, 1 recce pl, 1 armd pl, 1 amph aslt pl, 1 inf bn, 1 arty bty, 1 cbt engr pl), an aviation combat element (1 medium-lift sqn with attached atk hel, FGA ac and AD assets) and a composite log bn, with a combined total of about 2,200 personnel. Composition varies with mission requirements

FORCES BY ROLE

SPECIAL FORCES
(see USSOCOM)

MANOEUVRE

Reconnaissance
3 (MEF) recce coy

Amphibious
1 (1st) mne div (2 armd recce bn, 1 recce bn, 1 tk bn, 3 mne regt (4 mne bn), 1 amph aslt bn, 1 arty regt (3 arty bn, 1 MRL bn), 1 cbt engr bn, 1 EW bn, 1 int bn, 1 sigs bn)
1 (2nd) mne div (1 armd recce bn, 1 recce bn, 1 tk bn, 3 mne regt (3 mne bn), 1 amph aslt bn, 1 arty regt (2 arty bn), 1 cbt engr bn, 1 EW bn, 1 int bn, 1 sigs bn)
1 (3rd) mne div (1 recce bn, 1 inf regt (3 inf bn), 1 arty regt (2 arty bn), 1 cbt spt bn (1 armd recce coy, 1 amph aslt coy, 1 cbt engr coy), 1 EW bn, 1 int bn, 1 sigs bn)

COMBAT SERVICE SUPPORT
3 log gp

EQUIPMENT BY TYPE

ARMOURED FIGHTING VEHICLES
IFV 488 LAV-25
APC • APC (W) 207 LAV variants (66 CP; 127 log; 14 EW)
AAV 1,242: 1,200 AAV-7A1 (all roles); 42 ACV (in test)
AUV 4,929+: 1,725 *Cougar*; 2,500+ JLTV; 704 M-ATV

ENGINEERING & MAINTENANCE VEHICLES
AEV 42 M1 ABV
ARV 105: 60 AAVRA1; 45 LAV-R

MW 38 *Buffalo*; some *Husky* 2G
VLB ε30 M60 AVLB
ANTI-TANK/ANTI-INFRASTRUCTURE •
MSL
SP 106 LAV-AT
MANPATS FGM-148 *Javelin*; FGM-172B SRAW-MPV; TOW
ARTILLERY 1,452
TOWED 812: **105mm**: 331 M101A1; **155mm** 481 M777A2
MRL 227mm 40 M142 HIMARS
MOR 600: **81mm** 535 M252; **SP 81mm** 65 LAV-M; **120mm** (49 EFSS in store for trg)
UNMANNED AERIAL VEHICLES
ISR • Light 100 BQM-147 *Exdrone*
AIR DEFENCE • SAM • Point-defence FIM-92 *Stinger*

Marine Corps Aviation 34,700

3 active Marine Aircraft Wings (MAW) and 1 MCR MAW

FORCES BY ROLE
FIGHTER
1 sqn with F/A-18A++/C/C+ *Hornet*
3 sqn with F/A-18C *Hornet*
2 sqn with F/A-18D *Hornet*
FIGHTER/GROUND ATTACK
5 sqn with AV-8B *Harrier* II
5 sqn with F-35B *Lightning* II
1 sqn with F-35C *Lightning* II (forming)
COMBAT SEARCH & RESCUE/TRANSPORT
1 sqn with Beech A200/B200 *King Air* (UC-12F/M *Huron*); Beech 350 *King Air* (UC-12W *Huron*); Cessna 560 *Citation Ultra/Encore* (UC-35C/D); Gulfstream IV (C-20G)
TANKER
3 sqn with KC-130J *Hercules*
TRANSPORT
17 sqn with MV-22B *Osprey*
TRAINING
1 sqn with AV-8B *Harrier* II; TAV-8B *Harrier*
1 sqn with F/A-18C/D *Hornet*
2 sqn with F-35B *Lightning* II
1 sqn with MV-22B *Osprey*
1 hel sqn with AH-1Z *Viper*; UH-1Y *Venom*
1 hel sqn with CH-53E *Sea Stallion*
ATTACK HELICOPTER
7 sqn with AH-1Z *Viper*; UH-1Y *Venom*
TRANSPORT HELICOPTER
8 sqn with CH-53E *Sea Stallion*
1 (VIP) sqn with MV-22B *Osprey*; VH-3D *Sea King*; VH-60N *White Hawk*
ISR UAV
3 sqn with RQ-21A *Blackjack*
AIR DEFENCE
2 bn with M1097 *Avenger*; FIM-92 *Stinger*

EQUIPMENT BY TYPE
AIRCRAFT 362 combat capable
FGA 362: 95 F-35B *Lightning* II; 11 F-35C *Lightning* II; 10 F/A-18A++ *Hornet*; 70 F/A-18C *Hornet*; 1 F/A-18C+ *Hornet*; 50 F/A-18D *Hornet*; 109 AV-8B *Harrier* II; 16 TAV-8B *Harrier*
TKR 44 KC-130J *Hercules*
TPT 17: **Light** 16: 5 Beech A200/B200 *King Air* (UC-12F/M *Huron*); 5 Beech 350 *King Air* (C-12W *Huron*); 6 Cessna 560 *Citation Encore* (UC-35D); **PAX** 1 Gulfstream IV (C-20G)
TRG 3 T-34C *Turbo Mentor*
TILTROTOR • TPT 309 MV-22B *Osprey*
HELICOPTERS
ATK 141 AH-1Z *Viper*
TPT 285: **Heavy** 138: 137 CH-53E *Sea Stallion*; 1 CH-53K *King Stallion*; **Medium** 19: 8 VH-60N *White Hawk* (VIP tpt); 11 VH-3D *Sea King* (VIP tpt); **Light** 128 UH-1Y *Venom*
UNINHABITED AERIAL VEHICLES
ISR • Light 80 RQ-21A *Blackjack*
AIR DEFENCE
SAM • Point-defence FIM-92 *Stinger*; M1097 *Avenger*
AIR-LAUNCHED MISSILES
AAM • IR AIM-9M *Sidewinder*; **IIR** AIM-9X *Sidewinder* II; **SARH** AIM-7P *Sparrow*; **ARH** AIM-120C AMRAAM
ASM AGM-65E/F IR *Maverick*; AGM-114 *Hellfire*; AGM-176 *Griffin*; APKWS
AShM AGM-84D *Harpoon*
ARM AGM-88 HARM
LACM AGM-84E/H/K SLAM/SLAM-ER
BOMBS
Laser-guided GBU-10/12/16 *Paveway* II
INS/GPS guided GBU-31 JDAM; AGM-154A/C/C-1 JSOW

Reserve Organisations

Marine Corps Reserve 35,500

FORCES BY ROLE
MANOEUVRE
Reconnaissance
2 MEF recce coy
Amphibious
1 (4th) mne div (1 armd recce bn, 1 recce bn, 2 mne regt (3 mne bn), 1 amph aslt bn, 1 arty regt (2 arty bn, 1 MRL bn), 1 cbt engr bn, 1 int bn, 1 sigs bn)
COMBAT SERVICE SUPPORT
1 log gp

Marine Corps Aviation Reserve 12,000 reservists

FORCES BY ROLE
FIGHTER
1 sqn with F/A-18A++/C+ *Hornet*
TANKER
1 sqn with KC-130J *Hercules*
1 sqn with KC-130J/T *Hercules*
TRANSPORT
2 sqn with MV-22B *Osprey*
TRAINING
1 sqn with F-5F/N *Tiger* II
ATTACK HELICOPTER
2 sqn with AH-1Z *Viper*; UH-1Y *Venom*
TRANSPORT HELICOPTER
1 sqn with CH-53E *Sea Stallion*
ISR UAV
1 sqn with RQ-21A *Blackjack*

EQUIPMENT BY TYPE
AIRCRAFT 31 combat capable
FTR 12: 1 F-5F *Tiger* II; 11 F-5N *Tiger* II
FGA 19: 13 F/A-18A++ *Hornet*; 6 F/A-18C+ *Hornet*
TKR 21: 11 KC-130J *Hercules*; 10 KC-130T *Hercules*
TPT • Light 8: 2 Beech 350 *King Air* (UC-12W *Huron*); 2 Cessna 560 *Citation Ultra* (UC-35C); 4 Cessna 560 *Citation Encore* (UC-35D)
TILTROTOR • TPT 24 MV-22B *Osprey*
HELICOPTERS
ATK 12 AH-1Z *Viper*
TPT 29: **Heavy** 7 CH-53E *Sea Stallion*; **Light** 22 UH-1Y *Venom*
UNINHABITED AERIAL VEHICLES
ISR • Light 20 RQ-21A *Blackjack*

Marine Stand-by Reserve 700 reservists
Trained individuals available for mobilisation

US Coast Guard 41,450
9 districts (4 Pacific, 5 Atlantic)
EQUIPMENT BY TYPE
PATROL AND COASTAL COMBATANTS 342
PSOH 24: 1 *Alex Haley*; 13 *Famous*; 2 *Hamilton*; 8 *Legend* with 1 Mk 15 *Phalanx* Block 1B CIWS, 1 57mm gun (capacity 2 MH-65 hel)
PCO 52: 14 *Reliance* (with 1 hel landing platform); 38 *Sentinel* (Damen 4708)
PCC 19 *Island*
PBF 174 *Response Boat-Medium* (RB-M)
PBI 73 *Marine Protector*
LOGISTICS AND SUPPORT 65
ABU 52: 16 *Juniper*; 4 WLI; 14 *Keeper*; 18 WLR
AGB 12: 9 *Bay*; 1 *Mackinaw*; 1 *Healy*; 1 *Polar* (1 *Polar* in reserve)
AXS 1 *Eagle*

US Coast Guard Aviation
EQUIPMENT BY TYPE
AIRCRAFT
SAR 42: 11 HC-130H *Hercules*; 13 HC-130J *Hercules*; 10 HC-144A; 8 HC-144B
TPT 16: **Medium** 14 C-27J *Spartan*; **PAX** 2 Gulfstream V (C-37A)
HELICOPTERS
SAR 146: 44 MH-60T *Jayhawk*; 102 AS366G1 (MH-65C/D) *Dauphin* II

US Air Force (USAF) 331,400
Almost the entire USAF (plus active-force ANG and AFR) is divided into 10 Aerospace Expeditionary Forces (AEF), each on call for 120 days every 20 months. At least 2 of the 10 AEFs are on call at any one time, each with 10,000–15,000 personnel, 90 multi-role ftr and bbr ac, 31 intra-theatre refuelling aircraft and 13 aircraft for ISR and EW missions

Global Strike Command (GSC)
2 active air forces (8th & 20th); 8 wg
FORCES BY ROLE
SURFACE-TO-SURFACE MISSILE
9 ICBM sqn with LGM-30G *Minuteman* III
BOMBER
4 sqn with B-1B *Lancer*
2 sqn with B-2A *Spirit*
5 sqn (incl 1 trg) with B-52H *Stratofortress*
COMMAND & CONTROL
1 sqn with E-4B
TRANSPORT HELICOPTER
3 sqn with UH-1N *Iroquois*

Air Combat Command (ACC)
2 active air forces (9th & 12th); 12 wg. ACC numbered air forces provide the air component to CENTCOM, SOUTHCOM and NORTHCOM
FORCES BY ROLE
FIGHTER
3 sqn with F-22A *Raptor*
FIGHTER/GROUND ATTACK
4 sqn with F-15E *Strike Eagle*
3 sqn with F-16C/D *Fighting Falcon* (+6 sqn personnel only)
3 sqn with F-35A *Lightning* II
GROUND ATTACK
3 sqn with A-10C *Thunderbolt* II (+1 sqn personnel only)
ELECTRONIC WARFARE
1 sqn with EA-18G *Growler* (personnel only – USN aircraft)
2 sqn with EC-130H *Compass Call*
ISR
2 sqn with E-8C J-STARS (personnel only)
5 sqn with OC-135/RC-135/WC-135
2 sqn with U-2S
AIRBORNE EARLY WARNING & CONTROL
5 sqn with E-3 *Sentry*
COMBAT SEARCH & RESCUE
2 sqn with HC-130J *Combat King* II
2 sqn with HH-60G *Pave Hawk*
TRAINING
1 sqn with A-10C *Thunderbolt* II
1 sqn with E-3 *Sentry*
2 sqn with F-15E *Strike Eagle*
1 sqn with F-22A *Raptor*
1 sqn with RQ-4A *Global Hawk*; TU-2S
1 UAV sqn with MQ-9A *Reaper*
COMBAT/ISR UAV
9 sqn with MQ-9A *Reaper*
ISR UAV
2 sqn with EQ-4B/RQ-4B *Global Hawk*
2 sqn with RQ-170 *Sentinel*
1 sqn with RQ-180

Pacific Air Forces (PACAF)
Provides the air component of PACOM, and commands air units based in Alaska, Hawaii, Japan and South Korea. 3 active air forces (5th, 7th, & 11th); 8 wg
FORCES BY ROLE
FIGHTER
2 sqn with F-15C/D *Eagle*
2 sqn with F-22A *Raptor* (+1 sqn personnel only)
FIGHTER/GROUND ATTACK
5 sqn with F-16C/D *Fighting Falcon*
1 sqn with F-35A *Lightning* II

GROUND ATTACK
1 sqn with A-10C *Thunderbolt* II
AIRBORNE EARLY WARNING & CONTROL
2 sqn with E-3 *Sentry*
COMBAT SEARCH & RESCUE
1 sqn with HH-60G *Pave Hawk*
TANKER
1 sqn with KC-135R (+1 sqn personnel only)
TRANSPORT
1 sqn with B-737-200 (C-40B); Gulfstream V (C-37A)
1 sqn with C-17A *Globemaster* (+1 sqn personnel only)
1 sqn with C-130J-30 *Hercules*
1 sqn with Beech 1900C (C-12J); UH-1N *Huey*
TRAINING
1 (aggressor) sqn with F-16C/D *Fighting Falcon*

United States Air Forces Europe (USAFE)

Provides the air component to both EUCOM and AFRICOM. 1 active air force (3rd); 5 wg

FORCES BY ROLE
FIGHTER
1 sqn with F-15C/D *Eagle*
FIGHTER/GROUND ATTACK
2 sqn with F-15E *Strike Eagle*
3 sqn with F-16C/D *Fighting Falcon*
COMBAT SEARCH & RESCUE
1 sqn with HH-60G *Pave Hawk*
TANKER
1 sqn with KC-135R *Stratotanker*
TRANSPORT
1 sqn with C-130J-30 *Hercules*
2 sqn with Gulfstream V (C-37A); Learjet 35A (C-21A); B-737-700 (C-40B)

Air Mobility Command (AMC)

Provides strategic and tactical airlift, air-to-air refuelling and aeromedical evacuation. 1 active air force (18th); 12 wg and 1 gp

FORCES BY ROLE
TANKER
4 sqn with KC-10A *Extender*
1 sqn with KC-46A *Pegasus* (forming)
8 sqn with KC-135R/T *Stratotanker* (+2 sqn with personnel only)
TRANSPORT
1 VIP sqn with B-737-200 (C-40B); B-757-200 (C-32A)
1 VIP sqn with Gulfstream V (C-37A); Gulfstream 550 (C-37B)
1 VIP sqn with VC-25 *Air Force One*
2 sqn with C-5M *Super Galaxy*
8 sqn with C-17A *Globemaster* III (+1 sqn personnel only)
5 sqn with C-130J-30 *Hercules* (+1 sqn personnel only)
1 sqn with Learjet 35A (C-21A)

Air Education and Training Command

1 active air force (2nd), 10 active air wg and 1 gp

FORCES BY ROLE
TRAINING
1 sqn with C-17A *Globemaster* III
1 sqn with C-130J-30 *Hercules*
4 sqn with F-16C/D *Fighting Falcon*
4 sqn with F-35A *Lightning* II
1 sqn with KC-46A *Pegasus* (forming)
1 sqn with KC-135R *Stratotanker*
5 (flying trg) sqn with T-1A *Jayhawk*
10 (flying trg) sqn with T-6A *Texan* II
10 (flying trg) sqn with T-38C *Talon*
5 UAV sqn with MQ-9A *Reaper*

EQUIPMENT BY TYPE
SURFACE-TO-SURFACE MISSILE LAUNCHERS
ICBM • Nuclear 400 LGM-30G *Minuteman* III (1 Mk12A or Mk21 re-entry veh per missile)
AIRCRAFT 1,543 combat capable
BBR 139: 61 B-1B *Lancer*; 20 B-2A *Spirit*; 58 B-52H *Stratofortress* (46 nuclear capable)
FTR 261: 86 F-15C *Eagle*; 10 F-15D *Eagle*; 165 F-22A *Raptor*
FGA 1,000: 219 F-15E *Strike Eagle*; 439 F-16C *Fighting Falcon*; 111 F-16D *Fighting Falcon*; 231 F-35A *Lightning* II
ATK 143 A-10C *Thunderbolt* II
CSAR 16 HC-130J *Combat King* II
EW 13 EC-130H *Compass Call*
ISR 40: 2 E-9A; 4 E-11A; 2 OC-135B *Open Skies*; 26 U-2S; 4 TU-2S; 2 WC-135 *Constant Phoenix*
ELINT 22: 8 RC-135V *Rivet Joint*; 9 RC-135W *Rivet Joint*; 3 RC-135S *Cobra Ball*; 2 RC-135U *Combat Sent*
AEW&C 31: 10 E-3B *Sentry*; 1 E-3C *Sentry*; 20 E-3G *Sentry*
C2 4 E-4B
TKR 156: 126 KC-135R *Stratotanker*; 30 KC-135T *Stratotanker*
TKR/TPT 83: 56 KC-10A *Extender*; 27 KC-46A *Pegasus*
TPT 333: **Heavy** 182: 36 C-5M *Super Galaxy*; 146 C-17A *Globemaster* III; **Medium** 105 C-130J/J-30 *Hercules*; **Light** 23: 4 Beech 1900C (C-12J); 19 Learjet 35A (C-21A); **PAX** 23: 4 B-737-700 (C-40B); 4 B-757-200 (C-32A); 9 Gulfstream V (C-37A); 4 Gulfstream 550 (C-37B); 2 VC-25A *Air Force One*
TRG 1,126: 178 T-1A *Jayhawk*; 443 T-6A *Texan* II; 505 T-38A/C *Talon*
HELICOPTERS
CSAR 76: 74 HH-60G *Pave Hawk*; 2 HH-60W *Jolly Green* II
TPT • Light 62 UH-1N *Huey*
UNINHABITED AERIAL VEHICLES 261
CISR • Heavy 210 MQ-9A *Reaper*
ISR • Heavy 51: 3 EQ-4B; 31 RQ-4B *Global Hawk*; ε10 RQ-170 *Sentinel*; ε7 RQ-180
AIR DEFENCE
SAM • Point-defence FIM-92 *Stinger*
AIR-LAUNCHED MISSILES
AAM • IR AIM-9M *Sidewinder*; **IIR** AIM-9X *Sidewinder* II; **SARH** AIM-7M *Sparrow*; **ARH** AIM-120C/D AMRAAM
ASM AGM-65D/G *Maverick*; AGM-114K/M/N/R *Hellfire* II; AGM-130A; AGM-176 *Griffin*; APKWS
AShM AGM-158C LRASM
ALCM
Nuclear AGM-86B (ALCM)
Conventional AGM-158A JASSM; AGM-158B JASSM-ER
ARM AGM-88B/C HARM
EW MALD/MALD-J

BOMBS
Laser-guided GBU 10/12/16 *Paveway* II, GBU-24 *Paveway* III; GBU-28
INS/GPS guided GBU 31/32/38 JDAM; GBU-54 Laser JDAM; GBU-15 (with BLU-109 penetrating warhead or Mk84); GBU-39B Small Diameter Bomb (250lb); GBU-43B MOAB; GBU-57A/B MOP; Enhanced *Paveway* III

Reserve Organisations

Air National Guard 107,400 reservists

FORCES BY ROLE
BOMBER
1 sqn with B-2A *Spirit* (personnel only)
FIGHTER
5 sqn with F-15C/D *Eagle*
1 sqn with F-22A *Raptor* (+1 sqn personnel only)
FIGHTER/GROUND ATTACK
10 sqn with F-16C/D *Fighting Falcon*
1 sqn with F-35A *Lightning* II
GROUND ATTACK
4 sqn with A-10C *Thunderbolt* II
ISR
1 sqn with E-8C J-STARS
COMBAT SEARCH & RESCUE
3 sqn with HC-130J *Combat King* II
3 sqn with HH-60G *Pave Hawk*
TANKER
1 sqn with KC-46A *Pegasus* (forming)
16 sqn with KC-135R *Stratotanker* (+1 sqn personnel only)
3 sqn with KC-135T *Stratotanker*
TRANSPORT
1 sqn with B-737-700 (C-40C)
6 sqn with C-17A *Globemaster* (+2 sqn personnel only)
12 sqn with C-130H *Hercules*
1 sqn with C-130H/LC-130H *Hercules*
2 sqn with C-130J-30 *Hercules*
TRAINING
1 sqn with C-130H *Hercules*
1 sqn with F-15C/D *Eagle*
4 sqn with F-16C/D *Fighting Falcon*
1 sqn with MQ-9A *Reaper*
COMBAT/ISR UAV
10 sqn with MQ-9A *Reaper*

EQUIPMENT BY TYPE
AIRCRAFT 594 combat capable
FTR 157: 123 F-15C *Eagle;* 14 F-15D *Eagle;* 20 F-22A *Raptor*
FGA 352: 288 F-16C *Fighting Falcon*; 44 F-16D *Fighting Falcon*; 20 F-35A *Lightning* II
ATK 85 A-10C *Thunderbolt* II
CSAR 12 HC-130J *Combat King* II
ISR 13 E-8C J-STARS
ELINT 11 RC-26B *Metroliner*
TKR 172: 148 KC-135R *Stratotanker;* 24 KC-135T *Stratotanker*
TKR/TPT 9 KC-46A *Pegasus*
TPT 206: **Heavy** 50 C-17A *Globemaster* III; **Medium** 153: 123 C-130H *Hercules*; 20 C-130J/J-30 *Hercules*; 10 LC-130H *Hercules*; **PAX** 3 B-737-700 (C-40C)
HELICOPTERS • CSAR 18 HH-60G *Pave Hawk*
UNINHABITED AERIAL VEHICLES • CISR • Heavy 24 MQ-9A *Reaper*

Air Force Reserve Command 69,050 reservists

FORCES BY ROLE
BOMBER
1 sqn with B-52H *Stratofortress* (personnel only)
FIGHTER
2 sqn with F-22A *Raptor* (personnel only)
FIGHTER/GROUND ATTACK
2 sqn with F-16C/D *Fighting Falcon* (+1 sqn personnel only)
1 sqn with F-35A *Lightning* II (personnel only)
GROUND ATTACK
1 sqn with A-10C *Thunderbolt* II (+2 sqn personnel only)
ISR
1 (Weather Recce) sqn with WC-130J *Hercules*
AIRBORNE EARLY WARNING & CONTROL
1 sqn with E-3 *Sentry* (personnel only)
COMBAT SEARCH & RESCUE
1 sqn with HC-130J *Combat King* II
2 sqn with HH-60G *Pave Hawk*
TANKER
4 sqn with KC-10A *Extender* (personnel only)
1 sqn with KC-46A *Pegasus* (forming)
1 sqn with KC-46A *Pegasus* (personnel only)
6 sqn with KC-135R *Stratotanker* (+2 sqn personnel only)
TRANSPORT
1 (VIP) sqn with B-737-700 (C-40C)
2 sqn with C-5M *Super Galaxy* (+2 sqn personnel only)
3 sqn with C-17A *Globemaster* (+9 sqn personnel only)
6 sqn with C-130H *Hercules*
1 sqn with C-130J-30 *Hercules*
1 (Aerial Spray) sqn with C-130H *Hercules*
TRAINING
1 (aggressor) sqn with A-10C *Thunderbolt* II; F-15C/E *Eagle*; F-16 *Fighting Falcon*; F-22A *Raptor* (personnel only)
1 sqn with A-10C *Thunderbolt* II
1 sqn with B-52H *Stratofortress*
1 sqn with C-5M *Super Galaxy*
1 sqn with F-16C/D *Fighting Falcon*
5 (flying training) sqn with T-1A *Jayhawk*; T-6A *Texan* II; T-38C *Talon* (personnel only)
COMBAT/ISR UAV
2 sqn with MQ-9A *Reaper* (personnel only)
ISR UAV
1 sqn with RQ-4B *Global Hawk* (personnel only)

EQUIPMENT BY TYPE
AIRCRAFT 126 combat capable
BBR 18 B-52H *Stratofortress*
FGA 53: 49 F-16C *Fighting Falcon*; 4 F-16D *Fighting Falcon*
ATK 55 A-10C *Thunderbolt* II
CSAR 1 HC-130J *Combat King* II
ISR 10 WC-130J *Hercules* (Weather Recce)
TKR 70 KC-135R *Stratotanker*
TKR/TPT 4 KC-46A *Pegasus*

TPT 104: **Heavy** 42: 16 C-5M *Super Galaxy*; 26 C-17A *Globemaster* III; **Medium** 58: 48 C-130H *Hercules*; 10 C-130J-30 *Hercules*; **PAX** 4 B-737-700 (C-40C)
HELICOPTERS • CSAR 16 HH-60G *Pave Hawk*

Civil Reserve Air Fleet
Commercial ac numbers fluctuate
AIRCRAFT • TPT 517 international (391 long-range and 126 short-range); 36 national

Air Force Stand-by Reserve 16,850 reservists
Trained individuals for mobilisation

US Space Force 2,400
New service established December 2019, currently in the process of being stood up. Tasked with organising, training and equipping forces to protect US and allied space interests and to provide space capabilities to the joint Combatant Commands.

EQUIPMENT BY TYPE
SATELLITES see Space

US Special Operations Command (USSOCOM) 63,150; 6,550 (civilian)
Commands all active, reserve and National Guard Special Operations Forces (SOF) of all services based in CONUS

Joint Special Operations Command
Reported to comprise elite US SOF, including Special Forces Operations Detachment Delta ('Delta Force'), SEAL Team 6 and integral USAF support

US Army Special Operations Command 34,100
FORCES BY ROLE
SPECIAL FORCES
5 SF gp (4 SF bn, 1 spt bn)
1 ranger regt (3 ranger bn; 1 cbt spt bn)
COMBAT SUPPORT
1 civil affairs bde (5 civil affairs bn)
1 psyops gp (3 psyops bn)
1 psyops gp (4 psyops bn)
COMBAT SERVICE SUPPORT
1 (sustainment) log bde (1 sigs bn)
HELICOPTER
1 (160th SOAR) hel regt (4 hel bn)

EQUIPMENT BY TYPE
ARMOURED FIGHTING VEHICLES
APC • APC (W) 28: 16 M1126 *Stryker* ICV; 12 *Pandur*
AUV 640 M-ATV
ARTILLERY 20
MOR • 120mm 20 XM905 EMTAS
AIRCRAFT
TPT 12: **Medium** 7 C-27J *Spartan* (parachute training); **Light** 5 C-212 (parachute training)
HELICOPTERS
MRH 51 AH-6M/MH-6M *Little Bird*
TPT 140: **Heavy** 67 MH-47G *Chinook*; **Medium** 73 MH-60M *Black Hawk*
UNINHABITED AERIAL VEHICLES
CISR • Heavy 24 MQ-1C *Gray Eagle*
ISR • Light 29: 15 XPV-1 *Tern*; 14 XPV-2 *Mako*
TPT • Heavy 28 CQ-10 *Snowgoose*

Reserve Organisations

Army National Guard
FORCES BY ROLE
SPECIAL FORCES
2 SF gp (3 SF bn)

Army Reserve
FORCES BY ROLE
COMBAT SUPPORT
2 psyops gp
4 civil affairs comd HQ
8 civil affairs bde HQ
32 civil affairs bn (coy)

US Navy Special Warfare Command 9,850
FORCES BY ROLE
SPECIAL FORCES
8 SEAL team (total: 48 SF pl)
2 SEAL Delivery Vehicle team

Reserve Organisations

Naval Reserve Force
FORCES BY ROLE
SPECIAL FORCES
8 SEAL det
10 Naval Special Warfare det
2 Special Boat sqn
2 Special Boat unit
1 SEAL Delivery Vehicle det

US Marine Special Operations Command (MARSOC) 3,000
FORCES BY ROLE
SPECIAL FORCES
1 SF regt (3 SF bn)
COMBAT SUPPORT
1 int bn
COMBAT SERVICE SUPPORT
1 spt gp

Air Force Special Operations Command (AFSOC) 16,200
FORCES BY ROLE
GROUND ATTACK
2 sqn with AC-130J *Ghostrider*
1 sqn with AC-130W *Stinger* II
TRANSPORT
4 sqn with CV-22B *Osprey*
1 sqn with Do-328 (C-146A)
1 sqn with MC-130H *Combat Talon*
3 sqn with MC-130J *Commando* II
3 sqn with PC-12 (U-28A)
TRAINING
1 sqn with M-28 *Skytruck* (C-145A)
1 sqn with CV-22A/B *Osprey*
1 sqn with HC-130J *Combat King* II; MC-130J *Commando* II

1 sqn with Bell 205 (TH-1H *Iroquois*)
1 sqn with HH-60G *Pave Hawk*; UH-1N *Huey*

COMBAT/ISR UAV
3 sqn with MQ-9 *Reaper*

EQUIPMENT BY TYPE

AIRCRAFT 33 combat capable
ATK 33: 23 AC-130J *Ghostrider*; 10 AC-130W *Stinger* II
ISR 22 MC-12 *Javaman*
CSAR 3 HC-130J *Combat King* II
TPT 113: **Medium** 54: 13 MC-130H *Combat Talon* II; 41 MC-130J *Commando* II; **Light** 59: 19 Do-328 (C-146A); 5 M-28 *Skytruck* (C-145A); 35 PC-12 (U-28A)

TILT-ROTOR 51 CV-22A/B *Osprey*

HELICOPTERS
CSAR 3 HH-60G *Pave Hawk*
TPT • Light 34: 28 Bell 205 (TH-1H *Iroquois*); 6 UH-1N *Huey*

UNINHABITED AERIAL VEHICLES • CISR • Heavy 50 MQ-9 *Reaper*

Reserve Organisations

Air National Guard

FORCES BY ROLE

ELECTRONIC WARFARE
1 sqn with C-130J *Hercules*/EC-130J *Commando Solo*

ISR
1 sqn with MC-12W *Liberty*

TRANSPORT
1 flt with B-737-200 (C-32B)

EQUIPMENT BY TYPE

AIRCRAFT
EW 7 EC-130J *Commando Solo*
ISR 13 MC-12W *Liberty*
TPT 5: **Medium** 3 C-130J *Hercules*; **PAX** 2 B-757-200 (C-32B)

Air Force Reserve

FORCES BY ROLE

TRAINING
1 sqn with AC-130J *Ghostrider* (personnel only)
1 sqn with M-28 *Skytruck* (C-145A) (personnel only)

COMBAT/ISR UAV
1 sqn with MQ-9 *Reaper* (personnel only)

DEPLOYMENT

AFGHANISTAN: NATO • *Operation Resolute Support* ε4,000; 1 div HQ; 1 spec ops bn; 2 inf bde(-); 1 EOD bn; 1 (cbt avn) hel bde(-); 1 EW sqn with EC-130H *Compass Call*; 1 ISR unit with RC-12X *Guardrail*; 1 tpt sqn with C-130J-30 *Hercules*; 1 CSAR sqn with HH-60G *Pave Hawk*; 1 CISR UAV sqn with MQ-9A *Reaper*
US Central Command • *Operation Freedom's Sentinel* ε500

ARABIAN SEA: US Central Command • US Navy • 5th Fleet: 2 SSGN; 1 CGHM; 2 DDGHM; **Combined Maritime Forces •** TF 53: 1 AE; 2 AKE; 1 AOH; 3 AO; **Combined Maritime Forces •** CTF-150: 1 DDGHM

ARUBA: US Southern Command • 1 Forward Operating Location

ASCENSION ISLAND: US Strategic Command • 1 detection and tracking radar at Ascension Auxiliary Air Field

AUSTRALIA: US Pacific Command • 700; 1 SEWS at Pine Gap; 1 comms facility at Pine Gap; 1 SIGINT stn at Pine Gap; **US Strategic Command •** 1 detection and tracking radar at Naval Communication Station Harold E. Holt

BAHRAIN: US Central Command • 4,700; 1 HQ (5th Fleet); 1 ASW sqn with 5 P-8A *Poseidon*; 1 EP-3E *Aries* II; 2 AD bty with MIM-104E/F *Patriot* PAC-2/-3

BELGIUM: US European Command • 1,150

BRITISH INDIAN OCEAN TERRITORY: US Strategic Command • 300; 1 Spacetrack Optical Tracker at Diego Garcia; 1 ground-based electro-optical deep space surveillance system (GEODSS) at Diego Garcia
US Pacific Command • 1 MPS sqn (MPS-2 with equipment for one MEB) at Diego Garcia with 2 AKRH; 3 AKR; 1 AKEH; 1 ESD; 1 naval air base at Diego Garcia, 1 support facility at Diego Garcia

CAMEROON: US Africa Command • 300; MQ-1C *Gray Eagle*

CANADA: US Northern Command • 140

CENTRAL AFRICAN REPUBLIC: UN • MINUSCA 7

COLOMBIA: US Southern Command • 70

CUBA: US Southern Command • 800 (JTF-GTMO) at Guantanamo Bay

CURACAO: US Southern Command • 1 Forward Operating Location

DEMOCRATIC REPUBLIC OF THE CONGO: UN • MONUSCO 4

DJIBOUTI: US Africa Command • 4,000; 1 tpt sqn with C-130H/J-30 *Hercules*; 1 spec ops sqn with MC-130H/J; PC-12 (U-28A); 1 CSAR sqn with HH-60G *Pave Hawk*; 1 CISR UAV sqn with MQ-9A *Reaper*; 1 naval air base

EGYPT: MFO 454; elm 1 ARNG recce bn; 1 ARNG spt bn

EL SALVADOR: US Southern Command • 1 Forward Operating Location (Military, DEA, USCG and Customs personnel)

GERMANY: US Africa Command • 1 HQ at Stuttgart
US European Command • 37,100; 1 Combined Service HQ (EUCOM) at Stuttgart–Vaihingen

US Army 22,850

FORCES BY ROLE
1 HQ (US Army Europe (USAREUR)) at Wiesbaden; 1 SF gp; 1 recce bn; 1 mech bde(-); 1 MRL bde (2 MRL bn); 1 fd arty bn; 1 (cbt avn) hel bde; 1 (cbt avn) hel bde HQ; 1 int bde; 1 MP bde; 1 sigs bde; 1 spt bde; 1 ARNG SAM bde(-); 1 (APS) armd bde eqpt set

EQUIPMENT BY TYPE
M1A2 SEPv2 *Abrams;* M2A3/M3A3 *Bradley*; M1296 *Stryker Dragoon*, M109A6; M777A2; AH-64D/E *Apache*; CH-47F *Chinook*; UH-60M *Black Hawk*; HH-60M *Black Hawk*

US Navy 450
USAF 13,400

FORCES BY ROLE

1 HQ (US Air Force Europe (USAFE)) at Ramstein AB; 1 HQ (3rd Air Force) at Ramstein AB; 1 ftr wg at Spangdahlem AB with 1 ftr sqn with 24 F-16C/D *Fighting Falcon*; 1 tpt wg at Ramstein AB with 14 C-130J-30 *Hercules;* 2 Gulfstream V (C-37A); 5 Learjet 35A (C-21A); 1 B-737-700 (C-40B)

USMC 400

GREECE: US European Command • 400; 1 naval base at Makri; 1 naval base at Souda Bay; 1 air base at Iraklion

GREENLAND (DNK): US Strategic Command • 160; 1 AN/FPS-132 Upgraded Early Warning Radar and 1 Spacetrack Radar at Thule

GUAM: US Pacific Command • 9,000; 4 SSGN; 1 MPS sqn (MPS-3 with equipment for one MEB) with 2 AKRH; 4 AKR; 1 ESD; 1 AKEH; 1 tkr sqn with 12 KC-135R *Stratotanker*; 1 tpt hel sqn with MH-60S; 1 SAM bty with THAAD; 1 air base; 1 naval base

HONDURAS: US Southern Command • 370; 1 avn bn with CH-47F *Chinook*; UH-60 *Black Hawk*

ICELAND: NATO • Icelandic Air Policing; 4 F-15C *Eagle*

IRAQ: US Central Command • *Operation Inherent Resolve* 3,000; 1 AB bde(-); 1 EOD pl; 1 atk hel bn with AH-64D *Apache;* 4 AD bty with MIM-104E/F *Patriot* PAC-2/-3; 1 CISR UAV sqn with MQ-9A *Reaper*

ISRAEL: US Strategic Command • 1 AN/TPY-2 X-band radar at Mount Keren

ITALY: US European Command • 12,550

US Army 3,900; 1 AB bde(-)

US Navy 3,900; 1 HQ (US Navy Europe (USNAVEUR)) at Naples; 1 HQ (6th Fleet) at Gaeta; 1 ASW sqn with 4 P-8A *Poseidon* at Sigonella

USAF 4,550; 1 ftr wg with 2 ftr sqn with 21 F-16C/D *Fighting Falcon* at Aviano; 1 CSAR sqn with 8 HH-60G *Pave Hawk* at Aviano; 1 CISR UAV sqn with MQ-9A *Reaper* at Sigonella

USMC 200

JAPAN: US Pacific Command • 55,300

US Army 2,500; 1 corps HQ (fwd); 1 SF gp; 1 avn bn; 1 SAM bn

US Navy 20,100; 1 HQ (7th Fleet) at Yokosuka; 1 base at Sasebo; 1 base at Yokosuka

FORCES BY ROLE

3 FGA sqn at Iwakuni with 10 F/A-18E *Super Hornet*; 1 FGA sqn at Iwakuni with 10 F/A-18F *Super Hornet*; 2 EW sqn at Iwakuni/Misawa with 5 EA-18G *Growler*; 1 AEW&C sqn at Iwakuni with 5 E-2D *Hawkeye*; 2 ASW hel sqn at Atsugi with 12 MH-60R; 1 tpt hel sqn at Atsugi with 12 MH-60S

EQUIPMENT BY TYPE

1 CVN; 3 CGHM; 2 DDGHM; 5 DDGM; 1 LCC; 4 MCO; 1 LHA; 1 LPD; 2 LSD

USAF 12,700

FORCES BY ROLE

1 HQ (5th Air Force) at Okinawa – Kadena AB; 1 ftr wg at Misawa AB with (2 FGA sqn with 22 F-16C/D *Fighting Falcon*); 1 wg at Okinawa – Kadena AB with (2 ftr sqn with 27 F-15C/D *Eagle*; 1 tkr sqn with 15 KC-135R *Stratotanker*; 1 AEW&C sqn with 2 E-3B *Sentry*; 1 CSAR sqn with 10 HH-60G *Pave Hawk*); 1 tpt wg at Yokota AB with 10 C-130J-30 *Hercules*; 3 Beech 1900C (C-12J); 1 Spec Ops gp at Okinawa – Kadena AB with (1 sqn with 5 MC-130J *Commando* II; 1 sqn with 5 CV-22B *Osprey*); 1 ISR sqn with RC-135 *Rivet Joint*; 1 ISR UAV flt with 5 RQ-4A *Global Hawk*

USMC 20,000

FORCES BY ROLE

1 mne div; 1 mne regt HQ; 1 arty regt HQ; 1 recce bn; 1 mne bn; 1 amph aslt bn; 1 arty bn; 1 FGA sqn with 12 F/A-18C *Hornet*; 1 FGA sqn with 12 F/A-18D *Hornet*; 2 FGA sqn with 12 F-35B *Lightning* II; 1 tkr sqn with 15 KC-130J *Hercules*; 2 tpt sqn with 12 MV-22B *Osprey*

US Strategic Command • 1 AN/TPY-2 X-band radar at Shariki; 1 AN/TPY-2 X-band radar at Kyogamisaki

JORDAN: US Central Command • *Operation Inherent Resolve* 2,300: 1 FGA sqn with 12 F-15E *Strike Eagle*; 1 CISR UAV sqn with 12 MQ-9A *Reaper*; 2 AD bty with MIM-104E/F *Patriot* PAC-2/-3

KOREA, REPUBLIC OF: US Pacific Command • 31,050

US Army 22,500

FORCES BY ROLE

1 HQ (8th Army) at Seoul; 1 div HQ (2nd Inf) located at Pyeongtaek; 1 armd bde; 1 (cbt avn) hel bde; 1 MRL bde; 1 AD bde; 1 SAM bty with THAAD

EQUIPMENT BY TYPE

M1A2 SEPv2 *Abrams*; M2A2/M3A3 *Bradley*; M109A6; M270A1 MLRS; AH-64D/E *Apache*; CH-47F *Chinook*; UH-60L/M *Black Hawk*; MIM-104 *Patriot*; FIM-92A *Avenger*; 1 (APS) armd bde eqpt set

US Navy 350

USAF 8,000

FORCES BY ROLE

1 (AF) HQ (7th Air Force) at Osan AB; 1 ftr wg at Osan AB with (1 ftr sqn with 20 F-16C/D *Fighting Falcon*; 1 atk sqn with 24 A-10C *Thunderbolt* II); 1 ftr wg at Kunsan AB with (2 ftr sqn with 20 F-16C/D *Fighting Falcon*); 1 ISR sqn at Osan AB with U-2S

USMC 200

KUWAIT: US Central Command • 13,500; 1 armd bde(-); 1 ARNG (cbt avn) hel bde; 1 spt bde; 1 tpt sqn with 12 MV-22B *Osprey*; 1 CISR UAV sqn with MQ-9A *Reaper*; 3 AD bty with MIM-104E/F *Patriot* PAC-2/-3; 1 (APS) armd bde set; 1 (APS) inf bde set

LIBYA: UN • UNSMIL 1

MALI: UN • MINUSMA 9

MARSHALL ISLANDS: US Strategic Command • 1 detection and tracking radar at Kwajalein Atoll

MEDITERRANEAN SEA: US European Command • US Navy • 6th Fleet: 4 DDGM; 1 LCC

MIDDLE EAST: UN • UNTSO 3

NETHERLANDS: US European Command • 400

NIGER: US Africa Command • 800; 1 CISR sqn with MQ-9A *Reaper*

NORWAY: US European Command • 1,100; 1 (USMC) MEU eqpt set; 1 (APS) SP 155mm arty bn set

PERSIAN GULF: US Central Command • US Navy • 5th Fleet: 1 CVN; 1 CGHM; 1 DDGHM; 10 PCFG; 6 (Coast Guard) PCC
Combined Maritime Forces • CTF-152: 4 MCO; 1 ESB

PHILIPPINES: US Pacific Command • *Operation Pacific Eagle – Philippines* 200

POLAND: NATO • Enhanced Forward Presence 670; 1 mech bn with M1296 *Stryker Dragoon*; M777A2
US European Command • 4,500; 1 corps HQ (fwd); 1 div HQ (fwd); 1 armd bde; M1A2 SEPv2 *Abrams*; M3A3 *Bradley*; M2A3 *Bradley*; M109A6; 1 CISR UAV sqn with MQ-9A *Reaper*

PORTUGAL: US European Command • 250; 1 spt facility at Lajes

QATAR: US Central Command • 10,000: 1 ISR sqn with 4 RC-135 *Rivet Joint*; 1 ISR sqn with 4 E-8C JSTARS; 2 tkr sqn with 12 KC-135R/T *Stratotanker;* 1 tpt sqn with 4 C-17A *Globemaster*; 4 C-130H/J-30 *Hercules*; 2 AD bty with MIM-104E/F *Patriot* PAC-2/-3
US Strategic Command • 1 AN/TPY-2 X-band radar

ROMANIA: US European Command • 120

SAUDI ARABIA: US Central Command • 2,000; 1 FGA sqn with 12 F-16C *Fighting Falcon*; 2 AD bty with MIM-104E/F *Patriot* PAC-2/-3; 1 SAM bty with THAAD

SERBIA: NATO • KFOR • *Joint Enterprise* 660; elm 1 ARNG inf bde HQ; 1 ARNG recce bn; 1 hel flt with UH-60

SINGAPORE: US Pacific Command • 200; 1 log spt sqn; 1 spt facility

SOMALIA: US Africa Command • 700

SOUTH SUDAN: UN • UNMISS 7

SPAIN: US European Command • 3,300; 1 air base at Morón; 1 naval base at Rota

SYRIA: US Central Command • *Operation Inherent Resolve* 750; 1 armd inf coy; 1 mne bn(-)

THAILAND: US Pacific Command • 100

TURKEY: US European Command • 1,700; 1 tkr sqn with 14 KC-135; 1 air base at Incirlik; 1 support facility at Ankara; 1 support facility at Izmir
US Strategic Command • 1 AN/TPY-2 X-band radar at Kürecik

UKRAINE: JMTG-U 150 (trg mission)

UNITED ARAB EMIRATES: US Central Command • 5,000: 1 FGA sqn with 12 F-35A *Lightning* II; 1 ISR sqn with 4 U-2; 1 AEW&C sqn with 4 E-3B/G *Sentry;* 1 tkr sqn with 12 KC-10A; 1 ISR UAV sqn with RQ-4 *Global Hawk*; 2 AD bty with MIM-104E/F *Patriot* PAC-2/-3

UNITED KINGDOM: US European Command • 9,300

FORCES BY ROLE

1 ftr wg at RAF Lakenheath with 1 ftr sqn with 24 F-15C/D *Eagle*, 2 FGA sqn with 23 F-15E *Strike Eagle*; 1 ISR sqn at RAF Mildenhall with OC-135/RC-135; 1 tkr wg at RAF Mildenhall with 15 KC-135R/T *Stratotanker*; 1 spec ops gp at RAF Mildenhall with (1 sqn with 8 CV-22B *Osprey*; 1 sqn with 8 MC-130J *Commando* II)

US Strategic Command • 1 AN/FPS-132 Upgraded Early Warning Radar and 1 Spacetrack Radar at Fylingdales Moor

FOREIGN FORCES

Germany Air Force: trg units with 40 T-38 *Talon*; 69 T-6A *Texan* II; 24 *Tornado* IDS; • Missile trg at Fort Bliss (TX)
Netherlands 1 hel trg sqn with AH-64D *Apache*; CH-47D *Chinook*
Singapore Air Force: trg units with F-16C/D; 12 F-15SG; AH-64D *Apache*; 6+ CH-47D *Chinook* hel

Arms procurements and deliveries – North America

Significant events in 2020

APRIL — UTC–RAYTHEON MERGER

Raytheon Technologies (RTC) was formed by the merger of United Technologies Corporation (UTC) and Raytheon Company. A month before, the United States Department of Justice (DoJ) antitrust division expressed concern that the merger might harm competition in areas including GPS receivers. As a result of the DoJ's ruling that both needed to divest themselves of certain assets, UTC sold the GPS business of its subsidiary Collins Aerospace to BAE Systems for US$1.93bn. Raytheon sold to BAE its Airborne Tactical Radios firm for US$275m and its electro-optics technology business to AMERGINT Technologies Holdings Inc. In November, RTC announced further acquisitions and divestitures, including the sale of data-security firm Forcepoint, in order to focus on its defence and aerospace portfolio.

JULY — OMFV RESTRUCTURE

The third attempt to replace the *Bradley* infantry fighting vehicle (the first began in 2000) suffered a setback when the US Army cancelled its Request for Proposals (RFP) for the Optionally Manned Fighting Vehicle (OMFV) programme. This RFP had been launched in early 2019. Raytheon–Rheinmetall were unable to ship their *Lynx* prototype to the US in time and the *Griffin* model offered by General Dynamics Land Systems did not meet weight requirements. In April, the army restructured the programme. It is now organised in five phases and a draft request for preliminary digital designs (phase 1) was issued in July. The plan is that the final RFP for phase one leads to five contracts in June 2021, with low-rate production in 2028 and full-rate production in 2029.

JULY — CANADIAN FIGHTER REPLACEMENT

Boeing, Saab and Lockheed Martin submitted bids for Canada's Future Fighter Capability Project. The requirement is for 88 aircraft to replace the air force's CF-18 *Hornet* fighter aircraft and the project is valued at C$15–19bn (US$11.41–14.45bn). The aircraft entered into the competition are the F/A-18 E/F *Super Hornet* from Boeing, Saab's *Gripen* E, and the Lockheed Martin F-35 *Lightning* II. After the proposals have been evaluated, an agreement will be negotiated in 2022 with the first aircraft expected to be delivered in 2025. The programme has been ongoing for over a decade, and in recent years both BAE Systems and Dassault Aviation have withdrawn from the competition.

AUGUST — *IRON DOME* JV

Raytheon Technologies and Israel's Rafael established a joint venture, Raytheon Rafael Area Protection Systems, in order to build the *Iron Dome* medium-range surface-to-air missile system in the US. The site will mainly produce the *Tamir* interceptor and launcher, and the *Skyhunter* missile (the US variant of *Tamir*). The two companies have a long partnership in developing and producing the *Iron Dome* system, which has a claimed success rate of more than 90% since it entered Israeli service in March 2011. The first of two initial batteries was delivered to the US Army in September and the army plans to order more than a dozen more over the next five years.

AUGUST — MUSV PROTOTYPE

The US Navy has selected a team led by L3Harris Technologies to build a prototype for the Medium Unmanned Surface Vehicle (MUSV) programme, for delivery by 2023. The contract is worth an initial US$35m, though if options for nine more MUSVs are exercised, the contract value could increase to US$281.4m. To date, L3Harris Technologies has exported more than 100 uninhabited surface vehicles globally, and it was a key subcontractor on the DARPA *Sea Hunter* programme. A speech by then-secretary of defence Mark Esper in October outlined an ambition for the navy to have between 140 and 240 uninhabited or optionally crewed surface and subsurface vessels by 2045.

Table 2: Canada: armoured-vehicle procurement and upgrade programmes since 2005

Programme	Equipment	Type	Quantity	Value (US$)	Prime contractor(s)	First contract date	Deliveries
Armoured Patrol Vehicle	RG-31 *Nyala*	Protected patrol vehicle	75	$76.86m	GDLS Canada	Dec 2005	2006
Tank Replacement Project	*Leopard* 2A6M	Main battle tank	20	$605.19m	Netherlands government surplus	Dec 2007	2008–15
	Leopard 2A4M		20		German government surplus		
	Leopard 2A4		42		Krauss-Maffei Wegmann (KMW)		
	BPz-3 *Büffel*	Armoured-recovery-vehicle conversion	8		Rheinmetall; Rheinmetall Canada		
Lightweight Towed Howitzer	M777	155mm towed howitzer	25	$123.89m	BAE Systems Land & Armaments	May 2009	2010–11
Light Armoured Vehicle III Upgrade Project	LAV 6.0	Armoured personnel carrier (wheeled) upgrade	550	$1.35bn	GDLS Canada	Oct 2011	2013–19
Force mobility Enhancement	BPz-3 *Büffel*	Armoured-recovery-vehicle conversion	4	$356.23m	Flensburger Fahrzeugbau Gesellschaft (FFG); Rheinmetall; Krauss-Maffei Wegmann (KMW)	Mar 2012	2015–18
	Wisent 2	Armoured-engineer-vehicle conversion	18				
	Leopard 2	Main-battle-tank mobility enhancements	82				
Tactical Armoured Patrol Vehicle	TAPV	Reconnaissance vehicle	193	$603.89m	Textron Systems Canada	Jun 2012	2016–18
		Armoured utility vehicle	307				
Light Armoured Vehicle Reconnaissance Surveillance System	LAV 6.0	Reconnaissance-vehicle upgrade	66	$473.55m	GDLS Canada	Feb 2015	2020–21*
Armoured Combat Support Vehicles	LAV 6.0	Armoured personnel carrier (wheeled)	360	$1.51bn	GDLS Canada	Sep 2019	2020–25*
TOTAL				**$5.1bn**			

*Planned

Figure 4: US: Lockheed Martin Terminal High Altitude Area Defense (THAAD) ballistic-missile defence system

The Terminal (originally Theater) High Altitude Area Defense (THAAD) programme was devised in the late 1980s. Although Congress said that a deployable system should enter service in the mid-1990s, flight tests only began in 1995. While six consecutive failures left the programme's future in doubt, the first battery was eventually activated in 2008 and the system has now had 16 consecutive successful test intercepts.

The test failures in the 1990s led to the programme being reviewed and restructured. One review, by the US Government Accountability Office (GAO), concluded that the compressed test schedule gave too little time to detect and assess issues before and after launch.

The GAO also said that there was not enough time to uncover quality-control issues in the manufacturing process. Additionally, the 1992 development contract with the Lockheed Corporation included delivery of two prototype batteries intended to give the army an interim capability before fielding the production-standard system. (This is also a requirement in the Long Range Hypersonic Weapon contracts awarded in 2019.) This meant that there was a concurrent focus on operational as well as development considerations. Furthermore, though 'hit-to-kill' technology was seen as technically viable, the Department of Defense had underestimated its practical difficulty.

The programme was restructured after the fifth successive test failure in May 1998. The requirement for prototype batteries was removed, the flight-test schedule was extended and more ground tests were added. The contract was also renegotiated to include penalties and incentives. Lockheed Martin had earlier started its own restructuring process, which involved assigning significantly more staff to quality control. Two successful intercepts in 1999 (launches 10 and 11) gave the Army confidence to proceed to the engineering and manufacturing development phase in 2000. However, this measured approach meant that the next test launch did not take place until 2005. As of late 2020, seven THAAD batteries have entered service with the US Army and in tests the system has intercepted several types of ballistic-missile targets.

THAAD: flight-test timeline

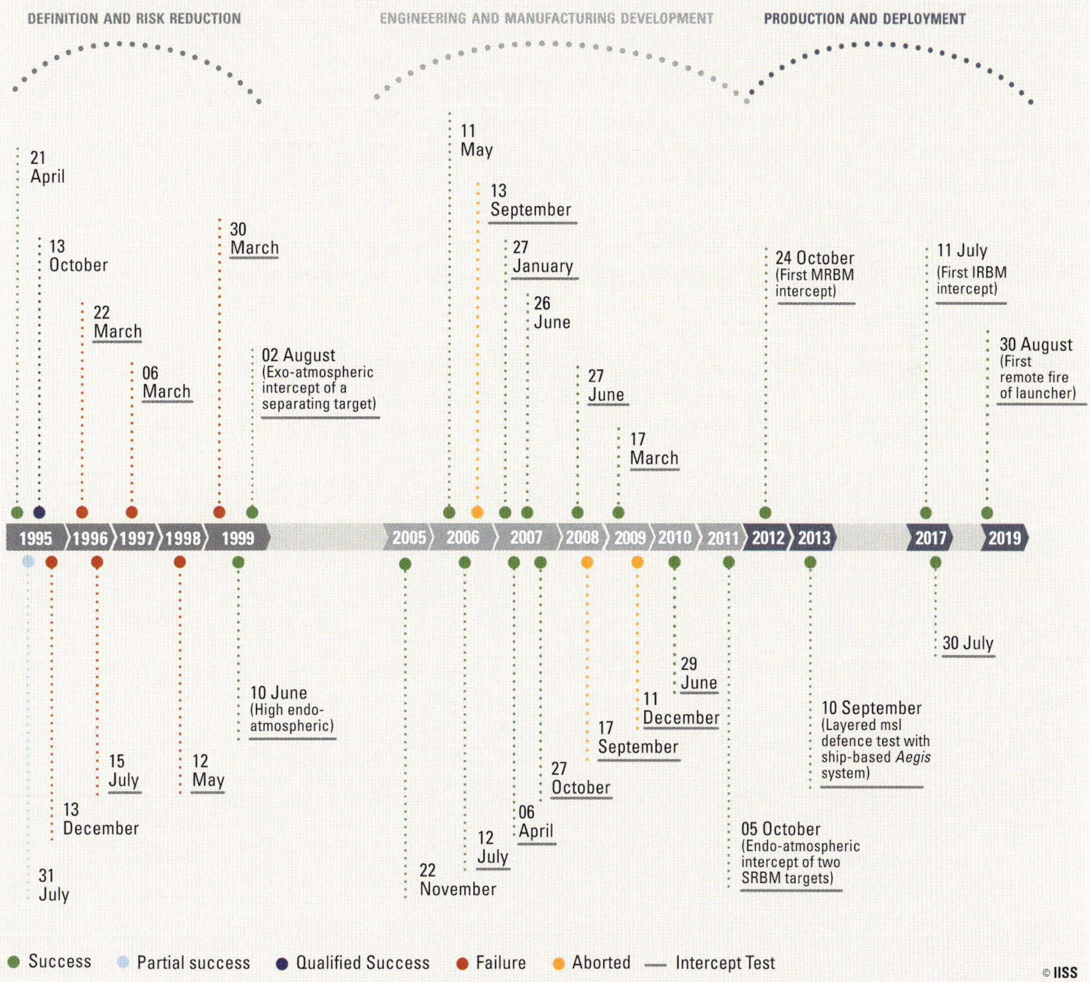

Chapter Four

Europe

- The US decision to withdraw nearly 12,000 of the (around 36,000) permanently assigned US personnel from Germany was confirmed on 29 July 2020. This decision discomfited European allies and would mean the withdrawal of the 2nd Cavalry Regiment which had only recently received the 30mm-cannon-armed *Stryker Dragoon* variant specifically for the European theatre.
- The EU's multi-year budget agreement, in July 2020, saw funding for the European Defence Fund, for military mobility and for the European Peace Facility lowered by almost 40%, 75% and just over 45% respectively, compared to the figures circulating at the beginning of the process. Setting and efficiently implementing priorities therefore becomes paramount.
- Germany's plan to replace its *Tornado* aircraft has highlighted the dual-capable aircraft (DCA) element of NATO's deterrent posture. Germany's *Tornados* are currently tasked with this, among other roles, and Belgium, Italy, the Netherlands and Turkey also operate aircraft for the DCA role, where aircraft are rated to carry US-owned free-fall nuclear bombs that are stored in each country.
- Europe lacks sufficient ground-based air defence capability to counter conventionally armed cruise and ballistic missiles. Russia has introduced ship and submarine-launched land-attack cruise missiles (LACM) with ranges of up to 2,500 kilometre (3M14 *Kalibr* (SS-N-30A *Sagaris*)), as well as the *Iskander*-M (SS-26 *Stone*/SSC-7 *Southpaw*) cruise- and ballistic-missile combination, and the air force's Kh-101 (AS-23 *Kodiak*) 3,500 km+ air-launched LACM. In response, some European states have ordered the US *Patriot* SAM system, while others are upgrading existing air defences.
- The Franco-German Major Ground Combat System (MGCS) project is designed to produce potentially revolutionary systems to succeed the *Leclerc* and *Leopard* 2 main battle tanks. The contract for a System Architecture Definition Study – Part 1 was signed in 2020, and Krauss-Maffei Wegmann (KMW), Nexter Systems and Rheinmetall AG are now working towards the MGCS demonstration phase. The MGCS may also have export potential, with the *Leopard* and *Leclerc* also serving in armies outside Europe.

Europe defence spending, 2020 – top 5

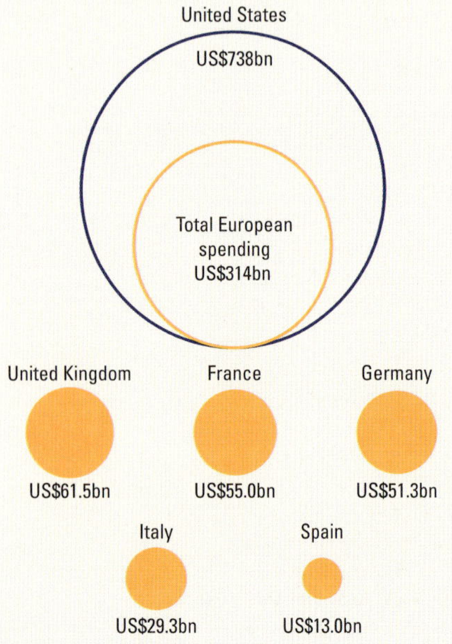

Active military personnel – top 10

(15,000 per unit)

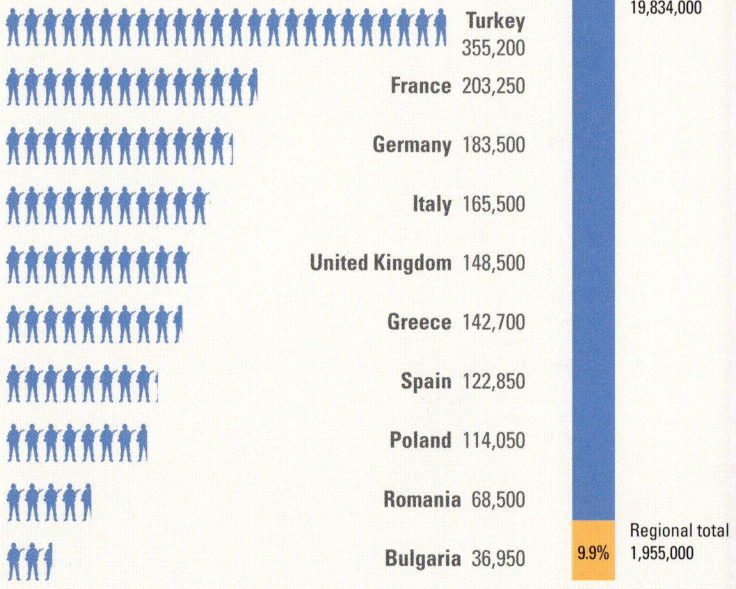

European operators of *Leclerc* and *Leopard* 2 tanks*

Country/Types	Active total
Austria *Leopard* 2A4	56
Denmark *Leopard* 2A5/2A7	44
Finland *Leopard* 2A4/2A6	100[a]
France *Leclerc*	222
Germany *Leopard* 2A5/A6/A7/A7V	245[b]
Greece *Leopard* 2A4/2A6HEL	353
Hungary *Leopard* 2A4HU	4 (trg)
Norway *Leopard* 2A4	36[c]
Poland *Leopard* 2A4/A5/2PL	247
Portugal *Leopard* 2A6	37
Spain *Leopard* 2A4/2E	327
Sweden *Leopard* 2A5 (Strv 122)	120
Switzerland *Leopard* 2 (Pz-87 *Leo*)	134
Turkey *Leopard* 2A4	316

[a] 100 more in store; [b] 78 more A4 in store; [c] 16 more in store

*Not including other variants such as engineering and recovery vehicles

Leclerc and *Leopard* 2 fleets by region

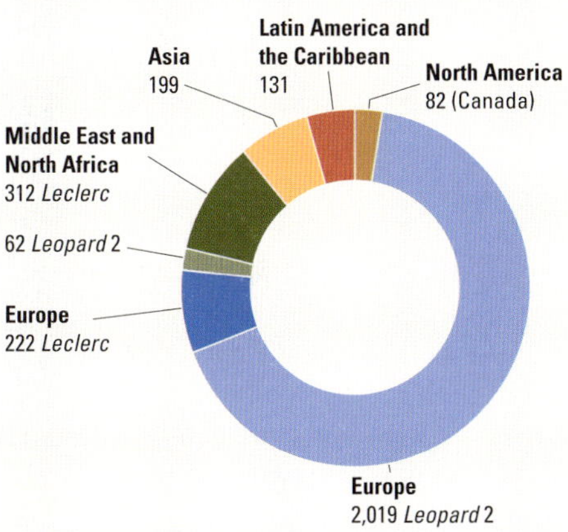

NATO tactical nuclear-weapons-capable delivery platforms

Current fleet	Future fleet
Belgium	
■ 45 F-16 AM *Fighting Falcon* ■ 9 F-16 BM *Fighting Falcon*	■ 34 F-35A *Lightning* II
Germany	
■ 68 *Tornado* IDS	to be decided
Italy	
■ 34 *Tornado* IDS ■ 12 F-35A *Lightning* II ■ 3 F-35B *Lightning* II	■ 90 F-35A/B *Lightning* II
Netherlands	
■ 61 F-16AM/BM *Fighting Falcon* ■ 12 F-35A *Lightning* II	■ 46 F-35A *Lightning* II
Turkey	
■ 260 F-16C/D *Fighting Falcon*	n.k.

All aircraft may be wired for the nuclear role, but not all will be tasked or trained for the nuclear role.

Europe: heavy transport aircraft

Country	Type	Total
France	A400M	17
Germany	A400M	34
Luxembourg	A400M	1
Spain	A400M	8
Turkey	A400M	9
United Kingdom	A400M *Atlas*/C-17A *Globemaster* III	28
Multinational	C-17A *Globemaster* III	3

Europe: medium transport aircraft

Country	Types	Total
Austria	C-130K *Hercules*	3
Belgium	C-130H *Hercules*	7
Bulgaria	C-27J *Spartan*	3
Denmark	C-130J-30 *Hercules*	4
France	C-130H/C-130H-30/C-130J *Hercules*/C-160R *Transall*	27
Germany	C-160D *Transall*	17
Greece	C-130B/C-130H *Hercules*/C-27J *Spartan*	23
Italy	C-130J/C-130J-30 *Hercules*/C-27J *Spartan*	30
Lithuania	C-27J *Spartan*	3
Netherlands	C-130H/C-130H-30 *Hercules*	4
Norway	C-130J-30 *Hercules*	4
Poland	C-130E *Hercules*	5
Portugal	C-130H/C-130H-30 *Hercules*	5
Romania	C-130B/C-130H *Hercules*/C-27J *Spartan*	12
Slovakia	C-27J *Spartan*	2
Spain	C-130H/C-130H-30 *Hercules*	5
Sweden	C-130H *Hercules* (Tp-84)	5
Turkey	C-130B/C-130E *Hercules*/C-160D *Transall*	30
United Kingdom	C-130J/C-130J-30 *Hercules*	14

Europe

For several weeks in early 2020, Europe was the epicentre of the global coronavirus pandemic. Italy, Spain and the United Kingdom were particularly badly affected. Governments across Europe had long considered the possibility that a pandemic might become a security risk. The 2015 UK National Security Strategy and Strategic Defence and Security Review considered that a major public-health crisis within the next five years caused by a pandemic was a Tier 1 national-security risk. The German government's 2016 White Paper on German Security Policy and the Future of the Bundeswehr suggested that pandemics could overwhelm public health systems and carry considerable economic costs.

While the analysis proved correct, levels of preparation were often found wanting and, when the crisis hit, uncomfortable questions were raised, for instance over the precise purpose of some military equipment, when national-security needs were measured by security of supply questions, or in stocks of personal protective equipment. Armed forces in many European countries contributed to the pandemic response by lending military assistance to civilian authorities (MACA). However, it became apparent that it would prove increasingly difficult to generate support for maintaining future defence budgets, let alone increasing them, as death tolls mounted in the first half of 2020 and the reality of shrinking government revenue became apparent, alongside ballooning public expenditure to finance the recovery for years to come.

Meanwhile, none of the security challenges, risks and threats that motivated defence concerns before the pandemic had disappeared: namely an assertive Russia with increasingly modern armed forces; a rising China looking to expand its global footprint; persistent instability on Europe's southern and south-eastern flanks; and terrorism and cyber threats. All these continued to demand policymakers' attention. At the 17 June 2020 meeting of the Council of the European Union, officials suggested that Europeans were 'facing an already challenging international environment in which the unprecedented COVID-19 pandemic risks amplifying existing global fragilities and tensions'.

It might appear that these factors would increase the demand for European military capabilities suitable for crisis-management, cooperative-security and collective-defence tasks. Indeed, Europeans are not structurally unable to defend themselves. However, the pandemic has made more difficult the political, financial and military choices governments need to make when generating defence capabilities. It has also highlighted the requirement to address wider questions of societal resilience and also to broaden the conception of defence. The latter theme featured prominently in preparatory work, during 2020, for the UK's integrated review. The plan was that this would better harness all levers of state power for future foreign, security and defence challenges, though its progress was unclear at the time of writing.

The pandemic triggered a number of MACA operations, with armed forces providing medical capacity and logistic support, helping to construct healthcare facilities, and in some countries carrying out internal- and border-security tasks. While it is not clear yet whether these tasks will permanently rise in priority for Europe's armed forces, there has been widespread support among European citizens for military support to authorities during civil emergencies. The pandemic also had an effect on ongoing crisis-management operations. In some cases, pre-deployment training was compressed, or personnel saw their tour of duty extended because of delays in organising rotations (for example due to quarantine regulations). Some personnel were withdrawn because the effects of the pandemic had reduced the need for their presence, or to assist with tasks at home. Recruit training was adversely affected in some countries, while ministries had to monitor closely the effect on readiness. The large NATO exercise *Defender* 2020 was the most prominent casualty, and this was reduced in size and scope. The pandemic has caused some delays in the defence-industrial supply chain. For instance, in April 2020 Joël Barre, head of France's defence procurement agency, the Direction générale de l'Armement (DGA), told parliament that some procurement programmes had suffered delays of one to two months. However, while it is true that defence ministries will need to redouble their efforts to make the case for defence spending, it will likely take until 2021–22 for the full economic effects of governments' pandemic responses to feed through to cuts in defence funding. Indeed, it is not unusual for there to be a lag in the effect that significant economic dislocations have on defence spending; this was the case after the 2008 financial crisis. Indeed, perhaps paradoxically, the immediate effect of the pandemic seems to have been that some governments – for example Germany and Denmark – have accelerated defence-investment spending in order to support local suppliers, and also that some states felt their threat perceptions sharpen such that they merited increased defence spending.

The coronavirus pandemic affected both the EU and NATO. European states are trying to position themselves in a world in which the dominant paradigm is once again great-power competition. The EU and NATO are, in this context, still the principal multilateral institutions used by European governments to pursue their security goals. However, the cohesion of these bodies continued to be challenged, for instance by continued transatlantic disagreements on issues ranging from defence spending to climate change and trade, simmering tensions between NATO members Greece and Turkey, and the still-unfolding effects of the UK's decision to leave the European Union, which took effect on 31 January 2020. As the EU reduced in size, NATO expanded, with North Macedonia becoming its 30th member state at the end of March 2020.

The US decision to withdraw nearly 12,000 of the (around 36,000) permanently assigned US personnel from Germany was confirmed on 29 July 2020. This was an example of how politics and military matters can intersect in unhelpful ways, with European allies feeling there was a lack of consultation before the decision was announced. When this did take place, the carefully constructed reasoning presented by then-US Secretary of Defense Mark Esper, the commander of US European Command (EUCOM), General Tod Wolters, and US National Security Advisor Robert O'Brien, was undermined by President Donald Trump, who said also on 29 July that Germany was 'delinquent', owed 'billions and billions of dollars to NATO' and that the troops would be withdrawn because 'we don't want to be the suckers anymore'. Irrespective of its rationale, the decision appears incoherent from an alliance-management point of view, with any benefits to NATO's deterrence posture spurious at best. For instance, while it might make some sense to consolidate units in Italy, and move the headquarters of US European Command closer to Supreme Headquarters Allied Powers Europe, in Belgium, other moves seem likely to reduce EUCOM's operational capabilities, such as the withdrawal of the 2nd Cavalry Regiment, which had only two years before started operating the 30mm-cannon-armed *Stryker Dragoon* variant specifically intended for the European theatre.

NATO: the vision question – again

On 3–4 December 2019, Alliance heads of state and government met in London. Expectations for the meeting were low and, after several NATO leaders had questioned the Alliance's political and military health throughout the year, the absence of open acrimony was considered a success. In London, NATO declared space an operational domain for NATO and signalled that it has begun to think about the implications of China's rise as a global security player, which, as the London declaration said, presented 'both opportunities and challenges' for NATO. Secretary General Jens Stoltenberg, meanwhile, was invited to start a reflection process to consider how the alliance could be strengthened, including politically, and to define a vision for NATO 2030.

The reflection process formally began in June 2020 and is expected to report in 2021. Before it started, Stoltenberg appointed, to assist in this effort, a ten-strong group of experts jointly chaired by former German defence minister Thomas de Maizière and Wess Mitchell, a former US assistant secretary of state for European and Eurasian affairs. They submitted their report in November. Stoltenberg defined three key elements guiding him in the task of defining NATO's 2030 vision: to ensure the Alliance remained militarily strong; that it would be more united; and that it would pursue a broader global approach. The military dimension of this task list effectively means continuing NATO's adaptation process, so Stoltenberg's review will likely have most effect on the political and global dimensions.

Stoltenberg suggested, during speeches in mid-2020, that it should be possible to widen the range of security issues discussed in NATO in order to help build early and systematic consensus among allies. Meanwhile, the coronavirus pandemic, international terrorism and China's rise underlined for Stoltenberg the requirement for a more global approach. Early in the reflection process, NATO's global approach was essentially portrayed as meaning closer engagement with like-minded countries (such as Australia, Japan, New Zealand and South Korea in the Asia-Pacific) rather than entailing some form of global presence. Previous attempts to give NATO a more global role have faced opposition from member states who consider this a dilution of NATO's core missions. At the same time, NATO has long sought to build stronger relations, in some cases including an operational dimension, with partners across the globe. As such, the precise benefit this proposal is seen as bringing to NATO remains unclear.

That said, the clearest focal point is perhaps China. Stoltenberg clarified that NATO would not view China as an adversary but suggested that China's rise would 'fundamentally' change the global balance of power. With China now projecting power internationally, including through military means, the areas where friction might occur in light of NATO members' interests are becoming apparent. These include China's exercise activities, its defence relationships (including in Africa), its Arctic interests, China's military-modernisation process and its growing military reach, and even its arms-export policies. The EU began to speak in March 2019 of China as 'a systemic rival promoting alternative models of governance'. NATO policymakers sense not only that China's challenge is coming 'closer to us', as Stoltenberg said in November, but that the US wishes to enrol European allies for what many decision-makers and specialists in Washington now consider to be the defining competition and confrontation of the future.

A key priority for NATO, since mid-2018, has been to meet the NATO Readiness Initiative, otherwise known as the 'Four Thirties', that requires NATO allies to be able to collectively commit 30 battalions, 30 air squadrons and 30 combat ships to NATO within 30 days. Although it was not directly referred to in the London declaration, the secretary general's 2019 annual report (published in March 2020) suggested that NATO allies had by December 2019 contributed all forces required by the 'Four Thirties' initiative.

DCA and the *Tornado* replacement in Germany

In April 2020, German's Ministry of Defence (MoD) said it wanted to replace its ageing fleet of *Tornado* aircraft with a mixed fleet of Eurofighter, Boeing F/A-18E/F *Super Hornet* and EA-18G *Growler* aircraft. The Bundeswehr began receiving *Tornado*s in 1981 and intends to retire the aircraft no later than 2030, with withdrawals beginning around 2025. The aircraft is currently used by the Luftwaffe in fighter/ground-attack and electronic-warfare roles. Importantly the *Tornado* is also earmarked for the nuclear-delivery task as part of NATO's dual-capable aircraft (DCA) element of its deterrent posture. DCA relates to the capacity of the air forces of several European NATO members – Belgium, Germany, Italy, the Netherlands, Turkey – to be able to carry US-owned free-fall nuclear bombs stored in each country. The F/A-18F Block III version of the *Super Hornet* would provide the air force with a platform to continue to perform the DCA role

The replacement package outlined by the MoD is, however, a recommendation and not a commitment. According to MoD statements, the government currently plans to introduce supporting documents to parliament in 2022 or 2023. Analysts judge that this timeline seems to assume that a firm decision by 2025 would provide a seamless continuation of all capabilities provided by *Tornado*. Germany will go to the polls in the third quarter of 2021 and the outcome of the election could have an impact on the project. Moreover, the air force's nuclear role is politically contentious.

In November 2020, a contract was signed for 38 new-build Eurofighter aircraft, under Project Quadriga, to replace the air force's Eurofighter Tranche 1 aircraft. Eight will be two-seat aircraft. The plan is to purchase an additional 55 Eurofighters, as well as 30 F/A-18Fs and 15 EA-18Gs to replace the *Tornado* fleet.

In 2017, the MoD considered the Boeing F-15, F/A-18, Lockheed Martin F-35 and Eurofighter as plausible successors to the *Tornado*, but in early 2019 announced that the F-15 and F-35 were no longer under consideration. This was primarily to align the replacement decision with Franco-German plans for a Future Combat Air System (FCAS), scheduled to enter service in the 2040s and to provide uninterrupted access to the capabilities provided by *Tornado*. While senior Luftwaffe personnel had voiced a preference for the F-35, that choice was judged to risk undermining the rationale for the Franco-German (and now Spanish) FCAS project and was rejected. The MoD has since declared that it considers the period from 2025 to 2040 as a bridging period before the arrival of FCAS. It also said that the ability to secure jobs, skills and technical knowledge in Germany and in the wider European defence-industrial base will be key factors in the decision-making process.

The annual report clarifies that the goal is not aimed at generating new forces but rather improving the readiness of existing forces. NATO's reporting on the readiness initiative is patchy, but it seems that a state's decision to commit forces is distinct from questions over how it will achieve and maintain the required readiness in case these troops are called up. As such, it is likely that the military effect of the readiness initiative is questionable so far. The UK, one of the countries that publicised some details about its contribution, in February 2020, suggested it had committed the UK carrier strike group and the Joint Expeditionary Force (JEF). Given that the JEF draws on contributions from eight other partners, including non-NATO members Finland and Sweden, this perhaps refers only to the UK elements of the JEF.

NATO has also been working on a Concept for Deterrence and Defence of the Euro-Atlantic Area and a NATO Warfighting Capstone Concept. These are intended to help clarify the requirements and priorities flowing from the new military strategy that was adopted by NATO in May 2019. A key concern in this context is to achieve a defence posture that is constantly responsive, rather than one that is activated only after threats come to light, sometimes with little or no warning. This thinking is closely linked to ideas around 'multi-domain integration', which seek to achieve higher and more persistent levels of coordination among and within military domains, as well as between the armed forces and other actors.

The EU: looking for ambition

On 1 December 2019, the new European Commission took office, led by former German defence minister Ursula von der Leyen, and with former Spanish foreign minister Josep Borrell Fontelles as the new EU high representative for foreign affairs and security policy. Ursula von der Leyen pledged that she would lead a 'geopolitical' Commission. Meanwhile, in an opinion piece in February 2020, Josep Borrell said that 'Europeans must deal with the world as it is, not as they wish it to be. And that means relearning the language of power and combining the European Union's resources in a way that maximizes their geopolitical impact.' Three processes, either under way or launched in 2020, will provide an indication of whether the EU will make progress in fulfilling its geopolitical aspirations.

The first is the so-called 'strategic compass'. This initiative goes back to a decision in June 2019 to examine the EU's security and defence priorities with a view to delivering additional political guidance. What is now referred to as the strategic compass is intended to provide objectives to help implement the EU's global agenda for foreign affairs (which was adopted in 2016); to generate a common understanding among EU member states of the threats to their security; and to provide additional and coherent impetus for the EU's recent security and defence initiatives. In June 2020, Borrell was tasked to present an assessment of the threats and challenges to European security by the end of the year, drawing on the EU's Intelligence and Situation Centre, which relies on civilian and military input from member states. The plan is that the strategic compass is drafted by late 2021 and finalised by 2022, and that it focuses on crisis management and resilience as well as military capabilities and partnerships. This process will be driven by EU member states, as they hold responsibility for the policy areas in question as well as the required capabilities, with Brussels-based institutions acting as drafting authorities. Even before the process began, more than a dozen EU member states used non-papers to define the political parameters of the process, to establish boundaries but also to signal commitment.

The second process relates to Permanent Structured Cooperation (PESCO), a defence-cooperation mechanism launched in 2017. PESCO was originally conceived – in the context of the Treaty of Lisbon (which entered into force in 2009) – as a coordinating structure for the most capable and willing states. However, PESCO now brings together 25 of the 27 EU member states, and 47 cooperation projects have been launched since 2017. Under PESCO, it is envisaged that investments in and planning for the development and operation of defence capabilities will evolve in an EU framework. Capabilities created by PESCO initiatives would be available for national and multinational purposes (including for NATO). A review of PESCO, mandated by the Council, was completed in November 2020 and concluded member states would need to put more effort into meeting their commitments. The review also suggested new PESCO projects would need to be linked closer to other processes such as the EU's Capability Development Plan (CDP), which is meant to set joint priorities, the Coordinated Annual Review on Defence (CARD), which should help to harmonise defence planning assumptions among member states and the future European Defence Fund (EDF), an EU-level budget for defence research and development and capability

Map 1 **EU and NATO member states: FDI-screening mechanisms**

The European Union adopted a foreign direct investment (FDI) screening regulation in March 2019, in response to concerns that certain investments could affect European security, risk critical infrastructure and harm Europe's future competitiveness. Following the coronavirus pandemic, questions over the security of critical supply chains intensified concerns about FDI. The regulation requires all EU member states to establish FDI-screening mechanisms by 11 October 2020. NATO, too, has become similarly concerned about the varying approaches to FDI among its members. However, nearly 18 months later, FDI in EU and NATO states remains subject to a varied patchwork of screening mechanisms, or in many cases no mechanism at all.

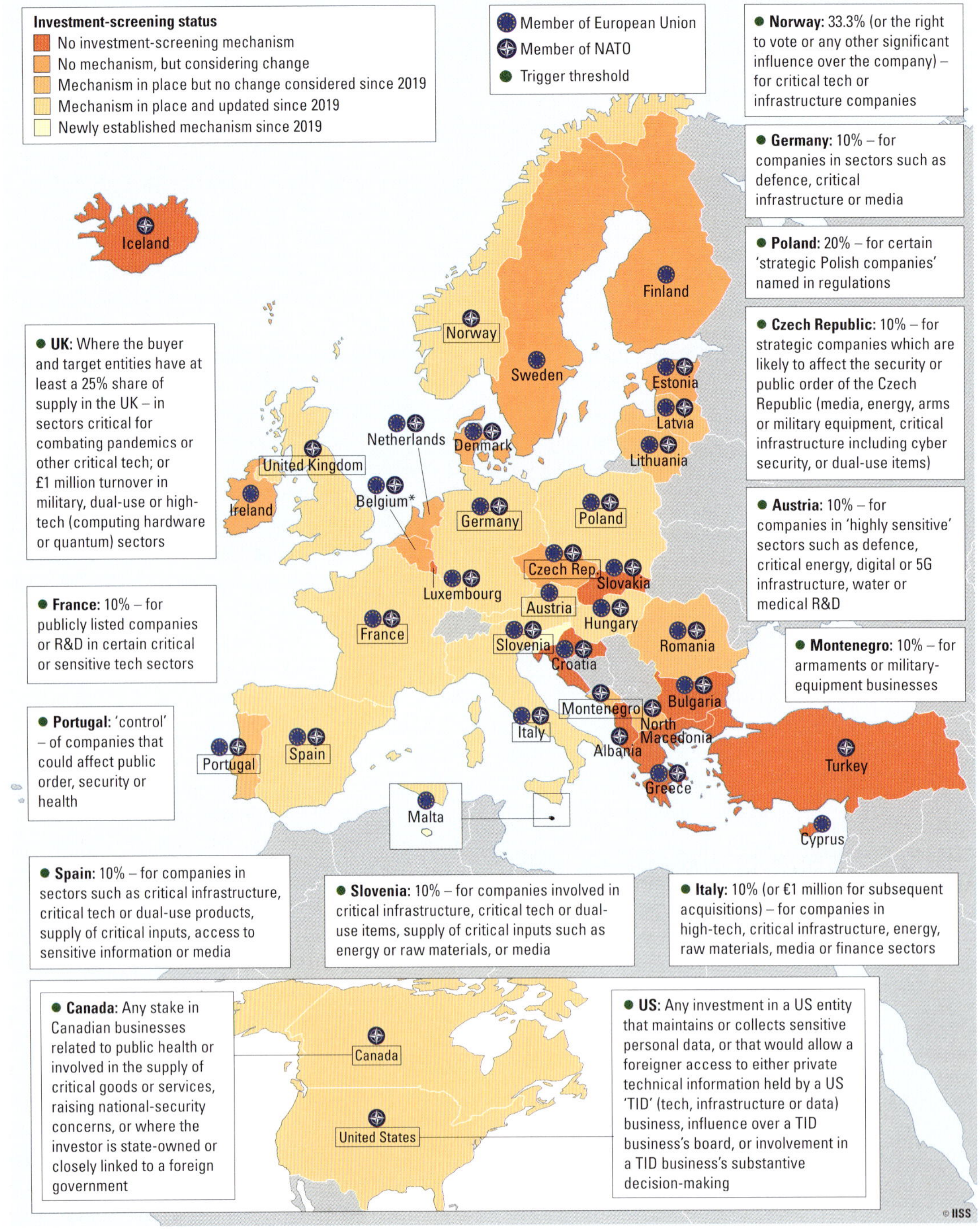

*Flanders region only – not Wallonia

development. At this point, the benefits of PESCO have been marginal – the PESCO review argued, projects that are nonperforming could be merged with related projects, where feasible, or could be closed if a revival is unlikely.

Related to this is the question of finances. On 21 July 2020, European leaders reached agreement on the outline of the next multi-year EU budget (and also the European recovery fund triggered by the coronavirus pandemic). As part of this settlement, which is still awaiting final approval, just over €7 billion (US$8bn) has been earmarked for the EDF, another €1.5bn (US$1.7bn) for military-mobility projects to facilitate the movement of military personnel and equipment around Europe, and €5bn (US$5.7bn) is reserved for the off-budget European Peace Facility (EPF), which is meant to help finance EU operations and those of partners. The budget agreement saw funding for the EDF, for military mobility and for the EPF lowered by almost 40%, 75% and just over 45% respectively, compared to the figures circulating at the beginning of the process. At reduced funding levels, setting priorities and implementing them efficiently becomes paramount. Hence, the strategic compass, the PESCO strategic review and providing an EU-level financial foundation are all increasingly linked directly to the European Union's ambition to be a geopolitical actor that is at least comfortable with the language of power. However, the time taken to agree on a united policy response to the ongoing challenges of the pandemic may complicate future EU defence-cooperation efforts. When combined with the fiscal overhang facing individual countries from their own policy response, this paints a bleak picture for defence investment and wider public spending.

DEFENCE ECONOMICS

Macroeconomics

The International Monetary Fund (IMF) predicts that economic growth in Europe will fall to -6.5% on average in 2020, after stalling in 2019, because of the contraction in activity caused by the coronavirus pandemic. With growth projected at just 4.9% in 2021, economic output in Europe may not recover to pre-pandemic levels until early 2022. Indeed, the contraction in 2020 is markedly larger than the 4.6% decline in output seen in 2009 after the 2007–08 financial crisis. Every country in the region is expected to experience negative economic growth in 2020.

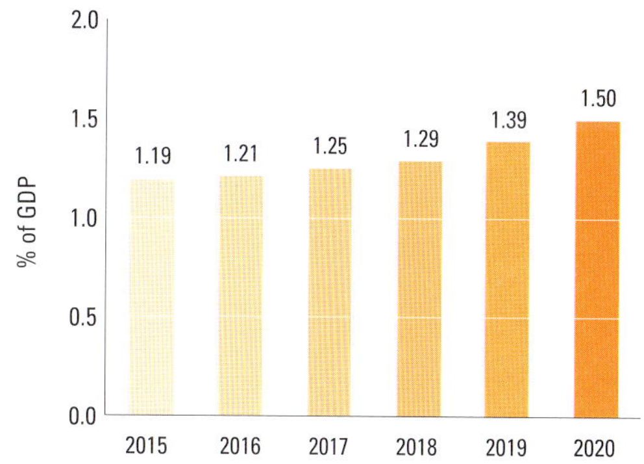

▲ Figure 5 **Europe regional defence spending** as % of GDP (average)

Therefore, the economic impact of the pandemic in Europe will be more immediate, wider reaching and far more severe than the impact of the financial crisis. European states are vulnerable to global repercussions due to their openness to trade and financial flows and their reliance on exports and tourism. Between April and June 2020, the IMF had downgraded the economic-growth projections for most major European economies: France's 2020 projection worsened from −7.2% to −12.5%; Germany's from −7.0% to −7.8%; Italy's from −9.1% to −12.8%; Spain's from −8.0% to −12.8%; and the UK's from −7.8% to −10.2%. By the IMF's October outlook, projections had improved to −9.8% for France, −6.0% for Germany, −10.7% for Italy and −9.8% for the UK while the Spanish projection was maintained at −12.8% for 2020.

Europe's economic outlook was muted even before the pandemic. Mounting global commercial uncertainties, weakening international demand and further uncertainty around the eventual trading relationship between the EU and UK, after the transition period ends in December 2020, led the IMF to project in October 2019 that EU economic growth would reach just 1.6% in 2020, with eurozone growth reaching just 1.4%. Furthermore, the short-term outlook to 2024 was equally subdued, with no sign that average annual growth would increase from the 2020 figure.

Euro-area economic output is projected to contract by 8.3% in 2020, with 2021 growth of 5.2% insufficient to see a full recovery in output. Meanwhile, the IMF projects that fiscal deficits will widen from 0.6% of GDP in 2019 to 10.1% of GDP in 2020 before recovering a little to 5.0% of GDP in 2021. Gross debt levels in the eurozone will grow from 84% of GDP to

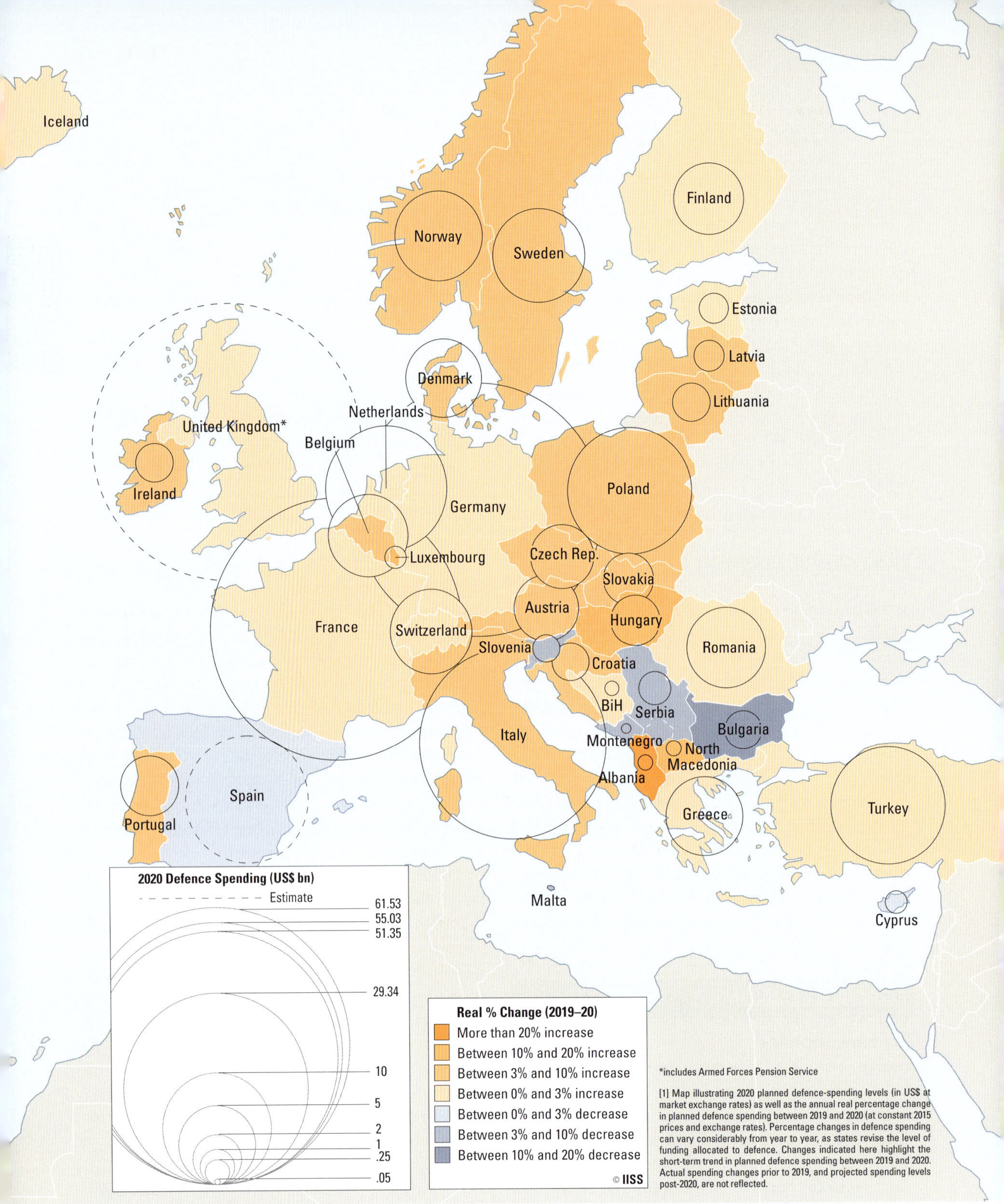

▲ Map 2 **Europe regional defence spending**[1]

Sub-regional groupings referred to in defence economics text: Central Europe (Austria, Czech Republic, Germany, Hungary, Poland, Slovakia and Switzerland), Northern Europe (Denmark, Estonia, Finland, Latvia, Lithuania, Norway and Sweden), Southern Europe (Cyprus, Greece, Italy, Malta, Portugal and Spain), Southeastern Europe (Bulgaria, Romania and Turkey), the Balkans (Albania, Bosnia-Herzegovina, Croatia, Montenegro, North Macedonia, Serbia and Slovenia) and Western Europe (Belgium, France, Iceland, Ireland, Luxembourg, the Netherlands and the United Kingdom).

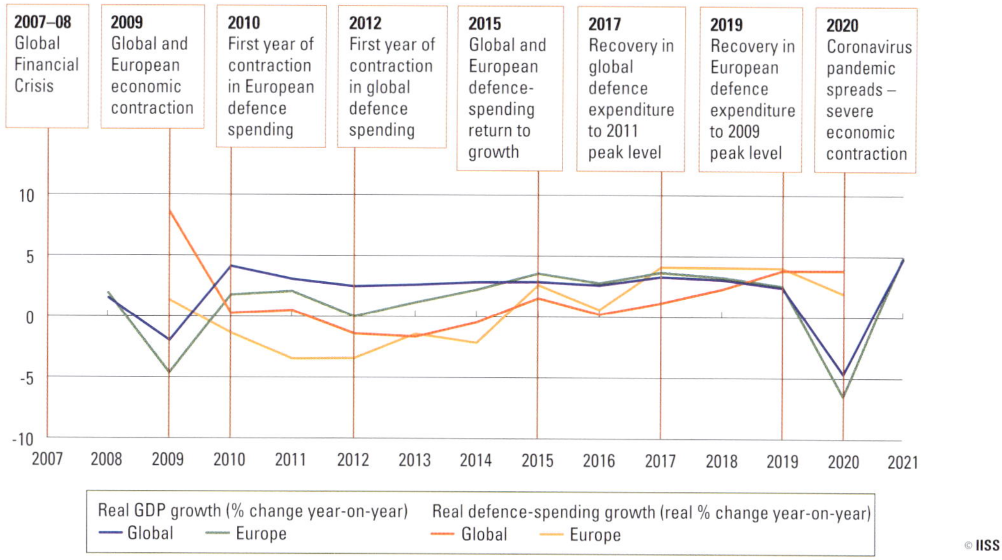

▲ Figure 6 **Global and European economic growth and defence-spending trends, 2007–21**

101% of GDP in 2020, dropping slightly to 100% of GDP in 2021.

This increase in borrowing and debt is driven by the expansive fiscal measures that countries in the region have adopted to try to support vulnerable sectors of society as unemployment soared and manufacturing ground to a halt. Of the approximately US$10 trillion spent globally on these measures by June 2020, one-third was spent by EU members.

Beyond spending at the national level, EU leaders agreed a €750 billion (US$857bn) economic-stimulus plan in July 2020, known as Next Generation EU. Together with revised proposals for the EU's €1.07tr (US$1.22tr) 2021–27 budget (the Multiannual Financial Framework or MFF), this plan was intended to help the EU rebuild and support investment in green and digital transitions. Of the total stimulus package, €390bn (US$446bn) will be distributed in the form of grants to member states while €360bn (US$411bn) will be distributed as loans.

Defence economics

Sizeable and sustained growth in European defence expenditure after 2017 saw total spending in real terms finally recover, in 2019, to the levels seen before the financial crisis. Growth of 4.1% resulted in regional spending of US$297bn in constant (2015) US dollars in 2019, surpassing the previous peak in 2009 of US$287bn. Driving this growth was a 9.8% real increase in the German defence budget, which accounted for over a third of the total regional increase in 2019. Indeed, without it, Europe's growth rate is less impressive in 2020, at just 2.0% in real terms.

Since 2015, European NATO members have steadily increased defence expenditure when measured as a proportion of GDP. This trend continued in 2020, though it reflects the economic impact of the coronavirus pandemic more than any heightened commitment to defence. Average spending increased from 1.25% of GDP in 2014 to 1.52% in 2019 and will rise to 1.64% of GDP in 2020.

Sub-regional trends shifted significantly this year. Spending in Southeastern Europe (Bulgaria, Romania and Turkey) increased by 4.9% in 2019 but fell by 2.4% in 2020 in real terms. The sub-regional trend tends to be dominated by movements in the much larger Turkish defence budget, though sharp increases in the Romanian and Bulgarian budgets in 2019 were sufficient to offset a 2.9% real-terms reduction in the Turkish defence budget that year. Romania's defence budget increased by 14.3% in 2019 while Bulgaria's defence budget surged by 118.2% in 2019 to fund the acquisition of new combat aircraft. Similarly, a 1.7% real increase in Turkey's defence budget in 2020 was offset by stagnating spending in Romania and a 46.5% real reduction in Bulgaria's budget to a more sustainable level of defence investment. Between 2008 and 2017, Bulgaria's defence budget ranged between US$600–800 million and accounted for just 1.31% of GDP on average. After increasing to US$1.0bn in

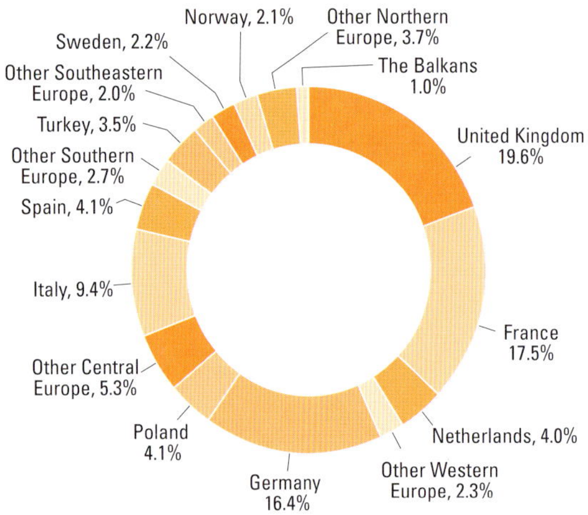

Other Western Europe – Belgium, Iceland, Ireland, Luxembourg
Other Central Europe – Austria, Czech Republic, Hungary, Slovakia, Switzerland
Other Northern Europe – Denmark, Estonia, Finland, Latvia, Lithuania
Other Southern Europe – Cyprus, Malta, Portugal
The Balkans – Albania, Bosnia-Herzegovina, Croatia, Montenegro, North Macedonia, Serbia, Slovenia
Other Southeastern Europe – Bulgaria, Romania

© IISS

▲ Figure 7 **Europe defence spending by country and sub-region, 2020**

2018, the 2019 surge increased spending to US$2.1bn or 3.05% of GDP, though this reduced to US$1.2bn (US$949m in constant 2015 US dollars) or 1.70% of GDP in 2020.

Meanwhile, after posting 16.2% real growth in 2019, defence spending in the Balkans stagnated in 2020 as every country apart from Albania implemented budget cuts or dramatically slowed the rate of their budget increase. The Croatian defence budget grew by 12.7% in real terms in 2019 but growth slowed to just 3.7% in 2020. Serbia, the other key defence spending state in the sub-region, implemented a 31.6% real increase in 2019, bringing the defence budget to US$799m in constant (2015) US dollars, within touching distance of the peak spending level of US$826m in 2010, before the effects of the financial crisis on defence funding became fully apparent. However, the defence budget contracted by 6.3% in real terms in 2020 to bring the total to US$748m (US$878m at current market exchange rates).

Central Europe also saw growth continue but slow markedly in 2020, down to 3.4% from 9.0% in 2019 in real terms. However, this trend is again distorted by shifts in one country, in this case Germany. Spending growth in the Czech Republic, Hungary and Poland was sustained but not sufficient to offset the lower rate of real growth in German spending of 1.9% in 2020 compared to 9.8% in 2019.

Growth also stalled in Northern Europe, falling from 7.3% in 2019 to 4.2% in 2020 in real terms as Norway and Sweden slowed budget increases. The slowdowns were implemented before the impact of the pandemic became clear. Sweden's defence budget growth decelerated mainly because the robust increase seen in 2019 was not repeated in 2020. In Norway, meanwhile, the slowdown reflected reduced funding required for payments related to the F-35 programme. Indeed, Norway intends to continue to increase core defence expenditure, by 2028, to a level that is Kr16.5bn (US$1.8bn) higher than 2020 levels, with the budget reaching 2% of GDP by that year.

Despite concerns regarding public spending in the wake of the pandemic, according to the 2021 budget released on 21 September 2020, Sweden intends to increase defence expenditure to SKr71.15bn (US$7.7bn) in 2021, a 9.8% real increase over the 2020 budget and in line with the plans outlined in the 2020 budget and 2020 Spring Fiscal Policy Bill. Similarly, Finland did not downgrade its defence-spending plans. The announcement of the 2021 budget in August 2020 saw the country move ahead with the HX fighter programme with a 44% real uplift planned for the defence budget.

The rate of real growth in West European defence expenditure slowed slightly in 2020, from 3.3% to 1.6%, but growth rates for the states with the largest budgets were in keeping with the three-year trend. The French budget has increased by 2.1% on average annually since 2017 and this trajectory continued in 2020, with an albeit milder 1.5% real increase over 2019 levels as the country implemented growth plans outlined in the Loi de Programmation Militaire (LPM) 2019–2025. The 2020/21 UK defence budget grew by 0.6% in real terms, benefiting from an uplift in the 2019 Spending Review. The financial settlement for defence will be bolstered from 2021, with the announcement in November 2020 of £16.5bn (US$21.1bn) in additional funding for the UK armed forces over the period from 2021–25.

Defence procurement and industry

The impact of the coronavirus pandemic on defence spending and procurement will become more apparent in 2021 as budgetary adjustments are made to account for extensive fiscal-stimulus packages. A group of scientific advisers from the Armament

Industry European Research (ARES) Group argued in April that defence should be included in stimulus packages and cuts to defence should be avoided as the region faces significant geopolitical challenges which may be exacerbated by the pandemic.

Several countries did implement measures to support defence industries and procurement in the wake of the pandemic. France announced a €600m (US$686m) programme of investment in June intended to accelerate key military aerospace programmes including the Avion Léger de Surveillance et de Renseignement (ALSR: light intelligence and surveillance aircraft) programme and the acquisition of three Airbus A330 aircraft for the air force. Meanwhile, uninhabited aerial vehicle (UAV) programmes for the navy will also be accelerated. Beyond core military services, orders for civil protection and the Gendarmerie will also be brought forward. The investment programme is part of a wider €15bn (US$17bn) rescue package to support the national aviation sector.

Germany announced a €130bn (US$149bn) stimulus package in June. €10bn (US$11.4bn) was earmarked for the acceleration of armaments programmes deemed to have a high domestic share of 'added value' (i.e. include a high work-share for local firms) and that were able to begin immediately in 2020 or 2021. Similarly, Norway brought forward the Kr500m (US$53m) upgrade of the navy's *Skjold*-class fast patrol craft and the Kr600m (US$64m) upgrade of the army's CV90 armoured vehicles.

The UK did not bring forward major programmes but instead ensured that cash flow was maintained and that the supply chain was supported through forward ordering and the adjustment of payment schedules. Despite improvements in affordability gaps in the ten-year Equipment Plan, concerns from the National Audit Office persisted in 2020. The Paliamentary body estimated in February 2020 that the disparity between funding and programme commitments ranged from £2.9bn (US$3.7bn) to £13.0bn (US$16.7bn) over the course of the plan. While the announcement of £16.5bn (US$21.1bn) in additional funding to 2024–25 will reduce this disparity, much of the new funding is already ring-fenced for emerging technologies (Artificial Intelligence, cyber-defence, autonomous systems, Space Command) or major programmes (such as the Future Combat Air System and expansion of the navy's surface fleet). Therefore it is likely that difficult programme decisions will still be required in the UK's Integrated Review in order to ensure affordability. Major programme adjustments were reportedly considered, including reducing the planned acquisition of Boeing's E-7 *Wedgetail* airborne early-warning aircraft from five to three. An overarching focus on resilience and new technology was apparent in the drive by the Ministry of Defence to fast-track new virtual-reality technology in an attempt to bolster the UK's response to future pandemics, natural disasters and terrorist attacks.

Meanwhile, the coronavirus pandemic led to delays in the acquisition process for Switzerland's fighter and ground-based air-defence requirement programme as well as Finland's HX fighter programme. In the former, it was announced in May 2020 that responses from industry to the second request for proposals would be delayed to November 2020, while in the latter Finland's Ministry of Defence announced in August that it would delay issuing best-and-final-offer requests until later in 2020 or early 2021. In both cases, however, the overarching programme timeline and contract expectations remain on track.

The effect of the coronavirus pandemic on the defence sector largely depends on firms' exposure to commercial aviation and on the success of government support initiatives. Airbus reported a 39% drop in revenue in the first half of 2020 (H1) and in June announced that it would reduce its headcount, which could lead to the loss of 15,000 jobs. Similarly, Rolls-Royce reported a 24% drop in H1 2020 revenue and announced in May the intention to reduce its global headcount by 9,000 posts, or 17%, as demand for engines plummeted as global commercial aviation ground to a halt.

FINLAND

Strategy and policy

Finland's defence policy is shaped by the country's recent history and its geopolitical environment, as well as the political preferences of governments in Helsinki. The role of history in shaping Finland's national identity is visible including in support for a nationwide and comprehensive defence system. Finland has never been occupied by a foreign state, though it fought three wars against the Soviet Union in the years after its independence in 1917. Finland's geopolitical environment is dominated by its proximity to Russia. Russia's key military concentrations on the Kola Peninsula run along the northernmost one-third of the shared border, while the former Russian capital of Saint Petersburg is situated close to southern Finland. The shared border between Finland and Russia, which effectively runs between these two strategic areas, is over 1,300 kilometres in length. Finland is non-aligned though it 'maintains the option to seek membership in a military alliance' and Finland is an active NATO partner. Meanwhile, there has been increased defence cooperation in recent years, particularly with its neighbours Sweden and Norway, as well as the United States. International cooperation has deepened after Russia's annexation of Crimea, and Finland has since increased its readiness, its military planning and exercises and also boosted its defence acquisitions, as part of a national-defence concept that includes territorial-defence structures and conscription.

The Government Report on Finnish Foreign and Security Policy, published by the prime minister's office in October 2020, states that one of the country's prime objectives is to avoid becoming a party to a military conflict. While the report says that there is no immediate military threat, it follows by saying that 'nonetheless, Finland must prepare for the use or the threat of use of military force against it'.

The government, and defence authorities, are also closely studying hybrid challenges, including what the October 2020 report terms 'hybrid influencing', as well as cyber challenges. Finland's nationwide comprehensive security concept is important in countering these challenges, and under this concept critical areas of society and infrastructure are supported by regular cooperation between the authorities, the business community and other relevant organisations, and citizens. The 2017 Security Strategy for Society outlines Finland's comprehensive security model, and states that national defence capability rests not only on military systems but also on the preparedness of society and whole-of-government cooperation as well as international defence cooperation. In practical terms, Finland's defence forces plan, prepare and exercise regularly together with police units as well as those of the Border Guard and the Customs authorities, among others. The country's comprehensive security model could, in sum, be described as a combination of military deterrence, societal resilience and defensive as well as offensive actions all designed to constrain adversary activities.

Defence cooperation

There is wide-ranging defence cooperation with Sweden, including bilateral operational planning, and as far as Finland is concerned there are no 'predetermined limitations' to 'deeper foreign and security policy cooperation and defence cooperation with Sweden'. These ties have deepened in recent years and the goal is to facilitate the use of national capabilities both in Finland and Sweden if required. As the October 2020 report puts it, 'the purpose of cooperation is to lay foundations for joint action and operations of Finland and Sweden under any circumstances'. This has resulted from a practical and commonly accepted notion that in a crisis both states would probably become involved in the same conflict. In the Nordic context, the Nordefco cooperation forum (comprising Denmark, Finland, Iceland, Norway and Sweden) has the goal of reducing national restrictions for the movement and storage of military units and materiel, and improving member states' interoperability, among other initiatives. From this viewpoint, Finland shares several security concerns with its neighbours, including Norway. In line with the aspirations of the Nordic Defence Cooperation vision for 2025, agreed at the Nordefco summit in 2018, a tripartite statement of intent was signed with Norway and Sweden in September 2020, setting targets for closer defence cooperation.

There has been a steady increase in cooperation with the US, particularly since the introduction of US-manufactured F-18 C/D *Hornet* fighter aircraft in the mid-1990s, through to the procurement of long-range air-to-ground missiles (Joint Air-to-Surface Standoff Missile, JASSM). A trilateral statement of intent was signed in 2018 with the US and Sweden to further deepen defence cooperation, based on earlier bilateral defence agreements in 2014 and 2016.

EU membership has been perceived by Finland's political leaders, and many of its citizens, as a source of enhanced security and defence. A sign of this approach is Finland's readiness to proceed with the solidarity and assistance clauses in the Lisbon Treaty (Articles 222 and 42). Finland has been preparing to provide and receive military assistance since the early 2000s by improving national facilities, and in 2017 modernised legislation to remove legal obstacles. Bilateral framework-style defence-cooperation agreements have also been signed with Denmark, Estonia, France, Germany, Norway, Poland, Sweden, the UK and the US. It is understood that the idea underpinning these activities is to promote arrangements for building an interoperable defence system, so leading to a more credible deterrence policy. Finland is also a participant in the UK-led Joint Expeditionary Force.

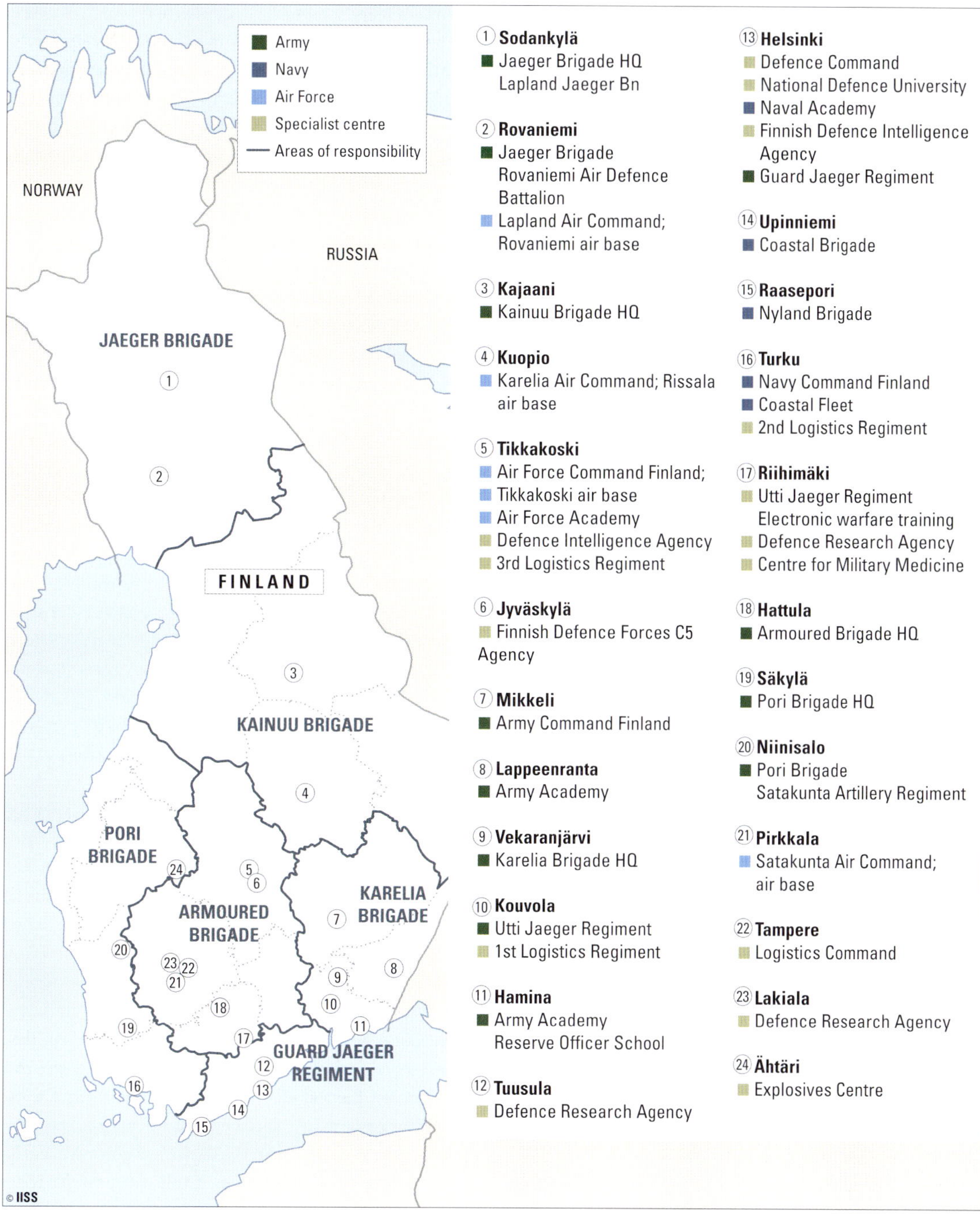

▲ Map 3 **Finland: key military units and locations**

Finland joined NATO's Partnership for Peace programme in 1994 and since then has engaged in political dialogue and practical cooperation with NATO. Finland is an Enhanced Opportunities partner, as is Sweden. According to an opinion poll on defence and security issues conducted by the Advisory Board for Defence Information (ABDI) and released in 2020, 64% of respondents say that Finland should not seek membership of the Alliance, with 94% supporting military cooperation with Nordic countries, particularly with Sweden. Cooperation with the EU is supported by some 88% and with the US by 57%. Some 57% were reported as being in favour of military cooperation with NATO, with this opposed by one-third of Finns. Nevertheless, key elements of the Finnish armed forces, especially operational naval and air-force units, are trained to meet NATO interoperability requirements. All the units declared available for crisis-response operations, NATO-led or otherwise, have fully met the evaluation requirements set for NATO standards. In many areas, the same requirements inform the training curriculum of units that are designated for national defence.

Armed forces

The government and defence ministry provide guidance on key defence policy issues, often including Finland's operating environment as well as capability objectives. The Defence Report to Parliament, published in February 2017, contains the most recent defence policy guidelines. The report recognises that there is a shorter early-warning period for a military crisis and a lower threshold for using force. It advocates higher readiness and deeper defence cooperation as well as the development of related national legislation. In short, defence-related demands have grown, and Finland's authorities judge that it must prepare for the use or threat of military force.

Particular attention is being paid to improving military readiness, together with intelligence, cyber-defence and long-range-strike systems. There are also plans to replace ageing capabilities. With military forces that are overwhelmingly based on reservists, fully mobilising Finland's wartime forces requires additional defence materiel as well as resources that are under the administrative control of other government departments.

With Finland's population numbering only five and a half million inhabitants and its territory the fifth largest in the EU, its system of compulsory conscription enables the wartime mobilisation of a sufficient number of units. Consequently, the defence structure in peacetime is designed for readiness-response tasks but also for training conscripts and in turn generating wartime units. The conscription system is widely supported by citizens, with 77% in favour, according to the ABDI poll. This system provides the majority of personnel for army units, though the air force and the navy are staffed by professional forces and held at a state of high readiness. A review of conscription is due to report in early 2021. However, the challenge from complex hybrid threats is creating new deterrence demands, and one challenge facing the armed forces is that the national mobilisation system is judged as vulnerable, particularly in the earliest phases of a crisis. Legislation to accelerate flexible mobilisation has been developed in a bid to address these concerns. High-level exercises are intended to boost readiness and contribute to the credibility of Finland's deterrence policy.

Equipment modernisation

The political requirement is for the armed forces to defend all territory, including the Åland Islands (which are demilitarised in peacetime), and the sparsely populated northern half of the country. This has resulted in a land-heavy defence system with 280,000-strong armed forces (at wartime strength), including mechanised and Jaeger brigades, armoured regiments, special-operations forces and helicopter battalions. Currently, the navy operates missile-capable vessels, minelayers and mine-countermeasures vessels. For coastal defence, coastal jaeger and amphibious brigades focus on conscript training, maritime surveillance and special operations. The air force's three fighter squadrons operate 62 F/A-18 C/D *Hornet*s and there are also units for reconnaissance, training and transport tasks, supported by surveillance systems.

Perhaps the major challenge for the armed forces stems from its equipment deficits, in terms of both quality and quantity. Defence funding increased after late 2016, and the army will also benefit from a share of additional funding, amounting to €150m (US$171m) annually from 2021, which will fund procurement, readiness and equipment recapitalisation, as well as joint capabilities. However, the principal defence-modernisation initiatives, dubbed strategic-capability projects, are the HX programme, designed to select a replacement for the *Hornet* combat aircraft (due to be phased out by 2030), and the Squadron 2020 project, intended to replace four *Rauma*-class guided-missile patrol boats and two *Hämeenmaa*-class minelayers (and the already retired minelayer *Pohjanmaa*) with a new class of four multi-role corvettes.

Military readiness in case of a surprise attack, is a concern that has risen in prominence after Russia's seizure of Crimea in 2014; overall, readiness is improving. There is also concern over Russia's dual-capable precision-strike systems, such as its ballistic and cruise missiles, which have been deployed within striking distance of Finland and other countries in the

Nordic–Baltic region. The defence authorities assess that new capabilities and tactics are needed to counter missile attacks that are beyond the capability of Finland's current air-defence system. Missile defence had been viewed by analysts as an area beyond the reach of Finland's defence funding, though in late October 2020 the authorities sent invitations to tender to five firms for a project to develop a high-altitude air-defence system.

Defence economics and industry

Traditionally, Finland's approach to defence has been threat-based, existential and has benefited from broad popular support. The latest government defence reports are designed to address previous reductions by raising total wartime strength and increasing slightly the defence budget for readiness and acquisitions. The 2021 budget, released in August 2020, proposed a €1.7 billion (US$1.9bn) increase for defence funding over 2020 levels, allocating €4.8bn (US$5.5bn) to the core Ministry of Defence (MoD) budget. The total defence budget will come to €5.2bn (US$5.9bn) for 2021 (including an estimated pensions cost of €361m (US$413m)), accounting for 2.1% of GDP, compared to €3.6bn (US$4.1bn) or 1.5% of GDP in 2020. The increase in 2021 is the result of additional funding for major acquisitions, primarily for Finland's two strategic capability projects, the air force's HX combat aircraft programme and the navy's Squadron 2020 programme.

The proposed 2021 budget includes €1.5bn (US$1.7bn – excluding VAT) in funding for the HX programme and €209m (US$239m – excluding VAT) for the Squadron 2020 programme. The total cost of the Squadron 2020 programme is estimated to be €1.3bn (US$1.5bn) to 2028 while the MoD has proposed €10bn (US$11.4bn) for the HX programme.

Finland maintains a specialised defence industry capable of maintenance and overhaul as well as defence-manufacturing capacity in certain areas; in some cases the defence ministry outsources maintenance to the private sector. Maintaining industrial capacity is a government priority, given concerns over security of supply. In its 2016 report on Securing the Finnish Technological and Industrial Base, Helsinki also noted Finland's dependence on the global defence market and stressed the need for international cooperation, because the technical sophistication of systems limits 'full autonomy in system support and maintenance', as well as the need to maintain the technical skills required by the defence authorities. The report listed four technology and engineering areas important for Finland's national-defence capabilities: command, control, communications, and computers (C4) and intelligence, surveillance, target acquisition and reconnaissance (ISTAR); materials technology and structural engineering; technology and engineering for multi-technology systems; and 'bio and chemical technologies and engineering'. Exports are a growing share of the revenue of Finland's defence firms, with Patria and Marine Alutech recently exporting land and naval platforms respectively, while a number of other firms provide services as subcontractors on other orders. The replacement aircraft under the HX programme is due to be selected in 2021, and some of the five bidders (Boeing with the F/A-18E/F *Super Hornet* and EA-18G *Growler*; Lockheed Martin with the F-35 *Lightning* II; Saab's *Gripen* E/F; Dassault's *Rafale*; and BAE leading the bid for the Eurofighter *Typhoon*) have submitted plans for technology transfer as part of their proposals. The government is keen to ensure local servicing and maintenance capacity and has set a requirement for industrial cooperation amounting to 30% of the HX acquisition price. Meanwhile, Finnish shipbuilding firm Rauma Marine Constructions has been selected to build the corvettes under the Squadron 2020 programme, under an overall contract with Saab. Perhaps the most well-known firm is Patria, headquartered in Helsinki and which manufactures defence materiel including armoured vehicles, artillery and patrol

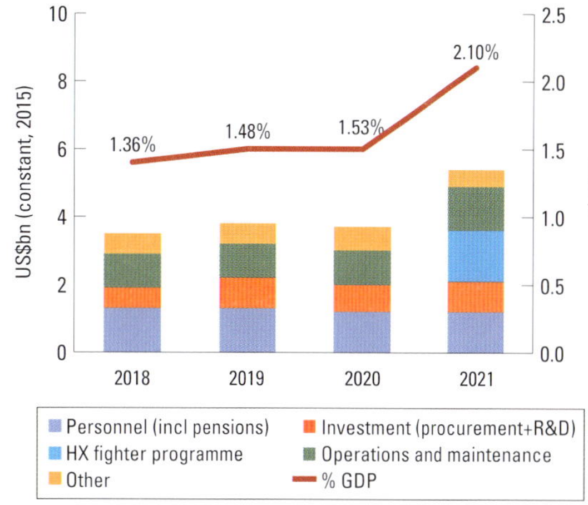

Source: 2018–20 Finnish government responses to IISS, 2021 based on 2021 Draft Budget, Finland Ministry of Finance

▲ Figure 8 **Finland: defence budget 2018–21** (constant US$bn)

craft, also exporting these to a range of states. Patria is owned by Finland's government (50.1%) and Kongsberg (49.9%). Kongsberg is itself half-owned by the Norwegian government. Nammo, which is owned in equal share by Patria and the Norwegian government, has five sites in Finland, three of which are involved in the manufacture of ammunition and related components.

UNITED KINGDOM

For much of 2020, the UK government was engaged in its Integrated Review of Security, Defence, Development and Foreign Policy. Work on this was disrupted by the coronavirus pandemic and it was expected, at the time of writing, that the review would be released in early 2021. Meanwhile, on 19 November 2020 the UK prime minister announced that UK defence spending would increase. Speaking in parliament, Prime Minister Boris Johnson said that an additional £16.5 billion (US$21.1bn) would be added to the defence budget over the next four years, this being beyond the existing commitment to implement 0.5% increases to the annual budget in real terms. The subsequent 'Spending Review 2020' confirmed this increase and provided details about the annual funding allocation. According to the spending review, the 2021–22 core defence budget will increase to £46.0bn (US$59.0bn), an 11.7% nominal increase over the 2020-21 budget of £41.2bn (US$52.8bn). Considerably lower nominal increases in subsequent years will bring the core budget to £47.6bn (US$61.0bn) by 2024–25. In real terms, the budget will be 9% higher by 2024–25, when compared to the 2019–20 budget, with real-terms growth averaging 1.8% over the period. The nominal increases in 2023–24 and 2024–25 actually translate to reductions in real terms.

The 2015 Strategic Defence and Security Review (SDSR) set ambitious targets for modernising UK military capability in the framework of a 'Joint Force 2025'. This was intended to provide deployed joint force including a carrier-task group, an army division and a large air group, though delivery of the full capability by 2025 seems unlikely. For example, a National Audit Office (NAO) review of the Royal Navy/Royal Air Force Carrier Strike capability suggests that the necessary complement of F-35 combat aircraft, airborne early-warning radar-equipped helicopters, and logistics ships, will not be delivered by 2025. And the Ministry of Defence (MoD) states that the SDSR requirement to deliver a new medium-weight strike brigade by 2025 will not be delivered in time, perhaps reducing the planned combat power of the Army's single heavy division by up to half.

In his speech, Johnson confirmed details about a number of projects and also noted additional areas of investment. The procurement of eight Type-26 and five Type-31 frigates was confirmed, as was a plan for support ships. New 'Type 32 frigates' were mentioned too and later in November, a ministerial answer in parliament, responding to a question from an opposition politician, indicated that these were envisaged to be platforms that would include 'autonomous systems'. While it is possible that the platform will be based on the Type-31, there is little detail publicly available, while the same parliamentary response indicated that the concept phase for the vessels had yet to be launched. Meanwhile, the prime minister said that the army would be 'reshaped for the age of networked warfare,' which raised questions over numbers of personnel and 'legacy' platforms, while the Royal Air Force would 'receive a new fighter system, harnessing artificial intelligence and drone technology'. The announcement confirmed other developments, including that a new Space Command would be set up within the RAF and a joint National Cyber Force, that succeeds existing initiatives and combines personnel from the Government Communications Headquarters, the MoD, the Secret Intelligence Service and the Defence Science and Technology Laboratory.

While the statement gave the defence ministry a firmer foundation for spending plans, the accumulated impact of cost escalation and delays to major programmes make it likely that some existing equipment modernisation plans will have to be curtailed, delayed or cancelled. Potential contenders for capability reductions could include the army's plan to modernise heavy armour, or the numbers of F-35 *Lightning* combat aircraft and *Wedgetail* airborne early warning and control aircraft. However, without the Integrated Review – which is intended to provide the strategic direction to enable capability choices – at the time of writing it remained difficult to assess the relative strategic value of the additional funding or, indeed, of the project announcements made by the prime minister.

Figure 9 'Global Combat Ship' frigate design

The Type-26 design originated in a United Kingdom requirement for a new high-end future surface combatant, principally to replace the Royal Navy's Type-23 frigates. The aim was to produce a platform able to operate both independently and as part of a task group to provide both general-purpose tasks and a sophisticated anti-submarine-warfare (ASW) capability. A major issue throughout the design and procurement process has been the high unit cost of such a platform, which led the UK to reduce its planned procurement from 13 to eight* vessels. The resurgence of great-power competition has helped the design's export prospects, because it has revived the requirement for high-end surface combatants such as the Type-26 and the Franco-Italian Multi-Mission Frigate FREMM and has renewed attention on ASW skills and capability. As a result, both Australia and Canada have chosen the Type-26 as the basis for their future surface-combatant requirements; these vessels will be built in local shipyards.

Type-26/FREMM customers

Type-26	FREMM
Australia	Egypt
Canada	France
UK	Italy
	Morocco
	United States

Basic dimensions/characteristics

- Full-load displacement up to 8,800 tonnes
- Length 149.9m
- Beam 20.8m
- Speed 27+ knots

Operating concepts

The UK variant is designed to operate as part of a mixed fleet including destroyers and aircraft carriers. The Australian and Canadian variants will be either their navies' sole or predominant major surface combatants. Their enhanced radar and anti-air-warfare capabilities reflect this and also indicate a different emphasis on operations. The sensor and equipment choices on these variants also reflect national priorities to support domestic defence industry.

Royal Navy variant

- Three 8-cell Mk 41 strike-length vertical launch systems (VLS) for possible land-attack cruise, anti-ship and anti-submarine missile fits
- Type 997 *Artisan* medium-range radar
- MBDA *Sea Ceptor* missiles for local area air defence
- Two 24-cell VLSs
- Medium-calibre Mk 45 127mm gun, with the potential for incorporating smart munitions
- Type 2087 active/passive towed array sonar (TAS) as principal ASW sensor

Common features on all variants

- Acoustically stealthy hull design, including rafted machinery installation
- Quiet combined diesel–electric or gas-turbine propulsion installation for ASW performance
- Type 2150 hull-mounted medium-frequency bow sonar
- Mission bay capable of accommodating boats, including for an embarked force; uninhabited or autonomous vehicles; and containers for humanitarian-assistance and disaster-relief supplies
- Aviation facilities including a flight deck able to accommodate heavy-lift helicopters up to the size of a CH-47 *Chinook*
- Hangar able to support medium-lift helicopters up to the size of the UK's *Merlin* and/or uninhabited air systems

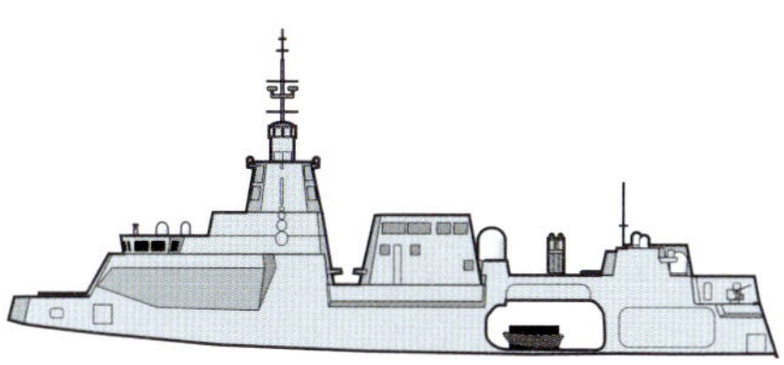

Royal Canadian Navy variant

- 15* hulls
- Replacing *Halifax*-class frigates and *Iroquois*-class destroyers
- SPY-7 active electronically-scanned radar
- Mk 41 VLS
- CMS 330 combat-management system with AEGIS
- Towed low-frequency active and passive sonar
- SM-2/ESSM/*Sea Ceptor* missiles
- Anti-ship missile and possible *Tomahawk* land-attack cruise missile fit
- Some equipment choices still to be finalised

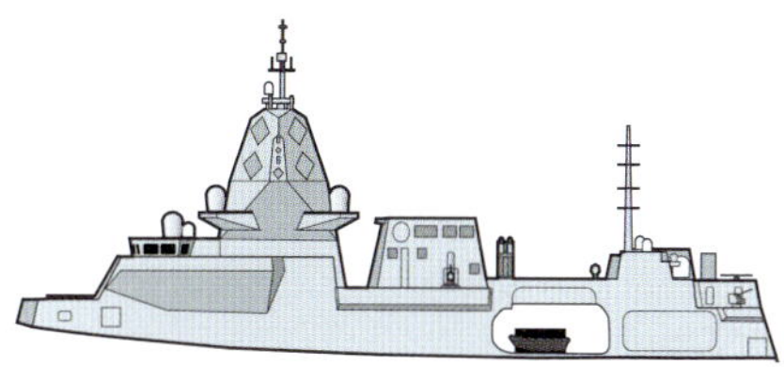

Royal Australian Navy variant

- Nine* hulls
- Replacing ANZAC-class frigates
- CEAFAR 2 active phased-array radar
- AEGIS combat-management system
- Mk 41 VLS with SM-2/ESSM missiles
- Type 2087 active/passive TAS
- Anti-ship missile fit

* As of October 2020

Albania ALB

Albanian Lek		2019	2020	2021
GDP	lek	1.68tr	1.57tr	
	US$	15.3bn	14.0bn	
per capita	US$	5,323	4,898	
Growth	%	2.2	-7.5	
Inflation	%	1.4	1.4	
Def exp [a]	lek	21.7bn	24.1bn	
	US$	197m	215m	
Def bdgt [b]	lek	15.7bn	20.9bn	18.9bn
	US$	143m	187m	
US$1=lek		109.87	111.88	

[a] NATO definition

[b] Excludes military pensions

Real-terms defence budget trend (US$m, constant 2015)

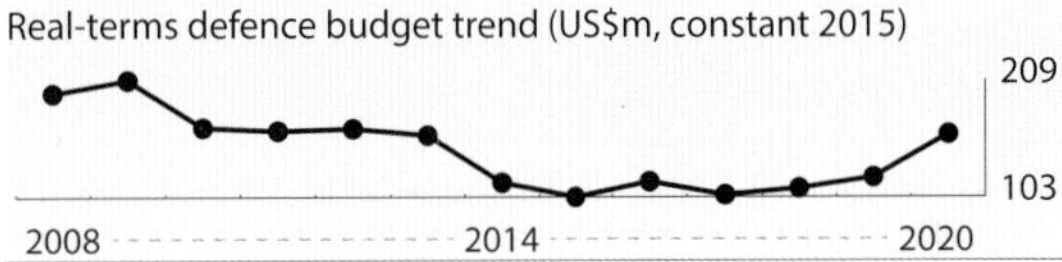

Population 3,074,579

Age	0–14	15–19	20–24	25–29	30–64	65 plus
Male	9.3%	3.6%	4.4%	4.8%	21.3%	6.1%
Female	8.3%	3.3%	4.1%	4.6%	23.4%	7.0%

Capabilities

Principal missions for Albania's armed forces include territorial defence, internal security, disaster-relief tasks, and small-scale peacekeeping or training deployments. Tirana is looking to improve the operational readiness of its infantry battalion in order to fulfil obligations to NATO, which it joined in 2009. Other priorities include improving border management and information sharing to prevent transnational crime and terrorism, establishing a cyber-security unit and developing air capabilities. Greece and Italy police Albania's airspace. Albania contributes to NATO, UN and EU missions but does not possess an independent expeditionary capability. Most Soviet-era equipment has been sold. Limited defence modernisation under the Long-term Development Plan 2016–25 is proceeding, including the purchase of helicopters and the installation of an airspace surveillance system. The navy is expected to receive upgrades to vessels that have been or still are deployed in the Aegean Sea. NATO plans to support the modernisation of Kuçova air base. Albania has little in the way of domestic defence industry, with no ability to design and manufacture modern military platforms. Nevertheless, the country has some publicly owned defence companies that are capable of producing small arms, explosives and ammunition.

ACTIVE 8,000 (Land Force 3,000 Naval Force 650 Air Force 550 Other 3,800)

ORGANISATIONS BY SERVICE

Land Force 3,000

FORCES BY ROLE

SPECIAL FORCES

1 SF bn

1 cdo bn

MANOEUVRE

Light

3 lt inf bn

COMBAT SUPPORT

1 mor bty

1 NBC coy

EQUIPMENT BY TYPE

ARMOURED FIGHTING VEHICLES

APC • PPV 40 *MaxxPro Plus*

ARTILLERY • **MOR** 93: **82mm** 81; **120mm** 12

Naval Force 650

EQUIPMENT BY TYPE

PATROL AND COASTAL COMBATANTS • **PBF** 5 *Archangel*

Coast Guard

EQUIPMENT BY TYPE

PATROL AND COASTAL COMBATANTS 22

PB 9: 4 *Iluria* (Damen Stan Patrol 4207); 3 Mk3 *Sea Spectre*; 2 (other)

PBR 13: 4 Type-227; 1 Type-246; 1 Type-303; 7 Type-2010

Air Force 550

EQUIPMENT BY TYPE

HELICOPTERS

TPT 27: **Medium** 4 AS532AL *Cougar*; **Light** 23: 1 AW109; 5 Bell 205 (AB-205); 7 Bell 206C (AB-206C); 8 Bo-105; 2 H145

Regional Support Brigade 700

FORCES BY ROLE

COMBAT SUPPORT

1 cbt spt bde (1 engr bn, 1 (rescue) engr bn, 1 CIMIC det)

Military Police

FORCES BY ROLE

COMBAT SUPPORT

1 MP bn

EQUIPMENT BY TYPE

ARMOURED FIGHTING VEHICLES

AUV 8 IVECO LMV

Logistics Brigade 1,200

FORCES BY ROLE

COMBAT SERVICE SUPPORT

1 log bde (1 tpt bn, 2 log bn)

DEPLOYMENT

AFGHANISTAN: NATO • *Operation Resolute Support* 99

BOSNIA-HERZEGOVINA: EU • EUFOR • *Operation Althea* 1

LATVIA: NATO • Enhanced Forward Presence 21; 1 EOD pl

MALI: EU • EUTM Mali 4

MEDITERRANEAN SEA: NATO • SNMG 2; 1 PB

SERBIA: NATO • KFOR 29

Austria AUT

Euro €		2019	2020	2021
GDP	€	399bn	379bn	
	US$	446bn	433bn	
per capita	US$	50,380	48,634	
Growth	%	1.6	-6.7	
Inflation	%	1.5	1.2	
Def bdgt [a]	€	2.89bn	3.04bn	3.08bn
	US$	3.24bn	3.47bn	
US$1=€		0.89	0.87	

[a] Includes military pensions

Real-terms defence budget trend (US$bn, constant 2015)

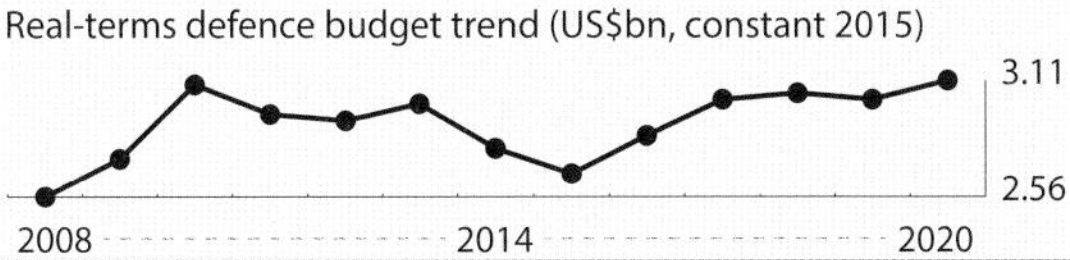

Population 8,859,449

Age	0–14	15–19	20–24	25–29	30–64	65 plus
Male	7.2%	2.5%	2.8%	3.3%	24.5%	8.7%
Female	6.8%	2.4%	2.7%	3.2%	24.7%	11.2%

Capabilities

Austria remains constitutionally non-aligned, but is an EU member and actively engaged in the CSDP. Defence-policy objectives are based on the 2013 National Security Strategy, the 2014 Defence Strategy and the 2017 Military Strategy, including providing military capabilities to maintain sovereignty and territorial integrity, to enable military assistance to the civil authorities and to participate in crisis-management missions abroad. A 2017 defence plan included structural changes at the operational and tactical command-and-control level and in 2020 a political debate about further Bundesheer reform was underway. As of summer 2020, reported areas of focus include cyber defence and providing military assistance to civilian authorities. Army brigades have specialised according to roles, such as rapid response, mechanised (heavy), air-mobile (light) and mountain warfare. While not a NATO member, Austria joined NATO's Partnership for Peace framework in 1995. A September 2019 defence ministry report defined recapitalisation requirements out to 2030. The level of ambition for crisis response is to be able to deploy and sustain a minimum (on average) of 1,100 troops. The September 2019 report also called for the *Typhoon* fleet to be upgraded rather than replaced. In July 2020, the defence ministry confirmed that it would retain the *Typhoon* and that its Saab 105Oe trainers would be withdrawn after 2021; they had been temporarily grounded in 2019 due to maintenance issues. There are plans to cooperate with Italy over helicopter procurement; the *Alouette* IIIs are due to retire in 2023. Austria's defence-industrial base is comprised of some 100 companies with significant niche capabilities and international ties in the areas of weapons and ammunitions, communications equipment and vehicles.

ACTIVE 22,050 (Land Forces 12,200 Air 2,750 Support 7,100)

Conscript liability 6 months recruit trg, 30 days reservist refresher trg for volunteers; 120–150 days additional for officers, NCOs and specialists. Authorised maximum wartime strength of 55,000

RESERVE 125,600 (Joint structured 33,900; Joint unstructured 91,700)

Some 12,000 reservists a year undergo refresher trg in tranches

ORGANISATIONS BY SERVICE

Land Forces 12,200

FORCES BY ROLE

MANOEUVRE

Armoured

1 (4th) armd inf bde (1 recce/SP arty bn, 1 tk bn, 2 armd inf bn, 1 spt bn)

Mechanised

1 (3rd) mech inf bde (1 recce/SP arty bn, 2 mech inf bn, 1 mot inf bn; 1 cbt engr bn, 1 spt bn)

Light

1 (7th) lt inf bde (1 recce bn, 3 inf bn, 1 cbt engr bn, 1 spt bn)

1 (6th) mtn inf bde (3 mtn inf bn, 1 cbt engr bn, 1 spt bn)

EQUIPMENT BY TYPE

ARMOURED FIGHTING VEHICLES

MBT 56 *Leopard* 2A4

AIFV 112 *Ulan*

APC 144

APC (T) 32 BvS-10

APC (W) 112: 78 *Pandur*; 34 *Pandur* EVO

AUV 216: 66 *Dingo* 2; 150 IVECO LMV

ENGINEERING & MAINTENANCE VEHICLES

ARV 49: 27 4KH7FA-SB *Greif* (11 more in store); 12 *Dingo* 2 ARV; 10 M88A1

NBC VEHICLES 12 *Dingo* 2 AC NBC

ANTI-TANK/ANTI-INFRASTRUCTURE

MSL • **MANPATS** *Bill* 2 (PAL 2000)

ARTILLERY 105

SP 155mm 48 M109A5ÖE

MOR 120mm 57 sGrW 86 (40 more in store)

Air Force 2,750

The Air Force is part of Joint Forces Comd and consists of 2 bde; Air Support Comd and Airspace Surveillance Comd

FORCES BY ROLE

FIGHTER

2 sqn with Eurofighter *Typhoon*

ISR

1 sqn with PC-6B *Turbo Porter*

TRANSPORT

1 sqn with C-130K *Hercules*

TRAINING

1 trg sqn with Saab 105Oe*

1 trg sqn with PC-7 *Turbo Trainer*

TRANSPORT HELICOPTER

2 sqn with Bell 212 (AB-212)

1 sqn with OH-58B *Kiowa*

1 sqn with S-70A *Black Hawk*
2 sqn with SA316/SA319 *Alouette* III
AIR DEFENCE
2 bn
1 radar bn

EQUIPMENT BY TYPE
AIRCRAFT 27 combat capable
FTR 15 Eurofighter *Typhoon* Tranche 1
TPT 11: **Medium** 3 C-130K *Hercules*; **Light** 8 PC-6B *Turbo Porter*
TRG 28: 12 PC-7 *Turbo Trainer*; 12 Saab 105Oe*; 4 DA40NG
HELICOPTERS
MRH 19 SA316/SA319 *Alouette* III
ISR 10 OH-58B *Kiowa*
TPT 32: **Medium** 9 S-70A-42 *Black Hawk*; **Light** 23 Bell 212 (AB-212)
AIR DEFENCE
SAM • Point-defence *Mistral*
GUNS 35mm 24 GDF-005 (6 more in store)
AIR-LAUNCHED MISSILES • AAM • IIR IRIS-T

Special Operations Forces

FORCES BY ROLE
SPECIAL FORCES
2 SF gp
1 SF gp (reserve)

Support 7,100

Support forces comprise Joint Services Support Command and several agencies, academies and schools

DEPLOYMENT

AFGHANISTAN: NATO • *Operation Resolute Support* 11

BOSNIA-HERZEGOVINA: EU • EUFOR • *Operation Althea* 296; 1 inf bn HQ; 2 inf coy; 1 hel unit

CYPRUS: UN • UNFICYP 3

LEBANON: UN • UNIFIL 182; 1 log coy

MALI: EU • EUTM Mali 13; **UN** • MINUSMA 1

MIDDLE EAST: UN • UNTSO 5

SERBIA: NATO • KFOR 322; 2 mech inf coy; 1 log coy

WESTERN SAHARA: UN • MINURSO 5

Belgium BEL

Euro €		2019	2020	2021
GDP	€	473bn	440bn	
	US$	530bn	503bn	
per capita	US$	46,237	43,814	
Growth	%	1.4	-8.3	
Inflation	%	1.2	0.6	
Def exp [a]	€	4.25bn	4.76bn	
	US$	4.76bn	5.44bn	
Def bdgt [b]	€	4.30bn	4.77bn	5.46bn
	US$	4.82bn	5.45bn	
US$1=€		0.89	0.87	

[a] NATO definition
[b] Includes military pensions

Real-terms defence budget trend (US$bn, constant 2015)

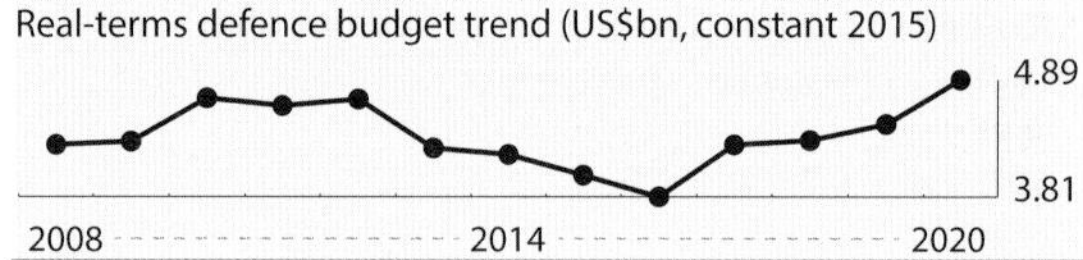

Population 11,720,716

Age	0–14	15–19	20–24	25–29	30–64	65 plus
Male	8.8%	2.8%	2.9%	3.2%	23.1%	8.4%
Female	8.4%	2.7%	2.8%	3.1%	22.9%	10.8%

Capabilities

In July 2016, the government published its strategic vision for defence up to 2030. Brussels intends to stabilise Belgium's defence effort and then provide for growth after 2020. It also envisages a reduced personnel component of around 25,000. However, a large number of impending service retirements means that a gradual increase in recruitment is planned. Belgium also continues to pursue high readiness levels and deployable niche capabilities. NATO, EU and UN membership are central to defence policy. Troops have been deployed to support police counter-terrorism efforts. Due to its limited force size, Belgium often collaborates with neighbours and has committed with Denmark and the Netherlands to form a composite combined special-operations command. Belgium can deploy forces for a small-scale overseas operation and maintains overseas deployments on EU and UN missions. The government is investing in short-term requirements related to aircraft readiness, personal equipment and land-forces vehicles. As part of the defence plan, the government envisages launching five investment projects for fighter aircraft, frigates, mine-countermeasures vessels (to be procured jointly with the Netherlands), UAVs and land-combat vehicles. The army has ordered French *Griffon* and *Jaguar* wheeled armoured vehicles for its mechanised brigade as well as the US JLTV. It is forming a multinational unit of A400M transport aircraft with Luxembourg. The air force has selected the F-35 to replace its F-16s and is to procure MQ-9B *Sky Guardian* UAVs. Belgium has an advanced, export-focused defence industry, focusing on components and subcontracting, though in FN Herstal it has one of the world's largest manufacturers of small arms.

ACTIVE 25,000 (Army 9,400 Navy 1,500 Air 5,250 Medical Service 1,300 Joint Service 7,550)

RESERVE 5,350

ORGANISATIONS BY SERVICE

Land Component 9,400

FORCES BY ROLE

SPECIAL FORCES

1 spec ops regt (1 SF gp, 1 cdo bn, 1 para bn, 1 sigs gp)

MANOEUVRE

Mechanised

1 mech bde (1 ISR bn; 3 mech bn; 2 lt inf bn; 1 arty bn; 2 engr bn; 2 sigs gp; 2 log bn)

COMBAT SUPPORT

1 EOD unit

1 MP coy

COMBAT SERVICE SUPPORT

1 log bn

EQUIPMENT BY TYPE

ARMOURED FIGHTING VEHICLES

ASLT 18 *Piranha* III-C DF90

RECCE 36 *Pandur Recce*

IFV 19 *Piranha* III-C DF30

APC • APC (W) 78: 64 *Piranha* III-C; 14 *Piranha* III-PC (CP)

AUV 655: 219 *Dingo* 2 (inc 52 CP); 436 IVECO LMV

ENGINEERING & MAINTENANCE VEHICLES

AEV 14: 6 Pionierpanzer 2 *Dachs*; 8 *Piranha* III-C

ARV 13: 4 *Pandur*; 9 *Piranha* III-C

VLB 4 *Leguan*

ANTI-TANK/ANTI-INFRASTRUCTURE

MSL • MANPATS *Spike*-MR

ARTILLERY 60

TOWED 105mm 14 LG1 MkII

MOR 46: **81mm** 14 Expal; **120mm** 32 RT-61

Naval Component 1,500

EQUIPMENT BY TYPE

PRINCIPAL SURFACE COMBATANTS • FRIGATES 2

FFGHM 2 *Leopold* I (ex-NLD *Karel Doorman*) with 2 quad lnchr with RGM-84 *Harpoon* AShM, 1 16-cell Mk 48 mod 1 VLS with RIM-7P *Sea Sparrow* SAM, 2 twin 324mm SVTT Mk 32 ASTT with Mk 46 LWT, 1 *Goalkeeper* CIWS, 1 76mm gun (capacity 1 med hel)

PATROL AND COASTAL COMBATANTS

PCC 2 *Castor* (FRA *Kermorvan* mod)

MINE WARFARE • MINE COUNTERMEASURES

MHC 5 *Flower* (*Tripartite*)

LOGISTICS AND SUPPORT 3

AGFH 1 *Godetia* (log spt/comd) (capacity 1 *Alouette* III)

AGOR 1 *Belgica* (owned by BELSPO, managed by RBINS)

AXS 1 *Zenobe Gramme*

Air Component 5,250

FORCES BY ROLE

FIGHTER/GROUND ATTACK/ISR

4 sqn with F-16AM/BM *Fighting Falcon*

SEARCH & RESCUE

1 sqn with NH90 NFH

TRANSPORT

1 sqn with A321; ERJ-135 LR (VIP); ERJ-145 LR (VIP); *Falcon* 7X

1 sqn (BEL/LUX) with A400M; C-130H *Hercules*

TRAINING

1 OCU sqn with F-16AM/BM *Fighting Falcon*

1 sqn with SF-260D/M

1 OCU unit with AW109

TRANSPORT HELICOPTER

2 sqn with AW109 (ISR)

EQUIPMENT BY TYPE

AIRCRAFT 54 combat capable

FTR 54: 45 F-16AM *Fighting Falcon*; 9 F-16BM *Fighting Falcon*

TPT 13: **Medium** 7 C-130H *Hercules*; **PAX** 6: 1 A321; 2 ERJ-135 LR (VIP); 1 ERJ-145 LR (VIP); 2 *Falcon* 7X (VIP)

TRG 32: 9 SF-260D; 23 SF-260M

HELICOPTERS

ASW 4 NH90 NFH (opcon Navy)

MRH 3 SA316B *Alouette* III (opcon Navy)

TPT 14: **Medium** 4 NH90 TTH; **Light** 10 AW109 (ISR) (7 more in store)

AIR-LAUNCHED MISSILES

AAM • IR AIM-9M *Sidewinder*; **IIR** AIM-9X *Sidewinder* II; **ARH** AIM-120B AMRAAM

BOMBS

Laser-guided: GBU-10/GBU-12 *Paveway* II; GBU-24 *Paveway* III

INS/GPS guided: GBU-31 JDAM; GBU-38 JDAM; GBU-54 Laser JDAM (dual-mode)

Medical Service 1,300

FORCES BY ROLE

COMBAT SERVICE SUPPORT

4 med unit

1 fd hospital

EQUIPMENT BY TYPE

ARMOURED FIGHTING VEHICLES

APC • APC (W) 10: 4 *Pandur* (amb); 6 *Piranha* III-C (amb)

AUV 10 *Dingo* 2 (amb)

DEPLOYMENT

AFGHANISTAN: NATO • *Operation Resolute Support* 92

BALTIC SEA: NATO • SNMCMG 1: 1 MHC

DEMOCRATIC REPUBLIC OF THE CONGO: UN • MONUSCO 1

FRANCE: NATO • Air Component 28 *Alpha Jet* located at Cazaux/Tours

IRAQ: *Operation Inherent Resolve* 5

LITHUANIA: NATO • Enhanced Forward Presence 1

MALI: EU • EUTM Mali 15; **UN •** MINUSMA 38

MIDDLE EAST: UN • UNTSO 1

NIGER: METT *Maradi* 100 (trg)

FOREIGN FORCES

United States US European Command: 1,150

Bosnia-Herzegovina BIH

Convertible Mark KM		2019	2020	2021
GDP	KM	34.7bn	32.3bn	
	US$	19.9bn	18.9bn	
per capita	US$	6,015	5,762	
Growth	%	2.7	-6.5	
Inflation	%	0.6	-0.8	
Def bdgt	KM	289m	288m	298m
	US$	165m	169m	
US$1=KM		1.75	1.71	

Real-terms defence budget trend (US$m, constant 2015)

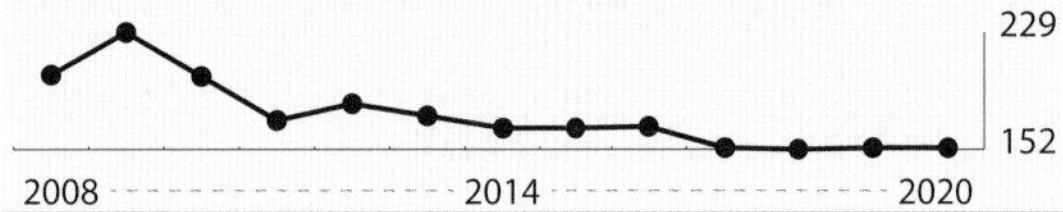

Population 3,835,586

Age	0–14	15–19	20–24	25–29	30–64	65 plus
Male	6.8%	2.4%	3.2%	3.2%	26.6%	6.5%
Female	6.4%	2.3%	3.0%	3.0%	26.9%	9.7%

Capabilities

The armed forces' primary goals are to defend territorial integrity and contribute to peacekeeping missions and civilian-support operations. Bosnia-Herzegovina joined NATO's Partnership for Peace in 2006 and a Membership Action Plan was presented in 2010. Its aspiration to join NATO has been delayed due to unresolved defence-property issues. The country is reforming its armed forces and modernising its equipment in accordance with its Defence Review, Development and Modernisation Plan for 2017–27 and its NATO aspirations. The armed forces are professional and represent all three ethnic groups. However, low salaries may negatively affect recruitment and retention. Bosnia-Herzegovina contributes to EU, NATO and UN missions and has deployed personnel to *Operation Resolute Support* in Afghanistan, but the armed forces have no capacity to independently deploy and self-sustain beyond national borders. The inventory comprises mainly ageing Soviet-era equipment, though some new helicopters have been procured from the US. Bosnia-Herzegovina has little in the way of a domestic defence industry, with only the capability to produce small arms, ammunition and explosives.

ACTIVE 10,500 (Armed Forces 10,500)

ORGANISATIONS BY SERVICE

Armed Forces 10,500

1 ops comd; 1 spt comd

FORCES BY ROLE

MANOEUVRE

Light

3 inf bde (1 recce coy, 3 inf bn, 1 arty bn)

COMBAT SUPPORT

1 cbt spt bde (1 tk bn, 1 engr bn, 1 EOD bn, 1 int bn, 1 MP bn, 1 CBRN coy, 1 sigs bn)

COMBAT SERVICE SUPPORT

1 log comd (5 log bn)

EQUIPMENT BY TYPE

ARMOURED FIGHTING VEHICLES

MBT 45 M60A3

APC • APC (T) 20 M113A2

ENGINEERING & MAINTENANCE VEHICLES

VLB MTU

MW *Bozena*

ANTI-TANK/ANTI-INFRASTRUCTURE • MSL

SP 60: 8 9P122 *Malyutka*; 9 9P133 *Malyutka*; 32 BOV-1; 11 M-92

MANPATS 9K11 *Malyutka* (AT-3 *Sagger*); 9K111 *Fagot* (AT-4 *Spigot*); 9K115 *Metis* (AT-7 *Saxhorn*); HJ-8; *Milan*

ARTILLERY 224

TOWED 122mm 100 D-30

MRL 122mm 24 APRA-40

MOR 120mm 100 M-75

Air Force and Air Defence Brigade 800

FORCES BY ROLE

HELICOPTER

1 sqn with Bell 205; Mi-8MTV *Hip*; Mi-17 *Hip* H

1 sqn with Mi-8 *Hip*; SA-341H/SA-342L *Gazelle* (HN-42/45M)

AIR DEFENCE

1 AD bn

EQUIPMENT BY TYPE

AIRCRAFT

FGA (7 J-22 *Orao* in store)

ATK (6 J-1 (J-21) *Jastreb*; 3 TJ-1(NJ-21) *Jastreb* all in store)

ISR (2 RJ-1 (IJ-21) *Jastreb** in store)

TRG (1 G-4 *Super Galeb* (N-62)* in store)

HELICOPTERS

MRH 13: 4 Mi-8MTV *Hip*; 1 Mi-17 *Hip* H; 1 SA-341H *Gazelle* (HN-42); 7 SA-342L *Gazelle* (HN-45M)

TPT 21: **Medium** 8 Mi-8 *Hip* **Light** 13 Bell 205 (UH-1H *Iroquois*)

TRG 1 Mi-34 *Hermit*

AIR DEFENCE

SAM

Short-range 20 2K12 *Kub* (SA-6 *Gainful*)

Point-defence 9K34 *Strela-3* (SA-14 *Gremlin*); 9K310 *Igla-1* (SA-16 *Gimlet*)

GUNS • TOWED 40mm 47: 31 L/60, 16 L/70

DEPLOYMENT

AFGHANISTAN: NATO • *Operation Resolute Support* 66

CENTRAL AFRICAN REPUBLIC: EU • EUTM RCA 2

DEMOCRATIC REPUBLIC OF THE CONGO: UN • MONUSCO 3

MALI: UN • MINUSMA 2

FOREIGN FORCES

Part of EUFOR – *Operation Althea* unless otherwise stated

Albania 1

Austria 296; 1 inf bn HQ; 2 inf coy

Bulgaria 11

Chile 16

Czech Republic 2
Greece 6
Hungary 47
Ireland 5
Italy 5
Macedonia, North 1
Poland 38
Romania 58
Slovakia 40
Slovenia 10
Spain 2
Switzerland 20
Turkey 205; 1 inf coy

Bulgaria BLG

Bulgarian Lev L		2019	2020	2021
GDP	L	119bn	116bn	
	US$	67.9bn	67.9bn	
per capita	US$	9,772	9,826	
Growth	%	3.4	-4.0	
Inflation	%	2.5	1.2	
Def exp [a]	L	3.77bn	2.14bn	
	US$	2.16bn	1.25bn	
Def bdgt [b]	L	3.61bn	1.98bn	2.28bn
	US$	2.07bn	1.16bn	
FMA (US)	US$	7.5m	0m	0m
US$1=L		1.75	1.71	

[a] NATO definition

[b] Excludes military pensions

Real-terms defence budget trend (US$m, constant 2015)

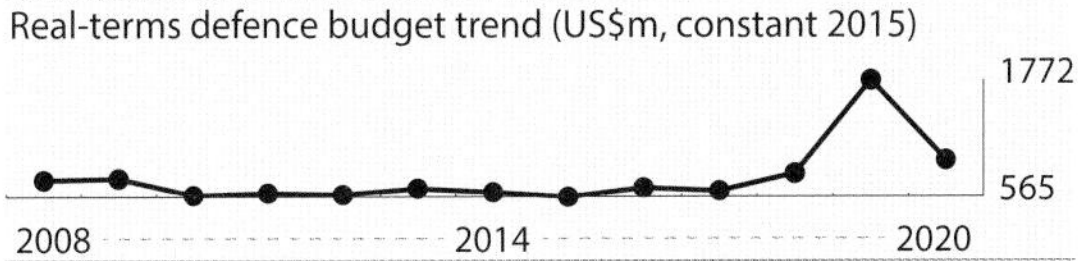

Population 6,966,899

Age	0–14	15–19	20–24	25–29	30–64	65 plus
Male	7.5%	2.5%	2.4%	2.9%	25.4%	8.1%
Female	7.1%	2.3%	2.2%	2.7%	25.1%	12.0%

Capabilities

The armed forces' main priority is defending state sovereignty and territorial integrity. Bulgaria conducted a review intended to generate a new armed forces development plan to 2032, which will replace the current development plan to 2020. The review emphasised building cyber-defence capabilities and developing a Cyber Defence Concept. Bulgaria signed a 10-year Defence Cooperation Roadmap with the US in 2020 to assist force development, including meeting NATO capability targets. There are several bilateral defence-cooperation agreements with regional states. Bulgaria's airspace is protected by NATO's Air Policing Mission, due to its limited numbers of combat aircraft. In an attempt to cope with personnel shortfalls, the military retirement age and salaries were increased. Training is prioritised for units intended for international operations and those with certain readiness levels declared to NATO and the EU. Bulgaria regularly trains and exercises with NATO partners and regional allies. The country contributes to NATO and EU missions but has little logistics-support capability. Despite long-term plans for reform, the armed forces still rely heavily on ageing Soviet-era equipment. In April 2020, a contract was signed for eight F-16 Block 70 fighter aircraft, to replace Bulgaria's MiG-29s. There are also plans for acquisitions to enable the formation of battalion battlegroups within its mechanised brigades. The navy is prioritising the procurement of a multi-purpose patrol vessel and the modernisation of its frigates to boost its presence in the Black Sea. Bulgaria's defence industry exports small arms but has limited capacity to design and manufacture platforms.

ACTIVE 36,950 (Army 17,000 Navy 4,450 Air 8,500 Central Staff 7,000)

RESERVE 3,000 (Joint 3,000)

ORGANISATIONS BY SERVICE

Army 17,000

FORCES BY ROLE

MANOEUVRE

Reconnaissance
1 recce bn

Mechanised
2 mech bde (4 mech inf bn, 1 SP arty bn, 1 cbt engr bn, 1 log bn, 1 SAM bn)

Light
1 mtn inf regt

COMBAT SUPPORT
1 arty regt (1 fd arty bn, 1 MRL bn)
1 engr regt (1 cbt engr bn, 1 ptn br bn, 1 engr spt bn)
1 NBC bn

COMBAT SERVICE SUPPORT
1 log regt

EQUIPMENT BY TYPE

ARMOURED FIGHTING VEHICLES
MBT 90 T-72M1/M2†
IFV 160: 90 BMP-1; 70 BMP-23
APC 120
APC (T) 100 MT-LB
APC (W) 20 BTR-60
AUV 44: 17 M1117 ASV; 27 Plasan *SandCat*

ENGINEERING & MAINTENANCE VEHICLES
AEV MT-LB
ARV T-54/T-55; MTP-1; MT-LB
VLB BLG67; TMM

ANTI-TANK/ANTI-INFRASTRUCTURE
MSL
SP 24 9P148 *Konkurs* (AT-5 *Spandrel*)
MANPATS 9K111 *Fagot* (AT-4 *Spigot*); 9K111-1 *Konkurs* (AT-5 *Spandrel*); (9K11 *Malyutka* (AT-3 *Sagger*) in store)
GUNS 126: **85mm** (150 D-44 in store); **100mm** 126 MT-12

ARTILLERY 176
SP 122mm 48 2S1
TOWED 152mm 24 D-20
MRL 122mm 24 BM-21
MOR 120mm ε80 *Tundza/Tundza Sani*

SURFACE-TO-SURFACE MISSILE LAUNCHERS
SRBM • Conventional 9K79 *Tochka* (SS-21 *Scarab*)

AIR DEFENCE
SAM • Point-defence 9K32 *Strela* (SA-7 *Grail*)‡; 24 9K33 *Osa* (SA-8 *Gecko*)
GUNS 400
SP 23mm ZSU-23-4
TOWED 23mm ZU-23-2; **57mm** S-60

Navy 4,450

EQUIPMENT BY TYPE
PRINCIPAL SURFACE COMBATANTS • FRIGATES 3
FFM 3 *Drazki* (ex-BEL *Wielingen*) with 1 octuple Mk 29 GMLS with RIM-7P *Sea Sparrow* SAM, 2 single 533mm ASTT with L5 mod 4 HWT, 1 sextuple Bofors ASW Rocket Launcher System 375mm A/S mor, 1 100mm gun (Fitted for but not with 2 twin lnchr with MM38 *Exocet* AShM)
PATROL AND COASTAL COMBATANTS 4
CORVETTES • FS 1 *Smeli* (ex-FSU *Koni*) with 2 RBU 6000 *Smerch* 2 A/S mor, 2 twin 76mm guns
PCFG 1 *Molnya*† (ex-FSU *Tarantul* II) with 2 twin lnchr with P-22 (SS-N-2C *Styx*) AShM, 2 AK630M CIWS, 1 76mm gun
PCT 2 *Reshitelni* (ex-FSU *Pauk* I) with 4 single 406mm TT, 2 RBU 1200 *Uragan* A/S mor, 1 76mm gun
MINE COUNTERMEASURES 7
MHC 1 *Tsibar* (*Tripartite* – ex-BEL *Flower*)
MSC 3 *Briz* (ex-FSU *Sonya*)
MSI 3 *Olya* (ex-FSU) (3 more non-operational)
AMPHIBIOUS • LANDING CRAFT
LCM 1 *Vydra* (capacity either 3 MBT or 200 troops)
LOGISTICS AND SUPPORT 8: 2 **AGS**; 2 **AOL**; 1 **ARS**; 2 **ATF**; 1 **AX**

Naval Aviation

EQUIPMENT BY TYPE
HELICOPTERS
ASW 2 AS565MB *Panther*
MRH 1 AS365N3+ *Dauphin 2*

Air Force 8,500

FORCES BY ROLE
FIGHTER/ISR
1 sqn with MiG-29/MiG-29UB *Fulcrum*
TRANSPORT
1 sqn with An-30 *Clank*; C-27J *Spartan*; L-410UVP-E; PC-12M
TRAINING
1 sqn with L-39ZA *Albatros**
1 sqn with PC-9M
ATTACK HELICOPTER
1 sqn with Mi-24D/V *Hind* D/E
TRANSPORT HELICOPTER
1 sqn with AS532AL *Cougar*; Bell 206 *Jet Ranger*; Mi-17 *Hip* H
EQUIPMENT BY TYPE
AIRCRAFT 24 combat capable
FTR 15: 12 MiG-29 *Fulcrum*†; 3 MiG-29UB *Fulcrum*†
FGA (Some MiG-21bis *Fishbed*/MiG-21UM *Mongol* B in store)
ATK 3: 2 Su-25K *Frogfoot* K; 1 Su-25UBK *Frogfoot* B
ISR 1 An-30 *Clank*
TPT 7: **Medium** 3 C-27J *Spartan*; **Light** 4: 1 An-2T *Colt*; 2 L-410UVP-E; 1 PC-12M
TRG 12: 6 L-39ZA *Albatros**; 6 PC-9M (basic)
HELICOPTERS
ATK 6 Mi-24D/V *Hind* D/E
MRH 5 Mi-17 *Hip* H
TPT 18: **Medium** 12 AS532AL *Cougar*; **Light** 6 Bell 206 *Jet Ranger*
UNINHABITED AERIAL VEHICLES • EW *Yastreb*-2S
AIR DEFENCE
SAM 12+
Long-range 12+: 12 S-200 (SA-5 *Gammon*); S-300P (SA-10 *Grumble*)
Short-range S-125 *Pechora* (SA-3 *Goa*); 2K12 *Kub* (SA-6 *Gainful*)
AIR-LAUNCHED MISSILES
AAM • IR R-3 (AA-2 *Atoll*)‡ R-73 (AA-11A *Archer*)
SARH R-27R (AA-10 *Alamo* A)
ASM Kh-29 (AS-14 *Kedge*); Kh-25 (AS-10 *Karen*)

Special Forces

FORCES BY ROLE
SPECIAL FORCES
1 spec ops bde (1 SF bn, 1 para bn)

DEPLOYMENT

AFGHANISTAN: NATO • *Operation Resolute Support* 160

BOSNIA-HERZEGOVINA: EU • EUFOR • *Operation Althea* 11

MALI: EU • EUTM Mali 5

SERBIA: NATO • KFOR 23

Croatia CRO

Croatian Kuna k		2019	2020	2021
GDP	k	400bn	368bn	
	US$	60.4bn	56.8bn	
per capita	US$	14,853	14,033	
Growth	%	2.9	-9.0	
Inflation	%	0.8	0.3	
Def exp [a]	k	6.63bn	6.85bn	
	US$	1.00bn	1.06bn	
Def bdgt [b]	k	6.94bn	7.27bn	7.49bn
	US$	1.05bn	1.12bn	
US$1=k		6.62	6.49	

[a] NATO definition

[b] Includes military pensions

Real-terms defence budget trend (US$m, constant 2015)

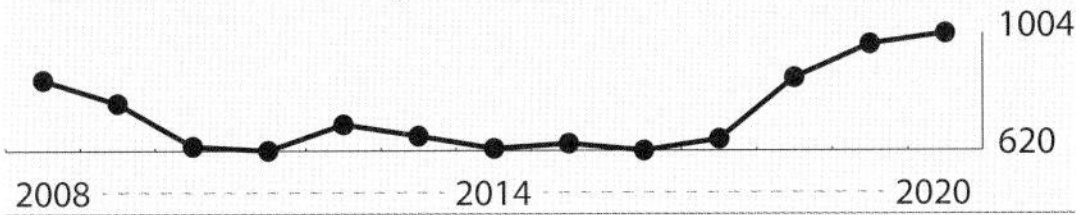

Population 4,227,746

Age	0–14	15–19	20–24	25–29	30–64	65 plus
Male	7.3%	2.5%	3.0%	3.0%	23.8%	8.6%
Female	6.9%	2.3%	2.9%	2.9%	24.4%	12.5%

Capabilities

Principal tasks for the armed forces include defending national sovereignty and territorial integrity as well as tackling terrorism and contributing to international peacekeeping missions. Croatia joined NATO in 2009, having reformed its armed forces to create a small professional force. There have been recent moves to improve conditions of service. Economic challenges have delayed further defence modernisation. Zagreb has defence-cooperation agreements with Bosnia-Herzegovina, Hungary and Romania, and personnel frequently train with regional and international allies. Croatia participates in EU and NATO missions, including NATO's Enhanced Forward Presence in Poland and Operation *Resolute Support* in Afghanistan. The inventory is almost entirely composed of ageing Soviet-era equipment. Modernisation objectives include the acquisition of UH-60M helicopters from US. The replacement process for Croatia's MiG-21 fighter aircraft is proceeding, after earlier US objections to the proposed sale of second-hand, upgraded Israeli F-16s. Croatia has a small defence industry, focused on small arms, ammunition, explosives and naval systems.

ACTIVE 15,200 (Army 10,750 Navy 1,300 Air 1,300 Joint 1,850) **Paramilitary 3,000**

Conscript liability Voluntary conscription, 8 weeks

RESERVE 18,350 (Army 18,350)

ORGANISATIONS BY SERVICE

Joint 1,850 (General Staff)

FORCES BY ROLE

SPECIAL FORCES

1 SF bn

Army 10,750

FORCES BY ROLE

MANOEUVRE

Armoured

1 armd bde (1 tk bn, 1 armd bn, 2 armd inf bn, 1 SP arty bn, 1 ADA bn, 1 cbt engr bn)

Mechanised

1 lt mech bde (2 mech inf bn, 2 mot inf bn, 1 fd arty bn, 1 ADA bn, 1 cbt engr bn)

Other

1 inf trg regt

COMBAT SUPPORT

1 arty/MRL regt

1 engr regt

1 NBC bn

1 sigs regt

COMBAT SERVICE SUPPORT

1 log regt

AIR DEFENCE

1 ADA regt

EQUIPMENT BY TYPE

ARMOURED FIGHTING VEHICLES

MBT 75 M-84

IFV 102 M-80

APC 198

APC (T) 14: 11 BTR-50; 3 OT M-60

APC (W) 132: 6 BOV-VP; 126 Patria AMV (incl variants)

PPV 52: 32 *Maxxpro Plus*; 20 RG-33 HAGA (amb)

AUV 172: 10 IVECO LMV; 162 M-ATV

ENGINEERING & MAINTENANCE VEHICLES

ARV 22: 12 JVBT-55A; 1 M-84AI; 1 WZT-2; 2 WZT-3; 6 *Maxxpro Recovery*

VLB 5 MT-55A

MW 4 MV-4

ANTI-TANK/ANTI-INFRASTRUCTURE • MSL

SP 20 BOV-1

MANPATS 9K11 *Malyutka* (AT-3 *Sagger*); 9K111 *Fagot* (AT-4 *Spigot*); 9K111-1 *Konkurs* (AT-5 *Spandrel*); 9K115 *Metis* (AT-7 *Saxhorn*)

ARTILLERY 167

SP 20: **122mm** 8 2S1 *Gvozdika*; **155mm** 12 PzH 2000

TOWED 122mm 20 D-30

MRL 122mm 27: 6 M91 *Vulkan*; 21 BM-21 *Grad*

MOR 100: **82mm** 54 LMB M96; **120mm** 46 M-75/UBM 52

AIR DEFENCE

SAM • Point-defence 9+: 9 9K35 *Strela*-10M3 (SA-13 *Gopher*); 9K310 *Igla*-1 (SA-16 *Gimlet*)

GUNS SP 20mm 33 BOV-3 SP

Navy 1,300

Navy HQ at Split

EQUIPMENT BY TYPE

PATROL AND COASTAL COMBATANTS 5

PCFG 1 *Končar* with 2 twin lnchr with RBS15B Mk I AShM, 1 AK630 CIWS, 1 57mm gun

PCG 4:

2 *Kralj* with 4 single lnchr with RBS15B Mk I AShM, 1 AK630 CIWS, 1 57mm gun (with minelaying capability)

2 *Vukovar* (ex-FIN *Helsinki*) with 4 single lnchr with RBS15B Mk I AShM, 1 57mm gun
MINE WARFARE • MINE COUNTERMEASURES 1
MHI 1 *Korcula*
AMPHIBIOUS • LANDING CRAFT 5:
LCT 2 *Cetina* (with minelaying capability)
LCVP 3: 2 Type-21; 1 Type-22
LOGISTICS AND SUPPORT • AKL 1
COASTAL DEFENCE • AShM 3 RBS15K

Marines

FORCES BY ROLE
MANOEUVRE
Amphibious
1 indep mne coy

Coast Guard

FORCES BY ROLE
Two divisions, headquartered in Split (1st div) and Pula (2nd div)
EQUIPMENT BY TYPE
PATROL AND COASTAL COMBATANTS • PB 5: 4 *Mirna*; 1 *Omiš*
LOGISTICS AND SUPPORT 3:
AAR 1 *Faust Vrancic* (YUG *Spasilac*)
AKL 1 PT-71
AX 1 *Andrija Mohorovicic* (POL Project 861)

Air Force and Air Defence 1,300

FORCES BY ROLE
FIGHTER/GROUND ATTACK
1 (mixed) sqn with MiG-21bis/UMD *Fishbed*
TRAINING
1 sqn with PC-9M; Z-242L
ISR HELICOPTER
1 hel sqn with Bell 206B *Jet Ranger* II; OH-58D *Kiowa Warrior*
TRANSPORT HELICOPTER
2 sqn with Mi-8MTV *Hip* H; Mi-8T *Hip* C; Mi-171Sh
EQUIPMENT BY TYPE
AIRCRAFT 11 combat capable
FGA 11: 8 MiG-21bis *Fishbed*; 3 MiG-21UMD *Fishbed*
TPT • Light (2 An-32 *Cline* in store)
TRG 21: 17 PC-9M; 4 Z-242L
HELICOPTERS
MRH 26: 11 Mi-8MTV *Hip* H; 15 OH-58D *Kiowa Warrior*
TPT 21: **Medium** 13: 3 Mi-8T *Hip* C; 10 Mi-171Sh; **Light** 8 Bell 206B *Jet Ranger* II
UNINHABITED AERIAL VEHICLES
ISR • Medium *Hermes* 450
AIR DEFENCE • SAM
Point-defence 9K31 *Strela*-1 (SA-9 *Gaskin*); 9K34 *Strela*-3 (SA-14 *Gremlin*); 9K310 *Igla*-1 (SA-16 *Gimlet*)
AIR-LAUNCHED MISSILES
AAM • IR R-3S (AA-2 *Atoll*)‡; R-60; R-60MK (AA-8 *Aphid*)
ASM AGM-114R *Hellfire*

Special Forces Command

FORCES BY ROLE
SPECIAL FORCES
2 SF gp

Paramilitary 3,000

Police 3,000 armed

DEPLOYMENT

AFGHANISTAN: NATO • *Operation Resolute Support* 76
INDIA/PAKISTAN: UN • UNMOGIP 9
LEBANON: UN • UNIFIL 1
POLAND: NATO • Enhanced Forward Presence 80; 1 MRL bty with M91 *Vulkan*
SERBIA: NATO • KFOR 37; 1 hel unit with Mi-8 *Hip*
WESTERN SAHARA: UN • MINURSO 10

Cyprus CYP

Euro €		2019	2020	2021
GDP	€	21.9bn	20.3bn	
	US$	24.6bn	23.2bn	
per capita	US$	28,049	26,240	
Growth	%	3.2	-6.4	
Inflation	%	0.6	-0.6	
Def bdgt	€	359m	355m	438m
	US$	402m	406m	
US$1=€		0.89	0.87	

Real-terms defence budget trend (US$m, constant 2015)

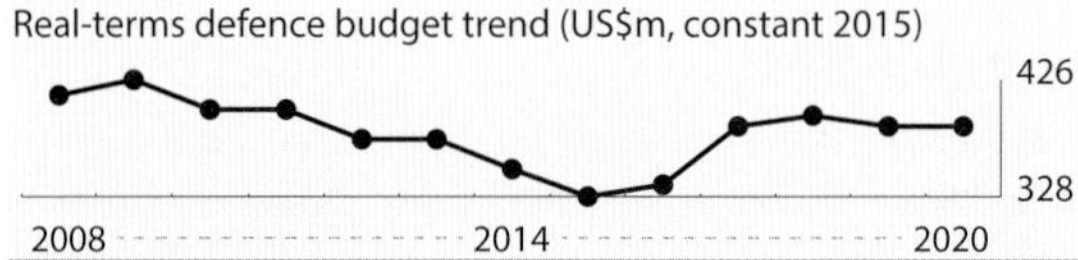

Population 1,266,676

Age	0–14	15–19	20–24	25–29	30–64	65 plus
Male	8.1%	2.8%	3.9%	4.7%	26.1%	5.6%
Female	7.6%	2.5%	3.1%	3.9%	24.5%	7.4%

Capabilities

The National Guard is focused on protecting the island's territorial integrity and sovereignty, and safeguarding Cyprus's EEZ. Its main objective is to deter any Turkish incursion, and to provide enough opposition until military support can be provided by Greece, its primary ally. Cyprus has been enhancing its defence cooperation with Greece, including on cyber defence. Nicosia has also pledged deeper military ties with Israel, while France has renewed and enhanced its defence-cooperation agreement with Cyprus. In 2018 Cyprus also signed a memorandum of understanding on enhancing defence and security cooperation with the UK. Having reduced conscript liability in 2016, Nicosia began recruiting additional contract-service personnel, as part of the effort to modernise and professionalise its forces. Cyprus exercises with several international partners, most notably France, Greece and Israel. External deployments have been limited to some officers joining EU and UN missions. Cyprus has little logistics capability to support operations abroad. Equipment comprises a mix of Soviet-era and modern European systems. The defence minister announced in 2019 the intention to introduce an eight-year military-modernisation programme that would also involve a higher defence budget. Cyprus has little in the way of a domestic defence industry, with no ability to design and manufacture modern equipment. However, the government is looking for opportunities to cooperate with defence firms in Greece.

ACTIVE 15,000 (National Guard 15,000)
Paramilitary 750
Conscript liability 14 months

RESERVE 50,000 (National Guard 50,000)
Reserve service to age 50 (officers dependent on rank; military doctors to age 60)

ORGANISATIONS BY SERVICE

National Guard 15,000 (incl conscripts)

FORCES BY ROLE
SPECIAL FORCES
1 comd (regt) (1 SF bn)
MANOEUVRE
Armoured
1 lt armd bde (2 armd bn, 1 armd inf bn)
Mechanised
4 (1st, 2nd, 6th & 7th) mech bde
Light
1 (4th) lt inf bde
2 (2nd & 8th) lt inf regt
COMBAT SUPPORT
1 arty comd (8 arty bn)
COMBAT SERVICE SUPPORT
1 (3rd) spt bde

EQUIPMENT BY TYPE
ARMOURED FIGHTING VEHICLES
MBT 134: 82 T-80U; 52 AMX-30B2
RECCE 72 EE-9 *Cascavel*
IFV 43 BMP-3
APC 295
APC (T) 168 *Leonidas*
APC (W) 127 VAB (incl variants)
ENGINEERING & MAINTENANCE VEHICLES
ARV 2+: 2 AMX-30D; BREM-80U
ANTI-TANK/ANTI-INFRASTRUCTURE
MSL
SP 33: 15 EE-3 *Jararaca* with *Milan*; 18 VAB with HOT
MANPATS *Milan*
RCL 106mm 144 M40A1
GUNS • TOWED 100mm 20 M-1944
ARTILLERY 444
SP 155mm 36: 12 Mk F3; 12 NORA B-52; 12 *Zuzana*
TOWED 84: **105mm** 72 M-56; **155mm** 12 TR-F-1
MRL 22: **122mm** 4 BM-21; **128mm** 18 M-63 *Plamen*
MOR 302: **81mm** 170 E-44 (70+ M1/M9 in store); **107mm** 20 M2/M30; **120mm** 112 RT61
AIR DEFENCE
SAM 22+
Medium-range 4 9K37M1 *Buk* M1-2 (SA-11 *Gadfly*)
Short-range 18: 12 *Aspide*; 6 9K331 *Tor*-M1 (SA-15 *Gauntlet*)
Point-defence *Mistral*
GUNS • TOWED 60: **20mm** 36 M-55; **35mm** 24 GDF-003 (with *Skyguard*)

Maritime Wing

FORCES BY ROLE
COMBAT SUPPORT
1 (coastal defence) AShM bty with MM40 *Exocet* AShM

EQUIPMENT BY TYPE
PATROL AND COASTAL COMBATANTS 6
PCC 2: 1 *Alasia* (ex-OMN *Al Mabrukha*) with 1 hel landing platform; 1 OPV 62 (ISR *Sa'ar* 4.5 derivative)
PBF 4: 2 Rodman 55; 2 *Vittoria*
COASTAL DEFENCE • **AShM** 3 MM40 *Exocet*

Air Wing

EQUIPMENT BY TYPE
HELICOPTERS
ATK 11 Mi-35P *Hind* E
MRH 7: 3 AW139 (SAR); 4 SA342L1 *Gazelle* (with HOT for anti-armour role)
TPT • Light 2 Bell 206L3 *Long Ranger*

Paramilitary 750+

Armed Police 500+

FORCES BY ROLE
MANOEUVRE
Other
1 (rapid-reaction) paramilitary unit

EQUIPMENT BY TYPE
ARMOURED FIGHTING VEHICLES
APC • APC (W) 2 VAB VTT
HELICOPTERS • **MRH** 4: 2 AW139; 2 Bell 412SP

Maritime Police 250

EQUIPMENT BY TYPE
PATROL AND COASTAL COMBATANTS 10
PBF 5: 2 *Poseidon*; 1 *Shaldag*; 2 *Vittoria*
PB 5 SAB-12

DEPLOYMENT

LEBANON: UN • UNIFIL 2

FOREIGN FORCES

Argentina UNFICYP 282; 2 inf coy; 1 hel flt
Australia UNFICYP 3
Austria UNFICYP 3
Brazil UNFICYP 3
Canada UNFICYP 1
Chile UNFICYP 12
Ghana UNFICYP 1
Greece Army: 950; ε200 (officers/NCOs seconded to Greek-Cypriot National Guard)
Hungary UNFICYP 14
India UNFICYP 1
Pakistan UNFICYP 3
Paraguay UNFICYP 12
Russia UNFICYP 1
Serbia UNFICYP 3

Slovakia UNFICYP 241; 1 inf coy; 1 engr pl
United Kingdom 2,260; 2 inf bn; 1 hel sqn with 4 Bell 412 *Twin Huey* • *Operation Inherent Resolve* (*Shader*) 500: 1 FGA sqn with 6 *Typhoon* FGR4; 1 *Sentinel* R1; 1 A330 MRTT *Voyager* KC3; 2 C-130J-30 *Hercules* • UNFICYP (*Operation Tosca*) 256: 1 inf coy

TERRITORY WHERE THE GOVERNMENT DOES NOT EXERCISE EFFECTIVE CONTROL

Data here represents the de facto situation on the northern section of the island. This does not imply international recognition as a sovereign state.

Capabilities

ACTIVE 3,000 (Army 3,000) **Paramilitary 150**
Conscript liability 15 months

RESERVE 15,000
Reserve liability to age 50

ORGANISATIONS BY SERVICE

Army ε3,000

FORCES BY ROLE
MANOEUVRE
Light
5 inf bn
7 inf bn (reserve)
EQUIPMENT BY TYPE
ANTI-TANK/ANTI-INFRASTRUCTURE
MSL • MANPATS *Milan*
RCL • **106mm** 36
ARTILLERY • MOR • **120mm** 73

Paramilitary

Armed Police ε150

FORCES BY ROLE
SPECIAL FORCES
1 (police) SF unit

Coast Guard

PATROL AND COASTAL COMBATANTS 6
PCC 5: 2 SG45/SG46; 1 *Rauf Denktash*; 2 US Mk 5
PB 1

FOREIGN FORCES

TURKEY
Army ε33,800
FORCES BY ROLE
1 corps HQ; 1 SF regt; 1 armd bde; 2 mech inf div; 1 mech inf regt; 1 arty regt; 1 avn comd
EQUIPMENT BY TYPE
ARMOURED FIGHTING VEHICLES
MBT ε342: ε300 M48A5T1; 42 *Leopard* 2A4
IFV 147 ACV AIFV
APC • **APC (T)** 492: 106 ACV AAPC (incl variants); 386 M113 (incl variants)
ANTI-TANK/ANTI-INFRASTRUCTURE
MSL
SP 60 ACV TOW
MANPATS *Milan*
RCL 106mm 219 M40A1
ARTILLERY 643
SP 155mm 194: 30 M44T; 144 M52T1; 20 T-155 *Firtina*
TOWED 84: **105mm** 36 M101A1; **155mm** 36 M114A2; **203mm** 12 M115
MRL 122mm 18 T-122
MOR 376: **81mm** 171; **107mm** 70 M30; **120mm** 135 HY-12
PATROL AND COASTAL COMBATANTS • **PB** 1
AIRCRAFT • **TPT** • **Light** 3 Cessna 185 (U-17)
HELICOPTERS • **TPT** 3 **Medium** 2 AS532UL *Cougar* **Light** 1 Bell 205 (UH-1H *Iroquois*)
AIR DEFENCE
SAM Point-defence FIM-92 *Stinger*
GUNS • **TOWED** 150: **20mm** 122: 44 Rh 202; 78 GAI-D01; **35mm** 28 GDF-003

Czech Republic CZE

Czech Koruna Kc		2019	2020	2021
GDP	Kc	5.75tr	5.57tr	
	US$	251bn	242bn	
per capita	US$	23,539	22,627	
Growth	%	2.3	-6.5	
Inflation	%	2.9	3.3	
Def exp [a]	Kc	66.7bn	75.5bn	
	US$	2.91bn	3.28bn	
Def bdgt [b]	Kc	66.7bn	75.5bn	85.4bn
	US$	2.91bn	3.28bn	
US$1=Kc		22.93	23.03	

[a] NATO definition
[b] Includes military pensions

Real-terms defence budget trend (US$bn, constant 2015)

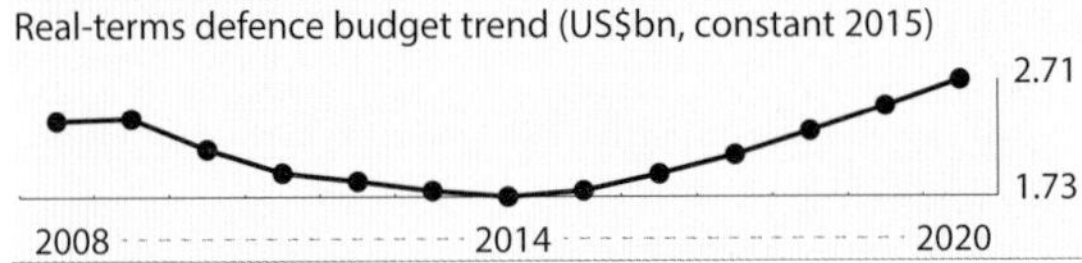

Population 10,702,498

Age	0–14	15–19	20–24	25–29	30–64	65 plus
Male	7.8%	2.4%	2.4%	3.1%	25.1%	8.5%
Female	7.4%	2.2%	2.2%	2.9%	24.3%	11.7%

Capabilities

The 2015 national-security strategy states that NATO is central to Czech security, while the 2017 defence strategy points to Russian assertiveness, an arc of instability to the south and southeast of Europe and information warfare, including cyber attacks, as core security challenges. The evolving security environment is discussed in the 'Long Term Perspective for Defence 2035' and the 'Czech Armed Forces Development Concept 2030', both published in 2019. Modernisation priorities include infantry fighting vehicles, self-propelled howitzers, multi-role helicopters, transport aircraft, short-range air-defence systems and UAVs. In February 2017, the Czech Republic signed a letter of intent with Germany to affiliate the 4th Czech Rapid Reaction Brigade with the 10th German Armoured Division under NATO's Framework Nations Concept. Recruitment and retention remains a challenge: in May 2020, the air-force chief outlined shortfalls in specialised trades such as air-traffic controllers, engineers and pilots. The armed forces are able to deploy on a variety of international crisis-management operations, including NATO's Enhanced Forward Presence, Baltic Air Policing and contribute to NATO's Very High Readiness Joint Task Force. There are plans to upgrade military training and simulation facilities by 2025. The government is trying to replace legacy equipment to modernise the armed forces and reduce dependence on Russia for spares and services. The defence-industrial base includes development and manufacturing capability, in particular small arms, vehicles, and training and light attack aircraft. The holding company Czechoslovak Group brings together several companies across the munitions, vehicles and aerospace sectors. In 2020, the government set up an agency to support the defence industry in government-to-government procurement activities.

ACTIVE 24,900 (Army 13,000 Air 5,850 Other 6,050)

ORGANISATIONS BY SERVICE

Army 13,000

FORCES BY ROLE

MANOEUVRE

Reconnaissance

1 ISR/EW regt (1 recce bn, 1 EW bn, 1 ISR UAV bn)

Armoured

1 (7th) mech bde (1 tk bn, 2 armd inf bn, 1 mot inf bn)

Mechanised

1 (4th) rapid reaction bde (2 mech inf bn, 1 mot inf bn, 1 AB bn)

COMBAT SUPPORT

1 (13th) arty regt (2 arty bn)

1 engr regt (2 engr bn, 1 EOD bn)

1 CBRN regt (2 CBRN bn)

COMBAT SERVICE SUPPORT

1 log regt (2 log bn, 1 maint bn)

Active Reserve

FORCES BY ROLE

COMMAND

14 (territorial defence) comd

MANOEUVRE

Armoured

1 armd coy

Light

14 inf coy (1 per territorial comd) (3 inf pl, 1 cbt spt pl, 1 log pl)

EQUIPMENT BY TYPE

ARMOURED FIGHTING VEHICLES

MBT 30 T-72M4CZ (89 T-72 in store)

RECCE 34 BPzV *Svatava*

IFV 247: 120 BMP-2; 127 *Pandur* II (incl variants); (98 BMP-1; 65 BMP-2 all in store)

APC • PPV 1 *Titus*

AUV 141: 21 *Dingo* 2; 120 IVECO LMV

ENGINEERING & MAINTENANCE VEHICLES

ARV 13+: 10 VPV-ARV (12 more in store); VT-55A; 3 VT-72M4

VLB 6 MT-55A (3 more in store)

MW *Bozena* 5; UOS-155 *Belarty*

NBC VEHICLES BRDM-2RCH

ANTI-TANK/ANTI-INFRASTRUCTURE

MSL • MANPATS 9K111-1 *Konkurs* (AT-5 *Spandrel*); FGM-148 *Javelin*; *Spike*-LR

RCL 84mm *Carl Gustaf*

ARTILLERY 96

SP 152mm 48 M-77 *Dana* (38 more in store)

MOR 48: **120mm** 40 M-1982; (45 more in store); **SP 120mm** 8 SPM-85

Air Force 5,850

Principal task is to secure Czech airspace. This mission is fulfilled within NATO Integrated Extended Air Defence System (NATINADS) and, if necessary, by means of the Czech national reinforced air-defence system. The air force also provides CAS for army SAR, and performs a tpt role

FORCES BY ROLE

FIGHTER/GROUND ATTACK

1 sqn with *Gripen* C/D

1 sqn with L-159 ALCA; L-159T1*

TRANSPORT

2 sqn with A319CJ; C295M; CL-601 *Challenger*; L-410FG/UVP-E *Turbolet*

TRAINING

1 sqn with L-159 ALCA; L-159T1*; L-159T2*

ATTACK HELICOPTER

1 sqn with Mi-24/Mi-35 *Hind* D/E

TRANSPORT HELICOPTER

1 sqn with Mi-17 *Hip* H; Mi-171Sh

1 sqn with Mi-8 *Hip*; Mi-17 *Hip* H; PZL W-3A *Sokol*

AIR DEFENCE

1 (25th) SAM regt (2 AD gp)

EQUIPMENT BY TYPE

AIRCRAFT 38 combat capable

FGA 14: 12 *Gripen* C; 2 *Gripen* D

ATK 16 L-159 ALCA

TPT 13: **Light** 10: 4 C295M; 2 L-410FG *Turbolet*; 4 L-410UVP-E *Turbolet*; **PAX** 3: 2 A319CJ; 1 CL-601 *Challenger*

TRG 8: 5 L-159T1*; 3 L-159T2*

HELICOPTERS

ATK 17: 7 Mi-24 *Hind* D; 10 Mi-35 *Hind* E

MRH 5 Mi-17 *Hip* H

TPT • Medium 30: 4 Mi-8 *Hip*; 16 Mi-171Sh; 10 PZL W3A *Sokol*

AIR DEFENCE • SAM

Point-defence 9K35 *Strela*-10 (SA-13 *Gopher*); 9K32 *Strela*-2‡ (SA-7 *Grail*) (available for trg RBS-70 gunners); RBS-70

AIR-LAUNCHED MISSILES

AAM • IR AIM-9M *Sidewinder*; **ARH** AIM-120C-5 AMRAAM

BOMBS

Laser-guided: GBU *Paveway*

Other Forces 6,050

FORCES BY ROLE

SPECIAL FORCES
- 1 SF gp

MANOEUVRE

Other
- 1 (presidential) gd bde (2 bn)
- 1 (honour guard) gd bn (2 coy)

COMBAT SUPPORT
- 1 int gp
- 1 (central) MP comd
- 3 (regional) MP comd
- 1 (protection service) MP comd

DEPLOYMENT

AFGHANISTAN: NATO • *Operation Resolute Support* 91

BOSNIA-HERZEGOVINA: EU • EUFOR • *Operation Althea* 2

CENTRAL AFRICAN REPUBLIC: UN • MINUSCA 3

DEMOCRATIC REPUBLIC OF THE CONGO: UN • MONUSCO 2

EGYPT: MFO 18; 1 C295M

IRAQ: *Operation Inherent Resolve* 30

LATVIA: NATO • Enhanced Forward Presence 60; 1 mor pl

LITHUAINA: NATO • Enhanced Forward Presence 35; 1 EW unit

MALI: EU • EUTM Mali 120; UN • MINUSMA 3

SERBIA: NATO • KFOR 9; UN • UNMIK 2

SYRIA/ISRAEL: UN • UNDOF 2

Denmark DNK

Danish Krone kr		2019	2020	2021
GDP	kr	2.31tr	2.23tr	
	US$	347bn	340bn	
per capita	US$	59,770	58,439	
Growth	%	2.3	-4.5	
Inflation	%	0.7	0.4	
Def exp [a]	kr	30.4bn	32.4bn	
	US$	4.56bn	4.94bn	
Def bdgt [b]	kr	30.4bn	32.2bn	33.8bn
	US$	4.56bn	4.91bn	
US$1=kr		6.67	6.56	

[a] NATO definition

[b] Includes military pensions

Real-terms defence budget trend (US$bn, constant 2015)

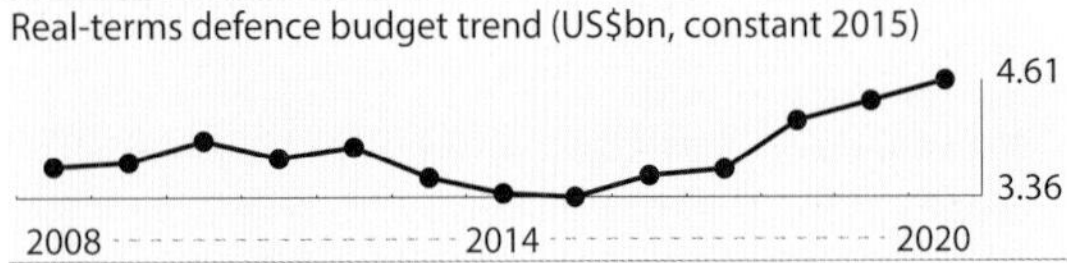

Population 5,869,410

Age	0–14	15–19	20–24	25–29	30–64	65 plus
Male	8.4%	3.0%	3.3%	3.5%	22.4%	9.2%
Female	8.0%	2.9%	3.1%	3.4%	22.1%	10.7%

Capabilities

Denmark's military capabilities remain compact but effective despite pressures on spending and deployments. In the foreign- and security-policy strategy 2019–20, released at the end of 2018, the government expressed concerns about Russia, instability in the Middle East and cyber attacks. Earlier, the government issued a defence agreement for 2018–23, envisaging increased defence spending to deal with a deteriorating security environment. In particular, it is intended to strengthen deterrence, cyber defence and Denmark's role in international operations, as well as support to civilian authorities. Denmark plans to set up a heavy brigade with ground-based air-defence capabilities and a light infantry battalion to support the police. Denmark also intends to strengthen naval air defence, as well as anti-submarine-warfare capabilities. Ties to NATO, NORDEFCO and other regional neighbours have increased. A joint declaration was signed in 2015 with other Nordic states, aimed at deterring Russia. Denmark has opted out of EU military cooperation under CSDP, but in late 2019 the government agreed an action plan to ensure the promotion of Danish interests in the EU's European Defence Fund, due to launch in 2021. The new defence agreement foresees retaining national service and that the annual conscript intake should rise. In June 2016, it was confirmed that the F-35A would replace the country's F-16AM/BM fleet. Industrial support from Terma, Denmark's largest defence company, may have been important to the F-35 procurement decision, as some key sub-components and composites are produced by the firm. Other acquisition priorities include an MLU for its *Leopard* 2A7s, EW equipment, UAVs and ASW capabilities. The defence-industrial base is focused on exports to Europe and North America and is mainly active in defence electronics and the design and manufacture of components and subsystems.

ACTIVE 15,400 (Army 8,000 Navy 2,250 Air 3,000 Joint 2,150)

Conscript liability 4–12 months, most voluntary

RESERVES 44,200 (Army 34,400 Navy 5,300 Air Force 4,500)

ORGANISATIONS BY SERVICE

Army 8,000

Div and a bde HQ transforming into operational formations

FORCES BY ROLE

COMMAND
- 1 (MND-N) div HQ

MANOEUVRE

Mechanised
- 1 (1st) mech bde (1 ISR bn, 3 mech inf bn, 1 SP arty bn, 1 cbt engr bn, 1 sigs bn, 1 log bn)
- 1 (2nd) mech bde (1 recce bn, 1 tk bn, 1 lt inf bn)

COMBAT SUPPORT
- 1 CBRN/construction bn
- 1 EOD bn
- 1 int bn
- 1 MP bn
- 2 sigs bn

COMBAT SERVICE SUPPORT
- 1 log bn
- 1 maint bn
- 1 spt bn

AIR DEFENCE
1 AD bn

EQUIPMENT BY TYPE

ARMOURED FIGHTING VEHICLES
MBT 44: 40 *Leopard* 2A5 (to be upgraded to 2A7); 4 *Leopard* 2A7
IFV 44 CV9035 MkIII
APC 344
APC (T) 125 M113 (incl variants)
APC (W) 219: 79 *Piranha* III (incl variants); ε140 *Piranha* V
AUV 120: 84 *Eagle* IV; 36 *Eagle* V

ENGINEERING & MAINTENANCE VEHICLES
AEV 6 *Wisent*
ARV 10 BPZ-2
VLB 9 *Biber*

ANTI-TANK/ANTI-INFRASTRUCTURE
RCL 84mm *Carl Gustaf*

ARTILLERY 32
SP 155mm 12 M109A3 (being replaced by CAESAR)
MOR • TOWED 120mm 20 Soltam K6B1

AIR DEFENCE • SAM • **Point-defence** FIM-92 *Stinger*

Navy 2,250

Three naval squadrons, headquartered at naval bases in Frederikshavn and Korsør

EQUIPMENT BY TYPE

PRINCIPAL SURFACE COMBATANTS 5
DESTROYERS • DDGHM 3 *Iver Huitfeldt* with 4 quad lnchr with RGM-84L *Harpoon* Block II AShM, 4 8-cell Mk 41 VLS (to be fitted with SAM), 2 12-cell Mk 56 VLS with RIM-162B ESSM SAM, 2 twin 324mm TT with MU90 LWT, 1 *Millennium* CIWS, 2 76mm guns (capacity 1 med hel)
FRIGATES • FFGHM 2 *Absalon* (flexible support ships) with 4 quad lnchr with RGM-84L *Harpoon* Block II AShM, 3 12-cell Mk 56 VLS with RIM-162B ESSM SAM, 2 twin 324mm TT with MU90 LWT, 2 *Millennium* CIWS, 1 127mm gun (capacity 2 AW101 *Merlin*; 2 LCP, 7 MBT or 40 vehicles; 130 troops)

PATROL AND COASTAL COMBATANTS 12
PSOH 4 *Thetis* 1 76mm gun (capacity 1 MH-60R *Seahawk*)
PSO 3 *Knud Rasmussen* with 1 76mm gun, 1 hel landing platform (ice-strengthened hull)
PCC 5 *Diana*

MINE WARFARE • MINE COUNTERMEASURES 2
MSD 2 *Holm*

LOGISTICS AND SUPPORT 12
ABU 2 *Gunnar Thorson* (primarily used for MARPOL duties)
AE 1 *Sleipner*
AGS 2 *Holm*
AKL 2 *Seatruck*
AX 1 *Soloven* (DNK *Flyvefisken*)
AXL 2 *Holm*
AXS 2 *Svanen*

Air Force 3,000

Tactical Air Command

FORCES BY ROLE

FIGHTER/GROUND ATTACK
2 sqn with F-16AM/BM *Fighting Falcon*

ANTI-SUBMARINE WARFARE
1 sqn with MH-60R *Seahawk*

SEARCH & RESCUE/TRANSPORT HELICOPTER
1 sqn with AW101 *Merlin*
1 sqn with AS550 *Fennec* (ISR)

TRANSPORT
1 sqn with C-130J-30 *Hercules*; CL-604 *Challenger* (MP/VIP)

TRAINING
1 unit with MFI-17 *Supporter* (T-17)

EQUIPMENT BY TYPE

AIRCRAFT 44 combat capable
FTR 44: 34 F-16AM *Fighting Falcon*; 10 F-16BM *Fighting Falcon* (30 operational)
TPT 8: **Medium** 4 C-130J-30 *Hercules*; **PAX** 4 CL-604 *Challenger* (MP/VIP)
TRG 27 MFI-17 *Supporter* (T-17)

HELICOPTERS
ASW 9 MH-60R *Seahawk*
SAR 8 AW101 *Merlin*
MRH 8 AS550 *Fennec* (ISR) (4 more non-operational)
TPT • Medium 6 AW101 *Merlin*

AIR-LAUNCHED MISSILES
AAM • IR AIM-9L *Sidewinder*; **IIR** AIM-9X *Sidewinder* II; **ARH** AIM-120B AMRAAM

BOMBS
Laser-guided EGBU-12/GBU-24 *Paveway* II/III
INS/GPS guided GBU-31 JDAM

Control and Air Defence Group

1 Control and Reporting Centre, 1 Mobile Control and Reporting Centre. 4 Radar sites

Special Operations Command

FORCES BY ROLE

SPECIAL FORCES
1 SF unit
1 diving unit

Reserves

Home Guard (Army) 34,400 reservists (to age 50)

2 (local) def region

Home Guard (Navy) 5,300 reservists (to age 50)

EQUIPMENT BY TYPE

PATROL AND COASTAL COMBATANTS 30
PB 30: 18 MHV800; 12 MHV900

Home Guard (Air Force) 4,500 reservists (to age 50)

EQUIPMENT BY TYPE

AIRCRAFT • TPT • **Light** 2 BN-2A *Islander*

Europe

DEPLOYMENT

AFGHANISTAN: NATO • *Operation Resolute Support* 140

ESTONIA: NATO • Enhanced Forward Presence 209; 1 armd inf coy

IRAQ: *Operation Inherent Resolve* 210; 1 SF gp; 1 trg team (DNK to assume leadership of NATO Mission Iraq Dec 2020)

KUWAIT: *Operation Inherent Resolve* 16

MALI: UN • MINUSMA 11

MIDDLE EAST: UN • UNTSO 10

PERSIAN GULF: EMASOH; 1 DDGHM

SERBIA: NATO • KFOR 35

UNITED ARAB EMIRATES: *Operation Inherent Resolve* 13

Estonia EST

Euro €		2019	2020	2021
GDP	€	28.1bn	26.7bn	
	US$	31.5bn	30.5bn	
per capita	US$	23,758	22,986	
Growth	%	5.0	-5.2	
Inflation	%	2.3	0.2	
Def Exp [a]	€	569m	615m	
	US$	637m	703m	
Def bdgt [b]	€	615m	627m	633m
	US$	688m	717m	
FMA (US)	US$	8m	5m	5m
US$1=€		0.89	0.87	

[a] NATO definition

[b] Includes military pensions

Real-terms defence budget trend (US$m, constant 2015)

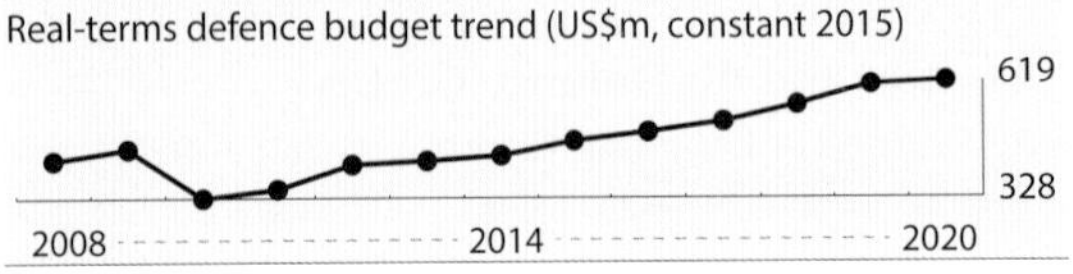

Population	1,228,624					
Age	**0–14**	**15–19**	**20–24**	**25–29**	**30–64**	**65 plus**
Male	8.3%	2.4%	2.2%	2.8%	24.0%	7.3%
Female	7.9%	2.2%	2.0%	2.5%	24.6%	13.7%

Capabilities

Estonia has small active armed forces and is reliant on NATO membership as a security guarantor. Security policy is predicated on the goals of ensuring sovereignty and territorial integrity, and there is concern over Russian security policy and military activity. The government's 2017–26 National Defence Development Plan (NDDP) reflects the worsening security environment in the Baltic region. The MoD publishes medium-term development plans on an annual basis. Covering a four-year period, these are intended to ensure that the NDDP's goals will be achieved within the planned time frame. The active armed forces are supplemented by a reserve component. In June 2018, Estonia joined the French-inspired European Intervention Force. A NATO battlegroup based in Estonia became operational in mid-2017 as part of the Alliance's Enhanced Forward Presence. Amari air base hosts a NATO Baltic Air Policing detachment. Estonia is also a member of the UK-led multinational Joint Expeditionary Force. Tallinn hosts NATO's Cybersecurity Centre of Excellence. The NDDP notes a desire to increase the annual conscript intake and active personnel numbers. There is limited organic capability to deploy beyond borders, though Estonian forces take part in EU, NATO and UN missions abroad on a small scale. The NDDP identifies the need for additional armoured mobility and armoured firepower, anti-armour weapons and increased munitions stocks. The country has a niche defence-industrial capability, including ship repair and digital systems.

ACTIVE 7,100 (Army 4,000 Navy 300 Air 500 Other 2,300)

Conscript liability 8 or 11 months (depending on specialisation; conscripts cannot be deployed)

RESERVE 17,500 (Joint 12,000; Territorial Defence 5,500)

ORGANISATIONS BY SERVICE

Army 1,500; 2,500 conscript (total 4,000)

4 def region. All units except one inf bn are reserve based

FORCES BY ROLE

MANOEUVRE

Mechanised

1 (1st) bde (1 recce coy, 1 armd inf bn; 2 mech inf bn, 1 arty bn, 1 AT coy, 1 cbt engr bn, 1 spt bn, 1 AD bn)

Light

1 (2nd) inf bde (1 recce coy, 4 inf bn, 2 arty bn, 1 AT coy, 1 cbt engr bn, 1 spt bn, 1 AD bn)

4 (territorial) inf bn

EQUIPMENT BY TYPE

ARMOURED FIGHTING VEHICLES

IFV 44 CV9035EE (incl 2 CP)

APC • **APC (W)** 136: 56 XA-180 *Sisu*; 80 XA-188 *Sisu*

ENGINEERING & MAINTENANCE VEHICLES

AEV 2 Pionierpanzer 2 *Dachs*

ARV 2 BPz-2

VLB 2 *Biber*

ANTI-TANK/ANTI-INFRASTRUCTURE

MSL • **MANPATS** FGM-148 *Javelin*; *Milan*; *Spike*-LR

RCL 84mm *Carl Gustaf*; **90mm** PV-1110

ARTILLERY 188

SP 155mm 2 K9 *Thunder*

TOWED 60: **122mm** 36 D-30 (H 63); **155mm** 24 FH-70

MOR 126: **81mm** 60 B455/NM 95/M252; **120mm** 66 2B11/M/41D

AIR DEFENCE

SAM • **Point-defence** *Mistral*

GUNS • **TOWED 23mm** ZU-23-2

Navy 200; 100 conscript (total 300)

EQUIPMENT BY TYPE

MINE WARFARE • MINE COUNTERMEASURES 4:

MCCS 1 *Tasuja* (ex-DNK *Lindormen*)

MHC 3 *Admiral Cowan* (ex-UK *Sandown*) (1 in refit)

Air Force 500

FORCES BY ROLE

TRANSPORT

1 sqn with An-2 *Colt*

TRANSPORT HELICOPTER

1 sqn with R-44 *Raven* II

EQUIPMENT BY TYPE

AIRCRAFT

TPT • Light 4: 2 An-2 *Colt*; 2 M-28 *Skytruck*

TRG 1+ L-39C *Albatros* (leased)

HELICOPTERS • TPT • Light 3 R-44 *Raven* II

Other 1,300; 800 conscript (total 2,100)

Includes Cyber Command, Support Command and Special Operations Forces

FORCES BY ROLE

SPECIAL FORCES

1 spec ops bn

COMBAT SUPPORT

2 MP coy

COMBAT SERVICE SUPPORT

1 log bn

Paramilitary

Border Guard

The Estonian Border Guard is subordinate to the Ministry of the Interior. Air support is provided by the Estonian Border Guard Aviation Corps

EQUIPMENT BY TYPE

PATROL AND COASTAL COMBATANTS 6

PCO 2: 1 *Kati*; 1 *Kindral Kurvits* (FIN *Tursas* derivative)

PCC 1 *Kou* (FIN *Silma*)

PB 3: 1 *Pikker*; 1 *Raju* (Baltic 4500WP); 1 *Valve*

AMPHIBIOUS • LANDING CRAFT • LCU 3

LOGISTICS & SUPPORT • AGF 1 *Balsam*

AIRCRAFT • TPT • Light 2: 1 Beech 305ER *King Air*; 1 Cessna 172R

HELICOPTERS • MRH 3 AW139

DEPLOYMENT

AFGHANISTAN: NATO • *Operation Resolute Support* 45

BALTIC SEA: NATO • SNMCMG 1: 1 MHC

IRAQ: *Operation Inherent Resolve* 10 • NATO Mission Iraq 5

LEBANON: UN • UNIFIL 1

MALI: *Operation Barkhane* 95; **EU** • EUTM Mali 10; **UN** • MINUSMA 3

MIDDLE EAST: UN • UNTSO 3

FOREIGN FORCES

All NATO Enhanced Forward Presence unless stated

Denmark 209; 1 armd inf coy

Germany NATO • Baltic Air Policing 6 Eurofighter *Typhoon*

United Kingdom 754; 1 armd regt HQ; 1 tk sqn; 1 armd inf coy (+); 1 cbt engr coy

Finland FIN

Euro €		2019	2020	2021
GDP	€	241bn	234bn	
	US$	269bn	268bn	
per capita	US$	48,810	48,461	
Growth	%	1.1	-4.0	
Inflation	%	1.1	0.7	
Def bdgt [a]	€	3.50bn	3.59bn	5.23bn
	US$	3.92bn	4.11bn	
US$1=€		0.89	0.87	

[a] Includes military pensions

Real-terms defence budget trend (US$bn, constant 2015)

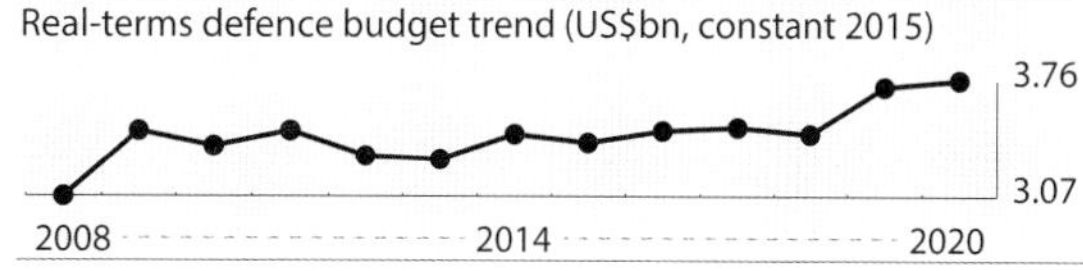

Population 5,571,665

Age	0–14	15–19	20–24	25–29	30–64	65 plus
Male	8.4%	2.7%	2.9%	3.2%	22.3%	9.8%
Female	8.0%	2.6%	2.7%	3.1%	21.8%	12.5%

Capabilities

Finland's armed forces are primarily focused on territorial defence. The country's long border with Russia has focused attention on Russia's military capabilities and plans. The 2017 Defence Report, due to be updated in 2021, argues that changes in the security environment have increased the demands on the armed forces and stresses that financial constraints are forcing trade-offs between long-term procurement plans and operational readiness. In October 2020, the government's report on foreign and security policy assessed a security environment that is rapidly changing and deteriorating An EU member state, Finland's principal multilateral defence relationships include NORDEFCO and the Northern Group, as well as strong bilateral cooperation with Sweden and the US; it is building close ties with NATO short of membership. In 2017, Finland joined a multinational cooperation programme for air-to-ground precision-guided munitions set up by a group of NATO member states. The country participates in UN peacekeeping missions and contributes to NATO operations. Legislation limits the number of personnel deployed on international crisis-management operations to an upper ceiling of 2,000 troops. In 2015, the air force launched the HX Fighter Programme to replace its F/A-18s. A request for quotations was issued in April 2018 and the replacement aircraft is expected to be selected in 2021. Under the Squadron 2020 programme, the navy will replace patrol boats and minelayers with corvette-sized vessels. Finland's defence industry consists largely of privately owned SMEs, concentrating on niche products for international markets, but it also features some internationally competitive larger companies producing wheeled armoured vehicles and turreted mortar systems.

ACTIVE 23,800 (Army 17,350 Navy 3,400 Air 3,050) Paramilitary 2,700

Conscript liability 165, 255 or 347 days (latter for NCOs, officers or those on 'especially demanding' duties)

RESERVE 216,000 (Army 170,000 Navy 20,000 Air 26,000) Paramilitary 11,500

18,000 reservists a year do refresher training: total obligation 80 days (150 for NCOs, 200 for officers) between conscript service and age 50 (NCOs and officers to age 60)

ORGANISATIONS BY SERVICE

Army 4,350; 13,000 conscript (total 17,350)

FORCES BY ROLE

Finland's army maintains a mobilisation strength of about 285,000. In support of this requirement, two conscription cycles, each for about 13,500 conscripts, take place each year. After conscript training, reservist commitment is to the age of 60. Reservists are usually assigned to units within their local geographical area. All service appointments or deployments outside Finnish borders are voluntary for all members of the armed services. All brigades are reserve based

Reserve Organisations 170,000

FORCES BY ROLE

SPECIAL FORCES
1 SF bn
MANOEUVRE
Armoured
2 armd BG (regt)
Mechanised
2 (Karelia & Pori Jaeger) mech bde
Light
3 (Jaeger) bde
6 lt inf bde
COMBAT SUPPORT
1 arty bde
1 AD regt
7 engr regt
3 sigs bn
COMBAT SERVICE SUPPORT
Some log unit
HELICOPTER
1 hel bn

EQUIPMENT BY TYPE

ARMOURED FIGHTING VEHICLES
MBT 100 *Leopard* 2A6 (100 *Leopard* 2A4 in store)
IFV 212: 110 BMP-2/-2MD; 102 CV9030FIN
APC 613
APC (T) 142: 40 MT-LBu; 102 MT-LBV
APC (W) 471: 260 XA-180/185 *Sisu*; 101 XA-202 *Sisu* (CP); 48 XA-203 *Sisu*; 62 AMV (XA-360)
ENGINEERING & MAINTENANCE VEHICLES
AEV 5 *Dachs*
ARV 36: 9 BPz-2; 15 MTP-LB; 12 VT-55A
VLB 27: 12 BLG-60M2; 6 *Leopard* 2S; 9 SISU *Leguan*
MW 6+: *Aardvark* Mk 2; KMT T-55; 6 *Leopard* 2R CEV; RA-140 DS
ANTI-TANK/ANTI-INFRASTRUCTURE
MSL • MANPATS NLAW; *Spike*-MR; *Spike*-LR
ARTILLERY 672
SP 122mm 49: 13 K9 *Thunder*; 36 2S1 *Gvozdika* (PsH 74)
TOWED 288: **122mm** 234 D-30 (H 63); **155mm** 54 K 83/GH-52 (K 98)
MRL 56: **122mm** 34 RM-70; **227mm** 22 M270 MLRS
MOR 279+: **81mm** Krh/71; **120mm** 261 Krh/92; **SP 120mm** 18 XA-361 AMOS
HELICOPTERS
MRH 7: 5 Hughes 500D; 2 Hughes 500E
TPT • Medium 20 NH90 TTH
UNINHABITED AERIAL VEHICLES
ISR
Medium 11 ADS-95 *Ranger*
AIR DEFENCE
SAM 60+
Short-range 44: 20 *Crotale* NG (ITO 90); 24 NASAMS II FIN (ITO 12)
Point-defence 16+: 16 ASRAD (ITO 05); FIM-92 *Stinger* (ITO 15); RBS 70 (ITO 05/05M)
GUNS 407+: **23mm** ItK 95/ZU-23-2 (ItK 61); **35mm** GDF-005 (ItK 88); **SP 35mm** 7 *Leopard* 2 ITK *Marksman*

Navy 1,400; 2,000 conscript (total 3,400)

FORCES BY ROLE

Naval Command HQ located at Turku; with two subordinate Naval Commands (Gulf of Finland and Archipelago Sea); 1 Naval bde; 3 spt elm (Naval Materiel Cmd, Naval Academy, Naval Research Institute)

EQUIPMENT BY TYPE

PATROL AND COASTAL COMBATANTS 20
PCGM 4 *Hamina* with 4 RBS15 Mk3 (MTO-85M) AShM, 1 8-cell VLS with *Umkhonto*-IR (ITO2004) SAM, 1 57mm gun
PBG 4 *Rauma* with 6 RBS15 Mk3 (MTO-85M) AShM
PBF 12 *Jehu* (U-700) (capacity 24 troops)
MINE WARFARE 8
MINE COUNTERMEASURES 3
MCC 3 *Katanpää* (ITA *Gaeta* mod)
MINELAYERS • ML 5:
2 *Hameenmaa* with 1 8-cell VLS with *Umkhonto*-IR (ITO2004) SAM, 2 RBU 1200 *Uragan* A/S mor, 1 57mm gun (can carry up to 100–120 mines)
3 *Pansio* with 50 mines
AMPHIBIOUS • LANDING CRAFT 51
LCM 1 *Kampela*
LCP 50
LOGISTICS AND SUPPORT 7
AG 3: 1 *Louhi*; 2 *Hylje*
AX 4: 3 *Fabian Wrede*; 1 *Lokki*

Coastal Defence

FORCES BY ROLE

MANOEUVRE
Amphibious
1 mne bde
COMBAT SUPPORT
1 cbt spt bde (1 AShM bty)

EQUIPMENT BY TYPE

COASTAL DEFENCE
AShM 4 RBS15K
ARTY • 130mm 30 K-53tk (static)
ANTI-TANK/ANTI-INFRASTRUCTURE
MSL • MANPATS *Spike* (used in AShM role)

Air Force 2,050; 1,000 conscript (total 3,050)

3 Air Comds: Satakunta (West), Karelia (East), Lapland (North)

FORCES BY ROLE

FIGHTER/GROUND ATTACK
2 sqn with F/A-18C/D *Hornet*

ISR
  1 (survey) sqn with Learjet 35A
TRANSPORT
  1 flt with C295M
  4 (liaison) flt with PC-12NG
TRAINING
  1 sqn with *Hawk* Mk50/51A/66* (air-defence and ground-attack trg)
  1 unit with G-115EA; L-70 *Vinka*

EQUIPMENT BY TYPE

AIRCRAFT 107 combat capable
  FGA 62: 55 F/A-18C *Hornet*; 7 F/A-18D *Hornet*
  ELINT 1 C295M
  TPT • Light 10: 2 C295M; 3 Learjet 35A (survey; ECM trg; tgt-tow); 5 PC-12NG
  TRG 99: 28 G-115EA; 29 *Hawk* Mk50/51A*; 16 *Hawk* Mk66*; 26 L-70 *Vinka*
AIR-LAUNCHED MISSILES
  AAM • IR AIM-9 *Sidewinder*; **IIR** AIM-9X *Sidewinder*
  ARH AIM-120C AMRAAM
  LACM Conventional AGM-158 JASSM
BOMBS
  INS/GPS-guided GBU-31 JDAM; AGM-154C JSOW

Paramilitary

Border Guard 2,700

Ministry of Interior. 4 Border Guard Districts and 2 Coast Guard Districts

FORCES BY ROLE

MARITIME PATROL
  1 sqn with Do-228 (maritime surv); AS332 *Super Puma*; Bell 412 (AB-412) *Twin Huey*; Bell 412EP (AB-412EP) *Twin Huey*;AW119KE *Koala*

EQUIPMENT BY TYPE

PATROL AND COASTAL COMBATANTS 44
  PSO 1 *Turva* with 1 hel landing platform
  PCC 2 *Tursas*
  PB 41
AMPHIBIOUS • LANDING CRAFT 6
  UCAC 6
AIRCRAFT • TPT • Light 2 Do-228
HELICOPTERS
  MRH 3: 1 Bell 412 (AB-412) *Twin Huey*; 2 Bell 412EP (AB-412EP) *Twin Huey*
  TPT 9: **Medium** 5 AS332 *Super Puma*; **Light** 4 AW119KE *Koala*

Reserve 11,500 reservists on mobilisation

DEPLOYMENT

AFGHANISTAN: NATO • *Operation Resolute Support* 60

IRAQ: *Operation Inherent Resolve* 80; 1 trg team; **NATO** • **NATO** Mission Iraq 5

LEBANON: UN • UNIFIL 198; 1 inf coy

MALI: EU • EUTM Mali 12; **UN** • MINUSMA 4

MIDDLE EAST: UN • UNTSO 15

SERBIA: NATO • KFOR 20

SOMALIA: EU • EUTM Somalia 10

France FRA

Euro €		2019	2020	2021
GDP	€	2.43tr	2.23tr	
	US$	2.72tr	2.55tr	
per capita	US$	41,897	39,257	
Growth	%	1.5	-9.8	
Inflation	%	1.3	0.5	
Def exp [a]	€	44.3bn	46.2bn	
	US$	49.6bn	52.8bn	
Def bdgt [b]	€	46.5bn	48.1bn	49.7bn
	US$	52.1bn	55.0bn	
US$1=€		0.89	0.87	

[a] NATO definition
[b] Includes pensions

Real-terms defence budget trend (US$bn, constant 2015)

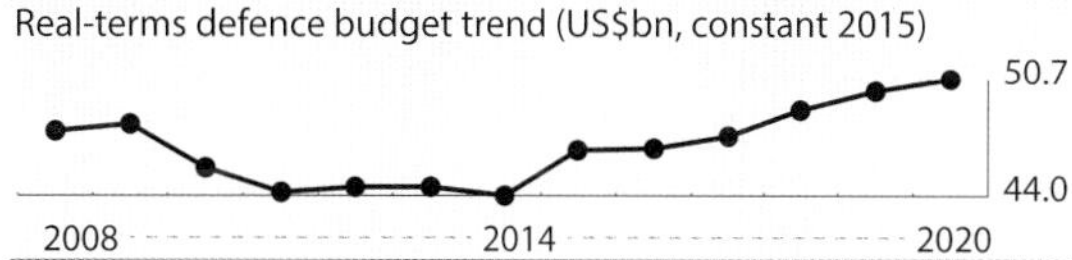

Population 67,848,156

Age	0–14	15–19	20–24	25–29	30–64	65 plus
Male	9.4%	3.1%	3.0%	2.9%	21.7%	8.9%
Female	9.0%	3.0%	2.9%	2.8%	21.8%	11.6%

Capabilities

France maintains globally deployed forces that are also engaged on enduring operations in Africa. The 2017 Strategic Review reiterated operational commitments in sub-Saharan Africa and the Middle East, as well as a continued presence in the Asia-Pacific. The Programme Budget Law for 2019–25 set out defence-budget increases to support these goals. France plays a leading military role in the EU, NATO and the UN. In 2018, Paris launched the European Intervention Initiative, joined by 11 other European countries, intended to foster a common strategic culture and develop the ability to jointly deploy quickly in case of crises. France is also expanding its capabilities in non-traditional domains, having set up a space command, developed a space strategy and formalised a cyber-offensive doctrine. Deployments abroad have demonstrated the ability to support expeditionary forces independently. In mid-2020, the army issued a strategy document, 'Operational Superiority 2030', which called for improved readiness in light of the risk of high-intensity conflict. However, these plans could be affected by the decision, in October, to deploy more troops on the domestic *Operation Sentinel* mission. Some strategic military air-transport requirements are dependent on allies and external contractors. The high operational tempo has increased the stress on equipment. The Programme Budget Law seeks to remedy this with a budget increase for maintenance and accelerated modernisation of multi-role tanker-transport and refuelling aircraft. France has a sophisticated defence industry, exemplified by companies such as Dassault, MBDA and Nexter, with most procurements undertaken domestically. France has called for increased European defence-industrial cooperation and aims to increase its exports to Europe. It is also seeking to invest in future technologies and supports start-ups and innovation in the defence domain.

ACTIVE 203,250 (Army 114,700 Navy 34,700 Air 40,450, Other Staffs 13,400) Paramilitary 100,500

RESERVE 41,050 (Army 24,900 Navy 6,250 Air 5,750 Other Staffs 4,150) Paramilitary 30,300

ORGANISATIONS BY SERVICE

Strategic Nuclear Forces

Navy 2,200

EQUIPMENT BY TYPE

SUBMARINES • STRATEGIC 4

SSBN 4 *Le Triomphant* with 16 M51 SLBM with 6 TN-75 nuclear warheads, 4 single 533mm TT with SM39 *Exocet* AShM/F17 mod 2 HWT

AIRCRAFT • FGA 20 *Rafale* M F3 with ASMPA msl

Air Force 1,800

Air Strategic Forces Command

FORCES BY ROLE

STRIKE

1 sqn with *Rafale* B with ASMPA msl

1 sqn with *Rafale* B with ASMPA msl (forming)

TANKER

1 sqn with C-135FR; KC-135 *Stratotanker*

EQUIPMENT BY TYPE

AIRCRAFT 20 combat capable

FGA 20 *Rafale* B

TKR/TPT 11 C-135FR

TKR 3 KC-135 *Stratotanker*

Paramilitary

Gendarmerie 40

Space

EQUIPMENT BY TYPE

SATELLITES 7

COMMUNICATIONS 3: 2 *Syracuse*-3 (designed to integrate with UK *Skynet* & ITA *Sicral*); 1 *Athena-Fidus* (also used by ITA)

ISR 4: 2 *Helios* (2A/2B); 2 *Pleiades*

Army 114,700

Regt and BG normally bn size

FORCES BY ROLE

COMMAND

1 corps HQ (CRR-FR)

2 div HQ

MANOEUVRE

Reconnaissance

1 recce regt

Armoured

1 (2nd) armd bde (2 tk regt, 3 armd inf regt, 1 SP arty regt, 1 engr regt)

1 (7th) armd bde (1 tk regt, 1 armd BG, 3 armd inf regt, 1 SP arty regt, 1 engr regt)

1 armd BG HQ (UAE)

Mechanised

1 (6th) lt armd bde (2 armd cav regt, 1 armd inf regt, 1 mech inf regt, 1 mech inf regt, 1 SP arty regt, 1 engr regt)

1 (FRA/GER) mech bde (1 armd cav regt, 1 mech inf regt)

1 mech regt HQ (Djibouti)

Light

1 (27th) mtn bde (1 armd cav regt, 3 mtn inf regt, 1 arty regt, 1 engr regt)

3 inf regt (French Guiana & French West Indies)

1 inf regt HQ (New Caledonia)

2 inf bn HQ (Côte d'Ivoire & Gabon)

Air Manoeuvre

1 (11th) AB bde (1 armd cav regt, 4 para regt, 1 arty regt, 1 engr regt, 1 spt regt)

1 AB regt (La Réunion)

Amphibious

1 (9th) amph bde (2 armd cav regt, 1 armd inf regt, 2 mech inf regt, 1 SP arty regt, 1 engr regt)

Other

4 SMA regt (French Guiana, French West Indies & Indian Ocean)

3 SMA coy (French Polynesia, Indian Ocean & New Caledonia)

COMBAT SUPPORT

1 MRL regt

2 engr regt

2 EW regt

1 int bn

1 CBRN regt

5 sigs regt

COMBAT SERVICE SUPPORT

5 tpt regt

1 log regt

1 med regt

3 trg regt

HELICOPTER

1 (4th) hel bde (3 hel regt)

ISR UAV

1 UAV regt

AIR DEFENCE

1 SAM regt

Special Operation Forces 2,200

FORCES BY ROLE

SPECIAL FORCES

2 SF regt

HELICOPTER

1 hel regt

Reserves 22,750 reservists

Reservists form 79 UIR (Reserve Intervention Units) of about 75 to 152 troops, for 'Proterre' – combined land projection forces bn, and 23 USR (Reserve Specialised Units) of about 160 troops, in specialised regt

EQUIPMENT BY TYPE

ARMOURED FIGHTING VEHICLES

MBT 222 *Leclerc*

ASLT 245 AMX-10RC

RECCE 1,458: 40 ERC-90D *Sagaie*; 1,418 VBL/VB2L

IFV 605: 500 VBCI VCI; 105 VBCI VPC (CP)

APC 2,530

APC (T) 49 BvS-10

APC (W) 2,468: 220 VBMR *Griffon*; 2,189 VAB; 59 VAB VOA (OP)

PPV 13 *Aravis*

ENGINEERING & MAINTENANCE VEHICLES

AEV 42 AMX-30EBG

ARV 47+: 30 AMX-30D; 17 *Leclerc* DNG; VAB-EHC

VLB 55: 27 EFA; 18 PTA; 10 SPRAT

MW 24+: AMX-30B/B2; 4 *Buffalo*; 20 *Minotaur*

NBC VEHICLES 40 VAB NRBC

ANTI-TANK/ANTI-INFRASTRUCTURE • MSL

SP 110 VAB *Milan*

MANPATS *Eryx*; FGM-148 *Javelin*; *Milan*; MMP

ARTILLERY 265+

SP 155mm 108: 32 AU-F-1; 76 CAESAR

TOWED 155mm 12 TR-F-1

MRL 227mm 13 M270 MLRS

MOR 132+: **81mm** LLR 81mm; **120mm** 132 RT-F-1

AIRCRAFT • TPT • Light 13: 5 PC-6B *Turbo Porter*; 5 TBM-700; 3 TBM-700B

HELICOPTERS

ATK 67: 29 *Tiger* HAP; 38 *Tiger* HAD

MRH 104: 18 AS555UN *Fennec*; 86 SA341F/342M *Gazelle* (all variants)

TPT 154: **Heavy** 8 H225M *Caracal* (CSAR); **Medium** 111: 24 AS532UL *Cougar*; 2 EC225LP *Super Puma*; 39 NH90 TTH; 46 SA330 *Puma*; **Light** 35 H120 *Colibri* (leased)

UNINHABITED AERIAL VEHICLES

ISR • Medium 23 SDTI (*Sperwer*)

AIR DEFENCE • SAM • Point-defence *Mistral*

Navy 35,100

EQUIPMENT BY TYPE

SUBMARINES 8

STRATEGIC • SSBN 4 *Le Triomphant* opcon Strategic Nuclear Forces with 16 M51 SLBM with 6 TN-75 nuclear warheads, 4 single 533mm TT with SM39 *Exocet* AShM/F17 mod 2 HWT

TACTICAL • SSN 4 *Rubis* (1 more severely damaged by fire) with 4 single 533mm TT with SM39 *Exocet* AShM/F17 mod 2 HWT

PRINCIPAL SURFACE COMBATANTS 22

AIRCRAFT CARRIERS • CVN 1 *Charles de Gaulle* with 4 8-cell *Sylver* A43 VLS with *Aster* 15 SAM, 2 sextuple *Sadral* lnchr with *Mistral* SAM (capacity 30 *Rafale* M FGA ac, 2 E-2C *Hawkeye* AEW&C ac, 8 AS365 *Dauphin*/NH90 NFH hel)

DESTROYERS • DDGHM 3:

1 *Cassard* with 2 quad lnchr with MM40 *Exocet* Block 2 AShM, 1 Mk 13 GMLS with SM-1MR Block VI SAM, 2 sextuple *Sadral* lnchr with *Mistral* SAM, 2 single 533mm ASTT with L5 mod 4 HWT, 1 100mm gun (capacity 1 AS565SA *Panther* ASW hel)

2 *Forbin* with 2 quad lnchr with MM40 *Exocet* Block 3 AShM, 4 8-cell *Sylver* A50 VLS with *Aster* 30 SAM, 2 8-cell *Sylver* A50 VLS with *Aster* 15 SAM, 2 twin 324mm ASTT with MU90 LWT, 2 76mm gun (capacity 1 NH90 NFH hel)

FRIGATES 18

FFGHM 12:

4 *Aquitaine* (FREMM ASM) with 2 8-cell *Sylver* A70 VLS with MdCN (SCALP Naval) LACM, 2 quad lnchr with MM40 *Exocet* Block 3 AShM, 2 8-cell *Sylver* A43 VLS with *Aster* 15 SAM, 2 twin 324mm B-515 ASTT with MU90 LWT, 1 76mm gun (capacity 1 NH90 NFH hel)

1 *Aquitaine* (FREMM ASM) with 2 8-cell *Sylver* A70 VLS with MdCN (SCALP Naval) LACM, 2 quad lnchr with MM40 *Exocet* Block 3 AShM, 2 8-cell *Sylver* A50 VLS with *Aster* 15 SAM (*Aster* 30 to be fitted), 2 twin 324mm B-515 ASTT with MU90 LWT, 1 76mm gun (capacity 1 NH90 NFH hel)

1 *Aquitaine* (FREMM ASM) with 2 8-cell *Sylver* A70 VLS with MdCN (SCALP Naval) LACM, 2 quad lnchr with MM40 *Exocet* Block 3 AShM, 2 8-cell *Sylver* A50 VLS with *Aster* 15 SAM/*Aster* 30 SAM, 2 twin 324mm B-515 ASTT with MU90 LWT, 1 76mm gun (capacity 1 NH90 NFH hel)

1 *Georges Leygues* (mod) with 2 quad lnchr with MM40 *Exocet* AShM, 1 octuple lnchr with *Crotale* SAM, 2 twin *Simbad* lnchr with *Mistral* SAM, 2 single 324mm ASTT with MU90 LWT, 1 100mm gun (capacity 2 hels)

5 *La Fayette* with 2 quad lnchr with MM40 *Exocet* Block 3 AShM, 1 octuple lnchr with *Crotale* SAM (space for fitting 2 8-cell VLS lnchr for *Aster* 15/30), 1 100mm gun (capacity 1 AS565SA *Panther* hel)

FFH 6 *Floreal* with 1 100mm gun (fitted for but not with 1 twin *Simbad* lnchr with *Mistral* SAM) (capacity 1 AS565SA *Panther* hel)

PATROL AND COASTAL COMBATANTS 20

FSM 6 *D'Estienne d'Orves* with 1 twin *Simbad* lnchr with *Mistral* SAM, 2 twin 533mm ASTT, 1 100mm gun

PSO 4 *d'Entrecasteaux* with 1 hel landing platform

PCO 6: 3 *La Confiance*, 1 *Lapérouse*; 1 *Le Malin*; 1 *Fulmar*

PCC 4: 1 *L'Audacieuse*; 3 *Flamant*

MINE WARFARE • MINE COUNTERMEASURES 17

MCD 4 *Vulcain*

MHC 3 *Antarès*

MHO 10 *Éridan*

AMPHIBIOUS

PRINCIPAL AMPHIBIOUS SHIPS 3

LHD 3 *Mistral* with 2 twin *Simbad* lnchr with *Mistral* SAM (capacity up to 16 NH90/SA330 *Puma*/AS532 *Cougar*/*Tiger* hel; 2 LCAC or 4 LCM; 13 MBTs; 50 AFVs; 450 troops)

LANDING CRAFT 38

LCT 4 EDA-R

LCM 9 CTM

LCVP 25

LOGISTICS AND SUPPORT 35

ABU 1 *Telenn Mor*

AG 2 *Chamois*

AGE 2: 1 *Corraline;* 1 *Lapérouse* (used as trials ships for mines and divers)
AGI 1 *Dupuy de Lome*
AGM 1 *Monge*
AGOR 2: 1 *Pourquoi pas?* (used 150 days per year by Ministry of Defence; operated by Ministry of Research and Education otherwise); 1 *Beautemps-beaupré*
AGS 3 *Lapérouse*
AORH 3 *Durance* with 3 twin *Simbad* lnchr with *Mistral* SAM (capacity 1 SA319 *Alouette* III/AS365 *Dauphin*/*Lynx*)
ATF 2 *Malabar*
ATS 4 *Loire* (BSAH)
AXL 10: 8 *Léopard;* 2 *Glycine*
AXS 4: 2 *La Belle Poule*; 2 other

Naval Aviation 6,500

FORCES BY ROLE
STRIKE/FIGHTER/GROUND ATTACK
2 sqn with *Rafale* M F3
1 sqn with *Rafale* M F3/F3-R
ANTI-SURFACE WARFARE
1 sqn with AS565SA *Panther*
ANTI-SUBMARINE WARFARE
2 sqn (forming) with NH90 NFH
MARITIME PATROL
2 sqn with *Atlantique* 2
1 sqn with *Falcon* 20H *Gardian*
1 sqn with *Falcon* 50MI
AIRBORNE EARLY WARNING & CONTROL
1 sqn with E-2C *Hawkeye*
SEARCH & RESCUE
1 sqn with AS365N/F *Dauphin* 2
TRAINING
1 sqn with EMB 121 *Xingu*
1 unit with SA319B *Alouette* III
1 unit with *Falcon* 10MER
1 unit with CAP 10M

EQUIPMENT BY TYPE
AIRCRAFT 54 combat capable
FGA 42: 41 *Rafale* M F3; 1 *Rafale* M F3-R
ASW 12 *Atlantique*-2 (10 more in store)
AEW&C 3 E-2C *Hawkeye*
SAR 4 *Falcon* 50MS
TPT 25: **Light** 10 EMB-121 *Xingu*; **PAX** 15: 6 *Falcon* 10MER; 5 *Falcon* 20H *Gardian*; 4 *Falcon* 50MI
TRG 6 CAP 10M
HELICOPTERS
ASW 24 NH90 NFH
MRH 39: 3 AS365F *Dauphin* 2; 6 AS365N *Dauphin* 2; 2 AS365N3; 16 AS565SA *Panther*; 12 SA319B *Alouette* III
AIR-LAUNCHED MISSILES
AAM • IR R-550 *Magic* 2; **IIR** *Mica* IR; **ARH** *Mica* RF
ASM AASM; AS-30L
AShM AM39 *Exocet*
LACM Nuclear ASMPA
BOMBS
Laser-guided: GBU-12 *Paveway* II

Marines 2,200

Commando Units 750

FORCES BY ROLE
MANOEUVRE
Reconnaissance
1 recce gp
Amphibious
2 aslt gp
1 atk swimmer gp
1 raiding gp
COMBAT SUPPORT
1 cbt spt gp
COMBAT SERVICE SUPPORT
1 spt gp

Fusiliers-Marin 1,450

FORCES BY ROLE
MANOEUVRE
Other
2 sy gp
7 sy coy

Reserves 6,000 reservists

Air Force 40,450

FORCES BY ROLE
STRIKE
1 sqn with *Rafale* B with ASMPA msl
1 sqn with *Rafale* B with ASMPA msl (forming)
SPACE
1 (satellite obs) sqn
FIGHTER
1 sqn with *Mirage* 2000-5
1 sqn with *Mirage* 2000B/C
FIGHTER/GROUND ATTACK
3 sqn with *Mirage* 2000D
1 (composite) sqn with *Mirage* 2000-5/D (Djibouti)
2 sqn with *Rafale* B/C
1 sqn with *Rafale* B/C (UAE)
ELECTRONIC WARFARE
1 flt with C-160G *Gabriel* (ESM); Beech 350ER *King Air*
AIRBORNE EARLY WARNING & CONTROL
1 (Surveillance & Control) sqn with E-3F *Sentry*
SEARCH & RESCUE/TRANSPORT
4 sqn with C-160R *Transall*; CN235M; SA330 *Puma*; AS555 *Fennec* (Djibouti, French Guiana, Gabon, Indian Ocean & New Caledonia)
TANKER
1 sqn with A330 MRTT
1 sqn with C-135FR; KC-135 *Stratotanker*
TANKER/TRANSPORT
1 sqn with C-160R *Transall*
TRANSPORT
1 sqn with A310-300; A330; A340-200 (leased)
1 sqn with A400M; KC-130J *Hercules*
1 sqn with C-130H/H-30 *Hercules*; C-160R *Transall*
1 sqn with C-130H/H-30/J-30 *Hercules*
2 sqn with CN235M

1 sqn with *Falcon* 7X (VIP); *Falcon* 900 (VIP); *Falcon* 2000
3 flt with TBM-700A
1 (mixed) gp with C-160 *Transall*; DHC-6-300 *Twin Otter*

TRAINING
1 OCU sqn with *Mirage* 2000D
1 OCU sqn with *Rafale* B/C
1 OCU sqn with SA330 *Puma*; AS555 *Fennec*
1 OCU unit with C-160 *Transall*
1 (aggressor) sqn with *Alpha Jet**
4 sqn with *Alpha Jet**
3 sqn with Grob G120A-F; TB-30 *Epsilon*
1 sqn with EMB-121

TRANSPORT HELICOPTER
2 sqn with AS555 *Fennec*
2 sqn with AS332C/L *Super Puma*; SA330 *Puma*; H225M

ISR UAV
1 sqn with MQ-9A *Reaper*

AIR DEFENCE
3 sqn with *Crotale* NG; SAMP/T
1 sqn with SAMP/T

EQUIPMENT BY TYPE

SATELLITES *see* Space
AIRCRAFT 279 combat capable
FTR 41: 35 *Mirage* 2000-5/2000C; 6 *Mirage* 2000B
FGA 166: 66 *Mirage* 2000D; 52 *Rafale* B; 48 *Rafale* C
ELINT 2 C-160G *Gabriel* (ESM)
AEW&C 4 E-3F *Sentry*
TKR 3 KC-135 *Stratotanker*
TKR/TPT 15: 3 A330 MRTT; 10 C-135FR; 2 KC-130J *Hercules*;
TPT 127: **Heavy** 17 A400M; **Medium** 27: 5 C-130H *Hercules*; 9 C-130H-30 *Hercules*; 2 C-130J-30 *Hercules*; 11 C-160R *Transall*; **Light** 72: 1 Beech 305ER *King Air*; 1 Beech 350i *King Air*; 19 CN235M-100; 8 CN235M-300; 5 DHC-6-300 *Twin Otter*; 23 EMB-121 *Xingu*; 15 TBM-700; **PAX** 11: 2 A310-300; 1 A330; 2 A340-200 (leased); 2 *Falcon* 7X; 2 *Falcon* 900 (VIP); 2 *Falcon* 2000
TRG 152: 72 *Alpha Jet**; 18 Grob G120A-F (leased); 25 TB-30 *Epsilon* (incl many in storage); 17 PC-21; 13 SR20 (leased); 7 SR22 (leased)

HELICOPTERS
MRH 37 AS555 *Fennec*
TPT 35: **Heavy** 10 H225M *Caracal*; **Medium** 25: 1 AS332C *Super Puma*; 4 AS332L *Super Puma*; 20 SA330B *Puma*

UNINHABITED AERIAL VEHICLES
CISR • Heavy 8 MQ-9A *Reaper*

AIR DEFENCE • SAM 64: **Long-range** 40 SAMP/T; **Short-range** 24 *Crotale* NG

AIR-LAUNCHED MISSILES
AAM • IR R-550 *Magic* 2; **IIR** *Mica* IR; **ARH** *Meteor*; *Mica* RF
ASM AASM; AS-30L; *Apache*
LACM
Nuclear ASMPA
Conventional SCALP EG

BOMBS • Laser-guided: GBU-12 *Paveway* II

Security and Intervention Brigade

FORCES BY ROLE
SPECIAL FORCES
3 SF gp
MANOEUVRE
Other
24 protection units
30 (fire fighting and rescue) unit

Reserves 5,700 reservists

Paramilitary 100,500

Gendarmerie 100,500; 30,300 reservists

EQUIPMENT BY TYPE
ARMOURED FIGHTING VEHICLES
APC • APC (W) 102: 82 VXB-170 (VBRG-170); 20 VAB
ARTILLERY • MOR 81mm some
PATROL AND COASTAL COMBATANTS 38
PB 38: 2 *Athos*; 4 *Géranium*; 24 VCSM; 8 VSMP
HELICOPTERS • TPT • Light 60: 25 AS350BA *Ecureuil*; 20 H135; 15 H145

DEPLOYMENT

ARABIAN SEA: Combined Maritime Forces • CTF-150: 1 FFH

BOSNIA-HERZEGOVINA: EU • EUFOR • *Operation Althea* 5

BURKINA FASO: *Operation Barkhane* 400; 1 SF gp; 1 C-130H; 1 DHC-6-300; 2 *Tiger*; 3 AS532UL *Cougar*; 2 H225M; 2 SA342 *Gazelle*

CENTRAL AFRICAN REPUBLIC: 160; **EU** • EUTM RCA 40 **UN** • MINUSCA 9

CHAD: *Operation Barkhane* 1,500; 1 mech inf BG; 1 FGA det with 4 *Mirage* 2000D; 1 tpt det with 1 C-130H; 2 CN235M; 1 UAV det with 1 MQ-9A *Reaper*

CÔTE D'IVOIRE: 950; 1 (Marine) inf bn; 1 AS555 *Fennec*; 2 SA330 *Puma*; 2 SA342 *Gazelle*

DEMOCRATIC REPUBLIC OF THE CONGO: UN • MONUSCO 4

DJIBOUTI: 1,450; 1 (Marine) combined arms regt with (2 recce sqn, 2 inf coy, 1 arty bty, 1 engr coy); 1 hel det with 2 SA330 *Puma*; 2 SA342 *Gazelle*; 1 LCM; 1 FGA sqn with 4 *Mirage* 2000-5; 1 SAR/tpt sqn with 1 CN235M; 2 SA330 *Puma*

EGYPT: MFO 1

FRENCH GUIANA: 2,100: 1 (Foreign Legion) inf regt; 1 (Marine) inf regt; 1 SMA regt; 2 PCO; 1 tpt sqn with 3 CN235M; 5 SA330 *Puma*; 4 AS555 *Fennec*; 3 gendarmerie coy; 1 AS350BA *Ecureuil*; 1 H145

FRENCH POLYNESIA: 900: 1 SMA coy; 1 naval HQ at Papeete; 1 FFGHM; 1 PSO; 1 PCO; 1 AFS; 3 *Falcon* 200 *Gardian*; 1 SAR/tpt sqn with 2 CN235M

FRENCH WEST INDIES: 1,000; 1 (Marine) inf regt; 2 SMA regt; 2 FFGHM; 1 AS565SA *Panther*; 1 SA319 *Alouette* III; 1

naval base at Fort de France (Martinique); 4 gendarmerie coy; 1 PCO; 1 PB; 2 AS350BA *Ecureuil*

GABON: 350; 1 inf bn

GERMANY: 2,000 (incl elm Eurocorps and FRA/GER bde); 1 (FRA/GER) mech bde (1 armd cav regt, 1 mech inf regt)

GULF OF GUINEA: *Operation Corymbe* 1 LHD

JORDAN: *Operation Inherent Resolve* (*Chammal*) 8 *Rafale* F3; 1 *Atlantique* 2

LA REUNION/MAYOTTE: 1,700; 1 (Marine) para regt; 1 (Foreign Legion) inf coy; 1 SMA regt; 1 SMA coy; 2 FFGHM; 1 PCO; 1 LCM; 1 naval HQ at Port-des-Galets (La Réunion); 1 naval base at Dzaoudzi (Mayotte); 1 Falcon 50M; 1 SAR/tpt sqn with 2 CN235M; 5 gendarmerie coy; 1 SA319 *Alouette* III

LEBANON: UN • UNIFIL 653; 1 mech inf bn(-); 1 maint coy; VBL; VBCI; VAB; *Mistral*

LITHUANIA: NATO • Enhanced Forward Presence: 275; 1 armd inf coy(+) with *Leclerc*; VBCI; VBL; VAB

MALI: *Operation Barkhane* 2,500; 1 mech inf BG; 1 log bn; 1 tpt unit with 1 CN235M; 1 PC-6B; 1 hel unit with 5 *Tiger*; 2 AS532 UL *Cougar*; 5 NH90 TTH; 4 SA342 *Gazelle*; **EU** • EUTM Mali 13; **UN** • MINUSMA 24

MEDITERRANEAN SEA: EU • EUNAVFOR MED • *Operation Irini* 1 FFGHM; **NATO** • SNMCMG 2: 1 MHO

NEW CALEDONIA: 1,450; 1 (Marine) mech inf regt; 1 SMA coy; 6 ERC-90F1 *Lynx*; 1 FFGHM; 1 PSO; 1 PCO; 1 base with 2 *Falcon* 200 *Gardian* at Nouméa; 1 tpt unit with 2 CN235 MPA; 2 SA330 *Puma;* 4 gendarmerie coy; 2 AS350BA *Ecureuil*

NIGER: *Operation Barkhane* 600; 1 FGA det with 4 *Mirage* 2000D; 1 tkr/tpt det with 1 C-135FR; 1 C-130J-30 *Hercules;* 1 UAV det with 3 MQ-9A *Reaper;* 1 ISR det with 1 *Atlantique*-2

PERSIAN GULF: EMASOH; 1 DDGHM

QATAR: *Operation Inherent Resolve* (*Chammal*) 1 E-3F *Sentry*

SAUDI ARABIA: 50 (radar det)

SENEGAL: 350; 1 *Falcon* 50MI

UNITED ARAB EMIRATES: 650: 1 armd BG (1 tk coy, 1 arty bty); *Leclerc*; CAESAR; • *Operation Inherent Resolve* (*Chammal*); 1 FGA sqn with 6 *Rafale* F3 • EMASOH; 1 *Atlantique*-2

WESTERN SAHARA: UN • MINURSO 3

FOREIGN FORCES

Belgium 28 *Alpha Jet* trg ac located at Cazaux/Tours
Germany 400 (GER elm Eurocorps)
Singapore 200; 1 trg sqn with 12 M-346 *Master*

Germany GER

Euro €		2019	2020	2021
GDP	€	3.45tr	3.31tr	
	US$	3.86tr	3.78tr	
per capita	US$	46,473	45,466	
Growth	%	0.6	-6.0	
Inflation	%	1.3	0.5	
Def exp [a]	€	46.9bn	51.5bn	
	US$	52.5bn	58.9bn	
Def bdgt [b]	€	43.2bn	44.9bn	46.8bn
	US$	48.4bn	51.3bn	
US$1=€		0.89	0.87	

[a] NATO definition
[b] Includes military pensions

Real-terms defence budget trend (US$bn, constant 2015)

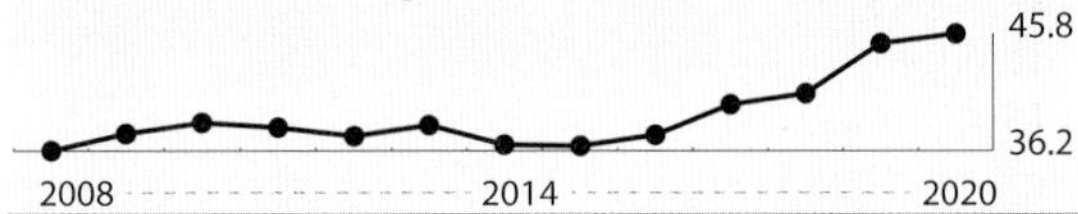

Population 80,159,662

Age	0–14	15–19	20–24	25–29	30–64	65 plus
Male	6.6%	2.4%	2.7%	2.9%	24.4%	10.2%
Female	6.3%	2.2%	2.6%	2.8%	24.2%	12.8%

Capabilities

The 2016 defence white paper committed Germany to a leadership role in European defence. It also emphasised the importance of NATO and the need for the armed forces to contribute to collective-defence tasks. The 2018 Konzeption der Bundeswehr underlines that collective- and territorial-defence tasks will drive military-modernisation efforts and are of equal standing with international crisis-management operations. The key implication for defence modernisation is that Germany will need to invest in readiness and return to fully equipping operational units, after having experimented in recent years with rotating equipment among units depending on their deployment or training demands. Germany is aligning its defence-planning process with capability goals derived from multinational guidance. Berlin has been a key sponsor of the NATO Framework Nations Concept and in the EU led the drive to implement Permanent Structured Cooperation on defence. Close military cooperation has been established, including the affiliation of units, with the Czech Republic, France, the Netherlands and Romania. The defence ministry has announced the objective of increasing authorised active force numbers. In July 2020, the government launched a new voluntary conscript initiative focused on homeland-security tasks to start in 2021 with 1,000 posts. Volunteers will serve for seven months plus five months as reservists over the course of six years. This is in addition to the existing voluntary conscript model that envisages between seven and 23 months of military service, and which has seen declining numbers of applications in recent years. The armed forces are struggling to improve readiness levels in light of increasing demands on NATO's eastern flank. In 2019, Germany was the lead nation for NATO's Very High Readiness Joint Task Force land component and is scheduled to provide this role again in 2023, with the earmarked units prioritised for modernisation and upgrades. Shortages of spare parts and maintenance problems are reported in all three services. Germany's defence-industrial base is able to design and manufacture equipment to meet requirements across all military domains, with strengths in land and naval systems. The government is pursuing a policy of closer defence-industrial cooperation in Europe.

ACTIVE 183,500 (Army 63,400 Navy 16,600 Air 27,600 Joint Support Service 27,800 Joint Medical Service 19,800 Cyber 13,300 Other 15,000)

Conscript liability Voluntary conscription only. Voluntary conscripts can serve up to 23 months

RESERVE 30,050 (Army 6,850 Navy 1,200 Air 3,400 Joint Support Service 11,950 Joint Medical Service 3,650 Cyber 1,150 Other 1,850)

ORGANISATIONS BY SERVICE

Space

EQUIPMENT BY TYPE

SATELLITES 7
- **COMMUNICATIONS** 2 COMSATBw (1 & 2)
- **ISR** 5 SAR-*Lupe*

Army 62,150

FORCES BY ROLE

COMMAND

elm 2 (1 GNC & MNC NE) corps HQ

MANOEUVRE

Armoured

1 (1st) armd div (1 (9th) armd bde (1 armd recce bn, 1 tk bn, 2 armd inf bn, 1 mech inf bn, 1 cbt engr bn, 1 spt bn); 1 (21st) armd bde (1 armd recce bn, 1 tk bn, 1 armd inf bn, 1 mech inf bn, 1 cbt engr bn, 1 spt bn); 1 (41st) mech inf bde (1 armd recce bn, 2 armd inf bn, 1 mech inf bn, 1 cbt engr bn, 1 sigs coy, 1 spt bn); 1 tk bn (for NLD 43rd Bde); 1 SP arty bn; 1 sigs coy)

1 (10th) armd div (1 (12th) armd bde (1 armd recce bn, 1 tk bn, 2 armd inf bn, 1 cbt engr bn, 1 sigs coy, 1 spt bn); 1 (37th) mech inf bde (1 armd recce bn, 2 tk bn, 2 armd inf bn, 1 engr bn, 1 sigs coy, 1 spt bn); 1 (23rd) mtn inf bde (1 recce bn, 3 mtn inf bn, 1 cbt engr bn, 1 spt bn); 1 SP arty bn; 1 SP arty trg bn; 2 mech inf bn (GER/FRA bde); 1 arty bn (GER/FRA bde); 1 cbt engr coy (GER/FRA bde); 1 spt bn (GER/FRA bde))

Air Manoeuvre

1 (rapid reaction) AB div (1 SOF bde (2 SOF bn); 1 AB bde (2 recce coy, 2 para regt, 2 cbt engr coy); 1 atk hel regt; 2 tpt hel regt; 1 sigs coy)

EQUIPMENT BY TYPE

ARMOURED FIGHTING VEHICLES
- **MBT** 245: 225 *Leopard* 2A5/A6; 19 *Leopard* 2A7; 1 *Leopard* 2A7V (78 *Leopard* 2A4 in store)
- **RECCE** 169 *Fennek* (incl 14 engr recce, 14 fires spt)
- **IFV** 710: 305 *Marder* 1A3/A4; 71 *Marder* 1A5; 334 *Puma*
- **APC** 728
  - **APC (T)** 112: 75 Bv-206S; 37 M113 (inc variants)
  - **APC (W)** 616: 257 *Boxer* (inc variants); 359 TPz-1 *Fuchs* (inc variants)
- **AUV** 683: 247 *Dingo* 2; 363 *Eagle* IV/V; 73 *Wiesel* 1 Mk20 (with 20mm gun)

ENGINEERING & MAINTENANCE VEHICLES
- **AEV** 57 *Dachs*
- **ARV** 120: 59 BPz-2 1; 61 BPz-3 *Büffel*
- **VLB** 57: 22 *Biber*; 5 *Leopard* 2 with *Leguan*; 30 M3
- **MW** 28: 4 *Fuchs* KAI; 24 *Keiler*

NBC VEHICLES 8 TPz-1 *Fuchs* NBC

ANTI-TANK/ANTI-INFRASTRUCTURE • MSL
- **SP** 102 *Wiesel* with TOW
- **MANPATS** *Milan*; *Spike*-LR (MELLS)

ARTILLERY 262
- **SP 155mm** 121 PzH 2000
- **MRL 227mm** 41 M270 MLRS
- **MOR** 100: **120mm** 70 Tampella; **SP 120mm** 30 M113 with Tampella

HELICOPTERS
- **ATK** 51 *Tiger*
- **TPT** 116: **Medium** 78 NH90; **Light** 38: 20 Bell 205 (UH-1D *Iroquois*); 13 H135; 5 H145 (SAR)

UNINHABITED AERIAL VEHICLES
- **ISR** 128: **Medium** 43 KZO; **Light** 85 LUNA

Navy 16,600

EQUIPMENT BY TYPE

SUBMARINES 6
- **SSK** 6 Type-212A (fitted with AIP) with 6 single 533mm TT with DM2A4 HWT

PRINCIPAL SURFACE COMBATANTS 10
- **DESTROYERS • DDGHM** 3 *Sachsen* (F124) with 2 quad lnchr with RGM-84C *Harpoon* Block 1B AShM, 4 8-cell Mk 41 VLS with SM-2 Block IIIA SAM/RIM-162B ESSM SAM, 2 21-cell Mk 49 GMLS with RIM-116 RAM SAM, 2 triple 324mm SVTT Mk 32 ASTT with MU90 LWT, 1 76mm gun (capacity 2 *Sea Lynx* Mk88A hel)
- **FRIGATES • FFGHM** 7:
  - 2 *Baden-Württemberg* (F125) with 2 quad lnchr with RGM-84C *Harpoon* Block 1B AShM, 2 21-cell Mk 49 GMLS with RIM-116C RAM Block 2 SAM, 1 127mm gun (capacity 2 NH90 hel)
  - 4 *Brandenburg* (F123) with 2 twin lnchr with MM38 *Exocet* AShM, 2 8-cell Mk 41 VLS with RIM-7P *Sea Sparrow* SAM, 2 Mk 49 GMLS with RIM-116 RAM SAM, 2 twin 324mm SVTT Mk 32 ASTT with Mk 46 LWT, 1 76mm gun (capacity 2 *Sea Lynx* Mk88A hel)
  - 1 *Bremen* (F122) with 2 quad lnchr with RGM-84C *Harpoon* Block 1B AShM, 1 octuple Mk 29 GMLS with RIM-7P *Sea Sparrow* SAM, 2 21-cell Mk 49 GMLS with RIM-116 RAM SAM, 2 twin 324mm SVTT Mk 32 ASTT with Mk 46 LWT, 1 76mm gun (capacity 2 *Sea Lynx* Mk88A hel)

PATROL AND COASTAL COMBATANTS • CORVETTES 5
- **FSGM** 5 *Braunschweig* (K130) with 2 twin lnchr with RBS15 Mk3 AShM, 2 21-cell Mk 49 GMLS with RIM-116 RAM SAM, 1 76mm gun, 1 hel landing platform

MINE WARFARE • MINE COUNTERMEASURES 23
- **MHO** 10: 7 *Frankenthal* (2 used as diving support); 3 *Frankenthal* (mod. MJ332CL)
- **MSO** 1 *Ensdorf*
- **MSD** 12 *Seehund*

AMPHIBIOUS • LANDING CRAFT 1
- **LCU** 1 Type-520

LOGISTICS AND SUPPORT 22
AG 4: 2 *Schwedeneck* (Type-748); 2 *Stollergrund* (Type-745)
AGI 3 *Oste* (Type-423)
AGOR 1 *Planet* (Type-751)
AOR 6 *Elbe* (Type-404) with 1 hel landing platform (2 specified for PFM support; 1 specified for SSK support; 3 specified for MHO/MSO support)
AORH 3 *Berlin* (Type-702) (fitted for but not with RIM-116 RAM SAM) (capacity 2 *Sea King* Mk41 hel)
AOT 2 *Rhön* (Type-704)
APB 2: 1 *Knurrhahn*; 1 *Ohre*
AXS 1 *Gorch Fock*

Naval Aviation 2,000

EQUIPMENT BY TYPE
AIRCRAFT 8 combat capable
ASW 8 AP-3C *Orion*
TPT • Light 2 Do-228 (pollution control)
HELICOPTERS
ASW 22 *Lynx* Mk88A
SAR 23: 19 *Sea King* Mk41; 4 NH90 NFH (*Sea Lion*)

Naval Special Forces Command

FORCES BY ROLE
SPECIAL FORCES
1 SF coy

Sea Battalion

FORCES BY ROLE
MANOEUVRE
Amphibious
1 mne bn

Air Force 27,750

FORCES BY ROLE
FIGHTER
3 wg (2 sqn with Eurofighter *Typhoon*)
FIGHTER/GROUND ATTACK
1 wg (2 sqn with *Tornado* IDS)
1 wg (2 sqn with Eurofighter *Typhoon* (multi-role))
ISR
1 wg (1 ISR sqn with *Tornado* ECR/IDS; 2 UAV sqn with *Heron*)
TANKER/TRANSPORT
1 (special air mission) wg (3 sqn with A310 MRTT; A319; A321; A340; A350; AS532U2 *Cougar* II; *Global* 5000; *Global* 6000)
TRANSPORT
1 wg (total: 1 sqn with C-160D *Transall*)
1 wg (3 sqn (forming) with A400M *Atlas*)
TRAINING
1 sqn located at Holloman AFB (US) with *Tornado* IDS
1 unit (ENJJPT) located at Sheppard AFB (US) with T-6A *Texan* II; T-38C *Talon*
1 hel unit located at Fassberg
TRANSPORT HELICOPTER
1 tpt hel wg (3 sqn with CH-53G/GA/GE/GS *Stallion*; 1 sqn with H145M)
AIR DEFENCE
1 wg (3 SAM gp) with MIM-104C/F *Patriot* PAC-2/3
1 AD gp with ASRAD *Ozelot*; C-RAM *Mantis* and trg unit
1 AD trg unit located at Fort Bliss (US) with MIM-104C/F *Patriot* PAC-2/3
3 (tac air ctrl) radar gp

Air Force Regiment

FORCES BY ROLE
MANOEUVRE
Other
1 sy regt

EQUIPMENT BY TYPE
AIRCRAFT 228 combat capable
FTR 140 Eurofighter *Typhoon*
ATK 68 *Tornado* IDS
ATK/EW 20 *Tornado* ECR*
ISR 1 A319CJ (Open Skies)
TKR/TPT 4 A310 MRTT
TPT 64: **Heavy** 34 A400M; **Medium** 17 C-160D *Transall*; **PAX** 13: 1 A321; 2 A340 (VIP); 1 A350 (VIP); 2 A319; 4 *Global* 5000; 3 *Global* 6000
TRG 109: 69 T-6A *Texan* II, 40 T-38C *Talon*
HELICOPTERS
MRH 15 H145M
TPT 73: **Heavy** 70 CH-53G/GA/GS/GE *Stallion*; **Medium** 3 AS532U2 *Cougar* II (VIP)
UNINHABITED AERIAL VEHICLES • ISR • Heavy 8 *Heron* 1
AIR DEFENCE
SAM 50
Long-range 30 MIM-104C/F *Patriot* PAC-2/PAC-3
Point-defence 20 ASRAD *Ozelot* (with FIM-92 *Stinger*)
GUNS 35mm 12 C-RAM *Mantis*
AIR-LAUNCHED MISSILES
AAM • IR AIM-9L/Li *Sidewinder*; **IIR** IRIS-T; **ARH** AIM-120B AMRAAM
LACM *Taurus* KEPD 350
ARM AGM-88B HARM
BOMBS
Laser-guided GBU-24 *Paveway* III; GBU-48 *Enhanced Paveway* II; GBU-54 JDAM

Joint Support Service 27,600

FORCES BY ROLE
COMBAT SUPPORT
3 MP regt
2 NBC bn
COMBAT SERVICE SUPPORT
6 log bn
1 spt regt

EQUIPMENT BY TYPE
ARMOURED FIGHTING VEHICLES
AUV 451: 206 *Dingo* 2; 245 *Eagle* IV/V
ENGINEERING & MAINTENANCE VEHICLES
ARV 35: 23 BPz-2; 12 BPz-3 *Büffel*
NBC VEHICLES 35 TPz-1 *Fuchs* A6/A7/A8 NBC

Joint Medical Services 19,900

FORCES BY ROLE

COMBAT SERVICE SUPPORT

4 med regt

EQUIPMENT BY TYPE

ARMOURED FIGHTING VEHICLES

APC • APC (W) 109: 72 *Boxer* (amb); 37 TPz-1 *Fuchs* (amb)

AUV 42 *Eagle* IV/V (amb)

Cyber & Information Command 13,150

FORCES BY ROLE

COMBAT SUPPORT

4 EW bn

6 sigs bn

DEPLOYMENT

AFGHANISTAN: NATO • *Operation Resolute Support* 1,300; 1 bde HQ; 1 recce bn; 1 hel flt with CH-53; 1 UAV flt with 3 *Heron* 1 UAV

BALTIC SEA: NATO • SNMCMG 1: 1 MHO

DJIBOUTI: EU • *Operation Atalanta* 1 AP-3C *Orion*

ESTONIA: NATO • Baltic Air Policing 6 Eurofighter *Typhoon*

FRANCE: 400 (incl GER elm Eurocorps)

IRAQ: *Operation Inherent Resolve* 150 (trg spt)

JORDAN: *Operation Inherent Resolve* 280; 4 *Tornado* ECR; 1 A400M

LEBANON: UN • UNIFIL 120; 1 FFGM

LITHUANIA: NATO • Enhanced Forward Presence 527; 1 armd inf bn HQ; 1 armd inf coy(+) with *Leopard* 2A6; *Marder*

MALI: EU • EUTM Mali 50; **UN •** MINUSMA 361; 1 sy coy; 1 int coy; 1 UAV sqn

MEDITERRANEAN SEA: EU • EU NAVFOR MED • *Operation Irini* 1 DDGHM

NIGER: *Operation Barkhane* 2 C-160 *Transall*

POLAND: 95 (GER elm MNC-NE)

SERBIA: NATO • KFOR 70

SOUTH SUDAN: UN • UNMISS 7

SUDAN: UN • UNAMID 1

UNITED STATES: Trg units with 40 T-38 *Talon*; 69 T-6A *Texan* II at Goodyear AFB (AZ)/Sheppard AFB (TX); NAS Pensacola (FL); Fort Rucker (AL); Missile trg at Fort Bliss (TX)

WESTERN SAHARA: UN • MINURSO 3

FOREIGN FORCES

France 2,000; 1 (FRA/GER) mech bde (1 armd cav regt, 1 mech inf regt)

United Kingdom 185

United States

US Africa Command: **Army**; 1 HQ at Stuttgart

US European Command: 37,100; 1 combined service HQ (EUCOM) at Stuttgart-Vaihingen

Army 22,850; 1 HQ (US Army Europe (USAREUR) at Wiesbaden; 1 div HQ (fwd); 1 SF gp; 1 recce bn; 1 mech bde(-); 1 fd arty bn; 1 MRL bde (2 MRL bn)1 (cbt avn) hel bde(-); 1 (cbt avn) hel bde HQ; 1 int bde; 1 MP bde; 1 sigs bde; 1 spt bde; 1 ARNG SAM bde(-); 1 (APS) armd bde eqpt set; M1A2 SEPv2 *Abrams*; M2A2 *Bradley*; *Stryker Dragoon*; M109A6; M119A3; M777A2; AH-64D *Apache*; CH-47F *Chinook*; UH-60L/M *Black Hawk*; HH-60M *Black Hawk*; M1097 *Avenger*

Navy 450

USAF 13,400; 1 HQ (US Airforce Europe (USAFE)) at Ramstein AB; 1 HQ (3rd Air Force) at Ramstein AB; 1 ftr wg at Spangdahlem AB with 1 ftr sqn with 24 F-16CJ *Fighting Falcon*; 1 airlift wg at Ramstein AB with 14 C-130J-30 *Hercules*; 2 Gulfstream V (C-37A); 5 Learjet 35A (C-21A); 1 B-737-700 (C-40B)

USMC 400

Greece GRC

Euro €		2019	2020	2021
GDP	€	187bn	170bn	
	US$	210bn	194bn	
per capita	US$	19,570	18,168	
Growth	%	1.9	-9.5	
Inflation	%	0.5	-0.6	
Def exp [a]	€	4.24bn	4.40bn	
	US$	4.74bn	5.03bn	
Def bdgt [b]	€	4.29bn	4.36bn	4.37bn
	US$	4.81bn	4.98bn	
US$1=€		0.89	0.87	

[a] NATO definition

[b] Includes military pensions

Real-terms defence budget trend (US$bn, constant 2015)

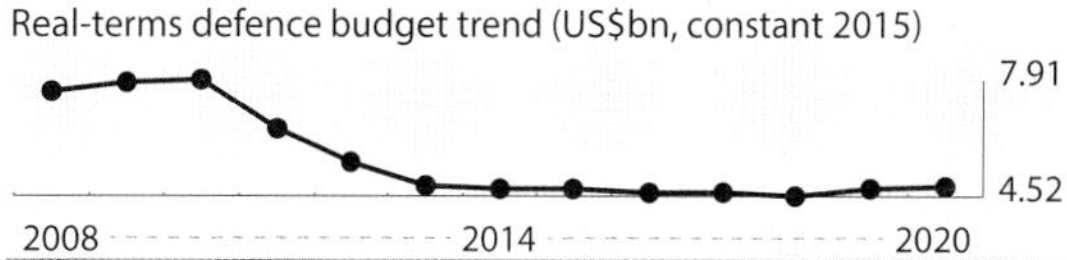

Population 10,607,051

Age	0–14	15–19	20–24	25–29	30–64	65 plus
Male	7.5%	2.7%	2.8%	2.8%	23.1%	10.0%
Female	7.0%	2.4%	2.5%	2.6%	24.3%	12.5%

Capabilities

Greece's 2014 National Military Strategy identifies safeguarding sovereignty and territorial integrity as principal defence objectives. The armed forces would also be expected to support Cyprus in the event of a conflict. The Force Structure 2013–27 document set out plans to make the armed forces more flexible, rapidly deployable and cost-effective. Greece is a NATO member and leads an EU battlegroup. In recent years, defence-cooperation agreements have been signed with Cyprus, Egypt and Israel. In 2018, talks began on an enhanced US presence in the country and elements of US Combat Aviation Brigades have deployed to Greece for training. The Mutual Defense Cooperation Agreement is the cornerstone of US–Greece defence cooperation and provides for a naval-support facility and an airfield at Souda Bay in Crete. The armed forces still include a number of conscripts but most personnel are regulars and Athens is looking to move to a fully professional force. However, financial difficulties and widespread misuse of the deferment process have slowed plans. Training levels are reportedly good, with a focus by the armed forces on joint operational training. Greece's deployments involve limited numbers of personnel and focus on the near abroad, although the country contributes to EU, NATO and UN missions. Greece is renewing its defence relationship with France and also with Egypt and Israel. An agreement to buy *Rafale* combat aircraft from France was announced, and Athens is looking to either replace or modernise its principal surface combatants. Greece has an extensive defence industry focused on the domestic market, capable of manufacturing and developing naval vessels, subsystems, ammunition and small arms.

ACTIVE 142,700 **(Army 93,500 Navy 16,000 Air 21,600 Joint 11,600) Paramilitary 4,000**

Conscript liability 9 months for all services

RESERVE 221,350 **(Army 181,500 Navy 5,850 Air 34,000)**

ORGANISATIONS BY SERVICE

Army 48,500; 45,000 conscripts (total 93,500)

Units are manned at 3 different levels – Cat A 85% fully ready, Cat B 60% ready in 24 hours, Cat C 20% ready in 48 hours (requiring reserve mobilisation). 3 military regions

FORCES BY ROLE

COMMAND

2 corps HQ (incl NRDC-GR)

1 armd div HQ

3 mech inf div HQ

1 inf div HQ

SPECIAL FORCES

1 SF comd

1 cdo/para bde

MANOEUVRE

Reconnaissance

4 recce bn

Armoured

4 armd bde (2 armd bn, 1 mech inf bn, 1 SP arty bn)

Mechanised

10 mech inf bde (1 armd bn, 2 mech bn, 1 SP arty bn)

Light

2 inf regt

Air Manoeuvre

1 air mob bde

1 air aslt bde

Amphibious

1 mne bde

COMBAT SUPPORT

2 MRL bn

3 AD bn (2 with I-*Hawk*, 1 with *Tor* M1)

3 engr regt

2 engr bn

1 EW regt

10 sigs bn

COMBAT SERVICE SUPPORT

1 log corps HQ

1 log div (3 log bde)

HELICOPTER

1 hel bde (1 hel regt with (2 atk hel bn), 2 tpt hel bn, 4 hel bn)

EQUIPMENT BY TYPE

ARMOURED FIGHTING VEHICLES

MBT 1,228: 170 *Leopard* 2A6HEL; 183 *Leopard* 2A4; 500 *Leopard* 1A4/5; 375 M48A5

RECCE 242 VBL

IFV 169 BMP-1

APC • APC (T) 2,130: 91 *Leonidas* Mk1/2; 1,852 M113A1/A2; 187 M577 (CP)

ENGINEERING & MAINTENANCE VEHICLES

ARV 262: 12 *Büffel*; 43 BPz-2; 94 M88A1; 113 M578

VLB 52: 34 M48/M60 AVLB; 10 *Biber*; 8 *Leopard* 1 with *Leguan*

MW *Giant Viper*

ANTI-TANK/ANTI-INFRASTRUCTURE

MSL

SP 558: 195 HMMWV with 9K135 *Kornet*-E (AT-14 *Spriggan*); 363 M901

MANPATS 9K111 *Fagot* (AT-4 *Spigot*); *Milan*; TOW

RCL 687+: **84mm** *Carl Gustaf*; **90mm** EM-67; **SP 106mm** 687 M40A1

ARTILLERY 3,518

SP 594: **155mm** 442: 418 M109A1B/A2/A3GEA1/A5; 24 PzH 2000; **175mm** 12 M107; **203mm** 140 M110A2

TOWED 459: **105mm** 229: 211 M101; 18 M-56; **155mm** 230 M114

MRL 145: **122mm** 109 RM-70; **227mm** 36 M270 MLRS

MOR 2,320: **81mm** 1,700; **107mm** 620 M30 (incl 231 SP)

SURFACE-TO-SURFACE MISSILE LAUNCHERS

SRBM • Conventional MGM-140A ATACMS (launched from M270 MLRS)

AIRCRAFT • TPT • Light 18: 1 Beech 200 *King Air* (C-12C) 2 Beech 200 *King Air* (C-12R/AP *Huron*); 15 Cessna 185 (U-17A/B)

HELICOPTERS

ATK 28: 19 AH-64A *Apache*; 9 AH-64D *Apache*

MRH 60 OH-58D *Kiowa Warrior*

TPT 139: **Heavy** 25: 19 CH-47D *Chinook*; 6 CH-47SD *Chinook*; **Medium** 14 NH90 TTH; **Light** 100: 86 Bell 205 (UH-1H *Iroquois*); 14 Bell 206 (AB-206) *Jet Ranger*

UNINHABITED AERIAL VEHICLES

ISR • Medium 4 *Sperwer*

AIR DEFENCE

SAM 155+

Medium-range 42 MIM-23B I-*Hawk*

Short-range 21 9K331 *Tor*-M1 (SA-15 *Gauntlet*)

Point-range 92+: 38 9K33 *Osa*-M (SA-8B *Gecko*); 54 ASRAD HMMWV; FIM-92 *Stinger*

GUNS • TOWED 727: **20mm** 204 Rh 202; **23mm** 523 ZU-23-2

National Guard 38,000 reservists

Internal security role

FORCES BY ROLE

MANOEUVRE

Light

1 inf div

Air Manoeuvre
1 para regt
COMBAT SUPPORT
8 arty bn
4 AD bn
HELICOPTER
1 hel bn

Navy 14,100; 1,900 conscript (total 16,000)

EQUIPMENT BY TYPE
SUBMARINES • SSK 11:
3 *Poseidon* (GER Type-209/1200) with 8 single 533mm TT with SUT HWT
1 *Poseidon* (GER Type-209/1200) (fitted with AIP technology) with 8 single 533mm TT with UGM-84C *Harpoon* Block 1B AShM/SUT HWT
3 *Glavkos* (GER Type-209/1100) with 8 single 533mm TT with UGM-84C *Harpoon* Block 1B AShM/SUT HWT
4 *Papanikolis* (GER Type-214) (fitted with AIP) with 8 single 533mm TT with UGM-84C *Harpoon* Block 1B AShM/SUT HWT
PRINCIPAL SURFACE COMBATANTS 13
FRIGATES • FFGHM 13:
4 *Elli* Batch I (ex-NLD *Kortenaer* Batch 2) with 2 quad lnchr with RGM-84C/G *Harpoon* Block 1B/G AShM, 1 octuple Mk 29 GMLS with RIM-7M/P *Sea Sparrow* SAM, 2 twin 324mm SVTT Mk 32 mod 9 ASTT with Mk 46 mod 5 LWT, 1 Mk 15 *Phalanx* CIWS, 1 76mm gun (capacity 2 Bell 212 (AB-212) hel or 1 S-70B *Seahawk* hel)
2 *Elli* Batch II (ex-NLD *Kortenaer* Batch 2) with 2 quad lnchr with RGM-84C/G *Harpoon* Block 1B/G AShM, 1 octuple Mk 29 GMLS with RIM-7M/P *Sea Sparrow* SAM, 2 twin 324mm SVTT Mk 32 mod 9 ASTT with Mk 46 mod 5 LWT, 1 Mk 15 *Phalanx* CIWS, 2 76mm gun (capacity 2 Bell 212 (AB-212) hel or 1 S-70B *Seahawk* hel)
3 *Elli* Batch III (ex-NLD *Kortenaer* Batch 2) with 2 quad lnchr with RGM-84C/G *Harpoon* Block 1B/G AShM, 1 octuple Mk 29 lnchr with RIM-7P *Sea Sparrow* SAM, 2 twin 324mm SVTT Mk 32 mod 9 ASTT with Mk 46 LWT, 1 Mk 15 *Phalanx* CIWS, 1 76mm gun (capacity 2 Bell 212 (AB-212) hel)
4 *Hydra* (GER MEKO 200) with 2 quad lnchr with RGM-84G *Harpoon* Block 1G AShM, 1 16-cell Mk 48 mod 2 VLS with RIM-162C ESSM SAM, 2 triple 324mm SVTT Mk 32 mod 5 ASTT with Mk 46 LWT, 2 Mk 15 *Phalanx* CIWS, 1 127mm gun (capacity 1 S-70B *Seahawk* ASW hel)
PATROL AND COASTAL COMBATANTS 34
PCGM 6 *Roussen* (*Super Vita*) with 2 quad lnchr with MM40 *Exocet* Block 3 AShM (of which 2 still fitted with Block 2), 1 21-cell Mk 49 GMLS with RIM-116 RAM SAM, 1 76mm gun
PCFG 8:
3 *Kavaloudis* (FRA *La Combattante* IIIB) with 2 twin lnchr with RGM-84C *Harpoon* Block 1B AShM, 2 single 533mm TT with SST-4 HWT, 2 76mm gun
1 *Laskos* (FRA *La Combattante* III) with 4 MM38 *Exocet* AShM, 2 single 533mm TT with SST-4 HWT, 2 76mm gun
3 *Laskos* (FRA *La Combattante* III) with 2 twin lnchr with RGM-84C *Harpoon* Block 1B AShM, 2 single 533mm TT with SST-4 HWT, 2 76mm gun
1 *Votsis* (ex-GER *Tiger*) with 2 twin lnchr with RGM-84C *Harpoon* AShM, 1 76mm gun
PCFT 2 *Kavaloudis* (FRA *La Combattante* IIIB) with 2 single 533mm TT with SST-4 HWT, 2 76mm gun
PCF 2 *Votsis* (ex-GER *Tiger*) with 1 76mm gun
PCO 8:
2 *Armatolos* (DNK *Osprey*) with 1 76mm gun
2 *Pirpolitis* with 1 76mm gun
4 *Machitis* with 1 76mm gun
PB 8: 4 *Andromeda* (NOR *Nasty*); 2 *Stamou*; 2 *Tolmi*
MINE WARFARE • MINE COUNTERMEASURES 3
MHO 3: 1 *Evropi* (ex-UK *Hunt*); 2 *Evniki* (ex-US *Osprey*)
AMPHIBIOUS
LANDING SHIPS • LST 5 *Chios* (capacity 4 LCVP; 300 troops) with 1 76mm gun, 1 hel landing platform
LANDING CRAFT 15
LCU 5
LCA 7
LCAC 3 *Kefallinia* (*Zubr*) with 2 AK630 CIWS (capacity either 3 MBT or 10 APC (T); 230 troops)
LOGISTICS AND SUPPORT 25
ABU 2
AG 2 *Pandora*
AGOR 1 *Naftilos*
AGS 2: 1 *Stravon*; 1 *Pytheas*
AOR 2 *Axios* (ex-GER *Luneburg*)
AORH 1 *Prometheus* (ITA *Etna*) with 1 Mk 15 *Phalanx* CIWS
AOT 4 *Ouranos*
AWT 6 *Kerkini*
AXS 5

Coastal Defence

EQUIPMENT BY TYPE
COASTAL DEFENCE • AShM 2 MM40 *Exocet*

Naval Aviation

FORCES BY ROLE
ANTI-SUBMARINE WARFARE
1 div with S-70B *Seahawk*; Bell 212 (AB-212) ASW
EQUIPMENT BY TYPE
AIRCRAFT 1 combat capable
ASW 1 P-3B *Orion* (4 P-3B *Orion* in store undergoing modernisation)
HELICOPTERS
ASW 18: 7 Bell 212 (AB-212) ASW; 11 S-70B *Seahawk*
AIR-LAUNCHED MISSILES
ASM AGM-114 *Hellfire*
AShM AGM-119 *Penguin*

Air Force 18,900; 2,700 conscripts (total 21,600)

Tactical Air Force

FORCES BY ROLE
FIGHTER/GROUND ATTACK
1 sqn with F-4E *Phantom* II
3 sqn with F-16CG/DG Block 30/50 *Fighting Falcon*

3 sqn with F-16CG/DG Block 52+ *Fighting Falcon*
2 sqn with F-16C/D Block 52+ ADV *Fighting Falcon*
1 sqn with *Mirage* 2000-5EG/BG Mk2
1 sqn with *Mirage* 2000EG/BG

AIRBORNE EARLY WARNING
1 sqn with EMB-145H *Erieye*

EQUIPMENT BY TYPE

AIRCRAFT 230 combat capable
FGA 230: 34 F-4E *Phantom* II; 69 F-16CG/DG Block 30/50 *Fighting Falcon*; 55 F-16CG/DG Block 52+; 30 F-16 C/D Block 52+ ADV *Fighting Falcon;* 19 *Mirage* 2000-5EG Mk2; 5 *Mirage* 2000-5BG Mk2; 16 *Mirage* 2000EG; 2 *Mirage* 2000BG
AEW 4 EMB-145AEW (EMB-145H) *Erieye*

AIR-LAUNCHED MISSILES
AAM • IR AIM-9L/P *Sidewinder*; R-550 *Magic* 2; **IIR** IRIS-T; *Mica* IR; **ARH** AIM-120B/C AMRAAM; *Mica* RF
ASM AGM-65A/B/G *Maverick*
LACM SCALP EG
AShM AM39 *Exocet*
ARM AGM-88 HARM

BOMBS
Electro-optical guided: GBU-8B HOBOS
Laser-guided: GBU-10/12/16 *Paveway* II; GBU-24 *Paveway* III; GBU-50 *Enhanced Paveway* II
INS/GPS-guided GBU-31 JDAM; AGM-154C JSOW

Air Defence

FORCES BY ROLE

AIR DEFENCE
6 sqn/bty with MIM-104A/B/D *Patriot/Patriot* PAC-1 SOJC/*Patriot* PAC-2 GEM
2 sqn/bty with S-300PMU1 (SA-20 *Gargoyle*)
12 bty with *Skyguard*/RIM-7 *Sparrow*/guns; *Crotale* NG/GR; *Tor*-M1 (SA-15 *Gauntlet*)

EQUIPMENT BY TYPE

AIR DEFENCE
SAM 81
Long-range 48: 36 MIM-104A/B/D *Patriot/Patriot* PAC-1 SOJC/PAC-2 GEM; 12 S-300PMU1 (SA-20 *Gargoyle*)
Short-range 33: 9 *Crotale* NG/GR; 4 9K331 *Tor*-M1 (SA-15 *Gauntlet*); 20 RIM-7M *Sparrow* with *Skygaurd*
GUNS 59: **20mm** some Rh-202; **30mm** 35+ *Artemis*-30; **35mm** 24 GDF-005 with *Skyguard*

Air Support Command

FORCES BY ROLE

SEARCH & RESCUE/TRANSPORT HELICOPTER
1 sqn with AS332C *Super Puma* (SAR/CSAR)
1 sqn with AW109; Bell 205A (AB-205A) (SAR); Bell 212 (AB-212 - VIP, tpt)

TRANSPORT
1 sqn with C-27J *Spartan*
1 sqn with C-130B/H *Hercules*
1 sqn with EMB-135BJ *Legacy*; ERJ-135LR; Gulfstream V

EQUIPMENT BY TYPE

AIRCRAFT
TPT 26: **Medium** 23: 8 C-27J *Spartan*; 5 C-130B *Hercules*; 10 C-130H *Hercules*; **Light** 2: 1 EMB-135BJ *Legacy*; 1 ERJ-135LR; **PAX** 1 Gulfstream V

HELICOPTERS
TPT 31: **Medium** 12 AS332C *Super Puma*; **Light** 19: 12 Bell 205A (AB-205A) (SAR); 4 Bell 212 (AB-212) (VIP, Tpt); 3 AW109

Air Training Command

FORCES BY ROLE

TRAINING
2 sqn with T-2C/E *Buckeye*
2 sqn with T-6A/B *Texan* II
1 sqn with P2002JF; T-41D

EQUIPMENT BY TYPE

AIRCRAFT • TRG 103: 12 P2002JF; 28 T-2C/E *Buckeye*; 20 T-6A *Texan* II; 25 T-6B *Texan* II; 18 T-41D

Paramilitary

Coast Guard and Customs 4,000

EQUIPMENT BY TYPE

PATROL AND COASTAL COMBATANTS 124
PCO 1 *Gavdos* (Damen 5009)
PCC 3
PBF 54
PB 66

AIRCRAFT • TPT • Light 4: 2 Cessna 172RG *Cutlass*; 2 TB-20 *Trinidad*

HELICOPTERS • SAR 3 AS365N3

DEPLOYMENT

AFGHANISTAN: NATO • *Operation Resolute Support* 11

BOSNIA-HERZEGOVINA: EU • EUFOR • *Operation Althea* 6

CYPRUS: Army 950 (ELDYK army); ε200 (officers/NCOs seconded to Greek-Cypriot National Guard) (total 1,150); 1 mech bde (1 armd bn, 2 mech inf bn, 1 arty bn); 61 M48A5 MOLF MBT; 80 *Leonidas* APC; 12 M114 arty; 6 M110A2 arty

IRAQ: NATO • NATO Mission Iraq 1

LEBANON: UN • UNIFIL 146; 1 FFGHM

MALI: EU • EUTM Mali 2

MEDITERRANEAN SEA: EU • EUNAVFOR MED • *Operation Irini*; 1 FFGHM; **NATO** • SNMG 2: 1 PCFG; 1 PCO

SERBIA: NATO • KFOR 111; 1 inf coy

FOREIGN FORCES

United States US European Command: 400; 1 naval base at Makri; 1 naval base at Souda Bay; 1 air base at Iraklion

Hungary HUN

Hungarian Forint f		2019	2020	2021
GDP	f	46.8tr	45.3tr	
	US$	161bn	150bn	
per capita	US$	16,470	15,373	
Growth	%	4.9	-6.1	
Inflation	%	3.4	3.7	
Def exp [a]	f	596bn	588bn	
	US$	2.05bn	1.95bn	
Def bdgt [b]	f	530bn	616bn	778bn
	US$	1.82bn	2.04bn	
US$1=f		290.68	301.79	

[a] NATO definition

[b] Includes military pensions

Real-terms defence budget trend (US$bn, constant 2015)

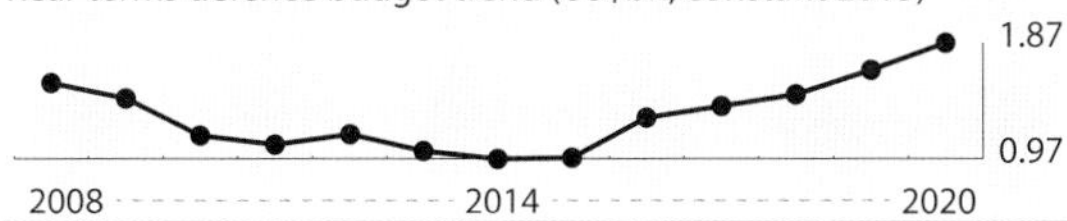

Population 9,771,827

Age	0–14	15–19	20–24	25–29	30–64	65 plus
Male	7.5%	2.6%	2.7%	3.2%	23.7%	7.9%
Female	7.1%	2.5%	2.6%	3.0%	24.4%	12.8%

Capabilities

Hungary published a new National Security Strategy in April 2020 and is reportedly working on a new National Military Strategy. The security strategy speaks of a deteriorating security environment, marked by great-power competition and an increasing military component. It also characterises mass migration as a key concern for Hungary. Hungary is implementing the Zrínyi 2026 national-defence and armed-forces modernisation plan. Hungary coordinates policy with the other member states of the Visegrád Group, including on defence, and hosts the NATO Centre of Excellence for Military Medicine. The armed forces participate in international crisis-management missions, notably in Afghanistan, the Balkans and Iraq, but have very limited organic capacity to deploy forces beyond national borders. Increasing migration pressure has involved the armed forces in internal border-control operations, assisting national police forces. Announced equipment-modernisation priorities focus on individual-soldier equipment and fixed- and rotary-wing aircraft. In 2019, the government announced that it had raised the level of ambition for the maximum number of Hungarian soldiers deployed on international missions from 1,000 to 1,200. While the air-force-related elements of Zrínyi 2026 had been a focus of attention, at the end of 2018 the ministry also initiated land procurements, including for main battle tanks and self-propelled artillery. Hungary's defence-industrial base is limited, though the defence ministry set up an inter-ministerial working group to boost domestic capacity in the small-arms sector. In 2020, a new defence procurement agency, reporting to the national armaments director, began its work and is intended to coordinate defence and security acquisitions.

ACTIVE 27,800 (Army 10,450 Air 5,750 Joint 11,600)
Paramilitary 12,000

RESERVE 20,000

ORGANISATIONS BY SERVICE

Hungary's armed forces have reorganised into a joint force

Land Component 10,450 (incl riverine element)

FORCES BY ROLE
SPECIAL FORCES
1 SF regt
MANOEUVRE
Reconnaissance
1 ISR regt
Mechanised
1 (5th) mech inf bde (3 mech inf bn, 1 cbt engr coy, 1 sigs coy, 1 log bn)
1 (25th) mech inf bde (1 tk bn; 2 mech inf bn, 1 arty bn, 1 AT bn, 1 log bn)
COMBAT SUPPORT
1 engr regt
1 EOD/rvn regt
1 CBRN bn
1 sigs regt
COMBAT SERVICE SUPPORT
1 log regt

EQUIPMENT BY TYPE
ARMOURED FIGHTING VEHICLES
MBT 48: 4 *Leopard* 2A4HU; 44 T-72M1
IFV 120 BTR-80A/AM
APC 272
APC (W) 260 BTR-80
PPV 12 *MaxxPro Plus*
ENGINEERING & MAINTENANCE VEHICLES
AEV 5 BAT-2
ARV 8 VT-55A
VLB 8 BLG-60; MTU; TMM
NBC VEHICLES 14 BTR-80M-NBC
ANTI-TANK/ANTI-INFRASTRUCTURE
MSL • MANPATS 9K111 *Fagot* (AT-4 *Spigot*); 9K111-1 *Konkurs* (AT-5 *Spandrel*)
ARTILLERY 31
TOWED 152mm 31 D-20
MOR 82mm
PATROL AND COASTAL COMBATANTS • **PBR** 2
MINE COUNTERMEASURES • **MSR** 3 *Nestin*

Air Component 5,750

FORCES BY ROLE
FIGHTER/GROUND ATTACK
1 sqn with *Gripen* C/D
TRANSPORT
1 sqn with A319; *Falcon* 7X
TRAINING
1 sqn with Z-143LSi; Z-242L; AS350 *Ecureuil*
ATTACK HELICOPTER
1 sqn with Mi-24V/P *Hind* E/F
TRANSPORT HELICOPTER
1 sqn with Mi-8 *Hip*; Mi-17 *Hip* H; H145M
AIR DEFENCE
1 SAM regt (9 bty with *Mistral*; 3 bty with 2K12 *Kub* (SA-6 *Gainful*))
1 radar regt

EQUIPMENT BY TYPE
AIRCRAFT 14 combat capable
FGA 14: 12 *Gripen* C; 2 *Gripen* D
TPT • PAX 4: 2 A319; 2 *Falcon* 7X
TRG 4: 2 Z-143LSi; 2 Z-242L
HELICOPTERS
ATK 8: 6 Mi-24V *Hind* E; 2 Mi-24P *Hind* F
MRH 12: 7 H145M (incl 2 SAR); 5 Mi-17 *Hip* H
TPT 5: **Medium** 3 Mi-8 *Hip* (10 in store); **Light** 2 AS350 *Ecureuil*
AIR DEFENCE
SAM • Point-defence 16 2K12 *Kub* (SA-6 *Gainful*); *Mistral*
AIR-LAUNCHED MISSILES
AAM • IR AIM-9 *Sidewinder*; **SARH** R-27 (AA-10 *Alamo* A); **ARH** AIM-120C AMRAAM
ASM AGM-65 *Maverick*; 3M11 *Falanga* (AT-2 *Swatter*); 9K114 *Shturm*-V (AT-6 *Spiral*)
BOMBS • Laser-guided *Paveway* II

Paramilitary 12,000

Border Guards 12,000 (to reduce)
Ministry of Interior
FORCES BY ROLE
MANOEUVRE
Other
1 (Budapest) paramilitary district (7 rapid reaction coy)
11 (regt/district) paramilitary regt
EQUIPMENT BY TYPE
ARMOURED FIGHTING VEHICLES
APC • APC (W) 68 BTR-80

DEPLOYMENT

AFGHANISTAN: NATO • *Operation Resolute Support* 89
BOSNIA-HERZEGOVINA: EU • *Operation Althea* 47
CYPRUS: UN • UNFICYP 14
IRAQ: *Operation Inherent Resolve* 170
LEBANON: UN • UNIFIL 4
MALI: EU • EUTM Mali 20
SERBIA: NATO • KFOR 397; 1 inf coy (KTM)
WESTERN SAHARA: UN • MINURSO 5

Iceland ISL

Icelandic Krona Kr		2019	2020	2021
GDP	Kr	2.97tr	2.83tr	
	US$	24.2bn	20.8bn	
per capita	US$	67,857	57,189	
Growth	%	1.9	-7.2	
Inflation	%	3.0	2.7	
Sy Bdgt [a]	Kr	6.75bn	7.07bn	5.53bn
	US$	55.0m	52.0m	
US$1=Kr		122.61	135.97	

[a] Coast Guard budget

Real-terms defence budget trend (US$m, constant 2015)

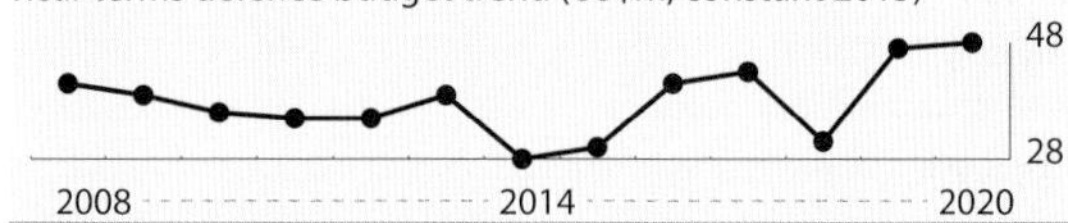

Population 350,734

Age	0–14	15–19	20–24	25–29	30–64	65 plus
Male	10.4%	3.2%	3.3%	3.6%	22.3%	7.3%
Female	9.9%	3.1%	3.3%	3.5%	22.0%	8.2%

Capabilities

Iceland is a NATO member but maintains only a coastguard service. In 2016, the country established a National Security Council to implement and monitor security policy. The coastguard controls the NATO Iceland Air Defence System, as well as a NATO Control and Reporting Centre that feeds into NATO air- and missile-defence and air-operations centres. Increased Russian air and naval activities in the Atlantic and close to NATO airspace have led to complaints from Iceland. Iceland considers its bilateral defence agreement with the US as an important pillar of its security policy and also participates in the security-policy dialogue of NORDEFCO. Iceland hosts NATO and regional partners for exercises, transits and naval task groups, as well as the Icelandic Air Policing mission. Despite there being no standing armed forces, Iceland makes financial contributions and on occasion deploys civilian personnel to NATO missions. In late 2016, following a joint declaration in June that year, the US Navy began operating P-8 *Poseidon* maritime-patrol aircraft from Keflavik air base. In summer 2019, it was reported that upgrades would include accommodation, dangerous-cargo handling facilities and runway extensions to enable tanker operations. The work is due to be completed in 2023.

ACTIVE NIL Paramilitary 250

ORGANISATIONS BY SERVICE

Paramilitary

Iceland Coast Guard 250
EQUIPMENT BY TYPE
PATROL AND COASTAL COMBATANTS 3
PSOH 2 *Aegir*
PSO 1 *Thor*
LOGISTICS AND SUPPORT • AGS 1 *Baldur*
AIRCRAFT • TPT • Light 1 DHC-8-300 (MP)
HELICOPTERS • TPT • Medium 2 H225 (leased)

FOREIGN FORCES

Icelandic Air Policing: Aircraft and personnel from various NATO members on a rotating basis

United States NATO • Icelandic Air Policing; 4 F-15C *Eagle*

Ireland IRL

Euro €		2019	2020	2021
GDP	€	356bn	349bn	
	US$	398bn	399bn	
per capita	US$	80,504	79,669	
Growth	%	5.9	-3.0	
Inflation	%	0.9	-0.2	
Def bdgt [a]	€	994m	1.04bn	1.07bn
	US$	1.11bn	1.19bn	
US$1=€		0.89	0.87	

[a] Includes military pensions and capital expenditure

Real-terms defence budget trend (US$bn, constant 2015)

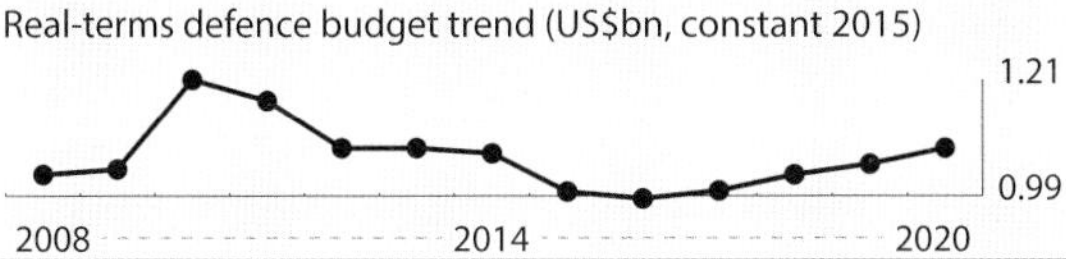

Population 5,176,569

Age	0–14	15–19	20–24	25–29	30–64	65 plus
Male	10.8%	3.1%	3.0%	3.0%	23.6%	6.4%
Female	10.3%	3.1%	2.9%	2.9%	23.4%	7.4%

Capabilities

The armed forces' core mission is defending the state against armed aggression, although a 2015 white paper broadened the scope of the national-security risk assessment beyond traditional military and paramilitary threats. It listed inter- and intra-state conflict, cyber attacks, terrorism, emergencies and natural disasters, among others. The army maintains substantial EOD capabilities. Ireland is active in EU defence cooperation and continues to contribute to multinational operations. Its forces are well trained for their roles. Ireland is also working to establish a specialist reserve with relevant professional qualifications. It has sufficient logistic capability to sustain its UN deployments but has no strategic-airlift capacity. After the 2015 white paper, Dublin identified a large number of defence projects to be completed over a ten-year period. Key priorities include a mid-life upgrade for the army's *Piranha* armoured personnel carriers, EOD robots and UAVs. Other stated priorities include new armoured vehicles, new aircraft, naval-vessel refits, training facilities and upgrades to military facilities. A 3% rise in the 2020 defence budget is intended to boost military staffing by improving pay and allowances, bases and infrastructure. Ireland has a small, specialist defence industry focused on areas including drivetrain technologies for land systems.

ACTIVE 8,750 (Army 7,100 Navy 950 Air 700)

RESERVE 4,050 (Army 3,850 Navy 200)

ORGANISATIONS BY SERVICE

Army 7,100

FORCES BY ROLE

SPECIAL FORCES

1 ranger coy

MANOEUVRE

Reconnaissance

1 armd recce sqn

Mechanised

1 mech inf coy

Light

1 inf bde (1 cav recce sqn, 4 inf bn, 1 arty regt (3 fd arty bty, 1 AD bty), 1 fd engr coy, 1 sigs coy, 1 MP coy, 1 tpt coy)

1 inf bde (1 cav recce sqn, 3 inf bn, 1 arty regt (3 fd arty bty, 1 AD bty), 1 fd engr coy, 1 sigs coy, 1 MP coy, 1 tpt coy)

EQUIPMENT BY TYPE

ARMOURED FIGHTING VEHICLES

RECCE 6 *Piranha* IIIH 30mm

APC 101

APC (W) 74: 56 *Piranha* III; 18 *Piranha* IIIH

PPV 27 RG-32M

ANTI-TANK/ANTI-INFRASTURCTURE

MSL • MANPATS FGM-148 *Javelin*

RCL 84mm *Carl Gustaf*

ARTILLERY 131

TOWED • 105mm 23: 17 L118 Light Gun; 6 L119 Light Gun

MOR 108: **81mm** 84 Brandt; **120mm** 24 Ruag M87

AIR DEFENCE

SAM • Point-defence RBS-70

GUNS • TOWED 40mm 32 L/70 with 8 *Flycatcher*

Reserves 3,850 reservists

FORCES BY ROLE

MANOEUVRE

Reconnaissance

1 (integrated) armd recce sqn

2 (integrated) cav sqn

Mechanised

1 (integrated) mech inf coy

Light

14 (integrated) inf coy

COMBAT SUPPORT

4 (integrated) arty bty

2 engr gp

2 MP coy

3 sigs coy

COMBAT SERVICE SUPPORT

2 med det

2 tpt coy

Naval Service 950

EQUIPMENT BY TYPE

PATROL AND COASTAL COMBATANTS 7

PSOH 1 *Eithne* with 1 57mm gun

PSO 6: 2 *Roisin* (of which 1 in refit) with 1 76mm gun; 4 *Samuel Beckett* with 1 76mm gun

PCO (2 *Orla* (ex-UK *Peacock*) (in reserve since mid-2019 due to crew shortage) with 1 76mm gun)

LOGISTICS AND SUPPORT • **AXS** 2

Air Corps 700

2 ops wg; 2 spt wg; 1 trg wg; 1 comms and info sqn

EQUIPMENT BY TYPE

AIRCRAFT

MP 2 CN235 MPA

TPT • Light 4: 1 Learjet 45 (VIP); 3 PC-12NG

TRG 8 PC-9M

HELICOPTERS:

MRH 6 AW139

TPT • Light 2 H135 (incl trg/medevac)

DEPLOYMENT

BOSNIA-HERZEGOVINA: EU • EUFOR • *Operation Althea* 5

DEMOCRATIC REPUBLIC OF THE CONGO: UN • MONUSCO 3

LEBANON: UN • UNIFIL 347; 1 mech inf bn(-)

MALI: EU • EUTM Mali 20

MIDDLE EAST: UN • UNTSO 12

SERBIA: NATO • KFOR 13

SYRIA/ISRAEL: UN • UNDOF 138; 1 inf coy

WESTERN SAHARA: UN • MINURSO 2

Italy ITA

Euro €		2019	2020	2021
GDP	€	1.79tr	1.62tr	
	US$	2.00tr	1.85tr	
per capita	US$	33,159	30,657	
Growth	%	0.3	-10.6	
Inflation	%	0.6	0.1	
Def exp [a]	€	21.0bn	22.8bn	
	US$	23.6bn	26.1bn	
Def bdgt [b]	€	24.2bn	25.7bn	
	US$	27.0bn	29.3bn	
US$1=€		0.89	0.87	

[a] NATO definition

[b] Includes military pensions

Real-terms defence budget trend (US$bn, constant 2015)

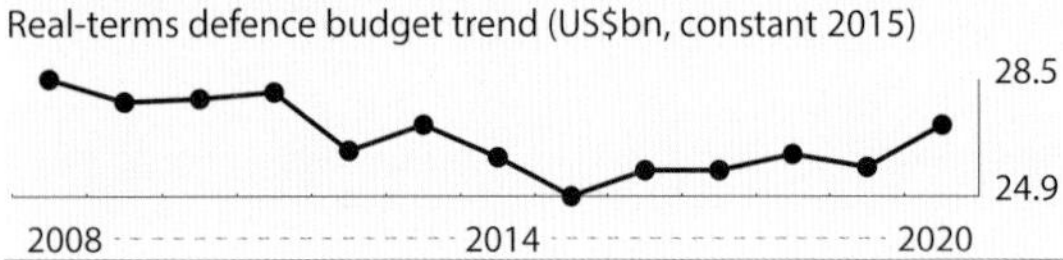

Population 62,402,659

Age	0–14	15–19	20–24	25–29	30–64	65 plus
Male	6.9%	2.4%	2.4%	2.6%	24.3%	9.5%
Female	6.6%	2.4%	2.4%	2.7%	25.2%	12.5%

Capabilities

Italy is concerned by security challenges in the Euro-Atlantic environment, as well as from Europe's southern flank. A defence white paper was issued in 2015. The latest three-year defence plan for 2020–22 outlined modernisation goals. The 2012 law governing personnel structures will remain but a changed strategic environment meant it needed to be updated. For instance, more personnel were needed in the defence industry. Italy has taken part in NATO's air-policing missions in the Baltic states, Iceland and Romania and since early 2017 has deployed to Latvia as part of the Enhanced Forward Presence. The EUNAVFOR-MED force is headquartered in Rome, while the US Navy 6th Fleet is based in Naples. The country takes part in and hosts NATO and other multinational exercises and continues to support NATO, EU and UN operations abroad. However, Italy is planning to gradually reduce its presence overseas to focus on Europe's southern flank. Italy's logistics capability is enabled by a fleet of medium transport aircraft and tankers. The white paper detailed capability-enhancement programmes including upgrades to main battle tanks and procurement of armoured fighting vehicles, counter-UAV systems and electronic-warfare capabilities. The expected retirement of much of the naval fleet has triggered a long-term replacement plan which includes the potential acquisition of two next-generation destroyers to replace the ageing *Luigi Durand de la Penne*-class vessels. F-35As have been ordered for the air force (and F-35Bs for both the air force and naval aviation). Italy has an advanced defence industry, producing equipment across all the domains, with particular strengths in shipbuilding and aircraft and helicopter manufacturing. The country hosts Europe's F-35 final assembly and check-out facility at Cameri. Italy takes part in European defence-industrial cooperation activities, including PESCO projects.

ACTIVE 165,500 **(Army 96,700 Navy 28,850 Air 39,950) Paramilitary 175,750**

RESERVES 18,300 **(Army 13,400 Navy 4,900)**

ORGANISATIONS BY SERVICE

Space

EQUIPMENT BY TYPE

SATELLITES 9

COMMUNICATIONS 4: 1 *Athena-Fidus* (also used by FRA); 3 *Sicral*

ISR 5: 4 *Cosmo* (*Skymed*); 1 OPSAT-3000

Army 96,700

Regt are bn sized

FORCES BY ROLE

COMMAND

1 (NRDC-ITA) corps HQ (1 spt bde, 1 sigs regt, 1 spt regt)

MANOEUVRE

Mechanised

1 (*Vittorio Veneto*) div (1 (*Ariete*) armd bde (1 cav regt, 2 tk regt, 1 armd inf regt, 1 SP arty regt, 1 cbt engr regt, 1 log regt); 1 (*Pozzuolo del Friuli*) cav bde (1 cav regt, 1 amph regt, 1 arty regt, 1 cbt engr regt, 1 log regt); 1 (*Folgore*) AB bde (1 cav regt, 3 para regt, 1 arty regt, 1 cbt engr regt, 1 log regt); 1 (*Friuli*) air mob bde (1 air mob regt, 2 atk hel regt))

1 (*Acqui*) div (1 (*Pinerolo*) mech bde (1 tk regt, 3 armd inf regt, 1 fd arty regt, 1 cbt engr regt, 1 log regt); 1 (*Granatieri*) mech bde (1 cav regt, 1 mech inf regt); 1 (*Garibaldi Bersaglieri*) mech bde (1 cav regt, 1 tk regt, 2 armd inf regt, 1 SP arty regt, 1 cbt engr regt, 1 log regt); 1 (*Aosta*) mech bde (1 cav regt, 1 armd inf regt, 2 mech inf regt, 1 fd arty regt, 1 cbt engr regt, 1 log regt); 1 (*Sassari*) lt mech bde (1 armd inf regt, 2 mech inf regt, 1 cbt engr regt, 1 log regt))

Mountain

1 (*Tridentina*) mtn div (2 mtn bde (1 cav regt, 3 mtn inf regt, 1 arty regt, 1 mtn cbt engr regt, 1 spt bn, 1 log regt))

COMBAT SUPPORT

1 arty comd (1 arty regt, 1 MRL regt, 1 NBC regt)
1 AD comd (3 SAM regt)
1 engr comd (2 engr regt, 1 ptn br regt, 1 CIMIC regt)
1 EW/sigs comd (1 EW/ISR bde (1 EW regt, 1 int regt, 1 STA regt); 1 sigs bde with (7 sigs regt))

COMBAT SERVICE SUPPORT

1 log comd (3 log regt, 4 med unit)

HELICOPTER

1 hel bde (3 hel regt)

EQUIPMENT BY TYPE

ARMOURED FIGHTING VEHICLES

MBT 200 C1 *Ariete*
ASLT 259 B1 *Centauro*
IFV 458: 200 VCC-80 *Dardo;* 258 VBM 8×8 *Freccia* (incl 20 CP and 36 with *Spike*-LR)
APC 640
APC (T) 156: 146 Bv-206S; 10 M113
APC (W) 428: 151 *Puma* 4×4; 277 *Puma* 6×6
PPV 56 VTMM *Orso* (incl 16 amb)
AUV 10 *Cougar*; IVECO LMV
AAV 15: 14 AAVP-7; 1 AAVC-7

ENGINEERING & MAINTENANCE VEHICLES

AEV 40 *Dachs*; M113
ARV 138: 137 BPz-2; 1 AAVR-7
VLB 64 *Biber*
MW 9: 6 *Buffalo*; 3 *Miniflail*

NBC VEHICLES 14: 5 VBR NBC; 9 VBR NBC Plus

ANTI-TANK/ANTI-INFRASTRUCTURE

MSL
SP 70 M901 with TOW-2A
MANPATS *Spike*

ARTILLERY 944

SP 155mm 69 PzH 2000
TOWED 188: **105mm** 25 Oto Melara Mod 56; **155mm** 163 FH-70
MRL 227mm 22 MLRS
MOR 665: **81mm** 283 Expal; **120mm** 325: 183 Brandt; 142 RT-61 (RT-F1) **SP 120mm** 57: 36 M106A1; 21 VBM 8×8 *Freccia*

AIRCRAFT • TPT • Light 6: 3 Do-228 (ACTL-1); 3 P-180 *Avanti*

HELICOPTERS

ATK 36 AW129CBT *Mangusta*
MRH 14 Bell 412 (AB-412) *Twin Huey*
TPT 137: **Heavy** 16 CH-47F *Chinook* (12 CH-47C *Chinook* in store); **Medium** 39 NH90 TTH (UH-90); **Light** 82: 8 AW109; 1 AW169; 32 Bell 205 (AB-205); 28 Bell 206 *Jet Ranger* (AB-206); 13 Bell 212 (AB-212)

AIR DEFENCE

SAM 52+
Long-range 20 SAMP/T
Short-range 32 *Aspide* with *Skyguard*
Point-defence FIM-92 *Stinger*

AIR-LAUNCHED MISSILES

ASM *Spike*-ER

Navy 28,850

EQUIPMENT BY TYPE

SUBMARINES • SSK 8:

4 *Pelosi* (imp *Sauro*, 3rd and 4th series) with 6 single 533mm TT with A184 mod 3 HWT
4 *Salvatore Todaro* (Type-212A) (fitted with AIP) with 6 single 533mm TT with *Black Shark* HWT

PRINCIPAL SURFACE COMBATANTS 18

AIRCRAFT CARRIERS • CVS 2:

1 *Cavour* with 4 8-cell *Sylver* A43 VLS with *Aster* 15 SAM, 2 76mm guns (capacity mixed air group of 20 AV-8B *Harrier* II; AW101 *Merlin*; NH90; Bell 212)
1 *G. Garibaldi* with 2 octuple *Albatros* lnchr with *Aspide* SAM, 2 triple 324mm ASTT with Mk 46 LWT (capacity mixed air group of 18 AV-8B *Harrier* II; AW101 *Merlin*; NH90; Bell 212)

DESTROYERS • DDGHM 4:

2 *Andrea Doria* with 2 quad lnchr with *Otomat* (*Teseo*) Mk2A AShM, 6 8-cell *Sylver* A50 VLS with *Aster* 15/*Aster* 30 SAM, 2 single 324mm B-515 ASTT with MU90 LWT, 3 76mm guns (capacity 1 AW101 *Merlin*/ NH90 hel)
2 *Luigi Durand de la Penne* (ex-*Animoso*) with 2 quad lnchr with *Otomat* (*Teseo*) Mk2A AShM/*Milas* A/S msl, 1 Mk 13 mod 4 GMLS with SM-1MR Block VI SAM, 1 octuple *Albatros* lnchr with *Aspide* SAM, 2 triple 324mm B-515 ASTT with Mk 46 LWT, 1 127mm gun, 3 76mm guns (capacity 1 NH90 or 2 Bell 212 (AB-212) hel)

FRIGATES • FFGHM 12:

4 *Bergamini* (GP) with 2 quad lnchr with *Otomat* (*Teseo*) Mk2A AShM, 2 8-cell *Sylver* A50 VLS with *Aster* 15/*Aster* 30 SAM, 2 triple 324mm B-515 ASTT with MU90 LWT, 1 127mm gun, 1 76mm gun (capacity 2 AW101/NH90 hel)
4 *Bergamini* (ASW) with 2 twin lnchr with *Otomat* (*Teseo*) Mk2A AShM, 2 twin lnchr with MILAS A/S msl, 2 8-cell *Sylver* A50 VLS with *Aster* 15/*Aster* 30 SAM, 2 triple 324mm B-515 ASTT with MU90 LWT, 2 76mm gun (capacity 2 AW101/NH90 hel)
4 *Maestrale* with 4 single lnchr with *Otomat* (*Teseo*) Mk2 AShM, 1 octuple *Albatros* lnchr with *Aspide* SAM, 2 triple 324mm SVTT Mk 32 ASTT with Mk 46 LWT, 1 127mm gun (capacity 1 NH90 or 2 Bell 212 (AB-212) hel)

PATROL AND COASTAL COMBATANTS 16

PSOH 10:
4 *Cassiopea* with 1 76mm gun (capacity 1 Bell 212 (AB-212) hel
4 *Comandante Cigala Fuligosi* with 1 76mm gun (capacity 1 Bell 212 (AB-212)/NH90 hel)
2 *Sirio* (capacity 1 Bell 212 (AB-212) or NH90 hel)
PB 6: 2 *Angelo Cabrini*; 4 *Esploratore*

MINE WARFARE • MINE COUNTERMEASURES 10

MHO 10: 8 *Gaeta*; 2 *Lerici*

AMPHIBIOUS

PRINCIPAL AMPHIBIOUS SHIPS • LHD 3:

2 *San Giorgio* (capacity 3-4 AW101/NH90/Bell 212; 3 LCM; 2 LCVP; 30 trucks; 36 APC (T); 350 troops)

1 *San Giusto* with 1 76mm gun (capacity 2 AW101 *Merlin*/NH90/Bell 212; 3 LCM; 2 LCVP; 30 trucks; 36 APC (T); 350 troops)
LANDING CRAFT 24: 15 **LCVP**; 9 **LCM**
LOGISTICS AND SUPPORT 53
ABU 5 *Ponza*
AFD 9
AGE 3: 1 *Leonardo* (coastal); 1 *Raffaele Rosseti*; 1 *Vincenzo Martellota*
AGI 1 *Elettra*
AGOR 1 *Alliance*
AGS 3: 1 *Ammiraglio Magnaghi* with 1 hel landing platform; 2 *Aretusa* (coastal)
AKSL 6 *Gorgona*
AORH 3: 1 *Etna* with 1 76mm gun (capacity 1 AW101/NH90/Bell 212 hel); 2 *Stromboli* with 1 76mm gun (capacity 1 AW101/NH90 hel)
AOT 4 *Panarea*
ARSH 1 *Anteo* (capacity 1 Bell 212 (AB-212) hel)
ATS 6 *Ciclope*
AWT 3: 1 *Bormida*; 2 *Simeto*
AXS 8: 1 *Amerigo Vespucci*; 5 *Caroly*; 1 *Italia*; 1 *Palinuro*

Naval Aviation 2,200

FORCES BY ROLE
FIGHTER/GROUND ATTACK
1 sqn with AV-8B *Harrier* II; TAV-8B *Harrier* II
ANTI-SUBMARINE WARFARE/TRANSPORT
5 sqn with AW101 ASW *Merlin*; Bell 212 ASW (AB-212AS); Bell 212 (AB-212); NH90 NFH
MARITIME PATROL
1 flt with P-180
AIRBORNE EARLY WANRING & CONTROL
1 flt with AW101 AEW *Merlin*

EQUIPMENT BY TYPE
AIRCRAFT 18 combat capable
FGA 18: 14 AV-8B *Harrier* II; 2 TAV-8B *Harrier* II; 2 F-35B *Lightning* II
MP 3 P-180
HELICOPTERS
ASW 47: 10 AW101 ASW *Merlin*; 9 Bell 212 ASW; 28 NH90 NFH (SH-90)
AEW 4 AW101 AEW *Merlin*
TPT 15: **Medium** 11: 8 AW101 *Merlin*; 3 NH90 MITT (MH-90); **Light** 4 Bell 212 (AB-212)
AIR-LAUNCHED MISSILES
AAM • IR AIM-9L *Sidewinder*; **ARH** AIM-120 AMRAAM
ASM AGM-65 *Maverick*
AShM *Marte* Mk 2/S

Marines 3,000

FORCES BY ROLE
MANOEUVRE
Amphibious
1 mne regt (1 recce coy, 2 mne bn, 1 log bn)
1 (boarding) mne regt (2 mne bn)
1 landing craft gp
Other
1 sy regt (3 sy bn)

EQUIPMENT BY TYPE
ARMOURED FIGHTING VEHICLES
APC (T) 27: 24 VCC-1; 3 VCC-2
AAV 17: 15 AAVP-7; 2 AAVC-7
ENGINEERING & MAINTENANCE VEHICLES
ARV 1 AAVR-7
ANTI-TANK/ANTI-INFRASTRUCTURE
MSL • MANPATS *Spike*
ARTILLERY
MOR 32: **81mm** 18 Brandt; **120mm** 10 Brandt; **SP 120mm** 4 M106
AIR DEFENCE • SAM • Point-defence FIM-92 *Stinger*

Air Force 39,950

FORCES BY ROLE
FIGHTER
4 sqn with Eurofighter *Typhoon*
FIGHTER/GROUND ATTACK
1 (SEAD/EW) sqn with *Tornado* ECR
2 sqn with *Tornado* IDS
1 sqn with F-35A/B *Lightning* II
FIGHTER/GROUND ATTACK/ISR
1 sqn with AMX *Ghibli*
MARITIME PATROL
1 sqn (opcon Navy) with ATR-72MP (P-72A)
TANKER/TRANSPORT
1 sqn with KC-767A
COMBAT SEARCH & RESCUE
1 sqn with AB-212 ICO; AW101 SAR (HH-101A)
SEARCH & RESCUE
1 wg with AW139 (HH-139A); Bell 212 (HH-212)
TRANSPORT
2 (VIP) sqn with A319CJ; AW139 (VH-139A); *Falcon* 50; *Falcon* 900 *Easy*; *Falcon* 900EX
2 sqn with C-130J/C-130J-30/KC-130J *Hercules*
1 sqn with C-27J *Spartan*
1 (calibration) sqn with P-180 *Avanti*/Gulfstream G550 CAEW
TRAINING
1 OCU sqn with Eurofighter *Typhoon*
1 sqn with MB-339PAN (aerobatic team)
1 sqn with MD-500D/E (NH-500D/E)
1 OCU sqn with *Tornado*
1 OCU sqn with AMX-T *Ghibli*
1 sqn with MB-339A
1 sqn with M-346
1 sqn with SF-260EA; 3 P2006T (T-2006A)
1 sqn with AW101 SAR (HH-101A); Bell 212 (HH-212)
ISR UAV
1 sqn with MQ-9A *Reaper*; RQ-1B *Predator*
AIR DEFENCE
2 bty with *Spada*

EQUIPMENT BY TYPE
AIRCRAFT 223 combat capable
FTR 94 Eurofighter *Typhoon*
FGA 52: 31 AMX *Ghibli*; 8 AMX-T *Ghibli*; 12 F-35A *Lightning* II; 1 F-35B *Lightning* II
ATK 34 *Tornado* IDS
ATK/EW 15 *Tornado* ECR*
MP 3 ATR-72MP (P-72A)
SIGINT 1 Beech 350 *King Air*

AEW&C 2 Gulfstream G550 CAEW
TKR/TPT 6: 4 KC-767A; 2 KC-130J *Hercules*
TPT 74: **Medium** 30: 8 C-130J *Hercules;* 10 C-130J-30 *Hercules;* 12 C-27J *Spartan;* **Light** 37: 17 P-180 *Avanti;* 20 S-208 (liaison); **PAX** 7: 2 A319CJ; 2 *Falcon* 50 (VIP); 2 *Falcon* 900 *Easy;* 1 *Falcon* 900EX (VIP)
TRG 111: 20 M-346; 21 MB-339A; 28 MB-339CD*; 16 MB-339PAN (aerobatics); 26 SF-260EA
HELICOPTERS
MRH 54: 13 AW139 (HH-139A/VH-139A); 2 MD-500D (NH-500D); 39 MD-500E (NH-500E)
CSAR 12 AW101 (HH-101A)
TPT • Light 29 Bell 212 (HH-212)/AB-212 ICO
UNINHABITED AERIAL VEHICLES • ISR • Heavy 12: 6 MQ-9A *Reaper;* 6 RQ-1B *Predator*
AIR DEFENCE • SAM • Short-range SPADA
AIR-LAUNCHED MISSILES
AAM • IR AIM-9L *Sidewinder;* **IIR** IRIS-T; **ARH** AIM-120B AMRAAM
ARM AGM-88 HARM
LACM SCALP EG/*Storm Shadow*
BOMBS
Laser-guided/GPS: Enhanced *Paveway* II; Enhanced *Paveway* III

Joint Special Forces Command (COFS)

Army

FORCES BY ROLE
SPECIAL FORCES
1 SF regt (9th *Assalto paracadutisti*)
1 STA regt
1 ranger regt (4th *Alpini paracadutisti*)
COMBAT SUPPORT
1 psyops regt
TRANSPORT HELICOPTER
1 spec ops hel regt

Navy (COMSUBIN)

FORCES BY ROLE
SPECIAL FORCES
1 SF gp (GOI)
1 diving gp (GOS)

Air Force

FORCES BY ROLE
SPECIAL FORCES
1 wg (sqn) (17th *Stormo Incursori*)

Paramilitary

Carabinieri

FORCES BY ROLE
SPECIAL FORCES
1 spec ops gp (GIS)

Paramilitary 175,750

Carabinieri 107,650

The Carabinieri are organisationally under the MoD. They are a separate service in the Italian Armed Forces as well as a police force with judicial competence

Mobile and Specialised Branch

FORCES BY ROLE
MANOEUVRE
Other
1 (mobile) paramilitary div (1 bde (1st) with (1 horsed cav regt, 11 mobile bn); 1 bde (2nd) with (1 (1st) AB regt, 2 (7th & 13th) mobile regt))
HELICOPTER
1 hel gp
EQUIPMENT BY TYPE
ARMOURED FIGHTING VEHICLES
APC • APC (T) 3 VCC-2
PATROL AND COASTAL COMBATANTS • PB 69
AIRCRAFT • TPT • Light: 1 P-180 *Avanti*
HELICOPTERS
MRH 24 Bell 412 (AB-412)
TPT • Light 19 AW109

Customs 68,100

(Servizio Navale Guardia Di Finanza)
EQUIPMENT BY TYPE
PATROL AND COASTAL COMBATANTS 177
PCF 1 *Antonio Zara*
PBF 144: 19 *Bigliani;* 22 *Corrubia;* 9 *Mazzei;* 62 V-2000; 32 V-5000/V-6000
PB 32: 24 *Buratti;* 8 *Meatini*
LOGISTICS AND SUPPORT • AX 1 *Giorgio Cini*
AIRCRAFT
MP 6: 4 ATR-42-500MP; 2 ATR-72-600 (P-72B)
TPT • Light 2 P-180 *Avanti*
HELICOPTERS
TPT • Light 30: 14 AW139; 1 AW169M; 15 Bell 412HP *Twin Huey*

DEPLOYMENT

AFGHANISTAN: NATO • *Operation Resolute Support* 800; 1 mech inf bde HQ; 1 mech inf regt(-); 1 hel regt(-); AW129 *Mangusta;* NH90; RQ-7

BOSNIA-HERZEGOVINA: EU • EUFOR • *Operation Althea* 5

CENTRAL AFRICAN REPUBLIC: EU • EUTM RCA 3

DJIBOUTI: 117

EGYPT: MFO 75; 3 PB

GULF OF ADEN & INDIAN OCEAN: EU • *Operation Atalanta* 1 FFGHM

INDIA/PAKISTAN: UN • UNMOGIP 2

IRAQ: *Operation Inherent Resolve* (*Prima Parthica*) 600; 1 inf regt; 1 trg unit; 1 hel sqn with 4 NH90; **NATO** • NATO Mission Iraq 46

KUWAIT: *Operation Inherent Resolve* (*Prima Parthica*) 250; 4 *Tornado* ECR; 2 MQ-9A *Reaper;* 1 KC-767A

LATVIA: NATO • Enhanced Forward Presence (*Baltic Guardian*) 200; 1 armd inf coy with C1 *Ariete;* VCC-80 *Dardo*

LEBANON: MIBIL 141; **UN** • UNIFIL 1,076; 1 mech bde HQ; 1 mech inf bn; 1 MP coy; 1 sigs coy; 1 hel bn

LIBYA: MIASIT 400; 1 inf coy; 1 CRBN unit; 1 trg unit

LITHUANIA: NATO • Baltic Air Policing: 4 Eurofighter *Typhoon*

MALI: EU • EUTM Mali 12; **UN** • MINUSMA 2

MEDITERRANEAN SEA: EU • EUNAVFOR MED 1 FFGHM; **NATO** • SNMCMG 2: 1 MHO

NIGER: MISIN 295; 1 inf coy; 1 engr unit; 1 CRBN unit; 1 med coy; 1 trg unit; 1 ISR unit

SERBIA: NATO • KFOR 628; 1 mtn inf BG HQ; 1 Carabinieri unit

SOMALIA: EU • EUTM Somalia 148

UNITED ARAB EMIRATES: 106; 1 tpt flt with 2 C-130J *Hercules*

WESTERN SAHARA: UN • MINURSO 2

FOREIGN FORCES

United States US European Command: 12,550
- Army 3,900; 1 AB bde(-)
- Navy 4,000; 1 HQ (US Navy Europe (USNAVEUR)) at Naples; 1 HQ (6th Fleet) at Gaeta; 1 ASW Sqn with 4 P-8A *Poseidon* at Sigonella
- USAF 4,550; 1 ftr wg with 2 ftr sqn with 21 F-16C/D *Fighting Falcon* at Aviano; 1 CSAR sqn with 8 HH-60G *Pave Hawk*; 1 CISR UAV sqn with MQ-9A *Reaper* at Sigonella
- USMC 200

Latvia LVA

Euro €		2019	2020	2021
GDP	€	30.5bn	28.9bn	
	US$	34.1bn	33.0bn	
per capita	US$	17,772	17,230	
Growth	%	2.2	-6.0	
Inflation	%	2.7	0.6	
Def exp [a]	€	618m	664m	
	US$	692m	759m	
Def bdgt [b]	€	634m	663m	709m
	US$	710m	758m	
FMA (US)	US$	8m	5m	5m
US$1=€		0.89	0.87	

[a] NATO definition

[b] Includes military pensions

Real-terms defence budget trend (US$m, constant 2015)

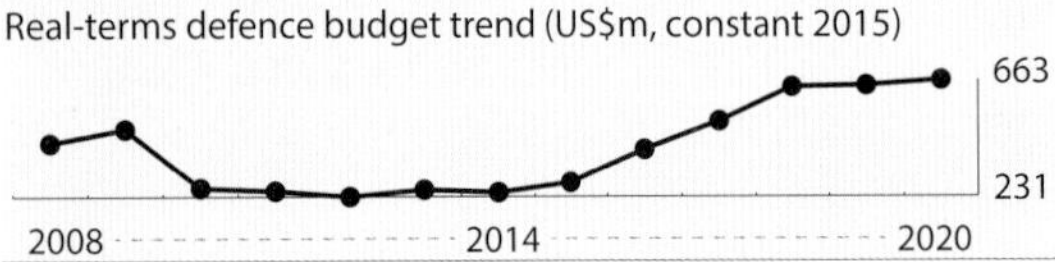

Population 1,881,232

Age	0–14	15–19	20–24	25–29	30–64	65 plus
Male	7.9%	2.4%	2.3%	3.1%	23.8%	6.8%
Female	7.4%	2.3%	2.1%	2.9%	25.4%	13.7%

Capabilities

Latvia has small armed forces focused on maintaining national sovereignty and territorial integrity but the country depends on NATO membership as a security guarantor. Russia is Latvia's overriding security concern. A National Armed Forces Development plan 2016–28 illustrated a capabilities-based planning process. Principal tasks were to improve early warning, detection and situational awareness, to increase combat readiness and to improve the ability to counter hybrid threats. An update to the State Defence Concept was announced in June 2020 to reflect shifts in Latvia's strategic environment. It identified challenges including new technologies, and low military spending in Europe and the consequent effect on capabilities and crisis response. It emphasised the importance of societal resilience and comprehensive defence as well as the significance of a NATO presence in the region. Latvia has no requirement and therefore no capacity to independently deploy and sustain forces beyond its national boundaries, although the armed forces have taken part in a range of NATO and EU missions. Capability-development projects include engineering, special operations, mechanised infantry, air defence, air surveillance and the National Guard. Acquisition requirements include self-propelled howitzers, CVR(T) and ATGW systems. Latvia has only a niche defence-industrial capability, with cyber security a focus.

ACTIVE 6,250 (Army 1,700 Navy 550 Air 500 Joint Staff 1,900 National Guard 1,100 Other 500)

RESERVE 11,000 (National Guard 8,000 Other 3,000)

ORGANISATIONS BY SERVICE

Joint 1,900

FORCES BY ROLE

SPECIAL FORCES
- 1 SF unit

COMBAT SUPPORT
- 1 MP bn

Army 1,700

FORCES BY ROLE

MANOEUVRE

Mechanised
- 1 mech inf bde (2 mech inf bn, 1 SP arty bn, 1 cbt spt bn HQ, 1 CSS bn HQ)

National Guard 1,100; 8,000 part-time (9,100 total)

FORCES BY ROLE

SPECIAL FORCES
- 1 SF unit

MANOEUVRE

Light
- 1 (2nd) inf bde (4 inf bn; 1 engr bn)
- 3 (1st, 3rd & 4th) inf bde (3 inf bn; 1 sy bn; 1 spt bn)

COMBAT SUPPORT
- 1 cyber unit
- 1 NBC coy
- 1 psyops pl

EQUIPMENT BY TYPE

ARMOURED FIGHTING VEHICLES
- **MBT** 3 T-55 (trg)
- **RECCE** 123 FV107 *Scimitar* (incl variants)

ANTI-TANK/ANTI-INFRASTRUCTURE
- **MANPATS** *Spike*-LR
- **RCL 84mm** *Carl Gustaf*; **90mm** Pvpj 1110

ARTILLERY 100
- **SP 155mm** 47 M109A5ÖE
- **TOWED 100mm** (23 K-53 in store)
- **MOR** 53: **81mm** 28 L16; **120mm** 25 M120

Navy 550 (incl Coast Guard)

Naval Forces Flotilla separated into an MCM squadron and a patrol-boat squadron. LVA, EST and LTU have set up a joint naval unit, BALTRON, with bases at Liepaja, Riga, Ventspils (LVA), Tallinn (EST), Klaipeda (LTU). Each nation contributes 1–2 MCMVs

EQUIPMENT BY TYPE

PATROL AND COASTAL COMBATANTS 5
- **PB** 5 *Skrunda* (GER *Swath*)

MINE WARFARE • MINE COUNTERMEASURES 5
- **MCCS** 1 *Vidar* (ex-NOR)
- **MHO** 4 *Imanta* (ex-NLD *Alkmaar/Tripartite*) (3 used in MCM role, 1 used for other duties)

LOGISTICS AND SUPPORT 1
- **AXL** 1 *Varonis* (comd and spt ship, ex-NLD)

Coast Guard

Under command of the Latvian Naval Forces

EQUIPMENT BY TYPE

PATROL AND COASTAL COMBATANTS 6
- **PB** 6: 1 *Astra*; 5 KBV 236 (ex-SWE)

Air Force 500

Main tasks are airspace control and defence, maritime and land SAR and air transportation

FORCES BY ROLE

TRANSPORT
- 1 (mixed) tpt sqn with An-2 *Colt*; Mi-17 *Hip* H; PZL Mi-2 *Hoplite*

AIR DEFENCE
- 1 AD bn
- 1 radar sqn (radar/air ctrl)

AIRCRAFT • TPT • Light 4 An-2 *Colt*

HELICOPTERS
- **MRH** 4 Mi-17 *Hip* H
- **TPT • Light** 2 PZL Mi-2 *Hoplite*

AIR DEFENCE
- **SAM • Point-defence** FIM-92 *Stinger*; RBS-70
- **GUNS • TOWED 40mm** 24 L/70

Paramilitary

State Border Guard

EQUIPMENT BY TYPE

PATROL AND COASTAL COMBATANTS 3
- **PB** 3: 1 *Valpas* (ex-FIN); 1 *Lokki* (ex-FIN); 1 *Randa*

HELICOPTERS
- **TPT • Light** 4: 2 Bell 206B (AB-206B) *Jet Ranger* II; 2 AW109E *Power*

DEPLOYMENT

AFGHANISTAN: NATO • *Operation Resolute Support* 24

BALTIC SEA: NATO • SNMCMG 1: 1 MHO

MALI: EU • EUTM Mali 3; **UN** • MINUSMA 1

FOREIGN FORCES

All NATO Enhanced Forward Presence unless stated

Albania 21; 1 EOD pl

Canada 527; 1 mech inf bn HQ; 1 mech inf coy(+); 1 cbt spt coy

Czech Republic 55; 1 mor pl

Italy 200; 1 armd inf coy

Montenegro 5

Poland 175; 1 tk coy

Slovakia 152; 1 mech inf coy

Slovenia 41; 1 mor pl

Spain 343; 1 armd inf coy(+); 1 cbt engr coy

Lithuania LTU

Euro €		2019	2020	2021
GDP	€	48.4bn	48.2bn	
	US$	54.2bn	55.1bn	
per capita	US$	19,482	19,883	
Growth	%	3.9	-1.8	
Inflation	%	2.2	1.3	
Def exp [a]	€	977m	1.03bn	
	US$	1.09bn	1.17bn	
Def bdgt [b]	€	948m	1.02bn	1.02bn
	US$	1.1bn	1.2bn	
FMA (US)	US$	8m	5m	5m
US$1=€		0.89	0.87	

[a] NATO definition

[b] Includes military pensions

Real-terms defence budget trend (US$m, constant 2015)

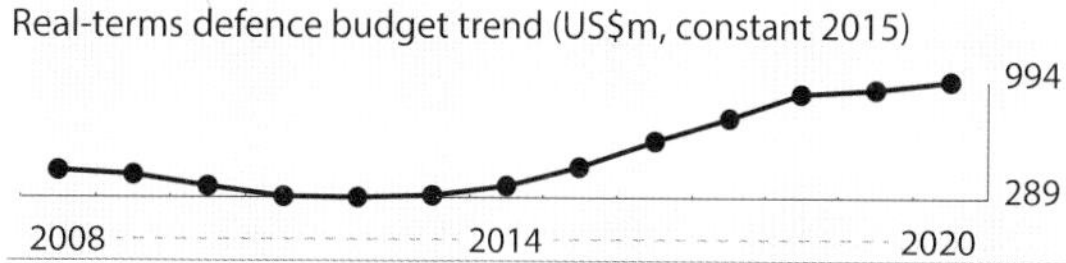

Population 2,731,464

Age	0–14	15–19	20–24	25–29	30–64	65 plus
Male	7.8%	2.5%	2.8%	3.2%	22.9%	7.0%
Female	7.4%	2.3%	2.6%	3.0%	25.0%	13.5%

Capabilities

Lithuania's small armed forces focus on maintaining territorial integrity and national sovereignty but the country relies on NATO membership for its security. Like the other Baltic states, it is reliant on NATO's air-policing deployment for a combat-aircraft capacity. Russia is the country's predominant security concern, and this shapes Lithuanian defence policy. In January 2017, Lithuania adopted a new National Security Strategy, reflecting the worsening regional security environment. Lithuania published a White Paper on Defence in 2017 and the National Defense System Development Programme in 2018, both of which establish long-term strategic objectives. There is a plan to improve combat readiness and the mobilisation system is being reformed. Compulsory military service was reintroduced in 2015. The age band for conscript service was changed in 2020 and the military reserve system was overhauled. A new law on Mobilisation and Host Nation Support is due to come into effect in January 2021. There is increased attention to communications security. Lithuania has a limited medium-airlift capability, for use in supporting its forces on multinational deployed operations. It takes an active part in NATO and EU operations. The country is purchasing the NASAMS SAM system to improve its ground-based air defences. Lithuania has a small defence-industrial base, with niche capabilities, for instance in helicopter support and maintenance.

ACTIVE 22,000 (Army 14,000 Navy 700 Air 1,450 Other 5,850) Paramilitary 15,150

Conscript liability 9 months, 18–23 years

RESERVE 7,100 (Army 7,100)

ORGANISATIONS BY SERVICE

Army 8,450; 5,550 active reserves (total 14,000)

FORCES BY ROLE
MANOEUVRE
Mechanised
1 (1st) mech bde (4 mech inf bn, 1 arty bn, 1 log bn)
Light
1 (2nd) mot inf bde (3 mot inf bn, 1 arty bn)
COMBAT SUPPORT
1 engr bn
COMBAT SERVICE SUPPORT
1 trg regt

EQUIPMENT BY TYPE
ARMOURED FIGHTING VEHICLES
IFV 22 *Boxer* (*Vilkas*) (in test; incl 2 trg)
APC • APC (T) 260: 234 M113A1; 26 M577 (CP)
ENGINEERING & MAINTENANCE VEHICLES
ARV 12: 8 BPz-2; 4 M113
ANTI-TANK/ANTI-INFRASTRUCTURE
MSL
SP 10 M1025A2 HMMWV with FGM-148 *Javelin*
MANPATS FGM-148 *Javelin*
RCL 84mm *Carl Gustaf*
ARTILLERY 91
SP 16 PzH 2000
TOWED 105mm 18 M101
MOR 57: **120mm** 42: 20 2B11; 22 M/41D; **SP 120mm** 15 M113 with Tampella
AIR DEFENCE • SAM • **Point-defence** GROM

Reserves

National Defence Voluntary Forces 5,550 active reservists

FORCES BY ROLE
MANOEUVRE
Other
6 (territorial) def unit

Navy 700

LVA, EST and LTU established a joint naval unit, BALTRON, with bases at Liepaja, Riga, Ventpils (LVA), Tallinn (EST), Klaipeda (LTU)

EQUIPMENT BY TYPE
PATROL AND COASTAL COMBATANTS 4
PCC 4 *Zemaitis* (ex-DNK *Flyvefisken*) with 1 76mm gun
MINE WARFARE • MINE COUNTERMEASURES 4
MHC 3: 1 *Sūduvis* (ex-GER *Lindau*); 2 *Skalvis* (ex-UK *Hunt*)
MCCS 1 *Jotvingis* (ex-NOR *Vidar*)
LOGISTICS AND SUPPORT • **AAR** 1 *Šakiai*

Air Force 1,350

FORCES BY ROLE
AIR DEFENCE
1 AD bn

EQUIPMENT BY TYPE
AIRCRAFT
TPT 6: **Medium** 3 C-27J *Spartan;* **Light** 3: 1 Cessna 172RG; 2 L-410 *Turbolet*
HELICOPTERS
MRH 3 AS365M3 *Dauphin* (SAR)
TPT • Medium 3 Mi-8 *Hip* (tpt/SAR)
AIR DEFENCE • SAM 4+
Medium-range 4 NASAMS III
Point-defence FIM-92 *Stinger*; RBS-70

Special Operation Force

FORCES BY ROLE
SPECIAL FORCES
1 SF gp (1 CT unit; 1 Jaeger bn, 1 cbt diver unit)

Logistics Support Command 1,400

FORCES BY ROLE
COMBAT SERVICE SUPPORT
1 log bn

Training and Doctrine Command 1,500

FORCES BY ROLE
COMBAT SERVICE SUPPORT
1 trg regt

Other Units 2,600

FORCES BY ROLE
COMBAT SUPPORT
1 MP bn

Paramilitary 15,150

Riflemen Union 12,000

State Border Guard Service 3,150

Ministry of Interior

EQUIPMENT BY TYPE
PATROL AND COASTAL COMBATANTS • **PB** 3: 1 *Lokki* (ex-FIN); 1 KBV 041 (ex-SWE); 1 *Bakauskas* (Baltic Patrol 2700)

AMPHIBIOUS • LANDING CRAFT • UCAC 2 *Christina* (*Griffon* 2000)
AIRCRAFT • TPT • Light 1 Cessna 172RG
HELICOPTERS • TPT • Light 5: 1 BK-117 (SAR); 2 H120 *Colibri*; 2 H135

DEPLOYMENT

AFGHANISTAN: NATO • *Operation Resolute Support* 25

BALTIC SEA: NATO • SNMCMG 1: 1 MCCS

CENTRAL AFRICAN REPUBLIC: EU • EUTM RCA 2

IRAQ: *Operation Inherent Resolve* 6; **NATO •** NATO Mission Iraq 9

MALI: EU • EUTM Mali 2; **UN •** MINUSMA 37

SERBIA: NATO • KFOR 1

UKRAINE: JMTG-U 26

FOREIGN FORCES

All NATO Enhanced Forward Presence unless stated
Belgium 1
Czech Republic 35; 1 EW unit
France 275; 1 amrd inf coy(+)
Germany 527; 1 armd inf bn HQ; 1 armd inf coy(+)
Italy NATO Baltic Air Policing: 4 Eurofighter *Typhoon*
Luxembourg 4
Netherlands 270; 1 armd inf coy
Norway 120; 1 armd inf coy

Luxembourg LUX

Euro €		2019	2020	2021
GDP	€	63.5bn	60.0bn	
	US$	71.1bn	68.6bn	
per capita	US$	115,839	109,602	
Growth	%	2.3	-5.8	
Inflation	%	1.7	0.4	
Def exp [a]	€	341m	388m	
	US$	382m	444m	
Def bdgt	€	305m	341m	348m
	US$	341m	390m	
US$1=€		0.89	0.87	

[a] NATO definition

Real-terms defence budget trend (US$m, constant 2015)

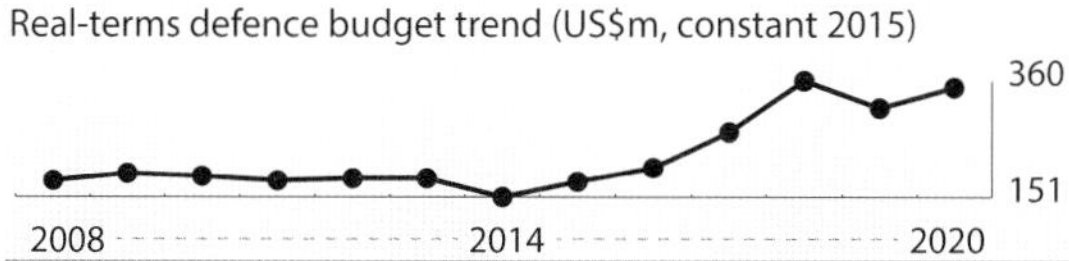

Population	628,381

Age	0–14	15–19	20–24	25–29	30–64	65 plus
Male	8.6%	2.8%	3.2%	3.6%	25.1%	6.9%
Female	8.1%	2.7%	3.0%	3.5%	23.9%	8.4%

Capabilities

Luxembourg maintains a limited military capability to participate in European collective security and crisis management. The 'Defence Guidelines for 2025 and Beyond' were published at the end of 2017. They contain strong statements of support for NATO and EU security policy and contributions to international missions. They also outline ambitious modernisation plans, including a reorganisation of the army, which will take on joint responsibilities, including for ISR, a new air component and a military-cyber cell. Luxembourg has contributed troops to the multinational battlegroup in Lithuania as part of NATO's Enhanced Forward Presence. It is part of the European Multi-Role Tanker Transport Fleet programme, partially funding one A330 MRTT, and has contributed its A400M to a multinational airlift squadron formed jointly with Belgium. The Belgian and Dutch air forces are responsible for policing Luxembourg's airspace. Sustaining the army's personnel strength depends on better recruiting and retention. A review is examining a specialised reserve of civilian experts. The defence guidelines envisage considerable equipment improvements and cooperative development of UAV capabilities with Belgium and the Netherlands. Ambitions for the new air component include tactical-airlift and medical-evacuation capabilities. There is a small but advanced space industry and some foreign defence firms have a presence, but the country is otherwise reliant on imports. A strategy for defence industry, innovation and research is to be developed as part of the new defence guidelines.

ACTIVE 410 (Army 410) **Paramilitary 600**

ORGANISATIONS BY SERVICE

Army 410

FORCES BY ROLE
MANOEUVRE
Reconnaissance
2 recce coy (1 to Eurocorps/BEL div, 1 to NATO pool of deployable forces)

EQUIPMENT BY TYPE
ARMOURED FIGHTING VEHICLES
AUV 48 *Dingo* 2
ANTI-TANK/ANTI-INFRASTRUCTURE
MSL • MANPATS NLAW; TOW
ARTILLERY • MOR 81mm 6+
AIRCRAFT • TPT • Heavy 1 A400M

Paramilitary 600

Gendarmerie 600

DEPLOYMENT

AFGHANISTAN: NATO • *Operation Resolute Support* 2

LITHUANIA: NATO • *Enhanced Forward Presence* 4

MALI: EU • EUTM Mali 2; **UN •** MINUSMA 2

MEDITERRANEAN SEA: EU • EUNAVFOR MED 2 *Merlin* IIIC (leased)

Macedonia, North MKD

Macedonian Denar d		2019	2020	2021
GDP	d	698bn	673bn	
	US$	12.7bn	12.5bn	
per capita	US$	6,109	6,019	
Growth	%	3.6	-5.4	
Inflation	%	0.8	0.9	
Def bdgt	d	8.31bn	10.1bn	12.6bn
	US$	151m	188m	
FMA (US)	US$	5m	0m	0m
US$1=d		54.93	53.80	

Real-terms defence budget trend (US$m, constant 2015)

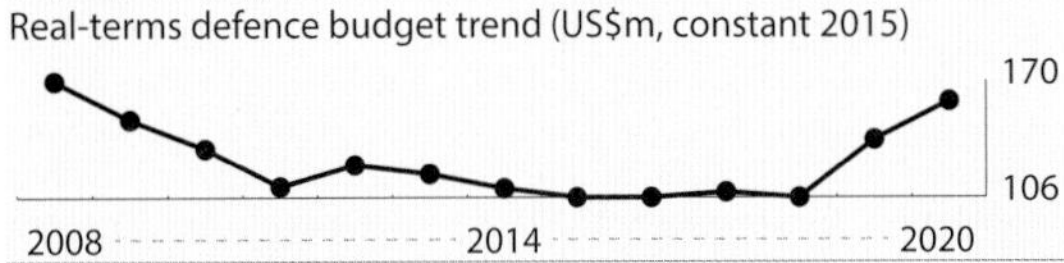

Population 2,125,971

Age	0 – 14	15–19	20–24	25–29	30–64	65 plus
Male	8.4%	3.0%	3.6%	3.7%	25.0%	6.2%
Female	7.8%	2.8%	3.3%	3.5%	24.7%	8.0%

Capabilities

The armed forces' primary goals are safeguarding the state's territorial integrity and sovereignty, as well as contributing to operations under the EU, NATO and UN umbrellas. North Macedonia formally became NATO's 30th member on 27 March 2020. A strategic defence review was released in 2018, which set out aims to optimise, reorganise and modernise the armed forces into a small, modern and flexible force. A 2019–2028 Defence Capability Development Plan (DCDP) consolidated long-term development goals aimed at developing collective-defence, cooperative-security and crisis-management capabilities. A Mid-term Defence Capabilities Development Plan, adopted in January 2020, is intended to help implement the DCDP. A new Defence Strategy, signed in March 2020, indicated development priorities as well as guidelines for improving defence planning. Work on MoD restructuring is under way. The armed forces are fully professional and the country aims to train all units, particularly those with deployable capability, to NATO standards. A number of units are earmarked for participation in NATO-led operations. Skopje contributes to EU, NATO and UN missions. Participation in international peacekeeping missions has increased logistics capability. The country has modest maritime and air wings and relies on Soviet-era equipment. Acquisition priorities include indirect fire-support, light armoured vehicles, cyber defence and multi-role helicopters. There is little in the way of a domestic defence industry, with no ability to design and manufacture modern equipment.

ACTIVE 8,000 (Army 8,000) **Paramilitary 7,600**

RESERVE 4,850

ORGANISATIONS BY SERVICE

Army 8,000

FORCES BY ROLE

SPECIAL FORCES
1 SF regt (1 SF bn, 1 Ranger bn)

MANOEUVRE
Mechanised
1 mech inf bde (1 tk bn, 4 mech inf bn, 1 arty bn, 1 engr bn, 1 NBC coy, 1 sigs coy)

COMBAT SUPPORT
1 MP bn
1 sigs bn

COMBAT SERVICE SUPPORT
1 log bde (3 log bn)

Reserves

FORCES BY ROLE

MANOEUVRE **Light**
1 inf bde

EQUIPMENT BY TYPE

ARMOURED FIGHTING VEHICLES
MBT 31 T-72A
IFV 11: 10 BMP-2; 1 BMP-2K (CP)
APC 200
APC (T) 46: 9 *Leonidas*; 27 M113; 10 MT-LB
APC (W) 154: 56 BTR-70; 12 BTR-80; 2 *Cobra*; 84 TM-170 *Hermelin*

ANTI-TANK/ANTI-INFRASTRUCTURE
MSL • MANPATS *Milan*
RCL 57mm; **82mm** M60A

ARTILLERY 131
TOWED 70: **105mm** 14 M-56; **122mm** 56 M-30 M-1938
MRL 17: **122mm** 6 BM-21; **128mm** 11
MOR 44: **120mm** 44

Marine Wing

EQUIPMENT BY TYPE

PATROL AND COASTAL COMBATANTS 2
PB 2 *Botica*

Aviation Brigade

FORCES BY ROLE

TRAINING
1 flt with Z-242; Bell 205 (UH-1H *Iroquois*); Bell 206B

ATTACK HELICOPTER
1 sqn with Mi-24V *Hind* E

TRANSPORT HELICOPTER
1 sqn with Mi-8MTV *Hip*; Mi-17 *Hip* H

AIR DEFENCE
1 AD bn

EQUIPMENT BY TYPE

AIRCRAFT
TPT • Light 1 An-2 *Colt*
TRG 5 Z-242

HELICOPTERS
ATK 2 Mi-24V *Hind* E (8: 2 Mi-24K *Hind* G2; 6 Mi-24V *Hind* E in store)
MRH 6: 4 Mi-8MTV *Hip*; 2 Mi-17 *Hip* H
TPT • Light 6: 2 Bell 205 (UH-1H *Iroquois*); 4 Bell 206B *Jet Ranger*

AIR DEFENCE
SAM • Point-defence 8+: 8 9K35 *Strela*-10 (SA-13 *Gopher*); 9K310 *Igla*-1 (SA-16 *Gimlet*)
GUNS 40mm 36 L/60

Paramilitary

Police 7,600 (some 5,000 armed)

incl 2 SF units

EQUIPMENT BY TYPE

ARMOURED FIGHTING VEHICLES
- **APC** • **APC (T)** M113; **APC (W)** BTR-80; TM-170 *Heimlin*
- **AUV** *Ze'ev*

HELICOPTERS
- **MRH** 1 Bell 412EP *Twin Huey*
- **TPT** 3: **Medium** 1 Mi-171; **Light** 2: 1 Bell 206B (AB-206B) *Jet Ranger* II; 1 Bell 212 (AB-212)

DEPLOYMENT

AFGHANISTAN: NATO • *Operation Resolute Support* 42

BOSNIA-HERZEGOVINA: EU • EUFOR • *Operation Althea* 12

CARIBBEAN: 1 PSOH

LEBANON: UN • UNIFIL 2

Malta MLT

Euro €		2019	2020	2021
GDP	€	13.4bn	12.5bn	
	US$	15.0bn	14.3bn	
per capita	US$	30,374	28,469	
Growth	%	4.9	-7.9	
Inflation	%	1.5	0.8	
Def bdgt [a]	€	74.7m	70.7m	71.8m
	US$	83.6m	80.8m	
US$1=€		0.89	0.87	

[a] Excludes military pensions

Real-terms defence budget trend (US$m, constant 2015)

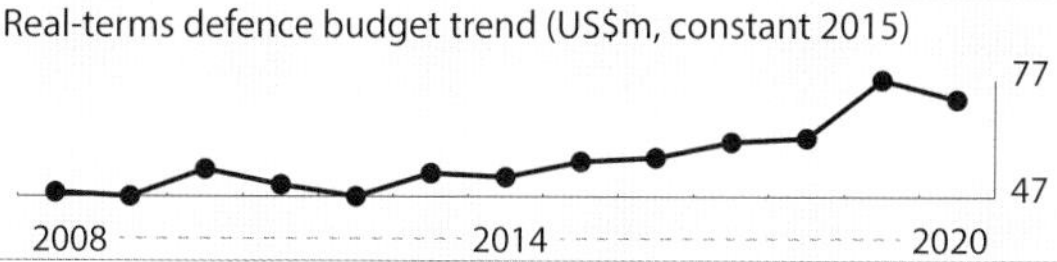

Population 457,267

Age	0–14	15–19	20–24	25–29	30–64	65 plus
Male	7.4%	2.4%	3.0%	3.7%	24.1%	9.8%
Female	7.0%	2.2%	2.8%	3.3%	22.9%	11.5%

Capabilities

The principal roles for the armed forces are maintaining external security and support for civil emergencies and to the police. There is also focus on maritime security in the Mediterranean. Malta is neutral but is a member of NATO's Partnership for Peace programme. The country also participates in bilateral and multilateral exercises. Although deployment capacity is limited, Malta has contributed to European missions. Italy has assisted Malta in meeting some security requirements, including air surveillance, while the European Internal Security Fund is funding some modernisation. Although there is some shipbuilding and ship-repair activity and a small aviation-maintenance industry, none are defence-specific and Malta relies on imports to equip its armed forces.

ACTIVE 1,700 (Armed Forces 1,700)

RESERVE 260 (Volunteer Reserve Force 110 Individual Reserve 150)

ORGANISATIONS BY SERVICE

Armed Forces of Malta 1,700

FORCES BY ROLE

SPECIAL FORCES
- 1 SF unit

MANOEUVRE
- **Light**
  - 1 (1st) inf regt (3 inf coy, 1 cbt spt coy)

COMBAT SUPPORT
- 1 (3rd) cbt spt regt (1 cbt engr sqn, 1 EOD sqn, 1 maint sqn)

COMBAT SERVICE SUPPORT
- 1 (4th) CSS regt (1 CIS coy, 1 sy coy)

EQUIPMENT BY TYPE

ARTILLERY • **MOR 81mm** L16

AIR DEFENCE • **GUNS 14.5mm** 1 ZPU-4

Maritime Squadron 500

Organised into 5 divisions: offshore patrol; inshore patrol; rapid deployment and training; marine engineering; and logistics

EQUIPMENT BY TYPE

PATROL AND COASTAL COMBATANTS 8
- **PCO** 1 *Emer*
- **PCC** 1 *Diciotti* (ITA *Saettia* mod) with 1 hel landing platform
- **PB** 6: 4 Austal 21m; 2 *Marine Protector*

LOGISTICS AND SUPPORT • **AAR** 2 *Cantiere Vittoria*

Air Wing

1 base party. 1 flt ops div; 1 maint div; 1 integrated log div; 1 rescue section

EQUIPMENT BY TYPE

AIRCRAFT
- **TPT** • **Light** 5: 3 Beech 200 *King Air* (maritime patrol); 2 BN-2B *Islander*
- **TRG** 3 *Bulldog* T MK1

HELICOPTERS
- **MRH** 6: 3 AW139 (SAR); 3 SA316B *Alouette* III

DEPLOYMENT

LEBANON: UN • UNIFIL 10

Montenegro MNE

Euro €		2019	2020	2021
GDP	€	4.91bn	4.32bn	
	US$	5.50bn	4.94bn	
per capita	US$	8,826	7,933	
Growth	%	3.6	-12.0	
Inflation	%	0.4	-0.1	
Def exp [a]	€	69.0m	90.0m	
	US$	77.3m	103m	
Def bdgt [b]	€	65.9m	64.7m	72.5m
	US$	73.7m	73.9m	
FMA (US)	US$	5m	0m	0m
US$1=€		0.89	0.87	

[a] NATO definition

[b] Includes military pensions

Real-terms defence budget trend (US$m, constant 2015)

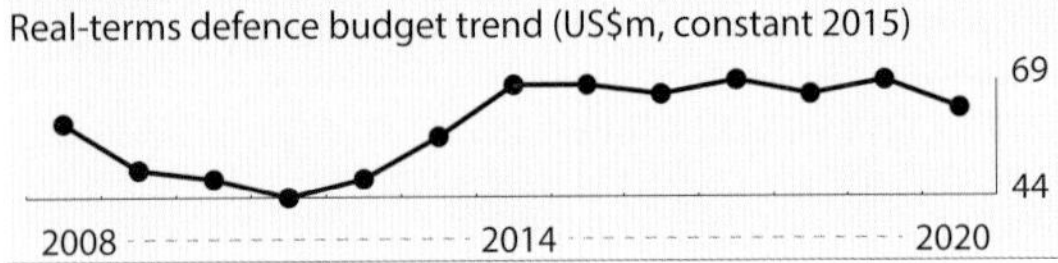

Population 609,859

Age	0–14	15–19	20–24	25–29	30–64	65 plus
Male	9.4%	3.3%	3.3%	3.1%	23.2%	6.9%
Female	8.7%	3.0%	3.1%	3.0%	23.8%	9.1%

Capabilities

According to its defence strategy, Montenegro intends to develop an integrated defence system, capable of defending and preserving independence, sovereignty and national territory. However, the principal concern of the authorities is integrating Montenegro into the collective security system of NATO as well as the EU. A NATO member since 2017, Montenegro has accepted NATO's capability targets and has been aligning its defence-planning process with NATO standards. Reform and professionalism of the armed forces has been slow, and developments have been focused on structural issues around improving recruitment, outflow and professional development. The armed forces are not designed to have an expeditionary capability, and as such have little logistics capability to support deployments beyond national borders. Nevertheless, personnel have deployed to EU, UN and NATO-led operations. Podgorica intends to replace ageing Soviet-era equipment. Procurement priorities include light and medium helicopters and light armoured vehicles as well as improved communications capacities according to NATO standards. Future plans include the formation of a SOF unit and an intelligence unit in the land forces. The country's defence industry is capable of producing small arms and ammunition.

ACTIVE 2,350 (Army 1,275 Navy 350 Air Force 225 Other 500) Paramilitary 10,100

ORGANISATIONS BY SERVICE

Army 1,275

FORCES BY ROLE

MANOEUVRE

Reconnaissance

1 recce coy

Light

1 mot inf bn

COMBAT SUPPORT

1 MP coy

1 sigs coy

COMBAT SERVICE SUPPORT

1 log bn

EQUIPMENT BY TYPE

ARMOURED FIGHTING VEHICLES

APC • APC (W) 8 BOV-VP M-86

AUV 20 JLTV

ANTI-TANK/ANTI-INFRASTRUCTURE

SP 9 BOV-1

MSL • MANPATS 9K111 *Fagot* (AT-4 *Spigot*); 9K111-1 *Konkurs* (AT-5 *Spandrel*)

ARTILLERY 135

TOWED 122mm 12 D-30

MRL 128mm 18 M-63/M-94 *Plamen*

MOR 105: **82mm** 73; **120mm** 32

Reserve

FORCES BY ROLE

MANOEUVRE

Light

2 inf bn

COMBAT SUPPORT

1 arty bn

Navy 350

1 Naval Cmd HQ with 4 operational naval units (patrol boat; coastal surveillance; maritime detachment; and SAR) with additional sigs, log and trg units with a separate coastguard element. Some listed units are in the process of decommissioning

EQUIPMENT BY TYPE

PATROL AND COASTAL COMBATANTS 4

PCF 2 *Rade Končar*†

PB 2 *Mirna* (Type-140) (Police units)

LOGISTICS AND SUPPORT 1

AXS 1 *Jadran*†

Air Force 225

Golubovci (Podgorica) air base under army command

FORCES BY ROLE

TRAINING

1 (mixed) sqn with G-4 *Super Galeb*; Utva-75 (none operational)

TRANSPORT HELICOPTER

1 sqn with SA341/SA342L *Gazelle*

EQUIPMENT BY TYPE

AIRCRAFT • TRG (4 G-4 *Super Galeb* non-operational; 4 Utva-75 non-operational)

HELICOPTERS

MRH 16: 1 Bell 412EP *Twin Huey*; 2 Bell 412EPI *Twin Huey*; 13 SA341/SA342L (HN-45M) *Gazelle*

TPT • Light 1 Bell 505 *Jet Ranger* X

Paramilitary ε10,100

Montenegrin Ministry of Interior Personnel ε6,000

Special Police Units ε4,100

DEPLOYMENT

AFGHANISTAN: NATO • *Operation Resolute Support* 32

LATVIA: NATO • Enhanced Forward Presence 5

SERBIA: NATO • KFOR 1

WESTERN SAHARA: UN • MINURSO 1

Multinational Organisations

Capabilities

The following represent shared capabilities held by contributors collectively rather than as part of national inventories

ORGANISATIONS BY SERVICE

NATO AEW&C Force

Based at Geilenkirchen (GER). Original participating countries (BEL, CAN, DNK, GER, GRC, ITA, NLD, NOR, PRT, TUR, US) have been subsequently joined by five more (CZE, ESP, HUN, POL, ROM)

FORCES BY ROLE

AIRBORNE EARLY WARNING & CONTROL

1 sqn with B-757 (trg); E-3A *Sentry* (NATO standard)

EQUIPMENT BY TYPE

AIRCRAFT

AEW&C 16 E-3A *Sentry* (NATO standard)

TPT • **PAX** 1 B-757 (trg)

NATO Alliance Ground Surveillance

Based at Sigonella (ITA)

EQUIPMENT BY TYPE

UNINHABITED AERIAL VEHICLES

ISR • **Heavy** 4 RQ-4D *Phoenix*

NATO Multinational Multi-Role Tanker Transport Fleet (MMF)

Based at Eindhoven (NLD). Six participating countries (BEL, CZE, GER, NLD, NOR & LUX)

EQUIPMENT BY TYPE

AIRCRAFT • **TKR/TPT** 2 A330 MRTT

Strategic Airlift Capability

Heavy Airlift Wing based at Papa air base (HUN). 12 participating countries (BLG, EST, FIN, HUN, LTU, NLD, NOR, POL, ROM, SVN, SWE, US)

EQUIPMENT BY TYPE

AIRCRAFT • **TPT** • **Heavy** 3 C-17A *Globemaster* III

Strategic Airlift Interim Solution

Intended to provide strategic-airlift capacity pending the delivery of A400M aircraft by leasing An-124s. 11 participating countries (BEL, CZE, FIN, FRA, GER, HUN, NOR, POL, SVK, SVN, SWE)

EQUIPMENT BY TYPE

AIRCRAFT • **TPT** • **Heavy** 2 An-124-100 (3 more available on 6–9 days' notice)

Netherlands NLD

Euro €		2019	2020	2021
GDP	€	810bn	775bn	
	US$	907bn	886bn	
per capita	US$	52,646	51,290	
Growth	%	1.7	-5.4	
Inflation	%	2.7	1.2	
Def exp [a]	€	11.0bn	11.1bn	
	US$	12.3bn	12.7bn	
Def bdgt [b]	€	10.8bn	11.0bn	11.6bn
	US$	12.1bn	12.6bn	
US$1=€		0.89	0.87	

[a] NATO definition

[b] Includes military pensions

Real-terms defence budget trend (US$bn, constant 2015)

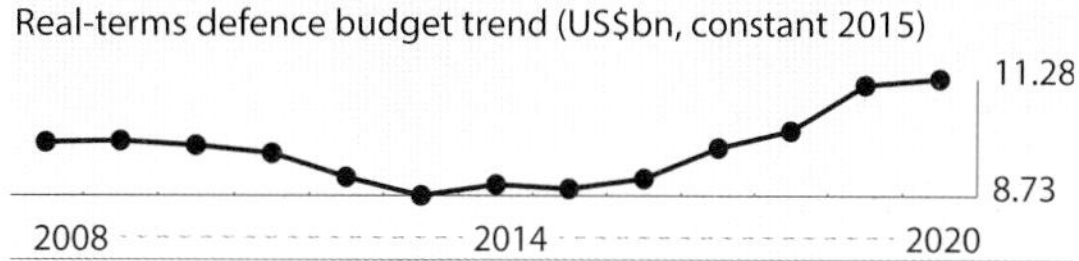

Population 17,280,397

Age	0–14	15–19	20–24	25–29	30–64	65 plus
Male	8.2%	3.0%	3.1%	3.2%	22.9%	9.0%
Female	7.9%	2.9%	3.0%	3.2%	22.9%	10.8%

Capabilities

The 2018 defence review tasks the armed forces with territorial defence and supporting national civil authorities with law enforcement, disaster relief and humanitarian assistance. The Defense Vision 2035 document was published in October 2020, which assessed the future security environment and indicated ten principles to guide the armed forces' response out to 2035. However, implementing the report would rely on adequate finances and political support. Dutch forces have increasingly integrated with NATO allies, particularly Germany. The army contributes to a Dutch–German tank battalion and its mechanised and air-mobile brigades are integral formations within German divisions. There is also cooperation and integration with the German armed forces in the air and naval domains. The Netherlands has air-policing agreements with France, Belgium and Luxembourg and is a member of the UK-led Joint Expeditionary Force. The Netherlands, Belgium and Denmark have committed to forming a composite special-operations command. Dutch forces are fully professional and well trained and the Netherlands can deploy and sustain a medium-scale force for a single operation, or a small-scale joint force for an extended period. The Netherlands makes significant contributions to NATO and EU military operations globally. The country has a modern European- and US-sourced equipment inventory. An agreement is in place with Belgium on the joint acquisition of new frigates and minehunters, while the air force is upgrading its *Patriot* missiles and *Apache* helicopters and acquiring F-35 combat aircraft and MQ-9 *Reaper* UAVs. The Netherlands is part of the programme for a multinational NATO unit of A330 transport/tanker aircraft. The country has an advanced domestic defence industry focusing on armoured vehicles, naval ships and air-defence systems, but also hosts a range of international aerospace-company subsidiaries. Damen Schelde Naval Shipbuilding exports frigates, corvettes and fast-attack craft, while DutchAero manufactures engine components for the F-35. The country also collaborates with Germany on the *Boxer* and *Fennek* armoured vehicles.

ACTIVE 33,600 (Army 15,350 Navy 7,350 Air 6,400 Other 4,500) Military Constabulary 6,500

RESERVE 6,000 (Army 3,900 Navy 1,100 Air 800 Other 200) Military Constabulary 300

Reserve liability to age 35 for soldiers/sailors, 40 for NCOs, 45 for officers

ORGANISATIONS BY SERVICE

Army 15,350

FORCES BY ROLE

COMMAND
elm 1 (1 GNC) corps HQ

SPECIAL FORCES
4 SF coy

MANOEUVRE

Reconnaissance
1 ISR bn (2 armd recce sqn, 1 EW coy, 2 int sqn, 1 UAV bty)

Mechanised
1 (43rd) mech bde (1 armd recce sqn, 2 armd inf bn, 1 engr bn, 1 maint coy, 1 med coy)
1 (13th) mech bde (1 recce sqn, 2 mech inf bn, 1 engr bn, 1 maint coy, 1 med coy)

Air Manoeuvre
1 (11th) air mob bde (3 air mob inf bn, 1 engr coy, 1 med coy, 1 supply coy, 1 maint coy)

COMBAT SUPPORT
1 SP arty bn (3 SP arty bty)
1 AD comd (1 AD sqn; 1 AD bty)
1 CIMIC bn
1 engr bn
2 EOD coy 1 (CIS) sigs bn 1 CBRN coy

COMBAT SERVICE SUPPORT
1 med bn
5 fd hospital
3 maint coy
2 tpt bn

Reserves 3,900 reservists

National Command
Cadre bde and corps tps completed by call-up of reservists (incl Territorial Comd)

FORCES BY ROLE

MANOEUVRE

Light
3 inf bn (could be mobilised for territorial def)

EQUIPMENT BY TYPE

ARMOURED FIGHTING VEHICLES
RECCE 197 *Fennek*
IFV 117 CV9035NL (32 more in store)
APC • APC (W) 200 *Boxer* (8 driver trg; 52 amb; 36 CP; 92 engr; 12 log)
AUV 248: 98 *Bushmaster* IMV; 150 *Fennek*

ENGINEERING & MAINTENANCE VEHICLES
AEV 10+: *Dachs*; 10 *Kodiak*
ARV 25+: BPz-2; 25 BPz-3 *Büffel*
VLB 22: 16 *Leopard* 1 with *Legaun*; 2 *Leopard* 2 with *Leguan*; 4 MLC70 with *Leguan*
MW *Bozena*

NBC VEHICLES 6 TPz-1 *Fuchs* NBC

ANTI-TANK/ANTI-INFRASTRUCTURE
MSL • MANPATS *Spike*-MR

ARTILLERY 119:
SP 155mm 18 PzH 2000 (38 more in store)
MOR 101: **81mm** 83 L16/M1; **120mm** 18 Brandt

AIR DEFENCE • SAM 42+
Long-range 18 MIM-104D/F *Patriot* PAC-2 GEM/PAC-3 (TMD capable)
Short-range 6 NASAMS II
Point-defence 18+: FIM-92 *Stinger*; 18 *Fennek* with FIM-92 *Stinger*

Navy 7,350 (incl Marines)

EQUIPMENT BY TYPE

SUBMARINES 4
SSK 4 *Walrus* with 4 single 533mm TT with Mk 48 ADCAP mod 7 HWT

PRINCIPAL SURFACE COMBATANTS 6
DESTROYERS • DDGHM 4:
3 *De Zeven Provinciën* with 2 quad lnchr with RGM-84C *Harpoon* Block 1B AShM, 5 8-cell Mk 41 VLS with SM-2 Block IIIA/RIM-162B ESSM SAM, 2 twin 324mm SVTT Mk 32 ASTT with Mk 46 LWT, 1 *Goalkeeper* CIWS, 1 127mm gun (capacity 1 NH90 hel)
1 *De Zeven Provinciën* with 2 quad lnchr with RGM-84C *Harpoon* Block 1B AShM, 5 8-cell Mk 41 VLS with SM-2 Block IIIA/RIM-162B ESSM SAM, 2 twin 324mm SVTT Mk 32 ASTT with Mk 46 LWT, 2 *Goalkeeper* CIWS, 1 127mm gun (capacity 1 NH90 hel)
FRIGATES • FFGHM 2 *Karel Doorman* with 2 quad lnchr with RGM-84C *Harpoon* Block 1B AShM, 1 16-cell Mk 48 mod 1 VLS with RIM-7P *Sea Sparrow* SAM, 2 twin 324mm SVTT Mk 32 ASTT with Mk 46 LWT, 1 *Goalkeeper* CIWS, 1 76mm gun (capacity 1 NH90 hel)

PATROL AND COASTAL COMBATANTS
PSOH 4 *Holland* with 1 76mm gun (capacity 1 NH90 hel)

MINE WARFARE • MINE COUNTERMEASURES 6
MHO 6 *Alkmaar* (*Tripartite*)

AMPHIBIOUS
PRINCIPAL AMPHIBIOUS SHIPS • LPD 2:
1 *Rotterdam* with 2 *Goalkeeper* CIWS (capacity 6 NH90/AS532 *Cougar* hel; either 6 LCVP or 2 LCM and 3 LCVP; either 170 APC or 33 MBT; 538 troops)

1 *Johan de Witt* with 2 *Goalkeeper* CIWS (capacity 6 NH90 hel or 4 AS532 *Cougar* hel; either 6 LCVP or 2 LCM and 3 LCVP; either 170 APC or 33 MBT; 700 troops)

LANDING CRAFT 17

LCU 5 LCU Mk II

LCVP 12 Mk5

LOGISTICS AND SUPPORT 7

AFSH 1 *Karel Doorman* with 2 *Goalkeeper* CIWS (capacity 6 NH90/AS532 *Cougar* or 2 CH-47F *Chinook* hel; 2 LCVP)

AGS 2 *Snellius*

AK 1 *Pelikaan*

AS 1 *Mercuur*

AXL 1 *Van Kingsbergen*

AXS 1 *Urania*

Marines 2,650

FORCES BY ROLE

SPECIAL FORCES

1 SF gp (1 SF sqn, 1 CT sqn)

MANOEUVRE

Amphibious

2 mne bn

1 amph aslt gp

COMBAT SERVICE SUPPORT

1 spt gp (coy)

EQUIPMENT BY TYPE

ARMOURED FIGHTING VEHICLES

APC • APC (T) 65 BvS-10 *Viking* (incl 20 CP)

ENGINEERING & MAINTENANCE VEHICLES

ARV 8: 4 BvS-10; 4 BPz-2

MED 4 BvS-10

ANTI-TANK/ANTI-INFRASTRUCTURE

MSL • MANPATS *Spike*-MR

ARTILLERY • MOR 81mm 12 L16/M1

AIR DEFENCE • SAM • Point-defence FIM-92 *Stinger*

Air Force 6,400

FORCES BY ROLE

FIGHTER/GROUND ATTACK

2 sqn with F-16AM/BM *Fighting Falcon*

1 sqn with F-16AM/BM *Fighting Falcon*; F-35A *Lightning* II

ANTI-SUBMARINE WARFARE/SEARCH & RESCUE

1 sqn with NH90 NFH

TANKER/TRANSPORT

1 sqn with C-130H/H-30 *Hercules*

1 sqn with KDC-10; Gulfstream IV

TRAINING

1 OEU sqn with F-35A *Lightning* II

1 sqn with PC-7 *Turbo Trainer*

1 hel sqn with AH-64D *Apache*; CH-47D *Chinook* (based at Fort Hood, TX)

ATTACK HELICOPTER

1 sqn with AH-64D *Apache*

TRANSPORT HELICOPTER

1 sqn with AS532U2 *Cougar* II; NH90 NFH

1 sqn with CH-47D/F *Chinook*

EQUIPMENT BY TYPE

AIRCRAFT 73 combat capable

FTR 61 F-16AM/BM *Fighting Falcon*

FGA 12 F-35A *Lightning* II (in test)

TKR 1 KDC-10

TPT 5: **Medium** 4: 2 C-130H *Hercules*; 2 C-130H-30 *Hercules*; **PAX** 1 Gulfstream IV

TRG 13 PC-7 *Turbo Trainer*

HELICOPTERS

ATK 28 AH-64D *Apache*

ASW 19 NH90 NFH (of which 8 not fitted with sonar)

TPT 28: **Heavy** 20: 11 CH-47D *Chinook*; 9 CH-47F *Chinook*; **Medium** 8 AS532U2 *Cougar* II

AIR-LAUNCHED MISSILES

AAM • IR AIM-9L/M *Sidewinder*; **IIR** AIM-9X *Sidewinder* II; **ARH** AIM-120B AMRAAM

ASM AGM-114K *Hellfire*; AGM-65D/G *Maverick*

BOMBS

Laser-guided GBU-10/GBU-12 *Paveway* II; GBU-24 *Paveway* III (all supported by LANTIRN)

INS/GPS guided GBU-39 Small Diameter Bomb

Paramilitary

Royal Military Constabulary 6,500

Subordinate to the Ministry of Defence, but performs most of its work under the authority of other ministries

FORCES BY ROLE

MANOEUVRE

Other

1 paramilitary comd (total: 28 paramilitary unit)

EQUIPMENT BY TYPE

ARMOURED FIGHTING VEHICLES

APC • APC (W) 24 YPR-KMar

DEPLOYMENT

AFGHANISTAN: NATO • *Operation Resolute Support* 160

BALTIC SEA: NATO • SNMCMG 1: 1 MHO

IRAQ: *Operation Inherent Resolve* 60; 2 trg unit; **NATO** • NATO Mission Iraq 2

LEBANON: UN • UNIFIL 1

LITHUANIA: NATO • Enhanced Forward Presence 270; 1 mech inf coy

MALI: EU • EUTM Mali 4; **UN** • MINUSMA 5

MIDDLE EAST: UN • UNTSO 13

SYRIA/ISRAEL: UN • UNDOF 2

UNITED STATES: 1 hel trg sqn with AH-64D *Apache*; CH-47D *Chinook* based at Fort Hood (TX)

FOREIGN FORCES

United States US European Command: 400

Norway NOR

Norwegian Kroner kr		2019	2020	2021
GDP	kr	3.55tr	3.44tr	
	US$	403bn	366bn	
per capita	US$	75,294	67,989	
Growth	%	1.2	-2.8	
Inflation	%	2.2	1.4	
Def exp [a]	kr	66.1bn	67.0bn	
	US$	7.51bn	7.13bn	
Def bdgt [b]	kr	58.9bn	61.0bn	64.5bn
	US$	6.69bn	6.49bn	
US$1=kr		8.80	9.39	

[a] NATO definition

[b] Includes military pensions

Real-terms defence budget trend (US$bn, constant 2015)

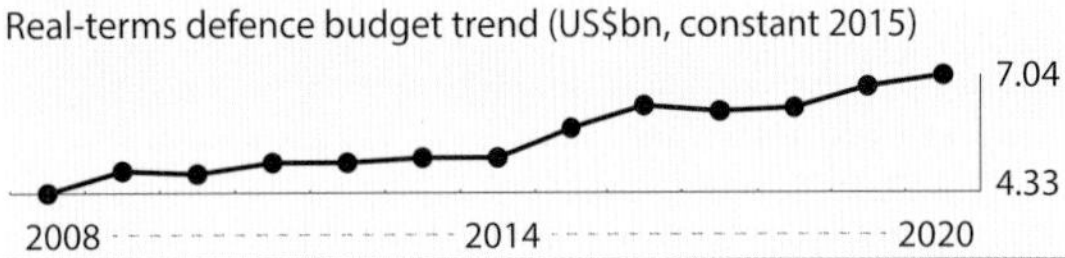

Population 5,467,439

Age	0–14	15–19	20–24	25–29	30–64	65 plus
Male	9.2%	3.0%	3.2%	3.5%	23.5%	8.1%
Female	8.8%	2.8%	3.1%	3.4%	22.2%	9.3%

Capabilities

Norway sustains small but well-equipped and highly trained armed forces. Territorial defence is at the heart of security policy. A new Long Term Defence Plan was published in October 2020, arguing that the security environment had deteriorated faster than expected. It envisages a gradual increase in personnel numbers and further measures to strengthen readiness and capability in the High North. A US Marine Corps contingent has deployed to Vaernes, on a rotational basis, since January 2017. In August 2018, this was extended for up to five years and a second location added at Setermoen. The US had planned to invest, through its European Deterrence Initiative (EDI), in infrastructure upgrades at Rygge air base to enable reinforcements in case of conflict. Cuts to EDI funding announced by the US in September 2019 put some of these investments in doubt. Norway is not an EU member, but it signed a cooperation agreement with the European Defence Agency in 2006. At any one time, around one-third of troops are conscripts. Senior officers reportedly expressed concerns in 2019 that Norway's force structure was too small for defence requirements. A report by the Norwegian chief of defence published in October 2019 argued that in order to address the challenging security situation, the number of combat units in all services should increase. Norway maintains a small presence in a range of international crisis-management missions. Equipment recapitalisation is ongoing. Norway's first F-35A arrived in late 2017 and the government announced that it would procure four submarines as part of a strategic partnership with Germany. Large procurements will stretch budgets, with the F-35 alone reportedly taking up 27% of all procurement spending between 2019 and 2026. In June 2018, it was announced that a planned upgrade to Norway's main-battle-tank fleet would be pushed to the mid-2020s; current budget forecasts foresee funding being made available for this purpose in 2025–26. Norway has an advanced and diverse defence-industrial base with a high percentage of SMEs and a mix of private and state-owned companies.

ACTIVE 23,350 (Army 8,400 Navy 4,250 Air 3,600 Central Support 6,450 Home Guard 650)

Conscript liability 19 months maximum. Conscripts first serve 12 months from 19–28, and then up to 4–5 refresher training periods until age 35, 44, 55 or 60 depending on rank and function. Conscription was extended to women in 2015

RESERVE 40,000 (Home Guard 40,000)

Readiness varies from a few hours to several days

ORGANISATIONS BY SERVICE

Army 3,800; 4,600 conscript (total 8,400)

The armoured infantry brigade – Brigade North – trains new personnel of all categories and provides units for international operations. At any time around one-third of the brigade will be trained and ready to conduct operations. The brigade includes one high-readiness armoured battalion (Telemark Battalion) with combat-support and combat-service-support units on high readiness

FORCES BY ROLE

MANOEUVRE

Reconnaissance

1 ISR bn

1 (GSV) bn (1 (border) recce coy, 1 ranger coy, 1 spt coy, 1 trg coy)

Armoured

1 armd inf bde (2 armd bn, 1 lt inf bn, 1 arty bn, 1 engr bn, 1 MP coy, 1 CIS bn, 1 spt bn, 1 med bn)

Light

1 lt inf bn (His Majesty The King's Guards)

EQUIPMENT BY TYPE

ARMOURED FIGHTING VEHICLES

MBT 36 *Leopard* 2A4 (16 more in store)

RECCE 46: 21 CV9030; 25 HMT *Extenda*

IFV 91: 76 CV9030N; 15 CV9030N (CP)

APC 390

APC (T) 315 M113 (incl variants)

APC (W) 75 XA-186 *Sisu*/XA-200 *Sisu*/XA-203 (amb)

AUV 140: 20 *Dingo* 2; 120 IVECO LMV (50 more in store)

ENGINEERING & MAINTENANCE VEHICLES

AEV 24+: 16 CV90 STING; 8 M113 AEV; NM109; 1 *Wisent*-2

ARV 12: 6 BPz-2; 6 *Wisent*-2

VLB 35: 26 *Leguan*; 9 *Leopard* 1

MW 9 910 MCV-2

NBC VEHICLES 6 TPz-1 *Fuchs* NBC

ANTI-TANK/ANTI-INFRASTRUCTURE

MANPATS FGM-148 *Javelin*

RCL 84mm *Carl Gustaf*

ARTILLERY 191

SP 155mm 48: 24 K9 *Thunder*; 24 M109A3GN (being withdrawn)

MOR 143: **81mm** 115 L16; **SP 81mm** 28: 16 CV9030; 12 M125A2

AIR DEFENCE

SAM • Medium-range NASAMS III

Navy 2,100; 2,150 conscripts (total 4,250)

Joint Command – Norwegian National Joint Headquarters. The Royal Norwegian Navy is organised into four elements under the command of the chief of staff of the Navy: the naval units (*Marinen*), the naval academy (KNM *Harald Haarfagre*), the navy medical branch and the Coast Guard (*Kystvakten*)

FORCES BY ROLE
MANOEUVRE
Reconnaissance
1 ISR coy (Coastal Rangers)
COMBAT SUPPORT
1 EOD pl

EQUIPMENT BY TYPE
SUBMARINES 6
SSK 6 *Ula* with 8 single 533mm TT with *SeaHake* (DM2A3) HWT
PRINCIPAL SURFACE COMBATANTS • FRIGATES 4
FFGHM 4 *Fridtjof Nansen* with *Aegis* C2 (mod), 2 quad lnchr with NSM AShM, 1 8-cell Mk 41 VLS with RIM-162A ESSM SAM, 2 twin 324mm ASTT with *Sting Ray* mod 1 LWT, 1 76mm gun (capacity 1 NH90 hel)
PATROL AND COASTAL COMBATANTS 12
PCFG 6 *Skjold* with 8 single lnchr with NSM AShM, 1 76mm gun
PBF 6 CB90N (capacity 20 troops)
MINE WARFARE • MINE COUNTERMEASURES 4
MSC 2 *Alta* with 1 twin *Simbad* lnchr with *Mistral* SAM
MHC 2 *Oksoy* with 1 twin *Simbad* lnchr with *Mistral* SAM
LOGISTICS AND SUPPORT 6
AGI 1 *Marjata* IV
AGS 2: 1 *HU Sverdrup* II; 1 *Eger* (*Marjata* III) with 1 hel landing platform
AORH 1 *Maud* (BMT *Aegir*) (capacity 2 med hel)
AXL 2 *Reine*

Coast Guard

EQUIPMENT BY TYPE
PATROL AND COASTAL COMBATANTS 13
PSOH 3 *Nordkapp* with 1 57mm gun (capacity 1 med tpt hel)
PSO 5: 3 *Barentshav*; 1 *Harstad*; 1 *Svalbard* with 1 57mm gun, 1 hel landing platform
PCC 5 *Nornen*

Air Force 2,400; 1,200 conscript (total 3,600)

Joint Command – Norwegian National HQ

FORCES BY ROLE
FIGHTER/GROUND ATTACK
1 sqn with F-16AM/BM *Fighting Falcon*
1 sqn with F-35A *Lightning* II (forming)
MARITIME PATROL
1 sqn with P-3C *Orion*; P-3N *Orion* (pilot trg)
ELECTRONIC WARFARE
1 sqn with *Falcon* 20C (EW, Flight Inspection Service)
SEARCH & RESCUE
1 sqn with *Sea King* Mk43B; AW101
TRANSPORT
1 sqn with C-130J-30 *Hercules*
TRAINING
1 sqn with MFI-15 *Safari*
TRANSPORT HELICOPTER
2 sqn with Bell 412SP *Twin Huey*
1 sqn with NH90 (forming)
AIR DEFENCE
1 bn with NASAMS III

EQUIPMENT BY TYPE
AIRCRAFT 68 combat capable
FTR 35: 30 F-16AM *Fighting Falcon*; 5 F-16BM *Fighting Falcon*
FGA 28 F-35A *Lightning* II
ASW 5: 4 P-3C *Orion*; 1 P-3N *Orion* (pilot trg)
EW 2 *Falcon* 20C
TPT • Medium 4 C-130J-30 *Hercules*
TRG 16 MFI-15 *Safari*
HELICOPTERS
ASW 8 NH90 NFH
SAR 19: 9 AW101; 10 *Sea King* Mk43B
MRH 18: 6 Bell 412HP; 12 Bell 412SP
AIR DEFENCE
SAM • Medium-range NASAMS III
AIR-LAUNCHED MISSILES
AAM • IR AIM-9L *Sidewinder*; **IIR** AIM-9X *Sidewinder* II; IRIS-T; **ARH** AIM-120B AMRAAM; AIM-120C AMRAAM
BOMBS
Laser-guided EGBU-12 *Paveway* II
INS/GPS guided JDAM

Special Operations Command (NORSOCOM)

FORCES BY ROLE
SPECIAL FORCES
1 (armed forces) SF comd (2 SF gp)
1 (navy) SF comd (1 SF gp)

Central Support, Administration and Command 6,150; 850 conscripts (total 7,000)

Central Support, Administration and Command includes military personnel in all joint elements and they are responsible for logistics and CIS in support of all forces in Norway and abroad

Home Guard 650 (40,000 reserves)

The Home Guard is a separate organisation, but closely cooperates with all services. The Home Guard is organised in 11 Districts with mobile Rapid Reaction Forces (3,000 troops in total) as well as reinforcements and follow-on forces (37,000 troops in total)

EQUIPMENT BY TYPE
PATROL AND COASTAL COMBATANTS • PB 11: 4 *Harek*; 2 *Gyda*; 5 *Alusafe* 1290

DEPLOYMENT

AFGHANISTAN: NATO • *Operation Resolute Support* 42

BALTIC SEA: NATO • SNMCMG 1: 1 MHC

EGYPT: MFO 3

IRAQ: *Operation Inherent Resolve* 60; 1 trg unit; **NATO** • NATO Mission Iraq 2

JORDAN: *Operation Inherent Resolve* 20

LITHUANIA: NATO • Enhanced Forward Presence 120; 1 armd inf coy; *Leopard* 2A4; CV9030

MALI: UN • MINUSMA 13

MIDDLE EAST: UN • UNTSO 10

SERBIA: NATO • KFOR 2

SOUTH SUDAN: UN • UNMISS 15

FOREIGN FORCES

United States US European Command: 1,100; 1 (USMC) MEU eqpt set; 1 (APS) 155mm SP Arty bn eqpt set

Poland POL

Polish Zloty z		2019	2020	2021
GDP	z	2.27tr	2.28tr	
	US$	592bn	581bn	
per capita	US$	15,601	15,304	
Growth	%	4.1	-3.6	
Inflation	%	2.3	3.3	
Def exp [a]	z	45.8bn	49.7bn	
	US$	11.9bn	12.7bn	
Def bdgt [b]	z	45.4bn	50.4bn	51.2bn
	US$	11.8bn	12.9bn	
US$1=z		3.84	3.92	

[a] NATO definition

[b] Includes military pensions

Real-terms defence budget trend (US$bn, constant 2015)

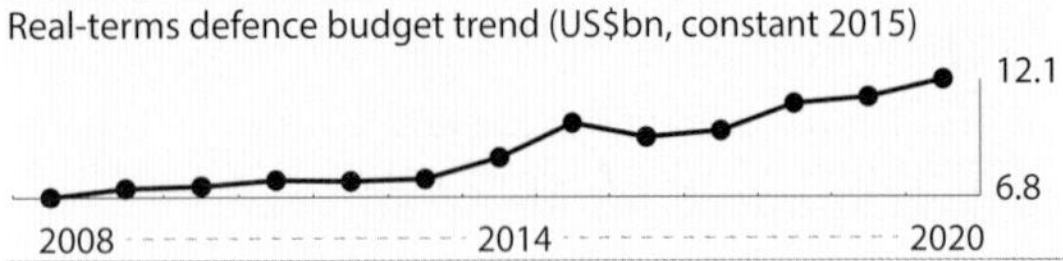

Population 38,282,325

Age	0–14	15–19	20–24	25–29	30–64	65 plus
Male	7.6%	2.4%	2.7%	3.3%	25.0%	7.5%
Female	7.2%	2.3%	2.5%	3.1%	25.3%	11.2%

Capabilities

Territorial defence and NATO membership are central pillars of Poland's defence policy. The primary focus of the 2017–32 defence concept is to prepare the armed forces to deter Russian aggression. Russia is characterised as a direct threat to Poland and to a stable international order. The government continues to pursue a goal of permanently stationing US troops in the country. A bilateral defence-cooperation agreement signed with the US in August 2020 provides for an increased enduring rotational presence of about 1,000 US personnel. Security and defence cooperation also takes place through the Visegrád Group. There are also defence ties through the Bucharest Nine, which brings together NATO's eastern-flank countries. The 2017–32 defence concept defines an ambition to restore divisions as tactical combat units, rather than administrative units. Recruitment is under way for the Territorial Defence Force, which was launched in 2017 and is intended to reach an end-strength of 53,000 by 2026. Poland has some capacity to independently deploy forces beyond national borders. Defence-acquisition reform is planned and proposals to set up a central armaments agency continue to circulate. Poland intends to build up its own anti-access/area-denial capacity and in the 2017 Defence Concept expressed an interest in research into emerging technologies. A technical-modernisation plan, covering the period 2021 to 2035, was released in October 2019, which extended the planning horizon from ten to 15 years. Warsaw continues plans to strengthen its domestic defence-industrial base, much of which is now consolidated in the state-owned holding company PGZ, using technology transfers and international partnering. Beyond PGZ, several international defence primes have subsidiaries in Poland.

ACTIVE 114,050 (Army 58,500 Navy 6,000 Air Force 14,300 Special Forces 3,150 Territorial 3,800 Joint 28,300) Paramilitary 75,400

ORGANISATIONS BY SERVICE

Army 58,500

FORCES BY ROLE

COMMAND

elm 1 (MNC NE) corps HQ

MANOEUVRE

Reconnaissance

3 recce regt

Armoured

1 (11th) armd cav div (2 armd bde, 1 mech bde, 1 arty regt)

Mechanised

1 (12th) mech div (2 mech bde, 1 (coastal) mech bde, 1 arty regt)

1 (16th) mech div (1 armd bde, 2 mech bde, 1 arty regt, 1 AT regt)

1 (18th) mech div (1 armd bde, 2 mech bde, 1 log regt)

Air Manoeuvre

1 (6th) AB bde (3 para bn)

1 (25th) air cav bde (2 air cav bn, 2 tpt hel bn, 1 (casevac) med unit)

COMBAT SUPPORT

2 engr regt

2 ptn br regt

2 chem def regt

COMBAT SUPPORT

2 log bde

HELICOPTER

1 (1st) hel bde (2 atk hel sqn with Mi-24D/V *Hind* D/E, 1 CSAR sqn with Mi-24V *Hind* E; PZL W-3PL *Gluszec*; 2 ISR hel sqn with Mi-2URP; 2 hel sqn with Mi-2)

AIR DEFENCE

3 AD regt

EQUIPMENT BY TYPE

ARMOURED FIGHTING VEHICLES

MBT 808: 137 *Leopard* 2A4 (being upgraded to 2PL); 105 *Leopard* 2A5; 5 *Leopard* 2PL (in test); 232 PT-91 *Twardy*; 329 T-72A/T-72M1

RECCE 407: 282 BRDM-2; 38 BWR-1 (being upgraded); 87 BRDM-2 R5

IFV 1,611: 1,252 BMP-1; 359 *Rosomak* IFV

APC 368
APC (T) 6 WDSz (OP)
APC (W) 332: 300 *Rosomak* APC (incl variants); 32 AWD RAK (arty CP)
PPV 30 *Maxxpro*
AUV 85: 40 *Cougar* (on loan from US); 45 M-ATV
ENGINEERING & MAINTENANCE VEHICLES
AEV 94+: IWT; 65 MT-LB AEV; 21 *Rosomak* WRT; 8 MID *Bizon*
ARV 122: 28 BPz-2; 68 MT-LB ARV; 26 WZT-3M
VLB 119: 4 *Biber*; 103 BLG67M2; 12 MS-20 *Daglezja*
MW 27: 17 *Bozena* 4; 6 ISM *Kroton*; 4 *Kalina* SUM
ANTI-TANK/ANTI-INFRASTRUCTURE
MSL • MANPATS 9K11 *Malyutka* (AT-3 *Sagger*); 9K111 *Fagot* (AT-4 *Spigot*); *Spike*-LR
ARTILLERY 719
SP 394: **122mm** 227 2S1 *Gvozdika*; **152mm** 111 M-77 *Dana*; **155mm** 56 *Krab*
MRL 122mm 179: 75 BM-21; 29 RM-70; 75 WR-40 *Langusta*
MOR 146: **120mm** 80: 15 2B11; 65 M120; **SP 120mm** 66 SMK120 RAK
HELICOPTERS
ATK 28 Mi-24D/V *Hind* D/E
MRH 64: 7 Mi-8MT *Hip*; 3 Mi-17 *Hip* H; 1 Mi-17AE *Hip* (aeromedical); 5 Mi-17-1V *Hip*; 16 PZL Mi-2URP *Hoplite*; 24 PZL W-3W/WA *Sokol*; 8 PZL W-3PL *Gluszec* (CSAR)
TPT 37: **Medium** 12: 6 Mi-8T *Hip*; 2 PZL W-3AE *Sokol* (aeromedical); 4 S-70i *Black Hawk*; **Light** 25 PZL Mi-2 *Hoplite*
AIR DEFENCE
SAM 227+
Short-range 20 2K12 *Kub* (SA-6 *Gainful*)
Point-defence 207+: 64 9K33 *Osa*-AK (SA-8 *Gecko*); 20 ZSU-23-4MP *Biala*; GROM; 81 ZUR-23-2KG *Jodek-G*; *Piorun*; 42 *Poprad*
GUNS 270
SP 23mm 2 ZSU-23-4
TOWED 23mm 268 ZU-23-2

Navy 6,000

EQUIPMENT BY TYPE
SUBMARINES 3
SSK 3:
2 *Sokół* (ex-NOR Type-207) with 8 single 533mm TT with Torped 613 HWT
1 *Orzeł* (ex-FSU *Kilo*) (on sea trials following extensive repairs) with 6 single 533mm TT each with 53-65KE HWT/TEST-71ME
PRINCIPAL SURFACE COMBATANTS • FRIGATES 2
FFH 2 *Pułaski* (ex-US *Oliver Hazard Perry*) (of which 1 used as training ship) with 2 triple 324mm SVTT Mk 32 ASTT with MU90 LWT, 1 Mk 15 *Phalanx* CIWS, 1 76mm gun (capacity 2 SH-2G *Super Seasprite* ASW hel)
PATROL AND COASTAL COMBATANTS 5
CORVETTES • FSM 1 *Kaszub* with 2 quad lnchr with 9K32 *Strela*-2 (SA-N-5 *Grail*) SAM, 2 twin 533mm ASTT with SET-53 HWT, 2 RBU 6000 *Smerch* 2 A/S mor, 1 76mm gun
PSO 1 *Ślązak* (MEKO A-100) with 1 76mm gun, 1 hel landing platform
PCFGM 3 *Orkan* (ex-GDR *Sassnitz*) with 1 quad lnchr with RBS15 Mk3 AShM, 1 quad lnchr (manual aiming) with 9K32 *Strela*-2M (SA-N-5 *Grail*) SAM, 1 AK630 CIWS, 1 76mm gun
MINE WARFARE • MINE COUNTERMEASURES 21
MCCS 1 *Kontradmiral Xawery Czernicki*
MCO 1 *Kormoran* II
MHO 2 *Krogulec*
MSI 17: 1 *Gopło*; 12 *Gardno*; 4 *Mamry*
AMPHIBIOUS 8
LANDING SHIPS • LSM 5 *Lublin* (capacity 9 tanks; 135 troops)
LANDING CRAFT • LCU 3 *Deba* (capacity 50 troops)
LOGISTICS AND SUPPORT 22
AGI 2 *Moma*
AGS 8: 2 *Heweliusz*; 4 *Wildcat* 40; 2 (coastal)
AORL 1 *Bałtyk*
AOL 1 *Moskit*
ARS 4: 2 *Piast*; 2 *Zbyszko*
ATF 4: 2 *Bolko* (B860); 2 H960
AX 1 *Wodnik* with 1 twin AK230 CIWS
AXS 1 *Iskra*
COASTAL DEFENCE • AShM 12 NSM

Naval Aviation 1,300

FORCES BY ROLE
ANTI SUBMARINE WARFARE/SEARCH & RESCUE
1 sqn with Mi-14PL *Haze* A; Mi-14PL/R *Haze* C
1 sqn with PZL W-3RM *Anakonda*; SH-2G *Super Seasprite*
MARITIME PATROL
1 sqn with An-28E/RM *Bryza*
TRANSPORT
1 sqn with An-28TD; M-28B TD *Bryza*
1 sqn with An-28TD; M-28B; PZL Mi-2 *Hoplite*; PZL W-3T/A

EQUIPMENT BY TYPE
AIRCRAFT
MP 10: 8 An-28RM *Bryza*; 2 An-28E *Bryza*
TPT • Light 4: 2 An-28TD *Bryza*; 2 M-28B TD *Bryza*
HELICOPTERS
ASW 8: 6 Mi-14PL *Haze*; 2 SH-2G *Super Seasprite*
SAR 8: 2 Mi-14PL/R *Haze* C; 4 PZL W-3RM *Anakonda*; 2 PZL W-3WA RM *Anakonda*
TPT • Light 7: 4 PZL Mi-2 *Hoplite*; 1 PZL W-3A; 2 PZL-W-3T

Air Force 14,300

FORCES BY ROLE
FIGHTER
2 sqn with MiG-29A/UB *Fulcrum*
FIGHTER/GROUND ATTACK
3 sqn with F-16C/D Block 52+ *Fighting Falcon*
FIGHTER/GROUND ATTACK/ISR
2 sqn with Su-22M-4 *Fitter*
SEARCH AND RESCUE
1 sqn with Mi-2; PZL W-3 *Sokol*

TRANSPORT
1 sqn with C-130E; M-28 *Bryza*
1 sqn with C295M; M-28 *Bryza*
TRAINING
1 sqn with PZL-130 *Orlik*
1 sqn with TS-11 *Iskra*
1 hel sqn with SW-4 *Puszczyk*
TRANSPORT HELICOPTER
1 (Spec Ops) sqn with Mi-17 *Hip* H
1 (VIP) sqn with Mi-8 *Hip*; W-3WA *Sokol*
AIR DEFENCE
1 bde with S-125 *Neva* SC (SA-3 *Goa*); S-200C *Vega* (SA-5 *Gammon*)

EQUIPMENT BY TYPE
AIRCRAFT 94 combat capable
FTR 28: 22 MiG-29A *Fulcrum*; 6 MiG-29UB *Fulcrum*
FGA 66: 36 F-16C Block 52+ *Fighting Falcon*; 12 F-16D Block 52+ *Fighting Falcon*; 12 Su-22M4 *Fitter*; 6 Su-22UM3K *Fitter*
TPT 47: **Medium** 5 C-130E *Hercules*; **Light** 39: 16 C295M; 10 M-28 *Bryza* TD; 13 M-28 *Bryza* PT; **PAX** 3: 2 Gulfstream G550; 1 B-737-800
TRG 54: 11 M-346; 28 PZL-130 *Orlik*; 15 TS-11 *Iskra*
HELICOPTERS
MRH 8 Mi-17 *Hip* H
TPT 65: **Medium** 29: 9 Mi-8 *Hip*; 10 PZL W-3 *Sokol*; 10 PZL W-3WA *Sokol* (VIP); **Light** 36: 14 PZL Mi-2 *Hoplite*; 22 SW-4 *Puszczyk* (trg)
AIR DEFENCE • SAM 18
Long-range 1 S-200C *Vega* (SA-5 *Gammon*)
Short-range 17 S-125 *Neva* SC (SA-3 *Goa*)
AIR-LAUNCHED MISSILES
AAM • IR R-60 (AA-8 *Aphid*); R-73 (AA-11A *Archer*); AIM-9 *Sidewinder*; R-27T (AA-10B *Alamo*); **IIR** AIM-9X *Sidwinder* II; **ARH** AIM-120C AMRAAM
ASM AGM-65J/G *Maverick*; Kh-25 (AS-10 *Karen*); Kh-29 (AS-14 *Kedge*)
LACM Conventional AGM-158 JASSM

Special Forces 3,150

FORCES BY ROLE
SPECIAL FORCES
3 SF units (GROM, FORMOZA & cdo)
COMBAT SUPPORT/
1 cbt spt unit (AGAT)
COMBAT SERVICE SUPPORT
1 spt unit (NIL)

Territorial Defence Forces 3,800 (plus 20,000 reservists)

FORCES BY ROLE
MANOEUVRE
Other
15 sy bde
2 sy bde (forming)

Paramilitary 75,400

Border Guards 14,300
Ministry of Interior

Maritime Border Guard 2,000
EQUIPMENT BY TYPE
PATROL AND COASTAL COMBATANTS 18
PCC 2 *Kaper*
PBF 6: 2 *Strażnik*; 4 IC16M
PB 10: 2 *Wisłoka*; 2 *Baltic* 24; 1 Project MI-6
AMPHIBIOUS • LANDING CRAFT
UCAC 2 *Griffon* 2000TDX

Prevention Units (Police) 61,100
Anti-terrorist Operations Bureau n.k.
Ministry of Interior

DEPLOYMENT

AFGHANISTAN: NATO • *Operation Resolute Support* 397

BOSNIA-HERZEGOVINA: EU • EUFOR • *Operation Althea* 38

CENTRAL AFRICAN REPUBLIC: EU • EUTM RCA 1

DEMOCRATIC REPUBLIC OF THE CONGO: UN • MONUSCO 2

IRAQ: *Operation Inherent Resolve* 150; **NATO • NATO** Mission Iraq 30

LATVIA: NATO • Enhanced Forward Presence 175; 1 tk coy

LEBANON: UN • UNIFIL 220; 1 mech inf coy

MIDDLE EAST: UN • UNTSO 1

ROMANIA: NATO • MNB-SE 220; 1 mech inf coy; *Rosomak*

SERBIA: NATO • KFOR 230; 1 inf coy; **UN** • UNMIK 1

SOUTH SUDAN: UN • UNMISS 1

UKRAINE: JMTG-U 40

WESTERN SAHARA: UN • MINURSO 1

FOREIGN FORCES

All NATO Enhanced Forward Presence unless stated
Croatia 80; 1 MRL bty with M91 *Vulkan*
Germany MNC-NE corps HQ: 95
Romania 102; 1 SP ADA bty
United Kingdom 140; 1 recce sqn
United States: 670; 1 mech bn with M1296 *Stryker Dragoon*; M777A2 • *Operation Atlantic Resolve* 4,500; 1 corps HQ (fwd); 1 div HQ (fwd); 1 armd bde; M1A2 SEPv2 *Abrams*; M3A3 *Bradley*; M2A3 *Bradley*; M109A6; 1 CISR UAV sqn with MQ-9A *Reaper*

Portugal PRT

Euro €		2019	2020	2021
GDP	€	212bn	194bn	
	US$	238bn	222bn	
per capita	US$	23,132	21,608	
Growth	%	2.2	-10.0	
Inflation	%	0.3	0.0	
Def exp [a]	€	2.95bn	3.19bn	
	US$	3.30bn	3.65bn	
Def bdgt	€	2.39bn	2.50bn	
	US$	2.67bn	2.86bn	
US$1=€		0.89	0.87	

[a] NATO definition

Real-terms defence budget trend (US$bn, constant 2015)

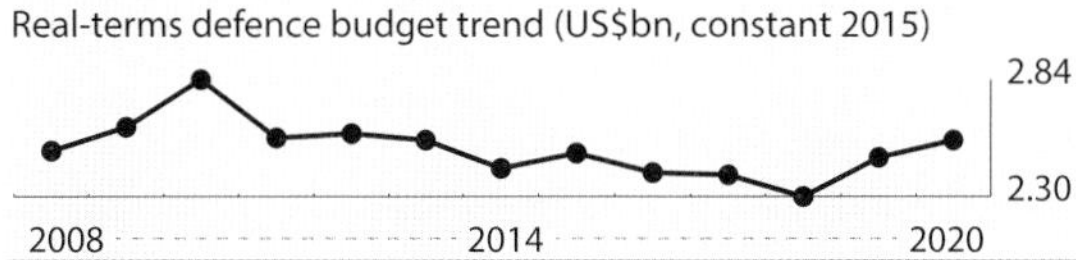

Population 10,302,674

Age	0–14	15–19	20–24	25–29	30–64	65 plus
Male	7.0%	2.8%	2.8%	2.7%	23.7%	8.3%
Female	6.6%	2.6%	2.7%	2.7%	25.4%	12.6%

Capabilities

Principal tasks for Portugal's all-volunteer armed forces are homeland defence, maritime security, multinational operations and responding to humanitarian disasters. The 2013 strategic review set out key defence tasks and envisaged a reduction in army strength and organisational changes to divide the services into immediate reaction forces, permanent defence forces and modular forces. Investment plans support Portugal's ambition to field rapid-reaction and maritime-surveillance capabilities for territorial defence and multinational operations. A new military programme law for 2019–30 was approved by parliament, funding the acquisition of five KC-390 aircraft, six offshore-patrol vessels, a replenishment tanker and a multi-purpose logistics ship, as well as cyber-defence and soldier-combat systems. Portugal hosts NATO's cyber-security academy and the country also contributes to EU military structures. There is a close relationship with former dependencies and with the US, which operates out of Lajes air base. All three services have programmes to modernise and sustain existing equipment platforms. There is an active defence industry, though principally in relation to shipbuilding, broader maintenance tasks and the manufacture of components, and small arms and light weapons.

ACTIVE 27,250 (Army 13,700 Navy 7,650 Air 5,900) Paramilitary 24,700

RESERVE 211,700 (Army 210,000 Navy 1,000, Air Force 700)

Reserve obligation to age 35

ORGANISATIONS BY SERVICE

Army 13,700

5 territorial comd (2 mil region, 1 mil district, 2 mil zone)

FORCES BY ROLE

SPECIAL FORCES

1 SF bn

MANOEUVRE

Reconnaissance

1 ISR bn

Mechanised

1 mech bde (1 recce sqn, 1 tk regt, 1 mech inf bn, 1 arty bn, 1 AD bty, 1 engr coy, 1 sigs coy, 1 spt bn)

1 (intervention) bde (1 recce regt, 2 mech inf bn, 1 arty bn, 1 AD bty, 1 engr coy, 1 sigs coy, 1 spt bn)

Air Manoeuvre

1 (rapid reaction) bde (1 cdo bn, 2 para bn, 1 arty bn, 1 AD bty, 1 engr coy, 1 sigs coy, 1 spt bn)

Other

1 (Azores) inf gp (2 inf bn, 1 AD bty)

1 (Madeira) inf gp (1 inf bn, 1 AD bty)

COMBAT SUPPORT

1 STA bty

1 engr bn (1 construction coy; 1 EOD unit; 1 ptn br coy; 1 CBRN coy)

1 EW coy

1 MP bn

1 psyops unit

1 CIMIC coy (joint)

1 sigs bn

COMBAT SERVICE SUPPORT

1 maint coy

1 log coy

1 tpt coy

1 med unit

AIR DEFENCE

1 AD bn

Reserves 210,000

FORCES BY ROLE

MANOEUVRE

Light

3 (territorial) def bde (on mobilisation)

EQUIPMENT BY TYPE

ARMOURED FIGHTING VEHICLES

MBT 37 *Leopard* 2A6

RECCE 16 VBL

IFV 30 *Pandur* II MK 30mm

APC 406

APC (T) 239: 176 M113A1; 14 M113A2; 49 M577A2 (CP)

APC (W) 167: 9 V-150 *Commando*; 12 V-200 *Chaimite*; 146 *Pandur* II (incl variants)

ENGINEERING & MAINTENANCE VEHICLES

AEV M728

ARV 13: 6 M88A1, 7 *Pandur* II ARV

VLB M48

ANTI-TANK/ANTI-INFRASTRUCTURE

MSL

SP 26: 17 M113 with TOW; 4 M901 with TOW; 5 *Pandur* II with TOW

MANPATS *Milan*; TOW

RCL • **84mm** *Carl Gustaf*; **106mm** 45 M40A1

ARTILLERY 320

SP 155mm 24: 6 M109A2; 18 M109A5

TOWED 62: **105mm** 39: 17 L119 Light Gun; 21 M101A1; **155mm** 24 M114A1
MOR 234: **81mm** 143; **SP 81mm** 12: 2 M125A1; 10 M125A2; **107mm** 11 M30; **SP 107mm** 18: 3 M106A1; 15 M106A2; **120mm** 50 Tampella
AIR DEFENCE
SAM • Point-defence 20+: 1 M48A2 *Chaparral*; 19 M48A3 *Chaparral*; FIM-92 *Stinger*
GUNS • TOWED 20mm 20 Rh 202

Navy 7,600 (incl 960 Marines)

EQUIPMENT BY TYPE
SUBMARINES 2
SSK 2 *Tridente* (GER Type-214) (fitted with AIP) with 8 533mm TT with UGM-84L *Harpoon* Block II AShM/*Black Shark* HWT
PRINCIPAL SURFACE COMBATANTS • FRIGATES 5
FFGHM 5:
2 *Bartolomeu Dias* (ex-NLD *Karel Doorman*) with 2 quad lnchr with RGM-84C *Harpoon* Block 1B AShM, 1 16-cell Mk 48 mod 1 VLS with RIM-7M *Sea Sparrow* SAM, 2 twin 324mm SVTT Mk 32 ASTT with Mk 46 LWT, 1 *Goalkeeper* CIWS, 1 76mm gun (capacity 1 *Lynx* Mk95 (*Super Lynx*) hel)
3 *Vasco Da Gama* with 2 quad lnchr with RGM-84C *Harpoon* Block 1B AShM, 1 octuple Mk 29 GMLS with RIM-7M *Sea Sparrow* SAM, 2 triple 324mm SVTT Mk 32 ASTT with Mk 46 LWT, 1 Mk 15 *Phalanx* Block 1B CIWS, 1 100mm gun (capacity 2 *Lynx* Mk95 (*Super Lynx*) hel)
PATROL AND COASTAL COMBATANTS 21
CORVETTES • FS 2:
1 *Baptista de Andrade* with 1 100mm gun, 1 hel landing platform
1 *Joao Coutinho* with 1 twin 76mm gun, 1 hel landing platform
PSO 4 *Viana do Castelo* with 1 hel landing platform
PCC 5: 1 *Cacine*; 4 *Tejo* (ex-DNK *Flyvisken*)
PBR 10: 5 *Argos*; 4 *Centauro*; 1 *Rio Minho*
LOGISTICS AND SUPPORT 11
AGS 4: 2 *D Carlos* I (ex-US *Stalwart*); 2 *Andromeda*
AORL 1 *Bérrio* (ex-UK *Rover*) with 1 hel landing platform (for medium hel)
AXS 6: 1 *Sagres*; 1 *Creoula*; 1 *Polar*; 2 *Belatrix*; 1 *Zarco*

Marines 950

FORCES BY ROLE
SPECIAL FORCES
1 SF det
MANOEUVRE
Light
1 lt inf bn
COMBAT SUPPORT
1 mor coy
1 MP coy
EQUIPMENT BY TYPE
ANTI-TANK/ANTI-INFRASTRUCTURE
MSL • MANPATS *Milan*; TOW
RCL • 84mm *Carl Gustaf*
ARTILLERY • MOR 30+: **81mm** some; **120mm** 30

Naval Aviation

EQUIPMENT BY TYPE
HELICOPTERS • ASW 5 *Lynx* Mk95 (*Super Lynx*)

Air Force 5,900

FORCES BY ROLE
FIGHTER/GROUND ATTACK
2 sqn with F-16AM/BM *Fighting Falcon*
MARITIME PATROL
1 sqn with P-3C *Orion*
ISR/TRANSPORT
1 sqn with C295M
COMBAT SEARCH & RESCUE
1 sqn with with AW101 *Merlin*
TRANSPORT
1 sqn with C-130H/C-130H-30 *Hercules*
1 sqn with *Falcon* 50
TRAINING
1 sqn with AW119 *Koala*
1 sqn with TB-30 *Epsilon*
EQUIPMENT BY TYPE
AIRCRAFT 35 combat capable
FTR 30: 26 F-16AM *Fighting Falcon*; 4 F-16BM *Fighting Falcon*
ASW 5 P-3C *Orion*
ISR: 7: 5 C295M (maritime surveillance), 2 C295M (photo recce)
TPT 13: **Medium** 5: 2 C-130H *Hercules*; 3 C-130H-30 *Hercules* (tpt/SAR); **Light** 5 C295M; **PAX** 3 *Falcon* 50 (tpt/VIP)
TRG 16 TB-30 *Epsilon*
HELICOPTERS
TPT 17: **Medium** 12 AW101 *Merlin* (6 SAR, 4 CSAR, 2 fishery protection); **Light** 5 AW119 *Koala*
AIR-LAUNCHED MISSILES
AAM • IR AIM-9L/I *Sidewinder*; **ARH** AIM-120C AMRAAM
ASM AGM-65A *Maverick*
AShM AGM-84A *Harpoon*
BOMBS
Laser-guided/GPS GBU-49 *Enhanced Paveway* II
INS/GPS guided GBU-31 JDAM

Paramilitary 24,700

National Republican Guard 24,700

EQUIPMENT BY TYPE
PATROL AND COASTAL COMBATANTS 32
PBF 12
PB 20
HELICOPTERS • MRH 7 SA315 *Lama*

DEPLOYMENT

AFGHANISTAN: NATO • *Operation Resolute Support* 188

CENTRAL AFRICAN REPUBLIC: EU • EUTM RCA 45; **UN** • MINUSCA 186; 1 AB coy

IRAQ: *Operation Inherent Resolve* 34

MALI: EU • EUTM Mali 11; **UN** • MINUSMA 63; 1 tpt flt with 1 C295M

NORTH SEA: NATO • SNMG 1: 1 FFGHM

SOMALIA: EU • EUTM Somalia 4

FOREIGN FORCES

United States US European Command: 250; 1 spt facility at Lajes

Romania ROM

New Lei		2019	2020	2021
GDP	lei	1.06tr	1.05tr	
	US$	250bn	249bn	
per capita	US$	12,887	12,813	
Growth	%	4.1	-4.8	
Inflation	%	3.8	2.9	
Def exp [a]	lei	19.5bn	24.3bn	
	US$	4.61bn	5.75bn	
Def bdgt [b]	lei	20.9bn	22.0bn	24.7bn
	US$	4.94bn	5.21bn	
FMA (US)	US$	12.5m	0m	0m
US$1=lei		4.24	4.22	

[a] NATO definition

[b] Includes military pensions

Real-terms defence budget trend (US$bn, constant 2015)

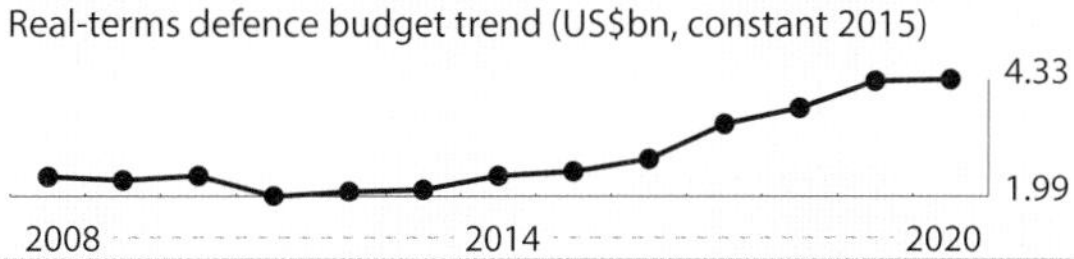

Population 21,302,893

Age	0–14	15–19	20–24	25–29	30–64	65 plus
Male	7.3%	2.6%	2.7%	2.9%	26.0%	7.1%
Female	6.9%	2.4%	2.6%	2.8%	26.2%	10.5%

Capabilities

Romania's armed forces are structured around territorial defence, support to NATO and EU missions and contributing to regional and global stability and security. According to the National Defence Strategy 2020–2024, principal security threats include Russia's increased presence in the Black Sea, hybrid warfare, cyber attacks, terrorism and the economic impact of the coronavirus pandemic. The government has stated the intention to strengthen operational capabilities and develop its partnerships and cooperation with other NATO and EU members, and there is an ongoing programme to modernise and upgrade the armed forces to NATO standards. Bucharest has signed defence-cooperation agreements with regional allies. There is a strategic partnership with the US. Romania hosts the *Aegis Ashore* ballistic-missile-defence system at Deveselu. There is broad training with NATO and regional allies and Romania contributes to EU and NATO missions. The inventory is mainly composed of Soviet-era equipment, which is seen as a factor limiting capability. Acquisition plans include armoured vehicles, air-defence radars, surface-to-air missiles and corvettes. The acquisition of additional second-hand F-16s will enhance Romania's air capabilities. The country's defence industry has struggled since 1989. Current production focuses on small arms and ammunition. However, Bucharest is looking to boost the industry through offset agreements and technology transfers.

ACTIVE 68,500 (Army 35,800 Navy 5,500 Air 10,700 Joint 16,500) **Paramilitary 57,000**

RESERVE 53,000 (Joint 53,000)

ORGANISATIONS BY SERVICE

Army 35,800

Readiness is reported as 70–90% for NATO-designated forces (1 div HQ, 1 mech bde, 1 inf bde & 1 mtn inf bde) and 40–70% for other forces

FORCES BY ROLE

COMMAND
2 div HQ (2nd & 4th)
elm 1 div HQ (MND-SE)

SPECIAL FORCES
1 SF bde (2 SF bn, 1 para bn, 1 log bn)

MANOEUVRE

Reconnaissance
1 recce bde
2 recce regt

Mechanised
5 mech bde (1 tk bn, 2 mech inf bn, 1 arty bn, 1 AD bn, 1 log bn)

Light
1 (MNB-SE) inf bde (3 inf bn, 1 arty bn, 1 AD bn, 1 log bn)
2 mtn inf bde (3 mtn inf bn, 1 arty bn, 1 AD bn, 1 log bn)

COMBAT SUPPORT
1 MRL bde (3 MRL bn, 1 STA bn, 1 log bn)
2 arty regt
1 engr bde (4 engr bn, 1 ptn br bn, 1 log bn)
2 engr bn
3 sigs bn
1 CIMIC bn
1 MP bn
3 CBRN bn

COMBAT SERVICE SUPPORT
3 spt bn

AIR DEFENCE
3 AD regt

EQUIPMENT BY TYPE

ARMOURED FIGHTING VEHICLES
MBT 377: 220 T-55AM; 103 TR-85; 54 TR-85 M1
IFV 142: 41 MLI-84 (incl CP); 101 MLI-84M *Jderul*
APC 785
APC (T) 76 MLVM
APC (W) 649: 69 B33 TAB *Zimbru*; 37 *Piranha* IIIC; 36 *Piranha* V; 354 TAB-71 (incl variants); 153 TAB-77 (incl variants)
PPV 60 *Maxxpro*
AUV 480 TABC-79 (incl variants)

ENGINEERING & MAINTENANCE VEHICLES
ARV 55: 3 MLI-84M TEHEVAC; 8 TERA-71L; 44 TERA-77L
VLB 43 BLG-67

NBC VEHICLES 109 RCH-84

ANTI-TANK/ANTI-INFRASTRUCTURE
MSL
SP 158: 12 9P122 *Malyutka* (AT-3 *Sagger*); 98 9P133 *Malyutka* (AT-3 *Sagger*); 48 9P148 *Konkurs* (AT-5 *Spandrel*)
MANPATS *Spike*-LR
GUNS
SP 100mm (23 SU-100 in store)
TOWED 100mm 218 M-1977
ARTILLERY 1,118
SP 122mm 40: 6 2S1; 34 Model 89
TOWED 447: **122mm** 96 (M-30) M-1938 (A-19); **152mm** 351: 247 M-1981; 104 M-1985
MRL 122mm 188: 134 APR-40; 54 LAROM
MOR 443: **SP 82mm** 177: 92 TAB-71AR; 85 TABC-79AR; **120mm** 266 M-1982
AIR DEFENCE
SAM 96
Short-range 48: 32 2K12 *Kub* (SA-6 *Gainful*); 16 9K33 *Osa* (SA-8 *Gecko*)
Point-defence 48 CA-95
GUNS 65+
SP 35mm 41 *Gepard*
TOWED 14.5mm ZPU-2; **35mm** 24 GDF-003; **57mm** S-60

Navy 5,500

EQUIPMENT BY TYPE
PRINCIPAL SURFACE COMBATANTS • FRIGATES 3
FFGH 1 *Marasesti* with 4 twin lnchr with P-22 (SS-N-2C *Styx*) AShM, 2 triple 533mm ASTT with 53–65 HWT, 2 RBU 6000 *Smerch* 2 A/S mor, 4 AK630M CIWS, 2 twin 76mm guns (capacity 2 SA-316 (IAR-316) *Alouette* III hel)
FFH 2 *Regele Ferdinand* (ex-UK Type-22), with 2 triple STWS Mk.2 324mm TT, 1 76mm gun (capacity 1 SA330 (IAR-330) *Puma*)
PATROL AND COASTAL COMBATANTS 24
CORVETTES 4
FSH 2 *Tetal* II with 2 twin 533mm ASTT with SET-53M HWT, 2 RBU 6000 *Smerch* 2 A/S mor, 2 AK630 CIWS, 1 76mm gun (capacity 1 SA316 (IAR-316) *Alouette* III hel)
FS 2 *Tetal* I with 2 twin 533mm ASTT with SET-53M HWT, 2 RBU 2500 *Smerch* 1 A/S mor, 2 AK230 CIWS, 2 twin 76mm guns
PCFG 3 *Zborul* with 2 twin lnchr with P-22 (SS-N-2C *Styx*) AShM, 2 AK630 CIWS, 1 76mm gun
PCFT 3 *Naluca* with 4 single 533mm ASTT
PCR 8: 5 *Brutar* II with 2 BM-21 MRL, 1 100mm gun; 3 *Kogalniceanu* with 2 BM-21 MRL, 2 100mm guns
PBR 6 VD141 (ex-MSR now used for river patrol)
MINE WARFARE 11
MINE COUNTERMEASURES 10
MSO 4 *Musca* with 2 RBU 1200 *Uragan* A/S mor, 2 AK230 CIWS
MSR 6 VD141
MINELAYERS • ML 1 *Corsar* with up to 120 mines, 2 RBU 1200 *Uragan* A/S mor, 2 AK230 CIWS
LOGISTICS AND SUPPORT 8
AE 2 *Constanta* with 2 RBU 1200 *Uragan* A/S mor, 2 AK230 CIWS, 2 twin 57mm guns
AGOR 1 *Corsar*
AGS 2: 1 *Emil Racovita*; 1 *Catuneanu*
AOL 1 *Tulcea*
ATF 1 *Grozavu*
AXS 1 *Mircea*

Naval Infantry

FORCES BY ROLE
MANOEUVRE
Light
1 naval inf regt
EQUIPMENT BY TYPE
ARMOURED FIGHTING VEHICLES
AUV 14: 11 ABC-79M; 3 TABC-79M

Air Force 10,700

FORCES BY ROLE
FIGHTER
2 sqn with MiG-21 *Lancer* C
FIGHTER/GROUND ATTACK
1 sqn with with F-16AM/BM *Fighting Falcon*
GROUND ATTACK
1 sqn with IAR-99 *Soim**
TRANSPORT
1 sqn with An-30 *Clank*; C-27J *Spartan*
1 sqn with C-130B/H *Hercules*
TRAINING
1 sqn with IAR-99 *Soim**
1 sqn with SA316B *Alouette* III (IAR-316B); Yak-52 (Iak-52)
TRANSPORT HELICOPTER
2 (multi-role) sqn with IAR-330 SOCAT *Puma*
2 sqn with SA330L/M *Puma* (IAR-330L/M)
AIR DEFENCE
1 AD bde
COMBAT SERVICE SUPPORT
1 engr spt regt
EQUIPMENT BY TYPE
AIRCRAFT 60 combat capable
FTR 16: 13 F-16AM *Fighting Falcon*; 3 F-16BM *Fighting Falcon*
FGA 24: 6 MiG-21 *Lancer* B; 18 MiG-21 *Lancer* C
ISR 2 An-30 *Clank*
TPT • Medium 12: 7 C-27J *Spartan*; 4 C-130B *Hercules*; 1 C-130H *Hercules*
TRG 32: 10 IAR-99*; 10 IAR-99C *Soim**; 12 Yak-52 (Iak-52)
HELICOPTERS
MRH 30: 22 IAR-330 SOCAT *Puma*; 8 SA316B *Alouette* III (IAR-316B)
TPT • Medium 24: 12 SA330L *Puma* (IAR-330L); 12 SA330M *Puma* (IAR-330M)
AIR DEFENCE • SAM 17
Long-range 4 MIM-104E/F *Patriot* PAC-2 GEM-T/PAC-3 MSE
Medium-range 13: 5 S-75M3 *Volkhov* (SA-2 *Guideline*); 8 MIM-23 *Hawk* PIP III
AIR-LAUNCHED MISSILES
AAM • IR AIM-9M *Sidewinder*; R-73 (AA-11A *Archer*); R-550 *Magic* 2; *Python* 3 **IIR** AIM-9X *Sidewinder* II; **ARH** AIM-120C AMRAAM
ASM *Spike*-ER

BOMBS

Laser-guided GBU-12 *Paveway*; GBU-54 Laser JDAM
INS/GPS guided GBU-38 JDAM

Paramilitary ε57,000

Gendarmerie ε57,000

Ministry of Interior

DEPLOYMENT

AFGHANISTAN: NATO • *Operation Resolute Support* 693; 1 inf bn

BOSNIA-HERZEGOVINA: EU • EUFOR • *Operation Althea* 58

CENTRAL AFRICAN REPUBLIC: EU • EUTM RCA 13

DEMOCRATIC REPUBLIC OF THE CONGO: UN • MONUSCO 6

IRAQ: *Operation Inherent Resolve* 10; **NATO** • **NATO** Mission Iraq 4

MALI: EU • EUTM Mali 1; **UN** • MINUSMA 124; 1 hel sqn

POLAND: NATO • Enhanced Forward Presence 102; 1 SP ADA bty with *Gepard*

SERBIA: NATO • KFOR 53; **UN** • UNMIK 1

SOMALIA: EU • EUTM Somalia 1

SOUTH SUDAN: UN • UNMISS 6

FOREIGN FORCES

Canada NATO Air Policing: 135; 6 F/A-18A *Hornet* (CF-18)
Poland NATO MNB-SE 220; 1 mech inf coy; *Rosomak*
United States US European Command: 120

Serbia SER

Serbian Dinar d		2019	2020	2021
GDP	d	5.41tr	5.48tr	
	US$	51.4bn	52.0bn	
per capita	US$	7,382	7,497	
Growth	%	4.2	-2.5	
Inflation	%	1.9	1.5	
Def bdgt	d	95.1bn	92.4bn	
	US$	903m	878m	
US$1=d		105.25	105.32	

Real-terms defence budget trend (US$m, constant 2015)

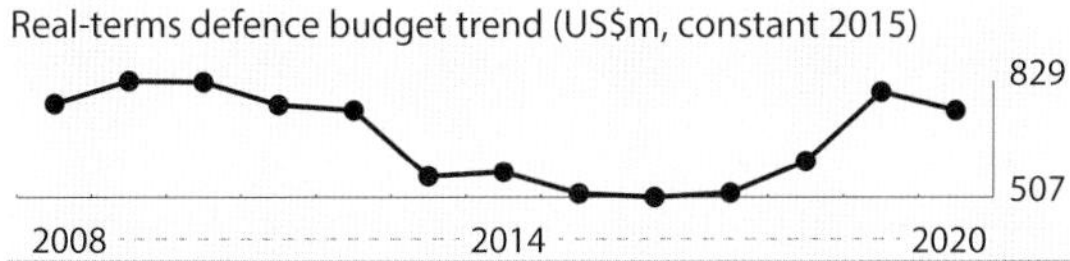

Population	7,012,165					
Age	0–14	15–19	20–24	25–29	30–64	65 plus
Male	7.2%	2.8%	2.9%	3.2%	24.3%	8.4%
Female	6.8%	2.7%	2.7%	3.0%	24.5%	11.6%

Capabilities

Serbia's armed forces focus on territorial defence, internal security and limited support to peacekeeping missions. According to the 2019 national-security strategy, key threats include separatism, ethnic and religious extremism, climate change and further international recognition of Kosovo. The armed forces are modernising to address long-term capability shortfalls and personnel shortages. Priorities include procurements; improving availability, maintenance and readiness levels; and bolstering air-defence systems. Serbia has agreed to deepen cooperation with NATO through an Individual Partnership Action Plan. Belgrade aspires to join the EU but not NATO. Serbia also maintains a close relationship with Russia, which in recent years has transferred military equipment to Serbia. The armed forces have reduced in size over the last decade, though annual recruitment goals are not met. The armed forces also lack skilled technicians to operate and maintain advanced systems and suffer from a shortage of pilots. Serbia mostly trains with its Balkan neighbours, Russia and NATO countries. Serbia contributes to EU, OSCE and UN peacekeeping missions. Serbia's defence industry focuses on missile and artillery systems, and small arms and ammunition, but the country is reliant on external suppliers for major platforms. Serbia continues to develop its defence industry with a focus on the aerospace industry.

ACTIVE 28,150 (Army 13,250 Air Force and Air Defence 5,100 Training Command 3,000 Guards 1,600 Other MoD 5,200) Paramilitary 3,700

Conscript liability 6 months (voluntary)

RESERVE 50,150

ORGANISATIONS BY SERVICE

Army 13,250

FORCES BY ROLE

SPECIAL FORCES

1 SF bde (1 CT bn, 1 cdo bn, 1 para bn)

MANOEUVRE

Mechanised

1 (1st) bde (1 tk bn, 2 mech inf bn, 1 inf bn, 1 SP arty bn, 1 MRL bn, 1 AD bn, 1 engr bn, 1 log bn)
3 (2nd, 3rd & 4th) bde (1 tk bn, 2 mech inf bn, 2 inf bn, 1 SP arty bn, 1 MRL bn, 1 AD bn, 1 engr bn, 1 log bn)

COMBAT SUPPORT

1 (mixed) arty bde (4 arty bn, 1 MRL bn, 1 spt bn)
2 ptn bridging bn
1 NBC bn
1 sigs bn
2 MP bn

Reserve Organisations

FORCES BY ROLE

MANOEUVRE

Light

8 (territorial) inf bde

EQUIPMENT BY TYPE

ARMOURED FIGHTING VEHICLES

MBT 214: 199 M-84; 13 T-72; 2 T-72MS
RECCE 76: 46 BRDM-2; 30 BRDM-2M
IFV 329: 323 M-80; 6 *Lazar-3*
APC 89
APC(T) 44: 12 BTR-50 (CP); 32 MT-LB (CP)

APC (W) 45: 39 BOV-VP M-86; 6 *Lazar-3*
AUV BOV M16 *Milos*
ENGINEERING & MAINTENANCE VEHICLES
AEV IWT
ARV M84A1; T-54/T-55
VLB MT-55; TMM
ANTI-TANK/ANTI-INFRASTRUCTURE
MSL
SP 48 BOV-1 (M-83) with 9K11 *Malyutka* (AT-3 *Sagger*)
MANPATS 9K11 *Malyutka* (AT-3 *Sagger*); 9K111 *Fagot* (AT-4 *Spigot*)
RCL 90mm M-79
ARTILLERY 449
SP 73+: **122mm** 67 2S1 *Gvozdika*; **155mm** 6+ B-52 NORA
TOWED 132: **122mm** 78 D-30; **130mm** 18 M-46; **152mm** 36 M-84 NORA-A
MRL 81: **128mm** 78: 18 M-63 *Plamen*; 60 M-77 *Organj*; **262mm** 3 M-87 *Orkan*
MOR 163: **82mm** 106 M-69; **120mm** 57 M-74/M-75
AIR DEFENCE
SAM 94+
Short-range 77 2K12 *Kub* (SA-6 *Gainful*);
Point-defence 17+: 12 9K31M *Strela*-1M (SA-9 *Gaskin*); 5 9K35M *Strela*-10M; 9K32M *Strela*-2M (SA-7B *Grail*)‡; *Šilo* (SA-16 *Gimlet*)
GUNS • TOWED 40mm 36 Bofors L/70
UNINHABITED AERIAL VEHICLES
CISR • Medium 6 CH-92A
AIR-LAUNCHED MISSILES
ASM FT-8C

River Flotilla

The Serbian–Montenegrin navy was transferred to Montenegro upon independence in 2006, but the Danube flotilla remained in Serbian control. The flotilla is subordinate to the Land Forces

EQUIPMENT BY TYPE
PATROL AND COASTAL COMBATANTS 5
PBR 5: 3 Type-20; 2 others
MINE WARFARE • MINE COUNTERMEASURES 4
MSI 4 *Nestin* with 1 quad lnchr with 9K32 *Strela*-2M (SA-N-5 *Grail*) SAM
AMPHIBIOUS • LANDING CRAFT
LCVP 5 Type-22
LOGISTICS AND SUPPORT 2
AGF 1 *Kozara*
AOL 1

Air Force and Air Defence 5,100

FORCES BY ROLE
FIGHTER
1 sqn with MiG-21bis *Fishbed*; MiG-29 *Fulcrum*
FIGHTER/GROUND ATTACK
1 sqn with IJ-22 *Orao* 1*; J-22 *Orao* 1
TRANSPORT
1 sqn with An-2; An-26; Yak-40 (Jak-40); 1 PA-34 *Seneca* V
TRAINING
1 sqn with G-4 *Super Galeb** (adv trg/light atk); SA341/342 *Gazelle*; *Lasta* 95; Utva-75 (basic trg)
ATTACK HELICOPTER
1 sqn with SA341H/342L *Gazelle*; (HN-42/45); Mi-24 *Hind*; Mi-35M *Hind*
TRANSPORT HELICOPTER
2 sqn with Mi-8 *Hip*; Mi-17 *Hip* H; Mi-17V-5 *Hip*
AIR DEFENCE
1 bde (5 bn (2 msl, 3 SP msl) with S-125 *Neva* (SA-3 *Goa*); 2K12 *Kub* (SA-6 *Gainful*); 9K32 *Strela*-2 (SA-7 *Grail*); 9K310 *Igla*-1 (SA-16 *Gimlet*))
2 radar bn (for early warning and reporting)
COMBAT SUPPORT
1 sigs bn
COMBAT SERVICE SUPPORT
1 maint bn

EQUIPMENT BY TYPE
AIRCRAFT 61 combat capable
FTR 13+: 2+ MiG-21bis *Fishbed*; 1+ MiG-21UM *Mongol* B; 7 MiG-29 *Fulcrum* (4 more non-operational, undergoing modernisation); 3 MiG-29UB *Fulcrum*
FGA 17 J-22 *Orao* 1
ISR 10 IJ-22R *Orao* 1*
TPT • Light 8: 1 An-2 *Colt*; 4 An-26 *Curl*; 2 Yak-40 (Jak-40); 1 PA-34 *Seneca* V
TRG 46: 21 G-4 *Super Galeb**; 11 Utva-75; 14 *Lasta* 95
HELICOPTERS
ATK 6: 2 Mi-24 *Hind*; 4 Mi-35M *Hind*
MRH 60: 5 H145M; 1 Mi-17 *Hip* H; 5 Mi-17V-5 *Hip*; 2 SA341H *Gazelle* (HI-42); 34 SA341H *Gazelle* (HN-42)/SA342L *Gazelle* (HN-45); 13 SA341H *Gazelle* (HO-42)/SA342L1 *Gazelle* (HO-45)
TPT • Medium 8 Mi-8T *Hip* (HT-40)
AIR DEFENCE
SAM 21+
Short-range 21: 6 S-125 *Pechora* (SA-3 *Goa*); 9 2K12 *Kub* (SA-6 *Gainful*); 6 96K6 *Pantsir*-S1 (SA-22 *Greyhound*)
Point-defence 9K32 *Strela*-2 (SA-7 *Grail*)‡; 9K310 *Igla*-1 (SA-16 *Gimlet*)
GUNS • TOWED 40mm 24 Bofors L/70
AIR-LAUNCHED MISSILES
AAM • IR R-60 (AA-8 *Aphid*)
ASM AGM-65 *Maverick*; A-77 *Thunder*

Guards 1,600

FORCES BY ROLE
MANOEUVRE
Other
1 (ceremonial) gd bde (1 gd bn, 1 MP bn, 1 spt bn)

Paramilitary 3,700

Gendarmerie 3,700

EQUIPMENT BY TYPE
ARMOURED FIGHTING VEHICLES
APC • APC (W) 24: 12 *Lazar-3*; 12 BOV-VP M-86
AUV BOV M16 *Milos*

DEPLOYMENT

CENTRAL AFRICAN REPUBLIC: EU • EUTM RCA 7; **UN** • MINUSCA 76; 1 med coy

CYPRUS: UN • UNFICYP 3

DEMOCRATIC REPUBLIC OF THE CONGO: UN • MONUSCO 1

LEBANON: UN • UNIFIL 178; 1 mech inf coy

MALI: EU • EUTM Mali 3

MIDDLE EAST: UN • UNTSO 1

SOMALIA: EU • EUTM Somalia 6

TERRITORY WHERE THE GOVERNMENT DOES NOT EXERCISE EFFECTIVE CONTROL

Data here represents the de facto situation in Kosovo. This does not imply international recognition as a sovereign state. In February 2008, Kosovo declared itself independent. Serbia remains opposed to this, and while Kosovo has not been admitted to the United Nations, a number of states have recognised Kosovo's self-declared status.

Kosovo Security Force 2,500; reserves 800

The Kosovo Security Force (KSF) was formed in January 2009 as a non-military organisation with responsibility for crisis response, civil protection and EOD. In 2017, a proposal by Pristina to establish an army was opposed by Russia, Serbia, the US and NATO. Legislation to this effect was passed by Pristina in October 2018. In December 2018, NATO said that should the KSF's mandate evolve, it would have to examine its level of engagement with the force. The KSF is armed with small arms and light vehicles only.

FOREIGN FORCES

All under Kosovo Force (KFOR) command unless otherwise specified
Albania 29
Armenia 41
Austria 322; 2 mech inf coy; 1 log coy
Bulgaria 23
Canada 5
Croatia 37; 1 hel flt with Mi-8
Czech Republic 9 • UNMIK 2 obs
Denmark 35
Finland 20
Germany 70
Greece 111; 1 inf coy
Hungary 397; 1 inf coy (KTM)
Ireland 13
Italy 628; 1 mtn inf BG HQ; 1 Carabinieri unit
Lithuania 1
Moldova 41 • UNMIK 1 obs
Montenegro 1
Norway 2
Poland 230; 1 inf coy • UNMIK 1 obs
Romania 53 • UNMIK 1 obs
Slovenia 229; 1 mot inf coy; 1 MP unit; 1 hel unit
Sweden 3
Switzerland 165; 1 inf coy; 1 engr pl; 1 hel flt with AS332
Turkey 311; 1 inf coy • UNMIK 1 obs
Ukraine 40 • UNMIK 2 obs
United Kingdom 21
United States 660; elm 1 ARNG inf bde HQ; 1 ARNG recce bn; 1 hel flt with UH-60

Slovakia SVK

Euro €		2019	2020	2021
GDP	€	94.2bn	89.1bn	
	US$	105bn	102bn	
per capita	US$	19,344	18,669	
Growth	%	2.4	-7.1	
Inflation	%	2.8	1.5	
Def exp [a]	€	1.61bn	1.61bn	
	US$	1.80bn	1.84bn	
Def bdgt	€	1.66bn	1.86bn	1.22bn
	US$	1.86bn	2.12bn	
US$1=€		0.89	0.87	

[a] NATO definition

Real-terms defence budget trend (US$m, constant 2015)

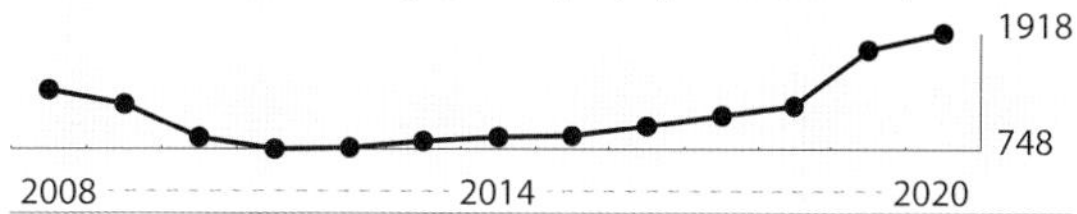

Population 5,440,602

Age	0–14	15–19	20–24	25–29	30–64	65 plus
Male	7.8%	2.4%	2.7%	3.3%	25.5%	6.7%
Female	7.4%	2.3%	2.6%	3.2%	25.7%	10.3%

Capabilities

Slovakia is trying to modernise its armed forces and replace obsolete equipment while contributing to international crisis-management missions. A defence white paper in September 2016 set out security priorities and a plan to increase defence capabilities. In 2017, the government approved a new defence strategy, a military strategy and a Long-Term Defence Development Plan. Work on a new security strategy and a new defence strategy began in July 2020 and was due to be completed by the end of the year. A NATO and EU member state, Slovakia cooperates closely with the Visegrád Group framework. Bratislava has signed an agreement to enable air policing and closer integration of air-defence capabilities. After amending the law on conscription in 2017, Slovakia began to implement its Active Reserves pilot project in order to help address shortfalls in specialist capacities, including in engineering. Results of the pilot project fell short of expectations, and Slovakia passed legislation in early 2018 to improve the training conditions for active reservists from mid-2018 onwards. Slovakia has committed to deploying a company-sized unit to NATO's Enhanced Forward Presence, serving with the Canadian-led battlegroup in Latvia, and has also contributed to EU operations and UN peacekeeping missions. Bratislava has begun to replace its small fighter and rotary-wing-transport fleets. Coinciding with the July 2018 NATO summit, the government announced it had selected the F-16, with delivery due between 2022 and 2024. There are also ambitions to replace land equipment and improve the level of technology in the armed forces. Part of

Slovakia's defence-industrial base is organised within the state-controlled holding company DMD Group, including KONSTRUKTA Defence, which produces land systems. Other companies focus on maintenance, repair and overhaul services.

ACTIVE 15,850 (Army 6,250 Air 3,950 Central Staff 2,550 Support and Training 3,100)

ORGANISATIONS BY SERVICE

Central Staff 2,550

FORCES BY ROLE
SPECIAL FORCES
1 (5th) spec ops bn

Army 6,250

FORCES BY ROLE
MANOEUVRE
Armoured
1 (2nd) armd bde (1 recce bn, 1 tk bn, 2 armd inf bn, 1 mot inf bn, 1 mixed SP arty bn)
Mechanised
1 (1st) mech bde (3 armd inf bn, 1 MRL bn, 1 engr bn, 1 NBC bn)
COMBAT SUPPORT
1 MP bn
COMBAT SERVICE SUPPORT
1 spt bde (2 log bn, 1 maint bn, 1 spt bn)
EQUIPMENT BY TYPE
ARMOURED FIGHITING VEHICLES
MBT 30 T-72M
RECCE 18 BPsVI
IFV 256: 148 BMP-1; 91 BMP-2; 17 BVP-M
APC 101+
APC (T) 72 OT-90
APC (W) 22: 7 OT-64; 15 *Tatrapan* (6×6)
PPV 7+ RG-32M
AUV IVECO LMV
ENGINEERING & MAINTENANCE VEHICLES
ARV MT-55; VT-55A; VT-72B; WPT-TOPAS
VLB AM-50; MT-55A
MW *Bozena*; UOS-155 *Belarty*
ANTI-TANK/ANTI-INFRASTRUCTURE
SP 9S428 with *Malyutka* (AT-3 *Sagger*) on BMP-1; 9P135 *Fagot* (AT-4 *Spigot*) on BMP-2; 9P148 *Konkurs* (AT-5 *Spandrel*) on BRDM-2
MANPATS 9K11 *Malyutka* (AT-3 *Sagger*); 9K111-1 *Konkurs* (AT-5 *Spandrel*)
RCL 84mm *Carl Gustaf*
ARTILLERY 49
SP 19: **152mm** 3 M-77 *Dana;* **155mm** 16 M-2000 *Zuzana*
MRL 30: **122mm** 4 RM-70; **122/227mm** 26 RM-70/85 MODULAR
AIR DEFENCE
SAM • Point-defence 9K310 *Igla*-1 (SA-16 *Gimlet*)

Air Force 3,950

FORCES BY ROLE
FIGHTER
1 sqn with MiG-29AS/UBS *Fulcrum*
TRANSPORT
1 flt with C-27J *Spartan*
1 flt with L-410FG/T *Turbolet*
TRANSPORT HELICOPTER
1 sqn with Mi-17 *Hip* H
1 sqn with UH-60M *Black Hawk*
TRAINING
1 sqn with L-39CM/ZAM *Albatros**
AIR DEFENCE
1 bde with 2K12 *Kub* (SA-6 *Gainful*); S-300PMU (SA-10 *Grumble*)
EQUIPMENT BY TYPE
AIRCRAFT 19 combat capable
FTR 11: 9 MiG-29AS *Fulcrum*; 2 MiG-29UBS *Fulcrum*;
TPT 5: **Medium** 2 C-27J *Spartan*; **Light** 3: 1 L-410FG *Turbolet*; 2 L-410T *Turbolet*; (4 L-410UVP *Turbolet* in store)
TRG 8: 6 L-39CM *Albatros**; 2 L-39ZAM *Albatros** (1 more in store)
HELICOPTERS
ATK (15: 5 Mi-24D *Hind* D; 10 Mi-24V *Hind* E all in store)
MRH 13 Mi-17 *Hip* H (incl 4 SAR)
TPT • Medium 9 UH-60M *Black Hawk*
AIR DEFENCE • SAM
Long-range S-300PMU (SA-10 *Grumble*)
Short-range 2K12 *Kub* (SA-6 *Gainful*)
AIR-LAUNCHED MISSILES
AAM • IR R-60 (AA-8 *Aphid*); R-73 (AA-11A *Archer*)
SARH R-27R (AA-10A *Alamo*)
ASM S5K/S5KO (57mm rockets); S8KP/S8KOM (80mm rockets)

DEPLOYMENT

AFGHANISTAN: NATO • *Operation Resolute Support* 20

BOSNIA-HERZEGOVINA: EU • EUFOR • *Operation Althea* 40

CENTRAL AFRICAN REPUBLIC: EU • EUTM RCA 4

CYPRUS: UN • UNFICYP 241; 2 inf coy; 1 engr pl

IRAQ: NATO • **NATO** Mission Iraq 42

LATVIA: NATO • Enhanced Forward Presence 152; 1 mech inf coy

MALI: EU • EUTM Mali 15

MIDDLE EAST: UN • UNTSO 2

Slovenia SVN

Euro €		2019	2020	2021
GDP	€	48.0bn	45.3bn	
	US$	53.7bn	51.8bn	
per capita	US$	25,992	25,039	
Growth	%	2.4	-6.7	
Inflation	%	1.6	0.5	
Def exp [a]	€	512m	437m	
	US$	573m	500m	
Def bdgt [b]	€	561m	530m	561m
	US$	628m	606m	
US$1=€		0.89	0.87	

[a] NATO definition

[b] Includes military pensions

Real-terms defence budget trend (US$m, constant 2015)

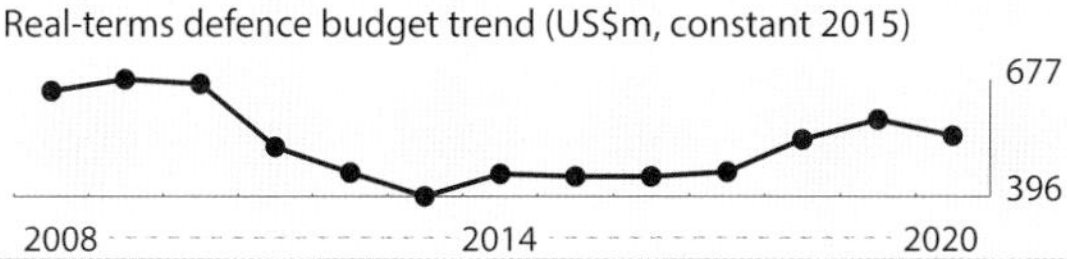

Population 2,102,678

Age	0–14	15–19	20–24	25–29	30–64	65 plus
Male	7.6%	2.3%	2.4%	2.7%	25.8%	9.2%
Female	7.2%	2.1%	2.2%	2.5%	24.0%	12.1%

Capabilities

Since joining NATO and the EU in 2004, territorial defence and the ability to take part in peace-support operations have been central to Slovenia's defence strategy. The defence ministry published a defence white paper in January 2020. This built on the Strategic Defence Review (SDR) of 2016, and acknowledged a deteriorating security environment in Europe. It described the core of Slovenia's combat capabilities as mechanised battalion-sized battle-groups, with the goal to form a first unit by 2027 and a second by 2030. The 2016 SDR had already defined the formation and equipment for two battalion-sized battlegroups by 2023 as its main goal. By 2035, active forces and the contract reserve are to grow to 8,000 and 2,000 personnel, respectively. The white paper indicates that higher defence expenditure will be required to meet modernisation goals. Slovenia acts as the framework nation for the NATO Mountain Warfare Centre of Excellence. Because its small air wing is not equipped to provide air policing, Italy and Hungary currently provide this capability under NATO arrangements. The country contributes to EU, NATO and UN operations and exercises with other member states. Recruitment and retention continues to be a challenge. Slovenia participates in NATO's Enhanced Forward Presence, where it contributes to the Canadian-led battlegroup in Latvia. Slovenia's defence industry relies heavily on exports for its revenue and focuses on individual-solider equipment, small arms and ammunition, and CBRN protection and detection.

ACTIVE 6,950 (Army 6,950)

RESERVE 1,200 (Army 1,200)

ORGANISATIONS BY SERVICE

Army 6,950

FORCES BY ROLE

Regt are bn sized

SPECIAL FORCES

1 SF unit (1 spec ops coy, 1 CSS coy)

MANOEUVRE

Mechanised

1 (1st) mech inf bde (1 mech inf regt, 1 mtn inf regt, 1 cbt spt bn (1 ISR coy, 1 arty bty, 1 engr coy, 1 MP coy, 1 CBRN coy, 1 sigs coy, 1 SAM bty))

1 (72nd) mech inf bde (2 mech inf regt, 1 cbt spt bn (1 ISR coy, 1 arty bty, 1 engr coy, 1 MP coy, 1 CBRN coy, 1 sigs coy, 1 SAM bty))

COMBAT SUPPORT

1 EW coy

COMBAT SERVICE SUPPORT

1 log bde (1 log regt, 1 maint regt (1 tk coy), 1 med regt)

Reserves

FORCES BY ROLE

MANOEUVRE

Mountain

2 inf regt (territorial – 1 allocated to each inf bde)

EQUIPMENT BY TYPE

ARMOURED FIGHTING VEHICLES

MBT 14 M-84 (trg role) (32 more in store)

APC 115+:

APC (W) 115: 85 *Pandur* 6×6 (*Valuk*); 30 Patria 8×8 (*Svarun*)

PPV *Cougar* 6×6 JERRV

ENGINEERING & MAINTENANCE VEHICLES

ARV VT-55A

VLB MT-55A

NBC VEHICLES 10 *Cobra* CBRN

ANTI-TANK/ANTI-INFRASTRUCTURE

MSL • MANPATS *Spike* MR/LR

ARTILLERY 68

TOWED • 155mm 18 TN-90

MOR 50+: **82mm** M-69; **120mm** 50 MN-9/M-74

AIR DEFENCE • **SAM • Point-defence** 9K338 *Igla*-S (SA-24 *Grinch*)

Army Maritime Element 130

FORCES BY ROLE

SPECIAL FORCES

1 SF unit

EQUIPMENT BY TYPE

PATROL AND COASTAL COMBATANTS 2

PCC 1 *Triglav* III (RUS *Svetlyak*)

PBF 1 *Super Dvora* MkII

Air Element 590

FORCES BY ROLE

TRANSPORT

1 sqn with *Falcon* 2000EX; L-410 *Turbolet*; PC-6B *Turbo Porter*;

TRAINING

1 unit with Bell 206 *Jet Ranger* (AB-206); PC-9M*; Z-143L; Z-242L

TRANSPORT HELICOPTER

1 sqn with AS532AL *Cougar*; Bell 412 *Twin Huey*

COMBAT SERVICE SUPPORT
1 maint sqn

EQUIPMENT BY TYPE

AIRCRAFT 9 combat capable
TPT 4: **Light** 3: 1 L-410 *Turbolet*; 2 PC-6B *Turbo Porter*
PAX 1 *Falcon* 2000EX
TRG 19: 9 PC-9M*; 2 Z-143L; 8 Z-242L

HELICOPTERS
MRH 8: 5 Bell 412EP *Twin Huey*; 2 Bell 412HP *Twin Huey*; 1 Bell 412SP *Twin Huey* (some armed)
TPT 8: **Medium** 4 AS532AL *Cougar*; **Light** 4 Bell 206 *Jet Ranger* (AB-206)

DEPLOYMENT

AFGHANISTAN: NATO • *Operation Resolute Support* 6

BOSNIA-HERZEGOVINA: EU • EUFOR • *Operation Althea* 10

IRAQ: *Operation Inherent Resolve* 6

LATVIA: NATO • Enhanced Forward Presence 41; 1 mor pl

LEBANON: UN • UNIFIL 6

MALI: EU • EUTM Mali 8

MIDDLE EAST: UN • UNTSO 2

SERBIA: NATO • KFOR 229; 1 mot inf coy; 1 MP unit; 1 hel unit

Spain ESP

Euro €		2019	2020	2021
GDP	€	1.25tr	1.09tr	
	US$	1.39tr	1.25tr	
per capita	US$	29,993	26,832	
Growth	%	2.0	-12.8	
Inflation	%	0.7	-0.2	
Def exp [a]	€	11.3bn	12.9bn	
	US$	12.6bn	14.8bn	
Def bdgt [b]	€	11.5bn	ε11.4bn	
	US$	12.9bn	ε13.0bn	
US$1=€		0.89	0.87	

[a] NATO definition
[b] Includes military pensions

Real-terms defence budget trend (US$bn, constant 2015)

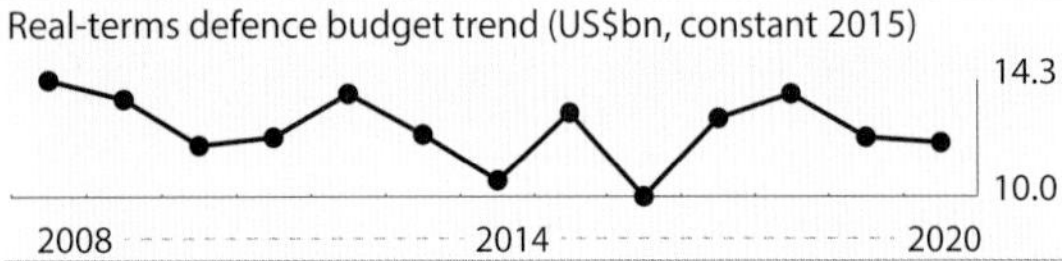

Population 50,015,792

Age	0–14	15–19	20–24	25–29	30–64	65 plus
Male	7.7%	2.6%	2.5%	2.7%	25.9%	7.9%
Female	7.3%	2.5%	2.3%	2.5%	25.4%	10.6%

Capabilities

The 2017 National Security Strategy indicated that Spain's defence policy was global in scope, though concerned with threats emanating from the Middle East and sub-Saharan Africa. The army reviewed its force structure in 2015, which resulted in a reorganisation into multipurpose brigades with heavy, medium and light capabilities, optimised for deployable operations. Spain is a member of NATO, continues to support NATO, EU and UN operations abroad, and hosts one of NATO's two Combined Air Operations Centres. The armed forces are well trained and there is a routine exercise programme for both domestic and multinational exercises. The country's equipment and logistic-support capability appears to be sufficient to meet its national commitments and contribution to NATO operations and exercises. In early 2018, Spain launched an equipment-modernisation plan, with funding for the modernisation of army *Chinook* helicopters, for the S-80 submarine programme and for military-communications satellites. Spain's defence industry manufactures across all domains and exports globally. Navantia is the principal, state-owned, shipbuilding firm. The industry is largely integrated within the European defence-industrial manufacturing base.

ACTIVE 122,850 (Army 71,300 Navy 20,350 Air 19,750 Joint 11,450) Paramilitary 75,800

RESERVE 14,900 (Army 8,800 Navy 3,150 Air 2,350 Other 600)

ORGANISATIONS BY SERVICE

Space

EQUIPMENT BY TYPE

SATELLITES 3
COMMUNICATIONS 2: 1 *Spainsat*; 1 *Xtar-Eur*
ISR 1 *Paz*

Army 71,300

The Land Forces High Readiness HQ Spain provides one NATO Rapid Deployment Corps HQ (NRDC-ESP)

FORCES BY ROLE

COMMAND
1 corps HQ (CGTAD/NRDC-ESP) (1 int regt, 1 MP bn)
2 div HQ

SPECIAL FORCES
1 comd (3 spec ops bn, 1 int coy, 1 sigs coy, 1 log bn)

MANOEUVRE

Reconnaissance
1 armd cav regt (2 armd recce bn)

Mechanised
2 (10th & 11th) mech bde (1 armd regt (1 armd recce bn, 1 tk bn), 1 mech inf regt (1 armd inf bn, 1 mech inf bn), 1 lt inf bn, 1 SP arty bn, 1 AT coy, 1 AD coy, 1 engr bn, 1 int coy, 1 NBC coy, 1 sigs coy, 1 log bn)
1 (12th) mech bde (1 armd regt (1 armd recce bn, 1 tk bn), 1 mech inf regt (1 armd inf bn, 1 mech inf bn), 1 mtn inf bn, 1 SP arty bn, 1 AT coy, 1 AD coy, 1 engr bn, 1 int coy, 1 NBC coy, 1 sigs coy, 1 log bn)
1 (1st) mech bde (1 armd regt (1 armd recce bn, 1 tk bn), 1 mech inf regt (1 armd inf bn, 1 mtn inf bn), 1 mtn inf bn, 1 SP arty bn, 1 AT coy, 1 AD coy, 1 engr bn, 1 int coy, 1 NBC coy, 1 sigs coy, 1 log bn)
2 (2nd/La Legion & 7th) lt mech bde (1 armd recce bn, 1 mech inf regt (2 mech inf bn), 1 lt inf bn, 1 fd arty bn, 1 AT coy, 1 AD coy, 1 engr bn, 1 int coy, 1 NBC coy, 1 sigs coy, 1 log bn)

Air Manoeuvre
1 (6th) bde (1 recce bn, 2 para bn, 1 lt inf bn, 1 fd arty bn, 1 AT coy, 1 AD coy, 1 engr bn, 1 int coy, 1 NBC coy, 1 sigs coy, 1 log bn)
Other
1 (Canary Islands) comd (1 lt inf bde (2 mech inf regt (1 mech inf bn), 1 lt inf regt (1 lt inf bn), 1 fd arty regt, 1 AT coy, 1 engr bn, 1 int coy, 1 NBC coy, 1 sigs coy, 1 log bn); 1 spt hel bn; 1 AD regt)
1 (Balearic Islands) comd (1 inf regt)
2 (Ceuta and Melilla) comd (1 recce regt, 1 mech inf bn, 1 inf bn, 1 arty regt, 1 engr bn, 1 sigs coy, 1 log bn)
COMBAT SUPPORT
1 arty comd (1 arty regt; 1 MRL regt; 1 coastal arty regt)
1 engr comd (2 engr regt, 1 bridging regt)
1 EW/sigs bde (2 EW regt, 3 sigs regt)
1 NBC regt
1 info ops regt (1 CIMIC bn; 1 Psyops bn)
1 int regt
COMBAT SERVICE SUPPORT
1 log bde (5 log regt; 1 tpt regt; 1 med gp (1 log unit, 2 med regt, 1 fd hospital unit))
HELICOPTER
1 hel comd (1 atk hel bn, 2 spt hel bn, 1 tpt hel bn, 1 sigs bn, 1 log unit (1 spt coy, 1 supply coy))
AIR DEFENCE
1 AD comd (3 SAM regt, 1 sigs unit)

EQUIPMENT BY TYPE
ARMOURED FIGHTING VEHICLES
MBT 327: 108 *Leopard* 2A4; 219 *Leopard* 2E
ASLT 84 B1 *Centauro*
RECCE 187 VEC-M1
IFV 225: 204 *Pizarro*; 21 *Pizarro* (CP)
APC 903
APC (T) 473: 20 Bv-206S; 453 M113 (incl variants)
APC (W) 320 BMR-600/BMR-600M1
PPV 110 RG-31
AUV 260 IVECO LMV
ENGINEERING & MAINTENANCE VEHICLES
AEV 26 CZ-10/25E
ARV 51: 16 *Leopard* REC; 5 BMR REC; 4 *Centauro* REC; 14 *Maxxpro* MRV; 12 M113
VLB 15 M60 AVLB
MW 6 *Husky* 2G
ANTI-TANK/ANTI-INFRASTRUCTURE
MSL • MANPATS *Spike*-LR; TOW
ARTILLERY 1,571
SP 155mm 95 M109A5
TOWED 287: **105mm** 223: 56 L118 Light Gun; 167 Model 56 pack howitzer; **155mm** 64 SBT 155/52 SIAC
MOR 1,189: **81mm** 777; **SP 81mm** 10 VAMTAC with *Cardom* 81mm; **120mm** 402
COASTAL DEFENCE • ARTY 155mm 19 SBT 155/52 APU SBT V07
HELICOPTERS
ATK 24: 6 *Tiger* HAP-E; 18 *Tiger* HAD-E
TPT 85: **Heavy** 17 CH-47D *Chinook* (HT-17D); **Medium** 47: 16 AS332B *Super Puma* (HU-21); 12 AS532UL *Cougar*; 6 AS532AL *Cougar*; 13 NH90 TTH; **Light** 21: 5 Bell 212 (HU.18); 16 H135 (HE.26/HU.26)
UAV • ISR • Medium 6: 2 *Searcher* MkII-J (PASI); 4 *Searcher* MkIII (PASI)
AIR DEFENCE
SAM 77+
Long-range 18 MIM-104C *Patriot* PAC-2
Medium-range 38 MIM-23B I-*Hawk* Phase III
Short-range 21: 8 NASAMS; 13 *Skyguard/Aspide*
Point-defence *Mistral*
GUNS • TOWED 35mm 67: 19 GDF-005; 48 GDF-007

Navy 20,350 (incl Naval Aviation and Marines)

EQUIPMENT BY TYPE
SUBMARINES 2
SSK 2 *Galerna* with 4 single 533mm TT with F17 mod 2 HWT
PRINCIPAL SURFACE COMBATANTS 11
DESTROYERS • DDGHM 5 *Alvaro de Bazan* with *Aegis* Baseline 5 C2, 2 quad lnchr with RGM-84F *Harpoon* Block 1D AShM, 6 8-cell Mk 41 VLS with SM-2 Block IIIA/RIM-162B ESSM SAM, 2 twin 324mm SVTT Mk 32 mod 9 ASTT with Mk 46 mod 5 LWT, 1 127mm gun (capacity 1 SH-60B *Seahawk* ASW hel)
FRIGATES • FFGH 6 *Santa Maria* with 1 Mk 13 GMLS with RGM-84C *Harpoon* Block 1B AShM, 2 triple 324mm SVTT Mk 32 ASTT with Mk 46 mod 5 LWT, 1 *Meroka* mod 2B CIWS, 1 76mm gun (capacity 2 SH-60B *Seahawk* ASW hel)
AMPHIBIOUS
PRINCIPAL AMPHIBIOUS SHIPS 3:
LHD 1 *Juan Carlos* I (capacity 18 hel or 10 AV-8B FGA ac; 4 LCM-1E; 42 APC; 46 MBT; 900 troops)
LPD 2 *Galicia* (capacity 6 Bell 212 or 4 SH-3D *Sea King* hel; 4 LCM or 2 LCM & 8 AAV; 130 APC or 33 MBT; 540 troops)
LANDING CRAFT • LCM 12 LCM 1E
LOGISTICS AND SUPPORT 2
AORH 2: 1 *Patino* (capacity 3 Bell 212 or 2 SH-3D *Sea King* hel); 1 *Cantabria* (capacity 3 Bell 212 or 2 SH-3D *Sea King* hel)

Maritime Action Force

EQUIPMENT BY TYPE
PATROL AND COASTAL COMBATANTS 23
PSOH 6 *Meteoro* (*Buques de Accion Maritima*) with 1 76mm gun
PSO 5: 3 *Alboran* each with 1 hel landing platform; 2 *Descubierta* with 1 76mm gun
PCO 4 *Serviola* with 1 76mm gun
PCC 3 *Anaga* with 1 76mm gun
PB 4: 2 P-101; 2 *Toralla*
PBR 1 *Cabo Fradera*
MINE WARFARE • MINE COUNTERMEASURES 6
MHO 6 *Segura*
LOGISTICS AND SUPPORT 29
AGI 1 *Alerta*
AGOR 2 (with ice-strengthened hull, for polar research duties in Antarctica)
AGS 3: 2 *Malaspina*; 1 *Castor*
AP 1 *Contramaestre Casado* with 1 hel landing platform
ASR 1 *Neptuno*

ATF 3: 1 *Mar Caribe*; 1 *Mahon*; 1 *La Grana*
AXL 10: 6 *Contramaestre*; 4 *Guardiamarina*
AXS 8

Naval Aviation 850

FORCES BY ROLE
FIGHTER/GROUND ATTACK
1 sqn with AV-8B *Harrier* II Plus
ANTI-SUBMARINE WARFARE
1 sqn with SH-60B/F *Seahawk*
TRANSPORT
1 (liaison) sqn with Cessna 550 *Citation* II; Cessna 650 *Citation* VII
TRAINING
1 sqn with Hughes 500MD8
1 flt with TAV-8B *Harrier*
TRANSPORT HELICOPTER
1 sqn with Bell 212 (HU-18)
1 sqn with SH-3D *Sea King*

EQUIPMENT BY TYPE
AIRCRAFT 13 combat capable
FGA 13: 12 AV-8B *Harrier* II Plus; 1 TAV-8B *Harrier* (on lease from USMC)
TPT • Light 4: 3 Cessna 550 *Citation* II; 1 Cessna 650 *Citation* VII
HELICOPTERS
ASW 20: 4 SH-3D *Sea King* (tpt); 12 SH-60B *Seahawk*; 4 SH-60F *Seahawk*
MRH 9 Hughes 500MD
TPT • Light 7 Bell 212 (HA-18)
AIR-LAUNCHED MISSILES
AAM • IR AIM-9L *Sidewinder*; **ARH** AIM-120 AMRAAM
ASM AGM-65G *Maverick*
AShM AGM-119 *Penguin*

Marines 5,350

FORCES BY ROLE
SPECIAL FORCES
1 spec ops bn
MANOEUVRE
Amphibious
1 mne bde (1 recce unit, 1 mech inf bn, 2 inf bn, 1 arty bn, 1 log bn)
Other
1 sy bde (5 mne garrison gp)

EQUIPMENT BY TYPE
ARMOURED FIGHTING VEHICLES
APC • APC (W) 34: 32 *Piranha* IIIC; 1 *Piranha* IIIC (amb); 1 *Piranha* IIIC EW (EW)
AAV 18: 16 AAV-7A1/AAVP-7A1; 2 AAVC-7A1 (CP)
ENGINEERING & MAINTENANCE VEHICLES
AEV 4 *Piranha* IIIC
ARV 3: 1 AAVR-7A1; 1 M88; 1 *Piranha* IIIC
ARTILLERY 30
SP 155mm 6 M109A2
TOWED 105mm 24 Model 56 pack howitzer
ANTI-TANK/ANTI-INFRASTRUCTURE
MSL • MANPATS *Spike*-LR; TOW-2
AIR DEFENCE • SAM • Point-defence *Mistral*

Air Force 19,750

The Spanish Air Force is organised in 3 commands – General Air Command, Combat Air Command and Canary Islands Air Command

FORCES BY ROLE
FIGHTER
2 sqn with Eurofighter *Typhoon*
FIGHTER/GROUND ATTACK
5 sqn with F/A-18A/B MLU *Hornet* (EF-18A/B MLU)
MARITIME PATROL
1 sqn with P-3A/M *Orion*
ISR
1 sqn with Beech C90 *King Air*
1 sqn with Cessna 550 *Citation* V; CN235 (TR-19A)
ELECTRONIC WARFARE
1 sqn with C-212 *Aviocar*; *Falcon* 20D
SEARCH & RESCUE
1 sqn with AS332B/B1 *Super Puma*; CN235 VIGMA
1 sqn with AS332B *Super Puma*; CN235 VIGMA; H215 (AS332C1) *Super Puma*
1 sqn with C-212 *Aviocar*; CN235 VIGMA
TANKER/TRANSPORT
1 sqn with KC-130H *Hercules*
TRANSPORT
1 VIP sqn with A310; *Falcon* 900
1 sqn with C-130H/H-30 *Hercules*; A400M
1 sqn with C-212 *Aviocar*
2 sqn with C295
1 sqn with CN235
TRAINING
1 OCU sqn with Eurofighter *Typhoon*
1 OCU sqn with F/A-18A/B (EF-18A/B MLU) *Hornet*
1 sqn with Beech F33C *Bonanza*
2 sqn with C-101 *Aviojet*
1 sqn with C-212 *Aviocar*
1 sqn with T-35 *Pillan* (E-26)
2 (LIFT) sqn with F-5B *Freedom Fighter*
1 hel sqn with H120 *Colibri*
1 hel sqn with S-76C
TRANSPORT HELICOPTER
1 sqn with AS332M1 *Super Puma*; AS532UL *Cougar* (VIP)
ISR UAV
1 sqn with MQ-9A *Reaper* (forming)

EQUIPMENT BY TYPE
AIRCRAFT 175 combat capable
FTR 88: 69 Eurofighter *Typhoon*; 19 F-5B *Freedom Fighter*
FGA 84: 20 F/A-18A *Hornet* (EF-18A); 52 EF-18A MLU; 12 EF-18B MLU
ASW 3 P-3M *Orion*
MP 8 CN235 VIGMA
ISR 2 CN235 (TR-19A)
EW 3: 1 C-212 *Aviocar* (TM.12D); 2 *Falcon* 20D
TKR 5 KC-130H *Hercules*
TPT 71: **Heavy** 8 A400M; **Medium** 5: 4 C-130H *Hercules*; 1 C-130H-30 *Hercules*; **Light** 51: 3 Beech C90 *King Air*; 15 Beech F33C *Bonanza*; 10 C-212 *Aviocar* (incl 9 trg); 12 C295; 8 CN235; 3 Cessna 560 *Citation* V (ISR); **PAX** 7: 2 A310; 5 *Falcon* 900 (VIP)
TRG 95: 59 C-101 *Aviojet*; 36 T-35 *Pillan* (E-26)

HELICOPTERS

TPT 38: **Medium** 16: 5 AS332B/B1 *Super Puma*; 4 AS332M1 *Super Puma*; 4 H215 (AS332C1) *Super Puma*; 2 AS532UL *Cougar* (VIP); 1 NH90 TTH; **Light** 22: 14 H120 *Colibri*; 8 S-76C

UNINHABITED AERIAL VEHICLES • CISR • Heavy 2 MQ-9A *Reaper*

AIR DEFENCE • SAM

Short-range *Skyguard/Aspide*

Point-defence *Mistral*

AIR-LAUNCHED MISSILES

AAM • IR AIM-9L/JULI *Sidewinder*; **IIR** IRIS-T; **SARH** AIM-7P *Sparrow*; **ARH** AIM-120B/C AMRAAM

ARM AGM-88B HARM

ASM AGM-65G *Maverick*

AShM AGM-84D *Harpoon*

LACM Taurus KEPD 350

BOMBS

Laser-guided: GBU-10/12/16 *Paveway* II; GBU-24 *Paveway* III; EGBU-16 *Paveway* II; BPG-2000

INS/GPS guided: GBU-38 JDAM

Emergencies Military Unit (UME) 3,500

FORCES BY ROLE

COMMAND

1 div HQ

MANOEUVRE

Other

5 Emergency Intervention bn

1 Emergency Support and Intervention regt

COMBAT SUPPORT

1 sigs bn

HELICOPTER

1 hel bn opcon Army

Paramilitary 75,800

Guardia Civil 75,800

17 regions, 54 Rural Comds

FORCES BY ROLE

SPECIAL FORCES

8 (rural) gp

MANOEUVRE

Other

15 (traffic) sy gp

1 (Special) sy bn

EQUIPMENT BY TYPE

PATROL AND COASTAL COMBATANTS 64

PSO 1 with 1 hel landing platform

PCC 2

PBF 34

PB 27

AIRCRAFT • TPT • Light 3: 2 CN235-300; 1 King Air 350i

HELICOPTERS

MRH 20: 2 AS653N3 *Dauphin*; 18 Bo-105ATH

TPT • Light 21: 8 BK-117; 13 H135

DEPLOYMENT

AFGHANISTAN: NATO • *Operation Resolute Support* 36

BOSNIA-HERZEGOVINA: EU • EUFOR **•** *Operation Althea* 3

CENTRAL AFRICAN REPUBLIC: EU • EUTM RCA 8

DJIBOUTI: EU • *Operation Atalanta* 1 P-3M *Orion*

GULF OF ADEN & INDIAN OCEAN: EU • *Operation Atalanta* 1 FFGHM

IRAQ: *Operation Inherent Resolve* 500; 1 trg unit; 1 hel unit with CH-47D *Chinook*; AS532 *Cougar*; **NATO • NATO** Mission Iraq 13

LATVIA: NATO • Enhanced Forward Presence 343; 1 armd inf coy(+); 1 cbt engr coy

LEBANON: UN • UNIFIL 630; 1 mech bde HQ; 1 mech inf bn(-); 1 engr coy; 1 sigs coy

MALI: EU • EUTM Mali 235

MEDITERRANEAN SEA: NATO • SNMG 2: 1 DDGHM; **NATO •** SNMCMG 2: 1 MHO

SENEGAL: *Operation Barkhane* 60; 2 C295M

SOMALIA: EU • EUTM Somalia 20

TURKEY: NATO • *Operation Active Fence* 150; 1 SAM bty with MIM-104C *Patriot* PAC-2

FOREIGN FORCES

United States US European Command: 3,300; 1 air base at Morón; 1 naval base at Rota

Sweden SWE

Swedish Krona Skr		2019	2020	2021
GDP	Skr	5.02tr	4.86tr	
	US$	531bn	529bn	
per capita	US$	51,404	50,339	
Growth	%	1.3	-4.7	
Inflation	%	1.6	0.8	
Def bdgt	Skr	60.3bn	64.8bn	71.2bn
	US$	6.37bn	7.05bn	
US$1=Skr		9.46	9.19	

Real-terms defence budget trend (US$bn, constant 2015)

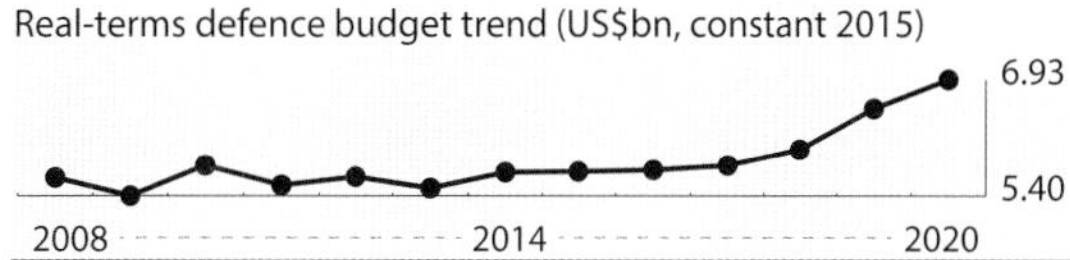

Population	10,202,491

Age	0–14	15–19	20–24	25–29	30–64	65 plus
Male	9.1%	2.8%	2.8%	3.5%	22.3%	9.6%
Female	8.6%	2.6%	2.6%	3.4%	21.8%	11.0%

Capabilities

Sweden's armed forces remain configured for territorial defence and there has been growing concern at Russian military activity in the Baltic area. There has also been a focus on increasing cooperation with neighbours and NATO in recent years. The 2021–25 defence bill, presented in October 2020, envisages increased spending and an expanded force structure with the re-establishment of five regiments and one air wing. Sweden also announced investments in what it calls civilian defence, effectively measures to enhance societal resilience and the ability to deal with civil emergencies. Sweden decided to relocate its service staffs from Stockholm to other locations in 2019 in order to provide better protection. There are plans to increase defence ties with the UK and the US. In July 2019, Sweden joined the UK-led *Tempest* future-combat-aircraft programme. Concerns over readiness levels have led to greater cooperation with NATO and NORDEFCO. In May 2018, Sweden, Finland and the US signed a statement of intent to develop closer cooperation on exercises and interoperability. Readiness, exercises and training, and intelligence, as well as cyber defence, are spending priorities. Amid recruitment challenges, Sweden announced in March 2017 that it would reinstate conscription from January 2018. Sweden has started to re-garrison the island of Gotland. Readiness challenges in the air force triggered a discussion about extending the service life of the JAS-39C *Gripen* Cs beyond their intended 2026 retirement date, not least since the air force was slated to receive a lower number of JAS-39Es than requested. In October 2020 it was confirmed by the government and industry that a number of *Gripen* Cs will be retained until at least 2030. In August 2018, Sweden proceeded with the acquisition of the *Patriot* air-defence system. The country's export-oriented defence industry is privately owned and capable of meeting most of the armed forces' equipment needs, including for advanced combat aircraft and conventional submarines.

ACTIVE 14,600 (Army 6,850 Navy 2,100 Air 2,700 Other 2,950) Voluntary Auxiliary Organisations 21,200

Conscript liability 4–11 months, depending on branch (selective conscription; 4,000 in total, gender neutral)

RESERVE 10,000

ORGANISATIONS BY SERVICE

Army 6,850

The army has been transformed to provide brigade-sized task forces depending on the operational requirement

FORCES BY ROLE

COMMAND
2 bde HQ

MANOEUVRE
Reconnaissance
1 recce bn
Armoured
5 armd bn
1 armd BG
Mechanised
1 mech bn
Light
1 mot inf bn
1 lt inf bn
Air Manoeuvre
1 AB bn
Other
1 sy bn

COMBAT SUPPORT
2 arty bn
2 engr bn
2 MP coy
1 CBRN coy

COMBAT SERVICE SUPPORT
1 tpt coy

AIR DEFENCE
2 AD bn

Reserves

FORCES BY ROLE

MANOEUVRE
Other
40 Home Guard bn

EQUIPMENT BY TYPE

ARMOURED FIGHTING VEHICLES
MBT 120 *Leopard* 2A5 (Strv 122)
IFV 411: 369 CV9040 (Strf 9040; incl 54 CP); 42 Epbv 90 (OP)
APC 1,064
APC (T) 389: 239 Pbv 302 (incl variants); 150 BvS-10 MkII
APC (W) 315: 34 XA-180 *Sisu* (Patgb 180); 20 XA-202 *Sisu* (Patgb 202); 148 XA-203 *Sisu* (Patgb 203); 113 Patria AMV (XA-360/Patgb 360)
PPV 360 RG-32M

ENGINEERING & MAINTENANCE VEHICLES
AEV 6 Pionierpanzer-3 *Kodiak* (Ingbv 120)
ARV 40: 14 Bgbv 120; 26 Bgbv 90
VLB 3 Brobv 120
MW 33+: *Aardvark* Mk2; 33 Area Clearing System

ANTI-TANK/ANTI-INFRASTRUCTURE
MSL • MANPATS NLAW; RBS-55
RCL 84mm *Carl Gustaf*

ARTILLERY 321
SP 155mm 27 *Archer*
MOR 294: **81mm** 201 M/86; **120mm** 81 M/41D **SP 120mm** 12 CV90 *Mjolnir* (Gkpbv 90)

AIR DEFENCE
SAM 8+
Medium-range MIM-23B *Hawk* (RBS-97)
Short-range 8 IRIS-T SLS (RBS-98); RBS-23 BAMSE
Point-defence RBS-70
GUNS • SP 40mm 30 Lvkv 90

Navy 1,250; 850 Amphibious (total 2,100)

EQUIPMENT BY TYPE

SUBMARINE • **SSK** 5:
1 *Gotland* (fitted with AIP) with 2 single 400mm TT with Torped 431 LWT/Torped 451 LWT, 4 single 533mm TT with Torped 613 HWT/Torped 62 HWT
2 *Gotland* mod (fitted with AIP) with 2 single 400mm TT with Torped 431 LWT/Torped 451 LWT, 4 single 533mm TT with Torped 613 HWT/Torped 62 HWT
2 *Sodermanland* (fitted with AIP) with 3 single 400mm TT with Torped 431 LWT/Torped 451 LWT, 6 single 533mm TT with Torped 613 HWT/Torped 62 HWT

PATROL AND COASTAL COMBATANTS 150
CORVETTES • FSG 5 *Visby* with 8 RBS15 Mk2 AShM, 4 single 400mm ASTT with Torped 45 LWT, 1 57mm gun, 1 hel landing platform

PCGT 4:
- 2 *Göteborg* with 4 twin lnchr with RBS15 Mk2 AShM, 4 single 400mm ASTT with Torped 431 LWT, 1 57mm gun
- 2 *Stockholm* with 4 twin lnchr with RBS15 Mk2 AShM, 4 single 400mm ASTT with Torped 431 LWT, 1 57mm gun

PBF 133: 100+ Combat Boat 90H (capacity 18 troops); 27 Combat Boat HS (capacity 18 troops); 6 Combat Boat 90HSM (capacity 18 troops)
PB 8 *Tapper* (Type 80)

MINE WARFARE • MINE COUNTERMEASURES 7
MCC 5 *Koster* (SWE *Landsort* mod)
MCD 2 *Spårö* (*Styrsö* mod)

AMPHIBIOUS • LANDING CRAFT 11
LCVP 8 *Trossbat*
LCAC 3 *Griffon* 8100TD

LOGISTICS AND SUPPORT 15
AG 2: 1 *Carlskrona* with 1 hel landing platform (former ML); 1 *Trosso* (spt ship for corvettes and patrol vessels but can also be used as HQ ship)
AGF 2 *Ledningsbåt* 2000
AGI 1 *Orion*
AKL 1 *Loke*
ARS 2: 1 *Belos* III; 1 *Furusund* (former ML)
AX 5 *Altair*
AXS 2: 1 *Falken*; 1 *Gladan*

Amphibious 850

FORCES BY ROLE
MANOEUVRE
Amphibious
1 amph bn

EQUIPMENT BY TYPE
ARTILLERY • MOR 81mm 12 M/86
COASTAL DEFENCE • AShM 8 RBS-17 *Hellfire*

Coastal Defence

FORCES BY ROLE
COASTAL DEFENCE
1 AShM bty with RBS-15

EQUIPMENT BY TYPE
COASTAL DEFENCE • AShM RBS-15

Air Force 2,700

FORCES BY ROLE
FIGHTER/GROUND ATTACK/ISR
6 sqn with JAS 39C/D *Gripen*
TRANSPORT/ISR/AEW&C
1 sqn with C-130H *Hercules* (Tp-84); KC-130H *Hercules* (Tp-84); Gulfstream IV SRA-4 (S-102B); S-100B/D *Argus*
TRAINING
1 unit with Sk-60
AIR DEFENCE
1 (fighter control and air surv) bn

EQUIPMENT BY TYPE
AIRCRAFT 96 combat capable
FGA 96 JAS 39C/D *Gripen*
ELINT 2 Gulfstream IV SRA-4 (S-102B)
AEW&C 3: 1 S-100B *Argus*; 2 S-100D *Argus*
TKR 1 KC-130H *Hercules* (Tp-84)
TPT 8: **Medium** 5 C-130H *Hercules* (Tp-84); **Light** 2 Saab 340 (OS-100A/Tp-100C); **PAX** 1 Gulfstream 550 (Tp-102D)
TRG 67 Sk-60W

UNINHABITED AERIAL VEHICLES
ISR • Medium 8 RQ-7 *Shadow* (AUV 3 *Örnen*)

AIR-LAUNCHED MISSILES
ASM AGM-65 *Maverick* (RB-75)
AShM RB-15F
AAM • IR AIM-9L *Sidewinder* (RB-74); **IIR** IRIS-T (RB-98); **ARH** AIM-120B AMRAAM (RB-99); *Meteor*

BOMBS
Laser-Guided GBU-12 *Paveway* II
INS/GPS guided GBU-39 Small Diameter Bomb

Armed Forces Hel Wing

FORCES BY ROLE
TRANSPORT HELICOPTER
3 sqn with AW109 (Hkp 15A); AW109M (Hkp-15B); NH90 TTH (Hkp-14) (SAR/ASW); UH-60M *Black Hawk* (Hkp-16)

EQUIPMENT BY TYPE
HELICOPTERS
TPT 53: **Medium** 33: 15 UH-60M *Black Hawk* (Hkp-16); 18 NH90 TTH (Hkp-14) (of which 9 configured for ASW); **Light** 20: 12 AW109 (Hkp-15A); 8 AW109M (Hkp-15B)

Special Forces

FORCES BY ROLE
SPECIAL FORCES
1 spec ops gp
COMBAT SUPPORT
1 cbt spt gp

Other 2,950

Includes staff, logisitics and intelligence personnel

FORCES BY ROLE
COMBAT SUPPORT
1 EW bn
1 psyops unit
COMBAT SERVICE SUPPORT
2 log bn
1 maint bn
4 med coy
1 tpt coy

DEPLOYMENT

AFGHANISTAN: NATO • *Operation Resolute Support* 13

CENTRAL AFRICAN REPUBLIC: EU • EUTM RCA 9

INDIA/PAKISTAN: UN • UNMOGIP 4

IRAQ: *Operation Inherent Resolve* 66; **NATO • NATO** Mission Iraq 1

KOREA, REPUBLIC OF: NNSC • 5 obs

MALI: EU • EUTM Mali 6; **UN** • MINUSMA 105; 1 int coy

MIDDLE EAST: UN • UNTSO 7

SERBIA: NATO • KFOR 3

SOMALIA: EU • EUTM Somalia 9

WESTERN SAHARA: UN • MINURSO 3

Switzerland CHE

Swiss Franc fr		2019	2020	2021
GDP	fr	700bn	663bn	
	US$	705bn	708bn	
per capita	US$	82,484	81,867	
Growth	%	1.2	-5.3	
Inflation	%	0.4	-0.8	
Def bdgt [a]	fr	5.31bn	5.37bn	5.45bn
	US$	5.34bn	5.74bn	
US$1=fr		0.99	0.94	

[a] Includes military pensions

Real-terms defence budget trend (US$bn, constant 2015)

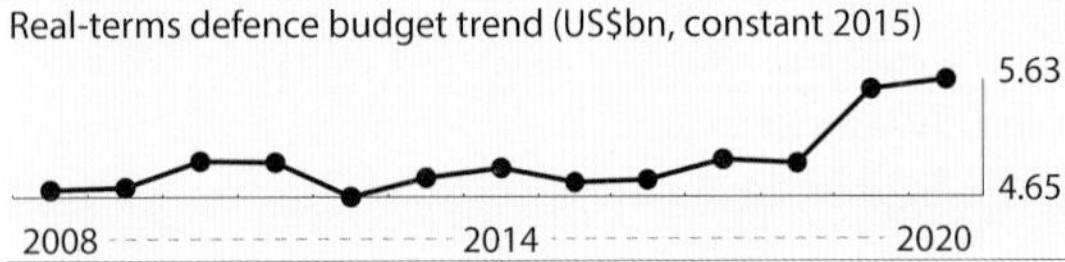

Population 8,403,994

Age	0–14	15–19	20–24	25–29	30–64	65 plus
Male	7.9%	2.5%	2.8%	3.1%	24.7%	8.3%
Female	7.4%	2.4%	2.7%	3.1%	24.6%	10.4%

Capabilities

The conscript-based armed forces are postured for territorial defence and limited participation in international peace-support operations. The government has begun to reduce its armed forces, reflecting an assessment that in the militia-based system not all personnel would realistically be available for active service. Switzerland is neutral and not a member of any alliances, although it joined NATO's Partnership for Peace programme in 1996 and on occasion contributes to NATO- and EU-led operations alongside its engagement in UN or OSCE missions. Switzerland does not participate in combat operations for peace-enforcement purposes and its deployments are limited in size. The 2016 armed-forces development plan emphasises improvements in readiness, training and equipment, and is meant to be implemented by the end of 2022. The approach to readiness is changing to a flexible model in which different units are called up for active service gradually and on different timelines. In July 2018 the government relaunched its attempt to procure a new combat aircraft, after plans to replace F-5 *Tiger* II combat aircraft with the *Gripen* were rejected by national referendum in May 2014. In September 2020, a corresponding referendum, which did not specify any aircraft type, passed with a small majority. The multi-stage selection process was expected to be completed by the end of 2020 but now has been pushed back into the second quarter of 2021. It includes replacement of the F/A-18 *Hornet*; the life of these platforms will be extended to 2030. In spring 2019, the government declared that the life-extension programme was behind schedule, further reducing availability. The fighter-acquisition programme was capped at CHF6 billion in May 2019 and has been separated from the procurement of ground-based air-defence systems. Previously both programmes were linked. Other priorities include upgrades to Switzerland's air-surveillance systems and transport helicopters and improved cyber-defence capabilities. Switzerland's defence industry has limited design and manufacturing capabilities, with recognised capacity in the land-vehicles sector, which has links to North American companies.

ACTIVE 19,550 (Armed Forces 19,550)

Conscript liability 260–600 compulsory service days depending on rank. 18 or 23 weeks' training (depending on branch) generally at age 20, followed by 6 refresher trg courses (3 weeks each). Alternative service available

RESERVE 123,450

Civil Defence 73,000 (51,000 Reserve)

ORGANISATIONS BY SERVICE

Armed Forces 3,100 active; 16,450 conscript (19,550 total)

Operations Command 72,600 on mobilisation

4 Territorial Regions. With the exception of military police all units are non-active

FORCES BY ROLE

COMMAND

4 regional comd

SPECIAL FORCES

2 SF bn

MANOEUVRE

Armoured

2 (1st & 11th) bde (1 recce bn, 1 tk bn, 2 armd inf bn, 1 SP arty bn, 1 engr bn, 1 sigs bn)

Mechanised

1 (4th) bde (2 recce bn, 2 SP arty bn, 1 ptn br bn)

Light

10 inf bn

7 mtn inf bn

1 mtn inf unit

COMBAT SUPPORT

4 engr bn

4 MP bn

1 NBC bn

1 int unit

COMBAT SUPPORT

4 engr rescue bn

EQUIPMENT BY TYPE

ARMOURED FIGHTING VEHICLES

MBT 134 *Leopard* 2 (Pz-87 *Leo*)

IFV 186: 154 CV9030CH; 32 CV9030 (CP)

APC 1,235

APC (T) 311 M113A2 (incl variants)

APC (W) 924 *Piranha* I/II/IIIC

AUV 441 *Eagle* I/II/III (incl 119 CP)

ENGINEERING & MAINTENANCE VEHICLES

AEV 12 *Kodiak*

ARV 25 *Büffel*

VLB 8 *Leopard* 2 with *Leguan*

MW 46: 26 Area Clearing System; 20 M113A2

NBC VEHICLES 12 *Piranha* IIIC CBRN
ANTI-TANK/ANTI-INFRASTRUCTURE
MSL • SP 106 *Piranha* I TOW-2
ARTILLERY 433
SP 155mm 133 M109 KAWEST
MOR • 81mm 300 Mw-72
PATROL AND COASTAL COMBATANTS 21
PB 21: 11 *Aquarius*; 10 *Watercat* 1250
AIR DEFENCE • SAM • Point-defence FIM-92 *Stinger*

Air Force 17,200 on mobilisation

FORCES BY ROLE
FIGHTER
2 sqn with F-5E/F *Tiger* II
3 sqn with F/A-18C/D *Hornet*
TRANSPORT
1 sqn with Beech 350 *King Air*; DHC-6 *Twin Otter*; PC-6 *Turbo Porter*; PC-12
1 VIP Flt with Beech 1900D; Cessna 560XL *Citation*; CL-604 *Challenger*; *Falcon* 900EX; PC-24
TRAINING
1 sqn with PC-7CH *Turbo Trainer*; PC-21
1 sqn with PC-9 (tgt towing)
1 OCU Sqn with F-5E/F *Tiger* II
TRANSPORT HELICOPTER
6 sqn with AS332M *Super Puma*; AS532UL *Cougar*; H135M
ISR UAV
1 sqn with *Hermes* 900
EQUIPMENT BY TYPE
AIRCRAFT 56 combat capable
FTR 26: 21 F-5E *Tiger* II; 5 F-5F *Tiger* II
FGA 30: 25 F/A-18C *Hornet*; 5 F/A-18D *Hornet*
TPT 24: **Light** 20: 1 Beech 350 *King Air*; 1 Beech 1900D; 1 Cessna 560XL *Citation*; 1 DHC-6 *Twin Otter*; 14 PC-6 *Turbo Porter*; 1 PC-6 (owned by armasuisse, civil registration); 1 PC-12 (owned by armasuisse, civil registration); **PAX** 4: 2 CL-604 *Challenger*; 1 *Falcon* 900EX; 1 PC-24 (VIP)
TRG 40: 27 PC-7CH *Turbo Trainer*; 5 PC-9; 8 PC-21
HELICOPTERS
MRH 20 H135M
TPT • Medium 25: 15 AS332M *Super Puma*; 10 AS532UL *Cougar*
UNINHABITED AERIAL VEHICLES
ISR • Medium 1 *Hermes* 900
AIR-LAUNCHED MISSILES • AAM • IR AIM-9P *Sidewinder*; **IIR** AIM-9X *Sidewinder* II; **ARH** AIM-120B/C-7 AMRAAM

Ground Based Air Defence (GBAD)

GBAD assets can be used to form AD clusters to be deployed independently as task forces within Swiss territory

EQUIPMENT BY TYPE
AIR DEFENCE
SAM • Point-defence 56+: 56 *Rapier*; FIM-92 *Stinger*
GUNS 35mm 27 GDF-003/-005 with *Skyguard*

Armed Forces Logistic Organisation 9,650 on mobilisation

FORCES BY ROLE
COMBAT SERVICE SUPPORT
1 log bde (6 log bn; 1 tpt bn; 6 med bn)

Command Support Organisation 11,150 on mobilisation

FORCES BY ROLE
COMBAT SERVICE SUPPORT
1 spt bde

Training Command 37,350 on mobilisation

COMBAT SERVICE SUPPORT
5 trg unit

Civil Defence 73,000 (51,000 Reserve)

(not part of armed forces)

DEPLOYMENT

BOSNIA-HERZEGOVINA: EU • EUFOR • *Operation Althea* 20

DEMOCRATIC REPUBLIC OF THE CONGO: UN • MONUSCO 1

INDIA/PAKISTAN: UN • UNMOGIP 4

KOREA, REPUBLIC OF: NNSC • 5 officers

MALI: UN • MINUSMA 5

MIDDLE EAST: UN • UNTSO 12

SERBIA: NATO • KFOR 165 (military volunteers); 1 inf coy; 1 engr pl; 1 hel flt with AS332M *Super Puma*

SOUTH SUDAN: UN • UNMISS 1

WESTERN SAHARA: UN • MINURSO 2

Europe

Turkey TUR

New Turkish Lira L		2019	2020	2021
GDP	L	4.32tr	4.49tr	
	US$	761bn	649bn	
per capita	US$	9,151	7,715	
Growth	%	0.9	-5.0	
Inflation	%	15.2	11.9	
Def exp [a]	L	79.4bn	89.0bn	
	US$	14.0bn	12.9bn	
Def bdgt [b]	L	68.6bn	76.3bn	
	US$	12.1bn	11.0bn	
US$1=L		5.68	6.91	

[a] NATO definition

[b] Includes funding for Undersecretariat of Defence Industries; Defence Industry Support Fund; TUBITAK Defense Industries R&D Institute (SAGE); and military pensions

Real-terms defence budget trend (US$bn, constant 2015)

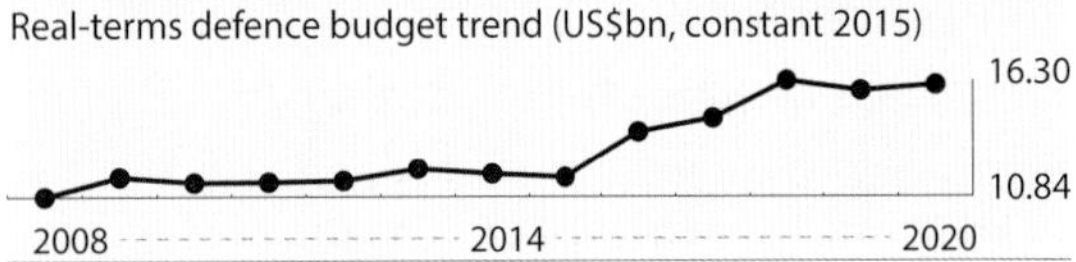

Population 82,017,514

Age	0–14	15–19	20–24	25–29	30–64	65 plus
Male	12.0%	4.1%	3.9%	3.8%	22.7%	3.7%
Female	11.4%	3.9%	3.8%	3.7%	22.4%	4.6%

Capabilities

Turkey has large, well-equipped armed forces that are primarily structured for national defence. Much recent activity has focused on internal security and cross-border operations in response to the continuing war in Syria and Libya. The Turkish Armed Forces 2033 Strategic Plan aims to modernise military equipment and the force structure. According to government officials, terrorism is the main security threat. Turkey is a NATO member and has provided access to its airspace and facilities for operations in Iraq and Syria. However, relationships with NATO allies have come under pressure after Ankara's decision to procure the Russian-made S-400 air-defence system, its operations in northern Syria, rising tensions with Greece in the eastern Mediterranean Sea, Turkey's role in Libya and its position over the late-2020 Azerbaijan–Armenia conflict. Following an attempted coup in July 2016, Ankara dismissed large numbers of officers from its armed forces. The armed forces train regularly, including with NATO allies. Turkish statements have indicated an intention to enhance its presence in Cyprus, possibly including a naval base in the northern part of the island. Equipment is mostly sourced from national firms. While key subcomponents are still often imported, a number of locally developed equipment designs are in production. Deliveries on the S-400 contract began in 2019 and as a result the US government terminated Turkey's participation in the F-35 programme. Turkey is also developing a domestic fighter aircraft but is dependent on collaboration with external defence companies for the implementation of this project. The president has authority over defence procurement and control over Turkey's top defence companies. Turkey has signed defence-cooperation agreements with a focus on exports and technology transfer, in an effort to boost its national defence industry and achieve defence-industrial autonomy.

ACTIVE 355,200 (Army 260,200 Navy 45,000 Air 50,000) **Paramilitary 156,800**

Conscript liability 12 months (5.5 months for university graduates; 21 days for graduates with exemption) (reducing to 6 months)

RESERVE 378,700 (Army 258,700 Navy 55,000 Air 65,000)

Reserve service to age 41 for all services

ORGANISATIONS BY SERVICE

Space

EQUIPMENT BY TYPE

SATELLITES • ISR 2 *Gokturk*-1/2

Army ε260,200 (incl conscripts)

FORCES BY ROLE

COMMAND
4 army HQ
9 corps HQ

SPECIAL FORCES
8 cdo bde
1 mtn cdo bde
1 cdo regt

MANOEUVRE

Armoured
1 (52nd) armd div (2 armd bde, 1 mech bde)
7 armd bde

Mechanised
2 (28th & 29th) mech div
14 mech inf bde

Light
1 (23rd) mot inf div (3 mot inf regt)
7 mot inf bde

COMBAT SUPPORT
2 arty bde
1 trg arty bde
6 arty regt
2 engr regt

AVIATION
4 avn regt
4 avn bn

EQUIPMENT BY TYPE

ARMOURED FIGHTING VEHICLES

MBT 2,378: 316 *Leopard* 2A4; 170 *Leopard* 1A4; 227 *Leopard* 1A3; 100 M60A1; 650 M60A3; 165 M60T (being upgraded); 750 M48A5 T2 (2,000 M48A5 T1 in store)

RECCE ε250 *Akrep*

IFV 645 ACV AIFV

APC 5,266

APC (T) 3,636: 823 ACV AAPC; 2,813 M113/M113A1/M113A2

PPV 1,630: 360 *Edjer Yaclin* 4×4; ε650 *Kirpi*; 320 *Kirpi*-2; ε300 *Vuran*

AUV 1,200: 800+ *Cobra*; ε400 *Cobra* II

ENGINEERING & MAINTENANCE VEHICLES

AEV 12+: AZMIM; 12 M48; M113A2T2

ARV 150: 12 *Leopard* 1; 105 M48T5; 33 M88A1

VLB 88: 36 *Leguan*; 52 Mobile Floating Assault Bridge

MW 4+: 4 *Husky* 2G; *Tamkar*; *Bozena*

ANTI-TANK/ANTI-INFRASTRUCTURE
MSL
SP 391: 365 ACV TOW; 26 *Kaplan*
MANPATS 9K135 *Kornet*-E (AT-14 *Spriggan*); *Eryx*; FGM-148 *Javelin*; *Milan*; OMTAS
RCL 106mm M40A1
ARTILLERY 7,833+
SP 1,080: **155mm** 825: ε150 M44T1; 365 M52T (mod); ε310 T-155 *Firtina*; **175mm** 36 M107; **203mm** 219 M110A2
TOWED 794+: **105mm** 75+ M101A1; **155mm** 557: 517 M114A1/M114A2; 40 *Panter*; **203mm** 162 M115
MRL 146+: **107mm** 48; **122mm** ε36 T-122; **227mm** 12 M270 MLRS; **302mm** 50+ TR-300 *Kasirga* (WS-1)
MOR 5,813+
SP 1,443+: **81mm**; **107mm** 1,264 M106; **120mm** 179
TOWED 4,370: **81mm** 3,792; **120mm** 578
SURFACE-TO-SURFACE MISSILE LAUNCHERS
SRBM • Conventional *Bora*; MGM-140A ATACMS (launched from M270 MLRS); J-600T *Yildrim* (B-611/CH-SS-9 mod 1)
AIRCRAFT
ISR 5 Beech 350 *King Air*
TPT • Light 8: 5 Beech 200 *King Air*; 3 Cessna 421
TRG 49: 45 Cessna T182; 4 T-42A *Cochise*
HELICOPTERS
ATK 89: 18 AH-1P *Cobra*; 12 AH-1S *Cobra*; 5 AH-1W *Cobra*; 4 TAH-1P *Cobra*; 9 T129A; 41 T129B
MRH 28 Hughes 300C
TPT 227+: **Heavy** 11 CH-47F *Chinook*; **Medium** 77+: 29 AS532UL *Cougar*; 48+ S-70A *Black Hawk*; **Light** 139: 12 Bell 204B (AB-204B); ε43 Bell 205 (UH-1H *Iroquois*); 64 Bell 205A (AB-205A); 20 Bell 206 *Jet Ranger*
UNINHABITED AERIAL VEHICLES
CISR • Medium 33 *Bayraktar* TB2
ISR • Heavy *Falcon* 600/*Firebee*; **Medium** CL-89; *Gnat*
LOITERING MUNITIONS *Harpy*
AIR DEFENCE
SAM • Point-defence 148+: 70 *Altigan* PMADS octuple *Stinger* lnchr, 78 *Zipkin* PMADS quad *Stinger* lnchr; FIM-92 *Stinger*
GUNS 1,382
SP 35mm ε20 *Korkut*
TOWED 1,362: **20mm** 439 GAI-D01/Rh-202; **35mm** 120 GDF-001/-003; **40mm** 803 L/60/L/70
AIR-LAUNCHED MISSILES
ASM *Mizrak*-U (UMTAS)
BOMBS
Laser-guided MAM-L; MAM-C

Navy ε45,000 (incl conscripts)

EQUIPMENT BY TYPE
SUBMARINES • SSK 12
4 *Atilay* (GER Type-209/1200) with 8 single 533mm TT with SST-4 HWT
4 *Gür* (GER Type-209/1400) with 8 single 533mm TT with UGM-84 *Harpoon* AShM/Mk 24 *Tigerfish* mod 2 HWT/*SeaHake* mod 4 (DM2A4) HWT
4 *Preveze* (GER Type-209/1400) (MLU ongoing) with 8 single 533mm TT with UGM-84 *Harpoon* AShM/Mk 24 *Tigerfish* mod 2 HWT/*SeaHake* mod 4 (DM2A4) HWT
PRINCIPAL SURFACE COMBATANTS • FRIGATES 16
FFGHM 16:
4 *Barbaros* (GER MEKO 200 mod) with 2 quad lnchr with RGM-84C *Harpoon* Block 1B AShM, 2 8-cell Mk 41 VLS with RIM-162B ESSM SAM, 2 triple 324mm SVTT Mk 32 ASTT with Mk 46 LWT, 3 *Sea Zenith* CIWS, 1 127mm gun (capacity 1 Bell 212 (AB-212) hel)
4 *Gabya* (ex-US *Oliver Hazard Perry*) with 1 Mk 13 GMLS with RGM-84C *Harpoon* Block 1B AShM/SM-1MR Block VI SAM, 1 8-cell Mk 41 VLS with RIM-162B ESSM SAM, 2 triple 324mm SVTT Mk 32 ASTT with Mk 46 LWT, 1 Mk 15 *Phalanx* Block 1B CIWS, 1 76mm gun (capacity 1 S-70B *Seahawk*/AB-212 ASW hel)
4 *Gabya* (ex-US *Oliver Hazard Perry*) with 1 Mk 13 GMLS with RGM-84C *Harpoon* Block 1B AShM/SM-1MR Block VI SAM, 2 triple 324mm SVTT Mk 32 ASTT with Mk 46 LWT, 1 Mk 15 *Phalanx* Block 1B CIWS, 1 76mm gun (capacity 1 S-70B *Seahawk*/AB-212 ASW hel)
4 *Yavuz* (GER MEKO 200TN) with 2 quad lnchr with RGM-84C *Harpoon* Block 1B AShM, 1 octuple Mk 29 GMLS with RIM-7M *Sea Sparrow* SAM, 2 triple 324mm SVTT Mk 32 ASTT with Mk 46 LWT, 3 *Sea Zenith* CIWS, 1 127mm gun (capacity 1 Bell 212 (AB-212) hel)
PATROL AND COASTAL COMBATANTS 45
CORVETTES 10:
FSGHM 4 *Ada* with 2 quad lnchr with RGM-84C *Harpoon* Block 1B AShM, 1 Mk 49 21-cell lnchr with RIM-116 SAM, 2 twin 324mm SVTT Mk 32 ASTT with Mk 46 LWT, 1 76mm gun (capacity 1 S-70B *Seahawk* hel)
FSG 6 *Burak* (ex-FRA *d'Estienne d'Orves*) with 2 single lnchr with MM38 *Exocet* AShM, 4 single 324mm ASTT with Mk 46 LWT, 1 Creusot-Loire Mk 54 A/S mor, 1 100mm gun (1 vessel with 1 76mm gun instead)
PCFG 19:
4 *Dogan* (GER Lurssen-57) with 2 quad lnchr with RGM-84C *Harpoon* Block 1B AShM, 1 76mm gun
9 *Kilic* with 2 quad lnchr with RGM-84C *Harpoon* Block 1B AShM, 1 76mm gun
4 *Rüzgar* (GER Lurssen-57) with 2 quad lnchr with RGM-84C *Harpoon* Block 1B AShM, 1 76mm gun
2 *Yildiz* with 2 quad lnchr with RGM-84C *Harpoon* Block 1B AShM, 1 76mm gun
PCC 16 *Tuzla*
MINE WARFARE • MINE COUNTERMEASURES 15
MHO 11: 5 *Engin* (FRA *Circe*); 6 *Aydin*
MSC 4 *Seydi* (US *Adjutant*)
AMPHIBIOUS
LANDING SHIPS • LST 5:
2 *Bayraktar* with 2 Mk 15 *Phalanx* Block 1B CIWS, 1 hel landing platform (capacity 20 MBT; 250 troops)
1 *Osman Gazi* with 1 Mk 15 *Phalanx* CIWS (capacity 4 LCVP; 17 tanks; 980 troops; 1 hel landing platform)
2 *Sarucabey* with 1 Mk 15 *Phalanx* CIWS (capacity 11 tanks; 600 troops; 1 hel landing platform)
LANDING CRAFT 30
LCT 21: 2 C-120/130; 11 C-140; 8 C-151
LCM 9: 1 C-310; 8 LCM 8

LOGISTICS AND SUPPORT 35
ABU 2: 1 AG5; 1 AG6 with 1 76mm gun
AGS 2: 1 *Cesme* (ex-US *Silas Bent*); 1 *Cubuklu*
AOR 2 *Akar* with 1 Mk 15 *Phalanx* CIWS, 1 hel landing platform
AOT 2 *Burak*
AOL 1 *Gurcan*
AP 1 *Iskenderun*
ASR 3: 1 *Alemdar* with 1 hel landing platform; 2 *Isin* II
ATF 9: 1 *Akbas*; 1 *Degirmendere*; 1 *Gazal*; 1 *Inebolu*; 5 *Onder*
AWT 3 *Sogut*
AXL 8
AX 2 *Pasa* (ex-GER *Rhein*)

Marines 3,000

FORCES BY ROLE
MANOEUVRE
Amphibious
1 mne bde (3 mne bn; 1 arty bn)

Naval Aviation

FORCES BY ROLE
ANTI-SUBMARINE WARFARE
2 sqn with Bell 212 ASW (AB-212 ASW); S-70B *Seahawk*
1 sqn with ATR-72-600; CN235M-100; TB-20 *Trinidad*
EQUIPMENT BY TYPE
AIRCRAFT 4 combat capable
ASW 4 ATR-72-600
MP 6 CN235M-100
TPT • Light 7: 2 ATR-72-600; 5 TB-20 *Trinidad*
HELICOPTERS
ASW 29: 11 Bell 212 ASW (AB-212 ASW); 18 S-70B *Seahawk*
UNINHABITED AERIAL VEHICLES 7
CISR 7: **Heavy** 3 ANKA-S; **Medium** 4 *Bayraktar* TB2

Air Force ε50,000

2 tac air forces (divided between east and west)
FORCES BY ROLE
FIGHTER/GROUND ATTACK
1 sqn with F-4E *Phantom* 2020
8 sqn with F-16C/D *Fighting Falcon*
ISR
1 sqn with F-16C/D *Fighting Falcon*
1 unit with *King Air* 350
AIRBORNE EARLY WARNING & CONTROL
1 sqn (forming) with B-737 AEW&C
EW
1 unit with CN235M EW
SEARCH & RESCUE
1 sqn with AS532AL/UL *Cougar*
TANKER
1 sqn with KC-135R *Stratotanker*
TRANSPORT
1 sqn with A400M; C-160D *Transall*
1 sqn with C-130B/E *Hercules*
1 (VIP) sqn with Cessna 550 *Citation* II (UC-35); Cessna 650 *Citation* VII; CN235M; Gulfstream 550
3 sqn with CN235M
10 (liaison) flt with Bell 205 (UH-1H *Iroquois*); CN235M
TRAINING
1 sqn with F-16C/D *Fighting Falcon*
1 sqn with F-5A/B *Freedom Fighter*; NF-5A/B *Freedom Fighter*
1 sqn with SF-260D
1 sqn with *Hurkus*-B; KT-IT
1 sqn with T-38A/M *Talon*
1 sqn with T-41D *Mescalero*
AIR DEFENCE
4 bn with S-400 (SA-21 *Growler*)
4 sqn with MIM-14 *Nike Hercules*
2 sqn with *Rapier*
8 (firing) unit with MIM-23 *Hawk*
MANOEUVRE
Air Manoeuvre
1 AB bde
EQUIPMENT BY TYPE
AIRCRAFT 306 combat capable
FTR 27: 17 NF-5A *Freedom Fighter*; 10 NF-5B *Freedom Fighter* (48 F-5s being upgraded as LIFT)
FGA 279: 19 F-4E *Phantom* 2020; 27 F-16C *Fighting Falcon* Block 30; 162 F-16C *Fighting Falcon* Block 50; 14 F-16C *Fighting Falcon* Block 50+; 8 F-16D Block 30 *Fighting Falcon*; 33 F-16D *Fighting Falcon* Block 50; 16 F-16D *Fighting Falcon* Block 50+
ISR 6: 5 Beech 350 *King Air*; 1 C-160D *Transall*
EW 2+ CN235M EW
AEW&C 4 B-737 AEW&C
TKR 7 KC-135R *Stratotanker*
TPT 89: **Heavy** 9 A400M; **Medium** 30: 6 C-130B *Hercules*; 12 C-130E *Hercules*; 12 C-160D *Transall*; **Light** 49: 2 Cessna 550 *Citation* II (UC-35 - VIP); 2 Cessna 650 *Citation* VII; 45 CN235M; **PAX** 1 Gulfstream 550
TRG 172: 4 *Hurkus*-B; 33 SF-260D; 70 T-38A/M *Talon*; 25 T-41D *Mescalero*; 40 KT-IT
HELICOPTERS
TPT 35: **Medium** 20: 6 AS532AL *Cougar* (CSAR); 14 AS532UL *Cougar* (SAR); **Light** 15 Bell 205 (UH-1H *Iroquois*)
UNINHABITED AERIAL VEHICLES
CISR • Heavy 19 ANKA-S
ISR 27: **Heavy** 9: 9 *Heron*; **Medium** 18 *Gnat* 750
AIR DEFENCE
SAM 32+
Long-range 32+ MIM-14 *Nike Hercules*; 32 S-400 (SA-21 *Growler*)
Medium-range MIM-23 *Hawk*
Point-defence *Rapier*
AIR-LAUNCHED MISSILES
AAM • IR AIM-9S *Sidewinder*; *Shafrir* 2(‡); **IIR** AIM-9X *Sidewinder* II; **SARH** AIM-7E *Sparrow*; **ARH** AIM-120A/B AMRAAM
ARM AGM-88A HARM
ASM AGM-65A/G *Maverick*; *Popeye* I
LACM Coventional AGM-84K SLAM-ER
BOMBS
Electro-optical guided GBU-8B HOBOS (GBU-15)
INS/GPS guided AGM-154A JSOW; AGM-154C JSOW
Laser-guided MAM-C; MAM-L; *Paveway* I; *Paveway* II

Paramilitary 156,800

Gendarmerie 152,100

Ministry of Interior; Ministry of Defence in war

FORCES BY ROLE

SPECIAL FORCES
1 cdo bde

MANOEUVRE

Other
1 (border) paramilitary div
2 paramilitary bde

EQUIPMENT BY TYPE

ARMOURED FIGHTING VEHICLES
RECCE 57+: *Akrep*; 57 *Ates*
APC 760+
APC (W) 560: 535 BTR-60/BTR-80; 25 *Condor*
PPV 200+: *Edjer Yaclin* 4×4; *Kirpi*; 200 *Kirpi* II
AUV *Cobra*; *Cobra* II; Otokar *Ural*

AIRCRAFT
ISR Some O-1E *Bird Dog*
TPT • Light 2 Do-28D

HELICOPTERS
ATK 6 T129B
MRH 19 Mi-17 *Hip* H
TPT 35: **Medium** 12 S-70A *Black Hawk*; **Light** 23: 8 Bell 204B (AB-204B); 6 Bell 205A (AB-205A); 8 Bell 206A (AB-206A) *Jet Ranger*; 1 Bell 212 (AB-212)

UNINHABITED AERIAL VEHICLES
CISR 24: **Heavy** 6 *Anka*-S; **Medium** 18 *Bayraktar* TB2

BOMBS
Laser-guided MAM-L; MAM-C

Coast Guard 4,700

EQUIPMENT BY TYPE

PATROL AND COASTAL COMBATANTS 104
PSOH 4 *Dost*
PBF 60
PB 40

AIRCRAFT • MP 3 CN235 MPA

HELICOPTERS • MRH 8 Bell 412EP (AB-412EP – SAR)

DEPLOYMENT

AFGHANISTAN: NATO • *Operation Resolute Support* 600; 1 mot inf bn(-)

AZERBAIJAN: Air Force: 6 F-16C/D *Fighting Falcon*; 1 CN235M

BOSNIA-HERZEGOVINA: EU • EUFOR • *Operation Althea* 205; 1 inf coy

CYPRUS (NORTHERN): ε33,800; 1 army corps HQ; 1 SF regt; 1 armd bde; 2 mech inf div; 1 mech inf regt; 1 arty regt; 1 avn comd; 287 M48A5T2; 147 ACV AIFV; 106 ACV AAPC (incl variants); 386 M113 (incl variants); 36 M101A1; 36 M114A2; 12 M115; 30 M44T; 144 M52T1; 18 T-122; 171 81mm mor; 70 M30; 135 HY-12; *Milan*; 60 ACV TOW; 219 M40A1; FIM-92 *Stinger*; 44 Rh 202; 78 GAI-D01; 16 GDF-003; 3 Cessna 185 (U-17); 2 AS532UL *Cougar*; 1 Bell 205 (UH-1H *Iroquois*); 1 PB

IRAQ: Army: 1,000; 1 cdo unit; **NATO • NATO** Mission Iraq up to 30

LEBANON: UN • UNIFIL 88; 1 PCFG

LIBYA: ε500; ACV-AAPC; *Kirpi*; 1 arty unit with T-155 *Firtina*; 1 AD unit with MIM-23B *Hawk*; Korkut; GDF-003; 1 CISR UAV unit with *Bayraktar* TB2

MEDITERRANEAN SEA: NATO • SNMG 2: 1 PCC • SNMCMG 2: 1 MHO

QATAR: Army: 300 (trg team); 1 mech inf coy; 1 arty unit; 12+ ACV AIFV/AAPC; 2 T-155 *Firtina*

SERBIA: NATO • KFOR 311; 1 inf coy
UN • UNMIK 1

SOMALIA: 200 (trg team); **UN** • UNSOM 1

SYRIA: *Operation Spring Shield* ε3,000; some cdo units; 3 armd BG; 1 SAM unit; 1 gendarmerie unit

FOREIGN FORCES

Spain *Active Fence*: 150; 1 SAM bty with MIM-104C *Patriot* PAC-2

United States US European Command: 1,700; 1 tkr sqn with 14 KC-135; 1 spt facility at Izmir; 1 spt facility at Ankara; 1 air base at Incirlik • US Strategic Command: 1 AN/TPY-2 X-band radar at Kürecik

Europe

United Kingdom UK

British Pound £		2019	2020	2021
GDP	£	2.22tr	2.06tr	
	US$	2.83tr	2.64tr	
per capita	US$	42,379	39,229	
Growth	%	1.5	-9.8	
Inflation	%	1.8	0.8	
Def exp [a]	£	46.5bn	48.4bn	
	US$	59.4bn	62.0bn	
Def bdgt [b]	£	46.4bn	ε48.0bn	
	US$	59.2bn	ε61.5bn	
US$1=£		0.78	0.78	

[a] NATO definition

[b] Includes total departmental expenditure limits; costs of military operations; Armed Forces Pension Service; and external income earned by the MoD

Real-terms defence budget trend (US$bn, constant 2015)

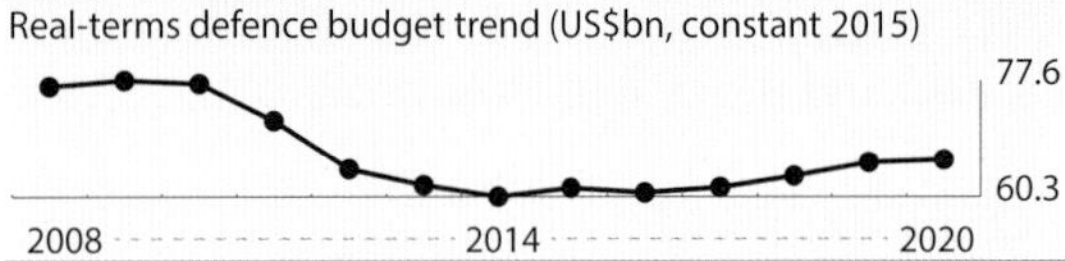

Population 65,761,117

Age	0–14	15–19	20–24	25–29	30–64	65 plus
Male	9.0%	2.8%	3.1%	3.4%	23.2%	8.3%
Female	8.6%	2.7%	3.0%	3.3%	22.6%	10.2%

Capabilities

The 2015 Strategic Defence and Security Review (SDSR) and 2018 National Security Capability Review highlighted a range of security challenges, including state-based threats and terrorism. UK defence policy is based on using the armed forces to reduce direct threats by projecting stability abroad. In 2020, the Ministry of Defence published an Integrated Operating Concept that aims to transform military capabilities for 'information-age warfare'. Principal defence priorities are counter-terrorism and contributing to the counter-ISIS coalition and NATO tasks, including in Afghanistan and in Eastern Europe. Strategic Command comprises key joint force elements, such as special-forces, defence-intelligence and military-cyber capabilities. Britain retains well-trained, rapidly deployable armed forces with a wide range of capabilities, combat ethos and sufficient strategic lift to deploy forces for small- and medium-scale operations. Equipment modernisation continues, but the defence budget is under pressure because of the fall in the value of the pound, cost growth of major equipment programmes, the difficulty of achieving savings targets and the likely effects of the pandemic on public spending. A defence-budget increase announced in September 2019 alleviated, but did not eliminate, these risks to the delivery of the modernised 'Joint Force 2025', while a greater budget increase, announced in November 2020, indicated that the budget will, in real terms, be 9% higher by 2024–25, when compared to the 2019–20 budget. Another risk is that personnel levels have declined to approximately 92% during the last year. The US is the country's closest military ally. There is also a close intelligence relationship with the 'Five Eyes' nations and a growing military partnership with France. The UK retains a small military footprint in Germany and leads the Combined Joint Expeditionary Force partnership. There is a naval base in Bahrain and a training and logistics base in Oman. The UK continues to support the FPDA in Southeast Asia. Expeditionary logistic capability meets policy requirements, but peacetime logistic support within the UK is dependent on contractors. The country's sophisticated defence industry is a world leader in defence exports but cannot meet all of the UK's requirements

ACTIVE 148,500 (Army 82,650 Navy 33,050 Air 32,800)

RESERVE 78,600 (Regular Reserve 39,450 (Army 26,550, Navy 6,050, Air 6,850); Volunteer Reserve 37,150 (Army 30,000, Navy 3,950, Air 3,200); Sponsored Reserve 2,000)

Includes both trained and those currently under training within the Regular Forces, excluding university cadet units

ORGANISATIONS BY SERVICE

Strategic Forces 1,000

Royal Navy

EQUIPMENT BY TYPE

SUBMARINES • STRATEGIC

SSBN 4: 4 *Vanguard* with 16 UGM-133A *Trident* II D-5/D-5LE nuclear SLBM, 4 533mm TT with *Spearfish* HWT (each boat will not deploy with more than 40 warheads, but each missile could carry up to 12 MIRV; some *Trident* D-5 capable of being configured for sub-strategic role)

MSL • SLBM • Nuclear 48 UGM-133A *Trident* II D-5 (fewer than 160 declared operational warheads)

Royal Air Force

EQUIPMENT BY TYPE

RADAR • STRATEGIC 1 Ballistic Missile Early Warning System (BMEWS) at Fylingdales Moor

Space

EQUIPMENT BY TYPE

SATELLITES • COMMUNICATIONS 8: 1 NATO-4B; 3 *Skynet*-4; 4 *Skynet*-5

Army 80,400; 3,100 Gurkhas (total 83,500)

Regt normally bn size. Many cbt spt and CSS regt and bn have reservist sub-units

FORCES BY ROLE

COMMAND

1 (ARRC) corps HQ

MANOEUVRE

Armoured

1 (3rd) armd div (3 armd inf bde (1 armd recce regt, 1 tk regt, 2 armd inf bn, 1 mech inf bn); 1 arty bde (2 SP arty regt, 1 MRL regt, 2 fd arty regt); 3 cbt engr regt; 1 sigs regt; 1 log bde (3 MP regt; 5 log regt; 3 maint regt; 3 med regt); 2 AD regt)

Light

1 (1st) lt inf div (2 (4th & 51st) inf bde (1 recce regt, 1 lt mech inf bn; 1 lt inf bn); 1 (7th) inf bde (1 recce regt, 3 lt inf bn); 1 (11th) inf bde (2 lt inf bn); 1 engr bde (1 cbt engr regt, 1 CBRN regt, 2 EOD regt, 1 (MWD) EOD search regt, 1 engr regt, 1 (air spt) engr regt, 1 log

regt); 1 log bde (2 log regt; 2 maint bn; 1 med regt); 1 log bde (3 log regt; 1 maint regt); 1 med bde (3 fd hospital))
1 (38th) inf bde (1 lt inf bn)
1 (160th) inf bde (2 lt inf bn)
2 inf bn (London)
1 inf bn (Brunei)
Air Manoeuvre
1 (16th) air aslt bde (1 recce pl, 2 para bn, 1 air aslt bn, 1 fd arty regt, 1 cbt engr regt, 1 log regt, 1 med regt)
COMBAT SUPPORT
1 (6th) cbt spt div (1 (Spec Inf Gp) inf bde(-) (5 inf bn(-)); 1 ISR bde (1 STA regt, 2 EW regt, 3 int bn, 1 ISR UAV regt); 1 (77th) info ops bde (3 info ops gp, 1 spt gp, 1 engr spt/log gp) 1 sigs bde (6 sigs regt); 1 sigs bde (2 sigs regt; 1 (ARRC) sigs bn)
1 (geographic) engr regt

Reserves

Army Reserve 27,450 reservists

The Army Reserve (AR) generates individuals, sub-units and some full units. The majority of units are subordinate to regular-formation headquarters and paired with one or more regular units

FORCES BY ROLE
MANOEUVRE
Reconnaissance
3 recce regt
Armoured
1 armd regt
Light
15 lt inf bn
Air Manoeuvre
1 para bn
COMBAT SUPPORT
3 arty regt
1 STA regt
1 MRL regt
3 engr regt
1 EOD regt
4 int bn
4 sigs regt
COMBAT SERVICE SUPPORT
11 log regt
3 maint regt
3 med regt
9 fd hospital
AIR DEFENCE
1 AD regt

EQUIPMENT BY TYPE
ARMOURED FIGHTING VEHICLES
MBT 227 *Challenger* 2
RECCE 617: 4 *Ajax* (in test); 197 *Jackal*; 110 *Jackal* 2; 130 *Jackal* 2A; 145 FV107 *Scimitar*; 31 *Scimitar* Mk2
IFV 388+: 388 FV510 *Warrior*; FV511 *Warrior* (CP); FV514 *Warrior* (OP); FV515 *Warrior* (CP)
APC 812
APC (T) 416: 6 *Ares* (in test); 1 *Athena* (CP – in test); 409 FV430 *Bulldog* (incl variants)
PPV 396 *Mastiff* (6×6)
AUV 1,238: 399 *Foxhound*; 252 FV103 *Spartan* (incl variants); 23 *Spartan* Mk2 (incl variants); 396 *Panther* CLV; 168 *Ridgback*
ENGINEERING & MAINTENANCE VEHICLES
AEV 88: 56 *Terrier*; 32 *Trojan*
ARV 288: 2 *Apollo* (in test); 2 *Atlas* (in test); 80 *Challenger* ARRV; 30 FV106 *Samson*; 28 *Samson* Mk2; 105 FV512 *Warrior*; 41 FV513 *Warrior*
MW 64 *Aardvark*
VLB 68: 35 M3; 33 *Titan*
NBC VEHICLES 8 TPz-1 *Fuchs* NBC
ANTI-TANK/ANTI-INFRASTRUCTURE • MSL
SP *Exactor*-2 (*Spike* NLOS)
MANPATS FGM-148 *Javelin*; NLAW
ARTILLERY 598
SP 155mm 89 AS90
TOWED 105mm 114 L118 Light Gun
MRL 227mm 35 M270B1 MLRS
MOR 81mm 360 L16A1
AMPHIBIOUS • LCM 3 Ramped Craft Logistic
AIR DEFENCE • SAM
Point-defence 74: 60 FV4333 *Stormer* with *Starstreak*; 14 *Rapier* FSC; *Starstreak* (LML)

Joint Helicopter Command

Tri-service joint organisation including Royal Navy, Army and RAF units

Army

FORCES BY ROLE
HELICOPTER
1 bde (1 atk hel regt (2 sqn with AH-64D *Apache*; 1 trg sqn with AH-64D *Apache*); 1 atk hel regt (2 sqn with AH-64D *Apache*); 1 regt (2 sqn with AW159 *Wildcat* AH1; 1 trg sqn with AW159 *Wildcat* AH1); 1 regt (1 sqn with SA341B *Gazelle* AH1); 1 (spec ops) sqn with AS365N3; SA341B *Gazelle* AH1; 1 flt with Bell 212 (Brunei); 1 flt with SA341B *Gazelle* AH1 (Canada); 1 ISR UAV regt; 1 maint regt)
TRAINING
1 hel regt (1 sqn with AH-64D *Apache*; 1 sqn with AS350B *Ecureuil*; 1 sqn with Bell 212; *Lynx* AH9A; SA341B *Gazelle* AH1)

Army Reserve

FORCES BY ROLE
HELICOPTER
1 hel regt (4 sqn personnel only)

Royal Navy

FORCES BY ROLE
ATTACK HELICOPTER
1 lt sqn with AW159 *Wildcat* AH1
TRANSPORT HELICOPTER
2 sqn with AW101 *Merlin* HC3/3A/4/4A

Royal Air Force

FORCES BY ROLE
TRANSPORT HELICOPTER
3 sqn with CH-47D/SD/F *Chinook* HC3/4/4A/6
2 sqn with SA330 *Puma* HC2

TRAINING
1 OCU sqn with CH-47D/SD/F *Chinook* HC3/4/4A/6; SA330 *Puma* HC2

EQUIPMENT BY TYPE

HELICOPTERS
ATK 40 AH-64D *Apache*
MRH 66: 5 AS365N3; 34 AW159 *Wildcat* AH1; 27 SA341B *Gazelle* AH1
TPT 122: **Heavy** 60: 8 CH-47D *Chinook* HC4; 30 CH-47D *Chinook* HC6A; 8 CH-47SD *Chinook* HC5; 14 CH-47F *Chinook* HC6; **Medium** 48: 25 AW101 *Merlin* HC3/3A/4/4A; 23 SA330 *Puma* HC2; **Light** 14: 9 AS350B *Ecureuil*; 5 Bell 212

UNINHABITED AERIAL VEHICLES • ISR • Medium 7 *Watchkeeper* (37+ more in store)

Royal Navy 32,450

EQUIPMENT BY TYPE

SUBMARINES 11
STRATEGIC • SSBN 4 *Vanguard*, opcon Strategic Forces with 16 UGM-133A *Trident* II D-5/D-5LE nuclear SLBM, 4 single 533mm TT with *Spearfish* HWT (each boat will not deploy with more than 40 warheads, but each missile could carry up to 12 MIRV; some *Trident* D-5 capable of being configured for sub-strategic role)
TACTICAL • SSN 7
3 *Trafalgar* with 5 single 533mm TT with UGM-109E *Tomahawk* Block IV LACM/*Spearfish* HWT
4 *Astute* with 6 single 533mm TT with UGM-109E *Tomahawk* Block IV LACM/*Spearfish* HWT

PRINCIPAL SURFACE COMBATANTS 21
AIRCRAFT CARRIERS 2:
CV 2 *Queen Elizabeth* with up to 3 Mk 15 *Phalanx* Block 1B CIWS (future capacity 24 F-35B *Lightning* II, 14 *Merlin* HM2/*Wildcat* HMA2/CH-47 *Chinook* hel) (1 in trials)
DESTROYERS 6:
DDGHM 3 *Daring* (Type-45) with 2 quad lnchr with RGM-84D *Harpoon* Block 1C AShM, 6 8-cell *Sylver* A50 VLS with *Aster* 15/30 (*Sea Viper*) SAM, 2 Mk 15 *Phalanx* Block 1B CIWS, 1 114mm gun (capacity 1 AW159 *Wildcat*/AW101 *Merlin* hel)
DDHM 3 *Daring* (Type-45) with 6 8-cell *Sylver* A50 VLS with *Aster* 15/30 (*Sea Viper*) SAM, 2 Mk 15 *Phalanx* Block 1B CIWS, 1 114mm gun (capacity 1 AW159 *Wildcat*/AW101 *Merlin* hel)
FRIGATES • FFGHM 13:
7 *Duke* (Type-23) with 2 quad lnchr with RGM-84D *Harpoon* Block 1C AShM, 1 32-cell VLS with *Sea Wolf* SAM, 2 twin 324mm ASTT with *Sting Ray* LWT, 1 114mm gun (capacity either 2 AW159 *Wildcat* or 1 AW101 *Merlin* hel)
6 *Duke* (Type-23) with 2 quad lnchr with RGM-84D *Harpoon* Block 1C AShM, 1 32-cell VLS with *Sea Ceptor* SAM, 2 twin 324mm ASTT with *Sting Ray* LWT, 1 114mm gun (capacity either 2 AW159 *Wildcat* or 1 AW101 *Merlin* hel)

PATROL AND COASTAL COMBATANTS 25
PSO 7: 3 *River* Batch 1; 4 *River* Batch 2 with 1 hel landing platform
PBI 18: 16 *Archer* (14 in trg role, 2 deployed to Gibraltar sqn); 2 *Scimitar*

MINE WARFARE • MINE COUNTERMEASURES 13
MCO 6 *Hunt* (incl 4 mod *Hunt*)
MHC 7 *Sandown*

AMPHIBIOUS
PRINCIPAL AMPHIBIOUS SHIPS 2
LPD 2 *Albion* with 2 Mk 15 *Phalanx* Block 1B CIWS (capacity 2 med hel; 4 LCU or 2 LCAC; 4 LCVP; 6 MBT; 300 troops) (of which 1 at extended readiness)

LOGISTICS AND SUPPORT 5
AGB 1 *Protector* with 1 hel landing platform
AGS 4: 1 *Scott*; 2 *Echo* (all with 1 hel landing platform); 1 *Magpie*

Royal Fleet Auxiliary

Support and miscellaneous vessels are mostly manned and maintained by the Royal Fleet Auxiliary (RFA), a civilian fleet owned by the UK MoD, which has approximately 1,900 personnel with type comd under Fleet Commander

AMPHIBIOUS • PRINCIPAL AMPHIBIOUS SHIPS 3:
LSD 3 *Bay* (capacity 4 LCU; 2 LCVP; 24 CR2 *Challenger* 2 MBT; 350 troops)

LOGISTICS AND SUPPORT 14
AOEH 4 *Tide* (capacity 1 AW159 *Wildcat*/AW101 *Merlin* hel)
AORH 3: 2 *Wave*; 1 *Fort Victoria* with 2 Mk 15 *Phalanx* Block 1B CIWS
AFSH 2 *Fort Rosalie*
AG 1 *Argus* (primary casualty-receiving ship with secondary aviation trg ship role)
AKR 4 *Point* (not RFA manned)

Naval Aviation (Fleet Air Arm) 4,900

FORCES BY ROLE

ANTI-SUBMARINE WARFARE
3 sqn with AW101 ASW *Merlin* HM2
2 sqn with AW159 *Wildcat* HMA2

AIRBORNE EARLY WARNING
1 sqn with *Merlin* HM2 *Crowsnest* (forming)

TRAINING
1 sqn with Beech 350ER *King Air*
1 sqn with G-115
1 sqn with *Hawk* T1

EQUIPMENT BY TYPE

AIRCRAFT
TPT • Light 4 Beech 350ER *King Air* (*Avenger*)
TRG 17: 5 G-115; 12 *Hawk* T1

HELICOPTERS
ASW 58: 28 AW159 *Wildcat* HMA2; 30 AW101 ASW *Merlin* HM2

Royal Marines 6,600

FORCES BY ROLE

MANOEUVRE
Amphibious
1 (3rd Cdo) mne bde (2 mne bn; 2 sy bn; 1 amph gp; 1 amph aslt sqn; 1 (army) arty regt; 1 (army) engr regt; 1 ISR gp (1 EW sqn; 1 cbt spt sqn; 1 sigs sqn; 1 log sqn), 1 log regt)
2 amph sqn

EQUIPMENT BY TYPE

ARMOURED FIGHTING VEHICLES

APC (T) 99 BvS-10 Mk2 *Viking* (incl 19 cabs with 81mm mor)

ANTI-TANK/ANTI-INFRASTUCTURE

MSL • MANPATS FGM-148 *Javelin*

ARTILLERY 39

TOWED 105mm 12 L118 Light Gun

MOR 81mm 27 L16A1

PATROL AND COASTAL COMBATANTS • PB 2 *Island*

AMPHIBIOUS • LANDING CRAFT 30

LCU 10 LCU Mk10 (capacity 4 *Viking* APC or 120 troops)

LCVP 16 LCVP Mk5B (capacity 35 troops)

UCAC 4 *Griffon* 2400TD

AIR DEFENCE • SAM • Point-defence *Starstreak*

Royal Air Force 32,500

FORCES BY ROLE

FIGHTER

2 sqn with *Typhoon* FGR4/T3

FIGHTER/GROUND ATTACK

3 sqn with *Typhoon* FGR4/T3

2 sqn with *Typhoon* FGR4/T3 (forming)

1 sqn with F-35B *Lightning* II

ANTI-SUBMARINE WARFARE

1 sqn with P-8A *Poseidon* (MRA Mk1) (forming)

ISR

1 sqn with *Sentinel* R1

1 sqn with *Shadow* R1

1 sqn with BN-2 *Defender/Islander*

ELINT

1 sqn with RC-135W *Rivet Joint*

AIRBORNE EARLY WARNING & CONTROL

1 sqn with E-3D *Sentry*

SEARCH & RESCUE

1 sqn with Bell 412EP *Griffin* HAR-2

TANKER/TRANSPORT

2 sqn with A330 MRTT *Voyager* KC2/3

TRANSPORT

1 (comms) sqn with AW109E/SP; BAe-146

1 sqn with A400M *Atlas*

1 sqn with C-17A *Globemaster*

3 sqn with C-130J/J-30 *Hercules*

TRAINING

1 OCU sqn with F-35B *Lightning* II (forming)

1 OCU sqn with *Typhoon*

1 OCU sqn with E-3D *Sentry*; *Sentinel* R1; RC-135W *Rivet Joint*

1 sqn with EMB-500 *Phenom* 100

1 sqn with *Hawk* T1/1A/1W*

2 sqn with *Hawk* T2

1 sqn with T-6C *Texan* II

2 sqn with G-115E *Tutor*

COMBAT/ISR UAV

2 sqn with MQ-9A *Reaper*

EQUIPMENT BY TYPE

AIRCRAFT 225 combat capable

FGA 162: 18 F-35B *Lightning* II; 138 *Typhoon* FGR4; 6 *Typhoon* T3

ASW 4 P-8A *Poseidon* (MRA Mk1)

ISR 11: 5 *Sentinel* R1; 6 *Shadow* R1

ELINT 3 RC-135W *Rivet Joint*

AEW&C 3 E-3D *Sentry*

TKR/TPT 10: 3 A330 MRTT *Voyager* KC2; 7 A330 MRTT Voyager KC3

TPT 58: **Heavy** 28: 20 A400M *Atlas*; 8 C-17A *Globemaster*; **Medium** 14: 1 C-130J *Hercules*; 13 C-130J-30 *Hercules*; **Light** 12: 9 BN-2T-4S *Defender*; 3 BN-2 *Islander* R1; **PAX** 4 BAe-146 CC2/C3

TRG 164: 5 EMB-500 *Phenom* 100; 86 G-115E *Tutor*; 28 *Hawk* T2*; 31 *Hawk* T1/1A/1W* (ε34 more in store); 14 T-6C *Texan* II

HELICOPTERS

MRH 5: 1 AW139; 4 Bell 412EP *Griffin* HAR-2

TPT • Light 3: 2 AW109E; 1 AW109SP

UNINHABITED AERIAL VEHICLES

CISR • Heavy 10 MQ-9A *Reaper*

AIR-LAUNCHED MISSILES

AAM • IR AIM-9L/L(I) *Sidewinder*; **IIR** ASRAAM; **ARH** AIM-120C-5 AMRAAM; *Meteor*

ASM AGM-114 *Hellfire*; *Brimstone*; *Dual-Mode Brimstone*; *Brimstone* II

LACM *Storm Shadow*

BOMBS

Laser/GPS-guided GBU-10 *Paveway* II; GBU-24 *Paveway* III; Enhanced *Paveway* II/III; *Paveway* IV

Royal Air Force Regiment

FORCES BY ROLE

MANOEUVRE

Other

6 sy sqn

Tri-Service Defence Helicopter School

FORCES BY ROLE

TRAINING

1 hel sqn with Bell 412EP *Griffin* HT1

2 hel sqn with AS350B *Ecureuil*

EQUIPMENT BY TYPE

HELICOPTERS

MRH 3 Bell 412EP *Griffin* HT1

TPT • Light 27: 25 AS350B *Ecureuil*; 2 AW109E

Volunteer Reserve Air Forces

(Royal Auxiliary Air Force/RAF Reserve)

MANOEUVRE

Other

5 sy sqn

COMBAT SUPPORT

2 int sqn

COMBAT SERVICE SUPPORT

1 med sqn

1 (air movements) sqn

1 (HQ augmentation) sqn

1 (C-130 Reserve Aircrew) flt

UK Special Forces

Includes Royal Navy, Army and RAF units

FORCES BY ROLE

SPECIAL FORCES

1 (SAS) SF regt
1 (SBS) SF regt
1 (Special Reconnaissance) SF regt
1 SF BG (based on 1 para bn)

AVIATION

1 wg (includes assets drawn from 3 Army hel sqn, 1 RAF tpt sqn and 1 RAF hel sqn)

COMBAT SUPPORT

1 sigs regt

Reserve

FORCES BY ROLE

SPECIAL FORCES

2 (SAS) SF regt

EQUIPMENT BY TYPE

ARMOURED FIGHTING VEHICLES

AUV 24 *Bushmaster* IMV

ANTI-TANK/ANTI-INFRASTRUCTURE • MSL

MANPATS FGM-148 *Javelin*; NLAW

DEPLOYMENT

AFGHANISTAN: NATO • *Operation Resolute Support* 950; 1 inf bn; 1 hel flt with 3 *Puma* HC2; **UN •** UNAMA 1

ARABIAN SEA: Combined Maritime Forces • 1 FFGHM

ASCENSION ISLAND: 20

ATLANTIC (NORTH)/CARIBBEAN: 1 AG

ATLANTIC (SOUTH): 1 PSO

BAHRAIN: 160; 1 naval base

BELIZE: BATSUB 12

BRITISH INDIAN OCEAN TERRITORY: 40; 1 navy/marine det

BRUNEI: 2,000; 1 (Gurkha) lt inf bn; 1 jungle trg centre; 1 hel flt with 3 Bell 212

CANADA: BATUS 370; 1 trg unit; 1 hel flt with SA341 *Gazelle* AH1

CYPRUS: 2,260; 2 inf bn; 1 SAR sqn with 4 Bell 412 *Griffin* HAR-2; 1 radar (on det); *Operation Shader* 500: 1 FGA sqn with 6 *Typhoon* FGR4; 1 *Sentinel* R1; 1 A330 MRTT *Voyager*; 2 C-130J-30 *Hercules*; **UN •** UNFICYP (*Operation Tosca*) 256; 1 inf coy

EGYPT: MFO 2

ESTONIA: NATO • Enhanced Forward Presence (*Operation Cabrit*) 754; 1 armd regt HQ; 1 tk sqn; 1 armd inf coy(+); 1 cbt engr coy

FALKLAND ISLANDS: 1,200: 1 inf coy(+); 1 sigs unit; 1 AD det with *Rapier*; 1 PSO; 1 ftr flt with 4 *Typhoon* FGR4; 1 tkr/tpt flt with 1 A330 MRTT *Voyager*; 1 A400M; 1 hel flt with 2 *Chinook*

GERMANY: 185

GIBRALTAR: 570 (incl Royal Gibraltar regt); 2 PBI

IRAQ: *Operation Shader* 400; 1 inf bn(-); 1 engr sqn(-)

KENYA: BATUK 350; 1 trg unit

KUWAIT: *Operation Shader* 50; 1 CISR UAV sqn with 8 MQ-9A *Reaper*

LIBYA: UN • UNSMIL (*Operation Tramal*) 1

MALI: *Operation Barkhane* 90; 1 hel flt with 3 *Chinook* HC5; **EU •** EUTM Mali 8; **UN •** MINUSMA (*Operation Newcombe*) 3

NEPAL: 60 (Gurkha trg org)

NIGERIA: 80 (trg team)

OMAN: 90

PERSIAN GULF: *Operation Kipion* 2 MCO; 2 MHC; 1 LSD

POLAND: NATO • Enhanced Forward Presence 140; 1 recce sqn

SAUDI ARABIA: 50 (radar det)

SERBIA: NATO • KFOR 21

SOMALIA: 65 (trg team); **EU •** EUTM Somalia 3; **UN •** UNSOM (*Operation Praiser*) 1; **UN •** UNSOS (*Operation Catan*) 10

SOUTH SUDAN: UN • UNMISS (*Operation Vogul*) 7

UKRAINE: *Operation Orbital* 53 (trg team)

UNITED ARAB EMIRATES: 200; 1 tpt/tkr flt with C-17A *Globemaster*; A400M *Atlas;* A330 MRTT *Voyager* (on rotation)

FOREIGN FORCES

United States

US European Command: 9,300; 1 ftr wg at RAF Lakenheath (1 ftr sqn with 24 F-15C/D *Eagle*, 2 FGA sqn with 23 F-15E *Strike Eagle*); 1 ISR sqn at RAF Mildenhall with OC-135/RC-135; 1 tkr wg at RAF Mildenhall with 15 KC-135R/T *Stratotanker*; 1 spec ops gp at RAF Mildenhall (1 sqn with 8 CV-22B *Osprey*; 1 sqn with 8 MC-130J *Commando* II) • US Strategic Command: 1 AN/FPS-132 Upgraded Early Warning Radar and 1 *Spacetrack* radar at Fylingdales Moor

Arms procurements and deliveries – Europe

Significant events in 2020

JANUARY — CROATIA FIGHTER RFP

Croatia issued a request for proposals (RFP) for 12 combat aircraft under its Multipurpose Fighter Aircraft (Višenamjenskog Borbenog Aviona (VBA)) programme, in order to replace its ageing MiG-21 *Fishbeds*. The RFP was issued to the US and Sweden for new-build aircraft (Lockheed Martin F-16s and Saab *Gripens*) and to France and Italy (for second-hand *Rafales* and *Eurofighters* respectively), as well as to Norway, Greece and Israel for second-hand F16s. This is Croatia's third attempt to replace its MiG-21s, which were upgraded in Ukraine in 2013–14. The first attempt fell through in November 2008 after the global financial crisis. US objections in January 2019 scuppered the second attempt to procure from Israel 12 F-16s for US$475m. This latest attempt was temporarily paused in April because of the coronavirus pandemic and was resumed in September 2020. Several offers have been received and a decision is planned in early 2021.

FEBRUARY — FCAS PHASE 1A

France and Germany signed an agreement to complete, within 18 months, initial elements of various systems under Phase 1A of the French-German-Spanish Future Combat Air System (FCAS) project. France and Germany will share the development costs worth a total of €155m (US$173.67m). Spain, which became a party to the initial phase in June 2019, plans to contribute €50m (US$56.02m) to the programme by the end of 2020. The test flight of the first demonstrator is planned for 2026, and service entry is projected in 2040 to replace the Dassault *Rafale* and the Eurofighter *Typhoon*. So far, Dassault, Airbus, Indra Sistemas, Safran, MTU Aero Engines, Thales and MBDA are involved in the project.

APRIL — MGCS AGREEMENT

Germany and France signed joint-development agreements for the Main Ground Combat System (MGCS) project. This is intended to replace, from the mid-2030s, Germany's *Leopard* 2 and France's *Leclerc* main battle tanks. The agreements set out the organisational and management structures for the project and also outline a 20-month System Architecture Definition Study constituting the project's first phase. Germany's Rheinmetall and Krauss-Maffei Wegmann and France's Nexter are leading on the development work.

JULY — TEMPEST GAINS NEW MEMBERS

The Team Tempest industrial partnership welcomed seven new members to its ranks, when Bombardier, Collins Aerospace, GE UK, GKN, Martin Baker, QinetiQ and Thales UK joined BAE Systems, Leonardo MW, Rolls Royce and MBDA UK. The UK hopes that Team Tempest will provide technology and designs to its Future Combat Air System Technology Initiative (FCAS TI) programme that aims to replace the *Typhoon* combat aircraft. It is expected to develop more than 60 technology prototypes and demonstrators costing £1.9bn (US$2.29bn) by 2025. Sweden has agreed to open negotiations with the UK on future combat air technologies, though it is not part of Team Tempest. Nonetheless, on the same day that Team Tempest expanded, Saab announced it would open an FCAS centre in the UK to act as a hub for work on the FCAS programme.

SEPTEMBER — SWISS FIGHTER VOTE

In a referendum, Swiss voters narrowly approved a plan to acquire new combat aircraft, by a margin of 0.3%. The Federal Council is due in 2021 to select an aircraft: either Airbus's Eurofighter *Typhoon*, Boeing's F/A-18E/F *Super Hornet*, Dassault's *Rafale* or Lockheed Martin's F-35A *Lightning* II. The acquisition is expected to cost up to fr6bn (US$6.21bn). The federal government's goal is to replace the air force's ageing fleet of 30 F/A-18C/D *Hornets* and 26 F-5 *Tigers* by 2030 with 30–40 new aircraft and for the prime contractor to award 60% of the total contract amount to Swiss suppliers and subcontractors. Six years previously, Swiss voters narrowly rejected acquiring 22 Saab *Gripen* combat aircraft for an estimated cost of fr3.13bn (US$3.41bn).

Both Romania's and Bulgaria's latest national-security strategies emphasise the need to improve military capabilities, within a NATO context, in order to counter Russia's increased presence in the Black Sea. Despite shared security concerns and requirements, Romania's defence budget is more than two times larger than Bulgaria's, and these funds are enabling Bucharest to acquire a wide range of equipment, much of which incorporates offset agreements and technology transfers.

In the mid-2000s, both countries could only acquire second-hand surface combatants from NATO allies or small vessels focusing on law enforcement. For example, 29 patrol boats for the Romanian border police, worth over €62m (US$85.92m), were partially financed through the EU's Schengen Facility in 2009–10. Separately, Romania also acquired two ex-Royal Navy Type-22 frigates. However, Russia's annexation of Crimea and the modernisation of Russia's Black Sea Fleet has led to closer alignment of both countries' naval procurement strategies. In July 2019, Romania selected a consortium of France's Naval Group and Romania's Santierul Naval Constanta to deliver four new *Gowind* 2500 frigates at a cost of €1.2bn (US$1.34bn). The ships will be built in Romania and deliveries are expected after 2022. Similarly, in November 2020, Bulgaria signed a contract with Germany's Lürssen Werft for two Modular Multi-role Patrol Boat (MMPK) corvettes for L984m (US$575.14m).

Though both countries have hitherto mainly focused on upgrading and modernising ageing Soviet equipment, they have recently begun to procure more modern equipment. Romania received the last of 12 surplus F-16s from Portugal in 2017, and in 2019 signed an agreement for an additional five. In April 2020, Bulgaria signed a contract with Lockheed Martin to procure a first batch of eight F-16Vs. Meanwhile, Romania is negotiating with German company Rheinmetall to design and manufacture 342 *Agilis* armoured personnel carriers (APC) through a locally based joint venture, and has also signed a contract with General Dynamics for 227 *Piranha* V APCs. Meanwhile, Bulgaria plans to acquire 150 new armoured vehicles in a programme worth over L1.46bn (US$884.38m). Although bids were submitted by ARTEC (*Boxer*), Patria (AMV), Nexter (VBCI) and General Dynamics (*Piranha* V), this programme was also halted due to COVID-19.

Table 3: Romania: defence procurement since 2010

Contract date	Equipment	Type	Quantity	Value	Prime contractor	Deliveries
Jan 2010	SNR-17	Patrol boat	5	€7.25m (US$9.61m)	Istanbul Shipyard	Nov 2010
Oct 2013	F-16AM/BM *Fighting Falcon*	Fighter/ground attack aircraft	12	€628m (US$834.07m)	Portuguese government surplus	Sep 2016–Sep 2017
Dec 2016	*Piranha* IIIC, *Piranha* IIIC CBRN, *Piranha* IIIC ARV	Armoured personnel carrier (wheeled)	12	lei187.77m (US$46.26m)	General Dynamics European Land Systems – Mowag	May 2018–ongoing
Nov 2017	MIM-104E *Patriot* PAC-2 GEM-T, MIM-104F *Patriot* PAC-3 MSE	Long-range surface-to-air missile launcher	28	US$3.9bn	Raytheon Technologies	Sep 2020–ongoing
Jan 2018	*Piranha* V	Armoured personnel carrier (wheeled)	227	€895m (US$1.06bn)	General Dynamics European Land Systems – Mowag	Jun 2020–ongoing
Aug 2018	M142 HIMARS	227mm multiple-rocket launcher	54	US$1.49bn	Lockheed Martin	Expected by Dec 2023
Dec 2019	F-16AM/BM *Fighting Falcon*	Fighter/ground attack aircraft	5	€280m (US$313.49m)	Portuguese government surplus	Aug 2020–ongoing

Table 4: Bulgaria: defence procurement since 2010

Contract date	Equipment	Type	Quantity	Value	Prime contractor	Deliveries
Oct 2014	M1117 ASV	Armoured utility vehicle	10	US$15.22m	Textron Marine & Land Systems	Jun 2015
Nov 2019	*Tsibar* (ex-NLD *Alkmaar*)	Oceangoing minehunter	2	€1.99m (US$2.23m)	Netherlands government surplus	Sep 2020
Apr 2020	F-16V *Fighting Falcon*	Fighter/ground attack aircraft	8	US$1.26bn	Lockheed Martin	Expected in 2023
Nov 2020	Modular Multi-role Patrol Boat (MMPK)	Corvette with surface-to-surface missiles, a hangar and surface-to-air missiles	2	L984m (US$575.14m)	Lürssen Werft	2025–26

Figure 10: **NHIndustries NH90 helicopter**

In the early 1980s, after NATO produced two studies for future helicopter designs, it was proposed that a single base helicopter design be used to produce a number of variants, primarily tactical transport helicopter (TTH) and ship-based NATO frigate helicopter (NFH) variants. A development contract was signed in 1992 with the NHIndustries consortium, which at the time comprised Eurocopter, Agusta and Stork Fokker. To date, around 400 helicopters have been delivered out of a current order book for 570, and the NH90 has been used operationally in Afghanistan and Mali. The programme experienced early problems. For example, after the production contract for core nations – France, Italy, Germany and the Netherlands – was signed in 2000, eight other countries then ordered more than 100 helicopters before a series-production helicopter had even been delivered – signing up for a product that was effectively still in the design phase. Some NH90 operators renegotiated their contracts to stagger deliveries in batches, at different levels of development, while fully operational designs were being developed. The early batches would then be upgraded to a final configuration. Meanwhile, the number of variants ordered by some countries reduced the economies of scale that might have resulted from agreement over common features. It also increased the amount of development work required across the programme.

NH90 TTH helicopter

NHIndustries shareholders

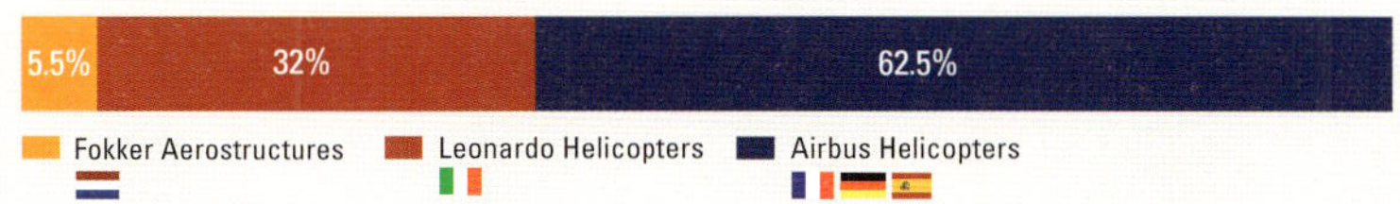

NH90 development, orders and delivery timeline

Country flag and quantity indicates number of NH90s ordered in that year.
Quantity in red font indicates order reduction/cancellation.

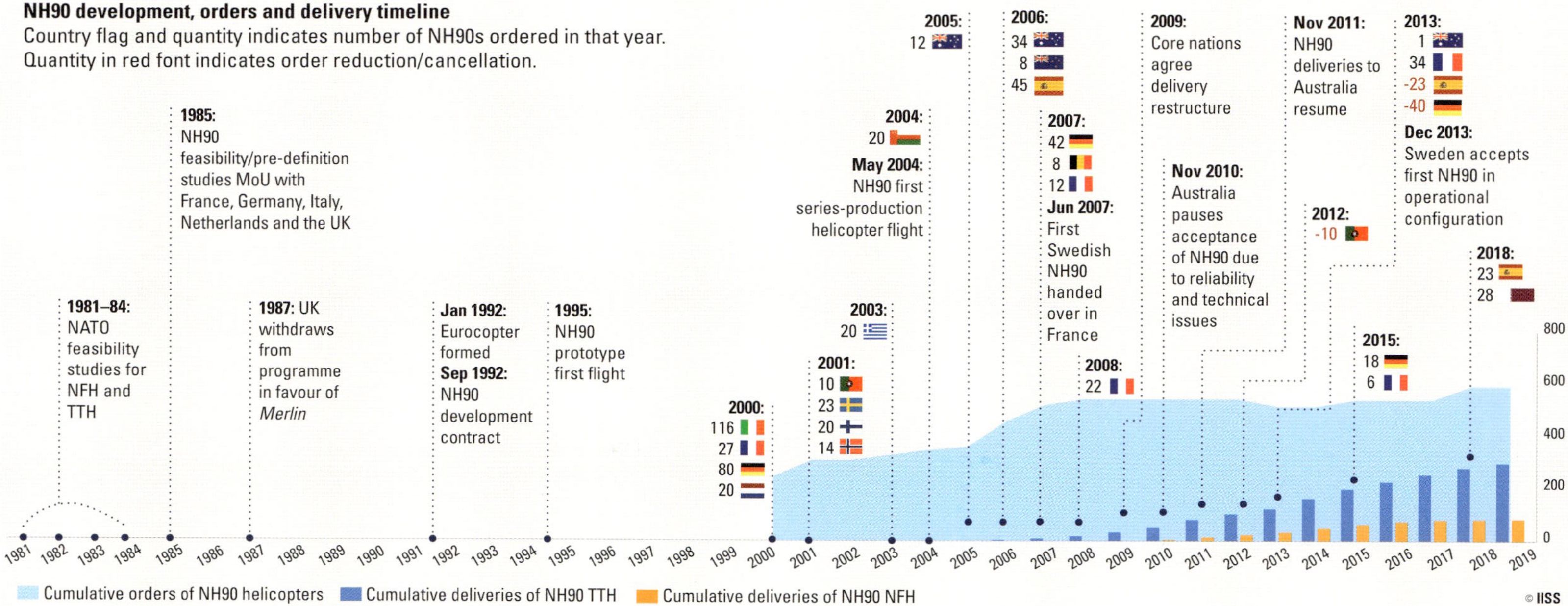

Europe

Chapter Five

Russia and Eurasia

- The Russian Aerospace Forces (VKS) will take delivery, in the fourth quarter of 2020, of the last of 50 Su-35S *Flanker* M fighter ground-attack aircraft ordered in 2015. This was the second large order for the Su-35S by the VKS, after a 2009 contract for 48 of the aircraft. A follow-on contract for a further 30 of the type was signed in August 2020. August 2020 also saw the conclusion of a 92-aircraft order for the Su-34 *Fullback* from 2012.
- The first Project 955A *Borey*-A ballistic-missile submarine, the *Knyaz Vladimir*, was handed over to the Northern Fleet in June 2020. The Russian Navy already has three Project 955 *Borey*-class (*Dolgorukiy*) submarines, while a further four hulls of the improved *Borey*-A have been laid down.
- Russia's Ground Forces continue to modernise their equipment, though at a slower pace than the other services. New armoured vehicles are arriving, slowly, while existing types are being modernised. The same is true of artillery, where the future *Koalitsiya* system is being developed alongside modernisation of older 2S19 *Msta* and 2S3 *Akatsiya* systems. The modernisation of older, large-calibre systems owes much to lessons from operations in Syria and Ukraine and requirements for heavier firepower against fortified positions.
- Azerbaijan's highly visible and successful use of both armed uninhabited aerial vehicles and loitering munitions in its late-2020 conflict with Armenia in Nagorno-Karabakh highlighted the importance for ground forces of mobile, layered air-defence systems capable of detecting and engaging such threats.
- Ukraine's Air Force Vision 2035, elements of which were released in May 2020, laid out ambitious recapitalisation goals for combat aircraft and ground-based air defence. Under the proposal nearly all Soviet-era combat-aircraft types are to be replaced within the inventory by 2030.
- Private military companies from Russia continue to operate abroad. Russian personnel have been deployed in the Central African Republic for some years, and troops from the Wagner Group have operated in Syria and most recently in Libya, in the latter supported by combat aircraft.

Russia real-terms total military expenditure, 2010–20 (US$bn, constant 2015)

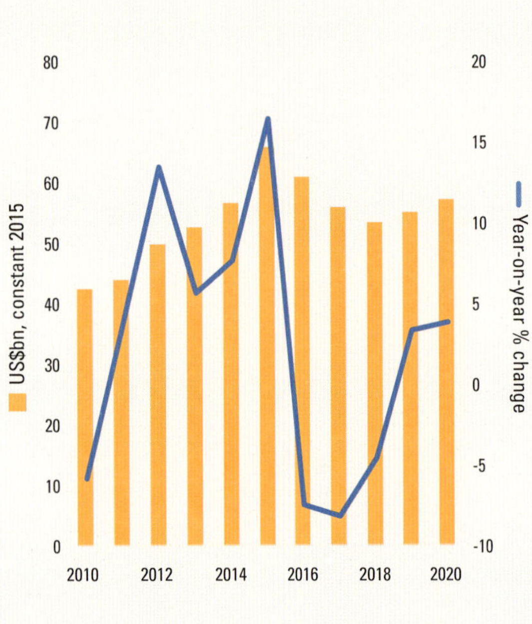

Active military personnel – top 10

(25,000 per unit)

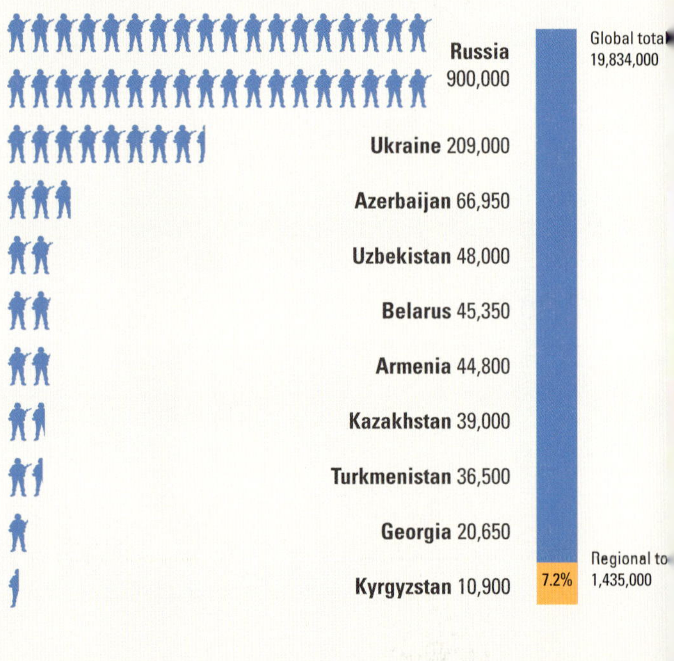

Russia's Aerospace Forces: Air-to-air weapons development timeline

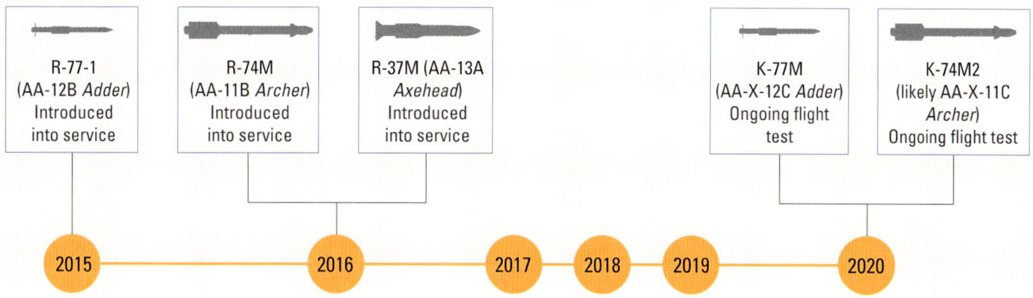

Note: K denotes a developmental AAM programme

Russia's *Borey*-class submarine: development timeline

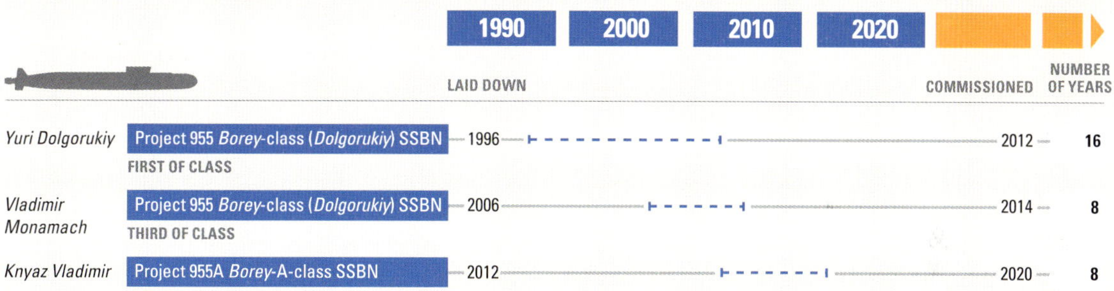

VKS FGA fleet total 2020

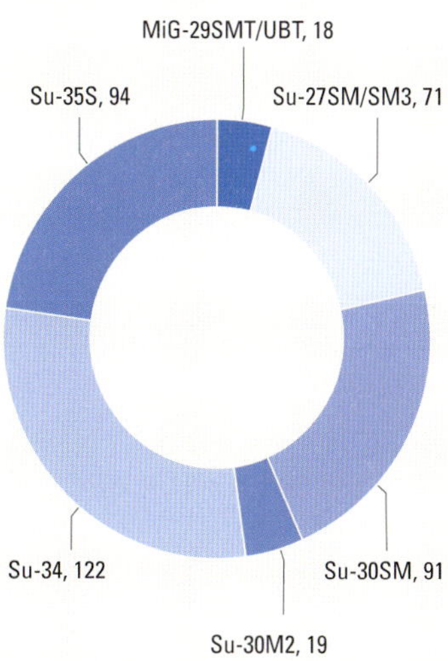

Russia's *Iskander* missile units, 2020

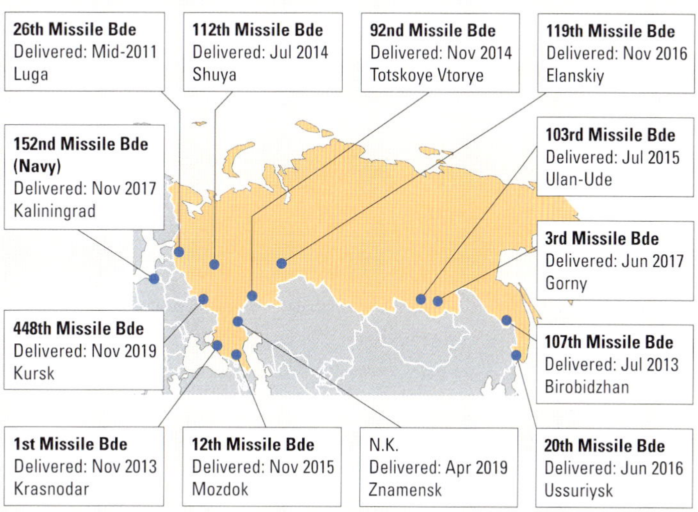

Russia and Eurasia

RUSSIA

2020 marks the end of a key period in Russia's post-Soviet military modernisation. It was intended that recapitalisation programmes under the State Armament Programme to 2020 (SAP 2020), signed in late 2010, would by 2020 lead to an improved inventory. Indeed, after Sergei Shoigu became defence minister in 2012 – not long after SAP 2020 began – the plan was that the armed forces would, by the end of the decade, field an inventory that comprised 70% 'modern' equipment. Although the Ground Forces lag significantly – their proportion of modern equipment is reduced by the sheer number of legacy armoured personnel carriers and armoured fighting vehicles – the other services have fared better, particularly the Strategic Rocket Forces and the air force, though the navy has benefited mostly from the introduction of more capable weapons rather than increased platform numbers.

The trajectory of the defence budget will be key to Russia's ability to continue delivering on its capability aspirations. The government has been forced to reduce defence expenditure by low oil prices and the effects of the coronavirus pandemic, which have left the federal budget with a record deficit. It is possible that expenditure on the current SAP 2027 may be cut by around 5% in 2021. Although it has not been announced which specific programmes will be pared back, Russian media reports indicate that the navy may suffer the most. That said, there remains a degree of pragmatism, in that the authorities have in the past proven willing to move some programmes to a new SAP, while others might be suspended until the fiscal situation improves. This flexibility has been reflected also in Russia's approach to modernising its force structures. If restructuring proved unattainable or sub-optimal it could be reversed or adjusted, either because of inefficiencies in the planned design or because of changed strategic priorities, such as with the revival in the Ground Forces of divisional-level formations.

Policy and strategy

A number of policy documents were due to be updated by the end of 2020. In November 2020, President Putin signed a new Defence Plan of the Russian Federation, covering the period 2021–25. This classified document is a principal means by which Russia sets out its threat perceptions and outlines its planned force dispositions and weapons developments. It is only the second such plan; the first was adopted in 2013 and was updated in 2015. Apart from references that the new plan will take into account 'changes in the military-political situation in the world', as President Putin put it in late 2019 when speaking to the defence ministry board, little has been disclosed about the document. The same can be said for the new version of Russia's National Security Strategy, which was expected to be released before the end of 2020 (the most recent doctrine dates from 2015). Officials have been focused on military activities by NATO forces, with representatives of the high command on occasion alleging not only that US and other NATO members' forces had increased the intensity of their operations but also that they were, in their exercises, practising airstrikes on Russia. Work has also started on a new State Armament Programme that will succeed SAP 2027. These documents are ordinarily reviewed every five years, which means that the new version should be adopted in 2023 and remain in force until 2030.

Russian officials continued in 2020 to assert that the United States was responsible for the demise of the Intermediate-range Nuclear Forces (INF) Treaty. Washington withdrew from the treaty in 2019, after it concluded – along with its NATO allies – that Russia had developed and deployed a ground-launched cruise missile (GLCM), the 9M729 (SSC-8 *Screwdriver*), in violation of the treaty. In October 2020, Moscow was still saying that it would not deploy INF-range GLCMs 'as long as US weapons of similar classes were not deployed'. Notwithstanding the allegations over the 9M729 – the US claimed that Russia had tested the system as far back as 2008 – in early 2019 Minister of Defence Shoigu acknowledged Russia's interest in these systems, when he said that during 2019–20 Russia needed 'to develop a land-based version of the sea-based *Kalibr* complex' and that 'in the same time frame [Russia has] to create a ground-based missile system with a long-range hypersonic missile'. Russian analysts suggest that the hypersonic system could be based on the naval 3M22 *Tsirkon* missile now in development.

After international concern over the collapse of the bilateral INF Treaty, and notwithstanding worry over the implications of the US decision to also withdraw from the Open Skies Treaty (due to be effective from November 2020), greater anxiety surrounded the fate of the New START (Strategic Arms Reduction Treaty) on limiting offensive strategic systems. The US wishes to widen the scope of new arms-control measures to include China, and also to include systems and technologies that are not covered by New START. Russia maintained its position that it respected China's stance on not engaging in trilateral disarmament talks, and also referred to frequently heard concerns about US missile-defence and long-range precision-strike systems. A series of strategic-stability talks took place in 2020, but at the time of writing there had been no agreement on extending the treaty, though Russia had in October offered to extend it by one year and at the same time jointly undertake a warhead freeze.

Despite pressure from the US, in August 2020 an agreement was reached with Turkey on the delivery of a second regimental set of the S-400 (SA-21 *Growler*) long-range surface-to-air missile system. Russian efforts to exploit the differences that emerged between Turkey and other NATO states in the wake of the S-400 deal have been hampered by a deterioration in relations with Ankara following the arrival of Turkish troops in the Syrian province of Idlib and in Libya, as well as the flare-up in the Nagorno-Karabakh conflict in September.

Although the coronavirus pandemic reduced Russia's contacts with its military allies in 2020, limited joint exercises still took place, including with the Chinese and Belarusian armies, though foreign participation was reduced in the strategic-level exercise *Kavkaz* 2020. An Airborne Forces (VDV) battalion took part in September in the two-week *Slavic Brotherhood* exercise in Belarus, against the backdrop of protests against President Alexander Lukashenko. Another exercise with Russian participation was held in Belarus only a month later. *Unbreakable Brotherhood* was, according to the Collective Security Treaty Organisation (CSTO), a peacekeeping exercise, though it was arguably also a demonstration by Russia that it was prepared to support the leader of an allied country including by the deployment of military force. September saw the first joint meeting of defence ministers from the Commonwealth of Independent States (CIS), Shanghai Cooperation Organisation (SCO) and the CSTO. Although no major agreements were reached at the meeting, it was an indication that Russia may intend to attach greater political–military importance to these alliances.

Structural changes

From 1 January 2021, the Northern Fleet will have the status of a full Military District, becoming Russia's fifth. Its area of responsibility will include the Northern Sea Route (as Russia terms the waters from its far east through to the north Atlantic) and the islands off Russia's Arctic coast as well as the north and northwest of Russia itself. The main change is administrative. The Fleet headquarters and its staff will have greater status and autonomy, but there have been no reports of adjustments to its size, structure or tasks – most of these reforms had already been undertaken after the Fleet was devolved from the Western Joint Strategic Command as a separate Strategic Command in 2014.

The formation of new military units has slowed, though infrastructure improvements continue. In recent years the Southern Military District has formed and relocated units close to the border with Ukraine. The defence minister said in March that the 150th Motor-Rifle Division (Novocherkassk, in Rostov oblast) was now fully staffed. Three motor-rifle divisions have been established near Ukraine since 2015, the 150th, as well as the 144th at Yelnya and the 3rd at Valyuki. The latter two are likely to be still missing at least a manoeuvre regiment each. It is also understood that most infrastructure development for the combat and support units of the 20th Combined Arms Army (headquartered at Voronezh) is now complete. Meanwhile, construction work is nearly complete at the Baltimor air base in Voronezh, in the Western Military District, which will house an Su-34 *Fullback* regiment (possibly the 47th mixed aviation regiment) as well as squadrons from the air force academy. Meanwhile, one tank and one motor-rifle regiment are being formed in the Eastern Military District, to complete the 127th Motor-Rifle Division, while there are reportedly plans to create new brigades of railway troops in order to build the second branch of the Baikal–Amur strategic railway line.

Russia has long aspired to improve its military infrastructure in the Arctic, and work on base construction is nearing completion in some cases.

Figure 11 **T-72 main battle tank (MBT): major upgrades**

The T-72 was originally envisaged as a cheaper and less complex alternative to the T-64 MBT. It incorporated many of the advances seen in the T-64, was easier to manufacture and was intended to replace older-model T-62s and T-55s in motor-rifle units. Over time, the T-72 became the backbone of Soviet, and later Russian, armoured formations. Originally entering service in 1973, the tank has had a series of upgrades in order to improve its serviceability, protection and offensive capability. It has been exported and also produced abroad under licence; globally, only the T-54/55 is more ubiquitous. The T-72B variant was the basis for Russia's most recent operational MBT design, the T-90. In various upgraded forms, the T-72 is likely to remain the mainstay of Russia's armoured forces for the foreseeable future, as it is unlikely in the near term that production of T-90 variants or the new T-14 *Armata* MBT will be able to – or perhaps are even intended to – replace the range of T-72 variants that remain in service.

T-72 (1973)

- 41 tonnes
- 780hp V-46 diesel engine

Firepower:

- 2A26M-2 125mm smoothbore gun
- Carousel autoloader with 22 ready rounds

Protection:

- Cast steel turret with composite armour in hull glacis

T-72A (1979)

- 41.5 tonnes
- Improved 780hp V-46-6 diesel engine

Firepower:

- Improved 2A46 125mm smoothbore gun
- Laser rangefinder

Protection

- Composite armour in hull glacis and turret
- Smoke grenade launchers

T-72B (1985)

- 44.5 tonnes
- Improved 840hp V-84-1 diesel engine

Firepower:

- Improved 2A46M 125mm smoothbore gun
- Improved autoloader to accommodate longer ammunition including 9M119 *Svir* gun-launched guided missile
- Improved fire-control systems and laser rangefinder

Protection:

- Improved composite armour in turret and hull
- *Kontakt*-1 explosive reactive armour

T-72B obr. 1989 (1989)

- 44.5 tonnes
- Improved 840hp V-84M diesel engine

Protection:

- Improved composite armour
- *Kontakt*-5 explosive reactive armour

Selected non-Russian variants

T-72AMT (Ukraine) (2020)

- Estimated 45–46 tonnes
- 840hp W-84-1 diesel engine

Firepower:

- 2A46M 125mm smoothbore gun
- Luch *Kombat* gun-launched guided missile

Protection:

- *Nozh* explosive reactive armour
- Bar armour fitted to engine compartment and rear of turret

PT-91 *Twardy* (Poland) (1995)

- 45.9 tonnes
- 850hp S-12U diesel engine

Firepower:

- 2A46 125mm smoothbore gun
- Improved fire-control systems including thermal gunner's sight

Protection:

- *Erawa* explosive reactive armour
- Laser warning receivers

Entry into service

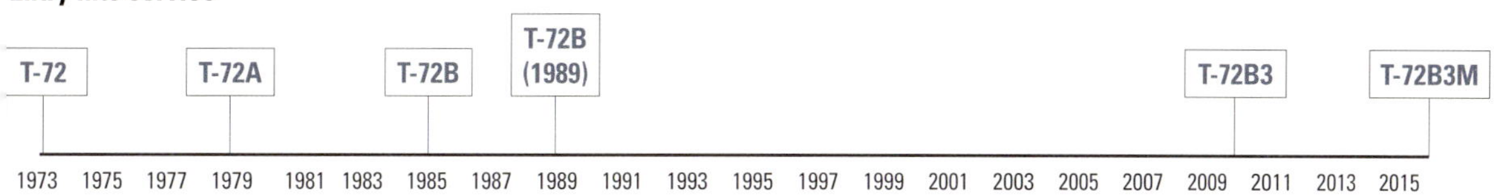

T-72B3 (2010)

- 45.6 tonnes

Firepower:

- Improved 2A46M-5 125mm smoothbore gun
- Autoloader modified to accommodate modern ammunition
- Improved fire-control systems including SOSNA-U thermal gunner's sight

Other:

- New tracks

T-72B3M (2016)

- 45.6 tonnes
- Improved V-92S2F 1,130hp engine
- Improved automatic transmission

Protection:

- New *Relikt* explosive reactive armour fitted in side skirts and to turret rear
- Bar armour fitted around engine compartment

An all-season runway was opened in April at the Nagurskoye air base on Alexandra Land, in the Franz Josef Land archipelago. The airfield was used during VDV exercises in 2020 by Il-76 *Candid* transport aircraft, indicating its current length, and the air force intends to extend the runway by a further kilometre in order to enable flights by strategic bombers. More broadly, Russia continues efforts to improve its air capabilities in the Arctic. In late 2019, a new (3rd) Air Defence Division was created in the Northern Fleet, and one of its regiments – equipped with the S-300PS (SA-10 *Grumble*) SAM system – was established in Tiksi (Republic of Yakutia).

Personnel

According to Shoigu there were 405,100 contract-service personnel serving in the armed forces by March 2020. Meanwhile, the number of conscripts has fallen to 225,000, leaving the overall size of the armed forces relatively unchanged.

The spring draft in 2020 took place under coronavirus restrictions, but there were still 135,000 recruits, the same number as in spring 2019. The Ministry of Defence (MoD) plans to increase further the number of contract personnel to 475,600 by 2027. However, the objective for contract personnel strength has fluctuated over the years. The aim in 2014 was 499,200, which was revised down to 425,000 before again being increased to the current figure. The autumn draft began on schedule in September, again under coronavirus restrictions, with the MoD saying that notices would this time go to 128,000 citizens.

Improving combat training has been an important part of the overall military modernisation programme. No-notice exercises and combat drills at various levels of command, including large-scale strategic-level exercises, have all been used to improve readiness. The MoD is also looking to improve synthetic training, and there are plans to deliver more sophisticated computer simulators for combat training. By 2024 the aim is that the first synthetic-training centre will be introduced to enable opposition-force exercises at company and battalion tactical group level.

Ground Forces

Equipment plans and deliveries for the Ground Forces focus on the same modernised systems delivered in the previous five years. A notable exception in 2020 was the delivery for trials of the first batch of eight 2S35 *Koalitsiya*-SV self-propelled guns to the Central Military District. Intended as a replacement for the 2S19 *Msta*-S family (now entering its third decade in service), the *Koalitsiya*-SV was originally planned to begin trials in the Western Military District in 2016. Deliveries have been repeatedly delayed, however, and in the interim deliveries of the latest modification of the *Msta*-S (2S19M2/2S33) to line units have continued at the rate of two battalion sets per year.

For tank units, the ten new-build T-90M main battle tanks from a 2017 contract (for 30 tanks, with 20 upgrades) were delivered to the 2nd Motor-Rifle Division in the Moscow region, while the nearby 4th Tank Division has begun taking delivery of a battalion set of T-80BVM tanks. However, the majority of modernised tanks being procured remain the T-72B3M upgrade, with both the Central and Western military districts seeing deliveries of the type in 2020. Motor-rifle units continued receiving a mix of new-build BMP-3s (the first deliveries under the 2018 contract) and modernised BMP-2Ms, the latter fitted with the *Berezhok* combat module. Meanwhile, the BTR-82A/AM also continues to be delivered in substantial numbers.

Russia's defence industry is still experiencing problems with the development of the new-generation vehicle platforms first shown at Moscow's 2015 Victory Day parade. Under the latest plans, the *Kurganets*-25 tracked infantry fighting vehicle and the *Bumerang* wheeled armoured personnel carrier will only complete state testing at the end of 2022. The *Armata* platform is also in test, and deliveries are not expected in 2021, even though assembly has already begun of the first trial batch of 132 tanks and support vehicles.

Air defence is being improved at all levels. The *Ptitselov* air defence system, based on the air-portable BMD-4M and reportedly with a 10 km-range, is reported to be still in development, as is a tracked version of the *Pantsir*-SM, with a reported range of up to 40km. Long-range S-300V4s (SA-23) are entering service, intended to protect combined-arms armies from air and missile threats.

In the VDV, the final battalion set of BMD-4Ms from the 2019 plan was handed over to the 76th Air Assault Division at Pskov in January 2020, with two more battalion sets handed over in March and June to the 7th and 76th divisions respectively. These deliveries bring the number of BMD-4M-equipped battalions in the VDV up to nine; regiments are each receiving two battalion sets of BMD-4M, likely one for each of the contract-manned BTGs every regiment is expected to generate. From 2021, these new-build vehicles are planned to be supplemented by the first of around 600 modernised BMD-2Ms. The VDV are continuing to receive tanks as part of the general move to heavier equipment, and a T-72B3M-equipped tank battalion was formed in the 7th Air Assault Division.

Navy

The first of the improved Project 955A *Borey*-A nuclear-powered ballistic-missile submarines, the *Knyaz Vladimir*, was commissioned into the Northern Fleet in June, adding to the three original Project 955 *Boreys* (*Dolgorukiy*). The next in the class, the *Knyaz Oleg*, was launched in July and three more were in various stages of construction, with orders for a further two expected. Meanwhile, the first of the improved Project 08851 *Yasen*-M-class of guided-missile-armed submarines, *Kazan*, continued trials. Seven more are now under construction. This submarine-building drive could see a significant number of vessels delivered during the rest of the decade. Meanwhile, the focus of conventionally-powered submarine development is currently on delivering six Project 06363 *Varshavyanka*-class (Improved *Kilo*) boats to the Pacific Fleet.

The second of the Project 22350 (*Gorshkov*) frigates commissioned in July, with the *Admiral Kasatonov* joining its sister ship, *Admiral Gorshkov*, in the Northern Fleet. These vessels are key to modernising the navy's blue-water surface-ship capability. Six more are being built, including four to a modified design; the last two were laid down in July.

The *Admiral Gorshkov* has been used as the naval test bed for the *Tsirkon* hypersonic missile system. A test in October was said to have been the first against a naval target, and the weapon was reported to have reached a speed of Mach 8. The potential incorporation of this system aboard Russian submarines and surface ships, in the latter half of this decade, would mark an important development in Russian naval strike capabilities.

July 2020 saw construction begin on a range of platforms. Perhaps the most significant were the two new Project 23900 large amphibious ships laid down at the Zaliv Shipyard in the Crimean city of Kerch. Russia had previously sought to introduce this capability through the acquisition of *Mistral* vessels from France, a deal that was halted after the annexation of Crimea.

Legacy platforms that completed lengthy refits included the Project 971 *Schuka*-B (*Akula* II) submarine *Vepr* and also the Project 1155 *Fregat*-class (*Udaloy* I) destroyer *Marshal Shaposhnikov*, which received an extensive weapons and equipment upgrade. The navy's ability to sustain its submarine and blue-water surface fleet inventories will depend on whether subsequent refits of vessels from these classes proceed.

There is little information on the progress of the *Poseidon* nuclear-powered and -armed uninhabited underwater vehicle, or of the submarines that are expected to deliver it. It was reported that the first of these, the *Belgorod* – based on a much-modified *Antey* (*Oscar* II) design and launched in April 2019 – could enter service for weapons tests before the end of 2020. Equally, the launch of a second vessel, the *Khabarovsk* – thought to be based on the *Borey*-class – was anticipated during 2020, but this may not now happen until 2021.

Aerospace Forces

The Aerospace Forces (VKS) faced a series of challenges in 2020, even if a number of capability milestones were met. 2019 saw the lowest number of combat-aircraft deliveries for nearly a decade, as several multi-year contracts were concluded: only two MiG-35 *Fulcrum*, ten Su-35S *Flanker* M and eight Su-34 *Fullback* aircraft were received. The crash of the first production-standard Su-57 *Felon* multi-role fighter in December 2019 resulted in further delay to the programme, while problems continued with the manufacture of the Il-76MD-90A heavy transport aircraft.

Service availability of the modern Yak-130 *Mitten* remains below target, and as a result the air force is having to continue using obsolescent types of training aircraft. An additional 25 Yak-130s have been ordered, while 20 DA-42T basic trainers are also being acquired. Across the training fleet, overall availability is around 55%, a figure that reflects the number of obsolescent L-39s that remain in service. Uninhabited aerial vehicle (UAV) and uninhabited combat aerial vehicle (UCAV) projects are also behind schedule. Both the *Altius* intelligence, surveillance and reconnaissance UAV and the S-70 *Okhotnik* UCAV are taking longer than was initially anticipated. A prototype of the S-70 entered the flight-test regime in 2019, but this appears to be a technology demonstrator.

With the delays to the *Felon*, the Su-35S *Flanker* M will remain the core of Russia's multi-role fighter fleet well into the 2030s, and potentially beyond. Firing trials were under way in 2020 to integrate the R-37M (AA-13 *Axehead*) long-range air-to-air missile (AAM) on the Su-35S; this would provide an extended-range engagement weapon for the aircraft. The R-37M is intended for use against high-value low-volume targets, such as tanker and airborne early warning and control aircraft, and also to engage cruise missiles. The R-37M was fielded first on the MiG-31BM *Foxhound* C interceptor in 2016.

The VKS's primary medium-range active radar-guided AAM is also being upgraded. The K-77M (K denoting a developmental missile) (AA-X-12C) is also being tested at the Aerospace Forces' 929th State Flight Test Centre at Akhtubinsk. The K-77M was seen on the Su-57's external wing pylons, though the missile is intended for internal carriage in the aircraft's two main weapons bays. The K-77M does not feature the R-77's lattice rear-fin design, reverting instead to a conventional blade design.

Along with what appeared to be a 'gap filling' order for 24 Su-34 *Fullbacks*, an order for improved Su-34Ms is expected. Additional types of weapon are being integrated onto the Su-34. The Kh-38 family of medium-range air-to-surface missiles is being added to the aircraft's weapons inventory. The Su-34 will also likely carry the Kh-69 medium-range land attack cruise missile. The design was previously associated with the Kh-59MK2 proposed variant of the Kh-59M (AS-18 *Kazoo*). The Kh-69 configuration is driven by the need for internal carriage, and the missile is designed so it can fit in the Su-57's main bays

As with the Su-57, the VKS's next-generation bomber is behind schedule, with existing types being upgraded or being newly manufactured in the case of the Tu-160 *Blackjack*. The Tupolev programme to meet the PAK DA (Future Aviation Complex – Long-range Aviation) is still delayed and the near-term emphasis is on Tu-160M production. The first new-build *Blackjack* airframe was being assembled at Kazan during the second half of 2020.

Russia continues also to reinvest in its space capabilities and there are indications that a suite of new counter-space capabilities are being developed. Russia has conducted multiple demonstrations of rendezvous and proximity operations on orbit, including in mid-2020 the reported injection into orbit of a projectile (catalogued as Object 45915) from one space vehicle (*Cosmos* 2543) that had itself been deployed from a larger satellite, *Cosmos* 2542. *Cosmos* 2542 was launched from a *Soyuz* rocket, while tests of anti-satellite capabilities have also been alleged, with a missile launched from a MiG-31 aircraft. Meanwhile, tests of a new direct-ascent anti-satellite missile have taken place since 2014, though this system has, to date, not been tested against a target in orbit.

Strategic Rocket Forces

The re-equipment programme of the Strategic Rocket Forces (RVSN) has been more successful than those of other services and branches. The 2020 target of 70% modernised equipment had been met before 2020. By mid-year it had reached 76%, according to RVSN commander Colonel-General Sergei Karakaev, and the plan was to reach 81% by the end of the year. To help meet this target, the RVSN planned to re-equip four more regiments with new missile systems in 2020. A continuing annual production of approximately 20 RS-24 *Yars* intercontinental ballistic missiles (ICBMs) would allow for three regiments to be re-equipped with this system – most likely the final silo-based regiment of the 28th Missile Division at Kozelsk, and two more mobile regiments from the 35th Missile Division at Barnaul. These latter formations are reportedly scheduled to receive the '*Yars-S*' modification of the RS-24, following the conversion of the 479th Missile Regiment in late 2019. There is little publicly available information about the differences, if any, between the standard *Yars* and *Yars-S*. The latter is reportedly intended to have a silo-based as well as a road-mobile version, but only the mobile system appears to have entered service at present.

The fourth and final regimental conversion will receive the *Avangard* hypersonic boost-glide vehicle system, still using the modified RS-18 (SS-19 mod 4) ICBM as a booster. The first regiment equipped with this system entered service with the 13th Missile Division at Dombarovsky in late 2019, and it is likely that the second regimental conversion will also be in the 13th Division, which is in the process of phasing out its ageing RS-20 (SS-18 *Satan*) silo-based ICBMs. These conversions would allow the re-equipment in 2021 of the final regiments equipped with the RS-12M *Topol* (SS-25 *Sickle*), so consolidating the road-mobile force on the later model RS-12M2 *Topol*-M, RS-24 *Yars* and *Yars*-S and effectively completing their modernisation. Seven to ten additional regiments of legacy silo-based ICBMs (a mix of RS-18 and RS-20 systems) would then remain in service to be replaced by the anticipated RS-28 *Sarmat* (SS-X-29) liquid-fuelled heavy ICBM.

DEFENCE ECONOMICS

Macroeconomics and defence spending

Russia's economy has shown only sluggish growth since 2015, reaching barely 1.5% on average. Military spending increased rapidly from 2012–15 with a surge of spending on the State Armament Programme (SAP), but then declined in both nominal and real terms from 2016–18, before a modest recovery in 2019. As a share of GDP, total military expenditure fell from almost 4.9% in 2015 to 3.8% in 2020.

The size of the annual State Defence Order (SDO) now appears to have stabilised at approximately 1,500 billion roubles (US$21.0bn). According to defence minister Sergei Shoigu, in 2020 r1,020bn (US$14.3bn) was to be allocated to the procurement of new weapons, some r290bn (US$4.1bn) to the repair and modernisation of existing systems and around r190bn (US$2.65bn) to research and development (R&D). This means that the SDO now comprises less than half of Russia's total spending on 'national defence'.

The coronavirus pandemic has seriously affected Russian economic performance, but not as severely as some European economies. At the end of August the Ministry of Economic Development forecast that GDP would decline in 2020 by 3.9% followed by modest recovery in 2021 and 3.4% growth in 2022, with the average annual inflation rate remaining at about 4%. The draft federal budget for 2021 and the planned budget for 2022 and 2023, approved on 16 September 2020, acknowledged the negative impact of the coronavirus pandemic on the Russian economy and the difficulties caused by the fall in the price of oil over the year. Government measures to revive the economy resulted in some readjustment to planned public expenditure to 2023. Spending on the 'national economy' and on social policy will increase. Of more concern is the growth in debt servicing, which will increase from 3.8% of GDP in 2019 to at least 7.2% in 2023.

Such movements create pressure on the defence budget, and before the draft budget was approved, the Ministry of Defence (MoD) was asked to plan for a 5% reduction in the SDO spending for 2021–23 as

▼ Map 4 **Russia and Eurasia regional defence spending**[1]

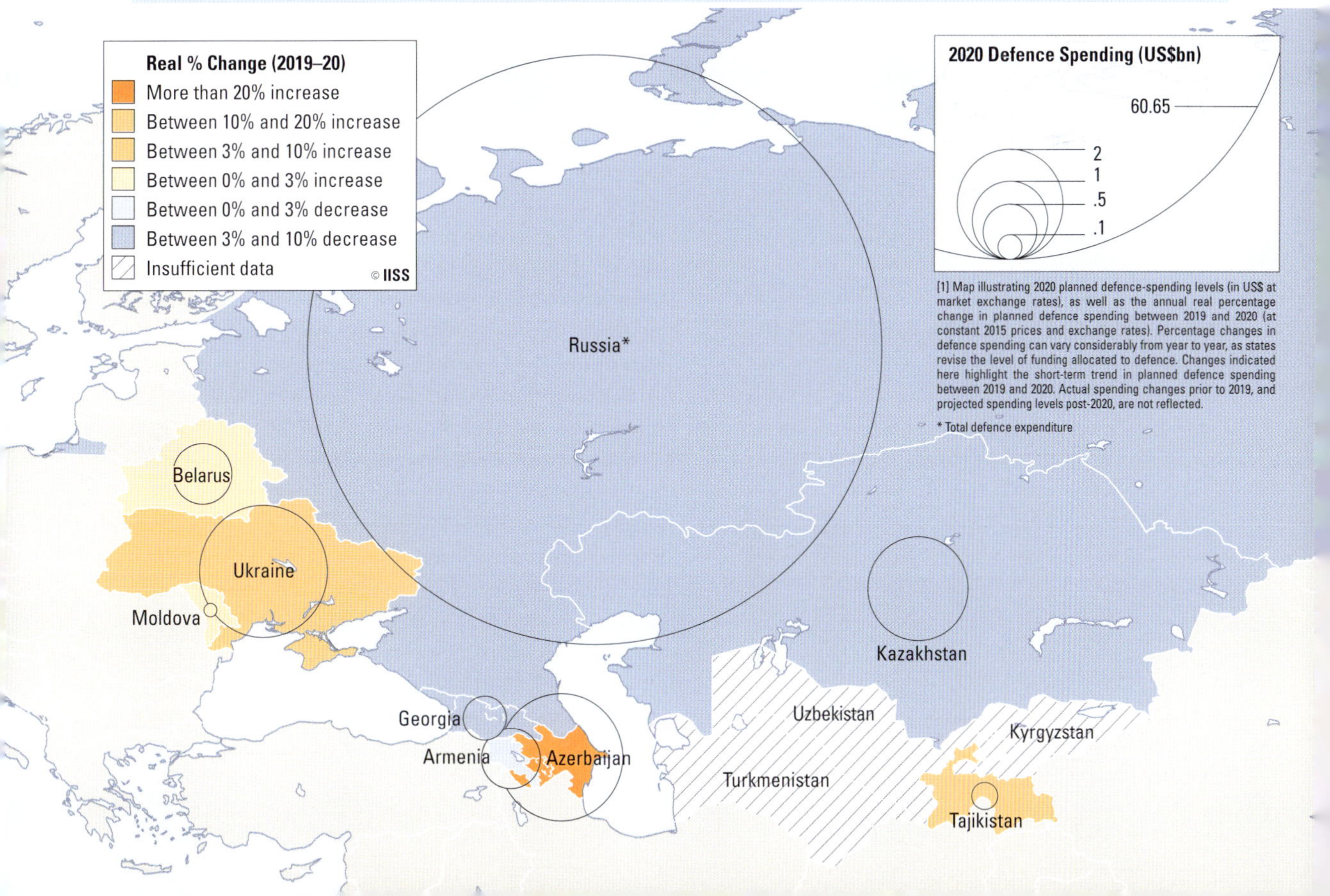

Table 5 **Russian defence expenditure as % of GDP**

Year	'National Defence'			Total military expenditure[1]			
	Roubles (trillion)	US$ (billion)	% of GDP	Roubles (trillion)	US$ (billion)[2]	% of GDP	% change real terms[3]
2020[4]	3.087	40.7	2.95	4.335	57.1	4.14	3.8
2019	2.998	39.2	2.72	4.211	55.0	3.83	3.3
2018	2.827	38.4	2.7	3.928	53.3	3.75	-4.6
2017	2.666[5]	40.2	2.9	3.704[5]	55.8	4.03	-8.2
2016	2.982[5]	47.3	3.48	3.831[5]	60.8	4.47	-7.5
2015	3.181	51.9	3.83	4.026	65.7	4.85	16.4
2014	2.479	43.4	3.14	3.224	56.5	4.08	7.6
2013	2.104	39.6	2.89	2.787	52.5	3.82	5.6
2012	1.812	35.9	2.66	2.505	49.7	3.68	13.4
2011	1.516	32.7	2.52	2.029	43.8	3.38	3.6

1. According to NATO definition. GDP as Rosstat's latest revision, April 2020. 2. Constant 2015 US$. 3. Calculated using annual GDP deflator. 4. Amended 2020 budget, March 2020. Totals based on draft budget as data of full final version not published, but the allocation to 'national defence' does not appear to have been changed. 5. Excluding a one-off payment to reduce accumulated debts of defence-industry enterprises under the scheme of state guaranteed credits. If this debt payment is included the total GDP share in 2016 rises to 5.4%, and 4.2% in 2017.

set out in the three-year budget adopted last year. Initial indications were that defence spending will be reduced only modestly, largely because efforts have been made to maintain the procurement levels planned under the SAP, notwithstanding the effect that the pandemic has had on the defence industry. The plan for military spending in the draft budget appears to be in line with these expectations. While funding for national defence – the core defence budget – is set to increase from r3.09tr (US$43.2bn) in 2020 to r3.26tr (US$45.6bn) by 2023, an inflation rate of 3% annually means that these nominal increases translate to modest cuts in real terms. Total military expenditure – that is, including wider spending items like military housing, pensions, health and social support – will increase from r4.33tr (US$60.6bn) to r4.73tr (US$66.2bn), though this still results in a cut in real terms. The 2021 budget and plan to 2023, signed into law by Putin in December 2020, was largely in line with the draft budget although total military spending will reach just r4.68tr (US$65.4bn) in 2023. Total military spending is set to fall from over 4.1% of GDP in 2020 to under 3.8% of GDP in 2023. Spending on new weapons might reduce if the MoD signals that it has reached its 70% target.

▼ Figure 12 **Russia: estimated total military expenditure** as % of GDP

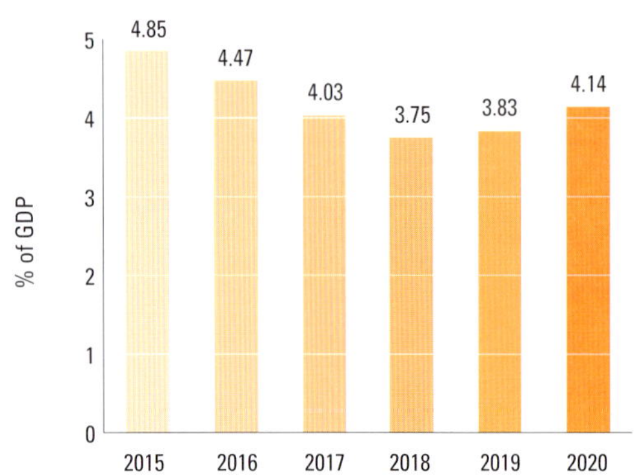

Weapons procurement

The key performance indicator of SAP 2020 was that the proportion of modern weapons in the armed forces should increase to 70%; by mid-2020 the ministry was saying that the total had reached 68.5%. The share by the end of the year was expected to be around 80% in the Strategic Rocket Forces, 78% in the air force, 63% in the navy and only around 50% in the Ground Forces – the figure here is dragged down by the large number of legacy armoured vehicles in service.

The successor programme, SAP 2027, was agreed by President Vladimir Putin in early 2018. Equipment recapitalisation will be focused towards maintaining the 70% modern share, and the serial procurement of systems that were developed and in some cases originally intended to begin their service-entry as part of SAP 2020. These include the S-350 (SA-28) air-defence system and the Su-57 multi-role fighter. *Armata* platforms may be acquired but not in large numbers, and modernisation of existing armour will remain a priority.

Defence industry

The most significant organisational change within the defence industry has been the absorption of the United Aviation Corporation (UAC) into the Rostec state corporation. Rostec now has the entire aircraft industry under its umbrella. UAC is being internally restructured, as is Russian Helicopters. Within UAC, specialised divisions are being created and some design bureaux are being amalgamated. A division is being formed for fixed-wing combat aircraft, and the Sukhoi and MiG bureaux are being merged and, although is larger, both brand names will be retained. Tupolev will bring together all strategic aviation, Ilyushin all transport aviation and Irkut will form the nucleus of a civil-aviation division. Similarly, a National Helicopter Centre will combine the Mil and Kamov design groups.

There is increasing attention to the task of diversifying into civil high-technology fields. In 2019, the civilian share of total defence-industry output reached 24% and efforts are now focused on reaching the target of 30% by 2025. A planned strategy for diversification to 2030 could form the basis of a new state programme.

Yuri Borisov, deputy prime minister responsible for oversight of the defence industry, and industry minister Denis Manturov have appeared confident that the coronavirus pandemic will have limited impact on the implementation of the SDO in 2020. Some defence-industry companies quickly organised the manufacture of medical equipment, including ventilators and personal protective equipment. Many designers, technical personnel and other white-collar personnel switched to working at home, while social distancing was introduced for workers on site. To help maintain production levels, advances of up to 80% have been introduced for work on SDO contracts.

However, the accumulation of debt is hampering the development of the defence industry. Companies have been obliged to resort to a large volume of bank credit at high interest rates. At a time of budget stringency there is little scope for resort to large-scale budget funding to provide debt relief. Instead, debts are being restructured, which means that resolution of the problem is delayed. Promsvyazbank now plays a key role in financing the activities of the defence industry. In December 2019 it became the designated state bank for servicing the SDO. In autumn 2020 it claimed to account for 60% of the total value of defence orders. It is now developing a centre of competence in order to assist diversification and will offer credit on preferential terms. According to Manturov, defence-industrial output increased by 38% after 2014 and labour productivity by almost 40%. He forecasts a 5–6% increase in output for 2020, notwithstanding the coronavirus pandemic.

Defence exports

In recent years the value of Russia's defence exports has been relatively stable, notwithstanding sanctions on some countries that opted to buy Russian weapons, and other measures adopted by the United States and other countries. Russia's own measure of export volumes relates to so-called military–industrial cooperation, which includes sales of weapons and other military equipment, components, licence sales, repairs, servicing and training.

Russia's principal defence exports remain combat aircraft and air-defence systems, in particular the S-400 now supplied to China and Turkey. Export to India has been agreed in principle and discussions have been reported with Turkey about a contract to purchase a second batch.

Russia does not import any end-product weapons but does build some systems domestically under licence and is dependent on external suppliers, notably Belarus, for some important equipment and subcomponents. Israeli *Searcher* II (*Forpost*) reconnaissance uninhabited aerial vehicles are assembled under licence in Ekaterinburg by Ural Civil Aviation Plant (UZGA), but a process of gradual localisation has been under way and most components are now sourced domestically. Since 2017 the same Ekaterinburg factory has been assembling a limited number of Austrian Diamond DA-42T light trainers for the air force. The most significant dependence on an external supplier remains the import from Belarus of heavy-duty transport vehicles for the *Yars* and other mobile strategic missiles, and also for the *Iskander* surface-to-surface missile. Russia has been trying for some time to develop a domestic substitute for these heavy wheeled transporters, but without success. However, transporters for other systems such as the S-350, S-400 and the new S-500 air-defence system are built at the Bryansk Automobile Plant (BAZ). Belarus is also an important supplier of electronic components and optical systems. Unrest in Belarus in the summer of 2020 is likely to intensify efforts for import substitution, after Russia ended its previous dependence on Ukraine for the supply of some defence articles.

UKRAINE

Ukraine's armed forces began a transformation process after Russia annexed Crimea in early 2014 and conflict began in the country's east. Personnel strength has increased and there is a growing reserve of combat-experienced personnel. Structural reforms are under way in command and control, training and education, and technical support. However, recapitalising the armed forces with modern weapons has not proven straightforward. Reforms are also targeting the procurement process. Some procurements have been successful but other capability areas are still dominated by Soviet systems. This legacy dominated Ukraine's defence environment until 2014. The armed forces had reduced in size after independence in 1992, their equipment inventory had shrunk, and combat training had also suffered.

Legislation and defence policy

The crisis caused by Russia's annexation of Crimea, and its subsequent military activity in Ukraine's east, spurred reform initiatives designed to improve the capability of the armed forces and Ukraine's defence and security institutions. A new National Security Strategy was approved in September 2020. This named threats to security including Russia's occupation of Crimea and its actions in Ukraine's east, and from other factors such as climate change and pandemics. It reiterated the aspiration to obtain EU and NATO membership as well as the need to boost deterrent capabilities – one of the identified threats was also from the still-low rate of armed forces' recapitalisation. Important in this regard will be updates to other documents, such as the Strategic Defence Bulletin, which was first published in 2016. This outlined reform plans and provided a guide to the armed forces strategic development. A more granular view is provided in the General Staff's ten-year vision for military development, which was published in 2020.

The reforms are intended to produce armed forces, and a defence sector, more aligned with Western military standards and that are capable of territorial defence. The plan is that they can also operate with NATO and EU armed forces, and engage in international deployments. As of early 2020, however, fully meeting NATO standards appeared to have been shelved in favour of achieving interoperability. The authorities also intend to establish a joint command structure and to improve defence policy, planning and administrative structures according to Western principles, and also to professionalise the armed forces and generate a military reserve.

Organisational changes

An important shift came with the decision to separate responsibilities between the posts of commander-in-chief (C-in-C) and the chief of general staff (CGS). The C-in-C is responsible for combat readiness and military taskings as well as capabilities subordinated from other organisations such as the National Guard or the State Border Service. The CGS, subordinate to the C-in-C, has a range of responsibilities including determining military capability and resource requirements, operational and modernisation planning, military training and support needs, and the allocation of equipment. Principal defence planning functions are held in the General Staff, which is also tasked with coordination between central government and local authorities with defence responsibilities. The joint-forces commander (currently a lieutenant-general rank) reports directly to the supreme commander (the president) and, through the Joint Operational Staff, commands subordinated elements of the armed forces, as well as peacekeeping contingents.

The General Staff's ten-year vision for military development sets out force structure aspirations and highlights development priorities. Particular priority is given to special-operations forces (SOF), artillery and missile troops, army aviation, the air force's air-defence units, and the navy's coastal missile and artillery forces. On current planning, the land forces will still be the largest service arm by 2030, comprising around 50% of total personnel strength. The main task of Land Forces Command, meanwhile, will be force generation, that is, deploying new formations on mobilisation. To help this process, a headquarters and training command analogous to the US Army's Training and Doctrine Command will be set up within Land Forces Command.

It is intended that the SOF remains relatively small, at around 2% of total strength, which, although the SOF is a priority in terms of force development, may reflect the influence of Western attitudes about retaining the selectiveness of such capabilities. Existing special-forces regiments are to transform into SOF Centres. (In 2020 the 35th Mixed Aviation Squadron was operationally subordinated to SOF.) The air force, navy and logisticians will comprise the remainder.

Other organisational developments are due to include fully staffing the 35th Naval Infantry Brigade and converting the 61st Motorised Infantry Brigade of the Reserve Corps into a *jager* infantry brigade specialised in operations in forests and swamps. The reserve structure changed in 2020. It consists of operational reserves, mobilisation reserves and civilian components. The operational reserve includes a first-line reserve made up of reservists and those subject to military service (primarily with combat experience in Donbas), intended to provide replacements for combat formations and for backfilling administrative posts. Reservists in this category are assigned to specific military units so these can be brought to wartime strength if required. The operational reserve has been divided in two, with a second tier becoming an expansion and a territorial reserve. The expansion reserve is intended to provide replacements for non-combat units, and the territorial reserve for Territorial Defence units. It is planned that the mobilisation reserve will provide personnel for military educational establishments and training centres and units. The civilian reserve, meanwhile, comprises citizens who volunteer for defence and national-security tasks. It appears that the plan is for the civilian reserve to include various non-state paramilitary formations such as the Ukrainian Volunteer Corps, the Ukrainian Volunteer Army, the Ukrainian Legion and others.

Territorial-defence brigades are part of the reserve system. The core of these units were set up in 2018, and the brigades replaced the territorial-defence detachments that were formed after 2014. In peacetime they consist of a relatively small cadre of officers administering the formation and its subordinate units, as well as reservists on contract. In wartime, or at a time of crisis, the brigades are filled to full strength by mobilised reservists, with the plan that they in effect become infantry brigades able to defend rear areas or protect borders. However, the plan has experienced some difficulties. For instance, analysts have indicated that setting up the organisational base within units, so that they can be effectively expanded under mobilisation conditions, has been hampered by a shortage of reservists on contract, as well as poor equipment and training deficiencies.

Equipment procurement

Equipment modernisation is also a priority, particularly the land forces' artillery and missile capabilities. This reflects military leaders' emphasis on developing long-range precision-strike capabilities. Soviet-era artillery systems are still being modernised (and others procured from states in Central Europe) but there is a design and production effort directed towards new weapons and munitions. Anti-armour capabilities have been a particular focus. Deliveries of *Javelin* missiles from the United States are accompanied by Ukrainian-made *Stugna*-P and *Korsar* anti-tank missiles. The *Shturm*-S vehicle is being modernised to carry the *Barrier*-V system with RK-2V missiles and the *Konkurs*-equipped vehicle, based on the BRDM-2, is being modernised into the *Amulet* variant, with RK-2 missiles from the equivalent *Stugna*-P system.

Munitions for both rocket and tube artillery are being developed and tested. The *Vilkha* precision MRL system, with new rockets based on the Soviet-era *Smerch* system, passed tests in 2020, as did the longer-range *Vilkha*-M. Large quantities of mortars have also been introduced. Test batches of *Bars*-8MMK mortar carriers have started to be delivered following the limited success of the *Molot* 120-mm mortar. Improved communications and reconnaissance, and the acquisition of both Ukrainian- and foreign-made UAVs, are also important for artillery capabilities. Army-aviation modernisation is primarily aimed at maintaining and increasing readiness levels. The acquisition of foreign helicopters has been discussed, and there is also focus on improving the quality and intensity of training for aircrew and related technical trades.

The air force is prioritising development of its missile-equipped air-defence forces, primarily through the modernisation of Soviet-era equipment. There are plans to increase inventory numbers by bringing back into service previously decommissioned assets, including S-300V (SA-12A *Gladiator*), 9K330 *Tor* (SA-15 *Gauntlet*), 2K12 *Kub* (SA-6 *Gainful*) and S-125 *Neva* (SA-3 *Goa*) surface-to-air missile systems.

The navy's priority is to improve its coastal-defence forces. Ukraine has developed the *Neptune* anti-ship missile system and has also tested the S-125 for use against surface targets.

Armoured vehicles have also been procured, including protected patrol vehicles. Soon after hostilities broke out in Ukraine's east, it was proposed that STREIT Group *Spartan* and *Cougar* light armoured vehicles be built at the KrAZ factory. However, most procurements are now of Ukrainian armoured vehicles made by Praktika and Ukrainskaya Bronetekhnika. Praktika has supplied the *Kazak*-2 armoured vehicle based on the IVECO Eurocargo lorry chassis to the armed forces, the National Guard and the State Border Service. Orders shifted then to the lighter *Kazak*-2M1 for the Airborne Assault Forces and Special Operations Forces under the 2020 state defence order. 2020 also saw the final delivery of a batch of 45 BTR-4s under a contract from 2016; the first 12 vehicles were delivered in December 2019.

Defence economics

Ukraine's defence budget growth has been volatile since 2008 – both in real and in nominal terms – with short-term surges interrupted by years in which spending plateaued. In 2011 and 2012, the budget grew by 34.1% and 23.8% respectively (17.3% and 14.4% in real terms) and those years were followed by, in 2013, a reduction of 10.3% in real terms. Then followed another, major, surge period with 79% and 78% nominal growth in 2014 and 2015 (53.7% and 30.0% in real terms) as the country reacted to Russia's annexation of Crimea. 2017 was another 'plateau' with spending dropping by 10.1% in real terms. There was a more moderate surge in 2018 and 2019 with real growth of 15.4% and 15.9%; as of 2020, spending is once again relatively flat. Nonetheless, the surge periods mean that Ukraine's budget has increased twelvefold between 2010 and 2020 and is almost three times greater in real terms. The defence budget grew from 1.1% in 2010 to 2.7%, as a proportion of GDP, in 2015. Spending has since been largely maintained at this level, although the drop in 2020 GDP means the defence budget grew to 3.1% of GDP from 2.8% in 2019 despite the budget remaining flat in nominal terms.

Ukraine's initial 2020 budget, released in September 2019, allocated 102.6 billion hryvnia (US$3.8bn) to the Ministry of Defence, a figure almost identical to the initial 2019 budget but lower in real terms than final 2019 expenditure of h108.4bn (US$4.0bn), a 12.7% reduction in real terms. As such the budget was revised upwards in November 2019 to h117.8bn (US$4.3bn), a 14.7% increase over the 2019 budget (5.4% in real terms). The 2021 MoD budget is projected to amount to h117.5bn (US$4.3bn), a figure in line with the revised 2020 budget but a 5.9% decrease in real terms. It is less likely that the budget will again be revised significantly upward in the context of a 7.2% contraction in real GDP in 2020 as a result of the coronavirus pandemic. However, in addition to the domestic budget, the US Congress approved a US$250 million military aid package for Ukraine in June covering air and maritime surveillance, cyber defence, naval and counter-artillery capabilities, and medical equipment.

Conflict and political instability

Moscow maintains its occupation of Crimea and continues to bolster its forces there and also support separatist forces in Ukraine's east. Political instability and conflict in 2020 in other places that were part of the former Soviet Union has prompted varying responses from Moscow. Protests in the wake of the 4 October parliamentary election in Kyrgyzstan led to the poll being annulled, troops being deployed and parliament electing an acting president. Protests also took place in Belarus after the disputed presidential elections on 9 August. On 27 August, President Putin announced he had created a 'law enforcement reserve' at the request of Minsk for use if the situation got 'out of control.' While there had at the time of writing been no overt Russian action in support of President Lukashenko, some analysts considered that advisers and possibly Russian security personnel had been in Belarus. In November, meanwhile, Russia brokered a ceasefire after fighting broke out in September between Armenian and Azerbaijani forces. Clashes lasting over six weeks resulted in Azerbaijan regaining territory lost since the 1990s surrounding Nagorno-Karabakh and regaining a swathe of territory in Nagorno-Karabakh itself. Following the ceasefire, Russian peacekeepers were deployed to the Lachin corridor now connecting Armenia with Nagorno-Karabakh. The conflict highlighted the effect of the differing military-modernisation paths followed by Armenia and Azerbaijan. Baku's budgets have enabled its armed forces to purchase modern systems in significant numbers. During the conflict, Azerbaijan's forces illustrated their ability to exploit tactical successes and integrate capabilities such as UAVs to help find, fix and strike at Armenian positions. These highlighted not just Azerbaijan's modernisation successes but also Armenia's long-standing problems in procuring and effectively operating modern equipment in sufficient numbers. Lessons from the conflict will, among others, relate to training, the integration of emerging capabilities like loitering munitions, better protection from top attack, the utility of precise rocket artillery as well as the importance of mobile layered air defences capable of detecting and engaging UAVs and loitering munitions.

Armenia ARM

Armenian Dram d		2019	2020	2021
GDP	d	6.57tr	6.34tr	
	US$	13.7bn	12.8bn	
per capita	US$	4,605	4,315	
Growth	%	7.6	-4.5	
Inflation	%	1.4	0.9	
Def bdgt [a]	d	309bn	307bn	
	US$	644m	621m	
US$1=d		480.44	494.84	

[a] Includes imported military equipment, excludes military pensions

Real-terms defence budget trend (US$m, constant 2015)

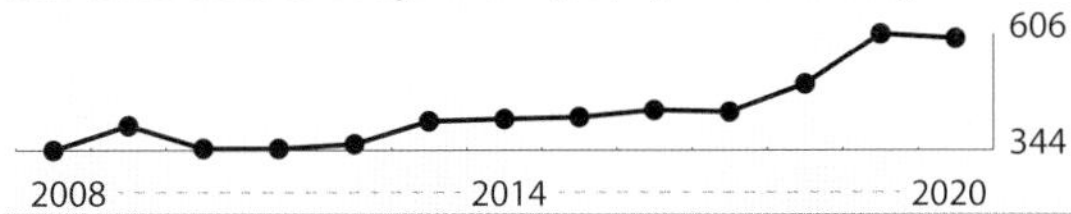

Population 3,021,324

Age	0–14	15–19	20–24	25–29	30–64	65 plus
Male	9.8%	2.9%	3.2%	3.9%	23.8%	5.1%
Female	8.8%	2.6%	2.9%	3.9%	25.6%	7.5%

Capabilities

The armed forces' main focus is territorial defence, and a resumption of armed conflict with Azerbaijan over Nagorno-Karabakh after September 2020 saw forces mobilised and combat losses to both equipment and personnel. Despite economic constraints, a 2018–24 modernisation plan outlined the goal of reorganising the command structure and modernising the equipment inventory. The programme includes sections on cyber- and information-domain capabilities. Armenia is a member of the CSTO and maintains close defence ties with Russia, centred on equipment-procurement, technical-advice and personnel-training programmes. Military doctrine remains influenced by Russian thinking. Armenia is also engaged in a NATO Individual Partnership Action Plan. Conscription continues, but there is a growing cohort of professional officers. The armed forces have deployed on NATO and UN missions in Afghanistan, Kosovo and Lebanon, providing learning opportunities. Personnel train regularly and take part in annual CSTO exercises and with Russia in bilateral drills. Equipment is mainly of Russian origin. Agreements have been reached in recent years to purchase modern Russian systems, though only in small numbers to date. Serviceability and maintenance of mainly ageing aircraft have been a problem for the air force. There is some capacity to manufacture defence equipment for the domestic market, including electro-optics, light weapons and UAVs, but Armenia is reliant on Russia for other equipment platforms and military systems.

ACTIVE 44,800 (Army 41,850 Air/AD Aviation Forces (Joint) 1,100 other Air Defence Forces 1,850) Paramilitary 4,300

Conscript liability 24 months

RESERVE

Some mobilisation reported, possibly 210,000 with military service within 15 years

ORGANISATIONS BY SERVICE

Army 22,900; 18,950 conscripts (total 41,850)

FORCES BY ROLE

SPECIAL FORCES
1 SF bde

MANOEUVRE

Mechanised
1 (1st) corps (1 recce bn, 1 tk bn, 2 MR regt, 1 maint bn)
1 (2nd) corps (1 recce bn, 1 tk bn, 2 MR regt, 1 lt inf regt, 1 arty bn)
1 (3rd) corps (1 recce bn, 1 tk bn, 4 MR regt, 1 lt inf regt, 1 arty bn, 1 MRL bn, 1 sigs bn, 1 maint bn)
1 (4th) corps (4 MR regt; 1 SP arty bn; 1 sigs bn)
1 (5th) corps (with 2 fortified areas) (1 MR regt)

Other
1 indep MR trg bde

COMBAT SUPPORT
1 arty bde
1 MRL bde
1 AT regt
1 AD bde
2 AD regt
2 (radiotech) AD regt
1 engr regt

SURFACE-TO-SURFACE MISSILE
1 SRBM regt

EQUIPMENT BY TYPE

Inventory totals represent pre-war figures. Fighting around Nagorno-Karabakh in late 2020 is likely to have resulted in significant combat losses but the breakdown of lost equipment between Armenian and Nagorno-Karabakh forces remains unclear at the time of writing.

ARMOURED FIGHTING VEHICLES
MBT 108: 3 T-54; 5 T-55; up to 100 T-72A/B
RECCE 12 BRM-1K (CP)
IFV 231: 75 BMP-1; 6 BMP-1K (CP); up to 150 BMP-2
APC • APC (W) 130: 8 BTR-60; 100 BTR-60 lookalike; 18 BTR-70; 4 BTR-80
AUV *Tigr*

ENGINEERING & MAINTENANCE VEHICLES
AEV MT-LB
ARV BREhM-D; BREM-1

ANTI-TANK/ANTI-INFRASTRUCTURE
MSL • SP 22+: 9 9P148 *Konkurs* (AT-5 *Spandrel*); 13 9P149 *Shturm* (AT-6 *Spiral*); 9K129 *Kornet*-E (AT-14 *Spriggan*)

ARTILLERY 231
SP 28: **122mm** some 2S1 *Gvozdika*; **152mm** 28 2S3 *Akatsiya*
TOWED 131: **122mm** 69 D-30; **152mm** 62: 26 2A36 *Giatsint*-B; 2 D-1; 34 D-20
MRL 60: **122mm** up to 50 BM-21 *Grad*; **273mm** 4 WM-80; **300mm** 6 9A52 *Smerch*
MOR 120mm 12 M120

SURFACE-TO-SURFACE MISSILE LAUNCHERS
SRBM • Conventional 14: 7+ 9K72 *Elbrus* (SS-1C *Scud* B); 3+ 9K79 *Tochka* (SS-21 *Scarab*); 4 9K720 *Iskander*-E

UNINHABITED AERIAL VEHICLES
ISR • Light *Krunk*

AIR DEFENCE
SAM
Medium-range 2K11 *Krug* (SA-4 *Ganef*); S-75 *Dvina* (SA-2 *Guideline*); 9K37M *Buk*-M1 (SA-11 *Gadfly*)
Short-range 2K12 *Kub* (SA-6 *Gainful*); S-125 *Pechora* (SA-3 *Goa*); 9K331MKM *Tor*-M2KM
Point-defence 9K33 *Osa* (SA-8 *Gecko*); 9K35M *Strela*-10; 9K310 *Igla*-1 (SA-16 *Gimlet*); 9K38 *Igla* (SA-18 *Grouse*); 9K333 *Verba* (SA-29 *Gizmo*); 9K338 *Igla*-S (SA-24 *Grinch*)
GUNS
SP 23mm ZSU-23-4
TOWED 23mm ZU-23-2

Air and Air Defence Aviation Forces 1,100

1 Air & AD Joint Command

FORCES BY ROLE
GROUND ATTACK
1 sqn with Su-25/Su-25UBK *Frogfoot*

EQUIPMENT BY TYPE
AIRCRAFT 17 combat capable
FGA 4 Su-30SM *Flanker* H
ATK 13: up to 12 Su-25 *Frogfoot*; 1 Su-25UBK *Frogfoot*
TPT 4: **Heavy** 3 Il-76 *Candid*; **PAX** 1 A319CJ
TRG 14: 4 L-39 *Albatros*; 10 Yak-52
HELICOPTERS
ATK 7 Mi-24P *Hind*
ISR 4: 2 Mi-24K *Hind*; 2 Mi-24R *Hind* (cbt spt)
MRH 10 Mi-8MT (cbt spt)
C2 2 Mi-9 *Hip* G (cbt spt)
TPT • Light 7 PZL Mi-2 *Hoplite*
AIR DEFENCE • SAM • Long-range S-300PT (SA-10 *Grumble*); S-300PS (SA-10 *Grumble*)

Paramilitary 4,300

Police

FORCES BY ROLE
MANOEUVRE
Other
4 paramilitary bn

EQUIPMENT BY TYPE
ARMOURED FIGHTING VEHICLES
RECCE 5 BRM-1K (CP)
IFV 45: 44 BMP-1; 1 BMP-1K (CP)
APC • APC (W) 24 BTR-60/BTR-70/BTR-152
ABCV 5 BMD-1

Border Troops

Ministry of National Security

EQUIPMENT BY TYPE
ARMOURED FIGHTING VEHICLES
RECCE 3 BRM-1K (CP)
IFV 35 BMP-1
APC • APC (W) 23: 5 BTR-60; 18 BTR-70
ABCV 5 BMD-1

DEPLOYMENT

AFGHANISTAN: NATO • *Operation Resolute Support* 121
LEBANON: UN • UNIFIL 33
SERBIA: NATO • KFOR 41

FOREIGN FORCES

Russia 3,500: 1 mil base with (1 MR bde; 74 T-72; 80 BMP-1; 80 BMP-2; 12 2S1; 12 BM-21); 1 ftr sqn with 18 MiG-29 *Fulcrum*; 1 hel sqn with 11 Mi-24P *Hind*; 4 Mi-8AMTSh *Hip*; 4 Mi-8MT *Hip*; 2 SAM bty with S-300V (SA-12 *Gladiator/Giant*); 1 SAM bty with *Buk*-M1-2 (SA-11 *Gadfly*)

Azerbaijan AZE

Azerbaijani New Manat m		2019	2020	2021
GDP	m	81.7bn	70.8bn	
	US$	48.0bn	41.7bn	
per capita	US$	4,814	4,125	
Growth	%	2.2	-4.0	
Inflation	%	2.7	3.0	
Def bdgt [a]	m	3.04bn	3.85bn	
	US$	1.79bn	2.27bn	
US$1=m		1.70	1.70	

[a] Official defence budget. Excludes a significant proportion of procurement outlays

Real-terms defence budget trend (US$bn, constant 2015)

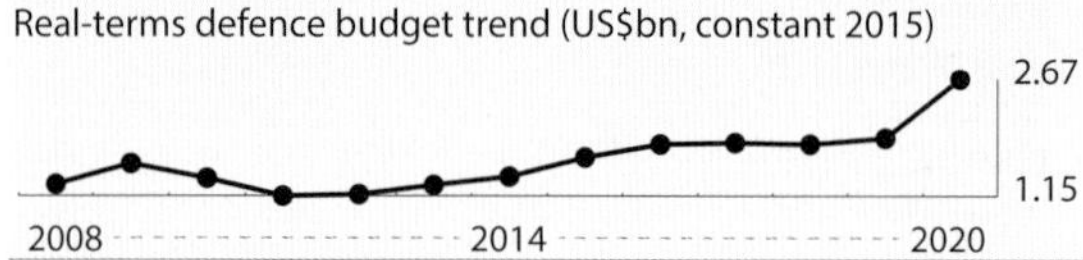

Population 10,205,810

Age	0–14	15–19	20–24	25–29	30–64	65 plus
Male	12.1%	3.4%	3.6%	4.7%	23.0%	2.8%
Female	10.7%	2.9%	3.2%	4.4%	24.6%	4.5%

Capabilities

The armed forces' principal focus is territorial defence, and there was a resumption of armed conflict with Armenia over Nagorno-Karabakh after September 2020. Defence cooperation persists with Russia and has, more recently, developed with Turkey. Azerbaijan maintains a defence relationship with NATO, concluding in 2019 a fifth cycle of its NATO Individual Partnership Action Plan. Readiness within Azerbaijan's conscript-based armed services varies between units. Azerbaijan has taken part in multilateral exercises and its forces have trained with Turkish troops in bilateral drills. The armed forces have little expeditionary capability though they contribute to NATO's *Resolute Support* mission in Afghanistan. Defence modernisation and procurement has been a focus in the past decade, intended to replace the ageing inventory of mainly Soviet-era equipment. Recent orders include for air-defence and artillery systems and wheeled and tracked armoured vehicles, predominantly of Russian origin. In recent years there has been significant procurement and industrial cooperation with Israel focused on UAVs and guided weapons. Azerbaijan's limited but growing defence-industrial capabilities are centred on the Ministry of Defence Industry, which manages and oversees the production of small arms and light weapons. While the country is reliant on external suppliers for major defence-equipment platforms and systems, some defence companies have started to export to foreign markets.

ACTIVE 66,950 (Army 56,850 Navy 2,200 Air 7,900)
Paramilitary 15,000
Conscript liability 18 months (12 for graduates)

RESERVE 300,000
Some mobilisation reported; 300,000 with military service within 15 years

ORGANISATIONS BY SERVICE

Army 56,850

FORCES BY ROLE
COMMAND
5 corps HQ
MANOEUVRE
Mechanised
4 MR bde
Light
19 MR bde
Other
1 sy bde
COMBAT SUPPORT
1 arty bde
1 arty trg bde
1 MRL bde
1 AT bde
1 engr bde
1 sigs bde
COMBAT SERVICE SUPPORT
1 log bde

EQUIPMENT BY TYPE
Inventory totals represent pre-war figures. Fighting in and around Nagorno-Karabakh in late 2020 is likely to have resulted in some combat losses as well as the capture of some equipment from Armenian and Nagorno-Karabakh forces.
ARMOURED FIGHTING VEHICLES
MBT 437: 95 T-55; 244 T-72A/AV/B; 98 T-90S
RECCE 15 BRM-1
IFV 215: 43 BMP-1; 33 BMP-2; 88 BMP-3; 7 BTR-80A; 44+ BTR-82A
APC 568
APC (T) 336 MT-LB
APC (W) 142: 10 BTR-60; 132 BTR-70
PPV 90: 45 *Marauder*; 45 *Matador*
AUV 65+: 35 *Cobra*; 30+ *Sand Cat*
ABCV 20 BMD-1
ENGINEERING & MAINTENANCE VEHICLES
AEV IMR-2; MT-LB
ARV BREM-L *Brelianka*
MW *Bozena*; GW-3 (minelayer)
ANTI-TANK/ANTI-INFRASTRUCTURE
SP 10 9P157-2 *Khrizantema*-S (AT-15 *Springer*); *Sand Cat* with *Spike*-ER
MSL • MANPATS 9K11 *Malyutka* (AT-3 *Sagger*); 9K111 *Fagot* (AT-4 *Spigot*); 9K111-1 *Konkurs* (AT-5 *Spandrel*); 9K115 *Metis* (AT-7 *Saxhorn*); *Spike*-LR
ARTILLERY 630
SP 102: **122mm** 46 2S1 *Gvozdika*; **152mm** 39: 6 2S3 *Akatsiya*; 18 2S19 *Msta*-S; 15 *Dana*-M1M; **155mm** 5 ATMOS 2000; **203mm** 12 2S7 *Pion*
TOWED 233: **122mm** 129 D-30; **130mm** 36 M-46; **152mm** 68: 44 2A36 *Giatsint*-B; 24 D-20
GUN/MOR 120mm 36: 18 2S9 NONA-S; 18 2S31 *Vena*
MRL 147: **122mm** 60+: 43 BM-21 Grad; 9+ IMI *Lynx*; 8 RM-70 *Vampir*; **128mm** 12 RAK-12; **220mm** 18 TOS-1A; **300mm** 36: 30 9A52 *Smerch*; 6+ *Polonez*; **302mm** 21 T-300 *Kasirga*
MOR 120mm 112: 5 *Cardom*; 107 M-1938 (PM-38) **SP 120mm** *Sand Cat* with *Spear*
SURFACE-TO-SURFACE MISSILE LAUNCHERS
SRBM • Conventional 6: 2 IAI LORA; ε4 9K79 *Tochka* (SS-21 *Scarab*)
UNINHABITED AERIAL VEHICLES
ISR • Medium 3 *Aerostar*
AIR DEFENCE
SAM
Medium-range 2K11 *Krug* (SA-4 *Ganef*)
Point-defence 9K33 *Osa* (SA-8 *Gecko*); 9K35 *Strela*-10 (SA-13 *Gopher*); 9K32 *Strela* (SA-7 *Grail*)‡; 9K34 *Strela*-3 (SA-14 *Gremlin*); 9K310 *Igla*-1 (SA-16 *Gimlet*); 9K338 *Igla*-S (SA-24 *Grinch*)
GUNS
SP 23mm ZSU-23-4
TOWED 23mm ZU-23-2

Navy 2,200

EQUIPMENT BY TYPE
PATROL AND COASTAL COMBATANTS 11
CORVETTES • FS 1 *Kusar* (ex-FSU *Petya* II) with 2 RBU 6000 *Smerch* 2 A/S mor, 2 twin 76mm gun
PSO 1 *Luga* (*Wodnik* 2) (FSU Project 888; additional trg role)
PCC 3: 2 *Petrushka* (FSU UK-3; additional trg role); 1 *Shelon* (ex-FSU Project 1388M)
PB 3: 1 *Araz* (ex-TUR AB 25); 1 *Bryza* (ex-FSU Project 722); 1 *Poluchat* (ex-FSU Project 368)
PBF 3 *Stenka*
MINE WARFARE • MINE COUNTERMEASURES 4
MHC 4: 2 *Korund* (Project 1258 (*Yevgenya*)); 2 *Yakhont* (FSU *Sonya*)
AMPHIBIOUS 6
LSM 3: 1 Project 770 (FSU *Polnochny A*) (capacity 6 MBT; 180 troops); 2 Project 771 (*Polnochny B*) (capacity 6 MBT; 180 troops)
LCM 3: 2 T-4 (FSU); 1 *Vydra*† (FSU) (capacity either 3 MBT or 200 troops)
LOGISTICS AND SUPPORT 3
AGS 1 (FSU Project 10470)
ATF 2 *Neftegaz* (Project B-92) (ex-Coast Guard)

Air Force and Air Defence 7,900

FORCES BY ROLE
FIGHTER
1 sqn with MiG-29 *Fulcrum* A; MiG-29UB *Fulcrum* B
FIGHTER/GROUND ATTACK
1 regt with Su-24 *Fencer*; Su-25 *Frogfoot*; Su-25UB *Frogfoot* B
TRANSPORT
1 sqn with An-12 *Cub*; Yak-40 *Codling*

TRAINING
1 sqn with L-39 *Albatros*
ATTACK/TRANSPORT HELICOPTER
1 regt with Ka-32 *Helix* C; Mi-8 *Hip*; Mi-24 *Hind*; PZL Mi-2 *Hoplite*

EQUIPMENT BY TYPE
AIRCRAFT 36 combat capable
FTR 15: 13 MiG-29 *Fulcrum* A; 2 MiG-29UB *Fulcrum* B
ATK 21: 2 Su-24 *Fencer*†; 16 Su-25 *Frogfoot*; 3 Su-25UB *Frogfoot* B
TPT 4: **Medium** 1 An-12 *Cub*; **Light** 3 Yak-40 *Codling*
TRG 15: 12 L-39 *Albatros*; 3+ *Super Mushshak*
HELICOPTERS
ATK 26 Mi-24 *Hind*
MRH: 20+ Mi-17-IV *Hip*
TPT 24: **Medium** 17: 1 Bell 412; 3 Ka-32 *Helix* C; 13 Mi-8 *Hip* **Light** 7 PZL Mi-2 *Hoplite*
UNINHABITED AERIAL VEHICLES
CISR • Medium *Bayraktar* TB2
ISR 16: **Heavy** 1 *Heron*; **Medium** 15: 4 *Aerostar*; 10+ *Hermes* 450; 1 *Hermes* 900
AIR DEFENCE • SAM
Long-range S-200 *Vega* (SA-5 *Gammon*); S-300PM/PMU2
Medium-range S-75 *Dvina* (SA-2 *Guideline*); 9K37M *Buk*-M1 (SA-11 *Gadfly*); *Buk*-MB (SA-11 mod.); S-125-2TM *Pechora*-2TM (SA-26); *Barak*-LRAD
Short-range *Abisr* (*Barak*-MRAD)
AIR-LAUNCHED MISSILES
AAM • IR R-60 (AA-8 *Aphid*); R-73 (AA-11A *Archer*) **IR/SARH** R-27 (AA-10 *Alamo*)
ASM *Barrier*-V

Paramilitary ε15,000

State Border Service ε5,000
Ministry of Internal Affairs

EQUIPMENT BY TYPE
ARMOURED FIGHTING VEHICLES
IFV 168 BMP-1/BMP-2
APC • APC (W) 19 BTR-60/70/80
ARTILLERY • **MRL 122mm** 3 T-122
HELICOPTERS • **ATK** 24 Mi-35M *Hind*
UNINHABITED AERIAL VEHICLES
ISR • Medium 3+: 3+ *Hermes* 450; *Hermes* 900
LOITERING MUNITIONS *Harop*

Coast Guard
The Coast Guard was established in 2005 as part of the State Border Service

EQUIPMENT BY TYPE
PATROL AND COASTAL COMBATANTS 19
PCG 6 *Sa'ar* 62 with 1 8-cell *Typhoon* MLS-NLOS lnchr with *Spike* NLOS SSM, 1 hel landing platform
PBF 9: 1 Project 205 (FSU *Osa* II); 6 *Shaldag* V; 2 Silver Ships 48ft
PB 4: 2 Baltic 150; 1 *Point* (US); 1 *Grif* (FSU *Zhuk*)
LOGISTICS AND SUPPORT 4
ARS 1 *Iva* (FSU *Vikhr*)
ATF 3 *Neftegaz* (Project B-92) (also used for patrol duties)

Internal Troops 10,000+
Ministry of Internal Affairs

EQUIPMENT BY TYPE
ARMOURED FIGHTING VEHICLES
APC • APC (W) 7 BTR-60/BTR-70/BTR-80

DEPLOYMENT

AFGHANISTAN: NATO • *Operation Resolute Support* 120

SOUTH SUDAN: UN • UNMISS 2

FOREIGN FORCES

Russia 817; 1 MR bn (peacekeeping)
Turkey 6 F-16C/D *Fighting Falcon*; 1 CN235

TERRITORY WHERE THE GOVERNMENT DOES NOT EXERCISE EFFECTIVE CONTROL

Nagorno-Karabakh was part of Azerbaijan, but mostly populated by ethnic Armenians. In 1988, when inter-ethnic clashes between Armenians and Azeris erupted in Azerbaijan, the local authorities declared their intention to secede and join Armenia. Baku rejected this and armed conflict erupted. A ceasefire was brokered in 1994; since then, Armenia controlled most of Nagorno-Karabakh. While Armenia provides political, economic and military support to Nagorno-Karabakh, the region has declared itself independent – although this has not been recognised by any other state, including Armenia. Baku claims Nagorno-Karabakh as part of Azerbaijan. Armed conflict over the disputed territory re-ignited in late 2020, leading to territorial realignment including the withdrawal of Armenian forces from territories surrounding Nagorno-Karabakh. Data presented here represents an assessment of the de facto situation.

Nagorno-Karabakh
Available estimates of military holdings in Nagorno-Karabakh have to be treated with caution due to the heavy level of attrition in the 2020 conflict. Some of the equipment listed may belong to Armenian forces.

EQUIPMENT BY TYPE
ARMOURED FIGHTING VEHICLES
MBT T-72
RECCE BRDM-2
IFV BMP-1; BMP-2
ANTI-TANK/ANTI-INFRASTRUCTURE
MSL
SP 9P148 *Konkurs* (AT-5 *Spandrel*); 9P149 *Shturm* (AT-6 *Spiral*)
MANPATS 9K111-1 *Konkurs* (AT-5 *Spandrel*)
RCL 73mm SPG-9
ARTILLERY
SP 122mm 2S1 *Gvozdika*; **152mm** 2S3 *Akatsiya*
TOWED 122mm D-30; **152mm** 2A36 *Giatsint*-B; D-20
MRL 122mm BM-21 *Grad*
MOR 120mm M-74/M-75
HELICOPTERS
ATK Mi-24 *Hind*
MRH Mi-8MT *Hip*

AIR DEFENCE
SAM
Medium-range 2K11 *Krug* (SA-4 *Ganef*); S-75 *Dvina* (SA-2 *Guideline*)
Short-range 2K12 *Kub* (SA-6 *Gainful*); S-125 *Pechora* (SA-3 *Goa*)
Point-defence 9K33 *Osa* (SA-8 *Gecko*); 9K310 *Igla*-1 (SA-16 *Gimlet*); 9K38 *Igla* (SA-18 *Grouse*)
GUNS
SP 23mm ZSU-23-4
TOWED 23mm ZU-23-2

Belarus BLR

Belarusian Ruble r		2019	2020	2021
GDP	r	132bn	138bn	
	US$	63.1bn	57.7bn	
per capita	US$	6,658	6,134	
Growth	%	1.2	-3.0	
Inflation	%	5.6	5.1	
Def bdgt	r	1.36bn	1.47bn	
	US$	651m	616m	
US$1=r		2.09	2.38	

Real-terms defence budget trend (US$m, constant 2015)

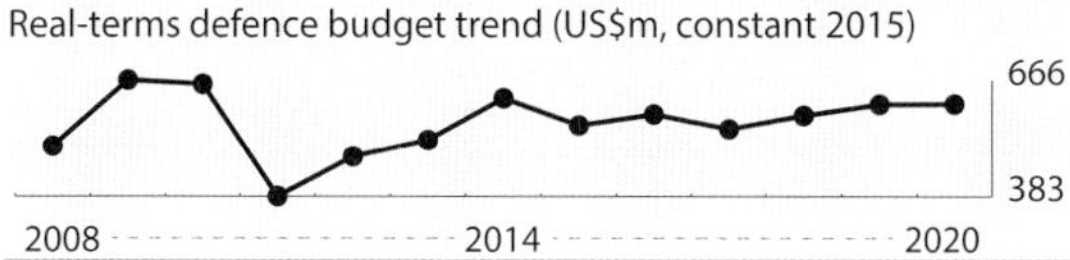

Population 9,477,918

Age	0–14	15–19	20–24	25–29	30–64	65 plus
Male	8.3%	2.4%	2.5%	3.3%	24.8%	5.2%
Female	7.8%	2.3%	2.4%	3.1%	27.2%	10.7%

Capabilities

Located between Russia and NATO European members, the main task of Belarus's armed forces is maintaining territorial integrity, though army units were visible in the government's attempt to curb civil protests during the latter part of 2020. The country's latest military doctrine was approved in July 2016 and identified as security challenges 'hybrid methods' and 'colour revolutions'. A new plan for the development of the armed forces until 2030 was approved in late 2019. The previous plan, dating from 2008, was due to be completed in 2020. Belarus is a member of the CSTO. Russia remains the country's principal defence partner, though Minsk has also looked to improve defence cooperation with China and Turkey. The forces remain conscript-based and train regularly with other CSTO partners. There has been increased emphasis on the training of territorial-defence troops to allow them to better operate with the regular forces. There is a small heavy-airlift fleet that could be supplemented by civil transport aircraft, and Minsk has a special-forces brigade trained for the air-assault role. There is no requirement to independently deploy and sustain the armed forces, but Belarus could do so as a part of the CSTO. Russia continues to be Minsk's main defence-equipment supplier. In recent years, Belarus has received air-defence systems and advanced combat-trainer/light-attack aircraft from Moscow. The local defence industry manufactures vehicles, guided weapons and electronic-warfare systems, among other equipment. However, there is no capacity to design or manufacture modern combat aircraft. The sector also undertakes upgrade work for foreign customers.

ACTIVE 45,350 (Army 10,700 Air 11,750 Special Operations Forces 5,900 Joint 17,000) Paramilitary 110,000

Conscript liability 18 months; 12 months for graduates (alternative service option)

RESERVE 289,500

(Joint 289,500 with mil service within last 5 years)

ORGANISATIONS BY SERVICE

Army 10,700

FORCES BY ROLE
COMMAND
2 comd HQ (West & North West)
MANOEUVRE
Mechanised
2 mech bde
2 mech bde(-)
COMBAT SUPPORT
2 arty bde
1 engr bde
1 engr regt

EQUIPMENT BY TYPE
ARMOURED FIGHTING VEHICLES
MBT 537: 517 T-72B; 20 T-72B3 mod
RECCE 145: 132 BRM-1; 13+ *Cayman* BRDM
IFV 932 BMP-2
APC • APC (T) 58 MT-LB
AUV 8 CS/VN3B mod; *Tigr*
ENGINEERING & MAINTENANCE VEHICLES
AEV BAT-2; IMR-2; MT-LB
VLB 24: 20 MTU-20; 4 MT-55A
MW UR-77
NBC VEHICLES BRDM-2RKhB; *Cayman* NRBC *Chimera*; RKhM-4; RKhM-K
ANTI-TANK/ANTI-INFRASTRUCTURE • MSL
SP 160: 75 9P148 *Konkurs*; 85 9P149 *Shturm*
MANPATS 9K111 *Fagot* (AT-4 *Spigot*); 9K111-1 *Konkurs* (AT-5 *Spandrel*); 9K115 *Metis* (AT-7 *Saxhorn*)
ARTILLERY 583
SP 333: **122mm** 125 2S1 *Gvozdika*; **152mm** 208: 125 2S3 *Akatsiya*; 71 2S5; 12 2S19 *Msta*-S
TOWED 152mm 72 2A65 *Msta*-B
MRL 164: **122mm** 128 BM-21 *Grad*; **220mm** 36 9P140 *Uragan*
MOR 120mm 14 2S12
AIR DEFENCE
SAM Point-defence 2K22 *Tunguska* (SA-19 *Grison*)
GUNS • SP 23mm ZU-23-2 (tch)

Air Force and Air Defence Forces 11,750

FORCES BY ROLE
FIGHTER
2 sqn with MiG-29/S/UB *Fulcrum*
GROUND ATTACK
2 sqn with Su-25K/UBK *Frogfoot* A/B
TRANSPORT
1 base with An-24 *Coke*; An-26 *Curl*; Il-76 *Candid*; Tu-134 *Crusty*

TRAINING
Some sqn with L-39 *Albatros*
ATTACK HELICOPTER
Some sqn with Mi-24 *Hind*
TRANSPORT HELICOPTER
Some (cbt spt) sqn with Mi-8 *Hip*; Mi-8MTV-5 *Hip*; Mi-26 *Halo*

EQUIPMENT BY TYPE

AIRCRAFT 72 combat capable
FTR 34: 28 MiG-29 *Fulcrum* A/MiG-29S *Fulcrum* C; 6 MiG-29UB *Fulcrum* B
FGA 4 Su-30SM *Flanker* H; (21 Su-27/UB *Flanker* B/C non-operational/stored)
ATK 22 Su-25K/UBK *Frogfoot* A/B
TPT 8: **Heavy** 2 Il-76 *Candid* (+9 civ Il-76 available for mil use); **Light** 6: 1 An-24 *Coke*; 4 An-26 *Curl*; 1 Tu-134 *Crusty*
TRG 12+: Some L-39 *Albatros*; 12 Yak-130 *Mitten**
HELICOPTERS
ATK 12 Mi-24 *Hind*
TPT 26: **Heavy** 6 Mi-26 *Halo*; **Medium** 20: 8 Mi-8 *Hip*; 12 Mi-8MTV-5 *Hip*
AIR-LAUNCHED MISSILES
AAM • IR R-60 (AA-8 *Aphid*); R-73 (AA-11A *Archer*)
SARH R-27R (AA-10 *Alamo* A)
ASM Kh-25 (AS-10 *Karen*); Kh-29 (AS-14 *Kedge*)
ARM Kh-58 (AS-11 *Kilter*) (likely WFU)

Air Defence

AD data from Uzal Baranovichi EW radar

FORCES BY ROLE

AIR DEFENCE
1 bde S-300PS (SA-10B *Grumble*)
3 regt with S-300PS (SA-10B *Grumble*)
1 bde with 9K37 *Buk* (SA-11 *Gadfly*); 9K331ME *Tor*-M2E (SA-15 *Gauntlet*)
1 regt with 9K331ME *Tor*-M2E (SA-15 *Gauntlet*)
2 regt with 9K33 *Osa* (SA-8 *Gecko*)

EQUIPMENT BY TYPE

AIR DEFENCE • SAM
Long-range S-300PS (SA-10B *Grumble*)
Medium-range 9K37 *Buk* (SA-11 *Gadfly*)
Short-range 17 9K331ME *Tor*-M2E (SA-15 *Gauntlet*)
Point-defence 9K33 *Osa* (SA-8 *Gecko*); 9K35 *Strela*-10 (SA-13 *Gopher*)

Special Operations Command 5,900

FORCES BY ROLE

SPECIAL FORCES
1 SF bde
MANOEUVRE
Mechanised
2 mech bde

EQUIPMENT BY TYPE

ARMOURED FIGHTING VEHICLES
APC • APC (W) 185: 32 BTR-70M1; 153 BTR-80
ARTILLERY 42
TOWED 122mm 24 D-30
GUN/MOR • TOWED 120mm 18 2B23 NONA-M1
ANTI-TANK/ANTI-INFRASTRUCTURE • MSL
MANPATS 9K111 *Fagot* (AT-4 *Spigot*); 9K111-1 *Konkurs* (AT-5 *Spandrel*); 9K115 *Metis* (AT-7 *Saxhorn*)

Joint 17,000 (Centrally controlled units and MoD staff)

FORCES BY ROLE

SURFACE-TO-SURFACE MISSILE
1 SRBM bde
COMBAT SUPPORT
1 arty bde
1 MRL bde
2 engr bde
1 EW unit
1 NBC regt
1 ptn bridging regt
2 sigs bde

EQUIPMENT BY TYPE

ARMOURED FIGHTING VEHICLES
APC • APC (T) 20 MT-LB
NBC VEHICLES BRDM-2RKhB; RKhM-4; RKhM-K
ARTILLERY 112
SP 152mm 36 2S5 *Giatsint*-S
TOWED 152mm 36 2A65 *Msta*-B
MRL 300mm 40: 36 9A52 *Smerch*; 4 *Polonez*
SURFACE-TO-SURFACE MISSILE LAUNCHERS
SRBM • Conventional 96: 36 9K79 *Tochka* (SS-21 *Scarab*); 60 9K72 *Elbrus* (SS-1C *Scud* B)

Paramilitary 110,000

State Border Troops 12,000

Ministry of Interior

Militia 87,000

Ministry of Interior

Internal Troops 11,000

DEPLOYMENT

LEBANON: UN • UNIFIL 7

Georgia GEO

Georgian Lari		2019	2020	2021
GDP	lari	50.0bn	49.9bn	
	US$	17.7bn	16.3bn	
per capita	US$	4,765	4,405	
Growth	%	5.1	-5.0	
Inflation	%	4.9	5.3	
Def bdgt	lari	875m	880m	900m
	US$	310m	288m	
FMA (US)	US$	35m	20m	20m
US$1=lari		2.82	3.06	

Real-terms defence budget trend (US$m, constant 2015)

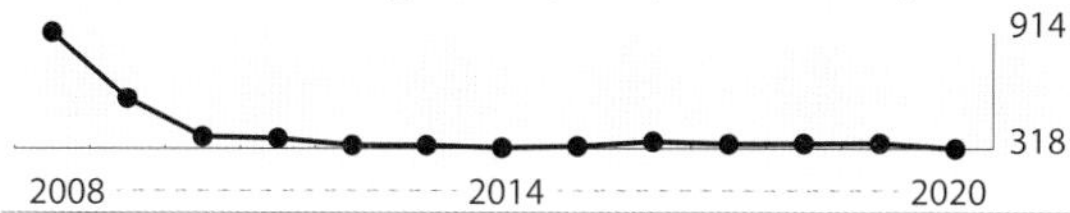

Population 4,930,030

Age	0–14	15–19	20–24	25–29	30–64	65 plus
Male	9.6%	2.7%	3.1%	3.8%	22.0%	6.6%
Female	8.8%	2.4%	2.7%	3.6%	24.4%	10.2%

Capabilities

Georgia's main security preoccupations concern Russian military deployments and the breakaway regions of Abkhazia and South Ossetia. A Strategic Defence Review 2017–20 was published in April 2017. This aimed at improving personnel structures, training facilities and equipment, and stressed a 'total defence' approach. It highlighted the importance of Georgia's reserves, and a ministerial document in 2020 indicated there may be more reliance on active and mobilised reserves. A National Security Strategy 2020–30 has been drafted. Security cooperation with the US includes the Georgia Defence Readiness Program (due to end in 2021). A combat training centre is being developed under the NATO–Georgia Joint Training and Evaluation Centre. Conscription was reinstated with revised terms and increased pay in early 2017. Forces take part in several NATO multinational exercises. Georgia's armed forces have limited expeditionary logistic capability. The backbone of the armed forces' military equipment is legacy Soviet-era systems with varying degrees of obsolescence. The Major Systems Acquisitions Strategy 2019–25 outlines efforts to procure new equipment in several areas, though funding availability will be key to meeting aspirations. There were plans to boost special-forces capacity in 2020 and there is a focus on boosting anti-armour and air-defence capability. The country has begun to develop a defence-industrial base, and this is intended mainly to support the armed forces. The State Military Scientific-Technical Center has demonstrated some maintenance, repair, overhaul and design capabilities for the production of light armoured vehicles.

ACTIVE 20,650 (Army 19,050 National Guard 1,600) **Paramilitary 5,400**

Conscript liability 12 months

ORGANISATIONS BY SERVICE

Army 15,000; 4,050 conscript (total 19,050)

FORCES BY ROLE

SPECIAL FORCES
1 SF bde

MANOEUVRE
Mechanised
1 mech inf bde
Light
3 inf bde
Amphibious
2 mne bn (1 cadre)

COMBAT SUPPORT
2 arty bde
1 engr bde
1 engr bn
1 sigs bn
1 SIGINT bn
1 MP bn

COMBAT SERVICE SUPPORT
1 med bn

EQUIPMENT BY TYPE

ARMOURED FIGHTING VEHICLES
MBT 123: 23 T-55AM2; 100 T-72B/SIM1
RECCE 36: 1 BRM-1K; 35 *Didgori-2*
IFV 71: 25 BMP-1; 46 BMP-2
APC 189+
APC (T) 69+: 3+ *Lazika*; 66 MT-LB
APC (W) 120+: 25 BTR-70; 19 BTR-80; 8+ *Didgori-1*; 3+ *Didgori-3*; 65 *Ejder*
AUV 10+: ATF *Dingo*; *Cobra*; 10 *Cougar*

ENGINEERING & MAINTENANCE VEHICLES
ARV IMR-2

ANTI-TANK/ANTI-INFRASTRUCTURE
MSL • MANPATS 9K111 *Fagot* (AT-4 *Spigot*); 9K113 *Konkurs* (AT-5 *Spandrel*); FGM-148 *Javelin*
GUNS • TOWED ε40: **85mm** D-44; **100mm** T-12

ARTILLERY 240
SP 67: **122mm** 20 2S1 *Gvozdika*; **152mm** 46: 32 M-77 *Dana*; 13 2S3 *Akatsiya*; 1 2S19 *Msta-S*; **203mm** 1 2S7 *Pion*
TOWED 71: **122mm** 58 D-30; **152mm** 13: 3 2A36 *Giatsint-B*; 10 2A65 *Msta-B*
MRL 122mm 37: 13 BM-21 *Grad*; 6 GradLAR; 18 RM-70
MOR 120mm 65: 14 2S12 *Sani*; 33 M-75; 18 M120

AIR DEFENCE • SAM
Short-range *Spyder*-SR
Point-defence *Grom*; *Mistral*-2; 9K32 *Strela*-2 (SA-7 *Grail*)‡; 9K35 *Strela*-10 (SA-13 *Gopher*); 9K36 *Strela*-3 (SA-14 *Gremlin*); 9K310 *Igla*-1 (SA-16 *Gimlet*)

Aviation and Air Defence Command 1,300 (incl 300 conscript)

1 avn base, 1 hel air base

EQUIPMENT BY TYPE

AIRCRAFT 3 combat capable
ATK 3 Su-25KM *Frogfoot* (6 Su-25 *Frogfoot* in store)
TPT • Light 9: 6 An-2 *Colt*; 1 Tu-134A *Crusty* (VIP); 2 Yak-40 *Codling*
TRG 9 L-29 *Delfin*

HELICOPTERS
ATK 6 Mi-24 *Hind*

TPT 29: **Medium** 17 Mi-8T *Hip*; **Light** 12 Bell 205 (UH-1H *Iroquois*)

UNINHABITED AERIAL VEHICLES

ISR • Medium 1+ *Hermes* 450

AIR DEFENCE • SAM

Medium-range 9K37 *Buk*-M1 (SA-11 *Gadfly*) (1–2 bn)

Point-defence 8 9K33 *Osa* AK (SA-8B *Gecko*) (two bty); 9K33 *Osa* AKM (6–10 updated SAM systems)

National Guard 1,600 active reservists opcon Army

FORCES BY ROLE

MANOEUVRE

Light

2 inf bde

Paramilitary 5,400

Border Police 5,400

EQUIPMENT BY TYPE

HELICOPTERS

TPT • Medium 3 Mi-8MTV-1 *Hip*

Coast Guard

HQ at Poti. The Navy was merged with the Coast Guard in 2009 under the auspices of the Georgian Border Police, within the Ministry of the Interior

EQUIPMENT BY TYPE

PATROL AND COASTAL COMBATANTS 24

PCC 2 *Ochamchira* (ex-US *Island*)

PBF 7: 4 Ares 43m; 1 *Kaan* 33; 1 *Kaan* 20; 1 Project 205P (*Stenka*)

PB 15: 1 *Akhmeta*; 2 *Dauntless*; 2 *Dilos* (ex-GRC); 1 *Kutaisi* (ex-TUR AB 25); 2 *Point*; 7 *Zhuk* (3 ex-UKR)

DEPLOYMENT

AFGHANISTAN: NATO • *Operation Resolute Support* 860; 1 inf bn

CENTRAL AFRICAN REPUBLIC: EU • EUTM RCA 35

MALI: EU • EUTM Mali 1

TERRITORY WHERE THE GOVERNMENT DOES NOT EXERCISE EFFECTIVE CONTROL

Following the August 2008 war between Russia and Georgia, the areas of Abkhazia and South Ossetia declared themselves independent. Data presented here represents the de facto situation and does not imply international recognition as sovereign states.

FOREIGN FORCES

Russia 7,000; 1 mil base at Gudauta (Abkhazia) with (1 MR bde; 40 T-90A; 120 BTR-82A; 18 2S3; 12 2S12; 18 BM-21; some S-300 SAM; some atk hel); 1 mil base at Djava/Tskhinvali (S. Ossetia) with (1 MR bde; 40 T-72; 120 BMP-2; 36 2S3; 12 2S12)

Kazakhstan KAZ

Kazakhstani Tenge t		2019	2020	2021
GDP	t	69.5tr	68.0tr	
	US$	182bn	166bn	
per capita	US$	9,750	8,782	
Growth	%	4.5	-2.7	
Inflation	%	5.2	6.9	
Def bdgt	t	610bn	591bn	
	US$	1.59bn	1.44bn	
US$1=t		382.75	410.36	

Real-terms defence budget trend (US$bn, constant 2015)

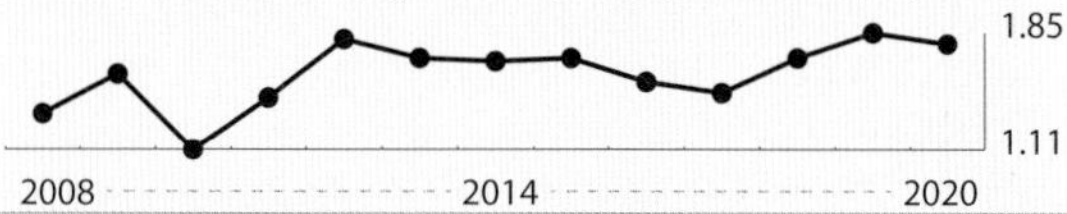

Population 19,091,949

Age	0–14	15–19	20–24	25–29	30–64	65 plus
Male	12.8%	3.3%	3.3%	4.1%	21.1%	3.0%
Female	13.4%	3.2%	3.2%	4.0%	23.2%	5.5%

Capabilities

The October 2017 military doctrine indicates a change in focus from countering violent extremism towards a wider concern for border security and hybrid threats. In the army, air-mobile units are held at the highest level of readiness. Kazakhstan entered a bilateral military agreement with Uzbekistan in September 2017 to cooperate on training and education, countering violent extremism and reducing militant movements in their region. There is a close defence relationship with Russia, reinforced by CSTO and SCO membership, and Moscow operates a radar station at Balkash. Kazakhstan takes part in regional and CSTO exercises, including anti-terror drills. In October 2019, in Saint Petersburg, Kazakhstan and four other Caspian littoral states signed a memorandum of understanding on military cooperation, among other discussions including on maritime security. By regional standards, the armed forces are relatively sizeable and well equipped, following the acquisition of significant amounts of new and upgraded materiel in recent years, primarily from Russia. Russia has supplied Kazakhstan with S-300PS self-propelled surface-to-air missile systems as part of a Joint Air-Defence Agreement, boosting its long-range air-defence capability. Kazakhstan is expanding its indigenous defence industry, and joint ventures and the production of rotary-wing and medium-lift fixed-wing aircraft are envisaged with European companies.

ACTIVE 39,000 (**Army 20,000 Navy 3,000 Air 12,000 MoD 4,000**) **Paramilitary 31,500**

Conscript liability 12 months (due to be abolished)

ORGANISATIONS BY SERVICE

Army 20,000

4 regional comd: Astana, East, West and Southern

FORCES BY ROLE

MANOEUVRE

Armoured

1 tk bde

Mechanised
3 mech bde
Air Manoeuvre
4 air aslt bde
COMBAT SUPPORT
3 arty bde
1 SSM unit
3 cbt engr bde
EQUIPMENT BY TYPE
ARMOURED FIGHTING VEHICLES
MBT 300 T-72BA
RECCE 100: 40 BRDM-2; 60 BRM-1
IFV 607: 500 BMP-2; 107 BTR-80A
APC 432
APC (T) 150 MT-LB
APC (W) 192: 2 BTR-3E; 190 BTR-80
PPV 90 *Arlan*
AUV 17+: 17 *Cobra; SandCat*
ENGINEERING & MAINTENANCE VEHICLES
AEV MT-LB
ANTI-TANK/ANTI-INFRASTRUCTURE
MSL
SP 3+: 3 BMP-T; HMMWV with 9K111-1 *Konkurs* (AT-5 *Spandrel*); 9P149 *Shturm* (MT-LB with AT-6 *Spiral*)
MANPATS 9K111 *Fagot* (AT-4 *Spigot*); 9K111-1 *Konkurs* (AT-5 *Spandrel*); 9K115 *Metis* (AT-7 *Saxhorn*)
GUNS 100mm 68 MT-12/T-12
ARTILLERY 611
SP 246: **122mm** 126: 120 2S1 *Gvozdika*; 6 *Semser*; **152mm** 120 2S3 *Akatsiya*
TOWED 150: **122mm** 100 D-30; **152mm** 50 2A65 *Msta*-B (**122mm** up to 300 D-30 in store)
GUN/MOR 120mm 25 2S9 NONA-S
MRL 127: **122mm** 100 BM-21 *Grad*; **220mm** 3 TOS-1A; **300mm** 24: 6 BM-30 *Smerch*; 18 IMI *Lynx* (with 50 msl) (**122mm** 100 BM-21 *Grad*; **220mm** 180 9P140 *Uragan* all in store)
MOR 63 **SP 120mm** 18 *Cardom*; **120mm** 45 2B11 *Sani*/M120
SURFACE-TO-SURFACE MISSILE LAUNCHERS
SRBM • Conventional 12 9K79 *Tochka* (SS-21 *Scarab*)

Navy 3,000

EQUIPMENT BY TYPE
PATROL AND COASTAL COMBATANTS 14
PCGM 3 *Kazakhstan* with 1 4-cell lnchr with 4 *Barrier*-VK SSM, 1 *Arbalet*-K lnchr with 4 9K38 *Igla* (SA-18 *Grouse*)
PCC 1 *Kazakhstan* with 1 122mm MRL
PBF 3 *Sea Dolphin*
PB 7: 3 *Archangel*; 1 *Dauntless*; 1 *Lashyn*; 1 *Turk* (AB 25); 1 Other
MINE WARFARE • MINE COUNTERMEASURES 1
MCC 1 *Alatau* (Project 10750E)
LOGISTICS AND SUPPORT • **AGS** 1 *Zhaik*

Coastal Defence

FORCES BY ROLE
MANOEUVRE
Mechanised
1 naval inf bde
EQUIPMENT BY TYPE
ARMOURED FIGHTING VEHICLES
IFV 70 BTR-82A

Air Force 12,000 (incl Air Defence)

FORCES BY ROLE
FIGHTER
1 sqn with MiG-29/MiG-29UB *Fulcrum* A/B
2 sqn with MiG-31B/MiG-31BM *Foxhound* A/C
FIGHTER/GROUND ATTACK
1 sqn with MiG-27 *Flogger* D; MiG-23UB *Flogger* C
1 sqn with Su-27/Su-27UB *Flanker* B/C
1 sqn with Su-27/Su-30SM *Flanker* B/H
GROUND ATTACK
1 sqn with Su-25 *Frogfoot*
TRANSPORT
1 unit with Tu-134 *Crusty*; Tu-154 *Careless*
1 sqn with An-12 *Cub*, An-26 *Curl*, An-30 *Clank*, An-72 *Coaler*, C295M
TRAINING
1 sqn with L-39 *Albatros*
ATTACK HELICOPTER
5 sqn with Mi-24V *Hind*
TRANSPORT HELICOPTER
Some sqn with Bell 205 (UH-1H *Iroquois*); H145; Mi-8 *Hip*; Mi-17V-5 *Hip*; Mi-171Sh *Hip*; Mi-26 *Halo*
AIR DEFENCE
Some regt with S-75M *Volkhov* (SA-2 *Guideline*); S-125 *Neva* (SA-3 *Goa*); S-300/S-300PS (SA-10 *Grumble*); 2K11 *Krug* (SA-4 *Ganef*); S-200 *Angara* (SA-5 *Gammon*); 2K12 *Kub* (SA-6 *Gainful*)
EQUIPMENT BY TYPE
AIRCRAFT 113 combat capable
FTR 45: 12 MiG-29 *Fulcrum* A; 2 MiG-29UB *Fulcrum* B; 31 MiG-31/MiG-31BM *Foxhound*
FGA 54: 12 MiG-27 *Flogger* D; 2 MiG-23UB *Flogger* C; 20 Su-27 *Flanker*; 4 Su-27UB *Flanker*; 16 Su-30SM *Flanker* H
ATK 14: 12 Su-25 *Frogfoot*; 2 Su-25UB *Frogfoot*
ISR 1 An-30 *Clank*
TPT 21: **Medium** 2 An-12 *Cub*; **Light** 18: 6 An-26 *Curl*, 2 An-72 *Coaler*; 8 C295; 2 Tu-134 *Crusty*; **PAX** 1 Tu-154 *Careless*
TRG 19: 17 L-39 *Albatros*; 2 Z-242L
HELICOPTERS
ATK 32: 20 Mi-24V *Hind* (some upgraded); 12 Mi-35M *Hind*
MRH 26: 20 Mi-17V-5 *Hip*; 6 Mi-171Sh *Hip*
TPT 16: **Heavy** 4 Mi-26 *Halo*; **Light** 12: 4 Bell 205 (UH-1H *Iroquois*); 8 H145
UNINHABITED AERIAL VEHICLES
CISR • Heavy 2 *Wing Loong* (GJ-1)
AIR DEFENCE • SAM
Long-range S-200 *Angara* (SA-5 *Gammon*); S-300 (SA-10 *Grumble*); 40+ S-300PS (SA-10B *Grumble*)
Medium-range 2K11 *Krug* (SA-4 *Ganef*); S-75M *Volkhov* (SA-2 *Guideline*)
Short-range 2K12 *Kub* (SA-6 *Gainful*); S-125 *Neva* (SA-3 *Goa*)
Point-defence 9K35 *Strela*-10 (SA-13 *Gopher*)

AIR-LAUNCHED MISSILES
AAM • IR R-60 (AA-8 *Aphid*); R-73 (AA-11A *Archer*); **IR/SARH** R-27 (AA-10 *Alamo*); **SARH** R-33 (AA-9A *Amos*); **ARH** R-77 (AA-12A *Adder* – on MiG-31BM)
ASM Kh-23 (AS-7 *Kerry*)‡; Kh-25 (AS-10 *Karen*); Kh-29 (AS-14 *Kedge*)
ARM Kh-27 (AS-12 *Kegler*); Kh-58 (AS-11 *Kilter*)

Paramilitary 31,500

National Guard ε20,000
Ministry of Interior
AIRCRAFT
TPT • Medium 1 Y-8F-200WA

State Security Service 2,500

Border Service ε9,000
Ministry of Interior
EQUIPMENT BY TYPE
AIRCRAFT 7: **Light** 6: 4 An-26 *Curl*; 1 An-74T; 1 An-74TK **PAX** 1 SSJ-100
HELICOPTERS • TPT • Medium 15: 1 Mi-171; 14 Mi-171Sh

Coast Guard
EQUIPMENT BY TYPE
PATROL AND COASTAL COMBATANTS 22
PBF 11: 1 *Aibar* (Project 0210); 8 FC-19; 2 *Saygak*
PB 11: 4 *Almaty*; 5 *Sardar*; 2 *Zhuk* (of which 1 may be non-operational)

DEPLOYMENT
LEBANON: UN • UNIFIL 124; 1 inf coy
WESTERN SAHARA: UN • MINURSO 6

Kyrgyzstan KGZ

Kyrgyzstani Som s		2019	2020	2021
GDP	s	590bn	561bn	
	US$	8.46bn	7.48bn	
per capita	US$	1,323	1,148	
Growth	%	4.5	-12.0	
Inflation	%	1.1	8.0	
Def bdgt	s	n.k	n.k	
	US$	n.k	n.k	
US$1=s		69.79	74.98	

Population 5,964,897

Age	0–14	15–19	20–24	25–29	30–64	65 plus
Male	15.6%	4.0%	4.0%	4.4%	18.8%	2.2%
Female	14.8%	3.8%	3.9%	4.3%	20.6%	3.6%

Capabilities
Kyrgyzstan has started to expand its ties with its neighbours on issues such as defence-industrial cooperation, though it remains generally dependent on Russian assistance for its defence requirements. Kyrgyzstan is a member of both the CSTO and the SCO. Moscow maintains a military presence, including a squadron of Su-25SM ground-attack aircraft at Kant air base, which it has leased since 2003. Plans were reported in 2020 to add air-defence systems and UAVs to the Russian air base. In 2020, Kyrgyzstan increased its annual fees, reportedly because Russian forces are using more land than outlined in the 2003 agreement. Talks are ongoing over a possible second Russian base. Joint training is held with regional countries, including on anti-terror drills, but combat readiness remains an issue. Kyrgyzstan has a limited capability to deploy externally, and personnel have been deployed to OSCE and UN missions. The armed forces possess ageing land equipment and limited air capabilities, relying instead on Russian support, training and deployments. There is little local defence industry, although in 2018 Kazakhstan and Kyrgyzstan discussed defence-industrial cooperation. Defence ties with India have increased and a joint working group has been formed on defence cooperation. Reports in 2019 that India might provide a credit line to enable Kyrgyzstan to make defence purchases have yet to transpire. Reports in 2020 indicated discussions with Russia over the transfer of air-defence equipment and helicopters.

ACTIVE 10,900 (Army 8,500 Air 2,400) **Paramilitary 9,500**
Conscript liability 18 months

ORGANISATIONS BY SERVICE

Army 8,500
FORCES BY ROLE
SPECIAL FORCES
1 SF bde
MANOEUVRE
Mechanised
2 MR bde
1 (mtn) MR bde
COMBAT SUPPORT
1 arty bde
1 AD bde
EQUIPMENT BY TYPE
ARMOURED FIGHTING VEHICLES
MBT 150 T-72
RECCE 39: 30 BRDM-2; 9 BRDM-2M
IFV 320: 230 BMP-1; 90 BMP-2
APC • APC (W) 55: 25 BTR-70; 20 BTR-70M; 10 BTR-80
ANTI-TANK/ANTI-INFRASTRUCTURE
MSL • MANPATS 9K11 *Malyutka* (AT-3 *Sagger*); 9K111 *Fagot* (AT-4 *Spigot*); 9K111-1 *Konkurs* (AT-5 *Spandrel*)
RCL 73mm SPG-9
GUNS 100mm 36: 18 MT-12/T-12; 18 M-1944
ARTILLERY 228
SP 122mm 18 2S1 *Gvozdika*
TOWED 123: **122mm** 107: 72 D-30; 35 M-30 (M-1938); **152mm** 16 D-1
GUN/MOR 120mm 12 2S9 NONA-S
MRL 21: **122mm** 15 BM-21; **220mm** 6 9P140 *Uragan*
MOR 120mm 54: 6 2S12; 48 M-120
AIR DEFENCE
SAM • Point-defence 9K32 *Strela*-2 (SA-7 *Grail*)‡
GUNS 48
SP 23mm 24 ZSU-23-4
TOWED 57mm 24 S-60

Air Force 2,400

FORCES BY ROLE

FIGHTER

1 regt with L-39 *Albatros**

TRANSPORT

1 regt with An-2 *Colt*; An-26 *Curl*

ATTACK/TRANSPORT HELICOPTER

1 regt with Mi-24 *Hind*; Mi-8 *Hip*

AIR DEFENCE

Some regt with S-125 *Pechora* (SA-3 *Goa*); S-75 *Dvina* (SA-2 *Guideline*); 2K11 *Krug* (SA-4 *Ganef*)

EQUIPMENT BY TYPE

AIRCRAFT 4 combat capable

TPT • Light 6: 4 An-2 *Colt*; 2 An-26 *Curl*

TRG 4 L-39 *Albatros**

HELICOPTERS

ATK 2 Mi-24 *Hind*

MRH 4 Mi-8MT *Hip*

TPT • Medium 8 Mi-8 *Hip*

AIR DEFENCE • SAM

Medium-range 2K11 *Krug* (SA-4 *Ganef*); S-75 *Dvina* (SA-2 *Guideline*)

Short-range S-125 *Pechora* (SA-3 *Goa*)

Paramilitary 9,500

Border Guards 5,000 (KGZ conscript, RUS officers)

Internal Troops 3,500

National Guard 1,000

DEPLOYMENT

SOUTH SUDAN: UN • UNMISS 2

SUDAN: UN • UNAMID 1; **UN** • UNISFA 1

FOREIGN FORCES

Russia ε500 Military Air Forces: 13 Su-25SM *Frogfoot*; 2 Mi-8 *Hip*

Moldova MDA

Moldovan Leu L		2019	2020	2021
GDP	L	210bn	205bn	
	US$	12.0bn	11.2bn	
per capita	US$	4,458	4,268	
Growth	%	3.6	-4.5	
Inflation	%	4.8	2.8	
Def bdgt	L	747m	770m	
	US$	42.5m	42.2m	
US$1=L		17.57	18.25	

Real-terms defence budget trend (US$m, constant 2015)

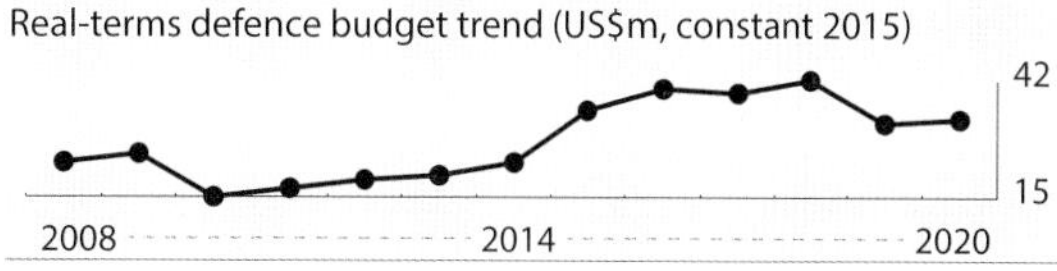

Population 3,364,496

Age	0–14	15–19	20–24	25–29	30–64	65 plus
Male	9.4%	2.9%	3.0%	3.6%	24.4%	5.6%
Female	8.9%	2.6%	2.8%	3.4%	24.9%	8.5%

Capabilities

The primary role of Moldova's armed forces is to maintain territorial integrity, though their size means they would be unable to offer more than token resistance to a determined adversary. The forces are constitutionally neutral. A National Defence Strategy for 2017–21 includes plans to strengthen border defence and airspace control and protection, and to improve the military-training system. Moldova continues to build relations with European states and with NATO. The country signed up to the NATO Defence Capacity Building Initiative in September 2014. Moldova is aiming to end mandatory conscription and develop professional armed forces. The Professional Army 2018–2021 programme was approved in June 2018. A draft government action plan (2019–20) circulated in August 2019 reportedly included closer ties with NATO and a continuing emphasis on professionalisation. A Long-Term Military Capabilities Development Plan was approved in March 2020, covering the period to 2030. The services exercise regularly with NATO states. Moldova has no requirement or capability to independently deploy and support its forces overseas. However, service members have deployed as part of KFOR. The country has no defence-industrial capabilities beyond the basic maintenance of front-line equipment. A permanent solution to the issue of the breakaway region of Transdniestr has yet to be agreed.

ACTIVE 5,150 (Army 3,250 Air 600 Logistic Support 1,300) Paramilitary 900

Conscript liability 12 months (3 months for university graduates)

RESERVE 58,000 (Joint 58,000)

ORGANISATIONS BY SERVICE

Army 1,300; 1,950 conscript (total 3,250)

FORCES BY ROLE

SPECIAL FORCES

1 SF bn

MANOEUVRE
Light
3 mot inf bde
1 lt inf bn
Other
1 gd bn
COMBAT SUPPORT
1 arty bn
1 engr bn
1 NBC coy
1 sigs bn

EQUIPMENT BY TYPE

ARMOURED FIGHTING VEHICLES
APC 163
APC (T) 69: 9 BTR-D; 60 MT-LB (variants)
APC (W) 94: 13 BTR-80; 81 TAB-71
ABCV 44 BMD-1
ANTI-TANK/ANTI-INFRASTRUCTURE
MSL • MANPATS 9K111 *Fagot* (AT-4 *Spigot*); 9K111-1 *Konkurs* (AT-5 *Spandrel*)
RCL 73mm SPG-9
GUNS 100mm 37 MT-12
ARTILLERY 221
TOWED 69: **122mm** 17 (M-30) M-1938; **152mm** 52: 21 2A36 *Giatsint*-B; 31 D-20
GUN/MOR • SP 120mm 9 2S9 NONA-S
MRL 220mm 11 9P140 *Uragan*
MOR 132: **82mm** 75 BM-37; **120mm** 57: 50 M-1989; 7 PM-38
AIR DEFENCE • **GUNS • TOWED** 39: **23mm** 28 ZU-23; **57mm** 11 S-60

Air Force 600 (incl 250 conscripts)

FORCES BY ROLE

TRANSPORT
1 sqn with An-2 *Colt*; Mi-8MTV-1/PS *Hip*; Yak-18
AIR DEFENCE
1 regt with S-125 *Neva* (SA-3 *Goa*)

EQUIPMENT BY TYPE

AIRCRAFT
TPT • Light 3: 2 An-2 *Colt*; 1 Yak-18
HELICOPTERS
TPT • Medium 6: 2 Mi-8PS *Hip*; 4 Mi-8MTV-1 *Hip*
AIR DEFENCE • **SAM • Short-range** 3 S-125 *Neva* (SA-3 *Goa*)

Paramilitary 900

OPON 900 (riot police)

Ministry of Interior

DEPLOYMENT

CENTRAL AFRICAN REPUBLIC: UN • MINUSCA 3
MALI: EU • EUTM Mali 3
SERBIA: NATO • KFOR 41; **UN** • UNMIK 1
SOUTH SUDAN: UN • UNMISS 5

FOREIGN FORCES

Russia ε1,500 (including 400 peacekeepers) 7 Mi-24 *Hind*/Mi-8 *Hip*
Ukraine 10 mil obs (Joint Peacekeeping Force)

Russia RUS

Russian Rouble r		2019	2020	2021
GDP	r	110tr	105tr	
	US$	1.70tr	1.46tr	
per capita	US$	11,601	9,972	
Growth	%	1.3	-4.1	
Inflation	%	4.5	3.2	
Def exp [a]	r	4.21tr	4.34tr	
	US$	65.1bn	60.6bn	
Def bdgt	r	3.00tr	3.09tr	3.20tr
	US$	46.4bn	43.2bn	
US$1=r		64.64	71.49	

[a] Calculated to be comparable with the NATO definition of defence expenditure

Real-terms defence expenditure trend (US$bn, constant)

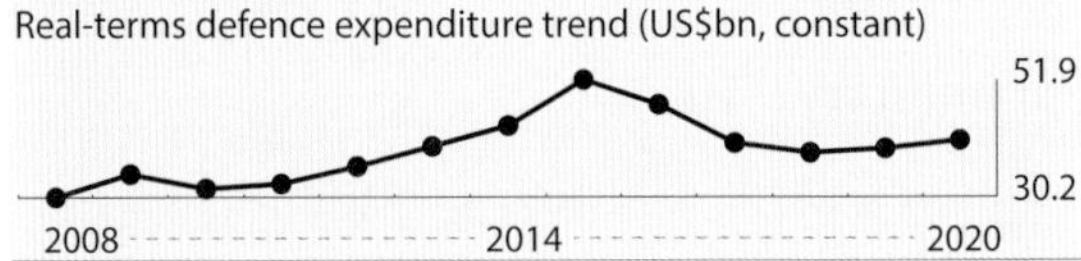

Population	141,722,205

Age	0–14	15–19	20–24	25–29	30–64	65 plus
Male	8.9%	2.5%	2.3%	2.9%	24.6%	5.0%
Female	8.4%	2.4%	2.2%	2.8%	27.3%	10.6%

Capabilities

Russia supports capable conventional military forces and retains the world's second-largest nuclear arsenal. The armed forces underpin an assertive foreign policy. Military aims are guaranteeing sovereignty and territorial integrity and maintaining and increasing Russia's influence in its near abroad and further afield. Russia is a leading member of both the CSTO and the SCO. The armed forces comprise a mix of volunteers and conscripts. Defence reforms launched in 2008 emphasised the shift from a conscript-based mass-mobilisation army to smaller, more professional ground forces. Morale has improved because of better pay, terms and conditions, and greater prestige associated with military service. The armed forces can independently deploy and sustain forces on a global scale, although likely only in modest size at extended distances. Its air-led intervention in Syria shows Russia can deploy, sustain and maintain a high operational tempo for a fixed- and rotary-wing air force, along with the required force-protection package for the main operating base. Russia continues to modernise its nuclear and conventional weapons. The 2020 State Armament Programme (SAP) has been broadly successful, although several of the more ambitious procurement goals were not met. The follow-on programme, SAP 2027, continues the emphasis on modernisation, though some aims are more modest. Russia can design, develop and manufacture advanced nuclear and conventional weaponry. Its defence-industrial base suffered from a lack of investment in the 1990s, and more recently from the loss of access to Ukrainian components. The defence-aerospace sector has been notably successful in terms of exports, particularly of combat aircraft and surface-to-air missile systems.

ACTIVE 900,000 (Army 280,000 Navy 150,000 Air 165,000 Strategic Rocket Force 50,000 Airborne 45,000 Special Operations Forces 1,000 Railway Forces 29,000 Command and Support 180,000) Paramilitary 554,000

Conscript liability 12 months (conscripts now can opt for contract service immediately, which entails a 24-month contract)

RESERVE 2,000,000 (all arms)

Some 2,000,000 with service within last 5 years; reserve obligation to age 50

ORGANISATIONS BY SERVICE

Strategic Deterrent Forces ε80,000 (incl personnel assigned from the Navy and Aerospace Forces)

Navy

EQUIPMENT BY TYPE

SUBMARINES • STRATEGIC • SSBN 11:

1 *Kalmar* (Project 667BDR (*Delta* III)) with 16 R-29RKU-02 *Stantsia*-02 (SS-N-18 *Stingray*) nuclear SLBM, 2 single 400mm TT with SET-72 LWT, 4 single 533mm TT with 53-65K HWT/SET-65K HWT/USET-80K *Keramika* HWT

6 *Delfin* (Project 667BDRM (*Delta* IV)) with 16 R-29RMU2 *Sineva*/R-29RMU2.1 *Layner* (SS-N-23 *Skiff*) nuclear SLBM, 4 single 533mm TT with 53-65K HWT/SET-65K HWT/USET-80K *Keramika* HWT

3 *Borey* (Project 955 (*Dolgorukiy*)) with 16 *Bulava* (SS-N-32) nuclear SLBM, 6 single 533mm TT with USET-80K *Keramika* HWT/UGST *Fizikov* HWT

1 *Borey*-A (Project 955A) with 6 *Bulava* (SS-N-32) nuclear SLBM, 6 single 533mm TT with USET-80K *Keramika* HWT/UGST *Fizikov* HWT

(1 *Akula* (Project 941UM (*Typhoon*)) in reserve for training and SLBM trials with capacity for 20 *Bulava* (SS-N-32) nuclear SLBM, 6 single 533mm TT with 53-65K HWT/SET-65K HWT/USET-80K *Keramika* HWT)

Strategic Rocket Forces 50,000

3 Rocket Armies operating silo and mobile launchers organised in 12 divs. Regt normally with 6 to 10 silos or 9 mobile launchers, and one control centre.

FORCES BY ROLE

SURFACE-TO-SURFACE MISSILE

2 ICBM regt with RS-12M *Topol* (SS-25 *Sickle*)

8 ICBM regt with RS-12M2 *Topol*-M (SS-27 mod 1)

2 ICBM regt with RS-18 (SS-19 *Stiletto*)

1 ICBM regt with RS-18 with Avangard HGV (SS-19 mod 4)

8 ICBM regt with RS-20 (SS-18 *Satan*)

14 ICBM regt with RS-24 *Yars* (SS-27 mod 2)

6 ICBM regt with *Yars*-S

EQUIPMENT BY TYPE

SURFACE-TO-SURFACE MISSILE LAUNCHERS

ICBM • Nuclear 336: 18 RS-12M *Topol* (SS-25 *Sickle*) (mobile single warhead); 60 RS-12M2 *Topol*-M (SS-27 mod 1) silo-based (single warhead); 18 RS-12M2 *Topol*-M (SS-27 mod 1) road mobile (single warhead); up to 20 RS-18 (SS-19 *Stiletto*) (mostly mod 3, 6 MIRV per msl) (being withdrawn); ε4 RS-18 with *Avangard* HGV (SS-19 mod 4 *Stiletto*); 46 RS-20 (SS-18 *Satan*) (mostly mod 5, 10 MIRV per msl); ε99 RS-24 *Yars* (SS-27 mod 2; ε3 MIRV per msl) road mobile; ε18 RS-24 *Yars* (SS-27 mod 2; ε3 MIRV per msl) silo-based; ε53 *Yars*-S (ε3 MIRV per msl) road mobile

Long-Range Aviation Command

FORCES BY ROLE

BOMBER

1 sqn with Tu-160/Tu-160 mod *Blackjack*

3 sqn with Tu-95MS/MS mod *Bear*

EQUIPMENT BY TYPE

AIRCRAFT

BBR 76: 9 Tu-160 *Blackjack* with Kh-55SM (AS-15B *Kent*) nuclear LACM; 7 Tu-160 mod *Blackjack* with Kh-55SM (AS-15B *Kent*)/Kh-102 (AS-23B *Kodiak*) nuclear LACM; 42 Tu-95MS *Bear* H with Kh-55SM (AS-15B *Kent*) nuclear LACM; 18 Tu-95MS mod *Bear* H with Kh-55SM (AS-15B *Kent*)/Kh-102 (AS-23B *Kodiak*) nuclear LACM

Space Command

EQUIPMENT BY TYPE

SATELLITES 109

COMMUNICATIONS 61: 4 *Blagovest*; 2 *Garpun*; 15 *Gonets*-M/M1 (dual-use); 3 Mod *Globus* (*Raduga*-1M); 5 *Meridian*; 3 *Raduga*; 21 *Rodnik* (*Strela*-3M); 8 *Strela*-3

EARLY WARNING 4 *Tundra*

NAVIGATION/POSITIONING/TIMING 28: 26 GLONASS; 2 *Parus*

ISR 11: 2 *Bars*-M; 3 GEO-IK-2; 1 *Kondor*; 1 *Kosmos*-2519; 2 *Persona*; 3 *Resurs*-P

ELINT/SIGINT 5: 4 *Liana* (*Lotos*-S); 1 *Tselina*-2

RADAR 12; Russia leases ground-based radar stations in Baranovichi (Belarus) and Balkhash (Kazakhstan). It also has radars on its own territory at Lekhtusi (St Petersburg); Armavir (Krasnodar); Olenegorsk (Murmansk); Mishelevka (Irkutsk); Kaliningrad; Pechora (Komi); Yeniseysk (Krasnoyarsk); Baranul (Altayskiy); Orsk (Orenburg); and Gorodets/Kovylkino (OTH)

Aerospace Defence Command

FORCES BY ROLE

AIR DEFENCE

2 AD div HQ

4 SAM regt with S-300PM1/PM2 (SA-20 *Gargoyle*)

5 SAM regt with S-400 (SA-21 *Growler*); 96K6 *Pantsir*-S1 (SA-22 *Greyhound*)

EQUIPMENT BY TYPE

AIR DEFENCE • SAM 222

Long-range 186: 90 S-300PM1/PM2 (SA-20 *Gargoyle*); 96 S-400 (SA-21 *Growler*)

Short-range 36 96K6 *Pantsir*-S1 (SA-22 *Greyhound*)

MISSILE DEFENCE 68 53T6 (AB-4A *Gazelle*)

RADAR 1 ABM engagement system located at Sofrino (Moscow)

Army ε280,000 (incl conscripts)

4 military districts (West (HQ St Petersburg), Centre (HQ Yekaterinburg), South (HQ Rostov-on-Don) & East (HQ Khabarovsk)), each with a unified Joint Strategic Command

FORCES BY ROLE

COMMAND

12 army HQ
1 corps HQ

SPECIAL FORCES

8 (Spetsnaz) SF bde
1 (Spetsnaz) SF regt

MANOEUVRE

Reconnaissance

2 recce bde

Armoured

1 (4th) tk div (1 armd recce bn, 2 tk regt, 1 MR regt, 1 arty regt, 1 AD regt)
1 (90th) tk div (1 armd recce bn, 2 tk regt, 1 MR regt, 1 arty regt)
2 tk bde (1 armd recce bn, 3 tk bn, 1 MR bn, 1 arty bn, 1 MRL bn, 2 AD bn, 1 engr bn, 1 EW coy, 1 NBC coy)
2 (3rd & 144th) MR div (1 armd recce bn, 1 tk regt, 2 MR regt, 1 arty regt)
1 (127th) MR div (1 tk bn, 2 MR regt)
1 (150th) MR div (1 armd recce bn, 2 tk regt, 1 MR regt; 1 arty regt, 1 AD regt)
13 (BMP) MR bde (1 armd recce bn, 1 tk bn, 3 armd inf bn, 2 arty bn, 1 MRL bn, 1 AT bn, 2 AD bn, 1 engr bn, 1 EW coy, 1 NBC coy)

Mechanised

1 (2nd) MR div (1 armd recce bn, 1 tk regt, 2 MR regt, 1 arty regt, 1 AD regt)
1 (42nd) MR div (1 armd recce bn, 3 MR regt, 1 arty regt)
8 (BTR/MT-LB) MR bde (1 recce bn; 1 tk bn; 3 mech inf bn; 2 arty bn; 1 MRL bn; 1 AT bn; 2 AD bn; 1 engr bn; 1 EW coy; 1 NBC coy)
2 MR bde (4–5 mech inf bn; 1 arty bn; 1 AD bn; 1 engr bn)
3 (lt/mtn) MR bde (1 recce bn; 2 mech inf bn; 1 arty bn)
1 (18th) MGA div (2 MGA regt; 1 arty regt; 1 tk bn; 2 AD bn)

SURFACE-TO-SURFACE MISSILE

12 SRBM/GLCM bde with 9K720 *Iskander*-M (SS-26 *Stone*/SSC-7 *Southpaw*) (3+ brigades also with 9M729 (SSC-8 *Screwdriver*))

COMBAT SUPPORT

9 arty bde
1 hy arty bde
4 MRL bde
4 engr bde
7 engr regt
1 ptn br bde
5 EW bde
5 NBC bde
10 NBC regt

COMBAT SERVICE SUPPORT

10 log bde

AIR DEFENCE

15 AD bde

EQUIPMENT BY TYPE

ARMOURED FIGHTING VEHICLES

MBT 2,840: 650 T-72B/BA; 850 T-72B3; 530 T-72B3M; 310 T-80BV/U; 140 T-80BVM; 350 T-90/T-90A; 10 T-90M; (10,200 in store: 7,000 T-72/T-72A/B; 3,000 T-80B/BV/U; 200 T-90)

RECCE 1,700: 1,000 BRDM-2/2A (1,000+ BRDM-2 in store); 700 BRM-1K (CP)

IFV 5,220: 500 BMP-1; 3,000 BMP-2; 600 BMP-3; 20+ BMP-3M; 100 BTR-80A; 1,000 BTR-82A/AM (8,500 in store: 7,000 BMP-1; 1,500 BMP-2)

APC 6,100+

APC (T) 3,500+: some BMO-T; 3,500 MT-LB (2,000 MT-LB in store)

APC (W) 2,600: 800 BTR-60 (all variants); 200 BTR-70 (all variants); 1,500 BTR-80; 100+ BPM-97 *Dozor* (4,000 BTR-60/70 in store)

PPV *Typhoon*-K

AUV 100+: 100+ *Tigr*/*Tigr*-M; some IVECO LMV

ENGINEERING & MAINTENANCE VEHICLES

AEV BAT-2; IMR; IMR-2; IMR-3; IRM; MT-LB

ARV BMP-1; BREM-1/64/K/L; BTR-50PK(B); M1977; MTP-LB; RM-G; T-54/55; VT-72A

VLB KMM; MT-55A; MTU; MTU-20; MTU-72; PMM-2

MW BMR-3M; GMX-3; MCV-2 (reported); MTK; MTK-2; UR-77

ANTI-TANK/ANTI-INFRASTRUCTURE

MSL

SP BMP-T with 9K120 *Ataka* (AT-9 *Spiral* 2); 9P149 with 9K114 *Shturm* (AT-6 *Spiral*); 9P149M with 9K132 *Shturm*-SM (AT-9 *Spiral*-2); 9P157-2 with 9K123 *Khrizantema* (AT-15 *Springer*); 9P163-3 with 9M133 *Kornet* (AT-14 *Spriggan*); 9K128-1 *Kornet*-T (AT-14 *Spriggan*)

MANPATS 9K111M *Fagot* (AT-4 *Spigot*); 9K111-1 *Konkurs* (AT-5 *Spandrel*); 9K115 *Metis* (AT-7 *Saxhorn*); 9K115-1 *Metis*-M (AT-13 *Saxhorn* 2); 9K115-2 *Metis*-M1 (AT-13 *Saxhorn* 2); 9K135 *Kornet* (AT-14 *Spriggan*)

RCL 73mm SPG-9

GUNS • TOWED 100mm 526 MT-12 (**100mm** 2,000 T-12/MT-12 in store)

ARTILLERY 4,684+

SP 1,938: **122mm** 150 2S1 *Gvozdika*; **152mm** 1,728: 800 2S3 *Akatsiya*; 100 2S5 *Giatsint*-S; 500 2S19/2S19M1 Msta-S; 320 2S19M2/2S33 *Msta*-SM; 8 2S35 *Koalitsiya*-SV (in test); **203mm** 60 2S7M *Malka* (4,260 in store: **122mm** 2,000 2S1 *Gvozdika*; **152mm** 2,000: 1,000 2S3 *Akatsiya*; 850 2S5 *Giatsint*-S; 150 2S19 *Msta*-S; **203mm** 260 2S7 *Pion*)

TOWED 150: **152mm** 150 2A65 *Msta*-B (12,415 in store: **122mm** 8,150: 4,400 D-30; 3,750 M-30 (M-1938); **130mm** 650 M-46; **152mm** 3,575: 1,100 2A36 *Giatsint*-B; 600 2A65 *Msta*-B; 1,075 D-20; 700 D-1 (M-1943); 100 M-1937 (ML-20); **203mm** 40 B-4M)

GUN/MOR 180+

SP 120mm 80+: 30 2S23 NONA-SVK; 50+ 2S34

TOWED 120mm 100 2B16 NONA-K

MRL 876+ **122mm** 730: 550 BM-21 *Grad*; 180 9K51M *Tornado-G*; **220mm** 206+: 200 9P140 *Uragan*; 6 9K512 *Uragan*-1M**;** some TOS-1A; **300mm** 120: 100 9A52 *Smerch;* 20 9K515 *Tornado*-S; (3,220 in store: **122mm** 2,420: 2,000

BM-21 *Grad*; 420 9P138; **132mm** 100 BM-13; **220mm** 700 9P140 *Uragan*)
MOR 1,540+: **82mm** 800+ 2B14; **120mm** 700 2S12 *Sani*; **240mm** 40 2S4 *Tulpan* (2,590 in store: **120mm** 1,900: 1,000 2S12 *Sani*; 900 M-1938 (PM-38); **160mm** 300 M-160; **SP 240mm** 390 2S4 *Tulpan*)
SURFACE-TO-SURFACE MISSILE LAUNCHERS
SRBM 150:
Dual-capable 150 9K720 *Iskander*-M (SS-26 *Stone*)
GLCM • Dual-capable Some 9M728 (SSC-7 *Southpaw*); some 9M729 (SSC-8 *Screwdriver*)
UNINHABITED AERIAL VEHICLES
ISR • Heavy Tu-243 *Reys*/Tu-243 *Reys* D (service status unclear); **Light** BLA-07; *Pchela*-1; *Pchela*-2
AIR DEFENCE
SAM 1,520+
Long-range S-300V (SA-12A/B *Gladiator*/*Giant*); S-300V4 (SA-23)
Medium-range 350: ε200 9K37M1-2 *Buk*-M1-2 (SA-11 *Gadfly*); ε90 9K317 *Buk*-M2 (SA-17 *Grizzly*); ε60 9K317M *Buk*-M3 (SA-27)
Short-range 120+ 9K331/9K331M/9K331MU *Tor*-M1/M2/M2U (SA-15 *Gauntlet*) (9M338 msl entering service)
Point-defence 1,050+: 250+ 2K22M *Tunguska* (SA-19 *Grison*); 400 9K33M3 *Osa*-AKM (SA-8B *Gecko*); 400 9K35M3 *Strela*-10 (SA-13 *Gopher*); 9K310 *Igla*-1 (SA-16 *Gimlet*); 9K34 *Strela*-3 (SA-14 *Gremlin*); 9K38 *Igla* (SA-18 *Grouse*); 9K333 *Verba* (SA-29 *Gizmo*); 9K338 *Igla*-S (SA-24 *Grinch*)
GUNS
SP 23mm ZSU-23-4
TOWED 23mm ZU-23-2; **57mm** S-60

Reserves

Cadre formations

FORCES BY ROLE
MANOEUVRE
Mechanised
13 MR bde

Navy ε150,000 (incl conscripts)

4 major fleet organisations (Northern Fleet, Pacific Fleet, Baltic Fleet, Black Sea Fleet) and Caspian Sea Flotilla

EQUIPMENT BY TYPE
SUBMARINES 49
STRATEGIC • SSBN 11:
1 *Kalmar* (Project 667BDR (*Delta* III)) with 16 R-29RKU-02 *Stantsia*-02 (SS-N-18 *Stingray*) nuclear SLBM, 2 single 400mm TT with SET-72 LWT, 4 single 533mm TT with 53-65K HWT/SET-65K HWT/USET-80K *Keramika* HWT
6 *Delfin* (Project 667BDRM (*Delta* IV)) with 16 R-29RMU2 *Sineva*/R-29RMU2.1 *Layner* (SS-N-23 *Skiff*) nuclear SLBM, 4 single 533mm TT with 53-65K HWT/SET-65K HWT/USET-80K *Keramika* HWT
3 *Borey* (Project 955 (*Dolgorukiy*)) with 16 *Bulava* (SS-N-32) nuclear SLBM, 6 single 533mm TT with USET-80K *Keramika* HWT/UGST *Fizikov* HWT
1 *Borey*-A (Project 955A) with 6 *Bulava* (SS-N-32) nuclear SLBM, 6 single 533mm TT with USET-80K *Keramika* HWT/UGST *Fizikov* HWT
(1 *Akula* (Project 941UM (*Typhoon*)) in reserve for training and SLBM trials with capacity for 20 *Bulava* (SS-N-32) nuclear SLBM, 6 single 533mm TT with 53-65K HWT/SET-65K HWT/USET-80K *Keramika* HWT)
TACTICAL 38
SSGN 7:
6 *Antey* (Project 949A (*Oscar* II)) (2 more non-operational, in long-term refit) with 24 single SM-225A lnchr with 3M45 *Granit* (SS-N-19 *Shipwreck*) dual-capable AShM, 2 single 650mm TT each with T-65 HWT/RPK-7 (SS-N-16 *Stallion*) ASW msl, 4 single 553mm TT with 53-65K HWT/SET-65K HWT/USET-80K *Keramika* HWT
1 *Yasen* (Project 885 (*Severodvinsk*)) with 8 4-cell SM-346 VLS with 3M14K (SS-N-30A *Sagaris*) dual-capable LACM/3M54K1 (SS-N-27) AShM/3M54K (SS-N-27B *Sizzler*) AShM/3M55 *Oniks* (SS-N-26 *Strobile*) AShM (3M54K/K1 operational status unclear); 10 single 533mm TT with USET-80K *Keramika* HWT/UGST *Fizikov* HWT
SSN 10:
2 *Kondor* (Project 945A (*Sierra* II)) with 4 single 533mm TT with TEST-71M HWT/USET-80K *Keramika* HWT (unclear if dual-capable 3M14 (SS-N-30A *Sagaris*) has replaced 3M10 *Granat* (SS-N-21 *Sampson*) nuclear LACM which is possibly withdrawn; AShM capability unconfirmed), 4 single 650mm TT with 65-73 HWT
3 *Schuka* (Project 671RTMK (*Victor* III)) with 4 single 533mm TT with 53-65K HWT/SET-65K HWT/USET-80K *Keramika* HWT (unclear if dual-capable 3M14 (SS-N-30A *Sagaris*) has replaced 3M10 *Granat* (SS-N-21 *Sampson*) nuclear LACM which is possibly withdrawn; AShM capability unconfirmed), 2 single 650mm TT with 65-73 HWT
3 *Schuka*-B (Project 971 (*Akula* I)) (5 more non-operational, return to service significantly delayed) with 4 single 533mm TT with 53-65K HWT/TEST-71M HWT/USET-80K *Keramika* HWT (unclear if dual-capable 3M14 (SS-N-30A *Sagaris*) has replaced 3M10 *Granat* (SS-N-21 *Sampson*) nuclear LACM which is possibly withdrawn; AShM capability unconfirmed), 4 single 650mm TT with 65-73 HWT/RPK-7 (SS-N-16 *Stallion*) ASW msl
2 *Schuka*-B (Project 971/09711 (*Akula* II)) with 4 single 533mm TT with 53-65K HWT/TEST-71M HWT/USET-80K *Keramika* HWT (unclear if dual-capable 3M14 (SS-N-30A *Sagaris*) has replaced 3M10 *Granat* (SS-N-21 *Sampson*) nuclear LACM which is possibly withdrawn; AShM capability unconfirmed), 4 single 650mm TT with 65-73 HWT/RPK-7 (SS-N-16 *Stallion*) ASW msl
(1 *Barracuda* (Project 945 (*Sierra* I)) (in reserve) with 6 single 533mm TT with TEST-71M HWT/USET-80K *Keramika* HWT (unclear if dual-capable 3M14 (SS-N-30A *Sagaris*) has replaced 3M10 *Granat* (SS-N-21 *Sampson*) nuclear LACM which is possibly withdrawn; AShM capability unconfirmed))

SSK 21:
12 *Paltus* (Project 877 (*Kilo*)) (2 more non-operational, in long-term refit) with 6 single 533mm TT with 53-65K HWT/TEST-71M HWT/USET-80K *Keramika* HWT
8 *Varshavyanka* (Project 06363 (Improved *Kilo*)) with 6 single 533mm TT with 3M14K *Kalibr*-PL (SS-N-30A) dual-capable LACM/3M54K (SS-N-27B *Sizzler*) AShM/3M54K1 (SS-N-27) AShM/53-65K HWT/TEST-71M HWT/USET-80K *Keramika* HWT (3M54K/K1 operational status unclear)
1 *Lada* (Project 677 (*Petersburg*)) (in test) with 6 single 533mm TT with 3M14K *Kalibr*-PL (SS-N-30A) dual-capable LACM/3M54K (SS-N-27B *Sizzler*) AShM/3M54K1 (SS-N-27) AShM/USET-80K *Keramika* HWT (3M54K/K1 operational status unclear)

PRINCIPAL SURFACE COMBATANTS 31
AIRCRAFT CARRIERS • CV 1 *Admiral Kuznetsov* (in extended refit) with 12 single SM-233A lnchr with 3M45 *Granit* (SS-N-19 *Shipwreck*) AShM, 24 8-cell 3S95 VLS with 3K95 *Kinzhal* (SA-N-9 *Gauntlet*) SAM, 2 RBU 12000 *Udav* 1 A/S mor, 8 3M87 *Kortik* CIWS with 9M311 SAM (CADS-N-1), 6 AK630M CIWS (capacity 18–24 Su-33 *Flanker* D Ftr ac; MiG-29KR/KUBR FGA ac; 15 Ka-27 *Helix* ASW hel, 2 Ka-31R *Helix* AEW hel)
CRUISERS 4:
CGHMN 1 *Orlan* (Project 11442 (*Kirov* I)) (1 other non-operational; undergoing extensive refit and planned to return to service in 2021) with 20 single SM-233 lnchr with 3M45 *Granit* (SS-N-19 *Shipwreck*) AShM, 6 6-cell B-203A VLS with S-300F *Fort* (SA-N-6 *Grumble*) SAM, 6 6-cell B-203A VLS with S-300FM *Fort*-M (SA-N-20 *Gargoyle*) SAM, 16 8-cell 3S95 VLS with 3K95 *Kinzhal* (SA-N-9 *Gauntlet*) SAM, 2 quintuple 533mm TT with RPK-6M *Vodopad*-NK (SS-N-16 *Stallion*) A/S msl, 1 RBU 6000 *Smerch* 2 A/S mor, 2 RBU 1000 *Smerch* 3 A/S mor, 6 3M87 *Kortik* CIWS with 9M311 SAM (CADS-N-1), 1 twin 130mm gun (capacity 3 Ka-27 *Helix* ASW hel)
CGHM 3 *Atlant* (Project 1164 (*Slava*)) with 8 twin SM-248 lnchr with 3M70 *Vulkan* (SS-N-12 mod 2 *Sandbox*) AShM, 8 octuple VLS with S-300F *Fort* (SA-N-6 *Grumble*) SAM/S-300FM *Fort* M (SA-N-20 *Gargoyle*) SAM, 2 twin ZIF-122 lnchr with 4K33 *Osa*-M (SA-N-4 *Gecko*) SAM, 2 quintuple 533mm PTA-53-1164 ASTT with SET-65K HWT, 2 RBU 6000 *Smerch* 2 A/S mor, 6 AK630 CIWS, 1 twin 130mm gun (capacity 1 Ka-27 *Helix* ASW hel)
DESTROYERS • DDGHM 11:
3 *Sarych* (Project 956 (*Sovremenny* I)) with 2 quad lnchr with 3M80 *Moskit* (SS-N-22 *Sunburn*) AShM, 2 twin 3S90 lnchr with 9M317 *Yezh* (SA-N-7B) SAM, 2 twin DTA-53-956 533mm TT with 53-65K HWT/SET-65K HWT, 2 RBU 1000 *Smerch* 3 A/S mor, 4 AK630 CIWS, 2 twin 130mm guns (capacity 1 Ka-27 *Helix* ASW hel)
6 *Fregat* (Project 1155 (*Udaloy* I)) with 2 quad lnchr with URK-5 *Rastrub*-B (SS-N-14 *Silex*) AShM/ASW, 8 8-cell 3S95 VLS with 3K95 *Kinzhal* (SA-N-9 *Gauntlet*) SAM, 2 quad 533mm ChTA-53-1155 ASTT with 53-65K HWT/SET-65K HWT, 2 RBU 6000 *Smerch* 2 A/S mor, 4 AK630 CIWS, 2 100mm guns (capacity 2 Ka-27 *Helix* ASW hel)
1 *Fregat* (Project 1155 (*Udaloy* I)) with 2 8-cll 3S14 UKSK VLS with with 3M14T *Kalibr*-NK (SS-N-30A *Sagaris*) dual-capable LACM/3M54T (SS-N-27B *Sizzler*) AShM/3M54T1 (SS-N-27) AShM/3M55 *Oniks* (SS-N-26 *Strobile*) AShM (3M54T/T1 operational status unclear), 2 quad lnchr with 3M24 *Uran* (SS-N-25 *Switchblade*) AShM, 2 quad 533mm ChTA-53-1155 ASTT with 53-65K HWT/SET-65K HWT, 2 RBU 6000 *Smerch* 2 A/S mor, 4 AK630 CIWS, 1 100mm gun (capacity 2 Ka-27 *Helix* ASW hel)
1 *Fregat* (Project 11551 (*Udaloy* II)) (in refit) with 2 quad lnchr with 3M80 *Moskit* (SS-N-22 *Sunburn*) AShM, 8 8-cell 3S95 VLS with 3K95 *Kinzhal* (SA-N-9 *Gauntlet*) SAM, 2 3M87 *Kortik* CIWS with 9M311 SAM (CADS-N-1), 2 RBU 6000 *Smerch* 2 A/S mor, 1 twin 130mm gun (capacity 2 Ka-27 *Helix* ASW hel)
(1 *Fregat* (Project 1155 (*Udaloy* I)) (in reserve) with 2 quad lnchr with URK-5 *Rastrub*-B (SS-N-14 *Silex*) AShM/ASW, 8 8-cell 3S95 VLS with 3K95 *Kinzhal* (SA-N-9 *Gauntlet*) SAM, 2 quad 533mm ChTA-53-1155 ASTT with 53-65K HWT/SET-65K HWT, 2 RBU 6000 *Smerch* 2 A/S mor, 4 AK630 CIWS, 2 100mm guns (capacity 2 Ka-27 *Helix* ASW hel))

FRIGATES 15
FFGHM 13:
3 Project 11356 (*Grigorovich*) with 1 8-cell 3S14 UKSK VLS with 3M14T *Kalibr*-NK (SS-N-30A *Sagaris*) dual-capable LACM/3M54T (SS-N-27B *Sizzler*) AShM/3M54T1 (SS-N-27) AShM/3M55 *Oniks* (SS-N-26 *Strobile*) AShM/91RT2 A/S msl (3M54T/T1 operational status unclear), 2 12-cell 3S90.1 VLS with 9M317 *Yezh* (SA-N-7B) SAM/9M317M *Yezh* (SA-N-7C) SAM, 2 twin DTA-53-11356 533mm TT with 53-65K HWT/SET-65K HWT, 1 RBU 6000 A/S mor, 2 AK630 CIWS, 1 100mm gun (capacity 1 Ka-27 *Helix* ASW hel)
2 *Jastreb* (Project 11540 (*Neustrashimyy*)) (of which 1 in refit) with 2 quad lnchr with 3M24 *Uran* (SS-N-25 *Switchblade*) AShM, 4 8-cell 3S95 VLS with 3K95 *Kinzhal* (SA-N-9 *Gauntlet*), 6 single 533mm ASTT with RPK-6M *Vodopad*-NK (SS-N-16 *Stallion*) A/S msl, 1 RBU 6000 *Smerch* 2 A/S mor, 2 3M87 *Kortik* CIWS with 9M311 SAM (CADS-N-1), 1 100mm gun (capacity 1 Ka-27 *Helix* ASW hel)
1 Project 20380 (*Steregushchiy* I) with 2 quad lnchr with 3M24 *Uran* (SS-N-25 *Switchblade)* AShM, 2 quad 324mm SM-588 ASTT with MTT LWT, 1 3M87 *Kortik*-M CIWS with 9M311 SAM (CADS-N-1), 2 AK630 CIWS, 1 100mm gun (capacity 1 Ka-27 *Helix* ASW hel)
5 Project 20380 (*Steregushchiy* II) with 2 quad lnchr with 3M24 *Uran* (SS-N-25 *Switchblade)* AShM, 3 4-cell 3S97 VLS with 3K96-3 *Redut* (SA-N-28) SAM, 2 quad 324mm SM-588 ASTT with MTT LWT, 2 AK630 CIWS, 1 100mm gun (capacity 1 Ka-27 *Helix* ASW hel)

2 Project 22350 (*Gorshkov*) with 2 8-cell 3S14 UKSK VLS with 3M14T *Kalibr*-NK (SS-N-30A *Sagaris*) dual-capable LACM/3M54T (SS-N-27B *Sizzler*) AShM/3M54T1 (SS-N-27) AShM/3M55 *Oniks* (SS-N-26 *Strobile*) AShM (3M54T/T1 operational status unclear), 4 8-cell 3S97 VLS with 3K96-2 *Poliment-Redut* (SA-N-28) SAM, 2 quad 324mm TT with MTT LWT, 2 3M89 *Palash* CIWS (CADS-N-2), 1 130mm gun (capacity 1 Ka-27 *Helix* ASW hel)

FFGM 2:

1 *Burevestnik* (Project 1135 (*Krivak* I))† with 1 quad lnchr with URK-5 *Rastrub*-B (SS-N-14 *Silex*) AShM/ASW, 1 twin ZIF-122 lnchr with *Osa*-M (SA-N-4 *Gecko*) SAM, 2 quad 533mm ChTA-53-1135 ASTT with 53-65K HWT/SET-65K HWT, 2 RBU 6000 *Smerch* 2 A/S mor, 2 twin 76mm guns

1 *Burevestnik* M (Project 1135M (*Krivak* II)) with 1 quad lnchr with URK-5 *Rastrub*-B (SS-N-14 *Silex*) AShM/ASW, 2 twin ZIF-122 lnchr with 4K33 *Osa*-M (SA-N-4 *Gecko* SAM), 2 quad 533mm ChTA-53-1135 ASTT with 53-65K HWT/SET-65K HWT, 2 RBU 6000 *Smerch* 2 A/S mor, 2 100mm guns

PATROL AND COASTAL COMBATANTS 125

CORVETTES 40

FSGM 12

8 *Buyan*-M (Project 21631 (*Sviyazhsk*)) with 1 8-cell 3S14 UKSK VLS with 3M14T *Kalibr*-NK (SS-N-30A *Sagaris*) dual-capable LACM/3M54T (SS-N-27B *Sizzler*) AShM/3M54T1 (SS-N-27) AShM/3M55 *Oniks* (SS-N-26 *Strobile*) AShM (3M54T/T1 operational status unclear), 2 sextuple 3M47 *Gibka* lnchr with *Igla*-1M (SA-N-10 *Grouse*) SAM, 1 AK630M-2 CIWS, 1 100mm gun

1 Project 11661K (*Gepard* I) with 2 quad lnchr with 3M24 *Uran* (SS-N-25 *Switchblade*) AShM, 1 twin ZIF-122 lnchr with 4K33 *Osa*-M (SA-N-4 *Gecko*) SAM, 2 AK630 CIWS, 1 76mm gun

1 Project 11661K (*Gepard* II) with 1 8-cell VLS with 3M14T *Kalibr*-NK (SS-N-30A *Sagaris*) dual-capable LACM/3M54T (SS-N-27B *Sizzler*) AShM/3M54T1 (SS-N-27) AShM/3M55 *Oniks* (SS-N-26 *Strobile*) AShM (3M54T/T1 operational status unclear), 1 3M89 *Palash* CIWS with 9M337 *Sosna*-R SAM (CADS-N-2), 1 76mm gun

2 *Sivuch* (Project 1239 (*Dergach*)) with 2 quad lnchr with 3M80 *Moskit* (SS-N-22 *Sunburn*) AShM, 1 twin ZIF-122 lnchr with 4K33AM *Osa*-MA2 (SA-N-4 *Gecko*) SAM, 2 AK630M CIWS, 1 76mm gun

FSG 2 *Karakurt* (Project 22800 (*Uragan*)) with 1 8-cell 3S14 VLS with 3M14T *Kalibr*-NK (SS-N-30A *Sagaris*) dual-capable LACM/3M54T (SS-N-27B *Sizzler*) AShM/3M54T1 (SS-N-27) AShM/3M55 *Oniks* (SS-N-26 *Strobile*) AShM (3M54T/T1 operational status unclear), 2 AK630M CIWS, 1 76mm gun

FSM 26:

2 *Albatros* (Project 1124 (*Grisha* III)) with 1 twin ZIF-122 lnchr with 4K33 *Osa*-M (SA-N-4 *Gecko*) SAM, 2 twin 533mm DTA-53-1124 ASTT, 2 RBU 6000 *Smerch* 2 A/S mor, 1 twin 57mm gun

18 *Albatros* (Project 1124M (*Grisha* V)) with 1 twin ZIF-122 lnchr with 4K33 *Osa*-M (SA-N-4 *Gecko*) SAM, 2 twin 533mm DTA-53-1124 ASTT, 1 RBU 6000 *Smerch* 2 A/S mor, 1 76mm gun

6 Project 1331M (*Parchim* II) with 2 quad lnchr with 9K32 *Strela*-2 (SA-N-5 *Grail*) SAM, 2 twin 533mm ASTT, 2 RBU 6000 *Smerch* 2 A/S mor, 1 AK630 CIWS, 1 76mm gun

PSOH 2 Project 22160 (*Bykov*) with 1 76mm gun (capacity 1 Ka-27 *Helix* ASW hel)

PCGM 11:

10 *Ovod*-1 (Project 1234.1 (*Nanuchka* III)) with 2 triple lnchr with P-120 *Malakhit* (SS-N-9 *Siren*) AShM, 1 twin ZIF-122 lnchr with 4K33 *Osa*-M (SA-N-4 *Gecko*) SAM, 1 AK630 CIWS, 1 76mm gun

1 *Ovod*-1 (Project 1234.1 (*Nanuchka* III)) with 4 quad lnchr with 3M24 *Uran* (SS-N-25 *Switchblade*) AShM, 1 twin lnchr with 4K33 *Osa*-M (SA-N-4 *Gecko*) SAM, 1 AK630 CIWS, 1 76mm gun

PCFG 23:

5 *Molnya* (*Tarantul* II) with 2 twin lnchr with P-22 *Termit*-R (SS-N-2D *Styx*) AShM, 2 AK630M CIWS, 1 76mm gun

17 *Molnya* (*Tarantul* III) with 2 twin lnchr with 3M80 *Moskit* (SS-N-22 *Sunburn*) AShM, 2 AK630M CIWS, 1 76mm gun

1 *Molnya* (*Tarantul* III) with 2 twin lnchr with 3M80 *Moskit* (SS-N-22 *Sunburn*) AShM, 1 3K89 *Palash* (CADS-N-2) CIWS, 1 76mm gun

PCM 3 *Buyan* (Project 21630 (*Astrakhan*)) with 1 sextuple lnchr with 3M47 *Gibka* lnchr with *Igla*-1M (SA-N-10 *Grouse*) SAM, 1 A-215 *Grad*-M 122mm MRL, 1 100mm gun

PCF 1 *Molnya* (*Tarantul* III) with 2 AK630M CIWS, 1 76mm gun

PBF 18: 16 *Raptor* (capacity 20 troops); 2 *Mangust*

PBR 4 *Shmel* with 1 17-cell BM-14 MRL, 1 76mm gun

PB 23 *Grachonok*

MINE WARFARE • MINE COUNTERMEASURES 41

MCC 3 *Alexandrit* (Project 12700)

MHI 7 *Sapfir* (Project 10750 (*Lida*)) with 1 AK630 CIWS

MHO 2 *Rubin* (Project 12660 (*Gorya*)) with 2 quad lnchr with 9K32 *Strela*-2 (SA-N-5 *Grail*) SAM, 1 AK630 CIWS, 1 76mm gun

MSC 21: 20 *Yakhont* (Project 1265 (*Sonya*)) with 4 AK630 CIWS (some with 2 quad lnchr with 9K32 *Strela*-2 (SA-N-5 *Grail*) SAM); 1 *Korund*-E (Project 1258E (*Yevgenya*))

MSO 8: 7 *Akvamaren*-M (Project 266M (*Natya*)); 1 *Agat* (Project 02668 (*Natya* II)) (all with 2 quad lnchr (manual aiming) with 9K32 *Strela*-2 (SA-N-5 *Grail*) SAM, 2 RBU 1200 *Uragan* A/S mor, 2 twin AK230 CIWS

AMPHIBIOUS

LANDING SHIPS • LST 20:

12 Project 775 (*Ropucha* I/II) with 2 twin 57mm guns (capacity either 10 MBT and 190 troops or 24 APC (T) and 170 troops)

3 Project 775M (*Ropucha* III) with 2 AK630 CIWS, 1 76mm gun (capacity either 10 MBT and 190 troops or 24 APC (T) and 170 troops)

4 *Tapir* (Project 1171 (*Alligator*)) with at least 2 twin lnchr with 9K32 *Strela*-2 (SA-N-5 *Grail*) SAM, 2 twin 57mm guns (capacity 20 tanks; 300 troops)

1 Project 11711 (*Gren*) with 1 AK630M-2 CIWS, 2 AK630M CIWS (capacity 1 Ka-29 *Helix* B hel; 13 MBT/36 AFV; 300 troops)

LANDING CRAFT 28

LCM 26: 9 *Akula* (Project 1176 (*Ondatra*)) (capacity 1 MBT); 5 *Dyugon* (Project 21820) (capacity 5 APC or 100 troops); 12 *Serna* (Project 11770) (capacity 2 APC or 100 troops)

LCAC 2 *Zubr* (Project 12322 (*Pomornik*)) with 2 22-cell 140mm MS-227 *Ogon'* MRL, 2 AK630 CIWS (capacity 230 troops; either 3 MBT or 10 APC(T))

LOGISTICS AND SUPPORT 276

SSAN 8:

2 *Halibut* (Project 18511 (*Paltus*))
3 *Kashalot* (Project 1910 (*Uniform*))
1 *Nelma* (Project 1851 (X-*Ray*))
1 *Orenburg* (*Delta* III Stretch)
1 *Podmoskovye* (Project 09787)
(1 non-operational *Losharik* (Project 10831 (*Norsub*-5)) reportedly damaged by fire in 2019)

SSA 1 *Sarov* (Project 20120)
ABU 12: 8 *Kashtan*; 4 Project 419 (*Sura*)
AE 8: 6 *Muna*; 1 *Dubnyak*; *Akademik Kovalev* (Project 20181) with 1 hel landing platform
AEM 2: 1 *Kalma*-3 (Project 1791R); 1 *Lama*
AFS 1 *Longvinik* (Project 23120)
AG 1 *Potok*
AGB 5: 1 *Dobrynya Mikitich*; 1 *Ilya Muromets*; 2 *Ivan Susanin*; 1 *Vladimir Kavraisky*
AGE 3: 2 *Seliger*; 1 *Tchusovoy*
AGHS 2 Project 23040G
AGI 14: 2 *Alpinist*; 2 *Dubridium* (Project 1826); 1 *Moma*; 7 *Vishnya*; 2 *Yuri Ivanov*
AGM 1 *Marshal Nedelin*
AGOR 7: 1 *Akademik Alexandrov* (Project 20183); 1 *Akademik Krylov*; 2 *Sibiriyakov*; 2 *Vinograd*; 1 *Yantar*
AGS 70: 8 *Biya*; 18 *Finik*; 7 *Kamenka*; 5 *Moma*; 9 *Onega*; 5 *Baklan* (Project 19920); 4 *Baklan* (Project 19920B); 4 *Vaygach*; 10 *Yug*
AGSH 1 *Samara*
AH 3 *Ob*†
AK 3: 2 *Irgiz*; 1 *Pevek*
AOL 10: 3 *Dubna*; 3 *Uda*; 4 *Altay* (mod)
AOR 4: 3 *Boris Chilikin*; 1 *Akademik Pashin* (Project 23130)
AORL 2: 1 *Kaliningradneft*; 1 *Olekma*
AOS 1 *Luza*
AR ε7 *Amur*
ARC 4: 3 *Emba*; 1 Improved *Klasma*
ARS 31: 1 *Kommuna*; 5 *Goryn*; 4 *Mikhail Rudnitsky*; 20 Project 23040; 1 *Zvezdochka* (Project 20180)
AS 3 Project 2020 (*Malina*)
ASR 2: 1 *Elbrus*; 1 *Igor Belousov*
ATF 54: 1 *Okhotsk*; 1 *Baklan*; ε3 *Katun*; 3 *Ingul*; 2 *Neftegaz*; 10 *Okhtensky*; 13 *Prometey*; 2 Project 23470 with 1 hel landing platform; 1 *Prut*; 4 *Sliva*; 14 *Sorum*
ATS 6 Project 22870
AWT 1 *Manych*
AXL 9: 7 *Petrushka*; 2 *Smolny* with 2 RBU 2500 *Smerch* 1 A/S mor, 2 twin 76mm guns

Naval Aviation ε31,000

FORCES BY ROLE

FIGHTER
1 regt with MiG-31B/BS/BM *Foxhound*
1 regt with Su-27/Su-27UB *Flanker*
1 regt with Su-33 *Flanker* D; Su-25UTG *Frogfoot*

FIGHTER/GROUND ATTACK
1 regt with MiG-29KR/KUBR *Fulcrum*
1 regt with MiG-31BM *Foxhound*; Su-24M/M2/MR *Fencer*

ANTI-SURFACE WARFARE/ISR
2 regt with Su-24M/MR *Fencer*; Su-30SM

ANTI-SUBMARINE WARFARE
3 sqn with Il-38/Il-38N *May**; Il-18D; Il-20RT *Coot* A; Il-22 *Coot* B
8 sqn with Ka-27/Ka-29 *Helix*
2 sqn with Tu-142MK/MZ/MR *Bear* F/J*
1 unit with Ka-31R *Helix*

MARITIME PATROL/TRANSPORT
1 sqn with An-26 *Curl*; Be-12 *Mail**; Mi-8 *Hip*

SEARCH & RESCUE/TRANSPORT
1 sqn with An-12PS *Cub*; An-26 *Curl*; Tu-134

TRANSPORT
1 sqn with An-12BK *Cub*; An-24RV *Coke*; An-26 *Curl*; An-72 *Coaler*; An-140
2 sqn with An-26 *Curl*; Tu-134

TRAINING
1 sqn with L-39 *Albatros*; Su-25UTG *Frogfoot*
1 sqn with An-140; Tu-134; Tu-154, Il-38 *May*

ATTACK/TRANSPORT HELICOPTER
1 sqn with Mi-24P *Hind*; Mi-8 *Hip*

TRANSPORT HELICOPTER
1 sqn with Mi-8 *Hip*

AIR DEFENCE
1 SAM regt with S-300PM1 (SA-20 *Gargoyle*)
1 SAM regt with S-300PM1 (SA-20 *Gargoyle*); S-300PS (SA-10B *Grumble*)
1 SAM regt with S-300PM1 (SA-20 *Gargoyle*); S-400 (SA-21 *Growler*); 96K6 *Pantsir*-S1 (SA-22 *Greyhound*)
1 SAM regt with S-300PS (SA-10B *Grumble*)
1 SAM regt with S-300PS (SA-10B *Grumble*); S-400 (SA-21 *Growler*); 96K6 *Pantsir*-S1 (SA-22 *Greyhound*)

EQUIPMENT BY TYPE

AIRCRAFT 219 combat capable
FTR 67: 10 MiG-31B/BS *Foxhound*; 22 MiG-31BM *Foxhound* C; 17 Su-33 *Flanker* D; 18 Su-27/Su-27UB *Flanker*
FGA 44: 19 MiG-29KR *Fulcrum*; 3 MiG-29KUBR *Fulcrum*; 22 Su-30SM
ATK 46: 41 Su-24M *Fencer*; 5 Su-25UTG *Frogfoot* (trg role)
ASW 44: 12 Tu-142MK/MZ *Bear* F; 10 Tu-142MR *Bear* J (comms); 15 Il-38 *May*; 7 Il-38N *May*
MP 7: 6 Be-12PS *Mail**; 1 Il-18D
ISR 12 Su-24MR *Fencer* E*
SAR 4: 3 An-12PS *Cub*; 1 Be-200ES
ELINT 4: 2 Il-20RT *Coot* A; 2 Il-22 *Coot* B
TPT 49: **Medium** 2 An-12BK *Cub*; **Light** 45: 1 An-24RV *Coke*; 24 An-26 *Curl*; 6 An-72 *Coaler*; 4 An-140; 9 Tu-134; 1 Tu-134UBL; **PAX** 2 Tu-154M *Careless*
TRG 4 L-39 *Albatros*

HELICOPTERS
ATK 8 Mi-24P *Hind*
ASW 63: 41 Ka-27PL *Helix*; 22 Ka-27M *Helix*
EW 8 Mi-8 *Hip* J
AEW 2 Ka-31R *Helix*
SAR 16 Ka-27PS *Helix* D
TPT 41: **Medium** 35: 27 Ka-29 *Helix*; 4 Mi-8T *Hip*; 4 Mi-8MT *Hip*; **Light** 6 Ka-226T
AIR DEFENCE • SAM
Long-range 136: 56 S-300PM1 (SA-20 *Gargoyle*); 40 S-300PS (SA-10B *Grumble*); 40 S-400 (SA-21 *Growler*)
Short-range 12 96K6 *Pantsir*-S1 (SA-22 *Greyhound*)
AIR-LAUNCHED MISSILES
AAM • IR R-27T/ET (AA-10B/D *Alamo*); R-60 (AA-8 *Aphid*); R-73 (AA-11A *Archer*); R-74M (AA-11B *Archer*); **ARH** R-77-1 (AA-12B *Adder*); **SARH** R-27R/ER (AA-10A/C *Alamo*); R-33 (AA-9A *Amos*)
ARM Kh-25MP (AS-12A *Kegler*); Kh-31P (AS-17A *Krypton*); Kh-58 (AS-11 *Kilter*)
ASM Kh-59 (AS-13 *Kingbolt*); Kh-59M (*AS-18 Kazoo*); Kh-29T (AS-14 *Kedge*)
AShM Kh-31A (AS-17B *Krypton*)

Naval Infantry (Marines) ε35,000

FORCES BY ROLE
COMMAND
3 corps HQ
SPECIAL FORCES
4 (OMRP) SF unit
11 (PDSS) cbt diver unit
MANOEUVRE
Reconnaissance
1 recce bde
Mechanised
3 MR bde
1 MR regt
6 naval inf bde
1 naval inf regt
SURFACE-TO-SURFACE MISSILE
1 SRBM/GLCM bde with 9K720 *Iskander*-M (SS-26 *Stone*/SSC-7 *Southpaw*)
COMBAT SUPPORT
2 arty bde
AIR DEFENCE
2 SAM regt with 9K33 *Osa* (SA-8 *Gecko*); *Strela*-1/*Strela*-10 (SA-9 *Gaskin*/SA-13 *Gopher*)
3 SAM regt with S-400 (SA-21 *Growler*); 96K6 *Pantsir*-S1 (SA-22 *Greyhound*)

EQUIPMENT BY TYPE
ARMOURED FIGHTING VEHICLES
MBT 330: 50 T-72B; 150 T-72B3; 30 T-72B3M; 50 T-80BV; 50 T-80BVM
IFV 1,100: 400 BMP-2; 700 BTR-82A
APC 400
APC (T) 300 MT-LB
APC (W) 100 BTR-80
ANTI-TANK/ANTI-INFRASTRUCTURE
MSL
SP 60 9P148 with 9K111-1 *Konkurs* (AT-5 *Spandrel*); 9P149 with 9K114 *Shturm* (AT-6 *Spiral*); 9P157-2 with 9K123 *Khrisantema* (AT-15 *Springer*)
MANPATS 9K111-1 *Konkurs* (AT-5 *Spandrel*); 9K135 *Kornet* (AT-14 *Spriggan*)
GUNS 100mm T-12
ARTILLERY 405
SP 181: **122mm** 95 2S1 *Gvozdika*; **152mm** 86: 50 2S3 *Akatsiya*; 36 2S19M1 *Msta-S*
TOWED 152mm 100: 50 2A36 *Giatsint*-B; 50 2A65 *Msta*-B
GUN/MOR 66
SP 120mm 42: 12 2S23 NONA-SVK; 30 2S9 NONA-S
TOWED 120mm 24 2B16 NONA-K
MRL 58: **122mm** 36 BM-21 *Grad*/*Tornado*-G; **220mm** 18 9P140 *Uragan*; **300mm** 4+ 9A52 *Smerch*
SURFACE-TO-SURFACE MISSILE LAUNCHER
SRBM • Dual-capable 12 9K720 *Iskander*-M (SS-26 *Stone*)
GLCM • Dual-capable Some 9M728 (SSC-7 *Southpaw*)
AIR DEFENCE
SAM
Long-range 64 S-400 (SA-21 *Growler*)
Short-range 30: 18 96K6 *Pantsir*-S1 (SA-22 *Greyhound*); 12+ *Tor*-M2DT
Point-defence 70+: 20 9K33 *Osa* (SA-8 *Gecko*); 50 9K31 *Strela*-1/9K35 *Strela*-10 (SA-9 *Gaskin*/SA-13 *Gopher*); 9K338 *Igla*-S (SA-24 *Grinch*)
GUNS 23mm 60 ZSU-23-4

Coastal Missile and Artillery Forces 2,000

FORCES BY ROLE
COASTAL DEFENCE
5 AShM bde
1 AShM regt

EQUIPMENT BY TYPE
COASTAL DEFENCE
ARTY • SP 130mm ε36 A-222 *Bereg*
AShM 92+: 40 3K60 *Bal* (SSC-6 *Sennight*); 52 3K55 *Bastion* (SSC-5 *Stooge*); some 4K44 *Redut* (SSC-1 *Sepal*); some 4K51 *Rubezh* (SSC-3 *Styx*)

Aerospace Forces ε165,000 (incl conscripts)

A joint CIS Unified Air Defence System covers RUS, ARM, BLR, KAZ, KGZ, TJK, TKM and UZB

FORCES BY ROLE
BOMBER
3 regt with Tu-22M3 *Backfire* C
3 sqn with Tu-95MS/MS mod *Bear*
1 sqn with Tu-160/Tu-160 mod *Blackjack*
FIGHTER
1 sqn with MiG-29/MiG-29UB *Fulcrum* (Armenia)
2 regt with MiG-31BM *Foxhound* C
1 regt with MiG-31BM *Foxhound* C; Su-35S *Flanker* M
1 regt with Su-27/Su-27SM/Su-27UB *Flanker* B/J/C; Su-30M2 *Flanker* G
2 regt with Su-30SM *Flanker* H
FIGHTER/GROUND ATTACK
1 regt with MiG-31BM *Foxhound* C; Su-27SM *Flanker* J; Su-30M2 *Flanker* G; Su-30SM *Flanker* H; Su-35S *Flanker M*
1 regt with Su-27SM *Flanker* J; Su-35S *Flanker* M
1 regt with Su-35S *Flanker* M; Su-30SM *Flanker* H
1 regt with Su-27SM3 *Flanker*; Su-30M2 *Flanker* G
1 regt with Su-25 *Frogfoot*; Su-30SM *Flanker* H

GROUND ATTACK
1 regt with Su-24M/M2 *Fencer*; Su-34 *Fullback*
1 regt with Su-24M *Fencer*; Su-25SM *Frogfoot*
3 regt with Su-25SM/SM3 *Frogfoot*
1 sqn with Su-25SM *Frogfoot* (Kyrgyzstan)
3 regt with Su-34 *Fullback*
GROUND ATTACK/ISR
1 regt with Su-24M/MR *Fencer*
ISR
3 sqn with Su-24MR *Fencer*
1 flt with An-30 *Clank*
AIRBORNE EARLY WARNING & CONTROL
1 sqn with A-50/A-50U *Mainstay*
TANKER
1 sqn with Il-78/Il-78M *Midas*
TRANSPORT
6 regt/sqn with An-12BK *Cub*; An-148-100E; An-26 *Curl*; Tu-134 *Crusty*; Tu-154 *Careless*; Mi-8 *Hip*
1 regt with An-124 *Condor*; Il-76MD *Candid*
1 regt with An-124 *Condor*; Il-76MD *Candid*; Il-76MD-90A *Candid*
1 regt with An-12BK *Cub*; Il-76MD *Candid*
1 sqn with An-22 *Cock*
3 regt with Il-76MD *Candid*
ATTACK/TRANSPORT HELICOPTER
1 bde with Ka-52A *Hokum* B; Mi-28N *Havoc* B; Mi-35 *Hind*; Mi-26 *Halo*; Mi-8MTV-5 *Hip*
1 bde with Ka-52A *Hokum* B; Mi-26 *Halo*; Mi-8 *Hip*
1 bde with Mi-28N *Havoc* B; Mi-35 *Hind*; Mi-26 *Halo*; Mi-8 *Hip*
2 regt with Ka-52A *Hokum* B; Mi-28N *Havoc* B; Mi-35 *Hind*; Mi-8 *Hip*
1 regt with Ka-52A *Hokum* B; Mi-24P *Hind*; Mi-8PPA *Hip*; Mi-8 *Hip*
1 regt with Ka-52A *Hokum* B; Mi-8 *Hip*
1 regt with Mi-28N *Havoc* B; Mi-35 *Hind*; Mi-8 *Hip*
1 regt with Mi-28N *Havoc* B; Mi-24P *Hind*; Mi-35 *Hind*; Mi-8 *Hip*
2 regt with Mi-24P *Hind*; Mi-8 *Hip*
2 sqn with Mi-24P *Hind*; Mi-8 *Hip*
AIR DEFENCE
9 AD div HQ
4 regt with 9K37M1-2 *Buk*-M1-2 (SA-11 *Gadfly*); 9K317 *Buk*-M2 (SA-17 *Grizzly*); S-300V (SA-12 *Gladiator/Giant*)
1 bde with S-300PS (SA-10B *Grumble*)
2 regt with S-300PS (SA-10B *Grumble*)
6 regt with S-300PM1/PM2 (SA-20 *Gargoyle*)
12 regt with S-400 (SA-21 *Growler*); 96K6 *Pantsir*-S1 (SA-22 *Greyhound*)

EQUIPMENT BY TYPE
AIRCRAFT 1,160 combat capable
BBR 137: 60 Tu-22M3 *Backfire* C; 1 Tu-22MR *Backfire*† (1 in overhaul); 42 Tu-95MS *Bear*; 18 Tu-95MS mod *Bear*; 9 Tu-160 *Blackjack*; 7 Tu-160 mod *Blackjack*
FTR 185: 70 MiG-29/MiG-29UB *Fulcrum*; 85 MiG-31BM *Foxhound*; 20 Su-27 *Flanker*; 10 Su-27UB *Flanker*
FGA 415: 16 MiG-29SMT *Fulcrum*; 2 MiG-29UBT *Fulcrum*; 47 Su-27SM *Flanker* J; 24 Su-27SM3 *Flanker*; 19 Su-30M2 *Flanker* G; 91 Su-30SM *Flanker* H; 122 Su-34 *Fullback*; 94 Su-35S *Flanker* M
ATK 264: 70 Su-24M/M2 *Fencer*; 40 Su-25 *Frogfoot*; 139 Su-25SM/SM3 *Frogfoot*; 15 Su-25UB *Frogfoot*
ISR 58: 4 An-30 *Clank*; up to 50 Su-24MR *Fencer**; 2 Tu-214ON; 2 Tu-214R
EW 3 Il-22PP
ELINT 31: 14 Il-20M *Coot* A; 5 Il-22 *Coot* B; 12 Il-22M *Coot* B
AEW&C 9: 5 A-50 *Mainstay*; 4 A-50U *Mainstay*
C2 8: 2 Il-80 *Maxdome*; 1 Il-82; 4 Tu-214SR; 1 Tu-214PU-SBUS
TKR 15: 5 Il-78 *Midas*; 10 Il-78M *Midas*
TPT 445: **Heavy** 123: 11 An-124 *Condor*; 4 An-22 *Cock*; 99 Il-76MD *Candid*; 3 Il-76MD-M *Candid*; 6 Il-76MD-90A *Candid* **Medium** 65 An-12BK *Cub*; **Light** 225: 114 An-26 *Curl*; 25 An-72 *Coaler*; 5 An-140; 27 L-410; 54 Tu-134 *Crusty*; **PAX** 32: 15 An-148-100E; 17 Tu-154 *Careless*
TRG 262: 35 DA42T; 118 L-39 *Albatros*; 109 Yak-130 *Mitten**
HELICOPTERS
ATK 394+: 127 Ka-52A *Hokum* B; 100 Mi-24D/V/P *Hind*; 90+ Mi-28N *Havoc* B; 13 Mi-28UB *Havoc*; 64+ Mi-35 *Hind*
EW 27: 20 Mi-8PPA *Hip*; 7 Mi-8MTRP-1 *Hip*
TPT 333: **Heavy** 33 Mi-26/Mi-26T *Halo*; **Medium** 300: 100 Mi-8/Mi-8MT *Hip*; 60 Mi-8AMTSh *Hip*; 5 Mi-8AMTSh-VA *Hip*; 135 Mi-8MTV-5-1 *Hip*
TRG 69: 19 Ka-226U; 50 *Ansat*-U
UNINHABITED AERIAL VEHICLES
ISR • Medium *Forpost (Searcher* II)
AIR DEFENCE
SAM 714:
Long-range 584: 160 S-300PS (SA-10B *Grumble*); 150 S-300PM1/PM2 (SA-20 *Gargoyle*); 20 S-300V (SA-12 *Gladiator/Giant*); 6 S-350 *Vityaz* (SA-28); 248 S-400 (SA-21 *Growler*)
Medium-range 80 9K37M1-2 *Buk*-M1-2/9K317 *Buk*-M2 (SA-11 *Gadfly*/SA-17 *Grizzly*)
Short-range 50 96K6 *Pantsir*-S1/S2 (SA-22 *Greyhound*)
COUNTERSPACE • DE • Laser *Peresvet*
AIR-LAUNCHED MISSILES
AAM • IR R-27T/ET (AA-10B/D *Alamo*); R-73 (AA-11A *Archer*); R-74M (AA-11B *Archer*); R-60T (AA-8 *Aphid*); **SARH** R-27R/ER (AA-10A/C *Alamo*); R-33 (AA-9A *Amos*); **ARH** R-77-1 (AA-12B *Adder*); R-37M (AA-13A *Axehead*); **PRH** R-27P/EP (AA-10E/F *Alamo*)
ARM Kh-25MP (AS-12A *Kegler*); Kh-25M (AS-12B *Kegler*); Kh-31P/PM (AS-17A/C *Krypton*); Kh-58 (AS-11 *Kilter*)
ASM Kh-29 (AS-14 *Kedge*); Kh-38; Kh-59 (AS-13 *Kingbolt*) Kh-59M (AS-18 *Kazoo*); *Kinzhal*; 9M114 *Kokon* (AT-6 *Spiral*); 9M120 *Ataka* (AT-9 *Spiral* 2); 9M120-1 *Vikhr* (AT-16 *Scallion*)
AShM Kh-22 (AS-4 *Kitchen*); Kh-31A/AM (AS-17B/D *Krypton*); Kh-32 (AS-4A mod); Kh-35U (AS-20 *Kayak*)
LACM
Nuclear Kh-55SM (AS-15B *Kent*); Kh-102 (AS-23B *Kodiak*)
Conventional Kh-101 (AS-23A *Kodiak*); Kh-555 (AS-22 *Kluge*)

BOMBS
Laser-guided KAB-500; KAB-1500L
TV-guided KAB-500KR; KAB-1500KR; KAB-500OD; UPAB 1500
INS/GLONASS-guided KAB-500S

Airborne Forces ε45,000

FORCES BY ROLE
SPECIAL FORCES
1 (AB Recce) SF bde
MANOEUVRE
Air Manoeuvre
1 AB div (1 tk bn, 3 para/air aslt regt, 1 arty regt, 1 AD regt)
1 AB div (1 tk bn, 2 para/air aslt regt, 1 para/air aslt bn, 1 arty regt, 1 AD regt)
2 AB div (2 para/air aslt regt, 1 arty regt, 1 AD regt)
1 indep AB bde
3 air aslt bde

EQUIPMENT BY TYPE
ARMOURED FIGHTING VEHICLES
MBT 160: 150 T-72B3; 10 T-72B3M
IFV 130 BTR-82AM
APC • APC (T) 808: 700 BTR-D; 108 BTR-MDM
AUV GAZ *Tigr*; UAMZ *Toros*
ABCV 1,289: 1,000 BMD-2; 289 BMD-4M
ENGINEERING & MAINTENANCE VEHICLES
ARV BREM-D; BREhM-D
ANTI-TANK/ANTI-INFRASTRUCTURE
MSL
SP 100 BTR-RD
MANPATS 9K111 *Fagot* (AT-4 *Spigot*); 9K113 *Konkurs* (AT-5 *Spandrel*); 9K115 *Metis* (AT-7 *Saxhorn*); 9K115-1 *Metis*-M (AT-13 *Saxhorn* 2); 9K135 *Kornet* (AT-14 *Spriggan*)
RCL 73mm SPG-9
GUNS • SP 125mm 36+ 2S25 *Sprut*-SD
ARTILLERY 600+
TOWED 122mm 150 D-30
GUN/MOR • SP 120mm 250: 220 2S9 NONA-S; 30 2S9 NONA-SM; (500 2S9 NONA-S in store)
MOR • TOWED 200+ **82mm** 150 2B14; **120mm** 50+ 2B23 NONA-M1
AIR DEFENCE
SAM • Point-defence 30+: 30 *Strela*-10MN; 9K310 *Igla*-1 (SA-16 *Gimlet*); 9K38 *Igla* (SA-18 *Grouse*); 9K333 *Verba* (SA-29 *Gizmo*); 9K338 *Igla*-S (SA-24 *Grinch*); 9K34 *Strela*-3 (SA-14 *Gremlin*)
GUNS • SP 23mm 150 BTR-ZD

Special Operations Forces ε1,000

FORCES BY ROLE
SPECIAL FORCES
2 SF unit

Railway Forces ε29,000

4 regional commands
FORCES BY ROLE
COMBAT SERVICE SUPPORT
10 (railway) tpt bde

Russian Military Districts

Western Military District

HQ at St Petersburg

Army

FORCES BY ROLE
COMMAND
3 army HQ
SPECIAL FORCES
2 (Spetsnaz) SF bde
MANOEUVRE
Reconnaissance
1 recce bde
Armoured
1 tk div
1 tk bde
2 MR div
Mechanised
1 MR div
3 MR bde
SURFACE-TO-SURFACE MISSILE
3 SRBM/GLCM bde with *Iskander*-M
COMBAT SUPPORT
2 arty bde
1 (hy) arty bde
1 MRL bde
1 engr bde
1 ptn br bde
1 EW bde
1 NBC bde
2 NBC regt
COMBAT SERVICE SUPPORT
2 log bde
AIR DEFENCE
4 AD bde

Reserves

FORCES BY ROLE
MANOEUVRE
Mechanised
2 MR bde

Northern Fleet

EQUIPMENT BY TYPE
SUBMARINES 26
STRATEGIC 8 **SSBN** (of which 1 in refit) (1 additional in reserve)
TACTICAL 18: 4 **SSGN**; 9 **SSN**; 5 **SSK**
PRINCIPAL SURFACE COMBATANTS 10: 1 **CV** (in refit); 1 **CGHMN**; 1 **CGHM**; 5 **DDGHM** (1 more in reserve); 2 **FFGHM**
PATROL AND COASTAL COMBATANTS 16: 6 **FSM**; 2 **PCGM**; 8 **PB**
MINE WARFARE • MINE COUNTERMEASURES 9: 1 **MHO**; 1 **MSO**; 7 **MSC**
AMPHIBIOUS 7: 5 **LST**; 2 **LCM**

Naval Aviation

FORCES BY ROLE
FIGHTER
1 regt with Su-33 *Flanker* D; Su-25UTG *Frogfoot*

FIGHTER/GROUND ATTACK
1 regt with MiG-29KR/KUBR *Fulcrum*
FIGHTER/GROUND ATTACK/ISR
1 regt with MiG-31BM *Foxhound* C; Su-24M/M2/MR *Fencer*
ANTI-SUBMARINE WARFARE
1 sqn with Il-38 *May*; Il-20RT *Coot* A; Tu-134
3 sqn with Ka-27/Ka-29 *Helix*
1 sqn with Tu-142MK/MZ/MR *Bear* F/J
AIR DEFENCE
5 SAM regt with S-300PS (SA-10B *Grumble*); S-300PM1 (SA-20 *Gargoyle*); S-400 (SA-21 *Growler*); 96K6 *Pantsir*-S1 (SA-22 *Greyhound*)

EQUIPMENT BY TYPE
AIRCRAFT
FTR 38: 20 MiG-31BM *Foxhound* C; 18 Su-33 *Flanker* D
FGA 25: 19 MiG-29KR *Fulcrum*; 4 MiG-29KUBR *Fulcrum*; 2 Su-30SM *Flanker* H
ATK 18: 13 Su-24M *Fencer*; 5 Su-25UTG *Frogfoot* (trg role)
ASW 21: 10 Il-38 *May*; 11 Tu-142MK/MZ/MR *Bear* F/J
ISR 4 Su-24MR *Fencer*
ELINT 3: 2 Il-20RT *Coot* A; 1 Il-22 *Coot* B
TPT 9: 8 An-26 *Curl*; 1 Tu-134
HELICOPTERS
ASW Ka-27 *Helix* A
TPT • Medium Ka-29 *Helix* B; Mi-8 *Hip*
AIR DEFENCE • SAM
Long-range S-300PS (SA-10B *Grumble*); S-300PM1 (SA-20 *Gargoyle*); S-400 (SA-21 *Growler*)
Short-range 96K6 *Pantsir*-S1 (SA-22 *Greyhound*)

Naval Infantry
FORCES BY ROLE
COMMAND
1 corps HQ
MANOEUVRE
Mechanised
2 MR bde
1 naval inf bde

Coastal Artillery and Missile Forces
FORCES BY ROLE
COASTAL DEFENCE
1 AShM bde

Baltic Fleet
EQUIPMENT BY TYPE
SUBMARINES • TACTICAL • **SSK** 1
PRINCIPAL SURFACE COMBATANTS 7: 1 **DDGHM**; 6 **FFGHM** (of which 1 in refit)
PATROL AND COASTAL COMBATANTS 32: 2 **FSGM**; 2 **FSG**; 6 **FSM**; 4 **PCGM**; 7 **PCFG**; 10 **PBF**; 1 **PB**
MINE WARFARE • MINE COUNTERMEASURES 12: 1 **MCC**; 5 **MSC**; 6 **MHI**
AMPHIBIOUS 13: 4 **LST**; 7 **LCM**; 2 **LCAC**

Naval Aviation
FORCES BY ROLE
FIGHTER
1 sqn with Su-27 *Flanker* B
ANTI-SURFACE WARFARE/ISR
1 sqn with Su-24M/MR *Fencer*; Su-30SM *Flanker* H
ANTI-SUBMARINE WARFARE
1 sqn with Ka-27/Ka-29 *Helix*
TRANSPORT
1 sqn with An-26 *Curl*; Tu-134 *Crusty*
ATTACK/TRANSPORT HELICOPTER
1 sqn with Mi-24P *Hind*; Mi-8 *Hip*
TRANSPORT HELICOPTER
1 sqn with Mi-8 *Hip*

EQUIPMENT BY TYPE
AIRCRAFT
FTR 18 Su-27/Su-27UB *Flanker* B/C
FGA 8 Su-30SM *Flanker* H
ATK 10 Su-24M *Fencer*
ISR 4 Su-24MR *Fencer*
TPT 8: 6 An-26 *Curl*; 2 Tu-134 *Crusty*
HELICOPTERS
ATK Mi-24P *Hind*
ASW Ka-27 *Helix*
TPT • Medium Ka-29 *Helix*; *Mi-8 Hip*

Naval Infantry
FORCES BY ROLE
COMMAND
1 corps HQ
MANOEUVRE
Mechanised
1 MR bde
1 MR regt
1 naval inf bde
SURFACE-TO-SURFACE MISSILE
1 SRBM/GLCM bde with *Iskander*-M
COMBAT SUPPORT
1 arty bde
AIR DEFENCE
3 SAM regt

Coastal Artillery and Missile Forces
FORCES BY ROLE
COASTAL DEFENCE
1 AShM regt

Military Air Force

6th Air Force & Air Defence Army
FORCES BY ROLE
FIGHTER
1 regt with Su-30SM *Flanker* H
1 regt with MiG-31BM *Foxhound* C; Su-35S *Flanker* M
1 regt with Su-27SM *Flanker* J; Su-35S *Flanker* M
GROUND ATTACK
1 regt with Su-34 *Fullback*
ISR
1 sqn with Su-24MR *Fencer* E
1 flt with A-30 *Clank*

TRANSPORT
1 regt with An-12 *Cub*; An-26 *Curl*; Tu-134 *Crusty*
ATTACK HELICOPTER
1 bde with Ka-52A *Hokum* B; Mi-28N *Havoc* B; Mi-35 *Hind*; Mi-26 *Halo*; Mi-8MTV-5 *Hip*
1 regt with Mi-24P/Mi-35 *Hind*; Mi-28N *Havoc* B; Mi-8 *Hip*
1 regt with Mi-24P *Hind*; Ka-52A *Hokum* B; Mi-8 *Hip*; Mi-8PPA *Hip*
AIR DEFENCE
3 SAM regt with S-300PM1/PM2 (SA-20 *Gargoyle*)
4 SAM regt with S-400 (SA-21 *Growler*); 96K6 *Pantsir*-S1 (SA-22 *Greyhound*)

EQUIPMENT BY TYPE
AIRCRAFT
FTR 30 MiG-31BM *Foxhound* C
FGA 98: 12 Su-27SM *Flanker* J; 24 Su-30SM *Flanker* H; 24 Su-34 *Fullback*; 38 Su-35S *Flanker* M
ISR 19: 4 An-30 *Clank*; 15 Su-24MR *Fencer* E*
TPT 12 An-12/An-26/Tu-134
HELICOPTERS
ATK 76+: 24 Ka-52A *Hokum* B; 16 Mi-24P *Hind*; 24 Mi-28N *Havoc* B; 12+ Mi-35 *Hind*
EW 10 Mi-8PPA *Hip*
TPT • Medium 50 Mi-8 *Hip*
AIR DEFENCE
SAM
Long-range S-300PM1/PM2 (SA-20 *Gargoyle*); S-400 (SA-21 *Growler*)
Short-range 96K6 *Pantsir*-S1 (SA-22 *Greyhound*)

Airborne Forces

FORCES BY ROLE
SPECIAL FORCES
1 (AB Recce) SF bde
MANOEUVRE
Air Manoeuvre
3 AB div

Central Military District

HQ at Yekaterinburg

Army

FORCES BY ROLE
COMMAND
2 army HQ
SPECIAL FORCES
2 (Spetsnaz) SF bde
MANOEUVRE
Armoured
1 tk div
3 MR bde
Mechanised
1 MR bde
2 (lt/mtn) MR bde
SURFACE-TO-SURFACE MISSILE
2 SRBM/GLCM bde with *Iskander*-M
COMBAT SUPPORT
2 arty bde
1 MRL bde
1 engr bde
1 EW bde
2 NBC bde
2 NBC regt
COMBAT SERVICE SUPPORT
2 log bde
AIR DEFENCE
3 AD bde

Reserves

FORCES BY ROLE
MANOEUVRE
Mechanised
3 MR bde

Military Air Force

14th Air Force & Air Defence Army

FORCES BY ROLE
FIGHTER
2 regt with MiG-31BM *Foxhound* C
GROUND ATTACK
1 regt with Su-34 *Fullback*
1 sqn with Su-25SM *Frogfoot* (Kyrgyzstan)
ISR
1 sqn with Su-24MR *Fencer* E
TRANSPORT
1 regt with An-12 *Cub*; An-26 *Curl*; Tu-134 *Crusty*; Tu-154; Mi-8 *Hip*
ATTACK/TRANSPORT HELICOPTER
1 bde with Mi-24P *Hind*; Mi-8 *Hip*
1 regt with Mi-24P *Hind*; Mi-8 *Hip*
1 sqn with Mi-24P *Hind*; Mi-8 *Hip* (Tajikistan)
AIR DEFENCE
2 regt with S-300PS (SA-10B *Grumble*)
1 bde with S-300PS (SA-10B *Grumble*)
1 regt with S-300PM1 (SA-20 *Gargoyle*)
3 regt with S-400 (SA-21 *Growler*); 96K6 *Pantsir*-S1 (SA-22 *Greyhound*)

EQUIPMENT BY TYPE
AIRCRAFT
FTR 40 MiG-31BM *Foxhound* C
FGA 24 Su-34 *Fullback*
ATK 13 Su-25SM *Frogfoot*
ISR 9 Su-24MR *Fencer* E
TPT 36 An-12 *Cub*/An-26 *Curl*/Tu-134 *Crusty*/Tu-154 *Careless*
HELICOPTERS
ATK 24 Mi-24 *Hind*
TPT 46: 6 Mi-26 *Halo*; 40 Mi-8 *Hip*
AIR DEFENCE
SAM
Long-range S-300PS (SA-10B *Grumble*); S-300PM1 (SA-20 *Gargoyle*); S-400 (SA-21 *Growler*)
Short-range 96K6 *Pantsir*-S1 (SA-22 *Greyhound*)

Airborne Troops

FORCES BY ROLE
MANOEUVRE
Air Manoeuvre
1 AB bde

Southern Military District

HQ located at Rostov-on-Don

Army

FORCES BY ROLE

COMMAND
3 army HQ

SPECIAL FORCES
3 (Spetsnaz) SF bde
1 (Spetsnaz) SF regt

MANOEUVRE

Reconnaissance
1 recce bde

Armoured
1 MR div
3 MR bde
1 MR bde (Armenia)
1 MR bde (South Ossetia)

Mechanised
1 MR div
1 MR bde
1 MR bde (Abkhazia)
1 (lt/mtn) MR bde

SURFACE-TO-SURFACE MISSILE
3 SRBM/GLCM bde with *Iskander*-M

COMBAT SUPPORT
2 arty bde
1 MRL bde
1 engr bde
1 EW bde
1 NBC bde
2 NBC regt

COMBAT SERVICE SUPPORT
2 log bde

AIR DEFENCE
4 AD bde

Black Sea Fleet

The Black Sea Fleet is primarily based in Crimea, at Sevastopol, Karantinnaya Bay and Streletskaya Bay

EQUIPMENT BY TYPE

SUBMARINES • TACTICAL 6 **SSK**
PRINCIPAL SURFACE COMBATANTS 6: 1 **CGHM**; 3 **FFGHM**; 2 **FFGM**
PATROL AND COASTAL COMBATANTS 35: 7 **FSGM**; 6 **FSM**; 2 **PSOH**; 1 **PCGM**; 5 **PCFG**; 6 **PB**; 8 **PBF**
MINE WARFARE • MINE COUNTERMEASURES 9: 2 **MCC**; 1 **MHO**; 5 **MSO**; 1 **MSC**
AMPHIBIOUS 10: 7 **LST**; 3 **LCM**

Naval Aviation

FORCES BY ROLE

FIGHTER

ANTI-SURFACE WARFARE/ISR
1 regt with Su-24M/MR *Fencer*; Su-30SM *Flanker* H

ANTI-SUBMARINE WARFARE
1 sqn with Ka-27 *Helix*
1 sqn with Mi-14 *Haze*

MARITIME PATROL/TRANSPORT
1 sqn with An-26 *Curl*; Be-12PS *Mail**; Mi-8

EQUIPMENT BY TYPE

AIRCRAFT
FGA 12 Su-30SM *Flanker* H
ATK 13 Su-24M *Fencer* D
ISR 4 Su-24MR *Fencer* E
MP 3 Be-12PS *Mail**
TPT 6 An-26

HELICOPTERS
ASW Ka-27 *Helix*
TPT • Medium Mi-8 *Hip* (MP/EW/Tpt)

Naval Infantry

FORCES BY ROLE

COMMAND
1 corps HQ

MANOEUVRE

Mechanised
2 naval inf bde

COMBAT SUPPORT
1 arty bde

AIR DEFENCE
1 SAM regt

Coastal Artillery and Missile Forces

FORCES BY ROLE

COASTAL DEFENCE
2 AShM bde

Caspian Sea Flotilla

EQUIPMENT BY TYPE

PATROL AND COASTAL COMBATANTS 15: 3 **FSGM**; 1 **PCFG**; 3 **PCM**; 3 **PB**; 1 **PBF**; 4 **PBR**
MINE WARFARE • MINE COUNTERMEASURES 3: 2 **MSC**; 1 **MHI**
AMPHIBIOUS 9 **LCM**

Naval Infantry

FORCES BY ROLE

MANOEUVRE

Mechanised
1 naval inf regt

Military Air Force

4th Air Force & Air Defence Army

FORCES BY ROLE

FIGHTER
1 regt with Su-30SM *Flanker* H
1 sqn with MiG-29 *Fulcrum* (Armenia)

FIGHTER/GROUND ATTACK
1 regt with Su-27/Su-27SM *Flanker* B/J; Su-30M2 *Flanker* G
1 regt with Su-27SM3 *Flanker*; Su-30M2 *Flanker* G

GROUND ATTACK
1 regt with Su-24M *Fencer*; Su-25SM *Frogfoot*
2 regt with Su-25SM/SM3 *Frogfoot*
1 regt with Su-34 *Fullback*

GROUND ATTACK/ISR
1 regt with Su-24M/MR *Fencer* D/E

TRANSPORT
1 regt with An-12 *Cub*/Mi-8 *Hip*

ATTACK/TRANSPORT HELICOPTER
1 bde with Mi-28N *Havoc* B; Mi-35 *Hind*; Mi-8 *Hip*; Mi-26 *Halo*
1 regt with Mi-28N *Havoc* B; Mi-35 *Hind*; Mi-8 *Hip*
2 regt with Ka-52A *Hokum* B; Mi-28N *Havoc* B; Mi-35 *Hind*; Mi-8AMTSh *Hip*
1 sqn with Mi-24P *Hind*; Mi-8 *Hip* (Armenia)
AIR DEFENCE
1 regt with 9K317 *Buk*-M2 (SA-17 *Grizzly*)
1 regt with S-300PM1 (SA-20 *Gargoyle*)
3 regt with S-400 (SA-21 *Growler*); 96K6 *Pantsir*-S1 (SA-22 *Greyhound*)

EQUIPMENT BY TYPE
AIRCRAFT
FTR 26: 12 MiG-29 *Fulcrum*; 14 Su-27 *Flanker* B
FGA 95: 12 Su-27SM *Flanker* J; 24 Su-27SM3 *Flanker*; 14 Su-30M2 *Flanker* G; 21 Su-30SM *Flanker* H; 24 Su-34 *Fullback*
ATK 109: 24 Su-24M *Fencer* D; 85 Su-25SM/SM3 *Frogfoot*
ISR 12 Su-24MR *Fencer* E
TPT 12 An-12 *Cub*
HELICOPTERS
ATK 117: 25 Ka-52A *Hokum* B; 44 Mi-28N *Havoc* B; 8 Mi-24P *Hind*; 40 Mi-35 *Hind*
TPT 72: **Heavy** 10 Mi-26 *Halo*; **Medium** 62 Mi-8 *Hip*
AIR DEFENCE • **SAM**
Long-range S-300PM1 (SA-20 *Gargoyle*); S-400 (SA-21 *Growler*)
Medium-range 9K317 *Buk*-M2 (SA-17 *Grizzly*)
Short-range 96K6 *Pantsir*-S1 (SA-22 *Greyhound*)

Airborne Forces

FORCES BY ROLE
MANOEUVRE
Air Manoeuvre
1 AB div
1 air aslt bde

Eastern Military District

HQ located at Khabarovsk

Army

FORCES BY ROLE
COMMAND
4 army HQ
SPECIAL FORCES
1 (Spetsnaz) SF bde
MANOEUVRE
Armoured
1 tk bde
1 MR div
5 MR bde
Mechanised
3 MR bde
1 MGA div
SURFACE-TO-SURFACE MISSILE
4 SRBM/GLCM bde with *Iskander*-M
COMBAT SUPPORT
3 arty bde
1 MRL bde
1 engr bde
1 EW bde
1 NBC bde
4 NBC regt
COMBAT SERVICE SUPPORT
4 log bde
AIR DEFENCE
5 AD bde

Reserves

FORCES BY ROLE
MANOEUVRE
Mechanised
8 MR bde

Pacific Fleet

EQUIPMENT BY TYPE
SUBMARINES 16
STRATEGIC 3 **SSBN**
TACTICAL 13: 3 **SSGN**; 1 **SSN**; 9 **SSK**
PRINCIPAL SURFACE COMBATANTS 8: 1 **CGHM**; 5 **DDGHM** (of which 1 in refit); 2 **FFGHM**
PATROL AND COASTAL COMBATANTS 27: 8 **FSM**; 4 **PCGM**; 10 **PCFG**; 5 **PB**
MINE WARFARE 8: 2 **MSO**; 6 **MSC**
AMPHIBIOUS 9: 4 **LST**; 5 **LCM**

Naval Aviation

FORCES BY ROLE
FIGHTER
1 sqn with MiG-31BS/BM *Foxhound* A/C
ANTI-SUBMARINE WARFARE
3 sqn with Ka-27/Ka-29 *Helix*
2 sqn with Il-38 *May*; Il-18D; Il-22 *Coot* B
1 sqn with Tu-142MK/MZ/MR *Bear* F/J*
TRANSPORT
2 sqn with An-12BK *Cub*; An-26 *Curl*; Tu-134

EQUIPMENT BY TYPE
AIRCRAFT
FTR 12 MiG-31B/BS *Foxhound* A
ASW 23: 11 Tu-142MK/MZ/MR *Bear* F/J; 12 Il-38 *May*
EW • ELINT 1 Il-22 *Coot* B
TPT 6: 2 An-12BK *Cub*; 3 An-26 *Curl*; 1 Tu-134
HELICOPTERS
ASW Ka-27 *Helix*
TPT • Medium Ka-29 *Helix*; Mi-8 *Hip*

Naval Infantry

FORCES BY ROLE
MANOEUVRE
Mechanised
2 naval inf bde
AIR DEFENCE
1 SAM regt

Coastal Artillery and Missile Forces

FORCES BY ROLE
COASTAL DEFENCE
2 AShM bde

Military Air Force

11th Air Force & Air Defence Army

FORCES BY ROLE

FIGHTER/GROUND ATTACK

1 regt with MiG-31BM *Foxhound* C; Su-27SM *Flanker* J; Su-30M2 *Flanker* G; Su-30SM *Flanker* H; Su-35S *Flanker* M

1 regt with Su-35S *Flanker* M; Su-30SM *Flanker* H

1 regt with Su-25 *Frogfoot*; Su-30SM *Flanker* H

GROUND ATTACK

1 regt with Su-24M/M2 *Fencer* D/D mod.; Su-34 *Fullback*

1 regt with Su-25SM *Frogfoot*

ISR

1 sqn with Su-24MR *Fencer* E

TRANSPORT

1 regt with An-12 *Cub*; An-26 *Curl*; Tu-134 *Crusty*/Tu-154 *Careless*

ATTACK/TRANSPORT HELICOPTER

1 bde with Ka-52A *Hokum* B; Mi-8 *Hip*; Mi-26 *Halo*

1 regt with Ka-52A *Hokum* B; Mi-8 *Hip*; Mi-26 *Halo*

1 regt with Mi-24P *Hind*; Mi-8 *Hip*

AIR DEFENCE

2 regt with 9K37M *Buk*-M1-2 (SA-11 *Gadfly*); S-300V (SA-12 *Gladiator*/*Giant*)

4 regt with S-400 (SA-21 *Growler*); 96K6 *Pantsir*-S1 (SA-22 *Greyhound*)

EQUIPMENT BY TYPE

AIRCRAFT

FTR 20 MiG-31BM *Foxhound* C

FGA 120: 23 Su-27SM *Flanker* J; 6 Su-30M2 *Flanker* G; 31 Su-30SM *Flanker* H; 26 Su-34 *Fullback*; 34 Su-35S *Flanker* M

ATK 102: 20 Su-24M *Fencer* D; 10 Su-24M2 *Fencer*; 72 Su-25/Su-25SM *Frogfoot*

ISR 12 Su-24MR *Fencer* E

TPT 24: 22 An-12 *Cub*/An-26 *Curl*; 1 Tu-134 *Crusty*; 1 Tu-154 *Careless*

HELICOPTERS

ATK 36: 24 Ka-52A *Hokum* B; 12 Mi-24P *Hind*

TPT 60: **Heavy** 4 Mi-26 *Halo*; **Medium** 56 Mi-8 *Hip*

AIR DEFENCE • SAM

Long-range S-300PS (SA-10B *Grumble*); S-300V (SA-12 *Gladiator*/*Giant*); S-400 (SA-21 *Growler*)

Medium-range 9K317 *Buk*-M1-2 (SA-11 *Gadfly*)

Short-range 96K6 *Pantsir*-S1 (SA-22 *Greyhound*)

Airborne Forces

FORCES BY ROLE

MANOEUVRE

Air Manoeuvre

2 air aslt bde

Paramilitary 554,000

Border Guard Service ε160,000

Subordinate to Federal Security Service

FORCES BY ROLE

10 regional directorates

MANOEUVRE

Other

7 frontier gp

EQUIPMENT BY TYPE

ARMOURED FIGHTING VEHICLES

IFV/APC (W) 1,000 BMP/BTR

ARTILLERY 90

SP 122mm 2S1 *Gvozdika*

GUN/MOR • SP 120mm 2S9 NONA-S

MOR 120mm 2S12 *Sani*

PATROL AND COASTAL COMBATANTS 197

PSOHM 2 *Nerey* (*Krivak* III) with 1 twin ZIF-122 lnchr with 4K33 *Osa*-M (SA-N-4 *Gecko*) SAM, 2 quad PTA-53-1135 533mm TT lnchr, 2 RBU 6000 *Smerch* 2 A/S mor, 1 100mm gun (capacity 1 Ka-27 *Helix* A ASW hel)

PSO 5: 4 *Komandor*; 1 *Okean* (Project 22100) with 1 76mm gun, 1 hel landing platform

PCO 26: 8 *Alpinist* (Project 503); 1 *Sprut*; 12 *Okhotnik* (Project 22460) with 1 AK630M CIWS, 1 hel landing platform; 5 *Purga* with 1 hel landing platform

PCC 34: 4 *Molnya* II (*Pauk* II); 4 *Svetlyak* (Project 10410); 16 *Svetlyak* (Project 10410) with 1 AK630M CIWS, 1 76mm gun; 8 *Svetlyak* (Project 10410) with 2 AK630M CIWS; 1 *Svetlyak* (Project 10410) with 1 AK630M CIWS; 1 *Yakhont*

PCR 1 *Slepen* (*Yaz*) with 1 AK630 CIWS, 2 100mm guns

PBF 87: 57 *Mangust*; 3 *Mirazh* (Project 14310); 4 *Mustang*-2 (Project 18623); 21 *Sobol*; 2 *Sokzhoi*

PBR 27: 4 *Ogonek*; 8 *Piyavka* with 1 AK630 CIWS; 15 *Moskit* (*Vosh*) with 1 AK630 CIWS, 1 100mm gun

PB 15: 3 *Gyuys* (Project 03050); 2 *Morzh* (Project 1496M; 10 *Lamantin* (Project 1496M1)

LOGISTICS AND SUPPORT 34

AE 1 *Muna*

AGB 2 *Ivan Susanin* (primarily used as patrol ships) with 2 AK630 CIWS, 1 76mm gun, 1 hel landing platform

AK 8 *Pevek*

AKSL 5 *Kanin*

AO 3: 1 *Ishim* (Project 15010); 2 *Envoron*

ATF 15: 14 *Sorum* (primarily used as patrol ships) with 2 AK230M CIWS; 1 *Sorum* (primarily used as patrol ship)

AIRCRAFT • TPT ε86: 70 An-24 *Coke*/An-26 *Curl*/An-72 *Coaler*/Il-76 *Candid*/Tu-134 *Crusty*/Yak-40 *Codling*; 16 SM-92

HELICOPTERS: ε200 Ka-28 (Ka-27) *Helix* ASW/Mi-24 *Hind* Atk/Mi-26 *Halo* Spt/Mi-8 *Hip* Spt

Federal Guard Service ε40,000–50,000

Org include elm of ground forces (mech inf bde and AB regt)

FORCES BY ROLE

MANOEUVRE

Mechanised

1 mech inf regt

Air Manoeuvre

1 AB regt

Other

1 (Presidential) gd regt

Federal Security Service Special Purpose Centre ε4,000

FORCES BY ROLE

SPECIAL FORCES

2 SF unit (Alfa and Vympel units)

National Guard ε340,000

FORCES BY ROLE

MANOEUVRE

Other

10 paramilitary div (2–5 paramilitary regt)
17 paramilitary bde (3 mech bn, 1 mor bn)
36 indep paramilitary rgt
90 paramilitary bn (incl special motorised units)

Aviation

8 sqn

COMBAT SUPPORT

1 arty regt

EQUIPMENT BY TYPE

ARMOURED FIGHTING VEHICLES

RECCE some BRDM-2A
IFV/APC (W) 1,650 BMP-2/BTR-70M/BTR-80/BTR-82A/BTR-82AM

ARTILLERY 35

TOWED 122mm 20 D-30
MOR 120mm 15 M-1938 (PM-38)

PATROL AND COASTAL COMBATANTS 3

PBF 2 BK-16 (Project 02510)
PB 1 *Grachonok*

AIRCRAFT

TPT 29: **Heavy** 9 Il-76 *Candid*; **Medium** 2 An-12 *Cub*; **Light** 18: 12 An-26 *Curl*; 6 An-72 *Coaler*

HELICOPTERS

TPT 71: **Heavy** 10 Mi-26 *Halo*; **Medium** 60+: 60 Mi-8 *Hip*; some Mi-8AMTSh *Hip*; **Light** 1 Ka-226T

DEPLOYMENT

ARMENIA: 3,500: 1 mil base with (1 MR bde; 74 T-72; 80 BMP-1; 80 BMP-2; 12 2S1; 12 BM-21); 1 ftr sqn with 18 MiG-29 *Fulcrum*; 1 hel sqn with 11 Mi-24P *Hind*; 4 Mi-8AMTSh *Hip*; 4 Mi-8MT *Hip*; 2 AD bty with S-300V (SA-12 *Gladiator/Giant*); 1 AD bty with *Buk*-M1-2 (SA-11 *Gadfly*)

AZERBAIJAN: 817; 1 MR bn (peacekeeping; to rise to 1,960)

BELARUS: 1 radar station at Baranovichi (*Volga* system; leased); 1 naval comms site

CENTRAL AFRICAN REPUBLIC: UN • MINUSCA 13

CYPRUS: UN • UNFICYP 3

DEMOCRATIC REPUBLIC OF THE CONGO: UN • MONUSCO 8

GEORGIA: 7,000; Abkhazia 1 mil base with (1 MR bde; 40 T-90A; 120 BTR-82A; 18 2S3; 12 2S12; 18 BM-21; some S-300 SAM; some atk hel); South Ossetia 1 mil base with (1 MR bde; 40 T-72; 120 BMP-2; 36 2S3; 12 2S12)

KAZAKHSTAN: 1 radar station at Balkash (*Dnepr* system; leased)

KYRGYZSTAN: ε500; 13 Su-25SM *Frogfoot*; 2 Mi-8 *Hip* spt hel

MEDITERRANEAN SEA: 2 SSK; 1 FFGHM; 1 FFGM; 1 AGI

MIDDLE EAST: UN • UNTSO 3

MOLDOVA/TRANSDNIESTR: ε1,500 (including 441 peacekeepers); 2 MR bn; 100 MBT/AIFV/APC; 7 Mi-24 *Hind*; some Mi-8 *Hip*

SOUTH SUDAN: UN • UNMISS 6

SUDAN: UN • UNISFA 2

SYRIA: 4,000: 1 inf BG; 3 MP bn; 1 engr unit; ε10 T-72B3/T-90; ε20 BTR-82A; *Typhoon*-K; *Tigr*; 12 2A65; 4 9A52 *Smerch*; 10 Su-24M *Fencer* D; 6 Su-34; 6 Su-35S *Flanker* M; 1 A-50U *Mainstay*; 1 Il-20M; 12 Mi-24P/Mi-35M *Hind*; 4 Mi-8AMTSh *Hip*; 1 AShM bty with 3K55 *Bastion*; 1 SAM bty with S-400; 1 SAM bty with *Pantsir*-S1/S2; air base at Latakia; naval facility at Tartus

TAJIKISTAN: 5,000; 1 (201st) mil base with (40 T-72B1; 60 BMP-2; 80 BTR-82A; 40 MT-LB; 18 2S1; 36 2S3; 6 2S12; 12 9P140 *Uragan*); 1 hel sqn with 4 Mi-24P *Hind*; 4 Mi-8MTV *Hip*; 2 Mi-8MTV-5-1 *Hip*

UKRAINE: Crimea: 28,000; 1 recce bde, 2 naval inf bde; 1 arty bde; 1 NBC regt; 40 T-72B3 MBT; 80 BMP-2 AIFV; 200 BTR-82A; 20 BTR-80 APC: 150 MT-LB; 18 2S1; 18 2A65; 12 BM-21; 8 9P140 *Uragan*; 1 AShM bde with 3K60 *Bal*; 3K55 *Bastion*; 1 FGA regt with Su-24M/MR; Su-30SM; 1 FGA regt with Su-27SM/SM3; Su-30M2; 1 FGA regt with Su-24M/Su-25SM; 1 atk/tpt hel regt; 1 ASW hel regt; 2 AD regt with S-400; *Pantsir*-S1; 1 Fleet HQ located at Sevastopol; 2 radar stations located at Sevastopol (*Dnepr* system) and Mukachevo (*Dnepr* system); Donetsk/Luhansk: 3,000 (reported)

WESTERN SAHARA: UN • MINURSO 10

Russia and Eurasia

Tajikistan TJK

Tajikistani Somoni Tr		2019	2020	2021
GDP	Tr	77.4bn	81.7bn	
	US$	8.12bn	7.90bn	
per capita	US$	873	834	
Growth	%	7.5	1.0	
Inflation	%	7.8	8.1	
Def bdgt [a]	Tr	ε847m	ε917m	
	US$	ε89m	ε89m	
US$1=Tr		9.53	10.34	

[a] Excl. budget for law enforcement

Real-terms defence budget trend (US$m, constant 2015)

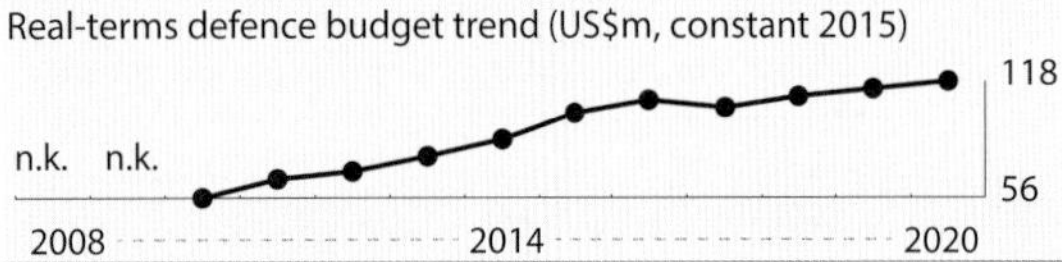

Population 8,873,669

Age	0–14	15–19	20–24	25–29	30–64	65 plus
Male	16.0%	4.7%	4.5%	4.6%	18.5%	1.5%
Female	15.4%	4.5%	4.4%	4.5%	19.3%	2.1%

Capabilities

The Tajik armed forces have little capacity to deploy other than token forces and most equipment is of Soviet-era origin. Regional security and terrorism remain key security concerns, given the border with Afghanistan. Tajikistan has been building its capability by hosting a CSTO counter-terrorism exercise, and by taking part in stability and counter-terror exercises organised by US CENTCOM. Tajikistan is a member of the CSTO and there is a large Russian military presence at the 201st military base. Reports in early 2019 indicated that there may be a Chinese military facility in eastern Tajikistan, though this remains unconfirmed by either Beijing or Dushanbe. In 2018, India and Tajikistan agreed to strengthen defence cooperation, in particular on counter-terrorism. Border deployments have been stepped up recently in response to regional security and terrorism concerns. In late 2016, a Military Cooperation Plan was signed with Russia. Moscow has indicated that Tajikistan is to receive military equipment, including aircraft. Some donations of personal equipment have been received from the US. Barring maintenance facilities, Tajikistan only has minimal defence-industrial capacity.

ACTIVE 8,800 (Army 7,300 Air Force/Air Defence 1,500) Paramilitary 7,500

Conscript liability 24 months

ORGANISATIONS BY SERVICE

Army 7,300

FORCES BY ROLE

MANOEUVRE

Mechanised

3 MR bde

Air Manoeuvre

1 air aslt bde

COMBAT SUPPORT

1 arty bde

AIR DEFENCE

1 SAM regt

EQUIPMENT BY TYPE

ARMOURED FIGHTING VEHICLES

MBT 40: 30 T-72; 3 T-72B1; 7 T-62

RECCE 10 BRDM-2M

IFV 23: 8 BMP-1; 15 BMP-2

APC 36

APC (W) 23 BTR-60/BTR-70/BTR-80

PPV 13 VP11

AUV 8 CS/VN3B mod

ARTILLERY 41

SP 122mm 3 2S1 *Gvozdika*

TOWED 122mm 13 D-30

MRL 122mm 15 BM-21 *Grad*

MOR 10+: **SP 82mm** CS/SS4; **120mm** 10

AIR DEFENCE

SAM

Medium-range S-75 *Dvina* (SA-2 *Guideline*); S-125 *Pechora*-2M (SA-26)

Point-defence 9K32 *Strela*-2 (SA-7 *Grail*)‡

GUNS • TOWED 23mm ZU-23M1

Air Force/Air Defence 1,500

FORCES BY ROLE

TRANSPORT

1 sqn with Tu-134A *Crusty*

ATTACK/TRANSPORT HELICOPTER

1 sqn with Mi-24 *Hind*; Mi-8 *Hip*; Mi-17TM *Hip* H

EQUIPMENT BY TYPE

AIRCRAFT

TPT • Light 1 Tu-134A *Crusty*

TRG 4+: 4 L-39 *Albatros*; some Yak-52

HELICOPTERS

ATK 4 Mi-24 *Hind*

TPT • Medium 11 Mi-8 *Hip*/Mi-17TM *Hip* H

Paramilitary 7,500

Internal Troops 3,800

National Guard 1,200

Emergencies Ministry 2,500

Border Guards

FOREIGN FORCES

China ε300 (trg)

Russia 5,000; 1 (201st) mil base with (40 T-72B1; 60 BMP-2; 80 BTR-82A; 40 MT-LB; 18 2S1; 36 2S3; 6 2S12; 12 9P140 *Uragan*); 1 hel sqn with 4 Mi-24P *Hind*; 4 Mi-8MTV *Hip*; 2 Mi-8MTV-5-1 *Hip*

Turkmenistan TKM

Turkmen New Manat T		2019	2020	2021
GDP	T	158bn	168bn	
	US$	45.2bn	48.0bn	
per capita	US$	7,724	8,074	
Growth	%	6.3	1.8	
Inflation	%	5.1	8.0	
Def exp	T	n.k	n.k	
	US$	n.k	n.k	
USD1=T		3.50	3.50	

Population 5,528,627

Age	0–14	15–19	20–24	25–29	30–64	65 plus
Male	12.9%	4.0%	4.3%	4.8%	21.1%	2.3%
Female	12.5%	3.9%	4.3%	4.8%	21.9%	3.0%

Capabilities

Turkmenistan has concerns over potential spillover from security challenges in Afghanistan, but its armed forces lack significant capabilities and equipment. Ashgabat has maintained a policy of neutrality since 1995 and confirmed this commitment in its 2016 military doctrine. This aimed to increase the armed forces' defensive capability in order to safeguard national interests and territorial integrity. Turkmenistan is not a member of the CSTO. While the ground forces are shifting from a Soviet-era divisional structure to a brigade system, progress is slow. The armed forces are largely conscript-based and reliant on Soviet-era equipment and doctrine, and the government has stated a requirement to improve conditions of service. Turkmenistan has participated in multinational exercises and is reported to have restarted joint exercises with Russia and Uzbekistan, but has limited capacity to deploy externally and maintains no international deployments. In October 2019, in Saint Petersburg, Turkmenistan and four other Caspian littoral states signed a memorandum of understanding on military cooperation, among other discussions including on maritime security. There are plans to strengthen the border guard with new equipment and facilities. Plans to bolster the naval forces have resulted in some procurements, leading to a modest improvement in the naval presence in the Caspian Sea. There has been limited procurement of systems such as UAVs, including from China. Barring maintenance facilities, Turkmenistan has little domestic defence industry, but is building a number of patrol vessels of Turkish design under licence.

ACTIVE 36,500 (Army 33,000 Navy 500 Air 3,000)
Paramilitary 5,000

Conscript liability 24 months

ORGANISATIONS BY SERVICE

Army 33,000

5 Mil Districts

FORCES BY ROLE

SPECIAL FORCES
1 spec ops regt

MANOEUVRE

Armoured
1 tk bde

Mechanised
1 (3rd) MR div (1 tk regt; 3 MR regt, 1 arty regt)
1 (22nd) MR div (1 tk regt; 1 MR regt, 1 arty regt)
4 MR bde
1 naval inf bde

Other
1 MR trg div

SURFACE-TO-SURFACE MISSILE
1 SRBM bde with SS-1 *Scud*

COMBAT SUPPORT
1 arty bde
1 (mixed) arty/AT regt
1 MRL bde
1 AT regt
1 engr regt

AIR DEFENCE
2 SAM bde

EQUIPMENT BY TYPE†

ARMOURED FIGHTING VEHICLES
MBT 654: 4 T-90S; 650 T-72/T-72UMG
RECCE 260+: 200 BRDM-2; 60 BRM-1; Nimr *Ajban*
IFV 1,038: 600 BMP-1/BMP-1M; 430 BMP-2; 4 BMP-3; 4 BTR-80A
APC 898+
APC (W) 870+: 120 BTR-60 (all variants); 300 BTR-70; 450 BTR-80
PPV 28+ *Kirpi*
AUV 12+: 8 Nimr *Ajban* 440A; 4+ *Cobra*
ABCV 8 BMD-1

ANTI-TANK/ANTI-INFRASTRUCTURE
MSL
SP 58+: 8 9P122 *Malyutka*-M (AT-3 *Sagger* on BRDM-2); 8 9P133 *Malyutka*-P (AT-3 *Sagger* on BRDM-2); 2 9P148 *Konkurs* (AT-5 *Spandrel* on BRDM-2); 36 9P149 *Shturm* (AT-6 *Spiral* on MT-LB); 4+ *Baryer* (on *Karakal*)
MANPATS 9K11 *Malyutka* (AT-3 *Sagger*); 9K111 *Fagot* (AT-4 *Spigot*); 9K111-1 *Konkurs* (AT-5 *Spandrel*); 9K115 *Metis* (AT-7 *Saxhorn*)
GUNS 100mm 60 MT-12/T-12

ARTILLERY 765
SP 122mm 40 2S1
TOWED 457: **122mm** 350 D-30; **130mm** 6 M-46; **152mm** 101: 17 D-1; 72 D-20; 6 2A36 *Giatsint*-B; 6 2A65 *Msta*-B
GUN/MOR 120mm 17 2S9 NONA-S
MRL 154+: **122mm** 88: 18 9P138; 70 BM-21 *Grad*; RM-70; **220mm** 60 9P140 *Uragan*; **300mm** 6 9A52 *Smerch*
MOR 97: **82mm** 31; **120mm** 66 M-1938 (PM-38)

SURFACE-TO-SURFACE MISSILE LAUNCHERS
SRBM • Conventional 16 SS-1 *Scud*

UNINHABITED AERIAL VEHICLES
CISR • Heavy CH-3A; WJ-600
ISR • Medium *Falco*

AIR DEFENCE
SAM
Short-range: FM-90; 2K12 *Kub* (SA-6 *Gainful*)
Point-defence 53+: 40 9K33 *Osa* (SA-8 *Gecko*); 13 9K35 *Strela*-10 mod (SA-13 *Gopher*); 9K38 *Igla* (SA-18 *Grouse*); 9K32M *Strela*-2M (SA-7 *Grail*)‡; 9K34 *Strela*-3 (SA-14 *Gremlin*); *Mistral* (reported); QW-2

GUNS 70
SP 23mm 48 ZSU-23-4
TOWED 22+: **23mm** ZU-23-2; **57mm** 22 S-60
AIR-LAUNCHED MISSILES
ASM CM-502KG; AR-1

Navy 500

EQUIPMENT BY TYPE
PATROL AND COASTAL COMBATANTS 4
PCFG 2 *Edermen* (RUS *Molnya*) with 4 quad lnchr with 3M24E *Uran*-E (SS-N-25 *Switchblade*) AShM, 2 AK630 CIWS, 1 76mm gun
PCGM 2 *Arkadag* (TUR *Tuzla*) with 2 twin lnchr with *Otomat* AShM, 2 twin *Simbad*-RC lnchr with *Mistral* SAM, 1 Roketsan ASW Rocket Launcher System A/S mor
LOGISTICS AND SUPPORT• AGHS 1 (Dearsan 41m)

Air Force 3,000

FORCES BY ROLE
FIGHTER
2 sqn with MiG-29A/S/UB *Fulcrum*
GROUND ATTACK
1 sqn with Su-25 *Frogfoot*
1 sqn with Su-25MK *Frogfoot*
TRANSPORT
1 sqn with An-26 *Curl*; Mi-8 *Hip*; Mi-24 *Hind*
TRAINING
1 unit with L-39 *Albatros*
AIR DEFENCE
Some sqn with S-75 *Dvina* (SA-2 *Guideline*); S-125 *Pechora* (SA-3 *Goa*); S-125 *Pechora*-2M (SA-26); S-200 *Angara* (SA-5 *Gammon*); FD-2000 (HQ-9); KS-1A (HQ-12)

EQUIPMENT BY TYPE
AIRCRAFT 55 combat capable
FTR 24: 22 MiG-29A/S *Fulcrum*; 2 MiG-29UB *Fulcrum*
ATK 31: 19 Su-25 *Frogfoot*; 12 Su-25MK *Frogfoot*
TPT • Light 3: 1 An-26 *Curl*; 2 An-74TK *Coaler*
TRG 2 L-39 *Albatros*
HELICOPTERS
ATK 10 Mi-24P *Hind* F
MRH 2+ AW139
TPT 11+: **Medium** 8: 6 Mi-8 *Hip*; 2 Mi-17V-V *Hip*; **Light** 3+ AW109
AIR-LAUNCHED MISSILES
AAM • IR R-60 (AA-8 *Aphid*); R-73 (AA-11A *Archer*)
AIR DEFENCE • SAM
Long-range S-200 *Angara* (SA-5 *Gammon*); FD-2000 (HQ-9)
Medium-range S-75 *Dvina* (SA-2 *Guideline*); S-125 *Pechora*-2M (SA-26); KS-1A (HQ-12)
Short-range S-125 *Pechora* (SA-3 *Goa*); S-125-2BM *Pechora*

Paramilitary 5,000

Federal Border Guard Service ε5,000

EQUIPMENT BY TYPE
PATROL AND COASTAL COMBATANTS 33
PCGM 8 *Arkadag* (TUR *Tuzla*) with 2 twin lnchr with *Otomat* AShM, 2 twin *Simbad*-RC lnchr with *Mistral* SAM, 1 Roketsan ASW Rocket Launcher System A/S mor
PBFG 6 *Nazya* (Dearsan 33) with 2 single lnchr with *Marte* Mk2/N AShM
PBF 18: 10 *Bars*-12; 5 *Grif*-T; 3 *Sobol*
PB 1 *Point*
AMPHIBIOUS • LCM 1 Dearsan LCM-1
HELICOPTERS
MRH 2 AW139
TPT 3+: **Medium** some Mi-8 *Hip*; **Light** 3 AW109

Ukraine UKR

Ukrainian Hryvnia h		2019	2020	2021
GDP	h	3.97tr	3.87tr	
	US$	155bn	142bn	
per capita	US$	3,707	3,425	
Growth	%	3.2	-7.2	
Inflation	%	7.9	3.2	
Def bdgt [a]	h	102bn	118bn	118bn
	US$	3.99bn	4.32bn	
FMA (US)	US$	115m	20.0m	115m
USD1=h		25.69	27.20	

[a] Including military pensions

Real-terms defence budget trend (US$bn, constant 2015)

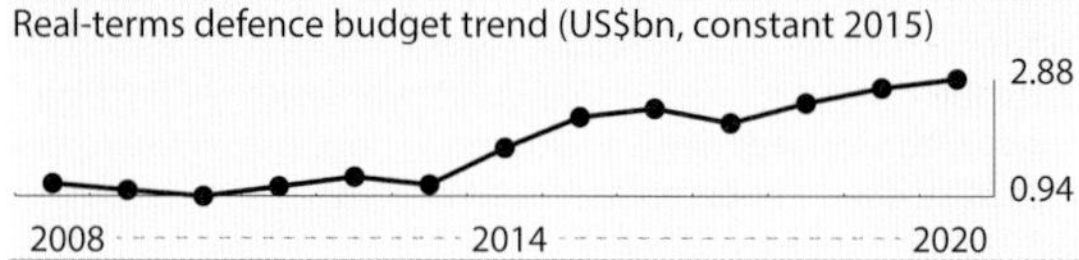

Population 43,922,939

Age	0–14	15–19	20–24	25–29	30–64	65 plus
Male	8.3%	2.3%	2.5%	3.2%	24.3%	5.7%
Female	7.8%	2.2%	2.3%	3.1%	26.9%	11.3%

Capabilities

Ukraine's overriding security concern is Russia's support for separatists in the east of the country and its occupation of Crimea. Sporadic exchanges of fire continued in 2020, and a further ceasefire, agreed in July 2020, had been breached a number of times by early November. Defence policy is centred on maintaining sovereignty and territorial integrity. Ukraine adopted a revised doctrine in 2015 that identified Moscow as a 'military adversary', called for 'comprehensive reform' of the security sector and revoked the country's 'nonbloc status'. In 2017, parliament identified NATO membership as a strategic goal, with the 'State Program for the Development of the Armed Forces until 2020' intended to support this goal. This aspiration was again noted in the 2020 National Security Strategy. As of early 2020, however, fully meeting NATO standards appeared to have been shelved in favour of achieving interoperability. This state programme is based on the Strategic Defence Bulletin. which was first published in 2016. There are two conscript intakes per year, though professionalisation is a long-term ambition. The armed forces participate in bilateral and multinational exercises. The equipment inventory still consists predominantly of Soviet-era weaponry. Sustaining and in some cases upgrading these systems is a near-term concern, and equipment replacements will be required over the coming decade in order to forestall potential problems with obsolescence. For example, Air Force Vision 2035 sets out aircraft-recapitalisation goals to replace all Russian types in service over the next 15 years. Funding this within the intended period may prove challenging. The country has a broad defence-manufacturing industry, operating in all sectors, though its capability remains shaped, and limited, by its Soviet heritage. Ukraine was a key provider of guided-weapons technologies in the Soviet Union. It retains the capability to build Soviet-era land systems and can maintain and modestly upgrade Soviet-era tactical combat aircraft.

ACTIVE 209,000 (Army 145,000 Navy 11,000 Air Force 45,000 Airborne 8,000 Special Operations Forces n.k.) Paramilitary 102,000

Conscript liability Army, Air Force 18 months, Navy 2 years. Minimum age for conscription raised from 18 to 20 in 2015

RESERVE 900,000 (Joint 900,000)

Military service within 5 years

ORGANISATIONS BY SERVICE

Army 145,000

4 regional HQ

FORCES BY ROLE

MANOEUVRE

Reconnaissance

5 recce bn

Armoured

4 tk bde

Mechanised

9 mech bde

2 mtn bde

Light

4 mot inf bde

1 lt inf bde

SURFACE-TO-SURFACE MISSILES

1 SSM bde

COMBAT SUPPORT

5 arty bde

1 MRL bde

2 MRL regt

1 engr regt

1 EW regt

1 EW bn

2 EW coy

1 CBRN regt

4 sigs regt

COMBAT SERVICE SUPPORT

3 maint regt

1 maint coy

HELICOPTERS

4 avn bde

AIR DEFENCE

4 AD regt

Reserves

FORCES BY ROLE

MANOEUVRE

Armoured

2 tk bde

Mechanised

2 mech bde

COMBAT SUPPORT

2 arty bde

EQUIPMENT BY TYPE

ARMOURED FIGHTING VEHICLES

MBT 858: 410 T-64BV; 210 T-64BV mod 2017; 100 T-64BM *Bulat*; 100 T-72AV/B1; 33 T-72AMT; 5 T-84 *Oplot*; (34 T-80; 500 T-72; 578 T-64; 20 T-55 all in store)

RECCE 547: 327 BRDM-2; 80 BRDM-2L1; 25 BRDM-2T; 115 BRM-1K (CP)

IFV 1,184: 213 BMP-1/BMP-1AK; 890 BMP-2; 4 BMP-3; 54+ BTR-3DA; 6+ BTR-3E1; 17+ BTR-4E *Bucephalus*

APC 622

APC (T) 60: 15 BTR-D; 45 MT-LB

APC (W) 322: 5 BTR-60; 215 BTR-70; 102 BTR-80

PPV 240: 40 *Kozak*-2 /-2M; 200 Varta

ABCV 30: 15 BMD-1, 15 BMD-2

AUV 50 *Novator*

ENGINEERING & MAINTENANCE VEHICLES

AEV 53 BAT-2; MT-LB

ARV 12+: 10 BREM-1; 2 BREM-M; BREM-2; BREM-64; T-54/T-55

VLB MTU-20

ANTI-TANK/ANTI-INFRASTRUCTURE

MSL

SP 9P149 with 9K114 *Shturm* (AT-6 *Spiral*)

MANPATS 9K111 *Fagot* (AT-4 *Spigot*); 9K113 *Konkurs* (AT-5 *Spandrel*); FGM-148 *Javelin*; *Stugna*-P; *Corsar*

GUNS 100mm ε500 MT-12/T-12

ARTILLERY 1,818

SP 607+: **122mm** 292 2S1 *Gvozdika*; **152mm** 302: 249 2S3 *Akatsiya*; 18 2S5 *Giatsint*-S; 35 2S19 *Msta*-S; **203mm** 13+ 2S7 *Pion* (up to 83 2S7 *Pion* in store)

TOWED 515+: **122mm** 75 D-30; **152mm** 440: 180 2A36 *Giatsint*-B; 130 2A65 *Msta*-B; 130+ D-20

GUN/MOR • 120mm • TOWED 2 2B16 NONA-K

MRL 354: **122mm** 203: 18 9P138; 185 BM-21 *Grad*; **220mm** 70 9P140 *Uragan*; **300mm** 81+: *Adler*; 81 9A52 *Smerch*

MOR 120mm 340: 190 2S12 *Sani*; 30 M-1938 (PM-38); 120 M120-15

SURFACE-TO-SURFACE MISSILE LAUNCHERS

SRBM • Conventional 90 9K79 *Tochka* (SS-21 *Scarab*)

HELICOPTERS

ATK ε35 Mi-24 *Hind*

MRH 1 *Lev*-1

TPT • Medium ε23 Mi-8 *Hip*

AIR DEFENCE

SAM 75+

Long-range Some S-300V (SA-12A *Gladiator*)

Short-range 6 9K330 *Tor*-M (SA-15 *Gauntlet*)

Point-defence 75+: 9K35 *Strela*-10 (SA-13 *Gopher*); 9K33 *Osa*-AKM (SA-8 *Gecko*); 75 2K22 *Tunguska* (SA-19 *Grison*)

GUNS

SP 23mm ZSU-23-4 *Shilka*

TOWED 23mm ZU-23-2; **57mm** S-60

AIR-LAUNCHED MISSILES • **ASM** *Barrier*-V

Navy 11,000 (incl Naval Aviation and Naval Infantry)

After Russia's annexation of Crimea, HQ shifted to Odessa. Several additional vessels remain in Russian possession in Crimea

2 Regional HQ

EQUIPMENT BY TYPE

PRINCIPAL SURFACE COMBATANTS • FRIGATES 1

FFHM 1 *Hetman Sagaidachny* (RUS *Krivak* III) with 1 twin lnchr with 4K33 *Osa*-M (SA-N-4 *Gecko*) SAM, 2 quad 533mm ASTT with SET-65 HWT/53-65K HWT, 2 RBU 6000 *Smerch* 2 A/S mor, 2 AK630M CIWS, 1 100mm gun (capacity 1 Ka-27 *Helix* ASW hel)

PATROL AND COASTAL COMBATANTS 12

CORVETTES • FS 1 *Albatros* (Project 1124P (*Grisha* II)) with 2 twin 533mm ASTT with SAET-60 HWT, 2 RBU 6000 *Smerch* 2 A/S mor, 2 57mm guns

PCC 2 *Slavyansk* (ex-US *Island*)

PHG 1 *Vekhr* (FSU *Matka*) with 2 single lnchr with P-15M/R *Termit*-M/R (SS-N-2C/D *Styx*) AShM, 1 AK630M CIWS, 1 76mm gun

PBG 7 *Gyurza*-M (Project 51855) with 2 *Katran*-M RWS with *Barrier* SSM

PB 1 *Grif*-M (FSU *Zhuk*)

MINE WARFARE • MINE COUNTERMEASURES 1

MHI 1 *Korund* (Project 1258 (*Yevgenya*))

AMPHIBIOUS

LANDING SHIPS • LSM 1 Project 773I (*Polnochny* C) (capacity 6 MBT; 180 troops)

LANDING CRAFT • LCM 1 *Akula* (*Ondatra*)

LOGISTICS AND SUPPORT 8

ABU 1 Project 419 (*Sura*)

AG 1 *Bereza*

AGI 1 *Muna*

AKL 1

AWT 1 *Sudak*

AXL 3 *Petrushka*

Naval Aviation ε1,000

EQUIPMENT BY TYPE

FIXED-WING AIRCRAFT

ASW (2 Be-12 *Mail* non-operational)

TPT • Light (2 An-26 *Curl* in store)

HELICOPTERS

ASW 7+: 4+ Ka-27 *Helix* A; 1 Mi-14PS *Haze* A; 2 Mi-14PL *Haze* C

TPT • Medium 1 Ka-29 *Helix*-B

TRG 1 Ka-226

Naval Infantry ε2,000

FORCES BY ROLE

MANOEUVRE

Reconnaissance

1 recce bn

Light

2 nav inf bde

EQUIPMENT BY TYPE

ARMOURED FIGHTING VEHICLES

MBT 69: 41 T-64BV; 28 T-80BV mod

IFV BMP-1

APC

APC (T) MT-LB

APC (W) BTR-60; BTR-80

PPV *Varta*

ANTI-TANK/ANTI-INFRASTRUCTURE

GUNS 100mm MT-12

ARTILLERY

SP 122mm 2S1 *Gvozdika*

TOWED 152mm 2A36 *Giatsint*-B

AIR DEFENCE

GUNS • SP 23mm ZSU-23-4

Coastal Defence ε1,500

FORCES BY ROLE

COASTAL DEFENCE

1 arty bde

1 MRL regt

EQUIPMENT BY TYPE

ARTILLERY

TOWED 152mm some D-20

MRL 220mm some 9P140 Uragan

Air Forces 45,000

3 Regional HQ

FORCES BY ROLE

FIGHTER

4 bde with MiG-29 *Fulcrum*; Su-27 *Flanker* B; L-39 *Albatros*

FIGHTER/GROUND ATTACK

2 bde with Su-24M *Fencer*; Su-25 *Frogfoot*

ISR

2 sqn with Su-24MR *Fencer* E*

TRANSPORT

3 bde with An-24; An-26; An-30; Il-76 *Candid*; Tu-134 *Crusty*

TRAINING

Some sqn with L-39 *Albatros*

TRANSPORT HELICOPTER

Some sqn with Mi-8; Mi-9; PZL Mi-2 *Hoplite*

AIR DEFENCE

6 bde with 9K37M *Buk*-M1 (SA-11 *Gadfly*); S-300P/PS/PT (SA-10 *Grumble*)

4 regt with 9K37M *Buk*-M1 (SA-11); S-300P/PS/PT (SA-10)

EQUIPMENT BY TYPE

AIRCRAFT ε125 combat capable

FTR 71: ε37 MiG-29 *Fulcrum*; ε34 Su-27 *Flanker* B

FGA ε14 Su-24M *Fencer* D

ATK ε31 Su-25 *Frogfoot*

ISR 12: 3 An-30 *Clank*; ε9 Su-24MR *Fencer* E*

TPT 29: **Heavy** 5 Il-76 *Candid*; **Medium** 1 An-70; **Light** ε23: 3 An-24 *Coke*; ε19 An-26 *Curl*; 1 Tu-134 *Crusty*

TRG ε31 L-39 *Albatros*

HELICOPTERS

C2 ε14 Mi-9

TPT 32: **Medium** ε30 Mi-8 *Hip*; **Light** 2 PZL Mi-2 *Hoplite*

UNINHABITED AERIAL VEHICLES

CISR • Medium 6 *Bayraktar* TB2

AIR DEFENCE • SAM 322:

Long-range 250 S-300P/PS/PT (SA-10 *Grumble*)

Medium-range 72 9K37M *Buk*-M1 (SA-11 *Gadfly*)
Short-range some S-125 *Pechora* (SA-3 *Goa*)
AIR-LAUNCHED MISSILES
AAM • IR R-60 (AA-8 *Aphid*); R-73 (AA-11A *Archer*)
SARH R-27 (AA-10A *Alamo*)
ASM Kh-25 (AS-10 *Karen*); Kh-29 (AS-14 *Kedge*); MAM-L
ARM Kh-25MP (AS-12A *Kegler*); Kh-58 (AS-11 *Kilter*); Kh-28 (AS-9 *Kyle*) (likely WFU)

Airborne Assault Troops ε8,000

FORCES BY ROLE
MANOEUVRE
Air Manoeuvre
1 AB bde
5 air aslt bde
1 air mob bde
COMBAT SUPPORT
1 SP arty bn
EQUIPMENT BY TYPE
ARMOURED FIGHTING VEHICLES
MBT 60 T-80BV mod
IFV some BTR-3E1; some BTR-4 *Bucephalus*
APC 180+
APC (T) 25 BTR-D
APC (W) 155+: 1 BTR-60; 2 BTR-70; 122 BTR-80; 30+ *Dozor*-B
ABCV 75: 30 BMD-1; 45 BMD-2
AUV KrAZ *Spartan*
ANTI-TANK/ANTI-INFRASTRUCTURE
MSL • MANPATS 9K111 *Fagot* (AT-4 *Spigot*); 9K111-1 *Konkurs* (AT-5 *Spandrel*)
ARTILLERY 142
SP 122mm 2S1 *Gvozdika*; **152mm** 18 2S3 *Akatsiya*
TOWED • 122mm 54 D-30
MRL 122mm 6 BM-21 *Grad*
GUN/MOR • SP • 120mm 40 2S9 NONA-S
MOR 120mm 24 2S12 *Sani*
AIR DEFENCE
SAM • Point-defence 9K35M *Strela*-10M
GUNS • SP 23mm some ZU-23-2 (truck mounted)

Special Operations Forces n.k.

FORCES BY ROLE
SPECIAL FORCES
2 SF regt

Paramilitary 102,000

National Guard ε60,000

Ministry of Internal Affairs; 5 territorial comd

FORCES BY ROLE
MANOEUVRE
Armoured
Some tk bn
Mechanised
Some mech bn
Light
Some lt inf bn
EQUIPMENT BY TYPE
ARMOURED FIGHTING VEHICLES
MBT T-64; T-64BV; T-64BM; T-72
IFV 83: BTR-3; 32+ BTR-3E1; ε50 BTR-4 *Bucephalus*; some BTR-4E; 1 BMP-2
APC 22+
APC (W) BTR-70; BTR-80
PPV 22+: Streit *Cougar*; Streit *Spartan*; 22 *Kozak*-2; *Varta*
AUV *Novator*
ANTI-TANK/ANTI-INFRASTRUCTURE
RCL 73mm some SPG-9
ARTILLERY
TOWED 122mm some D-30
MOR 120mm some
AIRCRAFT
TPT • Light 24: 20 An-26 *Curl*; 2 An-72 *Coaler*; 2 Tu-134 *Crusty*
HELICOPTERS • TPT 12: **Medium** 9: 2 H225M; 7 Mi-8 *Hip*; **Light** 3: 2 H125; 1 Mi-2MSB
AIR DEFENCE
SAM • Point-defence 9K38 *Igla* (SA-18 *Grouse*)
GUNS • SP 23mm some ZU-23-2 (truck mounted)

Border Guard ε42,000

FORCES BY ROLE
MANOEUVRE
Light
some mot inf gp
EQUIPMENT BY TYPE
ARMOURED FIGHTING VEHICLES
APC • PPV 17 *Kozak*-2

Maritime Border Guard

The Maritime Border Guard is an independent subdivision of the State Commission for Border Guards and is not part of the navy

EQUIPMENT BY TYPE
PATROL AND COASTAL COMBATANTS 21
PCT 1 *Molnya* (*Pauk* I) with 4 single 406mm TT, 2 RBU 1200 *Uragan* A/S mor, 1 76mm gun
PCC 4 *Tarantul* (*Stenka*)
PB 12: 11 *Zhuk*; 1 *Orlan*
PBR 4 *Shmel* with 1 76mm gun
LOGISTICS AND SUPPORT • AGF 1
AIRCRAFT • TPT Medium An-8 *Camp*; **Light** An-24 *Coke*; An-26 *Curl*; An-72 *Coaler*
HELICOPTERS • ASW: Ka-27 *Helix* A

DEPLOYMENT

AFGHANISTAN: NATO • *Operation Resolute Support* 13

CYPRUS: UN • UNFICYP 1

DEMOCRATIC REPUBLIC OF THE CONGO: UN • MONUSCO 259; 1 atk hel sqn

MALI: UN • MINUSMA 9

MOLDOVA: 10 obs

SERBIA: NATO • KFOR 40; **UN** • UNMIK 2

SOUTH SUDAN: UN • UNMISS 1

SUDAN: UN • UNISFA 3

FOREIGN FORCES

Canada *Operation Unifier* 200
Lithuania JMTG-U 26
Poland JMTG-U 40
United Kingdom *Operation Orbital* 53
United States JMTG-U 150

TERRITORY WHERE THE GOVERNMENT DOES NOT EXERCISE EFFECTIVE CONTROL

In late February 2014, Russian forces occupied Crimea. The region then requested to join the Russian Federation after a referendum in March, regarded as unconstitutional by the government in Kyiv. Months after Russia's annexation of Crimea, fighting began in Ukraine's Donetsk and Luhansk oblasts, with separatist forces there allegedly operating with Russian support. Conflict in the east has persisted since that date. The information displayed for these forces reflects equipment that has been observed as employed in support of the separatist cause in eastern Ukraine. Data presented here represents the de facto situation and does not imply international recognition.

EASTERN UKRAINE SEPARATIST FORCES

ORGANISATIONS BY SERVICE

Donetsk People's Republic ε20,000

FORCES BY ROLE
SPECIAL FORCES
2 (Spetsnaz) SF bn
MANOEUVRE
Reconnaissance
1 recce bn
Armoured
1 tk bn
Light
6 mot inf bde
COMBAT SUPPORT
1 arty bde
1 engr coy
1 EW coy
COMBAT SERVICE SUPPORT
1 log bn
AIR DEFENCE
1 AD bn

Luhansk People's Republic ε14,000

FORCES BY ROLE
MANOEUVRE
Reconnaissance
1 recce bn
Armoured
1 tk bn
Light
4 mot inf bde
COMBAT SUPPORT
1 arty bde
1 engr coy
1 EW coy
COMBAT SERVICE SUPPORT
1 log bn
AIR DEFENCE
1 AD bn

EQUIPMENT BY TYPE
ARMOURED FIGHTING VEHICLES
MBT T-64BV; T-64B; T-64BM†; T-72B1; T-72BA
RECCE BDRM-2
IFV BMP-1; BMP-2; BTR-4
APC
APC (T) BTR-D; MT-LB; GT-MU
APC (W) BTR-60; BTR-70; BTR-80
ABCV BMD-1, BMD-2
ANTI-TANK/ANTI-INFRASTRUCTURE
MSL 9K115 *Metis* (AT-7 *Saxhorn*); 9K135 *Kornet* (AT-14 *Spriggan*)
RCL 73mm SPG-9
GUNS 100mm MT-12
ARTILLERY
SP 122mm 2S1 *Gvozdika*; 152mm 2S3 *Akatsiya*; 2S19 *Msta*-S†; **203mm** 2S7 *Pion*
TOWED 122mm D-30; **152mm** 2A65 *Msta*-B
GUN/MOR
SP 120mm 2S9 NONA-S
TOWED 120mm 2B16 NONA-K
MRL 122mm BM-21 *Grad*
MOR 82mm 2B14; **120mm** 2B11 *Sani*
AIR DEFENCE
SAM
Short-range 9K331M *Tor*-M2 (SA-15 *Gauntlet*)
Point-defence 2K22 *Tunguska* (SA-19 *Grison*); 9K32M *Strela*-2M (SA-7B *Grail*); 9K33 *Osa* (SA-8 *Gecko*); 9K35 *Strela*-10 (SA-13 *Gopher*); 9K38 *Igla* (SA-18 *Grouse*); GROM
GUNS
SP 23mm ZU-23-2 (tch/on MT-LB)
TOWED 14.5mm ZPU-2; **57mm** S-60

FOREIGN FORCES

Russia Crimea: 28,000; 1 recce bde, 2 naval inf bde; 1 arty bde; 1 NBC bde; 40 T-72B3; 80 BMP-2; 200 BTR-82A; 20 BTR-80: 150 MT-LB; 18 2S1; 18 2A65; 12 BM-21 MRL; 8 9P140; 1 AShM bde with 3K60 Bal; 3K55 *Bastion*; 1 FGA regt with Su-24M/MR; Su-30SM; 1 FGA regt with Su-27SM/SM3; Su-30M2; 1 FGA regt with Su-24M/Su-25SM; 1 atk/tpt hel regt; 1 ASW hel regt; 1 AD regt with S-300PM; 1 AD regt with S-400; 1 Fleet HQ located at Sevastopol; 2 radar stations located at Sevastopol (*Dnepr* system) and Mukachevo (*Dnepr* system) • Donetsk/Luhansk: 3,000 (reported)

Uzbekistan UZB

Uzbekistani Som s		2019	2020	2021
GDP	s	512tr	598tr	
	US$	57.9bn	59.8bn	
per capita	US$	1,742	1,763	
Growth	%	5.6	0.7	
Inflation	%	14.5	13.0	
Def exp	s	n.k	n.k	
	US$	n.k	n.k	
US$1=s		8836.83	10008.30	

Population 30,565,411

Age	0–14	15–19	20–24	25–29	30–64	65 plus
Male	11.9%	4.0%	4.5%	5.0%	21.8%	2.5%
Female	11.3%	3.8%	4.3%	5.0%	22.5%	3.4%

Capabilities

Uzbekistan introduced a new military doctrine in early 2018, which highlighted increased concern over terrorism and the potential impact of conflicts including in Afghanistan. It noted a requirement for military modernisation. The doctrine also focuses on border security and hybrid-warfare concerns. Uzbekistan is a member of the SCO, but suspended its CSTO membership in 2012. It maintains bilateral defence ties with Moscow and in late 2018 a defence-cooperation agreement was reported with India. A protocol on military-education cooperation was ratified in Turkey in 2020. The armed forces are army-dominated and conscript-based. Uzbekistan has a limited capacity to deploy its forces externally and does not have any international deployments. A sizeable air fleet was inherited from the Soviet Union, but minimal recapitalisation in the intervening period has substantially reduced the active inventory. Logistical and maintenance shortcomings hinder aircraft availability. Uzbekistan is reliant on foreign suppliers for advanced military equipment, and procured equipment including military helicopters and armoured personnel carriers from Russia in 2019. Meetings took place with India in 2020 to advance defence cooperation (three defence-related MOUs were signed in 2019). A State Committee for the Defence Industry was established in late 2017 to organise domestic industry and defence orders. The 2018 doctrine calls for improvements to the domestic defence industry.

ACTIVE 48,000 (Army 24,500 Air 7,500 Joint 16,000)
Paramilitary 20,000

Conscript liability 12 months

ORGANISATIONS BY SERVICE

Army 24,500

4 Mil Districts; 2 op comd; 1 Tashkent Comd

FORCES BY ROLE

SPECIAL FORCES
1 SF bde

MANOEUVRE
Armoured
1 tk bde
Mechanised
11 MR bde
Air Manoeuvre
1 air aslt bde
1 AB bde
Mountain
1 lt mtn inf bde

COMBAT SUPPORT
3 arty bde 1 MRL bde

EQUIPMENT BY TYPE

ARMOURED FIGHTING VEHICLES
MBT 340: 70 T-72; 100 T-64B/MV; 170 T-62
RECCE 19: 13 BRDM-2; 6 BRM-1
IFV 270 BMP-2
APC 388
APC (T) 50 BTR-D
APC (W) 259: 24 BTR-60; 25 BTR-70; 210 BTR-80
PPV 79: 50 *Maxxpro+*; 24 *Ejder Yalcin*; 5 *Typhoon*-K
ABCV 129: 120 BMD-1; 9 BMD-2
AUV 11+: 7 *Cougar*; 4+ M-ATV; some *Tigr*-M

ENGINEERING & MAINTENANCE VEHICLES
ARV 20 *Maxxpro* ARV

ANTI-TANK/ANTI-INFRASTRUCTURE
MSL • MANPATS 9K11 *Malyutka* (AT-3 *Sagger*); 9K111 *Fagot* (AT-4 *Spigot*)
GUNS 100mm 36 MT-12/T-12

ARTILLERY 487+
SP 83+: **122mm** 18 2S1 *Gvozdika*; **152mm** 17+: 17 2S3 *Akatsiya*; 2S5 *Giatsint*-S (reported); **203mm** 48 2S7 *Pion*
TOWED 200: **122mm** 60 D-30; **152mm** 140 2A36 *Giatsint*-B
GUN/MOR 120mm 54 2S9 NONA-S
MRL 108: **122mm** 60: 36 BM-21 *Grad*; 24 9P138; **220mm** 48 9P140 *Uragan*
MOR 120mm 42: 5 2B11 *Sani*; 19 2S12 *Sani*; 18 M-120

AIR DEFENCE • SAM
Point-defence QW-18 (CH-SA-11)

Air Force 7,500

FORCES BY ROLE

FIGHTER
1 sqn with MiG-29/MiG-29UB *Fulcrum* A/B;
1 sqn with Su-27/Su-27UB *Flanker* B/C

GROUND ATTACK
1 sqn with Su-24 *Fencer*
1 sqn with Su-25/Su-25BM *Frogfoot*

ELINT/TRANSPORT
1 regt with An-12/An-12PP *Cub*; An-26/An-26RKR *Curl*

TRANSPORT
Some sqn with An-24 *Coke*; C295W; Tu-134 *Crusty*

TRAINING
1 sqn with L-39 *Albatros*

ATTACK/TRANSPORT HELICOPTER
1 regt with Mi-24 *Hind*; Mi-26 *Halo*; Mi-8 *Hip*;
1 regt with Mi-6 *Hook*; Mi-6AYa *Hook* C

EQUIPMENT BY TYPE

AIRCRAFT 41 combat capable
FTR 12 MiG-29/MiG-29UB *Fulcrum* A/B (18 more in store)
FGA 13 Su-27/Su-27UB *Flanker* B/C (11 more in store) (26 Su-17M (Su-17MZ)/Su-17UM-3 (Su-17UMZ) *Fitter* C/G non-operational)

ATK 16: 12 Su-25/Su-25BM *Frogfoot*; 4 Su-24 *Fencer*
EW/Tpt 26 An-12 *Cub* (med tpt)/An-12PP *Cub* (EW)
ELINT/Tpt 13 An-26 *Curl* (lt tpt)/An-26RKR *Curl* (ELINT)
TPT 7: **Heavy** 1 Il-76 *Candid*; **Light** 6: 1 An-24 *Coke*; 4 C295W; 1 Tu-134 *Crusty*
TRG 14 L-39 *Albatros*

HELICOPTERS
ATK 33: 29 Mi-24 *Hind*; 4 Mi-35M *Hind*
TPT 69: **Heavy** 9: 8 H225M *Caracal*; 1 Mi-26 *Halo*; **Medium** 52 Mi-8 *Hip*; **Light** 8 AS350 *Ecureuil*

AIR DEFENCE • **SAM** 45
Long-range S-200 *Angara* (SA-5 *Gammon*); FD-2000 (HQ-9)
Medium-range S-75 *Dvina* (SA-2 *Guideline*)
Short-range S-125 *Pechora* (SA-3 *Goa*)

AIR-LAUNCHED MISSILES
AAM • **IR** R-60 (AA-8 *Aphid*); R-73 (AA-11A *Archer*); **IR/SARH** R-27 (AA-10 *Alamo*)
ASM Kh-23 (AS-7 *Kerry*); Kh-25 (AS-10 *Karen*)
ARM Kh-25MP (AS-12A *Kegler*); Kh-28 (AS-9 *Kyle*); Kh-58 (AS-11 *Kilter*)

Paramilitary up to 20,000

Internal Security Troops up to 19,000
Ministry of Interior

National Guard 1,000
Ministry of Defence

Arms procurements and deliveries – Russia and Eurasia

Significant events in 2020

APRIL

BUMERANG OFFERED FOR EXPORT

Rosoboronexport launched an international marketing and sales campaign for the *Bumerang* 8×8 armoured personnel carrier developed by the Moscow-based Military Industrial Company. The firm hopes that the *Bumerang* will raise around US$1bn in export sales, though potential customers may have to wait for the vehicle to complete its development. *Bumerang* was first displayed in public at the 2015 Victory Day parade. Like other new vehicles first shown on that day, such the *Armata* and *Kurganets*, *Bumerang's* entry into service has been delayed and the vehicle has been redesigned to make it wider, in order to improve protection and buoyancy for river crossings. The plan is for state tests to be complete by the end of 2021.

MAY

UKROBORONPROM REFORM

Ukraine's government outlined plans to reform UkrOboronProm, the state-owned defence conglomerate. This will begin with management changes in 2021 before transferring its 137 current enterprises into six holding companies: missiles; aviation; aircraft repair; armoured vehicles; radars; and air-defence and marine systems. Many of these 137 are financially unstable and some are in parts of the country no longer under the control of the central government. After assessing each member company's financial stability, a restructuring model will be issued. This phase of the reform is planned to be complete by the end of 2022.

JULY

MiG–SUKHOI MERGER

Further details were announced about the planned 'merger' between MiG and Sukhoi, first announced in 2017. The plan is that after the merger, which is scheduled to be completed by the end of 2020, the new company will continue to produce aircraft under both brands. The merged companies will be tasked with developing a sixth-generation combat aircraft. The merger, which may be more a case of Sukhoi absorbing what is left of MiG, saves the latter from potential bankruptcy after the Russian government selected Sukhoi to build the Su-57 fighter. MiG's most modern fighter is the MiG-35, an updated version of the MiG-29, but this has not been ordered in large enough quantities to secure the firm's future, with only two prototypes and six production aircraft contracted so far for the Russian Air Force.

AUGUST

KPE FINANCIAL TROUBLE

A reduction in the number of *Arlan* 4×4 protected patrol vehicles planned to be ordered by the Kazakhstan Ministry of Defence (MoD) for 2020–21 raises the risk that the Kazakhstan Paramount Engineering (KPE) joint venture could face bankruptcy and closure at the end of 2020, leaving the firm with t9bn (US$23.43m) of debt and causing the loss of 200 jobs. According to the July 2017 memorandum of understanding between KPE and the MoD, Kazakhstan was to acquire 274 *Arlan* 4×4 patrol vehicles and 106 *Barys* 8x8 infantry fighting vehicles up to 2030. However, with Kazakhstan hard hit by the fall in oil revenues, as well as negative economic impacts from the coronavirus pandemic, the MoD amended the order from 30 to 11 *Arlans*. KPE is one of several efforts by the Kazakh government to boost local industrial capability by teaming with foreign companies.

AUGUST

NAVAL HELICOPTER R&D

Russian Helicopters was awarded a contract to begin development of a new naval helicopter. Design work for the *Minoga* programme will be completed by the recently formed Mil and Kamov National Centre for Helicopter Engineering. The first flight of the prototype is planned for 2025. The *Minoga* will replace the approximately 100 Ka-27 and Ka-29 *Helix* helicopters in Russian Navy service. The Ka-29 is an improved version of the Ka-27, but production ended in the early 1990s. Kumertau Aviation Production Enterprise does manufacture a civilian variant of the Ka-27, the Ka-32, meaning that there is potentially some capability to produce new Ka-27 helicopters if required. However, the several warships that Russia has exported within the past decade have not come with a naval helicopter as part of the package.

Table 6: Armenia: selected defence procurement since 2010

Contract Date	Equipment	Type	Quantity	Prime contractor	Deliveries
c. 2012	T-72B1	Main battle tank	35	Russian government surplus	2013
	n.k.	Armoured vehicle	110		
2015	9K720 *Iskander*-E	Short-range ballistic missile	≥2	KB Mashinostroyenia	2016
2015	9A52 *Smerch*	300mm multiple rocket launcher	6	Splav	2016
2015	9K135 *Kornet*-E (AT-14 *Spriggan*)	Man-portable anti-tank missiles	n.k.	KBP Instrument Design Bureau	2017
c. 2018	9K33 *Osa*-AK (SA-8 *Gecko*)	Point-defence surface-to-air missile system	≥4	Russian government surplus	2019
c. 2018	9K331MKM *Tor*-M2KM (SA-15 *Gauntlet*)	Short-range surface-to-air missile system	≥2	Almaz-Antey	2019
2019	Su-30SM *Flanker* H	Fighter/ground attack aircraft	4	Irkut Corporation	2019

Table 7: Azerbaijan: selected defence procurement since 2010

Contract date	Equipment	Type	Quantity	Prime contractor	Deliveries
c. 2010	T-72	Main battle tank	93	Belarusian government surplus	2011–12
2010	Mi-35M *Hind*	Attack helicopter	24	Rostvertol	2011–13
2010	Mi-17-1V	Multi-role helicopter	≥40	Kazan Helicopter Plant	2011–13
2011	T-90S	Main battle tank	≥100	UralVagonZavod	2012–15
	BMP-3	Infantry fighting vehicle	≥88	Kurganmashzavod	
	TOS-1A	220mm multiple rocket launcher	≥18	UralVagonZavod	
	9A52 *Smerch*	300mm multiple rocket launcher	≥18	Splav	
	2S19 *Msta*-S	152mm self-propelled howitzer	≥18	UralVagonZavod	
	2S31 *Vena*	120mm self-propelled gun/mortar	≥18	Kurganmashzavod	
2011	*Tufan* (*Sa'ar* 62)	Guided-missile patrol craft	6	Israel Shipyards	2015–19
	Shaldag V	Fast guided-missile patrol boat	6		
2012	*Abisr* (*Barak*-MRAD)	Short-range surface-to-air missile system	n.k.	Israel Aerospace Industries (IAI)	2016
2013	T-300 *Kasirga*	302mm multiple rocket launcher	21	Roketsan	2015–16
2013	BTR-82A	Infantry fighting vehicle	230	Arzamas Machine-Building Plant	2013–16
2014	9P157-2 *Khrizantema*-S (AT-15 *Springer*)	Self-propelled anti-tank missile system	n.k.	KB Mashinostroyenia	2017
2016	DANA M1M	152mm self-propelled howitzer	≥9	Excalibur Army	2017
	RM-70 *Vampir*	122mm multiple rocket launcher	≥8		
c. 2017	SPEAR	120mm self-propelled mortar	≥3	Elbit Systems	2018
c. 2017	*Hermes* 900	Medium ISR UAV	≥1	Elbit Systems	2018
c. 2017	*Barak*-LRAD	Self-propelled medium-range surface-to-air missile system	n.k.	Israel Aerospace Industries (IAI)	2018
c. 2017	*Polonez*	300mm multiple rocket launcher	≥6	Precise Electromechanics Factory (ZDEM)	2018
c. 2017	LORA	Short-range ballistic missile	≥2	Israel Aerospace Industries (IAI)	2018

Figure 13: Russia: attack-helicopter programmes

Russia is currently producing three types of attack helicopter for its Aerospace Forces (VKS): the Ka-52A *Hokum*-B, Mi-28 *Havoc* and Mi-35M *Hind*. In contrast, the United States is producing two types, with each being delivered to a different service, while China is also producing two (the WZ-10 and WZ-19 mainly to the army, with a WZ-10 variant to the air force). The Soviet Union began work on what was to become the Mi-24 *Hind* in 1967. A decade later it also issued a requirement for a new dedicated combat helicopter which led to the Mi-28 *Havoc* and the Kamov Ka-50 *Hokum*. Neither the Mi-28 or the Ka-50 entered series production during the Soviet era. Over 3,000 Mi-24s were built before 1991. Export orders fell in the 1990s, but manufacture of the Mi-35M after the mid-2000s has revitalised the only *Hind* production line, though on a substantially smaller scale than during the Cold War.

A requirement for a dedicated anti-tank helicopter, similar to the US AH-64 *Apache*, spurred the development of both the Ka-50 *Hokum*-A and the Mi-28 *Havoc*. Kamov's Ka-50 was built as a single-seater and featured the bureau's characteristic co-axial contra-rotating rotors, with the reduction in torque negating the need for a tail rotor. The Mi-28 featured a more traditional tandem cockpit for both pilots. A 1987 directive to develop versions capable of night operations was hindered by the Soviet collapse and subsequent economic problems. In the 1990s, both types were built in small quantities, although only the Ka-50 entered service. Development of the Mi-28N and the Ka-52A resumed in the 2000s with the addition of modern subsystems. Close to 100 Mi-28Ns and over 120 Ka-52As have now been delivered to the Russian armed forces with several dozen of each also delivered to export customers. An improved Mi-28NM and a navalised Ka-52K are under development.

A development contract for a new high-speed attack helicopter was awarded to Russian Helicopters in 2017 with the Mil design bureau reportedly taking the lead and a first flight planned to take place soon after 2025. In 2019, Kamov and Mil merged into the National Helicopter Centre and this new entity will now lead the work. Whether Russia will be able to maintain two separate attack-helicopter production lines will in all likelihood depend as much on the Russian economy and defence budget, as it does on identifying enough potential export customers.

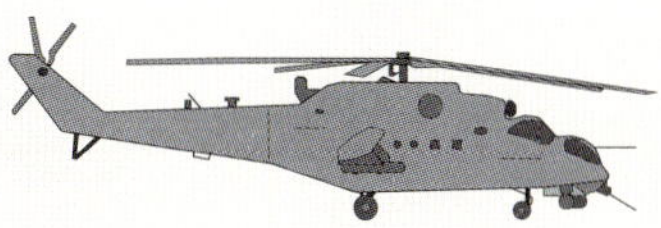

Mi-35M (*Hind*)
Designer: Mil
Manufacturer: Rostvertol
ISD Russia: 2011
Other operators:

- Azerbaijan
- Brazil
- Iraq
- Kazakhstan
- Mali
- Nigeria
- Pakistan
- Serbia
- Uzbekistan
- Venezuela

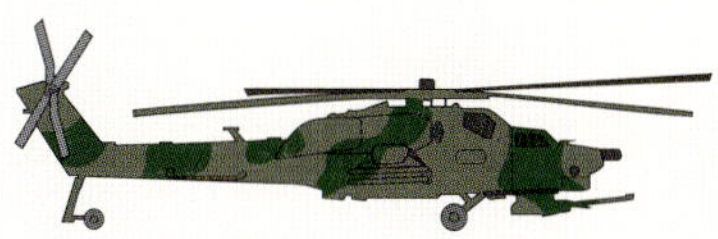

Mi-28N (*Havoc* B)
Designer: Mil
Manufacturer: Rostvertol
ISD Russia: 2009
Other operators:

- Algeria
- Iraq

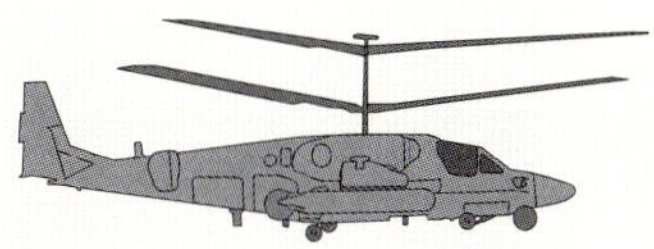

Ka-52A (*Hokum* B)
Designer: Kamov
Manufacturer: Progress Arsenyev
ISD Russia: 2009
Other operators:

- Egypt

Estimated Russian attack-helicopter deliveries for domestic customers

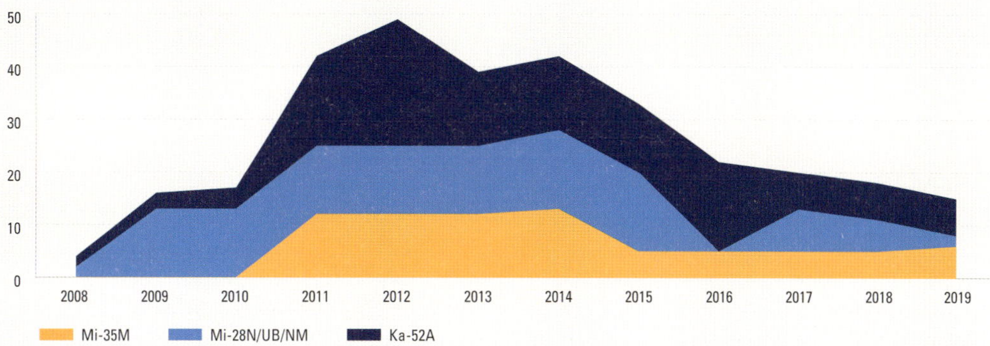

© IISS

Chapter Six

Asia

- Improving maritime surveillance and situational awareness remains a priority for some regional states, both by procuring new military platforms and, in some cases, by developing satellite capabilities. A year after Vietnam published its first defence white paper in ten years, Hanoi in April 2020 ordered a new satellite-based surveillance system. A number of other regional states are pursuing satellite capabilities; some are dual-use to meet civil and military surveillance purposes.
- China continued to exert pressure on regional states, with the People's Liberation Army Navy (PLAN) maintaining an 'over the horizon' presence and Beijing's maritime paramilitary forces in the lead, in some cases harassing and intimidating foreign vessels and also probing other states' exclusive economic zones.
- Australia's July 2020 'Defence Strategic Update' outlined plans to adjust Australia's defence posture to develop a more powerful and self-reliant deterrent. An accompanying force structure plan identified technology and capability areas for investment, including high-speed missile systems and 'capabilities to counter emerging space threats'.
- China's PLA continues to test its structures and capabilities. Exercises in the Eastern and Southern Theatre Commands focused primarily on what appeared to be Taiwan scenarios, including amphibious operations. These contained opposing-force drills, a development which has again been seen in China's brigade-level army exercises, indicating that the PLA continues its focus not only on integrated operations involving more than one theatre command, but also on more realistic drills.
- The PLA continues to boost its numbers of modern platforms, particularly in the air force and navy. The air force's fleet of heavy transport aircraft has effectively doubled in size since 2016, with Y-20As now outnumbering Il-76s in the inventory.

Asia defence spending, 2020 – top 5

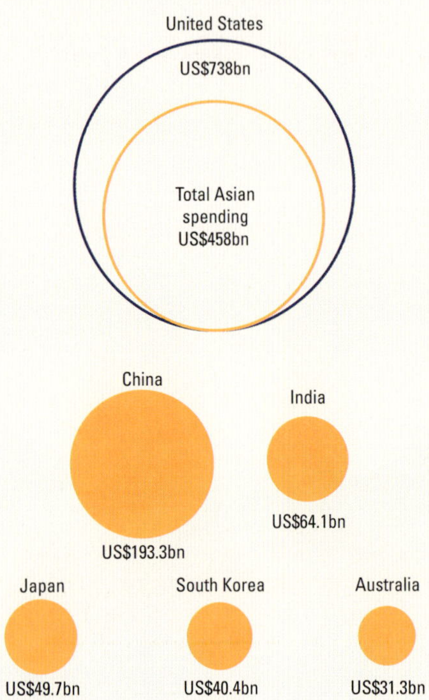

Active military personnel – top 10

(25,000 per unit)

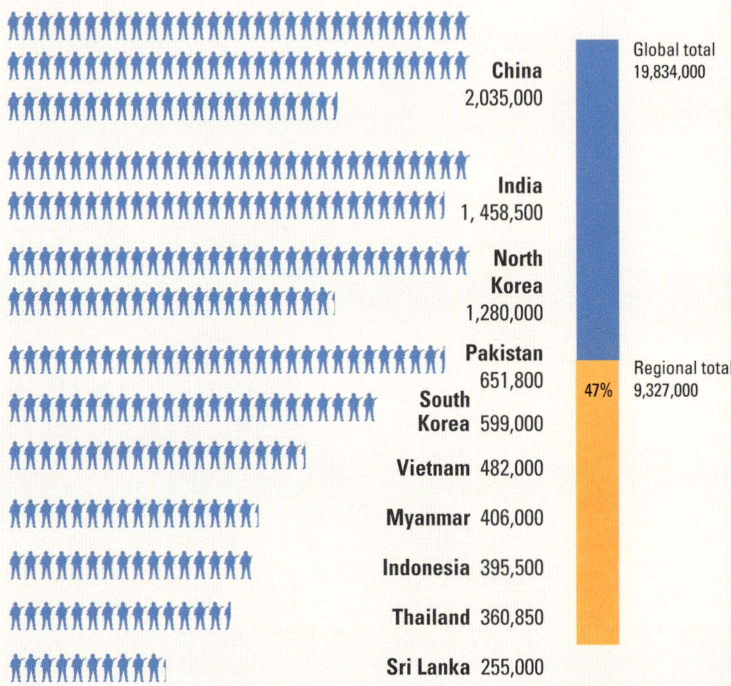

China's heavy airlift fleet, 2016–20

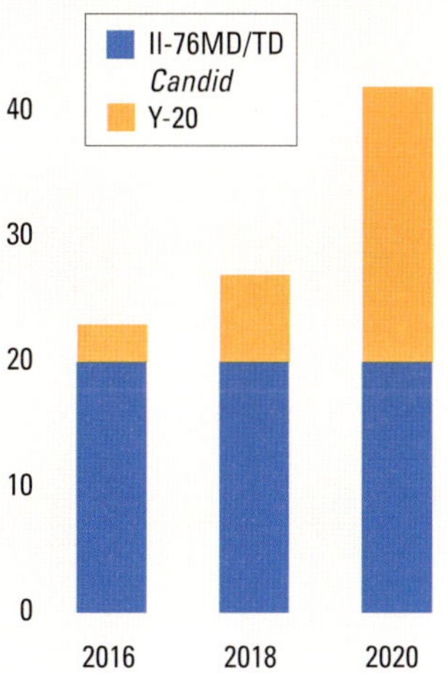

China's medium airlift fleet, 2016–20

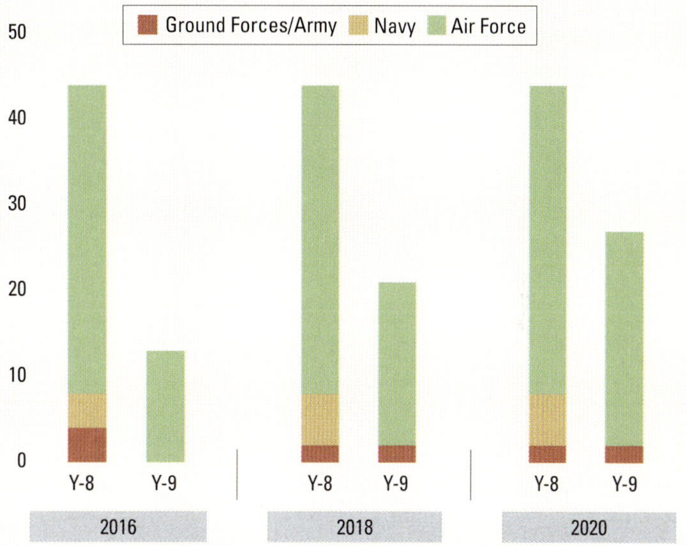

China's corvettes and offshore patrol ships 2016–20

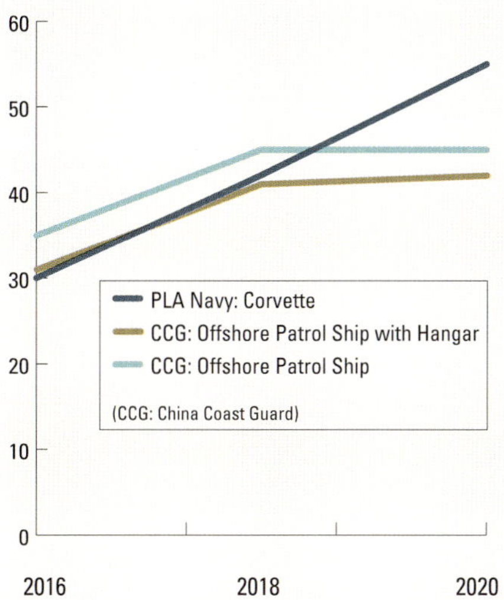

Australia, China and Japan: anti-submarine warfare (ASW) and maritime patrol aircraft (MPA) 2016–20

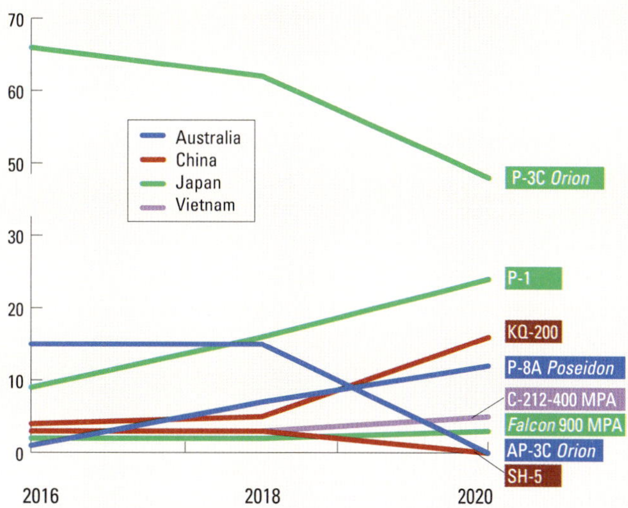

Asia

During 2020, Asia's security environment was dominated by China's drive to become the region's dominant power, the Trump administration's robust policies towards Beijing as well as its own allies and partners, North Korea's continuing missile and nuclear-weapons programme, and also sub-regional disputes and internal security challenges. However, the coronavirus outbreak in China at the start of the year, which quickly spread internationally and became a global pandemic, significantly affected the region's armed forces.

Asian armed forces and the coronavirus pandemic

Armed forces in many Asian states have supported governments' efforts to manage the consequences of the pandemic. In China, the People's Liberation Army (PLA) deployed medical units to support the public-health response to the initial outbreak in Wuhan and provided logistical support by transporting medical supplies and building temporary hospitals. In Malaysia, the army helped the police to enforce a nationwide 'movement control order' and the air force helped distribute supplies, while in Taiwan conscripts boosted the workforce producing face masks. In Singapore, the armed forces played an important part in the city-state's whole-of-government response, notably by packing face masks and delivering them to distribution points and also by temporarily housing in military accommodation large numbers of foreign workers displaced from dormitories with a high incidence of infection.

Activities related to the coronavirus response did not themselves significantly undermine the capacity of regional states' forces in performing their normal roles, though outbreaks in military establishments did temporarily affect operational readiness. The most public manifestation of this came in March and April, when sailors on a number of ships in the United States' 7th Fleet fell ill and the crew of the aircraft carrier USS *Theodore Roosevelt* was evacuated at Guam. Nonetheless, by June the US was able to send three carrier strike groups to sea in the Indo-Pacific region for the first time since 2017. This move was widely interpreted as a demonstration of continuing US military commitment to the region in the face of China's escalating geopolitical extroversion.

Fear of infection spreading among military personnel led many national defence establishments to curtail routine training and suspend national exercises. Although the annual US-led multilateral *Cobra Gold* exercise in Thailand proceeded in late February and early March, some land-based joint exercises were cancelled, including US drills with South Korea and the Philippines, while others were delayed or restructured. The pandemic also affected regional defence diplomacy. With international travel increasingly difficult or, in some cases, impossible during the second quarter of 2020, defence ministers, senior officials and military commanders turned to virtual communications in order to maintain their regional and international links.

In the medium to long term, the most important defence-related effect of the pandemic will – as in other regions – be the extent to which it leads governments to reduce military budgets, as they look to spend more on the health sector and on measures to support the wider economy. It is also widely anticipated that reduced economic growth in many countries will significantly constrain government spending, with the defence sector often carrying at least a proportionate burden of budget cuts. Even in the short term, the crisis affected some Asian states' defence spending. Indonesia, South Korea and Thailand had, by June, all announced cuts in their planned military budgets for 2020 as governments re-directed funds to crisis-relief measures.

China's operations challenge its neighbours

Strategic competition in the region intensified in 2020, with China apparently intent on asserting primacy in its maritime littoral. The People's Liberation Army Navy maintained an 'over-the-horizon' presence, and China's maritime paramilitary forces took the lead. China maintained pressure on its main South China Sea adversary, Vietnam, and in early April a vessel from the China Coast Guard (CCG) sank a Vietnamese fishing boat near the Paracel Islands. Indonesia and Malaysia are now effectively also 'front-line states', and vessels from the CCG and China's maritime militia have probed both nations' exclusive economic zones, using facilities on Chinese-occupied features in the Spratly Islands as forward bases. During August, four Chinese ballistic-missile tests ended in splashdowns between Hainan and the Paracel Islands. Some observers interpreted the tests of these missiles, which could perhaps be employed in war to target aircraft carriers and other large ships, as signalling to the US the potential cost if it intervened in a future South China Sea conflict.

China also kept up pressure in the East China Sea. In early May, CCG ships harassed a Japanese fishing boat inside the territorial waters of the Senkaku/Diaoyu islands (administered by Tokyo, but claimed by Beijing), leading to a stand-off with the Japan Coast Guard (JCG), which reported in late June that it had observed Chinese government vessels in waters close to the Senkaku/Diaoyu islands for a record 70 consecutive days. In mid-July, Tokyo's annual defence white paper accused China of 'relentlessly' attempting to change the status quo in the East China Sea, as well as trying to 'create a fait accompli' in the South China Sea. In early August, Japan's then-defence minister, Kono Taro, warned China that the Japan Self-Defense Force would 'act firmly when necessary' in support of the JCG.

China also stepped up efforts to intimidate Taiwan. From early February onwards, Chinese combat aircraft flew close to or within Taiwan's Air Defence Identification Zone, crossing the 'median line' in the Taiwan Strait numerous times. It was reported in September that on one occasion 19 Chinese aircraft entered Taiwan's airspace. That month, a spokesman for China's foreign ministry said that 'the so-called median line is non-existent'. During April, at a time when the coronavirus pandemic had undermined the operational readiness of the US Navy in the region, the Chinese aircraft carrier *Liaoning* and its escorts sailed close to Taiwan as part of a month-long deployment.

Some observers interpreted this as a demonstration not only to Taipei but also to Japan and Southeast Asian countries (particularly Vietnam) of Beijing's growing capacity for power projection. In early August, a report by Taipei's defence ministry on China's military capabilities claimed that the PLA's existing amphibious and logistical capabilities were insufficient for the task of invading Taiwan, notwithstanding the launch of China's second Type-075 (*Yushen*) amphibious assault ship in April. However, the report argued that Beijing could use long-range missile strikes, as well as sea and air blockades and the occupation of Taiwan's outlying islands, to undermine Taipei's resistance.

Emphasising the extent of China's geopolitical offensive, along with growing strategic tensions with India, from early May onwards there was a series of clashes on the two countries' Himalayan border. In mid-June, clashes between PLA and Indian troops reportedly led to casualties on both sides in the Galwan Valley between India's Ladakh territory and the Chinese-occupied Aksai Chin region. Both sides subsequently strengthened their military positions on the border despite talks aimed at de-escalation and, in early September, shots were exchanged across the Line of Actual Control, the notional demarcation line between Indian- and Chinese-controlled territory, for the first time in 45 years.

Korean Peninsula

While Pyongyang maintained its moratorium on intercontinental ballistic missile (ICBM) tests, which has been in place since early 2018, from May 2019 it resumed testing other types of missiles. Following the test of the new *Pukguksong*-3 submarine-launched ballistic missile (SLBM) in October 2019, other tests in October and November 2019 – and again in March 2020 – involved a mix of short-range ballistic missiles and projectiles launched from multiple-rocket launchers. North Korea simultaneously continued its nuclear-weapons programme. In August 2020, a report by a UN expert panel said that Pyongyang had 'probably developed' miniaturised nuclear warheads that could be fitted to ballistic missiles. Japan's 2020 defence white paper suggested that North Korea already had 'the ability to attack Japan with nuclear weapons' fitted to *Nodong* and *Scud*-ER ballistic missiles. Meanwhile, the 10 October 2020 parade marking the 75th anniversary of the Korean Workers Party showcased several apparently new systems, including what appeared to be a large ICBM and a *Pukguksong*-4 SLBM, as well as several previously unseen designs for a main battle tank, other armoured vehicles, and ground-based air-defence systems.

Deterring Pyongyang remained the focus of Seoul's defence policy. North Korea's demolition of the inter-Korean Joint Liaison Office in June underscored the hostility it had displayed towards Seoul since the failure of the second US–North

Korea summit in 2019. Meanwhile, Seoul continues to make capability-improvement efforts under its Defense Reform 2.0 programme. While there were relatively minor cuts to defence spending during 2020 because of the coronavirus pandemic, the defence-budget proposal for 2021–25, released in early August, increased by 3.5% the funding proposed a year earlier in the previous mid-term defence plan. Although marginally less was allocated for modernisation, over the coming decade South Korea's armed forces will still benefit from the arrival of new equipment, much of it supplied by the domestic defence industry. Efforts to expand naval capabilities are motivated at least partially by competition with Japan. In early October, South Korea's Joint Chiefs of Staff requested the acceleration of plans to design and build the 30–35,000-tonne LPX-II light aircraft carrier. This is expected to operate F-35B combat aircraft as well as helicopters when it enters service in the early 2030s. Amid the emphasis on re-equipping South Korea's armed forces, some observers expressed concern that the cancellation or alteration of national and bilateral US–South Korean exercises – resulting from the relative warming in 2018 of diplomatic relations with Pyongyang as well as more recently from the coronavirus pandemic – could have undermined their operational readiness.

Southeast Asian developments

Concerned by the deteriorating regional security environment, Singapore and Vietnam continued to demonstrate greater urgency than their Southeast Asian neighbours in developing their defence capabilities. The defence allocation in Singapore's February 2020 budget declined by 2.5% in nominal terms, when compared with the previous year, because of a slowdown in economic growth. However, major procurement projects were set to continue. Importantly, following a Letter of Request from Singapore in April 2019, in January 2020 the US Defense Security Cooperation Agency approved the potential sale to the city state of four F-35B combat aircraft, together with an option on a further eight. Soon afterwards, Singapore's Chief of Air Force, Major General Kelvin Khong, said that the first four F-35Bs would be based in the US for training and testing – an approach closely following that adopted by Singapore with previous purchases of US military aircraft. The revelation that Singapore will buy the short-take-off and vertical-landing version of the F-35 lent credence to long-standing speculation that it is keeping open the option of operating the aircraft from its navy's planned through-deck Joint Multi Mission Ships.

In late November 2019, Vietnam published its first defence white paper for a decade, against the backdrop of escalating pressure from Beijing on its maritime interests, seen particularly in the stand-off between Vietnamese and Chinese coastguard vessels at Vanguard Bank in the Spratly Islands earlier in the year. While still lacking details of military spending and capability-development efforts, unlike its predecessor the 2019 document several times referred explicitly to Vietnam's differences with China and the need for both sides to 'put more effort into maintaining stability'. It is evident that, from Hanoi's perspective, stability will depend in good measure on Vietnam bolstering its deterrent capabilities in the South China Sea. While long-expected orders for major equipment such as new combat and maritime-patrol aircraft have still failed to materialise – perhaps reflecting budgetary constraints – in April 2020 Vietnam ordered a Japanese-produced satellite-based surveillance system equipped with X-band synthetic aperture radar. While the manufacturer, NEC, has stressed the civil applications of this LOTUSat-1 earth-observation system, which is due to start operating in 2023, it seems almost certain to boost the armed forces' situational awareness, particularly in the South China Sea. In another potentially significant development, in September the Viettel Military Industry and Telecoms Group (owned by Vietnam's Ministry of National Defence) displayed a mock-up of a locally developed medium-sized uninhabited aerial vehicle (UAV) fitted with air-to-surface weapons and a forward upper fuselage bulge that could house a satellite datalink antenna.

In December 2019, Malaysia's first-ever defence white paper was tabled by the Pakatan Harapan government, a coalition of former opposition parties which came to power in 2018. The white paper described the 'three pillars' of Malaysia's defence strategy as 'concentric deterrence' (the armed forces' protection of national interests in 'core', 'extended' and 'forward' zones); 'comprehensive defence' (involving whole-of-government and whole-of-society support for the national defence effort); and 'credible partnerships' (involving engagement in regional and wider international defence cooperation). To provide a framework for assessing the country's geopolitical circumstances and defence requirements, it also promoted the idea of Malaysia as 'a maritime nation with continental roots' and as 'a bridging linchpin' between the Indian and Pacific oceans. However, while the paper also identified tensions in the South China Sea, terrorism and cyber threats as posing challenges for Malaysia's defence policy, it provided no detailed insights into future resource allocation or capability development. Moreover, in late February 2020 the governing coalition collapsed, leading to a new and less reform-minded administration. That said, whatever government was in power, resources for defence would likely be constrained, ambitious re-equipment programmes unlikely and external defence capabilities limited, despite incidents such as the month-long intrusion by a Chinese survey ship into Malaysia's Exclusive Economic Zone in April–May 2020. However, a priority inherited by the latest government was to salvage the troubled Littoral Combat Ship programme, under which the local Boustead Naval Shipyard (BNS) was supposed to build six vessels based on the *Gowind* 2500 corvette produced by France's Naval Group. In August, defence minister Ismail Sabri told parliament that the government was considering three options, two of which involved assigning either Naval Group or BNS to oversee completion of at least two ships; alternatively, the government might take control of the project.

Australian plans

In July 2020, Australia published a Defence Strategic Update, which said that the country's strategic environment had deteriorated more rapidly than was anticipated when the latest Defence White Paper was published in 2016. It outlined plans to adjust Australia's defence posture in order to develop a more powerful and self-reliant military deterrent. Although there were few direct references to China, it was clear that it was the main focus of Canberra's external security concerns. The document pointed to an increased prospect of war in Australia's region, claimed that a ten-year 'strategic warning time' could no longer be assumed and highlighted potential threats from 'grey-zone activities'. It also argued that threats to human security, such as pandemics and natural disasters, should be assigned higher priority in defence planning and highlighted three new strategic objectives to guide defence planning: to shape Australia's strategic environment, to deter actions against Australia's interests and to respond with 'credible military force' if necessary.

The implications of the Defence Strategic Update for the Australian Defence Force (ADF) will include a sharpened geographical focus on Australia's 'immediate region' ('from the north-eastern Indian Ocean through maritime and mainland Southeast Asia to Papua New Guinea and the South West

Brunei: defending the sultanate

Brunei Darussalam is Southeast Asia's smallest state in terms of population and second-smallest by area, but its government – in which Sultan Hassanal Bolkiah is Minister of Defence – places a high priority on defence. The well-trained, professional Royal Brunei Armed Forces (RBAF) comprise a brigade-sized army, a navy centred on four corvettes armed with *Exocet* anti-ship missiles, and an air force mainly operating support helicopters and short-range air-defence missiles. Similar to Singapore – with which it maintains close bilateral defence relations – Brunei emphasises diplomacy, deterrence and 'holistic defence' as key defence policy principles. In March 2020, the government's annual budget statement identified protecting national assets in the South China Sea – where Brunei has a large exclusive economic zone and claims an atoll, Louisa Reef – cyber defence and supporting efforts against the coronavirus pandemic as key immediate tasks for the RBAF. Although the defence budget increased significantly in 2018 (by 9.1%) and 2019 (19.7%), continuing economic stagnation resulting mainly from low oil and natural-gas prices meant there was a negligible increase in 2020 (2.7%), and no major equipment procurement is imminent. Brunei's Ministry of Defence has made a commitment to issue the country's first defence white paper in a decade during 2021, when the Sultanate will take over the annually rotating chairmanship of the Association of Southeast Asian Nations (ASEAN). In preparation for hosting the 2021 ASEAN Summit, in January–February 2020 Exercise *Setia Bersama* saw RBAF units test their operational readiness across the country, in close coordination with the police force and the paramilitary Gurkha Reserve Unit. Meanwhile, under a long-standing bilateral arrangement which has been extended to 2025, Brunei continues to host a British military presence of 2,000 personnel including a Gurkha infantry battalion, a helicopter-flight and a jungle-warfare school.

Pacific'); the need for greater 'self-reliance for delivering deterrent effects'; expanded capability for responding to grey-zone challenges; and enhanced lethality in high-intensity operations. While existing procurement will continue, the '2020 Force Structure Plan' that accompanied the Strategic Update highlighted emerging technology areas in which Australia would invest over the next decade, including remotely piloted or autonomous systems for air combat, strike, in-flight refuelling, surveillance, undersea warfare and land operations; 'capabilities to counter emerging space threats'; high-speed missile systems to provide 'more deterrence options'; and directed energy weapons. It also emphasised the development of a 'more durable supply chain' and strengthened defence-industrial capabilities to make the ADF more self-reliant. Crucially, the Strategic Update reiterated the '10-year funding model' set out in the 2016 White Paper, and envisaged the core defence budget (excluding pensions) increasing from A$42.2 billion (US$29bn) in 2020–21 to A$73.7bn (US$50.7bn) in 2029–30, with the proportion allocated to procurement increasing from 34% to 40%.

DEFENCE ECONOMICS

Macroeconomics

The Asia-Pacific region as a whole was able to avoid an economic contraction in the wake of the downturn after the 2008 financial crisis, but the effect of the coronavirus pandemic has proved so serious that the region was unable to repeat this performance in 2020. The IMF projects that real GDP growth will average –3.6% in 2020 before rebounding to 5.0% growth in 2021. The World Bank expects that the region's overall economic output will stagnate in 2020, contracting marginally by 0.5% and only supported by mild growth in China. In contrast, there was relatively robust performance after the 2008 crisis, with regional economic growth averaging 3.1% in 2009, even as global output declined. The situation was more positive even after the Asian financial crisis in 1997, with regional economic growth averaging 0.9% in 1998.

The World Bank expects East Asian GDP to expand only marginally in 2020, by 0.5%. This is enabled by the relatively stronger performance in China and Vietnam, though overall it is still the lowest growth rate in the sub-region since 1967. However, economic output in South Asia is expected to decline by 2.7% in 2020, the most severe contraction for over 40 years. The downgrade in the South Asian outlook is due to the contraction in investment consumption domestically in response to stringent lockdown measures while international demand for exports has plummeted and the tourism sector has been crippled by restrictions on travel.

Australia, Japan, Singapore and South Korea – the region's principal mature economies – entered recession in the second quarter of 2020, with GDP for each country projected to shrink by 4.2%, 5.3%, 6.0% and 1.9% respectively in 2020. Improved consumer spending in the second half of 2020 helped to limit the damage, though rising COVID-19 infection rates in Tokyo and another lockdown in Melbourne in July highlighted the fragility of any anticipated recovery.

However, as well as these immediate effects, the combination of an economic recession and emergency spending on pandemic relief and economic stimulus will widen budget deficits and require a fiscal response. In order to deal with these issues, most governments pursued various fiscal response measures including providing substantial liquidity support and credit facilities, cutting monetary-policy rates and pursuing asset-purchase programmes. Indonesia, South Korea and Thailand were among the first states globally to announce that parts of their budgets would be redirected towards tackling the crisis. The IMF projects that the fiscal deficits in China and India will widen in 2020 to 11.9% of GDP and 13.1% of GDP compared to deficits of 6.3% and 8.2% of GDP respectively in 2019. Nevertheless, the

▼ Figure 14 **Asia defence spending by country and sub-region, 2020**

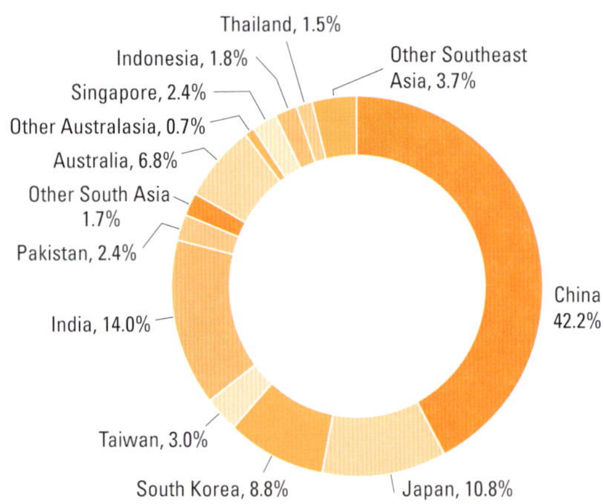

Note: analysis excludes North Korea and Laos due to insufficient data.

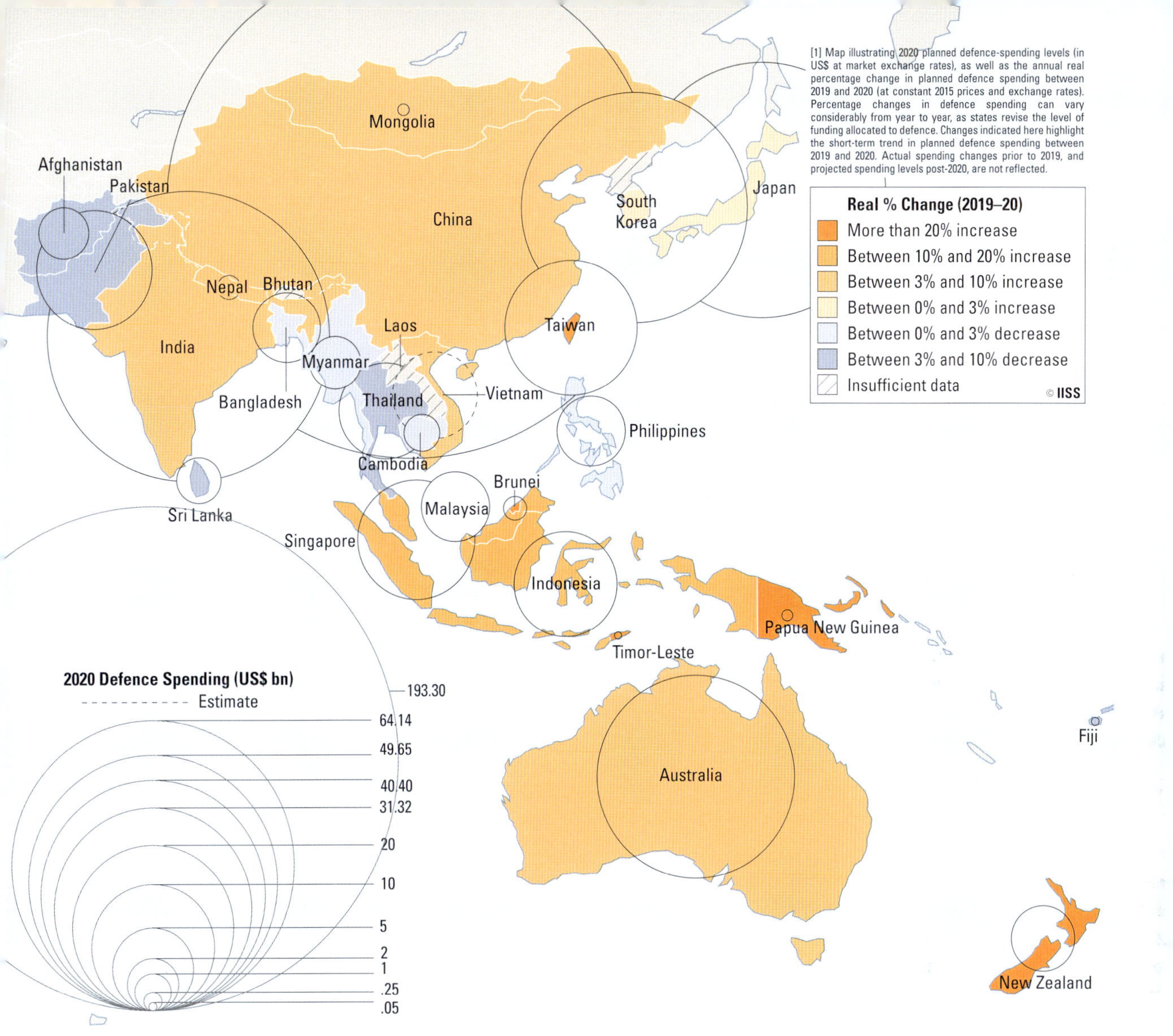

▲ Map 5 **Asia regional defence spending**[1]

generally lower levels of sovereign debt held by Asia-Pacific states when compared to other regions, and the anticipation that relatively robust economic growth will return in future, offers governments a degree of flexibility in their responses when compared to other regions.

In the second half of 2020, signs of a global economic recovery were a crucial development for the region given the extent to which its economies were integrated within global supply chains. However, the huge gains in economic output seen since the 1997 Asian financial crisis mean that the region will now be able to drive its own recovery in a way that it was unable to following previous economic crises. In this context, the region's economic recovery will likely benefit from the reduced prices for key imported commodities, such as energy and metals, in the wake of the pandemic.

Even with all this turbulence, the region's economic outlook is more positive than that of any other region. Importantly, trends over the past two decades have shown that this economic strength, and the increase in discretionary budgets that it enables, has underpinned rising military spending in the region.

Defence spending and procurement

Notwithstanding measures in some countries to reduce planned defence-budget increases, total Asia-Pacific defence spending still grew in real terms by 4.3% in 2020, only slightly lower than the figure of 4.7% real-terms growth seen in 2019. Total regional spending grew from US$419 billion in 2019 to US$437bn in constant (2015) dollars in 2020 (US$434bn to US$458bn in nominal terms). Of this US$18bn real increase, 51% could be attributed to increases in China, 15% to India, 13% to Taiwan and 6% to South Korea. Around two-thirds of the increase was driven by the growth in East Asian spending, while South Asia and Southeast Asia continue to underperform when compared to the regional average.

The extent of the GDP contraction forecast for Australia and New Zealand in 2020 (–4.2% and –6.1% respectively) appears to contrast with the increases in their real defence expenditure, of 4.3% and 21.3% respectively. The increase in New Zealand's budget from NZ$4.1bn (US$2.7bn) to NZ$5.1bn (US$3.3bn) was driven by continuing procurement programmes (including C-130J-30 *Super Hercules* transport aircraft and P-8A *Poseidon* maritime-patrol aircraft) while Australia has indicated its intention to strengthen its deterrent posture, with supporting defence investments, as outlined in its 2020 Defence Strategic Update.

China's continued defence-spending increases have the potential to influence budget decisions by some regional states. Indeed, although the 6.4% increase announced by Beijing (a 5.2% rise in real terms) might be one of its slowest rates of growth since the early 1990s, the US$9bn (constant 2015 dollars) increase was still more than the combined real increase of all other regional states in 2020. Indeed, as other Asian capitals considered budgetary adjustments to deal with the costs of the coronavirus pandemic, the strong increase in China's 2020 defence budget may be interpreted as confirmation that military modernisation remains a priority for Beijing, coming as it did at a time of great economic uncertainty.

There are, however, statements of intent elsewhere. Perhaps the clearest indication of the importance given to regional security concerns despite fiscal pressures came in the form of Australia's Defence Strategic Update, released in June. The document outlined plans to increase Australia's defence budget by 9% in 2020 and to maintain growth at an average rate of 6.7% each year over the next decade. The result is that, under current government projections, Australia's defence budget will rise from A$39.1bn (US$26.9bn) in the 2019–20 budget to A$73.7bn (US$50.7bn) by 2029–30, with a total of A$220bn (US$151bn) being invested in defence equipment over the period. While changes in government over the intervening period mean that attaining, or even maintaining, these targets cannot be guaranteed, the strategic update demonstrated continued political determination to invest in Australia's defence despite the effects of the coronavirus pandemic.

In South Korea, the 2020 defence budget, as initially approved, was to see spending rise to 50.2 trillion won (US$41.9 bn), marking a 7.4% real increase over 2019. Then in April, as COVID-19

▼ Figure 15 **China defence budget compared with rest of Asia total, 2008–20,** US$bn, constant 2015

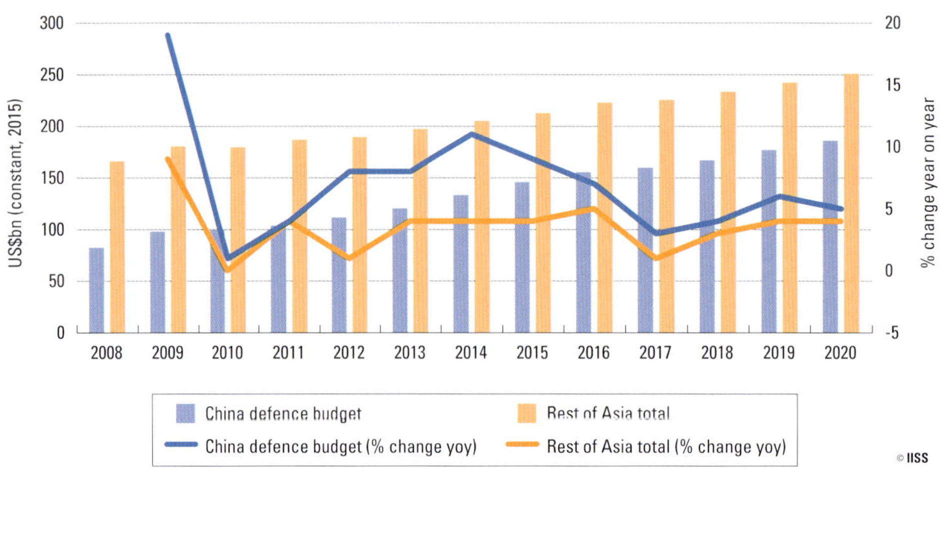

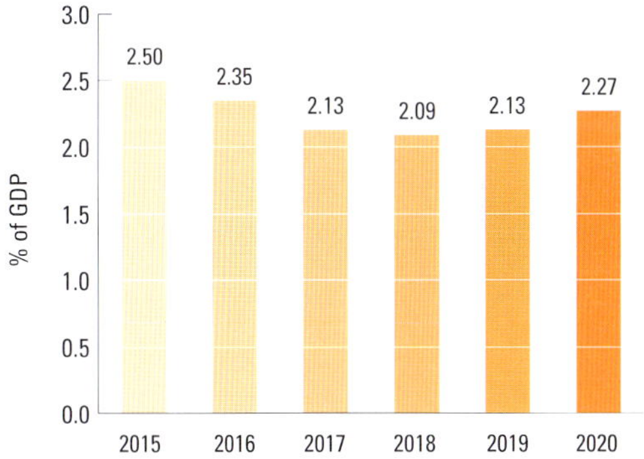

▲ Figure 16 **Asia regional defence spending** as % of GDP (average)

cases surged, Seoul announced an emergency relief fund, with ₩900bn (US$752 million) diverted from the defence budget. Plans to divert a further ₩300bn (US$251m) were announced in June in order to fund further stimulus, bringing the final 2020 defence budget to ₩48.4tr (US$40.4bn). However, despite these downward adjustments, Korean defence spending remained around 2.7% higher in real terms than in 2019. Furthermore, in August the Ministry of National Defense announced an ambitious update to its mid-term defence plan, outlining plans to increase defence spending by 35% by 2025. Indeed, the 2021 budget that was announced in September allocated ₩52.9tr (US$44.2bn) to defence, 8.4% higher in real terms than the final 2020 budget. Mirroring wider regional trends, budget cuts in Korea over the course of 2020 have slowed the growth in defence spending but have not caused a year-on-year reduction, while plans persist for longer-term spending increases.

Taiwan's defence spending continued to rise in 2020 with a 5.1% real increase in the budget. Defence spending in the country had declined in real terms between 2012 and 2015, though President Tsai Ing-wen's government has since sought to boost defence resources. A key element of this strategy has been the re-introduction of special budgets which can be approved by the government to augment defence spending in order to support major defence procurements. Through the 1990s and early 2000s this represented a key source of procurement spending for Taiwan, though the reluctance of allies to clear transfers of military equipment saw such funding largely discontinued. The approval in November 2019 of a NT$247bn (US$8.3bn) special budget or the acquisition of 66 Lockheed Martin F-16V Block 70 combat aircraft represents an important reversal of recent trends.

Budgetary conditions in Southeast Asia have seen a more immediate correction resulting from the challenges created by the coronavirus pandemic. Indonesia announced that defence spending for 2020 would be reduced from Rp131tr (US$9.0bn) to Rp122tr (US$8.4bn), with the funding reallocated to stimulate the post-pandemic economy. The final figure for Indonesia's defence spending in 2020 fell even further to Rp118tr (US$8.1bn), although this is still a 7.3% real increase over 2019 levels.

In Thailand, the defence budget for 2020 was also reduced, by b18bn (US$59m) to b214.0bn (US$7.0bn), to aid the government's response to the crisis, resulting in a 5.3% real reduction in the budget compared to 2019. The proposed 2021 defence budget of just b214bn (US$7.0bn) maintains the budget at 2020 levels in nominal terms but is a 0.7% cut in real terms. Singapore reduced its 2020 defence budget by a more marginal 0.5% in real terms. Minister for Defence Dr Ng Eng Hen announced in June that the armed forces were 'taking measures to cut cost where we can' in anticipation that Singapore 'will have to tighten our belts for this year, even the next, even 2022, until the economy fully recovers'.

In India, measures adopted in April and May suggest that the coronavirus pandemic will have a significant effect on the defence budget, particularly on capital investment. In April, among wider restrictions on government expenditure, the Ministry of Finance announced that it would cap defence spending at 20% of the annual total for the first quarter of the 2020 fiscal year (encompassing April, May and June). With personnel expenses ostensibly ring-fenced from the worst of the cuts, this placed pressure on procurement budgets, particularly new acquisition projects. This development has the potential to divert funding away from capital investment and military modernisation and towards personnel and operating costs.

Defence industry

Despite the economic downturn, it has been apparent over the past 12 months that regional states remain committed to supporting national defence industries. Even though some regional states might have reduced their defence budgets, most have nonetheless ensured continuing support for projects benefiting local industry. Asia's defence sector has developed over the past decade, and governments in the region have sought to ensure a steady stream of investment in order to stimulate the wider economy.

India's Prime Minister Narendra Modi has been a particularly strong advocate of such policies and has maintained efforts to boost self-sufficiency in defence equipment despite budgetary constraints. An order for 83 locally made *Tejas* Mk 1A combat aircraft worth Rs390bn (US$5.3bn) was approved in March, followed in July and August by clearance to procure 12 additional Su-30MKI fighters and 70 indigenously developed HTT-40 basic training aircraft from Hindustan Aeronautics Limited. In support of wider indigenisation policies, India announced a list of 101 defence-equipment items that can no longer be imported from abroad including offshore-patrol vessels, land-attack cruise missiles, towed artillery and short-range maritime-patrol aircraft. Nevertheless, new procurement guidelines issued by the government underlined the importance of foreign assistance in supporting the longer-term goal of self-sufficiency, as the limit on foreign direct investment in local defence manufacturers was increased from 49% to 74%.

Australia also intends to boost the local defence sector. In August the Department of Defence emphasised that an update to the government's procurement rules was 'aimed at forging new opportunities for Australia's defence industries'. The changes provide guidance on engaging local suppliers within larger procurement contracts, and on assessing economic benefits and 'value for money' in such deals. The department's 2020 Defence Strategic Update further emphasised ambitions in this area, with the intent to leverage an 'unprecedented programme of investment' to 'build a resilient and internationally competitive defence industrial base'.

As the capabilities of local defence suppliers increase, regional governments are increasingly looking to export markets to boost sales. Both Japan and South Korea have put measures in place to improve coordination between government agencies and departments in providing support for exports. South Korea also announced plans to encourage local firms to develop components for locally-made platforms in order to enable import substitution and higher levels of indigenous content, while Japan's defence ministry plans to explore potential collaboration with foreign companies in export markets.

CHINA

The People's Liberation Army (PLA) was due to accomplish a key modernisation goal in 2020. By the end of this year, the Central Military Commission (CMC) had directed the PLA to achieve full mechanisation, the first of three military-related goals identified in President Xi Jinping's October 2017 address to the 19th Congress of the Chinese Communist Party. However, in 2019, China's latest defence white paper – *China's National Defense in the New Era* – said the PLA had yet to complete this task and was also 'in urgent need of improving its informatisation'. Other targets were to complete 'basic modernisation' of the PLA by 2035 and, by the middle of the century, to have transformed the PLA into 'world-class' forces. The communiqué issued at the Fifth Plenum of the Chinese Communist Party's 19th Central Committee, in October 2020, continued the call to 'accelerate mechanisation, informatisation and intelligent integration'. Mechanisation is generally held to refer to ambitions to replace legacy equipment; informatisation to efforts at improving technical sophistication (the US DoD has likened it to 'net-centric' capability); while intelligent integration likely refers to the premise that military systems will be improved by the integration of automation, big data and artificial intelligence.

Accelerating military modernisation

An important development for military modernisation was the announcement at the Fifth Plenum that modernisation should be accelerated and that the PLA must meet the goal of 'building' the army by 2027, the year which marks the PLA's centenary. This is the first time that this goal has been mentioned, though it does not seem to replace the goal of achieving 'basic modernisation' by 2035. Beijing remains intent on funding that latter goal. Despite the effect of the coronavirus pandemic and cuts to central budget lines such as education and foreign affairs, the defence budget increased by 6.4%. Moreover, 2020 marks the end of the 13th Five Year Special Plan for Science and Technology Military–Civil Fusion (MCF) Development, which aimed to establish top-level design of China's MCF strategy and eliminate existing stovepipes between the civilian and military innovation systems. The 14th Five Year Plan is set to come into force in January 2021 and will reflect China's continued development of emerging high technologies, drive Chinese innovation efforts and also increase investment in basic research.

Training and exercises

The PLA continued to put its reformed structure and new capabilities to the test in 2020. Exercises in the Eastern and Southern theatre commands focused primarily on Taiwan scenarios, particularly following Taiwan's overwhelming vote, in January 2020, to re-elect President Tsai Ing-wen. For example, in May 2020, the PLA issued Notices to Mariners that effectively cordoned off a large area of the Bohai Sea for a series of opposing-force live-fire exercises. These drills included training for the seizure of islands and air-defence and anti-missile manoeuvres, possibly aimed at scenarios related to an amphibious operation on Taiwan's western shore (or at least, for messaging purposes, indicating preparations for such an eventuality). In mid-September, exercises crossing the median line of the Taiwan Strait became more frequent. On 18 and 19 September, the PLA sent 37 fighters, bombers and anti-submarine warfare aircraft into the northern and southern sections of Taiwan's Air-Defence Identification Zone (ADIZ) in two exercises. PLA drills at the northern and southern ends of the Taiwan Strait have become more common in 2020, as have air incursions across the median line. Indeed, Taipei claimed that the likelihood of misperception had risen, after the Chinese government in 2020 said that it did not recognise a median line in the Taiwan Strait. Meanwhile, exercises involving all of China's military theatre commands continued for a second year in 2020.

Border clashes, and the coronavirus pandemic

The military stand-off between the PLA and Indian forces in May, along the disputed western border with China, resulted in a series of scuffles and ultimately fatalities. The PLA subsequently seems to have reinforced its border regiments in these areas with additional operational reserves. Imagery also showed that the PLA Army mobilised additional conventional combat forces, including tank companies and towed artillery batteries deployed in existing Chinese positions north and east of Gogra. While fatalities are rare on the border, transgressions are not and fall within a more regular pattern that occurs each spring.

By the end of May, the PLA's role in China's national response to the coronavirus pandemic had become apparent. The pandemic proved to be the Joint Logistics Support Force's first real test since its inception in 2016. It coordinated the response of thousands of military medical personnel from all branches of the armed forces in 19 cities, organised air, road and rail transports of supplies and personnel, and directed assistance by militia.

While the coronavirus pandemic affected PLA personnel, the PLA's leadership was reluctant to disclose details and quickly moved to prevent further transmission. Joint training and exercises were reportedly put on hold, though large single-service activities went ahead. For instance, the Navy conducted anti-submarine warfare drills in the South China Sea and East China Sea. Meanwhile, both state-owned enterprises and small and medium-sized enterprises received support from central government to overcome any financial impact on their business from the pandemic.

PLA Army

Although the army has never publicly defined how it will judge whether 'basic mechanisation' is complete by the end of 2020, it is likely that this includes both structural reorganisation to better enable combined-arms operations, as well as the replacement of its sizeable number of ageing and obsolescent armoured vehicles and artillery systems.

The higher-level reorganisation in the army after 2015, and in its combat and combat-support formations from 2017, is now essentially complete, suggesting that the structural aspect of the 2020 targets has been achieved. In terms of modern equipment, however, the army apparently remains somewhat short of full recapitalisation.

In the group armies' combined arms brigades, modern armoured-vehicle designs now comprise between 66% and 75% of the total inventory. By late 2020, 25 out of 32 heavy brigades were equipped with modern main battle tanks or light tanks, and 17 also had the ZBD-04/04A family of tracked infantry fighting vehicles. For the medium-weight forces, the situation appears broadly similar, with at least 14 of the 18 brigades equipped with the ZBL-08 family of wheeled armoured vehicles.

The delivery of wheeled artillery and air-defence systems has been a focus in 2020. The PCL-181 (or PCL-18-1) 155mm truck-mounted gun/howitzer, displayed in the 2019 parade, is now in service with five artillery brigades in the Eastern, Southern and Central theatre armies as well as with the artillery brigade in Tibet. A new 122mm truck-mounted system is also now in service with a combined arms brigade in Tibet. Mounting artillery systems on truck chassis allows for faster movement on roads when compared to towed systems, while they are also easier to transport by air or rail than their tracked equivalents. This will better enable these brigades to quickly re-deploy both within and between theatre commands, which is a key modernisation objective.

Similarly, the army has also now begun to adopt both a new wheeled version of the HQ-17 (CH-SA-15) surface-to-air missile system (the HQ-17A) and a wheeled self-propelled anti-aircraft gun with a medium-weight combined arms brigade in the Northern Theatre Command. As these systems are introduced more widely amongst other medium-weight brigades, they will be able to give these formations a mobile air-defence capability that is comparable to their heavy, tracked equivalents without unduly limiting their strategic or operational mobility.

The army's aviation branch continues to work on modifications to its WZ-10 attack-helicopter fleet and improvements to its land-attack capability, including testing an improved missile design, similar to the export model *Blue Arrow* 21, in mid-2020. Production of the Z-20 medium transport helicopter continues for the two dedicated air-assault brigades, and the army is now introducing into service a further upgraded variant of its Z-8 heavy-lift helicopter (reportedly designated Z-8L).

In addition to improving organisation and equipment, more realistic training has been a key component of the army's modernisation effort. Brigade-level 'opposing force' training exercises, such as *Stride* or *Firepower*, which were just beginning to increase in 2019 following the restructuring in 2017 and 2018, may have temporarily been curtailed again by the coronavirus pandemic. By September 2020, only two iterations of the *Stride* series had taken place, both at the Qingtongxia combined arms training base on the edge of the Gobi Desert. This was followed in mid-September by a single iteration of the *Firepower* series, also conducted at Qingtongxia. However, other individual formation training exercises continued, as did army participation in what now appears to be an annual national-level multi-theatre-command exercise series.

PLA Rocket Force

The PLA Rocket Force (PLARF) continues to equip the new missile brigades it has established from 2017, as well as re-equip some of its existing formations with new or upgraded designs. Of the PLARF systems displayed during the Chinese Communist Party's 2019 anniversary parade, it remains unclear whether either the DF-17 HGV system or the DF-41 (CH-SS-X-20) intercontinental ballistic missile (ICBM) has yet achieved initial operating capability. Both the road-mobile DF-26 intermediate-range ballistic missile (IRBM) and the DF-31A(G) ICBM designs, however, have already entered service in substantial numbers.

In its 2020 report on the PLA, the US Department of Defense estimated that the Rocket Force now has around 200 launchers for IRBMs – probably all for variants of the DF-26. This is greater than the PLARF's total number of medium-range ballistic-missile launchers, which have traditionally formed the bulk of their nuclear and conventional theatre missile inventory, and likely reflects the growing emphasis the PLA places on being able to hold targets in the second island chain at risk. Two hundred launchers would allow the PLARF to equip ten missile brigades (out of 39 in total) exclusively with IRBM variants, or more if they mix intermediate- and medium-range ballistic missile systems within brigades. The DF-26 is believed to have nuclear, conventional land-attack and anti-ship variants, which would give the design utility across the spectrum of PLARF mission types.

PLA Navy

In August 2020, the first of the new Type-075 (*Yushen*) 30–35,000-tonne large amphibious ships (LHDs) embarked on sea trials. This first vessel was rapidly followed by a second launched in April 2020, and a third is under construction. These ships should add considerably to the PLAN's ability to project power for a range of contingencies. While the new PLA Marine Corps aviation brigade continues to work up, the provision of helicopters to support amphibious operations in China's littoral remains primarily the responsibility of army units.

China's second aircraft carrier, commissioned in December 2019 and named *Shandong*, continued its trials, including operations with J-15 combat aircraft. This is China's first locally built aircraft carrier. The PLAN demonstrated the ability to deploy both the *Shandong* and the PLAN's first carrier, the *Liaoning*, at sea at the same time, raising the question of when this might be translated into a true two-carrier operational capability and what options this could present for Beijing. A third, larger, vessel is also under construction, and Beijing's longer-term carrier ambitions remain a source of considerable debate.

▼ Figure 17 **The PLA's Su-27 *Flanker* family**

China's armed forces have taken delivery of more than 600 Sukhoi Su-27 *Flanker* variants since the first Su-27SK aircraft were ordered from Russia in 1991. The *Flanker* family is central to air-force and naval-aviation air combat capabilities and will remain so well into the next decade, if not beyond. The basic design continues to provide a platform for further developments, including a dedicated electronic-warfare (EW) variant. China's defence aerospace sector also benefited from the *Flanker* relationship, in developing the assembly and manufacturing skills required. Key to this was the 1996 licence assembly leading to a licence production agreement with Shenyang Aircraft Corporation. The latest domestic variants of the *Flanker*, such as the J-16, are arguably more advanced than their Russian counterparts. The J-16, for example, is fitted with an active electronically scanned array radar, while the Su-35 *Flanker* M is equipped with a less-capable passive electronically scanned array.

Russian build, Russian systems

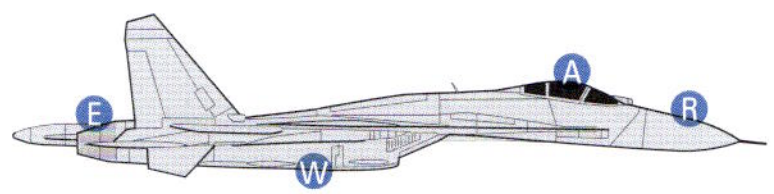

Su-27SK *Flanker* B (and Su-27UBK two-seat trainer)
Status: Deliveries from 1992

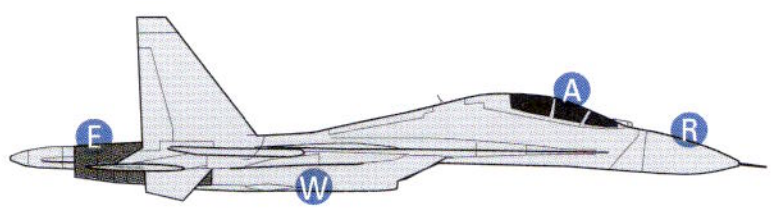

Su-30MKK *Flanker* G
Status: Deliveries from late 2000

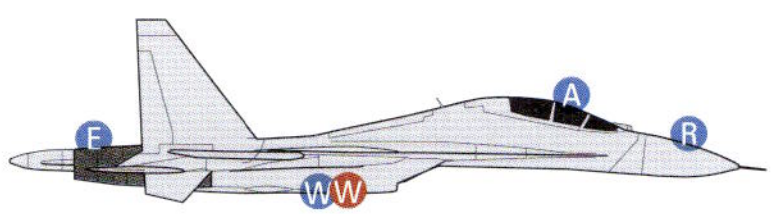

Su-30MK2 (PLA Navy) *Flanker* G
Status: Deliveries 2004

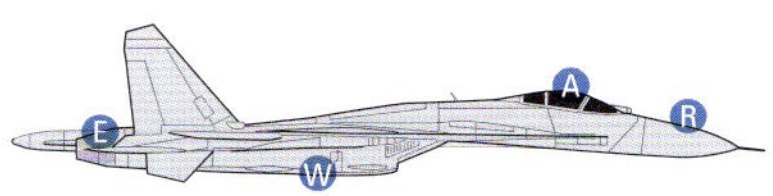

Su-35 *Flanker* M
Status: Deliveries from late 2016

Chinese final assembly, Russian systems

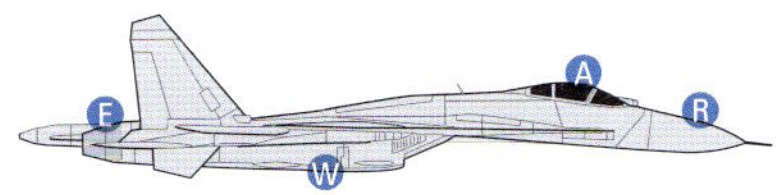

J-11
Status: Delivery from 1999 licence assembly

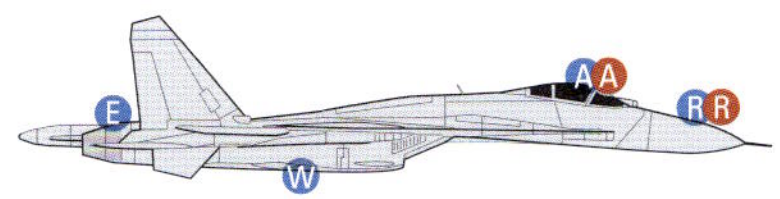

J-11A
Status: Delivery from 2001, with avionics and radar upgraded with external assistance

Chinese build, Chinese systems

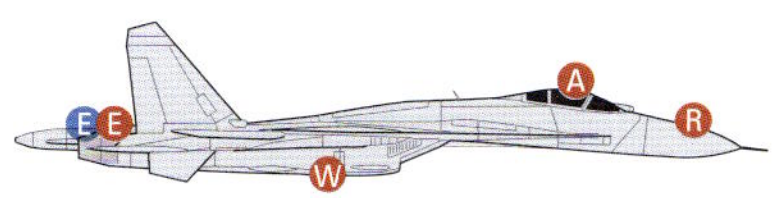

J-11B *Flanker* L (and J-11BS two-seat trainer)
Status: Deliveries from 2007 (PLA Navy 2010)

Chinese build, Chinese systems (continued)

J-11D
E W A R
Status: Uncertain

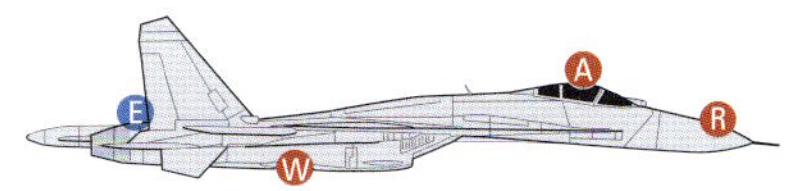

J-15 (PLA Navy)
Status: Deliveries from 2013

J-15D
E W A R
Two-seat EW version of J-15
Status: Flight testing ongoing

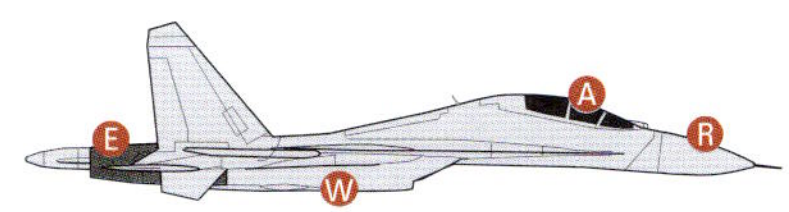

J-16
Status: Deliveries from 2016

J-16D
EW version of J-16
Status: Service entry pending

● Russian build, ●● Chinese upgrade, ● Chinese build. Ⓔ Engines Ⓦ Weapons Ⓐ Avionics Ⓡ Radar All deliveries are to the PLA Air Force unless otherwise specified.

In the meantime, the third production batch of J-15 aircraft appears to have entered service in 2020, presumably to form an air group for the *Shandong*.

China launched the eighth Type-055 (*Renhai*) cruiser and the 25th Type-052D/Type-052D mod (*Luyang* III/*Luyang* III mod) destroyer in mid-2020. With no apparent sign that more of these vessels are in production, it is unclear whether this represents the end of a prodigious phase of shipbuilding output for these surface combatants. Alternatively, particularly in the case of the Type-055s, it could indicate a pause either between procurement cycles between five-year plans or to prepare for a new, possibly improved batch incorporating further technical developments. At the same time more legacy platforms, including the last two Type-051 mod (*Luda* IV) destroyers, disappeared from the fleet. Only a handful of Type-053H1/H1G (*Jianghu* I/I Upgrade) frigates now remain in service from the PLAN's pre-1990 surface combat fleet, relegated to coastal-patrol taskings.

There was evidence in the first half of 2020 pointing to the service-entry of the fifth and sixth nuclear-powered ballistic-missile submarines of the Type-094 (*Jin*)-class, even as analysts debated the extent to which they are conducting operational patrols.

Generally, the PLAN and China's other maritime security agencies, including the coastguard, appeared to maintain and even increase assertive deployments in and around the South China Sea as the coronavirus pandemic unfolded. The carrier *Liaoning* was also deployed close to Taiwan and in the East China Sea.

PLA Air Force

New evidence emerged to suggest the air force is close to again becoming a part of Beijing's nuclear-deterrent forces. In October 2020 imagery emerged of the Xian H-6N variant of the long-serving H-6 (Tu-16 *Badger*) that is now entering service carrying a very large air-launched ballistic system semi-recessed under the fuselage. It is possible that this weapon is what the US intelligence community has designated the CH-AS-X-13 air-launched ballistic missile. This weapon may be dual capable. The H-6N can also carry cruise missiles on each of its six underwing pylons. Development of a replacement for the H-6 family, possibly known as the H-20, is understood to be continuing. This aircraft is assumed to be a low-observable design, very likely with a nuclear-weapons delivery role.

PLA Air Force (PLAAF) combat production currently seems set to continue with the trio of J-20, J-16 and J-10C types. At the same time, older types continue to be replaced, and the remaining Cold War-era J-7 and J-8 designs now account for only around 25% of the PLAAF's 1,500-strong tactical-combat aircraft inventory.

Additional numbers of Shenyang J-16 two-seat fighter/ground-attack aircraft were delivered, as were a limited number of the Chengdu J-20A multi-role fighters. Domestic Chinese reports from August 2020 suggested that a Chengdu J-10A-equipped brigade (the 26th Brigade at Huiyang, Guangzhou province) was the first to convert to the J-16. The two-seat J-16 is a more capable multi-role platform than the single-seat J-10A *Firebird*, although the conversion will also require the unit to introduce weapons-system officers to occupy the rear cockpit. An electronic-warfare derivative of the J-16, the J-16D, is also being developed, and this version of the aircraft is being flight tested.

An upgraded variant of the J-20, notionally the J-20B, is also reportedly in development. This variant has improvements resulting from the initial testing and fielding of the J-20A and may become the first version to be fielded in operationally significant numbers. The air force also continued to take delivery of special-mission platforms, with the KJ-500 airborne early-warning and control aircraft (based on the Y-9) emerging as the most numerous. An improved endurance version of the aircraft fitted with an in-flight refuelling probe is also in development. The Y-9 continues to be used as the platform for a variety of signals-intelligence roles. The PLAAF's fleet of the Xian Y-20 heavy-transport aircraft also continued to grow and there are now more Y-20As than Il-76s, meaning that the air force has effectively doubled its heavy-transport fleet in the last four years The Y-20 is also likely to form the basis of a dedicated tanker and possibly an airborne early-warning and control platform to replace the KJ-2000 *Mainring*.

PLAAF training continues to aim to complement the continuing improvement in the equipment inventory, and the shift from a homeland defence to a regional power-projection role. Training 'exercises' were notable in the region around Taiwan, where the air force has, along with other service arms, increased activity amid growing tensions with Taipei. These have included reportedly mixed sorties of J-10, J-11, J-16 fighter and H-6 bomber aircraft, and there has also been evidence of coordination between air-force units and naval-aviation anti-submarine warfare units.

DEFENCE ECONOMICS

The coronavirus-pandemic crisis and escalating restrictions by the US on technology exports threatened to hinder the growth of China's defence economy in 2020. Nonetheless, China's defence-industrial base and its associated components continued to perform well both in financial as well as technological terms. It will be important to maintain this momentum in order to meet the demands for more-capable and increasingly indigenous defence, technological and industrial capabilities, not least after China's leaders again called for an acceleration in its military-modernisation drive, during the Fifth Plenum meeting in October 2020.

The coronavirus pandemic

As China locked down in January 2020 following the coronavirus outbreak in Wuhan in Hubei province, the defence industry followed the rest of the economy in curtailing its activities. This shutdown lasted around a month, before production lines and research-and-development (R&D) work began to gradually resume from mid-February, in order to fulfil the People's Liberation Army's (PLA) acquisition priorities as well as to meet the goals of the 13th Five Year Plan (2016–20) and the 2006–20 Medium and Long-Term Science and Technology Development Plan.

The effect of the pandemic on the defence industry varied by sector and by company. The listed arms of China Shipbuilding Industry Group Co. Ltd (CSIC) and China South Industries Group Corporation (CSGC) each suffered a 28% drop in their incomes in the first half of 2020, with profits of the latter declining by 82% over the same period in 2019. That said, even before the pandemic there had been reports of overcapacity in the global shipbuilding sector, while CSGC's revenues had reduced since 2016, a development in part attributed to weak performance in the civilian car manufacturing industry. By contrast, China North Industries Group Corporation (NORINCO), another major state-owned ordnance conglomerate, has performed well since 2010 and only saw its income and profits fall by 8.9% and 7.3% respectively in the first half of 2020.

The impact on the aerospace sector was more modest. Aviation Industry Corp. of China (AVIC), the country's monopoly defence-aviation supplier, posted a 4.8% fall in income in the first half of 2020, while profits were unchanged. The space and missile sectors were largely unaffected, although China Aerospace Science and Industry Corporation Ltd (CASIC) saw single-digit reductions in its income and profits in the first half of 2020. One reason for these uneven economic repercussions is that the balance between civilian and military business varies across defence sectors. Civilian items comprise around two-thirds of the overall annual value of output produced by the defence industry, and the proportion is higher in the shipbuilding, ordnance and nuclear sectors when compared to the more defence-focused aviation, space and missile domains. Civilian business operations have been harder hit by the coronavirus pandemic. As China's economy rebounded in the second half of 2020, the defence industry looked to recover lost ground and to continue the run of year-on-year increases in revenues and profits that stretch back to the early 2000s.

The government sought to protect the PLA and the defence economy by only marginally trimming back the rate of growth in the official defence budget, to 6.7% in 2020, compared to an annual average increase of around 7.2% between 2014 and 2019. In real terms, the growth rate in Chinese defence expenditure slowed from 5.9% in 2019 to 5.2% in 2020. The 2020 defence budget, including funding for local militias, came to US$193.3 billion (US$186bn in constant 2015 dollars), although total expenditure is estimated to be much higher if foreign-weapons purchases, military R&D funding and the People's Armed Police central budget is included. As the defence economy has access to sources of funding beyond the official defence budget – such as the capital markets, arms sales and special projects – this sector may effectively be shielded from volatility in the general economy and significant reductions in state allocations.

Growing security and technology frictions with the US

While the effects of the coronavirus pandemic on China's defence economy appear so far to be limited, tighter US controls over the export of critical high-technology goods poses a more serious and long-term threat to China's dual-use strategic, and high-technology and industrial base. These are critical sources for the local defence industry and for the PLA. While China has been subject to a long-running arms embargo by the US and European Union since Tiananmen Square in 1989, this has not covered commercial and dual-use technologies (the EU had a dual-use export control list, but this was in practice unevenly applied). As a result, China has been able to improve its domestic science, technology and innovation systems by accessing advanced Western technologies and scientific research. This has particularly benefited the electronics, naval, space and missile sectors.

In recent years, China's defence and national science, technology and innovation systems have stepped up efforts to broaden its innovation model from one of absorbing foreign technologies to also emphasising fundamental and basic research so it can produce original work. This has meant a dual-track strategy of building up a high-end domestic R&D base, as well as tapping into advanced foreign innovation ecosystems, of which the US academic and commercial start-up communities have been a prominent area of focus.

China has faced increased scrutiny and higher barriers to entry to the US market in the wake of reviews conducted by the Committee on Foreign Investment in the United States (CFIUS). The US is now clamping down on China's access to its science and technology systems through a diverse range of measures. These include tightening controls on Chinese investment in high-technology and sensitive economic areas, cutting off sales and the transfer of technology products to a growing list of Chinese technology firms suspected of having close ties to the PLA, and prosecuting US-based academic researchers who have been recruited to work for Chinese science talent programmes run by Chinese authorities. These initiatives include the China Initiative, launched in November 2018 by the US Department of Justice, which targets Chinese talent-recruitment and academic-espionage activities, and restrictions on semi-conductor-related exports since 2019. This crackdown has strong bipartisan political support within the US Congress, which means that the momentum for further tightening, or decoupling of the US and Chinese science and technology systems and supply chains, can be expected to gather pace well into the 2020s.

China's response, and long-term planning

The authorities in Beijing have been assessing how to respond to the US stifling of technology flows to China. The issue has become more pressing as deliberations move to a close over finalising the details of the 14th Five Year Plan (FYP) and new 15 Year Medium and Long-Term Science and Technology Development Plan (MLP), both due to start in 2021.

The 5th Plenum of the 19th Chinese Communist Party Central Committee, held in October 2020, reviewed and approved guidance for the 14th Five Year Plan and development objectives out to 2035. The plenum communiqué pointed out that the world was seeing significant change and that 'instability and uncertainty has increased significantly'. In response, the plenum said that the pace of defence modernisation should be accelerated, and that the task of 'building' the army should be achieved by 2027 – the PLA's centenary year. This was the first time that this 'centenary' target has been publicly disclosed, and while the communiqué offered no detailed explanation of what the deadline meant, it pointed out the need for the PLA to improve its 'strategic ability to defend national sovereignty, security, and development interests'. Official Chinese media outlets suggested that 'strategic ability' referred to the military capabilities to successfully carry out military operations across the Taiwan Strait and in the South China Sea. In addition to the 2027 objective, the plenum put forward a number of other military goals that included strengthening 'war preparations' and promoting concurrent improvement in national defence and economic development.

At a Politburo study session on 30 July 2020 devoted to defence issues and military modernisation, President Xi Jinping discussed the importance of coordinating the military version of the 14th FYP with national development plans. He stressed that it was imperative that the defence establishment 'adhere to the strategic basis of indigenous innovation, strengthen basic research and original innovation, accelerate breakthroughs in key core technologies, accelerate the development of strategic, cutting-edge, and disruptive technologies, accelerate the implementation of major strategic projects in defence technology and weaponry, and continuously improve the scientific and technological content of our military construction'. Key emerging technologies of interest to the PLA include big data, blockchain, artificial intelligence, and quantum computing and communications.

Senior Chinese technology advisors have said that the US technology restrictions provide a road map for China as to where to focus its development efforts. Bai Chunli, President of the Chinese Academy of Sciences (CAS), China's premier scientific R&D institution, said in September 2020 that to avoid 'being strangled' technologically, China should prioritise the development of core capabilities such as photolithography machines, high-end chips and basic algorithms. CAS launched a strategic pilot programme in 2018 with the goal of achieving technological breakthroughs in cutting-edge scientific problems related to national security and economic competitiveness.

The central government's defence-industrial regulatory agency, the State Administration for Science, Technology and Industry for National Defence (SASTIND), has sought to influence this planning process, but has seen its bureaucratic authority and independence steadily eroded since the end of the 1990s. Since the mid-2010s, SASTIND has established two special committees – the National Defense Science and Technology Development Strategy Committee and a Science and Technology Committee – in order to draw up long-term defence-technology and -industrial development plans.

The Science and Technology (S&T) Committee was established in late 2018. One of its key tasks has been to provide assistance in building a robust defence innovation system, addressing issues such as cultivating human talent, enhancing the governance regime and project evaluation. The S&T Committee is chaired by a SASTIND deputy director and its membership is dominated by corporate representatives, including the heads of many of the major state defence conglomerates, as well as scientists and engineers from CAS, defence universities, the China Academy of Engineering Physics, government ministries and PLA representatives.

Military–civil fusion

Details of how these initiatives will be implemented are likely to become even more opaque as the US and other countries pay increasing attention to Chinese strategies and plans and seek to counter them. Military–civil fusion (MCF), which refers to the seamless integration of the civilian and defence economies, is a high-profile strategic venture that received significant leadership attention after the early 2010s. MCF remains a top priority in Beijing's plans to restructure China's defence economy so that it becomes increasingly fused into the civilian economy and is better able to leverage commercial innovation and commercial resources. Chinese authorities have said that the beginning of the 2020s marks the start of the next phase of the MCF development strategy, moving from a preparatory transition stage to 'deep implementation'. A key MCF goal is to allow private civilian firms to provide materiel for the PLA. Around 1,800 companies have so far been granted the necessary permits, although most provide non-war-fighting services such as food, clothing, fuel and other logistics supplies.

While MCF is primarily focused on domestic priorities, China's authorities are looking to leverage MCF capabilities to support the export of defence, dual-use and strategic high-technology products. The uninhabited aerial vehicle (UAV) and surveillance industry is being promoted in this way. Sales of Chinese military, dual-use and commercial UAVs have increased since the early 2010s.

The Chinese government has also sought to use its Belt and Road Initiative (BRI) to support MCF-related exports, such as the sale of nuclear power plants and related equipment as well as aviation and aerospace equipment, and the construction of a BRI Space Information Corridor. This corridor initiative was started in 2016, and the objective of this ten-year project is to provide an integrated information system for countries along the Belt and Road served by Chinese weather, communications, navigation, data-relay and remote-sensing satellites. Alongside this is China's development of a Digital Silk Road project in which Chinese firms are looking to build a comprehensive digital information infrastructure.

INDONESIA

Indonesia's armed forces have been engaged on a long-term modernisation strategy since 2010, in a bid to produce more modern and mobile forces able to more effectively tackle a growing range of security challenges. Some expected that this modernisation process, and Indonesia's defence establishment, would have been reinvigorated by the appointment by President Joko Widodo of his former rival Prabowo Subianto as Minister of Defence in October 2019. The defence minister, a retired lieutenant general, was a former commander of the army's special forces. Nonetheless, while Indonesia's defence policies and capability-development plans remain ambitious in scope, over the past year they have stayed largely underdeveloped. This reflects the defence ministry's need to balance competing, and perhaps even contradictory, priorities.

The defence ministry inherited several policies and priorities from previous administrations, including completion of the 2010 plan to develop a Minimum Essential Force (MEF) by 2024, as well as the 2012 law that mandated the development of domestic defence industries. At the same time, defence minister Prabowo came into office looking to modernise the armed forces' ageing equipment inventory, even if it meant restarting or sidelining pre-existing plans. Meanwhile, he has had to carefully manage the defence ministry's relationship with the Indonesian National Defence Force (Tentara Nasional Indonesia, or TNI). While the defence minister does not have operational command over the TNI (this is vested in the TNI commander, as instructed by the president), he is responsible for overall strategic policy and management of the defence budget. As a result, defence policymaking needs to take into account the TNI's policy priorities, particularly regarding its ongoing organisational expansion, as well as its operational demands, including counter-terrorism, maritime security and the response to the coronavirus pandemic.

Strategy and policy

Defence minister Prabowo issued the MoD's latest State Defence Policy document on 20 January 2020. These documents have recently been issued every two years and this latest one is, at the time of writing, the only major defence policy guidance that Prabowo's defence ministry has issued. It is designed to advance, between 2020–24, the plan to develop a modern defence establishment and a professional TNI.

The 2020 document outlines four strategic objectives: protection of territorial integrity, sovereignty and society; the development of an integrated and modern 'Total People's Defence and Security' system; the creation of a 'resource management system for national defence' and improved defence management. Twenty-eight policy targets identified within these objectives include creating an Air Defence Identification Zone; boosting satellite-based remote

▼ Map 6 **Indonesia: key military units and locations**

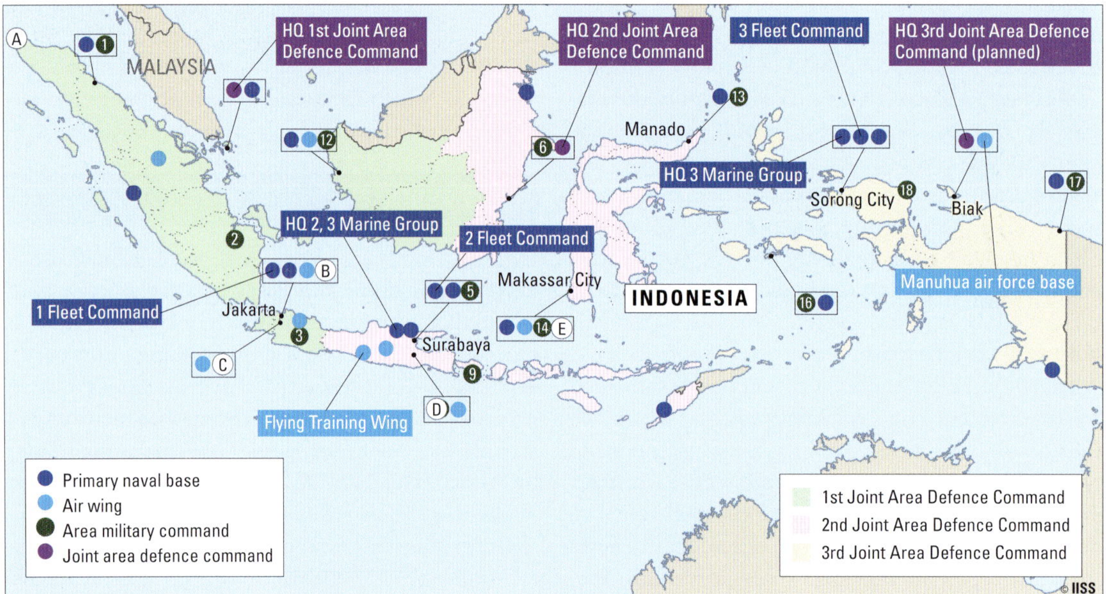

❶ HQ KODAM 1 (Bukit Barisan)
Including: 1 x inf bde, 1 x raider inf bn

❷ HQ KODAM 2 (Sriwijaya)
Including: 1 x raider inf bn

❸ HQ KODAM 3 (Siliwangi)
Including: 1 x inf bde, 1 x raider inf bn

❺ HQ KODAM 5 (Brawijaya)
Including: 1 x mech inf bde, 1 x raider inf bn

❻ HQ KODAM 6 (Mulawarman)
Including: 1 x inf bde, 1 x raider inf bn

❾ HQ KODAM 9 (Udayana)
Including: 1 x inf bde, 1 x raider inf bn, 1 x mech inf bn

⓬ HQ KODAM 12 (Tanjungpura)
Including: 1 x inf bde

⓭ HQ KODAM 13 (Merdeka)
Including: 1 x inf bde, 1 x raider inf bn

⓮ HQ KODAM 14 (Hasanuddin)
Including: 1 x inf bde, 1 x raider inf bn

⓰ HQ KODAM 16 (Pattimura)
Including: 1 x inf bde, 1 x raider inf bn

⓱ HQ KODAM 17 (Cendrawasih)
Including: 1 x raider inf bn

⓲ HQ KODAM 18 (Kasuari)
Including: 1 x inf bde, 1 x inf bn

Ⓐ HQ KODAM (Iskandar Muda)
Including: 1 x special raider inf bde, 1 x raider bn

Ⓑ HQ KODAM (Jaya)
Including: 1 x mech inf bde, 1 x AD regt

Ⓒ HQ 1 Inf Div (KOSTRAD)
Including: 1 x inf bde, 1 x AB bde, 1 x arty regt

Ⓓ HQ 2 Inf Div
Including: 1 x mech inf bde, 1 x inf bde, 1 x AB bde, 1 x arty regt

Ⓔ HQ 3 Inf Div (KOSTRAD)
Including: 1 x AB bde

sensing and surveillance; developing a tri-service TNI presence in key outer islands; boosting land, naval, air and cyber capabilities; deploying missile-defence systems; securing strategic choke points; and establishing 100 battalion-level reserve units as well as decentralised defence logistics depots. Some of these targets were in earlier plans and have been established for the past decade, while others are more recent additions. The reserve units, for example, were mandated by Law No. 23 of 2019, concerning the establishment of the resource-management system for national defence.

Overall, the document comprises a mix of legacy policies, long-established defence strategy and doctrine, and urgent modernisation requirements for ageing equipment. Indeed, the deterioration of Indonesia's strategic environment in recent years, especially recurring maritime incidents with China, has created political support for defence policymakers to implement both pre-existing plans and new objectives. These factors increase the likelihood that the 2020 State Defence Policy will encapsulate the defence ministry's ambitions under defence minister Prabowo until 2024, if not also beyond that date.

Moreover, the document elaborates the reasons behind various international defence engagements and activities over the past year. Since late 2019, the defence minister has travelled to and met senior officials and ministers from almost a dozen countries – including China, France, India and Turkey – to discuss proposals for defence procurement and technological cooperation. This flurry of activity was partially influenced by concern over possible sanctions by the United States, under the Countering America's Adversaries Through Sanctions Act (CAATSA), if Indonesia were to follow through on its plan to obtain Su-35 combat aircraft from Russia. That said, these efforts to find additional defence suppliers might have long-term consequences. As things stand, Indonesia's inventory of major systems comes from around two dozen different suppliers. Further increasing supplier diversity would likely exacerbate inefficiencies as the TNI would have to resolve interoperability and training problems as well as various maintenance, repair and overhaul concerns. It remains unclear how the defence ministry would address these challenges.

As well as seeking alternative sources of defence equipment, the commitment of Prabowo's defence ministry to 'complete' the MEF blueprint by 2024 also appears to have driven this increase in international engagement. While the 2020 State Defence Policy does not explicitly mention the MEF, it has taken some of the MEF's remaining capability-development plans and framed them as ongoing efforts to fulfil the TNI's 'essential force' development. However, the 2020 State Defence Policy might be the nucleus of a new long-term strategic-planning blueprint for the next 20 years.

The armed forces

Under Prabowo, the defence ministry has taken the lead in shaping Indonesia's capability-development plans, while the TNI General Headquarters has focused on addressing various organisational challenges, from doctrinal development to personnel and career management. However, the process of post-authoritarian defence transformation, following the fall of President Suharto in 1998, remains far from complete. Following efforts at de-politicisation in the early 2000s and technological-modernisation plans under the MEF after 2010, the armed forces have only recently focused on improving personnel, infrastructure and organisation.

The TNI leadership has also had to confront increasing internal pressures, relating to the career management of its senior mid-ranking and high-ranking officers. Problems of 'promotional logjams', where too many officers are available for too few posts, have been particularly challenging for the Indonesian Army (Tentara Nasional Indonesia Angkatan Darat, TNI-AD). Internal studies show, for example, that between 2011 and 2017, the TNI-AD had on average a 'surplus' of some 30 generals and about 330 colonels per year – officers who had to wait between one and three years before they could be posted to an available position corresponding to their rank and experience.

To address this problem, the TNI leadership has gradually expanded the armed forces' organisational structure, including reviving defunct units or commands, creating new ones, and 'upgrading' the ranks required for key staff and command positions. This expansion has taken place across both the TNI's territorial as well as its combat forces, although most are within the army.

The army created or upgraded at least 127 high-ranking posts between 2013 and 2019. During that period the number of established three-star, two-star and one-star posts increased by 300%, 164% and 162% respectively. The broader TNI leadership also created or upgraded at least 145 high-ranking posts between 2011 and 2019 (including 57 upgrades from colonel-rank posts to one-star posts). Between these years, the number of established three-star, two-star and one-star generals went up by 86%, 33% and 106% respectively.

Career management problems therefore provide the organisational context underlying the establishment of new military units and commands over the past two years. These include the three Joint Area Defence Commands (KOGABWILHAN), the Special Forces Command, the 3rd Infantry Division of the Strategic Army Reserve Command (KOSTRAD) and others. The army has also expanded and upgraded over a dozen of its territorial units and infantry battalions. The TNI is responsible for organisational expansion, and the TNI commander has said that these commands were designed to address the requirements of future warfare, strengthen the forces' geographical focus and their interoperability, and boost deterrence and crisis-management capacity.

The TNI is also continuing to gradually reformulate its doctrinal documents, formalising what it calls its 'parent doctrine' in June 2018. Dubbed 'Tri Dharma Eka Karma' or TRIDEK (roughly translated as 'dedicated and unified tri-service struggle'), the document provides a foundation for the TNI's force management, deployment and employment. As a philosophical and fundamental strategy, it is intended as a springboard for future doctrinal documents, from single-service to operational and tactical doctrines.

The TNI has been testing and revising its doctrines over the past two decades. Some of the current drafts are temporary in nature as it is envisaged that the revision process could take years. Indeed, single-service doctrines might be revised in future as the various doctrinal precepts including on Joint Operations and Military Operations Other Than War are still in their development and testing stages. Detailed Rules of Engagement documents will also likely follow the finalisation of these doctrines. These challenges suggest that the creation of new units and commands is not just intended to address changes in Indonesia's operational environment, and the TNI leadership remains concerned with addressing pressing organisational challenges, even as it tries to meet contemporary operational challenges.

The navy (Tentara Nasional Indonesia Angkatan Laut, TNI-AL) remains concerned with various strategic and day-to-day operational challenges, ranging from illegal, unregulated and unreported fishing to piracy and illegal incursions into the country's territorial waters. In responding to these relatively low-intensity challenges, the TNI-AL sometimes competes, rather than cooperates, with the Coast Guard, Marine Police, and fisheries ministry enforcement units. The need for capabilities suitable for use in humanitarian-assistance and disaster-relief tasks has led the navy to procure four *Makassar*-class landing-platform dock vessels, to a South Korean design but with two of the four built locally. At the higher end of the capability spectrum, introducing advanced German-designed and South Korean-built Type 209/1400 submarines (with one launched at a new facility in Surabaya) will boost the TNI-AL's capacity for sea-denial.

The army remains involved in addressing the internal conflict in Papua and West Papua as well as the ongoing *Operation Tinombala*, a joint military–police counter-terrorism operation in Central Sulawesi since 2016. Recently, thousands of army personnel have also been deployed to assist the government's coronavirus response; army and police personnel have been deployed to assist localised government lockdown efforts. Hundreds of army officers have also been assigned as deputy chiefs to local pandemic-management task forces.

The air force (Tentara Nasional Indonesia Angkatan Udara (TNI-AU) remains concerned with managing airspace violations as well as its broader management, which includes plans to take over the parts of the Flight Information Region shared with Singapore around Batam and the Natuna islands. Although the air force's modernisation focus has been on developing its combat-aircraft development force, it has – like the navy – developed its mobility capability. Focus on HA/DR requirements is one influence, but so too is Indonesia's sheer geographic scale. Current programmes include procurement of C295 light-transport aircraft, while the air force has indicated interest in buying two A400M heavy-transport aircraft.

Defence economics and industry

Indonesia's government seems to be committed to sustaining the level of defence-spending growth seen since 2008. Amid broader fiscal demands arising from the coronavirus pandemic, the defence budget for 2021 is nonetheless projected to amount to Rp137tr (US$9.4 billion), representing a 12% real increase from the final 2020 budget of Rp122tr (US$8.4bn) and a more sizeable 16% real-terms increase on final 2020 spending, which amounted to Rp118tr (US$8.1bn).

Indeed, notwithstanding cuts in 2016 and 2018, Indonesia's defence budget has grown substantially since 2008, with annual growth averaging 8.5% in real terms (13.0% in nominal terms). Nonetheless, over this period the defence budget has peaked at 0.9% of GDP in 2015 and with annual allocations averaging 0.7% between 2008 and 2020. The 2021 budget increase is sizeable, but defence spending will still only account for 0.8% of GDP.

The costs allocated for procurement, research and development, operations, and exercises or training hovers around 30–40% of total expenditure, and most of the budget is devoted to personnel and routine expenditures. The small budget share allocated to modernisation has exacerbated 'economy-of-scale' challenges for the navy and air force, who, for major capabilities, have to settle for a small number of complex platforms insufficient to boost operational and capability development in the long term.

Since Law No. 16 on defence-industrial development was passed in 2012, the government has been legally required to engage in various offset policies for major procurement projects, especially transfer of knowledge and technology. An example of this can be seen in the collaboration with Daewoo Shipbuilding and Marine Engineering (DSME) to build and deliver three Type-209/1400 submarines for the navy. The first two boats, KRI *Nagapasa* (pennant number 403) and KRI *Ardadedali* (404), were built in South Korea and delivered in 2017 and 2018, respectively. The third, KRI *Alugoro* (405), has been locally built at PT PAL Indonesia in Surabaya and is currently undergoing sea trials. Indeed, there has also been increased focus on boosting domestic defence industries. Indonesia's PINDAD, for example, has worked with Turkey's FNSS to develop the *Harimau Hitam* medium tank. The MoD has also tasked PT PINDAD to boost the production of small- and medium-calibre munitions. It was reported in July 2020 that the defence ministry had ordered 4bn small-arms rounds, the largest contract in PINDAD's history. Overall, the basic framework for increasing domestic companies' share in arms procurement is in place, even if the MoD and the TNI are likely to continue acquiring major, complex platforms from overseas.

These broad patterns have remained in place over the past year, despite the fiscal and security challenges posed by the coronavirus pandemic. Any further budgetary adjustment would more likely be intended to provide additional funds to cover the armed forces' costs related to managing the pandemic. It does not appear that the defence budget as a whole will be significantly affected. Indeed, despite the downward revisions to the 2020 defence budget over the course of the year, final expenditure was still higher than 2019 levels. As the government relies on the security forces to manage the pandemic response, Indonesia's political leaders are unlikely to reduce their defence budget and potentially risk domestic insecurity.

That said, one recent priority could affect these broad patterns – the possible creation of a new capital city in Kalimantan that President Widodo had announced before the coronavirus pandemic. It is estimated that the armed forces would need to spend nearly US$10bn on the move. The costs include moving major headquarters; creating new regional commands, naval and air bases; setting up missile-defence systems; and procuring other assets. Indeed, providing adequate security for the new capital would necessitate moving significant forces to Kalimantan (the northwest of the island contains Malaysian territory, and also Brunei). However, the pandemic has temporarily halted these plans, and their long-term prospects remain unclear.

Afghanistan AFG

New Afghan Afghani Afs		2019	2020	2021
GDP	Afs	1.47tr	1.47tr	
	US$	18.9bn	19.0bn	
per capita	US$	507	499	
Growth	%	3.9	-5.0	
Inflation	%	2.3	5.4	
Def bdgt [a]	Afs	153bn	155bn	
	US$	1.96bn	2.01bn	
US$1=Afs		77.86	77.13	

[a] Security expenditure. Includes expenditure on Ministry of Defence, Ministry of Interior, Ministry of Foreign Affairs, National Security Council and the General Directorate of National Security. Also includes donor funding.

Real-terms defence budget trend (US$bn, constant 2015)

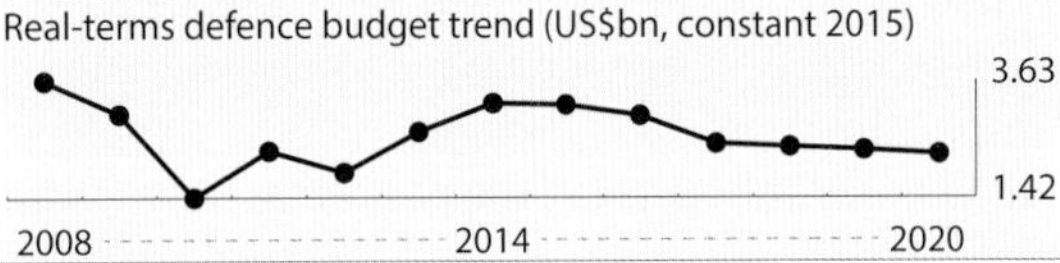

Population 36,643,815

Age	0–14	15–19	20–24	25–29	30–64	65 plus
Male	20.6%	5.4%	5.4%	4.4%	13.6%	1.2%
Female	20.0%	5.2%	5.2%	4.3%	13.2%	1.4%

Capabilities

The Afghan National Defence and Security Forces (ANDSF) are focused on counter-insurgency and counter-terrorist operations against the Taliban and other groups, and reports suggested that already high losses and high levels of desertion further increased in 2020. There was a reported 22% personnel shortage in mid-2019, and there are problems in retaining key specialists including pilots and special-operations troops. Army and police special forces bear the brunt of offensive operations and intelligence-led strike operations against insurgent networks, while regular forces are held in a defensive posture, mainly restricted to bases and outposts and at risk from Taliban attack. NATO advisers remain embedded in the defence and interior ministries, although the ANDSF are now responsible for the majority of training, albeit with NATO support. Western troop numbers continue to reduce. Efforts are under way to improve leadership, intelligence, logistics and coordination and an ANDSF Road Map contains plans to improve capability but depends on international support and improvements in maintenance. Afghan forces' logistics are organised for internal deployments, but remain of limited effectiveness. Talks were under way between the government and Taliban in 2020, and the year saw government forces under pressure in rural areas and Taliban control expand.

ACTIVE 178,800 (Army 171,500 Air Force 7,300)
Paramilitary 99,000

ORGANISATIONS BY SERVICE

Afghan National Army (ANA) 171,500

5 regional comd

FORCES BY ROLE

SPECIAL FORCES

1 spec ops div (1 (National Mission) SF bde (1 SF gp; 1 mech inf bn); 2 cdo bde (1 mech inf coy, 4 cdo bn); 1 mech bde (2 mech inf bn); 1 mech bde (3 mech inf bn); 2 cdo bn; 2 mech bn)

MANOEUVRE

Light

1 (201st) corps (3 inf bde (4 inf bn, 1 sy coy, 1 cbt spt bn, 1 CSS bn), 1 inf bde (3 inf bn, 1 sy coy, 1 cbt spt bn, 1 CSS bn), 1 sy bde, 1 engr bn, 1 int bn, 2 MP coy, 1 sigs bn)
1 (203rd) corps (2 inf bde (5 inf bn, 1 sy coy, 1 cbt spt bn, 1 CSS bn), 3 inf bde (4 inf bn, 1 sy coy, 1 cbt spt bn, 1 CSS bn), 1 sy bde, 1 engr bn, 1 int bn, 2 MP coy, 1 sigs bn)
1 (205th) corps (4 inf bde (4 inf bn, 1 sy coy, 1 cbt spt bn, 1 CSS bn), 1 sy bde, 1 engr bn, 1 int bn, 2 MP coy, 1 sigs bn)
1 (207th) corps (3 inf bde (4 inf bn, 1 sy coy, 1 cbt spt bn, 1 CSS bn), 1 sy bde, 1 engr bn, 1 int bn, 2 MP coy, 1 sigs bn)
1 (209th) corps (2 inf bde (4 inf bn, 1 sy coy, 1 cbt spt bn, 1 CSS bn), 1 sy bde, 1 engr bn, 1 int bn, 2 MP coy, 1 sigs bn)
1 (215th) corps (3 inf bde (4 inf bn, 1 sy coy, 1 cbt spt bn, 1 CSS bn), 1 inf bde (2 inf bn, 1 cbt spt bn, 1 CSS bn), 1 sy bde, 1 engr bn, 1 int bn, 2 MP coy, 1 sigs bn)
1 (217th) corps (3 inf bde (4 inf bn, 1 sy coy, 1 cbt spt bn, 1 CSS bn), 1 sy bde)
1 (111st Capital) div (1 inf bde (1 tk bn, 1 mech inf bn, 2 inf bn, 1 sy coy, 1 cbt spt bn, 1 CSS bn), 1 inf bde (4 inf bn, 1 sy coy, 1 cbt spt bn, 1 CSS bn), 1 int bn)

EQUIPMENT BY TYPE

ARMOURED FIGHTING VEHICLES

MBT 20 T-55/T-62 (24 more in store†)
APC 1,013
APC (T) 173 M113A2†
APC (W) ε640 MSFV (inc variants)
PPV 200 *Maxxpro*

ENGINEERING & MAINTENANCE VEHICLES

ARV 20 *Maxxpro* ARV
MW *Bozena*

ARTILLERY 775

TOWED 109: **122mm** 85 D-30†; **155mm** 24 M114A1†
MOR 82mm 666: 521 2B14†; 105 M-69†; 40 M252†

Afghan Air Force (AAF) 7,300

Including Special Mission Wing

EQUIPMENT BY TYPE

AIRCRAFT 34 combat capable

ISR 10 Cessna AC-208 *Combat Caravan**
TPT 45: **Medium** 4 C-130H *Hercules*; **Light** 41: 23 Cessna 208B; 18 PC-12 (Special Mission Wing)
TRG 24 EMB-314 *Super Tucano** (of which 6 in the US for trg)

HELICOPTERS

ATK 5 Mi-35 *Hind*

MRH 110: 3 *Cheetal*; 53 MD-530F (11 armed); 54 Mi-17 *Hip* H (incl 30 Special Mission Wing hel)
TPT • Medium 42 UH-60A+ *Black Hawk*
AIR-LAUNCHED MISSILES
AAM AGR-20A APKWS
BOMBS
Laser-guided GBU-12 *Paveway* II; GBU-58 *Paveway* II

Paramilitary 99,000

Afghan National Police 99,000

Under control of Interior Ministry. Includes Afghan Uniformed Police (AUP), Afghan Border Police (ABP), Public Security Police (PSP), Police Special Forces (GCPSU) and Afghan Anti-Crime Police (AACP)

FOREIGN FORCES

All *Operation Resolute Support* unless otherwise specified
Albania 99
Armenia 121
Australia 200; 1 SF unit; 1 sy unit; 1 sigs unit
Austria 11
Azerbaijan 120
Belgium 92
Bosnia-Herzegovina 66
Bulgaria 160
Croatia 76
Czech Republic 91
Denmark 140
Estonia 45
Finland 60
Georgia 860; 1 lt inf bn
Germany 1,300; 1 bde HQ; 1 recce bn; 1 hel flt with CH-53G *Stallion*; 1 ISR UAV flt with *Heron* UAV
Greece 11
Hungary 89
India Indo-Tibetan Border Police 335 (facilities protection)
Italy 800; 1 mech inf bde HQ; 1 mech inf regt(-); 1 hel regt(-) with AW129 *Mangusta*; NH90; RQ-7
Latvia 24
Lithuania 25
Luxembourg 2
Macedonia, North 42
Mongolia 233
Montenegro 32
Netherlands 160
New Zealand 9
Norway 42
Poland 397
Portugal 188
Romania 693; 1 inf bn
Slovakia 20
Slovenia 6
Spain 36
Sweden 13
Turkey 600; 1 mot inf bn(-)
Ukraine 13
United Kingdom 950; 1 inf bn; 1 hel flt with 3 SA330 *Puma* HC2 • UNAMA 1
United States 4,000; 1 div HQ; 1 spec ops bn; 2 inf bde(-); 1 EOD bn; 1 cbt avn bde(-); 1 ISR unit with RC-12X *Guardrail*; 1 EW sqn with EC-130H *Compass Call*, 1 tpt sqn with C-130J-30 *Hercules*, 1 CSAR sqn with HH-60G *Pave Hawk*; 1 CISR UAV sqn with MQ-9A *Reaper* • *Operation Freedom's Sentinel* 500

Australia AUS

Australian Dollar A$		2019	2020	2021
GDP	A$	1.99tr	1.94tr	
	US$	1.39tr	1.33tr	
per capita	US$	54,348	51,885	
Growth	%	1.8	-4.2	
Inflation	%	1.6	0.7	
Def bdgt [a]	A$	43.0bn	45.5bn	45.5bn
	US$	29.9bn	31.3bn	
US$1=A$		1.44	1.45	

[a] Includes pensions

Real-terms defence budget trend (US$bn, constant 2015)

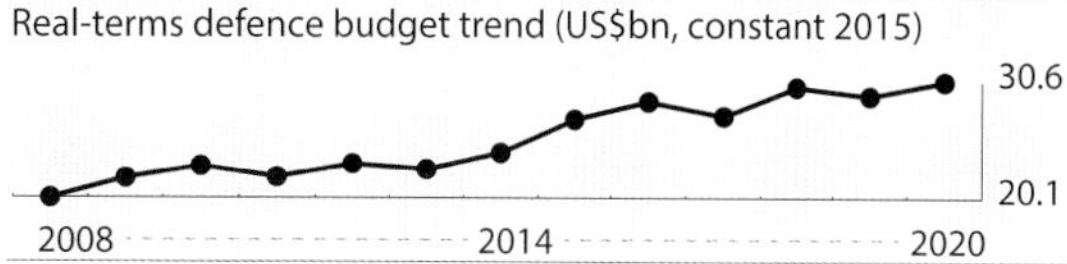

Population	25,466,459

Age	0–14	15–19	20–24	25–29	30–64	65 plus
Male	9.6%	3.1%	3.6%	3.8%	22.2%	7.3%
Female	9.1%	2.9%	3.3%	3.7%	22.8%	8.6%

Capabilities

The Australian Defence Force (ADF) is capable, well trained and well equipped. It also has considerable recent operational experience. In March 2016, the government published Australia's third defence white paper in seven years. This identified China's growing regional role, regional military modernisation and inter-state rivalry as among the influences shaping defence policy. A 'Strategic Update' to the white paper, published in July 2020, outlined plans to adjust Australia's defence posture in order to develop a more powerful and self-reliant military deterrent. The document pointed to an increased prospect of war in Australia's region, claimed that a ten-year 'strategic warning time' could no longer be assumed, and highlighted potential threats from 'grey-zone activities'. The country's primary ally is the US, but it is also forging closer defence ties with India, Japan and South Korea, while remaining committed to the Five Power Defence Arrangements and close defence relations with New Zealand. Strategic air- and sealift platforms give the ADF considerable capability to move and sustain deployments overseas. Australia is modernising its navy and locally building submarines, frigates and patrol ships based on European designs. Combat-air, maritime-patrol and armoured-vehicle capabilities are also being boosted, and more closely integrating Australia's armed forces – and its modern platforms – is becoming a priority. Australia imports most of its significant defence equipment but possesses a growing defence industry. Its largest naval shipbuilders are ASC and Austal, whose US subsidiary, Austal USA, builds vessels for the US Navy.

ACTIVE 58,600 (Army 29,500 Navy 14,700 Air 14,400)

RESERVE 30,100 (Army 18,750 Navy 6,300 Air 5,050)

Integrated units are formed from a mix of reserve and regular personnel. All ADF operations are now controlled by Headquarters Joint Operations Command (HQJOC)

ORGANISATIONS BY SERVICE

Space

EQUIPMENT BY TYPE

SATELLITES • COMMUNICATIONS 1 *Optus* C1 (dual use for civil/mil comms)

Army 29,500

FORCES BY ROLE

COMMAND

1 (1st) div HQ (1 sigs regt)

MANOEUVRE

Mechanised

3 mech inf bde (1 armd cav regt, 1 mech inf bn, 1 lt mech inf bn, 1 arty regt, 1 cbt engr regt, 1 sigs regt, 1 CSS bn)

Amphibious

1 (2nd RAR) amph bn

Aviation

1 (16th) avn bde (1 regt (2 ISR hel sqn), 1 regt (3 tpt hel sqn), 1 regt (2 spec ops hel sqn, 1 avn sqn))

COMBAT SUPPORT

1 (6th) cbt spt bde (1 STA regt (1 STA bty, 2 UAV bty, 1 CSS bty), 1 AD/FAC regt (integrated), 1 engr regt (2 construction sqn, 1 EOD sqn), 1 EW regt, 1 int bn)

COMBAT SERVICE SUPPORT

1 (17th) CSS bde (3 log bn, 3 med bn, 1 MP bn)

Special Operations Command

FORCES BY ROLE

SPECIAL FORCES

1 (SAS) SF regt

1 (SF Engr) SF regt

2 cdo regt

COMBAT SUPPORT

3 sigs sqn (incl 1 reserve sqn)

COMBAT SERVICE SUPPORT

1 CSS sqn

Reserve Organisations 18,750 reservists

FORCES BY ROLE

COMMAND

1 (2nd) div HQ

MANOEUVRE

Reconnaissance

3 (regional force) surv unit (integrated)

Light

1 (4th) inf bde (1 recce regt, 2 inf bn, 1 engr regt, 1 spt bn)

1 (5th) inf bde (1 recce bn, 4 inf bn, 1 engr regt, 2 spt bn)

1 (9th) inf bde (1 recce sqn, 2 inf bn, 1 spt bn)

1 (11th) inf bde (1 recce regt, 3 inf bn, 1 engr regt, 1 spt bn)

1 (13th) inf bde (1 recce sqn, 2 inf bn, 1 spt bn)

COMBAT SUPPORT

1 arty regt

1 sigs regt

COMBAT SERVICE SUPPORT

1 trg bde

EQUIPMENT BY TYPE

ARMOURED FIGHTING VEHICLES

MBT 59 M1A1 *Abrams*

RECCE 1 *Boxer* CRV (in test)

IFV 253 ASLAV-25 (all variants)

APC • APC (T) 431 M113AS4

AUV 1,120: 1,020 *Bushmaster* IMV; 100 *Hawkei*

ENGINEERING & MAINTENANCE VEHICLES

ARV 51: 15 ASLAV-F; 17 ASLAV-R; 19 M88A2

VLB 5 *Biber*

MW 20: 12 *Husky*; 8 MV-10

ANTI-TANK/ANTI-INFRASTRUCTURE

MSL • MANPATS FGM-148 *Javelin*

RCL • 84mm *Carl Gustaf*

ARTILLERY 239

TOWED 155mm 54 M777A2

MOR 81mm 185

AIR DEFENCE • SAM • Point-defence RBS-70

AMPHIBIOUS 15 LCM-8 (capacity either 1 MBT or 200 troops)

HELICOPTERS

ATK 22 *Tiger*

TPT 85: **Heavy** 10 CH-47F *Chinook*; **Medium** 75: 41 NH90 TTH (MRH90 TTH); 34 S-70A *Black Hawk*

UNINHABITED AERIAL VEHICLES

ISR • Medium 15 RQ-7B *Shadow* 200

AIR-LAUNCHED MISSILES

ASM AGM-114M *Hellfire*

Navy 14,700

Fleet Comd HQ located at Sydney. Naval Strategic Comd HQ located at Canberra

EQUIPMENT BY TYPE

SUBMARINES 6

SSK 6 *Collins* with 6 single 533mm TT with UGM-84C *Harpoon* Block 1B AShM/Mk 48 ADCAP mod 7 HWT

PRINCIPAL SURFACE COMBATANTS 11

DESTROYERS • DDGHM 3 *Hobart* with *Aegis* Baseline 8.1 C2, 2 quad lnchr with RGM-84L *Harpoon* Block II AShM, 6 8-cell Mk 41 VLS with SM-2 Block IIIB SAM/RIM-162A ESSM SAM, 2 twin 324mm SVTT Mk 32 mod 9 ASTT with MU90 LWT/Mk 54 LWT, 1 MK 15 *Phalanx* Block 1B CIWS, 1 127mm gun (capacity 1 MH-60R *Seahawk*)

FRIGATES • FFGHM 8 *Anzac* (GER MEKO 200) with 2 quad lnchr with RGM-84L *Harpoon* Block II AShM, 1 8-cell Mk 41 VLS with RIM-162B ESSM SAM, 2 triple 324mm SVTT Mk 32 mod 5 ASTT with MU90 LWT, 1 127mm gun (capacity 1 MH-60R *Seahawk* ASW hel)

PATROL AND COASTAL COMBATANTS 15

PCO 15: 13 *Armidale* (*Bay* mod); 2 *Cape* (leased)

MINE WARFARE • MINE COUNTERMEASURES 4
MHC 4 *Huon*
AMPHIBIOUS
PRINCIPAL AMPHIBIOUS SHIPS 3
LHD 2 *Canberra* (capacity 18 hel; 4 LCM-1E; 110 veh; 12 M1 *Abrams* MBT; 1,000 troops)
LSD 1 *Choules* (ex-UK *Bay*) (capacity 1 med hel; 24 MBT; 350 troops)
LANDING CRAFT • LCM 12 LCM-1E
LOGISTICS AND SUPPORT 12
AGHS 2 *Leeuwin* with 1 hel landing platform
AGS 4 *Paluma*
AOR 1 *Sirius*
AX 1 *Sycamore* (capacity 1 med hel) (operated by private company, Teekay Shipping; multi-role aviation training vessel)
AXS 1 *Young Endeavour*
The following vessels are operated by a private company, DMS Maritime:
ASR 2: 1 *Besant*; 1 *Stoker*
AXL 1 *Seahorse Mercator*

Naval Aviation 1,450

FORCES BY ROLE
ANTI SUBMARINE WARFARE
1 sqn with NH90 (MRH90)
1 sqn with MH-60R *Seahawk*
TRAINING
1 OCU sqn with MH-60R *Seahawk*
1 sqn with H135
EQUIPMENT BY TYPE
HELICOPTERS
ASW 24 MH-60R *Seahawk*
TPT 21: **Medium** 6 NH90 (MRH90); **Light** 15 H135
AIR-LAUNCHED MISSILES
ASM AGM-114M *Hellfire*

Clearance Diving Branch

FORCES BY ROLE
SPECIAL FORCES
2 diving unit

Air Force 14,400

FORCES BY ROLE
FIGHTER/GROUND ATTACK
2 sqn with F/A-18A/B *Hornet*
1 sqn with F/A-18F *Super Hornet*
1 sqn (forming) with F-35A *Lightning* II
ANTI SUBMARINE WARFARE
1 sqn with P-8A *Poseidon*
ELECTRONIC WARFARE
1 sqn with EA-18G *Growler*
ISR
1 (FAC) sqn with PC-21
1 sqn with AP-3C *Orion*
AIRBORNE EARLY WARNING & CONTROL
1 sqn with B-737-700 *Wedgetail* (E-7A)
TANKER/TRANSPORT
1 sqn with A330 MRTT (KC-30A)
TRANSPORT
1 VIP sqn with B-737BBJ; *Falcon* 7X
1 sqn with C-17A *Globemaster* III
1 sqn with C-27J *Spartan*
1 sqn with C-130J-30 *Hercules*
TRAINING
1 OCU with F-35A *Lightning* II
1 sqn with Beech 350 *King Air*
2 sqn with PC-21
2 (LIFT) sqn with *Hawk* MK127*
EQUIPMENT BY TYPE
AIRCRAFT 145 combat capable
FGA 89: 28 F/A-18A *Hornet*; 7 F/A-18B *Hornet*; 24 F/A-18F *Super Hornet*; 30 F-35A *Lightning* II (in test)
ASW 12 P-8A *Poseidon*
EW 13: 2 AP-3C *Orion* mod; 11 EA-18G *Growler**
AEW&C 6 B-737-700 *Wedgetail* (E-7A)
TKR/TPT 7 A330 MRTT (KC-30A)
TPT 47: **Heavy** 8 C-17A *Globemaster* III; **Medium** 22: 10 C-27J *Spartan*; 12 C-130J-30 *Hercules*; **Light** 12 Beech 350 *King Air*; **PAX** 5: 2 B-737BBJ (VIP); 3 *Falcon* 7X (VIP)
TRG 82: 33 *Hawk* Mk127*; 49 PC-21
AIR-LAUNCHED MISSILES
AAM • IIR AIM-9X *Sidewinder* II; ASRAAM; **ARH** AIM-120B/C-5/C-7 AMRAAM
ARM AGM-88B HARM; AGM-88E AARGM
AShM AGM-84A *Harpoon*
LACM Conventional AGM-158A JASSM
BOMBS
Laser-guided *Paveway* II/IV; Laser JDAM
INS/GPS-guided AGM-154C JSOW; JDAM; JDAM-ER

DEPLOYMENT

AFGHANISTAN: NATO • ISAF *Operation Resolute Support* (*Operation Highroad*) 200; 1 SF unit; 1 sy unit; 1 sigs unit

CYPRUS: UN • UNFICYP 3

EGYPT: MFO (*Operation Mazurka*) 27

IRAQ: *Operation Inherent Resolve* (*Okra*) 110; 1 SF gp; **NATO** • NATO Mission Iraq 2

MALAYSIA: 120; 1 inf coy (on 3-month rotational tours); 1 P-8A *Poseidon* (on rotation)

MIDDLE EAST: UN • UNTSO (*Operation Paladin*) 12

PHILIPPINES: *Operation Augury* 100 (trg team)

SOUTH SUDAN: UN • UNMISS (*Operation Aslan*) 10

UNITED ARAB EMIRATES: *Operation Accordion* 400: 1 tpt det with 2 C-130J-30 *Hercules*

FOREIGN FORCES

Singapore 230: 1 trg sqn at Pearce with PC-21 trg ac; 1 trg sqn at Oakey with 12 AS332 *Super Puma*; AS532 *Cougar*
United States US Pacific Command: 700; 1 SEWS at Pine Gap; 1 comms facility at NW Cape; 1 SIGINT stn at Pine Gap • US Strategic Command: 1 detection and tracking radar at Naval Communication Station Harold E. Holt

Bangladesh BGD

Bangladeshi Taka Tk		2019	2020	2021
GDP	Tk	25.4tr	27.7tr	
	US$	303bn	318bn	
per capita	US$	1,816	1,888	
Growth	%	8.2	3.8	
Inflation	%	5.5	5.6	
Def bdgt	Tk	307bn	321bn	344bn
	US$	3.65bn	3.68bn	
FMA (US)	US$	1.5m	0m	0m
US$1=Tk		84.04	87.31	

Real-terms defence budget trend (US$bn, constant 2015)

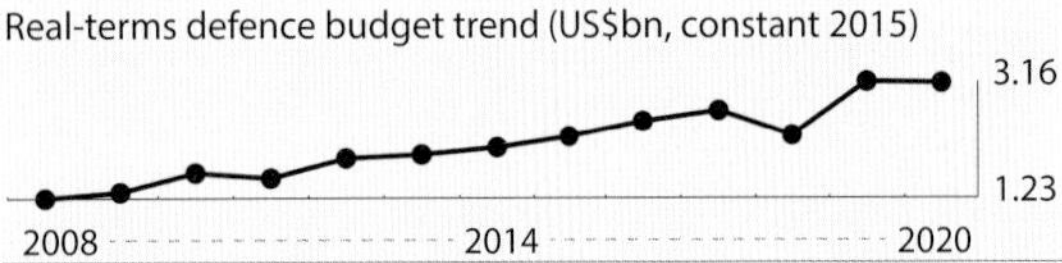

Population 162,650,853

Age	0–14	15–19	20–24	25–29	30–64	65 plus
Male	13.5%	4.6%	4.7%	4.1%	19.0%	3.2%
Female	13.0%	4.5%	4.7%	4.4%	20.6%	3.6%

Capabilities

Bangladesh has limited military capability, which is optimised for border and domestic security, including domestic disaster relief. A defence-modernisation plan is under way, called Forces 2030, although acquisitions have been limited. Bangladesh has relied on Chinese and Russian aid and credit to overcome its limited procurement funding. It has increased defence collaboration with India. The country has a long record of UN peacekeeping deployments. A major naval-recapitalisation and -expansion programme, including local manufacture of patrol boats, is under way to better protect the country's large EEZ. Airlift capability has improved with the addition of ex-UK C-130Js. Substantial efforts have also been made to strengthen the shipbuilding industry and work has begun on a new submarine-support facility. The armed forces reportedly retain extensive commercial interests, including in real estate, banks and other businesses.

ACTIVE 163,050 **(Army 132,150 Navy 16,900 Air 14,000) Paramilitary 63,900**

ORGANISATIONS BY SERVICE

Army 132,150

FORCES BY ROLE

COMMAND
- 10 inf div HQ

SPECIAL FORCES
- 1 cdo bde (2 cdo bn)

MANOEUVRE

Armoured
- 1 armd bde
- 3 indep armd regt

Light
- 25 inf bde
- 2 (composite) bde

COMBAT SUPPORT
- 10 arty bde
- 1 engr bde
- 1 sigs bde

AVIATION
- 1 avn regt (1 avn sqn; 1 hel sqn)

AIR DEFENCE
- 1 AD bde

EQUIPMENT BY TYPE

ARMOURED FIGHTING VEHICLES

MBT 276: 174 Type-59/-59G(BD); 58 Type-69/-69G; 44 Type-90-II (MBT-2000)
LT TK 8 Type-62
RECCE 8+ BOV M11
APC 514
APC (T) 134 MT-LB
APC (W) 330 BTR-80
PPV 50 *Maxxpro*
AUV 17 *Cobra*

ENGINEERING & MAINTENANCE VEHICLES

AEV MT-LB
ARV 3+: T-54/T-55; Type-84; 3 Type-654
VLB MTU

ANTI-TANK/ANTI-INFRASTRUCTURE

MSL • MANPATS 9K115-2 *Metis* M1 (AT-13 *Saxhorn*-2)
RCL 106mm 238 M40A1

ARTILLERY 889+

SP 155mm 18 NORA B-52
TOWED 363+: **105mm** 170 Model 56 pack howitzer; **122mm** 131: 57 Type-54/54-1 (M-30); 20 Type-83; 54 Type-96 (D-30), **130mm** 62 Type-59-1 (M-46)
MRL 122mm 36+ WS-22
MOR 472: **81mm** 11 M29A1; **82mm** 366 Type-53/type-87/M-31 (M-1937); **120mm** 95 AM-50/UBM 52

AMPHIBIOUS • **LANDING CRAFT** 3: 1 **LCT**; 2 **LCVP**

AIRCRAFT • **TPT** • **Light** 7: 1 C295; 5 Cessna 152; 1 PA-31T *Cheyenne*

HELICOPTERS

MRH 2 AS365N3 *Dauphin*
TPT 6: **Medium** 3 Mi-171Sh **Light** 3 Bell 206L-4

AIR DEFENCE

SAM
Short-range FM-90
Point-defence FN-16 (CH-SA-14); QW-2
GUNS • TOWED 174: **35mm** 8 GDF-009 (with *Skyguard*-3); **37mm** 132 Type-65/74; **57mm** 34 Type-59 (S-60)

Navy 16,900

EQUIPMENT BY TYPE

SUBMARINES 2

SSK 2 *Nabajatra* (ex-PRC Type-035G (*Ming*)) with 8 single 533mm TT

PRINCIPAL SURFACE COMBATANTS • **FRIGATES** 6

FFGHM 3:
1 *Bangabandhu* (ROK modified *Ulsan*) with 2 twin lnchr with *Otomat* Mk2 AShM, 1 octuple FM-90N (CH-SA-N-4) SAM, 2 triple ILAS-3 (B-515) 324mm TT with A244/S LWT, 1 76mm gun (capacity 1 AW109E hel)

2 *Umar Farooq* (ex-PRC Type-053H3 (*Jiangwei* II)) with 2 quad lnchr with YJ-83 AShM, 1 octuple GMLS with HHQ-7 (CH-SA-N-4) SAM, 2 FQF 3200 A/S mor, 1 twin 100mm gun (capacity 1 hel)

FFG 3:

2 *Abu Bakr* (ex-PRC Type-053H2 (*Jianghu* III)) with 2 twin lnchr with C-802A AShM, 2 RBU 1200 *Uragan* A/S mor, 2 twin 100mm gun

1 *Osman* (ex-PRC Type-053H1 (*Jianghu* I)) with 2 quad lnchr with C-802 (CH-SS-N-6) AShM, 2 RBU 1200 *Uragan* A/S mor, 2 twin 100mm gun

PATROL AND COASTAL COMBATANTS 46

CORVETTES 6

FSGM 4 *Shadhinota* (PRC C13B) with 2 twin lnchr with C-802 (CH-SS-N-6) AShM, 1 octuple lnchr with FL-3000N (HHQ-10) (CH-SA-N-17) SAM, 1 76mm gun, 1 hel landing platform

FSG 2 *Bijoy* (ex-UK *Castle*) (of which 1 damaged in 2020 Beirut port explosion) with 2 twin lnchr with C-704 AShM, 1 76mm gun, 1 hel landing platform

PSOH 2 *Somudra Joy* (ex-US *Hero*) with 1 76mm gun, hel landing platform

PCFG 4 *Durdarsha* (ex-PRC *Huangfeng*) with 4 single lnchr with HY-2 (CH-SS-N-2 *Safflower*) AShM

PCG 2 *Durjoy* with 2 twin lnchr with C-704 AShM, 1 76MM gun

PCO 8: 1 *Madhumati* (*Sea Dragon*) with 1 57mm gun; 5 *Kapatakhaya* (ex-UK *Island*); 2 *Durjoy* with 2 triple 324mm ASTT, 1 76mm gun

PCC 8: 2 *Meghna* with 1 57mm gun (fishery protection); 1 *Nirbhoy* (ex-PRC *Hainan*) with 4 RBU 1200 *Uragan* A/S mor; 2 twin 57mm gun; 5 *Padma*

PBFG 5 *Durbar* (PRC *Hegu*) with 2 single lnchr with SY-1 (CH-SS-N-1 *Scrubbrush*) AShM

PBF 4 *Titas* (ROK *Sea Dolphin*)

PB 7: 1 *Barkat* (ex-PRC *Shanghai* III); 2 *Karnaphuli*; 1 *Salam* (ex-PRC *Huangfen*); 3 *Shaheed Daulat* (PRC *Shanghai* II)

MINE WARFARE • MINE COUNTERMEASURES 5

MSO 5: 1 *Sagar*; 4 *Shapla* (ex-UK *River*)

AMPHIBIOUS

LANDING SHIPS • LSL 1

LANDING CRAFT 14

LCU 4 (of which 2†)

LCT 2

LCM 5 *Darshak* (*Yuchin*)

LCVP 3†

LOGISTICS AND SUPPORT 11

AG 1

AGHS 2: 1 *Agradoot*; 1 *Anushandhan*

AGS 2 *Darshak*

AOR 2 (coastal)

AOT 1 *Khan Jahangir Ali*

AR 1†

ATF 1†

AX 1 *Shaheed Ruhul Amin*

Naval Aviation

EQUIPMENT BY TYPE

AIRCRAFT • TPT • Light 2 Do-228NG (MP)

HELICOPTERS • TPT • Light 2 AW109E *Power*

Special Warfare and Diving Command 300

Air Force 14,000

FORCES BY ROLE

FIGHTER

1 sqn with MiG-29/MiG-29UB *Fulcrum*

FIGHTER/GROUND ATTACK

1 sqn with F-7MB/FT-7B *Airguard*

1 sqn with F-7BG/FT-7BG *Airguard*

1 sqn with F-7BGI/FT-7BGI *Airguard*

GROUND ATTACK

1 sqn with Yak-130 *Mitten**

TRANSPORT

1 sqn with An-32 *Cline*

1 sqn with C-130B/J *Hercules*

1 sqn with L-410UVP

TRAINING

1 sqn with K-8W *Karakorum**; L-39ZA *Albatros**

1 sqn with PT-6

TRANSPORT HELICOPTER

1 sqn with AW139; Mi-17 *Hip* H; Mi-17-1V *Hip* H; Mi-171Sh

1 sqn with Mi-17 *Hip* H; Mi-17-1V *Hip* H; Mi-171Sh

1 sqn with Bell 212

1 trg sqn with Bell 206L *Long Ranger*; AW119 *Koala*

EQUIPMENT BY TYPE

AIRCRAFT 88 combat capable

FTR 53: 9 F-7MB *Airguard*; 11 F-7BG *Airguard*; 12 F-7BGI *Airguard*; 5 FT-7B *Airguard*; 4 FT-7BG *Airguard*; 4 FT-7BGI *Airguard*; 6 MiG-29 *Fulcrum*; 2 MiG-29UB *Fulcrum* B

TPT 14: **Medium** 7: 4 C-130B *Hercules*; 3 C-130J *Hercules*; **Light** 7: 3 An-32 *Cline*†; 3 L-410UVP; 1 C295W

TRG 65: 15 K-8W *Karakorum**; 7 L-39ZA *Albatros**; 30+ PT-6; 13 Yak-130 *Mitten**

HELICOPTERS

MRH 16: 2 AW139 (SAR); 12 Mi-17 *Hip* H; 2 Mi-17-1V *Hip* H (VIP)

TPT 19: **Medium** 11 Mi-171Sh; **Light** 8: 2 Bell 206L *Long Ranger*; 4 Bell 212; 2 AW119 *Koala*

AIR-LAUNCHED MISSILES

AAM • IR R-73 (AA-11A *Archer*); PL-5; PL-7; **SARH** R-27R (AA-10A *Alamo*)

Paramilitary 63,900

Ansars 20,000+

Security Guards

Rapid Action Battalions 5,000

Ministry of Home Affairs

FORCES BY ROLE

MANOEUVRE

Other

14 paramilitary bn

Border Guard Bangladesh 38,000

FORCES BY ROLE
MANOEUVRE
Amphibious
1 rvn coy
Other
54 paramilitary bn

Coast Guard 900

EQUIPMENT BY TYPE
PATROL AND COASTAL COMBATANTS 23
PSO 4 *Syed Nazrul* (ex-ITA *Minerva*) with 1 hel landing platform
PCC 5 *Sobuj Bangla* (*Padma* mod)
PB 9: 1 *Ruposhi Bangla;* 4 *Shaheed Daulat;* 2 *Shetgang;* 2 *Sonadia*
PBR 5 *Pabna*

DEPLOYMENT

CENTRAL AFRICAN REPUBLIC: UN • MINUSCA 1,299; 1 cdo coy; 1 inf bn; 1 med coy; 1 hel coy

DEMOCRATIC REPUBLIC OF THE CONGO: UN • MONUSCO 1,703; 1 inf bn; 1 engr coy; 1 avn coy; 1 hel coy

LEBANON: UN • UNIFIL 117; 1 FSGM

MALI: UN • MINUSMA 1,310; 1 inf bn; 1 engr coy; 2 sigs coy; 1 tpt coy

SOUTH SUDAN: UN • UNMISS 1,618; 1 inf bn; 2 rvn coy; 2 engr coy

SUDAN: UN • UNAMID 4; **UN** • UNISFA 3

WESTERN SAHARA: UN • MINURSO 23; 1 fd hospital

Brunei BRN

Brunei Dollar B$		2019	2020	2021
GDP	B$	18.4bn	14.8bn	
	US$	13.5bn	10.6bn	
per capita	US$	29,314	23,117	
Growth	%	3.9	0.1	
Inflation	%	-0.4	0.3	
Def bdgt	B$	590m	606m	
	US$	432m	437m	
US$1=B$		1.36	1.39	

Real-terms defence budget trend (US$m, constant 2015)

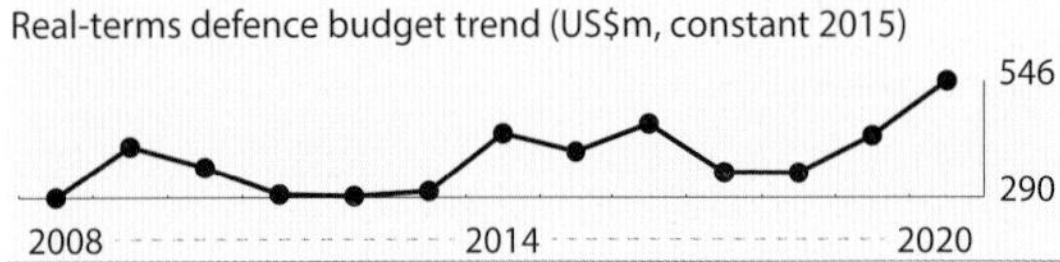

Population 464,478

Age	0–14	15–19	20–24	25–29	30–64	65 plus
Male	11.6%	3.8%	4.3%	4.4%	22.1%	2.9%
Female	10.9%	3.7%	4.4%	4.8%	24.3%	3.0%

Capabilities

The Royal Brunei Armed Forces are professional and well trained. The 2011 defence white paper set out missions such as ensuring territorial integrity and upholding the constitution. In March 2020, the government's budget statement identified protecting national assets in the South China Sea – where Brunei has a large EEZ and claims an atoll, Louisa Reef – cyber defence and supporting efforts against the coronavirus pandemic as key immediate tasks. C4ISR capabilities are being improved to offset the forces' relatively small size, and the white paper advocates procurements including to strengthen airspace control and harden C4 systems. Brunei plans to develop a fully mechanised battalion. Under a long-standing bilateral arrangement, extended to 2025, Brunei continues to host a British military presence including a Gurkha infantry battalion, and helicopter-flight and jungle-warfare schools. Brunei is a member of ASEAN and has a close relationship with Singapore, for whom it hosts a permanent training base, and the white paper advocates participation in regional exercises. The next edition of the document is expected to be released in May 2021. Brunei does not have the ability to deploy abroad without assistance, but has maintained a small deployment to UNIFIL in Lebanon since 2008 and to the IMT in southern Philippines since 2004. Brunei has no domestic defence industry and imports its military equipment. In 2010, the Centre of Science and Technology Research and Development was established to lead on defence-technology research, manage defence procurements, and provide engineering and support services to the armed forces.

ACTIVE 7,200 (Army 4,900 Navy 1,200 Air 1,100)
Paramilitary 400–500

RESERVE 700 (Army 700)

ORGANISATIONS BY SERVICE

Army 4,900

FORCES BY ROLE
MANOEUVRE
Light
3 inf bn
COMBAT SUPPORT
1 cbt spt bn (1 armd recce sqn, 1 engr sqn)

Reserves 700

FORCES BY ROLE
MANOEUVRE
Light
1 inf bn

EQUIPMENT BY TYPE
ARMOURED FIGHTING VEHICLES
LT TK 20 FV101 *Scorpion* (incl FV105 *Sultan* CP)
APC • APC (W) 45 VAB
ENGINEERING & MAINTENANCE VEHICLES
ARV 2 *Samson*
ARTILLERY • **MOR 81mm** 24

Navy 1,200

FORCES BY ROLE
SPECIAL FORCES
1 SF sqn

EQUIPMENT BY TYPE
PATROL AND COASTAL COMBATANTS 9
CORVETTES • FSG 4 *Darussalam* with 2 twin lnchr with MM40 *Exocet* Block 2 AShM, 1 57mm gun, 1 hel landing platform
PCC 4 *Ijtihad*
PBF 1 *Mustaed*
AMPHIBIOUS • LANDING CRAFT • LCM 4: 2 *Teraban*; 2 *Cheverton Loadmaster*

Air Force 1,100

FORCES BY ROLE
MARITIME PATROL
1 sqn with CN235M
TRAINING
1 sqn with Bell 206B *Jet Ranger* II
TRANSPORT HELICOPTER
1 sqn with Bell 214 (SAR)
1 sqn with Bo-105
1 sqn with S-70i *Black Hawk*
AIR DEFENCE
1 sqn with *Rapier*
1 sqn with *Mistral*
EQUIPMENT BY TYPE
AIRCRAFT
MP 1 CN235M
HELICOPTERS
TPT 21: **Medium** 13: 1 Bell 214 (SAR); 12 S-70i *Black Hawk*; **Light** 8: 2 Bell 206B *Jet Ranger* II; 6 Bo-105 (armed, 81mm rockets)
AIR DEFENCE • SAM • Point-defence *Rapier*; *Mistral*

Paramilitary 400–500

Gurkha Reserve Unit 400–500

FORCES BY ROLE
MANOEUVRE
Light
2 inf bn(-)

DEPLOYMENT

LEBANON: UN • UNIFIL 30
PHILIPPINES: IMT 8

FOREIGN FORCES

Singapore 1 trg camp with infantry units on rotation; 1 trg school; 1 hel det with AS332 *Super Puma*
United Kingdom 2,000; 1 Gurkha bn; 1 jungle trg centre; 1 hel flt with 3 Bell 212

Cambodia CAM

Cambodian Riel r		2019	2020	2021
GDP	r	109tr	108tr	
	US$	26.7bn	26.3bn	
per capita	US$	1,620	1,572	
Growth	%	7.0	-2.8	
Inflation	%	2.0	2.5	
Def bdgt [a]	r	4.24tr	4.24tr	
	US$	1.0bn	1.0bn	
US$1=r		4061.27	4108.94	

[a] Defence and security budget

Real-terms defence budget trend (US$m, constant 2015)

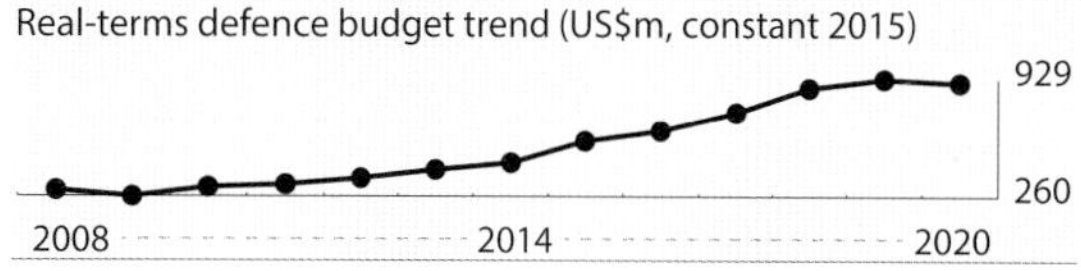

Population 16,926,984

Age	0–14	15–19	20–24	25–29	30–64	65 plus
Male	15.3%	4.4%	4.1%	4.7%	18.4%	1.7%
Female	14.9%	4.4%	4.3%	4.9%	19.9%	2.9%

Capabilities

Despite their name, which reflects Cambodia's formal status as a constitutional monarchy, the Royal Cambodian Armed Forces (RCAF) are essentially the modern manifestation of the armed forces of the former People's Republic of Kampuchea, and were established in 1979 following Vietnam's invasion. Cambodia faces no direct external military threats, besides border clashes with Thailand which last occurred in 2011. Security concerns regard mainly transnational threats that can generate instability, such as drug trafficking. Cambodia's most important international links are with the Chinese and Vietnamese armed forces. Training ties have developed with China in recent years, and exercises have grown in scale. Skirmishes on the border with Thailand since 2008 provided little indication of capacity for high-intensity combat. Cambodia lacks significant resources for personnel training, which is partly financed by Chinese military assistance. The RCAF has an excessive number of senior officers, while many formations and units appear to be of only nominal status. Cambodia has contributed personnel to UN peacekeeping missions. The armed forces rely largely on equipment donations and second-hand procurements, including from China and South Korea. Cambodia has no domestic defence industry, with no ability to design and manufacture modern equipment for its armed forces.

ACTIVE 124,300 (Army 75,000 Navy 2,800 Air 1,500 Provincial Forces 45,000) Paramilitary 67,000
Conscript liability 18 months service authorised but not implemented since 1993

ORGANISATIONS BY SERVICE

Army ε75,000

6 Military Regions (incl 1 special zone for capital)
FORCES BY ROLE
SPECIAL FORCES
1 (911th) AB/SF Bde

MANOEUVRE
Light
2 (2nd & 3rd Intervention) inf div (3 inf bde)
5 (Intervention) indep inf bde
8 indep inf bde
Other
1 (70th) sy bde (4 sy bn)
17 (border) sy bn
COMBAT SUPPORT
2 arty bn
4 fd engr regt
COMBAT SERVICE SUPPORT
1 (construction) engr regt
2 tpt bde
AIR DEFENCE
1 AD bn

EQUIPMENT BY TYPE
ARMOURED FIGHTING VEHICLES
MBT 200+: 50 Type-59; 150+ T-54/T-55
LT TK 20+: Type-62; 20 Type-63
RECCE 4+ BRDM-2
IFV 70 BMP-1
APC 230+
APC (T) M113
APC (W) 230: 200 BTR-60/BTR-152; 30 OT-64
ENGINEERING & MAINTENANCE VEHICLES
ARV T-54/T-55
MW *Bozena*; RA-140 DS
ANTI-TANK/ANTI-INFRASTRUCTURE
RCL 82mm B-10; **107mm** B-11
ARTILLERY 433+
TOWED 400+: **76mm** ZIS-3 (M-1942)/**122mm** D-30/**122mm** M-30 (M-1938)/**130mm** Type-59-I
MRL 33+: **107mm** Type-63; **122mm** 13: 8 BM-21; 5 RM-70; **132mm** BM-13-16 (BM-13); **140mm** 20 BM-14-16 (BM-14)
MOR 82mm M-37; **120mm** M-43; **160mm** M-160
AIR DEFENCE
SAM • Point-defence FN-6; FN-16 (reported)
GUNS • TOWED 14.5mm ZPU-1/ZPU-2/ZPU-4; **37mm** M-1939; **57mm** S-60

Navy ε2,800 (incl 1,500 Naval Infantry)

EQUIPMENT BY TYPE
PATROL AND COASTAL COMBATANTS 15
PBF 4 Project 205P (*Stenka*)
PB 9: 4 (PRC 46m); 3 (PRC 20m); 2 *Shershen*
PBR 2 *Kaoh Chhlam*
AMPHIBIOUS • LANDING CRAFT • LCU 1
LOGISTICS AND SUPPORT • AFDL 1

Naval Infantry 1,500

FORCES BY ROLE
MANOEUVRE
Light
1 (31st) nav inf bde
COMBAT SUPPORT
1 arty bn

Air Force 1,500

FORCES BY ROLE
ISR/TRAINING
1 sqn with P-92 *Echo*
TRANSPORT
1 VIP sqn (reporting to Council of Ministers) with An-24RV *Coke*; AS350 *Ecureuil*; AS355F2 *Ecureuil* II
1 sqn with BN-2 *Islander*; Y-12 (II)
TRANSPORT HELICOPTER
1 sqn with Mi-17 *Hip* H; Mi-8 *Hip*; Z-9

EQUIPMENT BY TYPE
AIRCRAFT
TPT • Light 9: 2 MA60; 5 P-92 *Echo* (pilot trg/recce); 2 Y-12 (II) (2 An-24RV *Coke*; 1 BN-2 *Islander* in store)
TRG (5 L-39C *Albatros** in store)
HELICOPTERS
MRH 14: 3 Mi-17 *Hip* H; 11 Z-9
TPT 4: **Heavy** (2 Mi-26 *Halo* in store); **Light** 4: 2 AS350 *Ecureuil;* 2 AS355F2 *Ecureuil* II

Provincial Forces 45,000+

Reports of at least 1 inf regt per province, with varying numbers of inf bn (with lt wpn)

Paramilitary

Police 67,000 (including gendarmerie)

DEPLOYMENT

CENTRAL AFRICAN REPUBLIC: UN • MINUSCA 214; 1 engr coy

LEBANON: UN • UNIFIL 184; 1 EOD coy

MALI: UN • MINUSMA 297: 2 engr coy(-); 1 EOD coy

SOUTH SUDAN: UN • UNMISS 79; 1 MP unit

SUDAN: UN • UNAMID 1; **UN** • UNISFA 1

China, People's Republic of PRC

Chinese Yuan Renminbi Y		2019	2020	2021
GDP	Y	99.5tr	102tr	
	US$	14.7tr	15.2tr	
per capita	US$	10,522	10,839	
Growth	%	6.1	1.9	
Inflation	%	2.9	2.9	
Def exp	Y	1.58tr	n.k	
	US$	234bn	n.k	
Def bdgt [a]	Y	1.22tr	1.30tr	
	US$	181bn	193bn	
US$1=Y		6.75	6.73	

[a] Central Expenditure budget – includes local militia funding

Real-terms defence budget trend (US$bn, constant 2015)

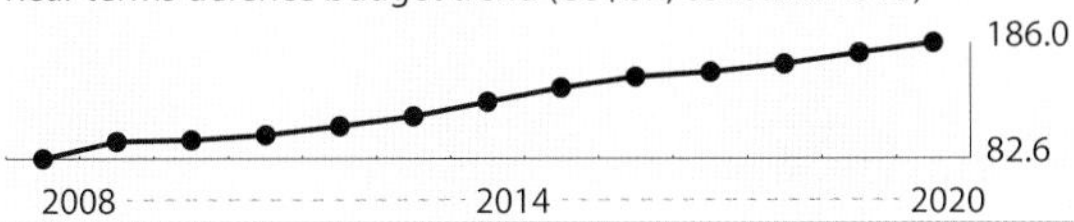

Population 1,401,880,342

Age	0–14	15–19	20–24	25–29	30–64	65 plus
Male	9.3%	2.9%	3.3%	3.8%	26.2%	5.9%
Female	8.0%	2.4%	2.8%	3.5%	25.4%	6.5%

Capabilities

China's People's Liberation Army (PLA) is the world's largest armed force, with an increasingly advanced equipment inventory. Its operational effectiveness, however, remains hampered by training and doctrine issues. China's 2019 defence white paper did not significantly alter the strategic direction laid out in the 2015 edition, and was focused more on updating the progress of PLA modernisation efforts. A major restructuring process is now mostly complete and the Strategic Support Force continues to develop China's cyber, space and information-dominance capabilities. China does not maintain any formal alliances, but it does have a number of key defence relationships with regional states and through its membership of the SCO. It has also worked to develop defence ties with several African states. Improving readiness for combat operations is a key objective of the current reforms; the PLA currently lacks any significant recent combat experience and its training has traditionally suffered from over-scripted and unrealistic exercises. Though these weaknesses are acknowledged and possibly being addressed, it is unclear how effective the newly established structures will be at generating and controlling high-intensity combined-arms capabilities. In the short term, changes to roles and requirements for officers may have had a detrimental effect on morale within the PLA, as well as on its overall level of readiness. The requirement for out-of-area operations is relatively new for the PLA; the navy is the only service to have experience in extended deployments, assisted by its support base in Djibouti. Major platform inventories in all the services comprise a mix of modern, older and obsolescent designs. The reduction in overall force size as part of the restructuring process has seen some older equipment designs withdrawn from service. China has an extensive defence-industrial base, capable of producing advanced equipment across all domains, although questions persist over quality and reliability.

ACTIVE 2,035,000 (Ground Forces 965,000 Navy 260,000 Air Force 395,000 Strategic Missile Forces 120,000 Strategic Support Force 145,000 Other 150,000) Paramilitary 500,000

Conscript liability Selective conscription; all services 24 months

RESERVE ε510,000

ORGANISATIONS BY SERVICE

Strategic Missile Forces 120,000+

People's Liberation Army Rocket Force

The People's Liberation Army Rocket Force (formerly the Second Artillery Force) organises and commands its own troops to launch nuclear counter-attacks with strategic missiles and to conduct operations with conventional missiles. Organised as launch bdes subordinate to 6 army-level msl bases. Org varies by msl type

FORCES BY ROLE

SURFACE-TO-SURFACE MISSILE

1 ICBM bde with DF-4
3 ICBM bde with DF-5A/B
1 ICBM bde with DF-31
2 ICBM bde with DF-31A
2 ICBM bde with DF-31A(G)
2 ICBM bde with DF-41 (forming)
6 IRBM bde with DF-26
2 MRBM bde with DF-16
2 MRBM bde with DF-17 with HGV (forming)
6 MRBM bde with DF-21A/E
2 MRBM bde with DF-21C/D
3 SRBM bde with DF-11A/DF-15B
2 GLCM bde with CJ-10/CJ-10A/CJ-100
5 SSM bde (forming)

EQUIPMENT BY TYPE

SURFACE-TO-SURFACE MISSILE LAUNCHERS

ICBM • Nuclear 104: ε10 DF-4 (CH-SS-3); ε20 DF-5A/B (CH-SS-4 Mod 2/3); ε8 DF-31 (CH-SS-10 Mod 1); ε24 DF-31A (CH-SS-10 Mod 2); ε24 DF-31A(G) (CH-SS-10 Mod 3); ε18 DF-41 (CH-SS-20) (entering service)

IRBM • Dual-capable 110+ DF-26

MRBM 186: **Nuclear** ε80 DF-21A/E (CH-SS-5 Mod 2/6); **Conventional** 106: ε36 DF-16 (CH-SS-11 Mod 1/2); ε16 DF-17 with HGV (entering service); ε24 DF-21C (CH-SS-5 Mod 4); ε30 DF-21D (CH-SS-5 Mod 5 – ASBM)

SRBM • Conventional 189: ε108 DF-11A (CH-SS-7 Mod 2); ε81 DF-15B (CH-SS-6 Mod 3)

GLCM • Conventional 108: ε54 CJ-10/CJ-10A (CH-SSC-9 mod 1/2); ε54 CJ-100 (entering service)

Navy

EQUIPMENT BY TYPE

SUBMARINES • STRATEGIC 6

SSBN 6 Type-094 (*Jin*) with up to 12 JL-2 (CH-SS-N-14) strategic SLBM, 6 single 533mm TT with Yu-6 HWT

Defensive

EQUIPMENT BY TYPE

RADAR • STRATEGIC: 4+ large phased array radars; some detection and tracking radars

Space

EQUIPMENT BY TYPE

SATELLITES 132

COMMUNICATIONS 9: 2 *Shen Tong*-1; 3 *Shen Tong*-2; 2 *Feng Huo*-1; 2 *Feng Huo*-2

NAVIGATION/POSITIONING/TIMING 45: 3 *Beidou*-2(M); 5 *Beidou*-2(G); 7 *Beidou*-2(IGSO); 24 *Beidou*-3(M); 3 *Beidou*-3(G); 3 *Beidou*-3(ISGO)

METEOROLOGY/OCEANOGAPHY 8: 2 *Yunhai*-1; 6 *Yunhai*-2

ISR 29: 2 *Jianbing*-5; 4 *Jianbing*-6; 3 *Jianbing*-7; 5 *Jianbing*-9; 4 *Jianbing*-10; 3 *Jianbing*-11/-12; 4 LKW; 2 *Tianhui*-2; 2 ZY-1

ELINT/SIGINT 41: 9 *Jianbing*-8; 8 *Shijian* 6 (4 pairs – reported ELINT/SIGINT role); 7 *Shijian* 11 (reported ELINT/SIGINT role); 15 *Yaogan*-30; 2 *Yaogan*-32

COUNTERSPACE • **MSL** SC-19 (reported)

Army ε965,000

FORCES BY ROLE

COMMAND

13 (Group) army HQ

SPECIAL FORCES

15 spec ops bde

MANOEUVRE

Armoured

27 (cbd arms) armd bde

1 hy mech inf div (1 armd regt, 2 mech inf regt, 1 arty regt, 1 AD regt)

Mechanised

1 (high alt) mech inf div (1 armd regt, 2 mech inf regt, 1 arty regt, 1 AD regt)

23 (cbd arms) mech inf bde

1 indep mech inf regt

Light

3 (high alt) mot inf div (1 armd regt, 2 mot inf regt, 1 arty regt, 1 AD regt)

24 (cbd arms) inf bde

Air Manoeuvre

2 air aslt bde

Amphibious

6 amph aslt bde

Other

1 (OPFOR) armd bde

1 mech gd div (1 armd regt, 2 mech inf regt, 1 arty regt, 1 AD regt)

1 sy gd div (4 sy regt)

16 (border) sy bde

15 (border) sy regt

1 (border) sy gp

COMBAT SUPPORT

15 arty bde

14 engr/NBC bde

1 engr regt

COMBAT SERVICE SUPPORT

13 spt bde

COASTAL DEFENCE

19 coastal arty/AShM bde

AVIATION

1 mixed avn bde

HELICOPTER

12 hel bde

TRAINING

4 hel trg regt

AIR DEFENCE

15 AD bde

Reserves

The People's Liberation Army Reserve Force is being restructured, and the army component reduced. As a result some of the units below may have been re-roled or disbanded

FORCES BY ROLE

MANOEUVRE

Armoured

2 armd regt

Light

18 inf div

4 inf bde

3 indep inf regt

COMBAT SUPPORT

3 arty div

7 arty bde

15 engr regt

1 ptn br bde

3 ptn br regt

10 chem regt

10 sigs regt

COMBAT SERVICE SUPPORT

9 log bde

1 log regt

AIR DEFENCE

17 AD div

8 AD bde

8 AD regt

EQUIPMENT BY TYPE

ARMOURED FIGHTING VEHICLES

MBT 5,650: 600 ZTZ-59/-59-II; 600 ZTZ-59D; 200 ZTZ-79; 300 ZTZ-88A/B; 1,000 ZTZ-96; 1,500 ZTZ-96A; 600 ZTZ-99; 600 ZTZ-99A; 250 ZTQ-15

LT TK 350: 250 ZTD-05; 100 ZTS-63A

ASLT 950 ZTL-11

IFV 6,700: 400 ZBD-04; 1,900 ZBD-04A; 2,000 ZBL-08; 600 ZBD-86; 650 ZBD-86A; 550 ZSL-92; 600 ZSL-92B

APC 3,950

APC (T) 2,700: 750 ZSD-63; 200 ZSD-63C; 1,750 ZSD-89

APC (W) 1,250: 700 ZSL-92A; 500 ZSL-10; 50 ZSL-93

AAV 600 ZBD-05

AUV Dongfeng Mengshi; *Tiger* 4×4

ENGINEERING & MAINTENANCE VEHICLES

ARV Type-73; Type-84; Type-85; Type-97; Type-654

VLB KMM; MTU; TMM; Type-84A

MW Type-74; Type-79; Type-81-II; Type-84

ANTI-TANK/ANTI-INFRASTRUCTURE

MSL

SP 1,000: 450 HJ-8 (veh mounted); 100 HJ-10; 450 ZSL-02B

MANPATS HJ-73D; HJ-8A/C/E; HJ-11

RCL 3,966: **75mm** PF-56; **82mm** PF-65 (B-10); PF-78; **105mm** PF-75; **120mm** PF-98
GUNS 1,788
SP 480: **100mm** 250 PTL-02; **120mm** 230 PTZ-89
TOWED • 100mm 1,308 PT-73 (T-12)/PT-86
ARTILLERY 9,204+
SP 2,350: **122mm** 1,680: 500 PLZ-89; 350 PLZ-07A; 150 PLZ-07B; 300 PCL-09; 350 PLL-09; 30 PCL-161; **152mm** 150 PLZ-83A/B; **155mm** 520: 320 PLZ-05; 200 PCL-181; (400 in store: **122mm** 200 PLZ-89; **152mm** 200 PLZ-83A)
TOWED 1,234: **122mm** 500 PL-96 (D-30); **130mm** 234 PL-59 (M-46)/PL-59-I; **152mm** 500 PL-66 (D-20); (4,400 in store: **122mm** 2,800 PL-54-1 (M-1938)/PL-83/PL-60 (D-74)/PL-96 (D-30); **152mm** 1,600 PL-54 (D-1)/PL-66 (D-20))
GUN/MOR 120mm 1,250: 450 PLL-05; 800 PPZ-10
MRL 1,570+ **107mm** PH-63; **122mm** 1,375: 550 PHL-81/PHL-90; 350 PHL-11; 375 PHZ-89; 100 PHZ-11; **300mm** 175 PHL-03; **370mm** 20+ PHL-19; (700 in store: **122mm** 700 PHL-81)
MOR 2,800: **82mm** PP-53 (M-37)/PP-67/PP-82/PP-87; **SP 82mm** PCP-001; **100mm** PP-89
COASTAL DEFENCE
AShM HY-1 (CH-SSC-2 *Silkworm*); HY-2 (CH-SSC-3 *Seersucker*); HY-4 (CH-SSC-7 *Sadsack*); YJ-62
PATROL AND COASTAL COMBATANTS 25
PB 25: 9 *Huzong*; 16 *Shenyang*
AMPHIBIOUS • LANDING CRAFT • LCM 255: 3+ *Yugong*; 50+ *Yunnan* II; 100+ *Yupen*; 2+ *Yutu*; approx. 100 *Yuwei*
LOGISTICS AND SUPPORT 22
AK 6+ *Leizhuang*
AKR 1 *Yunsong* (capacity 1 MBT; 1 med hel)
ARC 1
AOT 11: 1 *Fuzhong*; 8 *Fubing*; 2 *Fulei*
ATF 2 *Huntao*
AX 1 *Haixun* III
AIRCRAFT • TPT 6: **Medium** 4: 2 Y-8; 2 Y-9; **Light** 2 Y-7
HELICOPTERS
ATK 270+: 150 WZ-10; 120+ WZ-19
MRH 351: 22 Mi-17 *Hip* H; 3 Mi-17-1V *Hip* H; 38 Mi-17V-5 *Hip* H; 25 Mi-17V-7 *Hip* H; 8 SA342L *Gazelle*; 21 Z-9A; 31 Z-9W; 10 Z-9WA; 193 Z-9WZ
TPT 412: **Heavy** 111: 9 Z-8A; 96 Z-8B; 6 Z-8L; **Medium** 233: 50 Mi-8T *Hip*; 140 Mi-171; 19 S-70C2 (S-70C) *Black Hawk*; 24+ Z-20; **Light** 68: 53 AS350 *Ecureuil*; 15 H120 *Colibri*
UNINHABITED AERIAL VEHICLES
ISR • Heavy BZK-005; BZK-009 (reported); **Medium** BZK-006 (incl variants); BZK-007; BZK-008
LOITERING MUNITIONS *Harpy*
AIR DEFENCE
SAM 614+
Medium-range 200 HQ-16A/B (CH-SA-16)
Short-range 414: 24 9K331 *Tor*-M1 (SA-15 *Gauntlet*); 30 HQ-6D (CH-SA-6); 200 HQ-7A/B (CH-SA-4); 140 HQ-17 (CH-SA-15); 20+ HQ-17A
Point-defence HN-5A/B (CH-SA-3); FN-6 (CH-SA-10); QW-1 (CH-SA-7); QW-2
GUNS 7,396+
SP 396: **25mm** 270 PGZ-04A; **30mm** some PGL-19; **35mm** 120 PGZ-07; **37mm** 6 PGZ-88
TOWED 7,000+: **25mm** PG-87; **35mm** PG-99 (GDF-002); **37mm** PG-55 (M-1939)/PG-65/PG-74; **57mm** PG-59 (S-60); **100mm** PG-59 (KS-19)
AIR-LAUNCHED MISSILES
AAM • IR TY-90
ASM AKD-8; AKD-9; AKD-10

Navy ε260,000

The PLA Navy is organised into five service arms: submarine, surface, naval aviation, coastal defence and marine corps, as well as other specialised units. There are three fleets, one each in the Eastern, Southern and Northern theatre commands

EQUIPMENT BY TYPE
SUBMARINES 59
STRATEGIC • SSBN 6 Type-094 (*Jin*) with up to 12 JL-2 (CH-SS-N-14) strategic SLBM, 6 single 533mm TT with Yu-6 HWT
TACTICAL 53
SSN 6:
2 Type-093 (*Shang* I) with 6 single 533mm TT with YJ-82 (CH-SS-N-7) AShM or YJ-18 (CH-SS-N-13) AShM/Yu-3 HWT/Yu-6 HWT
4 Type-093A (*Shang* II) with 6 single 533mm TT with YJ-82 (CH-SS-N-7) AShM or YJ-18 (CH-SS-N-13) AShM/Yu-3 HWT/Yu-6 HWT
(3 Type-091 (*Han*) in reserve with 6 single 533mm TT with YJ-82 (CH-SS-N-7) AShM/Yu-3 HWT)
SSK 46:
2 Project 877 (*Kilo*) with 6 single 533mm TT with TEST-71ME HWT/53-65KE HWT
2 Project 636 (Improved *Kilo*) with 6 single 533mm TT with TEST-71ME HWT/53-65KE HWT
8 Project 636M (Improved *Kilo*) with 6 single 533mm TT with TEST-71ME HWT/53-65KE HWT/3M54E *Klub*-S (SS-N-27B *Sizzler*) AShM
4 Type-035B (*Ming*) with 8 single 533mm TT with Yu-3 HWT/Yu-4 HWT
12 Type-039(G) (*Song*) with 6 single 533mm TT with YJ-82 (CH-SS-N-7) AShM or YJ-18 (CH-SS-N-13) AShM/Yu-3 HWT/Yu-6 HWT
4 Type-039A (*Yuan*) (fitted with AIP) with 6 533mm TT with YJ-82 (CH-SS-N-7) AShM or YJ-18 (CH-SS-N-13) AShM/Yu-3 HWT/Yu-6 HWT
14 Type-039B (*Yuan*) (fitted with AIP) with 6 533mm TT with YJ-82 (CH-SS-N-7) AShM or YJ-18 (CH-SS-N-13) AShM/Yu-3 HWT/Yu-6 HWT
(10 Type-035(G) (*Ming*) in reserve with 8 single 533mm TT with Yu-3 HWT/Yu-4 HWT)
SSB 1 Type-032 (*Qing*) (SLBM trials)
PRINCIPAL SURFACE COMBATANTS 80
AIRCRAFT CARRIERS • CV 2:
1 Type-001 (*Kuznetsov*) with 3 18-cell GMLS with HHQ-10 (CH-SA-N-17) SAM, 2 RBU 6000 *Smerch* 2 A/S mor, 3 H/PJ-11 CIWS (capacity 18–24 J-15 ac; 17 Ka-28/Ka-31/Z-8S/Z-8JH/Z-8AEW hel)
1 Type-002 (*Kuznetsov* mod) with 3 18-cell GMLS with HHQ-10 (CH-SA-N-17) SAM, 2 RBU 6000 *Smerch* 2 A/S mor, 3 H/PJ-11 CIWS (capacity 32 J-15 ac; 12 Ka-28/Ka-31/Z-8S/Z-8JH/Z-8AEW hel)

CRUISERS • CGHM 1 Type-055 (*Renhai*) with 14 8-cell VLS (8 fore, 6 aft) with YJ-18A (CH-SS-N-13) AShM/HHQ-9B (CH-SA-N-21) SAM/Yu-8 A/S msl, 1 24-cell GMLS with HHQ-10 (CH-SA-N-17) SAM, 2 triple 324mm TT with Yu-7 LWT, 1 H/PJ-11 CIWS, 1 130mm gun (capacity 2 med hel)

DESTROYERS 31

DDGHM 29:

1 *Hangzhou* (Project 956E (*Sovremenny* I)) (in refit) with 2 quad lnchr with 3M80E *Moskit*-E (SS-N-22A *Sunburn*) AShM, 2 single 3S90E lnchr with 9M38E M-22E *Shtil* (SA-N-7 *Gadfly*) SAM, 2 twin 533mm DTA-53-956 ASTT with SET-65KE HWT/53-65KE HWT, 2 RBU 1000 *Smerch* 3 A/S mor, 4 AK630 CIWS, 2 twin 130mm gun (capacity 1 Z-9C/Ka-28 *Helix* A hel)

2 *Hangzhou* (Project 956EM (*Sovremenny* II)) with 2 quad lnchr with 3M80MVE *Moskit*-E (SS-N-22B *Sunburn*) AShM, 2 single 3S90E lnchr with 9M38E M-22E *Shtil* (SA-N-7 *Gadfly*) SAM, 2 twin 533mm DTA-53-956 ASTT with SET-65KE HWT/53-65KE HWT, 2 RBU 1000 *Smerch* 3 A/S mor, 2 *Kashtan* (CADS-N-1) CIWS, 1 twin 130mm gun (capacity 1 Z-9C/Ka-28 *Helix* A hel)

1 *Hangzhou* (Project 956E (*Sovremenny* III)) with 2 quad lnchr with YJ-12A AShM, 4 8-cell H/AJK-16 VLS with HHQ-16 (CH-SA-N-16) SAM/Yu-8 A/S msl, 2 triple 324mm ASTT with Yu-7 LWT, 4 AK630M CIWS, 2 twin 130mm gun (capacity 1 Z-9C/Ka-28 *Helix* A hel)

1 Type-051B (*Luhai*) with 4 quad lnchr with YJ-12A AShM, 4 8-cell H/AJK-16 VLS with HHQ-16 (CH-SA-N-16) SAM/Yu-8 A/S msl, 2 triple 324mm ASTT with Yu-7 LWT, 2 H/PJ-11 CIWS, 1 twin 100mm gun (capacity 2 Z-9C/Ka-28 *Helix* A hel)

2 Type-052 (*Luhu*) with 4 quad lnchr with YJ-83 AShM, 1 octuple lnchr with HHQ-7 (CH-SA-N-4) SAM, 2 triple 324mm ASTT with Yu-7 LWT, 2 FQF 2500 A/S mor, 2 H/PJ-12 CIWS, 1 twin 100mm gun (capacity 2 Z-9C hel)

2 Type-052B (*Luyang*) with 4 quad lnchr with YJ-83 AShM, 2 single 3S90E lnchr with 9M317E *Shtil*-1 (SA-N-7B) SAM, 2 triple 324mm ASTT with Yu-7 LWT, 2 H/PJ-12 CIWS, 1 100mm gun (capacity 1 Ka-28 *Helix* A hel)

6 Type-052C (*Luyang* II) with 2 quad lnchr with YJ-62 AShM, 8 8-cell VLS with HHQ-9 (CH-SA-N-9) SAM (CH-SA-N-9), 2 triple 324mm TT with Yu-7 LWT, 2 H/PJ-12 CIWS, 1 100mm gun (capacity 2 Ka-28 *Helix* A hel)

10 Type-052D (*Luyang* III) with 8 octuple VLS with YJ-18A (CH-SS-N-13) AShM/HHQ-9B (CH-SA-N-21) SAM/Yu-8 A/S msl, 1 24-cell GMLS with HHQ-10 (CH-SA-N-17) SAM, 2 triple 324mm TT with Yu-7 LWT, 1 H/PJ-12 CIWS, 1 130mm gun (capacity 2 Ka-28 *Helix* A hel)

2 Type-052D (*Luyang* III) with 8 octuple VLS with YJ-18A (CH-SS-N-13) AShM/HHQ-9B (CH-SA-N-21) SAM/Yu-8 A/S msl, 1 24-cell GMLS with HHQ-10 (CH-SA-N-17) SAM, 2 triple 324mm TT with Yu-7 LWT, 1 H/PJ-11 CIWS, 1 130mm gun (capacity 2 Ka-28 *Helix* A hel)

2 Type-052D mod (*Luyang* III mod) with 8 octuple VLS with YJ-18A (CH-SS-N-13) AShM/HHQ-9B (CH-SA-N-21) SAM/Yu-8 A/S msl, 1 24-cell GMLS with HHQ-10 (CH-SA-N-17) SAM, 2 triple 324mm TT with Yu-7 LWT, 1 H/PJ-11 CIWS, 1 130mm gun (capacity 2 Z-9/Z-20 hel)

DDGM 2 Type-051C (*Luzhou*) with 2 quad lnchr with YJ-83 AShM; 6 6-cell B-204 VLS with S-300FM *Rif*-M (SA-N-20 *Gargoyle*) SAM, 2 H/PJ-12 CIWS, 1 100mm gun, 1 hel landing platform

FRIGATES 46

FFGHM 40:

4 Type-053H3 (*Jiangwei* II) with 2 quad lnchr with YJ-83 AShM, 1 octuple lnchr with HHQ-7 (CH-SA-N-4) SAM, 2 RBU 1200 A/S mor, 1 twin 100mm gun (capacity 1 Z-9C hel)

4 Type-053H3 (*Jiangwei* II upgrade) with 2 quad lnchr with YJ-83 AShM, 1 8-cell GMLS with HHQ-10 (CH-SA-N-17) SAM, 2 RBU 1200 A/S mor, 1 twin 100mm gun (capacity 1 Z-9C hel)

2 Type-054 (*Jiangkai*) with 2 quad lnchr with YJ-83 AShM, 1 octuple lnchr with HHQ-7 (CH-SA-N-4) SAM, 2 triple 324mm TT with Yu-7 LWT, 2 RBU 1200 A/S mor, 4 AK630 CIWS, 1 100mm gun (capacity 1 Ka-28 *Helix* A/Z-9C hel)

16 Type-054A (*Jiangkai* II) with 2 quad lnchr with YJ-83 AShM, 4 8-cell VLS with Yu-8 A/S msl/HHQ-16 (CH-SA-N-16) SAM, 2 triple 324mm TT with Yu-7 LWT, 2 FQF 3200 A/S mor, 2 H/PJ-12 CIWS, 1 76mm gun (capacity 1 Ka-28 *Helix* A/Z-9C hel)

14 Type-054A (*Jiangkai* II) with 2 quad lnchr with YJ-83 AShM, 4 8-cell VLS with Yu-8 A/S msl/HHQ-16 (CH-SA-N-16) SAM, 2 triple 324mm TT with Yu-7 LWT, 2 FQF 3200 A/S mor, 2 H/PJ-11 CIWS, 1 76mm gun (capacity 1 Ka-28 *Helix* A/Z-9C hel)

FFG 6:

2 Type-053H1 (*Jianghu* I) with 2 twin lnchr with HY-2 (CH-SS-N-2 *Safflower*) AShM, 2 RBU 1200 *Uragan* A/S mor, 1 twin 100mm gun (capacity 1 Z-9C hel)

4 Type-053H1G (*Jianghu* I Upgrade) with 2 quad lnchr with YJ-83 AShM, 2 RBU 1200 *Uragan* A/S mor, 2 twin 100mm gun

PATROL AND COASTAL COMBATANTS ε179

CORVETTES • FSGM 55:

22 Type-056 (*Jiangdao*) with 2 twin lnchr with YJ-83 AShM, 1 8-cell GMLS with HHQ-10 (CH-SA-N-17) SAM, 2 triple 324mm ASTT with Yu-7 LWT, 1 76mm gun, 1 hel landing platform

33 Type-056A (*Jiangdao*) with 2 twin lnchr with YJ-83 AShM, 1 8-cell GMLS with HHQ-10 (CH-SA-N-17) SAM, 2 triple 324mm ASTT with Yu-7 LWT, 1 76mm gun, 1 hel landing platform

PCFG ε60 Type-022 (*Houbei*) with 2 quad lnchr with YJ-83 AShM, 1 H/PJ-13 CIWS

PCG 24: 6 Type-037-II (*Houjian*) with 2 triple lnchr with YJ-8 (CH-SS-N-4) AShM; 18 Type-037-IG (*Houxin*) with 2 twin lnchr with YJ-8 (CH-SS-N-4) AShM

PCC 8+:

some Type-037 (*Hainan*) with ε4 RBU 1200 *Uragan* A/S mor, 2 twin 57mm gun

some Type-037-I (*Haijiu*) with 4 RBU 1200 *Uragan* A/S mor, 1 twin 57mm gun
ε8 Type-037-IS (*Haiqing*) with 2 FQF-3200 A/S mor
PB ε32 Type-062-1 (*Shanghai* III)
MINE WARFARE • MINE COUNTERMEASURES 56:
MCO 19: 4 Type-081 (*Wochi*); 8 Type-081A (*Wochi* mod); 7 Type-082II (*Wozang*)
MSC 16: 4 Type-082 (*Wosao* I); 12 Type-082-II (*Wosao* II)
MSD 21 Type-529 (*Wonang*) (operated by *Wozang* MCO)
AMPHIBIOUS
PRINCIPAL AMPHIBIOUS SHIPS • LPD 6 Type-071 (*Yuzhao*) with 4 AK630 CIWS, 1 76mm gun (capacity 4 *Yuyi* LCAC plus supporting vehicles; 800 troops; 60 armoured vehs; 4 hel)
LANDING SHIPS 49
LST 28:
4 Type-072-IIG (*Yukan*) (capacity 2 LCVP; 10 tk; 200 troops)
9 Type-072-II/III (*Yuting* I) (capacity 10 tk; 250 troops; 2 hel)
9 Type-072A (*Yuting* II) (capacity 4 LCVP; 10 tk; 250 troops)
6 Type-072B (*Yuting* II) (capacity 4 LCVP; 10 tk; 250 troops)
LSM 21:
1 Type-073-II (*Yudeng*) with 1 twin 57mm gun (capacity 5 tk or 500 troops)
10 Type-073A (*Yunshu*) (capacity 6 tk)
7 Type-074 (*Yuhai*) (capacity 2 tk; 250 troops)
3 Type-074 (mod)
LANDING CRAFT 55
LCU 11 Type-074A (*Yubei*) (capacity 10 tanks or 150 troops)
LCM ε30 Type-067A (*Yunnan*)
LCAC 14: 10+ Type-726 (*Yuyi*); 4 *Zubr*
LOGISTICS AND SUPPORT 154
ABU 1 Type-744A
AFS 2: 1 Type-904 (*Dayun*); 1 Type-904A (*Danyao* I)
AFSH 2 Type-904B (*Danyao* II)
AG 7: 6 *Kanhai*; 1 *Kanwu*
AGB 2 Type-272 (*Yanrao*) with 1 hel landing platform
AGE 8: 2 Type-909 (*Dahua*) with 1 hel landing platform (weapons test platform); 1 *Kantan*; 3 Type-636 (*Shupang*); 1 Type-904I (*Yanqian*); 1 *Yuting* I (naval rail gun test ship)
AGI 19: 1 *Dadie*; 1 Type-815 (*Dongdiao*) with 1 hel landing platform; 9 Type-815A (*Dongdiao*) with 1 hel landing platform; 8 FT-14
AGM 4 Type-718 (*Yuan Wang*) (space and missile tracking)
AGOR 2 *Dahua*
AGS 8 Type-636A (*Shupang*) with 1 hel landing platform
AH 8: 5 *Ankang*; 1 Type-920 (*Anwei*); 2 *Qiongsha* (hospital conversion)
AOEH 2 Type-901 (*Fuyu*) with 2 H/PJ-13 CIWS
AORH 10: 2 Type-903 (*Fuchi*); 7 Type-903A (*Fuchi* mod); 1 *Fusu*
AOT 22: 4 *Fubai*; 16 Type-632 (*Fujian*); 2 *Fuxiao*
AP 4: 2 *Daguan*; 2 *Darong*
ARC 2 *Youlan*
ARS 18: 1 *Dadao*; 1 *Dadong*; 1 Type-922III (*Dalang* II); 3 Type-922IIIA (*Dalang* III); 3 *Dasan*; 4 *Datuo*; 2 *Dazhou*; 3 *Hai Jiu* 101 with 1 hel landing platform
ASR 6: 3 Type-926 (*Dalao*); 3 Type-925 (*Dajiang*) (capacity 2 Z-8)
ATF 14: ε11 *Hujiu*; 3 *Tuqiang*
AWT 8: 4 *Fujian*; 3 *Fushi*; 1 *Jinyou*
AX 4:
1 Type-0891A (*Dashi*) with 2 hel landing platforms
1 *Daxin* with 2 FQF 1200 A/S mor, 1 57mm gun, 1 hel landing platform
1 Type-927 (*Qi Ji Guang*) with 1 76mm gun, 1 hel landing platform
1 *Yudao*
ESD 1 *Donghaidao*
COASTAL DEFENCE • AShM 72 YJ-12/YJ-62 (3 regt)

Naval Aviation 26,000

FORCES BY ROLE
Naval aviation fighter/ground-attack units adopted brigade structure in 2017
BOMBER
2 regt with H-6DU/G/J
FIGHTER/GROUND ATTACK
1 bde with J-10A/S *Firebird*; Su-30MK2 *Flanker* G
1 bde with J-11B/BS *Flanker* L
1 bde with J-11B/BS *Flanker* L; JH-7A *Flounder*
1 bde with J-8F *Finback*; JH-7A *Flounder*
1 regt with J-15 *Flanker*
GROUND ATTACK
1 bde with JH-7 *Flounder*
ANTI-SUBMARINE WARFARE
2 regt with KQ-200
ELINT/ISR/ASW
1 regt with Y-8JB/X; Y-9JZ; KQ-200
AIRBORNE EARLY WARNING & CONTROL
3 regt with Y-8J; KJ-200; KJ-500
TRANSPORT
1 regt with Y-7H; Y-8C; CRJ-200/700
TRAINING
1 regt with CJ-6A
1 regt with HY-7
2 regt with JL-8
1 regt with JL-9G
1 regt with JL-9
1 regt with JL-10
1 regt with Z-9C
HELICOPTER
1 regt with Ka-27PS; Ka-28; Ka-31
1 regt with AS365N; Z-9C/D; Z-8J/JH
1 regt with Y-7G; Z-8; Z-8J; Z-8S; Z-9C/D
AIR DEFENCE
2 SAM bde with HQ-9; HQ-9B: HQ-6A

EQUIPMENT BY TYPE
AIRCRAFT 426 combat capable
BBR 45: 27 H-6G/G mod; 18 H-6J
FTR 24 J-8F *Finback*
FGA 153: 16 J-10A *Firebird*; 7 J-10S *Firebird*; 72 J-11B/BS *Flanker* L; 34 J-15 *Flanker*; 24 Su-30MK2 *Flanker* G
ATK 120: 48 JH-7; 72 JH-7A *Flounder*
ASW 16+ KQ-200

Asia

ELINT 13: 4 Y-8JB *High New* 2; 3 Y-8X; 6 Y-9JZ
AEW&C 24: 6 KJ-200 *Moth*; 14 KJ-500; 4 Y-8J *Mask*
TKR 5 H-6DU
TPT 38: **Medium** 6 Y-8C; **Light** 28: 20 Y-5; 2 Y-7G; 6 Y-7H; **PAX** 4: 2 CRJ-200; 2 CRJ-700
TRG 118: 38 CJ-6; 12 HY-7; 16 JL-8*; 28 JL-9*; 12 JL-9G*; 12 JL-10*

HELICOPTERS
ASW 32: 14 Ka-28 *Helix* A; 14 Z-9C; 4 Z-18F
AEW 10+: 9 Ka-31; 1+ Z-18 AEW
MRH 18: 7 AS365N; 11 Z-9D
SAR 11: 3 Ka-27PS; 4 Z-8JH; 2 Z-8S; 2 Z-9S
TPT 38: **Heavy** 30: 8 SA321 *Super Frelon*; 9 Z-8; 13 Z-8J; **Medium** 8 Mi-8 *Hip*

UNINHABITED AERIAL VEHICLES
ISR Heavy BZK-005; **Medium** BZK-007

AIR DEFENCE • SAM
Long-range 32: 16 HQ-9 (CH-SA-9); 16 HQ-9B (CH-SA-21)
Short-range HQ-6A (CH-SA-6)

AIR-LAUNCHED MISSILES
AAM • IR PL-5; PL-8; PL-9; R-73 (AA-11A *Archer*); **IR/SARH** R-27 (AA-10 *Alamo*); **SARH** PL-11; **ARH** R-77 (AA-12A *Adder*); PL-12
ASM KD-88
AShM Kh-31A (AS-17B *Krypton*); YJ-12; YJ-61; YJ-8K; YJ-83K; YJ-9
ARM Kh-31P (AS-17A *Krypton*); YJ-91

BOMBS
Laser-guided: LS-500J
TV-guided: KAB-500KR; KAB-1500KR

Marines ε35,000

FORCES BY ROLE

SPECIAL FORCES
1 spec ops bde

MANOEUVRE
Mechanised
1 mne bde
Light
3 mne bde
Amphibious
2 mne bde

HELICOPTER
1 bde (forming) with Z-8C

EQUIPMENT BY TYPE

ARMOURED FIGHTING VEHICLES
MBT some ZTQ-15
LT TK 73 ZTD-05
ASLT 2+ ZTL-11
IFV 10+ ZBL-08
AAV 152 ZBD-05

ANTI-TANK/ANTI-INFRASTRUCTURE
MSL • MANPATS HJ-73; HJ-8
RCL 120mm Type-98

ARTILLERY 40+
SP 122mm 40+: 20+ PLZ-07; 20+ PLZ-89
MRL 107mm PH-63
MOR 82mm

HELICOPTERS
TPT • Heavy 5 Z-8C

AIR DEFENCE • SAM • Point-defence HN-5 (CH-SA-3)

Air Force 395,000

FORCES BY ROLE

BOMBER
1 regt with H-6M
2 regt with H-6H
4 regt with H-6K
1 bde with H-6N (forming)

FIGHTER
2 bde with J-7 *Fishcan*
4 bde with J-7E *Fishcan*
4 bde with J-7G *Fishcan*
2 bde with J-8F/H *Finback*
2 bde with J-11A/Su-27UBK *Flanker*
4 bde with J-11A/J-11B/Su-27UBK *Flanker*
3 bde with J-11B/BS *Flanker* L

FIGHTER/GROUND ATTACK
6 bde with J-10A/S *Firebird*
1 bde with J-10A/C/S *Firebird*
2 bde with J-10B/S *Firebird*
1 bde with J-10B/C/S *Firebird*
2 bde with J-10C/S *Firebird*
1 bde with Su-35 *Flanker* M; Su-30MKK *Flanker* G
5 bde with J-16 *Flanker*
2 bde with Su-30MKK *Flanker* G
1 bde with J-20A (forming)

GROUND ATTACK
6 bde with JH-7A *Flounder*

ELECTRONIC WARFARE
4 regt with Y-8CB/DZ/G/XZ; Y-9G/XZ

ISR
1 regt with JZ-8F *Finback**
1 bde with JZ-8F *Finback**

AIRBORNE EARLY WARNING & CONTROL
1 regt with KJ-500
1 regt with KJ-200 *Moth*; KJ-2000; Y-8T

SEARCH & RESCUE
4 bde with Y-5; Mi-171E; Z-8
1 regt with Y-5; Mi-171E; Z-8

TANKER
1 bde with H-6U

TRANSPORT
1 (VIP) regt with A319; B-737; CRJ-200/700
1 (VIP) regt with Tu-154M; Tu-154M/D
1 regt with Il-76MD/TD *Candid*
1 regt with Il-76MD *Candid*; Il-78 *Midas*
1 regt with Y-7
2 regt with Y-9
1 regt with Y-8C/Y-20
1 regt with Y-20

TRAINING
5 bde with CJ-6/6A/6B; Y-5
6 bde with J-7; JJ-7A
13 bde with JL-8; JL-9; JL-10
1 trg bde with Y-7; Y-8C

TRANSPORT HELICOPTER
1 (VIP) regt with AS332 *Super Puma*; H225

ISR UAV
2 bde with GJ-1; GJ-2

AIR DEFENCE
1 SAM div (3 SAM regt)
24 SAM bde

EQUIPMENT BY TYPE

AIRCRAFT 2,367 combat capable

BBR 176: ε12 H-6A (trg role); ε60 H-6H/M; ε100 H-6K; 4+ H-6N

FTR 517: 100 J-7 *Fishcan*; 120 J-7E *Fishcan*; 120 J-7G *Fishcan*; 50 J-8F/H *Finback*; 95 J-11; 32 Su-27UBK *Flanker*

FGA 866+: 220 J-10A *Firebird*; 55 J-10B *Firebird*; 120+ J-10C *Firebird*; 70 J-10S *Firebird*; 130 J-11B/BS *Flanker* L; 150+ J-16 *Flanker*; 24+ J-20A; 73 Su-30MKK *Flanker* G; 24 Su-35 *Flanker* M

ATK 140 JH-7A *Flounder*

EW 19: 4 Y-8CB *High New* 1; 2 Y-8DZ; 6 Y-8G *High New* 3; 2 Y-8XZ *High New* 7; 3 Y-9G; 2 Y-9XZ

ELINT 4 Tu-154M/D *Careless*

ISR 48: 24 JZ-8 *Finback**; 24 JZ-8F *Finback**

AEW&C 19: 4 KJ-200 *Moth*; 11 KJ-500; 4 KJ-2000

C2 5: 2 B-737; 3 Y-8T *High New* 4

TKR 13: 10 H-6U; 3 Il-78 *Midas*

TPT 238+ **Heavy** 42+: 20 Il-76MD/TD *Candid*; 22+ Y-20; **Medium** 55+: 30 Y-8C; 25+ Y-9; **Light** 111: 70 Y-5; 41 Y-7/Y-7H; **PAX** 30: 3 A319; 9 B-737 (VIP); 5 CRJ-200; 5 CRJ-700; 8 Tu-154M *Careless*

TRG 1,012+: 400 CJ-6/-6A/-6B; 12+ HY-7; 50 JJ-7*; 150 JJ-7A*; 350 JL-8*; 30 JL-9*; 40+ JL-10*

HELICOPTERS

MRH 22: 20 Z-9; 2 Mi-17V-5 *Hip* H

TPT 31+: **Heavy** 18+ Z-8; **Medium** 13+: 6+ AS332 *Super Puma* (VIP); 3 H225 (VIP); 4+ Mi-171

UNINHABITED AERIAL VEHICLES

CISR • Heavy 12+ GJ-1; GJ-2; GJ-11 (in test)

ISR • Heavy 14+: 12+ EA-03; 2+ WZ-8

AIR DEFENCE

SAM 846+

Long-range 552+: 180 HQ-9 (CH-SA-9); 80 HQ-9B (CH-SA-21); 100+ HQ-22; 32 S-300PMU (SA-10 *Grumble*); 64 S-300PMU1 (SA-20 *Gargoyle*); 64 S-300PMU2 (SA-20 *Gargoyle*); 32 S-400 (SA-21B *Growler*)

Medium-range 190: ε40 HQ-2/-2A/-2B (CH-SA-1); 150 HQ-12 (CH-SA-12)

Short-range 104+: 50+ HQ-6A (CH-SA-6); 24 HQ-6D (CH-SA-6); ε30 HQ-7 (CH-SA-4)

GUNS • TOWED • 57mm PG-59 (S-60)

AIR-LAUNCHED MISSILES

AAM • IR PL-5B/C; PL-8; R-73 (AA-11A *Archer*); **IIR** PL-10; **IR/SARH** R-27 (AA-10 *Alamo*); **SARH** PL-11; **ARH** PL-12; PL-15; R-77 (AA-12A *Adder*); R-77-1 (RVV-SD) (AA-12B *Adder*)

ASM AKD-9; AKD-10; KD-88; Kh-29 (AS-14 *Kedge*); Kh-59M (AS-18 *Kazoo*)

AShM Kh-31A (AS-17B *Krypton*)

ARM Kh-31P (AS-17A *Krypton*); YJ-91 (Domestically produced Kh-31P variant)

ALCM • Conventional CJ-20; YJ(KD)-63

BOMBS

Laser-guided: LS-500J; LT-2

TV-guided: KAB-500KR; KAB-1500KR

Airborne Corps

FORCES BY ROLE

SPECIAL FORCES

1 spec ops bde

MANOEUVRE

Air Manoeuvre

6 AB bde

COMBAT SERVICE SUPPORT

1 spt bde

TRANSPORT

1 bde with Y-5; Y-7; Y-8; Y-12

HELICOPTER

1 regt with WZ-10K; Z-8KA; Z-9WZ

EQUIPMENT BY TYPE

ARMOURED FIGHTING VEHICLES

ABCV 180 ZBD-03

APC • APC (T) 4 ZZZ-03 (CP)

AUV CS/VN3 mod

ANTI-TANK/ANTI-INFRASTRUCTURE

SP some HJ-9

ARTILLERY 162+

TOWED 122mm ε54 PL-96 (D-30)

MRL 107mm ε54 PH-63

MOR 54+: **82mm** some; **100mm** 54

AIRCRAFT • TPT 40: **Medium** 6 Y-8; **Light** 34: 20 Y-5; 2 Y-7; 12 Y-12D

HELICOPTERS

ATK 8 WZ-10K

CSAR 8 Z-8KA

MRH 12 Z-9WZ

AIR DEFENCE

SAM • Point-defence QW-1 (CH-SA-7)

GUNS • TOWED 25mm 54 PG-87

Strategic Support Force ε175,000

At the end of 2015, a new Strategic Support Force was established by drawing upon capabilities previously exercised by the PLA's 3rd and 4th departments and other central functions. It reports to the Central Military Commission and is responsible for the PLA's space and cyber capabilities

Theatre Commands

In early 2016, the previous seven military regions were consolidated into five new theatre commands

Eastern Theatre Command

Eastern Theatre Ground Forces

71st Group Army

(1 spec ops bde, 4 armd bde, 1 mech inf bde, 1 inf bde, 1 arty bde, 1 engr/NBC bde bde, 1 spt bde, 1 hel bde, 1 AD bde)

72nd Group Army

(1 spec ops bde, 1 armd bde, 1 mech inf bde, 2 inf bde, 2 amph bde, 1 arty bde, 1 engr/NBC bde, 1 spt bde, 1 hel bde, 1 AD bde)

73rd Group Army

(1 spec ops bde, 1 armd bde, 1 mech inf bde, 2 inf bde, 2 amph bde, 1 arty bde, 1 engr/NBC bde, 1 spt bde, 1 hel bde, 1 AD bde)

Eastern Theatre Navy

Coastal defence from south of Lianyungang to Dongshan (approx. 35°10′N to 23°30′N), and to

seaward; HQ at Ningbo; support bases at Fujian, Zhoushan, Ningbo
18 **SSK**; 12 **DDGHM**; 17 **FFGHM**; 2 **FFG**; 23 **FSGM**; ε30 **PCFG/PCG**; ε22 **MCMV**; 2 **LPD**; ε22 **LST/M**

Eastern Theatre Navy Aviation

1st Naval Aviation Division
(1 AEW&C regt with KJ-500: 1 ASW regt with KQ-200)
Other Forces
(1 bbr regt with H-6DU/G/J; 1 FGA bde with JH-7; 1 FGA bde with Su-30MK2; J-10A; 1 hel regt with Ka-27PS; Ka-28; Ka-31)

Eastern Theatre Air Force

10th Bomber Division
(1 bbr regt with H-6H; 1 bbr regt with H-6K; 1 bbr regt with H-6M)
26th Special Mission Division
(1 AEW&C regt with KJ-500; 1 AEW&C regt with KJ-200/KJ-2000/Y-8T)
Fuzhou Base
(1 ftr bde with J-7E; 1 ftr bde with J-11A/B; 1 FGA bde with J-16; 1 FGA bde with Su-30MKK; 2 SAM bde)
Shanghai Base
(1 ftr bde with J-7E; 1 ftr bde with J-8F/H; 1 ftr bde with J-11B; 1 FGA bde with J-10A; 1 FGA bde with J-16; 1 FGA bde with J-20A; 2 atk bde with JH-7A; 1 trg bde with J-7/JJ-7A; 2 SAM bde)
Other Forces
(1 ISR bde with JZ-8F; 1 SAR bde; 1 Flight Instructor Training Base with CJ-6; JL-8; JL-9; JL-10)

Other Forces

Marines
(2 mne bde)

Southern Theatre Command

Southern Theatre Ground Forces

74th Group Army
(1 spec ops bde, 1 armd bde, 1 mech inf bde, 2 inf bde, 2 amph bde, 1 arty bde, 1 engr/NBC bde, 1 spt bde, 1 hel bde, 1 AD bde)
75th Group Army
(1 spec ops bde, 4 armd bde, 1 mech inf bde, 1 inf bde, 1 air aslt bde, 1 arty bde, 1 engr/NBC bde, 1 spt bde, 1 AD bde)
Other Forces
(1 (composite) inf bde (Hong Kong); 1 hel sqn (Hong Kong), 1 AD bn (Hong Kong))

Southern Theatre Navy

Coastal defence from Dongshan (approx. 23°30′N) to VNM border, and to seaward (including Paracel and Spratly islands); HQ at Zhanjiang; support bases at Yulin, Guangzhou
6 **SSBN**; 2 **SSN**; 13 **SSK**; 1 **CV**; 10 **DDGHM**; 12 **FFGHM**; 2 **FFG**; 22 **FSGM**; ε38 **PCFG/PCG**; ε16 **MCMV**; 4 **LPD**; ε21 **LST/M**

Southern Theatre Navy Aviation

3rd Naval Aviation Division
(1 ASW regt with KQ-200; 1 AEW&C regt with KJ-500)
Other Forces
(1 bbr regt with H-6DU/G/J; 1 FGA bde with J-11B; 1 FGA bde with J-11B; JH-7A; 1 tpt/hel regt with Y-7G; Z-8; Z-8J; Z-8S; Z-9C/D; 1 SAM bde)

Southern Theatre Air Force

8th Bomber Division
(2 bbr regt with H-6K)
20th Special Mission Division
(3 EW regt with Y-8CB/DZ/G/XZ; Y-9G/XZ)
Kunming Base
(1 FGA bde with J-10A; 1 FGA bde with J-10C; 1 trg bde with JJ-7A; 1 SAM bde)
Nanning Base
(1 ftr bde with J-11A; 1 FGA bde with J-10A; 1 FGA bde with J-10B/C; 1 FGA bde with J-16; 1 FGA bde with Su-35; 1 FGA bde with Su-30MKK; 1 atk bde with JH-7A; 3 SAM bde)
Other Forces
(1 tkr bde with H-6U; 1 SAR bde; 1 UAV bde)

Other Forces

Marines
(1 spec ops bde; 2 mne bde)

Western Theatre Command

Western Theatre Ground Forces

76th Group Army
(1 spec ops bde, 3 armd bde, 1 mech inf bde, 2 inf bde, 1 arty bde, 1 engr/NBC bde, 1 spt bde, 1 hel bde, 1 AD bde)
77th Group Army
(1 spec ops bde, 1 armd bde, 2 mech inf bde; 3 inf bde, 1 arty bde, 1 engr/NBC bde, 1 spt bde, 1 hel bde, 1 AD bde)
Xinjiang Military District
(1 spec ops bde, 1 (high alt) mech div, 3 (high alt) mot div, 1 mech inf regt, 1 arty bde, 1 AD bde, 1 engr regt, 1 hel bde)
Xizang Military District
(1 spec ops bde; 1 mech inf bde; 2 inf bde; 1 arty bde, 1 AD bde, 1 engr/NBC bde, 1 hel bde)

Western Theatre Air Force

4th Transport Division
(2 tpt regt with Y-9; 1 tpt regt with Y-20A)
Lanzhou Base
(1 ftr bde with J-11A/B; 1 ftr bde with J-7; 1 ftr bde with J-7E; 1 FGA bde with J-16; 1 SAM bde)
Urumqi Base
(1 ftr bde with J-8F/H; 1 ftr bde with J-11A/B; 1 atk bde with JH-7A; 2 SAM bde)
Lhasa Base
(1 SAM bde)
Xi'an Flying Academy
(1 trg bde with JJ-7A; 1 trg bde with JL-9A; 2 trg bde with JL-8; 1 trg bde with Y-7; Y-8)
Other Forces
(1 SAR regt)

Northern Theatre Command

Northern Theatre Ground Forces

78th Group Army
(1 spec ops bde, 4 armd bde, 1 mech inf bde, 1 inf bde, 1 arty bde, 1 engr/NBC bde, 1 spt bde, 1 hel bde, 1 AD bde)
79th Group Army
(1 spec ops bde, 2 armd bde, 3 mech inf bde, 1 inf bde, 1 arty bde, 1 engr/NBC bde, 1 spt bde, 1 hel bde, 1 AD bde)
80th Group Army
(1 spec ops bde, 1 armd bde; 1 mech inf bde, 4 inf bde, 1 arty bde, 1 engr/NBC bde, 1 spt bde, 1 hel bde, 1 AD bde)

Northern Theatre Navy

Coastal defence from the DPRK border (Yalu River) to south of Lianyungang (approx 35°10′N), and to seaward; HQ at Qingdao; support bases at Lushun, Qingdao.
4 **SSN**; 16 **SSK**; 1 **CV**; 1 **CGHM**; 7 **DDGHM**; 2 **DDGM**; 11 **FFGHM**; 2 **FFG**; 10 **FSGM**; ε18 **PCFG/PCG**; ε18 **MCMV**; ε7 **LST/M**

Northern Theatre Navy Aviation

2nd Naval Air Division
(1 EW/ISR/ASW regt with KQ-200; Y-8JB/X; Y-9JZ; 1 AEW&C regt with Y-8J; KJ-200; KJ-500)
Other Forces
(1 FGA regt with J-15; 1 FGA bde with JH-7A; J-8F; ; 1 hel regt with AS365N; Z-8J/JH; Z-9C/D1 tpt regt with Y-7H/Y-8C/CRJ-200/CRJ-700; 1 trg regt with CJ-6A; 2 trg regt with JL-8; 1 trg regt with HY-7; 1 trg regt with JL-9G; 1 trg regt with JL-9; 1 trg regt with JL-10)

Northern Theatre Air Force

16th Special Mission Division
(1 EW regt with Y-8CB/G; 1 ISR regt with JZ-8F)
Dalian Base
(1 ftr bde with J-7; 1 ftr bde with J-7E; 2 ftr bde with J-11B; 1 FGA bde with J-10A/C; 1 FGA bde with J-10B; 1 FGA bde with J-16; 1 atk bde with JH-7A; 1 trg bde with JJ-7A; 3 SAM bde)
Jinan Base
(1 ftr bde with J-7G; 1 FGA bde with J-10A; 1 atk bde with JH-7A; 2 SAM bde)
Harbin Flying Academy
(1 trg bde with CJ-6; Y-5; 1 trg bde with H-6; HY-7; 2 trg bde with JL-8; 1 trg bde with JL-9)
Other Forces
(1 SAR bde)

Other Forces

Marines
(2 mne bde; 1 hel bde)

Central Theatre Command

Central Theatre Ground Forces

81st Group Army
(1 spec ops bde, 2 armd bde, 1 (OPFOR) armd bde, 2 mech inf bde, 1 inf bde, 1 arty bde, 1 engr/NBC bde, 1 spt bde, 1 avn bde, 1 AD bde)
82nd Group Army
(1 spec ops bde, 2 armd bde, 2 mech bde, 2 inf bde, 1 arty bde, 1 engr/NBC bde, 1 spt bde, 1 hel bde, 1 AD bde)
83rd Group Army
(1 spec ops bde, 1 armd bde, 5 mech inf bde, 1 air aslt bde, 1 arty bde, 1 engr/NBC bde, 1 spt bde, 1 AD bde)
Other Forces
(1 hy mech inf div, 2 (Beijing) gd div)

Central Theatre Air Force

13th Transport Division
(1 tpt regt with Y-8C/Y-20A; 1 tpt regt with Il-76MD/TD; 1 tpt regt with Il-76MD; Il-78)
34th VIP Transport Division
(1 tpt regt with A319; B-737; CRJ200/700; 1 tpt regt with Tu-154M; Tu-154M/D; 1 tpt regt with Y-7; 1 hel regt with AS332; H225)
36th Bomber Division
(1 bbr regt with H-6K; 1 bbr regt with H-6H)
Datong Base
(3 ftr bde with J-7E/G; 1 ftr bde with J-11A/B; 2 FGA bde with J-10A; 1 FGA bde with J-10C; 1 SAM div; 4 SAM bde)
Wuhan Base
(2 ftr bde with J-7E/G; 1 ftr bde with J-11A; 1 FGA bde with J-10B; 1 trg bde with J-7/JJ-7A; 3 SAM bde)
Shijiazhuang Flying Academy
(3 trg bde with JL-8; 1 trg bde with JL-8; JL-10)
Airborne Corps
(6 AB bde; 1 tpt bde; 1 hel regt)
Other Forces
(1 bbr bde with H-6N; 1 SAR bde)

Paramilitary 500,000+ active

People's Armed Police ε500,000

In 2018 the People's Armed Police (PAP) divested its border-defence, firefighting, gold, forest, hydropower and security-guard units. In addition to the forces listed below, PAP also has 32 regional commands, each with one or more mobile units

FORCES BY ROLE
MANOEUVRE
Other
1 (1st Mobile) paramilitary corps (3 SF regt; 9 (mobile) paramilitary units; 1 engr/CBRN unit; 1 hel unit)
1 (2nd Mobile) paramilitary corps (2 SF unit; 9 (mobile) paramilitary units; 1 engr/CBRN unit; 1 hel unit)

China Coast Guard (CCG)

In 2018 the CCG was moved from the authority of the State Oceanic Administration to that of the People's Armed Police. The CCG is currently reorganising its pennant-number system, making it problematic to assess the number of vessels that entered service since 2019.

EQUIPMENT BY TYPE
PATROL AND COASTAL COMBATANTS 524
PSOH 42:
2 *Zhaotou* with 1 76mm gun (capacity 2 med hel)

7 Type-054 mod (*Zhaoduan*) with 1 76mm gun (capacity 1 med hel)
3 Type-053H2G (*Jiangwei* I) (capacity 1 med hel) (ex-PLAN)
4 *Shuoshi* II (capacity 1 med hel)
2 *Shucha* I (capacity 1 med hel)
10 *Shucha* II (capacity 1 med hel)
12 *Zhaoyu* (capacity 1 med hel)
1 *Zhaochang* (capacity 1 med hel)
1 *Zhongyang* (capacity 1 med hel)

PSO 45:
9 Type-718B (*Zhaojun*) with 1 76mm gun, 1 hel landing platform
1 Type-922 (*Dalang* I) (ex-PLAN)
1 Type-625C (*Hai Yang*) (ex-PLAN)
1 Type-053H (*Jianghu* I) (ex-PLAN)
1 Type-636A (*Kanjie*) with 1 hel landing platform (ex-PLAN)
6 *Shusheng* with 1 hel landing platform
3 *Shuwu*
3 *Tuzhong* (ex-PLAN)
1 Type-918 (*Wolei*) (ex-PLAN)
1 *Xiang Yang Hong* 9 (ex-PLAN)
4 *Zhaolai* with 1 hel landing platform
14 *Zhaotim*

PCO 33: 4 Type-056 mod (*Zhaogao*) with 1 hel landing platform; 1 *Shuke* I; 4 *Shuke* II; 14 *Shuke* III; 3 *Shuyou*; 4 *Zhaodai*; 3 *Zhaoming*

PCC 104: 25+ Type-618B-II; 45 *Hailin* I/II; 1 *Shuzao* II; 14 *Shuzao* III; 10 *Zhongeng*; 2 *Zhongmel*; 7 *Zhongsui*

PB/PBF 300+

AMPHIBIOUS • LANDING SHIPS 2
LST 2 Type-072-II (*Yuting* I) (ex-PLAN; used as hospital vessels and island supply)

LOGISTICS AND SUPPORT 27
AG 6: 5+ *Kaobo*; 1 *Shutu*
AGB 1 Type-071 (*Yanbing*) (ex-PLAN)
AGOR 9: 4 *Haijian*; 3 *Shuguang* 04 (ex-PLAN); 2 *Xiang Yang Hong* 9
ATF 11

AIRCRAFT
MP 1+ MA60H
TPT • Light Y-12 (MP role)

HELICOPTERS
TPT • Light Z-9

Maritime Militia

Composed of full- and part-time personnel. Reports to PLA command and trains to assist PLAN and CCG in a variety of military roles. These include ISR, maritime law enforcement, island supply, troop transport and supporting sovereignty claims. The Maritime Militia operates a variety of civilian vessels including fishing boats and oil tankers.

DEPLOYMENT

DEMOCRATIC REPUBLIC OF THE CONGO: UN • MONUSCO 226; 1 engr coy; 1 fd hospital

DJIBOUTI: 240; 1 mne coy(-); 1 med unit; 2 ZTL-11; 8 ZBL-08; 1 LPD; 1 ESD

GULF OF ADEN: 1 DDGHM; 1 FFGHM; 1 AORH

LEBANON: UN • UNIFIL 419; 2 engr coy; 1 med coy

MALI: UN • MINUSMA 426; 1 engr coy; 1 fd hospital

MIDDLE EAST: UN • UNTSO 4

SOUTH SUDAN: UN • UNMISS 1,050; 1 inf bn; 1 engr coy; 1 fd hospital

SUDAN: UN • UNAMID 367; 1 engr coy

TAJIKISTAN: ε300 (trg)

WESTERN SAHARA: UN • MINURSO 9

Fiji FJI

Fijian Dollar F$		2019	2020	2021
GDP	F$	11.7bn	9.1bn	
	US$	5.41bn	3.93bn	
per capita	US$	6,043	4,371	
Growth	%	-1.3	-21.0	
Inflation	%	1.8	-1.3	
Def bdgt	F$	121m	112m	95m
	US$	55.9m	48.5m	
US$1=F$		2.16	2.32	

Real-terms defence budget trend (US$m, constant 2015)

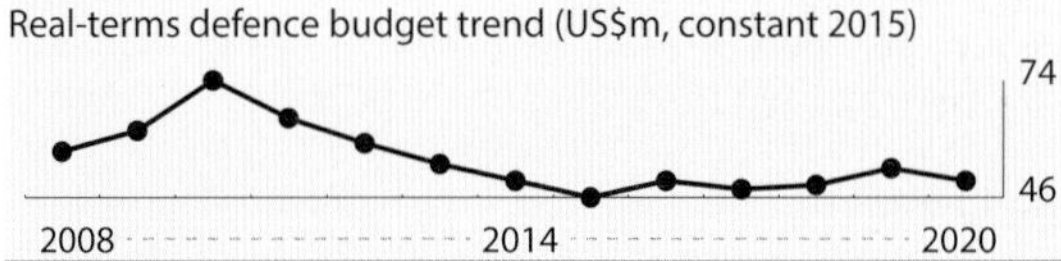

Population	935,974

Age	0–14	15–19	20–24	25–29	30–64	65 plus
Male	13.7%	4.0%	3.9%	4.0%	21.8%	3.4%
Female	13.1%	3.8%	3.8%	3.8%	20.8%	4.0%

Capabilities

The Republic of Fiji Military Forces (RFMF) are an infantry-dominated defence force with a small naval element. The RFMF has intervened heavily in Fiji's domestic politics, and between 2006 and 2014 after a third coup, democracy was effectively suspended. Guidelines issued in 2018 emphasised the need to confront non-traditional threats such as climate change, terrorism and transnational crime. The RFMF is constructing a deployable-force headquarters, funded by Australia, which will administer and train all peacekeeping and HA/DR forces. International peacekeeping operations are an important source of revenue. Fiji's principal defence relationships are with Australia and New Zealand, with whom the RFMF regularly conducts training and maritime patrols. Defence relations with China, South Korea and the US are growing, with all three countries providing training or donating equipment. The RFMF is attempting to improve the quality of senior NCOs and to raise standards across the rest of the force. Previously, personnel were sent overseas to receive this level of training. Fiji has no significant defence industry and is only able to carry out basic equipment maintenance domestically. Significant upgrade and maintenance work is usually conducted in Australia.

ACTIVE 4,040 (Army 3,700 Navy 340)

RESERVE ε6,000
(to age 45)

ORGANISATIONS BY SERVICE

Army 3,700 (incl 300 recalled reserves)

FORCES BY ROLE
SPECIAL FORCES
1 spec ops coy
MANOEUVRE
Light
3 inf bn
COMBAT SUPPORT
1 arty bty
1 engr bn
COMBAT SUPPORT
1 log bn

Reserves 6,000

FORCES BY ROLE
MANOEUVRE
Light
3 inf bn

EQUIPMENT BY TYPE
ARMOURED FIGHTING VEHICLES
AUV 10 *Bushmaster* IMV
ARTILLERY 12
MOR 81mm 12

Navy 340

EQUIPMENT BY TYPE
PATROL AND COASTAL COMBATANTS 4:
PCO 1 *Guardian* (AUS *Bay* mod)
PB 3: 1 *Kula* (AUS *Pacific*); 2 *Levuka*
LOGISTICS AND SUPPORT • AGHS 1 *Kacau*

DEPLOYMENT

EGYPT: MFO 170; elm 1 inf bn

IRAQ: UN • UNAMI 167; 2 sy unit

LEBANON: UN • UNIFIL 1

MIDDLE EAST: UN • UNTSO 2

SOUTH SUDAN: UN • UNMISS 4

SYRIA/ISRAEL: UN • UNDOF 137; 1 inf bn(-); elm 1 log bn

India IND

Indian Rupee Rs		2019	2020	2021
GDP	Rs	203tr	191tr	
	US$	2.87tr	2.59tr	
per capita	US$	2,098	1,877	
Growth	%	4.2	-10.3	
Inflation	%	4.8	4.9	
Def bdgt [a]	Rs	4.31tr	4.71tr	
	US$	60.8bn	64.1bn	
US$1=Rs		70.90	73.49	

[a] Includes defence civil estimates, which include military pensions

Real-terms defence budget trend (US$bn, constant 2015)

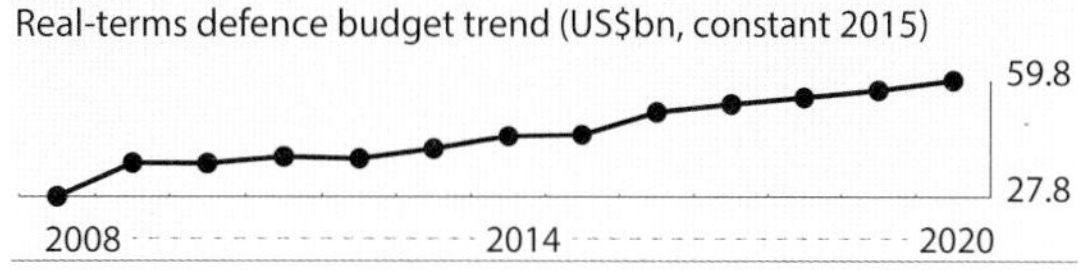

Population 1,326,093,247

Age	0–14	15–19	20–24	25–29	30–64	65 plus
Male	14.0%	4.7%	4.6%	4.4%	21.1%	3.2%
Female	12.4%	4.1%	4.1%	3.9%	20.1%	3.5%

Capabilities

India continues to modernise its armed forces, though progress in some areas remains slow. The armed forces are orientated against both Pakistan and China, and violence on the Western frontier with China in 2020 raised tensions. There is growing focus on Indian Ocean security. Large numbers of paramilitary forces remain employed in the internal-security role. Army doctrine issued in late 2018 identified requirements including for 'integrated battle groups' and improved cyber, information-warfare and electronic-warfare capabilities. A Joint Armed Forces Doctrine was issued in 2017, much of which was consistent with similar US and NATO doctrines. It set out doctrine for Indian nuclear command and control, and envisaged an 'emerging triad' of space, cyber and special-operations capabilities complementing conventional land, sea and air capabilities. India continues to develop its nuclear capabilities. In 2020 the first Chief of Defence Staff was appointed, which may improve high-level coordination of military planning. Foreign defence cooperation continues to grow, including with the US. Recent imports of foreign equipment have primarily been from the US and France, although India is also interested in Russian equipment, such as air-defence systems. Indian personnel participate in numerous bilateral and multilateral exercises, and the country is one of the main troop contributors to UN peacekeeping operations. However, the overall capability of India's large conventional forces is limited by inadequate logistics, maintenance and shortages of ammunition, spare parts and maintenance personnel. Though modernisation continues, many equipment projects have seen delays and cost overruns, particularly indigenous systems. The government's 'Make in India' policy aims to strengthen the defence-industrial base.

ACTIVE 1,458,500 (Army 1,237,000 Navy 69,050 Air 139,850 Coast Guard 12,600) Paramilitary 1,585,950

RESERVE 1,155,000 (Army 960,000 Navy 55,000 Air 140,000) Paramilitary 941,000

Army first-line reserves (300,000) within 5 years of full-time service, further 500,000 have commitment to age 50

ORGANISATIONS BY SERVICE

Strategic Forces Command

Strategic Forces Command (SFC) is a tri-service command established in 2003. The commander-in-chief of SFC, a senior three-star military officer, manages and administers all strategic forces through separate army and air-force chains of command

FORCES BY ROLE

SURFACE-TO-SURFACE MISSILE
- 1 SRBM gp with *Agni* I
- 1 MRBM gp with *Agni* II
- 1 IRBM gp (reported forming) with *Agni* III
- 2 SRBM gp with SS-250 *Prithvi* II

EQUIPMENT BY TYPE

SURFACE-TO-SURFACE MISSILE LAUNCHERS 54
- **ICBM • Nuclear** *Agni* V (in test)
- **IRBM • Nuclear** *Agni* III (entering service); *Agni* IV (in test)
- **MRBM • Nuclear** ε12 *Agni* II
- **SRBM • Nuclear** 42: ε12 *Agni* I; ε30 SS-250 *Prithvi* II; some SS-350 *Dhanush* (naval testbed)

SUBMARINES • STRATEGIC • SSBN 1 *Arihant* with 4 1-cell VLS with K-15 *Sagarika* SLBM, 6 533mm TT

AIR-LAUNCHED MISSILES • ALCM • Nuclear *Nirbhay* (likely nuclear capable; in development)

Some Indian Air Force assets (such as *Mirage* 2000H or Su-30MKI) may be tasked with a strategic role

Space

EQUIPMENT BY TYPE

SATELLITES 21
- **NAVIGATION, POSITIONING, TIMING**: 7 IRNSS
- **COMMUNICATIONS**: 2 GSAT-7/-7A
- **ISR** 11: 8 *Cartosat*; 3 RISAT
- **ELINT/SIGINT** 1 EMISAT

Army 1,237,000

6 Regional Comd HQ (Northern, Western, Central, Southern, Eastern, Southwestern), 1 Training Comd (ARTRAC)

FORCES BY ROLE

COMMAND
- 4 (strike) corps HQ
- 10 (holding) corps HQ

SPECIAL FORCES
- 8 SF bn

MANOEUVRE

Armoured
- 3 armd div (2–3 armd bde, 1 arty bde (2 arty regt))
- 8 indep armd bde

Mechanised
- 6 (RAPID) mech inf div (1 armd bde, 2 mech inf bde, 1 arty bde)
- 2 indep mech bde

Light
- 15 inf div (2–5 inf bde, 1 arty bde)
- 1 inf div (forming)
- 7 indep inf bde
- 12 mtn div (3-4 mtn inf bde, 1 arty bde)
- 2 indep mtn bde

Air Manoeuvre
- 1 para bde

SURFACE-TO-SURFACE MISSILE
- 1 SRBM gp with *Agni* I
- 1 MRBM gp with *Agni* II
- 1 IRBM gp (reported forming) with *Agni* III
- 2 SRBM gp with SS-250 *Prithvi* II
- 3 GLCM regt with PJ-10 *Brahmos*

COMBAT SUPPORT
- 3 arty div (2 arty bde, 1 MRL bde)
- 2 indep arty bde
- 4 engr bde

HELICOPTER
- 23 hel sqn

AIR DEFENCE
- 8 AD bde

Reserve Organisations

Reserves 300,000 reservists (first-line reserve within 5 years full-time service); **500,000 reservists** (commitment until age 50) (total 800,000)

Territorial Army 160,000 reservists (only 40,000 regular establishment)

FORCES BY ROLE

MANOEUVRE

Light
- 42 inf bn

COMBAT SUPPORT
- 6 (Railway) engr regt
- 2 engr regt
- 1 sigs regt

COMBAT SERVICE SUPPORT
- 6 ecological bn

EQUIPMENT BY TYPE

ARMOURED FIGHTING VEHICLES
- **MBT** 3,640: 122 *Arjun*; 2,418 T-72M1; ε1,100 T-90S (ε1,100 various models in store)
- **RECCE** *Ferret* (used for internal-security duties along with some indigenously built armd cars)
- **IFV** 3,100: 700 BMP-1; 2,400 BMP-2 *Sarath* (incl some BMP-2K CP)
- **APC** 336+
  - **APC (W)** 157+ OT-64
  - **PPV** 179: 165 *Casspir*; 14+ *Yukthirath* MPV

ENGINEERING & MAINTENANCE VEHICLES
- **AEV** BMP-2; FV180
- **ARV** 730+: T-54/T-55; 156 VT-72B; 222 WZT-2; 352 WZT-3
- **VLB** AM-50; BLG-60; BLG T-72; *Kartik*; MTU-20; MT-55; *Sarvatra*
- **MW** 24 910 MCV-2

ANTI-TANK/ANTI-INFRASTRUCTURE
- **MSL**
  - **SP** 110 9P148 *Konkurs* (AT-5 *Spandrel*)
  - **MANPATS** 9K113 *Konkurs* (AT-5 *Spandrel*); *Milan* 2
- **RCL** 3,000+: **84mm** *Carl Gustaf*; **106mm** 3,000+ M40A1 (10 per inf bn)

ARTILLERY 9,809+
SP 155mm ε80 K9 *Vajra*-T
TOWED 2,995+: **105mm** 1,350+: 600+ IFG Mk1/Mk2/Mk3; up to 700 LFG; 50 M-56; **122mm** 520 D-30; **130mm** ε600 M-46 (500 in store) **155mm** 525: ε300 FH-77B; ε200 M-46 (mod); 25 M777A2
MRL 214: **122mm** ε150 BM-21/LRAR **214mm** 36 *Pinaka*; **300mm** 28 9A52 *Smerch*
MOR 6,520+: **81mm** 5,000+ E1; **120mm** ε1,500 AM-50/E1; **SP 120mm** E1; **160mm** 20 M-58 Tampella
SURFACE-TO-SURFACE MISSILE LAUNCHERS
IRBM • Nuclear some *Agni*-III (entering service)
MRBM • Nuclear ε12 *Agni*-II
SRBM • Nuclear 42: ε12 *Agni*-I; ε30 250 *Prithvi* II
GLCM • Conventional 15 PJ-10 *Brahmos*
HELICOPTERS
MRH 320+: 79 *Dhruv*; 12 *Lancer*; 50+ *Rudra*; 119 SA315B *Lama* (*Cheetah*); 60 SA316B *Alouette* III (*Chetak*)
UNINHABITED AERIAL VEHICLES
ISR • Medium 25: 13 *Nishant*; 12 *Searcher* Mk I/II
AIR DEFENCE
SAM 680+
Medium-range *Akash*
Short-range 180 2K12 *Kub* (SA-6 *Gainful*)
Point-defence 500+: 50+ 9K33 *Osa* (SA-8B *Gecko*); 200 9K31 *Strela*-1 (SA-9 *Gaskin*); 250 9K35 *Strela*-10 (SA-13 *Gopher*); 9K310 *Igla*-1 (SA-16 *Gimlet*); 9K38 *Igla* (SA-18 *Grouse*)
GUNS 2,395+
SP 155+: **23mm** 75 ZSU-23-4; ZU-23-2 (truck-mounted); **30mm** 20-80 2S6 *Tunguska*
TOWED 2,240+: **20mm** Oerlikon (reported); **23mm** 320 ZU-23-2; **40mm** 1,920 L40/70

Navy 69,050 (incl 7,000 Naval Avn and 1,200 Marines)

Fleet HQ New Delhi. Commands located at Mumbai, Vishakhapatnam, Kochi & Port Blair

EQUIPMENT BY TYPE
SUBMARINES 16
STRATEGIC • SSBN 1 *Arihant* with 4 1-cell VLS with K-15 *Sagarika* SLBM, 6 533mm TT
TACTICAL 15
SSN 1 *Chakra* (ex-RUS *Schuka*-B (Project 971I (*Akula* II))) (RUS lease agreement) with 4 single 533mm TT with 3M14E *Klub*-S (SS-N-30B) LACM/3M54E1/E *Klub*-S (SS-N-27A/B) (*Klub*-S AShM variant unclear) AShM, 4 single 650mm TT with 65-73 HWT
SSK 14:
4 *Shishumar* (GER T-209/1500) with 8 single 533mm TT with SUT mod 1 HWT
1 *Sindhughosh* (FSU *Kilo*) with 6 single 533mm TT with 53-65KE HWT/TEST-71ME HWT/SET-65E HWT
7 *Sindhughosh* (FSU *Kilo*) with 6 single 533mm TT with 3M54E1/E *Klub*-S (SS-N-27A/B) (*Klub*-S AShM variant unclear) AShM/53-65KE HWT/TEST-71ME HWT/SET-65E HWT
2 *Kalvari* (FRA *Scorpène*) with 6 533mm TT with SM39 *Exocet* Block 2 AShM

PRINCIPAL SURFACE COMBATANTS 28
AIRCRAFT CARRIERS • CV 1 *Vikramaditya* (ex-FSU *Kiev* mod) with 3 8-cell VLS with *Barak*-1 SAM, 4 AK630M CIWS (capacity: 12 MiG-29K/KUB *Fulcrum* FGA ac; 6 Ka-28 *Helix* A ASW hel/Ka-31 *Helix* B AEW hel)
DESTROYERS 10
DDGHM 6:
2 *Delhi* with 4 quad lnchr with 3M24E *Uran*-E (SS-N-25 *Switchblade*) AShM, 2 single 3S90E lnchr with 9M38E M-22E *Shtil* (SA-N-7 *Gadfly*) SAM, 4 8-cell VLS with *Barak*-1 SAM, 5 single 533mm ASTT with SET-65E HWT/*Varunastra* HWT, 2 RBU 6000 *Smerch* 2 A/S mor; 2 AK630 CIWS, 1 100mm gun (capacity either 2 *Dhruv* hel/*Sea King* Mk42A ASW hel)
1 *Delhi* with 4 quad lnchr with 3M24E *Uran*-E (SS-N-25 *Switchblade*) AShM, 2 single 3S90E lnchr with 9M38E M-22E *Shtil* (SA-N-7 *Gadfly*) SAM, 5 single 533mm ASTT with SET-65E HWT/*Varunastra* HWT, 2 RBU 6000 *Smerch* 2 A/S mor; 2 AK630 CIWS, 1 100mm gun (capacity either 2 *Dhruv* hel/*Sea King* Mk42A ASW hel)
3 *Kolkata* with 2 8-cell UVLM VLS with *Brahmos* AShM, 4 8-cell VLS with *Barak*-8 SAM; 2 twin 533mm TT with SET-65E HWT/*Varanustra* HWT, 2 RBU 6000 *Smerch* 2 A/S mor, 4 AK630M CIWS, 1 76mm gun (capacity 2 *Dhruv*/*Sea King* Mk42B hel)
DDGM 4:
1 *Rajput* (FSU *Kashin*) with 2 twin lnchr with P-27 *Termit*-R (SS-N-2D *Styx*) AShM, 2 twin ZIF-101 lnchr with 4K91 M-1 *Volnya* (SA-N-1 *Goa*) SAM, 5 single 533mm PTA-51-61ME ASTT with SET-65E HWT/*Varanustra* HWT, 2 RBU 6000 *Smerch* 2 A/S mor, 2 AK630M CIWS, 1 76mm gun (capacity Ka-28 *Helix* A hel)
1 *Rajput* (FSU *Kashin*) with 2 twin lnchr with *Brahmos* AShM, 2 single lnchr with P-27 *Termit*-R (SS-N-2D *Styx*) AShM, 2 twin ZIF-101 lnchr with 4K91 M-1 *Volnya* (SA-N-1 *Goa*) SAM, 5 single 533mm ASTT with SET-65E HWT/*Varanustra* HWT, 2 RBU 6000 *Smerch* 2 A/S mor, 4 AK630M CIWS, 1 76mm gun (capacity 1 Ka-28 *Helix* A hel)
2 *Rajput* (FSU *Kashin*) with 1 8-cell UVLM VLS with *Brahmos* AShM, 2 twin lnchr with P-27 *Termit*-R (SS-N-2D *Styx*) AShM, 2 8-cell VLS with *Barak*-1 SAM, 1 twin ZIF-101 lnchr with 4K91 M-1 *Volnya* (SA-N-1 *Goa*) SAM, 5 single 533mm ASTT with SET-65E HWT/*Varanustra* HWT, 2 RBU 6000 *Smerch* 2 A/S mor, 4 AK630M CIWS, 1 76mm gun (capacity 1 Ka-28 *Helix* A hel)
FRIGATES 17
FFGHM 13: 3 *Brahmaputra* (of which 1 in refit) with 4 quad lnchr with 3M24E *Uran*-E (SS-N-25 *Switchblade*) AShM, 3 8-cell VLS with *Barak*-1 SAM, 2 triple ILAS-3 (B-515) 324mm ASTT with A244 LWT, 4 AK630M CIWS, 1 76mm gun (capacity 2 SA316B *Alouette* III (*Chetak*)/*Sea King* Mk42 ASW hel) (of which 1 non-operational)
1 *Godavari* with 4 single lnchr with P-27 *Termit*-R (SS-N-2D *Styx*) AShM, 1 8-cell VLS with *Barak*-1

SAM, 2 triple 324mm ILAS-3 (B-515) ASTT with A244 LWT, 4 AK630 CIWS, 1 76mm gun (capacity 2 SA316B *Alouette* III (*Chetak*)/*Sea King* Mk42 ASW hel)

3 *Shivalik* with 1 8-cell 3S14E VLS with 3M54TE *Klub*-N (SS-N-27B *Sizzler*) AShM/*Brahmos* AShM, 4 8-cell VLS with *Barak*-1 SAM, 1 single 3S90E lnchr with 9M317E *Shtil*-1 (SA-N-7B) SAM, 2 triple 324mm ILAS-3 (B-515) ASTT, 2 RBU 6000 *Smerch* 2 A/S mor, 2 AK630M CIWS, 1 76mm gun (capacity 1 *Sea King* Mk42B ASW hel)

3 *Talwar* I with 1 8-cell 3S14E VLS with 3M54TE *Klub*-N (SS-N-27B *Sizzler*) AShM, 1 single 3S90E lnchr with 9M317E *Shtil*-1 (SA-N-7B) SAM, 2 twin 533mm DTA-53-11356 ASTT with SET-65E HWT/*Varunastra* HWT, 2 RBU 6000 *Smerch* 2 A/S mor, 2 *Kashtan* (CADS-N-1) CIWS, 1 100mm gun (capacity 1 *Dhruv*/Ka-28 *Helix* A ASW hel)

3 *Talwar* II with 1 8-cell UVLM VLS with *Brahmos* AShM, 1 single 3S90E lnchr with 9M317E *Shtil*-1 (SA-N-7B) SAM, 2 twin 533mm DTA-53-11356 ASTT with SET-65E HWT/*Varunastra* HWT, 2 RBU 6000 *Smerch* 2 A/S mor, 2 AK630M CIWS, 1 100mm gun (capacity 1 *Dhruv*/Ka-28 *Helix* A ASW hel)

FFH 4 *Kamorta* with 2 twin 533mm ITTL ASTT with *Varunastra* HWT, 2 RBU 6000 *Smerch* 2 A/S mor, 2 AK630 CIWS, 1 76mm gun (capacity 1 *Dhruv*/Ka-28 *Helix* A ASW hel)

PATROL AND COASTAL COMBATANTS 170

CORVETTES • FSGM 8:

4 *Khukri* with 2 twin lnchr with P-27 *Termit*-R (SS-N-2D *Styx*) AShM, 2 twin lnchr (manual aiming) with 9K32M *Strela*-2M (SA-N-5 *Grail*) SAM, 2 AK630M CIWS, 1 76mm gun, 1 hel landing platform (for *Dhruv*/SA316 *Alouette* III (*Chetak*))

4 *Kora* with 4 quad lnchr with 3M24E *Uran*-E (SS-N-25 *Switchblade*) AShM, 1 quad lnchr (manual aiming) with 9K32M *Strela*-2M (SA-N-5 *Grail*) SAM, 2 AK630M CIWS, 1 76mm gun, 1 hel landing platform (for *Dhruv*/SA316 *Alouette* III (*Chetak*))

PSOH 10: 4 *Saryu* with 2 AK630M CIWS, 1 76mm gun (capacity 1 *Dhruv*); 6 *Sukanya* with 4 RBU 2500 A/S mor (capacity 1 SA316 *Alouette* III (*Chetak*))

PCFGM 8:

6 *Veer* (FSU *Tarantul*) with 4 single lnchr with P-27 *Termit*-R (SS-N-2D *Styx*) AShM, 2 quad lnchr (manual aiming) with 9K32M *Strela*-2M (SA-N-5 *Grail*), 2 AK630M CIWS, 1 76mm gun

2 *Prabal* (mod *Veer*) each with 4 quad lnchr with 3M24E *Uran*-E (SS-N-25 *Switchblade*) AShM, 1 quad lnchr (manual aiming) with 9K32M *Strela*-2M (SA-N-5 *Grail*) SAM, 2 AK630M CIWS, 1 76mm gun

PCMT 3 *Abhay* (FSU *Pauk* II) with 1 quad lnchr (manual aiming) with 9K32M *Strela*-2M (SA-N-5 *Grail*) SAM, 2 twin 533mm DTA-53 ASTT with SET-65E, 2 RBU 1200 *Uragan* A/S mor, 1 AK630M CIWS, 1 76mm gun

PCC 15: 4 *Bangaram*; 10 *Car Nicobar*; 1 *Trinkat* (SDB Mk5)

PCF 4 *Tarmugli* (*Car Nicobar* mod)

PBF 122: 9 Immediate Support Vessel (Rodman 78); 13 Immediate Support Vessel (Craftway); 15 Plascoa 1300 (SPB); 5 *Super Dvora*; 79 Solas Marine Interceptor

AMPHIBIOUS

PRINCIPAL AMPHIBIOUS VESSELS • LPD 1 *Jalashwa* (ex-US *Austin*) with 1 Mk 15 *Phalanx* CIWS (capacity up to 6 med spt hel; either 9 LCM or 4 LCM and 2 LCAC; 4 LCVP; 930 troops)

LANDING SHIPS 8

LSM 3 *Kumbhir* (FSU *Polnochny* C) (capacity 5 MBT or 5 APC; 160 troops)

LST 5: 2 *Magar* (capacity 15 MBT or 8 APC or 10 trucks; 500 troops); 3 *Magar* mod (capacity 11 MBT or 8 APC or 10 trucks; 500 troops)

LANDING CRAFT 11

LCT 7 LCU Mk-IV (capacity 1 *Arjun* MBT/2 T-90 MBT/4 IFV/160 troops)

LCM 4 LCM 8 (for use in *Jalashwa*)

LOGISTICS AND SUPPORT 42

AFD 2: 1 FDN-1; 1 FDN-2

AGOR 1 *Sagardhwani* with 1 hel landing platform

AGHS 8: 1 *Makar*; 7 *Sandhayak*

AO 4 GSL 1,000T Fuel Barge

AOL 10: 1 *Ambika*; 2 *Poshak* II; 7 *Purak*

AOR 1 *Jyoti* with 1 hel landing platform

AORH 3: 1 *Aditya* (based on *Deepak* (1967) Bremer Vulkan design); 2 *Deepak* with 4 AK630 CIWS

AP 3 *Nicobar* with 1 hel landing platform

ASR 1

ATF 1

AWT 3 *Ambuda* II

AX 1 *Tir*

AXS 4: 2 *Mhadei*; 2 *Tarangini*

Naval Aviation 7,000

FORCES BY ROLE

FIGHTER/GROUND ATTACK

2 sqn with MiG-29K/KUB *Fulcrum*

ANTI-SUBMARINE WARFARE

1 sqn with Ka-28 *Helix* A

1 sqn with *Sea King* Mk42B

MARITIME PATROL

4 sqn with BN-2 *Islander*; Do-228-101

1 sqn with Do-228

1 sqn with Il-38SD *May*

1 sqn with P-8I *Neptune*

AIRBORNE EARLY WARNING & CONTROL

1 sqn with Ka-31 *Helix* B

SEARCH & RESCUE

1 sqn with SA316B *Alouette* III (*Chetak*); *Sea King* Mk42C

1 sqn with *Dhruv*

TRANSPORT

1 sqn with HS-748M (HAL-748M)

TRAINING

1 sqn with Do-228

1 sqn with HJT-16 *Kiran* MkI/II, *Hawk* Mk132*

1 hel sqn with *Sea King* Mk42B

TRANSPORT HELICOPTER

1 sqn with UH-3H *Sea King*

ISR UAV
3 sqn with *Heron*; *Searcher* MkII

EQUIPMENT BY TYPE

AIRCRAFT 73 combat capable
FTR 43 MiG-29K/KUB *Fulcrum*
ASW 13: 5 Il-38SD *May*; 8 P-8I *Neptune*
MP 13+ Do-228-101
TPT 37:
Light 27: 17 BN-2 *Islander*; 10 Do-228
PAX 10 HS-748M (HAL-748M)
TRG 29: 6 HJT-16 *Kiran* MkI; 6 HJT-16 *Kiran* MkII; 17 *Hawk* Mk132*

HELICOPTERS
ASW 30: 12 Ka-28 *Helix* A; 18 *Sea King* Mk42B
MRH 57: 10 *Dhruv*; 24 SA316B *Alouette* III (*Chetak*); 23 SA319 *Alouette* III
AEW 11 Ka-31 *Helix* B
TPT • Medium 11: 5 *Sea King* Mk42C; up to 6 UH-3H *Sea King*

UNINHABITED AERIAL VEHICLES
ISR 10: **Heavy** 4 *Heron;* **Medium** 6 *Searcher* Mk II

AIR-LAUNCHED MISSILES
AAM • IR R-550 *Magic*/*Magic* 2; R-73 (AA-11A *Archer*) **IR/SARH** R-27 (AA-10 *Alamo*); **ARH**: R-77 (AA-12A *Adder*)
AShM AGM-84 *Harpoon* (on P-8I ac); Kh-35 (AS-20 *Kayak*; on *May* ac); *Sea Eagle* (possibly WFU)

Marines ε1,200 (Additional 1,000 for SPB duties)

After the Mumbai attacks, the Sagar Prahari Bal (SPB), with 80 PBF, was established to protect critical maritime infrastructure

FORCES BY ROLE

SPECIAL FORCES
1 (marine) cdo force

MANOEUVRE
Amphibious
1 amph bde

Air Force 139,850

5 regional air comds: Western (New Delhi), Southwestern (Gandhinagar), Eastern (Shillong), Central (Allahabad), Southern (Trivandrum). 2 support comds: Maintenance (Nagpur) and Training (Bangalore)

FORCES BY ROLE

FIGHTER
3 sqn with MiG-29 *Fulcrum*; MiG-29UB *Fulcrum*

FIGHTER/GROUND ATTACK
4 sqn with *Jaguar* IB/IS
6 sqn with MiG-21 *Bison*
3 sqn with *Mirage* 2000E/ED/I/IT (2000H/TH – secondary ECM role)
1 sqn with *Rafale* DH/EH (forming)
11 sqn with Su-30MKI *Flanker*
2 sqn with *Tejas*

ANTI SURFACE WARFARE
1 sqn with *Jaguar* IM

ISR
1 unit with Gulfstream IV SRA-4

AIRBORNE EARLY WARNING & CONTROL
1 sqn with Il-76TD *Phalcon*

TANKER
1 sqn with Il-78 *Midas*

TRANSPORT
1 sqn with C-130J-30 *Hercules*
1 sqn with C-17A *Globemaster* III
5 sqn with An-32/An-32RE *Cline*
1 (comms) sqn with B-737; B-737BBJ; EMB-135BJ
4 sqn with Do-228; HS-748
1 sqn with Il-76MD *Candid*
1 flt with HS-748

TRAINING
1 OCU sqn with Su-30MKI *Flanker*

ATTACK HELICOPTER
1 sqn with Mi-25 *Hind*; Mi-35 *Hind*
1 sqn with Mi-25 *Hind*; Mi-35 *Hind*; AH-64E *Apache Guardian*

TRANSPORT HELICOPTER
5 sqn with *Dhruv*
7 sqn with Mi-17/Mi-17-1V *Hip* H
12 sqn with Mi-17V-5 *Hip* H
2 sqn with SA316B *Alouette* III (*Chetak*)
1 flt with Mi-26 *Halo*
2 flt with SA315B *Lama* (*Cheetah*)
2 flt with SA316B *Alouette* III (*Chetak*)

ISR UAV
5 sqn with *Heron*; *Searcher* MkII

AIR DEFENCE
25 sqn with S-125 *Pechora* (SA-3B *Goa*)
6 sqn with 9K33 *Osa*-AK (SA-8B *Gecko*)
2 sqn with *Akash*
10 flt with 9K38 *Igla*-1 (SA-18 *Grouse*)

EQUIPMENT BY TYPE

AIRCRAFT 768 combat capable
FTR 61: 54 MiG-29 *Fulcrum* (incl 12+ MiG-29UPG); 7 MiG-29UB *Fulcrum* B
FGA 491: 112 MiG-21 *Bison*; 38 MiG-21U/UM *Mongol*; 39 *Mirage* 2000E/I (2000H); 10 *Mirage* 2000ED/IT (2000TH); 8 *Rafale* DH; 5 *Rafale* EH; 263 Su-30MKI *Flanker* H; 16 *Tejas*
ATK 115: 28 *Jaguar* IB; 79 *Jaguar* IS; 8 *Jaguar* IM
ISR 3 Gulfstream IV SRA-4
AEW&C 5: 2 EMB-145AEW *Netra* (1 more in test); 3 Il-76TD *Phalcon*
TKR 6 Il-78 *Midas*
TPT 243: **Heavy** 28: 11 C-17A *Globemaster* III; 17 Il-76MD *Candid*; **Medium** 10 C-130J-30 *Hercules*; **Light** 141: 47 An-32; 55 An-32RE *Cline*; 35 Do-228; 4 EMB-135BJ; **PAX** 64: 1 B-707; 4 B-737; 3 B-737BBJ; 56 HS-748
TRG 308: 101 *Hawk* Mk132*; 90 HJT-16 *Kiran* MkI/IA; 42 HJT-16 *Kiran* MkII; 75 PC-7 *Turbo Trainer* MkII

HELICOPTERS
ATK 39: 22 AH-64E *Apache Guardian*; 17 Mi-25/Mi-35 *Hind*
MRH 389: 60 *Dhruv*; 35 Mi-17 *Hip* H; 45 Mi-17-1V *Hip* H; 148 Mi-17V-5 *Hip* H; 59 SA315B *Lama* (*Cheetah*); 39 SA316B *Alouette* III (*Chetak*); 3+ *Rudra*
TPT • Heavy 16: 15 CH-47F *Chinook*; 1+ Mi-26 *Halo*

UNINHABITED AERIAL VEHICLES
ISR • Heavy 9 *Heron*; **Medium** some *Searcher* MkII

LOITERING MUNITIONS *Harop*
AIR DEFENCE • SAM
Medium-range *Akash*
Short-range S-125 *Pechora* (SA-3B *Goa*); *Spyder*-SR
Point-defence 9K33 *Osa*-AK (SA-8B *Gecko*); 9K38 *Igla* (SA-18 *Grouse*)
AIR-LAUNCHED MISSILES
AAM • IR R-60 (AA-8 *Aphid*); R-73 (AA-11A *Archer*) R-550 *Magic*; **IIR** *Mica IR*; **IR/SARH** R-27 (AA-10 *Alamo*); **SARH** Super 530D **ARH** R-77 (AA-12A *Adder*); *Mica* RF
AShM AGM-84 *Harpoon*; AM39 *Exocet*; Kh-31A (AS-17B *Krypton*); *Sea Eagle*†
ASM AGM-114L/R *Hellfire*; Kh-29 (AS-14 *Kedge*); Kh-59 (AS-13 *Kingbolt*); Kh-59M (AS-18 *Kazoo*); AS-30; *Popeye* II (*Crystal Maze*)
ARM Kh-25MP (AS-12A *Kegler*); Kh-31P (AS-17A *Krypton*)
ALCM • Nuclear *Nirbhay* (likely nuclear capable; in development)
BOMBS
INS/SAT guided *Spice*
Laser-guided *Paveway* II

Coast Guard 12,600

EQUIPMENT BY TYPE
PATROL AND COASTAL COMBATANTS 132
PSOH 21: 2 *Sankalp* (capacity 1 *Chetak*/*Dhruv* hel); 4 *Samar* with 1 76mm gun (capacity 1 *Chetak*/*Dhruv* hel); 7 *Samarth*; 5 *Vikram* (capacity 1 *Dhruv* hel); 3 *Vishwast* (capacity 1 *Dhruv* hel)
PSO 3 *Samudra Prahari* with 1 hel landing platform
PCC 44: 20 *Aadesh*; 8 *Rajshree* (Flight I); 4 *Rajshree* (Flight II) 5 *Rani Abbakka*; 7 *Sarojini Naidu*
PBF 63: 6 C-154; 2 C-141; 11 C-143; 44 C-401
PB 1 *Priyadarshini*
AMPHIBIOUS • UCAC 18: 6 H-181 (*Griffon* 8000TD); 12 H-187 (*Griffon* 8000TD)
AIRCRAFT • MP 23 Do-228-101
HELICOPTERS • MRH 21: 4 *Dhruv*; 17 SA316B *Alouette* III (*Chetak*)

Paramilitary 1,585,950

Rashtriya Rifles 65,000

Ministry of Defence. 15 sector HQ

FORCES BY ROLE
MANOEUVRE
Other
65 paramilitary bn

Assam Rifles 63,750

Ministry of Home Affairs. Security within northeastern states, mainly army-officered; better trained than BSF

FORCES BY ROLE
Equipped to roughly same standard as an army inf bn
COMMAND
7 HQ
MANOEUVRE
Other
46 paramilitary bn

EQUIPMENT BY TYPE
ARTILLERY • MOR 81mm 252

Border Security Force 257,350

Ministry of Home Affairs

FORCES BY ROLE
MANOEUVRE
Other
186 paramilitary bn

EQUIPMENT BY TYPE
Small arms, lt arty, some anti-tank weapons
ARTILLERY • MOR 81mm 942+
AIRCRAFT • TPT some (air spt)
HELICOPTERS • MRH 2 Mi-17V-5 *Hip*

Central Industrial Security Force 144,400 (lightly armed security guards)

Ministry of Home Affairs. Guards public-sector locations

Central Reserve Police Force 313,650

Ministry of Home Affairs. Internal-security duties, only lightly armed, deployable throughout the country

FORCES BY ROLE
MANOEUVRE
Other
236 paramilitary bn
10 (rapid action force) paramilitary bn
10 (CoBRA) paramilitary bn
6 (Mahila) paramilitary bn (female)
2 sy gp
COMBAT SUPPORT
5 sigs bn

Defence Security Corps 31,000

Provides security at Defence Ministry sites

Indo-Tibetan Border Police 89,450

Ministry of Home Affairs. Tibetan border security SF/ guerrilla-warfare and high-altitude-warfare specialists

FORCES BY ROLE
MANOEUVRE
Other
56 paramilitary bn

National Security Guards 12,000

Anti-terrorism contingency deployment force, comprising elements of the armed forces, CRPF and Border Security Force

Railway Protection Forces 70,000

Sashastra Seema Bal 76,350

Guards the borders with Nepal and Bhutan

Special Frontier Force 10,000

Mainly ethnic Tibetans

Special Protection Group 3,000

Protection of ministers and senior officials

State Armed Police 450,000

For duty primarily in home state only, but can be moved to other states. Some bn with GPMG and army-standard infantry weapons and equipment

FORCES BY ROLE

MANOEUVRE

Other

144 (India Reserve Police) paramilitary bn

Reserve Organisations

Civil Defence 500,000 reservists

Operate in 225 categorised towns in 32 states. Some units for NBC defence

Home Guard 441,000 reservists (547,000 authorised str)

In all states except Arunachal Pradesh and Kerala; men on reserve lists, no trg. Not armed in peacetime. Used for civil defence, rescue and firefighting provision in wartime; 6 bn (created to protect tea plantations in Assam)

DEPLOYMENT

AFGHANISTAN: 335 (Indo-Tibetan Border Police paramilitary: facilities protection)

CYPRUS: UN • UNFICYP 1

DEMOCRATIC REPUBLIC OF THE CONGO: UN • MONUSCO 1,867; 2 inf bn; 1 med coy

LEBANON: UN • UNIFIL 775; 1 inf bn; 1 med coy

MIDDLE EAST: UN • UNTSO 2

SOMALIA: UN • UNSOM 1

SOUTH SUDAN: UN • UNMISS 2,378; 2 inf bn; 1 engr coy; 1 sigs coy; 1 fd hospital

SUDAN: UN • UNISFA 3

SYRIA/ISRAEL: UN • UNDOF 159; 1 MP coy(-); 1 log bn(-)

WESTERN SAHARA: UN • MINURSO 1

FOREIGN FORCES

Total numbers for UNMOGIP mission in India and Pakistan

Chile 2

Croatia 9

Italy 2

Korea, Republic of 7

Mexico 1

Philippines 5

Sweden 4

Switzerland 4

Thailand 4

Uruguay 3

Indonesia IDN

Indonesian Rupiah Rp		2019	2020	2021
GDP	Rp	15834tr	15919tr	
	US$	1.12tr	1.09tr	
per capita	US$	4,197	4,038	
Growth	%	5.0	-1.5	
Inflation	%	2.8	2.1	
Def bdgt	Rp	107tr	122tr	137tr
	US$	7.60bn	8.37bn	
FMA (US)	US$	14m	0m	0m
US$1=Rp		14135.67	14621.24	

Real-terms defence budget trend (US$bn, constant 2015)

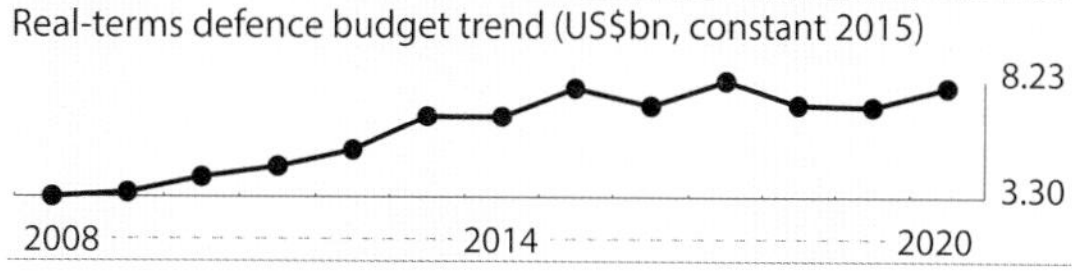

Population 267,026,366

Age	0–14	15–19	20–24	25–29	30–64	65 plus
Male	12.2%	4.3%	4.2%	3.9%	22.0%	3.4%
Female	11.7%	4.1%	4.1%	3.8%	21.8%	4.4%

Capabilities

Indonesia's TNI is the largest armed force in Southeast Asia. It has traditionally been concerned primarily with internal security and counter-insurgency. All three services are based on regional commands. The army remains the dominant service and is deployed operationally in West Papua, central Sulawesi and elsewhere. A modernisation plan adopted in 2010 called for the establishment by 2024 of a 'Minimum Essential Force' including strengthened naval and air capabilities. The 2015 defence white paper outlined Indonesia's 'Global Maritime Fulcrum' policy and advocated building up maritime, satellite and UAV capabilities. Some of these objectives were reflected in the 2020 State Defence Policy document. In 2018, Indonesia expanded its forces in the eastern areas of the country and stood up a third naval fleet command and a third air-force command to organise existing units in that area. Indonesia also created a new army reserve division and a third marines group, both to be stationed in the east. An ASEAN member, Indonesia has no formal defence alliances but there are a number of defence-cooperation agreements. China has supplied some military equipment, including UAVs. The armed forces have contributed to UN and other international peacekeeping operations. Indonesia regularly exercises with Australian and US armed forces and those of Southeast Asian states. Indonesia's inventory comprises equipment from diverse international sources, and the country uses technology-transfer agreements to develop its national defence industry. Indonesia has a number of public and private defence companies that provide services and equipment across the domains.

ACTIVE 395,500 (Army 300,400 Navy 65,000 Air 30,100) Paramilitary 280,000

Conscription liability 24 months selective conscription authorised (not required by law)

RESERVE 400,000

Army cadre units; numerical str n.k., obligation to age 45 for officers

ORGANISATIONS BY SERVICE

Army ε300,400

Mil Area Commands (KODAM)

15 comd (I, II, III, IV, V, VI, IX, XII, XIII, XIV, XVI, XVII, XVIII, Jaya & Iskandar Muda)

FORCES BY ROLE

MANOEUVRE

Mechanised

3 armd cav bn
8 cav bn
1 mech inf bde (1 cav bn, 3 mech inf bn)
1 mech inf bde (3 mech inf bn)
3 indep mech inf bn

Light

1 inf bde (3 cdo bn)
1 inf bde (2 cdo bn, 1 inf bn)
1 inf bde (1 cdo bn, 2 inf bn)
2 inf bde (3 inf bn)
3 inf bde (1 cdo bn, 1 inf bn)
3 inf bde (2 inf bn)
24 indep inf bn
20 indep cdo bn

COMBAT SUPPORT

11 fd arty bn
11 cbt engr bn

COMBAT SERVICE SUPPORT

4 construction bn

AVIATION

1 composite avn sqn

HELICOPTER

1 hel sqn with Bo-105; Bell 205A; Bell 412; AH-64E *Apache Guardian*
1 hel sqn Mi-35P *Hind*; Mi-17V-5 *Hip* H

AIR DEFENCE

1 AD regt (2 ADA bn, 1 SAM unit)
8 ADA bn
3 SAM unit

Special Forces Command (KOPASSUS)

FORCES BY ROLE

SPECIAL FORCES

3 SF gp (total: 2 cdo/para unit, 1 CT unit, 1 int unit)

Strategic Reserve Command (KOSTRAD)

FORCES BY ROLE

COMMAND

3 div HQ

MANOEUVRE

Armoured

2 tk bn

Mechanised

1 mech inf bde (3 mech inf bn)

Light

2 inf bde (3 cdo bn)
1 inf bde (2 inf bn)

Air Manoeuvre

3 AB bde (3 AB bn)

COMBAT SUPPORT

2 arty regt (1 SP arty bn; 1 MRL bn; 1 fd arty bn)
1 fd arty bn
2 cbt engr bn

AIR DEFENCE

3 AD bn

EQUIPMENT BY TYPE

ARMOURED FIGHTING VEHICLES

MBT 103: 42 *Leopard* 2A4; 61 *Leopard* 2RI
LT TK 350: 275 AMX-13 (partially upgraded); 15 PT-76; 60 FV101 *Scorpion*-90
RECCE 142: 55 *Ferret* (13 upgraded); 69 *Saladin* (16 upgraded); 18 VBL
IFV 64: 22 *Black Fox*; 42 *Marder* 1A3
APC 834+
APC (T) 267: 75 AMX-VCI; 34 BTR-50PK; 15 FV4333 *Stormer*; 143 M113A1-B
APC (W) 567+: 350 *Anoa*; some *Barracuda*; 40 BTR-40; 45 FV603 *Saracen* (14 upgraded); 100 LAV-150 *Commando*; 32 VAB-VTT
PPV some *Casspir*
AUV 39: 14 APR-1; 3 *Bushmaster*; 22 *Commando Ranger*

ENGINEERING & MAINTENANCE VEHICLES

AEV 4: 3 *Leopard* 2; 1 M113A1-B-GN
ARV 15+: 2 AMX-13; 6 AMX-VCI; 3 BREM-2; 4 BPz-3 *Buffel*; *Stormer*; T-54/T-55
VLB 20: 10 AMX-13; 4 *Leguan*; 4 M3; 2 *Stormer*

ANTI-TANK/ANTI-INFRASTRUCTURE

MSL • MANPATS FGM-148 *Javelin*; SS.11; *Milan*; 9K11 *Malyutka* (AT-3 *Sagger*)
RCL 90mm M67; **106mm** M40A1
RL 89mm LRAC

ARTILLERY 1,238+

SP 87: **105mm** 20 AMX Mk61; **155mm** 67: 49 CAESAR; 18 M109A4
TOWED 133+: **105mm** 110+: some KH-178; 60 M101; 50 M-56; **155mm** 23: 5 FH-88; 18 KH-179
MRL 127mm 63 ASTROS II Mk6
MOR 955: **81mm** 800; **120mm** 155: 75 Brandt; 80 UBM 52

AMPHIBIOUS

LANDING SHIPS • LST 2 ADRI LI with 1 hel landing platform (capacity 8 MBT; 500 troops)
LANDING CRAFT • LCU 17: 1 ADRI XXXII; 4 ADRI XXXIII; 1 ADRI XXXIX; 1 ADRI XL; 3 ADRI XLI; 2 ADRI XLIV; 2 ADRI XLVI; 2 ADRI XLVIII; 1 ADRI L

AIRCRAFT • TPT • Light 9: 1 BN-2A *Islander*; 6 C-212 *Aviocar* (NC-212); 2 *Turbo Commander* 680

HELICOPTERS

ATK 14: 6 Mi-35P *Hind*; 8 AH-64E *Apache Guardian*
MRH 45: 12 H125M *Fennec*; 17 Bell 412 *Twin Huey* (NB-412); 16 Mi-17V-5 *Hip* H
TPT • Light 29: 7 Bell 205A; 20 Bo-105 (NBo-105); 2 H120 *Colibri*
TRG up to 19 Hughes 300C

AIR DEFENCE

SAM • Point-defence 95+: 2 *Kobra* (with 125 GROM-2 msl); *Starstreak*; TD-2000B (*Giant Bow* II); 51 *Rapier*; 42 RBS-70; QW-3
GUNS • TOWED 411: **20mm** 121 Rh 202; **23mm** *Giant Bow*; **40mm** 90 L/70; **57mm** 200 S-60

AIR-LAUNCHED MISSILES

ASM AGM-114 *Hellfire*

Navy ε65,000 (including Marines and Aviation)

Three fleets: East (Sorong), Central (Surabaya) and West (Jakarta). Two Forward Operating Bases at Kupang (West Timor) and Tahuna (North Sulawesi)

EQUIPMENT BY TYPE

SUBMARINES • SSK 4:

2 *Cakra* (Type-209/1300) with 8 single 533mm TT with SUT HWT

2 *Nagapasa* (Type-209/1400) with 8 single 533mm TT with *Black Shark* HWT

PRINCIPAL SURFACE COMBATANTS 7

FRIGATES 7:

FFGHM 5:

1 *Ahmad Yani* (ex-NLD *Van Speijk*) with 2 twin-cell VLS with 3M55E *Yakhont* (SS-N-26 *Strobile*) AShM; 2 twin *Simbad* lnchr (manual) with *Mistral* SAM, 2 triple 324mm SVTT Mk 32 ASTT with Mk 46 LWT, 1 76mm gun (capacity 1 Bo-105 (NBo-105) hel)

2 *Ahmad Yani* (ex-NLD *Van Speijk*) with 2 twin lnchr with C-802 (CH-SS-N-6) AShM, 2 twin *Simbad* lnchr (manual) with *Mistral* SAM, 2 triple 324mm SVTT Mk 32 ASTT with Mk 46 LWT, 1 76mm gun (capacity 1 Bo-105 (NBo-105) hel)

2 *R.E. Martadinata* (SIGMA 10514) with 2 quad lnchr with MM40 *Exocet* Block 3 AShM, 2 6-cell CLA VLS with VL MICA SAM, 2 triple 324mm ILAS-3 (B-515) ASTT with A244/S LWT, 1 *Millennium* CIWS, 1 76mm gun (capacity 1 med hel)

FFHM 2 *Ahmad Yani* (ex-NLD *Van Speijk*) with 2 twin *Simbad* lnchr (manual) with *Mistral* SAM, 2 triple 324mm ASTT with Mk 46 LWT, 1 76mm gun (capacity 1 Bo-105 (NBo-105) hel)

PATROL AND COASTAL COMBATANTS 122

CORVETTES 24

FSGM 7:

3 *Bung Tomo* with 2 quad lnchr with MM40 *Exocet* Block 2 AShM, 1 18-cell VLS with *Sea Wolf* SAM, 2 triple 324mm ASTT, 1 76mm gun (capacity: 1 Bo-105 hel)

4 *Diponegoro* (SIGMA 9113) with 2 twin lnchr with MM40 *Exocet* Block 2 AShM, 2 quad *Tetral* lnchr with *Mistral* SAM, 2 triple 324mm ILAS-3 (B-515) ASTT with MU90 LWT, 1 76mm gun, 1 hel landing platform

FSGH 1 *Nala* with 2 twin lnchr with MM38 *Exocet* AShM, 1 twin Bofors ASW Rocket Launcher System 375mm A/S mor, 1 120mm gun (capacity 1 lt hel)

FS 16:

2 *Fatahillah* with 2 triple 324mm SVTT Mk 32 ASTT with Mk 46 LWT, 1 twin 375mm A/S mor, 1 120mm gun

14 *Kapitan Pattimura* (GDR *Parchim* I) with 4 single 400mm ASTT, 2 RBU 6000 *Smerch* 2 A/S mor, 1 twin 57mm gun

PCFG 3 *Mandau* with 4 single lnchr with MM38 *Exocet* AShM, 1 57mm gun

PCG 4:

2 *Sampari* (KCR-60M)with 2 twin lnchr for C-705 AShM

2 *Todak* with 2 single lnchr with C-802 (CH-SS-N-6), 1 57mm gun

PCT 2 *Andau* with 2 single 533mm TT with SUT, 1 57mm gun

PCC 13: 4 *Kakap* with 1 hel landing platform; 2 *Pandrong*; 3 *Pari*; 2 *Sampari* (KCR-60M) with 1 NG-18 CIWS; 2 *Todak* with 1 57mm gun

PBG 8: 2 *Clurit* with 2 single lnchr with C-705 AShM, 1 AK630 CIWS; 6 *Clurit* with 2 single lnchr with C-705 AShM

PBF 4 Combat Boat AL D-18

PB 64: 2 *Badau* (ex-BRN *Waspada*); 9 *Boa*; 1 *Cucut* (ex-SGP *Jupiter*); 4 *Kobra*; 1 *Krait*; 8 *Sibarau*; 22 *Sinabang* (KAL 28); 4 *Tarihu*; 9 *Tatihu* (PC-40); 4 *Viper*

MINE WARFARE • MINE COUNTERMEASURES 8

MCO 2 *Pulau Rengat*

MSC 6 *Pulau Rote* (ex-GDR *Wolgast*)

AMPHIBIOUS

PRINCIPAL AMPHIBIOUS VESSELS • LPD 6:

1 *Dr Soeharso* (ex-*Tanjung Dalpele*) (capacity 2 LCU/ LCVP; 13 tanks; 500 troops; 2 AS332L *Super Puma*) (used in AH role)

4 *Makassar* (capacity 2 LCU or 4 LCVP; 13 tanks; 500 troops; 2 AS332L *Super Puma*)

1 *Semarang* (IDN *Makassar* mod) (capacity 2 LCM; 3 hels; 28 vehs; 650 troops) (used in AH role)

LANDING SHIPS • LST 16

1 *Teluk Amboina* (capacity 16 tanks; 800 troops)

1 *Teluk Bintuni* (capacity 10 MBT)

9 *Teluk Gilimanuk* (ex-GDR *Frosch*)

1 *Teluk Lada* with 1 hel landing platform (capacity 4 LCVP; 470 troops; 15 APC; 10 MBT)

4 *Teluk Semangka* (capacity 17 tanks; 200 troops)

LANDING CRAFT 54

LCM 20

LCU 4

LCVP 30

LOGISTICS AND SUPPORT 20

AGF 1 *Multatuli* with 1 hel landing platform

AGOR 2 *Rigel*

AGOS 1 *Leuser*

AGHS 1

AGS 2 *Pulau Rote* (ex-GDR *Wolgast*)

AKSL 3

AORLH (1 *Arun* (ex-UK *Rover*) damaged at sea 2018, non-operational and in repair)

AOR 2: 1 *Bontang* with 1 hel landing platform; 1 *Tarakan* with 1 hel landing platform

AOT 1 *Sorong*

AP 3: 1 *Tanjung Kambani* (troop transport) with 1 hel landing platform; 2 *Karang Pilang* (troop transport)

ATF 1

AXS 3

Naval Aviation ε1,000

EQUIPMENT BY TYPE

AIRCRAFT

MP 28: 3 C212-200; 5 CN235-220 (MPA); 14 N-22B *Searchmaster* B; 6 N-22SL *Searchmaster* L

TPT • Light 33: 1 Beech 350i *King Air* (VIP transport); 8 Beech G36 *Bonanza*; 2 Beech G38 *Baron*; 17 C-212-200 *Aviocar*; 3 TB-9 *Tampico*; 2 TB-10

HELICOPTERS
ASW 11 AS565MBe *Panther*
MRH 4 Bell 412 (NB-412) *Twin Huey*
CSAR 4 H225M *Caracal*
TPT 15: **Medium** 3 AS332L *Super Puma* (NAS322L); **Light** 12: 3 H120 *Colibri*; 9 Bo-105 (NBo-105)

Marines ε20,000

FORCES BY ROLE

SPECIAL FORCES
1 SF bn
MANOEUVRE
Amphibious
2 mne gp (1 cav regt, 3 mne bn, 1 arty regt, 1 cbt spt regt, 1 CSS regt)
1 mne gp (forming)
1 mne bde (3 mne bn)

EQUIPMENT BY TYPE

ARMOURED FIGHTING VEHICLES
LT TK 65: 10 AMX-10 PAC 90; 55 PT-76†
RECCE 21 BRDM-2
IFV 114: 24 AMX-10P; 22 BMP-2; 54 BMP-3F; 2 BTR-4; 12 BTR-80A
APC 103: • **APC (T)** 100 BTR-50P; **APC (W)** 3 BTR-4M
AAV 15: 10 LVTP-7A1; 5 M113 *Arisgator*
ARTILLERY 71+
TOWED 50: **105mm** 22 LG1 MK II; **122mm** 28 M-38
MRL 122mm 21: 4 PHL-90B; 9 RM-70; 8 RM-70 *Vampir*
MOR 81mm some
AIR DEFENCE • GUNS • 40mm 5 L/60/L/70; **57mm** S-60

Air Force 30,100

3 operational comd (East, Central and West) plus trg comd

FORCES BY ROLE

FIGHTER
1 sqn with F-16A/B/C/D *Fighting Falcon*
FIGHTER/GROUND ATTACK
1 sqn with F-16C/D *Fighting Falcon*
1 sqn with Su-27SK *Flanker*; Su-30MK *Flanker*
1 sqn with Su-27SKM *Flanker*; Su-30MK2 *Flanker*
2 sqn with *Hawk* Mk109*/Mk209*
1 sqn with T-50i *Golden Eagle**
GROUND ATTACK
1 sqn with EMB-314 (A-29) *Super Tucano**
MARITIME PATROL
1 sqn with B-737-200
1 sqn with CN235M-220 MPA; CN235M-110
TANKER/TRANSPORT
1 sqn with C-130B/KC-130B *Hercules*
TRANSPORT
1 VIP sqn with B-737-200; C-130H/H-30 *Hercules*; L-100-30; F-27-400M *Troopship*; F-28-1000/3000
1 sqn with C-130H/H-30 *Hercules*; L-100-30
1 sqn with C-130H *Hercules*
1 sqn with C-212 *Aviocar* (NC-212)
1 sqn with C295M
TRAINING
1 sqn with Grob 120TP
1 sqn with KT-1B
TRANSPORT HELICOPTER
2 sqn with H225M; AS332L *Super Puma* (NAS332L); SA330J/L *Puma* (NAS330J/L)
1 VIP sqn with AS332L *Super Puma* (NAS332L); SA330SM *Puma* (NAS300SM)
1 sqn with H120 *Colibri*
ISR UAV
1 sqn with *Aerostar*

EQUIPMENT BY TYPE

Only 45% of ac op
AIRCRAFT 108 combat capable
FTR 9: 7 F-16A *Fighting Falcon*; 2 F-16B *Fighting Falcon* (8 F-5E *Tiger* II; 4 F-5F *Tiger* II non-operational)
FGA 40: 19 F-16C *Fighting Falcon*; 5 F-16D *Fighting Falcon*; 2 Su-27SK *Flanker*; 3 Su-27SKM *Flanker*; 2 Su-30MK *Flanker* F; 9 Su-30MK2 *Flanker* G
MP 8: 3 B-737-200; 2 CN235M-220 MPA
ISR 1 C295M
TKR 1 KC-130B *Hercules*
TPT 51: **Medium** 19: 4 C-130B *Hercules*; 7 C-130H *Hercules*; 6 C-130H-30 *Hercules*; 2 L-100-30; **Light** 23: 9 C295; 9 C-212 *Aviocar* (NC-212); 5 CN235M-110; **PAX** 9: 1 B-737-200; 3 B-737-400; 1 B-737-500; 1 B-737-800BBJ; 1 F-28-1000; 2 F-28-3000
TRG 103: 15 EMB-314 (A-29) *Super Tucano**; 30 Grob 120TP; 7 *Hawk* Mk109*; 22 *Hawk* Mk209*; 14 KT-1B; 15 T-50i *Golden Eagle**
HELICOPTERS
TPT 36: **Heavy** 6 H225M (CSAR); **Medium** 18: 9 AS332 *Super Puma* (NAS332L) (VIP/CSAR); 1 SA330SM *Puma* (NAS330SM) (VIP); 4 SA330J *Puma* (NAS330J); 4 SA330L *Puma* (NAS330L); **Light** 12 H120 *Colibri*
UNINHABITED AERIAL VEHICLES
CISR • Heavy CH-4B (in test)
ISR • Medium *Aerostar*
AIR-LAUNCHED MISSILES
AAM • IR AIM-9P *Sidewinder*; R-73 (AA-11A *Archer*); **IR/SARH** R-27 (AA-10 *Alamo*)
ARH R-77 (AA-12A *Adder*)
ASM AGM-65G *Maverick*; Kh-59M (AS-18 *Kazoo*); Kh-59T (AS-14B *Kedge*)
ARM Kh-31P (AS-17A *Krypton*)

Special Forces (Paskhasau)

FORCES BY ROLE

SPECIAL FORCES
3 (PASKHASAU) SF wg (total: 6 spec ops sqn)
4 indep SF coy

EQUIPMENT BY TYPE

AIR DEFENCE
SAM • Point *Chiron*; QW-3
GUNS • TOWED 35mm 6 Oerlikon *Skyshield*

Paramilitary 280,000+

Police ε280,000 (including 14,000 police 'mobile bde' (BRIMOB) org in 56 coy, incl CT unit (Gegana))

EQUIPMENT BY TYPE

ARMOURED FIGHTING VEHICLES
APC (W) 34 *Tactica*

AIRCRAFT • TPT • Light 6: 2 Beech 18; 2 C-212 *Aviocar* (NC-212); 1 C295; 1 *Turbo Commander* 680

HELICOPTERS

MRH 1 Bell 412EP

TPT • Light 22: 3 Bell 206 *Jet Ranger*; 19 Bo-105 (NBo-105)

KPLP (Coast and Seaward Defence Command)

Responsible to Military Sea Communications Agency

EQUIPMENT BY TYPE

PATROL AND COASTAL COMBATANTS 36

PCO 6: 1 *Arda Dedali*; 3 *Chundamani*; 1 *Kalimasada*; 2 *Trisula*

PB 30: 4 *Golok* (SAR); 5 *Kujang*; 6 *Rantos*; 15 (various)

LOGISTICS AND SUPPORT • ABU 1 *Jadayat*

Bakamla (Maritime Security Agency)

EQUIPMENT BY TYPE

PATROL AND COASTAL COMBATANTS 10

PSO 4: 3 *Pulau Nipah* with 1 hel landing platform; 1 *Tanjung Datu* with 1 hel landing platform

PB 6 *Bintang Laut* (KCR-40 mod)

Reserve Organisations

Kamra People's Security ε40,000

Report for 3 weeks' basic training each year; part-time police auxiliary

DEPLOYMENT

CENTRAL AFRICAN REPUBLIC: UN • MINUSCA 212; 1 engr coy

DEMOCRATIC REPUBLIC OF THE CONGO: UN • MONUSCO 1,033; 1 inf bn; 3 engr coy

LEBANON: UN • UNIFIL 1,259; 1 mech inf bn; 1 MP coy; 1 FSGHM

MALI: UN • MINUSMA 6

PHILIPPINES: IMT 9

SOUTH SUDAN: UN • UNMISS 4

SUDAN: UN • UNAMID 3; **UN** • UNISFA 4

WESTERN SAHARA: UN • MINURSO 2

Japan JPN

Japanese Yen ¥		2019	2020	2021
GDP	¥	554tr	526tr	
	US$	5.08tr	4.91tr	
per capita	US$	40,256	39,048	
Growth	%	0.7	-5.3	
Inflation	%	0.5	-0.1	
Def bdgt	¥	5.26tr	5.32tr	5.33tr
	US$	48.2bn	49.7bn	
US$1=¥		109.01	107.19	

Real-terms defence budget trend (US$bn, constant 2015)

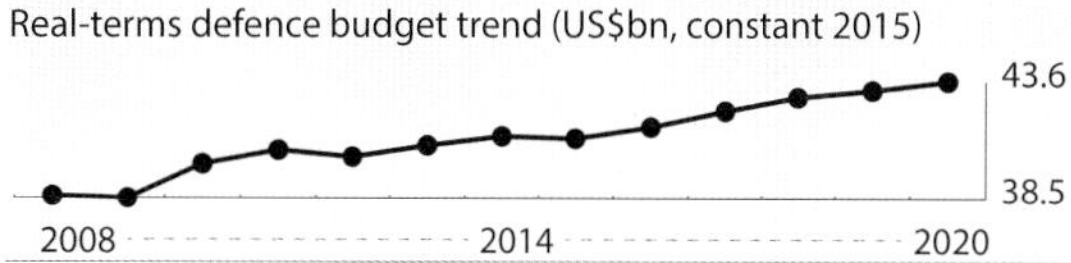

Population 125,507,472

Age	0–14	15–19	20–24	25–29	30–64	65 plus
Male	6.4%	2.4%	2.6%	2.5%	21.7%	12.8%
Female	6.1%	2.2%	2.3%	2.4%	22.3%	16.4%

Capabilities

Japan's concerns over its regional security environment have heightened, as evidenced in its 2020 Defense White Paper. It was reported that updates might be made by the end of 2020 to the 2013 National Security Strategy, the 2018 National Defense Program Guidelines and the 2019–23 Mid-Term Defense Program. These principally relate to an emerging security challenge from China and an established concern over North Korea. This has stimulated defence-budget increases and defence-policy and legislative reforms to enable Japan to play a more active international security role and strengthen the Japan Self-Defense Force (JSDF). While the JSDF's offensive capacity remains weak, the navy has strengths in anti-submarine warfare and air defence. In 2018, a Ground Component Command was created. An Amphibious Rapid Deployment Brigade was also created, tasked mainly with the defence of remote islands. There are plans to convert the *Izumo* helicopter carriers into aircraft carriers. In 2020, Japan launched a Space Operations Squadron to boost space situational-awareness capabilities. It also laid out plans to set up electromagnetic-warfare units and to expand the cyber-defence unit. Japan's alliance with the US remains the cornerstone of its defence policy, reflected by continued US basing, the widespread use of US equipment across all three services and regular training with US forces. Due to their defensive mandate, JSDF deployments are mostly for peacekeeping purposes. The ongoing military-procurement drive has focused for the first time on power projection, mobility and ISR. Japan is considering options to boost its ballistic-missile-defence capability. Budget documents also note research on a hypersonic glide body and new anti-ship missiles. Japan has an advanced defence-industrial base, which produces modern equipment for the JSDF.

ACTIVE 247,150 (Ground Self-Defense Force 150,700 Maritime Self-Defense Force 45,350 Air Self-Defense Force 46,900 Central Staff 4,200) Paramilitary 14,350

RESERVE 56,000 (General Reserve Army (GSDF) 46,000 Ready Reserve Army (GSDF) 8,100 Navy 1,100 Air 800)

ORGANISATIONS BY SERVICE

Space

EQUIPMENT BY TYPE

SATELLITES 11

COMMUNICATIONS 2: 1 *Kirameki*-1; 1 *Kirameki*-2

ISR 9 IGS

Ground Self-Defense Force 150,700

FORCES BY ROLE

COMMAND

5 army HQ (regional comd)

SPECIAL FORCES

1 spec ops unit (bn)

MANOEUVRE

Armoured

1 (7th) armd div (1 armd recce sqn, 3 tk regt, 1 armd inf regt, 1 hel sqn, 1 SP arty regt, 1 AD regt, 1 cbt engr bn, 1 sigs bn, 1 NBC bn, 1 log regt)

1 indep tk bn

Mechanised

1 (2nd) inf div (1 armd recce sqn, 1 tk regt, 3 inf regt, 1 hel sqn, 1 SP arty regt, 1 AT coy, 1 ADA bn, 1 cbt engr bn, 1 sigs bn, 1 NBC bn, 1 log regt)

1 (4th) inf div (1 armd recce bn, 3 inf regt, 1 inf coy, 1 hel sqn, 1 AT coy, 1 SAM bn, 1 cbt engr bn, 1 sigs bn, 1 NBC bn, 1 log regt)

1 (6th) inf div (1 recce sqn, 1 mech inf regt; 3 inf regt, 1 hel sqn, 1 SAM bn, 1 cbt engr bn, 1 sigs bn, 1 NBC bn, 1 log regt)

1 (9th) inf div (1 armd recce sqn, 1 tk bn, 3 inf regt, 1 hel sqn, 1 SAM bn, 1 cbt engr bn, 1 sigs bn, 1 NBC bn, 1 log regt)

1 (5th) inf bde (1 armd recce sqn, 1 tk bn, 3 inf regt, 1 hel sqn, 1 SP arty bn, 1 SAM coy, 1 cbt engr coy, 1 sigs coy, 1 NBC coy, 1 log bn)

1 (11th) inf bde (1 armd recce sqn, 1 tk sqn, 3 inf regt, 1 hel sqn, 1 SP arty bn, 1 SAM coy, 1 cbt engr coy, 1 sigs coy, 1 NBC coy, 1 log bn)

Light

2 (1st & 3rd) inf div (1 recce sqn, 1 tk bn, 3 inf regt, 1 hel sqn, 1 fd arty bn, 1 SAM bn, 1 cbt engr bn, 1 sigs bn, 1 NBC bn, 1 log regt)

1 (10th) inf div (1 recce sqn, 1 tk bn, 3 inf regt, 1 hel sqn, 1 fd arty regt, 1 SAM bn, 1 cbt engr bn, 1 sigs bn, 1 NBC bn, 1 log regt)

1 (8th) inf div (1 recce sqn, 3 inf regt, 1 hel sqn, 1 SAM bn, 1 cbt engr bn, 1 sigs bn, 1 NBC bn, 1 log regt)

1 (13th) inf bde (1 recce sqn, 1 tk coy, 3 inf regt, 1 hel sqn, 1 fd arty bn, 1 SAM coy, 1 cbt engr coy, 1 NBC coy, 1 sigs coy, 1 log bn)

1 (14th) inf bde (1 recce sqn, 2 inf regt, 1 hel sqn, 1 SAM coy, 1 cbt engr coy, 1 NBC coy, 1 sigs coy, 1 log bn)

1 (15th) inf bde (1 recce sqn, 1 inf regt, 1 avn sqn, 1 AD regt, 1 cbt engr coy, 1 NBC coy, 1 sigs coy, 1 log bn)

Air Manoeuvre

1 (1st) AB bde (3 AB bn, 1 fd arty bn, 1 cbt engr coy, 1 sigs coy, 1 log bn)

1 (12th) air mob inf bde (1 recce sqn, 3 inf regt, 1 avn sqn, 1 fd arty bn, 1 SAM coy, 1 cbt engr coy, 1 NBC coy, 1 sigs coy, 1 log bn)

Amphibious

1 amph bde (1 recce coy, 2 amph regt, 1 amph aslt bn, 1 log bn)

COMBAT SUPPORT

1 (1st) arty bde (1 SP arty regt (2 SP arty bn, 1 MRL bn); 1 SP arty regt (1 SP arty bn, 1 MRL bn); 3 AShM regt)

1 (Northwestern Army) arty bde (1 fd arty regt (4 fd arty bn); 1 AShM regt)

1 (Western Army) arty bde (1 fd arty regt (4 fd arty bn); 1 MRL bn; 1 AShM regt)

1 (Central Army) fd arty bn

4 engr bde

1 engr unit

1 EW bn

5 int bn

1 MP bde

1 sigs bde

COMBAT SERVICE SUPPORT

5 log unit (bde)

5 trg bde

HELICOPTER

2 sqn with MV-22B *Osprey* (forming)

HELICOPTER

1 hel bde (5 tpt hel sqn; 1 VIP tpt hel bn)

5 hel gp (1 atk hel bn, 1 hel bn)

AIR DEFENCE

2 SAM bde (2 SAM gp)

2 SAM gp

EQUIPMENT BY TYPE

ARMOURED FIGHTING VEHICLES

MBT 580: 99 Type-10; 140 Type-74; 341 Type-90

ASLT 109 Type-16 MCV

RECCE 111 Type-87

IFV 68 Type-89

APC 795

APC (T) 226 Type-73

APC (W) 569: 204 Type-82; 365 Type-96

AAV 52 AAV-7

AUV 8 *Bushmaster*

ENGINEERING & MAINTENANCE VEHICLES

ARV 70: 4 Type-11; 36 Type-78; 30 Type-90

VLB 22 Type-91

NBC VEHICLES 49: 30 Chemical Reconnaissance Vehicle; 19 NBC Reconnaissance Vehicle

ANTI-TANK/ANTI-INFRASTRUCTURE

MSL

SP 37 Type-96 MPMS

MANPATS Type-79 *Jyu*-MAT; Type-87 *Chu*-MAT; Type-01 LMAT

RCL • 84mm *Carl Gustaf*

ARTILLERY 1,583

SP 179: **155mm** 143: 136 Type-99; 7 Type-19 **203mm** 36 M110A2

TOWED 155mm 220 FH-70

MRL 227mm 60 M270 MLRS

MOR 1,124: **81mm** 660 L16 **120mm** 440; **SP 120mm** 24 Type-96

COASTAL DEFENCE • AShM 90: 30 Type-12; 60 Type-88
AIRCRAFT • TPT • Light 7 Beech 350 *King Air* (LR-2)
TILTROTOR • TPT 3 MV-22B *Osprey*
HELICOPTERS
ATK 101: 52 AH-1S *Cobra*; 12 AH-64D *Apache*; 37 OH-1
TPT 250: **Heavy** 53: 18 CH-47D *Chinook* (CH-47J); 35 CH-47JA *Chinook*; **Medium** 43: 3 H225 *Super Puma* MkII+ (VIP); 40 UH-60L *Black Hawk* (UH-60JA); **Light** 154: 124 Bell 205 (UH-1J); 30 Enstrom 480B (TH-480B)
AIR DEFENCE
SAM 327+
Medium-range 163: 43 Type-03 *Chu*-SAM; 120 MIM-23B I-*Hawk*
Short-range 5 Type-11 *Tan*-SAM
Point-defence 159+: 46 Type-81 *Tan*-SAM; 113 Type-93 *Kin*-SAM; Type-91 *Kei*-SAM
GUNS • SP 35mm 52 Type-87

Maritime Self-Defense Force 45,350

Surface units organised into 4 Escort Flotillas with a mix of 8 warships each. Bases at Yokosuka, Kure, Sasebo, Maizuru, Ominato. SSK organised into two flotillas with bases at Kure and Yokosuka

EQUIPMENT BY TYPE

SUBMARINES • SSK 22:
11 *Oyashio* (of which 2 in trg role) with 6 single 533mm TT with UGM-84C *Harpoon* Block 1B AShM/Type-89 HWT
11 *Soryu* (of which 10 fitted with AIP and 1 fitted with lithium-ion fuel battery) with 6 single 533mm TT with UGM-84C *Harpoon* Block 1B AShM/Type-89 HWT
PRINCIPAL SURFACE COMBATANTS 51
AIRCRAFT CARRIERS • CVH 4:
2 *Hyuga* with 2 8-cell Mk 41 VLS with ASROC/RIM-162B ESSM SAM, 2 triple 324mm HOS-303 ASTT with Mk 46/Type-97 LWT, 2 Mk 15 *Phalanx* Block 1B CIWS (normal ac capacity 3 SH-60 *Seahawk* ASW hel; plus additional ac embarkation up to 7 SH-60 *Seahawk* or 7 MCH-101)
2 *Izumo* with 2 11-cell Mk 15 SeaRAM lnchr with RIM-116 SAM, 2 Mk 15 *Phalanx* Block 1B CIWS (normal ac capacity 7 SH-60 *Seahawk* ASW hel; plus additional ac embarkation up to 5 SH-60 *Seahawk*/MCH-101 hel)
CRUISERS • CGHM 3:
2 *Atago* with *Aegis* Baseline 9 C2, 2 quad lnchr with SSM-1B (Type-90) AShM, 12 8-cell Mk 41 VLS (8 fore, 4 aft) with SM-2 Block IIIA/B SAM/SM-3 Block IA/IB SAM/ASROC A/S msl, 2 triple 324mm HOS-302 ASTT with Mk 46 LWT, 2 Mk 15 *Phalanx* Block 1B CIWS, 1 127mm gun (capacity 1 SH-60 *Seahawk* ASW hel)
1 *Maya* (*Atago* mod) with *Aegis* Baseline 9 C2, w quad lnchr with SSM-1B (Type-90) AShM), 12 8-cell Mk 41 VLS (8 fore, 4 aft) with SM-2 Block IIIA/B SAM/SM-3 Block IA/IB SAM/Type-07 A/S msl, 2 triple 324mm HOS-303 ASTT with Mk 46 LWT, 2 Mk 15 *Phalanx* Block 1B CIWS, 1 127mm gun (capacity 1 SH-60 *Seahawk* ASW hel)
DESTROYERS 38
DDGHM 32:
8 *Asagiri* with 2 quad lnchr with RGM-84C *Harpoon* Block 1B AShM, 1 octuple Mk 29 lnchr with RIM-7M *Sea Sparrow* SAM, 2 triple 324mm HOS-302 ASTT with Mk 46 LWT, 1 octuple Mk 112 lnchr with ASROC, 2 Mk 15 *Phalanx* CIWS, 1 76mm gun (capacity 1 SH-60 *Seahawk* ASW hel)
4 *Akizuki* with 2 quad lnchr with SSM-1B (Type-90) AShM, 4 8-cell Mk 41 VLS with ASROC/RIM-162B ESSM SAM, 2 triple 324mm HOS-303 ASTT with Type-97 LWT, 2 Mk 15 *Phalanx* Block 1B CIWS, 1 127mm gun (capacity 1 SH-60 *Seahawk* ASW hel)
2 *Asahi* (*Akizuki* mod) with 2 quad lnchr with SSM-1B (Type-90) AShM, 4 8-cell Mk 41 VLS with RIM-162B ESSM SAM/Type-07 A/S msl, 2 triple 324mm HOS-303 ASTT with Type-12 LWT, 2 Mk 15 *Phalanx* Block 1B CIWS, 1 127mm gun (capacity 1 SH-60 *Seahawk* ASW hel)
4 *Hatsuyuki* (of which 2 in trg role) with 2 quad lnchr with RGM-84C *Harpoon* Block 1B AShM, 1 octuple Mk 29 lnchr with RIM-7F/M *Sea Sparrow* SAM, 2 triple 324mm HOS-301 ASTT with Mk 46 LWT, 1 octuple Mk 112 lnchr with ASROC A/S msl, 2 Mk 15 *Phalanx* CIWS, 1 76mm gun (capacity 1 SH-60 *Seahawk* ASW hel)
9 *Murasame* with 2 quad lnchr with SSM-1B (Type-90) AShM, 1 16-cell Mk 48 mod 0 VLS with RIM-162C ESSM SAM, 2 triple 324mm HOS-302 ASTT with Mk 46 LWT, 2 8-cell Mk 41 VLS with ASROC, 2 Mk 15 *Phalanx* CIWS, 2 76mm gun (capacity 1 SH-60 *Seahawk* ASW hel)
5 *Takanami* (improved *Murasame*) with 2 quad lnchr with SSM-1B (Type-90) AShM, 4 8-cell Mk 41 VLS with RIM-162B ESSM SAM/ASROC A/S msl, 2 triple 324mm HOS-302 ASTT with Mk 46 LWT, 2 Mk 15 *Phalanx* Block 1B CIWS, 1 127mm gun (capacity 1 SH-60 *Seahawk* ASW hel)
DDGM 6:
2 *Hatakaze* with 2 quad lnchr with RGM-84C *Harpoon* Block 1B AShM, 1 Mk 13 GMLS with SM-1MR Block VI SAM, 2 triple 324mm HOS-301 ASTT with Mk 46 LWT, 1 octuple Mk 112 lnchr with ASROC, 2 Mk 15 *Phalanx* CIWS, 2 127mm gun, 1 hel landing platform
4 *Kongou* with *Aegis* Baseline 5 C2, 2 quad lnchr with RGM-84C *Harpoon* Block 1B AShM, 12 8-cell Mk 41 VLS (of which 2 only 5-cell and fitted with reload crane) with SM-2 Block IIIA/B SAM/SM-3 Block IA SAM/ASROC A/S msl, 2 triple 324mm HOS-302 ASTT with Mk 46 LWT, 2 Mk 15 *Phalanx* Block 1B CIWS, 1 127mm gun
FRIGATES • FFG 6 *Abukuma* with 2 quad lnchr with RGM-84C *Harpoon* Block 1B AShM, 2 triple 324mm HOS-301 ASTT with Mk 46 LWT, 1 octuple Mk 112 lnchr with ASROC A/S msl, 1 Mk 15 *Phalanx* CIWS, 1 76mm gun
PATROL AND COASTAL COMBATANTS 6
PBFG 6 *Hayabusa* with 4 SSM-1B (Type-90) AShM, 1 76mm gun

MINE WARFARE • MINE COUNTERMEASURES 24
MCCS 4:
1 *Ieshima*
1 *Uraga* with 1 76mm gun, 1 hel landing platform (for MCH-101 hel)
1 *Uraga* with 1 hel landing platform (for MCH-101)
1 *Uwajima*
MSC 18: 3 *Hirashima*; 11 *Sugashima*; 1 *Uwajima*; 3 *Enoshima*
MSO 2 *Awaji*
AMPHIBIOUS
PRINCIPAL AMPHIBIOUS SHIPS • LHD 3 *Osumi* with 2 Mk 15 *Phalanx* CIWS (capacity for 2 CH-47 hel) (capacity 10 Type-90 MBT; 2 LCAC(L) ACV; 330 troops)
LANDING CRAFT 8
LCM 2 LCU-2001
LCAC 6 LCAC(L) (capacity either 1 MBT or 60 troops)
LOGISTICS AND SUPPORT 18
AGBH 1 *Shirase* (capacity 2 AW101 *Merlin* hel)
AGEH 1 *Asuka* (wpn trials) with 1 8-cell Mk 41 VLS (capacity 1 SH-60 *Seahawk* hel)
AGOS 2 *Hibiki* with 1 hel landing platform
AGS 3: 1 *Futami*; 1 *Nichinan*; 1 *Shonan*
AOE 5: 2 *Mashu* (capacity 1 med hel); 3 *Towada* with 1 hel landing platform
ARC 1 *Muroto*
ASR 2: 1 *Chihaya* with 1 hel landing platform; 1 Chiyoda II with 1 hel landing platform
AX 3:
1 *Kashima* with 2 triple 324mm HOS-301 ASTT, 1 76mm gun, 1 hel landing platform
1 *Kurobe* with 1 76mm gun (trg spt ship)
1 *Tenryu* (trg spt ship); with 1 76mm gun (capacity: 1 med hel)

Naval Aviation ε9,800

7 Air Groups

FORCES BY ROLE
ANTI SUBMARINE/SURFACE WARFARE
5 sqn with SH-60B (SH-60J)/SH-60K *Seahawk*
MARITIME PATROL
1 sqn with P-1
1 sqn with P-1; P-3C *Orion*
2 sqn with P-3C *Orion*
ELECTRONIC WARFARE
1 sqn with EP-3 *Orion*
MINE COUNTERMEASURES
1 sqn with MCH-101
SEARCH & RESCUE
1 sqn with *Shin Meiwa* US-2
2 sqn with UH-60J *Black Hawk*
TRANSPORT
1 sqn with AW101 *Merlin* (CH-101); Beech 90 *King Air* (LC-90); KC-130R *Hercules*
TRAINING
1 sqn with Beech 90 *King Air* (TC-90)
1 sqn with P-3C *Orion*
1 sqn with T-5J
1 hel sqn with H135 (TH-135); OH-6DA; SH-60B (SH-60J) *Seahawk*

EQUIPMENT BY TYPE
AIRCRAFT 72 combat capable
ASW 72: 24 P-1; 48 P-3C *Orion*
ELINT 5 EP-3C *Orion*
SAR 5 *Shin Meiwa* US-2
TPT 24: **Medium** 6 C-130R *Hercules*; **Light** 18: 5 Beech 90 *King* Air (LC-90); 13 Beech 90 *King Air* (TC-90) (trg)
TRG 30 T-5J
HELICOPTERS
ASW 80: 17 SH-60B *Seahawk* (SH-60J); 63 SH-60K *Seahawk*
MCM 10 MCH-101
SAR 12 UH-60J *Black Hawk*
TPT 18: **Medium** 3 AW101 *Merlin* (CH-101); **Light** 15 H135 (TH-135) (trg)
AIR-LAUNCHED MISSILES
AShM ASM-1C (Type-90)

Air Self-Defense Force 46,900

7 cbt wg

FORCES BY ROLE
FIGHTER
7 sqn with F-15J *Eagle*
1 sqn with F-4EJ (F-4E) *Phantom* II
3 sqn with Mitsubishi F-2
1 sqn with F-35A *Lightning* II
ELECTRONIC WARFARE
1 sqn with Kawasaki EC-1; YS-11EA
ELINT
1 sqn with YS-11EB
AIRBORNE EARLY WARNING & CONTROL
2 sqn with E-2C/D *Hawkeye*
1 sqn with E-767
SEARCH & RESCUE
1 wg with U-125A *Peace Krypton*; UH-60J *Black Hawk*
TANKER
1 sqn with KC-767J
TRANSPORT
1 (VIP) sqn with B-777-300ER
1 sqn with C-1; Gulfstream IV (U-4)
1 sqn with C-2
1 sqn with C-130H *Hercules*; KC-130H *Hercules*
Some (liaison) sqn with Gulfstream IV (U-4); T-4*
TRAINING
1 (aggressor) sqn with F-15J *Eagle*
TEST
1 wg with F-15J *Eagle*; T-4*
TRANSPORT HELICOPTER
4 flt with CH-47JA *Chinook*

EQUIPMENT BY TYPE
AIRCRAFT 534 combat capable
FTR 201: 156 F-15J *Eagle*; 45 F-15DJ *Eagle*
FGA 134: 64 F-2A; 27 F-2B; 26 F-4E *Phantom* II (F-4EJ); 17 F-35A *Lightning* II
EW 3: 1 Kawasaki EC-1; 2 YS-11EA
ELINT 1 RC-2
ISR 4 YS-11EB
AEW&C 21: 13 E-2C *Hawkeye*; 4 E-2D *Hawkeye*; 4 E-767
SAR 26 U-125A *Peace Krypton*
TKR 6: 2 KC-130H *Hercules*; 4 KC-767J

TPT 56: **Medium** 36: 14 C-130H *Hercules*; 11 C-1; 11 C-2; **PAX** 20: 2 B-777-300ER (VIP); 13 Beech T-400; 5 Gulfstream IV (U-4)
TRG 248: 199 T-4*; 49 T-7
HELICOPTERS
SAR 39 UH-60J *Black Hawk*
TPT • Heavy 15 CH-47JA *Chinook*
AIR-LAUNCHED MISSILES
AAM • IR AAM-3 (Type-90); AIM-9L *Sidewinder*; **IIR** AAM-5 (Type-04); **SARH** AIM-7 *Sparrow*; **ARH** AAM-4 (Type-99); AIM-120C5/C7 AMRAAM (limited numbers)
AShM ASM-1 (Type-80); ASM-2 (Type-93)
BOMBS
INS/SAT guided GBU-38 JDAM; GBU-54 Laser JDAM

Air Defence
Ac control and warning. 4 wg; 28 radar sites
FORCES BY ROLE
AIR DEFENCE
6 SAM gp (total: 24 SAM bty with MIM-104D/F *Patriot* PAC-2/3)
1 AD gp with Type-81 *Tan*-SAM; M167 *Vulcan*
EQUIPMENT BY TYPE
AIR DEFENCE
SAM 120+
Long-range 120 MIM-104D/F *Patriot* PAC-2 GEM/PAC-3
Point-defence Type-81 *Tan*-SAM
GUNS • TOWED 20mm M167 *Vulcan*

Paramilitary 14,350

Coast Guard 14,350
Ministry of Land, Transport, Infrastructure and Tourism (no cbt role)
EQUIPMENT BY TYPE
PATROL AND COASTAL COMBATANTS 379
PSOH 17: 2 *Mizuho* (capacity 2 hels); 1 *Mizuho* II (capacity 2 hels); 3 *Shikishima* (capacity 2 hels); 1 *Shunko* (capacity 2 hels); 1 *Soya* (capacity 1 hel) (icebreaking capability); 9 *Tsugaru* (*Soya* mod) (capacity 1 hel)
PSO 46: 9 *Hateruma* with 1 hel landing platform; 3 *Hida* with 1 hel landing platform; 6 *Iwami*; 1 *Izu* with 1 hel landing platform; 1 *Kojima* (trg) with 1 hel landing platform; 2 *Kunigami* with 1 hel landing platform; 1 *Miura* with 1 hel landing platform; 1 *Miyako* with 1 hel landing platform; 6 *Ojika* with 1 hel landing platform; 16 *Taketomi* with 1 hel landing platform
PCO 16: 3 *Aso*; 8 *Katori*; 5 *Teshio*
PCC 26: 4 *Amami*; 22 *Tokara*
PBF 51: 24 *Hayagumo*; 4 *Mihashi*; 15 *Raizan*; 2 *Takatsuki*; 6 *Tsuruugi*
PB 56: 4 *Asogiri*; 4 *Hamagumo*; 11 *Hayanami*; 15 *Katonami*; 1 *Matsunami*; 2 *Natsugiri*; 9 *Shimoji*; 10 *Yodo*
PBI 167: 2 *Hakubai*; 1 *Hayagiku*; 164 *Himegiku*
LOGISTICS AND SUPPORT 17
ABU 1 *Teshio*
AGS 13: 6 *Hamashio*; 1 *Jinbei*; 2 *Meiyo*; 1 *Peiyo*; 1 *Shoyo*; 1 *Takuyo*; 1 *Tenyo*
AX 3
AIRCRAFT
MP 3 *Falcon* 2000MSA
SAR 4 Saab 340B
TPT 25: **Light** 23: 5 Cessna 172; 9 Beech 350 *King Air* (LR-2); 9 DHC *Dash-7* (Bombardier 300) (MP); **PAX** 2 Gulfstream V (MP)
HELICOPTERS
MRH 5 Bell 412 *Twin Huey*
SAR 11 S-76D
TPT 33: **Medium** 8: 2 AS332 *Super Puma*; 6 H225 *Super Puma*; **Light** 25: 18 AW139; 4 Bell 505 *Jet Ranger* X; 3 S-76C

DEPLOYMENT

ARABIAN SEA & GULF OF ADEN: Combined Maritime Forces • CTF-151: 2 DDGHM

DJIBOUTI: 170; 2 P-3C *Orion*

SOUTH SUDAN: UN • UNMISS 6

FOREIGN FORCES

United States
US Pacific Command: 55,300
Army 2,500; 1 corps HQ (fwd); 1 SF gp; 1 avn bn; 1 SAM bn
Navy 20,100; 1 CVN; 3 CGHM; 2 DDGHM; 5 DDGM; 1 LCC; 4 MCO; 1 LHA; 1 LPD; 2 LSD; 3 FGA sqn with 10 F/A-18E *Super Hornet*; 1 FGA sqn with 10 F/A-18F *Super Hornet*; 2 ASW sqn with 6 P-8A *Poseidon*; 1 ASW flt with 2 P-3C *Orion*; 2 EW sqn with 5 EA-18G *Growler*; 1 AEW&C sqn with 5 E-2D *Hawkeye*; 2 ASW hel sqn with 12 MH-60R *Seahawk*; 1 tpt hel sqn with MH-60S *Knight Hawk*; 1 base at Sasebo; 1 base at Yokosuka
USAF: 12,700; 1 HQ (5th Air Force) at Okinawa–Kadena AB; 1 ftr wg at Misawa AB (2 ftr sqn with 22 F-16C/D *Fighting Falcon*); 1 ftr wg at Okinawa–Kadena AB (2 ftr sqn with 27 F-15C/D *Eagle*; 1 tkr sqn with 15 KC-135R *Stratotanker*; 1 AEW sqn with 2 E-3B *Sentry*; 1 CSAR sqn with 10 HH-60G *Pave Hawk*); 1 tpt wg at Yokota AB with 10 C-130J-30 *Hercules*; 2 Beech 1900C (C-12J); 1 spec ops gp at Okinawa–Kadena AB with (1 sqn with 5 MC-130J *Commando* II; 1 sqn with 5 CV-22B *Osprey*); 1 ISR sqn with RC-135 *Rivet Joint*; 1 ISR UAV flt with 5 RQ-4A *Global Hawk*
USMC 20,000; 1 mne div; 1 mne regt HQ; 1 arty regt HQ; 1 recce bn; 1 mne bn; 1 amph aslt bn; 1 arty bn; 1 FGA sqn at Iwakuni with 12 F/A-18C *Hornet*; 1 FGA sqn at Iwakuni with 12 F/A-18D *Hornet*; 2 FGA sqn at Iwakuni with 12 F-35B *Lightning* II; 1 tkr sqn at Iwakuni with 15 KC-130J *Hercules*; 2 tpt sqn at Futenma with 12 MV-22B *Osprey*
US Strategic Command: 1 AN/TPY-2 X-band radar at Shariki; 1 AN/TPY-2 X-band radar at Kyogamisaki

Korea, Democratic People's Republic of DPRK

North Korean Won		2019	2020	2021
GDP	US$			
per capita	US$			
Def exp	won			
	US$			
US$1=won				

*definitive economic data not available

Population 25,643,466

Age	0–14	15–19	20–24	25–29	30–64	65 plus
Male	10.4%	3.6%	3.8%	4.0%	23.4%	3.4%
Female	10.0%	3.5%	3.8%	4.0%	23.8%	6.3%

Capabilities

Renewed diplomacy has reduced overall tensions on the Korean Peninsula since 2018, though the prospect of limiting Pyongyang's nuclear ambitions remains uncertain. Aware of the qualitative inferiority of its conventional forces, North Korea has invested in asymmetric capabilities, particularly the development of nuclear weapons and ballistic-missile delivery systems. The 2018 moratorium covering nuclear- and long-range-missile tests technically remains in place, though testing of systems below such thresholds resumed in 2019. A number of new tactical missiles and guided rockets have been demonstrated, as has a new medium-range and possibly nuclear-capable SLBM system. Pyongyang's continued ambitions were further indicated with additional new ICBM and SLBM designs, both paraded in October 2020. North Korea remains diplomatically isolated. While foreign defence cooperation is restricted by international pressure and sanctions, Pyongyang has nonetheless often found ways to develop military ties. Official conscription for both men and women is often extended, sometimes indefinitely. Training is focused on fighting a short intensive war on the peninsula, but the armed forces' overall effectiveness in a modern conflict against technologically superior opposition is unclear. Internal exercises are conducted regularly, but those shown are staged and are not necessarily representative of wider operational capability. North Korea's conventional forces remain reliant on increasingly obsolete equipment, with older Soviet-era and Chinese-origin equipment supplemented by a growing number of indigenous designs and upgrades. Overall effectiveness and serviceability of some equipment remains in doubt but there is local maintenance, repair and overhaul capacity. Local defence-industrial capacity includes the manufacture of light arms, armoured vehicles, artillery and missile systems. North Korea has exported weaponry in the past. It is unclear whether the country would have had the capability to indigenously develop some of the technical advances it has demonstrated, including in rocket propulsion.

ACTIVE 1,280,000 (Army 1,100,000 Navy 60,000 Air 110,000 Strategic Forces 10,000) Paramilitary 189,000

Conscript liability Army 5–12 years, Navy 5–10 years, Air Force 3–4 years, followed by compulsory part-time service to age 40. Thereafter service in the Worker/Peasant Red Guard to age 60

RESERVE ε600,000 (Armed Forces ε600,000), Paramilitary 5,700,000

Reservists are assigned to units (see also Paramilitary)

ORGANISATIONS BY SERVICE

Strategic Forces ε10,000

North Korea's ballistic missiles and obsolete H-5 (Il-28) bombers could be used to deliver nuclear warheads or bombs. At present, however, there is no conclusive evidence to verify that North Korea has successfully produced a warhead or bomb capable of being delivered by these systems

EQUIPMENT BY TYPE (ε)

SURFACE-TO-SURFACE MISSILE LAUNCHERS

ICBM 8+: *Hwasong*-13/*Hwasong*-13 mod/*Hwasong*-14 (in test); *Hwasong*-15 (in test); *Hwasong*-16 (status uncertain)

IRBM *Hwasong*-12 (in test)

MRBM ε10 *Nodong* mod 1/mod 2 (ε90+ msl); some *Scud*-ER; *Pukgusong*-2 (in test); *Hwasong*-10 (*Musudan*) (in test)

SBRM 30+ *Hwasong*-5 (SS-1C *Scud*-B)/*Hwasong*-6 (SS-1D *Scud*-C) (ε200+ msl); some *Scud* (mod) (in test)

Army ε1,100,000

FORCES BY ROLE

COMMAND

2 mech corps HQ
10 inf corps HQ
1 (Capital Defence) corps HQ

MANOEUVRE

Armoured

1 armd div
15 armd bde

Mechanised

4 mech div

Light

27 inf div
14 inf bde

COMBAT SUPPORT

1 arty div
21 arty bde
9 MRL bde
5–8 engr river crossing/amphibious regt
1 engr river crossing bde

Special Purpose Forces Command 88,000

FORCES BY ROLE

SPECIAL FORCES

8 (Reconnaissance General Bureau) SF bn

MANOEUVRE

Reconnaissance

17 recce bn

Light

9 lt inf bde
6 sniper bde

Air Manoeuvre

3 AB bde
1 AB bn
2 sniper bde

Amphibious

2 sniper bde

Reserves 600,000

FORCES BY ROLE

MANOEUVRE

Light

40 inf div

18 inf bde

EQUIPMENT BY TYPE (ε)

ARMOURED FIGHTING VEHICLES

The Korean People's Army displayed a number of new armoured-vehicle designs at a parade in 2020, but it is unclear if any of them have entered operational service

MBT 3,500+ T-34/T-54/T-55/T-62/Type-59/*Chonma*/*Pokpoong*/*Songun*

LT TK 560+: 560 PT-76; M-1985

IFV 32 BTR-80A

APC 2,500+

APC (T) BTR-50; Type-531 (Type-63); VTT-323

APC (W) 2,500 BTR-40/BTR-60/M-1992/1/BTR-152/M-2010 (6×6)/M-2010 (8×8)

ANTI-TANK/ANTI-INFRASTRUCTURE

MSL

SP 9K11 *Malyutka* (AT-3 *Sagger*); M-2010 ATGM

MANPATS 2K15 *Shmel* (AT-1 *Snapper*); 9K111 *Fagot* (AT-4 *Spigot*); 9K113 *Konkurs* (AT-5 *Spandrel*)

RCL 82mm 1,700 B-10

ARTILLERY 21,600+

SP/TOWED 8,600:

SP 122mm M-1977; M-1981; M-1985; M-1991; **130mm** M-1975; M-1981; M-1991; **152mm** M-1974; M-1977; M-2018; **170mm** M-1978; M-1989

TOWED 122mm D-30; D-74; M-1931/37; **130mm** M-46; **152mm** M-1937; M-1938; M-1943

GUN/MOR 120mm (reported)

MRL 5,500: **107mm** Type-63; VTT-323 107mm; **122mm** BM-11; M-1977 (BM-21); M-1985; M-1992; M-1993; VTT-323 122mm; **200mm** BMD-20; **240mm** BM-24; M-1985; M-1989; M-1991; **300mm** some M-2015 (KN-SS-X-09) (in test); **600mm** some M-2019 (in test)

MOR 7,500: **82mm** M-37; **120mm** M-43; **160mm** M-43

SURFACE-TO-SURFACE MISSILE LAUNCHERS

SBRM 24+: 24 FROG-3/5/7; some *Toksa* (SS-21B *Scarab* mod); some (*Iskander* lookalike); some (ATACMS lookalike)

AIR DEFENCE

SAM

Point-defence 9K35 *Strela*-10 (SA-13 *Gopher*); 9K310 *Igla*-1 (SA-16 *Gimlet*); 9K32 *Strela*-2 (SA-7 *Grail*)‡

GUNS 11,000+

SP 14.5mm M-1984; **23mm** M-1992; **37mm** M-1992; **57mm** M-1985

TOWED 11,000: **14.5mm** ZPU-1/ZPU-2/ZPU-4; **23mm** ZU-23; **37mm** M-1939; **57mm** S-60; **85mm** M-1939 *KS*-12; **100mm** KS-19

Navy ε60,000

EQUIPMENT BY TYPE

SUBMARINES 71

SSB 1 *Gorae* (*Sinpo*-B) (SLBM trials) with 1 *Pukguksong*-1 SLBM

SSK ε20 Type-033 (*Romeo*) with 8 single 533mm TT with SAET-60 HWT

SSC ε40 (some *Sang-O* some with 2 single 533mm TT with 53–65E HWT; some *Sang-O* II with 4 single 533mm TT with 53–65E HWT)

SSW ε10† (some *Yugo* some with 2 single 406mm TT; some *Yeono* some with 2 single 533mm TT)

PRINCIPAL SURFACE COMBATANTS 2

FRIGATES • FFG 2:

1 *Najin* with 2 single lnchr with P-20 (SS-N-2A *Styx*) AShM, 2 RBU 1200 *Uragan* A/S mor, 2 100mm gun, 2 twin 57mm gun

1 *Najin* with 2 twin lnchr with *Kumsong*-3 (KN-SS-N-2 *Stormpetrel*) AShM, 2 RBU 1200 *Uragan* A/S mor, 2 100mm gun, 2 twin 57mm gun (operational status unclear)

PATROL AND COASTAL COMBATANTS 383+

CORVETTES • FS 5: 4 *Sariwon* with 2 twin 57mm gun; 1 *Tral* with 1 85mm gun

PCG 10 *Soju* (FSU Project 205 mod (*Osa*)) with 4 single lnchr with P-20 (SS-N-2A *Styx*) AShM

PCC 18:

6 Type-037 (*Hainan*) with 4 RBU 1200 A/S mor, 2 twin 57mm gun

7 *Taechong* I with 2 RBU 1200 *Uragan* A/S mor, 1 85mm gun, 1 twin 57mm gun

5 *Taechong* II with 2 RBU 1200 *Uragan* A/S mor, 1 100mm gun, 1 twin 57mm gun

PBFG 25+:

4 *Huangfeng* (Type-021) with 4 single lnchr with P-15 *Termit* (SS-N-2) AShM, 2 twin AK230 CIWS

6 *Komar* with 2 single lnchr with P-20 (SS-N-2A *Styx*) AShM

8 Project 205 (*Osa* I) with 4 single lnchr with P-20 (SS-N-2A *Styx*) AShM, 2 twin AK230 CIWS

6 *Sohung* (*Komar* mod) with 2 single lnchr with P-20 (SS-N-2A *Styx*) AShM

1+ *Nongo* with 2 twin lnchr with *Kumsong*-3 (KN-SS-N-2 *Stormpetrel*) AShM, 2 30mm CIWS (operational status unknown)

PBF 229: 54 *Chong-Jin* with 1 85mm gun; 142 *Ku Song*/*Sin Hung*/*Sin Hung* (mod); 33 *Sinpo*

PB 96: 59 *Chaho*; 6 *Chong-Ju* with 2 RBU 1200 *Uragan* A/S mor, 1 85mm gun; 13 Type-062 (*Shanghai* II); 18 SO-1 with 4 RBU 1200 *Uragan* A/S mor, 2 twin 57mm gun

MINE WARFARE • MINE COUNTERMEASURES 20

MSC 20: 15 *Yukto* I; 5 *Yukto* II

AMPHIBIOUS

LANDING SHIPS • LSM 10 *Hantae* (capacity 3 tanks; 350 troops)

LANDING CRAFT 255

LCM 25

LCPL approx. 95 *Nampo* (capacity 35 troops)

UCAC 135 *Kongbang* (capacity 50 troops)

LOGISTICS AND SUPPORT 23:

AGI 14 (converted fishing vessels)

AS 8 (converted cargo ships)

ASR 1 *Kowan*

Coastal Defence

FORCES BY ROLE

COASTAL DEFENCE

2 AShM regt with HY-1/*Kumsong*-3 (6 sites, some mobile launchers)

EQUIPMENT BY TYPE
COASTAL DEFENCE
ARTY 130mm M-1992; SM-4-1
AShM HY-1; *Kumsong*-3
ARTILLERY • TOWED 122mm M-1931/37; **152mm** M-1937

Air Force 110,000

4 air divs. 1st, 2nd and 3rd Air Divs (cbt) responsible for N, E and S air-defence sectors respectively; 8th Air Div (trg) responsible for NE sector. The AF controls the national airline

FORCES BY ROLE
BOMBER
3 lt regt with H-5; Il-28 *Beagle*
FIGHTER
1 regt with MiG-15 *Fagot*
6 regt with J-5; MiG-17 *Fresco*
4 regt with J-6; MiG-19 *Farmer*
5 regt with J-7; MiG-21F-13/PFM *Fishbed*
1 regt with MiG-21bis *Fishbed*
1 regt with MiG-23ML/P *Flogger*
1 regt with MiG-29A/S/UB *Fulcrum*
GROUND ATTACK
1 regt with Su-25K/UBK *Frogfoot*
TRANSPORT
Some regt with An-2 *Colt*/Y-5 (to infiltrate 2 air-force sniper brigades deep into ROK rear areas); Il-62M *Classic*
TRAINING
Some regt with CJ-6; FT-2; MiG-21U/UM
TRANSPORT HELICOPTER
Some regt with Hughes 500D/E; Mi-8 *Hip*; Mi-17 *Hip* H; Mil-26 *Halo*; PZL Mi-2 *Hoplite*; Mi-4 *Hound*; Z-5
AIR DEFENCE
19 bde with S-125 *Pechora* (SA-3 *Goa*); S-75 *Dvina* (SA-2 *Guideline*); S-200 *Angara* (SA-5 *Gammon*); 9K36 *Strela*-3 (SA-14 *Gremlin*); 9K310 *Igla*-1 (SA-16 *Gimlet*); 9K32 *Strela*-2 (SA-7 *Grail*)‡; *Pongae*-5

EQUIPMENT BY TYPE
AIRCRAFT 545 combat capable
BBR 80 Il-28 *Beagle*/H-5†
FTR 401+: MiG-15 *Fagot*; 107 MiG-17 *Fresco*/J-5; 100 MiG-19 *Farmer*/J-6; 120 MiG-21F-13 *Fishbed*/J-7; MiG-21PFM *Fishbed*; 46 MiG-23ML *Flogger*; 10 MiG-23P *Flogger*; 18+ MiG-29A/S/UB *Fulcrum*
FGA 30 MiG-21bis *Fishbed* (18 Su-7 *Fitter* in store)
ATK 34 Su-25K/UBK *Frogfoot*
TPT 205: **Heavy** 3 Il-76 (operated by state airline); **Light** ε200 An-2 *Colt*/Y-5; **PAX** 2 Il-62M *Classic* (VIP)
TRG 215+: 180 CJ-6; 35 FT-2; some MiG-21U/UM
HELICOPTERS
MRH 80 Hughes 500D/E†
TPT 206: **Heavy** 4 Mi-26 *Halo*; **Medium** 63: 15 Mi-8 *Hip*/Mi-17 *Hip* H; 48 Mi-4 *Hound*/Z-5; **Light** 139 PZL Mi-2 *Hoplite*
UNINHABITED AERIAL VEHICLES
ISR • Medium some (unidentified indigenous type); **Light** *Pchela*-1 (*Shmel*) (reported)
AIR DEFENCE • **SAM** 350+
Long-range 38 S-200 *Angara* (SA-5 *Gammon*)
Medium-range 179+: some *Pongae*-5 (status unknown); 179+ S-75 *Dvina* (SA-2 *Guideline*)
Short-range 133 S-125 *Pechora* (SA-3 *Goa*)
Point-defence 9K32 *Strela*-2 (SA-7 *Grail*)‡; 9K36 *Strela*-3 (SA-14 *Gremlin*); 9K310 *Igla*-1 (SA-16 *Gimlet*)
AIR-LAUNCHED MISSILES
AAM • IR R-3 (AA-2 *Atoll*)‡; R-60 (AA-8 *Aphid*); R-73 (AA-11A *Archer*); PL-5; PL-7; **SARH** R-23/24 (AA-7 *Apex*); R-27R/ER (AA-10 A/C *Alamo*)
ASM Kh-23 (AS-7 *Kerry*)‡; Kh-25 (AS-10 *Karen*); Kh-29L (AS-14A *Kedge*)

Paramilitary 189,000 active

Security Troops 189,000 (incl border guards, public-safety personnel)

Ministry of Public Security

Worker/Peasant Red Guard ε5,700,000 reservists

Org on a province/town/village basis; comd structure is bde–bn–coy–pl; small arms with some mor and AD guns (but many units unarmed)

Korea, Republic of ROK

South Korean won ₩		2019	2020	2021
GDP	₩	1919tr	1900tr	
	US$	1.65tr	1.59tr	
per capita	US$	31,846	30,644	
Growth	%	2.0	-1.9	
Inflation	%	0.4	0.5	
Def bdgt	₩	46.7tr	48.4tr	52.9tr
	US$	40.1bn	40.4bn	
US$1=₩		1165.36	1197.39	

Real-terms defence budget trend (US$bn, constant 2015)

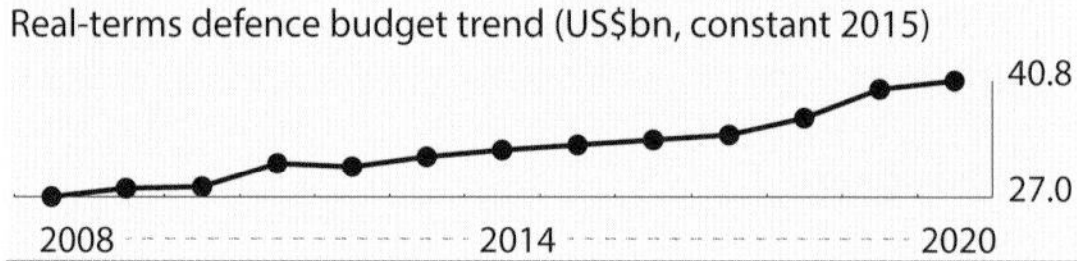

Population	51,835,110					
Age	**0–14**	**15–19**	**20–24**	**25–29**	**30–64**	**65 plus**
Male	6.6%	2.5%	3.3%	3.8%	27.0%	6.9%
Female	6.2%	2.3%	3.0%	3.3%	26.1%	9.0%

Capabilities

South Korea's defence policy remains focused on North Korea, notwithstanding diplomatic contact since 2018. Seoul has looked to recapitalise conventional military systems to ensure a qualitative edge over Pyongyang. It continues to prioritise acquiring new capabilities to respond to North Korea's military and nuclear threat. The three-axis approach formerly known as 'Kill Chain', 'Korea Air and Missile Defense' and 'Korea Massive Punishment and Retaliation', was reportedly rebranded in early 2019. The Defense Reform 2.0 project announced in 2018 sets out ambitions to modernise and restructure the armed forces, placing emphasis on new technologies. The established alliance with the US is a major element of defence strategy, though the planned transfer of wartime operational control of forces to Seoul is now 'conditions based' with no firm date set. A large number of US military personnel and equipment remain stationed in South Korea including the THAAD missile-defence system. South Korea's forces are some of the best equipped and trained in the region. South Korea has demonstrated the capacity to support small international deployments, including contributions to UN missions and counter-piracy operations in the Arabian Sea. The inventory increasingly comprises modern systems. South Korea has developed a broad range of domestic defence industries which are capable of supplying the majority of military requirements. However, there is still reliance on the US in areas such as frontline combat aircraft. Local defence industries are finding growing export success.

ACTIVE 599,000 (Army 464,000 Navy 70,000 Air 65,000) Paramilitary 13,500

Conscript liability Service period reducing from Oct 2018, by three months for the army and marines (now 18 months), and the navy (now 20 months); and by two months for the air force (now 22 months)

RESERVE 3,100,000

Reserve obligation of three days per year. First Combat Forces (Mobilisation Reserve Forces) or Regional Combat Forces (Homeland Defence Forces) to age 33

Reserve Paramilitary 3,000,000

Being reorganised

ORGANISATIONS BY SERVICE

Army 464,000

FORCES BY ROLE

COMMAND

8 corps HQ

1 (Capital Defence) comd HQ

SPECIAL FORCES

1 (Special Warfare) SF comd

6 SF bde

1 indep SF gp

2 cdo bde

6 cdo regt

2 indep cdo bn

MANOEUVRE

Armoured

6 armd bde

2 (8th and Capital) mech inf div (1 recce bn, 1 armd bde, 1 armd inf bde; 1 mech inf bde, 1 arty bde, 1 engr bn)

1 (11th) mech inf div (1 recce bn, 3 armd inf bde, 1 arty bde, 1 engr bn)

2 tk bn

Mechanised

1 (30th) mech inf div (1 recce bn, 2 mech inf bde, 1 arty bde, 1 engr bn)

Light

15 inf div (1 recce bn, 1 tk bn, 3 inf regt, 1 arty regt (4 arty bn), 1 engr bn)

2 indep inf bde

Air Manoeuvre

1 air aslt bde

Other

5 sy regt

SURFACE-TO-SURFACE MISSILE

3 SSM bn

COMBAT SUPPORT

1 MRL bde (3 MRL bn)

6 engr bde

5 engr gp

1 CBRN defence bde

8 sigs bde

COMBAT SERVICE SUPPORT

4 log spt comd

HELICOPTER

1 (army avn) comd

AIR DEFENCE

1 ADA bde

5 ADA bn

Reserves

FORCES BY ROLE

COMMAND

1 army HQ

MANOEUVRE

Light

24 inf div

EQUIPMENT BY TYPE

ARMOURED FIGHTING VEHICLES

MBT 2,221: 1,000 K1/K1E1; 484 K1A1/K1A2; 100 K2; 597 M48A5; 40 T-80U

IFV 540: ε500 K21; 40 BMP-3

APC 2,490

APC (T) 2,260: 1,700 KIFV; 420 M113; 140 M577 (CP)

APC (W) 220; 20 BTR-80; 200 K806/K808

PPV 10 *MaxxPro*

ENGINEERING & MAINTENANCE VEHICLES

AEV 207 M9

ARV 238+: 200 K1; K21 ARV; K288A1; M47; 38 M88A1

VLB 56 K1

ANTI-TANK/ANTI-INFRASTRUCTURE

MSL

SP *Hyeongung*

MANPATS 9K115 *Metis* (AT-7 *Saxhorn*); *Hyeongung*; TOW-2A

RCL 75mm; 90mm M67; **106mm** M40A2

GUNS 58

SP 90mm 50 M36

TOWED 76mm 8 M18 *Hellcat* (AT gun)

ARTILLERY 12,237+

SP 2,403+: **105mm** ε50 K105A1; **155mm** 2,340: ε1,300 K9/K9A1 *Thunder;* 1,040 M109A2 (K55/K55A1); **175mm** some M107; **203mm** 13 M110

TOWED 3,500+: **105mm** 1,700 M101/KH-178; **155mm/203mm** 1,800+ KH-179/M114/M115

MRL 334: **130mm** 156 K136 *Kooryong*; **227mm** 58: 48 M270 MLRS; 10 M270A1 MLRS; **239mm** ε120 K239 *Cheonmu*

MOR 6,000: **81mm** KM29 (M29); **107mm** M30

SURFACE-TO-SURFACE MISSILE LAUNCHERS

SRBM • Conventional 30+: 30 *Hyonmu* IIA/IIB; MGM-140A/B ATACMS (launched from M270/M270A1 MLRS)

GLCM • Conventional *Hyonmu* III

HELICOPTERS
ATK 96: 60 AH-1F/J *Cobra*; 36 AH-64E *Apache*
MRH 175: 130 Hughes 500D; 45 MD-500
TPT 236+: **Heavy** 37: 31 CH-47D *Chinook*; 6 MH-47E *Chinook*; **Medium** 187+: 100+ KUH-1 *Surion*; 87 UH-60P *Black Hawk*; **Light** 12 Bo-105
AIR DEFENCE
SAM • Point-defence *Chiron*; *Chun Ma* (*Pegasus*); FIM-92 *Stinger*; *Javelin*; *Mistral*; 9K310 *Igla*-1 (SA-16 *Gimlet*)
GUNS 477+
SP 317: **20mm** ε150 KIFV *Vulcan* SPAAG; **30mm** 167 K30 *Biho*
TOWED 160: **20mm** 60 M167 *Vulcan*; **35mm** 20 GDF-003; **40mm** 80 L/60/L/70; M1
AIR-LAUNCHED MISSILES
ASM AGM-114R1 *Hellfire*

Navy 70,000 (incl marines)

Three separate fleet elements: 1st Fleet Donghae (East Sea/Sea of Japan); 2nd Fleet Pyeongtaek (West Sea/Yellow Sea); 3rd Fleet Busan (South Sea/Korea Strait); independent submarine command; three additional flotillas (incl SF, mine-warfare, amphibious and spt elements) and 1 Naval Air Wing (3 gp plus spt gp)

EQUIPMENT BY TYPE
SUBMARINES • SSK 18:
6 *Chang Bogo* I (GER Type-209/1200; KSS-1) with 8 single 533mm TT with SUT HWT/K731 *White Shark* HWT
3 *Chang Bogo* I (GER Type 209/1200; KSS 1) with 8 single 533mm TT with UGM-84 *Harpoon* AShM/SUT HWT/K731 *White Shark* HWT
9 *Chang Bogo* II (GER Type-214; KSS-2) (fitted with AIP) with 8 single 533mm TT with *Hae Sung* III LACM/*Hae Sung* I AShM/SUT HWT/K731 *White Shark* HWT
PRINCIPAL SURFACE COMBATANTS 23
CRUISERS • CGHM 3:
3 *Sejong* (KDD-III) with *Aegis* Baseline 7 C2, 6 8-cell K-VLS with *Hae Sung* II LACM/*Red Shark* A/S msl, 4 quad lnchr with *Hae Sung* I AShM, 10 8-cell Mk 41 VLS (6 fore, 4 aft) with SM-2 Block IIIA/B SAM, 1 21-cell Mk 49 GMLS with RIM-116 RAM SAM, 2 triple 324mm SVTT Mk 32 ASTT with K745 *Blue Shark* LWT, 1 *Goalkeeper* CIWS, 1 127mm gun (capacity 2 *Lynx* Mk99/AW159 *Wildcat* hels)
DESTROYERS • DDGHM 6 *Chungmugong Yi Sun-Sin* (KDD-II) with 2 8-cell K-VLS with *Hae Sung* II LACM/*Red Shark* A/S msl, 2 quad lnchr with RGM-84 *Harpoon* AShM/*Hae Sung* I AShM, 4 8-cell Mk 41 VLS with SM-2 Block IIIA/B SAM, 1 21-cell Mk 49 GMLS with RIM-116 RAM SAM, 2 triple 324mm SVTT Mk 32 ASTT with Mk 46 LWT, 1 *Goalkeeper* CIWS, 1 127mm gun (capacity 1 *Lynx* Mk99/AW159 *Wildcat* hel)
FRIGATES 14
FFGHM 10:
3 *Gwanggaeto Daewang* (KDD-I) with 2 quad lnchr with RGM-84 *Harpoon* AShM, 2 8-cell Mk 48 mod 2 VLS with RIM-7P *Sea Sparrow* SAM, 2 triple 324mm SVTT Mk 32 ASTT with Mk 46 LWT, 2 *Goalkeeper* CIWS, 1 127mm gun (capacity 1 *Lynx* Mk99/AW159 *Wildcat* hel)
6 *Incheon* with 2 quad lnchr with TSLM LACM/*Hae Sung* I AShM, 1 21-cell Mk 49 lnchr with RIM-116 RAM SAM, 2 triple 324mm KMk. 32 ASTT with K745 *Blue Shark* LWT, 1 Mk 15 *Phalanx* Block 1B CIWS, 1 127 mm gun (capacity 1 *Lynx* Mk99/AW159 *Wildcat* hel)
1 *Daegu* (*Incheon* Batch II) with 2 8-cell K-VLS with *Hae Sung* II LACM/TSLM LACM/*Haegung* (K-SAAM) SAM/*Red Shark* A/S msl, 2 quad lnchr with TSLM LACM/*Hae Sung* I AShM, 2 triple 324mm KMk. 32 ASTT with K745 *Blue Shark* LWT, 1 Mk 15 *Phalanx* Block 1B CIWS, 1 127mm gun (capacity 1 *Lynx* Mk99/AW159 *Wildcat* hel)
FFG 4 *Ulsan* with 2 quad lnchr with RGM-84 *Harpoon* AShM, 2 triple 324mm SVTT Mk 32 ASTT with Mk 46 LWT, 2 76mm gun
PATROL AND COASTAL COMBATANTS ε84
CORVETTES • FSG 12:
6 *Po Hang* (Flight IV) with 2 twin lnchr with RGM-84 *Harpoon* AShM, 2 triple 324mm ASTT with Mk 46 LWT, 2 76mm gun
6 *Po Hang* (Flight V) with 2 twin lnchr with *Hae Sung* I AShM, 2 triple 324mm KMk. 32 ASTT with K745 *Blue Shark* LWT, 2 76mm gun
PCFG 22: 18 *Gumdoksuri* with 2 twin lnchr with *Hae Sung* I AShM, 1 76mm gun; 4 *Chamsuri* II with 1 12-cell 130mm MRL, 1 76mm gun
PBF ε50 *Sea Dolphin*
MINE WARFARE 11
MINE COUNTERMEASURES 9
MHO 6 *Kan Kyeong*
MSO 3 *Yang Yang*
MINELAYERS • ML 2:
1 *Won San* with 2 triple 324mm SVTT Mk 32 ASTT with Mk 46 LWT/K745 *Blue Shark* LWT, 1 76mm gun, 1 hel landing platform
1 *Nampo* (MLS-II) with 1 4-cell K-VLS VLS with *Haegung* (K-SAAM) SAM, 2 triple KMk. 32 triple 324mm ASTT with K745 *Blue Shark* LWT, 1 76mm gun (capacity 1 med hel)
AMPHIBIOUS
PRINCIPAL AMPHIBIOUS SHIPS 5
LHD 1 *Dokdo* with 1 Mk 49 GMLS with RIM-116 RAM SAM, 2 *Goalkeeper* CIWS (capacity 2 LCAC; 10 tanks; 700 troops; 10 UH-60 hel)
LPD 4 *Cheonwangbong* (LST-II) (capacity 3 LCM; 2 MBT; 8 AFV; 300 troops; 2 med hel)
LANDING SHIPS • LST 4 *Go Jun Bong* with 1 hel landing platform (capacity 20 tanks; 300 troops)
LANDING CRAFT 26
LCU 8 *Mulgae* I
LCT 3 *Mulgae* II
LCM 10 LCM-8
LCAC 5: 3 *Tsaplya* (capacity 1 MBT; 130 troops); 2 LSF-II (capacity 150 troops or 1 MBT & 24 troops)
LOGISTICS AND SUPPORT 11
AG 1 *Sunjin* (trials spt)
AOEH 1 *Soyangham* (AOE-II) with 1 Mk 15 *Phalanx* Block 1B CIWS (capacity 1 med hel)
AORH 3 *Chun Jee*
ARS 1 *Cheong Hae Jin*

ATS 2 *Tongyeong*
AX 3: 1 *Hansando* with 2 triple 324mm KMk. 32 ASTT with K745 *Blue Shark* LWT, 1 76mm gun (fitted for but not with K-VLS) (capacity 2 med hels; 300 students); 2 MTB

Naval Aviation

EQUIPMENT BY TYPE
AIRCRAFT 16 combat capable
ASW 16: 8 P-3C *Orion;* 8 P-3CK *Orion*
TPT • Light 5 Cessna F406 *Caravan* II
HELICOPTERS
ASW 31: 11 *Lynx* Mk99; 12 *Lynx* Mk99A; 8 AW159 *Wildcat*
TPT 15: **Medium** 8 UH-60P *Black Hawk* **Light** 7 Bell 205 (UH-1H *Iroquois*)

Marines 29,000

FORCES BY ROLE
SPECIAL FORCES
1 SF regt
MANOEUVRE
Amphibious
2 mne div (1 recce bn, 1 tk bn, 3 mne regt, 1 amph bn, 1 arty regt, 1 engr bn)
1 mne bde
COMBAT SUPPORT
Some cbt spt unit
EQUIPMENT BY TYPE
ARMOURED FIGHTING VEHICLES
MBT 100: 50 K1A1; 50 M48
AAV 166 AAV-7A1
ANTI-TANK/ANTI-INFRASTUCTURE • MSL
SP *Spike* NLOS
MANPATS *Hyeongung*
ARTILLERY
SP 155mm K9/K9A1 *Thunder*
TOWED 105mm KH-178; **155mm** KH-179
MRL 130mm K136 *Kooryong*
MOR 81mm KM29 (M29)
COASTAL DEFENCE • AShM RGM-84A *Harpoon* (truck mounted)
HELICOPTERS • TPT • Medium 5+ MUH-1 *Surion*
AIR DEFENCE
GUNS • Towed • 20mm M167 *Vulcan* (direct fire role)

Naval Special Warfare Flotilla

Air Force 65,000

4 Comd (Ops, Southern Combat, Logs, Trg)
FORCES BY ROLE
FIGHTER/GROUND ATTACK
1 sqn with F-4E *Phantom* II
6 sqn with F-5E/F *Tiger* II
3 sqn with F-15K *Eagle*
10 sqn with F-16C/D *Fighting Falcon* (KF-16C/D)
2 sqn with FA-50 *Fighting Eagle*
ISR
1 wg with KO-1
SIGINT
1 sqn with Hawker 800RA/XP
SEARCH & RESCUE
2 sqn with AS332L *Super Puma;* Bell 412EP; HH-47D *Chinook;* HH-60P *Black Hawk;* Ka-32 *Helix* C
TRANSPORT
1 VIP sqn with B-737-300; B-747; CN235-220; S-92A *Superhawk;* VH-60P *Black Hawk* (VIP)
3 sqn (incl 1 Spec Ops) with C-130H/H-30/J-30 *Hercules*
2 sqn with CN235M-100/220
TRAINING
2 sqn with F-5E/F *Tiger* II
1 sqn with F-16C/D *Fighting Falcon*
4 sqn with KT-1
1 sqn with Il-103
3 sqn with T-50/TA-50 *Golden Eagle**
TRANSPORT HELICOPTER
1 sqn with UH-60P *Black Hawk* (Spec Ops)
ISR
1 sqn with RQ-4B *Global Hawk* (forming)
AIR DEFENCE
3 AD bde (total: 2 SAM bn with MIM-23B I-*Hawk;* 5 SAM bn with *Cheongung;* 2 SAM bn with MIM-104E *Patriot* PAC-2 GEM-T)
EQUIPMENT BY TYPE
AIRCRAFT 579 combat capable
FTR 174: 142 F-5E *Tiger* II; 32 F-5F *Tiger* II
FGA 325: 30 F-4E *Phantom* II; 59 F-15K *Eagle;* 118 F-16C *Fighting Falcon* (KF-16C); 44 F-16D *Fighting Falcon* (KF-16D); 24 F-35A *Lightning* II; 50 FA-50 *Fighting Eagle*
AEW&C 4 B-737 AEW
ISR 24: 4 Hawker 800RA; 20 KO-1
SIGINT 6: 4 Hawker 800SIG; 2 *Falcon* 2000 (COMINT/SIGINT)
TKR/TPT 4 A330 MRTT
TPT 38: **Medium** 16: 8 C-130H *Hercules;* 4 C-130H-30 *Hercules;* 4 C-130J-30 *Hercules;* **Light** 20: 12 CN235M-100; 8 CN235M-220 (incl 2 VIP); **PAX** 2: 1 B-737-300; 1 B-747-400
TRG 183: 83 KT-1; 49 T-50 *Golden Eagle**; 9 T-50B *Black Eagle** (aerobatics); 22 TA-50 *Golden Eagle**; ε20 KT-100
HELICOPTERS
SAR 16: 5 HH-47D *Chinook;* 11 HH-60P *Black Hawk*
MRH 3 Bell 412EP
TPT • Medium 30: 2 AS332L *Super Puma;* 8 Ka-32 *Helix* C; 3 S-92A *Super Hawk;* 7 UH-60P *Black Hawk;* 10 VH-60P *Black Hawk* (VIP)
UNINHABITED AERIAL VEHICLES • ISR 7+: **Heavy** 4 RQ-4B *Global Hawk;* **Medium** 3+: some *Night Intruder;* 3 *Searcher*
LOITERING MUNITIONS 100 *Harpy*
AIR DEFENCE • SAM 168
Long-range 48 MIM-104 *Patriot* PAC-2 GEM-T/PAC-3 CRI
Medium-range 120: 72 *Cheongung* (KM-SAM); 48 MIM-23B I-*Hawk*
AIR-LAUNCHED MISSILES
AAM • IR AIM-9 *Sidewinder;* **IIR** AIM-9X *Sidewinder* II; **SARH** AIM-7 *Sparrow;* **ARH** AIM-120B/C-5/7 AMRAAM
ASM AGM-65A *Maverick;* AGM-130
AShM AGM-84L *Harpoon* Block II; AGM-142 *Popeye*

Asia

ARM AGM-88 HARM
ALCM AGM-84H SLAM-ER; KEPD-350 *Taurus*
BOMBS • **Laser-guided** *Paveway* II

Paramilitary 13,500 active

Civilian Defence Corps 3,000,000 reservists (to age 50)

Coast Guard 13,500

Part of the Ministry of Maritime Affairs and Fisheries. Five regional headquarters with 19 coastguard stations and one guard unit

EQUIPMENT BY TYPE
PATROL AND COASTAL COMBATANTS 84
PSOH 15: 1 *Lee Cheong-ho* with 1 76mm gun; 1 *Sambongho*; 13 *Tae Pung Yang* with 1 med hel
PSO 21: 3 *Han Kang* with 1 76mm gun, 1 hel landing platform; 5 *Han Kang* II with 1 76mm gun, 1 hel landing pllatform; 12 *Jaemin* with 1 hel landing platform; 1 *Sumjinkang*
PCO 23 *Tae Geuk*
PCC 21: 15 *Hae Uri*; 2 *Hae Uri* II; 4 others
PB ε4 (various)
AMPHIBIOUS • **LANDING CRAFT** 8
UCAC 8: 1 BHT-150; 4 *Griffon* 470TD; 3 *Griffon* 8000TD
AIRCRAFT
MP 5: 1 C-212-400 MP; 4 CN235-110 MPA
TPT • **PAX** 1 CL-604
HELICOPTERS
MRH 7: 5 AS565MB *Panther*; 1 AW139; 1 Bell 412SP
SAR 2 S-92
TPT • **Medium** 10: 8 Ka-32 *Helix* C; 2 KUH-1 *Surion*

DEPLOYMENT

ARABIAN SEA & GULF OF ADEN: Combined Maritime Forces • CTF-151: 1 DDGHM

INDIA/PAKISTAN: UN • UNMOGIP 7

LEBANON: UN • UNIFIL 409; 1 mech inf coy; 1 engr coy; 1 sigs coy; 1 maint coy

SOUTH SUDAN: UN • UNMISS 272; 1 engr coy

UNITED ARAB EMIRATES: 170 (trg activities at UAE Spec Ops School)

WESTERN SAHARA: UN • MINURSO 4

FOREIGN FORCES

Sweden NNSC: 5 obs
Switzerland NNSC: 5 obs
United States US Pacific Command: 31,050
Army 22,500; 1 HQ (8th Army) at Yongsan; 1 div HQ at Pyeongtaek; 1 armd bde with M1A2 SEPv2 *Abrams*; M2A2/M3A3 *Bradley*; M109A6; 1 (cbt avn) hel bde with AH-64D/E *Apache*; CH-47 *Chinook*; UH-60L/M *Black Hawk*; 1 MRL bde with M270A1 MLRS; 1 AD bde with MIM-104 *Patriot*/FIM-92A *Avenger*; 1 SAM bty with THAAD; 1 (APS) armd bde eqpt set
Navy 350
USAF 8,000; 1 HQ (7th Air Force) at Osan AB; 1 ftr wg at Kunsan AB (2 ftr sqn with 20 F-16C/D *Fighting Falcon*); 1 ftr wg at Osan AB (1 ftr sqn with 20 F-16C/D *Fighting Falcon*, 1 atk sqn with 24 A-10C *Thunderbolt* II); 1 ISR sqn at Osan AB with U-2S
USMC 200

Laos LAO

New Lao Kip		2019	2020	2021
GDP	kip	164tr	177tr	
	US$	19.1bn	18.7bn	
per capita	US$	2,661	2,567	
Growth	%	5.2	0.2	
Inflation	%	3.3	6.5	
Def exp	kip	n.k.	n.k.	
	US$	n.k.	n.k.	
US$1=kip		8582.18	9474.02	

Population 7,447,396

Age	0–14	15–19	20–24	25–29	30–64	65 plus
Male	15.8%	5.3%	5.0%	4.6%	17.1%	1.9%
Female	15.4%	5.3%	5.1%	4.7%	17.7%	2.3%

Capabilities

The Lao People's Armed Forces (LPAF) have considerable military experience from the Second Indo-China War and the 1988 border war with Thailand. They are closely linked to the ruling Communist Party and their primary role is internal security. A lack of financial resources has limited defence spending and military procurement for two decades. Contacts continue with the Chinese and Vietnamese armed forces, while there is strong defence cooperation with Russia. Laos has also participated in ADMM-Plus military exercises. Training support has been provided by countries including Russia and Vietnam. The LPAF have participated in regional exercises with neighbouring countries but have made no international deployments and have little capacity for sustained operations. Laos still operates Soviet-era military equipment, and relies on Russian supplies, as illustrated by ongoing deliveries of training aircraft, armoured reconnaissance vehicles and main battle tanks. The country lacks a traditional defence-industrial base and maintenance capacity is limited, reflected in a support contract for a Russian firm to maintain part of the rotary-wing fleet.

ACTIVE 29,100 (Army 25,600 Air 3,500) **Paramilitary 100,000**
Conscript liability 18 months minimum

ORGANISATIONS BY SERVICE

Space

EQUIPMENT BY TYPE
SATELLITES • **ISR** 1 LaoSat-1

Army 25,600

FORCES BY ROLE
4 mil regions

MANOEUVRE

Armoured

1 armd bn

Light

5 inf div

7 indep inf regt

65 indep inf coy

COMBAT SUPPORT

5 arty bn

1 engr regt

2 (construction) engr regt

AIR DEFENCE

9 ADA bn

EQUIPMENT BY TYPE

ARMOURED FIGHTING VEHICLES

MBT 25: 15 T-54/T-55; 10 T-72B1

LT TK 10 PT-76

RECCE BRDM-2M

IFV 10+ BMP-1

APC • APC (W) 50: 30 BTR-40/BTR-60; 20 BTR-152

AUV Dongfeng Mengshi 4×4; ZYZ-8002 (CS/VN3)

ENGINEERING & MAINTENANCE VEHICLES

ARV T-54/T-55

VLB MTU

ANTI-TANK/ANTI-INFRASTRUCTURE • **RCL 57mm** M18/A1; **75mm** M20; **106mm** M40; **107mm** B-11

ARTILLERY 62+

TOWED 62: **105mm** 20 M101; **122mm** 20 D-30/M-30 M-1938; **130mm** 10 M-46; **155mm** 12 M114

MOR 81mm; **82mm**; **107mm** M-1938/M2A1; **120mm** M-43

AIR DEFENCE

SAM

Short-range S-125 *Pechora* (SA-3 *Goa*); *Yitian* (CH-SA-13)

Point-defence 9K32M *Strela*-2M (SA-7 *Grail*)‡; 9K35 *Strela*-10 (SA-13 *Gopher*); 9K310 *Igla*-1 (SA-16 *Gimlet*)

GUNS

SP 23mm ZSU-23-4

TOWED 14.5mm ZPU-1/ZPU-4; **23mm** ZU-23; **37mm** M-1939; **57mm** S-60

Army Marine Section ε600

EQUIPMENT BY TYPE

PATROL AND COASTAL COMBATANTS • **PBR** some

AMPHIBIOUS • **LCM** some

Air Force 3,500

FORCES BY ROLE

TRANSPORT

1 regt with MA60; MA600; Mi-17 *Hip* H

EQUIPMENT BY TYPE

AIRCRAFT 4 combat capable

TPT • Light 5: 1 An-74TK *Coaler*; 2 MA60; 2 MA600

TRG 4 Yak-130 *Mitten**

HELICOPTERS

MRH 15: 6 Mi-17 *Hip* H; 5 Mi-17V-5 *Hip*; 4 Z-9A

TPT 4: **Medium** 1 Ka-32T *Helix* C; **Light** 3 SA360 *Dauphin*

Paramilitary

Militia Self-Defence Forces 100,000+

Village 'home guard' or local defence

Malaysia MYS

Malaysian Ringgit RM		2019	2020	2021
GDP	RM	1.51tr	1.41tr	
	US$	365bn	336bn	
per capita	US$	11,193	10,192	
Growth	%	4.3	-6.0	
Inflation	%	0.7	-1.1	
Def bdgt	RM	13.9bn	15.6bn	
	US$	3.36bn	3.72bn	
US$1=RM		4.14	4.19	

Real-terms defence budget trend (US$bn, constant 2015)

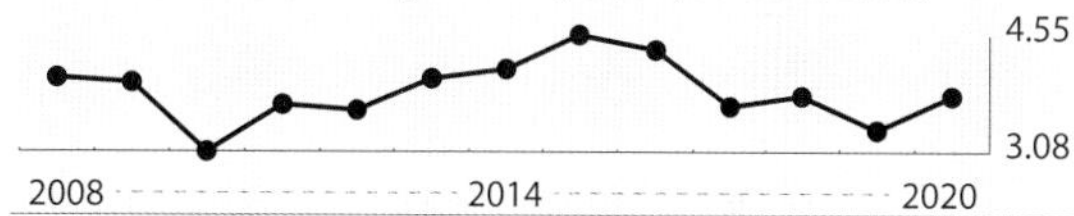

Population 32,652,083

Age	0–14	15–19	20–24	25–29	30–64	65 plus
Male	13.8%	4.4%	4.1%	3.9%	21.2%	3.3%
Female	13.0%	4.2%	4.0%	3.8%	20.7%	3.6%

Capabilities

Substantial modernisation programmes over the past 30 years have developed the Malaysian armed forces' capacity for external defence. The government tabled its first-ever defence white paper in December 2019, which described 'three pillars' of Malaysia's defence strategy as 'concentric deterrence' (the armed forces' protection of national interests in 'core', 'extended' and 'forward' zones); 'comprehensive defence' (involving whole-of-government and whole-of-society support for the national-defence effort); and 'credible partnerships' (involving engagement in regional and wider international defence cooperation). While the paper also identified other challenges, including tensions in the South China Sea and cyber threats, it provided no detailed insights into future resource allocation or capability development. However, it is likely that budgetary constraints will continue to constrain defence resources. Malaysian forces regularly participate in ADMM-Plus, Five Power Defence Arrangements and other exercises with regional and international partners, including the US. Malaysia has invested in synthetic military-training aids. In 2017, Malaysia began trilateral joint maritime patrols and joint Sulu Sea air patrols with Indonesia and the Philippines. However, much of Malaysia's military equipment is ageing and in several cases non-operational. However, addressing capability gaps is a high priority, particularly in air and maritime surveillance, and investments in new equipment have been made in recent years. For example, 22 Squadron was established in 2015 to operate Malaysia's new A400M transport aircraft. In recent decades, Malaysia has maintained a small defence industry providing maintenance, repair and overhaul services.

ACTIVE 113,000 **(Army 80,000 Navy 18,000 Air 15,000) Paramilitary 22,500**

RESERVE 51,600 **(Army 50,000, Navy 1,000 Air Force 600) Paramilitary 244,700**

ORGANISATIONS BY SERVICE

Army 80,000

2 mil region

FORCES BY ROLE

COMMAND

5 div HQ

SPECIAL FORCES

1 SF bde (3 SF bn)

MANOEUVRE

Armoured

1 tk regt

Mechanised

4 armd regt

1 mech inf bde (4 mech bn, 1 cbt engr sqn)

Light

1 inf bde (6 inf bn, 1 arty regt)

4 inf bde (3 inf bn, 1 arty regt)

2 inf bde (3 inf bn)

1 inf bde (2 inf bn, 1 arty regt)

1 inf bde (2 inf bn)

1 inf bde (forming)

Air Manoeuvre

1 (Rapid Deployment Force) AB bde (1 lt tk sqn, 4 AB bn, 1 lt arty regt, 1 engr sqn)

Other

2 (border) sy bde (5 bn)

COMBAT SUPPORT

9 arty regt

1 STA regt

1 MRL regt

1 cbt engr sqn

3 fd engr regt (total: 7 cbt engr sqn, 3 engr spt sqn)

1 construction regt

1 int unit

4 MP regt

1 sigs regt

HELICOPTER

1 hel sqn

1 tpt sqn with S-61A-4 *Nuri* (forming)

AIR DEFENCE

3 ADA regt

EQUIPMENT BY TYPE

ARMOURED FIGHTING VEHICLES

MBT 48 PT-91M *Twardy*

LT TK 21 *Scorpion*-90

RECCE 74 SIBMAS (some†)

IFV 136: 31 ACV300 *Adnan* (25mm *Bushmaster*); 13 ACV300 *Adnan* AGL; 46 AV8 *Gempita* IFV25; 46 AV8 *Gempita* IFV30 (incl 10 with *Ingwe* ATGM)

APC 620

APC (T) 265: 149 ACV300 *Adnan* (incl 69 variants); 13 FV4333 *Stormer* (upgraded); 63 K200A; 40 K200A1

APC (W) 326: 26 AV8 *Gempita* APC (incl 13 CP; 3 sigs); 300 *Condor* (incl variants)

PPV 29: 9 IAG *Guardian*; 20 *Lipanbara*

ENGINEERING & MAINTENANCE VEHICLES

AEV 3 MID-M

ARV 47+: *Condor*; 15 ACV300; 4 K288A1; 22 SIBMAS; 6 WZT-4

VLB 5+: *Leguan*; 5 PMCz-90

NBC VEHICLES K216A1

ANTI-TANK/ANTI-INFRASTRUCTURE • MSL

SP 8 ACV300 *Baktar Shikan*

MANPATS 9K115 *Metis* (AT-7 *Saxhorn*); 9K115-2 *Metis*-M (AT-13 *Saxhorn* 2); *Eryx*; *Baktar Shihan* (HJ-8); SS.11

RCL 84mm *Carl Gustaf*

ARTILLERY 418

TOWED 128: **105mm** 106: 6 LG1 MkIII; 100 Model 56 pack howitzer; **155mm** 22 G-5

MRL 36 ASTROS II (equipped with 127mm SS-30)

MOR 254: **81mm** 232; **SP 81mm** 14: 4 K281A1; 10 ACV300-S; **SP 120mm** 8 ACV-S

AMPHIBIOUS • LANDING CRAFT

LCA 165 Damen Assault Craft 540 (capacity 10 troops)

HELICOPTERS • TPT 12: **Medium** 2 S-61A-4 *Nuri*; **Light** 10 AW109

AIR DEFENCE

SAM • Point-defence 15+: 15 *Jernas* (*Rapier* 2000); *Anza*-II; HY-6 (FN-6); 9K38 *Igla* (SA-18 *Grouse*); *Starstreak*

GUNS 52+

SP 20mm K263

TOWED 52: **35mm** 16 GDF-005; **40mm** 36 L40/70

Reserves

Territorial Army

Some paramilitary forces to be incorporated into a re-organised territorial organisation

FORCES BY ROLE

MANOEUVRE

Mechanised

4 armd sqn

Light

16 inf regt (3 inf bn)

Other

5 (highway) sy bn

COMBAT SUPPORT

5 arty bty

2 fd engr regt

1 int unit

3 sigs sqn

COMBAT SUPPORT

4 med coy

5 tpt coy

Navy 18,000

3 Regional Commands: MAWILLA 1 (Kuantan), MAWILLA 2 (Sabah) and MAWILLA 3 (Langkawi). A fourth is being formed (Bintulu)

EQUIPMENT BY TYPE

SUBMARINES 2

SSK 2 *Tunku Abdul Rahman* (FRA *Scorpène*) with 6 single 533mm TT with SM39 *Exocet* AShM/*Black Shark* HWT

PRINCIPAL SURFACE COMBATANTS • FRIGATES 2

FFGHM 2 *Lekiu* with 2 quad lnchr with MM40 *Exocet* Block 2 AShM, 1 16-cell VLS with *Sea Wolf* SAM, 2 triple

324mm ILAS-3 (B-515) ASTT with A244/S LWT, 1 57mm gun (capacity 1 *Super Lynx* 300 hel)
PATROL AND COASTAL COMBATANTS 46
CORVETTES 8
FSG 2 *Kasturi* with 2 quad lnchr with MM40 *Exocet* Block 2 AShM, 2 triple 324mm ILAS-3 (B-515) ASTT with A244/S LWT, 1 57mm gun, 1 hel landing platform
FSH 6 *Kedah* (GER MEKO 100) with 1 76mm gun, 1 hel landing platform (fitted for but not with MM40 *Exocet* AShM & RAM SAM)
PCFM 4 *Laksamana* with 1 *Albatros* quad lnchr with *Aspide* SAM, 1 76mm gun
PCF 4 *Perdana* (FRA *Combattante* II) with 1 57mm gun
PCC 1 *Keris* (Littoral Mission Ship)
PB 4 *Handalan* (SWE *Spica*-M) with 1 57mm gun
PBF 17 *Tempur* (SWE CB90)
PB 8: 6 *Jerong* (Lurssen 45) with 1 57mm gun; 2 *Sri Perlis*
MINE WARFARE • MINE COUNTERMEASURES 4
MCO 4 *Mahamiru* (ITA *Lerici*)
LOGISTICS AND SUPPORT 12
AFS 2: 1 *Mahawangsa* with 2 57mm guns, 1 hel landing platform; 1 *Sri Indera Sakti* with 1 57mm gun, 1 hel landing platform
AG 2 *Bunga Mas Lima* with 1 hel landing platform
AGS 1 *Perantau*
AP 2 *Sri Gaya*
ASR 1 *Mega Bakti*
ATF 1
AX 2 *Gagah Samudera* with 1 hel landing platform
AXS 1

Naval Aviation 160

EQUIPMENT BY TYPE
HELICOPTERS
ASW 6 *Super Lynx* 300
MRH 6 AS555 *Fennec*
AIR-LAUNCHED MISSILES • AShM *Sea Skua*

Special Forces

FORCES BY ROLE
SPECIAL FORCES
1 (mne cdo) SF unit

Air Force 15,000

1 air op HQ, 2 air div, 1 trg and log comd, 1 Intergrated Area Def Systems HQ

FORCES BY ROLE
FIGHTER/GROUND ATTACK
1 sqn with F/A-18D *Hornet*
1 sqn with Su-30MKM *Flanker*
2 sqn with *Hawk* Mk108*/Mk208*
MARITIME PATROL
1 sqn with Beech 200T
TANKER/TRANSPORT
2 sqn with KC-130H *Hercules*; C-130H *Hercules*; C-130H-30 *Hercules*; Cessna 402B
TRANSPORT
1 sqn with A400M *Atlas*
1 (VIP) sqn with A319CT; AW109; BD700 *Global Express*; F-28 *Fellowship*; *Falcon* 900
1 sqn with CN235M-220
TRAINING
1 unit with PC-7
TRANSPORT HELICOPTER
4 (tpt/SAR) sqn with H225M *Super Cougar*; S-61A-4 *Nuri*; S-61N; S-70A *Black Hawk*
AIR DEFENCE
1 sqn with *Starburst*
SPECIAL FORCES
1 (Air Force Commando) unit (airfield defence/SAR)

EQUIPMENT BY TYPE
AIRCRAFT 43 combat capable
FTR (8 F-5E *Tiger* II; 3 F-5F *Tiger* II; 8 MiG-29 *Fulcrum* (MiG-29N); 2 MiG-29UB *Fulcrum* B (MIG-29NUB) in store)
FGA 26: 8 F/A-18D *Hornet* (some serviceability in doubt); 18 Su-30MKM (some serviceability in doubt)
ISR 3 Beech 200T (2 RF-5E *Tigereye** in store)
TKR 4 KC-130H *Hercules*
TKR/TPT 4 A400M *Atlas*
TPT 30: **Medium** 10: 2 C-130H *Hercules*; 8 C-130H-30 *Hercules*; **Light** 16: 7 CN235M-220 (incl 1 VIP); 9 Cessna 402B (2 modified for aerial survey); **PAX** 4: 1 A319CT; 1 BD700 *Global Express*; 1 F-28 *Fellowship*; 1 *Falcon* 900
TRG 71: 5 *Hawk* Mk108*; 12 *Hawk* Mk208*; 7 MB-339C; 30 PC-7; 17 PC-7 Mk II *Turbo Trainer*
HELICOPTERS
TPT 41: **Heavy** 12 H225M *Super Cougar*; **Medium** 28: 24 S-61A-4 *Nuri*; 2 S-61N; 2 S-70A *Black Hawk*; **Light** 1 AW109
AIR DEFENCE • SAM • Point-defence *Starstreak*
AIR-LAUNCHED MISSILES
AAM • IR AIM-9 *Sidewinder*; R-73 (AA-11A *Archer*); **IIR** AIM-9X *Sidewinder* II; **IR/SARH** R-27 (AA-10 *Alamo*); **SARH** AIM-7 *Sparrow*; **ARH** AIM-120C AMRAAM; R-77 (AA-12A *Adder*)
ASM AGM-65 *Maverick*; Kh-29T (AS-14B *Kedge*); Kh-29L (AS-14A *Kedge*); Kh-31P (AS-17A *Krypton*); Kh-59M (AS-18 *Kazoo*)
ARM Kh-31P (AS-17A *Krypton*);
AShM AGM-84D *Harpoon*; Kh-31A (AS-17B *Krypton*)
BOMBS
Electro-optical guided KAB-500KR; KAB-500OD
Laser-guided *Paveway* II

Paramilitary ε22,500

Police–General Ops Force 18,000

FORCES BY ROLE
COMMAND
5 bde HQ
SPECIAL FORCES
1 spec ops bn
MANOEUVRE
Other
19 paramilitary bn
2 (Aboriginal) paramilitary bn
4 indep paramilitary coy

EQUIPMENT BY TYPE
ARMOURED FIGHTING VEHICLES
APC • APC (W) AT105 *Saxon*
AUV ε30 SB-301

Malaysian Maritime Enforcement Agency (MMEA) ε4,500

Controls 5 Maritime Regions (Northern Peninsula; Southern Peninsula; Eastern Peninsula; Sarawak; Sabah), subdivided into a further 18 Maritime Districts. Supported by one provisional MMEA Air Unit

EQUIPMENT BY TYPE

PATROL AND COASTAL COMBATANTS 127

PSO 4: 1 *Arau* (ex-JPN *Nojima*) with 1 hel landing platform; 2 *Langkawi* with 1 57mm gun, 1 hel landing platform; 1 *Pekan* (ex-JPN *Ojika*) with 1 hel landing platform

PCC 3 *Bagan Datuk*

PBF 57: 18 *Penggalang* 17 (TUR MRTP 16); 2 *Penggalang* 18; 6 *Penyelamat* 20; 16 *Penggalang* 16; 15 *Tugau*

PB 63: 15 *Gagah*; 4 *Malawali*; 2 *Nusa*; 3 *Nusa 28*; 1 *Peninjau*; 7 *Ramunia*; 2 *Rhu*; 4 *Semilang*; 8 *Icarus* 1650; 10 *Pengawal*; 4 *Penyelamat*; 2 *Perwira*; 1 *Sugut*

LOGISTICS AND SUPPORT • AX 1 *Marlin*

AIRCRAFT • MP 2 Bombardier 415MP

HELICOPTERS

SAR 3 AW139

MRH 3 AS365 *Dauphin*

Area Security Units 3,500 reservists

(Auxiliary General Ops Force)

FORCES BY ROLE

MANOEUVRE

Other

89 paramilitary unit

Border Scouts 1,200 reservists

in Sabah, Sarawak

People's Volunteer Corps 240,000 reservists (some 17,500 armed)

RELA

DEPLOYMENT

DEMOCRATIC REPUBLIC OF THE CONGO: UN • MONUSCO 2

LEBANON: UN • UNIFIL 827; 1 mech inf bn

PHILIPPINES: IMT 11

SUDAN: UN • UNISFA 2

WESTERN SAHARA: UN • MINURSO 7

FOREIGN FORCES

Australia 130; 1 inf coy (on 3-month rotational tours); 1 P-8A *Poseidon* (rotational)

Mongolia MNG

Mongolian Tugrik t		2019	2020	2021
GDP	t	36.9tr	38.4tr	
	US$	13.9bn	13.4bn	
per capita	US$	4,202	3,990	
Growth	%	5.1	-2.0	
Inflation	%	7.3	5.0	
Def bdgt	t	261bn	295bn	
	US$	98.1m	103m	
FMA (US)	US$	2.6m	0m	0m
US$1=t		2663.51	2871.01	

Real-terms defence budget trend (US$m, constant 2015)

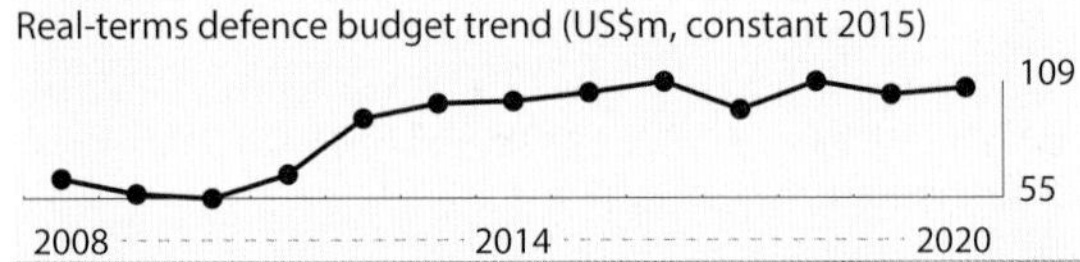

Population 3,168,026

Age	0–14	15–19	20–24	25–29	30–64	65 plus
Male	13.7%	3.6%	3.9%	4.2%	21.3%	1.9%
Female	13.2%	3.5%	3.9%	4.3%	23.4%	2.9%

Capabilities

Mongolia's latest defence-policy document, from 2015, stresses the importance of peacekeeping and anti-terrorist capabilities. The country has no formal military alliances, but pursues defence ties and bilateral training with regional states and others including India, Turkey and the US. Mongolia hosts the annual *Khaan Quest* multinational peacekeeping-training exercises. The country's main exercise partners are India and Russia, with each country running regular bilateral exercises. Mongolia's most significant deployment is to the UN peacekeeping mission in South Sudan. The generally Soviet-era inventory has been supplemented by deliveries of second-hand Russian weapons. Barring maintenance facilities, there is no significant defence-industrial base.

ACTIVE 9,700 (Army 8,900 Air 800) **Paramilitary 7,500**

Conscript liability 12 months for males aged 18–25

RESERVE 137,000 (Army 137,000)

ORGANISATIONS BY SERVICE

Army 5,600; 3,300 conscript (total 8,900)

FORCES BY ROLE

MANOEUVRE

Mechanised

1 MR bde

Light

1 (rapid deployment) lt inf bn (2nd bn to form)

Air Manoeuvre

1 AB bn

COMBAT SUPPORT

1 arty regt

EQUIPMENT BY TYPE

ARMOURED FIGHTING VEHICLES

MBT 420: 370 T-54/T-55; 50 T-72A

RECCE 120 BRDM-2
IFV 310 BMP-1
APC • APC (W) 210: 150 BTR-60; 40 BTR-70M; 20 BTR-80
ENGINEERING & MAINTENANCE VEHICLES
ARV T-54/T-55
ANTI-TANK/ANTI-INFRASTRUCTURE
GUNS • TOWED 200: **85mm** D-44/D-48; **100mm** M-1944/MT-12
ARTILLERY 570
TOWED ε300: **122mm** D-30/M-30 (M-1938); **130mm** M-46; **152mm** ML-20 (M-1937)
MRL 122mm 130 BM-21
MOR 140: **120mm**; **160mm**; **82mm**
AIR DEFENCE
SAM Medium-range 2+ S-125 *Pechora*-2M (SA-26)
GUNS • TOWED 23mm ZU-23-2

Air Force 800

FORCES BY ROLE
FIGHTER
1 sqn (forming) with MiG-29UB *Fulcrum* B
TRANSPORT
1 sqn with An-24 *Coke*; An-26 *Curl*
ATTACK/TRANSPORT HELICOPTER
1 sqn with Mi-8 *Hip*; Mi-171
AIR DEFENCE
2 regt with S-60/ZPU-4/ZU-23
EQUIPMENT BY TYPE
AIRCRAFT 2 combat capable
FTR 2 MiG-29UB *Fulcrum* B
TPT • Light 3: 2 An-24 *Coke*; 1 An-26 *Curl*
HELICOPTERS
TPT • Medium 12: 10 Mi-8 *Hip*; 2 Mi-171
AIR DEFENCE • GUNS • TOWED 150: **14.5mm** ZPU-4; **23mm** ZU-23; **57mm** S-60

Paramilitary 7,500 active

Border Guard 1,300; 4,700 conscript (total 6,000)

Internal Security Troops 400; 800 conscript (total 1,200)

FORCES BY ROLE
MANOEUVRE
Other
4 gd unit

Construction Troops 300

DEPLOYMENT

AFGHANISTAN: NATO • *Operation Resolute Support* 233

DEMOCRATIC REPUBLIC OF THE CONGO: UN • MONUSCO 2

SOUTH SUDAN: UN • UNMISS 865; 1 inf bn

SUDAN: UN • UNAMID 1; UN • UNISFA 5

WESTERN SAHARA: UN • MINURSO 3

Myanmar MMR

Myanmar Kyat K		2019	2020	2021
GDP	K	105tr	114tr	
	US$	68.6bn	70.9bn	
per capita	US$	1,299	1,333	
Growth	%	6.5	2.0	
Inflation	%	8.6	6.1	
Def bdgt	K	3.26tr	3.39tr	3.41tr
	US$	2.13bn	2.11bn	
US$1=K		1529.87	1607.00	

Real-terms defence budget trend (US$bn, constant 2015)

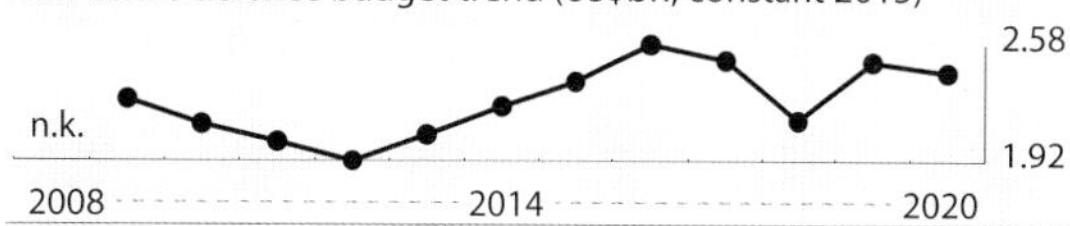

Population 56,590,071

Age	0–14	15–19	20–24	25–29	30–64	65 plus
Male	13.3%	4.3%	4.3%	4.1%	20.7%	2.6%
Female	12.7%	4.2%	4.3%	4.2%	21.9%	3.4%

Capabilities

Since the country's independence struggle in the 1940s, Myanmar's large, army-dominated Tatmadaw (armed forces) has been intimately involved in domestic politics. Even though the National League for Democracy (NLD) won the November 2015 election, the armed forces remain politically powerful. A defence white paper published in 2015 said that ending conflicts with domestic armed groups was a priority. It also gives a 'state-building' role to the Tatmadaw, legitimising continued intervention in the country's politics. In its counter-insurgency operations, the Tatmadaw has been accused by international organisations of human-rights violations, and concerns increased after military actions against the Rohingya minority in 2017. China and Russia are key defence-cooperation partners, including bilateral military exercises and the provision of weapons. Due to long-running domestic conflicts, the Tatmadaw has experience with counter-insurgency operations and jungle warfare. Although there have been small deployments to UN missions, the Tatmadaw remains essentially an internally focused force. Since the 1990s, the armed forces have attempted to develop limited conventional-warfare capabilities, and have brought into service new armoured vehicles, air-defence weapons, artillery, combat aircraft and ships procured mainly from China and Russia. In 2020, Myanmar's navy began operating its first submarine, a formerly Indian *Kilo*-class boat, though training and support arrangements were unclear. There is limited defence-industrial capacity. The Aircraft Production and Maintenance Base has assembled Chinese K-8 trainer aircraft. Myanmar also has growing shipbuilding capabilities, notably through the Naval Dockyard in Yangon, which launched patrol and utility vessels in 2018.

ACTIVE 406,000 (Army 375,000 Navy 16,000 Air 15,000) **Paramilitary 107,000**
Conscript liability 24–36 months

ORGANISATIONS BY SERVICE

Army ε375,000

14 military regions, 7 regional op comd

FORCES BY ROLE

COMMAND

20 div HQ (military op comd)

10 inf div HQ

34+ bde HQ (tactical op comd)

MANOEUVRE

Armoured

10 armd bn

Light

100 inf bn (coy)

337 inf bn (coy) (regional comd)

COMBAT SUPPORT

7 arty bn

37 indep arty coy

6 cbt engr bn

54 fd engr bn

40 int coy

45 sigs bn

AIR DEFENCE

7 AD bn

EQUIPMENT BY TYPE

ARMOURED FIGHTING VEHICLES

MBT 185+: 10 T-55; 50 T-72S; 25+ Type-59D; 100 Type-69-II

LT TK 105 Type-63 (ε60 serviceable)

ASLT 24 PTL-02 mod

RECCE 87+: 12+ EE-9 *Cascavel*; 45 *Ferret*; 30 Mazda; MAV-1

IFV 10+ BTR-3U

APC 431+

APC (T) 331: 26 MT-LB; 250 Type-85; 55 Type-90

APC (W) 90+: 20 Hino; 40 Humber *Pig*; 30+ Type-92

PPV 10 MPV

ENGINEERING & MAINTENANCE VEHICLES

ARV Type-72

VLB MT-55A

ANTI-TANK/ANTI-INFRASTRUCTURE

RCL 84mm *Carl Gustaf*; **106mm** M40A1

GUNS • TOWED 60: **57mm** 6-pdr; **76mm** 17-pdr

ARTILLERY 422+

SP 155mm 42: 30 NORA B-52; 12 SH-1

TOWED 264+: **105mm** 132: 36 M-56; 96 M101; **122mm** 100 D-30; **130mm** 16 M-46; **140mm**; **155mm** 16 Soltam M-845P

MRL 36+: **107mm** 30 Type-63; **122mm** BM-21 *Grad* (reported); Type-81; **240mm** 6+ M-1985 mod

MOR 80+: **82mm** Type-53 (M-37); **120mm** 80+: 80 Soltam; Type-53 (M-1943)

SURFACE-TO-SURFACE MISSILE LAUNCHERS

SRBM • Conventional some *Hwasong*-6 (reported)

AIR DEFENCE

SAM 4+

Medium-range 4+: 4 KS-1A (HQ-12); S-125 *Pechora*-2M (SA-26); 2K12 *Kvadrat*-M (SA-6 *Gainful*)

Point-defence Some 2K22 *Tunguska* (SA-19 *Grison*); HN-5 *Hong Nu/Red Cherry* (reported); 9K310 *Igla*-1 (SA-16 *Gimlet*)

GUNS 46

SP 57mm 12 Type-80

TOWED 34: **37mm** 24 Type-74; **40mm** 10 M1

Navy ε16,000

EQUIPMENT BY TYPE

SUBMARINES 1

SSK 1 *Min Ye Thein Kha Thu* (ex-IND *Sindhughosh* (Project 877EKM (*Kilo*))) with 6 single 533mm TT

PRINCIPAL SURFACE COMBATANTS • FRIGATES 5

FFGHM 2 *Kyansitthar* with 2 twin lnchr with C-802 (CH-SS-N-6) AShM, 1 sextuple lnchr with MANPAD SAM, 2 RDC-32 A/S mor, 4 AK630 CIWS, 1 76mm gun (capacity 1 med hel)

FFG 3:

1 *Aung Zeya* with 2 quad lnchr with DPRK AShM (possibly 3M24 derivative), 4 AK630 CIWS, 1 76mm gun, 1 hel landing platform

2 *Mahar Bandoola* (ex-PRC Type-053H1 (*Jianghu* I)) with 2 quad lnchr with C-802 (CH-SS-N-6) AShM, 2 RBU 1200 *Uragan* A/S mor, 2 twin 100mm gun

PATROL AND COASTAL COMBATANTS 79

CORVETTES 3

FSGHM 1 *Tabinshwethi* (*Anawrahta* mod) with 2 twin lnchr with C-802 (CH-SS-N-6), 1 sectuple lnchr with unknown MANPADs, 2 RBU 1200 *Uragan* A/S mor, 1 76mm gun (capacity 1 med hel)

FSG 2 *Anawrahta* with 2 twin lnchr with C-802 (CH-SS-N-6) AShM, 2 RDC-32 A/S mor, 1 76mm gun, 1 hel landing platform

PSOH 1 *Inlay* with 1 twin 57mm gun

PCG 7: 6 Type-037-IG (*Houxin*) with 2 twin lnchr with C-801 (CH-SS-N-4) AShM; 1 FAC(M) mod with 2 twin lnchr with C-802 (CH-SS-N-6) AShM, 1 AK630 CIWS

PCO 2 *Indaw*

PCC 11: 2 *Admirable* (ex-US); 9 Type-037 (*Hainan*) with 4 RBU 1200 *Uragan* A/S mor, 2 twin 57mm gun

PBG 4 *Myanmar* with 2 single lnchr with C-801 (CH-SS-N-4) AShM

PBF 5: 1 Type-201; 4 *Super Dvora* Mk III

PB 32: 3 PB-90; 6 PGM 401; 6 PGM 412; 14 *Myanmar*; 3 *Swift*

PBR 14: 4 *Sagu*; 9 Y-301†; 1 Y-301 (Imp)

AMPHIBIOUS

PRINCIPAL AMPHIBIOUS VESSELS • LPD 1:

1 *Moattama* (ROK *Makassar*) (capacity 2 LCVP; 2 hels; 13 tanks; 500 troops)

LANDING CRAFT 21: **LCU** 5; **LCM** 16

LOGISTICS AND SUPPORT 13

ABU 1

AGHS 2: 1 *Innya*; 1 (near shore)

AGS 1

AH 2

AK 1

AKSL 5

AP 1 *Chindwin*

Naval Infantry 800

FORCES BY ROLE

MANOEUVRE

Light

1 inf bn

Air Force ε15,000

FORCES BY ROLE

FIGHTER

4 sqn with F-7 *Airguard*; FT-7; JF-17 *Thunder*; MiG-29 *Fulcrum*; MiG-29SE/SM *Fulcrum*; MiG-29UB *Fulcrum*

GROUND ATTACK

2 sqn with A-5C *Fantan*

TRANSPORT

1 sqn with F-27 *Friendship*; FH-227; PC-6AB *Turbo Porter*

TRAINING

2 sqn with G-4 *Super Galeb**; PC-7 *Turbo Trainer**; PC-9*

1 (trg/liaison) sqn with Cessna 550 *Citation* II; Cessna 180 *Skywagon*; K-8 *Karakorum**

TRANSPORT HELICOPTER

4 sqn with Bell 205; Bell 206 *Jet Ranger*; Mi-17 *Hip* H; Mi-35P *Hind*; PZL Mi-2 *Hoplite*; PZL W-3 *Sokol*; SA316 *Alouette* III

EQUIPMENT BY TYPE

AIRCRAFT 165 combat capable

FTR 63: 21 F-7 *Airguard*; 10 FT-7; 11 MiG-29 *Fulcrum*; 6 MiG-29SE *Fulcrum*; 10 MiG-29SM *Fulcrum*; 5 MiG-29UB *Fulcrum*

FGA 6: 4 JF-17 *Thunder* (FC-1 Block 2); 2 JF-17B *Thunder* (FC-1 Block 2)

ATK 22 A-5C *Fantan*

TPT 20: **Medium** 5: 4 Y-8D; 1 Y-8F-200W **Light** 16: 3 Beech 1900D; 4 Cessna 180 *Skywagon*; 1 Cessna 550 *Citation* II; 3 F-27 *Friendship*; 5 PC-6A/B *Turbo Porter*; **PAX** 1+ FH-227

TRG 88: 11 G-4 *Super Galeb**; 20 Grob G120; 24+ K-8 *Karakorum**; 12 PC-7 *Turbo Trainer**; 9 PC-9*; 12 Yak-130 *Mitten**

HELICOPTERS

ATK 12 Mi-35P *Hind*

MRH 23: 3 AS365; 11 Mi-17 *Hip* H; 9 SA316 *Alouette* III

TPT 45: **Medium** 10 PZL W-3 *Sokol*; **Light** 35: 12 Bell 205; 6 Bell 206 *Jet Ranger*; 17 PZL Mi-2 *Hoplite*

UNINHABITED AERIAL VEHICLES

CISR • **Heavy** 4 CH-3

AIR-LAUNCHED MISSILES

AAM • **IR** PL-5; R-73 (AA-11A *Archer*); PL-5E-II; **IR/SARH** R-27 (AA-10 *Alamo*); **ARH** PL-12

AShM C-802A

Paramilitary 107,000

People's Police Force 72,000

People's Militia 35,000

Nepal NPL

Nepalese Rupee NR		2019	2020	2021
GDP	NR	3.46tr	3.69tr	
	US$	30.7bn	32.2bn	
per capita	US$	1,079	1,116	
Growth	%	7.1	0.0	
Inflation	%	4.6	6.4	
Def bdgt	NR	44.9bn	50.1bn	49.2bn
	US$	398m	437m	
FMA (US)	US$	1.7m	0m	0m
US$1=NR		112.88	114.63	

Real-terms defence budget trend (US$m, constant 2015)

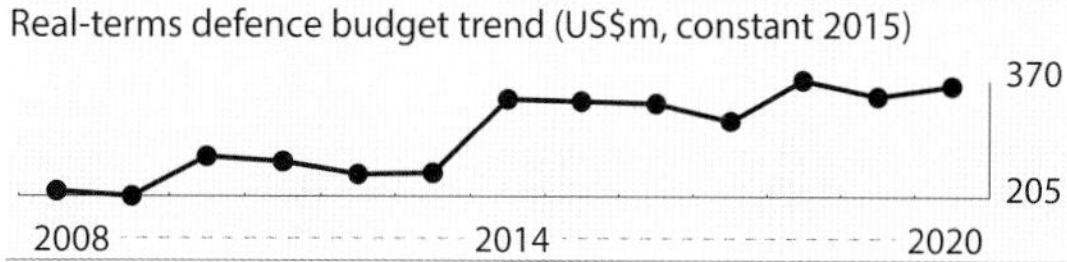

Population 30,327,877

Age	0–14	15–19	20–24	25–29	30–64	65 plus
Male	14.9%	5.4%	5.4%	4.5%	15.9%	2.8%
Female	13.4%	4.9%	5.2%	5.0%	19.6%	2.9%

Capabilities

The principal role of Nepal's armed forces is maintaining territorial integrity, but they have also traditionally focused on internal security and humanitarian relief. Nepal has a history of deploying contingents to UN peacekeeping operations. Training support is provided by several countries, including China, India and the US. Following a 2006 peace accord with the Maoist People's Liberation Army, Maoist personnel underwent a process of demobilisation or integration into the armed forces. Gurkhas continue to be recruited by the British and Indian armed forces and the Singaporean police. The small air wing provides a limited transport and support capacity but mobility remains a challenge, in part because of topography. Nepal's logistic capability appears to be sufficient for internal-security operations; however, its contingents on UN peacekeeping operations appear to largely depend on contracted logistic support. Modernisation plans include a very limited increase in the size of its air force. Barring maintenance capacities there is no defence-industrial base, and Nepal is dependent on foreign suppliers for modern equipment.

ACTIVE 96,600 (Army 96,600) **Paramilitary 15,000**

ORGANISATIONS BY SERVICE

Army 96,600

FORCES BY ROLE

COMMAND

6 inf div HQ

1 (valley) comd

SPECIAL FORCES

1 bde (1 SF bn, 1 AB bn, 1 cdo bn, 1 ranger bn, 1 mech inf bn)

MANOEUVRE

Light

16 inf bde (total: 62 inf bn; 32 indep inf coy)

COMBAT SUPPORT
4 arty regt
5 engr bn
AIR DEFENCE
2 AD regt
4 indep AD coy
EQUIPMENT BY TYPE
ARMOURED FIGHTING VEHICLES
RECCE 40 *Ferret*
APC 253
APC (W) 13: 8 OT-64C; 5 WZ-551
PPV 240: 90 *Casspir*; 150 MPV
ARTILLERY 92+
TOWED 105mm 22: 8 L118 Light Gun; 14 pack howitzer (6 non-operational)
MOR 70+: **81mm**; **120mm** 70 M-43 (est 12 op)
AIR DEFENCE • GUNS • TOWED 32+: **14.5mm** 30 Type-56 (ZPU-4); **37mm** (PRC); **40mm** 2 L/60

Air Wing 320
EQUIPMENT BY TYPE†
AIRCRAFT • TPT • **Light** 5: 1 BN-2T *Islander*; 1 CN235M-220; 3 M-28 *Skytruck*
HELICOPTERS
MRH 12: 2 *Dhruv*; 2 *Lancer*; 3 Mi-17-1V *Hip* H; 2 Mi-17V-5 *Hip*; 1 SA315B *Lama (Cheetah)*; 2 SA316B *Alouette* III
TPT 3: **Medium** 1 SA330J *Puma*; **Light** 2 AS350B2 *Ecureuil*

Paramilitary 15,000

Armed Police Force 15,000
Ministry of Home Affairs

DEPLOYMENT

CENTRAL AFRICAN REPUBLIC: UN • MINUSCA 737; 1 inf bn; 1 MP pl

DEMOCRATIC REPUBLIC OF THE CONGO: UN • MONUSCO 898; 1 inf bn; 1 engr coy

IRAQ: UN • UNAMI 77; 1 sy unit

LEBANON: UN • UNIFIL 870; 1 mech inf bn

LIBYA: UN • UNISMIL 229; 2 sy coy

MALI: UN • MINUSMA 198; 1 EOD coy

MIDDLE EAST: UN • UNTSO 3

SOUTH SUDAN: UN • UNMISS 1,737; 2 inf bn

SUDAN: UN • UNAMID 5; **UN** • UNISFA 8

SYRIA/ISRAEL: UN • UNDOF 357; 2 mech inf coy

WESTERN SAHARA: UN • MINURSO 4

FOREIGN FORCES

United Kingdom 60 (Gurkha trg org)

New Zealand NZL

New Zealand Dollar NZ$		2019	2020	2021
GDP	NZ$	311bn	301bn	
	US$	205bn	194bn	
per capita	US$	41,667	38,675	
Growth	%	2.2	-6.1	
Inflation	%	1.6	1.7	
Def bdgt	NZ$	4.05bn	5.06bn	4.62bn
	US$	2.67bn	3.25bn	
US$1=NZ$		1.52	1.55	

Real-terms defence budget trend (US$bn, constant 2015)

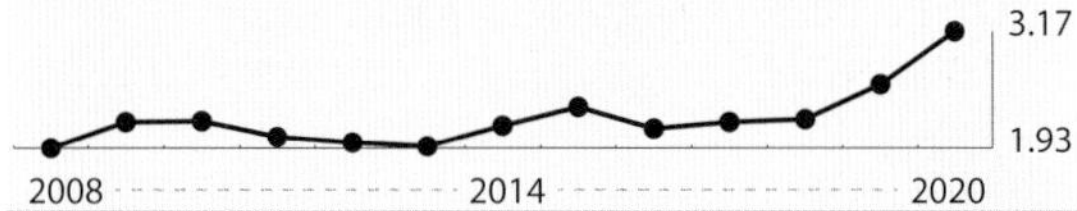

Population 4,925,477

Age	0–14	15–19	20–24	25–29	30–64	65 plus
Male	10.1%	3.2%	3.4%	3.7%	22.3%	7.3%
Female	9.5%	3.1%	3.2%	3.6%	22.3%	8.3%

Capabilities

New Zealand has a strong military tradition. The New Zealand Defence Force (NZDF) is well trained and has operational experience. The June 2016 defence white paper forecast a range of challenges likely to affect the country's security in the period to 2040, including rising tension in the South and East China seas. The white paper indicated investments including in improved maritime air-surveillance capability and cyber-support capability for deployed operations, but said that until 2030 defence spending was expected to remain pegged at around 1% of GDP. New Zealand's closest defence partner is Australia but the country has revived defence relations with the US. The 2016 Defence Capability Plan outlined plans including deliveries of new frigates in the late 2020s. In 2018, it was announced that four P-8 *Poseidon* maritime-patrol aircraft would be procured, while at-sea replenishment capability was revived in 2020 with the commissioning of HMNZS *Aotearoa*. A new 2019 Defence Capability Plan detailed plans to expand the army to 6,000 personnel by 2035, as well as to acquire a sealift vessel in the late 2020s. Replacement of the ANZAC frigates, both of which are being upgraded, has now been postponed until the 2030s. New Zealand has a small defence industry consisting of numerous private companies and subsidiaries of larger North American and European companies. These companies are able to provide some maintenance, repair and overhaul capability but significant work is contracted to foreign companies.

ACTIVE 9,400 (Army 4,650 Navy 2,250 Air 2,500)

RESERVE 2,850 (Army 2,050 Navy 500 Air Force 300)

ORGANISATIONS BY SERVICE

Army 4,650
FORCES BY ROLE
SPECIAL FORCES
1 SF regt

MANOEUVRE

Light

1 inf bde (1 armd recce regt, 2 lt inf bn, 1 arty regt (2 arty bty), 1 engr regt(-), 1 MP coy, 1 sigs regt, 2 log bn)

EQUIPMENT BY TYPE

ARMOURED FIGHTING VEHICLES

IFV 93 NZLAV-25

ENGINEERING & MAINTENANCE VEHICLES

AEV 7 NZLAV

ARV 3 LAV-R

ANTI-TANK/ANTI-INFRASTRUCTURE

MSL • MANPATS FGM-148 *Javelin*

RCL 84mm *Carl Gustaf*

ARTILLERY 60

TOWED 105mm 24 L118 Light Gun

MOR 81mm 36

Reserves

Territorial Force 2,050 reservists

Responsible for providing trained individuals for augmenting deployed forces

FORCES BY ROLE

COMBAT SERVICE SUPPORT

3 (Territorial Force Regional) trg regt

Navy 2,250

Fleet based in Auckland. Fleet HQ at Wellington

EQUIPMENT BY TYPE

PRINCIPAL SURFACE COMBATANTS • FRIGATES 2

FFHM 2 *Anzac* (GER MEKO 200) with 1 8-cell Mk 41 VLS with RIM-7M *Sea Sparrow* SAM, 2 triple SVTT Mk 32 324mm TT with Mk 46 mod 5 LWT, 1 Mk 15 *Phalanx* Block 1B CIWS, 1 127mm gun (capacity 1 SH-2G(I) *Super Seasprite* ASW hel) (both vessels in refit in Canada since 2018)

PATROL AND COASTAL COMBATANTS 4

PSOH 2 *Otago* (capacity 1 SH-2G(I) *Super Seasprite* ASW hel) (ice-strengthened hull)

PCC 2 *Lake*

AMPHIBIOUS • LANDING CRAFT 2

LCM 2 (operated off HMNZS *Canterbury*)

LOGISTICS AND SUPPORT • 3

AGHS 1 *Manawanui* with 1 hel landing platform

AKRH 1 *Canterbury* (capacity 4 NH90 tpt hel; 1 SH-2G(I) *Super Seasprite* ASW hel; 2 LCM; 16 NZLAV; 20 trucks; 250 troops)

AORH 1 *Aotearoa* (capacity 1 NH90/SH-2G(I) hel)

Air Force 2,500

FORCES BY ROLE

MARITIME PATROL

1 sqn with P-3K2 *Orion*

TRANSPORT

1 sqn with B-757-200 (upgraded); C-130H *Hercules* (upgraded)

ANTI-SUBMARINE/SURFACE WARFARE

1 (RNZAF/RNZN) sqn with SH-2G(I) *Super Seasprite*

TRAINING

1 sqn with T-6C *Texan* II

1 sqn with Beech 350 *King Air* (leased)

TRANSPORT HELICOPTER

1 sqn with AW109LUH; NH90

EQUIPMENT BY TYPE

AIRCRAFT 6 combat capable

ASW 6 P-3K2 *Orion*

TPT 11: **Medium** 5 C-130H *Hercules* (upgraded); **Light** 4 Beech 350 *King Air* (leased); **PAX** 2 B-757-200 (upgraded)

TRG 11 T-6C *Texan* II

HELICOPTERS

ASW 8 SH-2G(I) *Super Seasprite*

TPT 13: **Medium** 8 NH90; **Light** 5 AW109LUH

AIR-LAUNCHED MISSILES • AShM AGM-119 *Penguin* Mk2 mod7

DEPLOYMENT

AFGHANISTAN: NATO • *Operation Resolute Support* 9

EGYPT: MFO 26; 1 trg unit; 1 tpt unit

MIDDLE EAST: UN • UNTSO 7

SOUTH SUDAN: UN • UNMISS 3

Pakistan PAK

Pakistani Rupee Rs		2019	2020	2021
GDP	Rs	38.0tr	41.7tr	
	US$	276bn	303bn	
per capita	US$	1,349	1,482	
Growth	%	1.9	-0.4	
Inflation	%	6.7	10.7	
Def bdgt [a]	Rs	1.40tr	1.48tr	1.65tr
	US$	10.2bn	10.8bn	
FMA (US)	US$	80m	0m	0m
US$1=Rs		137.52	137.52	

[a] Includes defence allocations to the Public Sector Development Programme (PSDP), including funding to the Defence Division and the Defence Production Division.

Real-terms defence budget trend (US$bn, constant 2015)

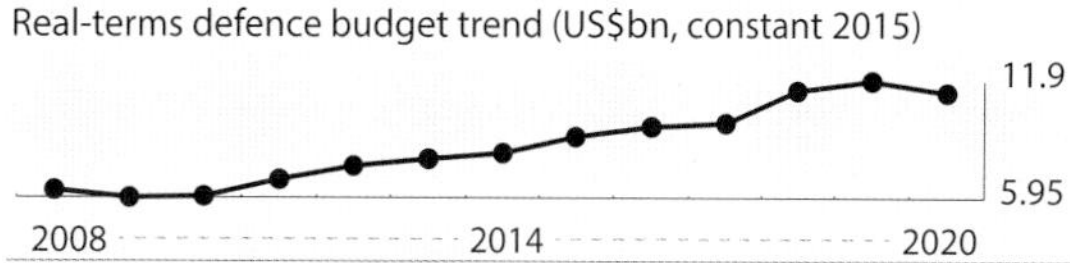

Population	233,500,636					
Age	0–14	15–19	20–24	25–29	30–64	65 plus
Male	18.4%	5.2%	4.7%	4.2%	16.4%	2.1%
Female	17.6%	5.0%	4.4%	4.0%	15.7%	2.4%

Capabilities

The armed forces have considerable domestic political influence and are the dominant voice on defence and security policy. Pakistan's nuclear and conventional forces have traditionally been oriented and structured against a prospective threat from India. Since 2008, counter-insurgency and counter-terrorism have been the forces' main effort. Although an army-led counter-terrorism operation has improved domestic security, terrorist attacks continue. There continue to be exchanges of fire with Indian forces in the disputed territory of Kashmir. The armed forces have a major role in disaster relief. China is Pakistan's main defence partner, with all three services employing a large amount of Chinese equipment. Military cooperation with the US is limited by sanctions aiming to improve cooperation on counter-terrorism. Recruitment is good, retention is high and the forces are well trained. The army and air force have considerable operational experience from a decade of counter-insurgency operations in Pakistan's tribal areas. Funds have been directed towards improving security on the border with Afghanistan. Major investment in military nuclear programmes continues, including the testing of a nuclear-capable sea-launched cruise missile. The navy plans to increase surface combatants, patrol vessels, submarines (in collaboration with China), maritime-patrol aircraft and UAVs. This is to both improve combat capability and the protection of sea-based nuclear weapons. The air force is modernising its inventory while improving its precision-strike and ISR capabilities. The indigenous defence industry exports platforms, weapons and ammunition and there is considerable defence-industrial collaboration with China.

ACTIVE 651,800 (Army 560,000 Navy 21,800 Air 70,000) Paramilitary 291,000

ORGANISATIONS BY SERVICE

Strategic Forces

Operational control rests with the National Command Authority. The Strategic Plans Directorate (SPD) manages and commands all of Pakistan's military nuclear capability. The SPD also commands a reportedly 25,000-strong military security force responsible for guarding military nuclear infrastructure

Army Strategic Forces Command 12,000–15,000

Commands all land-based strategic nuclear forces

EQUIPMENT BY TYPE

SURFACE-TO-SURFACE MISSILE LAUNCHERS 60+

MRBM • Nuclear ε30 *Ghauri/Ghauri* II (*Hatf*-5)/ *Shaheen*-2 (*Hatf*-6 – in test); *Shaheen*-3 (in test)

SRBM • Nuclear 30+: ε30 *Ghaznavi* (*Hatf*-3 – PRC M-11)/*Shaheen*-1 (*Hatf*-4); some *Abdali* (*Hatf*-2); some *Nasr* (*Hatf*-9)

GLCM • Nuclear *Babur* (*Hatf*-7); *Ra'ad* (*Hatf*-8 – in test)

Air Force

1–2 sqn of F-16A/B or *Mirage* 5 may be assigned a nuclear-strike role

Army 560,000

FORCES BY ROLE

COMMAND

9 corps HQ

1 (Northern) comd

SPECIAL FORCES

2 SF gp (total: 4 SF bn)

MANOEUVRE

Armoured

2 armd div

7 indep armd bde

Mechanised

2 mech inf div

1 indep mech bde

Light

18 inf div

5 indep inf bde

4 (Northern Command) inf bde

Other

1 sy div (1 more div forming)

COMBAT SUPPORT

1 arty div

14 arty bde

7 engr bde

AVIATION

1 VIP avn sqn

4 avn sqn

HELICOPTER

3 atk hel sqn

2 ISR hel sqn

2 SAR hel sqn

2 tpt hel sqn

1 spec ops hel sqn

AIR DEFENCE

1 AD comd (3 AD gp (total: 8 AD bn))

EQUIPMENT BY TYPE

ARMOURED FIGHTING VEHICLES

MBT 2,467: 300 *Al-Khalid* (MBT 2000); ε80 *Al-Khalid* I; 315 T-80UD; ε500 *Al-Zarrar*; 400 Type-69; 268 Type-85-IIAP; 4+ VT-4; ε600 ZTZ-59

APC 3,545

APC (T) 3,200: 2,300 M113A1/A2/P; ε200 *Talha*; 600 VCC-1/VCC-2; ε100 ZSD-63

APC (W) 120 BTR-70/BTR-80

PPV 225 *Maxxpro*

AUV 10 *Dingo* 2

ENGINEERING & MAINTENANCE VEHICLES

ARV 262+: 175 Type-70/Type-84 (W653/W653A); *Al-Hadeed*; 52 M88A1; 35 *Maxxpro* ARV; T-54/T-55

VLB M47M; M48/60

MW *Aardvark* Mk II

ANTI-TANK/ANTI-INFRASTRUCTURE

MSL

SP M901 TOW; ε30 *Maaz* (HJ-8 on *Talha* chassis)

MANPATS HJ-8; TOW

RCL 75mm Type-52; **106mm** M40A1 **RL 89mm** M20

GUNS 85mm 200 Type-56 (D-44)

ARTILLERY 4,595+

SP 498: **155mm** 438: 200 M109A2; ε115 M109A5; 123 M109L; **203mm** 60 M110/M110A2

TOWED 1,659: **105mm** 329: 216 M101; 113 M-56; **122mm** 570: 80 D-30 (PRC); 490 Type-54 (M-1938); **130mm** 410 Type-59-I; **155mm** 322: 144 M114; 148 M198; ε30 *Panter*; **203mm** 28 M115

MRL 88+: **107mm** Type-81; **122mm** 52+: 52 *Azar* (Type-83); some KRL-122; **300mm** 36 A100

MOR 2,350+: **81mm**; **120mm** AM-50

SURFACE-TO-SURFACE MISSILE LAUNCHERS

MRBM • Nuclear ε30 *Ghauri/Ghauri* II (*Hatf*-5); some *Shaheen*-2 (*Hatf*-6 – in test); *Shaheen*-3 (in test)

SRBM 135+: **Nuclear** 30+: ε30 *Ghaznavi* (*Hatf*-3 – PRC M-11)/*Shaheen*-1 (*Hatf*-4); some *Abdali* (*Hatf*-2); some *Nasr* (*Hatf*-9); **Conventional** 105 *Hatf*-1

GLCM • Nuclear some *Babur* (*Hatf*-7)

AIRCRAFT
TPT • Light 13: 1 Beech 350 *King Air*; 3 Cessna 208B; 1 Cessna 421; 1 Cessna 550 *Citation*; 1 Cessna 560 *Citation*; 2 *Turbo Commander* 690; 4 Y-12(II)
TRG 87 MFI-17B *Mushshak*
HELICOPTERS
ATK 42: 38 AH-1F/S *Cobra* with TOW; 4 Mi-35M *Hind* (1 Mi-24 *Hind* in store)
MRH 115+: 10 H125M *Fennec*; 7 AW139; 26 Bell 412EP *Twin Huey*; 38+ Mi-17 *Hip* H; 2 Mi-171E *Hip*; 12 SA315B *Lama*; 20 SA319 *Alouette* III
TPT 76: **Medium** 36: 31 SA330 *Puma*; 4 Mi-171; 1 Mi-172; **Light** 40: 17 H125 *Ecureuil* (SAR); 5 Bell 205 (UH-1H *Iroquois*); 5 Bell 205A-1 (AB-205A-1); 13 Bell 206B *Jet Ranger* II
TRG 10 Hughes 300C
UNINHABITED AERIAL VEHICLES
ISR • Light *Bravo*; *Jasoos*; *Vector*
AIR DEFENCE
SAM 27+
Medium-range 27 LY-80 (CH-SA-16)
Short-range FM-90 (CH-SA-4)
Point-defence M113 with RBS-70; *Anza*-II; FN-6 (CH-SA-10); *Mistral*; QW-18 (CH-SA-11); RBS-70
GUNS • TOWED 1,933: **14.5mm** 981; **35mm** 248 GDF-002/GDF-005 (with 134 *Skyguard* radar units); **37mm** 310 Type-55 (M-1939)/Type-65; **40mm** 50 L/60; **57mm** 144 Type-59 (S-60); **85mm** 200 Type-72 (M-1939) KS-12

Navy 21,800 (incl ε3,200 Marines)

EQUIPMENT BY TYPE
SUBMARINES 8
SSK 5:
2 *Hashmat* (FRA *Agosta* 70) with 4 single 533mm ASTT with UGM-84 *Harpoon* AShM/F-17P HWT
3 *Khalid* (FRA *Agosta* 90B) (of which 2 fitted with AIP) with 4 single 533mm ASTT with SM39 *Exocet* AShM/ *SeaHake* mod 4 (DM2A4) HWT
SSW 3 MG110 (SF delivery) each with 2 single 533mm TT with F-17P HWT
PRINCIPAL SURFACE COMBATANTS • FRIGATES 8
FFGHM 4 *Sword* (F-22P) with 2 quad lnchr with C-802A AShM, 1 octuple lnchr with FM-90N (CH-SA-N-4) SAM, 2 triple 324mm ASTT with ET-52C (A244/S) LWT, 2 RDC-32 A/S mor, 1 Type 730B (H/PJ-12) CIWS, 1 76mm gun (capacity 1 Z-9C *Haitun* hel)
FFGH 2:
1 *Alamgir* (ex-US *Oliver Hazard Perry*) with 2 quad lnchr with RGM-84 *Harpoon* AShM, 2 triple 324mm ASTT with Mk 46 LWT, 1 Mk 15 *Phalanx* CIWS, 1 76mm gun
1 *Tariq* (ex-UK *Amazon*) with 2 quad lnchr with RGM-84 *Harpoon* AShM, 2 triple 324mm ASTT with Mk 46 LWT, 1 Mk 15 *Phalanx* Block 1B CIWS, 1 114mm gun (capacity 1 hel)
FFHM 2 *Tariq* (ex-UK *Amazon*) with 1 sextuple lnchr with LY-60N SAM, 2 triple 324mm ASTT with Mk 46 LWT, 1 Mk 15 *Phalanx* Block 1B CIWS, 1 114mm gun (capacity 1 hel)

PATROL AND COASTAL COMBATANTS 18
CORVETTES • FSH 1 *Yarmook* (Damen OPV 1900) (fitted for but not with 2 quad lnchr for AShM) (capacity 1 hel)
PCG 3: 2 *Azmat* (FAC(M)) with 2 quad lnchr with C-802A AShM, 1 AK630 CIWS; 1 *Azmat* (FAC(M)) with 2 triple lnchr with C-602 AShM, 1 AK630 CIWS
PBG 4: 2 *Jalalat* with 2 twin lnchr with C-802 (CH-SS-N-6) AShM; 2 *Jurrat* with 2 twin lnchr with C-802 (CH-SS-N-6) AShM
PBF 4: 2 *Kaan* 15; 2 *Zarrar* (33)
PB 6: 1 *Larkana*; 1 *Rajshahi*; 4 M16 Fast Assault Boat
MINE WARFARE • MINE COUNTERMEASURES 3
MCC 3 *Munsif* (FRA *Eridan*)
AMPHIBIOUS • LANDING CRAFT 8
LCM 2
LCAC 2 *Griffon* 8100TD
UCAC 4 *Griffon* 2000
LOGISTICS AND SUPPORT 9
AGS 2: 1 *Behr Masa*; 1 *Behr Paima*
AOL 2 *Madagar*
AOR 1 *Moawin* II (Fleet Tanker) with 1 hel landing platform
AORH 1 *Fuqing* with 1 Mk 15 *Phalanx* CIWS (capacity 1 SA319 *Alouette* III hel)
AOT 2 *Gwadar*
AXS 1

Asia

Marines ε3,200

FORCES BY ROLE
SPECIAL FORCES
1 cdo gp
MANOEUVRE
Amphibious
3 mne bn
AIR DEFENCE
1 AD bn

Naval Aviation

EQUIPMENT BY TYPE
AIRCRAFT 9 combat capable
ASW 9: 7 P-3B/C *Orion*; 2 ATR-72-500
MP 6 F-27-200 MPA
TPT 3: **Light** 2 ATR-72-500; **PAX** 1 Hawker 850XP
HELICOPTERS
ASW 11: 4 *Sea King* Mk45; 7 Z-9C *Haitun*
MRH 6 SA319B *Alouette* III
SAR 1 *Sea King* (ex-HAR3A)
TPT • Medium 1 *Sea King* (ex-HC4)
AIR-LAUNCHED MISSILES • AShM AM39 *Exocet*

Coastal Defence

FORCES BY ROLE
COASTAL Defence
1 AShM regt with *Zarb* (YJ-62)
EQUIPMENT BY TYPE
COASTAL DEFENCE • AShM *Zarb* (YJ-62)

Air Force 70,000

3 regional comds: Northern (Peshawar), Central (Sargodha), Southern (Masroor). The Composite Air Tpt Wg, Combat Cadres School and PAF Academy are Direct Reporting Units

FORCES BY ROLE

FIGHTER

3 sqn with F-7PG/FT-7PG *Airguard*
1 sqn with F-16A/B MLU *Fighting Falcon*
1 sqn with F-16A/B ADF *Fighting Falcon*
1 sqn with *Mirage* IIID/E (IIIOD/EP)

FIGHTER/GROUND ATTACK

1 sqn with JF-17 *Thunder* (FC-1 Block 1)
3 sqn with JF-17 *Thunder* (FC-1 Block 2)
1 sqn with JF-17B *Thunder* (FC-1 Block 2)
1 sqn with F-16C/D Block 52 *Fighting Falcon*
3 sqn with *Mirage* 5 (5PA)

ANTI-SURFACE WARFARE

1 sqn with *Mirage* 5PA2/5PA3 with AM-39 *Exocet* AShM

ELECTRONIC WARFARE/ELINT

1 sqn with *Falcon* 20F

AIRBORNE EARLY WARNING & CONTROL

1 sqn with Saab 2000; Saab 2000 *Erieye*
1 sqn with ZDK-03 **SEARCH & RESCUE**
1 sqn with Mi-171Sh (SAR/liaison)
6 sqn with SA316 *Alouette* III
1 sqn with AW139

TANKER

1 sqn with Il-78 *Midas*

TRANSPORT

1 sqn with C-130B/E *Hercules*; L-100-20
1 sqn with CN235M-220
1 VIP sqn with B-707; Cessna 560XL *Citation Excel*; CN235M-220; F-27-200 *Friendship*; *Falcon* 20E; Gulfstream IVSP
1 (comms) sqn with EMB-500 *Phenom* 100; Y-12 (II)

TRAINING

1 OCU sqn with F-7P/FT-7P *Skybolt*
1 OCU sqn with *Mirage* III/*Mirage* 5
1 OCU sqn with F-16A/B MLU *Fighting Falcon*
2 sqn with K-8 *Karakorum**
2 sqn with MFI-17
2 sqn with T-37C *Tweet*

AIR DEFENCE

1 bty with HQ-2 (SA-2 *Guideline*); 9K310 *Igla*-1 (SA-16 *Gimlet*)
6 bty with *Crotale*
10 bty with SPADA 2000

EQUIPMENT BY TYPE

AIRCRAFT 413 combat capable

FTR 151: 46 F-7PG *Airguard*; 20 F-7P *Skybolt*; 23 F-16A MLU *Fighting Falcon*; 21 F-16B MLU *Fighting Falcon*; 9 F-16A ADF *Fighting Falcon*; 4 F-16B ADF *Fighting Falcon*; 21 FT-7; 5 FT-7PG; 2 *Mirage* IIIB

FGA 214: 12 F-16C Block 52 *Fighting Falcon*; 6 F-16D Block 52 *Fighting Falcon*; 49 JF-17 *Thunder* (FC-1 Block 1); 61 JF-17 *Thunder* (FC-1 Block 2); 12 JF-17B *Thunder*; 7 *Mirage* IIID (*Mirage* IIIOD); 30 *Mirage* IIIE (IIIEP); 25 *Mirage* 5 (5PA)/5PA2; 2 *Mirage* 5D (5DPA)/5DPA2; 10 *Mirage* 5PA3 (ASuW)

ISR 10 *Mirage* IIIR* (*Mirage* IIIRP)

ELINT 2 *Falcon* 20F

AEW&C 10: 6 Saab 2000 *Erieye*; 4 ZDK-03

TKR 4 Il-78 *Midas*

TPT 35: **Medium** 16: 5 C-130B *Hercules*; 10 C-130E *Hercules*; 1 L-100-20; **Light** 14: 2 Cessna 208B; 1 Cessna 560XL *Citation Excel*; 4 CN235M-220; 4 EMB-500 *Phenom* 100; 1 F-27-200 *Friendship*; 2 Y-12 (II); **PAX** 5: 1 B-707; 1 *Falcon* 20E; 2 Gulfstream IVSP; 1 Saab 2000

TRG 142: 38 K-8 *Karakorum**; 80 MFI-17B *Mushshak*; 24 T-37C *Tweet*

HELICOPTERS

MRH 19: 15 SA316 *Alouette* III; 4 AW139

TPT • Medium 4 Mi-171Sh

UNINHABITED AERIAL VEHICLES

CISR • Heavy CH-3 (*Burraq*); CH-4 (reported)

ISR • Medium *Falco*

AIR DEFENCE • SAM 190+

Medium-range 6 HQ-2 (CH-SA-1)

Short-range 184: 144 *Crotale*; ε40 SPADA 2000

Point-defence 9K310 *Igla*-1 (SA-16 *Gimlet*)

AIR-LAUNCHED MISSILES

AAM • IR AIM-9L/P *Sidewinder*; *U-Darter*; PL-5; PL-5E-II; **SARH** Super 530; **ARH** PL-12; AIM-120C AMRAAM

ASM AGM-65 *Maverick*; *Raptor* II

AShM AM39 *Exocet*

ARM MAR-1

ALCM • Nuclear *Ra'ad*

BOMBS

INS/SAT-guided FT-6 (REK)

Laser-guided *Paveway* II

Paramilitary 291,000 active

Airport Security Force 9,000

Government Aviation Division

Pakistan Coast Guards

Ministry of Interior

EQUIPMENT BY TYPE

PATROL AND COASTAL COMBATANTS 5:

PBF 4

PB 1

Frontier Corps 70,000

Ministry of Interior

FORCES BY ROLE

MANOEUVRE

Reconnaissance

1 armd recce sqn

Other

11 paramilitary regt (total: 40 paramilitary bn)

EQUIPMENT BY TYPE

ARMOURED FIGHTING VEHICLES

APC (W) 45 UR-416

Maritime Security Agency ε2,000

FORCES BY ROLE

MARITIME PATROL

1 sqn with BN-2T *Defender*

EQUIPMENT BY TYPE
PATROL AND COASTAL COMBATANTS 19
- **PSO** 1 *Kashmir*
- **PCC** 10: 4 *Barkat*; 4 *Hingol*; 2 *Sabqat* (ex-US *Island*)
- **PBF** 5
- **PB** 3 *Guns*

AIRCRAFT • TPT • Light 3 BN-2T *Defender*

National Guard 185,000
Incl Janbaz Force; Mujahid Force; National Cadet Corps; Women Guards

Pakistan Rangers 25,000
Ministry of Interior

DEPLOYMENT

ARABIAN SEA & GULF OF ADEN: Combined Maritime Forces • CTF-151: 1 FFGHM

CENTRAL AFRICAN REPUBLIC: UN • MINUSCA 1,252; 1 inf bn; 2 engr coy; 1 hel sqn

CYPRUS: UN • UNFICYP 3

DEMOCRATIC REPUBLIC OF THE CONGO: UN • MONUSCO 1,970; 2 inf bn; 1 hel sqn

MALI: UN • MINUSMA 142; 1 hel sqn

SOMALIA: UN • UNSOS 1

SOUTH SUDAN: UN • UNMISS 11

SUDAN: UN • UNAMID 907; 1 inf bn; 2 fd hospital; **UN** • UNISFA 2

WESTERN SAHARA: UN • MINURSO 10

FOREIGN FORCES

Figures represent total numbers for UNMOGIP mission in India and Pakistan

Chile 2
Croatia 9
Italy 2
Korea, Republic of 7
Mexico 1
Philippines 5
Sweden 4
Switzerland 4
Thailand 4
Uruguay 3

Papua New Guinea PNG

Papua New Guinea Kina K		2019	2020	2021
GDP	K	84.0bn	81.3bn	
	US$	24.8bn	23.3bn	
per capita	US$	2,884	2,652	
Growth	%	4.9	-3.3	
Inflation	%	3.7	3.4	
Def bdgt	K	267m	329m	
	US$	78.8m	94.2m	
US$1=K		3.39	3.49	

Real-terms defence budget trend (US$m, constant 2015)

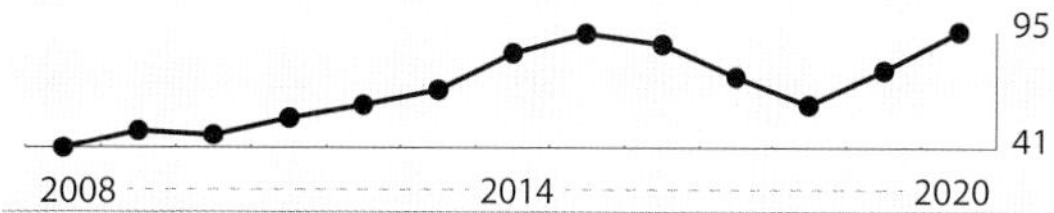

Population 7,259,456

Age	0–14	15–19	20–24	25–29	30–64	65 plus
Male	16.3%	5.2%	4.9%	4.3%	18.0%	2.3%
Female	15.7%	5.1%	4.7%	4.2%	17.1%	2.4%

Capabilities

Since independence in 1975, the Papua New Guinea Defence Force (PNGDF) has suffered from underfunding and lack of capacity to perform its core roles. After personnel reductions in the 2000s, the government made efforts in the next decade to revive defence capability. A 2013 defence white paper identified core roles including defending the state and civil-emergency assistance, but noted that 'defence capabilities have deteriorated to the extent that we have alarming gaps in our land, air and maritime borders'. The white paper called for strengthening defence capability on an ambitious scale, with long-term plans calling for a 'division-sized force' of 10,000 personnel by 2030. The PNGDF continues to receive substantial external military assistance from Australia but also from China, which has donated equipment. In late 2018, plans to build a joint US–Australia–Papua New Guinea naval base at Lombrum were announced. The PNGDF is not able to deploy outside of the country without outside assistance and there have only been small PNGDF deployments to UN peacekeeping missions. The PNGDF is receiving four of the *Guardian*-class patrol boats that Australia is donating to small Pacific Ocean nations. These will replace the four *Pacific*-class boats Australia donated in the 1980s. Papua New Guinea has no significant defence industry, though there is some local maintenance capacity.

ACTIVE 3,600 (Army 3,300 Maritime Element 200 Air 100)

ORGANISATIONS BY SERVICE

Army ε3,300

FORCES BY ROLE
SPECIAL FORCES
- 1 spec ops unit

MANOEUVRE
- **Light**
  - 2 inf bn

COMBAT SUPPORT
1 engr bn
1 EOD unit
1 sigs sqn

EQUIPMENT BY TYPE

ARTILLERY • MOR 3+: **81mm** Some; **120mm** 3

Maritime Element ε200

1 HQ located at Port Moresby

EQUIPMENT BY TYPE

PATROL AND COASTAL COMBATANTS 4
PCO 1 *Guardian* (AUS *Bay* mod)
PB 3 *Rabaul* (AUS *Pacific*)

AMPHIBIOUS • LANDING SHIPS • LCT 3 *Salamaua* (ex-AUS *Balikpapan*) (of which 1 in trg role)

Air Force ε100

FORCES BY ROLE

TRANSPORT
1 sqn with CN235M-100; IAI-201 *Arava*

TRANSPORT HELICOPTER
1 sqn with Bell 205 (UH-1H *Iroquois*)†

EQUIPMENT BY TYPE

AIRCRAFT • TPT • Light 3: 1 CN235M-100 (1 more in store); 2 IAI-201 *Arava*

HELICOPTERS • TPT • Light 3: 2 Bell 412 (leased); 1 Bell 212 (leased) (2 Bell 205 (UH-1H *Iroquois*) non-operational)

Philippines PHL

Philippine Peso P		2019	2020	2021
GDP	P	19.5tr	18.3tr	
	US$	377bn	367bn	
per capita	US$	3,512	3,373	
Growth	%	6.0	-8.3	
Inflation	%	2.5	2.4	
Def bdgt [a]	P	183bn	182bn	209bn
	US$	3.54bn	3.65bn	
FMA (US)	US$	40.0m	45.9m	40.0m
US$1=P		51.80	49.80	

[a] Excludes military pensions

Real-terms defence budget trend (US$bn, constant 2015)

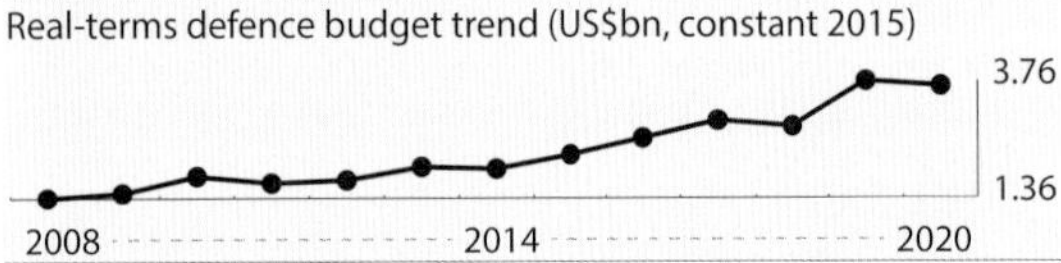

Population	109,180,815

Age	0–14	15–19	20–24	25–29	30–64	65 plus
Male	16.5%	5.1%	4.7%	4.2%	17.7%	2.0%
Female	15.9%	4.9%	4.5%	4.0%	17.7%	2.9%

Capabilities

Despite modest increases in defence funding in recent years, mainly in response to rising tensions in the South China Sea, the capabilities and procurement plans of the Armed Forces of the Philippines (AFP) remain limited. The National Defense Strategy 2018–22 identified policy priorities including ensuring sovereignty and territorial integrity, and internal stability. Organisational changes include the establishment in 2018 of an AFP Special Operations Command to command all special-forces units. There are also plans to upgrade anti-submarine-warfare capabilities and create a space agency with a national-security mandate. Although President Duterte announced in 2016 a 'separation' from the US and the pursuit of closer relations with China, he described the US as an important security ally, especially in support of counter-terrorism, in September 2017. The Philippines is an ASEAN member. In 2017 it began trilateral joint maritime patrols and joint Sulu Sea patrols with Indonesia and Malaysia to counter regional terrorist activity. The armed forces continue to be deployed on internal-security duties in the south, where Manila faces continuing challenges from insurgent groups. The second phase (2018–22) of the 'second horizon' AFP modernisation programme was approved in 2018. Acquisition plans include armoured vehicles, air-surveillance radars, OPVs and, in the long term, light and medium airlift and maritime-patrol aircraft. The Philippine Aerospace Development Corporation has assembled a variety of small helicopters and aircraft for the AFP, as well as providing maintenance, repair and overhaul services for military aircraft.

ACTIVE 143,100 (Army 101,000 Navy 24,500 Air 17,600) Paramilitary 12,300

RESERVE 131,000 (Army 100,000 Navy 15,000 Air 16,000) Paramilitary 50,000 (to age 49)

ORGANISATIONS BY SERVICE

Army 101,000

5 Area Unified Comd (joint service), 1 National Capital Region Comd

FORCES BY ROLE

SPECIAL FORCES
1 spec ops comd (1 ranger regt, 1 SF regt, 1 CT regt)

MANOEUVRE

Mechanised
1 armd div (2 mech bde (total: 3 lt armd sqn; 7 armd cav tp; 4 mech inf bn; 1 cbt engr coy; 1 avn regt; 1 cbt engr coy, 1 sigs coy))

Light
1 div (4 inf bde; 1 arty bn, 1 int bn, 1 sigs bn)
9 div (3 inf bde; 1 arty bn, 1 int bn, 1 sigs bn)
1 bde (1 mech inf bn; 2 inf bn, 1 arty bn)

Other
1 (Presidential) gd gp

COMBAT SUPPORT
1 arty regt HQ
1 MRL bty (forming)
5 engr bde

SURFACE-TO-SURFACE MISSILE
1 SSM bty (forming)

AIR DEFENCE
1 AD bty

EQUIPMENT BY TYPE

ARMOURED FIGHTING VEHICLES
LT TK 7 FV101 *Scorpion*
IFV 54: 2 YPR-765; 34 M113A1 FSV; 18 M113A2 FSV

APC 387
APC (T) 168: 6 ACV300; 42 M113A1; 120 M113A2 (some with *Dragon* RWS)
APC (W) 219: 73 LAV-150 *Commando*; 146 *Simba*
ENGINEERING & MAINTENANCE VEHICLES
ARV ACV-300; *Samson*; M578; 4 M113 ARV
ANTI-TANK-ANTI-INFRASTRUCTURE • RCL 75mm M20; 90mm M67; 106mm M40A1
ARTILLERY 260+
TOWED 220: 105mm 204 M101/M102/Model 56 pack howitzer; 155mm 16: 10 M114/M-68; 6 Soltam M-71
MOR 40+: 81mm M29; 107mm 40 M30; 120mm some *Cardom*
AIRCRAFT
TPT • Light 4: 1 Beech 80 *Queen Air*; 1 Cessna 170; 1 Cessna 172; 1 Cessna P206A
HELICOPTERS
TPT • Light 2 R-44 *Raven* II
UNINHABITED AERIAL VEHICLES • ISR • Medium *Blue Horizon*

Navy 24,500

EQUIPMENT BY TYPE
PRINCIPAL SURFACE COMBATANTS • FRIGATES 1
FFGHM 1 *Jose Rizal* (HDF-3000) with 2 quad lnchr with *Hae Sung* I AShM, 2 twin *Simbad*-RC lnchr with *Mistral* SAM, 2 triple 324mm SEA TLS ASTT with K745 *Blue Shark* LWT, 1 76mm gun (fitted for but not with 1 8-cell VLS) (capacity 1 AW159 *Wildcat*)
PATROL AND COASTAL COMBATANTS 63
CORVETTES • FS 1 *Conrado Yap* (ex-ROK *Po Hang* (Flight III)) with 2 triple 324mm SVTT Mk 32 ASTT, 2 76mm gun
PSOH 3 *Del Pilar* (ex-US *Hamilton*) with 1 76mm gun (capacity 1 Bo 105)
PCF 1 *General Mariano Alvares* (ex-US *Cyclone*)
PCO 7: 3 *Emilio Jacinto* (ex-UK *Peacock*) with 1 76mm gun; 3 *Miguel Malvar* (ex-US) with 1 76mm gun; 1 *Rizal* (ex-US *Auk*) with 2 76mm gun
PBFG 3 MPAC Mk3 with 1 *Typhoon* MLS-ER quad lnchr with *Spike*-ER SSM
PBF 12: 3 *Tomas Batilo* (ex-ROK *Chamsuri*); 6 MPAC Mk1/2; 3 MPAC Mk3 (to be fitted with *Spike*-ER SSM)
PB 30: 22 *Jose Andrada*; 2 *Kagitingan*; 2 *Point* (ex-US); 4 *Swift* Mk3 (ex-US)
PBR 6 Silver Ships
AMPHIBIOUS
PRINCIPAL AMPHIBIOUS SHIPS • LPD 2:
2 *Tarlac* (IDN *Makassar*) (capacity 2 LCVP; 2 hels; 13 tanks; 500 troops)
LANDING SHIPS • LST 4:
2 *Bacolod City* (US *Besson*) with 1 hel landing platform (capacity 32 tanks; 150 troops)
2 LST-1/542 (ex-US) (capacity 16 tanks; 200 troops)
LANDING CRAFT 11
LCM 2: 1 *Manobo*; 1 *Tagbanua* (capacity 100 tons; 200 troops)
LCT 5 *Ivatan* (ex-AUS *Balikpapan*)
LCU 4: 3 LCU Mk 6 (ex-US); 1 *Mulgae* I (ex-RoK)
LOGISTICS AND SUPPORT 6
AGOR 1 *Gregorio Velasquez* (ex-US *Melville*)
AO 1 *Lake Caliraya*
AOL 1
AP 1
AWT 2

Naval Aviation

EQUIPMENT BY TYPE
AIRCRAFT • TPT • Light 10: 3 BN-2A *Defender*; 2 Cessna 177 *Cardinal*; 5 Beech 90 *King Air* (TC-90)
HELICOPTERS
ASW 2 AW159 *Wildcat*
TPT 13: Medium 4 Mi-171Sh; Light 9: 3 AW109; 2 AW109E; 4 Bo-105

Marines 8,300

FORCES BY ROLE
SPECIAL FORCES
1 (force recon) spec ops bn
MANOEUVRE
Amphibious
4 mne bde (total: 12 mne bn)
COMBAT SUPPORT
1 CSS bde (6 CSS bn)
EQUIPMENT BY TYPE
ARMOURED FIGHTING VEHICLES
APC • APC (W) 42: 19 LAV-150 *Commando*; 23 LAV-300
AAV 67: 8 AAV-7A1; 4 LVTH-6†; 55 LVTP-7
ARTILLERY 37+
TOWED 37: 105mm 31: 23 M101; 8 M-26; 155mm 6 Soltam M-71
MOR 107mm M30

Naval Special Operations Group

FORCES BY ROLE
SPECIAL FORCES
1 SEAL unit
1 diving unit
10 naval spec ops unit
1 special boat unit
COMBAT SUPPORT
1 EOD unit

Air Force 17,600

FORCES BY ROLE
FIGHTER
1 sqn with FA-50PH *Fighting Eagle**
GROUND ATTACK
1 sqn with OV-10A/C *Bronco**
ISR
1 sqn with *Turbo Commander* 690A
SEARCH & RESCUE
4 (SAR/Comms) sqn with Bell 205 (UH-1M *Iroquois*); AUH-76
TRANSPORT
1 sqn with C-130B/H/T *Hercules*
1 sqn with N-22B *Nomad*; N-22SL *Searchmaster*; C-212 *Aviocar* (NC-212i)

1 sqn with F-27-200 MPA; F-27-500 *Friendship*
1 VIP sqn with F-28 *Fellowship*

TRAINING
1 sqn with SF-260F/TP
1 sqn with T-41B/D/K *Mescalero*
1 sqn with S-211*

ATTACK HELICOPTER
1 sqn with MD-520MG

TRANSPORT HELICOPTER
1 sqn with AUH-76
1 sqn with W-3 *Sokol*
4 sqn with Bell 205 (UH-1H *Iroquois*)
1 (VIP) sqn with Bell 412EP *Twin Huey*; S-70A *Black Hawk* (S-70A-5)

EQUIPMENT BY TYPE

AIRCRAFT 39 combat capable
FGA 12 FA-50PH *Fighting Eagle*
MP 3: 1 C-130T MP mod; 1 F-27-200 MPA; 1 N-22SL *Searchmaster*
ISR 11: 2 Cessna 208B *Grand Caravan*; 9 OV-10A/C *Bronco**
TPT 15: **Medium** 4: 1 C-130B *Hercules*; 2 C-130H *Hercules*; 1 C-130T *Hercules* **Light** 9: 3 C295; 1 C295M; 1 F-27-500 *Friendship*; 1 N-22B *Nomad*; 1 *Turbo Commander* 690A; 2 C-212 *Aviocar* (NC-212i); **PAX** 2: 1 F-28 *Fellowship* (VIP); 1 Gulfstream G280
TRG 45: 6 EMB-314 *Super Tucano**; 12 S-211*; 7 SF-260F; 10 SF-260TP; 10 T-41B/D/K *Mescalero*

HELICOPTERS
ATK 2 AH-1F *Cobra*
MRH 39: 8 W-3 *Sokol*; 2 AUH-76; 8 AW109E; 8 Bell 412EP *Twin Huey*; 2 Bell 412HP *Twin Huey*; 11 MD-520MG
TPT 32: **Medium** 1 S-70A *Black Hawk* (S-70A-5); **Light** 31: 11 Bell 205 (UH-1D); 20 Bell 205 (UH-1H *Iroquois*) (25 more non-operational)

UNINHABITED AERIAL VEHICLES
ISR • Medium 5: 2 *Blue Horizon* II; 3 *Hermes* 900

AIR-LAUNCHED MISSILES
AAM • IR AIM-9L *Sidewinder*
ASM AGM-65D *Maverick*; AGM-65G2 *Maverick*

Paramilitary 12,300

Coast Guard 12,300

EQUIPMENT BY TYPE

Rodman 38 and Rodman 101 owned by Bureau of Fisheries and Aquatic Resources

PATROL AND COASTAL COMBATANTS 87
PSOH 1 *Gabriela Silang* (OCEA OPV 270)
PCO 5: 4 *San Juan* with 1 hel landing platform; 1 *Balsam*
PCC 2 *Tirad*
PB 68: 4 *Boracay* (FPB 72 Mk II); 3 *De Haviland*; 4 *Ilocos Norte*; 1 *Palawan*; 12 PCF 50 (US *Swift* Mk1/2); 10 PCF 46; 10 PCF 65 (US *Swift* Mk3); 4 Rodman 38; 10 Rodman 101; 10 *Parola* (MRRV)
PBR 11

LOGISTICS AND SUPPORT • ABU 1 *Corregidor*
AIRCRAFT • TPT • Light 2 BN-2 *Islander*
HELICOPTERS • TPT • Light 4: 2 Bo-105; 2 H145

Citizen Armed Force Geographical Units
50,000 reservists

FORCES BY ROLE

MANOEUVRE
Other 56 militia bn (part-time units which can be called up for extended periods)

DEPLOYMENT

CENTRAL AFRICAN REPUBLIC: UN • MINUSCA 3

INDIA/PAKISTAN: UN • UNMOGIP 5

SOUTH SUDAN: UN • UNMISS 2

FOREIGN FORCES

Australia *Operation Augury* 100
Brunei IMT 8
Indonesia IMT 9
Malaysia IMT 11
United States US Pacific Command: *Operation Pacific Eagle – Philippines* 200

Singapore SGP

Singapore Dollar S$		2019	2020	2021
GDP	S$	508bn	468bn	
	US$	372bn	337bn	
per capita	US$	65,234	58,484	
Growth	%	0.7	-6.0	
Inflation	%	0.6	-0.4	
Def bdgt	S$	15.5bn	15.1bn	
	US$	11.3bn	10.9bn	
US$1=S$		1.36	1.39	

Real-terms defence budget trend (US$bn, constant 2015)

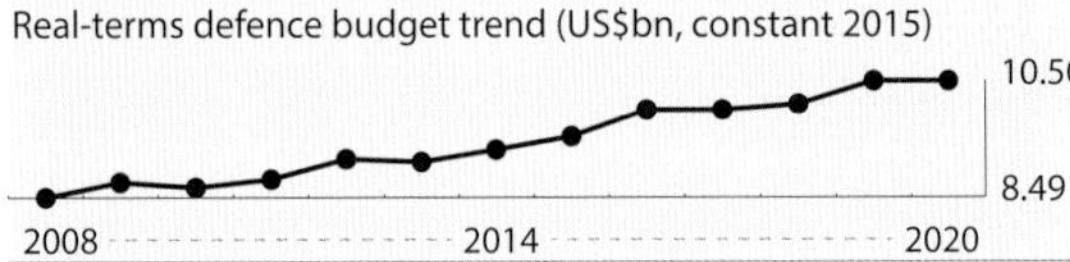

Population 6,209,660

Age	0–14	15–19	20–24	25–29	30–64	65 plus
Male	6.6%	3.0%	4.4%	5.1%	24.8%	5.0%
Female	6.2%	3.0%	4.7%	5.4%	25.9%	5.9%

Capabilities

The Singapore Armed Forces (SAF) are the best equipped in South-east Asia. They are organised essentially along Israeli lines, with the air force and navy staffed mainly by professional personnel while, apart from a small core of regulars, the much larger army is based on conscripts and reservists. Although there are no publicly available defence-policy documents, it is widely presumed that the SAF's primary role is to deter attacks on the city state or interference with its vital interests – particularly its sea lines of communication – by potential regional adversaries. There is an additional focus on counter-terrorist operations. With an ageing population and declining conscript cohort, there is a significant personnel challenge, which the defence ministry is looking to address by lean staffing and increased use of technology. There is routine overseas training, and plans have been announced to further improve domestic training areas. The SAF also engages extensively in bilateral and multilateral exercises with regional and international partners. Singaporean forces have gradually become more involved in multinational operations, and deployed to support the US-led air offensive against ISIS. While such deployments have provided some operational experience, and training standards and operational readiness are high, the army's reliance on conscripts and reservists limits its capacity for sustained operations abroad. Equipment modernisation continues, with plans to acquire F-35 combat aircraft, new domestically produced armoured fighting vehicles, multi-role combat vessels and multi-mission ships, with a view to retaining Singapore's military edge over other Southeast Asian powers. There is a small but sophisticated defence industry. ST Engineering group manufactures several types of armoured vehicles and corvettes for the SAF.

ACTIVE 51,000 (Army 41,000 Navy 4,000 Air 6,000) Paramilitary 7,400

Conscription liability 22–24 months

RESERVE 252,500 (Army 240,000 Navy 5,000 Air 7,500)

Annual trg to age 40 for army other ranks, 50 for officers

ORGANISATIONS BY SERVICE

Army 41,000 (including 26,000 conscripts)

FORCES BY ROLE

COMMAND
- 3 (combined arms) div HQ
- 1 (rapid reaction) div HQ
- 4 armd bde HQ
- 9 inf bde HQ
- 1 air mob bde HQ
- 1 amph bde HQ

SPECIAL FORCES
- 1 cdo bn

MANOEUVRE

Reconnaissance
- 3 lt armd/recce bn

Armoured
- 1 armd bn

Mechanised
- 6 mech inf bn

Light
- 2 (gds) inf bn

Other
- 2 sy bn

COMBAT SUPPORT
- 2 arty bn
- 1 STA bn
- 2 engr bn
- 1 EOD bn
- 1 ptn br bn
- 1 int bn
- 2 ISR bn
- 1 CBRN bn
- 3 sigs bn

COMBAT SERVICE SUPPORT
- 3 med bn
- 2 tpt bn
- 3 spt bn

Reserves

Activated units form part of divisions and brigades listed above; 1 op reserve div with additional armd & inf bde; People's Defence Force Comd (homeland defence) with 12 inf bn

FORCES BY ROLE

SPECIAL FORCES
- 1 cdo bn

MANOEUVRE

Reconnaissance
- 6 lt armd/recce bn

Mechanised
- 6 mech inf bn

Light
- ε56 inf bn

COMBAT SUPPORT
- ε12 arty bn
- ε8 engr bn

EQUIPMENT BY TYPE

ARMOURED FIGHTING VEHICLES

MBT 96+ *Leopard* 2SG

LT TK 372: 22 AMX-10 PAC 90; ε350 AMX-13 SM1

IFV 622+: 22 AMX-10P; 250 *Bionix* IFV-25; 250 *Bionix* IFV-40/50; ε50 *Hunter* AFV; 50+ M113A1/A2 (some with 40mm AGL, some with 25mm gun)

APC 1,655+

APC (T) 1,100+: 700+ M113A1/A2; 400+ ATTC *Bronco*

APC (W) 415: 250 LAV-150 *Commando*/V-200 *Commando*; 135 *Terrex* ICV; 30 V-100 *Commando*

PPV 140: 74 *Belrex*; 15 *MaxxPro Dash*; 51 *Peacekeeper*

ENGINEERING & MAINTENANCE VEHICLES

AEV 94: 18 CET; 54 FV180; 14 *Kodiak*; 8 M728

ARV *Bionix*; *Büffel*; LAV-150; LAV-300

VLB 72+: *Bionix*; LAB 30; *Leguan*; M2; 60 M3; 12 M60

MW 910-MCV-2; *Trailblazer*

ANTI-TANK/ANTI-INFRASTRUCTURE

MSL • MANPATS *Milan*; *Spike*-SR; *Spike*-MR

RCL 90+: **84mm** *Carl Gustaf*; **106mm** 90 M40A1

ARTILLERY 798+

SP 155mm 54 SSPH-1 *Primus*

TOWED 88: **105mm** (37 LG1 in store); **155mm** 88: 18 FH-2000; ε18 *Pegasus*; 52 FH-88

MRL 227mm 18 M142 HIMARS

MOR 638+

SP 90+: **81mm**; **120mm** 90: 40 on *Bronco*; 50 on M113

TOWED 548: **81mm** 500 **120mm** 36 M-65; **160mm** 12 M-58 Tampella

UNINHABITED AERIAL VEHICLES • **ISR** • **Light** *Skylark*

Navy 4,000 (incl 1,000 conscripts)

EQUIPMENT BY TYPE

SUBMARINES • **SSK** 4:

2 *Archer* (ex-SWE *Västergötland*) (fitted with AIP) with 3 single 400mm TT with Torped 431, 6 single 533mm TT with *Black Shark* HWT

2 *Challenger* (ex-SWE *Sjoormen*) with 2 single 400mm TT with Torped 431, 4 single 533mm TT with Torped 613

PRINCIPAL SURFACE COMBATANTS • FRIGATES 6
FFGHM 6 *Formidable* with 2 quad lnchr with RGM-84 *Harpoon* AShM, 4 8-cell *Sylver* A43 VLS with *Aster* 15 SAM, 2 triple 324mm ILAS-3 (B-515) ASTT with A244/S LWT, 1 76mm gun (capacity 1 S-70B *Sea Hawk* hel)
PATROL AND COASTAL COMBATANTS 24
CORVETTES • FSM 8 *Independence* (Littoral Mission Vessel) with 1 12-cell CLA VLS with VL MICA, 1 76mm gun, 1 hel landing platform
PCGM 6 *Victory* with 2 quad lnchr with RGM-84 *Harpoon* AShM, 2 8-cell VLS with *Barak*-1 SAM, 2 triple 324mm ILAS-3 (B-515) ASTT with A244/S LWT, 1 76mm gun
PCO 2 *Fearless* with 1 76mm gun (can be fitted with 2 sextuple *Sadral* lnchr with *Mistral* SAM)
PBF 8: 2 SMC Type 1; 6 SMC Type 2
MINE WARFARE • MINE COUNTERMEASURES 4
MCC 4 *Bedok*
AMPHIBIOUS
PRINCIPAL AMPHIBIOUS SHIPS • LPD 4 *Endurance* with 2 twin *Simbad* lnchr with *Mistral* SAM, 1 76mm gun (capacity 2 hel; 4 LCVP; 18 MBT; 350 troops)
LANDING CRAFT • LCVP 23: ε17 FCEP; 6 FCU
LOGISTICS AND SUPPORT 2
ASR 1 *Swift Rescue*
AX 1

Naval Diving Unit

FORCES BY ROLE
SPECIAL FORCES
1 SF gp
1 (diving) SF gp
COMBAT SUPPORT
1 EOD gp

Air Force 6,000 (incl 3,000 conscripts)

5 comds

FORCES BY ROLE
FIGHTER/GROUND ATTACK
2 sqn with F-15SG *Eagle*
2 sqn with F-16C/D *Fighting Falcon* (some used for ISR with pods)
ANTI-SUBMARINE WARFARE
1 sqn with S-70B *Seahawk*
MARITIME PATROL/TRANSPORT
1 sqn with F-50
AIRBORNE EARLY WARNING & CONTROL
1 sqn with G550-AEW
TANKER
1 sqn with A330 MRTT
TANKER/TRANSPORT
1 sqn with KC-130B/H *Hercules*; C-130H *Hercules*
TRAINING
1 (aggressor) sqn with F-15SG *Eagle*; F-16C/D *Fighting Falcon*
1 (FRA-based) sqn with M-346 *Master*
4 (US-based) units with AH-64D *Apache*; CH-47D *Chinook*; F-15SG; F-16C/D
1 (AUS-based) sqn with PC-21
1 hel sqn with H120 *Colibri*
ATTACK HELICOPTER
1 sqn with AH-64D *Apache*
TRANSPORT HELICOPTER
1 sqn with CH-47SD *Super D Chinook*
2 sqn with AS332M *Super Puma*; AS532UL *Cougar*
ISR UAV
1 sqn with *Hermes* 450
2 sqn with *Heron* 1
AIR DEFENCE
1 AD bn with *Mistral* (opcon Army)
3 AD bn with RBS-70; 9K38 *Igla* (SA-18 *Grouse*); *Mechanised Igla* (opcon Army)
1 ADA sqn with Oerlikon
1 AD sqn with SAMP/T
1 AD sqn with *Spyder*-SR
1 radar sqn with radar (mobile)
1 radar sqn with LORADS
MANOEUVRE
Other
4 (field def) sy sqn

EQUIPMENT BY TYPE
AIRCRAFT 105 combat capable
FGA 100: 40 F-15SG *Eagle*; 20 F-16C Block 52 *Fighting Falcon*; 20 F-16D Block 52 *Fighting Falcon*; 20 F-16D Block 52+ *Fighting Falcon* (incl reserves)
MP 5 F-50 *Maritime Enforcer**
AEW&C 4 G550-AEW
TKR 1 KC-130H *Hercules*
TKR/TPT 10: 6 A330 MRTT; 4 KC-130B *Hercules*
TPT 9: **Medium** 5 C-130H *Hercules* (2 ELINT); **PAX** 4 F-50
TRG 31: 12 M-346 *Master*; 19 PC-21
HELICOPTERS
ATK 19 AH-64D *Apache*
ASW 8 S-70B *Seahawk*
TPT 51: **Heavy** 16: 6 CH-47D *Chinook*; 10 CH-47SD *Super D Chinook*; **Medium** 30: 18 AS332M *Super Puma* (incl 5 SAR); 12 AS532UL *Cougar*; **Light** 5 H120 *Colibri* (leased)
UNINHABITED AERIAL VEHICLES
ISR 17+: **Heavy** 8+ *Heron* 1; **Medium** 9+ *Hermes* 450
AIR DEFENCE
SAM 4+
Long-range 4+ SAMP/T
Short-range *Spyder*-SR
Point-defence 9K38 *Igla* (SA-18 *Grouse*); *Mechanised Igla*; *Mistral*; RBS-70
GUNS 34
SP 20mm GAI-C01
TOWED 34 **20mm** GAI-C01; **35mm** 34 GDF (with 25 *Super-Fledermaus* fire-control radar)
AIR-LAUNCHED MISSILES
AAM • IR AIM-9P/S *Sidewinder*; *Python* 4 (reported); **IIR** AIM-9X *Sidewinder* II; **SARH** AIM-7P *Sparrow*; **ARH** (AIM-120C5/7 AMRAAM in store in US)
ASM: AGM-65B/G *Maverick*; AGM-114K/L *Hellfire*; AGM-154A/C JSOW
AShM AGM-84 *Harpoon*; AM39 *Exocet*
BOMBS
INS/GPS guided GBU-31 JDAM
Laser-guided *Paveway* II

Paramilitary 7,400 active

Civil Defence Force 5,600 (incl conscripts); 500 auxiliaries (total 6,100)

Singapore Gurkha Contingent 1,800

Under the Police

FORCES BY ROLE

MANOEUVRE

Other

6 paramilitary coy

DEPLOYMENT

AUSTRALIA: 2 trg schools – 1 with 12 AS332 *Super Puma*/AS532 *Cougar* (flying trg) located at Oakey; 1 with PC-21 (flying trg) located at Pearce. Army: prepositioned AFVs and heavy equipment at Shoalwater Bay training area

BRUNEI: 1 trg camp with inf units on rotation; 1 hel det with AS332 *Super Puma*

FRANCE: 200: 1 trg sqn with 12 M-346 *Master*

KUWAIT: *Operation Inherent Resolve* 11

TAIWAN: 3 trg camp (incl inf and arty)

THAILAND: 1 trg camp (arty, cbt engr)

UNITED STATES: Trg units with F-16C/D; 12 F-15SG; AH-64D *Apache*; 6+ CH-47D *Chinook*

FOREIGN FORCES

United States US Pacific Command: 200; 1 naval spt facility at Changi naval base; 1 USAF log spt sqn at Paya Lebar air base

Sri Lanka LKA

Sri Lankan Rupee Rs		2019	2020	2021
GDP	Rs	15.0tr	15.0tr	
	US$	84.0bn	81.1bn	
per capita	US$	3,852	3,698	
Growth	%	2.3	-4.6	
Inflation	%	4.3	4.7	
Def bdgt	Rs	298bn	294bn	
	US$	1.67bn	1.59bn	
FMA (US)	US$	0.5m	0m	0m
US$1=Rs		178.77	184.98	

Real-terms defence budget trend (US$bn, constant 2015)

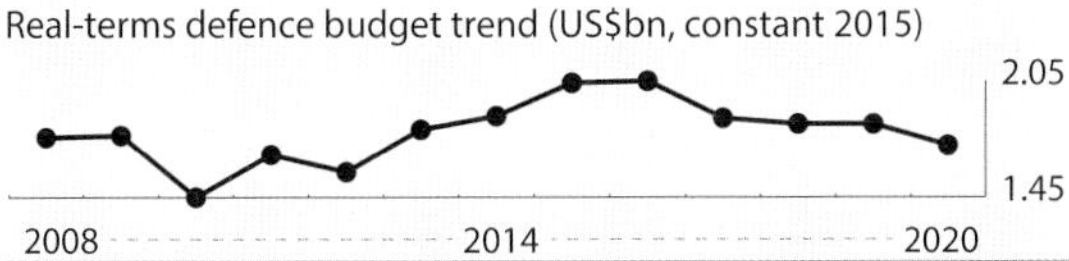

Population 22,889,201

Age	0–14	15–19	20–24	25–29	30–64	65 plus
Male	11.8%	3.9%	3.5%	3.6%	21.5%	4.5%
Female	11.3%	3.8%	3.4%	3.5%	23.0%	6.2%

Capabilities

Since the defeat of the Tamil Tigers, the armed forces have reoriented to a peacetime internal-security role. Support has been provided by China, in an indication of a growing military-to-military relationship. The US has eased its long-standing military trade restrictions and Japan has stated an intention to increase maritime cooperation. Sri Lanka has little capacity for force projection beyond its national territory but has sent small numbers of troops on UN missions. The navy's littoral capability, based on fast-attack and patrol boats, has been strengthened with the acquisition of offshore-patrol vessels, while the US has gifted a former US coastguard cutter and China has gifted a frigate. The army is reducing in size and there appears to have been little spending on new equipment since the end of the civil war. Sri Lanka is looking to begin a series of procurements to fill key capability gaps but ambitions are limited by budget constraints. Beyond maintenance facilities and limited fabrication, such as at Sri Lanka's shipyards, there is no defence-industrial base.

ACTIVE 255,000 (Army 177,000 Navy 50,000 Air 28,000) Paramilitary 62,200

RESERVE 5,500 (Army 1,100 Navy 2,400 Air Force 2,000) Paramilitary 30,400

ORGANISATIONS BY SERVICE

Army 113,000; 64,00 active reservists (recalled) (total 177,000)

Regt are bn sized

FORCES BY ROLE

COMMAND

7 region HQ

21 div HQ

SPECIAL FORCES

1 indep SF bde

MANOEUVRE

Reconnaissance

3 armd recce regt

Armoured

1 armd bde(-)

Mechanised

1 mech inf bde

Light

60 inf bde

1 cdo bde

Air Manoeuvre

1 air mob bde

COMBAT SUPPORT

7 arty regt

1 MRL regt

8 engr regt

6 sigs regt

EQUIPMENT BY TYPE

ARMOURED FIGHTING VEHICLES

MBT 62 T-55A/T-55AM2

RECCE 15 *Saladin*

IFV 62+: 13 BMP-1; 49 BMP-2; WZ-551 20mm

APC 211+
APC (T) 30+: some Type-63; 30 Type-85; some Type-89
APC (W) 181: 25 BTR-80/BTR-80A; 31 *Buffel*; 20 WZ-551; 105 *Unicorn*
ENGINEERING & MAINTENANCE VEHICLES
ARV 16 VT-55
VLB 2 MT-55
ANTI-TANK/ANTI-INFRASTRUCTURE
MANPATS HJ-8
RCL 40: **105mm** ε10 M-65; **106mm** ε30 M40
GUNS 85mm 8 Type-56 (D-44)
ARTILLERY 908
TOWED 96: **122mm** 20; **130mm** 30 Type-59-I; **152mm** 46 Type-66 (D-20)
MRL 122mm 28: 6 KRL-122; 22 RM-70
MOR 784: **81mm** 520; **82mm** 209; **120mm** 55 M-43
UNINHABITED AERIAL VEHICLES
ISR • Medium 1 *Seeker*

Navy ε37,000; ε13,000 active reserves (total 50,000)

Seven naval areas

EQUIPMENT BY TYPE
PRINCIPAL SURFACE COMBATANTS • FRIGATES 1
FFH 1 *Parakramabahu* (ex-PRC Type-053H2G (*Jiangwei* I)) with 1 twin 100mm gun (capacity 1 med hel)
PATROL AND COASTAL COMBATANTS 120
PSOH 4: 1 *Gajabahu* (ex-US *Hamilton*) with 1 76mm gun (capacity 1 med hel); 1 *Sayura* (ex-IND *Sukanya*); 2 *Sayurala* (IND *Samarth*)
PCO 2: 1 *Samudura* (ex-US *Reliance*); 1 *Sagara* (IND *Vikram*) with 1 hel landing platform
PCC 3: 1 *Jayasagara*; 2 *Nandimithra* (ISR *Sa'ar* 4) with 1 76mm gun
PBF 74: 26 *Colombo*; 6 *Shaldag*; 4 *Super Dvora* Mk II; 6 *Super Dvora* Mk III; 5 *Trinity Marine*; 27 *Wave Rider*
PB 11: 2 *Mihikatha* (ex-AUS *Bay*); 2 *Prathapa* (PRC mod *Haizhui*); 3 *Ranajaya* (PRC *Haizhui*); 1 *Ranarisi* (PRC mod *Shanghai* II); 3 *Weeraya* (PRC *Shanghai* II)
PBR 26
AMPHIBIOUS
LANDING SHIPS • LSM 1 *Shakthi* (PRC *Yuhai*) (capacity 2 tanks; 250 troops)
LANDING CRAFT 8
LCM 2
LCU 2 *Yunnan*
UCAC 1 M 10 (capacity 56 troops)
LOGISTICS AND SUPPORT 3: 2 **AP**; 1 **AX**

Marines ε500

FORCES BY ROLE
MANOEUVRE
Amphibious
1 mne bn

Special Boat Service ε100

Reserve Organisations

Sri Lanka Volunteer Naval Force (SLVNF) 13,000 active reservists

Air Force 28,000 (incl SLAF Regt)

FORCES BY ROLE
FIGHTER
1 sqn with F-7BS/G; FT-7
FIGHTER/GROUND ATTACK
1 sqn with *Kfir* C-2
1 sqn with K-8 *Karakorum**
TRANSPORT
1 sqn with An-32B *Cline*; C-130K *Hercules*; Cessna 421C *Golden Eagle*
1 sqn with Beech B200 *King Air*; Y-12 (II)
TRAINING
1 wg with PT-6, Cessna 150L
ATTACK HELICOPTER
1 sqn with Mi-24V *Hind* E; Mi-35P *Hind*
TRANSPORT HELICOPTER
1 sqn with Mi-17 *Hip* H; Mi-171Sh
1 sqn with Bell 206A/B (incl basic trg), Bell 212
1 (VIP) sqn with Bell 212; Bell 412 *Twin Huey*
ISR UAV
1 sqn with *Blue Horizon* II
1 sqn with *Searcher* MkII
MANOEUVRE
Other
1 (SLAF) sy regt

EQUIPMENT BY TYPE
AIRCRAFT 13 combat capable
FTR 5: 3 F-7GS; 2 FT-7 (3 F-7BS; 1 F-7GS non-operational)
FGA 1 *Kfir* C-2 (2 *Kfir* C-2; 1 *Kfir* C-7; 2 *Kfir* TC-2; 6 MiG-27M *Flogger* J; 1 MiG-23UB *Flogger* C non-operational)
TPT 20: **Medium** 2 C-130K *Hercules*; **Light** 18: 3 An-32B *Cline*; 6 Cessna 150L; 1 Cessna 421C *Golden Eagle*; 6 Y-12 (II); 2 Y-12 (IV)
TRG 13: 7 K-8 *Karakorum**; 6 PT-6
HELICOPTERS
ATK 11: 6 Mi-24P *Hind*; 3 Mi-24V *Hind* E; 2 Mi-35V *Hind*
MRH 18: 6 Bell 412 *Twin Huey* (VIP); 2 Bell 412EP (VIP); 10 Mi-17 *Hip* H
TPT 16: **Medium** 4 Mi-171Sh; **Light** 12: 2 Bell 206A *Jet Ranger*; 2 Bell 206B *Jet Ranger*; 8 Bell 212
UNINHABITED AERIAL VEHICLES
ISR • Medium 2+: some *Blue Horizon* II; 2 *Searcher* MkII
AIR DEFENCE • GUNS • TOWED 27: **40mm** 24 L/40; **94mm** 3 (3.7in)
AIR-LAUNCHED MISSILES
AAM • IR PL-5E

Paramilitary ε62,200

Home Guard 13,000

National Guard ε15,000

Police Force 30,200; 1,000 (women) (total 31,200) 30,400 reservists

Ministry of Defence Special Task Force 3,000

Anti-guerrilla unit

Coast Guard n/k

Ministry of Defence

EQUIPMENT BY TYPE

PATROL AND COASTAL COMBATANTS 28

PCO 1 *Suraksha* (ex-IND *Vikram*) with 1 hel landing platform

PBF 22: 2 *Dvora*; 4 *Super Dvora* Mk I; 3 *Killer* (ROK); 10 (Inshore Patrol Craft); 3 (Fast Patrol Craft)

PB 4: 2 Simonneau Type-508; 2 *Samudra Raksha*

PBR 1

DEPLOYMENT

CENTRAL AFRICAN REPUBLIC: UN • MINUSCA 110; 1 hel sqn

LEBANON: UN • UNIFIL 146; 1 inf coy

MALI: UN • MINUSMA 249; 1 sy coy

SOUTH SUDAN: UN • UNMISS 164; 1 fd hospital; 1 hel sqn

WESTERN SAHARA: UN • MINURSO 1

Taiwan (Republic of China) ROC

New Taiwan Dollar NT$		2019	2020	2021
GDP	NT$	18.9tr	18.9tr	
	US$	611bn	636bn	
per capita	US$	25,873	26,910	
Growth	%	2.7	0.0	
Inflation	%	0.5	-0.1	
Def bdgt	NT$	341bn	411bn	367bn
	US$	11.0bn	13.9bn	
US$1=NT$		30.93	29.66	

Real-terms defence budget trend (US$bn, constant 2015)

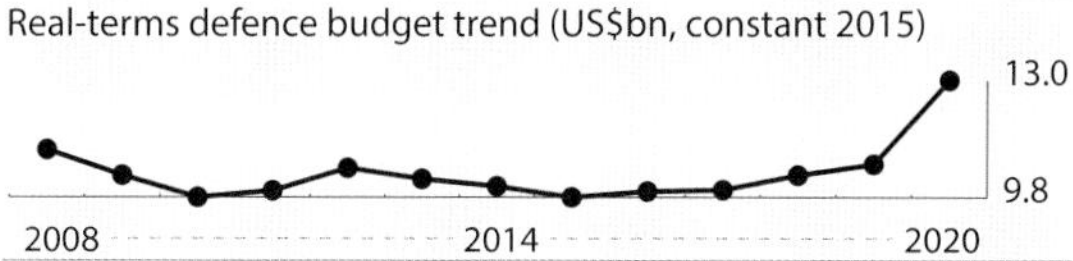

Population 23,603,049

Age	0–14	15–19	20–24	25–29	30–64	65 plus
Male	6.4%	2.6%	3.3%	3.5%	26.4%	7.1%
Female	6.0%	2.5%	3.2%	3.4%	27.0%	8.6%

Capabilities

Taiwan's security policy is dominated by its relationship with China and its attempts to sustain a credible military capability. Taiwan's current focus is on air defence and deterrence in coastal areas, on both sides of the island. The 2019 'National Defense Report' highlighted the importance of coastal defence. The armed forces are well trained and exercise regularly. Demographic pressure has influenced plans for force reductions and a shift towards an all-volunteer force. Taiwan's main security partnership is with the US. The Taiwan Relations Act from 1979 states that 'the United States shall provide Taiwan with arms of a defensive character'. In 2019, the United States approved the transfer of new F-16C/D Block 70 combat aircraft to Taiwan. Nevertheless, Taipei maintains an interest in the F-35. In late 2020 the US issued notifications regarding the possible sale of MQ-9B UAVs, HIMARS launchers and a variety of long-range precision-strike missiles which, if these sales proceeded, would significantly boost Taiwan's military capabilities. Nonetheless, Taiwan is modernising its existing holdings and developing its domestic defence industry through increased funding and the development of new weapons programmes. The government launched a new defence-industrial policy in 2019, aimed at further strengthening independent defence-manufacturing capacities. Taiwan's defence-industrial base has strengths in aerospace, shipbuilding and missiles.

ACTIVE 163,000 (Army 88,000 Navy 40,000 Air 35,000) Paramilitary 11,800

Conscript liability (19–40 years) 12 months for those born before 1993; four months for those born after 1994 (alternative service available)

RESERVE 1,657,000 (Army 1,500,000 Navy 67,000 Air Force 90,000)

Some obligation to age 30

ORGANISATIONS BY SERVICE

Space

EQUIPMENT BY TYPE

SATELLITES • **ISR** 1 *Formosat-5*

Army 88,000 (incl ε5,000 MP)

FORCES BY ROLE

COMMAND

3 corps HQ

5 defence comd HQ

SPECIAL FORCES/HELICOPTER

1 SF/hel comd (2 spec ops gp, 2 hel bde)

MANOEUVRE

Armoured

4 armd bde

Mechanised

3 mech inf bde

Light

6 inf bde

COMBAT SUPPORT

3 arty gp

3 engr gp

3 CBRN gp

3 sigs gp

COASTAL DEFENCE

1 AShM bn

Reserves

FORCES BY ROLE

MANOEUVRE

Light

21 inf bde

EQUIPMENT BY TYPE

ARMOURED FIGHTING VEHICLES

MBT 565: 200 M60A3; 100 M48A5; 265 M48H *Brave Tiger*

LT TK ε100 M41A3/D

IFV 257: 225 CM-25 (M113 with 20–30mm cannon); 32 CM-34 *Yunpao*

APC 1,318

APC (T) 650 M113

APC (W) 668: 368 CM-32 *Yunpao*; 300 LAV-150 *Commando*

ENGINEERING & MAINTENANCE VEHICLES
AEV 18 M9
ARV CM-27/A1; 37 M88A1
VLB 22 M3; M48A5
NBC VEHICLES 48+: BIDS; 48 K216A1; KM453
ANTI-TANK/ANTI-INFRASTRUCTURE
MSL
SP TOW
MANPATS FGM-148 *Javelin*; TOW
RCL 500+: **90mm** M67; **106mm** 500+: 500 M40A1; Type-51
ARTILLERY 2,093
SP 488: **105mm** 100 M108; **155mm** 318: 225 M109A2/A5; 48 M44T; 45 T-69; **203mm** 70 M110
TOWED 1,060+: **105mm** 650 T-64 (M101); **155mm** 340+: 90 M59; 250 T-65 (M114); M44; XT-69; **203mm** 70 M115
MRL 223: **117mm** 120 *Kung Feng* VI; **126mm** 103: 60 *Kung Feng* III/*Kung Feng* IV; 43 RT 2000 *Thunder*
MOR 322+
SP 162+: **81mm** 72+: M29; 72 M125; **107mm** 90 M106A2
TOWED 81mm 160 M29; T-75; **107mm** M30; **120mm** K5; XT-86
COASTAL DEFENCE
ARTY 54: **127mm** ε50 US Mk32 (reported); **240mm** 4 M1
AShM *Ching Feng*
HELICOPTERS
ATK 96: 67 AH-1W *Cobra*; 29 AH-64E *Apache*
MRH 38 OH-58D *Kiowa Warrior*
TPT 38: **Heavy** 8 CH-47SD *Super D Chinook*; **Medium** 30 UH-60M *Black Hawk*
TRG 29 TH-67 *Creek*
UNINHABITED AERIAL VEHICLES
ISR • Light *Mastiff* III
AIR DEFENCE
SAM • Point-defence 76+: 74 M1097 *Avenger*; 2 M48 *Chaparral*; FIM-92 *Stinger*
GUNS
SP 40mm M42
TOWED 40mm L/70

Navy 40,000

EQUIPMENT BY TYPE
SUBMARINES • SSK 4:
2 *Hai Lung* with 6 single 533mm TT with UGM-84L *Harpoon* Block II AShM/SUT HWT
2 *Hai Shih*† (ex-US *Guppy* II (used in trg role) with 10 single 533mm TT (6 fwd, 4 aft) with SUT HWT
PRINCIPAL SURFACE COMBATANTS 26
DESTROYERS • DDGHM 4 *Keelung* (ex-US *Kidd*) with 2 quad lnchr with RGM-84L *Harpoon* Block II AShM, 2 twin Mk 26 GMLS with SM-2 Block IIIA SAM, 2 triple 324mm SVTT Mk 32 ASTT with Mk 46 LWT, 2 Mk 15 *Phalanx* Block 1B CIWS, 2 127mm gun (capacity 1 S-70 ASW hel)
FRIGATES 22
FFGHM 21:
8 *Cheng Kung* (US *Oliver Hazard Perry* mod) with 2 quad lnchr with *Hsiung Feng* II/III AShM, 1 Mk 13 GMLS with SM-1MR Block VI SAM, 2 triple 324mm SVTT Mk 32 ASTT with Mk 46 LWT, 1 Mk 15 *Phalanx* Block 1B CIWS, 1 76mm gun (capacity 2 S-70C ASW hel)
2 *Meng Chuan* (ex-US *Oliver Hazard Perry*) with 1 Mk13 GMLS with RGM-84 *Harpoon* AShM/SM-1MR Block VI SAM, 2 triple 324mm SVTT Mk 32 ASTT with Mk 46 LWT, 1 Mk 15 *Phalanx* Block 1B CIWS, 1 76mm gun (capacity 2 S-70C ASW hel)
5 *Chin Yang* (ex-US *Knox*) with 1 octuple Mk 16 lnchr with RGM-84C *Harpoon* Block 1B AShM/ASROC A/S msl, 2 triple lnchr with SM-1MR Block VI SAM, 2 twin lnchr with SM-1MR Block VI SAM, 2 twin 324mm SVTT Mk 32 ASTT with Mk 46 LWT, 1 Mk 15 *Phalanx* Block 1B CIWS, 1 127mm gun (capacity 1 MD-500 hel)
6 *Kang Ding* with 2 quad lnchr with *Hsiung Feng* II AShM, 1 quad lnchr with *Sea Chaparral* SAM, 2 triple 324mm SVTT Mk 32 ASTT with Mk 46 LWT, 1 Mk 15 *Phalanx* Block 1B CIWS, 1 76mm gun (capacity 1 S-70C ASW hel)
FFGH 1 *Chin Yang* (ex-US *Knox*) with 1 octuple Mk 112 lnchr with RGM-84C *Harpoon* Block 1B AShM, 2 twin 324mm SVTT Mk 32 ASTT with Mk 46 LWT, 1 Mk 15 *Phalanx* Block 1B CIWS, 1 127mm gun (capacity 1 MD-500 hel)
PATROL AND COASTAL COMBATANTS 44
PCFG 1 *Tuo Jiang* (*Hsun Hai*) with 4 twin lnchr with *Hsiung Feng* II AShM, 4 twin lnchr with *Hisung Feng* III AShM, 2 triple 324mm SVTT Mk 32 ASTT, 1 Mk 15 *Phalanx* Block 1B CIWS; 1 76mm gun
PCG 11:
1 *Jin Chiang* with 1 twin lnchr with *Hsiung Feng* II AShM
4 *Jin Chiang* with 2 twin lnchr with *Hsiung Feng* II AShM, 1 76mm gun
6 *Jin Chiang* with 1 twin lnchr with *Hsiung Feng* III AShM, 1 76mm gun
PCC 1 *Jin Chiang* (test platform)
PBG 31 *Kwang Hua* with 2 twin lnchr with *Hsiung Feng* II AShM
MINE WARFARE • MINE COUNTERMEASURES 8
MHC 6: 4 *Yung Feng*; 2 *Yung Jin* (ex-US *Osprey*)
MSO 2 *Yung Yang* (ex-US *Aggressive*)
COMMAND SHIPS • LCC 1 *Kao Hsiung*
AMPHIBIOUS
PRINCIPAL AMPHIBIOUS SHIPS • LSD 1 *Shiu Hai* (ex-US *Anchorage*) with 2 Mk 15 *Phalanx* CIWS, 1 hel landing platform (capacity either 2 LCU or 18 LCM; 360 troops)
LANDING SHIPS
LST 8:
6 *Chung Hai* (capacity 16 tanks; 200 troops)
2 *Chung Ho* (ex-US *Newport)* with 1 Mk 15 *Phalanx* CIWS , 1 hel landing platform (capacity 3 LCVP, 23 AFVs, 400 troops)
LANDING CRAFT 47
LCM ε35 (various)
LCU 12 LCU 1610 (capacity 2 M60A3 or 400 troops) (minelaying capability)
LOGISTICS AND SUPPORT 10
AGOR 1 *Ta Kuan*

AOEH 1 *Panshih* with 1 quad lnchr with *Sea Chaparral* SAM, 2 Mk 15 *Phalanx* CIWS (capacity 3 med hel)
AOE 1 *Wu Yi* with 1 quad lnchr with *Sea Chaparral* SAM, 1 hel landing platform
ARS 2: 1 *Da Hu* (ex-US *Diver*); 1 *Da Juen* (ex-US *Bolster*)
ATF 5 *Ta Tung* (ex-US *Cherokee*)

Marines 10,000

FORCES BY ROLE
MANOEUVRE
Amphibious
3 mne bde
COMBAT SUPPORT
Some cbt spt unit

EQUIPMENT BY TYPE
ARMOURED FIGHTING VEHICLES
AAV 202: 52 AAV-7A1; 150 LVTP-5A1
ENGINEERING & MAINTENANCE VEHICLES
ARV 2 AAVR-7
ANIT-TANK/ANTI-INFRASTRUCTURE
RCL 106mm
ARTILLERY • TOWED 105mm; **155mm**

Naval Aviation

FORCES BY ROLE
ANTI SUBMARINE WARFARE
2 sqn with S-70C *Seahawk* (S-70C *Defender*)
1 sqn with MD-500 *Defender*
ISR UAV
1 bn with *Chung Shyang* II

EQUIPMENT BY TYPE
HELICOPTERS
ASW 20 S-70C *Seahawk* (S-70C *Defender*)
MRH 10 MD-500 *Defender*
UNINHABITED AERIAL VEHICLES • ISR • Medium ε28 *Chung Shyang* II

Air Force 35,000

FORCES BY ROLE
FIGHTER
3 sqn with *Mirage* 2000-5E/D (2000-5EI/DI)
FIGHTER/GROUND ATTACK
3 sqn with F-5E/F *Tiger* II
6 sqn with F-16A/B *Fighting Falcon*
5 sqn with F-CK-1A/B/C/D *Ching Kuo*
ANTI-SUBMARINE WARFARE
1 sqn with P-3C *Orion*
ELECTRONIC WARFARE
1 sqn with C-130HE *Tien Gian*
ISR
1 sqn with RF-5E *Tigereye*
AIRBORNE EARLY WARNING & CONTROL
1 sqn with E-2T *Hawkeye*
SEARCH & RESCUE
1 sqn with H225; UH-60M *Black Hawk*
TRANSPORT
2 sqn with C-130H *Hercules*
1 (VIP) sqn with B-727-100; B-737-800; Beech 1900; F-50; S-70C *Black Hawk*
TRAINING
1 sqn with AT-3A/B *Tzu-Chung**
1 sqn with Beech 1900
1 (basic) sqn with T-34C *Turbo Mentor*

EQUIPMENT BY TYPE
AIRCRAFT 478 combat capable
FTR 280: 86 F-5E/F *Tiger* II (some in store); up to 139 F-16A/B *Fighting Falcon*; 9 *Mirage* 2000-5D (2000-5DI); 46 *Mirage* 2000-5E (2000-5EI)
FGA 131: 127 F-CK-1C/D *Ching Kuo*; 4 F-16V *Fighting Falcon*
ASW 12 P-3C *Orion*
EW 1 C-130HE *Tien Gian*
ISR 7 RF-5E *Tigereye*
AEW&C 6 E-2T *Hawkeye*
TPT 33: **Medium** 19 C-130H *Hercules*; **Light** 10 Beech 1900; **PAX** 4: 1 B-737-800; 3 F-50
TRG 97: 55 AT-3A/B *Tzu-Chung**; 42 T-34C *Turbo Mentor*
HELICOPTERS
TPT • Medium 17: 3 H225; 14 UH-60M *Black Hawk*
AIR-LAUNCHED MISSILES
AAM • IR AIM-9J/P *Sidewinder*; R-550 *Magic* 2; *Shafrir*; *Sky Sword* I; **IIR** AIM-9X *Sidewinder* II MICA IR; **ARH** *Mica* RF; **ARH** AIM-120C AMRAAM; *Sky Sword* II
ASM AGM-65A *Maverick*
AShM AGM-84 *Harpoon*
ARM *Sky Sword* IIA
LACM Conventional *Wan Chien*
BOMBS • Laser-guided *Paveway* II

Air Defence and Missile Command

FORCES BY ROLE
SURFACE-TO-SURFACE MISSILE
1 GLCM bde (2 GLCM bn with *Hsiung Feng* IIE)
AIR DEFENCE
1 (792) SAM bde (1 SAM bn with *Tien Kung* III; 2 ADA bn)
2 (793 & 794) SAM bde (1 SAM bn with Tien Kung II; 1 SAM bn with MIM-104F *Patriot* PAC-3; 1 SAM bn with MIM-23 *Hawk*)
1 (795) SAM bde (1 SAM bn with MIM-104F *Patriot* PAC-3; 2 ADA bn)

EQUIPMENT BY TYPE
SURFACE-TO-SURFACE MISSILE LAUNCHERS
GLCM • Conventional ε12 *Hsiung Feng* IIE
AIR DEFENCE
SAM 202+
Long-range 122+: 72+ MIM-104F *Patriot* PAC-3; ε50 *Tien Kung* II
Medium-range 50 MIM-23 *Hawk*
Short-range 30 RIM-7M *Sparrow* with *Skyguard*
Point-defence *Antelope*
GUNS • 20mm some T-82; **35mm** 20+ GDF-006 with *Skyguard*
MISSILE DEFENCE *Tien Kung* III

Paramilitary 11,800

Coast Guard 11,800

EQUIPMENT BY TYPE
PATROL AND COASTAL COMBATANTS 167

PSOH 4: 2 *Tainan*; 2 *Yilan*
PSO 6: 4 *Miaoli* with 1 hel landing platform; 2 *Ho Hsing*
PCO 13: 2 *Kinmen*; 2 *Mou Hsing*; 3 *Shun Hu 7*; 4 *Taichung*; 2 *Taipei*
PBF ε57 (various)
PB 87: 1 *Shun Hu 6*; ε86 (various)

FOREIGN FORCES

Singapore 3 trg camp (incl inf and arty)

Thailand THA

Thai Baht b		2019	2020	2021
GDP	b	16.9tr	15.6tr	
	US$	544bn	509bn	
per capita	US$	7,807	7,295	
Growth	%	2.4	-7.1	
Inflation	%	0.7	-0.4	
Def bdgt	b	227bn	214bn	215bn
	US$	7.32bn	6.99bn	
US$1=b		31.05	30.60	

Real-terms defence budget trend (US$bn, constant 2015)

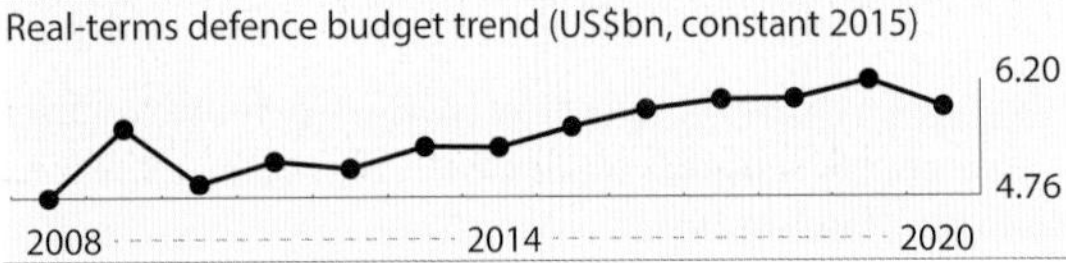

Population 68,977,400

Age	0–14	15–19	20–24	25–29	30–64	65 plus
Male	8.4%	3.1%	3.5%	3.8%	25.0%	5.2%
Female	8.0%	3.0%	3.4%	3.7%	26.2%	6.7%

Capabilities

Thailand has large, well-funded armed forces and its air force is one of the best equipped and trained in Southeast Asia. The Vision 2026 defence-modernisation plan, approved by the defence council in October 2017, outlines the armed forces' capability improvements over the next decade. Thailand is considered a major non-NATO ally by the US, while the country has developed deeper defence ties with China since the military coup in 2014. The armed forces regularly take part in international military exercises. Personnel remain deployed on the UNMISS mission to South Sudan. The military-modernisation effort includes the development of submarines, anti-submarine-warfare capabilities and a surface-ship procurement programme. The armoured vehicle fleet has been recapitalised with deliveries from China and Ukraine. The arrival of Saab 340 AEW aircraft, *Gripen* combat aircraft and a command-and-control system has improved air capability. In January 2020, the Royal Thai Air Force (RTAF) issued a white paper which detailed further acquisition and upgrade requirements through the 2020s, including fighters, tactical-transport and VIP aircraft. The RTAF launched a space-operations centre in August 2019, a priority identified in the National Strategy 2018–37 development programme. Under its Defence Industry Masterplan, the government indicates that expanding Thailand's presently limited defence sector can be an important way of developing military capability. The government is making efforts to reform defence procurement and offsets by expanding the role of its Defence Technology Institute.

ACTIVE 360,850 (Army 245,000 Navy 69,850 Air 46,000) Paramilitary 93,700
Conscription liability 24 months

RESERVE 200,000 Paramilitary 45,000

ORGANISATIONS BY SERVICE

Army 130,000; ε115,000 conscript (total 245,000)

Cav, lt armd, recce and tk sqn are bn sized

FORCES BY ROLE
COMMAND
4 (regional) army HQ
3 corps HQ
SPECIAL FORCES
1 SF div
1 SF regt
MANOEUVRE
Armoured
1 (3rd) mech cav div (2 tk regt (2 tk sqn); 1 sigs bn; 1 maint bn; 1 hel sqn)
Mechanised
1 (1st) mech cav div (1 armd recce sqn; 2 mech cav regt (3 mech cav sqn); 1 indep mech cav sqn; 1 sigs bn; 1 maint bn; 1 hel sqn)
1 (2nd) mech cav div (1 armd recce sqn; 2 (1st & 5th) mech cav regt (1 tk sqn, 2 mech cav sqn); 1 (4th) mech cav regt (3 mech cav sqn); 1 sigs bn; 1 maint bn; 1 hel sqn)
1 (2nd) mech inf div (1 armd recce sqn; 1 tk bn; 3 mech inf regt (3 mech inf bn); 1 arty regt (4 arty bn); 1 engr bn; 1 sigs bn)
1 (11th) mech inf div (2 mech inf regt (3 mech inf bn); 1 engr bn; 1 sigs bn)
Light
1 (1st) inf div (1 lt armd sqn; 1 ranger regt (3 ranger bn); 1 arty regt (4 arty bn); 1 engr bn; 1 sigs bn)
1 (3rd) inf div (3 inf regt (3 inf bn); 1 arty regt (3 arty bn); 1 engr bn; 1 sigs bn)
1 (4th) inf div (1 lt armd sqn; 2 inf regt (3 inf bn); 1 arty regt (3 arty bn); 1 engr bn; 1 sigs bn)
1 (5th) inf div (1 lt armd sqn; 3 inf regt (3 inf bn); 1 arty regt (4 arty bn); 1 engr bn; 1 sigs bn)
1 (6th) inf div (2 inf regt (3 inf bn); 1 arty regt (4 arty bn); 1 engr bn; 1 sigs bn)
1 (7th) inf div (2 inf regt (3 inf bn); 1 arty regt (2 arty bn); 1 engr bn; 1 sigs bn)
1 (9th) inf div (1 mech cav sqn; 3 inf regt (3 inf bn); 1 arty regt (3 arty bn); 1 engr bn; 1 sigs bn)
1 (15th) inf div (1 mech cav sqn; 3 inf regt (3 inf bn); 1 engr bn; 1 sigs bn)
COMBAT SUPPORT
1 arty div (1 arty regt (1 SP arty bn; 2 fd arty bn); 1 arty regt (1 MRL bn; 2 fd arty bn))
1 engr div
COMBAT SERVICE SUPPORT
4 economic development div
HELICOPTER
Some hel flt

ISR UAV
1 UAV bn with *Hermes* 450; *Searcher* II
AIR DEFENCE
1 ADA div (6 bn)

EQUIPMENT BY TYPE
ARMOURED FIGHTING VEHICLES
MBT 370: 53 M60A1; 125 M60A3; 105 M48A5; 49 T-84 *Oplot*; 38 VT-4; (50 Type-69 in store)
LT TK 194: 24 M41; 104 *Scorpion* (50 in store); 66 *Stingray*
RECCE 32 S52 *Shorland*
IFV 202: 168 BTR-3E1; 34 VN-1
APC 1,190
APC (T) 880: *Bronco*; 430 M113A1/A3; 450 Type-85
APC (W) 210: 9 BTR-3K (CP); 6 BTR-3C (amb); 18 *Condor*; 142 LAV-150 *Commando*; 35 M1126 *Stryker* ICV
PPV 100 REVA
ENGINEERING & MAINTENANCE VEHICLES
ARV 58+: 2 BREM-84 *Atlet*; 13 BTR-3BR; 22 M88A1; 6 M88A2; 10 M113; 5 Type-653; WZT-4
VLB Type-84
MW *Bozena*; *Giant Viper*
ANTI-TANK/ANTI-INFRASTRUCTURE
MSL
SP 30+: 18+ M901A5 (TOW); 12 BTR-3RK
MANPATS M47 *Dragon*
RCL 180: **75mm** 30 M20; **106mm** 150 M40
ARTILLERY 2,567
SP 155mm 42: 16 ATMOS-2000; 6 CAESAR; 20 M109A5
TOWED 525: **105mm** 296: 24 LG1 MkII; 12 M-56; 200 M101A1; 60 L119 Light Gun; (12 M102; 32 M618A2 in store); **155mm** 229: 90 GHN-45 A1; 118 M198; 21 M-71 (48 M114 in store)
MRL 68: **122mm** 4 SR-4; **130mm** 60 PHZ-85; **302mm** 4: 1 DTI-1 (WS-1B); 3 DTI-1G (WS-32)
MOR 1,932+: SP **81mm** 39: 18 BTR-3M1; 21 M125A3; SP **107mm** M106A3; **SP 120mm** 26: 8 BTR-3M2; 6+ Elbit *Spear*; 12 M1064A3; 1,867 **81mm/107mm/120mm**
AIRCRAFT
TPT • Light 22: 2 Beech 200 *King Air*; 2 Beech 1900C; 1 C-212 *Aviocar*; 1 C295W; 3 Cessna 182T *Skylane*; 9 Cessna A185E (U-17B); 2 ERJ-135LR; 2 *Jetstream* 41
TRG 33: 11 MX-7-235 *Star Rocket*; 22 T-41B *Mescalero*
HELICOPTERS
ATK 7 AH-1F *Cobra*
MRH 17: 8 AS550 *Fennec*; 2 AW139; 7 Mi-17V-5 *Hip* H
TPT 216: **Heavy** 5 CH-47D *Chinook*; **Medium** 12: 9 UH-60L *Black Hawk*; 3 UH-60M *Black Hawk*; **Light** 199: 93 Bell 205 (UH-1H *Iroquois*); 27 Bell 206 *Jet Ranger*; 52 Bell 212 (AB-212); 16 Enstrom 480B; 6 H145M (VIP tpt); 5 UH-72A *Lakota*
TRG 53 Hughes 300C
UNINHABITED AERIAL VEHICLES
ISR • Medium 4+: 4 *Hermes* 450; *Searcher*; *Searcher* II
AIR DEFENCE
SAM 8+
Short-range *Aspide*
Point-defence 8+: 8 *Starstreak*; 9K338 *Igla*-S (SA-24 *Grinch*)
GUNS 184
SP 54: **20mm** 24 M163 *Vulcan*; **40mm** 30 M1/M42 SP
TOWED 138: **20mm** 24 M167 *Vulcan*; **35mm** 8 GDF-007 with *Skyguard* 3; **37mm** 52 Type-74; **40mm** 48 L/70; **57mm** ε6 Type-59 (S-60) (18+ more non-operational)

Navy 44,000 (incl Naval Aviation, Marines, Coastal Defence); 25,850 conscript (total 69,850)

EQUIPMENT BY TYPE
PRINCIPAL SURFACE COMBATANTS 8
AIRCRAFT CARRIERS • CVH 1 *Chakri Naruebet* with 3 sextuple *Sadral* lnchr with *Mistral* SAM (capacity 6 S-70B *Seahawk* ASW hel)
FRIGATES 7
FFGHM 3:
2 *Naresuan* with 2 quad lnchr with RGM-84 *Harpoon* AShM, 1 8 cell Mk 41 VLS with RIM-162B ESSM SAM, 2 triple SVTT Mk 32 324mm TT with Mk 46 LWT, 1 127mm gun (capacity 1 *Super Lynx* 300 hel)
1 *Bhumibol Adulyadej* (DW3000F) with 2 quad lnchr with RGM-84L *Harpoon* Block II AShM, 1 8-cell Mk 41 VLS with RIM-162B ESSM SAM, 2 triple 324mm SEA TLS ASTT with Mk 54 LWT, 1 Mk 15 *Phalanx* Block 1B CIWS, 1 76mm gun (capacity 1 med hel)
FFG 4:
2 *Chao Phraya* (trg role) with 4 twin lnchr with C-802A AShM, 2 RBU 1200 *Uragan* A/S mor, 2 twin 100mm gun
2 *Chao Phraya* with 4 twin lnchr with C-802A AShM, 2 RBU 1200 *Uragan* A/S mor, 1 twin 100mm gun, 1 hel landing platform
PATROL AND COASTAL COMBATANTS 90
CORVETTES 7:
FSGM 2 *Rattanakosin* with 2 twin lnchr with RGM-84 *Harpoon* AShM, 1 octuple *Albatros* lnchr with *Aspide* SAM, 2 triple 324mm SVTT Mk 32 ASTT with *Stingray* LWT, 1 76mm gun
FSG 1 *Krabi* (UK *River* mod) with 2 twin lnchr with RGM-84L *Harpoon* Block II AShM, 1 76mm gun
FS 4:
1 *Makut Rajakumarn* with 2 triple 324mm ASTT, 2 114mm gun
1 *Pin Klao* (ex-US *Cannon*) (trg role) with 2 triple 324mm SVTT Mk 32 ASTT, 3 76mm gun
2 *Tapi* with 2 triple 324mm SVTT Mk 32 ASTT with Mk 46 LWT, 1 76mm gun
PSO 1 *Krabi* (UK *River* mod) with 1 76mm gun
PCFG 4: 3 *Prabparapak* with 2 single lnchr with *Gabriel* I AShM, 1 triple lnchr with *Gabriel* I AShM, 1 57mm gun; 1 *Ratcharit* with 2 twin lnchr with MM38 *Exocet* AShM, 1 76mm gun
PCT 3 *Khamronsin* with 2 triple 324mm ASTT with *Stingray* LWT, 1 76mm gun
PCOH 2 *Pattani* (1 in trg role) with 1 76mm gun
PCO 4: 3 *Hua Hin* with 1 76mm gun; 1 M58 Patrol Gun Boat with 1 76mm gun
PCC 9: 3 *Chon Buri* with 2 76mm gun; 6 *Sattahip* with 1 76mm gun
PBF 4 M18 Fast Assault Craft (capacity 18 troops)

PB 56: 1 T-11 (US PGM-71); 3 T-81; 6 T-91; 3 M36 Patrol Boat; 13 T-213; 1 T-227; 23 M21 Patrol Boat; 3 T-991; 3 T-994

MINE WARFARE • MINE COUNTERMEASURES 17
MCCS 1 *Thalang*
MCO 2 *Lat Ya*
MCC 2 *Bang Rachan*
MSR 12: 7 T1; 5 T6

AMPHIBIOUS
PRINCIPAL AMPHIBIOUS SHIPS • LPD 1 *Angthong* (SGP *Endurance*) with 1 76mm gun (capacity 2 hel; 19 MBT; 500 troops)
LANDING SHIPS 2
LST 2 *Sichang* with 2 hel landing platform (capacity 14 MBT; 300 troops)
LANDING CRAFT 14
LCU 9: 3 *Man Nok*; 2 *Mataphun* (capacity either 3–4 MBT or 250 troops); 4 *Thong Kaeo*
LCM 2
UCAC 3 *Griffon* 1000TD

LOGISTICS AND SUPPORT 13
ABU 1 *Suriya*
AGOR 1 *Sok*
AGS 2
AOL 6: 1 *Matra* with 1 hel landing platform; 3 *Proet*; 1 *Prong*; 1 *Samui*
AOR 1 *Chula*
AORH 1 *Similan* (capacity 1 hel)
AWT 1

Naval Aviation 1,200

EQUIPMENT BY TYPE
AIRCRAFT 3 combat capable
ASW 2 P-3A *Orion* (P-3T)
ISR 9 *Sentry* O-2-337
MP 1 F-27-200 MPA*
TPT • Light 15: 7 Do-228-212; 2 ERJ-135LR; 2 F-27-400M *Troopship*; 3 N-24A *Searchmaster*; 1 UP-3A *Orion* (UP-3T)
HELICOPTERS
ASW 8: 6 S-70B *Seahawk*; 2 *Super Lynx* 300
MRH 2 MH-60S *Knight Hawk*
TPT 18: **Medium** 2 Bell 214ST (AB-214ST); **Light** 16: 6 Bell 212 (AB-212); 5 H145M; 5 S-76B
AIR-LAUNCHED MISSILES • AShM AGM-84 *Harpoon*

Marines 23,000

FORCES BY ROLE
COMMAND
1 mne div HQ
MANOEUVRE
Reconnaissance
1 recce bn
Mechanised
1 mech bn
Light
2 inf regt (total: 6 bn)
Amphibious
1 amph aslt bn
COMBAT SUPPORT
1 arty regt (3 fd arty bn, 1 ADA bn)

EQUIPMENT BY TYPE
ARMOURED FIGHTING VEHICLES
IFV 14 BTR-3E1
APC (W) 24 LAV-150 *Commando*
AAV 33 LVTP-7
ENGINEERING & MAINTENANCE VEHICLES
ARV 1 AAVR-7
ANTI-TANK/ANTI-INFRASTRUCTURE
MSL
SP 10 M1045A2 HMMWV with TOW
MANPATS M47 *Dragon*; TOW
RCL • SP 106mm M40A1
ARTILLERY • TOWED 48: **105mm** 36 M101A!; **155mm** 12 GC-45
AIR DEFENCE
SAM Point-defence QW-18
GUNS 12.7mm 14

Naval Special Warfare Command

Air Force ε46,000

4 air divs, one flying trg school

FORCES BY ROLE
FIGHTER
2 sqn with F-5E/5F *Tiger* II
3 sqn with F-16A/B *Fighting Falcon*
FIGHTER/GROUND ATTACK
1 sqn with *Gripen* C/D
GROUND ATTACK
1 sqn with *Alpha Jet**
1 sqn with AU-23A *Peacemaker*
1 sqn with L-39ZA *Albatros**; T-50TH *Golden Eagle**
ELINT/ISR
1 sqn with DA42 MPP *Guardian*
AIRBORNE EARLY WARNING & CONTROL
1 sqn with Saab 340B; Saab 340 *Erieye*
TRANSPORT
1 (Royal Flight) sqn with A319CJ; A340-500; B-737-800
1 sqn with ATR-72; BAe-748
1 sqn with BT-67
1 sqn with C-130H/H-30 *Hercules*
TRAINING
1 sqn with L-39ZA *Albatros**
1 sqn with CT-4A/B *Airtrainer*; T-41D *Mescalero*
1 sqn with CT-4E *Airtrainer*
1 sqn with PC-9
TRANSPORT HELICOPTER
1 sqn with Bell 205 (UH-1H *Iroquois*)
1 sqn with Bell 412 *Twin Huey*; S-92A

EQUIPMENT BY TYPE
AIRCRAFT 151 combat capable
FTR 78: 1 F-5B *Freedom Fighter*; 20 F-5E *Tiger* II; 2 F-5F *Tiger* II (F-5E/F being upgraded); 1 F-5TH(E) *Tiger* II; 1 F-5TH(F) *Tiger* II; 38 F-16A *Fighting Falcon*; 15 F-16B *Fighting Falcon*
FGA 11: 7 *Gripen* C; 4 *Gripen* D
ATK 16 AU-23A *Peacemaker*
ISR 5 DA42 MPP *Guardian*
AEW&C 2 Saab 340 *Erieye*
ELINT 2 Saab 340 *Erieye* (COMINT/ELINT)

TPT 42: **Medium** 14: 6 C-130H *Hercules*; 6 C-130H-30 *Hercules*; 2 Saab 340B; **Light** 21: 3 ATR-72; 3 Beech 200 *King Air*; 8 BT-67; 1 *Commander* 690; 6 DA42M; **PAX** 7: 1 A319CJ; 1 A320CJ; 1 A340-500; 1 B-737-800; 3 SSJ-100-95LR (1 A310-324 in store)
TRG 113: 16 *Alpha Jet**; 13 CT-4A *Airtrainer*; 6 CT-4B *Airtrainer*; 20 CT-4E *Airtrainer*; 26 L-39ZA *Albatros**; 21 PC-9; 7 T-41D *Mescalero*; 4 T-50TH *Golden Eagle**

HELICOPTERS
MRH 11: 2 Bell 412 *Twin Huey*; 2 Bell 412SP *Twin Huey*; 1 Bell 412HP *Twin Huey*; 6 Bell 412EP *Twin Huey*
CSAR 8 H225M *Super Cougar*
TPT 20: **Medium** 3 S-92A *Super Hawk*; **Light** 17 Bell 205 (UH-1H *Iroquois*)

UNINHABITED AERIAL VEHICLES • ISR • Light U-1

AIR-LAUNCHED MISSILES
AAM • IR AIM-9P/S *Sidewinder*; *Python* 3; **IIR** IRIS-T; *Python* 5 (reported); **ARH** AIM-120 AMRAAM; *Derby* (reported)
ASM AGM-65 *Maverick*
AShM RBS15F

BOMBS
Laser-guided *Paveway* II
INS/GPS-guided GBU-38 JDAM

Royal Security Command

FORCES BY ROLE
MANOEUVRE
Light
2 inf regt (3 inf bn)

Paramilitary ε93,700

Border Patrol Police 20,000

Marine Police 2,200

EQUIPMENT BY TYPE
PATROL AND COASTAL COMBATANTS 98
PCO 1 *Srinakrin*
PCC 2 *Hameln*
PB 49: 2 *Chasanyabadee*; 3 *Cutlass*; 2 *Ratayapibanbancha* (*Reef Ranger*); 1 *Sriyanont*; 41 (various)
PBR 46

National Security Volunteer Corps 45,000 – Reserves

Police Aviation 500

EQUIPMENT BY TYPE
AIRCRAFT 6 combat capable
ATK 6 AU-23A *Peacemaker*
TPT 16: **Light** 15: 2 CN235; 8 PC-6 *Turbo-Porter*; 3 SC-7 3M *Skyvan*; 2 Short 330UTT; **PAX** 1 F-50

HELICOPTERS
MRH 12: 6 Bell 412 *Twin Huey*; 6 Bell 429
TPT • Light 61: 27 Bell 205A; 14 Bell 206 *Jet Ranger*; 20 Bell 212 (AB-212)

Provincial Police 50,000 (incl ε500 Special Action Force)

Thahan Phran (Hunter Soldiers) 21,000

Volunteer irregular force
FORCES BY ROLE
MANOEUVRE
Other
22 paramilitary regt (total: 275 paramilitary coy)

DEPLOYMENT

INDIA/PAKISTAN: UN • UNMOGIP 4
SOUTH SUDAN: UN • UNMISS 266; 1 engr coy
SUDAN: UN • UNAMID 3

FOREIGN FORCES

United States US Pacific Command: 300

Timor-Leste TLS

US$		2019	2020	2021
GDP	US$	1.62bn	1.92bn	
per capita	US$	1,252	1,456	
Growth	%	3.1	-6.8	
Inflation	%	0.9	0.9	
Def bdgt	US$	31.3m	41.7m	

Real-terms defence budget trend (US$m, constant 2015)

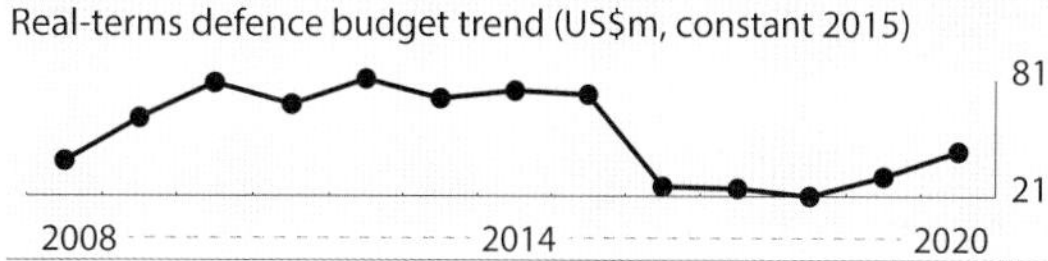

Population 1,383,723

Age	0–14	15–19	20–24	25–29	30–64	65 plus
Male	20.5%	5.6%	4.7%	3.8%	13.3%	2.0%
Female	19.4%	5.4%	4.6%	4.0%	14.5%	2.1%

Capabilities

The small Timor-Leste Defence Force (F-FDTL) has been afflicted by funding, personnel and morale challenges since it was established in 2001. The F-FDTL was reconstituted in the wake of fighting in 2006, but is still a long way from meeting the ambitious force-structure goals set out in the Force 2020 plan published in 2007. In 2016, the government published a Strategic Defence and Security Concept (SDSC). This outlined the roles of the F-FDTL as including the protection of the country from external threats and combating violent crime. However, this parallel internal-security role has sometimes brought it into conflict with the national police force. The SDSC also stated that the F-FDTL needs to improve its naval capabilities, owing to the size of Timor-Leste's exclusive economic zone. The origins of the F-FDTL in the Falintil national resistance force, and continuing training and doctrinal emphasis on low-intensity infantry tactics, mean that the force provides a deterrent to invasion. In 2017, Portugal and Timor-Leste signed a defence-cooperation agreement up to 2022. The F-FDTL sometimes receives training from Australian and US personnel. Australia is also donating two *Guardian*-class patrol vessels as part of its Pacific Patrol Boat Replacement programme; these are due to arrive in 2023. Maintenance capacity is unclear and the country has no traditional defence industry.

ACTIVE 2,280 (Army 2,200 Naval Element 80)

ORGANISATIONS BY SERVICE

Army 2,200

Training began in January 2001 with the aim of deploying 1,500 full-time personnel and 1,500 reservists. Authorities are engaged in developing security structures with international assistance

FORCES BY ROLE

MANOEUVRE

Light

2 inf bn

COMBAT SUPPORT

1 MP pl

COMBAT SERVICE SUPPORT

1 log spt coy

Naval Element 80

EQUIPMENT BY TYPE

PATROL AND COASTAL COMBATANTS 7

PB 7: 2 *Albatros*; 2 *Dili* (ex-ROK); 2 *Shanghai* II; 1 *Kamenassa* (ex-ROK *Chamsuri*)

Air Component

EQUIPMENT BY TYPE

AIRCRAFT • TPT • Light 1 Cessna 172

Vietnam VNM

Vietnamese Dong d		2019	2020	2021
GDP	d	7654tr	7972tr	
	US$	330bn	341bn	
per capita	US$	3,416	3,498	
Growth	%	7.0	1.6	
Inflation	%	2.8	3.8	
Def bdgt	d	ε121tr	ε133tr	
	US$	ε5.22bn	ε5.68bn	
FMA (US)	US$	12.0m	45.0m	10.9m
US$1=d		23227.32	23406.12	

Real-terms defence budget trend (US$bn, constant 2015)

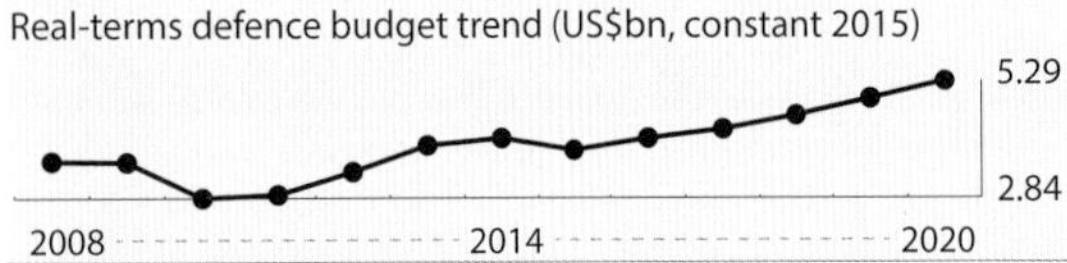

Population	98,721,275

Age	0–14	15–19	20–24	25–29	30–64	65 plus
Male	11.9%	4.0%	3.9%	4.5%	23.1%	2.7%
Female	10.7%	3.6%	3.7%	4.2%	23.4%	4.2%

Capabilities

Vietnam has a stronger military tradition and its armed forces have more operational experience than any of its neighbours. Its defence efforts and armed forces also benefit from broad popular support, particularly in the context of tensions with China over conflicting claims in the South China Sea. Vietnam adopted a new Law on National Defence in 2018 and issued a defence white paper in 2019. The latter several times referred to Vietnam's differences with China and the need for both sides to 'put more effort into maintaining stability'. It is evident, from Hanoi's perspective, that stability will depend in good measure on Vietnam bolstering its deterrent capabilities in the South China Sea. While Russia continues to be the dominant defence supplier, Washington lifted its arms embargo on Vietnam in 2016, and New Delhi and Seoul are understood to be seeking inroads into Vietnam's defence market. Recapitalisation efforts have focused on the navy and air force, mainly with a view to disputes in the Spratly Islands. Long-expected orders for major equipment such as new combat aircraft and maritime-patrol aircraft have failed to materialise, though in April 2020 Vietnam ordered a Japanese-produced satellite-based surveillance system. In addition, the coastguard's capabilities are being enhanced. Vietnam is developing its limited defence-industrial capacities, with the launch of a defence-focused subsidiary to state-owned Viettel Military Industry and Telecoms Group. Called Viettel High Technology Industries Corporation, it will focus on defence electronics and communications.

ACTIVE 482,000 (Army 412,000 Navy 40,000 Air 30,000) Paramilitary 40,000

Conscript liability 2 years army and air defence, 3 years air force and navy, specialists 3 years, some ethnic minorities 2 years

RESERVES Paramilitary 5,000,000

ORGANISATIONS BY SERVICE

Space

EQUIPMENT BY TYPE

SATELLITES • ISR 1 VNREDSat

Army ε412,000

8 Mil Regions (incl capital)

FORCES BY ROLE

COMMAND

4 corps HQ

SPECIAL FORCES

1 SF bde (1 AB bde, 1 demolition engr regt)

MANOEUVRE

Armoured

9 armd bde

Mechanised

2 mech inf div

Light

23 inf div

SURFACE-TO-SURFACE MISSILE

1 SRBM bde

COMBAT SUPPORT

13 arty bde

1 arty regt

11 engr bde

1 engr regt

1 EW unit

3 sigs bde

2 sigs regt

COMBAT SERVICE SUPPORT

9 economic construction div

1 log regt

1 med unit

1 trg regt

AIR DEFENCE

11 AD bde

Reserve

FORCES BY ROLE

MANOEUVRE

Light

9 inf div

EQUIPMENT BY TYPE

ARMOURED FIGHTING VEHICLES

MBT 1,379: 45 T-34; 850 T-54/T-55; 70 T-62; 64 T-90S; 350 Type-59;

LT TK 620: 300 PT-76; 320 Type-62/Type-63

RECCE 100 BRDM-1/BRDM-2

IFV 300 BMP-1/BMP-2

APC 1,380+

APC (T) 280+: Some BTR-50; 200 M113 (to be upgraded); 80 Type-63

APC (W) 1,100 BTR-40/BTR-60/BTR-152

ENGINEERING & MAINTENANCE VEHICLES

AEV IMR-2

ARV BREM-1M

VLB TMM-3

ANTI-TANK/ANTI-INFRASTRUCTURE

MSL • MANPATS 9K11 *Malyutka* (AT-3 *Sagger*); 9M14 mod

RCL 75mm Type-56; **82mm** Type-65 (B-10); **87mm** Type-51

GUNS

SP 100mm SU-100; **122mm** SU-122

TOWED 100mm T-12 (arty); M-1944

ARTILLERY 3,040+

SP 30+: **122mm** 2S1 *Gvozdika*; **152mm** 30 2S3 *Akatsiya*; **175mm** M107

TOWED 2,300: **105mm** M101/M102; **122mm** D-30/Type-54 (M-1938)/Type-60 (D-74); **130mm** M-46; **152mm** D-20; **155mm** M114

MRL 710+: **107mm** 360 Type-63; **122mm** 350 BM-21 *Grad*; **140mm** BM-14

MOR 82mm; **120mm** M-1943; **160mm** M-1943

SURFACE-TO-SURFACE MISSILE LAUNCHERS

SRBM • Coventional *Scud*-B/C

AIR DEFENCE

SAM • Point-defence 9K32 *Strela*-2 (SA-7 *Grail*)‡; 9K310 *Igla*-1 (SA-16 *Gimlet*); 9K38 *Igla* (SA-18 *Grouse*)

GUNS 12,000

SP 23mm ZSU-23-4

TOWED 14.5mm/30mm/37mm/57mm/85mm/100mm

Navy ε40,000 (incl ε27,000 Naval Infantry)

EQUIPMENT BY TYPE

SUBMARINES 8

SSK 6 *Hanoi* (RUS Project 636.1 (Improved *Kilo*)) with 6 533mm TT with 3M14E *Klub*-S (SS-N-30B) LACM/3M54E1/E *Klub*-S (SS-N-27A/B) AShM (*Klub*-S AShM variant unclear)/53-65KE HWT/TEST-71ME HWT

SSW 2 *Yugo* (DPRK)

PATROL AND COASTAL COMBATANTS 75

CORVETTES 12:

FSGM 5:

1 BPS-500 with 2 quad lnchr with 3M24E *Uran*-E (SS-N-25 *Switchblade*) AShM, 1 9K32 *Strela*-2M (SA-N-5 *Grail*) SAM (manually operated), 2 twin 533mm TT, 1 RBU 1600 A/S mor, 1 AK630 CIWS, 1 76mm gun

2 *Dinh Tien Hoang* (RUS *Gepard* 3.9 (Project 11661E)) with 2 quad lnchr with 3M24E *Uran*-E (SS-N-25 *Switchblade*) AShM, 1 3M89E *Palma* (*Palash*) CIWS with *Sosna*-R SAM (CADS-N-2), 2 AK630M CIWS, 1 76mm gun, 1 hel landing platform

2 *Tran Hung Dao* (RUS *Gepard* 3.9 (Project 11661E)) with 2 quad lnchr with 3M24E *Uran*-E (SS-N-25 *Switchblade*), 1 3M89E *Palma* (*Palash*) CIWS with *Sosna*-R SAM (CADS-N-2), 2 twin 533mm TT with SET-53M HWT, 2 AK630M CIWS, 1 76mm gun, 1 hel landing platform

FSG 1 *Po Hang* (Flight III) (ex-ROK) with 2 quad lnchr with 3M24E *Uran*-E (SS-N-25 *Switchblade*) AShM, 2 76mm guns

FS 6:

3 Project 159A (ex-FSU *Petya* II) with 1 quintuple 406mm ASTT, 4 RBU 6000 *Smerch* 2 A/S mor, 2 twin 76mm gun

2 Project 159AE (ex-FSU *Petya* III) with 1 triple 533mm ASTT with SET-53ME HWT, 4 RBU 2500 *Smerch* 1 A/S mor, 2 twin 76mm gun

1 *Po Hang* (Flight III) (ex-ROK) with 2 76mm guns

PCFGM 12:

4 Project 1241RE (*Tarantul* I) with 2 twin lnchr with P-15 *Termit*-R (SS-N-2D *Styx*) AShM, 1 quad lnchr with 9K32 *Strela*-2M (SA-N-5 *Grail*) SAM (manually operated), 2 AK630M CIWS, 1 76mm gun

8 Project 12418 (*Tarantul* V) with 4 quad lnchr with 3M24E *Uran*-E (SS-N-25 *Switchblade*) AShM, 1 quad lnchr with 9K32 *Strela*-2M (SA-N-5 *Grail*) SAM (manually operated), 2 AK630M CIWS, 1 76mm gun

PCO 7: 1 Project FC264; 6 TT-400TP with 2 AK630M CIWS, 1 76mm gun

PCC 6 *Svetlyak* (Project 1041.2) with 1 AK630M CIWS, 1 76mm gun

PBFG 8 Project 205 (*Osa* II) with 4 single lnchr with P-20U (SS-N-2B *Styx*) AShM

PBFT 1+ *Shershen*† (FSU) with 4 single 533mm TT

PH 2 *Shtorm* (ex-FSU Project 206M (*Turya*))† with 1 twin 57mm gun

PHT 3 *Shtorm* (ex-FSU Project 206M (*Turya*))† with 4 single 533mm TT with 53-65KE HWT, 1 twin 57mm gun

PB 20: 14 *Zhuk* (*Grif*-M)†; 4 *Zhuk* (mod); 2 TP-01

PBR 4 *Stolkraft*

MINE WARFARE • MINE COUNTERMEASURES 13

MSO 2 *Akvamaren* (Project 266 (*Yurka*))

MSC 4 *Sonya* (Project 1265 (*Yakhont*))

MHI 2 *Korund* (Project 1258 (*Yevgenya*))

MSR 5 K-8

AMPHIBIOUS

LANDING SHIPS 7

LST 2 *Tran Khanh Du* (ex-US LST 542) with 1 hel landing platform (capacity 16 Lt Tk/APC; 140 troops)

LSM 5:

1 *Polnochny* A (capacity 6 Lt Tk/APC; 200 troops)

2 *Polnochny* B (capacity 6 Lt Tk/APC; 200 troops)

2 *Nau Dinh*

LANDING CRAFT • LCM 13

8 LCM 6 (capacity 1 Lt Tk or 80 troops)

4 LCM 8 (capacity 1 MBT or 200 troops)
1 VDN-150

LOGISTICS AND SUPPORT 17
AGS 1 *Tran Dai Nia* (Damen Research Vessel 6613)
AH 1 *Khanh Hoa* (*Truong Sa* mod)
AKSL 10+
AP 1 *Truong Sa*
AT 2
AWT 1
AXS 1 *Le Quy Don*

Naval Infantry ε27,000

EQUIPMENT BY TYPE
ARMOURED FIGHTING VEHICLES
LT TK PT-76; Type-63
APC • APC (W) BTR-60

Coastal Defence

FORCES BY ROLE
COASTAL DEFENCE
3 AShM bde
1 coastal arty bde

EQUIPMENT BY TYPE
COASTAL DEFENCE • AShM 4K44 *Redut* (SSC-1B *Sepal*); 4K51 *Rubezh* (SSC-3 *Styx*); K-300P *Bastion*-P (SSC-5 *Stooge*)
ARTILLERY • MRL 160mm AccuLAR-160; **306mm** EXTRA

Navy Air Wing

FORCES BY ROLE
ASW/SAR
1 regt with H225; Ka-28 (Ka-27PL) *Helix* A; Ka-32 *Helix* C

EQUIPMENT BY TYPE
AIRCRAFT • TPT • Light 6 DHC-6-400 *Twin Otter*
HELICOPTERS
ASW 10 Ka-28 *Helix* A
TPT • Medium 4: 2 H225; 2 Ka-32 *Helix* C

Air Force 30,000

3 air div, 1 tpt bde

FORCES BY ROLE
FIGHTER/GROUND ATTACK
3 regt with Su-22M3/M4/UM *Fitter* (some ISR)
1 regt with Su-27SK/Su-27UBK *Flanker*
1 regt with Su-27SK/Su-27UBK *Flanker*; Su-30MK2 *Flanker*
2 regt with Su-30MK2 *Flanker*
TRANSPORT
2 regt with An-2 *Colt*; An-26 *Curl*; Bell 205 (UH-1H *Iroquois*); Mi-8 *Hip*; Mi-17 *Hip H*; M-28 *Bryza*; C295M
TRAINING
1 regt with L-39 *Albatros*
1 regt with Yak-52
ATTACK/TRANSPORT HELICOPTER
2 regt with Mi-8 *Hip*; Mi-17 *Hip* H; Mi-171; Mi-24 *Hind*
AIR DEFENCE
6 AD div HQ
2 SAM regt with S-300PMU1 (SA-20 *Gargoyle*)
2 SAM regt with *Spyder*-MR
3 SAM regt with S-75 *Dvina* (SA-2 *Guideline*)
4 SAM regt with S-135-2TM *Pechora* (SA-26)
5 ADA regt

EQUIPMENT BY TYPE
AIRCRAFT 72 combat capable
FGA 72: 26 Su-22M3/M4/UM *Fitter* (some ISR); 6 Su-27SK *Flanker*; 5 Su-27UBK *Flanker* B; 35 Su-30MK2 *Flanker* G
TPT • Light 24: 6 An-2 *Colt*; 12 An-26 *Curl*; 3 C295M; 1 M-28 *Bryza*; 2 C-212 *Aviocar* (NC-212i)
TRG 47: 17 L-39 *Albatros*; 30 Yak-52
HELICOPTERS
MRH 6 Mi-17 *Hip* H
TPT 28: **Medium** 17: 14 Mi-8 *Hip*; 3 Mi-171; **Light** 11 Bell 205 (UH-1H *Iroquois*)
AIR DEFENCE
SAM 12+:
Long-range 12 S-300PMU1 (SA-20 *Gargoyle*)
Medium-range S-75 *Dvina* (SA-2 *Guideline*); S-125-2TM *Pechora* (SA-26), *Spyder*-MR
Short-range 2K12 *Kub* (SA-6 *Gainful*);
Point-defence 9K32 *Strela*-2 (SA-7 *Grail*)‡; 9K310 *Igla*-1 (SA-16 *Gimlet*)
GUNS 37mm; **57mm**; **85mm**; **100mm**; **130mm**
AIR-LAUNCHED MISSILES
AAM • IR R-60 (AA-8 *Aphid*); R-73 (AA-11A *Archer*); **IR/SARH** R-27 (AA-10 *Alamo*); **ARH** R-77 (AA-12A *Adder*)
ASM Kh-29L/T (AS-14 *Kedge*); Kh-59M (AS-18 *Kazoo*)
AShM Kh-31A (AS-17B *Krypton*)
ARM Kh-28 (AS-9 *Kyle*); Kh-31P (AS-17A *Krypton*)

Paramilitary 40,000+ active

Border Defence Corps ε40,000

Coast Guard

EQUIPMENT BY TYPE
PATROL AND COASTAL COMBATANTS 72+
PSOH 1 *Hamilton* (ex-US) with 1 76mm gun (capacity 1 med hel)
PSO 4 DN2000 (Damen 9014)
PCO 13+: 1 *Mazinger* (ex-ROK); 9 TT-400; 3+ other
PCC 2 *Hae Uri* (ex-ROK)
PBF 26: 24 MS-50S; 2 *Shershen*
PB 26: 1 MS-50; 12 TT-200; 13 TT-120
LOGISTICS AND SUPPORT 5
AFS 1
ATF 4 Damen Salvage Tug
AIRCRAFT • MP 5 C-212-400 MPA

Local Forces ε5,000,000 reservists

Incl People's Self-Defence Force (urban units) and People's Militia (rural units); comprises static and mobile cbt units, log spt and village protection pl; some arty, mor and AD guns; acts as reserve

DEPLOYMENT

CENTRAL AFRICAN REPUBLIC: UN • MINUSCA 6
SOUTH SUDAN: UN • UNMISS 68; 1 fd hospital

Arms procurements and deliveries – Asia

Significant events in 2020

FEBRUARY — INDONESIA OFFSETS AND CONSOLIDATIONS

Indonesia announced plans to ensure that local manufacturing capacity was being developed in line with Law 16 (enacted in 2012), which mandates the government to engage in various offset policies, especially technology and knowledge transfer. To boost the 2020–24 defence-industrial development plan, within the framework of Law 16, in August the government announced plans to localise the production of defence components by up to 50% by 2024 by consolidating four state-owned companies: electronics company PT Len Industri, aerospace company PT Dirgantara Indonesia, armoured-vehicle producer PT Pindad and munitions manufacturer PT Dahana.

FEBRUARY — INDIA–ISRAEL COOPERATION CONTINUES

A number of collaboration agreements were agreed between India and Israel. Bharat Electronics Limited and Israel Aerospace Industries (IAI) signed a memorandum of understanding (MoU) covering technical support and maintenance for India's air-defence systems. Hindustan Aeronautics Limited (HAL) and Dynamatic Technologies Limited signed an agreement with IAI to collaborate on uninhabited aerial vehicle (UAV) production in India. Elbit Systems and HAL signed an MoU to develop and produce vertical take-off and landing UAVs for the Indian armed forces. India already operates a range of Israeli-designed equipment, including the *Barak*-8 surface-to-air missile system and the *Heron* UAV.

AUGUST — KOREAN CARRIER PLANS

In August, the Republic of Korea announced a Mid-Term Defense Plan for 2021–25 which included plans for the Korean Light Aircraft Carrier project (formerly known as the LPX-II project). The project will begin in 2021, and the aim is that the carrier enters service in 2033. In October 2019, the Defense Acquisition Program Administration awarded Hyundai Heavy Industries (HHI) a contract for a conceptual design of a platform of some 30,000 tons' displacement. This design is due to be completed by the end of 2020. Daewoo Shipbuilding & Marine Engineering (DSME) has previously proposed an LPX-II design with a displacement of 70,000 tons. The acquisition cost estimated by HHI is ₩3.1tr (US$2.64bn), while DSME estimates ₩5.4tr (US$4.6bn).

SEPTEMBER — INDIA INDIGENOUS CONTENT

India's new Defence Acquisition Procedure (DAP), formerly known as Defence Procurement Procedure (DPP), was released after a fourth amendment by the government. This replaces the last DPP issued in 2016. The DAP builds on efforts to further increase 'indigenous manufacturing and [reduce] timelines for procurement of defence equipment'. The plan is to increase, from 10% to about 50%, the amount of local content in various procurement categories, in order to support the government's 'Make in India' initiative. However, equipment acquired through government-to-government agreements, like the *Apache* attack helicopter and the *Rafale* fighter, will no longer require offsets, though they will still be a feature of competitions. The government hopes that the new policy will simplify the acquisition procedure, although it is unclear that the latest legislation will do anything to speed up the decision-making process, a key issue for Indian procurement. Also included in the legislation is an embargo on the importation of 101 pieces of military equipment, including artillery and transport aircraft; India has imported both recently.

OCTOBER — JAPAN F-X FIGHTER PROGRAMME

Japan's Ministry of Defense (JMoD) announced that it had contracted Mitsubishi Heavy Industries (MHI) to lead the development of the F-X indigenous fighter-jet programme. F-X is intended to replace Japan's 90 F-2 fighter aircraft from the mid-2030s. MHI built the F-2, a Japanese version of the F-16, in the 1990s and 2000s, and more recently conducts final assembly of F-35 aircraft at its facility at Nagoya. The first flight of the MHI X-2 Advanced Technology Demonstrator aircraft, which will inform much of the F-X development work, took place in April 2016; which likely informed the decision to award the contract to MHI. In August the JMoD invited foreign companies to participate in the programme, and there are ongoing negotiations with the UK and US governments about their potential involvement. As of late 2020, the total amount allocated for the F-X project has now reached ¥28bn (US$258.78m) including ¥11.1bn (US$102.58m) which will be spent in the conceptual-design process. A first F-X prototype is planned to be built by 2024 with full-rate production to start from 2031.

Figure 18: China: PLAA mechanised formation equipment modernisation

Since 2017, the People's Liberation Army Army (PLAA) has been transforming its manoeuvre forces into combined arms brigades in order to improve its ability to conduct joint operations. These brigades can be categorised by assigned equipment as either 'heavy', with tanks and other tracked armoured vehicles, 'medium', with wheeled armoured vehicles, or 'light', typically with 4x4s for increased mobility. However, acquisition of the vehicles that are now at the core of these formations began in the mid-2000s, when the PLA was creating its first modern mechanised force as part of its overall plan to achieve basic mechanisation by 2020. Approximately 2,500 ZBD-04 and improved ZBD-04A tracked infantry fighting vehicles have been produced to date and several variants have been fielded. Perhaps inspired by the US Army's *Stryker* family, the ZBL-08 wheeled armoured vehicle entered service in 2009 and has since been developed into at least 13 other variants. These vehicles are significant improvements on their predecessors. The ZBL-08 is replacing the ZSL-92, which was designed in the 1980s and is more lightly armed and armoured than its replacement. Likewise, the ZBD-04A will likely replace the ZBD-86, which was first produced in the 1980s and is a copy of the 1960s-era Soviet BMP-1.

Heavy mechanised units

ZBD-04

Manufacturer: Beijing North Vehicle Group

ISD: mid-2000s (ZBD-04)
late 2000s (ZBD-04A)

Quantity produced to date:
- 400 ZBD-04
- 1,900 ZBD-04A

Replacing: ZBD-86 and ZSD-63

ZBD-04A

Observed variants:
- Infantry fighting vehicle (base)
- Command post
- 120mm gun/mortar (PPZ-10)
- HJ-10 anti-tank missile launcher
- Reconnaissance
- Armoured recovery vehicle
- Ambulance

Medium mechanised units

ZBL-08

Manufacturer: Inner Mongolia First Machinery Group

ISD: 2009

Quantity produced to date:
- 2,000 ZBL-08
- 950 ZTL-11
- 500 ZSL-10
- 350 PLL-09

Replacing: ZSL-92A and ZSL-93

Exports:
- Gabon
- Nigeria
- Thailand
- Venezuela

ZTL-11

Observed variants of ZBL-08:
- Infantry fighting vehicle (base)
- Wheeled asslt gun (ZTL-11)
- Armoured personnel carrier (ZSL-10)
- 122mm howitzer (PLL-09)
- Command post
- Reconnaissance
- SATCOM
- Armoured engineering vehicle
- Armoured recovery vehicle
- Mine warfare
- Vehicle launched bridge
- Ambulance
- Mine clearance
- Electronic warfare

Figure 19: India: major surface-combatant programmes at state-owned shipyards

In 2005, the Indian Navy outlined a plan to develop a 160-ship navy. Several surface-combatant programmes approved in previous years formed a key part of the planned expansion. Although most of those vessels have now entered into service, the projects were dogged by significant delays and higher-than-anticipated costs. Indian procurement programmes have to be approved by a government committee before production contracts are signed, and the project cost is estimated for the committee. However, a 2010 government auditor report found the navy's estimates had not been rigorous enough. For example, the cost estimate for the Project 15A destroyer was approved before the weapons and sensor fit had been finalised. Costs for the Project 15A programme, as well as for the Project 17 and Project 28 vessels, were estimated based on previous-year prices. This meant that the estimates were significantly lower than actual costs. Disagreements over scheduling and labour costs meant that the actual production contracts were signed after the first platform was supposed to have been delivered. Moreover, changes to platform designs, weapons and sensors, after production had already begun, added to the delay and the cost.

Although it remains unclear how much procurement procedure has improved since these ships entered service, production contracts for the follow-on Project 15B destroyers and 17A frigates – signed in 2011 and 2015 respectively – appear more comparable with the cost of other complex warships. While Project 17A appears to be on schedule, Project 15B is not, in part due to a fire on the lead ship in 2019 as well as delays in acquiring critical subsystems.

Project 15A *Kolkata*-class destroyer

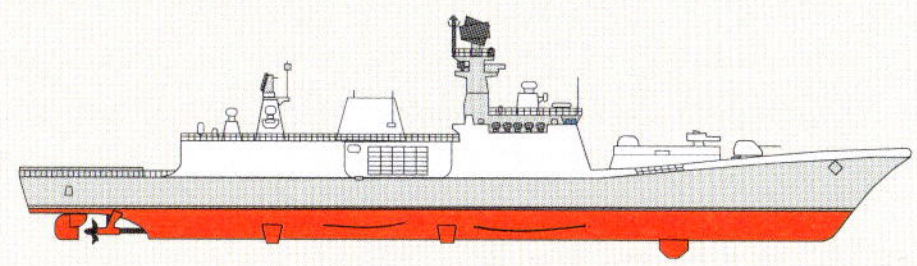

Project 17 *Shivalik*-class frigate

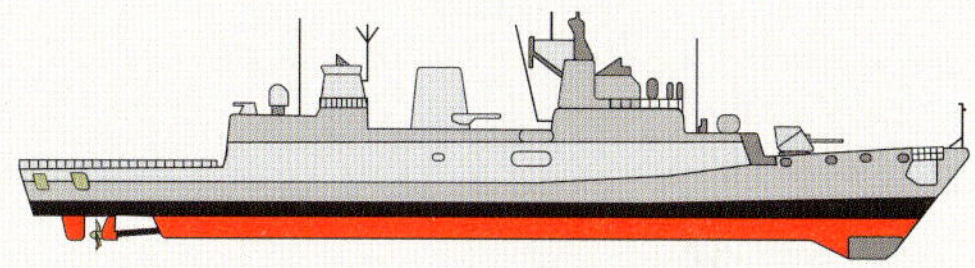

Project 28 *Kamorta*-class frigate

Indian surface-combatant construction

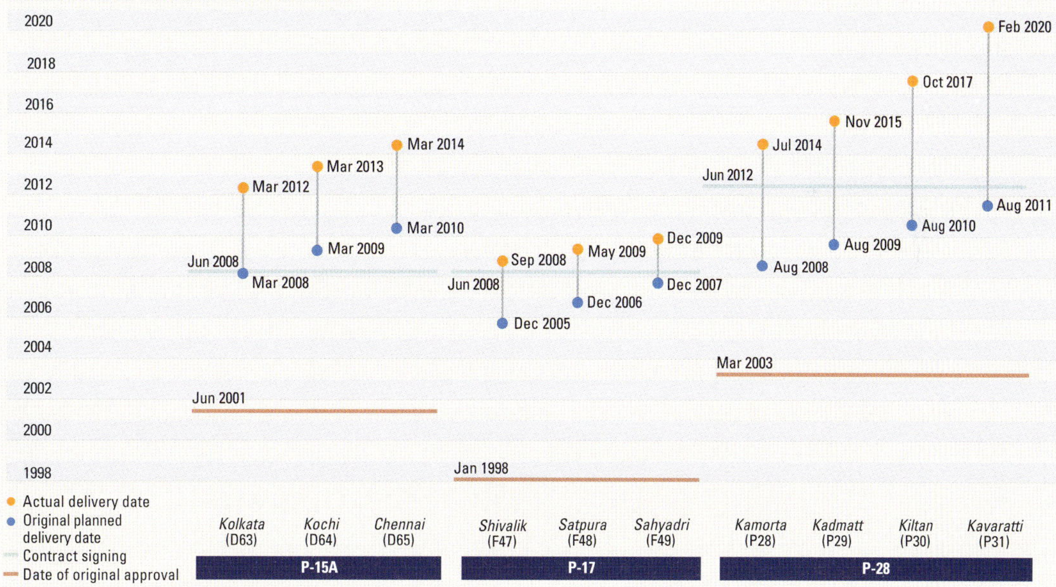

Indian Navy cost estimates (ten million (crore) rupees)

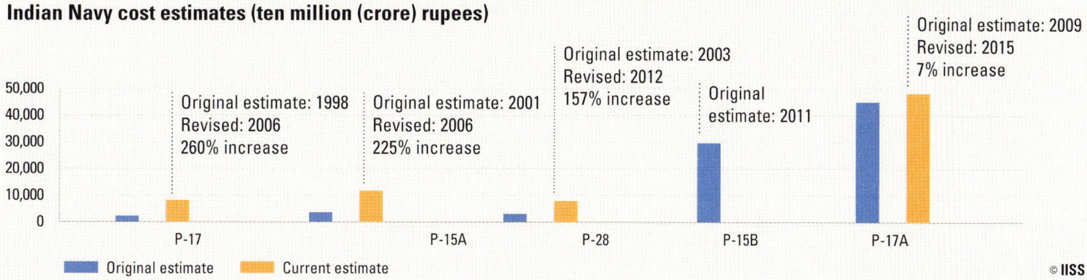

Chapter Seven

Middle East and North Africa

- Turkey's armed forces continued operations in Syria, conducting airstrikes and ground operations in early 2020 to halt government advances in Idlib province. Turkey's military capabilities were also deployed in Libya, supporting the Tripoli-based recognised government. Meanwhile, General Khalifa Haftar's forces obtained support from Egypt, Russia and the UAE. External actors helped improve the military potential of warring parties, and their sustained involvement indicated growing operational capacity.
- Conflict grinds on in Yemen. The Houthis conducted fewer missile and rocket attacks against Saudi Arabia, but they displayed increasing military competence, such as through the use of uninhabited aerial vehicles and cruise missiles sourced from Iran.
- The UN embargo on conventional-arms sales to Iran ended in October 2020, though other embargoes remain in place. Initial modernisation may focus on air defences and missile capabilities, as economic challenges make it unlikely that Iran will comprehensively recapitalise an inventory that for the air force and navy, for example, generally relies on combat platforms acquired in the 1970s.
- The UAE–Israel normalisation agreement heralded a possible acceleration in defence and technological cooperation between the two states and also seemed to unlock greater defence cooperation with the US. Shortly afterwards, defence sales from the US were mooted including F-35s, though Washington's intent to maintain Israel's Qualitative Military Edge raised questions not just about the capability of the sensors and weapons in any such sale, but also about the level of sophistication likely in current and future US military sales to Israel.
- Middle East regional missile-defence concerns were again underscored when Iran struck two US bases in Iraq in January 2020 using short-range ballistic missiles. Saudi Arabia, meanwhile, continued to be the target of missile attacks from opposition forces in Yemen.

Saudi Arabia, real-terms defence budget trend, 2010–20 (US$bn, constant 2015)*

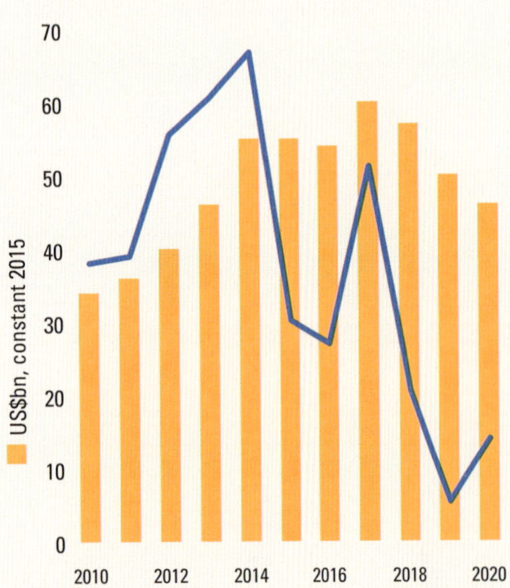

Note: Defence budget only – excludes security expenditure (Ministry of the Interior etc.)

Active military personnel – top 10
(25,000 per unit)

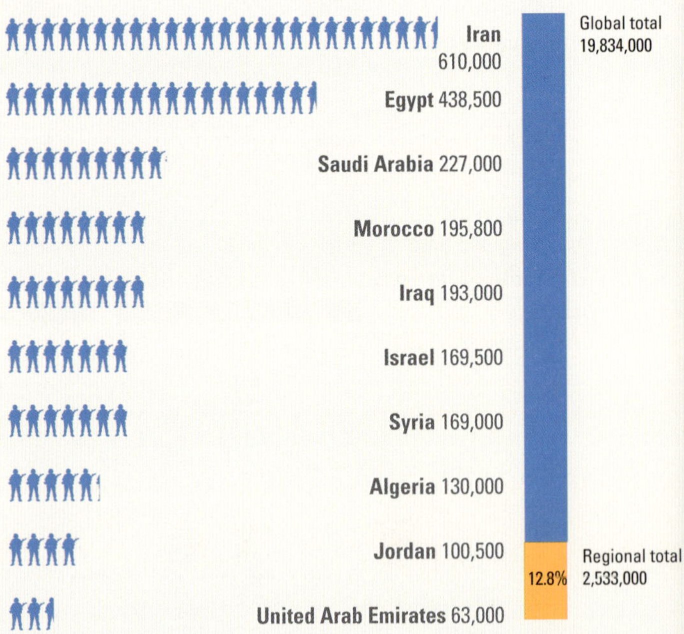

GCC combat aircraft orders and deliveries

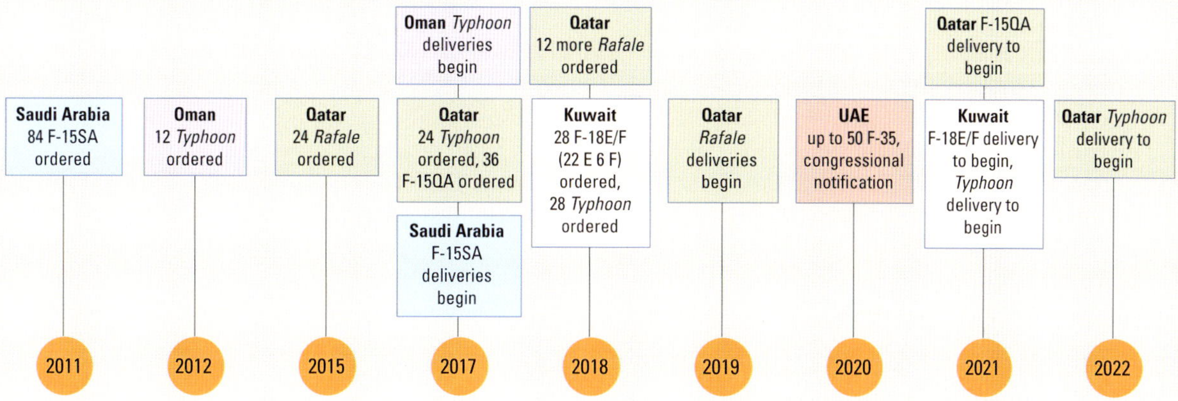

Selected GCC air and missile defence

	System	Status	Launcher quantity
Bahrain	*Patriot* PAC-3	DSCA request	up to 9
Kuwait	*Patriot* PAC-2	Being upgraded to PAC-3	40
Qatar	*Patriot* PAC-3	Deliveries ongoing	10+
	THAAD	DSCA request	up to 12
Saudi Arabia	*Patriot* PAC-2/-3	In service	108
	THAAD	On order	44
United Arab Emirates	*Patriot* PAC-3	In service	n.k.
	THAAD	In service	12

Middle East and North Africa: selected amphibious capabilities

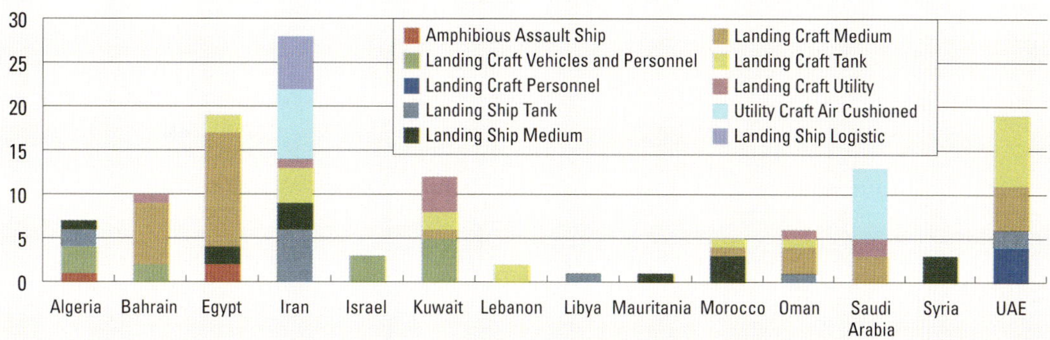

Middle East and North Africa

The Middle East and North Africa saw continued conflict in 2020, including overt intervention by regional powers and the deployment of increasingly sophisticated weapons systems on key battlefields.

Turkey's military interventions

In October 2019, Turkey conducted an operation in northeast Syria targeting the Democratic Union Party (PYD), a Kurdish organisation affiliated with the Kurdistan Workers' Party (PKK), which ended in the seizure of a large pocket of territory in Syria. Turkey deployed its own troops as well as Syrian militia and also used airpower against the Kurdish forces.

In February 2020, Turkey clashed for the first time with Syrian government forces that were conducting an operation to seize the rebel-held pocket of Idlib. After losing a number of soldiers in an airstrike, Turkey launched an air and artillery campaign, and deployed personnel alongside Syrian militia partners, in a bid to halt and reverse the Syrian operation. Airstrikes from Turkish aircraft and uninhabited aerial vehicles (UAVs) destroyed a number of armoured vehicles and troop concentrations, while the deployment of air defences prevented the Syrian and Russian air forces from providing cover to ground forces. By March, the Syrian campaign had ground to a halt and a Russian–Turkish ceasefire came into effect. At the time of writing, the ceasefire was still in place, with joint Russian–Turkish patrols along the front line, though its durability was also partly the result of enduring equipment and personnel shortages among the Syrian government forces.

Turkey also deployed military capabilities to Libya in early 2020, where its intervention helped

Libya: increasing involvement by foreign forces

A growing number of foreign forces delivered military assistance to warring parties in Libya in 2020. This came despite renewed calls – at the 19 January 2020 Berlin Conference on Libya – for a ceasefire and that the UN arms embargo should be implemented. In January, Turkey deployed personnel to Tripoli to provide training and support to forces loyal to the Government of National Accord (GNA). Also, the US Department of Defense's Inspector General reported that Turkey had used charter flights to move around 3,800 Syrian combatants into Libya. Meanwhile, Turkish naval vessels have been deployed on several occasions to shield cargo ships from inspection by the EU Naval Force Mediterranean's IRINI mission. There are reports that Turkey has delivered equipment to GNA forces including armoured personnel carriers, self-propelled artillery pieces and 35mm anti-aircraft guns.

Similarly, the United Arab Emirates has bolstered its military support for General Khalifa Haftar's Libyan Arab Armed Forces in 2020. A number of UAE-registered transport aircraft (An-124 *Condors*, Il-76 *Candids* and C-17A *Globemasters*) have been observed and reports have asserted that the UAE used LAAF-controlled Benina airport and Al-Khadim air base as well as Sidi Barrani air base in Egypt to deliver military support to Haftar's forces. In June, Egyptian President Abdel Fattah al-Sisi reportedly said Egypt had a 'legitimate right' to intervene in Libya. Although the weight of Egypt's army deployments remain in the east of the country, the Egyptian Air Force has renovated its forward desert base at Habata close to the Libyan border, deploying in 2020 several *Rafale*, *Mirage* 2000 and MiG-29M *Fulcrum* combat aircraft.

Moreover, 2020 has seen growing Russian involvement. By June, US AFRICOM reported the presence in Libya of 3,000 combatants associated with the Wagner Group, a Russian private military company. Between May and July, AFRICOM released satellite imagery indicating that at least 14 MiG-29 combat aircraft had been flown from Russia to Syria, where Russian markings were reportedly painted over in an attempt to conceal their origin. The aircraft were then flown into Libya, in contravention of the UN arms embargo. Other Russian-origin equipment identified by AFRICOM included at least two Su-24 combat aircraft, one Il-76 heavy transport aircraft and at least one 96K6 *Pantsir* (SA-22 *Greyhound*) short-range surface-to-air missile system at the LAAF-controlled Al-Khadim air base, south of Sirte.

the beleaguered Tripoli-based Government of National Accord and its allied militias to repel the forces of General Khalifa Haftar, which had been besieging the capital since April 2019. Haftar had obtained military assistance from Egypt, but also most significantly from Russia and the United Arab Emirates (UAE). The support from these states had included air support, armed UAVs, personnel from private military companies, anti-tank and anti-aircraft systems, and training. Russian and UAE aircraft and UAVs were spotted at airfields in Libya. Turkey reciprocated by reportedly deploying advisers and advanced weapons systems, notably air defences and armed UAVs, as well as Syrian rebel fighters. This influx led to an uneasy stalemate on an axis running from Sirte to Jufra. Though it was forbidden by UN resolutions, the persistent provision of weaponry had the effect of frustrating efforts at conflict mediation, as did the increasingly complex mosaic of forces in Libya, both local and foreign, and state and non-state actors. Meanwhile, though their deployments may have been modest in scale, regional actors were able to sustain their presence, pointing to yet more progress in the development of their military capabilities.

Regional complexities

Fighting in Yemen between the Houthi forces and the anti-Houthi coalition remained intense, notably in Ma'rib province where the former made considerable gains. Meanwhile, Saudi Arabia reduced the scale of its military operations, while there was a draw-down in the UAE's contribution. Moreover, fissures appeared in the anti-Houthi coalition, leading to fighting between secessionists and government forces in Aden and elsewhere. While the Houthis conducted fewer missile and rocket attacks against Saudi Arabia, they remained intent on seizing territory and displayed increasing military competence, such as through the use of UAVs and more precise rockets. Meanwhile, the Yemeni population have had to grapple with the effects not just of continuing conflict, a collapsed economy and persistent food shortages, but also the coronavirus pandemic.

In the Gulf and Iraq, escalating tension between the US and Iran turned into direct confrontation. Throughout 2019, Iran had conducted operations designed to increase the cost to US partners, particularly Saudi Arabia and the UAE, in supporting the US strategy of maximum pressure on Iran. These actions involved targeting and seizing tankers and shooting down a US UAV in the Gulf. In September 2019, an Iranian attack combining UAVs and missiles hit the Abqaiq oil-processing facility and the Khurais oil-field in Saudi Arabia, taking offline for several weeks around 50% of Saudi oil production. This attack demonstrated not only Iranian skill and technological ability, but also the challenge of delivering effective air defence in the region. On this occasion, the US declined to retaliate against Iran, leading some observers in the Gulf to question the reliability of the US as a security guarantor.

The most dangerous escalation in the region happened in Iraq. In response to rocket attacks by Iranian-backed militias against US facilities, the US mounted a large-scale strike against militia compounds. In turn, militias mounted an attack that killed a US national and besieged the US embassy in Baghdad. In early January, the Trump administration ordered a UAV strike against Major General Qasem Soleimani, the head of the Islamic Revolutionary Guard Corps (IRGC) Quds Force and the architect of Iran's regional influence. Although it generated fears of wider attacks, if anything the killing of Soleimani served to illustrate the constraints on each actor: Iran sought revenge but without risking all-out war; likewise, the US showed no interest in a direct conflict. A few days after Soleimani's killing, Iran conducted a retaliatory missile strike against US facilities located at Iraq's al-Asad air base. While around a hundred personnel were injured, no troops were killed and the damage was limited. This allowed both Iran and the US to de-escalate. That said, later in 2020, Iranian pressure increased against US patrols and facilities in Iraq, in the form of harassment and attacks by Iraqi militias. Iran's apparent intention was to press the US to downsize its presence in Iraq, ultimately leading to its withdrawal.

The coronavirus pandemic compelled several states to mobilise their armed and security forces as part of their response to the crisis. In many Middle Eastern countries, the armed forces played numerous internal roles, from closing borders and enforcing curfews and lockdowns to logistics support and the supply of humanitarian aid. Military hospitals were also enrolled in efforts to manage the crisis.

More broadly, several developments appeared to herald new regional defence trends. The coronavirus pandemic had a significant effect on Middle Eastern economies, with the region's GDP expected to shrink by 10.2% on average in 2020, according to the IMF in

October. While this might push the region's leaders to reallocate elements of their budgets, the effect on defence budgets may be relatively limited, depending on threat perceptions. For instance, though Saudi Arabia's budget had already been reduced (as the country sought to reduce its fiscal deficit), the Kingdom nonetheless insisted that its defence-modernisation and -industrialisation programmes would proceed as planned. Indeed, no major cancellation of weapons purchases was recorded, and Saudi Arabia seemed intent on upgrading its air defences after the September 2019 Abqaiq attacks. The surprise UAE–Israel normalisation agreement, announced in August, suggested that defence and technological cooperation between the two states would accelerate. Then on 10 November, the US Defense Security Cooperation Agency announced three possible defence sales to the UAE. A common factor that has hitherto hindered procurements of this sort is that they risk eroding Israel's Qualitative Military Edge in the region. When it announced the potential sales, including F-35 combat aircraft, the US reiterated this guarantee.

News that the US would reconsider its adherence to the Missile Technology Control Regime (MTCR) revived hopes in Middle Eastern states that the US would begin sales of armed UAVs; the announcement of a possible MQ-9B sale to the UAE was the first manifestation of Washington's changing policy towards the region. Several countries, such as the UAE, Algeria and Jordan, had turned to China to obtain such capabilities in the past, but complaints about performance and technology surfaced over the years. However, a major concern for the US and its regional allies was the lifting of the UN embargo on arms sales to Iran on 18 October. US attempts to prolong the embargo at the UN Security Council were rebuffed. Both Russia and China appeared ready to supply weapons systems to Iran, but low oil prices and Iran's economic difficulties seemed likely to prevent a quick and comprehensive rearmament programme. Instead, Iran was believed likely to focus on building up its missiles (such as anti-ship systems) and its air defences. That said, Iran's inventory does, more broadly, require recapitalisation. For instance, the air force – like the navy – generally relies on combat platforms originally acquired in the 1970s, but in the absence of funds to procure full fleet replacements, short-term priorities may include upgrades and additional focus on improving weapons packages.

DEFENCE ECONOMICS

Macroeconomics

As 2020 began, the outlook for the region's economies was already dampened by low oil prices that had, since 2016, persisted at levels well below the price needed to balance the budgets (the fiscal-breakeven price) of the region's major economies. Regional real GDP growth was already slow, falling from 2.1% in 2018 to 1.5% in 2019. In 2020, the region therefore just encountered an additional problem when the coronavirus pandemic negatively affected domestic output, while the subsequent drop in global demand caused the oil price to fall to even lower levels. According to the IMF's October 2020 World Economic Outlook, the region's economies are expected to contract by 10.2% in 2020 on average. However, the IMF predicts that there will be an improvement as early as 2021, with projected growth estimated at around 6.7%, mostly due to improving economic prospects for non-Gulf Cooperation Council oil exporters.

However, economic uncertainty will likely continue for as long as the coronavirus pandemic persists, or at least until there are reliable plans for countering the virus. Moreover, further mitigation and containment measures could have an additional effect on economic health, and also potentially exacerbate volatility in global oil markets. There has been a significant fall in oil prices in 2020, amid a collapse in demand, after 'lockdown' measures designed to hinder transmission restricted travel, and a price war between Russia and Saudi Arabia in the first quarter. From an average of US$65 a barrel in 2019, prices fell to an average of just above US$40 a barrel in 2020, hitting a low point in April at just over an average of US$20 a barrel. Although the US Energy Information Administration (EIA) estimates that worldwide crude prices will recover to US$49 a barrel in 2021, this projection is subject to a high level of uncertainty as the pandemic continues to drag down global economic growth.

The economies of GCC states experienced some of the sharpest contractions seen in the region in 2020: their budget deficits are expected to range from 10.6% of GDP in Saudi Arabia to 18.7% in Oman, with only Qatar expected to remain in surplus. Crucially, even accounting for the current EIA projections that there will be a moderate increase in oil prices in 2021, the projected price of US$49 a barrel would not be high enough to balance the budgets in most of these states. According to the IMF, Saudi Arabia's fiscal-

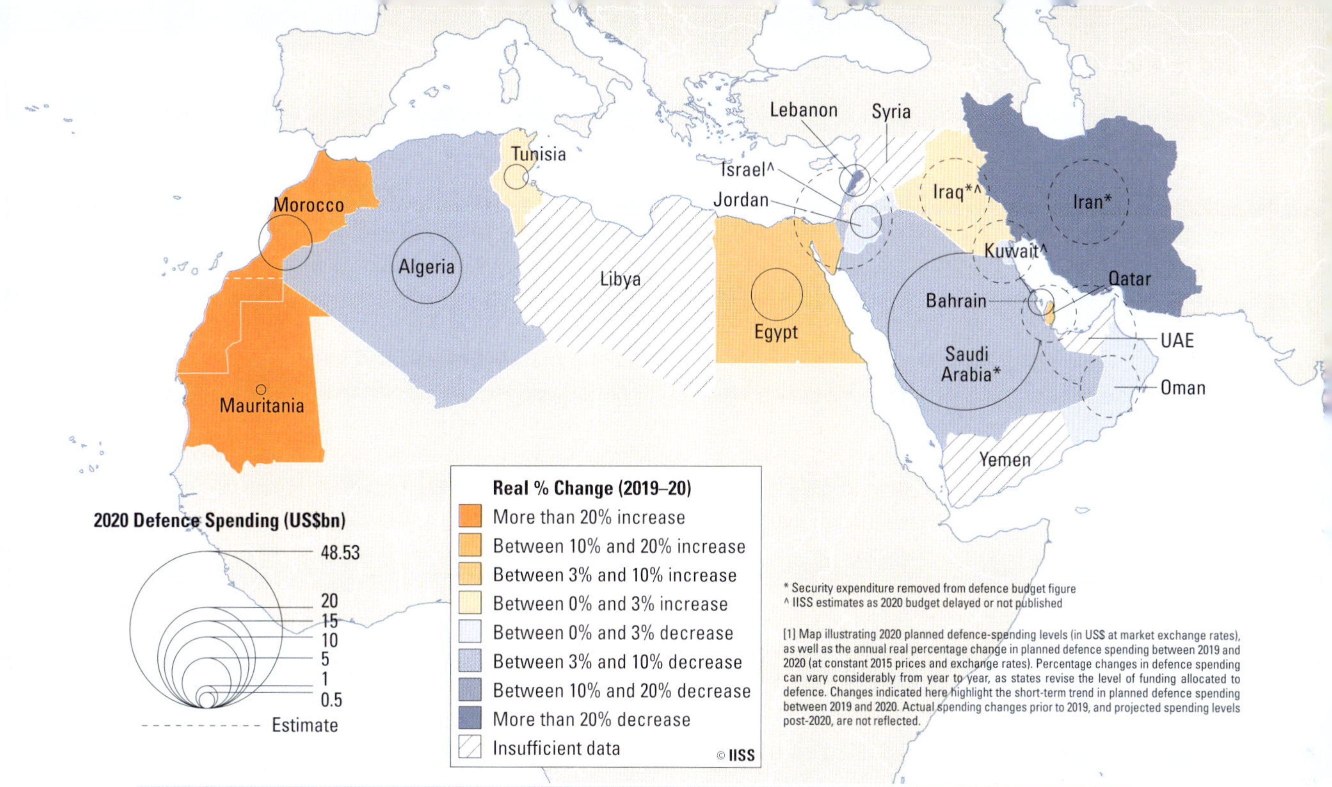

▲ Map 7 **Middle East and North Africa regional defence spending**[1]

breakeven oil price is estimated to be US$78 a barrel in 2020 and US$68 a barrel in 2021. Of greater concern are Bahrain and Oman, where an oil price of US$93 a barrel and US$105 a barrel is needed, respectively, to balance their budgets in 2020. Non-GCC oil exporters will also experience a sharp decline in output in 2020. For example, the October 2020 IMF World Economic Outlook estimated that the economies in Algeria and Iraq would contract by 5.5% and 12.1% respectively in 2020. Completing this gloomy picture, the Libyan economy, crippled by years of conflict, is expected to contract by 67% while Lebanon's economy could also contract by 25%, as the August blast in the port of Beirut exacerbated the country's economic and political crisis.

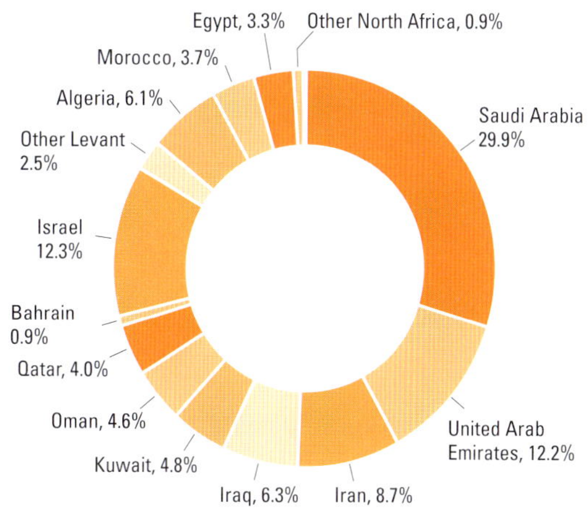

Note: Analysis excludes Libya, Palestinian Authority, Syrian Arab Republic and Yemen. © IISS

▲ Figure 20 **Middle East and North Africa defence spending by country and sub-region, 2020**

Defence-budget trajectories

These economic challenges will influence the trajectory of regional defence budgets, as will uncertainty over the pace of economic recovery. In the short term, budgets may remain relatively stable because 2020 allocations have already been set, and because defence has historically been a priority area for regional governments. That said, it is possible that the pace at which procurements and spending plans are implemented could be influenced by the coronavirus pandemic as well as the effect that low oil prices will have on funding allocations. For GCC countries, another effect might be to slow the distribution of financial aid and military assistance, for example to Egypt.

In the longer run, it is possible that the coronavirus pandemic could impel regional leaders to reallocate elements of broader state budgets in order to tackle the vulnerabilities exposed by the pandemic, perhaps spurring investments in the social and healthcare

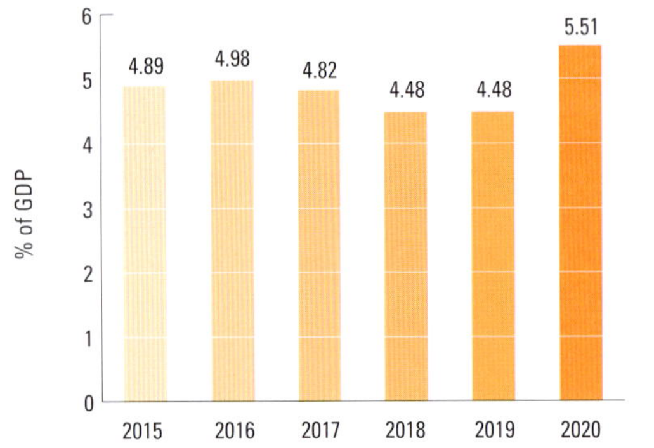

▲ Figure 21 **Middle East and North Africa regional defence spending** as % of GDP (average)

sectors beyond the emergency funds announced in 2020. That said, this might not affect defence budgets as profoundly as might be expected, particularly if policymakers also adopted a broader definition of national security. For instance, this could lead states to reprioritise elements within defence budgets, rather than cut them. In this scenario, regional states could sharpen focus on capabilities including cyber security and artificial intelligence rather than conventional platforms. Nonetheless, where an urgent operational requirement or specific threat emerges, it is likely that regional states will continue to prioritise the funding required.

However, regional states' defence budgets had already contracted in 2019. For instance, Saudi Arabia's defence budget fell from US$58.1 billion in 2018 to US$50.9bn in 2019 in nominal terms (the total defence and security budget fell from US$82.9bn to US$80.5bn), while Iran's defence budget fell from US$18.5bn to US$17.2bn, a 35% reduction when measured in real terms, due to inflation hitting 41% in 2019. Before the economic ramifications of the pandemic became clear, Saudi Arabia had already announced a cut to the 2020 defence budget, down to US$48.5bn (US$77.3bn for defence and security), as the country sought to reduce its fiscal deficit. Iran's defence budget also fell in 2020 to an estimated US$14.1bn, a 26% reduction from 2019 in real terms. Conversely, North Africa witnessed an upward trend in 2019 and most of the non-oil-dependent countries there have managed to avoid implementing cuts in 2020. In June 2019, Morocco, for example, announced that its defence budget would be increased by 21% in real terms in 2020 to US$6bn (US$5.5bn in constant 2015 dollars), building on the more mild growth seen between 2015 and 2019 where the defence budget increased from US$4.3bn to US$4.8bn (US$4.4bn to US$4.6bn in constant 2015 dollars). While Algeria implemented a 3.4% real cut to the 2020 defence budget, increases were evident in Egypt, Tunisia and Mauritania.

Defence procurement and industry

The Middle East and North Africa is home to a number of major importers of defence equipment and has routinely seen orders of significant size and cost. Procurement trends in the Arabian Peninsula, Egypt and Morocco indicate that air-power remains a continued priority, notably the modernisation of combat aircraft and rotary-wing fleets. Examples include the order by the United Arab Emirates (UAE) for 26 AH-64E *Apache* attack helicopters and Morocco's Foreign Military Sales (FMS) contract for 24 AH-64E *Apaches* in 2020 and the FMS notifications for sales/refurbishment of *Apache* helicopters to Qatar in 2019 and to Egypt in 2020. At the end of 2019, Egyptian and US military officers also formally reopened the fighter base at Cairo West Airport, which now operates the Block 52 F-16C/D. The UAE's growing focus on intelligence, surveillance and reconnaissance (ISR) capabilities is exemplified by its contract for two *GlobalEye* airborne early-warning and control aircraft based on Bombardier *Global* 6000 business jets; the UAE signed a support and maintenance agreement with Saab in July 2020. The priority given to airpower reflects the growing force-projection ambitions of regional powers. Recent years have seen some regional air forces add heavy transport aircraft to their inventories, and an increasing number of operational deployments in, for instance, the conflicts in Yemen and also in Libya. One issue that had stymied regional states' procurement ambitions for the latest-generation military equipment from the United States was Washington's determination to maintain Israel's Qualitative Military Edge (QME), though for some states this situation may now have changed after the normalisation of relations between some Gulf states and Israel in late 2020. November saw the US Defense Sales Cooperation Agency announce the possible sale to the UAE of F-35A *Lightning* II combat aircraft as well as a range of other equipment. The first was for munitions, sustainment and support including AIM-120C8 AMRAAM, AGM-154C Joint Stand-off Attack Weapons, AGM-154E Joint Stand-off Attack Weapons–Extended Range and AGM-88E Advanced

Anti-Radiation Guided Missiles. The second announcement covered the possible sale to the UAE of up to 50 F-35A Joint Strike Fighters and related equipment, while the third announcement noted the possible sale of up to 18 weapons-ready MQ-9B *Sky Guardian* uninhabited aerial vehicles. If all options were exercised on these deals, and they were approved by the US authorities, they would cost an estimated US$10bn, US$10.4bn and US$2.97bn respectively. Given that Washington said at the same time that it was determined to retain Israel's QME, the capability of the proposed weapons and sensors remains unclear in relation to Israeli systems. Nonetheless, for regional states facing an Iran that has just emerged from a UN conventional-arms embargo and is – at least in theory, and notwithstanding economic challenges – able to begin recapitalising its ageing conventional capabilities, the arrival of advanced platforms like these would significantly boost their combat capabilities.

For Iran, the UN conventional-arms embargo might have lifted, but other embargos remain. Moreover, Tehran's economic problems mean it is unlikely to rush to procure equipment in large numbers to replace ageing inventories. Moreover, Iran has for some years prioritised the development of asymmetric capabilities. That said, over time Iran's conventional platforms will need to be replaced, including land platforms and its US-origin combat aircraft, the last of which would have been delivered by 1978, and also its surface combatants (its main platforms are the 1960s-designed *Alvand*-class).

There is growing interest in boosting regional maritime security capacities. In 2018, Saudi Arabia placed orders for five *Avante* 2200 frigates from Spain as well as for 39 HSI-32 patrol craft from France. In 2019, the German Bundestag approved the sale of six frigates to Egypt while the UAE signed a deal for two *Gowind* 2500 frigates from France. Importantly, it is planned that at least one of the frigates and 18 of the HSI-32 patrol craft will be built in Egypt and Saudi Arabia respectively. (The UAE is also due to build 24 B-250 light attack aircraft contracted from local firm Calidus in 2019; the aircraft was reportedly designed by Brazilian firm Novaer.) All these point to continued interest in developing the local defence industry in these countries. For the UAE as well as Saudi Arabia and, to a lesser extent, Egypt, local production ambitions are linked to economic diversification plans and – perhaps in the long term – sovereign production capabilities, though establishing or securing the supporting design and supplier base would be an additional task.

EGYPT

Ten years after a wave of popular uprisings swept across the Arab world, upending several regimes, Egypt's armed forces have instead deepened their grip on political and economic life. President Abdel Fattah al-Sisi assumed power in 2013, after the sitting president Muhammad Morsi – elected the year before – was ousted by the armed forces in the wake of widespread protests against him. Sisi had been appointed as the minister of defence by Morsi and had previously been head of military intelligence.

President Sisi has sought to boost national prestige by reinvigorating the armed forces, by procuring advanced weapons and awarding major infrastructure contracts. His analogous, if not immediate, predecessors, Hosni Mubarak and Anwar Sadat, gave the Egyptian armed forces a significant role in the national economy, but also used political parties and other elites to support their rule. President Sisi, by contrast, has appeared to place his faith in fellow military officers, appointing them to key civilian posts and entrusting the armed forces with large infrastructure projects, such as construction of the future seat of Egypt's government, the New Administrative Capital, east of Cairo. That said, President Sisi consolidated power by effectively sidelining prominent potential rivals from within the defence establishment. In 2020, he ratified a parliamentary law that effectively bans any retired military officers from running for elected office without military approval.

However, Egypt's military modernisation is also a response to a regional security environment in which Cairo perceives itself encircled by threats and powerful competitors. Within Egypt itself, there is an ongoing low-intensity conflict in the Sinai Peninsula between the army and an Islamic State affiliate. To the west, continuing instability in Libya has stoked Egyptian fears about the spread of Islamism and Turkey's role in supporting the Tripoli-based Government of National Accord (GNA). Egypt, in contrast, has expressed support for the Libyan Arab Armed Forces, based in the east under General Khalifa Haftar.

A long-running dispute continues between Egypt, Ethiopia and Sudan over the waters of the Nile, as Ethiopia moves toward completion of the

Grand Ethiopian Renaissance Dam. In the eastern Mediterranean, competition over maritime boundaries related to natural-gas exploration has brought Egypt closer to Cyprus, Greece and Israel, while exacerbating its rivalry with Turkey. Meanwhile, while Israel may be Egypt's primary competitor, Israeli–Egyptian relations have become closer than ever under Sisi, as the two countries cooperate to contain Hamas in the Gaza Strip. However, this cooperation masks continued Egyptian concern over the growing technological sophistication of Israel's armed forces.

One aspect of Cairo's response to these challenges has been to advocate a brand of Egyptian militaristic nationalism that stresses Egyptian independence, while another has been an attempt to make Egypt more relevant in the age of renewed great-power competition, notably by diversifying its weapons suppliers. This strategy has succeeded in reviving military ties to Russia, but it has not markedly changed Egypt's defence relationship with the United States.

While Egypt has moved ahead with purchases of new Russian weapons, including the US$2 billion purchase of 24 Sukhoi Su-35 *Flanker* M fighter jets, its relationship with the US remains complex; President Sisi's relative closeness to Israel and his struggle against Sinai insurgents have won him support in the Republican Party, while his crackdown on dissent has left many Democrats questioning continued US military aid to Egypt. Western European countries also have expressed misgivings over Egypt's human-rights record, though not enough to cancel various arms deals.

Policy and strategy

Egypt's recent defence acquisitions have raised questions over the degree to which future defence policy will continue to be largely defensive in nature, or whether Egypt is developing capabilities to project power beyond its borders. Analysts have perceived a sense of defensiveness and caution in Egypt's military doctrine, based on the legacies of European colonialism, the 1956 Suez invasion, Egypt's ill-fated deployment of troops to North Yemen in the 1960s, and its defeat by Israel in 1967 and 1973. At its core, the Egyptian Armed Forces perceives itself as the protector of the homeland, the Suez Canal and Egypt's offshore natural-gas fields.

Moreover, Egypt has continued to demonstrate caution in using force abroad. The last major deployment of troops was its 35,000-plus troop contribution to *Operation Desert Storm* in Iraq in 1991. More recently, Egypt has avoided the turmoil in nearby Arab states, such as Syria. In 2015, Egypt made a minimal contribution to the Saudi-led intervention in Yemen by sending four vessels and F-16s to enforce the coalition's blockade, though it did not deploy any ground troops. Also in 2015, Egyptian fighter jets conducted air-strikes in Libya in response to the murder of 21 Egyptian Coptic Christians by Islamic State militants.

Egypt has also supported US-led coalition operations in the Middle East, though its role has been generally hesitant due to both a lack of enthusiasm for foreign entanglements and misgivings over certain US policies. After the terrorist attacks of 11 September 2001, Egypt endorsed the US administration's war on terror, but objected to Washington's subsequent regional democratisation agenda. After the launch of *Operation Enduring Freedom* in Afghanistan, Egypt opened a field hospital in Bagram, Afghanistan. Egypt was not an active member of the international coalition during *Operation Iraqi Freedom*, but it did provide US forces with overflight rights and expedited passage through the Suez Canal. A decade later, Egypt played little role in *Operation Inherent Resolve* against the Islamic State, arguing that it was preoccupied with its own militancy problem in the Sinai Peninsula.

However, as the war in Libya has developed in a way inimical to Egyptian interests, Cairo's response has become somewhat of a litmus test for a more forward-leaning military doctrine. In January 2020, the armed forces held a large-scale exercise, which was perhaps as much a message that Egypt was ready to intervene if its interests in Libya were jeopardised, as it was a combat drill aimed at coastal-defence and Sinai contingencies. Then, in summer 2020, as Turkish support for the Tripoli-based GNA helped swing the balance of fighting in Libya and the GNA's forces pushed eastward, President Sisi issued a clear warning to halt their advances. To further demonstrate resolve, Egypt held a second military exercise, simulating an amphibious assault. The exercises, which albeit were carefully choreographed, featured some of the advanced weapons procured in the last five years, including the *Mistral*-class of helicopter carrier.

That said, capability gaps could hinder swift movement beyond Egypt's borders. For example, if Egypt were to launch an amphibious assault in Libya or elsewhere, its *Mistral* ships lack dedicated air defence, with imagery in 2020 purportedly showing *Avenger* short-range air-defence systems secured to the deck for this role, while fleet defence could also prove problematic with the current number of frigates. The *Mistrals* themselves are unable to embark the maximum number (16) of rotary aircraft until Egypt acquires helicopters with folding rotors. Moreover, some analysts say that if Egypt were to act on concerns over events in Libya solely using air-power, it could face challenges in sustaining a prolonged campaign. Egypt only recently unveiled footage showing its first-ever aerial-refuelling capability using under-wing pods, and it is unclear how prepared the air force may be to carry out such operations on a more frequent and wider scale.

The armed forces

Army

The army is considered the heart of the Egyptian Armed Forces. With just under 700,000 active and reserve personnel, it remains oriented toward defeating a comparably armed conventional opponent. Most of the army remains postured against Israel, with seven of 13 armoured and mechanised divisions deployed along the Suez Canal as part of the 2nd and 3rd armies. By contrast, only two divisions are understood to be normally assigned to the Libyan Border (the 21st Armoured Division at Sidi Barrani and the 33rd Mechanised Division at El Salloum). The US has been a key source of equipment and training. Between 1992 and 2015, Egypt acquired over 1,100 M1A1 *Abrams* main battle tanks through United States military assistance (there has been co-production at the Egyptian Tank Plant outside Cairo). Over the past year, there have been unsubstantiated reports that Egypt also plans to procure 400–500 T-90 main battle tanks from Russia.

Conscription and professionalisation remain sensitive topics. Continued conscription has led to a skills-gap between conscript and professional soldiers. Unskilled conscripts are generally assigned relatively menial roles in military companies, while officers receive training at a combination of Egyptian military academies and foreign-partnership programmes. Recent military training has emphasised combined-arms operations, joint training with foreign armed forces and additional focus on counter-terrorism, but observers have said that the organisational culture within the armed forces can impede reform. It has been said that decision-making is too hierarchical, providing little flexibility for independent decision-making by junior officers. Moreover, according to US analysts of the Egyptian armed forces, training scenarios can be too rigid and pre-determined.

Ongoing military operations in the Sinai Peninsula have helped the army gain counter-insurgency experience, though after several years, the armed forces have only been able to contain the spread of militancy, not eliminate it. The army continues to deal with terrorist activity in the Sinai Peninsula by launching intermittent campaigns that succeed in dispersing the terrorist presence in populated areas, but those soldiers left behind when the main force leaves often become targets. That said, there are signs that the government is addressing the root causes of militancy in the Sinai by attempting to improve the livelihoods of Northern Sinai Bedouin tribes by building water-treatment plants, factories and housing.

Navy

There has been considerable attention on improving naval capabilities. The two *Mistral*-class helicopter carriers (produced by Naval Group) were bought from France after their sale to Russia was cancelled following Russia's annexation of Crimea. Egypt also acquired a FREMM multi-mission frigate and issued a contract in July 2014 for four *Gowind* 2500 frigates, again from France's Naval Group; the first of the four, built in France, has been commissioned while the remaining three will be built at the Alexandria Shipyard in Egypt.

Egypt has purchased four Type-209/1400 submarines from a subsidiary of Germany's ThyssenKrupp Marine Systems, which are to eventually replace Egypt's Chinese-made Type-033 (*Romeo*) boats that it acquired in the early 1980s (and that have since been upgraded with US systems including *Harpoon* missiles). A deal for Germany to supply Egypt with perhaps six MEKO A-200 corvettes was approved by Germany's Bundestag in April 2019, but the current status is unclear. Meanwhile, it was reported in 2020 that Egypt had purchased two FREMM frigates from Italy and that discussions were underway on other potential sales.

Air Force

Egypt's Air Force is generally considered by outside specialists to be the most professional and technically capable branch of the armed forces. While there has been much media coverage of Egypt's procurement of 24 *Rafale* multi-role fighters from France and around 50 Mig-29M/M2 fighter/ground attack aircraft from Russia, the US comprises the largest single source of Egypt's fixed-wing inventory, including around 200 F-16-variant combat aircraft. Egypt's diversification strategy is particularly challenging for the air force, which has to balance different training and maintenance procedures for European, Russian and US platforms.

Reports since 2019 of Egyptian plans to purchase Russian Su-35 fighter/ground attack aircraft were confirmed in 2020. In May 2020, it was reported in the Russian media that the Gagarin Aircraft Manufacturing plant in Komsomolsk-on-Amur had started production of the aircraft under a contract signed in 2018; five aircraft were observed at Novosibirsk in July, reportedly on their delivery flight. It had been held that this move could trigger US sanctions under the 2017 Countering America's Adversaries Through Sanctions Act (CAATSA), though the formal response from Washington was unclear at the time of writing.

Egypt's fleet of helicopters are a mix of European, Russian and US models. There are 45 AH-64D *Apache* attack helicopters in the inventory, and in May 2020 the US announced the approval of a potential deal, worth some US$2.3bn, to refurbish them to AH-64E standard with improved engines and navigation aids. Egypt also has purchased 46 standard Ka-52A *Hokum* B Russian attack helicopters, as well as perhaps a dozen older Russian Mi-24 *Hind* attack helicopters. When Egypt conducted naval exercises in mid-2020, one *Mistral* helicopter carrier was seen with both *Apaches* and Ka-52s on its flight deck.

Defence economics and industry

The armed forces have been deeply involved in the Egyptian economy for decades. Though there is little defence-budget transparency in Egypt, it is assessed that final expenditure on defence amounted to E£66bn (US$4.1bn) in the 2019–20 financial year. Including FMF allocations of US$1.3bn brings total defence expenditure to US$5.4bn which accounts for just 1.4% of GDP, well below the regional average of 4.5% in 2019 and 5.5% in 2020. Compared to the levels spent in 2009, the regional average has increased from 4.2% of GDP while Egypt's allocation is now markedly lower than the 2.7% of GDP it spent that year. It is assumed by outside observers that the armed forces' various enterprises are profitable enough to make the armed forces mostly self-funded. This self-sufficiency derives from its main corporate branches, the National Service Projects Organisation, the Arab Organisation for Industrialisation and the National Organisation for Military Production. Combined, these quasi-state entities control numerous companies operating in sectors such as food processing, agriculture, heavy industry, mining, government contracting, construction, transport, pharmaceuticals, tourism and retail.

Although Egypt held its first-ever defence exhibition in 2018 to showcase its domestically produced armaments, Egypt's defence-industry is overall geared more toward the manufacture of foreign items or reproduction of licensed products rather than innovation and design. For instance, Egyptian factories and shipyards have assembled tanks and ships originally produced in Western states. For decades, Egyptian military factories also have produced artillery pieces, mortars and small arms. In recent years, three different indigenously designed armoured vehicles have been built, the *Temsah*-3, the ST-100 and the *Fahd*-300. That said, one report has asserted that Egyptian attempts to move up the value chain and produce higher-end products will run into systemic problems in military companies relating to deficiencies in management, training and marketing.

CONTRACTOR SUPPORT TO GULF DEFENCE ESTABLISHMENTS

Contractors have become an important means of delivering services and capability in advanced armed forces. They can be used for roles including the provision of base and support functions that do not need to be entirely delivered by uniformed personnel or that can be delivered more cheaply or more reliably. Contractors have also been used for specialist roles where skills are either developing or are in short supply, as well as for various non-deployable functions. The firms that operate in this space often rely heavily on recruitment of retired military personnel, including from Western states.

The delivery of military support by contractors is nothing new in the Gulf region. This tends to be delivered either directly by contractors, or by them in support of armed forces. Oman has benefited from contract support from the UK armed forces, and BAE Systems, since the 1970s. Meanwhile, the major aircraft capability programmes in Saudi Arabia delivered by the US and the UK are extensively supported by contractors, though this is declining as Saudi industrial capability grows in these programmes. Vinnell Arabia provides platform-oriented support (training, logistics, engineering) to the Saudi Arabian National Guard (SANG) as an element of a government-to-government agreement with the United States. However, as budgets are pressured and defence ministries are having to become more accountable, and with the catalyst of the conflict in Yemen since March 2015, the value of this support has been acknowledged by states and is being slowly increased, reflecting the need to consider through-life- as well as initial purchase-costs.

The United Arab Emirates illustrates the evolving use of contractors in the Middle East, where experienced personnel who have retired from foreign armed forces have been hired to bolster and accelerate operational capability, as can be seen in the Presidential Guard and the Joint Helicopter Command. In both cases, former Australian and US military personnel are filling mainstream posts with a view to improving internal processes. The aim is to accelerate Emirati personnel-development capacities such that the country is able to, in time, create self-sustaining personnel-development structures. Some of these contractors have executive authority as opposed to being advisers. Moreover, contractors are used elsewhere in the UAE. For instance, personnel from France, the UK and US currently augment staff at training centres, such as the Staff College and Desert Warfare School, through direct delivery, and over time it is hoped that these contractors will step back to an advisory and mentoring role. There is also extensive use of contractor support within the UAE's defence industry, such as engineering support for air and land platforms at AMMROC and Al Taif, while a domestic engineering capability is developed. As with other Gulf neighbours, the UAE Ministry of Defence (MoD) has traditionally presided over disparate single services, and there has been extensive work since 2010 to develop defence institutions, drawing on support from major consulting firms including Booz Allen Hamilton, AT Kearney and PWC, who in turn rely heavily on retired military personnel from Australia, Canada, the UK and US.

In contrast, Saudi Arabia presents a larger challenge given its size and also the need to meet the ambitions of the Kingdom's Vision 2030 reform programme. That said, little specific information has been made public about Riyadh's defence-reform ambitions, other than the desire to develop national defence industry and to 'bring onshore' more defence production. Niche elements of the Saudi armed forces have benefited from contractor support delivered largely to the Royal Saudi Air Force (RSAF) through government-to-government agreements involving BAE Systems and US prime contractors designed to deliver platform-oriented maintenance, repair and overhaul, and technical and other training, while national capability is developed. This has led to the recruitment of a number of former UK and US military and defence-industry personnel.

Operations in Yemen illustrated a gap between these elements of the RSAF and the rest of Saudi Arabia's MoD. Notwithstanding criticism of other aspects of RSAF operations in Yemen, a clear lesson from those operations has been that historic platform-centric arrangements have had only limited impact on overall military capability. This patchwork picture of contractor support may, however, change following the establishment, by Crown Prince Mohammed bin Salman, of an MoD Transformation Management Office (TMO). The head of this organisation was brought in from Saudi Aramco to develop what is, over time, intended to be a top-down reform programme. Progress has overall been slow for reasons including ongoing operations in Yemen; insufficient experience among senior personnel; problems in delegating decision-making; and, arguably, the pressure of multiple consultants and contractors looking to engage without sight of an overall plan. Although contractor support will not resolve these challenges quickly, agreements have been signed with the UK and US to provide senior, often ex-military, advisers to advise and assist. As elements of these programmes progress to the delivery stage, it is likely that increasing numbers of contractors will be deployed as advisers and in delivery roles working alongside Saudi counterparts.

In an environment where regional budgets might remain pressured by low oil revenues, it is likely that the balance of contractor support requirements will lean towards defence transformation and the development of effective military forces. Analysts contend that it is unlikely that the UAE's unique approach, of putting non-nationals in uniform and giving them executive roles, will become widespread, and perhaps there is more likely to be continued contractor support for platform delivery and integration, as well as to support wider defence reforms intended to pursue the force development that has been neglected at the expense of 'trophy' procurements. That said, these will likely be programmes with 20–30-year timescales, dependent on the ability of education systems (that are also in need of reform) delivering individuals with the necessary skills, and also dependent on soldiers and officers progressing through structured careers that give them the experience needed for senior positions.

OFFICER EDUCATION IN GULF ARMED FORCES

Gulf states are working to boost the quality of their military leaders as part of their attempts to improve overall military capability. An important part of this process is to enhance the system of officer education. Modern armed forces expend substantial effort on education and training throughout an officer's career in order to improve skills and deliver better leaders. Courses punctuate the careers of military leaders as they rise through the ranks. The ideal is to achieve meritocratic advancement, with the best candidate appointed to each post.

Gulf states face particular challenges in improving officer education, not least because of specific factors relating to religion and tribe. These continue to affect the selection of officers in the Gulf and their further education, even though there have been recent attempts to promote the broader concept of service to the nation. There are Shia officers in the armed forces of both Bahrain and Kuwait, but analysts have detected a 'glass ceiling' when it comes to promotion. Moreover, even in Gulf countries where Shia are in the minority they can still be marginalised in government service. Oman has taken a different course. Most Omani Muslims are Ibadi but the Sunni predominate in Dhofar. In the wake of the rebellion there in the 1960s and 1970s, Sultan Qaboos drew Dhofaris into the Omani establishment. It is common for Dhofari Sunnis to hold senior roles in the Armed Forces and the Royal Office. Even if religion plays a part in Gulf states' officer selection, the tribe is more important.

Tribe is important in Gulf societies because of proximity to ruling families. Loyalty was and is rewarded with and secured by government employment. With the armed forces seen by many as simply another element of government service, military service has – according to analysts – often been perceived more as a job than as a profession, except by the best officers.

That said, there have in recent years been efforts to harness nationalism in a bid to boost *esprit de corps* and professionalism, not least in the wake of expeditionary operations. This is easier in countries where the indigenous population is homogeneous or where the government has looked to build national spirit through government service. Indeed, a measure of how this can succeed is that the Omani Chief of Staff is the son of one of the rebels from the Jebel Akhdar rebellion that ended in 1959. The United Arab Emirates (UAE) and Qatar recently took a different approach and have introduced a form of national service. The overall motivations have been to help to tie the nation together by instilling discipline, improving physical fitness, raising the population's awareness of security challenges to the state and improving the rewards from a career in the armed forces. While these initiatives have been introduced too recently to have had a notable effect on the professionalism of the officer corps, they have had an effect in generating a more mature appreciation of military risk and sacrifice among the wider population.

Institutions and courses

The six countries of the Gulf Cooperation Council (GCC) (Bahrain, Kuwait, Oman, Qatar, Saudi Arabia and the UAE) approach initial officer training in similar ways. Each has one or more military academies. Oman, for example, has a single-service academy for each of the services, whereas Kuwait has one academy for land and air cadets with maritime cadets being trained abroad, mainly in France. Courses vary in duration from one to three years. Degrees are highly valued in the Gulf, so most courses either award or contribute to a degree. Abu Dhabi University, for example, awards a BSc in Military Science that can be earned by Land Forces and Presidential Guard officers after two years at the Zayed Military Academy followed by one year of distance learning as newly commissioned officers.

In common with many Gulf higher-educational establishments, rote learning is the norm in military academies. Critical thinking and the concept of mission command are given less priority than in similar Western institutions. Foreign staff serve as faculty members in almost all academies. For instance, staff from Bangladesh serve in Kuwait, UK nationals in Bahrain, and UAE and UK personnel are on 'loan service' arrangements in Oman. Course standards are not comparable to Western academies for a number of reasons. For instance, academies tend to be relatively unpopular postings, and unit commanders are reluctant to release their best officers as instructors.

This has the effect of hindering the feedback loop that proves so useful in Western academies, such as the UK's Royal Military Academy Sandhurst (RMAS), West Point in the US, France's Ecole d'état-major and Australia's Command and Staff College. In Western states, there is a tradition of posting high-calibre experienced officers (as well as non-commissioned officers) to instruct cadets.

Currently, the regional alumni of the UK's RMAS include Kuwait's Chief of Staff, Bahrain's Commander in Chief, Qatar's Chief of Staff (COS), Oman's COS and the UAE Deputy Supreme Commander. While some cadets are sent by Saudi Arabia, the other GCC states have a quota of places. According to analysts, the cadets sent by Oman, the UAE and, more recently, Bahrain are of good quality but often find the RMAS course more challenging than their British counterparts because of their relative youth. In each country, cadets are chosen either by a selection process or by the involvement of senior leadership. Tribal and familial links remain important: the first female cadet from Bahrain was the granddaughter of Bahrain's late prime minister, while in 2020 there have been cadets from Abu Dhabi with the name Al Nahyan, and from Dubai with the name Al Maktoum. In both Emirates this is the family name of the respective ruling families.

However, the abilities and motivations of the officers that emerge from international and regional academies can be variable. Gulf officers enter armed forces where, for the most part, promotion is based on time served, and service is for life with a generous pension on retirement. Moreover, career planning is often difficult because officers can remain in post for long periods. Currently, the Chiefs of Staff of the GCC countries are all long-serving – with the exception of Kuwait and Saudi Arabia. Retaining loyal officers in post for extended periods may be reasonable when loyalty is a priority, but doing so risks adversely affecting the development of the next generation of leaders.

Command and staff courses

Officers undertake a range of courses throughout their career. The most relevant for command and staff purposes are intermediate, advanced and higher-level courses. Most Gulf countries run short courses for young officers before formal staff training begins at the rank of major. Oman's structure is the most systematic. Modelled on that of the UK, it covers all three levels and is, according to observers, highly competitive. For example, graduates from the one-year command and staff course at Oman's Bait Al Falaj Joint Command and Staff College receive their certificates according to their grade in front of the COS of the Sultan's Armed Forces and the three Service Chiefs, making it apparent who came top, as well as who came bottom. This course is well regarded by outside military specialists, who assert that the UAE's is likely to be in the same bracket.

In 2014, Qatar contracted services firm Serco to run its command and staff course at the Joaan Bin Jassim Joint Command and Staff College. However, in 2017 the contract in Qatar was not renewed. Analysts have reported that this owed much to faculty reports about difficulties that students were having with the course, which was set at the military-strategic level – including that English was the language of instruction – and that the decision coincided with the arrival of a new minister of state for defence affairs. Meanwhile, Kuwait's Mubarak al-Abdullah Joint Command and Staff College stresses its English-language tuition and reportedly focuses on inculcating critical thinking and instruction at the operational and strategic levels. That said, overseas institutions remain the preferred location for career-development courses at the intermediate and advanced levels, with UK and US institutions generally preferred. There is strong competition for places. The UAE, for example, sends three students each year to the UK's Advanced Command and Staff Course at Shrivenham, all of whom will first have completed the Emirati staff course.

Regional states do maintain the capacity to provide some higher-level courses. For instance, Bahrain has a course for one-star officers; Oman and the UAE both maintain National Defence Colleges; Saudi Arabia has a War College for officers at colonel rank; while Kuwait recently added a short Senior Leaders' Development Programme for its most promising colonels and brigadiers. Nonetheless, the preference for many regional officers looking to advance at the higher level is still to attend a US War College or the UK's Royal College of Defence Studies.

The curricula of the region's officer-education courses tend to mirror the format seen in other international armed forces. However, the value of the military education delivered by these institutions is undermined, analysts attest, by a focus on format rather than content and continued emphasis on rote learning rather than critical thinking. For officer selection, while there remains value in drawing a loyal officer corps from a select group of families, recent efforts to highlight the benefit of service to the nation could provide a more durable and stronger motivation. Moreover, the willingness to honour those killed in expeditionary campaigns as martyrs – as the UAE and Bahrain have done – will serve to illustrate the commitment required of officers. In time, it is possible that this will diminish the attitude that military service is little more than government service, but in uniform; and the net result could be that more military posts are filled by the most able individuals.

Algeria ALG

Algerian Dinar D		2019	2020	2021
GDP	D	20.2tr	18.3tr	
	US$	169bn	147bn	
per capita	US$	3,898	3,331	
Growth	%	0.8	-5.5	
Inflation	%	2.0	3.5	
Def bdgt	D	1.23tr	1.23tr	
	US$	10.3bn	9.92bn	
US$1=D		119.35	124.01	

Real-terms defence budget trend (US$bn, constant 2015)

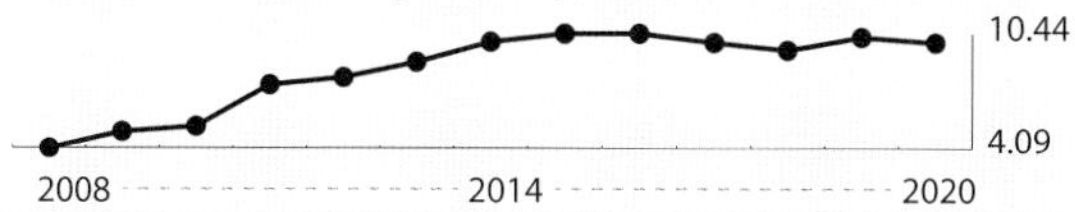

Population 42,972,878

Age	0–14	15–19	20–24	25–29	30–64	65 plus
Male	15.1%	3.6%	3.5%	4.2%	21.2%	2.9%
Female	14.4%	3.5%	3.4%	4.0%	20.8%	3.3%

Capabilities

The armed forces are among the most capable and best equipped in North Africa. Their primary roles are territorial integrity, internal security and regional stability. In April 2019, the army was instrumental in ending President Abdelaziz Bouteflika's two decades in power, after widespread demonstrations, and it retains a key political role. Algeria is part of the African Union's North African Regional Capability Standby Force, hosting the force's logistics base in Algiers. Algeria discusses with its neighbours regional security challenges like counter-terrorism, and has particularly close security cooperation with Tunisia. The conscript-based force exercises regularly, with training appearing to be of a relatively good standard. There is an ongoing attempt to make the armed forces more professional, which was reflected in the reduction of conscription liability from 18 to 12 months in 2014. The armed forces' logistics capabilities appear sufficient to support internal deployments. The army's and air force's inventories consist of a core of modern, primarily Russian-sourced equipment, though China has also supplied equipment, including self-propelled artillery. Algiers has continued investing in fixed-wing combat-aircraft recapitalisation and the navy has invested in its submarine and frigate fleet. Local industry, and the services, are capable of equipment maintenance. However, while Algeria is largely dependent on foreign suppliers for new equipment, it has in recent years made significant investments towards developing a domestic defence industry. This has led to a number of joint ventures with foreign partners, such as with Italy's Leonardo and Germany's Rheinmetall.

ACTIVE 130,000 (Army 110,000 Navy 6,000 Air 14,000) **Paramilitary 187,200**

Conscript liability 12 months

RESERVE 150,000 (Army 150,000) to age 50

ORGANISATIONS BY SERVICE

Army 35,000; 75,000 conscript (total 110,000)

FORCES BY ROLE

6 Mil Regions

MANOEUVRE

Armoured

2 (1st & 8th) armd div (3 tk regt; 1 mech regt, 1 arty gp)
2 indep armd bde

Mechanised

2 (12th & 40th) mech div (1 tk regt; 3 mech regt, 1 arty gp)
4 indep mech bde

Light

1 indep mot bde

Air Manoeuvre

1 AB div (4 para regt; 1 SF regt)

COMBAT SUPPORT

2 arty bn
1 AT regt
4 engr bn

AIR DEFENCE

7 AD bn

EQUIPMENT BY TYPE

ARMOURED FIGHTING VEHICLES

MBT 1,495: 270 T-55AMV; 300 T-62; 325 T-72M1/M1M; 600+ T-90SA
RECCE 134: 44 AML-60; 26 BRDM-2; 64 BRDM-2M with 9M133 *Kornet* (AT-14 *Spriggan*)
IFV 980: ε220 BMP-2; 760 BMP-2M with 9M133 *Kornet* (AT-14 *Spriggan*)
APC 1,107+
APC (T) VP-6
APC (W) 1,105: 250 BTR-60; 150 BTR-80; 150 OT-64; 55 M3 Panhard; ε400 *Fuchs* 2; 100 *Fahd*
PPV 2 *Marauder*
AUV Nimr *Ajban*; Nimr *Ajban* LRSOV

ENGINEERING & MAINTENANCE VEHICLES

AEV IMR-2
ARV BREM-1
VLB MTU-20
MW M58 MICLIC

ANTI-TANK/ANTI-INFRASTRUCTURE

SP 28 9P163-3 *Kornet*-EM (AT-14 *Spriggan*)
MSL • MANPATS 9K11 *Malyutka* (AT-3 *Sagger*); 9K111 *Fagot* (AT-4 *Spigot*); 9K111-1 *Konkurs* (AT-5 *Spandrel*); 9K115-2 *Metis*-M1 (AT-13 *Saxhorn*-2); 9K135 *Kornet*-E (AT-14 *Spriggan*); Luch *Skif*; *Milan*
RCL 180: **82mm** 120 B-10; **107mm** 60 B-11
GUNS 100mm 10 T-12

ARTILLERY 1,127

SP 224: **122mm** 140 2S1 *Gvozdika*; **152mm** 30 2S3 *Akatsiya*; **155mm** ε54 PLZ-45
TOWED 393: **122mm** 345: 160 D-30 (incl some truck mounted SP); 25 D-74; 100 M-1931/37; 60 M-30; **130mm** 10 M-46; **152mm** 20 M-1937 (ML-20); **155mm** 18 PLL-01
MRL 180: **122mm** 48 BM-21 *Grad*; **140mm** 48 BM-14; **220mm** 36: 18+ SR5; ε18 TOS-1A; **240mm** 30 BM-24; **300mm** 18 9A52 *Smerch*
MOR 330+: **82mm** 150 M-37; **120mm** 120 M-1943; W86; **SP 120mm** SM4; W86 (SP); **160mm** 60 M-1943

SURFACE-TO-SURFACE MISSILE LAUNCHERS

SRBM 12 *Iskander*-E

AIR DEFENCE
SAM 106+
Short-range 38 96K6 *Pantsir*-S1 (SA-22 *Greyhound*); *Pantsir*-SM
Point-defence 68+: ε48 9K33M *Osa* (SA-8B *Gecko*); ε20 9K31 *Strela*-1 (SA-9 *Gaskin*); 9K32 *Strela*-2 (SA-7A/B *Grail*)‡; QW-2
GUNS ε425
SP 23mm ε225 ZSU-23-4
TOWED 200: **14.5mm** 100: 60 ZPU-2; 40 ZPU-4; **23mm** 100 ZU-23-2

Navy ε6,000

EQUIPMENT BY TYPE
SUBMARINES • SSK 6:
2 *Paltus* (FSU Project 877 (*Kilo*)) with 6 single 533mm TT with Test-71ME HWT
4 *Varshavyanka* (RUS Project 636.1 (Improved *Kilo*)) with 6 single 533mm TT with 3M14E *Klub*-S (SS-N-30B) LACM/3M54E1/E *Klub*-S (SS-N-27A/B) AShM (*Klub*-S AShM variant unclear)/TEST-71ME HWT
PRINCIPAL SURFACE COMBATANTS • FRIGATES 5
FFGHM 5:
3 *Adhafer* (C28A) with 2 quad lnchr with C-802A AShM, 1 octuple lnchr with FM-90 (CH-SA-N-4) SAM, 2 triple 324mm ASTT, 2 Type-730B (H/PJ-12) CIWS, 1 76mm gun (capacity 1 hel)
2 *Erradii* (MEKO 200AN) with 2 octuple lnchrs with RBS15 Mk3 AShM, 4 8-cell VLS with *Umkhonto*-IR SAM, 2 twin 324mm TT with MU90 LWT, 1 127mm gun (capacity 1 *Super Lynx* 300)
PATROL AND COASTAL COMBATANTS 28
CORVETTES • FS 3 *Mourad Rais* (FSU Project 1159 (*Koni*)) with 2 twin 533mm TT, 2 RBU 6000 *Smerch* 2 A/S mor, 2 twin 76mm gun
PCGM 3 *Rais Hamidou* (FSU Project 1234E (*Nanuchka* II)) with 4 twin lnchr with 3M24E *Uran*-E (SS-N-25 *Switchblade*) AShM, 1 twin lnchr with 4K33 *Osa*-M (SA-N-4 *Gecko*) SAM, 1 AK630 CIWS, 1 twin 57mm gun
PCG 4: 3 *Djebel Chenoua* with 2 twin lnchr with C-802 (CH-SS-N-6) AShM, 1 AK630 CIWS, 1 76mm gun; 1 *Rais Hassen Barbiar* (*Djebel Chenoua* mod) with 2 twin lnchr with C-802 (CH-SS-N-6) AShM, 1 Type-730 (H/PJ-12) CIWS, 1 76mm gun
PBFG 9 Project 205 (ex-FSU *Osa* II) (or which 3†) with 4 single lnchr with P-20U (SS-N-2B *Styx*) AShM
PB 9 *Kebir* with 1 76mm gun
MINE WARFARE • MINE COUNTERMEASURES 1
MCC 1 *El-Kasseh* (ITA *Gaeta* mod)
AMPHIBIOUS
PRINCIPAL AMPHIBIOUS SHIPS • LHD 1 *Kalaat Beni Abbes* with 1 8-cell *Sylver* A50 VLS with *Aster* 15 SAM, 1 76mm gun (capacity 5 med hel; 3 LCVP; 15 MBT; 350 troops)
LANDING SHIPS 3:
LSM 1 *Polnochny* B with 1 twin AK230 CIWS (capacity 6 MBT; 180 troops)
LST 2 *Kalaat beni Hammad* (capacity 7 MBT; 240 troops) with 1 med hel landing platform
LANDING CRAFT • LCVP 3
LOGISTICS AND SUPPORT 3
AGS 1 *El Idrissi*
AX 1 *Daxin* with 2 AK230 CIWS, 1 76mm gun, 1 hel landing platform
AXS 1 *El Mellah*

Naval Infantry

FORCES BY ROLE
MANOEUVRE
Amphibious
1 naval inf bn

Naval Aviation

EQUIPMENT BY TYPE
HELICOPTERS
MRH 9: 3 AW139 (SAR); 6 *Super Lynx* 300
SAR 9: 5 AW101 SAR; 4 *Super Lynx* Mk130

Coastal Defence

EQUIPMENT BY TYPE
COASTAL DEFENCE
AShM 4K51 *Rubezh* (SSC-3 *Styx*)

Coast Guard ε500

EQUIPMENT BY TYPE
PATROL AND COASTAL COMBATANTS 68
PBF 6 *Baglietto* 20
PB 62: 6 *Baglietto Mangusta*; 12 *Jebel Antar*; 34 *Deneb*; 4 *El Mounkid*; 6 *Kebir* with 1 76mm gun
LOGISTICS AND SUPPORT 9
AR 1 *El Mourafek*
ARS 3 *El Moundjid*
AXL 5 *El Mouderrib* (PRC *Chui-E*) (2 more in reserve†)

Air Force 14,000

FORCES BY ROLE
FIGHTER
1 sqn with MiG-25PDS/RU *Foxbat*
4 sqn with MiG-29S/UB *Fulcrum*
FIGHTER/GROUND ATTACK
3 sqn with Su-30MKA *Flanker* H
GROUND ATTACK
2 sqn with Su-24M/MK *Fencer* D
ELINT
1 sqn with Beech 1900D
MARITIME PATROL
2 sqn with Beech 200T/300 *King Air*
ISR
1 sqn with Su-24MR *Fencer* E*; MiG-25RBSh *Foxbat* D*
TANKER
1 sqn with Il-78 *Midas*
TRANSPORT
1 sqn with C-130H/H-30 *Hercules*; L-100-30
1 sqn with C295M
1 sqn with Gulfstream IV-SP; Gulfstream V
1 sqn with Il-76MD/TD *Candid*
TRAINING
2 sqn with Z-142

1 sqn with Yak-130 *Mitten**
2 sqn with L-39C/ZA *Albatros*
1 hel sqn with PZL Mi-2 *Hoplite*

ATTACK HELICOPTER
3 sqn with Mi-24 *Hind* (one re-equipping with Mi-28NE *Havoc*)

TRANSPORT HELICOPTER
1 sqn with AS355 *Ecureuil*
5 sqn with Mi-8 *Hip*; Mi-17 *Hip* H
1 sqn with Ka-27PS *Helix* D; Ka-32T *Helix*

ISR UAV
1 sqn with *Seeker* II

AIR DEFENCE
3 ADA bde
3 SAM regt with S-125 *Neva* (SA-3 *Goa*); 2K12 *Kub* (SA-6 *Gainful*); S-300PMU2 (SA-20 *Gargoyle*)

EQUIPMENT BY TYPE

AIRCRAFT 133 combat capable
FTR 34: 11 MiG-25PDS/RU *Foxbat*; 23 MiG-29S/UB *Fulcrum*
FGA 43 Su-30MKA *Flanker* H
ATK 33 Su-24M/MK *Fencer* D
ISR 7: 4 MiG-25RBSh *Foxbat* D*; 3 Su-24MR *Fencer* E*
TKR 6 Il-78 *Midas*
TPT 65: **Heavy** 11: 3 Il-76MD *Candid* B; 8 Il-76TD *Candid*; **Medium** 16: 8 C-130H *Hercules*; 6 C-130H-30 *Hercules*; 2 L-100-30; **Light** 32: 3 Beech C90B *King Air*; 5 Beech 200T *King Air*; 6 Beech 300 *King Air*; 12 Beech 1900D (electronic surv); 5 C295M; 1 F-27 *Friendship*; **PAX** 6: 1 A340; 4 Gulfstream IV-SP; 1 Gulfstream V
TRG 99: 36 L-39ZA *Albatros*; 7 L-39C *Albatros*; 16 Yak-130 *Mitten**; 40 Z-142

HELICOPTERS
ATK 44: 30 Mi-24 *Hind*; 14+ Mi-28NE/UB *Havoc*
SAR 3 Ka-27PS *Helix* D
MRH 85: 8 AW139 (SAR); 3 Bell 412EP; 74 Mi-8 *Hip* (med tpt)/Mi-17 *Hip* H
TPT 62: **Heavy** 14 Mi-26T2 *Halo*; **Medium** 4 Ka-32T *Helix*; **Light** 44: 8 AW119KE *Koala*; 8 AS355 *Ecureuil*; 28 PZL Mi-2 *Hoplite*

UNINHABITED AERIAL VEHICLES
CISR • Heavy CH-3; CH-4; *Yabhon United*-30
ISR • Medium *Seeker* II; *Yabhon Flash*-20

AIR DEFENCE • SAM
Long-range S-300PMU2 (SA-20 *Gargoyle*)
Medium-range 9K317 *Buk*-M2E (SA-17 *Grizzly*); S-125 *Pechora*-M (SA-3 *Goa*)
Short-range 2K12 *Kvadrat* (SA-6 *Gainful*)

AIR-LAUNCHED MISSILES
AAM • IR R-60 (AA-8 *Aphid*); R-73 (AA-11A *Archer*); **IR/SARH** R-40/46 (AA-6 *Acrid*); R-23/24 (AA-7 *Apex*); R-27 (AA-10 *Alamo*); **ARH** R-77 (AA-12A *Adder*);
ASM Kh-25 (AS-10 *Karen*); Kh-29 (AS-14 *Kedge*); Kh-59ME (AS-18 *Kazoo*); ZT-35 *Ingwe*; 9M120 *Ataka* (AT-9 *Spiral*-2)
AShM Kh-31A (AS-17B *Krypton*)
ARM Kh-25MP (AS-12A *Kegler*); Kh-31P (AS-17A *Krypton*)

Paramilitary ε187,200

Gendarmerie 20,000
Ministry of Defence control; 6 regions

EQUIPMENT BY TYPE

ARMOURED FIGHTING VEHICLES
RECCE AML-60
APC • APC (W) 210: 100 TH-390 *Fahd*; 110 Panhard M3

HELICOPTERS • TPT • Light 12+: 12 AW109; Some PZL Mi-2 *Hoplite*

National Security Forces 16,000
Directorate of National Security. Small arms

Republican Guard 1,200

EQUIPMENT BY TYPE

ARMOURED FIGHTING VEHICLES
RECCE AML-60
APC • APC (T) M3 half-track

Legitimate Defence Groups ε150,000
Self-defence militia, communal guards (60,000)

DEPLOYMENT

DEMOCRATIC REPUBLIC OF THE CONGO: UN • MONUSCO 2

Bahrain BHR

Bahraini Dinar D		2019	2020	2021
GDP	D	14.5bn	13.0bn	
	US$	38.6bn	34.6bn	
per capita	US$	25,998	22,878	
Growth	%	1.8	-4.9	
Inflation	%	1.0	0.0	
Def bdgt [a]	D	530m	528m	
	US$	1.41bn	1.41bn	
US$1=D		0.38	0.38	

[a] Excludes funds allocated to the Ministry of the Interior and the National Security Agency

Real-terms defence budget trend (US$m, constant 2015)

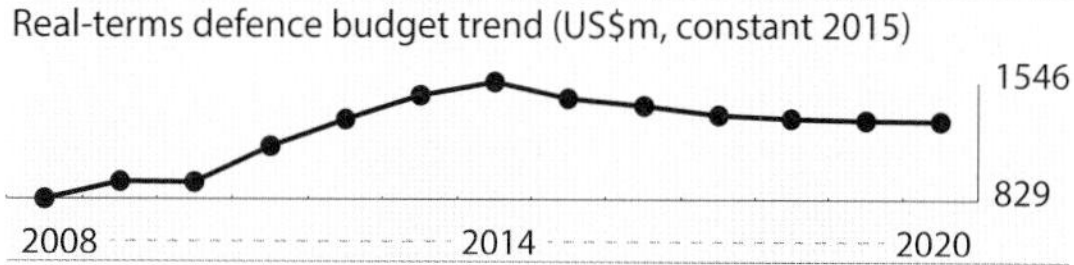

Population 1,505,003

Age	0–14	15–19	20–24	25–29	30–64	65 plus
Male	9.4%	3.6%	5.0%	6.3%	34.6%	1.7%
Female	9.1%	3.1%	3.5%	3.9%	18.3%	1.6%

Capabilities

Bahrain's armed forces are capable and well equipped. The country occupies a key strategic position between regional rivals Iran and Saudi Arabia. The principal roles of the armed forces are territorial defence and internal-security support. Bahrain is a member of the GCC. Its most critical security relationship is with Saudi Arabia, but it also has a strong defence relationship with the US and it has been a US major non-NATO ally since 2002. The US 5th Fleet is headquartered in Bahrain, as is the US-led Combined Maritime Forces and the UK-led *Operation Sentinel* international maritime security construct. In 2018, the UK reopened a permanent naval facility in Bahrain. Military service is voluntary and personnel are relatively well trained. Despite their small size, the armed forces have carried out a number of limited expeditionary deployments to support coalition operations, including in support of the Saudi-led intervention in Yemen. For a period in 2017–18, Bahrain commanded Combined Task Force 151 – the first time a GCC nation had assumed command of a CTF outside the Arabian Gulf. As part of a major air-force modernisation, Bahrain is in the process of acquiring F-16V fighters and upgrading its existing F-16C/Ds to that configuration, as well as acquiring the *Patriot* air- and missile-defence system. It is also upgrading its naval capabilities. The armed forces have organic maintenance support, but there is little in the way of a defence-industrial base beyond the limited maintenance support provided by the Arab Shipbuilding and Repair Yard.

ACTIVE 8,200 (Army 6,000 Navy 700 Air 1,500)
Paramilitary 11,260

ORGANISATIONS BY SERVICE

Army 6,000

FORCES BY ROLE
SPECIAL FORCES
1 SF bn
MANOEUVRE
Armoured
1 armd bde(-) (1 recce bn, 2 armd bn)
Mechanised
1 inf bde (2 mech bn, 1 mot bn)
Light
1 (Amiri) gd bn
COMBAT SUPPORT
1 arty bde (1 hvy arty bty, 2 med arty bty, 1 lt arty bty, 1 MRL bty)
1 engr coy
COMBAT SERVICE SUPPORT
1 log coy
1 tpt coy
1 med coy
AIR DEFENCE
1 AD bn (1 ADA bty, 2 SAM bty)

EQUIPMENT BY TYPE
ARMOURED FIGHTING VEHICLES
MBT 180 M60A3
RECCE 22 AML-90
IFV 67: 25 YPR-765 PRI; 42 AIFV-B-C25
APC 303+
APC (T) 303: 300 M113A2; 3 AIFV-B
APC (W) *Arma* 6×6
AUV M-ATV
ENGINEERING & MAINTENANCE VEHICLES
ARV 53 *Fahd* 240
ANTI-TANK/ANTI-INFRASTRUCTURE
MSL
SP 5 AIFV-B-*Milan*; HMMWV with BGM-71A TOW; 9P163-3 *Kornet*-EM (AT-14 *Spriggan*)
MANPATS BGM-71A TOW; *Kornet*-EM
RCL 31: **106mm** 25 M40A1; **120mm** 6 MOBAT
ARTILLERY 175
SP 82: **155mm** 20 M109A5; **203mm** 62 M110A2
TOWED 36: **105mm** 8 L118 Light Gun; **155mm** 28 M198
MRL 13: **220mm** 4 SR5; **227mm** 9 M270 MLRS
MOR 44: **81mm** 32: 12 L16; 20 EIMOS; **SP 120mm** 12 M113A2
SURFACE-TO-SURFACE MISSILE LAUNCHERS
SRBM • Conventional MGM-140A ATACMS (launched from M270 MLRS)
AIR DEFENCE
SAM 13+
Medium-range 6 MIM-23B I-*Hawk*
Short-range 7 *Crotale*
Point-defence 9K338 *Igla*-S (SA-24 *Grinch*) (reported); FIM-92 *Stinger*; RBS-70
GUNS 24: **35mm** 12 GDF-003/-005; **40mm** 12 L/70

Navy 700

EQUIPMENT BY TYPE
PRINCIPAL SURFACE COMBATANTS • FRIGATES 1
FFGHM 1 *Sabha* (ex-US *Oliver Hazard Perry*) with 1 Mk 13 GMLS with SM-1MR Block VI SAM/RGM-84C *Harpoon* Block 1B AShM, 2 triple 324mm SVTT Mk 32 ASTT with Mk 46 LWT, 1 Mk 15 *Phalanx* Block 1B CIWS, 1 76mm gun (capacity 1 Bo-105 hel)
PATROL AND COASTAL COMBATANTS 13
PSO 1 *Al Zubara* (ex-UK *River* (OPV) Batch 1 (mod)) with 1 hel landing platform
PCFG 4 *Ahmed el Fateh* (GER Lurssen 45m) with 2 twin lnchr with MM40 *Exocet* AShM, 1 76mm gun
PCG 2 *Al Manama* (GER Lurssen 62m) with 2 twin lnchr with MM40 *Exocet* AShM, 2 76mm guns, 1 hel landing platform
PB 4: 2 *Al Jarim* (US *Swift* FPB-20); 2 *Al Riffa* (GER Lurssen 38m)
PBF 2 Mk V SOC
AMPHIBIOUS • LANDING CRAFT 9
LCM 7: 1 *Loadmaster*; 4 *Mashtan*; 2 *Dinar* (ADSB 42m)
LCVP 2 *Sea Keeper*

Naval Aviation

EQUIPMENT BY TYPE
HELICOPTERS • TPT • Light 2 Bo-105

Air Force 1,500

FORCES BY ROLE
FIGHTER
2 sqn with F-16C/D *Fighting Falcon*
FIGHTER/GROUND ATTACK
1 sqn with F-5E/F *Tiger* II
TRANSPORT
1 (Royal) flt with B-727; B-747; BAe-146; Gulfstream II; Gulfstream IV; Gulfstream 450; Gulfstream 550; S-92A
TRAINING
1 sqn with *Hawk* Mk129*
1 sqn with T-67M *Firefly*
ATTACK HELICOPTER
2 sqn with AH-1E/F *Cobra*; TAH-1P *Cobra*

TRANSPORT HELICOPTER
1 sqn with Bell 212 (AB-212)
1 sqn with UH-60M *Black Hawk*
1 (VIP) sqn with Bo-105; S-70A *Black Hawk*; UH-60L *Black Hawk*

EQUIPMENT BY TYPE
AIRCRAFT 38 combat capable
FTR 12: 8 F-5E *Tiger* II; 4 F-5F *Tiger* II
FGA 20: 16 F-16C Block 40 *Fighting Falcon*; 4 F-16D Block 40 *Fighting Falcon*
TPT 12: **Medium** 2 C-130J *Hercules*; **PAX** 10: 1 B-727; 2 B-747; 1 Gulfstream II; 1 Gulfstream IV; 1 Gulfstream 450; 1 Gulfstream 550; 3 BAe-146
TRG 9: 6 *Hawk* Mk129*; 3 T-67M *Firefly*
HELICOPTERS
ATK 28: 16 AH-1E *Cobra*; 12 AH-1F *Cobra*
TPT 27: **Medium** 13: 3 S-70A *Black Hawk*; 1 S-92A (VIP); 1 UH-60L *Black Hawk*; 8 UH-60M *Black Hawk*; **Light** 14: 11 Bell 212 (AB-212); 3 Bo-105
TRG 6 TAH-1P *Cobra*
AIR-LAUNCHED MISSILES
AAM • IR AIM-9P *Sidewinder*; **SARH** AIM-7 *Sparrow*; **ARH** AIM-120B/C AMRAAM
ASM AGM-65D/G *Maverick*; some TOW
BOMBS
Laser-guided GBU-10/12 *Paveway* II

Paramilitary ε11,260

Police 9,000

Ministry of Interior

EQUIPMENT BY TYPE
ARMOURED FIGHTING VEHICLES
RECCE 8 S52 *Shorland*
APC • PPV Otokar ISV;
AUV *Cobra*
HELICOPTERS
MRH 2 Bell 412 *Twin Huey*
ISR 2 Hughes 500
TPT • Light 1 Bo-105

National Guard ε2,000

FORCES BY ROLE
MANOEUVRE
Other
3 paramilitary bn

EQUIPMENT BY TYPE
ARMOURED FIGHTING VEHICLES
APC • APC (W) *Arma* 6×6; *Cobra*

Coast Guard ε260

Ministry of Interior
PATROL AND COASTAL COMBATANTS 55
PBF 26: 2 *Ares* 18; 3 *Response Boat-Medium* (RB-M); 4 *Jaris*; 6 *Saham*; 6 *Fajr*; 5 *Jarada*
PB 29: 6 *Haris*; 1 *Al Muharraq*; 10 *Deraa* (of which 4 *Halmatic* 20, 2 *Souter* 20, 4 Rodman 20); 10 *Saif* (of which 4 *Fairey Sword*, 6 *Halmatic* 160); 2 *Hawar*
AMPHIBIOUS • LANDING CRAFT • LCU 1 *Loadmaster* II

DEPLOYMENT

SAUDI ARABIA: *Operation Restoring Hope* 250; 1 SF gp; 1 arty gp; 6 F-16C *Fighting Falcon*

FOREIGN FORCES

United Kingdom Air Force 160: 1 naval base
United States US Central Command 4,700; 1 HQ (5th Fleet); 1 ASW sqn with 5 P-8A *Poseidon*; EP-3E *Aries* II; 2 AD bty with MIM-104E/F *Patriot* PAC-2/3

Egypt EGY

Egyptian Pound E£		2019	2020	2021
GDP	E£	5.32tr	5.82tr	
	US$	302bn	362bn	
per capita	US$	3,044	3,561	
Growth	%	5.6	3.5	
Inflation	%	13.9	5.7	
Def bdgt	E£	59.0bn	66.0bn	
	US$	3.35bn	4.11bn	
FMA (US)	US$	1.30bn	1.30bn	1.30bn
US$1=E£		17.60	16.08	

Real-terms defence budget trend (US$bn, constant 2015)

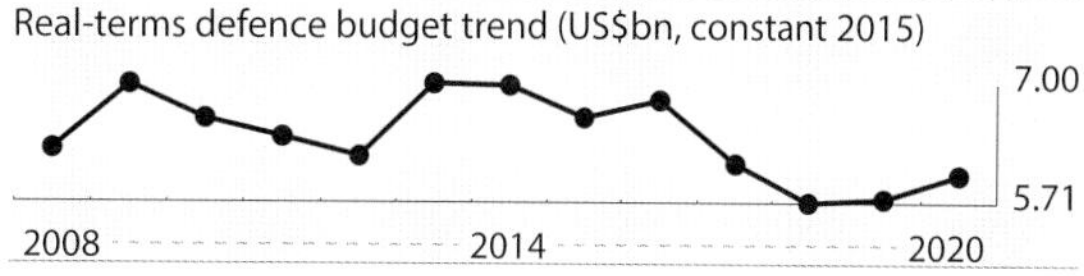

Population 104,124,440

Age	0–14	15–19	20–24	25–29	30–64	65 plus
Male	17.4%	4.6%	4.7%	4.6%	17.7%	2.1%
Female	16.2%	4.3%	4.5%	4.4%	17.2%	2.3%

Capabilities

Egypt's armed forces are the largest in the region and are principally focused on territorial integrity and internal security, including combating ISIS-affiliated groups in northern Sinai. The armed forces remain deeply involved in the civilian economy and retain a central role in Egyptian politics. Egypt and the US maintain a strong strategic partnership, which has seen significant US equipment deliveries and ongoing foreign-military-aid payments. Defence relations with Russia have developed, particularly regarding procurements. Operational experience will have been bolstered by counter-insurgency operations in the Sinai since 2011 and contributions to several UN deployments. Training is supplemented by regular involvement in a number of multinational exercises. Egypt has a developing capacity to deploy independently beyond its borders. It contributes to UN missions, has intervened militarily in Libya and sent combat aircraft to support the Saudi-led coalition in Yemen. The navy's two *Mistral*-class amphibious ships have bolstered the capacity to deploy regionally. The armed forces' inventory primarily comprises obsolete Soviet-era systems and newer Western equipment. However, the armed forces are undertaking an extensive equipment-recapitalisation programme, which is seeing the delivery of Russian multi-role fighters, attack helicopters and SAM systems. Combat aircraft have also been sourced from France and armed UAVs from China. Naval recapitalisation includes German-built submarines and French-designed frigates.

Egypt has an established domestic defence industry, supplying equipment for both the armed forces and export markets, ranging from small arms to armoured vehicles. There is a history of licensed and co-production with foreign companies, including the local assembly of M1A1 main battle tanks from US-supplied kits and the production of frigates with French assistance.

ACTIVE 438,500 (Army 310,000 Navy 18,500 Air 30,000 Air Defence Command 80,000) Paramilitary 397,000

Conscription liability 12–36 months (followed by refresher training over a period of up to 9 years)

RESERVE 479,000 (Army 375,000 Navy 14,000 Air 20,000 Air Defence Command 70,000)

ORGANISATIONS BY SERVICE

Space

EQUIPMENT BY TYPE

SATELLITES • ISR 1 *Egyptsat*-A

Army 90,000–120,000; 190,000–220,000 conscript (total 310,000)

FORCES BY ROLE

SPECIAL FORCES

5 cdo gp
1 counter-terrorist unit
1 spec ops unit

MANOEUVRE

Armoured

4 armd div (2 armd bde, 1 mech bde, 1 arty bde)
4 indep armd bde
1 Republican Guard bde

Mechanised

8 mech div (1 armd bde, 2 mech bde, 1 arty bde)
4 indep mech bde

Light

1 inf div
2 indep inf bde

Air Manoeuvre

2 air mob bde
1 para bde

SURFACE-TO-SURFACE MISSILE

1 SRBM bde with FROG-7
1 SRBM bde with 9K72 Elbrus (SS-1C *Scud*-B)

COMBAT SUPPORT

15 arty bde
6 engr bde (3 engr bn)
2 spec ops engr bn
6 salvage engr bn
24 MP bn
18 sigs bn

COMBAT SERVICE SUPPORT

36 log bn
27 med bn

EQUIPMENT BY TYPE

ARMOURED FIGHTING VEHICLES

MBT 2,480: 1,130 M1A1 *Abrams*; 300 M60A1; 850 M60A3; 200 T-62 (840 T-54/T-55; 300 T-62 all in store)

RECCE 412: 300 BRDM-2; 112 *Commando Scout*

IFV 690: 390 YPR-765 25mm; 300 BMP-1

APC 5,244+

APC (T) 2,700: 2,000 M113A2/YPR-765 (incl variants); 500 BTR-50; 200 OT-62

APC (W) 1,560: 250 BMR-600P; 250 BTR-60; 410 *Fahd*-30/TH 390 *Fahd*; 650 *Walid*

PPV 984+: 535 *Caiman*; some REVA III; some REVA V LWB; 360 RG-33L; 89 RG-33 HAGA (amb)

AUV 95+: *Panthera* T6; 95+ *Sherpa Light Scout*

ENGINEERING & MAINTENANCE VEHICLES

ARV 367+: *Fahd* 240; BMR 3560.55; 12 *Maxxpro* ARV; 220 M88A1; 90 M88A2; M113 ARV; 45 M578; T-54/55 ARV

VLB KMM; MTU; MTU-20

MW *Aardvark* JFSU Mk4

ANTI-TANK/ANTI-INFRASTRUCTURE • MSL

SP 352+: 52 M901, 300 YPR-765 PRAT; HMMWV with TOW-2

MANPATS 9K11 *Malyutka* (AT-3 *Sagger*) (incl BRDM-2); HJ-73; Luch *Corsar* (reported); *Milan*; *Stugna*-P (reported); TOW-2

ARTILLERY 4,468

SP 492+: **122mm** 124+: 124 SP 122; D-30 mod; **130mm** M-46 mod; **155mm** 368: 164 M109A2; 204 M109A5

TOWED 962: **122mm** 526: 190 D-30M; 36 M-1931/37; 300 M-30; **130mm** 420 M-46; **155mm** 16 GH-52

MRL 450: **122mm** 356: 96 BM-11; 60 BM-21; 50 *Sakr*-10; 50 *Sakr*-18; 100 *Sakr*-36; **130mm** 36 K136 *Kooryong*; **140mm** 32 BM-14; **227mm** 26 M270 MLRS; **240mm** (48 BM-24 in store)

MOR 2,564: **81mm** 50 M125A2; **82mm** 500; **SP 107mm** 100: 65 M106A1; 35 M106A2; **120mm** 1,848: 1,800 M-1943; 48 Brandt; **SP 120mm** 36 M1064A3; **160mm** 30 M-160

SURFACE-TO-SURFACE MISSILE LAUNCHERS

SRBM • Conventional 42+: 9 FROG-7; 24 *Sakr*-80; 9 9K72 *Elbrus* (SS-1C *Scud*-B)

UNINHABITED AERIAL VEHICLES

ISR • Medium R4E-50 *Skyeye*; ASN-209

AIR DEFENCE

SAM 45+

Point-defence 45 *Sinai*-23 with *Ayn al-Saqr*; *Ayn al-Saqr*; FIM-92 *Stinger*; 9K38 *Igla* (SA-18 *Grouse*); 9K338 *Igla*-S (SA-24 *Grinch*) (reported)

GUNS 860

SP 160: **23mm** 120 ZSU-23-4; **57mm** 40 ZSU-57-2

TOWED 700: **14.5mm** 300 ZPU-4; **23mm** 200 ZU-23-2; **57mm** 200 S-60

Navy ε8,500 (incl 2,000 Coast Guard); 10,000 conscript (total 18,500)

EQUIPMENT BY TYPE

SUBMARINES • SSK 7

4 Type-033 (PRC *Romeo*) with 8 single 533mm TT with UGM-84C *Harpoon* Block 1B AShM/Mk 37 HWT
3 Type-209/1400 with 8 single 533mm TT with UGM-84L *Harpoon* Block II AShM/*SeaHake* mod 4 (DM2A4) HWT

PRINCIPAL SURFACE COMBATANTS • FRIGATES 8

FFGHM 6:

4 *Alexandria* (ex-US *Oliver Hazard Perry*) with 1 Mk 13 GMLS with RGM-84C *Harpoon* Block 1B AShM/

SM-1MR Block VI SAM, 2 triple 324mm ASTT with Mk 46 LWT, 1 Mk 15 *Phalanx* CIWS, 1 76mm gun (capacity 2 SH-2G *Super Seasprite* ASW hel)

1 *El Fateh* (*Gowind* 2500) with 2 quad lnchrs with MM40 *Exocet* Block 3 AShM, 1 16-cell CLA VLS with VL MICA SAM, 2 triple 324mm ASTT with MU90 LWT, 1 76mm gun (capacity 1 med hel)

1 *Tahya Misr* (FRA *Aquitaine* (FREMM)) with 2 quad lnchr with MM40 *Exocet* Block 3 AShM, 2 8-cell *Sylver* A43 VLS with *Aster* 15 SAM, 2 twin 324mm B-515 ASTT with MU90 LWT, 1 76mm gun (capacity 1 med hel)

FFGH 2 *Damyat* (ex-US *Knox*) with 1 octuple Mk 16 GMLS with RGM-84C *Harpoon* Block 1B AShM/ASROC, 2 twin 324mm SVTT Mk 32 TT with Mk 46 LWT, 1 Mk 15 *Phalanx* CIWS, 1 127mm gun (capacity 1 SH-2G *Super Seasprite* ASW hel)

PATROL AND COASTAL COMBATANTS 63

CORVETTES 5

FSGM 2 *Abu Qir* (ESP *Descubierta*) (of which 1†) with 2 quad lnchr with RGM-84C *Harpoon* Block 1B AShM, 1 octuple *Albatros* lnchr with *Aspide* SAM, 2 triple 324mm SVTT Mk 32 ASTT with *Sting Ray* LWT, 1 twin 375mm Bofors ASW Rocket Launcher System A/S mor, 1 76mm gun

FSG 2 *Najim Al Zaffer* (PRC Type-053HE (*Jianghu* I)) with 2 twin lnchr with HY-2 (CH-SS-N-2 *Safflower*) AShM, 4 RBU 1200 A/S mor, 2 twin 57mm guns

FS 1 *Shabab Misr* (ex-RoK *Po Hang*) with 2 76mm guns

PCFGM 4 *Ezzat* (US *Ambassador* Fast Missile Craft) with 2 quad lnchr with RGM-84L *Harpoon* Block II AShM, 1 21-cell Mk49 lnchr with RAM Block 1A SAM, 1 Mk15 Mod 21 Block 1B *Phalanx* CIWS 1 76mm gun

PCFG 8:

1 Project 12418 (RUS *Tarantul* IV) with 2 twin lnchr with 3M80E *Moskit* (SS-N-22A *Sunburn*), 2 AK630 CIWS, 1 76mm gun

6 *Ramadan* with 4 single lnchr with *Otomat* Mk2 AShM, 1 76mm gun

1 *Tiger* with 2 twin lnchr with RGM-84 *Harpoon* AShM, 1 76mm gun

PCF 4 *Tiger* with 1 76mm gun

PCC 5 *Al-Nour* (ex-PRC *Hainan*) (3 more in reserve†) with 2 triple 324mm TT, 4 RBU 1200 A/S mor, 2 twin 57mm guns

PBFGM 8 Project 205 (ex-YUG *Osa* I) (of which 3†) with 4 single lnchr with P-20 (SS-N-2A *Styx*) AShM, 1 9K32 *Strela*-2 (SA-N-5 *Grail*) SAM (manual aiming)

PBFG 9:

4 Type-024 (PRC *Hegu*) (2 additional vessels in reserve) with 2 single lnchr with SY-1 (CH-SS-N-1 *Scrubbrush*) AShM

5 *October* (FSU *Komar*) (of which 1†) with 2 single lnchr with *Otomat* Mk2 AShM (1 additional vessel in reserve)

PBFM 4 *Shershen* (FSU) with 1 9K32 *Strela*-2 (SA-N-5 *Grail*) SAM (manual aiming), 1 12-tube BM-24 MRL

PBF 10:

6 *Kaan* 20 (TUR MRTP 20)

4 Project 205 (ex-FIN *Osa* II)

PB 6:

4 Type-062 (ex-PRC *Shanghai* II)

2 *Shershen* (FSU) (of which 1†) with 4 single 533mm TT, 1 8-tube BM-21 MRL

MINE WARFARE • MINE COUNTERMEASURES 14

MHC 5: 2 *Al Siddiq* (ex-US *Osprey*); 3 *Dat Assawari* (US Swiftships)

MSI 2 *Safaga* (US Swiftships)

MSO 7: 3 *Assiout* (FSU T-43); 4 *Aswan* (FSU *Yurka*)

AMPHIBIOUS

PRINCIPAL AMPHIBIOUS SHIPS • LHD 2 *Gamal Abdel Nasser* (FRA *Mistral*) (capacity 16 med hel; 2 LCT or 4 LCM; 13 MBTs; 50 AFVs; 450 troops)

LANDING SHIPS • LSM 2 *Polnochny* A (FSU) (capacity 6 MBT; 180 troops)

LANDING CRAFT 15:

LCT 2 EDA-R

LCM 13: 4 CTM NG; 9 *Vydra* (FSU) (capacity either 3 MBT or 200 troops)

LOGISTICS AND SUPPORT 24

AE 1 *Halaib* (ex-GER *Westerwald*)

AKR 3 *Al Hurreya*

AOT 7 *Ayeda* (FSU *Toplivo*) (1 more in reserve)

AR 1 *Shaledin* (ex-GER *Luneberg*)

ARS 2 *Al Areesh*

ATF 5 *Al Maks*† (FSU *Okhtensky*)

AX 5: 1 *El Fateh*† (ex-UK 'Z' class); 1 *El Horriya* (also used as the presidential yacht); 1 *Al Kousser*; 1 *Intishat*; 1 other

Coastal Defence

Army tps, Navy control

EQUIPMENT BY TYPE

COASTAL DEFENCE

ARTY **100mm**; **130mm** SM-4-1; **152mm**

AShM 4K87 (SSC-2B *Samlet*); *Otomat* MkII

Naval Aviation

All aircraft operated by Air Force

AIRCRAFT • TPT • Light 4 Beech 1900C (maritime surveillance)

UNINHABITED AERIAL VEHICLES

ISR • Light 2 S-100 *Camcopter*

Coast Guard 2,000

EQUIPMENT BY TYPE

PATROL AND COASTAL COMBATANTS 89

PBF 14: 6 *Crestitalia*; 5 *Swift Protector*; 3 *Peterson*

PB 75: 5 *Nisr*; 12 *Sea Spectre* MkIII; 25 Swiftships; 21 *Timsah*; 3 Type-83; 9 *Peterson*

Air Force 30,000 (incl 10,000 conscript)

FORCES BY ROLE

FIGHTER

1 sqn with F-16A/B *Fighting Falcon*

8 sqn with F-16C/D *Fighting Falcon*

1 sqn with J-7

3 sqn with MiG-21 *Fishbed*/MiG-21U *Mongol* A

2 sqn with *Mirage* 5D/E

1 sqn with *Mirage* 2000B/C

FIGHTER/GROUND ATTACK

1 sqn with *Mirage* 5E2

1 sqn (forming) with *Rafale* DM/EM
1 sqn (forming) with MiG-29M/M2 *Fulcrum*

ANTI-SUBMARINE WARFARE
1 sqn with SH-2G *Super Seasprite*

MARITIME PATROL
1 sqn with Beech 1900C

ELECTRONIC WARFARE
1 sqn with Beech 1900 (ELINT); *Commando* Mk2E (ECM)

ELECTRONIC WARFARE/TRANSPORT
1 sqn with C-130H/VC-130H *Hercules*

AIRBORNE EARLY WARNING
1 sqn with E-2C *Hawkeye*

SEARCH & RESCUE
1 unit with AW139

TRANSPORT
1 sqn with An-74TK-200A
1 sqn with C-130H/C-130H-30 *Hercules*
1 sqn with C295M
1 sqn with DHC-5D *Buffalo*
1 sqn with B-707-366C; B-737-100; Beech 200 *Super King Air*; *Falcon* 20; Gulfstream III; Gulfstream IV; Gulfstream IV-SP

TRAINING
1 sqn with *Alpha Jet**
1 sqn with DHC-5 *Buffalo*
3 sqn with EMB-312 *Tucano*
1 sqn with Grob 115EG
ε6 sqn with K-8 *Karakorum**
1 sqn with L-39 *Albatros*; L-59E *Albatros**

ATTACK HELICOPTER
1 sqn with Mi-24V
2 sqn with AH-64D *Apache*
1 sqn with Ka-52A *Hokum* B
2 sqn with SA-342K *Gazelle* (with HOT)
1 sqn with SA-342L *Gazelle*

TRANSPORT HELICOPTER
1 sqn with CH-47C/D *Chinook* 1 sqn with Mi-8
1 sqn with Mi-8/Mi-17-V1 *Hip*
1 sqn with S-70 *Black Hawk*; UH-60A/L *Black Hawk*

UAV
Some sqn with R4E-50 *Skyeye*; *Wing Loong* (GJ-1)

EQUIPMENT BY TYPE

AIRCRAFT 585 combat capable
FTR 62: 26 F-16A *Fighting Falcon*; 6 F-16B *Fighting Falcon*; ε30 J-7
FGA 320: 138 F-16C *Fighting Falcon*; 37 F-16D *Fighting Falcon*; 2 *Mirage* 2000B; 15 *Mirage* 2000C; 36 *Mirage* 5D/E; 12 *Mirage* 5E2; ε40 MiG-21 *Fishbed*/MiG-21U *Mongol* A; 16+ MiG-29M/M2 *Fulcrum*; 16 *Rafale* DM; 8 *Rafale* EM
ELINT 2 VC-130H *Hercules*
ISR 12: ε6 AT-802 *Air Tractor**; 6 *Mirage* 5R (5SDR)*
AEW&C 7 E-2C *Hawkeye*
TPT 82: **Heavy** 2 Il-76MF *Candid*; **Medium** 24: 21 C-130H *Hercules*; 3 C-130H-30 *Hercules*; **Light** 45: 3 An-74TK-200A; 1 Beech 200 *King Air*; 4 Beech 1900 (ELINT); 4 Beech 1900C; 24 C295M; 9 DHC-5D *Buffalo* (being withdrawn) **PAX** 11: 1 B-707-366C; 3 *Falcon* 20; 2 Gulfstream III; 1 Gulfstream IV; 4 Gulfstream IV-SP
TRG 329: 36 *Alpha Jet**; 54 EMB-312 *Tucano*; 74 Grob 115EG; 120 K-8 *Karakorum**; 10 L-39 *Albatros*; 35 L-59E*

HELICOPTERS
ATK 88: 45 AH-64D *Apache*; ε30 Ka-52A *Hokum* B; ε13 Mi-24V *Hind* E
ASW 10 SH-2G *Super Seasprite* (opcon Navy)
ELINT 4 *Commando* Mk2E (ECM)
MRH 72: 2 AW139 (SAR); 65 SA342K *Gazelle* (some with HOT); 5 SA342L *Gazelle* (opcon Navy)
TPT 96: **Heavy** 19: 3 CH-47C *Chinook*; 16 CH-47D *Chinook*; **Medium** 77: 2 AS-61; 24 *Commando* (of which 3 VIP); 40 Mi-8T *Hip*; 3 Mi-17-1V *Hip*; 4 S-70 *Black Hawk* (VIP); 4 UH-60L *Black Hawk* (VIP)
TRG 17 UH-12E

UNINHABITED AERIAL VEHICLES
CISR • Heavy 4+ *Wing Loong* (GJ-1)
ISR • Medium R4E-50 *Skyeye*

AIR LAUNCHED MISSILES
AAM • IR R-3 (AA-2 *Atoll*)‡; AIM-9M/P *Sidewinder*; R-73 (AA-11A *Archer*); R-550 *Magic*; 9M39 *Igla*-V; **IIR** *Mica* IR; **ARH** *Mica* RF; R-77 (AA-12 *Adder*); **SARH** AIM-7F/M *Sparrow*; R-530
ASM AASM; AGM-65A/D/F/G *Maverick*; AGM-114F/K *Hellfire*; AS-30L; HOT; AKD-10 (LJ-7); 9M120 *Ataka* (AT-9 *Spiral*-2)
LACM SCALP EG (reported)
AShM AGM-84L *Harpoon* Block II; AM39 *Exocet*;
ARM *Armat*; Kh-25MP (AS-12A *Kegler*)

BOMBS
Laser-guided GBU-10/12 *Paveway* II

Air Defence Command 80,000 conscript; 70,000 reservists (total 150,000)

FORCES BY ROLE

AIR DEFENCE
5 AD div (geographically based) (total: 12 SAM bty with M48 *Chaparral*, 12 radar bn, 12 ADA bde (total: 100 ADA bn), 12 SAM bty with MIM-23B I-*Hawk*, 14 SAM bty with *Crotale*, 18 AD bn with RIM-7M *Sea Sparrow* with *Skyguard*/GDF-003 with *Skyguard*, 110 SAM bn with S-125 *Pechora*-M (SA-3A *Goa*); 2K12 *Kub* (SA-6 *Gainful*); S-75M *Volkhov* (SA-2 *Guideline*))

EQUIPMENT BY TYPE

AIR DEFENCE
SAM 928
Long-range S-300V4 (SA-23)
Medium-range 612+: 40+ *Buk*-M1-2/M2E (SA-11/SA-17); 78+ MIM-23B I-*Hawk*; 282 S-75M *Volkhov* (SA-2 *Guideline*); 212+ S-125 *Pechora*-M (SA-3A *Goa*)
Short-range 180+: 56+ 2K12 *Kub* (SA-6 *Gainful*); 10 9K331 *Tor*-M1 (SA-15 *Gauntlet*); 10+ 9K331ME *Tor*-M2E (SA-15 *Gauntlet*); 24+ *Crotale*; 80 RIM-7M *Sea Sparrow* with *Skyguard*
Point-defence 136+: 50 M1097 *Avenger*; 50+ M48 *Chaparral*; 36+ *Sinai*-23 with *Ayn al-Saqr*
GUNS 910
SP • 23mm 230 ZSU-23-4 *Shilka*
TOWED 680: **35mm** 80 GDF-005 with *Skyguard*; **57mm** 600 S-60

Paramilitary ε397,000 active

Central Security Forces ε325,000

Ministry of Interior; includes conscripts
ARMOURED FIGHTING VEHICLES
APC • APC (W) *Walid*
AUV *Sherpa Light Scout*

National Guard ε60,000

Lt wpns only
FORCES BY ROLE
MANOEUVRE
Other
8 paramilitary bde (cadre) (3 paramilitary bn)
EQUIPMENT BY TYPE
ARMOURED FIGHTING VEHICLES APC • APC (W) 250 *Walid*

Border Guard Forces ε12,000

Ministry of Interior; lt wpns only
FORCES BY ROLE
MANOEUVRE
Other
18 Border Guard regt

DEPLOYMENT

CENTRAL AFRICAN REPUBLIC: UN • MINUSCA 1,000; 1 inf bn; 1 tpt coy

DEMOCRATIC REPUBLIC OF THE CONGO: UN • MONUSCO 10

MALI: UN • MINUSMA 1,149; 1 SF coy; 1 sy bn; 1 MP coy

SAUDI ARABIA: *Operation Restoring Hope* 6 F-16C *Fighting Falcon*

SOUTH SUDAN: UN • UNMISS 4

SUDAN: UN • UNAMID 161; 1 inf coy

WESTERN SAHARA: UN • MINURSO 21

FOREIGN FORCES

Australia MFO (*Operation Mazurka*) 27
Canada MFO 68
Colombia MFO 275; 1 inf bn
Czech Republic MFO 18; 1 C295M
Fiji MFO 170; elm 1 inf bn
France MFO 1
Italy MFO 75; 3 PB
New Zealand MFO 26; 1 trg unit; 1 tpt unit
Norway MFO 3
United Kingdom MFO 2
United Arab Emirates ε300: 12 F-16E/F *Fighting Falcon* (reported); *Wing Loong* I UAV; *Wing Loong* II UAV
United States MFO 454; elm 1 ARNG recce bn; 1 ARNG spt bn (1 EOD coy, 1 medical coy, 1 hel coy)
Uruguay MFO 41 1 engr/tpt unit

Iran IRN

Iranian Rial r		2019	2020	2021
GDP	r	24515tr	30153tr	
	US$	584bn	611bn	
per capita	US$	7,010	7,257	
Growth	%	-6.5	-5.0	
Inflation	%	41.0	30.5	
Def bdgt [a]	r	ε722tr	ε694tr	
	US$	ε17.2bn	ε14.1bn	
US$1=r		41999.99	49377.25	

[a] Excludes Law Enforcement Forces (NAJA)

Real-terms defence budget trend (US$bn, constant 2015)

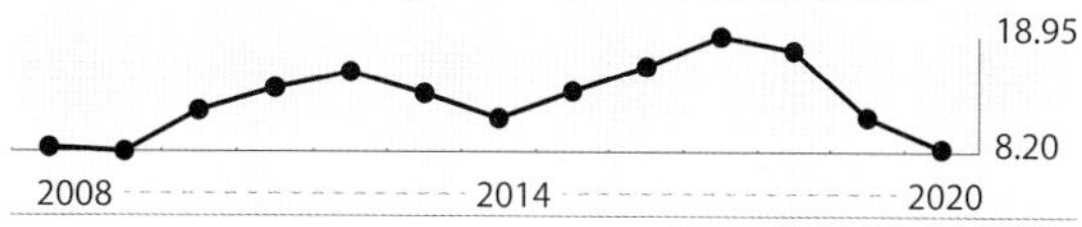

Population 84,923,314

Age	0–14	15–19	20–24	25–29	30–64	65 plus
Male	12.3%	3.5%	3.4%	4.6%	24.2%	2.7%
Female	11.8%	3.3%	3.2%	4.4%	23.5%	3.1%

Capabilities

Iran is a major regional military power, due to a combination of asymmetric and other strategies, despite significant handicaps to its conventional capabilities because of international sanctions and restrictions on arms imports. Iran exerts significant military influence via a range of regional allies and proxies, chiefly through the Islamic Revolutionary Guard Corps (IRGC). This is, in effect, a form of expeditionary capability which is likely to continue despite the killing in a US airstrike of influential commander Qasem Soleimani. It has also developed a ballistic-missile inventory. Iran has a key relationship with Syria and ties with Russia, including for defence sales. It has developed significant influence in weaker regional states like Lebanon, Iraq and Yemen through a network of non-state groups, such as Hizbullah and Houthi forces. The armed forces are large by regional standards. The IRGC's Quds Force is a principal element of Iran's military power abroad. An attack in Saudi Arabia in 2019 offers an indication that Iran has developed the capacity to conduct complex strikes utilising land-attack missiles and UAVs. The regular navy has limited power-projection capabilities, while the IRGC navy is responsible for maritime security close to home. The armed forces struggle with an ageing inventory of primary combat equipment that ingenuity and asymmetric-warfare techniques can only partially offset. The expiry of the UN arms embargo could allow greater access to new weaponry although available funding may remain limited. China and Russia are potentially major suppliers. In regional terms, Iran has a well-developed defence-industrial base, which has displayed the capacity to support and sustain equipment. Key sectors include missiles and guided weapons, but the defence industry is still incapable of meeting the need for major weapons systems.

ACTIVE 610,000 (Army 350,000 Islamic Revolutionary Guard Corps 190,000 Navy 18,000 Air 37,000 Air Defence 15,000) Paramilitary 40,000

Armed Forces General Staff coordinates two parallel organisations: the regular armed forces and the Islamic Revolutionary Guard Corps

Conscript liability 18–21 months (reported, with variations depending on location in which service is performed)

RESERVE 350,000 (Army 350,000, ex-service volunteers)

ORGANISATIONS BY SERVICE

Army 130,000; 220,000 conscript (total 350,000)

FORCES BY ROLE
5 corps-level regional HQ
COMMAND
1 cdo div HQ
4 armd div HQ
2 mech div HQ
4 inf div HQ
SPECIAL FORCES
1 cdo div (3 cdo bde)
6 cdo bde
1 SF bde
MANOEUVRE
Armoured
8 armd bde
Mechanised
14 mech bde
Light
12 inf bde
Air Manoeuvre
1 AB bde
Aviation
Some avn gp
COMBAT SUPPORT
5 arty gp

EQUIPMENT BY TYPE
Totals incl those held by IRGC Ground Forces. Some equipment serviceability in doubt
ARMOURED FIGHTING VEHICLES
MBT 1,513+: 480 T-72S; 150 M60A1; 75+ T-62; 100 *Chieftain* Mk3/Mk5; 540 T-54/T-55/Type-59/*Safir-74*; 168 M47/M48
LT TK 80+: 80 *Scorpion*
RECCE 35 EE-9 *Cascavel*
IFV 610+: 210 BMP-1; 400 BMP-2 with 9K111 *Fagot* (AT-4 *Spigot*); BMT-2 *Cobra*
APC 640+
APC (T) 340: 140 *Boragh* with 9K111 *Fagot* (AT-4 *Spigot*); 200 M113
APC (W) 300+: 300 BTR-50/BTR-60; *Rakhsh*
ENGINEERING & MAINTENANCE VEHICLES
ARV 20+: BREM-1 reported; 20 *Chieftain* ARV; M578; T-54/55 ARV reported
VLB 15: 15 *Chieftain* AVLB
MW *Taftan* 1
ANTI-TANK/ANTI-INFRASTRUCTURE
MSL • MANPATS 9K11 *Malyutka* (AT-3 *Sagger*/*I-Raad*); 9K111 *Fagot* (AT-4 *Spigot*); 9K111-1 *Konkurs* (AT-5 *Spandrel*/*Towsan*-1); *Dehleavieh* (*Kornet*); *Saeqhe* 1; *Saeqhe* 2; *Toophan*; *Toophan* 2
RCL 200+: **75mm** M20; **82mm** B-10; **106mm** ε200 M40; **107mm** B-11
ARTILLERY 6,798+
SP 292+: **122mm** 60+: 60 2S1 *Gvozdika*; *Raad*-1 (*Thunder* 1); **155mm** 150+: 150 M109A1; *Raad*-2 (*Thunder* 2); **170mm** 30 M-1978; **175mm** 22 M107; **203mm** 30 M110
TOWED 2,030+; **105mm** 150: 130 M101A1; 20 M-56; **122mm** 640: 540 D-30; 100 Type-54 (M-30); **130mm** 985 M-46; **152mm** 30 D-20; **155mm** 205: 120 GHN-45; 70 M114; 15 Type-88 WAC-21; **203mm** 20 M115
MRL 1,476+: **107mm** 1,300: 700 Type-63; 600 HASEB *Fadjr* 1; **122mm** 157: 7 BM-11; 100 BM-21 *Grad*; 50 *Arash*/*Hadid*/*Noor*; **240mm** 19+: ε10 *Fadjr* 3; 9 M-1985; **330mm** *Fadjr* 5
MOR 3,000: **81mm**; **82mm**; **107mm** M30; **120mm** M-65
SURFACE-TO-SURFACE MISSILE LAUNCHERS
SRBM • Conventional ε30 CH-SS-8 (175 msl); *Shahin*-1/*Shahin*-2; *Nazeat*; *Oghab*
AIRCRAFT • TPT 17 **Light** 16: 10 Cessna 185; 2 F-27 *Friendship*; 4 *Turbo Commander* 690; **PAX** 1 *Falcon* 20
HELICOPTERS
ATK 50 AH-1J *Cobra*
TPT 167: **Heavy** ε20 CH-47C *Chinook*; **Medium** 69: 49 Bell 214; 20 Mi-171; **Light** 78: 68 Bell 205A (AB-205A); 10 Bell 206 *Jet Ranger* (AB-206)
UNINHABITED AERIAL VEHICLES
CISR • Medium *Mohajer* 6
ISR • Medium *Ababil* 2; *Ababil* 3; *Mohajer* 3; *Mohajer* 4; **Light** *Mohajer* 2
AIR DEFENCE
SAM
Short-range FM-80
Point-defence 9K36 *Strela*-3 (SA-14 *Gremlin*); 9K32 *Strela*-2 (SA-7 *Grail*)‡; *Misaq* 1 (QW-1 *Vanguard*); *Misaq* 2 (QW-18); 9K338 *Igla-S* (SA-24 *Grinch*) (reported); HN-5A
GUNS 1,122
SP 180: **23mm** 100 ZSU-23-4; **57mm** 80 ZSU-57-2
TOWED 942+: **14.5mm** ZPU-2; ZPU-4; **23mm** 300 ZU-23-2; **35mm** 92 GDF-002; **37mm** M-1939; **40mm** 50 L/70; **57mm** 200 S-60; **85mm** 300 M-1939
BOMBS
Laser-guided *Qaem*
Electro-optical guided *Qaem*

Islamic Revolutionary Guard Corps 190,000

Islamic Revolutionary Guard Corps Ground Forces 150,000

Controls Basij paramilitary forces. Lightly manned in peacetime. Primary role: internal security; secondary role: external defence, in conjunction with regular armed forces

FORCES BY ROLE
COMMAND
31 provincial corps HQ (2 in Tehran)
SPECIAL FORCES
3 spec ops div
1 AB bde
MANOEUVRE
Armoured
2 armd div
3 armd bde

Light
8+ inf div
5+ inf bde

Islamic Revolutionary Guard Corps Naval Forces 20,000+ (incl 5,000 Marines)

FORCES BY ROLE
COMBAT SUPPORT
Some arty bty
Some AShM bty with HY-2 (CH-SSC-3 *Seersucker*) AShM

EQUIPMENT BY TYPE
In addition to the vessels listed, the IRGC operates a substantial number of patrol boats with a full-load displacement below 10 tonnes, including *Boghammar*-class vessels and small *Bavar*-class wing-in-ground effect air vehicles
PATROL AND COASTAL COMBATANTS 126
PBFG 56:
5 C14 with 2 twin lnchr with C-701 (*Kosar*)/C-704 (*Nasr*) AShM
10 Mk13 with 2 single lnchr with C-704 (*Nasr*) AShM, 2 single 324mm TT
10 *Thondor* (PRC *Houdong*) with 2 twin lnchr with C-802A (*Ghader*) AShM, 2 AK230 CIWS
25 *Peykaap* II (IPS-16 mod) with 2 single lnchr with C-701 (*Kosar*) AShM/C-704 (*Nasr*), 2 single 324mm TT
6 *Zolfaghar* (*Peykaap* III/IPS-16 mod) with 2 single lnchr with C-701 (*Kosar*)/C-704 (*Nasr*) AShM
PBFT 15 *Peykaap* I (IPS -16) with 2 single 324mm TT
PBF 35: 15 *Kashdom* II; 10 *Tir* (IPS-18); ε10 *Pashe* (MIG-G-1900)
PB ε20 *Ghaem*
AMPHIBIOUS
LANDING SHIPS • LST 3 *Hormuz* 24 (*Hejaz* design for commercial use)
LANDING CRAFT • LCT 2 *Hormuz* 21 (minelaying capacity)
LOGISTICS AND SUPPORT • AP 3 *Naser*
COASTAL DEFENCE • AShM C-701 (*Kosar*); C-704 (*Nasr*); C-802 (*Noor*); HY-2 (CH-SSC-3 *Seersucker*)
HELICOPTERS
MRH 5 Mi-171 *Hip*
TPT • Light some Bell 206 (AB-206) *Jet Ranger*

Islamic Revolutionary Guard Corps Marines 5,000+

FORCES BY ROLE
MANOEUVRE
Amphibious
1 mne bde

Islamic Revolutionary Guard Corps Aerospace Force 15,000

Controls Iran's strategic-missile force

FORCES BY ROLE
MISSILE
ε1 bde with *Shahab*-1/-2; *Qiam*-1
ε1 bn with *Shahab*-3

EQUIPMENT BY TYPE
SURFACE-TO-SURFACE MISSILE LAUNCHERS
MRBM • Conventional up to 50: *Shahab*-3 (mobile & silo); some *Ghadr*-1 (in test); some *Emad*-1 (in test); some *Sajjil*-2 (in devt); some *Khorramshahr* (in devt)
SRBM • Conventional up to 100: some *Fateh*-110; some *Fateh*-313; Some *Khalij Fars* (*Fateh*-110 mod ASBM); some *Shahab*-1/-2; some *Qiam*-1; some *Zelzal*; some *Zolfagar* (IR-SS-1)
GLCM • Conventional some *Ya'ali* (*Quds-1*)
UNINHABITED AERIAL VEHICLES
CISR • Heavy *Shahed* 129
ISR • Medium *Ababil* 3; *Mohajer* 4; *Shahed* 123
AIR DEFENCE
SAM
Medium-range *Ra'ad*/*3rd Khordad*; *Talash*/*15th Khordad*
Point-defence *Misaq 1* (QW-1 *Vanguard*); *Misaq 2* (QW-18)
BOMBS
Laser-guided *Sadid*
Electro-optical guided *Sadid*

Islamic Revolutionary Quds Force 5,000

Navy 18,000

HQ at Bandar Abbas

EQUIPMENT BY TYPE
In addition to the vessels listed, the Iranian Navy operates a substantial number of patrol boats with a full-load displacement below 10 tonnes
SUBMARINES • TACTICAL 17
SSK 1 *Taregh* (RUS *Paltus* (Project 877EKM (*Kilo*))) (2 more non-operational) with 6 single 533mm TT
SSC 1 *Fateh* with 4 single 533mm TT with C-704 (*Nasr*-1) AShM/*Valfajar* HWT
SSW 15: 14 *Ghadir* with 2 single 533mm TT with *Valfajar* HWT (additional vessels in build); 1 *Nahang*
PATROL AND COASTAL COMBATANTS 66
CORVETTES 6
FSGM 2 *Jamaran* (UK Vosper Mk 5 derivative – 1 more undergoing sea trials) with 2 twin lnchr with C-802 (*Noor*) (CH-SS-N-6) AShM, 2 single lnchr with SM-1 SAM, 2 triple 324mm SVTT Mk 32 ASTT, 1 76mm gun, 1 hel landing platform
FSG 4:
2 *Alvand* (UK Vosper Mk 5) with 2 twin lnchr with C-802 (CH-SS-N-6) AShM, 2 triple 324mm SVTT Mk 32 ASTT, 1 114mm gun
1 *Alvand* (UK Vosper Mk 5) with 2 twin lnchr with C-802 (CH-SS-N-6) AShM, 2 triple 324mm SVTT Mk 32 ASTT, 1 AK630M CIWS, 1 114mm gun
1 *Bayandor* (US PF-103) with 2 twin lnchr with C-802 (CH-SS-N-6) AShM, 2 triple 324mm SVTT Mk 32 ASTT, 1 76mm gun
PCFG 14+ *Kaman* (FRA *Combattante* II) with 1 twin lnchr with C-802 (*Noor*) (CH-SS-N-6) AShM, 1 76mm gun
PBG 9:
3 *Hendijan* with 2 twin lnchr with C-802 (*Noor*) (CH-SS-N-6) AShM

3 *Kayvan* with 2 single lnchr with C-704 (*Nasr*) AShM
3 *Parvin* with 2 single lnchr with C-704 (*Nasr*) AShM
PBFT 3 *Kajami* (semi-submersible) with 2 324mm TT
PBF 1 MIL55
PB 33: 9 C14; 8 *Hendijan*; 6 MkII; 10 MkIII
AMPHIBIOUS
LANDING SHIPS 12
LST 3 *Hengam* with 1 hel landing platform (capacity 9 tanks; 225 troops)
LSM 3 *Farsi* (ROK) (capacity 9 tanks; 140 troops)
LSL 6 *Fouque*
LANDING CRAFT 11
LCT 2
LCU 1 *Liyan 110*
UCAC 8: 2 *Wellington* Mk 4; 4 *Wellington* Mk 5; 2 *Tondar* (UK *Winchester*)
LOGISTICS AND SUPPORT 18
AE 2 *Delvar*
AFD 2 *Dolphin*
AG 1 *Hamzah* with 2 single lnchr with C-802 (*Noor*) (CH-SS-N-6) AShM
AK 3 *Delvar*
AORH 3: 2 *Bandar Abbas*; 1 *Kharg* with 1 76mm gun
AWT 5: 4 *Kangan*; 1 *Delvar*
AX 2 *Kialas*
COASTAL DEFENCE • AShM C-701 (*Kosar*); C-704 (*Nasr*); C-802 (*Noor*); C-802A (*Ghader*); *Ra'ad* (reported)

Marines 2,600

FORCES BY ROLE
MANOEUVRE
Amphibious
2 mne bde

Naval Aviation 2,600

EQUIPMENT BY TYPE
AIRCRAFT
TPT 16: **Light** 13: 5 Do-228; 4 F-27 *Friendship*; 4 *Turbo Commander* 680; **PAX** 3 *Falcon* 20 (ELINT)
HELICOPTERS
ASW ε10 SH-3D *Sea King*
MCM 3 RH-53D *Sea Stallion*
TPT • Light 17: 5 Bell 205A (AB-205A); 2 Bell 206 *Jet Ranger* (AB-206); 10 Bell 212 (AB-212)

Air Force 18,000

FORCES BY ROLE
Serviceability probably about 60% for US ac types and about 80% for PRC/Russian ac. Includes IRGC AF equipment
FIGHTER
1 sqn with F-7M *Airguard*; JJ-7*
2 sqn with F-14 *Tomcat*
2 sqn with MiG-29A/UB *Fulcrum*
FIGHTER/GROUND ATTACK
1 sqn with *Mirage* F-1E; F-5E/F *Tiger* II
5 sqn with F-4D/E *Phantom* II
3 sqn with F-5E/F *Tiger* II
1 sqn (forming) with Su-22M4 *Fitter K*; Su-22UM-3K *Fitter* G
GROUND ATTACK
1 sqn with Su-24MK *Fencer* D
MARITIME PATROL
1 sqn with P-3F *Orion*
ISR
1 (det) sqn with RF-4E *Phantom* II*
SEARCH & RESCUE
Some flt with Bell 214C (AB-214C)
TANKER/TRANSPORT
1 sqn with B-707; B-747; B-747F
TRANSPORT
1 sqn with B-707; *Falcon* 50; L-1329 *Jetstar*; Bell 412
2 sqn with C-130E/H *Hercules*
1 sqn with F-27 *Friendship*; *Falcon* 20
1 sqn with Il-76 *Candid*; An-140 (Iran-140 *Faraz*)
TRAINING
1 sqn with Beech F33A/C *Bonanza*
1 sqn with F-5B *Freedom Fighter*
1 sqn with PC-6
1 sqn with PC-7 *Turbo Trainer*
Some units with EMB-312 *Tucano*; MFI-17 *Mushshak*; TB-21 *Trinidad*; TB-200 *Tobago*
TRANSPORT HELICOPTER
1 sqn with CH-47 *Chinook*
Some units with Bell 206A *Jet Ranger* (AB-206A); *Shabaviz* 2-75; *Shabaviz* 2061

EQUIPMENT BY TYPE
AIRCRAFT 333 combat capable
FTR 183+: 20 F-5B *Freedom Fighter*; 55+ F-5E/F *Tiger* II; 24 F-7M *Airguard*; up to 43 F-14 *Tomcat*; 35 MiG-29A/UB *Fulcrum*; up to 6 *Azarakhsh* (reported)
FGA 88: 62 F-4D/E *Phantom* II; 10 *Mirage* F-1E; up to 6 *Saegheh* (reported); up to 7 Su-22M4 *Fitter K*; 3+ Su-22UM-3K *Fitter* G
ATK 39: 29 Su-24MK *Fencer* D; 7 Su-25K *Frogfoot* (status unknown); 3 Su-25UBK *Frogfoot* (status unknown)
ASW 3 P-3F *Orion*
ISR: 6+ RF-4E *Phantom* II*
TKR/TPT 3: ε1 B-707; ε2 B-747
TPT 116: **Heavy** 12 Il-76 *Candid*; **Medium** ε19 C-130E/H *Hercules*; **Light** 75: 11 An-74TK-200; 5 An-140 (Iran-140 *Faraz*); 10 F-27 *Friendship*; 1 L-1329 *Jetstar*; 10 PC-6B *Turbo Porter*; 8 TB-21 *Trinidad*; 4 TB-200 *Tobago*; 3 *Turbo Commander* 680; 14 Y-7; 9 Y-12; **PAX** 10: ε1 B-707; 1 B-747; 4 B-747F; 1 *Falcon* 20; 3 *Falcon* 50
TRG 141: 25 Beech F33A/C *Bonanza*; 15 EMB-312 *Tucano*; 14 JJ-7*; 25 MFI-17 *Mushshak*; 12 *Parastu*; 15 PC-6; 35 PC-7 *Turbo Trainer*
HELICOPTERS
MRH 2 Bell 412
TPT 34+: **Heavy** 2+ CH-47 *Chinook*; **Medium** 30 Bell 214C (AB-214C); **Light** 2+: 2 Bell 206A *Jet Ranger* (AB-206A); some *Shabaviz* 2-75 (indigenous versions in production); some *Shabaviz* 2061
AIR-LAUNCHED MISSILES
AAM • IR PL-2A‡; PL-7; R-60 (AA-8 *Aphid*); R-73 (AA-11A *Archer*); AIM-9J *Sidewinder*; **IR/SARH** R-27 (AA-10 *Alamo*); **SARH** AIM-7E-2 *Sparrow*; **ARH** AIM-54 *Phoenix*†
ASM AGM-65A *Maverick*; Kh-25 (AS-10 *Karen*); Kh-25ML (AS-10 *Karen*); Kh-29L/T (AS-14A/B *Kedge*)

AShM C-801K
ARM Kh-58 (AS-11 *Kilter*)
BOMBS
Electro-optical guided GBU-87/B *Qassed*

Air Defence Force 12,000

New service branch formed mid-2019

FORCES BY ROLE
AIR DEFENCE
16 bn with MIM-23B I-*Hawk*/*Shahin*
4 bn with S-300PMU2 (SA-20 *Gargoyle*)
5 sqn with FM-80 (*Crotale*); *Rapier*; S-75M *Volkhov* (SA-2 *Guideline*); S-200 *Angara* (SA-5 *Gammon*); 9K331 *Tor*-M1 (SA-15 *Gauntlet*)

EQUIPMENT BY TYPE
AIR DEFENCE
SAM 546+:
Long-range 42+: 10 S-200 *Angara* (SA-5 *Gammon*); 32 S-300PMU2 (SA-20 *Gargoyle*); *Bavar*-373
Medium-range 195+: 150+ MIM-23B I-*Hawk*/*Shahin*; 45 S-75 *Dvina* (SA-2 *Guideline*); *Talash*/*15th Khordad*
Short-range 279: 250 FM-80 (*Crotale*); 29 9K331 *Tor*-M1 (SA-15 *Gauntlet*)
Point-defence 30+: 30 *Rapier*; *Misaq 1* (QW-1 *Vanguard*); *Misaq 2* (QW-18)
GUNS • TOWED 23mm ZU-23-2; **35mm** GDF-002

Paramilitary 40,000–60,000

Law-Enforcement Forces 40,000–60,000 (border and security troops); 450,000 on mobilisation (incl conscripts)

Part of armed forces in wartime

EQUIPMENT BY TYPE
PATROL AND COASTAL COMBATANTS • PB ε90
AIRCRAFT • TPT • Light 2+: 2 An-140; some Cessna 185/Cessna 310
HELICOPTERS • TPT • Light ε24 AB-205 (Bell 205)/AB-206 (Bell 206) *Jet Ranger*

Basij Resistance Force ε600,000 on mobilisation

Paramilitary militia with claimed membership of 12.6 million; ε600,000 combat capable

FORCES BY ROLE
MANOEUVRE
Other
2,500 militia bn(-) (claimed, limited permanent membership)

DEPLOYMENT

GULF OF ADEN AND SOMALI BASIN: Navy: 1 FSG; 1 LST
SYRIA: 1,500

Iraq IRQ

Iraqi Dinar D		2019	2020	2021
GDP	D	272tr	211tr	
	US$	230bn	178bn	
per capita	US$	5,884	4,438	
Growth	%	4.4	-12.1	
Inflation	%	-0.2	0.8	
Def bdgt [a]	D	11.6tr	ε12.1tr	
	US$	9.9bn	ε10.3bn	
FMA (US$)	US$	250m	0m	0m
US$1=D		1182.00	1182.00	

[a] Excludes Ministry of the Interior and National Security Council budget

Real-terms defence budget trend (US$bn, constant 2015)

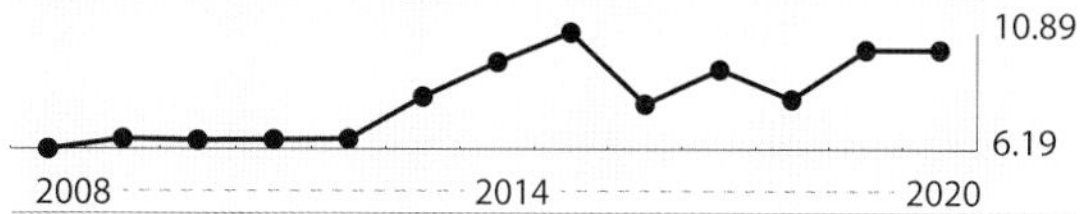

Population 38,872,655

Age	0–14	15–19	20–24	25–29	30–64	65 plus
Male	18.9%	5.4%	4.7%	3.9%	16.0%	1.5%
Female	18.1%	5.2%	4.5%	3.9%	16.1%	1.9%

Capabilities

The armed forces' capabilities and morale have improved since the collapse of several divisions in the face of the ISIS advance in the north in 2014. The recapture of Mosul in 2017 demonstrated incremental growth in capability, as Iraq's armed forces adapted to fight ISIS in urban areas. The future of the Kurdish Peshmerga forces and the Popular Mobilisation Units militias, particularly questions over their integration into a national-security framework, remains an issue for Baghdad. The government's most critical security relationship has been with the US, although this was strained after the US strike that killed Iranian commander Qasem Soleimani. The level of attrition among Iraqi forces has caused concern, particularly among the well-regarded Counter-Terrorism Service, which has been used as a spearhead force. The US has been engaged in a training effort and in late 2018 NATO established a new training and capacity-building mission. Iraqi forces appear to have improved their ability to conduct complex operations. However, there has been a reliance on US air support and coalition ISR assistance, suggesting continuing capability limitations in this area. The US and other international forces drew down their presence in 2020 citing increased confidence in the capabilities of Iraqi forces. Significant logistical shortcomings remain, including logistics support and intelligence integration. Internal political frictions, revived US–Iran tensions and efforts to rein in corruption add to concerns over the cohesion and reliability of the armed forces and associated PMU militias. The inventory comprises Soviet-era and Russian equipment combined with newer European- and US-sourced platforms. Barring military maintenance facilities, the Iraqi defence industry has only a limited ability to manufacture light weapons and ammunition.

ACTIVE 193,000 (Army 180,000 Navy 3,000 Air 5,000 Air Defence 5,000) Paramilitary 148,000

ORGANISATIONS BY SERVICE

Army ε180,000

Includes Counter Terrorism Service

FORCES BY ROLE

SPECIAL FORCES

3 SF bde

1 ranger bde (3 ranger bn)

MANOEUVRE

Armoured

1 (9th) armd div (2 armd bde, 2 mech bde, 1 engr bn, 1 sigs regt, 1 log bde)

Mechanised

3 (5th, 8th & 10th) mech div (4 mech inf bde, 1 engr bn, 1 sigs regt, 1 log bde)

1 (7th) mech div (2 mech inf bde, 1 inf bde, 1 engr bn, 1 sigs regt, 1 log bde)

Light

1 (6th) mot div (3 mot inf bde, 1 inf bde, 1 engr bn, 1 sigs regt, 1 log bde)

1 (14th) mot div (2 mot inf bde, 3 inf bde, 1 engr bn, 1 sigs regt, 1 log bde)

1 (1st) inf div (2 inf bde)

1 (11th) inf div (3 lt inf bde, 1 engr bn, 1 sigs regt, 1 log bde)

1 (15th) inf div (5 inf bde)

1 (16th) inf div (2 inf bde)

1 (17th Cdo) inf div (4 inf bde, 1 engr bn, 1 sigs regt, 1 log bde)

1 inf bde

Other

1 (PM SF) sy div (3 inf bde)

HELICOPTER

1 atk hel sqn with Mi-28NE *Havoc*

1 atk hel sqn with Mi-35M *Hind*

1 sqn with Bell 205 (UH-1H *Huey* II)

3 atk hel sqn with Bell T407; H135M

3 sqn with Mi-17 *Hip* H; Mi-171Sh

1 ISR sqn with SA342M *Gazelle*

2 trg sqn with Bell 206; OH-58C *Kiowa*

1 trg sqn with Bell 205 (UH-1H *Huey* II)

1 trg sqn with Mi-17 *Hip*

EQUIPMENT BY TYPE

ARMOURED FIGHTING VEHICLES

MBT 391+: ε100 M1A1 *Abrams;* 168+ T-72M/M1; ε50 T-55; 73 T-90S

RECCE 453: ε400 *Akrep*; 18 BRDM 2; 35 EE-9 *Cascavel*;

IFV 650: ε400 BMP-1; ε90 BMP-3M; ε60 BTR-4 (inc variants); 100 BTR-80A

APC 1,592+

APC (T) 900: ε500 M113A2/*Talha*; ε400 MT-LB

PPV 692+: 12 *Barracuda*; 250 *Caiman*; *Gorets*-M; ε400 ILAV *Badger; Mamba*; 30 *Maxxpro*

AUV M-ATV

ENGINEERING & MAINTENANCE VEHICLES

ARV 222+: 180 BREM; 35+ M88A1/2; 7 *Maxxpro* ARV; T-54/55 ARV; Type-653; VT-55A

NBC VEHICLES 20 *Fuchs* NBC

ANTI-TANK/ANTI-INFRASTRUCTURE

MSL • MANPATS 9K135 *Kornet* (AT-14 *Spriggan*) (reported)

ARTILLERY 1,061+

SP 48+: **152mm** 18+ Type-83; **155mm** 30: 6 M109A1; 24 M109A5

TOWED 60+: **130mm** M-46/Type-59; **152mm** D-20; Type-83; **155mm** ε60 M198

MRL 3+: **122mm** some BM-21 *Grad*; **220mm** 3+ TOS-1A

MOR 950+: **81mm** ε500 M252; **120mm** ε450 M120; **240mm** M-240

HELICOPTERS

ATK 35: 11 Mi-28NE *Havoc*; 4 Mi-28UB *Havoc*; 20+ Mi-35M *Hind*

MRH 63+: 4+ SA342 *Gazelle*; 17 Bell IA407; 23 H135M; ε19 Mi-17 *Hip* H/Mi-171Sh

ISR 10 OH-58C *Kiowa*

TPT • Light 44: 16 Bell 205 (UH-1H *Huey* II); 10 Bell 206B3 *Jet Ranger*; ε18 Bell T407

UNINHABITED AERIAL VEHICLES • CISR Heavy 10 CH-4†

AIR-LAUNCHED MISSILES • ASM 9K114 *Shturm* (AT-6 *Spiral*); AGR-20A APKWS; AR-1; *Ingwe*

Navy 3,000

EQUIPMENT BY TYPE

PATROL AND COASTAL COMBATANTS 32

PCO 2 *Al Basra* (US *River Hawk)*

PCC 4 *Fateh* (ITA *Diciotti*)

PB 20: 12 Swiftships 35; 5 *Predator* (PRC 27m); 3 *Al Faw*

PBR 6: 2 Type-200; 4 Type-2010

Marines 1,000

FORCES BY ROLE

MANOEUVRE

Amphibious

2 mne bn

Air Force ε5,000

FORCES BY ROLE

FIGHTER/GROUND ATTACK

1 sqn with F-16C/D *Fighting Falcon*

GROUND ATTACK

1 sqn with Su-25/Su-25K/Su-25UBK *Frogfoot*

1 sqn with L-159A; L-159T1

ISR

1 sqn with CH-2000 *Sama*; SB7L-360 *Seeker*

1 sqn with Cessna 208B *Grand Caravan*; Cessna AC-208B *Combat Caravan**

1 sqn with Beech 350 *King Air*

TRANSPORT

1 sqn with An-32B *Cline*

1 sqn with C-130E/J-30 *Hercules*

TRAINING

1 sqn with Cessna 172, Cessna 208B

1 sqn with *Lasta*-95

1 sqn with T-6A

1 sqn with T-50IQ *Golden Eagle**

EQUIPMENT BY TYPE

AIRCRAFT 90 combat capable

FGA 34: 26 F-16C *Fighting Falcon*; 8 F-16D *Fighting Falcon*;
ATK 30: 10 L-159A; 1 L-159T1; ε19 Su-25/Su-25K/Su-25UBK *Frogfoot*
ISR 10: 2 Cessna AC-208B *Combat Caravan**; 2 SB7L-360 *Seeker*; 6 Beech 350ER *King Air*
TPT 29: **Medium** 15: 3 C-130E *Hercules*; 6 C-130J-30 *Hercules*; 6 An-32B *Cline* (of which 2 combat capable); **Light** 14: 1 Beech 350 *King Air*; 5 Cessna 208B *Grand Caravan*; 8 Cessna 172
TRG 57+: 8 CH-2000 *Sama*; 10+ *Lasta*-95; 15 T-6A; 24 T-50IQ *Golden Eagle**

AIR-LAUNCHED MISSILES

AAM • IR AIM-9L *Sidewinder*; AIM-9M *Sidewinder*
ASM AGM-114 *Hellfire*

BOMBS

Laser-guided GBU-12 *Paveway* II
INS/GPS-guided FT-9

Air Defence Command ε5,000

FORCES BY ROLE

AIR DEFENCE

1 bn with 96K6 *Pantsir*-S1 (SA-22 *Greyhound*)
1 bn with M1097 *Avenger*
1 bn with 9K338 *Igla*-S (SA-24 *Grinch*)
1 bn with ZU-23-2; S-60

EQUIPMENT BY TYPE

AIR DEFENCE

SAM 24+
Short-range 24 96K6 *Pantsir*-S1 (SA-22 *Greyhound*)
Point-defence M1097 *Avenger*; 9K338 *Igla*-S (SA-24 *Grinch*)
GUNS • TOWED 23mm ZU-23-2; **57mm** S-60

Paramilitary ε148,000

Iraqi Federal Police ε36,000

Territorial Interdiction Force ε12,000

FORCES BY ROLE

MANOEUVRE

Other
4 sy bde

Militias ε100,000

Popular Mobilisation Units include: Badr Organisation; Kataib Hizbullah; Kataib Imam Ali; Kataib Sayyid al-Shuhada

FOREIGN FORCES

Australia *Operation Inherent Resolve (Okra)* 110 • NATO Mission Iraq 2
Belgium *Operation Inherent Resolve* (*Valiant Phoenix*) 5
Canada *Operation Inherent Resolve* (*Impact*) 100; 1 SF gp; 1 med unit • NATO Mission Iraq 250; 1 hel flt with 3 Bell 412 (CH-146 *Griffon*)
Czech Republic *Operation Inherent Resolve* 60
Denmark *Operation Inherent Resolve* 210; 1 SF gp; 1 trg team
Estonia *Operation Inherent Resolve* 10 • NATO Mission Iraq 5
Fiji UNAMI 167; 2 sy unit
Finland *Operation Inherent Resolve* 80; 1 trg unit • NATO Mission Iraq 5
Germany *Operation Inherent Resolve* 150; some trg units
Greece NATO Mission Iraq 1
Hungary *Operation Inherent Resolve* 170 • NATO Mission Iraq 1
Italy *Operation Inherent Resolve* (*Prima Parthica*) 600; 1 inf regt; 1 trg unit; 1 hel sqn with 4 NH90 • NATO Mission Iraq 46
Lithuania *Operation Inherent Resolve* 6 • NATO Mission Iraq 9
Nepal UNAMI 77; 1 sy unit
Netherlands *Operation Inherent Resolve* 60; 2 trg units • NATO Mission Iraq 2
Norway *Operation Inherent Resolve* 60; 1 trg unit • NATO Mission Iraq 10
Poland *Operation Inherent Resolve* 150 • NATO Mission Iraq 65
Portugal *Operation Inherent Resolve* 34
Romania *Operation Inherent Resolve* 10 • NATO Mission Iraq 4
Slovakia NATO Mission Iraq 42
Slovenia *Operation Inherent Resolve* 6
Spain *Operation Inherent Resolve* 500; 1 trg units; 1 hel unit • NATO Mission Iraq 13
Sweden *Operation Inherent Resolve* 66 • NATO Mission Iraq 1
Turkey Army 1,000; 1 cdo unit • NATO Mission Iraq up to 30
United Kingdom *Operation Inherent Resolve* (*Shader*) 400; 1 inf bn(-); 1 engr sqn(-)
United States *Operation Inherent Resolve* 3,000; 1 AB bde(-); 1 EOD pl; 1 atk hel bn with AH-64D *Apache*; 4 AD bty with MIM-104E/F *Patriot* PAC-2/-3; 1 CISR UAV sqn with MQ-9A *Reaper*

Israel ISR

New Israeli Shekel NS		2019	2020	2021
GDP	NS	1.41tr	1.33tr	
	US$	395bn	383bn	
per capita	US$	43,603	41,560	
Growth	%	3.4	-5.9	
Inflation	%	0.8	-0.5	
Def bdgt [a]	NS	59.9bn	ε57.6bn	
	US$	16.8bn	ε16.6bn	
FMA (US)	US$	3.30bn	3.30bn	3.30bn
US$1=NS		3.56	3.46	

[a] Excludes Foreign Military Financing

Real-terms defence budget trend (US$bn, constant 2015)

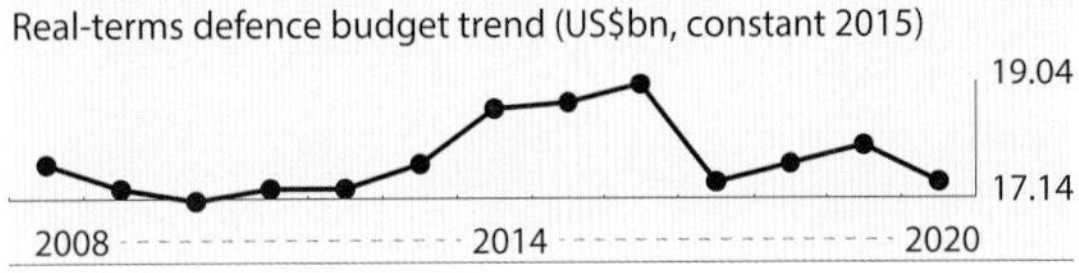

Population 8,675,475

Age	0–14	15–19	20–24	25–29	30–64	65 plus
Male	13.7%	4.2%	3.8%	3.6%	19.6%	5.4%
Female	13.1%	4.0%	3.7%	3.4%	19.0%	6.6%

Capabilities

The Israel Defense Forces (IDF) are organised for territorial defence, short-term interventions in neighbouring states and limited regional power projection. Israel is widely believed to possess a nuclear-weapons capability. Following the 2015 Plan *Gideon* a new multi-year Plan *Tnufa* was adopted by the IDF. It seeks to improve areas of relative superiority such as technology and intelligence, to ensure swifter and more decisive operations against future threats. However, budgetary challenges may affect the plan's implementation. The US remains Israel's key defence partner, as well as a significant source of funding, and is instrumental in several of the IDF's equipment programmes, particularly in missile defence and combat aviation. Israel also maintains discreet ties with a number of Arab states and has recently sought to normalise relations with some Gulf states. Personnel quality and training are generally high, despite the IDF's continuing reliance on national service. Ground-forces training is being overhauled, with new training centres under construction. The IDF has no requirement for out-of-area deployments and its logistics capabilities are limited to sustaining operations within Israel itself or in immediately neighbouring territories. The largely asymmetric nature of the threats faced by the IDF in recent years has focused modernisation efforts on force-protection, missile-defence and precision-strike capabilities. Israel maintains a broad defence-industrial base, with world-class capabilities in several areas, notably armoured vehicles, uninhabited systems, guided weapons and cyber security.

ACTIVE 169,500 (Army 126,000 Navy 9,500 Air 34,000) Paramilitary 8,000

Conscript liability Officers 48 months, other ranks 32 months, women 24 months (Jews and Druze only; Christians, Circassians and Muslims may volunteer)

RESERVE 465,000 (Army 400,000 Navy 10,000 Air 55,000)

Annual trg as cbt reservists to age 40 (some specialists to age 54) for male other ranks, 38 (or marriage/pregnancy) for women

ORGANISATIONS BY SERVICE

Strategic Forces

Israel is widely believed to have a nuclear capability – delivery means include F-15I and F-16I ac, *Jericho* 2 IRBM and, reportedly, *Dolphin*/*Tanin*-class SSKs with LACM

FORCES BY ROLE

SURFACE-TO-SURFACE MISSILE

3 IRBM sqn with *Jericho* 2

EQUIPMENT BY TYPE

SURFACE-TO-SURFACE MISSILE LAUNCHERS

IRBM • Nuclear: ε24 *Jericho* 2

Strategic Defences

FORCES BY ROLE

AIR DEFENCE

3 bty with *Arrow* 2 ATBM with *Green Pine*/*Super Green Pine* radar and *Citrus Tree* command post
10 bty with *Iron Dome* (incl reserve bty)
6 bty with MIM-104C *Patriot* PAC-2
2 bty with *David's Sling*

Space

EQUIPMENT BY TYPE

SATELLITES 10

COMMUNICATIONS 3 *Amos*
ISR 7: 1 EROS; 5 *Ofeq* (5, 7, 9, 10 & 16); 1 TecSAR-1 (*Polaris*)

Army 26,000; 100,000 conscript (total 126,000)

Organisation and structure of formations may vary according to op situations. Equipment includes that required for reserve forces on mobilisation

FORCES BY ROLE

COMMAND

3 (regional comd) corps HQ
2 armd div HQ
1 (Multidimensional) div HQ
5 (territorial) inf div HQ
1 (home defence) comd HQ

SPECIAL FORCES

1 spec ops bde (3 spec ops unit)

MANOEUVRE

Reconnaissance
1 indep recce bn
Armoured
3 armd bde (1 armd recce coy, 3 armd bn, 1 AT coy, 1 cbt engr bn)
1 (Multidimensional) armd inf/ISR bn
Mechanised
3 mech inf bde (3 mech inf bn, 1 cbt spt bn, 1 sigs coy)
1 mech inf bde (4 mech inf bn, 1 cbt spt bn)
1 indep mech inf bn
Light
2 indep inf bn
Air Manoeuvre
1 para bde (3 para bn, 1 cbt spt bn, 1 sigs coy)
Other
1 armd trg bde (3 armd bn)
1 (Border Protection) sy bde (4 sy bn; 2 ISR bn)

COMBAT SUPPORT
3 arty bde
1 engr bde (3 engr bn, 3 EOD coy)
1 CBRN bn
1 int bde (3 int bn)
1 SIGINT unit
2 MP bn

Reserves 400,000+ on mobilisation

FORCES BY ROLE
COMMAND
3 armd div HQ
1 AB div HQ
MANOEUVRE
Armoured
9 armd bde
Mechanised
8 mech inf bde
Light
16 (territorial/regional) inf bde
Air Manoeuvre
4 para bde
Mountain
1 mtn inf bn
COMBAT SUPPORT
5 arty bde
COMBAT SERVICE SUPPORT
6 log unit

EQUIPMENT BY TYPE
ARMOURED FIGHTING VEHICLES
MBT 490: ε160 *Merkava* MkIII; ε330 *Merkava* MkIV (ε660 *Merkava* MkIII; ε220 *Merkava* MkIV all in store)
APC • APC (T) 1,360: ε260 *Namer*; ε100 *Achzarit* (modified T-55 chassis); 500 M113A2; ε400 *Nagmachon* (*Centurion* chassis); *Nakpadon* (5,000 M113A1/A2 in store)
AUV *Ze'ev*
ENGINEERING & MAINTENANCE VEHICLES
AEV D9R; *Namer*; *Puma*
ARV *Centurion* Mk2; *Nemmera*; M88A1; M113 ARV
VLB *Alligator* MAB; M48/60; MTU
NBC VEHICLES ε8 TPz-1 *Fuchs* NBC
ANTI-TANK/ANTI-INFRASTRUCTURE • MSL
SP M113 with *Spike*; *Tamuz* (*Spike* NLOS)
MANPATS IMI MAPATS; *Spike* SR/MR/LR/ER
ARTILLERY 530
SP 250: **155mm** 250 M109A5 (**155mm** 30 M109A2; **175mm** 36 M107; **203mm** 36 M110 all in store)
TOWED (**155mm** 171: 40 M-46 mod; 50 M-68/M-71; 81 M-839P/M-845P all in store)
MRL 30: **227mm** 30 M270 MLRS; **306mm** IMI *Lynx* (**160mm** 50 LAR-160; **227mm** 18 M270 MLRS; **290mm** 20 LAR-290 all in store)
MOR 250: **81mm** 250 (**81mm** 1,100; **120mm** 650; **160mm** 18 Soltam M-66 all in store)
SURFACE-TO-SURFACE MISSILE LAUNCHERS
IRBM • Nuclear ε24 *Jericho* 2
SRBM • Dual-capable (7 *Lance* in store)
AIR DEFENCE • SAM • Point-defence *Machbet*; FIM-92 *Stinger*

Navy 7,000; 2,500 conscript (total 9,500)

EQUIPMENT BY TYPE
SUBMARINES 5
SSK 5:
3 *Dolphin* (GER HDW design) with 6 single 533mm TT with UGM-84C *Harpoon* Block 1B AShM/*SeaHake* (DM2A3) HWT/*SeaHake* mod 4 (DM2A4) HWT/*Kaved* HWT, 4 single 650mm TT with with dual-capable LACM (reported)
2 *Tanin* (GER HDW design) (fitted with AIP) with 6 single 533mm TT with UGM-84C *Harpoon* Block 1B AShM/*SeaHake* (DM2A3) HWT/*SeaHake* mod 4 (DM2A4) HWT/*Kaved* HWT, 4 single 650mm TT with with dual-capable LACM (reported)
PATROL AND COASTAL COMBATANTS 45
CORVETTES • FSGHM 3:
2 *Eilat* (*Sa'ar* 5) with 2 quad lnchr with RGM-84 *Harpoon* AShM/*Gabriel* V AShM, 4 8-cell VLS with *Barak*-1 SAM (being upgraded to *Barak*-8), 2 triple 324mm TT with Mk 46 LWT, 1 Mk 15 *Phalanx* CIWS (capacity 1 AS565SA *Panther* ASW hel)
1 *Eilat* (*Sa'ar* 5) with 2 quad lnchr with RGM-84 *Harpoon* AShM/*Gabriel* V AShM, 4 8-cell VLS with *Barak*-8 SAM, 2 triple 324mm TT with Mk 46 LWT, 1 Mk 15 *Phalanx* CIWS (capacity 1 AS565SA *Panther* ASW hel)
PCGM 8 *Hetz* (*Sa'ar* 4.5) with 2 quad lnchr with RGM-84 *Harpoon* AShM (can also be fitted with up to 6 single lnchr with *Gabriel* II AShM), 2 8-cell VLS with *Barak*-1 SAM, (can be fitted with 2 triple 324mm Mk32 TT with Mk46 LWT), 1 Mk 15 *Phalanx* CWIS, 1 76mm gun
PBF 34: 5 *Shaldag*; 3 *Stingray*; 9 *Super Dvora* Mk I (SSM & TT may be fitted); 4 *Super Dvora* Mk II (SSM & TT may be fitted); 6 *Super Dvora* Mk II-I (SSM & TT may be fitted); 4 *Super Dvora* Mk III (SSM & TT may be fitted); 3 *Super Dvora* Mk III (SSM may be fitted)
AMPHIBIOUS • LANDING CRAFT • LCVP 3 *Manta*
LOGISTICS AND SUPPORT • AG 1 *Bat Yam* (ex-GER Type-745)

Naval Commandos ε300

FORCES BY ROLE
SPECIAL FORCES
1 cdo unit

Air Force 34,000

Responsible for Air and Space Coordination

FORCES BY ROLE
FIGHTER & FIGHTER/GROUND ATTACK
1 sqn with F-15A/B/D *Eagle* (*Baz*)
1 sqn with F-15B/C/D *Eagle* (*Baz*)
1 sqn with F-15I *Ra'am*
5 sqn with F-16C/D *Fighting Falcon* (*Barak*)
4 sqn with F-16I *Fighting Falcon* (*Sufa*)
2 sqn with F-35I *Adir*
ANTI-SUBMARINE WARFARE
1 sqn with AS565SA *Panther* (missions flown by IAF but with non-rated aircrew)
ELECTRONIC WARFARE
1 sqn with RC-12D *Guardrail*; Beech A36 *Bonanza* (*Hofit*); Beech 200 *King Air*; Beech 200T *King Air*; Beech 200CT *King Air*

AIRBORNE EARLY WARNING & CONTROL
1 sqn with Gulfstream G550 *Eitam*; Gulfstream G550 *Shavit*
TANKER/TRANSPORT
1 sqn with C-130E/H *Hercules*; KC-130H *Hercules*
1 sqn with C-130J-30 *Hercules*
1 sqn with KC-707
TRAINING
1 OPFOR sqn with F-16C/D *Fighting Falcon* (*Barak*)
1 sqn with M-346 *Master* (*Lavi*)
ATTACK HELICOPTER
1 sqn with AH-64A *Apache* (*Peten*)
1 sqn with AH-64D *Apache* (*Sarat*)
TRANSPORT HELICOPTER
2 sqn with CH-53D *Sea Stallion*
2 sqn with S-70A *Black Hawk*; UH-60A *Black Hawk*
1 medevac unit with CH-53D *Sea Stallion*
UAV
1 ISR sqn with *Hermes* 450
1 ISR sqn with *Heron* (*Shoval*); *Heron* TP (*Eitan*)
1 ISR sqn with *Heron* (*Shoval*) (MP role)
AIR DEFENCE
3 bty with *Arrow* 2/3
10 bty with *Iron Dome*
6 bty with MIM-104C *Patriot* PAC-2
2 bty with *David's Sling*
SPECIAL FORCES
1 SF wg (2 SF unit, 1 CSAR unit, 1 int unit)

EQUIPMENT BY TYPE
AIRCRAFT 334 combat capable
FGA 304: 16 F-15A *Eagle* (*Baz*); 6 F-15B *Eagle* (*Baz*); 17 F-15C *Eagle* (*Baz*); 19 F-15D *Eagle* (*Baz*); 25 F-15I *Ra'am*; ε50 F-16C *Fighting Falcon* (*Barak*); 49 F-16D *Fighting Falcon* (*Barak*); 98 F-16I *Fighting Falcon* (*Sufa*); 24 F-35I *Adir*
ISR 6 RC-12D *Guardrail*
ELINT 4: 1 EC-707; 3 Gulfstream G550 *Shavit*
AEW 4: 2 B-707 *Phalcon*; 2 Gulfstream G550 *Eitam* (1 more on order)
TKR/TPT 10: 4 KC-130H *Hercules*; 6 KC-707
TPT 65: **Medium** 18: 5 C-130E *Hercules*; 6 C-130H *Hercules*; 7 C-130J-30 *Hercules*; **Light** 47: 3 AT-802 *Air Tractor*; 9 Beech 200 *King Air*; 8 Beech 200T *King Air*; 5 Beech 200CT *King Air*; 22 Beech A36 *Bonanza* (*Hofit*)
TRG 67: 17 Grob G-120; 30 M-346 *Master* (*Lavi*)*; 20 T-6A
HELICOPTERS
ATK 43: 26 AH-64A *Apache* (*Peten*); 17 AH-64D *Apache* (*Sarat*)
ASW 7 AS565SA *Panther* (missions flown by IAF but with non-rated aircrew)
ISR 12 OH-58B *Kiowa*
TPT 80: **Heavy** 25 CH-53D *Sea Stallion*; **Medium** 49: 39 S-70A *Black Hawk*; 10 UH-60A *Black Hawk*; **Light** 6 Bell 206 *Jet Ranger*
UNINHABITED AERIAL VEHICLES
ISR 3+: **Heavy** 3+: *Heron* (*Shoval*); 3 *Heron* TP (*Eitan*); RQ-5A *Hunter*; **Medium** *Hermes* 450; *Hermes* 900 (22+ *Searcher* MkII in store); (an unknown number of ISR UAVs are combat capable)
LOITERING MUNITIONS *Harop*; *Harpy*
AIR DEFENCE
SAM 40+:
Long-range MIM-104C *Patriot* PAC-2
Medium-range some *David's Sling*
Short-range up to 40 *Iron Dome*
Point-defence *Machbet*
GUNS • TOWED 20mm M167 *Vulcan*
MISSILE DEFENCE • SAM 24 *Arrow* 2/*Arrow* 3;
AIR-LAUNCHED MISSILES
AAM • IR AIM-9 *Sidewinder*; *Python* 4; **IIR** *Python* 5; **ARH** AIM-120C AMRAAM
ASM AGM-114 *Hellfire*; AGM-62B *Walleye*; AGM-65 *Maverick*; *Delilah* AL; *Popeye* I/*Popeye* II; *Spike* NLOS
BOMBS
IIR guided *Opher*
Laser-guided *Griffin*; *Lizard*; *Paveway* II
INS/GPS guided GBU-31 JDAM; GBU-39 Small Diameter Bomb (*Barad Had*); *Spice*

Airfield Defence 3,000 active (15,000 reservists)

Paramilitary ε8,000

Border Police ε8,000

FOREIGN FORCES

UNTSO unless specified. UNTSO figures represent total numbers for mission
Argentina 3
Australia 12
Austria 5
Belgium 1
Bhutan 4 • UNDOF 3
Canada 6
Chile 3
China 4
Czech Republic UNDOF 2
Denmark 10
Estonia 3
Fiji 2 • UNDOF 137; 1 inf bn(-); elm 1 log bn
Finland 15
Gambia 1
Ghana UNDOF 2
India 2 • UNDOF 159; 1 MP coy(-); 1 log bn(-)
Ireland 12 • UNDOF 138; 1 inf coy
Nepal 3 • UNDOF 357; 2 mech inf coy; 1 engr pl
Netherlands 13 • UNDOF 1
New Zealand 7
Norway 10
Poland 1
Russia 3
Serbia 1
Slovakia 2
Slovenia 2
Sweden 7
Switzerland 12
United States 3 • US Strategic Command; 1 AN/TPY-2 X-band radar at Mount Keren
Uruguay UNDOF 191; 1 mech inf coy

Jordan JOR

Jordanian Dinar D		2019	2020	2021
GDP	D	31.6bn	30.2bn	
	US$	44.6bn	42.6bn	
per capita	US$	4,426	4,174	
Growth	%	2.0	-5.0	
Inflation	%	0.7	-0.3	
Def bdgt [a]	D	1.20bn	1.22bn	
	US$	1.69bn	1.72bn	
FMA (US)	US$	425m	350m	500m
US$1=D		0.71	0.71	

[a] Excludes expenditure on public order and safety

Real-terms defence budget trend (US$bn, constant 2015)

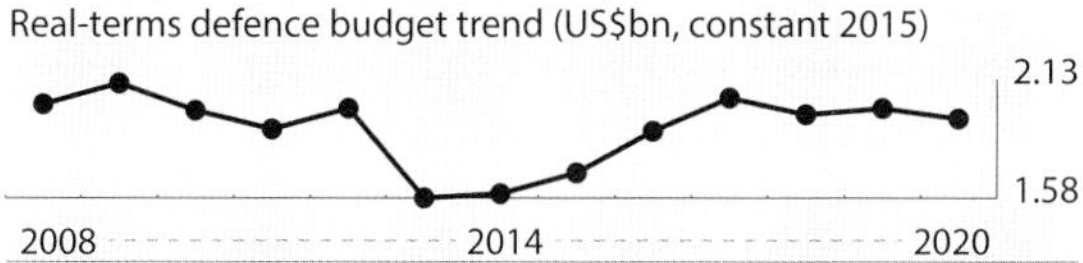

Population 10,820,644

Age	0–14	15–19	20–24	25–29	30–64	65 plus
Male	17.0%	5.3%	5.1%	4.8%	18.7%	1.8%
Female	16.1%	4.9%	4.4%	4.0%	16.1%	1.9%

Capabilities

The Jordanian armed forces are structured to provide border security and an armoured response to conventional threats. Their well-regarded operational capability belies their moderate size and ageing equipment inventory. There are ongoing efforts to restructure, review modernisation requirements and increase efficiency. There has been no recent public statement of defence policy, but regional instability is a prime concern, principally the ongoing war in Syria. Jordan is a major non-NATO ally of the US and there is a close bilateral defence relationship. The country has developed a bespoke special-forces training centre and has hosted training for numerous state and non-state military forces. Personnel are well trained, particularly aircrew and special forces, who are highly regarded internationally. Jordanian forces are able to independently deploy regionally and have participated in ISAF operations in Afghanistan and in coalition air operations over Syria and Yemen. Jordan's inventory largely comprises older systems. Although the state-owned King Abdullah II Design and Development Bureau (KADDB) has demonstrated a vehicle-upgrade capacity, the army has largely recapitalised its armoured-vehicle fleet with second-hand armour from European countries. KADDB produces some light armoured vehicles for domestic use, but the company currently has little export profile.

ACTIVE 100,500 (Army 86,000 Navy 500 Air 14,000) Paramilitary 15,000

RESERVE 65,000 (Army 60,000 Joint 5,000)

ORGANISATIONS BY SERVICE

Army 86,000

FORCES BY ROLE

SPECIAL FORCES
1 (Royal Guard) SF gp (1 SF regt, 1 SF bn, 1 CT bn)
1 (AB) SF bde (3 SF bn)

MANOEUVRE
Armoured
2 armd bde
1 armd inf bn
Mechanised
8 mech bde
Light
2 lt inf bde
Air Manoeuvre
1 (QRF) AB bde (1 SF bn, 2 AB bn)

COMBAT SUPPORT
4 arty bde
4 AD bde
1 MRL bn
1 engr bn

COMBAT SERVICE SUPPORT
1 log bn

EQUIPMENT BY TYPE

ARMOURED FIGHTING VEHICLES
MBT 266: ε80 FV4034 *Challenger* 1 (*Al Hussein*) (being withdrawn); 4+ *Leclerc*; 182 M60A3 (ε300 FV4034 *Challenger* 1 (*Al Hussein*) in store)
ASLT 80 B1 *Centauro* (61 more in store)
IFV 720: 13 AIFV-B-C25; 50 *Marder* 1A3; 321 *Ratel*-20; 336 YPR-765 PRI
APC 923+
APC (T) 729: 370 M113A1/A2 Mk1J; 269 M577A2 (CP); 87 YPR-765 PRCO (CP); 3 AIFV-B
PPV 194: 25 *Marauder*; 25 *Matador*; 100 *MaxxPro*; 44 *Nomad/Thunder*
AUV 35 *Cougar*

ENGINEERING & MAINTENANCE VEHICLES
ARV 85+: *Al Monjed*; 5 BPz-1; FV4204 *Chieftain* ARV; 32 M88A1; 30 M578; 18 YPR-806
MW 12 *Aardvark* Mk2

ANTI-TANK/ANTI-INFRASTRUCTURE • MSL
SP 115: 70 M901; 45 AIFV-B-*Milan*
MANPATS FGM-148 *Javelin*; TOW/TOW-2A; 9K135 *Kornet* (AT-14 *Spriggan*); Luch *Corsar*; *Stugna*-P

ARTILLERY 1,393+
SP 506: **155mm** 358 M109A1/A2; **203mm** 148 M110A2
TOWED 94: **105mm** 66: 54 M102; 12 M119A2; **155mm** 28: 10 M1/M59; 18 M114; **203mm** (4 M115 in store)
MRL 16+: **227mm** 12 M142 HIMARS; **273mm** 4+ WM-80
MOR 777: **81mm** 359; **SP 81mm** 50; **107mm** 50 M30; **120mm** 300 Brandt **SP 120mm** 18 *Agrab* Mk2

AIR DEFENCE
SAM • **Point-defence** 140+: 92 9K35 *Strela*-10 (SA-13 *Gopher*); 48 9K33 *Osa*-M (SA-8 *Gecko*); 9K36 *Strela*-3 (SA-14 *Gremlin*); 9K310 *Igla*-1 (SA-16 *Gimlet*); 9K38 *Igla* (SA-18 *Grouse*); 9K338 *Igla*-S (SA_24 *Grinch*)
GUNS • **SP 35mm** 60 *Gepard*

Navy ε500

EQUIPMENT BY TYPE

PATROL AND COASTAL COMBATANTS 9
PBF 2 Response Boat-Medium (RB-M)
PB 7: 4 *Abdullah* (US *Dauntless*); 3 *Al Hussein* (UK Vosper 30m)

Marines

FORCES BY ROLE

MANOEUVRE

Amphibious

1 mne unit

Air Force 14,000

FORCES BY ROLE

FIGHTER/GROUND ATTACK

2 sqn with F-16AM/BM *Fighting Falcon*

ISR

1 sqn with AT-802U *Air Tractor*

1 sqn with Cessna 208B

TRANSPORT

1 sqn with C-130E *Hercules*

TRAINING

1 OCU with F-16AM/BM *Fighting Falcon*

1 sqn with PC-21

1 sqn with Grob 120TP

1 hel sqn with R-44 *Raven* II

ATTACK HELICOPTER

2 sqn with AH-1F *Cobra* (with TOW)

TRANSPORT HELICOPTER

1 sqn with AS332M *Super Puma;* UH-60A *Black Hawk*

1 sqn with UH-60A *Black Hawk*

1 sqn with UH-60M *Black Hawk*

1 sqn with Mi-26T2 *Halo*

1 (Royal) flt with VH-60M *Black Hawk;* AW139

ISR UAV

1 sqn with S-100 *Camcopter*

AIR DEFENCE

2 bde with MIM-23B Phase III I-*Hawk*

EQUIPMENT BY TYPE

AIRCRAFT 57 combat capable

FGA 47: 33 F-16AM *Fighting Falcon*; 14 F-16BM *Fighting Falcon*

ATK (2 AC235 in store, offered for sale)

ISR 10 AT-802U *Air Tractor**

TPT 10: **Medium** 3 C-130E *Hercules* (1 C-130B *Hercules*; 4 C-130H *Hercules* in store); **Light** 7: 5 Cessna 208B; 2 M-28 *Skytruck* (2 C295M in store, offered for sale)

TRG 27: 15 Grob 120TP; 12 PC-21; (12 *Hawk* Mk63* in store, offered for sale)

HELICOPTERS

ATK 12 AH-1F *Cobra* (17 more in store, offered for sale)

MRH 14: 3 AW139; 11 H135M (Tpt/SAR) (6 MD-530F in store, offered for sale)

TPT 49: **Heavy** 4 Mi-26T2 *Halo*; **Medium** 33: 10 AS332M *Super Puma* (being WFU); 8 UH-60A *Black Hawk*; 12 UH-60M *Black Hawk*; 3 VH-60M *Black Hawk*; (8 UH-60L in store, offered for sale); **Light** 12 R-44 *Raven* II; (13 Bell 205 (UH-1H *Iroquois*) in store, offered for sale)

UNINHABITED AERIAL VEHICLES

CISR • Heavy (some CH-4B in store, offered for sale)

ISR • Light up to 10 S-100 *Camcopter*

AIR DEFENCE

SAM • Medium-range 24 MIM-23B Phase III I-*Hawk*

GUNS • TOWED 40mm 22 L/70 (with *Flycatcher* radar)

AIR-LAUNCHED MISSILES

AAM • IR AIM-9J/N/P *Sidewinder;* **SARH** AIM-7 *Sparrow;* **ARH** AIM-120C AMRAAM

ASM AGM-65D/G *Maverick;* BGM-71 TOW

BOMBS

Laser-guided GBU-10/12 *Paveway* II

Paramilitary ε15,000 active

Gendarmerie ε15,000 active

3 regional comd

FORCES BY ROLE

SPECIAL FORCES

2 SF unit

MANOEUVRE

Other

10 sy bn

EQUIPMENT BY TYPE

ARMOURED FIGHTING VEHICLES

APC • APC (W) 25+: AT105 *Saxon* (reported); 25+ EE-11 *Urutu*

AUV AB2 *Al-Jawad*

DEPLOYMENT

CENTRAL AFRICAN REPUBLIC: UN • MINUSCA 9

DEMOCRATIC REPUBLIC OF THE CONGO: UN • MONUSCO 10

MALI: UN • MINUSMA 64

SAUDI ARABIA: *Operation Restoring Hope* 6 F-16C *Fighting Falcon*

SOUTH SUDAN: UN • UNMISS 5

SUDAN: UN • UNAMID 4

WESTERN SAHARA: UN • MINURSO 2

FOREIGN FORCES

France *Operation Inherent Resolve* (*Chammal*) 8 *Rafale* F3; 1 *Atlantique* 2

Germany *Operation Inherent Resolve* 280; 4 *Tornado* ECR; 1 A400M

Norway *Operation Inherent Resolve* 20

United States Central Command: *Operation Inherent Resolve* 2,300; 1 FGA sqn with 12 F-15E *Strike Eagle*; 1 CISR sqn with 12 MQ-9A *Reaper*; 2 AD bty with MIM-104E/F *Patriot* PAC-2/-3

Kuwait KWT

Kuwaiti Dinar D		2019	2020	2021
GDP	D	40.9bn	32.8bn	
	US$	135bn	109bn	
per capita	US$	28,500	22,252	
Growth	%	0.4	-8.1	
Inflation	%	1.1	1.0	
Def bdgt [a]	D	2.37bn	ε2.34bn	
	US$	7.86bn	ε7.76bn	
US$1=D		0.30	0.30	

[a] Includes National Guard

Real-terms defence budget trend (US$bn, constant 2015)

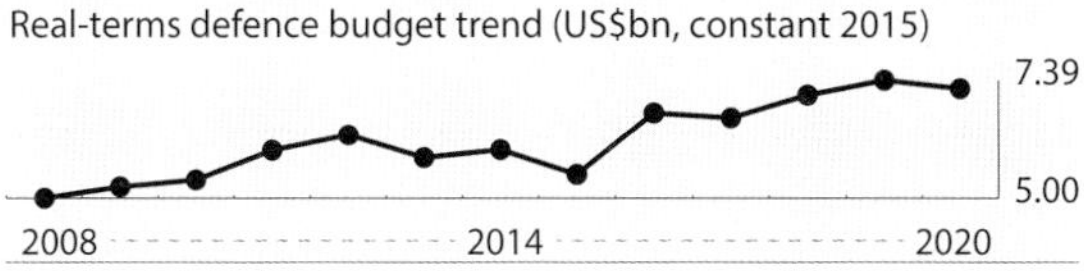

Population 2,993,706

Age	0–14	15–19	20–24	25–29	30–64	65 plus
Male	12.7%	3.2%	5.0%	7.2%	28.8%	1.3%
Female	11.6%	3.0%	3.7%	4.3%	17.6%	1.6%

Capabilities

Kuwait's small but capable armed forces have benefited considerably from the presence on Kuwaiti territory of sizeable US forces. The primary responsibility is territorial defence, through a strategy of sufficient readiness to provide a holding force until the mobilisation of friendly forces. Kuwait is a member of the GCC and has a bilateral defence-cooperation agreement with the US, which provides for a range of joint activities and mentoring, and the stationing and pre-positioning of significant numbers of US personnel and supplies of equipment. Since 2004, Kuwait has been designated a US major non-NATO ally. There is limited logistic-support capacity, although heavy-airlift and airborne-tanking assets grant a limited airborne-expeditionary capability. The equipment inventory includes a range of modern European- and US-sourced platforms, including advanced air-defence missile batteries for which it is seeking improvements. Kuwait is recapitalising its combat-aircraft fleet with the F/A-18E/F *Super Hornet* and Eurofighter *Typhoon*, which together will significantly enhance its air-combat capabilities. It is in the process of upgrading its fleet of main battle tanks and is seeking additional attack helicopters, and possibly new missile-armed multi-mission fast attack craft for the navy. Kuwait lacks a domestic defence-industrial base and is reliant on imports, albeit with offset requirements to help stimulate the country's wider industrial sector.

ACTIVE 17,500 (Army 11,500 Navy 2,000 Air 2,500 Emiri Guard 1,500) Paramilitary 7,100

Conscript liability 12 months, males 18–35 years

RESERVE 23,700 (Joint 23,700)

Reserve obligation to age 40; 1 month annual trg

ORGANISATIONS BY SERVICE

Army 11,500

FORCES BY ROLE

SPECIAL FORCES

1 SF unit

MANOEUVRE

Reconnaissance

1 mech/recce bde

Armoured

3 armd bde

Mechanised

2 mech inf bde

Light

1 cdo bde

Other

1 (Amiri) gd bde

COMBAT SUPPORT

1 arty bde

1 engr bde

1 MP bn

COMBAT SERVICE SUPPORT

1 log gp

1 fd hospital

Reserve

FORCES BY ROLE

MANOEUVRE

Mechanised

1 bde

EQUIPMENT BY TYPE

ARMOURED FIGHTING VEHICLES

MBT 293: 218 M1A2 *Abrams*; 75 M-84AB (75 more in store)

IFV 537: 76 BMP-2; 122 BMP-3; 103 BMP-3M; 236 *Desert Warrior*† (incl variants)

APC 260

APC (T) 260: 230 M113A2; 30 M577 (CP)

APC (W) (40 TH 390 *Fahd* in store)

AUV 118 *Sherpa Light Scout*

ENGINEERING & MAINTENANCE VEHICLES

ARV 19+: 19 M88A1/2; Type-653A; *Warrior*

MW *Aardvark* Mk2

NBC VEHICLES 12 *Fuchs*-2 NBC

ARTY 211

SP 155mm 106: 37 M109A3; 18 Mk F3; 51 PLZ-45 (18 AU-F-1 in store)

MRL 300mm 27 9A52 *Smerch*

MOR 78: **81mm** 60; **107mm** 6 M30; **120mm** ε12 RT-F1

ANTI-TANK/ANTI-INFRASTRUCTURE

MSL

SP 74: 66 HMMWV TOW; 8 M901

MANPATS 9K135 *Kornet* (AT-14 *Spriggan*); TOW-2

RCL 84mm *Carl Gustaf*

AIR DEFENCE

SAM • Point-defence *Starburst*; FIM-92 *Stinger*

Navy ε2,000 (incl 500 Coast Guard)

EQUIPMENT BY TYPE

PATROL AND COASTAL COMBATANTS 20

PCFG 2:

1 *Al Sanbouk* (GER Lurssen TNC-45) with 2 twin lnchr with MM40 *Exocet* AShM, 1 76mm gun

1 *Istiqlal* (GER Lurssen FPB-57) with 2 twin lnchr with MM40 *Exocet* AShM, 1 76mm gun
PBF 10 *Al Nokatha* (US Mk V *Pegasus*)
PBG 8 *Um Almaradim* (FRA P-37 BRL) with 2 twin lnchr with *Sea Skua* AShM
AMPHIBIOUS • LANDING CRAFT 8
LCT 2 *Assafar* (ADSB 64m)
LCM 1 *Abhan* (ADSB 42m)
LCVP 5 ADSB 16m
LOGISTICS AND SUPPORT • AG 1 *Sawahil* with 1 hel landing platform

Air Force 2,500

FORCES BY ROLE
FIGHTER/GROUND ATTACK
2 sqn with F/A-18C/D *Hornet*
TRANSPORT
1 sqn with C-17A *Globemaster* III; KC-130J *Hercules*; L-100-30
TRAINING
1 unit with EMB-312 *Tucano**; *Hawk* Mk64*
ATTACK HELICOPTER
1 sqn with AH-64D *Apache*
1 atk/trg sqn with SA342 *Gazelle* with HOT
TRANSPORT HELICOPTER
1 sqn with AS532 *Cougar*; SA330 *Puma*; S-92
EQUIPMENT BY TYPE
AIRCRAFT 66 combat capable
FGA 39: 31 F/A-18C *Hornet*; 8 F/A-18D *Hornet*
TKR 3 KC-130J *Hercules*
TPT 5: **Heavy** 2 C-17A *Globemaster* III; **Medium** 3 L-100-30
TRG 27: 11 *Hawk* Mk64*; 16 EMB-312 *Tucano**
HELICOPTERS
ATK 16 AH-64D *Apache*
MRH 13 SA342 *Gazelle* with HOT
TPT • Medium 13: 3 AS532 *Cougar*; 7 SA330 *Puma*; 3 S-92
AIR-LAUNCHED MISSILES
AAM • IR AIM-9L *Sidewinder*; R-550 *Magic*; **SARH** AIM-7F *Sparrow*; **ARH** AIM-120C7 AMRAAM
ASM AGM-65G *Maverick*; AGM-114K *Hellfire*; HOT
AShM AGM-84D *Harpoon* Block IC

Air Defence Command

FORCES BY ROLE
AIR DEFENCE
1 SAM bde (7 SAM bty with MIM-104D *Patriot* PAC-2 GEM)
1 SAM bde (6 SAM bty with *Skyguard/Aspide*)
EQUIPMENT BY TYPE
AIR DEFENCE
SAM 52
Long-range 40 MIM-104D *Patriot* PAC-2 GEM
Short-range 12 *Aspide* with *Skyguard*
GUNS • TOWED 35mm 12+ Oerlikon GDF

Emiri Guard 1,500

FORCES BY ROLE
MANOEUVRE
Other
1 (Emiri) gd bde

Paramilitary ε7,100 active

National Guard ε6,600 active

FORCES BY ROLE
SPECIAL FORCES
1 SF bn
MANOEUVRE
Reconnaissance
1 armd car bn
Other
3 security bn
COMBAT SUPPORT
1 MP bn
EQUIPMENT BY TYPE
ARMOURED FIGHTING VEHICLES
RECCE 20 VBL
IFV ε150 *Pandur* (incl variants)
APC 67+
APC (W) 27+: 5+ *Desert Chameleon*; 22 S600 (incl variants)
PPV 40 Otokar ISV
AUV 120 *Sherpa Light Scout*
ENGINEERING & MAINTENANCE VEHICLES
ARV *Pandur*

Coast Guard 500

EQUIPMENT BY TYPE
PATROL AND COASTAL COMBATANTS 32
PBF 12 *Manta*
PB 20: 3 *Al Shaheed*; 4 *Inttisar* (Austal 31.5m); 3 *Kassir* (Austal 22m); 10 *Subahi*
AMPHIBIOUS • LANDING CRAFT
LCU 4: 2 *Al Tahaddy*; 1 *Saffar*; 1 other
LOGISTICS AND SUPPORT • AG 1 *Sawahil*

DEPLOYMENT

SAUDI ARABIA: *Operation Restoring Hope* 4 F/A-18A *Hornet*

FOREIGN FORCES

Canada *Operation Inherent Resolve (Impact)* 2 C-130J-30 *Hercules* (CC-130J)
Denmark *Operation Inherent Resolve* 16
Italy *Operation Inherent Resolve (Prima Parthica)* 250; 4 *Tornado* ECR; 2 MQ-9A *Reaper*; 1 KC-767A
Singapore *Operation Inherent Resolve* 11
United Kingdom *Operation Inherent Resolve (Shader)* 50; 1 CISR UAV sqn with 8 MQ-9A *Reaper*
United States Central Command: 13,500; 1 armd bde(-); 1 ARNG (cbt avn) hel bde; 1 spt bde; 3 AD bty with MIM-104E/F *Patriot* PAC-2/3; 1 CISR UAV sqn with MQ-9A *Reaper*; 1 (APS) armd bde eqpt set; 1 (APS) inf bde eqpt set

Lebanon LBN

Lebanese Pound LP		2019	2020	2021
GDP	LP	79.2tr	93.7tr	
	US$	52.5bn	18.7bn	
per capita	US$	7,661	2,745	
Growth	%	-6.9	-25.0	
Inflation	%	2.9	85.5	
Def bdgt	LP	2.91tr	2.90tr	
	US$	1.93bn	1.92bn	
FMA (US)	US$	105.0m	50.0m	50.0m
US$1=LP		1507.50	1507.50	

Real-terms defence budget trend (US$bn, constant 2015)

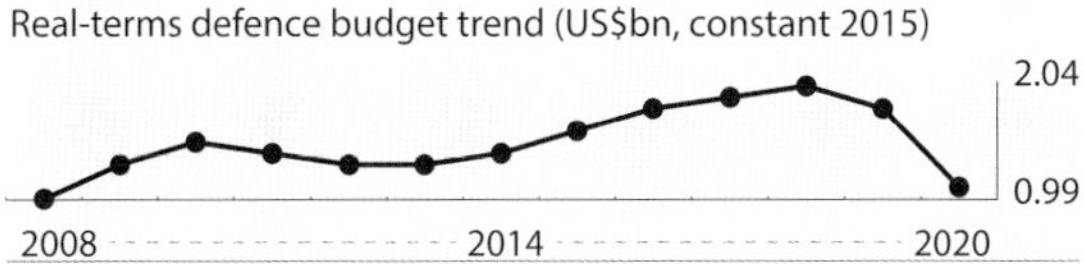

Population 5,469,612

Age	0–14	15–19	20–24	25–29	30–64	65 plus
Male	10.6%	3.8%	3.8%	4.1%	24.2%	3.4%
Female	10.1%	3.7%	3.7%	3.9%	24.1%	4.5%

Capabilities

Lebanon's Armed Forces (LAF) are focused on internal and border security. However, the LAF's ability to fulfil its missions remains under strain from Hizbullah's position in national politics and from the spillover effects of the Syrian conflict. The political and economic aftermath of the 2020 blast at Beirut port is expected to further delay plans for a national-defence strategy. Training and material support are received from the US, as well as from France, Italy and the UK. Personnel quality and capability is relatively high for the region, and US special-operations personnel continue to provide operational advice and assistance. LAF operations against ISIS have demonstrated improved capability. The LAF has no requirement for extraterritorial deployment and only minimal capability to do so. It remains dependent on foreign support to replace and modernise its ageing equipment inventory. Barring limited organic maintenance facilities, Lebanon has no significant domestic defence industry.

ACTIVE 60,000 (Army 56,600 Navy 1,800 Air 1,600) Paramilitary 20,000

ORGANISATIONS BY SERVICE

Army 56,600

FORCES BY ROLE

5 regional comd (Beirut, Bekaa Valley, Mount Lebanon, North, South)

SPECIAL FORCES

1 cdo regt

MANOEUVRE

Armoured

1 armd regt

Mechanised

11 mech inf bde

Air Manoeuvre

1 AB regt

Amphibious

1 mne cdo regt

Other

1 Presidential Guard bde

6 intervention regt 4 border sy regt

COMBAT SUPPORT

2 arty regt

1 cbt spt bde (1 engr regt, 1 AT regt, 1 sigs regt; 1 log bn)

1 MP gp

COMBAT SERVICE SUPPORT

1 log bde

1 med gp

1 construction regt

EQUIPMENT BY TYPE

MBT 334: 92 M48A1/A5; 10 M60A2; 185 T-54; 47 T-55

RECCE 55 AML

IFV 56: 24 AIFV-B-C25; 32 M2A2 *Bradley*

APC 1,378

APC (T) 1,274 M113A1/A2 (incl variants)

APC (W) 96: 86 VAB VCT; 10 VBPT-MR *Guarani*

PPV 8 *Maxxpro*

ENGINEERING & MAINTENANCE VEHICLES

ARV 3 M88A1; M113 ARV; T-54/55 ARV (reported)

VLB MTU-72 reported

MW *Bozena*

ARTILLERY 641

SP 155mm 12 M109A2

TOWED 313: **105mm** 13 M101A1; **122mm** 35: 9 D-30; 26 M-30; **130mm** 15 M-46; **155mm** 250: 18 M114A1; 218 M198; 14 Model-50

MRL 122mm 11 BM-21

MOR 305: **81mm** 134; **82mm** 112; **120mm** 59: 29 Brandt; 30 M120

ANTI-TANK/ANTI-INFRASTRUCTURE

MSL

SP 35 VAB with HOT

MANPATS *Milan*; TOW

RCL 106mm 113 M40A1

UNINHABITED AERIAL VEHICLES

ISR • Medium 8 *Mohajer* 4

AIR DEFENCE

SAM • Point-defence 9K32 *Strela*-2M (SA-7B *Grail*)‡

GUNS • TOWED 77: **20mm** 20; **23mm** 57 ZU-23-2

Navy 1,800

EQUIPMENT BY TYPE

PATROL AND COASTAL COMBATANTS 13

PCC 1 *Trablous*

PBF 1

PB 11: 1 *Aamchit* (ex-GER *Bremen*); 1 *Al Kalamoun* (ex-FRA *Avel Gwarlarn*); 7 *Tripoli* (ex-UK *Attacker/Tracker* Mk 2); 1 *Naquora* (ex-GER *Bremen*); 1 *Tabarja* (ex-GER *Bergen*)

AMPHIBIOUS • LANDING CRAFT

LCT 2 *Sour* (ex-FRA EDIC – capacity 8 APC; 96 troops)

Air Force 1,600

4 air bases

FORCES BY ROLE

GROUND ATTACK

1 sqn with Cessna AC-208 *Combat Caravan**

1 sqn with EMB-314 *Super Tucano**

ATTACK HELICOPTER

1 sqn with SA342L *Gazelle*

TRANSPORT HELICOPTER

4 sqn with Bell 205 (UH-1H *Iroquois/Huey* II)

1 sqn with SA330/IAR330SM *Puma*

1 trg sqn with R-44 *Raven* II

EQUIPMENT BY TYPE

AIRCRAFT 9 combat capable

ISR 3 Cessna AC-208 *Combat Caravan**

TRG 9: 3 *Bulldog*; 6 EMB-314 *Super Tucano**

HELICOPTERS

MRH 9: 1 AW139; 8 SA342L *Gazelle* (5 SA342L *Gazelle*; 5 SA316 *Alouette* III; 1 SA318 *Alouette* II all non-operational)

TPT 38: **Medium** 13: 3 S-61N (fire fighting); 10 SA330/IAR330 *Puma*; **Light** 25: 18 Bell 205 (UH-1H *Iroquois*); 3 Bell 205 (UH-1H *Huey* II); 4 R-44 *Raven* II (basic trg) (11 Bell 205; 7 Bell 212 all non-operational)

AIR LAUNCHED MISSILES

ASM AGM-114 *Hellfire*; AGR-20A APKWS

Paramilitary ε20,000 active

Internal Security Force ε20,000

Ministry of Interior

FORCES BY ROLE

Other Combat Forces

1 (police) judicial unit

1 regional sy coy

1 (Beirut Gendarmerie) sy coy

EQUIPMENT BY TYPE

ARMOURED FIGHTING VEHICLES

APC • **APC (W)** 60 V-200 *Chaimite*

Customs

EQUIPMENT BY TYPE

PATROL AND COASTAL COMBATANTS 7

PB 7: 5 *Aztec*; 2 *Tracker*

FOREIGN FORCES

Unless specified, figures refer to UNTSO and represent total numbers for the mission

Argentina 3

Armenia UNIFIL 33

Australia 12

Austria 5 • UNIFIL 182: 1 log coy

Bangladesh UNIFIL 117: 1 FSGM

Belarus UNIFIL 7

Belgium 1

Bhutan 4

Brazil UNIFIL 221: 1 FFGHM

Brunei UNIFIL 30

Cambodia UNIFIL 184: 1 EOD coy

Canada 6 (*Operation Jade*)

Chile 3

China, People's Republic of 4 • UNIFIL 419: 2 engr coy; 1 med coy

Colombia UNIFIL 1

Croatia UNIFIL 1

Cyprus UNIFIL 2

Denmark 10

El Salvador UNIFIL 52: 1 inf pl

Estonia 3 • UNIFIL 1

Fiji 2 • UNIFIL 1

Finland 15 • UNIFIL 198; 1 inf coy

France UNIFIL 653: 1 mech inf bn(-); VBL; VBCI; VAB; *Mistral*

Gambia 1

Germany UNIFIL 120: 1 FFGM

Ghana UNIFIL 864: 1 mech inf bn

Greece UNIFIL 146: 1 FFGHM

Guatemala UNIFIL 1

Hungary UNIFIL 4

India 3 • UNIFIL 775: 1 inf bn; 1 med coy

Indonesia UNIFIL 1,259: 1 mech inf bn; 1 MP coy; 1 FSGHM

Ireland 12 • UNIFIL 347: 1 mech inf bn(-)

Italy MIBIL 140; UNIFIL 1,076: 1 mech bde HQ; 1 mech inf bn; 1 MP coy; 1 sigs coy; 1 hel bn

Kazakhstan UNIFIL 124; 1 inf coy

Kenya UNIFIL 3

Korea, Republic of UNIFIL 409: 1 mech inf coy; 1 engr coy; 1 sigs coy; 1 maint coy

Macedonia, North UNIFIL 2

Malaysia UNIFIL 827: 1 mech inf bn

Malta UNIFIL 10

Nepal 3 • UNIFIL 870: 1 mech inf bn

Netherlands 13 • UNIFIL 1

New Zealand 7

Nigeria UNIFIL 1

Norway 10

Peru UNIFIL 1

Poland 1 • UNIFIL 220; 1 mech inf coy

Russia 3

Serbia 1 • UNIFIL 178; 1 mech inf coy

Sierra Leone UNIFIL 3

Slovakia 2

Slovenia 2 • UNIFIL 6

Spain UNIFIL 630: 1 mech bde HQ; 1 mech inf bn(-); 1 engr coy; 1 sigs coy

Sri Lanka UNIFIL 146: 1 inf coy

Sweden 7

Switzerland 12

Tanzania UNIFIL 120: 1 MP coy

Turkey UNIFIL 88: 1 PCFG

United States 3

Uruguay UNIFIL 2

Libya LBY

Libyan Dinar D		2019	2020	2021
GDP	D	55.2bn	30.2bn	
	US$	39.8bn	21.8bn	
per capita	US$	6,055	3,282	
Growth	%	9.9	-66.7	
Inflation	%	4.6	22.3	
Def bdgt	D	n.k.	n.k.	
	US$	n.k.	n.k.	
US$1=D		1.39	1.39	

Population 6,890,535

Age	0–14	15–19	20–24	25–29	30–64	65 plus
Male	17.2%	4.1%	3.7%	3.6%	20.7%	1.9%
Female	16.5%	3.9%	3.5%	3.4%	19.3%	2.2%

Capabilities

Prime Minister Fayez al-Sarraj's internationally recognised Government of National Accord (GNA) in the west of Libya continues to confront a mix of semi-regular military units, tribal militias and armed civilians based around General Haftar's Libyan Arab Armed Forces (LAAF, also known as the Libyan National Army (LNA)) in the east. Forces affiliated to both have relatively low levels of training. The presence in these formations of units from the former Gadhafi-era army has bolstered their military capability. Meanwhile, the GNA-affiliated forces have since 2016 benefited from several military advisory and training programmes, including EUNAVFOR–MED maritime-security training for the Libyan Navy and Coast Guard. Foreign-military involvement has increased in 2020. Turkey delivered military support to the GNA, while forces from the UAE as well as from the Wagner Group (a Russian private military company) are also in-country, both supporting Haftar's forces. There are also reports of Syrian combatants paid to fight for both sides. LAAF troops have combat experience from fighting ISIS in the eastern coastal region and they have allegedly received training and combat support from external actors in the region. Equipment is mainly of Russian or Soviet origin, including items from the former Libyan armed forces, and suffers from varying degrees of obsolescence. The country has no domestic defence-industrial capability.

Forces loyal to the Government of National Accord (Tripoli-based)

ACTIVE n.k.

ORGANISATIONS BY SERVICE

Ground Forces n.k.

EQUIPMENT BY TYPE

ARMOURED FIGHTING VEHICLES
MBT T-55; T-72
IFV BMP-2
APC
APC (T) ACV-AAPC; Steyr 4K-7FA
APC (W) *Mbombe-6*
PPV *Al-Wahsh*; *Kirpi-2*; *Vuran*
AUV Lenco *Bearcat* G3; Nimr *Ajban*
ENGINEERING & MAINTENANCE VEHICLES
ARV *Centurion* 105 AVRE
ANTI-TANK/ANTI-INFRASTRUCTURE
MSL • SP 9P157-2 *Khrizantema-S* (AT-15 *Springer*)
ARTILLERY
SP 155mm *Palmaria*
TOWED 122mm D-30
AIR DEFENCE
SAM • Point-defence QW-18 (CH-SA-11)
GUNS • SP 14.5mm ZPU-2 (on tch); **23mm** ZU-23-2 (on tch)

Navy n.k.

A number of intact naval vessels remain in Tripoli, although serviceability is questionable

EQUIPMENT BY TYPE

PATROL AND COASTAL COMBATANTS 3+
CORVETTES • FSGM (1 *Al Hani* (ex-FSU Project 1159 (*Koni*)) in Malta for refit since 2013 with 2 twin lnchr with P-22 (SS-N-2C *Styx*) AShM, 1 twin lnchr with 4K33 *Osa-M* (SA-N-4 *Gecko*) SAM, 2 twin 406mm ASTT, 1 RBU 6000 *Smerch* 2 A/S mor, 2 AK230 CIWS, 2 twin 76mm gun)
PBFG 1 *Sharaba* (FRA *Combattante* II) with 4 single lnchr with *Otomat* Mk2 AShM, 1 76mm gun†
PB 2+ PV30
AMPHIBIOUS
LANDING SHIPS • LST 1 *Ibn Harissa* (capacity 1 hel; 11 MBT; 240 troops)
LOGISTICS AND SUPPORT 2
AFD 1
ARS 1 *Al Munjed* (YUG *Spasilac*)†

Air Force n.k.

EQUIPMENT BY TYPE

AIRCRAFT 3+ combat capable
FGA 2 MiG-23BN
ATK 1 J-21 *Jastreb*†
TRG 10+: 3 G-2 *Galeb**; ε5 L-39ZO*; 1+ SF-260ML*
HELICOPTERS
ATK Mi-24 *Hind*
TPT • Medium Mi-17 *Hip*
AIR-LAUNCHED MISSILES • **AAM • IR** R-3 (AA-2 *Atoll*)‡; R-60 (AA-8 *Aphid*); R-24 (AA-7 *Apex*)

Paramilitary n.k.

Coast Guard n.k.

EQUIPMENT BY TYPE

PATROL AND COASTAL COMBATANTS 9
PCC 1 Damen Stan 2909 with 1 sextuple 122mm MRL
PBF 5: 4 *Bigliani*; 1 *Fezzan* (ex-ITA *Corrubia*)
PB 3: 1 *Burdi* (Damen Stan 1605); 1 *Hamelin*; 1 *Ikrimah* (FRA RPB 20)

FOREIGN FORCES

Italy MIASIT 400; 1 inf coy; 1 CBRN unit; 1 trg unit
Nepal UNSMIL 229; 2 sy coy
Turkey ε500; ACV-AAPC; *Kirpi*; 1 arty unit with T-155 *Firtina*; 1 AD unit with MIM-23B *Hawk*; *Korkut*; GDF-003; 1 CISR UAV unit with *Bayraktar* TB2

United Kingdom UNSMIL 1
United States UNSMIL 1

TERRITORY WHERE THE RECOGNISED AUTHORITY DOES NOT EXERCISE EFFECTIVE CONTROL

Data here represents the de facto situation. This does not imply international recognition

ACTIVE n.k.

ORGANISATIONS BY SERVICE

Libyan Arab Armed Forces n.k.

EQUIPMENT BY TYPE
ARMOURED FIGHTING VEHICLES
MBT T-55; T-62; T-72
RECCE BRDM-2; EE-9 *Cascavel*
IFV BMP-1; *Ratel*-20
APC
APC (T) M113
APC (W) *Al-Mared*; BTR-60PB; *Mbombe*-6; Nimr *Jais*; *Puma*
PPV *Al-Wahsh*; *Caiman*; Streit *Spartan*; Streit *Typhoon*; *Vuran*
AUV *Panthera* T6; *Panthera* F9
ANTI-TANK/ANTI-INFRASTRUCTURE
MSL
SP 10 9P157-2 *Khrizantema*-S (status unknown)
MANPATS 9K11 *Malyutka* (AT-3 *Sagger*); 9K111 *Fagot* (AT-4 *Spigot*); 9K111-1 *Konkurs* (AT-5 *Spandrel*); *Milan*
RCL some: **106mm** M40A1; **84mm** *Carl Gustaf*
ARTILLERY
SP 122mm 2S1 *Gvozdika*
TOWED 122mm D-30
MRL 107mm Type-63; **122mm** BM-21 *Grad*
MOR M106
AIR DEFENCE
SAM
Short-range 2K12 *Kvadrat* (SA-6 *Gainful*)
Point-defence 9K338 *Igla*-S (SA-24 *Grinch*)
GUNS • SP 14.5mm ZPU-2 (on tch); **23mm** ZSU-23-4 *Shilka*; ZU-23-2 (on tch)

Navy n.k.

EQUIPMENT BY TYPE
PATROL AND COASTAL COMBATANTS 7+
PB: 7+: 2 *Burdi* (Damen Stan 1605); 1 *Burdi* (Damen Stan 1605) with 1 73mm gun; 2 *Ikrimah* (FRA RPB20); 1 *Hamelin*; 1+ PV30
LOGISTICS AND SUPPORT • AFD 1

Air Force n.k.

EQUIPMENT BY TYPE
AIRCRAFT 22 combat capable
FGA 11: ε10 MiG-21MF *Fishbed*; 1 Su-22UM3 *Fitter* G
TRG 12: ε10 L-39ZO *Albatros**; 1+ MiG-21UM *Mongol* B; 1 SF-260ML*
HELICOPTERS
ATK Mi-24/35 *Hind*
TPT Medium Mi-8/Mi-17 *Hip*
AIR-LAUNCHED MISSILES • AAM • IR R-3 (AA-2 *Atoll*)‡; R-60 (AA-8 *Aphid*)

FOREIGN FORCES

United Arab Emirates UH-60M; *Pantsir*-S1
Wagner Group 3,000; Ural-432007 PPV; *Tigr*; *Pantsir*-S1; 12 MiG-29; 4 Su-24M

Mauritania MRT

Mauritanian Ouguiya OM		2019	2020	2021
GDP	OM	279bn	285bn	
	US$	7.60bn	7.43bn	
per capita	US$	1,873	1,791	
Growth	%	5.9	-3.2	
Inflation	%	2.3	3.9	
Def bdgt	OM	5.91bn	7.53bn	
	US$	161m	196m	
US$1=OM		36.69	38.34	

Real-terms defence budget trend (US$m, constant 2015)

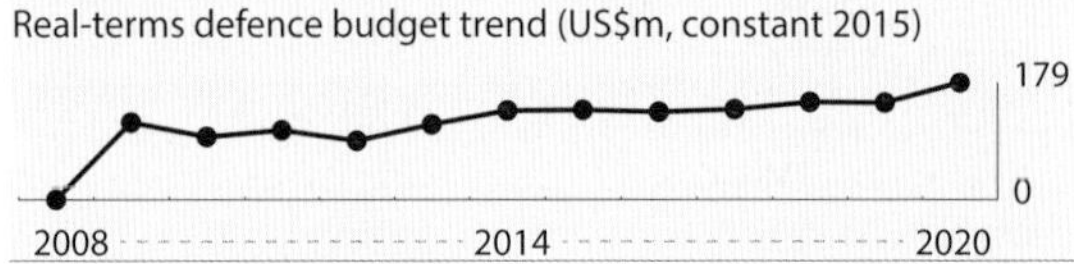

Population	4,005,475

Age	0–14	15–19	20–24	25–29	30–64	65 plus
Male	18.9%	5.2%	4.5%	3.9%	14.1%	1.7%
Female	18.7%	5.3%	4.8%	4.3%	16.6%	2.3%

Capabilities

The country's small and modestly equipped armed forces are tasked with maintaining territorial integrity and internal security. In light of the regional threat from extremist Islamist groups, border security is also a key role for the armed forces, which are accustomed to counter-insurgency operations in the desert. The country is a member of the G5 Sahel group. Mauritania's armed forces take part in the US-led special-operations *Flintlock* training exercise. The country also benefits from training with French armed forces. Deployment capabilities are limited to neighbouring countries without external support, but the armed forces have demonstrated mobility and sustainability in desert regions. Mauritania has a limited and ageing equipment inventory, which hampers operational capability. Despite some recent acquisitions, including small ISR aircraft, aviation resources are insufficient considering the size of the country. Naval equipment is geared toward coastal-surveillance missions and China's donation of a landing ship has helped establish a basic sealift capability. There is no domestic defence industry.

ACTIVE 15,850 (Army 15,000 Navy 600 Air 250)
Paramilitary 5,000
Conscript liability 24 months

ORGANISATIONS BY SERVICE

Army 15,000

FORCES BY ROLE
6 mil regions
MANOEUVRE
Reconnaissance
1 armd recce bn
Armoured
1 armd bn
Light
7 mot inf bn
8 (garrison) inf bn
Air Manoeuvre
1 cdo/para bn
Other
2 (camel corps) bn
1 gd bn
COMBAT SUPPORT
3 arty bn
4 ADA bty
1 engr coy

EQUIPMENT BY TYPE
ARMOURED FIGHTING VEHICLES
MBT 35 T-54/T-55
RECCE 70: 20 AML-60; 40 AML-90; 10 *Saladin*
APC • APC (W) 32: 5 FV603 *Saracen*; 7 *Bastion* APC; ε20 Panhard M3
AUV 12 *Cobra*
ENGINEERING & MAINTENANCE VEHICLES
ARV T-54/55 ARV reported
ANTI-TANK/ANTI-INFRASTRUCTURE
MSL • MANPATS *Milan*
RCL • 106mm ε90 M40A1
ARTILLERY 180
TOWED 80: **105mm** 36 HM-2/M101A1; **122mm** 44: 20 D-30; 24 D-74
MRL 10: **107mm** 4 Type-63; **122mm** 6 Type-81
MOR 90: **81mm** 60; **120mm** 30 Brandt
AIR DEFENCE
SAM • Point-defence ε4 SA-9 *Gaskin* (reported); 9K32 *Strela-2* (SA-7 *Grail*)‡
GUNS • TOWED 82: **14.5mm** 28: 16 ZPU-2; 12 ZPU-4; **23mm** 20 ZU-23-2; **37mm** 10 M-1939; **57mm** 12 S-60; **100mm** 12 KS-19

Navy ε600

EQUIPMENT BY TYPE
PATROL AND COASTAL COMBATANTS 17
PCO 1 *Voum-Legleita*
PCC 7: 1 *Abourbekr Ben Amer* (FRA OPV 54); 1 *Arguin*; 2 *Conejera*; 1 *Limam El Hidrami* (PRC); 2 *Timbédra* (PRC *Huangpu* Mod)
PB 9: 1 *El Nasr*† (FRA *Patra*); 4 *Mandovi*; 2 *Saeta*-12; 2 *Megsem Bakkar* (FRA RPB20 – for SAR duties)
AMPHIBIOUS • LANDING SHIPS 1
LSM 1 *Nimlane* (PRC)

Fusiliers Marins

FORCES BY ROLE
MANOEUVRE
Amphibious
1 mne unit

Air Force 250

EQUIPMENT BY TYPE
AIRCRAFT 2 combat capable
ISR 2 Cessna 208B *Grand Caravan*
TPT 9: **Light** 8: 2 BN-2 *Defender*; 1 C-212; 1 CN235; 2 PA-31T *Cheyenne* II; 2 Y-12(II); **PAX** 1 BT-67 (with sensor turret)
TRG 9: 3 EMB-312 *Tucano*; 2 EMB-314 *Super Tucano**; 4 SF-260E
HELICOPTERS • **MRH** 3: 1 SA313B *Alouette* II; 2 Z-9

Paramilitary ε5,000 active

Gendarmerie ε3,000

Ministry of Interior
FORCES BY ROLE
MANOEUVRE
Other
6 regional sy coy
EQUIPMENT BY TYPE
PATROL AND COASTAL COMBATANTS • **PB** 2 Rodman 55M

National Guard 2,000

Ministry of Interior

Customs

EQUIPMENT BY TYPE
PATROL AND COASTAL COMBATANTS • **PB** 2: 1 *Dah Ould Bah* (FRA *Amgram* 14); 1 *Yaboub Ould Rajel* (FRA RPB18)

DEPLOYMENT

CENTRAL AFRICAN REPUBLIC: UN • MINUSCA 452; 1 inf bn(-)

MALI: UN • MINUSMA 7

SOMALIA: UN • UNSOS 1

Middle East and North Africa

Morocco MOR

Moroccan Dirham D		2019	2020	2021
GDP	D	1.14tr	1.07tr	
	US$	119bn	112bn	
per capita	US$	3,332	3,121	
Growth	%	2.2	-7.0	
Inflation	%	0.2	0.2	
Def bdgt [a]	D	46.3bn	56.6bn	58.6bn
	US$	4.82bn	5.96bn	
FMA (US)	US$	10.0m	0.0m	0.0m
US$1=D		9.62	9.50	

[a] Includes autonomous defence spending (SEGMA) and Treasury funding for 'Acquisitions and Repair of Equipment for Royal Armed Forces'

Real-terms defence budget trend (US$bn, constant 2015)

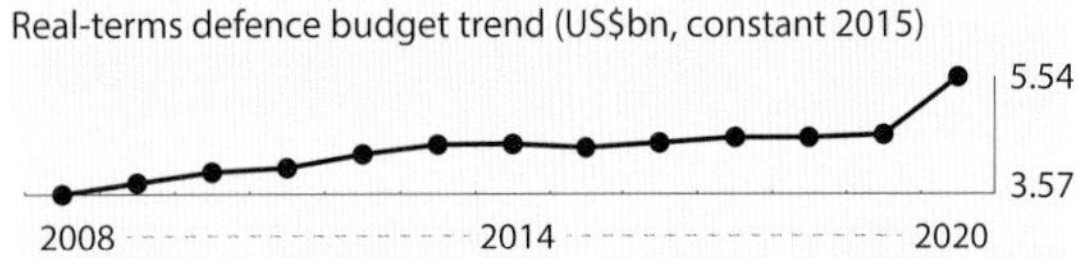

Population 35,561,654

Age	0–14	15–19	20–24	25–29	30–64	65 plus
Male	13.8%	4.3%	4.0%	3.8%	20.5%	3.4%
Female	13.2%	4.2%	4.0%	3.9%	21.1%	3.7%

Capabilities

Regional security challenges rank highly for Morocco's armed forces. Despite the UN-brokered 1991 ceasefire between Morocco and the Polisario Front, the conflict in Western Sahara remains unresolved. There was uncertainty over the future of the ceasefire in late 2020 after the Polisario leader reportedly said the group had ended the ceasefire. Morocco maintains defence ties with France and the US, receiving military training and equipment from both. There is also close cooperation with NATO, and in 2016 Morocco was granted access to the Alliance's Interoperability Platform in order to strengthen the defence and security sectors and bring the armed forces up to NATO standards. In 2017, Morocco rejoined the African Union. The armed forces have also gained experience from UN peacekeeping deployments and multinational exercises. Conscription was reintroduced in early 2019. The armed forces have some capacity to deploy independently within the region and on UN peacekeeping missions in sub-Saharan Africa, although they lack heavy sealift and airlift capabilities. Morocco has also recently deployed overseas in a combat role, contributing F-16 aircraft to the Saudi-led coalition intervention in Yemen from 2015 to early 2019. The inventory primarily comprises ageing French and US equipment. However, there are plans to re-equip all the services and to invest significantly in the navy. Morocco has also launched two Earth-observation satellites, aboard European rockets, meeting some surveillance requirements. Morocco does not yet have an established domestic defence industry and relies on imports and donations for major defence equipment. However, its relative stability has attracted Western defence companies, such as Airbus, Safran and Thales, to establish aerospace manufacturing and servicing facilities in the country.

ACTIVE 195,800 (Army 175,000 Navy 7,800 Air 13,000) **Paramilitary 50,000**

Conscript liability 12 months for men aged 19–25

RESERVE 150,000 (Army 150,000)

Reserve obligation to age 50

ORGANISATIONS BY SERVICE

Space

EQUIPMENT BY TYPE

SATELLITES • ISR 2 *Mohammed* VI

Army 175,000

FORCES BY ROLE

2 comd (Northern Zone, Southern Zone)

MANOEUVRE

Armoured

1 armd bde

11 armd bn

Mechanised

3 mech inf bde

Mechanised/Light

8 mech/mot inf regt (2–3 bn)

Light

1 lt sy bde

3 (camel corps) mot inf bn

35 lt inf bn

4 cdo unit

Air Manoeuvre

2 para bde

2 AB bn

Mountain

1 mtn inf bn

COMBAT SUPPORT

11 arty bn

7 engr bn

AIR DEFENCE

1 AD bn

Royal Guard 1,500

FORCES BY ROLE

MANOEUVRE

Other

1 gd bn

2 cav sqn

EQUIPMENT BY TYPE

ARMOURED FIGHTING VEHICLES

MBT 656: 222 M1A1SA *Abrams*; 220 M60A1 *Patton*; 120 M60A3 *Patton*; 40 T-72B; 54 Type-90-II (MBT-2000); (ε200 M48A5 *Patton* & ε60 T-72B in store)

LT TK 116: 5 AMX-13; 111 SK-105 *Kuerassier*

ASLT 80 AMX-10RC

RECCE 284: 38 AML-60-7; 190 AML-90; 40 EBR-75; 16 *Eland*

IFV 238: 10 AMX-10P; 30 *Ratel* Mk3-20; 30 *Ratel* Mk3-90; 45 VAB VCI; 123 YPR-765

APC 1,225

APC (T) 905: 400 M113A1/A2; 419 M113A3; 86 M577A2 (CP)

APC (W) 320 VAB VTT

ENGINEERING & MAINTENANCE VEHICLES

ARV 85+: 10 *Greif*; 55 M88A1; M578; 20 VAB-ECH

ANTI-TANK/ANTI-INFRASTRUCTURE

MSL

SP 80 M901

MANPATS 9K11 *Malyutka* (AT-3 *Sagger*); HJ-8L; M47 *Dragon*; *Milan*; TOW
RCL 106mm 350 M40A1
GUNS • SP 36: **90mm** 28 M56; **100mm** 8 SU-100
ARTILLERY 2,319
SP 357: **105mm** 5 AMX Mk 61; **155mm** 292: 84 M109A1/A1B; 43 M109A2; 4 M109A3; 1 M109A4; 70 M109A5; 90 Mk F3; **203mm** 60 M110
TOWED 118: **105mm** 50: 30 L118 Light Gun; 20 M101; **130mm** 18 M-46; **155mm** 50: 30 FH-70; 20 M114
MRL 47: **122mm** 35 BM-21 *Grad*; **300mm** 12+ PHL-03
MOR 1,797: **81mm** 1,100 Expal model LN; **SP 107mm** 36 M106A2; **120mm** 550 Brandt; **SP 120mm** 110: 20 (VAB APC); 91 M1064A3
UNINHABITED AERIAL VEHICLES
ISR • Medium R4E-50 *Skyeye*
AIR DEFENCE
SAM 67+
Medium-range 18 *Tianlong*-50
Short-range DK-9 (CH-SA-5)
Point-defence 49+: 12 2K22M *Tunguska*-M (SA-19 *Grison*); 37 M48 *Chaparral*; 9K38 *Igla* (SA-18 *Grouse*)
GUNS 390
SP 20mm 60 M163 *Vulcan*
TOWED 330: **14.5mm** 200: 150–180 ZPU-2; 20 ZPU-4; **20mm** 40 M167 *Vulcan*; **23mm** 75–90 ZU-23-2; **35mm** some PG-99

Navy 7,800 (incl 1,500 Marines)

EQUIPMENT BY TYPE
PRINCIPAL SURFACE COMBATANTS • FRIGATES 3
FFGHM 1 *Mohammed* VI (FRA FREMM) with 2 quad lnchr with MM40 *Exocet* Block 3 AShM, 2 8-cell *Sylver* A43 VLS with *Aster* 15 SAM, 2 triple 324mm ILAS-3 (B-515) ASTT with MU90 LWT, 1 76mm gun (capacity 1 AS565SA *Panther*)
FFGH 2 *Mohammed* V (FRA *Floreal*) with 2 single lnchr with MM38 *Exocet* AShM, 1 76mm gun (fitted for but not with *Simbad* SAM) (capacity 1 AS565SA *Panther*)
PATROL AND COASTAL COMBATANTS 53
CORVETTES 4
FSGHM 3:
2 *Sultan Moulay Ismail* (NLD SIGMA 9813) with 2 twin lnchr with MM40 *Exocet* Block 2/3 AShM, 1 12-cell CLA VLS with VL MICA SAM, 2 triple 324mm ILAS-3 (B-515) ASTT with MU90 LWT, 1 76mm gun (capacity 1 AS565SA *Panther*)
1 *Tarik ben Ziyad* (NLD SIGMA 10513) with 2 twin lnchr with MM40 *Exocet* Block 2/3 AShM, 1 12-cell CLA VLS with VL MICA SAM, 2 triple 324mm ILAS-3 (B-515) ASTT with MU90 LWT, 1 76mm gun (capacity 1 AS565SA *Panther*)
FSM 1 *Lt Col Errhamani* (ESP *Descubierto*) with 1 octuple *Albatros* lnchr with *Aspide* SAM, 2 triple 324mm ASTT with Mk46 LWT, 1 76mm gun
PSO 1 *Bin an Zaran* (OPV 70) with 1 76mm gun
PCG 4 *Cdt El Khattabi* (ESP *Lazaga* 58m) with 4 single lnchr with MM38 *Exocet* AShM, 1 76mm gun
PCO 5 *Rais Bargach* (under control of fisheries dept)
PCC 12:
4 *El Hahiq* (DNK *Osprey* 55, incl 2 with customs)
6 *LV Rabhi* (ESP 58m B-200D)
2 *Okba* (FRA PR-72) each with 1 76mm gun
PB 27: 6 *El Wacil* (FRA P-32); 10 VCSM (RPB 20); 10 Rodman 101; 1 other (UK *Bird*)
AMPHIBIOUS
LANDING SHIPS • LSM 3 *Ben Aicha* (FRA *Champlain* BATRAL) with 1 hel landing platform (capacity 7 tanks; 140 troops)
LANDING CRAFT 2:
LCT 1 *Sidi Ifni*
LCM 1 CTM (FRA CTM-5)
LOGISTICS AND SUPPORT 9
AG 1 Damen 3011
AGHS 1 *Dar Al Beida* (FRA BHO2M)
AGOR 1 *Abou Barakat Albarbari*† (ex-US *Robert D. Conrad*)
AGS 1 Stan 1504
AK 2
AX 1 *Essaouira*
AXS 2

Marines 1,500

FORCES BY ROLE
MANOEUVRE
Amphibious
2 naval inf bn

Naval Aviation

EQUIPMENT BY TYPE
AIRCRAFT • MP 2 Beech 350ER *King Air*
HELICOPTERS • ASW/ASUW 3 AS565SA *Panther*

Air Force 13,000

FORCES BY ROLE
FIGHTER/GROUND ATTACK
2 sqn with F-5E/F-5F *Tiger* II
3 sqn with F-16C/D *Fighting Falcon*
1 sqn with *Mirage* F-1C (F-1CH)
1 sqn with *Mirage* F-1E (F-1EH)
ELECTRONIC WARFARE
1 sqn with EC-130H *Hercules*; *Falcon* 20 (ELINT)
MARITIME PATROL
1 flt with Do-28
TANKER/TRANSPORT
1 sqn with C-130/KC-130H *Hercules*
TRANSPORT
1 sqn with CN235
1 VIP sqn with B-737BBJ; Beech 200/300 *King Air*; *Falcon* 50; Gulfstream II/III/V-SP/G550
TRAINING
1 sqn with *Alpha Jet**
1 sqn T-6C
ATTACK HELICOPTER
1 sqn with SA342L *Gazelle* (some with HOT)
TRANSPORT HELICOPTER
1 sqn with Bell 205A (AB-205A); Bell 206 *Jet Ranger* (AB-206); Bell 212 (AB-212)
1 sqn with CH-47D *Chinook*
1 sqn with SA330 *Puma*

EQUIPMENT BY TYPE

AIRCRAFT 90 combat capable

FTR 22: 19 F-5E *Tiger* II; 3 F-5F *Tiger* II

FGA 49: 15 F-16C *Fighting Falcon*; 8 F-16D *Fighting Falcon*; 15 *Mirage* F-1C (F-1CH); 11 *Mirage* F-1E (F-1EH)

ELINT 1 EC-130H *Hercules*

TKR/TPT 2 KC-130H *Hercules*

TPT 47: **Medium** 17: 4 C-27J *Spartan*; 13 C-130H *Hercules*; **Light** 19: 4 Beech 100 *King Air*; 2 Beech 200 *King Air*; 1 Beech 200C *King Air*; 2 Beech 300 *King Air*; 3 Beech 350 *King Air*; 5 CN235; 2 Do-28; **PAX** 11: 1 B-737BBJ; 2 *Falcon* 20; 2 *Falcon* 20 (ELINT); 1 *Falcon* 50 (VIP); 1 Gulfstream II (VIP); 1 Gulfstream III; 1 Gulfstream V-SP; 2 Gulfstream G550

TRG 80: 12 AS-202 *Bravo*; 19 *Alpha Jet**; 2 CAP-10; 24 T-6C *Texan*; 9 T-34C *Turbo Mentor*; 14 T-37B *Tweet*

HELICOPTERS

MRH 19 SA342L *Gazelle* (7 with HOT, 12 with cannon)

TPT 76: **Heavy** 10 CH-47D *Chinook*; **Medium** 24 SA330 *Puma*; **Light** 42: 24 Bell 205A (AB-205A); 11 Bell 206 *Jet Ranger* (AB-206); 3 Bell 212 (AB-212); 4 Bell 429

AIR-LAUNCHED MISSILES

AAM • **IR** AIM-9J *Sidewinder*; R-550 *Magic*; *Mica* IR; **IIR** AIM-9X *Sidewinder* II; **ARH** AIM-120C7 AMRAAM; *Mica* RF

ASM AASM; AGM-65 *Maverick*; HOT

ARM AGM-88B HARM

BOMBS

Laser-guided *Paveway* II; GBU-54 Laser JDAM

INS/GPS-guided GBU-31 JDAM

Paramilitary 50,000 active

Gendarmerie Royale 20,000

FORCES BY ROLE

MANOEUVRE

Air Manoeuvre

1 para sqn

Other

1 paramilitary bde

4 (mobile) paramilitary gp

1 coast guard unit

TRANSPORT HELICOPTER

1 sqn

EQUIPMENT BY TYPE

PATROL AND COASTAL COMBATANTS • **PB** 15 Arcor 53

AIRCRAFT • **TRG** 2 R-235 *Guerrier*

HELICOPTERS

MRH 14: 3 SA315B *Lama*; 2 SA316 *Alouette* III; 3 SA318 *Alouette* II; 6 SA342K *Gazelle*

TPT 8: **Medium** 6 SA330 *Puma*; **Light** 2 SA360 *Dauphin*

Force Auxiliaire 30,000 (incl 5,000 Mobile Intervention Corps)

Customs/Coast Guard

EQUIPMENT BY TYPE

PATROL AND COASTAL COMBATANTS

PB 36: 4 *Erraid*; 18 *Arcor* 46; 14 (other SAR craft)

DEPLOYMENT

CENTRAL AFRICAN REPUBLIC: UN • MINUSCA 763; 1 inf bn

DEMOCRATIC REPUBLIC OF THE CONGO: UN • MONUSCO 1,372; 1 inf bn; 1 fd hospital

SOUTH SUDAN: UN • UNMISS 1

Oman OMN

Omani Rial R		2019	2020	2021
GDP	R	29.4bn	24.0bn	
	US$	76.3bn	62.3bn	
per capita	US$	18,198	14,423	
Growth	%	-0.8	-10.0	
Inflation	%	0.1	1.0	
Def bdgt [a]	R	2.90bn	2.88bn	
	US$	7.53bn	7.48bn	
US$1=R		0.38	0.38	

[a] Excludes security funding

Real-terms defence budget trend (US$bn, constant 2015)

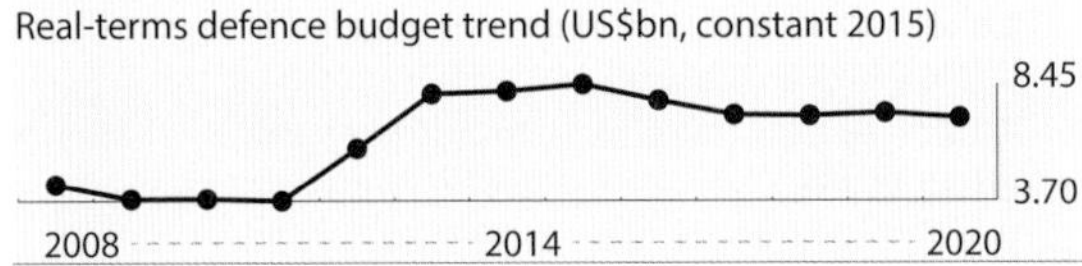

Population 3,634,689

Age	0–14	15–19	20–24	25–29	30–64	65 plus
Male	15.5%	4.2%	4.9%	5.8%	21.9%	1.8%
Female	14.7%	3.9%	4.3%	4.5%	16.6%	1.9%

Capabilities

Oman maintains small but capable armed forces, though the precise trajectory of defence policy and investment levels under Oman's new ruler remains unclear. The armed forces' principal task is ensuring territorial integrity. There is a particular focus on maritime security, given Oman's long coastline. With relatively limited resources, it maintains modern but modest equipment. Oman is a member of the GCC and also has close defence and security ties with the UK and the US. There is a particular emphasis on training, especially with the UK. In late 2018, the two countries announced a new joint training base in Oman, and a joint defence agreement was signed in February 2019. Recent deployment experience beyond Oman has been limited, but there are some logistics assets, particularly in the naval domain. The port of Duqm is being developed into a major potential logistics hub for partners, and in September 2020 the UK announced a further investment in its logistics hub. Oman is recapitalising its core inventory with the procurement of air and naval systems, including combat aircraft and patrol and high-speed support vessels. Oman has very limited indigenous defence-industrial capacity, but it has begun local production of various types of ammunition.

ACTIVE 42,600 (Army 25,000 Navy 4,200 Air 5,000 Foreign Forces 2,000 Royal Household 6,400) Paramilitary 4,400

ORGANISATIONS BY SERVICE

Army 25,000

FORCES BY ROLE
(Regt are bn size)
MANOEUVRE
Armoured
1 armd bde (2 armd regt, 1 recce regt)
Light
1 inf bde (5 inf regt, 1 arty regt, 1 fd engr regt, 1 engr regt, 1 sigs regt)
1 inf bde (3 inf regt, 2 arty regt)
1 indep inf coy (Musandam Security Force)
Air Manoeuvre
1 AB regt
COMBAT SERVICE SUPPORT
1 tpt regt
AIR DEFENCE
1 ADA regt (2 ADA bty)

EQUIPMENT BY TYPE
ARMOURED FIGHTING VEHICLES
MBT 117: 38 *Challenger* 2; 6 M60A1 *Patton*; 73 M60A3 *Patton*
LT TK 37 FV101 *Scorpion*
RECCE 149: 13 FV105 *Sultan* (CP); 12 *Pars* III 6×6; 124 VBL
IFV 72 *Pars* III 8×8
APC 262
APC (T) 10 FV4333 *Stormer*
APC (W) 252: 15 *Pars* III 6×6 (incl 10 CP; 1 trg); 47 *Pars* III 8×8 (38 CP; 8 amb; 1 trg); 175 *Piranha* (incl variants); 15 AT-105 *Saxon*
AUV 6 FV103 *Spartan*
ENGINEERING & MAINTENANCE VEHICLES
AEV 6 *Pars* III AEV
ARV 19: 4 *Challenger* ARV; 2 M88A1; 8 *Pars* III ARV; 2 *Piranha* ARV; 3 *Samson*
ARTILLERY 245
SP 155mm 24 G-6
TOWED 108: **105mm** 42 L118 Light Gun; **122mm** 30 D-30; **130mm** 24: 12 M-46; 12 Type-59-I; **155mm** 12 FH-70
MOR 113: **81mm** 69; **107mm** 20 M30; **120mm** 12 Brandt; **SP 120mm** 12 *Pars* III
ANTI-TANK/ANTI-INFRASTRUCTURE • MSL
SP 8 VBL with TOW
MANPATS FGM-148 *Javelin*; *Milan*; TOW/TOW-2A
AIR DEFENCE
SAM • Point-defence 8 *Mistral* 2; *Javelin*; 9K32 *Strela-2* (SA-7 *Grail*)‡
GUNS 26: **23mm** 4 ZU-23-2; **35mm** 10 GDF-005 (with *Skyguard*); **40mm** 12 L/60 (Towed)

Navy 4,200

EQUIPMENT BY TYPE
PRINCIPAL SURFACE COMBATANTS • FRIGATES 3
FFGHM 3 *Al-Shamikh* with 2 twin lnchr with MM40 *Exocet* Block 3 AShM, 2 6-cell CLA VLS with VL MICA SAM, 1 76mm gun
PATROL AND COASTAL COMBATANTS 10
CORVETTES • FSGM 2:
2 *Qahir Al Amwaj* with 2 quad lnchr with MM40 *Exocet* AShM, 1 octuple lnchr with *Crotale* SAM, 1 76mm gun, 1 hel landing platform
PCFG 1 *Dhofar* with 2 quad lnchr with MM40 *Exocet* AShM, 1 76mm gun
PCO 4 *Al Ofouq* with 1 76mm gun, 1 hel landing platform
PCC 3 *Al Bushra* (FRA P-400) with 1 76mm gun
AMPHIBIOUS
LANDING SHIPS • LST 1 *Nasr el Bahr*† with 1 hel landing platform (capacity 7 tanks; 240 troops) (in refit since 2017)
LANDING CRAFT 5: 1 **LCU**; 1 **LCT**; 3 **LCM**
LOGISTICS AND SUPPORT 8
AGS 1 *Al Makhirah*
AK 1 *Al Sultana*
AP 2 *Shinas* (commercial tpt – auxiliary military role only) (capacity 56 veh; 200 tps)
AX 1 *Al-Mabrukah*
AXS 1 *Shabab Oman* II
EPF 2 *Al Mubshir* (High Speed Support Vessel 72) with 1 hel landing platform (capacity 260 troops)

Air Force 5,000

FORCES BY ROLE
FIGHTER/GROUND ATTACK
2 sqn with F-16C/D Block 50 *Fighting Falcon*
1 sqn with *Hawk* Mk103; *Hawk* Mk203; *Hawk* Mk166
1 sqn with *Typhoon*
MARITIME PATROL
1 sqn with C295MPA
TRANSPORT
1 sqn with C-130H/J/J-30 *Hercules*
1 sqn with C295M
TRAINING
1 sqn with MFI-17B *Mushshak*; PC-9*; Bell 206 (AB-206) *Jet Ranger*
TRANSPORT HELICOPTER
4 (med) sqn; Bell 212 (AB-212); NH-90; *Super Lynx* Mk300 (maritime/SAR)
AIR DEFENCE
2 sqn with *Rapier*; *Blindfire*; S713 *Martello*

EQUIPMENT BY TYPE
AIRCRAFT 63 combat capable
FGA 35: 17 F-16C Block 50 *Fighting Falcon*; 6 F-16D Block 50 *Fighting Falcon*; 12 *Typhoon*
MP 4 C295MPA
TPT 12: **Medium** 6: 3 C-130H *Hercules*; 2 C-130J *Hercules*; 1 C-130J-30 *Hercules* (VIP); **Light** 4 C295M; **PAX** 2 A320-300
TRG 43: 4 *Hawk* Mk103*; 7 *Hawk* Mk166; 12 *Hawk* Mk203*; 8 MFI-17B *Mushshak*; 12 PC-9*
HELICOPTERS
MRH 15 *Super Lynx* Mk300 (maritime/SAR)
TPT 26+ **Medium** 20 NH90 TTH; **Light** 6: 3 Bell 206 (AB-206) *Jet Ranger*; 3 Bell 212 (AB-212)
AIR DEFENCE • SAM
Short-range NASAMS
Point-defence 40 *Rapier*

MSL
AAM • IR AIM-9/M/P *Sidewinder*; **IIR** AIM-9X *Sidewinder* II; **ARH** AIM-120C7 AMRAAM
ASM AGM-65D/G *Maverick*
AShM AGM-84D *Harpoon*
BOMBS
Laser-guided EGBU-10 *Paveway* II; EGBU-12 *Paveway* II
INS/GPS guided GBU-31 JDAM

Royal Household 6,400
(incl HQ staff)
FORCES BY ROLE
SPECIAL FORCES
2 SF regt

Royal Guard Brigade 5,000
FORCES BY ROLE
MANOEUVRE
Other
1 gd bde (1 armd sqn, 2 gd regt, 1 cbt spt bn)
EQUIPMENT BY TYPE
ARMOURED FIGHTING VEHICLES
ASLT 9 *Centauro* MGS (9 VBC-90 in store)
IFV 14 VAB VCI
APC • APC (W) ε50 Type-92
ANTI-TANK/ANTI-INFRASTRUCTURE
MSL • MANPATS *Milan*
ARTILLERY • MRL 122mm 6 Type-90A
AIR DEFENCE
SAM • Point-defence *Javelin*
GUNS • SP 9: **20mm** 9 VAB VDAA

Royal Yacht Squadron 150
EQUIPMENT BY TYPE
LOGISTICS AND SUPPORT 3
AP 1 *Fulk Al Salamah* (also veh tpt) with up to 2 AS332 *Super Puma* hel

Royal Flight 250
EQUIPMENT BY TYPE
AIRCRAFT • TPT • PAX 7: 1 747-400; 1 747-8; 1 B-747SP; 1 A319; 1 A320; 2 Gulfstream IV
HELICOPTERS • TPT • Medium 6 EC225LP *Super Puma*

Paramilitary 4,400 active

Tribal Home Guard 4,000
org in teams of ε100

Police Coast Guard 400
EQUIPMENT BY TYPE
PATROL AND COASTAL COMBATANTS 32
PCO 2 *Haras*
PBF 3 *Haras* (US Mk V *Pegasus*)
PB 27: 3 Rodman 101; 1 *Haras* (SWE CG27); 3 *Haras* (SWE CG29); 14 Rodman 58; 1 D59116; 5 *Zahra*

Police Air Wing
EQUIPMENT BY TYPE
AIRCRAFT • TPT • Light 4: 1 BN-2T *Turbine Islander*; 2 CN235M; 1 Do-228
HELICOPTERS • TPT • Light 5: 2 Bell 205A; 3 Bell 214ST (AB-214ST)

FOREIGN FORCES
United Kingdom 90

Palestinian Territories PT

New Israeli Shekel NS		2019	2020	2021
GDP	US$			
per capita	US$			
Growth	%			
Inflation	%			
US$1=NS				

Population 4,818,255

Age	0–14	15–19	20–24	25–29	30–64	65 plus
Male	19.6%	5.7%	5.0%	4.4%	14.4%	1.6%
Female	18.6%	5.5%	4.9%	4.4%	14.3%	1.7%

Capabilities
The Palestinian Territories remain effectively divided between the Palestinian Authority-run West Bank and Hamas-run Gaza. Each organisation controls its own security forces, principally the National Security Forces (NSF) in the West Bank and the Izz al-Din al-Qassam Brigades in Gaza. Both have generally proved effective at maintaining internal security in their respective territories. The Palestinian Authority has received support from the EU, Jordan and the US. NSF battalions, as well as the Presidential Guard and Civil Police, conduct US-funded internal-security training at the Jordanian International Police Training Center. A small number of Izz al-Din al-Qassam Brigades personnel are claimed by Israel to have received military training in Iran and Syria. None of the Palestinian security organisations conduct external military deployments, and they lack a formal military-logistics structure. Both Hamas and the Palestinian Authority lack heavy military equipment, although the former has retained a substantial arsenal of improvised rocket and mortar capabilities, as well as some portable guided weapons. No formal defence industry exists, although Hamas is able to acquire light or improvised weapons, either smuggled into Gaza or of local construction.

ACTIVE 0 Paramilitary n.k.
Precise personnel-strength figures for the various Palestinian groups are not known

ORGANISATIONS BY SERVICE
There is little available data on the status of the organisations mentioned below. Following internal fighting in June 2007, Gaza has been under the de facto control of Hamas, while the West Bank is controlled by the Palestinian Authority. In October 2017, both sides agreed a preliminary reconciliation deal on control of Gaza.

Paramilitary

Palestinian Authority n.k.

Presidential Security ε3,000

Special Forces ε1,200

Police ε9,000

National Security Force ε10,000

FORCES BY ROLE
MANOEUVRE
Other
9 paramilitary bn

Preventative Security ε4,000

Civil Defence ε1,000

The al-Aqsa Brigades n.k.

Profess loyalty to the Fatah group that dominates the Palestinian Authority

Hamas n.k.

Izz al-Din al-Qassam Brigades ε15,000–20,000

FORCES BY ROLE
COMMAND
6 bde HQ (regional)
MANOEUVRE
Other
1 cdo unit (Nukhba)
27 paramilitary bn
100 paramilitary coy
COMBAT SUPPORT Some engr units
COMBAT SERVICE SUPPORT
Some log units
EQUIPMENT BY TYPE
ANTI-TANK/ANTI-INFRASTRUCTURE • MSL • MANPATS 9K11 *Malyutka* (AT-3 *Sagger*) (reported); *Dehlavieh* (*Kornet*) (reported)
ARTILLERY
MRL • *Qassam* rockets (multiple calibres); **122mm** *Grad*
MOR some (multiple calibres)

Martime Police ε600

Qatar QTR

Qatari Riyal R		2019	2020	2021
GDP	R	640bn	538bn	
	US$	176bn	148bn	
per capita	US$	62,919	52,751	
Growth	%	0.8	-4.5	
Inflation	%	-0.6	-2.2	
Def bdgt [a]	R	ε23.3bn	ε23.5bn	
	US$	ε6.40bn	ε6.47bn	
US$1=R		3.64	3.64	

[a] Defence-budget estimate derived from analysis of public-spending breakdown, procurement-programme announcements, force size and structure, and operational tempo

Population 2,444,174

Age	0–14	15–19	20–24	25–29	30–64	65 plus
Male	6.5%	2.4%	6.0%	10.5%	51.1%	0.8%
Female	6.4%	1.6%	1.8%	2.6%	9.9%	0.4%

Capabilities

Qatar is attempting to transform its military capabilities and regional defence standing based on significant equipment acquisitions. The diplomatic crisis with several of its GCC neighbours brought Qatar and Turkey closer together in their limited but significant defence cooperation, which includes a small Turkish military presence in-country. The crisis appears not to have affected the significant Qatar–US military relationship, including the presence of forces from the US and other Western states at Al-Udeid air base, and the key US-run coalition air-operations centre. The pressure on personnel requirements is increasing due to Qatar's acquisition programme. The scale of the equipment plan suggests that Qatar will need significant foreign help to integrate and operate its new capabilities. The Italian Navy is supporting training for new Italian-built vessels and a joint Qatar–UK Eurofighter *Typhoon* squadron has begun training flights. Qatar is acquiring platforms with potentially significant power-projection capability. Combat-aircraft procurements will, when combined, dramatically increase the size of the air force, and it is in terms of air capabilities that there are the most questions about Qatar's ability to develop the necessary infrastructure, maintenance and personnel. Coastal-defence missiles are being acquired, while an AN/FPS-132 early-warning radar is being installed. Qatar currently has a limited indigenous defence-industrial capability, including in ship repair.

ACTIVE 16,500 (Army 12,000 Navy 2,500 Air 2,000)
Paramilitary up to 5,000

Conscript liability 12 months, males 18–35 years. Voluntary national service for women

ORGANISATIONS BY SERVICE

Space

EQUIPMENT BY TYPE
SATELLITES • COMMUNICATIONS 1 *Es'hail-2*

Army 12,000 (including Emiri Guard)

FORCES BY ROLE
SPECIAL FORCES
1 SF coy

Middle East and North Africa

MANOEUVRE
Armoured
1 armd bde (1 tk bn, 1 mech inf bn, 1 mor sqn, 1 AT bn)
Mechanised
3 mech inf bn
1 (Emiri Guard) bde (3 mech regt)
COMBAT SUPPORT
1 SP arty bn
1 fd arty bn

EQUIPMENT BY TYPE
ARMOURED FIGHTING VEHICLES
MBT 62 *Leopard* 2A7+
ASLT 48: 12 AMX-10RC; 36 *Piranha* II 90mm
RECCE 56: 32 *Fennek*; 8 V-150 *Chaimite*; 16 VBL
IFV 40 AMX-10P
APC 340
APC (T) 30 AMX-VCI
APC (W) 160 VAB
PPV 150 *Ejder Yalcin*; *Kirpi*-2; RG-31
AUV 14+: 14 *Dingo* 2; NMS
ENGINEERING & MAINTENANCE VEHICLES
AEV 6 *Wisent* 2
ARV 3: 1 AMX-30D; 2 *Piranha*
ANTI-TANK/ANTI-INFRASTRUCTURE
MSL
SP 24 VAB VCAC HOT; *Ejder Yalcin* with *Kornet;* NMS with *Kornet*
MANPATS FGM-148 *Javelin*; *Milan*; *Kornet*-EM
RCL 84mm *Carl Gustaf*
ARTILLERY 89+
SP 155mm 24 PzH 2000
TOWED 155mm 12 G-5
MRL 8+: **107mm** PH-63; **122mm** 2+ (30-tube); **127mm** 6 ASTROS II Mk3
MOR 45: **81mm** 26 L16; **SP 81mm** 4 VAB VPM 81; **120mm** 15 Brandt
SURFACE-TO-SURFACE MISSILE LAUNCHERS
SRBM • Conventional 2 BP-12A (CH-SS-14 mod 2)
AIR DEFENCE
SAM • Point-defence NMS with *Igla*

Navy 2,500 (incl Coast Guard)

EQUIPMENT BY TYPE
PATROL AND COASTAL COMBATANTS 11
PCFGM 4 *Barzan* (UK *Vita*) with 2 quad lnchr with MM40 *Exocet* Block 3 AShM, 1 sextuple *Sadral* lnchr with *Mistral* SAM, 1 *Goalkeeper* CIWS, 1 76mm gun
PCFG 3 *Damsah* (FRA *Combattante* III) with 2 quad lnchr with MM40 *Exocet* AShM, 1 76mm gun
PBF 3 MRTP 16
PB 1 MRTP 34

Coast Guard

EQUIPMENT BY TYPE
PATROL AND COASTAL COMBATANTS 12
PBF 4 DV 15
PB 8: 4 *Crestitalia* MV-45; 3 *Halmatic* M160; 1 other

Coastal Defence

FORCES BY ROLE
COASTAL DEFENCE
1 bty with 3 quad lnchr with MM40 *Exocet* AShM
EQUIPMENT BY TYPE
COASTAL DEFENCE • AShM 12 MM40 *Exocet* AShM

Air Force 2,000

FORCES BY ROLE
FIGHTER/GROUND ATTACK
1 sqn with *Alpha Jet**
1 sqn with *Mirage* 2000ED; *Mirage* 2000D
1 sqn with *Rafale* DQ/EQ
TRANSPORT
1 sqn with C-17A *Globemaster* III; C-130J-30 *Hercules*
1 sqn with A340; B-707; B-727; *Falcon* 900
ATTACK HELICOPTER
1 ASuW sqn with *Commando* Mk3 with *Exocet*
1 sqn with SA341 *Gazelle;* SA342L *Gazelle* with HOT
TRANSPORT HELICOPTER
1 sqn with *Commando* Mk2A; *Commando* Mk2C
1 sqn with AW139
EQUIPMENT BY TYPE
AIRCRAFT 33 combat capable
FGA 27: 9 *Mirage* 2000ED; 3 *Mirage* 2000D; 5 *Rafale* DQ; 18 *Rafale* EQ
TPT 18: **Heavy** 8 C-17A *Globemaster* III; **Medium** 4 C-130J-30 *Hercules*; **PAX** 6: 1 A340; 2 B-707; 1 B-727; 2 *Falcon* 900
TRG 27: 6 *Alpha Jet**; 21 PC-21
HELICOPTERS
ATK 24 AH-64E *Apache*
ASuW 8 *Commando* Mk3
MRH 34: 21 AW139 (incl 3 for medevac); 2 SA341 *Gazelle*; 11 SA342L *Gazelle*
TPT 5: **Medium** 4: 3 *Commando* Mk2A; 1 *Commando* Mk2C; **Light** 1 H125 *Ecureuil* (trg config)
UNINHABITED AERIAL VEHICLES
CISR • Medium 6 *Bayraktar* TB2
AIR DEFENCE • SAM
Long-range MIM-104E/F *Patriot* PAC-2 GEM-T/PAC-3
Point-defence FIM-92 *Stinger*; FN-6 (CH-SA-10); *Mistral*
AIR-LAUNCHED MISSILES
AAM • IR R-550 *Magic* 2; **ARH** *Mica* RF
ASM *Apache*; AGM-114R *Hellfire*; AGR-20A APKWS; HOT
AShM AM39 *Exocet*

Paramilitary up to 5,000 active

Internal Security Force up to 5,000

FOREIGN FORCES

Turkey 300 (trg team); 1 mech coy; 1 arty unit
United States US Central Command: 10,000; CAOC; 1 ISR sqn with 4 RC-135 *Rivet Joint;* 1 ISR sqn with 4 E-8C JSTARS; 1 tkr sqn with 24 KC-135R/T *Stratotanker;* 1 tpt

sqn with 4 C-17A *Globemaster*; 4 C-130H/J-30 *Hercules*; 2 AD bty with MIM-104E/F *Patriot* PAC-2/3 • US Strategic Command: 1 AN/TPY-2 X-band radar

Saudi Arabia SAU

Saudi Riyal R		2019	2020	2021
GDP	R	2.97tr	2.55tr	
	US$	793bn	681bn	
per capita	US$	23,266	19,587	
Growth	%	0.3	-5.4	
Inflation	%	-2.1	3.6	
Def bdgt [a]	R	191bn	182bn	
	US$	50.9bn	48.5bn	
US$1=R		3.75	3.75	

[a] Military budget only – excludes security budget

Real-terms defence budget trend (US$bn, constant 2015)

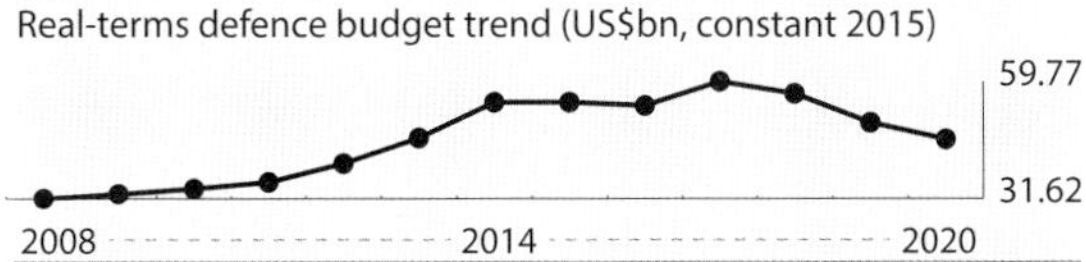

Population 34,173,498

Age	0–14	15–19	20–24	25–29	30–64	65 plus
Male	12.7%	4.1%	3.9%	4.5%	29.4%	1.9%
Female	12.2%	3.9%	3.5%	3.8%	18.4%	1.7%

Capabilities

The Saudi armed forces are among the best equipped in the region, and the Kingdom has displayed an increasing willingness to use them as part of a more assertive foreign policy. Principal roles are securing territorial integrity, internal security and regional stability. Saudi Arabia's defence posture continues to emphasise the deployment of airpower. Saudi Arabia is the leading member of the GCC. However, its most critical defence relationship is with the US. Riyadh also has significant security relationships with France and the UK, though recently there has been diversification of defence relationships, including with China. Significant training support is supplied by the US and, to a lesser extent, the UK. The armed forces continue to gain combat experience from their involvement in the conflict in Yemen, though the operation has exposed areas of comparative weakness and capability gaps, especially in the application of precision airpower, air–ground coordination, and in logistics support. Meanwhile, the cruise-missile and UAV attacks on Saudi oil infrastructure in September 2019 exposed further capability weaknesses. Saudi Arabia's conduct of its campaign in Yemen and other associated policies have increased the controversy surrounding arms sales to the country. Equipment recapitalisation continues, with orders for combat aircraft, corvettes and multi-mission surface combatants despite concerns about austerity. There is a modest domestic defence-industrial base, mainly in the assembly and overhaul of land systems. Riyadh has declared an intention to spend 50% of its defence outlays locally as part of its Vision 2030 initiative, and established the state-owned Saudi Arabian Military Industries to oversee local defence production.

ACTIVE 227,000 **(Army 75,000 Navy 13,500 Air 20,000 Air Defence 16,000 Strategic Missile Forces 2,500 National Guard 100,000) Paramilitary 24,500**

ORGANISATIONS BY SERVICE

Army 75,000

FORCES BY ROLE

MANOEUVRE

Armoured

4 armd bde (1 recce coy, 3 tk bn, 1 mech bn, 1 fd arty bn, 1 AD bn, 1 AT bn, 1 engr coy, 1 log bn, 1 maint coy, 1 med coy)

Mechanised

5 mech bde (1 recce coy, 1 tk bn, 3 mech bn, 1 fd arty bn, 1 AD bn, 1 AT bn, 1 engr coy, 1 log bn, 1 maint coy, 1 med coy)

Light

2 lt inf bde

Other

1 (Al-Saif Al-Ajrab) gd bde
1 (Royal Guard) gd regt (3 lt inf bn)

Air Manoeuvre

1 AB bde (2 AB bn, 3 SF coy)

Aviation

1 comd (3 hel gp)

COMBAT SUPPORT

3 arty bde

EQUIPMENT BY TYPE

MBT 960: 140 AMX-30; ε450 M1A2/A2S *Abrams*; ε370 M60A3 *Patton*
RECCE 300 AML-60/AML-90
IFV 794: 380 AMX-10P; 380 M2A2 *Bradley*; 34 VAB Mk3
APC 1,340
APC (T) 1,190 M113A4 (incl variants)
APC (W) 150 Panhard M3; (ε40 AF-40-8-1 *Al-Fahd* in store)
AUV 1,200+: 100 *Didgori* (amb); 1,000+ M-ATV; *Al-Shibl* 2; 100 Sherpa *Light Scout*; Terradyne *Gurkha*

ENGINEERING & MAINTENANCE VEHICLES

AEV 15 M728
ARV 275+: 8 ACV ARV; AMX-10EHC; 55 AMX-30D; *Leclerc* ARV; 122 M88A1; 90 M578
VLB 10 AMX-30
MW *Aardvark* Mk2

NBC VEHICLES 10 TPz-1 *Fuchs* NBC

ANTI-TANK/ANTI-INFRASTRUCTURE

MSL
SP 290+: 90+ AMX-10P (HOT); 200 VCC-1 ITOW; M-ATV with *Milan*
MANPATS *Hyeongung*; Luch *Corsar* (reported); Luch *Skif* (reported); *Stugna*-P (reported); TOW-2A
RCL 84mm *Carl Gustaf*; **90mm** M67; **106mm** M40A1

ARTILLERY 833

SP 155mm 224: 60 AU-F-1; 110 M109A1B/A2; 54 PLZ-45
TOWED 172: **105mm** 62 LG1; (100 M101/M102 in store); **155mm** 110: 50 M114; 60 M198; **203mm** (8 M115 in store)
MRL 70: **127mm** 60 ASTROS II Mk3; **220mm** 10 TOS-1A
MOR 367: **SP 81mm** 70; **SP 107mm** 150 M30; **120mm** 147: 110 Brandt; 37 M12-1535; **SP 120mm** 2R2M

HELICOPTERS

ATK 35: 11 AH-64D *Apache*; 24 AH-64E *Apache*

MRH 21: 6 AS365N *Dauphin* 2 (medevac); 15 Bell 406CS *Combat Scout*
TPT • Medium 58: 22 UH-60A *Black Hawk* (4 medevac); 36 UH-60L *Black Hawk*
AIR DEFENCE • SAM
Short-range *Crotale*
Point-defence FIM-92 *Stinger*

Navy 13,500

Navy HQ at Riyadh; Eastern Fleet HQ at Jubail; Western Fleet HQ at Jeddah

EQUIPMENT BY TYPE
PRINCIPAL SURFACE COMBATANTS • FRIGATES 6
FFGHM 6:
3 *Al Riyadh* (FRA *La Fayette* mod) with 2 quad lnchr with MM40 *Exocet* Block 2 AShM, 2 8-cell *Sylver* A43 VLS with *Aster* 15 SAM, 4 single 533mm TT with F17P HWT, 1 76mm gun (capacity 1 AS365N *Dauphin* 2 hel)
3 *Madina* (FRA F-2000) (1 more damaged in 2017 and non-operational) with 2 quad lnchr with *Otomat* Mk2 AShM, 1 octuple lnchr with *Crotale* SAM, 4 single 533mm TT with F17P HWT, 1 100mm gun (capacity 1 AS365N *Dauphin* 2 hel)
PATROL AND COASTAL COMBATANTS 45
CORVETTES • FSG 4 *Badr* (US *Tacoma*) with 2 quad lnchr with RGM-84C *Harpoon* Block 1B AShM, 2 triple 324mm ASTT with Mk 46 LWT, 1 Mk 15 *Phalanx* CIWS, 1 76mm gun
PCFG 9 *Al Siddiq* (US 58m) with 2 twin lnchr with RGM-84C *Harpoon* Block 1B AShM, 1 Mk 15 *Phalanx* CIWS, 1 76mm gun
PBF 13 HSI 32
PB 19: 17 (US) *Halter Marine* 24m; 2 *Plascoa* 2200
MINE WARFARE • MINE COUNTERMEASURES 3
MHC 3 *Al Jawf* (UK *Sandown*)
AMPHIBIOUS • LANDING CRAFT 5
LCU ε2 *Al Qiaq* (US LCU 1610) (capacity 120 troops)
LCM 3 LCM 6 (capacity 80 troops)
LOGISTICS AND SUPPORT 1
AORH 1 *Boraida* (mod FRA *Durance*) (1 more non-operational and in drydock since 2017) (capacity either 2 AS365F *Dauphin* 2 hel or 1 AS332C *Super Puma*)

Naval Aviation

EQUIPMENT BY TYPE
HELICOPTERS
MRH 34: 6 AS365N *Dauphin* 2; 15 AS565; 13 Bell 406CS *Combat Scout*
TPT • Medium 12 AS332B/F *Super Puma*
AIR-LAUNCHED MISSILES
AShM AM39 *Exocet*; AS-15TT

Marines 3,000

FORCES BY ROLE
SPECIAL FORCES
1 spec ops regt with (2 spec ops bn)

EQUIPMENT BY TYPE
ARMOURED FIGHTING VEHICLES
RECCE *Bastion Patsas*
APC • APC (W) 135 BMR-600P

Air Force 20,000

FORCES BY ROLE
FIGHTER
4 sqn with F-15C/D *Eagle*
FIGHTER/GROUND ATTACK
3 sqn with F-15S/SA *Eagle*
3 sqn with *Typhoon*
GROUND ATTACK
3 sqn with *Tornado* IDS; *Tornado* GR1A
AIRBORNE EARLY WARNING & CONTROL
1 sqn with E-3A *Sentry*
1 sqn with Saab 2000 *Erieye*
ELINT
1 sqn with RE-3A/B; Beech 350ER *King Air*
TANKER
1 sqn with KE-3A
TANKER/TRANSPORT
1 sqn with KC-130H/J *Hercules*
1 sqn with A330 MRTT
TRANSPORT
3 sqn with C-130H *Hercules*; C-130H-30 *Hercules*; CN-235; L-100-30HS (hospital ac)
2 sqn with Beech 350 *King Air* (forming)
TRAINING
1 OCU sqn with F-15SA *Eagle*
3 sqn with *Hawk* Mk65*; *Hawk* Mk65A*; *Hawk* Mk165*
1 sqn with *Jetstream* Mk31
1 sqn with MFI-17 *Mushshak*; SR22T
2 sqn with PC-9; PC-21
TRANSPORT HELICOPTER
4 sqn with AS532 *Cougar* (CSAR); Bell 212 (AB-212); Bell 412 (AB-412) *Twin Huey* (SAR)

EQUIPMENT BY TYPE
AIRCRAFT 443 combat capable
FTR 81: 56 F-15C *Eagle*; 25 F-15D *Eagle*
FGA 215: up to 67 F-15S *Eagle* (being upgraded to F-15SA configuration); 77 F-15SA *Eagle*; 71 *Typhoon*
ATK 66 *Tornado* IDS
ISR 14+: 12 *Tornado* GR1A*; 2+ Beech 350ER *King Air*
AEW&C 7: 5 E-3A *Sentry*; 2 Saab 2000 *Erieye*
ELINT 2: 1 RE-3A; 1 RE-3B
TKR/TPT 15: 6 A330 MRTT; 7 KC-130H *Hercules*; 2 KC-130J *Hercules*
TKR 7 KE-3A
TPT 47+: **Medium** 36: 30 C-130H *Hercules*; 3 C-130H-30 *Hercules*; 3 L-100-30; **Light** 11+: 10+ Beech 350 *King Air*; 1 *Jetstream* Mk31
TRG 188: 24 *Hawk* Mk65* (incl aerobatic team); 16 *Hawk* Mk65A*; 29 *Hawk* Mk165*; 20 MFI-17 *Mushshak*; 20 PC-9; 55 PC-21; 24 SR22T
HELICOPTERS
MRH 15 Bell 412 (AB-412) *Twin Huey* (SAR)
TPT 30: **Medium** 10 AS532 *Cougar* (CSAR); **Light** 20 Bell 212 (AB-212)
UNINHABITED AERIAL VEHICLES
CISR • Heavy some *Wing Loong* 1 (GJ-1) (reported); some CH-4
ISR • Medium some *Falco*

AIR-LAUNCHED MISSILES
AAM • IR AIM-9P/L *Sidewinder*; **IIR** AIM-9X *Sidewinder* II; IRIS-T; **SARH** AIM-7 *Sparrow*; AIM-7M *Sparrow*; **ARH** AIM-120C AMRAAM
ASM AGM-65 *Maverick*; AR-1
AShM AGM-84L *Harpoon* Block II
ARM ALARM
ALCM *Storm Shadow*
BOMBS
Laser-guided GBU-10/12 *Paveway* II; *Paveway* IV
INS/GPS-guided GBU-31 JDAM; FT-9

Royal Flt

EQUIPMENT BY TYPE
AIRCRAFT • **TPT** 24: **Medium** 8: 5 C-130H *Hercules*; 3 L-100-30; **Light** 3: 1 Cessna 310; 2 Learjet 35; **PAX** 13: 1 A340; 1 B-737-200; 2 B-737BBJ; 2 B-747SP; 4 BAe-125-800; 2 Gulfstream III; 1 Gulfstream IV
HELICOPTERS • **TPT** 3+: **Medium** 3: 2 AS-61; 1 S-70 *Black Hawk*; **Light** some Bell 212 (AB-212)

Air Defence Forces 16,000

FORCES BY ROLE
AIR DEFENCE
6 bn with MIM-104D/F *Patriot* PAC-2 GEM/PAC-3
17 bty with *Shahine*/AMX-30SA
16 bty with MIM-23B I-*Hawk*

EQUIPMENT BY TYPE
AIR DEFENCE
SAM 817+
Long-range 108 MIM-104D/F *Patriot* PAC-2 GEM/PAC-3
Medium-range 128 MIM-23B I-*Hawk*
Short-range 181: 40 *Crotale*; 141 *Shahine*
Point-defence 400+: 400 M1097 *Avenger*; *Mistral*
GUNS 218
SP • 20mm 90 M163 *Vulcan*
TOWED 128: **35mm** 128 GDF Oerlikon; **40mm** (150 L/70 in store)

Strategic Missile Forces 2,500

EQUIPMENT BY TYPE
MSL • TACTICAL
IRBM 10+ DF-3 (CH-SS-2) (service status unclear)
MRBM Some DF-21 (CH-SS-5 – variant unclear) (reported)

National Guard 73,000 active; 27,000 (tribal levies) (total 100,000)

FORCES BY ROLE
MANOEUVRE
Mechanised
5 mech bde (1 recce coy, 3 mech inf bn, 1 SP arty bn, 1 cbt engr coy, 1 sigs coy, 1 log bn)
Light
5 inf bde (3 combined arms bn, 1 arty bn, 1 log bn)
3 indep lt inf bn
Other
1 (Special Security) sy bde (3 sy bn)
1 (ceremonial) cav sqn
COMBAT SUPPORT
1 MP bn

EQUIPMENT BY TYPE
ARMOURED FIGHTING VEHICLES
ASLT 235: 204 LAV-AG (90mm); 31 LAV 6.0 (105mm)
IFV 1,035: ε635 LAV-25; ε400 LAV 6.0 (incl variants)
APC 778
APC (W) 514: 116 LAV-A (amb); 30 LAV-AC (ammo carrier); 296 LAV-CC (CP); 72 LAV-PC
PPV 264 *Aravis*; some *Arive*
ENGINEERING & MAINTENANCE VEHICLES
AEV 58 LAV-E
ARV 111 LAV-R; V-150 ARV
MW MV5; MV10
ANTI-TANK/ANTI-INFRASTRUCTURE
MSL
SP 182 LAV-AT
MANPATS TOW-2A; M47 *Dragon*
RCL • 106mm M40A1
ARTILLERY 363+
SP 155mm up to 136 CAESAR
TOWED 108: **105mm** 50 M102; **155mm** 58 M198
MOR 119+: **81mm** some; **120mm** 119 LAV-M
HELICOPTERS
ATK 12 AH-64E *Apache*
MRH 35: 23 AH-6i *Little Bird*; 12 MD530F (trg role)
TPT • Medium ε50 UH-60M *Black Hawk*
AIR DEFENCE
SAM 73
Short-range 5 VL MICA
Point-defence 68 MPCV
GUNS • TOWED • 20mm 30 M167 *Vulcan*
AIR-LAUNCHED MISSILES
ASM AGM-114R *Hellfire* II

Paramilitary 24,500+ active

Border Guard 15,000

FORCES BY ROLE
Subordinate to Ministry of Interior. HQ in Riyadh. 9 subordinate regional commands
MANOEUVRE
Other
Some mobile def (long-range patrol/spt) units
2 border def (patrol) units
12 infrastructure def units
18 harbour def units
Some coastal def units
COMBAT SUPPORT
Some MP units

EQUIPMENT BY TYPE
ARMOURED FIGHTING VEHICLES
APC • PPV *Caprivi* Mk1/Mk3
PATROL AND COASTAL COMBATANTS 106
PCC 13 CSB 40
PBF 85: 4 *Al Jouf*; 2 *Sea Guard*; 79 *Plascoa* FIC 1650
PB 8: 6 Damen Stan Patrol 2606; 2 *Al Jubatel*
AMPHIBIOUS • LANDING CRAFT • **UCAC** 8: 5 *Griffon* 8000; 3 other
LOGISTICS AND SUPPORT 4: 1 **AXL**; 3 **AO**

Facilities Security Force 9,000+

Subordinate to Ministry of Interior

General Civil Defence Administration Units

EQUIPMENT BY TYPE

HELICOPTERS • TPT • Medium 10 Boeing *Vertol* 107

Special Security Force 500

EQUIPMENT BY TYPE

ARMOURED FIGHTING VEHICLES

APC • APC (W) UR-416

AUV *Gurkha* LAPV

DEPLOYMENT

YEMEN: *Operation Restoring Hope* 2,500; 2 armd BG; M60A3; M2A2 *Bradley*; M113A4; M-ATV; 2+ MIM-104D/F *Patriot* PAC-2/3

FOREIGN FORCES

Bahrain *Operation Restoring Hope* 250; 1 SF gp; 1 arty gp; 6 F-16C *Fighting Falcon*

Egypt *Operation Restoring Hope* 6 F-16C *Fighting Falcon*

France 50 (radar det)

Jordan *Operation Restoring Hope* 6 F-16AM *Fighting Falcon*

Kuwait *Operation Restoring Hope* 4 F/A-18A *Hornet*

Sudan *Operation Restoring Hope* 3 Su-24 *Fencer*

United Arab Emirates *Operation Restoring Hope* 12 F-16E *Fighting Falcon*

United Kingdom 50 (radar det)

United States US Central Command: 2,000; 1 FGA sqn with 12 F-16C *Fighting Falcon*; 2 SAM bty with MIM-104 *Patriot* PAC-2/-3; 1 SAM bty with THAAD

Syria SYR

Syrian Pound S£		2019	2020	2021
GDP	S£			
	US$			
per capita	US$			
Growth	%			
Inflation	%			
Def exp	S£			
	US$			
US$1=S£				

Population 19,398,448

Age	0–14	15–19	20–24	25–29	30–64	65 plus
Male	17.1%	4.9%	4.7%	3.9%	17.1%	2.1%
Female	16.3%	4.8%	4.9%	4.0%	17.7%	2.4%

Capabilities

The protracted civil war has significantly depleted the combat capabilities of the Syrian armed forces and transformed them into an irregularly structured militia-style organisation focused on internal security. There is no published defence doctrine or white paper, the conflict instead dictating ad hoc requirements. Opposition groups maintain de facto control over parts of the country. Most formal pre-war structures and formations exist in name only, as resources have been channelled into the irregular network of military organisations that form the regime's most effective military capabilities. Russia is the regime's principal ally and has provided essential combat support and assistance, as well as replacement equipment. Russia is also involved in efforts to reconstitute the army's pre-war divisions. Iran and Hizbullah also continue to assist in the provision and training of militias and other ground forces. Overall levels of training remain poor but combat experience has improved proficiency in select regular and irregular military formations. The armed forces lack the requisite capabilities for external deployment, although they remain able to redeploy moderate numbers of formations and capabilities within the country. Logistics support for major internal operations away from established bases remains a challenge. Before the civil war, Syria did not have a major domestic defence industry, although it possessed facilities to overhaul and maintain its existing systems. It did, however, possess some capacity in focused areas, such as ballistic missiles and chemical weapons. International efforts to verify destruction of chemical-weapons stockpiles and production facilities continue.

ACTIVE 169,000 (Army 130,000 Navy 4,000 Air 15,000 Air Defence 20,000) Paramilitary 100,000

Conscript liability 30 months (there is widespread avoidance of military service)

ORGANISATIONS BY SERVICE

Army ε130,000

FORCES BY ROLE

The Syrian Arab Army combines conventional formations, special forces and auxiliary militias. The main fighting units are the 4th Armoured Division, the Republican Guard, the Special Forces (including Tiger Forces) and the brigades assigned to the 5th Assault Corps; they receive the most attention and training. Most other formations are under-strength, at an estimated 500–1,000 personnel in brigades and regiments, but Russia has been assisting in the reconstruction and re-equipment of some divisions.

COMMAND

4 corps HQ

1 (5th Assault) corps HQ

SPECIAL FORCES

2 SF div (total: 11 SF regt; 1 tk regt)

MANOEUVRE

Armoured

1 (4th) armd div (1 SF regt, 2 armd bde, 2 mech bde, 1 arty regt, 1 SSM bde (3 SSM bn with *Scud*-B/C))

3 armd div (being reconstituted)

2 armd div(-)

Mechanised

1 (Republican Guard) mech div (3 mech bde, 2 sy regt, 1 arty regt)

1 mech div (being reconstituted)

2 mech div(-)

8 mech bde (assigned to 5th Assault Corps)

2 indep inf bde(-)

Amphibious

1 mne unit

COMBAT SUPPORT

2 arty bde

2 AT bde
1 SSM bde (3 SSM bn with FROG-7)
1 SSM bde (3 SSM bn with 9K79)

EQUIPMENT BY TYPE

Attrition during the civil war has severely reduced equipment numbers for almost all types. It is unclear how much remains available for operations

ARMOURED FIGHTING VEHICLES
MBT T-55A; T-55AM; T-55AMV; T-62; T-62M; T-72; T-72AV; T-72B; T-72B3; T-72M1; T-90
RECCE BRDM-2
IFV BMP-1; BMP-2; BTR-82A
APC
APC (T) BTR-50
APC (W) BTR-152; BTR-60; BTR-70; BTR-80
APC IVECO LMV
ENGINEERING & MAINTENANCE VEHICLES
ARV BREM-1 reported; T-54/55
VLB MTU; MTU-20
MW UR-77
ANTI-TANK/ANTI-INFRASTRUCTURE • MSL
SP 9P133 *Malyutka*-P (BRDM-2 with AT-3C *Sagger*); 9P148 *Konkurs* (BRDM-2 with AT-5 *Spandrel*)
MANPATS 9K111 *Fagot* (AT-4 *Spigot*); 9K111-1 *Konkurs* (AT-5 *Spandrel*); 9K115 *Metis* (AT-7 *Saxhorn*); 9K115-2 *Metis*-M (AT-13 *Saxhorn* 2); 9K135 *Kornet* (AT-14 *Spriggan*); *Milan*
ARTILLERY
SP 122mm 2S1 *Gvozdika*; D-30 (mounted on T-34/85 chassis); **130mm** M-46 (truck–mounted); **152mm** 2S3 *Akatsiya*
TOWED 122mm D-30; M-30 (M1938); **130mm** M-46; **152mm** D-20; ML-20 (M-1937); **180mm** S-23
GUN/MOR 120mm 2S9 NONA-S
MRL 107mm Type-63; **122mm** BM-21 *Grad*; **140mm** BM-14; **220mm** 9P140 *Uragan*; **300mm** 9A52 *Smerch*; **330mm** some (also improvised systems of various calibres)
MOR 82mm some; **120mm** M-1943; **160mm** M-160; **240mm** M-240
SURFACE-TO-SURFACE MISSILE LAUNCHERS
SRBM • Conventional *Scud*-B/C/D; *Scud* lookalike; 9K79 *Tochka* (SS-21 *Scarab*); *Fateh*-110/M-600
UNINHABITED AERIAL VEHICLES
ISR • Medium *Mohajer* 3/4; **Light** *Ababil*
AIR DEFENCE
SAM
Medium-range 9K37 *Buk* (SA-11 *Gadfly*); 9K317 *Buk*-M2 (SA-17 *Grizzly*)
Short-range 96K6 *Pantsir*-S1 (SA-22 *Greyhound*)
Point-defence 9K31 *Strela*-1 (SA-9 *Gaskin*); 9K33 *Osa* (SA-8 *Gecko*); 9K35 *Strela*-10 (SA-13 *Gopher*); 9K32 *Strela*-2 (SA-7 *Grail*)‡; 9K38 *Igla* (SA-18 *Grouse*); 9K36 *Strela*-3 (SA-14 *Gremlin*); 9K338 *Igla-S* (SA-24 *Grinch*)
GUNS
SP 23mm ZSU-23-4; **57mm** ZSU-57-2; S-60 (on 2K12 chassis)
TOWED 23mm ZU-23-2; **37mm** M-1939; **57mm** S-60; **100mm** KS-19

Navy ε4,000

Some personnel are likely to have been drafted into other services

EQUIPMENT BY TYPE

PATROL AND COASTAL COMBATANTS 31:
CORVETTES • FS 1 Project 159AE (*Petya* III)† with 1 triple 533mm ASTT with SAET-60 HWT, 4 RBU 2500 *Smerch* 1 A/S mor, 2 twin 76mm gun
PBFG 22:
16 Project 205 (*Osa* I/II)† with 4 single lnchr with P-22 (SS-N-2C *Styx*) AShM
6 *Tir* with 2 single lnchr with C-802 (CH-SS-N-6) AShM
PB 8 *Zhuk*†
MINE WARFARE • MINE COUNTERMEASURES 7
MHC 1 Project 1265 (*Sonya*) with 2 quad lnchr with 9K32 *Strela*-2 (SA-N-5 *Grail*)‡ SAM, 2 AK630 CIWS
MSO 1 *Akvamaren*-M (FSU Project 266M (*Natya*)) with 2 quad lnchr with 9K32 *Strela*-2 (SA-N-5 *Grail*)‡ SAM
MSI 5 *Korund* (Project 1258 (*Yevgenya*))
AMPHIBIOUS • LANDING SHIPS • LSM 3 *Polnochny* B (capacity 6 MBT; 180 troops)
LOGISTICS AND SUPPORT • AX 1 *Al Assad*

Coastal Defence

FORCES BY ROLE

COASTAL DEFENCE
1 AShM bde with P-35 (SSC-1B *Sepal*); P-15M *Termit*-R (SSC-3 *Styx*); C-802; K-300P *Bastion* (SSC-5 *Stooge*)

EQUIPMENT BY TYPE

COASTAL DEFENCE • AShM P-35 (SSC-1B *Sepal*); P-15M *Termit*-R (SSC-3 *Styx*); C-802; K-300P *Bastion* (SSC-5 *Stooge*)

Naval Aviation

All possibly non-operational after vacating base for Russian deployment

EQUIPMENT BY TYPE

HELICOPTERS • ASW 10: 4 Ka-28 *Helix* A; 6 Mi-14 *Haze*

Air Force ε15,000(-)

FORCES BY ROLE

FIGHTER
2 sqn with Mig-23MF/ML/MLD/UM *Flogger*
2 sqn with MiG-29A/UB/SM *Fulcrum*
FIGHTER/GROUND ATTACK
4 sqn with MiG-21MF/bis *Fishbed*; MiG-21U *Mongol* A
2 sqn with MiG-23BN/UB *Flogger*
GROUND ATTACK
4 sqn with Su-22M3/M4 *Fitter* J/K
1 sqn with Su-24MK *Fencer* D
1 sqn with L-39ZA/ZO *Albatros**
TRANSPORT
1 sqn with An-24 *Coke*; An-26 *Curl*; Il-76 *Candid*
1 sqn with *Falcon* 20; *Falcon* 900
1 sqn with Tu-134B-3
1 sqn with Yak-40 *Codling*
ATTACK HELICOPTER
3 sqn with Mi-24D/P *Hind* D/F
2 sqn with SA342L *Gazelle*

Middle East and North Africa

TRANSPORT HELICOPTER
6 sqn with Mi-8 *Hip*/Mi-17 *Hip* H

EQUIPMENT BY TYPE

Heavy use of both fixed- and rotary-wing assets has likely reduced readiness and availability to very low levels. It is estimated that no more than 30–40% of the inventory is operational

AIRCRAFT 184 combat capable
FTR 55: ε25 MiG-23MF/ML/MLD/UM *Flogger*; ε30 MiG-29A/SM/UB *Fulcrum*
FGA 79: ε50 MiG-21MF/bis *Fishbed* J/L; 9 MiG-21U *Mongol* A; ε20 MiG-23BN/UB *Flogger*
ATK 30: 20 Su-22M3/M4 *Fitter* J/K; ε10 Su-24MK *Fencer* D
TPT 23: **Heavy** 3 Il-76 *Candid*; **Light** 13: 1 An-24 *Coke*; 6 An-26 *Curl*; 2 PA-31 *Navajo*; 4 Yak-40 *Codling*; **PAX** 7: 2 *Falcon* 20; 1 *Falcon* 900; 4 Tu-134B-3
TRG 20+: ε20 L-39ZA/ZO *Albatros**; some MBB-223 *Flamingo*†
HELICOPTERS
ATK 20+: ε20 Mi-24D *Hind* D; some Mi-24P *Hind* F
MRH 40: ε20 Mi-17 *Hip* H; ε20 SA342L *Gazelle*
TPT • Medium ε10 Mi-8 *Hip*
AIR-LAUNCHED MISSILES
AAM • IR R-60 (AA-8 *Aphid*); R-73 (AA-11 *Archer*); **IR/SARH**; R-23/24 (AA-7 *Apex*); R-27 (AA-10 *Alamo*); **ARH**; R-77 (AA-12A *Adder*)
ASM Kh-25 (AS-10 *Karen*); Kh-29T/L (AS-14 *Kedge*); HOT
ARM Kh-31P (AS-17A *Krypton*)

Air Defence Command ε20,000(-)

FORCES BY ROLE

AIR DEFENCE
4 AD div with S-125 *Pechora* (SA-3 *Goa*); 2K12 *Kub* (SA-6 *Gainful*); S-75 *Dvina* (SA-2 *Guideline*)
3 AD regt with S-200 *Angara* (SA-5 *Gammon*); S-300PMU2 (SA-20 *Gargoyle*)

EQUIPMENT BY TYPE

AIR DEFENCE • SAM
Long-range S-200 *Angara* (SA-5 *Gammon*); 24 S-300PMU2 (SA-20 *Gargoyle*)
Medium-range S-75 *Dvina* (SA-2 *Guideline*)
Short-range 2K12 *Kub* (SA-6 *Gainful*); S-125 *Pechora* (SA-3 *Goa*)
Point-defence 9K32 *Strela*-2/2M (SA-7A/B *Grail*)‡

Paramilitary ε100,000

National Defence Force ε50,000

An umbrella of disparate regime militias performing a variety of roles, including territorial control

Other Militias ε50,000

Numerous military groups fighting for the Assad regime, including Afghan, Iraqi, Pakistani and sectarian organisations. Some receive significant Iranian support

FOREIGN FORCES

Hizbullah 7,000–8,000
Iran 1,500
Russia 4,000: 1 inf BG; 3 MP bn; 1 engr unit; ε10 T-72B3/T-90; ε20 BTR-82A; 12 2A65; 4 9A52 *Smerch*; 10 Su-24M *Fencer*; 6 Su-34; 6 Su-35S; 1 A-50U; 1 Il-20M; 12 Mi-24P/Mi-35M *Hind*; 4 Mi-8AMTSh *Hip*; 1 AShM bty with 3K55 *Bastion* (SSC-5 *Stooge*); 1 SAM bty with S-400 (SA-21 *Growler*); 1 SAM bty with *Pantsir*-S1/S2; air base at Latakia; naval facility at Tartus

TERRITORY WHERE THE GOVERNMENT DOES NOT EXERCISE EFFECTIVE CONTROL

Data here represents the de facto situation for selected armed opposition groups and their observed equipment

Syrian Democratic Forces ε50,000

A coalition of predominantly Kurdish rebel groups in de facto control of much of northeastern Syria. Kurdish forces from the YPG/J (People's Protection Units/Women's Protection Units) provide military leadership and main combat power, supplemented by Arab militias and tribal groups.

EQUIPMENT BY TYPE

ARMOURED FIGHTING VEHICLES
MBT T-55; T-72 (reported)
IFV BMP-1
APC • PPV *Guardian*
AUV M-ATV
ANTI-TANK/ANTI-INFRASTRUCTURE
MSL • MANPATS 9K111-1 *Konkurs*
RCL 73mm SPG-9; **90mm** M-79 *Osa*
ARTILLERY
MRL 122mm BM-21 *Grad*; 9K132 *Grad*-P
MOR 82mm 82-BM-37; M-1938; **120mm** M-1943; improvised mortars of varying calibre
AIR DEFENCE • GUNS
SP 14.5mm ZPU-4 (tch); ZPU-2 (tch); ZPU-1 (tch); 1 ZPU-2 (tch/on T-55); **23mm** ZSU-23-4 *Shilka*; ZU-23-2 (tch); **57mm** S-60
TOWED 14.5mm ZPU-2; ZPU-1; **23mm** ZU-23-2

Syrian National Army & National Front for Liberation ε70,000

In late 2019 the Syrian National Army (SNA) and the National Front for Liberation (NLF) began to merge under the SNA umbrella. The SNA formed in late 2017 from Syrian Arab and Turkmen rebel factions operating under Turkish command in the Aleppo governate and northwestern Syria, including Afrin province. The NLF is a coalition of surviving Islamist and nationalist rebel factions formed in 2018 operating in northwestern Syria, particularly in and around Idlib.

EQUIPMENT BY TYPE

ARMOURED FIGHTING VEHICLES
MBT T-54; T-55; T-62
IFV BMP-1
ANTI-TANK/ANTI-INFRASTRUCTURE
MSL • MANPATS 9K11 *Malyutka* (AT-3 *Sagger*); 9K111 *Fagot* (AT-4 *Spigot*); 9K113 *Konkurs* (AT-5 *Spandrel*); 9K115 *Metis* (AT-7 *Saxhorn*); 9K115-2 *Metis*-M (AT-13

Saxhorn 2); 9K135 *Kornet* (AT-14 *Spriggan*); BGM-71 TOW; *Milan*
RCL 73mm SPG-9; **82mm** B-10
ARTILLERY
TOWED 122mm D-30
MRL 107mm Type-63; **122mm** 9K132 *Grad*-P; BM-21 *Grad*; *Grad* (6-tube tech)
MOR 82mm 2B9 *Vasilek*; improvised mortars of varying calibre
AIR DEFENCE
SAM • Point-defence MANPADS some
GUNS
SP 14.5mm ZPU-4 (tch); ZPU-2 (tch); ZPU-1 (tch); **23mm** ZU-23-2 (tch); ZSU-23-4 *Shilka*; **57mm** AZP S-60
TOWED 14.5mm ZPU-1; ZPU-2; ZPU-4; **23mm** ZU-23-2

Hayat Tahrir al-Sham (HTS) ε15,000

HTS was formed by Jabhat Fateh al-Sham (formerly known as Jabhat al-Nusra) in January 2017 by absorbing other hardline groups. It is designated a terrorist organisation by the US government.

EQUIPMENT BY TYPE
ANTI-TANK/ANTI-INFRASTRUCTURE
MSL • MANPATS 9K11 *Malyutka* (AT-3 *Sagger*); 9K113 *Konkurs* (AT-5 *Spandrel*); 9K115-2 *Metis*-M (AT-13 *Saxhorn* 2); 9K135 *Kornet* (AT-14 *Spriggan*)
RCL 73mm SPG-9; **106mm** M-40
ARTILLERY
MRL 107mm Type-63
MOR 120mm some; improvised mortars of varying calibres
AIR DEFENCE
SAM
Point-defence 9K32M *Strela*-2M (SA-7B *Grail*)‡
GUNS
SP 14.5mm ZPU-1; ZPU-2; **23mm** ZU-23-2; **57mm** S-60

Guardians of Religion (Huras al-Din) ε4,000

An al-Qaeda-affiliated group operating in Idlib province. It is designated a terrorist organisation by the US government.

FOREIGN FORCES

Turkey *Operation Spring Shield* ε3,000; 3 armd BG; some cdo units; 1 gendarmerie unit
United States *Operation Inherent Resolve* 750; 1 armd inf coy; 1 mne bn(-)

Tunisia TUN

Tunisian Dinar D		2019	2020	2021
GDP	D	114bn	111bn	
	US$	38.8bn	39.2bn	
per capita	US$	3,293	3,295	
Growth	%	1.0	-7.0	
Inflation	%	6.7	5.8	
Def bgt	D	2.93bn	3.24bn	
	US$	999m	1.14bn	
FMA (US)	US$	85.0m	40.0m	40.0m
US$1=D		2.93	2.83	

Real-terms defence budget trend (US$m, constant 2015)

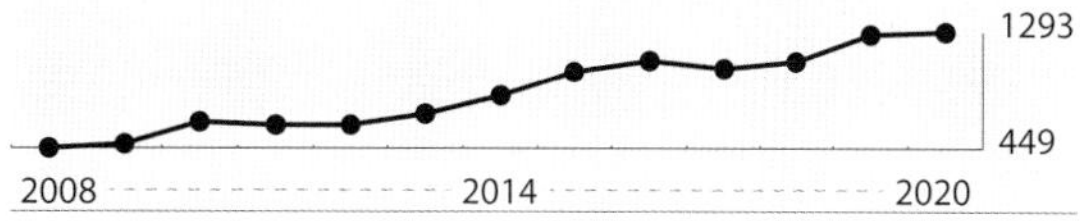

Population 11,721,177

Age	0–14	15–19	20–24	25–29	30–64	65 plus
Male	13.1%	3.3%	3.2%	3.7%	22.2%	4.2%
Female	12.2%	3.2%	3.2%	3.8%	23.3%	4.7%

Capabilities

Ensuring territorial sovereignty and internal security are the main tasks of the armed forces, which have limited capacities but are undergoing a modernisation process. The civil war in Libya and Islamist terrorist groups operating from there continue to pose a security concern. In the light of terrorist attacks, the armed forces are engaged in counter-terrorism operations, and have been tasked with securing sensitive industrial sites. Designated a major non-NATO ally by the US in 2015, Tunisia benefits from defence and security cooperation with US AFRICOM and also with France. In 2019, Tunisia sent a transport aircraft to support the UN's MINUSMA peacekeeping mission in Mali and reportedly plans to deploy an infantry battalion and military police to the mission. The country has received training from Algeria and is a member of the Saudi-led Islamic Military Counter Terrorism Coalition. The armed forces are involved in multinational exercises, notably those led by the US. The country is also strengthening its intelligence capabilities. Overall military capability is limited by the ageing equipment inventory, although Tunisia has been the recipient of surplus US systems, including armed utility helicopters. The country has limited defence-industrial capabilities.

ACTIVE 35,800 (Army 27,000 Navy 4,800 Air 4,000)
Paramilitary 12,000
Conscript liability 12 months selective

ORGANISATIONS BY SERVICE

Army 5,000; 22,000 conscript (total 27,000)

FORCES BY ROLE
SPECIAL FORCES
1 SF bde
1 (Sahara) SF bde

MANOEUVRE
Reconnaissance
1 recce regt
Mechanised
3 mech bde (1 armd regt, 2 mech inf regt, 1 arty regt, 1 AD regt, 1 engr regt, 1 sigs regt, 1 log gp)
COMBAT SUPPORT
1 engr regt

EQUIPMENT BY TYPE
ARMOURED FIGHTING VEHICLES
MBT 84: 30 M60A1; 54 M60A3
LT TK 48 SK-105 *Kuerassier*
RECCE 60: 40 AML-90; 20 FV601 *Saladin*
APC 425+
APC (T) 140 M113A1/A2
APC (W) 110 Fiat 6614
PPV 175+: 4 *Bastion* APC: 71 *Ejder Yalcin*; 100+ *Kirpi*
ENGINEERING & MAINTENANCE VEHICLES
ARV 11: 5 *Greif*; 6 M88A1
ANTI-TANK/ANTI-INFRASTRUCTURE • MSL
SP 35 M901 ITV TOW
MANPATS *Milan*; TOW
ARTILLERY 276
TOWED 115: **105mm** 48 M101A1/A2; **155mm** 67: 12 M114A1; 55 M198
MOR 161: **81mm** 95; **SP 107mm** 48 M106; **120mm** 18 Brandt
AIR DEFENCE
SAM • Point-defence 26 M48 *Chaparral*; RBS-70
GUNS 112
SP 40mm 12 M42
TOWED • 20mm 100 M-55

Navy ε4,800

EQUIPMENT BY TYPE
PATROL AND COASTAL COMBATANTS 35
PSO 4 *Jugurtha* (Damen Stan MSOPV 1400) (of which 2 with 1 hel landing platform)
PCFG 3 *La Galite* (FRA *Combattante* III) with 2 quad lnchr with MM40 *Exocet* AShM, 1 76mm gun
PCC 3 *Bizerte* (FRA P-48)
PCFT 6 *Albatros* (GER Type-143B) with 2 single 533mm TT, 2 76mm guns
PBF 2 20m Fast Patrol Boat
PB 17: 3 *Istiklal*; 3 *Utique* (mod PRC *Haizhui* II); 5 *Joumhouria*; 6 V Series
LOGISTICS AND SUPPORT 7:
ABU 3: 2 *Tabarka* (ex-US *White Sumac*); 1 *Sisi Bou Said*
AGE 1 *Hannibal*
AGS 1 *Khaireddine* (ex-US *Wilkes*)
AWT 1 *Ain Zaghouan* (ex-ITA *Simeto*)
AX 1 *Salambo* (ex-US *Conrad*, survey)

Air Force 4,000

FORCES BY ROLE
FIGHTER/GROUND ATTACK
1 sqn with F-5E/F-5F *Tiger* II
TRANSPORT
1 sqn with C-130B/H/J-30 *Hercules*; G.222; L-410 *Turbolet*
1 liaison unit with S-208A
TRAINING
2 sqn with L-59 *Albatros**; MB-326B; SF-260
1 sqn with MB-326K; MB-326L
TRANSPORT HELICOPTER
2 sqn with AS350B *Ecureuil*; AS365 *Dauphin* 2; AB-205 (Bell 205); SA313; SA316 *Alouette* III; UH-1H *Iroquois*; UH-1N *Iroquois*
1 sqn with HH-3E

EQUIPMENT BY TYPE
AIRCRAFT 23 combat capable
FTR 11: 9 F-5E *Tiger* II; 2 F-5F *Tiger* II
ATK 3 MB-326K
ISR 12 *Maule* MX-7-180B
TPT 18: **Medium** 13: 5 C-130B *Hercules*; 1 C-130H *Hercules*; 2 C-130J-30 *Hercules*; 5 G.222; **Light** 5: 3 L-410 *Turbolet*; 2 S-208A
TRG 30: 9 L-59 *Albatros**; 4 MB-326B; 3 MB-326L; 14 SF-260
HELICOPTERS
MRH 34: 1 AS365 *Dauphin* 2; 6 SA313; 3 SA316 *Alouette* III; 24 OH-58D *Kiowa Warrior*
SAR 11 HH-3E
TPT 39: **Medium** 8 UH-60M *Black Hawk*; **Light** 31: 6 AS350B *Ecureuil*; 15 Bell 205 (AB-205); 8 Bell 205 (UH-1H *Iroquois*); 2 Bell 212 (UH-1N *Iroquois*)
AIR-LAUNCHED MISSILES
AAM • IR AIM-9P *Sidewinder*
ASM AGM-114R *Hellfire*

Paramilitary 12,000

National Guard 12,000

Ministry of Interior

EQUIPMENT BY TYPE
ARMOURED FIGHTING VEHICLES
ASLT 2 EE-11 *Urutu* FSV
APC 29+
APC (W) 16 EE-11 *Urutu* (anti-riot); VAB Mk3
PPV 13 Streit *Typhoon*
AUV IVECO LMV
PATROL AND COASTAL COMBATANTS 24
PCC 6 *Rais el Blais* (ex-GDR *Kondor* I)
PBF 7: 4 *Gabes; 3 Patrouiller*
PB 11: 5 *Breitla* (ex-GDR *Bremse*); 4 Rodman 38; 2 *Socomena*
HELICOPTERS
MRH 8 SA318 *Alouette* II/SA319 *Alouette* III
TPT • Light 3 Bell 429

DEPLOYMENT

CENTRAL AFRICAN REPUBLIC: UN • MINUSCA 2

DEMOCRATIC REPUBLIC OF THE CONGO: UN • MONUSCO 7

MALI: UN • MINUSMA 85; 1 tpt flt with C-130J-30

SOUTH SUDAN: UN • UNMISS 2

United Arab Emirates UAE

Emirati Dirham D		2019	2020	2021
GDP	D	1.55tr	1.30tr	
	US$	421bn	354bn	
per capita	US$	39,180	31,948	
Growth	%	1.7	-6.6	
Inflation	%	-1.9	-1.5	
Def bdgt [a]	D	ε71.0bn	ε72.8bn	
	US$	ε19.3bn	ε19.8bn	
US$1=D		3.67	3.67	

[a] Defence-budget estimate derived from central MoD expenditure and a proportion of the Federal Services section of the Abu Dhabi budget

Population 9,992,083

Age	0–14	15–19	20–24	25–29	30–64	65 plus
Male	7.5%	2.1%	2.2%	3.9%	54.8%	1.5%
Female	7.0%	1.8%	1.8%	2.5%	14.5%	0.4%

Capabilities

The UAE's armed forces are arguably the best trained and most capable among the GCC states. Its diplomatic recognition of Israel reflects and may affect its defence and security posture, including potentially greater access to US-supplied weapons systems. In recent years, there has been a growing willingness to take part in operations and project power and influence, including sending an F-16 detachment to Afghanistan and involvement in the conflict in Libya. It also joined the Saudi-led intervention in Yemen, although it has subsequently drawn down its forces. Attacks on tankers off the UAE coast and on oil infrastructure have added to defence concerns. A new defence agreement with the US came into force in May 2019. The UAE hosts a French base and is diversifying its security relationships, including with China, India and Japan. The UAE's involvement in the Yemen campaign has offered combat lessons, not least in limited amphibious operations. This operation demonstrated the country's developing approach to the use of force and there are signs of an acceptance of military risk. The UAE has an advanced inventory of modern equipment across the domains and is taking steps to upgrade its airborne ISR capabilities. In 2016, the UAE began to receive US-manufactured THAAD ballistic-missile-defence batteries. The country continues to develop its defence-industrial base and in 2019 announced a new state-owned defence group, EDGE, that will include some of the UAE's leading defence firms and also absorb the existing defence-industry groupings EAIG, EDIC and Tawazun Holdings. The UAE remains reliant on external providers for major weapons systems.

ACTIVE 63,000 (Army 44,000 Navy 2,500 Air 4,500 Presidential Guard 12,000)

Conscript liability 16–24 months, males 18–30 years dependent on education level. Voluntary service enrolment for women

ORGANISATIONS BY SERVICE

Space

EQUIPMENT BY TYPE

SATELLITES • COMMUNICATIONS 3 *Yahsat*

Army 44,000

FORCES BY ROLE

MANOEUVRE

Armoured

2 armd bde

Mechanised

2 mech bde

Light

1 inf bde

COMBAT SUPPORT

1 arty bde (3 SP arty regt)

1 engr gp

EQUIPMENT BY TYPE

ARMOURED FIGHTING VEHICLES

MBT 303: 45 AMX-30; 258 *Leclerc*

LT TK 76 FV101 *Scorpion*

RECCE 73: 49 AML-90; 24 VBL

IFV 445: 390 BMP-3; 55 *Rabdan*

APC 1,656

APC (T) 136 AAPC (incl 53 engr plus other variants)

APC (W) 185: 45 AMV 8×8 (one with BMP-3 turret); 120 EE-11 *Urutu*; 20 VAB

PPV 1,335: ε460 *Caiman*; ε680 *Maxxpro* LWB; 150 Nimr *Hafeet* 630A (CP); 45 Nimr *Hafeet (Amb)*

AUV 650 M-ATV; Nimr *Ajban*; Nimr *Jais*

ENGINEERING & MAINTENANCE VEHICLES

AEV 53+: 53 ACV-AESV; *Wisent*-2

ARV 158: 8 ACV-AESV Recovery; 4 AMX-30D; 85 BREM-L; 46 *Leclerc* ARV; 15 *Maxxpro* ARV

NBC VEHICLES 32 TPz-1 *Fuchs* NBC

ANTI-TANK/ANTI-INFRASTRUCTURE

MSL

SP 135: 20 HOT; 115 Nimr *Ajban* 440A with *Kornet*-E

MANPATS FGM-148 *Javelin*; *Milan*; TOW

RCL 84mm *Carl Gustaf*

ARTILLERY 613+

SP 155mm 181: 78 G-6; 85 M109A3; 18 Mk F3

TOWED 93: **105mm** 73 L118 Light Gun; **130mm** 20 Type-59-I; **155mm** 6 AH-4

MRL 88+: **122mm** 50+: 48 Firos-25 (est 24 op); 2 *Jobaria*; Type-90 (reported); **227mm** 32 M142 HIMARS; **300mm** 6 9A52 *Smerch*

MOR 251: **81mm** 134: 20 Brandt; 114 L16; **120mm** 21 Brandt; **SP 120mm** 96 RG-31 MMP *Agrab* Mk2

SURFACE-TO-SURFACE MISSILE LAUNCHERS

SRBM • Conventional 6 *Scud*-B (up to 20 msl); MGM-140A/B ATACMS (launched from M142 HIMARS)

UNINHABITED AERIAL VEHICLES

ISR • Medium *Seeker* II

AIR DEFENCE

SAM • Point-defence *Mistral*

Navy 2,500

EQUIPMENT BY TYPE

PATROL AND COASTAL COMBATANTS 43

CORVETTES 7

FSGHM 6 *Baynunah* with 2 quad lnchr with MM40 *Exocet* Block 3 AShM, 1 8-cell Mk 56 VLS with RIM-162 ESSM SAM, 1 21-cell Mk 49 GMLS with RIM-116C RAM Block 2 SAM, 1 76mm gun

Middle East and North Africa

FSGM 1 *Abu Dhabi* with 2 twin lnchr with MM40 *Exocet* Block 3 AShM, 1 76mm gun
PCFGM 2 *Mubarraz* (GER Lurssen 45m) with 2 twin lnchr with MM40 *Exocet* AShM, 1 sextuple *Sadral* lnchr with *Mistral* SAM, 1 76mm gun
PCGM 4:
2 *Muray Jib* (GER Lurssen 62m) with 2 quad lnchr with MM40 *Exocet* Block 2 AShM, 1 octuple lnchr with *Crotale* SAM, 1 *Goalkeeper* CIWS, 1 76mm gun, 1 hel landing platform
2 *Ganthoot* with 2 twin lnchr with MM40 *Exocet* Block 3 AShM, 2 3-cell VLS with VL-MICA SAM, 1 76mm gun, 1 hel landing platform
PCFG 6 *Ban Yas* (GER Lurssen TNC-45) with 2 twin lnchr with MM40 *Exocet* Block 3 AShM, 1 76mm gun
PBFG 12 *Butinah* (*Ghannatha* mod) with 4 single lncher with *Marte* Mk2/N AShM
PBF 12: 6 *Ghannatha* with 1 120mm NEMO mor (capacity 42 troops); 6 *Ghannatha* (capacity 42 troops)
MINE WARFARE • MINE COUNTERMEASURES 2
MHO 2 *Al Murjan* (ex-GER *Frankenthal* Type-332)
AMPHIBIOUS
LANDING SHIPS • LST 2 *Alquwaisat* with 1 hel landing platform
LANDING CRAFT 17
LCM 5: 3 *Al Feyi* (capacity 56 troops); 2 (capacity 40 troops and additional vehicles)
LCP 4 Fast Supply Vessel (multipurpose)
LCT 8: 1 *Al Shareeah* (LSV 75m) with 1 hel landing platform; 7 ADSB 64m
LOGISTICS AND SUPPORT 3:
AFS 2 *Rmah* with 4 single 533mm TT
AX 1 *Al Semeih* with 1 hel landing platform

Air Force 4,500

FORCES BY ROLE
FIGHTER/GROUND ATTACK
3 sqn with F-16E/F Block 60 *Fighting Falcon*
3 sqn with *Mirage* 2000-9DAD/EAD/RAD
AIRBORNE EARLY WARNING AND CONTROL
1 flt with Saab 340 *Erieye*
SEARCH & RESCUE
2 flt with AW109K2; AW139
TANKER
1 flt with A330 MRTT
TRANSPORT
1 sqn with C-17A *Globemaster*
1 sqn with C-130H/H-30 *Hercules*; L-100-30
1 sqn with CN235M-100
TRAINING
1 sqn with Grob 115TA
1 sqn with *Hawk* Mk102*
1 sqn with PC-7 *Turbo Trainer*
1 sqn with PC-21
TRANSPORT HELICOPTER
1 sqn with Bell 412 *Twin Huey*
EQUIPMENT BY TYPE
AIRCRAFT 156 combat capable
FGA 137: 54 F-16E Block 60 *Fighting Falcon* (*Desert Eagle*); 24 F-16F Block 60 *Fighting Falcon* (13 to remain in US for trg); 15 *Mirage* 2000-9DAD; 44 *Mirage* 2000-9EAD
MP 4 DHC-8 *Dash* 8 MPA
ISR 7 *Mirage* 2000 RAD*
SIGINT 1 *Global* 6000
AEW&C 4: 2 *GlobalEye*; 2 Saab 340 *Erieye*
TPT/TKR 3 A330 MRTT
TPT 22: **Heavy** 7 C-17 *Globemaster* III; **Medium** 6: 3 C-130H *Hercules*; 1 C-130H-30 *Hercules*; 2 L-100-30; **Light** 13: 4 C295W; 5 CN235
TRG 79: 12 Grob 115TA; 12 *Hawk* Mk102*; 30 PC-7 *Turbo Trainer*; 25 PC-21
HELICOPTERS
MRH 21: 12 AW139; 9 Bell 412 *Twin Huey*
TPT • Light 4: 3 AW109K2; 1 Bell 407
UNINHABITED AERIAL VEHICLES
CISR • Heavy *Wing Loong* I; *Wing Loong* II
ISR • Heavy RQ-1E *Predator* XP
AIR-LAUNCHED MISSILES
AAM • IR AIM-9L *Sidewinder*; R-550 *Magic*; **IIR** AIM-9X *Sidewinder* II; **IIR/ARH** *Mica*; **ARH** AIM-120B/C AMRAAM
ASM AGM-65G *Maverick*; *Hakeem* 1/2/3 (A/B)
ARM AGM-88C HARM
ALCM *Black Shaheen* (*Storm Shadow*/SCALP EG variant)
BOMBS
INS/SAT guided *Al Tariq*
Laser-guided GBU-12/58 *Paveway* II

Air Defence

FORCES BY ROLE
AIR DEFENCE
2 AD bde (3 bn with MIM-23B I-*Hawk*; MIM-104F *Patriot* PAC-3)
3 (short range) AD bn with *Crotale*; *Mistral*; *Rapier*; RB-70; *Javelin*; 9K38 *Igla* (SA-18 *Grouse*); 96K6 *Pantsir*-S1
2 SAM bty with THAAD
EQUIPMENT BY TYPE
AIR DEFENCE
SAM 42+
Medium-range MIM-23B I-*Hawk*; MIM-104F *Patriot* PAC-3
Short-range 42+: *Crotale*; 42 96K6 *Pantsir*-S1
Point-defence 9K38 *Igla* (SA-18 *Grouse*); RBS-70; *Rapier*; *Mistral*
GUNS • Towed 35mm GDF-005
MISSILE DEFENCE 12 THAAD

Presidential Guard Command 12,000

FORCES BY ROLE
SPECIAL FORCES
1 SF bn
1 spec ops bn
MANOEUVRE
Reconaissance
1 recce sqn
Mechanised
1 mech bde (1 tk bn, 4 mech inf bn, 1 AT coy, 1 cbt engr coy, 1 CSS bn)

Amphibious
1 mne bn

EQUIPMENT BY TYPE

ARMOURED FIGHTING VEHICLES
MBT 50 *Leclerc*
IFV 290: 200 BMP-3; 90 BTR-3U *Guardian*

ANTI-TANK/ANTI-INFRASTRUCTURE
MSL • SP HMMWV with 9M133 *Kornet*

Joint Aviation Command

FORCES BY ROLE

GROUND ATTACK
1 sqn with *Archangel*; AT802 *Air Tractor*

ANTI-SURFACE/ANTI-SUBMARINE WARFARE
1 sqn with AS332F *Super Puma*; AS565 *Panther*

TRANSPORT
1 (Spec Ops) gp with AS365F *Dauphin* 2; H125M *Fennec*; AW139; Bell 407MRH; Cessna 208B *Grand Caravan*; CH-47C/F *Chinook*; DHC-6-300/400 *Twin Otter*; UH-60L/M *Black Hawk*

ATTACK HELICOPTER
1 gp with AH-64D *Apache*

EQUIPMENT BY TYPE

AIRCRAFT 37 combat capable
ATK 23 *Archangel*
ISR ε6 AT802 *Air Tractor**
TPT • Light 15: 2 Beech 350 *King Air*; 8 Cessna 208B *Grand Caravan**; 1 DHC-6-300 *Twin Otter*; 4 DHC-6-400 *Twin Otter*

HELICOPTERS
ATK 28 AH-64D *Apache*
ASW 7 AS332F *Super Puma* (5 in ASuW role)
MRH 53+: 4 AS365F *Dauphin* 2 (VIP); 9 H125M *Fennec*; 7 AS565 *Panther*; 3 AW139 (VIP); 20 Bell 407MRH; 4 SA316 *Alouette* III; 6+ UH-60M *Black Hawk* (ABH)
TPT 66: **Heavy** 22 CH-47F *Chinook*; **Medium** 44: 11 UH-60L *Black Hawk*; up to 33 UH-60M *Black Hawk*

AIR-LAUNCHED MISSILES
ASM AGM-114 *Hellfire*; *Cirit*; *Hydra-70*; HOT
AShM AS-15TT; AM39 *Exocet*

Paramilitary

Critical Infrastructure and Coastal Protection Agency (CICPA)

Ministry of Interior

EQUIPMENT BY TYPE

PATROL AND COASTAL COMBATANTS 114
PSO 1 *Al Wtaid*
PCM 2 *Arialah* (Damen Sea Axe 6711) with 1 11-cell Mk 15 SeaRAM GMLS with RIM-116C RAM Block 2 SAM, 1 57mm gun, 1 hel landing platform
PBF 58: 6 *Baglietto* GC23; 3 *Baglietto* 59; 15 DV-15; 34 MRTP 16
PB 53: 2 *Protector*; 16 (US Camcraft 65); 5 (US Camcraft 77); 6 Watercraft 45; 12 *Halmatic Work*; 12 *Al Saber*

DEPLOYMENT

EGYPT: ε300 12 F-16E/F *Fighting Falcon* (reported); *Wing Loong* I UAV; *Wing Loong* II UAV

ERITREA: *Operation Restoring Hope* ε500; 2 FSGHM; 4 Bell 407MRH; 2 CH-47F *Chinook*; 4 UH-60M *Black Hawk*; 1 C-130H; 1 DHC-6-400; 1 DHC-8

LIBYA: UH-60M; *Pantsir*-S1

SAUDI ARABIA: *Operation Restoring Hope* 12 F-16E *Fighting Falcon*

FOREIGN FORCES

Australia 400; 1 tpt det with 2 C-130J-30 *Hercules*
Denmark *Operation Inherent Resolve* 13
France 650: 1 armd BG (1 tk coy, 1 armd inf coy; 1 aty bty); *Leclerc*; VBCI; CAESAR; 6 *Rafale*; • EMASOH; 1 *Atlantique*-2
Italy 106; 1 tpt flt with 2 C-130J *Hercules*
Korea, Republic of 170 (trg activities at UAE Spec Ops School)
United Kingdom 200; 1 tkr/tpt flt with C-17A *Globemaster*; C-130J *Hercules*; A330 MRTT *Voyager*
United States 5,000; 1 FGA sqn with 12 F-35A *Lightning* II; 1 ISR sqn with 4 U-2S; 1 AEW&C sqn with 4 E-3B/G *Sentry*; 1 tkr sqn with 12 KC-10A; 1 ISR UAV sqn with RQ-4 *Global Hawk*; 2 AD bty with MIM-104E/F *Patriot* PAC-2/3

Yemen, Republic of YEM

Yemeni Rial R		2019	2020	2021
GDP	R	12.6tr	15.2tr	
	US$	22.6bn	20.9bn	
per capita	US$	713	645	
Growth	%	2.1	-5.0	
Inflation	%	10.0	26.4	
Def bdgt	R	n.k	n.k	
	US$	n.k	n.k	
US$1=R		558.59	725.55	

Population 29,884,405

Age	0–14	15–19	20–24	25–29	30–64	65 plus
Male	19.2%	5.8%	5.1%	4.4%	14.8%	1.3%
Female	18.5%	5.6%	4.9%	4.3%	14.5%	1.6%

Capabilities

Yemen continues to be afflicted by a conflict that is, according to the UN, the world's worst humanitarian crisis. There appears to be little prospect that any of the competing forces will be able to gain a decisive upper hand. UN peace efforts have faltered as the government has continued to struggle with Houthi rebels in the north and the secessionist Southern Transitional Council. The government of President Abdu Rabbu Mansour Hadi appears to exercise limited control over the forces nominally loyal to it, while the proxy forces supposedly allied to the government and supported by the members of the Saudi-led coalition answer to those member states rather than Yemeni military authorities. The rebel

Middle East and North Africa

Houthi forces, who are assumed to receive material support from Iran, are largely tribal-based militias, along with some elements of the Yemeni armed forces who were loyal to the former Saleh regime. Al-Qaeda affiliates also appear to be active in the country. Government forces tend to be under-equipped and poorly paid compared to the proxy groups supported by the Saudi-led coalition. The Houthi rebel forces benefited from the training and capabilities of Yemeni armed forces previously loyal to former president Saleh. The Saudi-led coalition continues to provide air support for the Hadi government, although the UAE has largely drawn down its forces. The conflict appears to have been sustained by a combination of large existing stockpiles of weapons and ammunition and external supplies, despite UN embargoes. There is no domestic defence industry, barring some limited maintenance and workshop facilities.

ACTIVE 40,000 (Goverment forces 40,000)

ORGANISATIONS BY SERVICE

Government forces ε40,000 (incl militia)

President Hadi's government is nominally supported by parts of the Yemeni armed forces, as well as a number of militia organisations in southern and eastern Yemen. However, the government's ability to exercise direct control over most of these forces is limited. The re-emergence of secessionist sentiment in the south, and the seizure of Aden in 2019 by forces loyal to the Southern Transitional Council (STC), complicated further the Hadi government's attempt to exert political and military authority. Though an accord (the Riyadh Agreement) was signed in late 2019, its future was unclear as of late 2020. This political uncertainty, which also relates to the future of the STC's military forces (which have been termed the 'Security Belt' militia), not only indicates a reduction in coherent government military capacity but will also make it more difficult for the government to effectively focus its energies on presenting a more unified military challenge to the Houthi rebels.

FORCES BY ROLE

MANOEUVRE

Mechanised

up to 20 bde(-)

EQUIPMENT BY TYPE

ARMOURED FIGHTING VEHICLES

MBT Some M60A1; T-34†; T-54/55; T-62; T-72

RECCE some BRDM-2

IFV BMP-2; BTR-80A; *Ratel*-20

APC

APC (W) BTR-60

PPV Streit *Cougar*; Streit *Spartan*

AUV M-ATV

ANTI-TANK/ANTI-INFRASTRUCTURE

MSL • MANPATS 9K11 *Malyutka* (AT-3 *Sagger*); M47 *Dragon*; TOW

GUNS • SP 100mm SU-100†

ARTILLERY • SP 122mm 2S1 *Gvozdika*

AIRCRAFT • ISR 6 AT-802 *Air Tractor**

AIR DEFENCE • GUNS • TOWED 14.5mm ZPU-4; **23mm** ZU-23-2

DEPLOYMENT

MALI: UN • MINUSMA 2

FOREIGN FORCES

All *Operation Restoring Hope* unless stated

Saudi Arabia 2,500: 2 armd BG; M60A3; M2A2 *Bradley*; M113A4; M-ATV; AH-64 *Apache*; 2+ MIM-104D/F *Patriot* PAC-2/3

Sudan 650; 1 mech BG; T-72AV; BTR-70M *Kobra* 2

TERRITORY WHERE THE GOVERNMENT DOES NOT EXERCISE EFFECTIVE CONTROL

Insurgent forces 20,000 (incl Houthi and tribes)

The Houthi-run de facto administration has controlled northern Yemen since 2015 and is supported by a combination of Houthi tribal militias and elements of the Yemeni armed forces previously loyal to former president Ali Abdullah Saleh. Following a break between the Houthis and Saleh in late 2017, and the latter's death, his former forces have further split between those that remained affiliated with the Houthis and those who have joined Saleh's son and nephew to fight against them.

FORCES BY ROLE

MANOEUVRE

Mechanised

up to 20 bde(-)

EQUIPMENT BY TYPE

ARMOURED FIGHTING VEHICLES

MBT T-55; T-72

IFV BMP-2; BTR-80A

APC • APC (W) Some BTR-40; BTR-60

AUV M-ATV

ANTI-TANK/ANTI-INFRASTRUCTURE

MSL • MANPATS M47 *Dragon*; 9K111-1 *Konkurs* (AT-5B *Spandrel*/*Towsan*-1); 9K115 *Metis* (AT-7 *Saxhorn*); *Dehlavieh* (*Kornet*)

RCL 82mm B-10

SURFACE-TO-SURFACE MISSILE LAUNCHERS

SRBM • Conventional 9K79 *Tochka* (SS-21 *Scarab*); 9K72 *Elbrus* (SS-1C *Scud*-B); *Hwasong*-5 (SS-1C *Scud*-B); *Hwasong*-6 (SS-1D *Scud*-C); *Borkan*-1 (extended-range *Scud* derivative); *Borkan*-2H (*Qiam*-1); *Qaher*-1 (converted S-75 SAM)

GLCM • Conventional *Quds*-1; *Quds*-2 (reported)

COASTAL DEFENCE • AShM C-801; C-802

UNINHABITED AERIAL VEHICLES

ISR • Medium *Qasef*-1; *Qasef*-2K; *Sammad*-1; *Sammad*-2; *Sammad*-3 (many of these systems have been fitted with a warhead payload to function as a form of improvised missile)

AIR DEFENCE • GUNS • TOWED 20mm M167 *Vulcan*; **23mm** ZU-23-2

Arms procurements and deliveries – Middle East and North Africa

Significant events in 2020

APRIL — CANADA SALES TO SAUDI ARABIA RESUME

Canada lifted the arms embargo on Saudi Arabia that had been in place since 2018, to allow the export of General Dynamics Land Systems Canada (GDLS-C)'s LAV wheeled armoured vehicle. The Canadian Commercial Corporation had imposed a moratorium on the 2014 contract to deliver 900 LAVs for US$13.4bn due to concerns over the Saudi-led military campaign in Yemen. However, the Canadian government issued export permits in 2016 after a re-negotiation reduced the total number to 777 for US$10.7bn. Further details of production status and delivery schedules have not been disclosed but Saudi Arabia will – through a newly established payment schedule – proceed with regular payment for products delivered. The UK has also resumed the approval of defence sales to Saudi Arabia, though fellow NATO member Germany has not.

MAY — OSHKOSH JV IN SAUDI ARABIA

Saudi Arabia's Al Tadrea Manufacturing Company and the United States' Oshkosh Defense have established a joint venture known as Oshkosh Al Tadrea Manufacturing to produce Oshkosh's armoured vehicles in Riyadh, starting with the Family of Medium Tactical Vehicles (FMTV) trucks. The joint venture will involve technology transfer to Saudi Arabia and, for Oshkosh, will help improve market access in the region. The company has already had some success in Saudi Arabia with the M-ATV armoured utility vehicle and the Heavy Equipment Transporter System (HETS) logistics vehicles.

JULY — MOROCCO DEFENCE INDUSTRY

The Moroccan Council of Ministers approved Law No. 10.20 which aims to encourage the development of the local defence industry. The law recommends that Moroccan companies collaborate with foreign defence companies, but requires that the majority of the capital in the enterprise be Moroccan. A national supervisory authority will be created to authorise activity and manufacturing processes. Morocco is heavily reliant on imports of equipment from the US and has recently signed a number of high-profile deals for new fighter aircraft and attack helicopters. However, the country also has a high youth unemployment rate and this legislation is likely part of a wider drive to diversify the economy.

JULY — EDGE AMMROC ACQUISITION

The UAE's Edge group announced a conditional agreement to acquire the 40% (US$307m) share of Advanced Military Maintenance Repair and Overhaul Center (AMMROC), currently owned by Lockheed Martin and Sikorsky. As the region's only authorised C-130 *Hercules* service centre, AMMROC has a large amount of potential customers with Gulf neighbours Bahrain, Iraq, Kuwait, Oman, Qatar and Saudi Arabia all operating the type. AMMROC also offers maintenance, repair and overhaul (MRO) services for F-16 combat aircraft and *Black Hawk* helicopters, which are widely in service in the region. The acquisition of the remainder of AMMROC follows the launch of the Edge conglomerate in late 2019 which brought together most of the UAE's defence industry under one roof.

Table 8: Jordan: defence procurement since 2010

Contract date	Equipment	Type	Quantity	Value (US$)	Prime contractor(s)	Deliveries
2010	M1A42 HIMARS	227mm multiple rocket launcher	12	$100.4m	Lockheed Martin	c. 2012
2010	YPR-765	Infantry fighting vehicle	336	n.k.	Netherlands government surplus	2011–16
	YPR-765	Armoured personnel carrier (T)	87			
	M109	152mm self-propelled artillery	121			
2010	UH-60M *Black Hawk*	Medical transport helicopter	2	$26.3m	Sikorsky	2013
2011	F-16AM/BM	Fighter/ground attack aircraft	9	$44.58m	Belgian government surplus	2011
c. 2012	AT-802U *Air Tractor*	Intelligence, surveillance and reconnaissance aircraft	6	Donation	UAE government surplus	2013
c. 2013	AH-1 *Cobra*	Attack helicopter	~16	Donation	Israeli government surplus	2014
c. 2013	WM-80	273mm multiple rocket launcher	≥2	n.k.	NORINCO	2014
2013	M-28 *Skytruck*	Light transport aircraft	2	n.k.	PZL Mielec	2014–15
2013	F-16AM *Fighting Falcon*	Fighter/ground attack aircraft	13	$86.35m	Netherlands government surplus	2017
	F-16BM *Fighting Falcon*		2			
2013	*Gepard*	35mm self-propelled anti-aircraft gun	60	$27.89m	Netherlands government surplus	2014–16
	L/70	40mm towed anti-aircraft gun	22			
c. 2014	*Hawk* Mk63	Training aircraft	13	Donation	UAE government surplus	2015
c. 2014	UH-60A *Black Hawk*	Medical transport helicopter	8	Donation	US government surplus	2015–16
2014	B1 *Centauro*	Wheeled assault gun	24	Donation	Italian government surplus	2015
c. 2015	CH-4B	Heavy CISR UAV	6	n.k.	CASC	2016
2015	B1 *Centauro*	Wheeled assault gun	117	$6.19m	Italian government surplus	c. 2016–18
2015	PC-21	Training aircraft	12	n.k.	Pilatus	2017–19
2015	UH-60M *Black Hawk*	Medical transport helicopter	1	$17.2m	Sikorsky	c. 2015
2015	AT-802U *Air Tractor*	Intelligence, surveillance and reconnaissance aircraft	4	$9.78m	L-3 Communications	2016
c. 2016	*Marder* 1A3	Infantry fighting vehicle	50	Donation	German government surplus	2016–19
2016	G 120TP	Training aircraft	14	n.k.	Grob Aircraft	2017
2016	Mi-26T2 *Halo*	Heavy transport helicopter	4	n.k.	Russian Helicopters	2018–ongoing
2016	UH-60M *Black Hawk*	Medical transport helicopter	12	$185m	Sikorsky	2017–18
c. 2017	G 120TP	Training aircraft	2	Donation	German government surplus	2018
c. 2017	*Agrab* Mk2	120mm self-propelled mortar	18	Donation	UAE government surplus	2018
2017	RB-M	Fast patrol boat	2	n.k.	Fincantieri Marinette Marine	2019
c. 2018	M119	105mm towed artillery	12	Donation	US government surplus	2019
2018	B1 *Centauro*	Wheeled assault gun upgrade	80	n.k.	Star Defence Logistics & Engineering	2019–ongoing
c. 2019	*Nomad*	Protected patrol vehicle	44	Donation	Qatari government surplus	2020–ongoing
	Thunder	Protected patrol vehicle				

Figure 22: United Arab Emirates: defence-equipment exports since 2010

In December 2014, the UAE began consolidating a number of state-owned companies under the Emirates Defense Industries Company (EDIC), as part of an effort to develop local defence-industrial capability. EDIC later became one of 25 firms merged into the new EDGE Group in November 2019. Defence exports form part of the UAE's broader industrial strategy, and export customers so far can be found among the UAE's neighbours, India, Russia and Turkmenistan, as well as in sub-Saharan Africa. The equipment exported, or donated, has been of relatively low technical sophistication, consisting largely of different types of armoured 4x4s. Arguably the most significant export to date has been that of *Yabhon* UAVs to Algeria. Imagery of these systems has indicated that these are armed, although it is unclear if they were exported with this capability or if this was subsequently added. If the former, then the UAE joins an exclusive club of nations (including China, Turkey and the US) that have exported this category of UAV. The UAE's shipyards have recently enjoyed some export success. While landing craft, of a similar type to that sold to Oman in the 2000s, were delivered to Kuwait in 2018, the UAE has yet to export the types of corvettes or patrol vessels that have been built for local use.

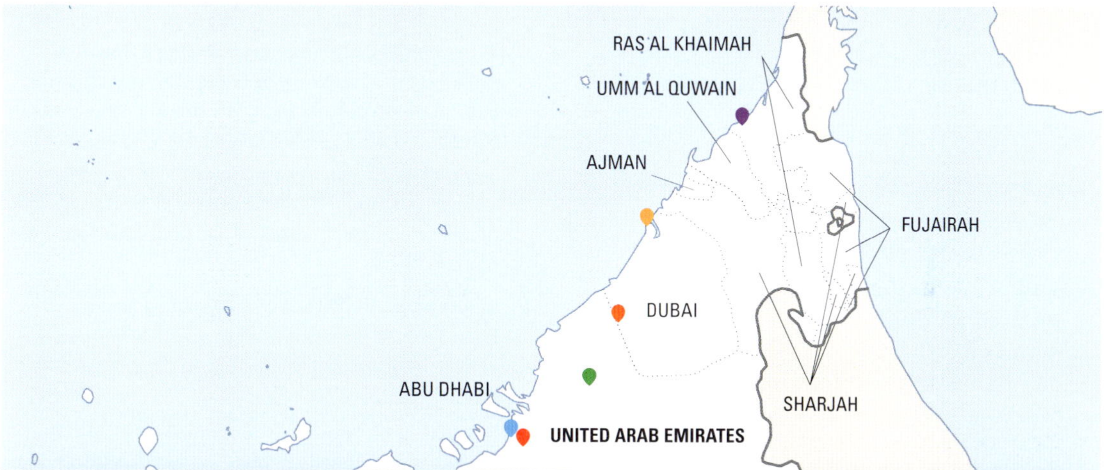

Company	Recipient country	Contract Date	Equipment	Type	Quantity	Deliveries
Adcom Systems	Russia	ε2013	*Yabhon United*-40	Heavy CISR UAV	≥2	ε2016
	Algeria	ε2013	*Yabhon United*-40 (Algeria-54/El Djazair-54)**	Heavy CISR UAV	≥2	ε2018
	Algeria	ε2015	*Yabhon United*-20 (Algeria-55/El Djazair-55)**	Heavy CISR UAV	≥2	ε2018
Abu Dhabi Ship Building Company (ADSB)*	India	Nov 2010	Immediate Support Vessel (Rodman)	Fast patrol boat	4 (23)***	2014–15
	Kuwait	Jan 2015	*Assafar* (ADSB 64m) *Abhan* (ADSB 42m) ADSB 16m	Landing craft	8	2017–18
Albwardy Damen	Nigeria	2019	LST 100	Landing ship tank	1	Expected by March 2022
Minerva Special Purpose Vehicles (MSPV)	India	n.k.	*Panthera* T6	Armoured utility vehicle	n.k.	ε2016
	Cameroon	n.k.	*Panthera* T6	Armoured utility vehicle	≥6	ε2018
	Madagascar	n.k.	*Panthera* T4	Armoured utility vehicle	≥6	ε2020
	Egypt	2013	*Panthera* T6	Armoured utility vehicle	1,200 (2,500)****	2014–19
NIMR Automotive*	Algeria	Jul 2012	*Ajban* 4x4**	Armoured utility vehicle	≥200	n.k.
	Turkmenistan	n.k.	*Ajban* 440A	Armoured utility vehicle	≥8	ε2016
	Turkmenistan	ε2016	*Ajban* LRSOV	Armoured utility vehicle	n.k.	ε2017
Streit Group	Libya*****	2012	*Cougar* *Spartan* *Cobra*	Armoured utility vehicle	n.k.	2012–13

*Part of EDGE Group
**Algerian licensed production
***Four ISVs (Roadman) built by ADSB; Five ISVs (Roadman) built by Spain's Rodman Polyships SAU; 14 ISVs (Craftway) built by India's SHM Shipcare
****1,200 *Panthera* T6 built by MSPV; 1,300 *Panthera* T6 built by Egypt's Eagles Defence International Systems (EDIS)
*****Production and final assembly sites are unclear, but exported from the UAE

Chapter Eight

Latin America and the Caribbean

- Armed forces across the region were tasked with responding to the coronavirus pandemic, adding to challenges including law-and-order responses for some. They were deployed to assist the public-health response by providing healthcare facilities, employing military-logistics capabilities in order to support health agencies, and enforcing lockdowns.
- The lack of any significant and sustained growth in defence spending and low levels of funding allocated to defence investment (procurement and research and development) among the major regional players constrains capability development in Latin America.
- Mexico's naval modernisation continues, with the entry into service of the *Benito Juárez*-class (formerly *Reformador*) light frigate. Construction plans for 2020–24 include two 1,680-ton *Oaxaca*-class Offshore Patrol Vessels (OPVs), four *Teotihuacan*-class (Damen Stan Patrol 4207) coastal-patrol vessels, five *Polaris* II-class interceptor craft and an *Isla Maria Madre*-class (Damen Fast Crew Supplier 5009) logistics supply vessel.
- Brazil's defence modernisation continues. The Island of Madeira submarine base (at the Itaguaí Naval Complex) was inaugurated in July 2020 and is scheduled to be complete in 2022. One nuclear- and four conventionally powered submarines will be based at the facility. The first conventionally powered boat, *Riachuelo*, is due to be commissioned in December 2020. The navy's Strategic Plan 2040 outlined the acquisition of *Tamandaré*-class frigates, coastal-patrol vessels and OPVs, among other projects. The first *Gripen* E combat aircraft arrived in Brazil in September 2020 under the October 2014 contract for 28 F-39E and eight F-39F *Gripens*, and local production of aerostructures is under way at Saab's purpose-built plant in Brazil.
- In Argentina, force-modernisation plans have been stymied by economic problems. Armoured-vehicle recapitalisation remains moribund. Naval-modernisation plans include four OPVs, and the plan to revive the *Super Etendard* fleet with five former French *Super Etendard Modernisé* aircraft is progressing slowly. Air-force priorities are fleet maintenance and rebuilding operational capability, as well as inventory upgrades, such as a replacement for the *Skyhawk* and (already retired) *Mirage* fleets.

Latin America and the Caribbean defence spending, 2020 – top 5,
including US foreign military financing

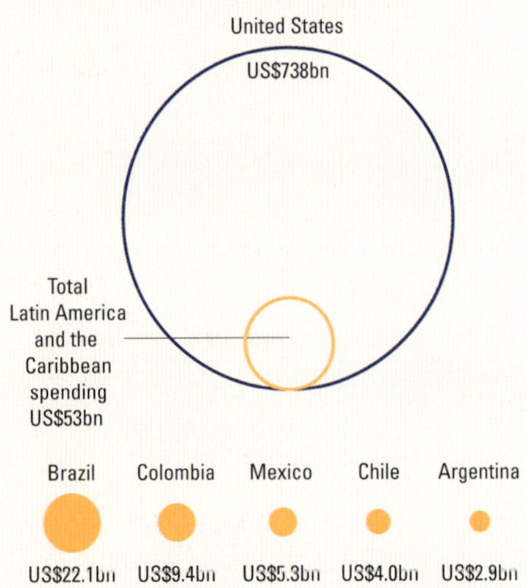

Active military personnel – top 10
(15,000 per unit)

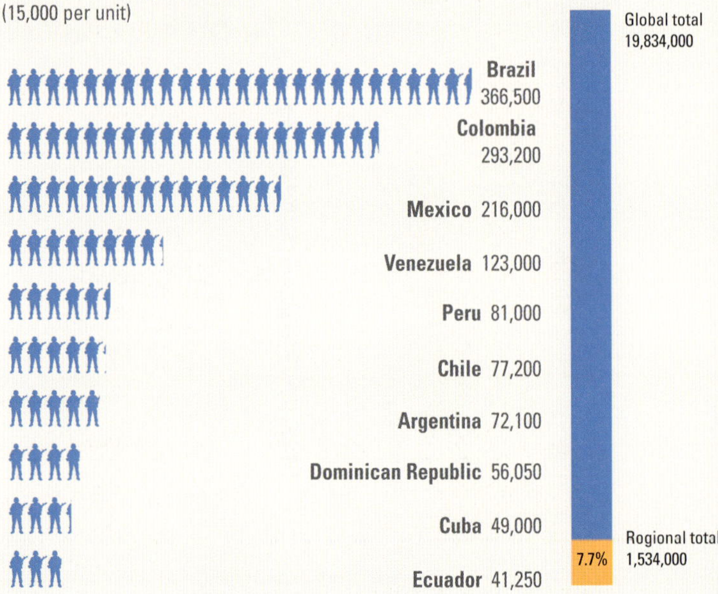

Latin American states' Antarctic bases

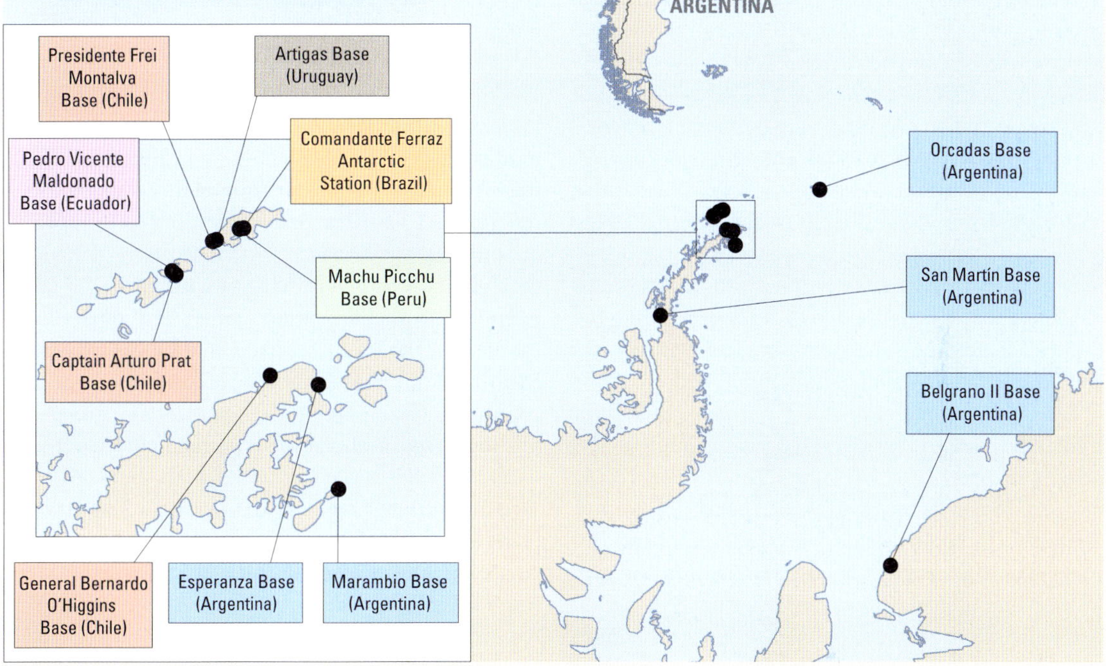

Latin American armed forces' coronavirus response

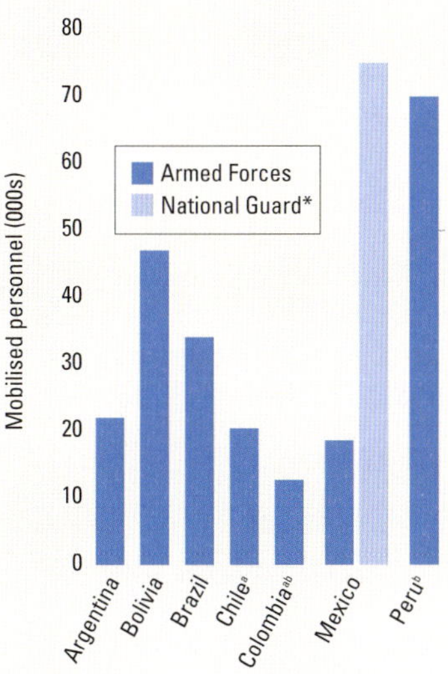

[a]Deployed; [b]Colombia and Peru deployed 76,392 and mobilised 80,000 police personnel respectively; *mobilised

Regional crewed and uninhabited fixed-wing ISR aviation platforms 2016–20

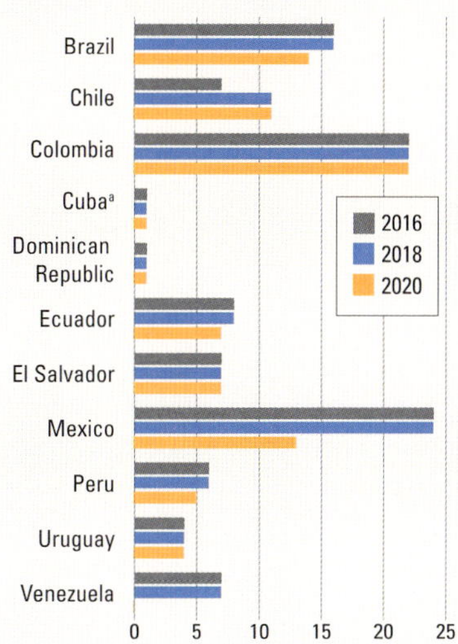

[a]Serviceability in doubt

Regional frigate inventories 2016–20

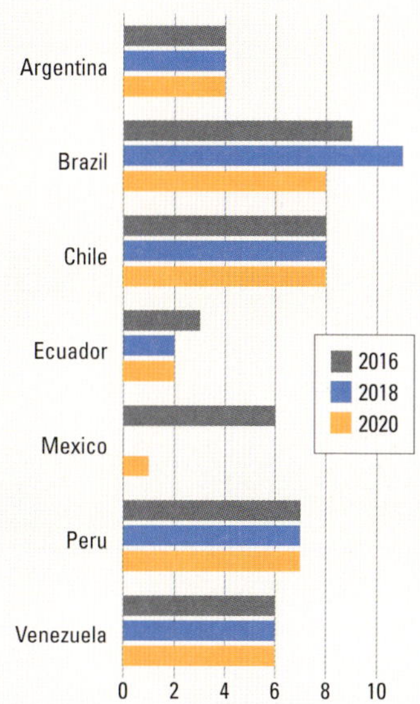

Latin America and the Caribbean

The effects of the coronavirus pandemic have dominated Latin America's security and defence environment in 2020. Armed forces across the region were deployed to assist with the public-health response by providing healthcare facilities, employing military-logistics capabilities in order to support health agencies and enforcing lockdowns. However, these activities have come at considerable cost, with defence spending in many countries likely to adjust, in future, due to the expense of the pandemic response. At the same time, regional armed forces and security services continued to focus on combating the trade in illegal drugs, tackling localised insurgencies and tasks related to maintaining public order.

Mexico continues to face perhaps the most challenging security environment in the region. The heavily armed Cártel de Jalisco Nueva Generación has expanded from the Pacific coast region to establish a foothold in other parts of the country, and was held responsible for an assassination attempt in June that targeted Mexico City's chief of security. Nevertheless, many drug-related security challenges stem from the activities of smaller groups and are often also connected with other forms of criminality such as fuel theft across the country. Piracy against offshore oil installations and commercial vessels in the Gulf of Mexico reportedly increased fourfold from 2017–19.

Mexico's armed forces are engaged in a significant reorganisation. Not only have large numbers of military personnel been assigned to the new National Guard, where they undertake a variety of internal security functions, but President Andrés Manuel López Obrador (AMLO) has increased the role of the military in public life, notably by tasking it with the transformation of its main air base into Mexico City's new airport, construction of sections of a railway in southeastern Mexico, as well as management of the country's ports and customs facilities, the latter intended to better target corruption and illegal trafficking. That said, it is likely that the armed forces' main roles for the remainder of AMLO's term will continue to be engaging the drug cartels, protecting offshore assets and providing humanitarian support.

In 2020 the air force brought three of its F-5E *Tiger* IIs and one F-5F back to operational service and plans to bring back the remaining F-5Es and one F-5F. Analysts understand that SEDENA, the department responsible for the army and air force, intends these to remain in service until 2028–30, while a replacement programme is expected to begin in 2024. The air force has taken delivery of its full fleet of T-6C+ *Texan* IIs, while retaining some armed Pilatus PC-7s. Rather than new acquisitions, the air force is understood to be focusing on the maintenance and repair of its existing inventory.

Mexico's National Guard

The National Guard was officially re-established in June 2019 as a 'national law enforcement agency' under direct control of the Secretariat of Security and Civilian Protection (which was established in November 2018), following a February 2019 constitutional amendment. The 'new' force is comprised of over 90,000 personnel as of late 2020. Of the total, 25,000 were absorbed from the former Federal Police and other personnel have temporarily been moved from the army and navy.

The army's recently expanded Military Police Corps, composed of 12 light infantry brigades, along with 12,000 marines and naval police and some 14,000 additional troops from various units, were seconded to the National Guard. While the new National Guard is at least on paper a 'civilian police force', in practice it is commanded by an army lieutenant-general and the military personnel within it retain their rank and benefits and are paid by either the Secretariat of National Defense (SEDENA) or the Secretariat of the Navy (SEMAR). The National Guard took over most of the Federal Police's former air wing, including its *Black Hawk* helicopter squadron, the fleet of short- and medium-range transport aircraft plus a sizeable number of small uninhabited aerial vehicles.

Naval modernisation is proceeding with the entry into service of the *Benito Juárez*-class (formerly *Reformador*) light frigate. This is the largest modern warship built in Mexico, and the over-2,500 tonne Damen Sigma 10514 vessel is armed with RGM-84 *Harpoon* missiles; plans to build a further seven ships have been put on hold until 2025–30. The navy outlined its construction plans for 2020–24, which include the local construction of two 1,680-ton *Oaxaca*-class offshore patrol vessels, four *Tenochtitlan*-class (Damen Stan 4207) coastal patrol vessels, five *Polaris* II-class interceptor craft and an *Isla Maria Madre*-class (Damen Fast Crew Supplier 5009) logistics supply vessel. While the army has highlighted the need to begin to procure new armoured vehicles as well as new light artillery, the administration's austerity programme – plus the effect of the coronavirus pandemic – may curtail these plans. This is perhaps one reason for reports that the army is looking to modernise its fleet of VCR TT and ECR-90F1 Panhard 6x6 armoured fighting vehicles.

The fourth anniversary of the landmark peace agreement between the government of Colombia and the Fuerzas Armadas Revolucionarias de Colombia (FARC) guerrilla group took place on 24 November 2020. However, armed activity by dissident insurgent groups continues to pose a security challenge and has led to continued offensive operations by all services, as well as missions by them designed to combat organised crime, deliver humanitarian support and also assist in the process of reintegrating former FARC rebels. Colombia continues to have close military ties with the United States, and June 2020 saw the deployment to Colombia of a company from the US Army's 1st Security Force Assistance Brigade in a counter-narcotics role. In July, the navy published the Naval Development Plan 2042, which envisages the construction of five (out of an originally projected eight) new surface combatants under the 2015 Plataformas Estratégicas de Superficie (PES) requirement, along with four new diesel–electric submarines, among other capabilities. The overarching programme is termed the Fleet Construction and Optimisation Plan (PROCYON). A key aspect of the plan is to further develop national shipbuilding capacity, primarily the shipyards operated by the Corporación de Ciencia y Tecnología para el Desarrollo de la Industria Naval Marítima y Fluvial (COTECMAR), as technology transfer is a key component of the fleet-modernisation plans. The air force is also planning to modernise its fighter fleet, currently comprised of *Kfir* fighter/ground-attack aircraft, with the Eurofighter *Typhoon*, F-16 and *Gripen* all mentioned as possible replacements. However, budgetary factors mean that these may be purchased second-hand.

Chile's armed forces were deployed internally on law-and-order-related tasks after protests broke out in the country in 2019, which were interrupted in 2020 by the coronavirus pandemic. The government enacted a special constitutional decree on 18 March 2020, which was extended by 90 days on 11 September, designed to give authorities special powers to tackle the pandemic. The armed forces have assisted public-health authorities in responding to the pandemic, including medical and logistics support as well as helping enforce lockdowns and curfews. Internal deployments continued to northern border regions under Decree 265, which was extended until 13 August 2021 and authorises the use of the armed forces to combat drug trafficking and organised crime. Meanwhile, the army continued to implement aspects of the Strategic Development Plan to 2026, and it was announced that the army's brigade-sized special-forces, aviation and intelligence units would combine under a new Special Operations Command. The navy commissioned in April two *Adelaide*-class vessels formerly operated by the Royal Australian Navy. The ships arrived in Chile mid-year and are equipped with *Harpoon* and SM-2 Block IIIA missiles. They were bought to replace the two former Dutch frigates (with *Harpoon* and SM-1MR missiles) that were retired in late 2019. There are also plans to further expand the shipbuilding capacities of the Astilleros y Maestranzas de la Armada (ASMAR) shipyards. In late 2020 it was announced that as part of an expansion in national space capabilities, Chile would replace its existing earth-observation satellite and that two further satellites would be manufactured, including one in Chile.

Brazil's armed forces remain focused on their military-modernisation plans and on fulfilling the range of deals signed in recent years, including for the *Gripen* combat aircraft under the F-X2 programme, for conventional- and nuclear-powered submarine capabilities, and for *Tamandaré* frigates. Government plans to develop the defence sector are also progressing. However, the armed forces remain involved in tackling a range of internal security challenges, including in urban areas as part of government measures designed to address domestic criminality.

DEFENCE ECONOMICS

Macroeconomics

The economic outlook for Latin America has deteriorated considerably over the course of 2020, reflecting the impact of the coronavirus pandemic and the costly measures implemented to counter its effects. In addition, the reliance of regional states on commodity exports increased their exposure to the global collapse in demand and to the resulting fall in commodity prices. According to International Monetary Fund (IMF) data, economic growth in Latin America will average –8% in 2020. While the region's largest economy, Brazil, is only projected to contract by 5.8% in 2020, severe downturns in Argentina (–11.8%), Ecuador (–11%), Mexico (–9%) and Peru (–13.9%) dragged down the 2020 average. Meanwhile, Venezuela's economic woes continued in 2020 as the oil price collapsed to below US$20/barrel in April; it is projected to average just over US$40/barrel in 2020.

The World Bank assesses that, in addition to the reliance on commodities, the abrupt slowdown in China and the US disrupted supply chains in Mexico and Brazil; there was also a reduction in trade and remittance channels as US activity slowed. The collapse of international tourism, meanwhile, particularly affected Mexico and the Caribbean.

Even before the coronavirus pandemic, regional growth was at risk from mounting regional political uncertainty. While in 2019 demonstrations were primarily evident in Bolivia, Chile and Nicaragua, there have also been protests in 2020 in Argentina, Brazil and Panama. Unrest has been driven by political dissatisfaction, social tensions and economic issues already present in the region but have in some cases been heightened by the pandemic. Several countries, including Argentina, Brazil, El Salvador and Colombia, have pursued monetary-policy measures including liquidity provision, interest-rate cuts and asset purchases, as well as fiscal stimulus including social assistance, support for small businesses and additional healthcare spending.

▼ Figure 23 **Latin America and the Caribbean defence spending by country and sub-region, 2020**

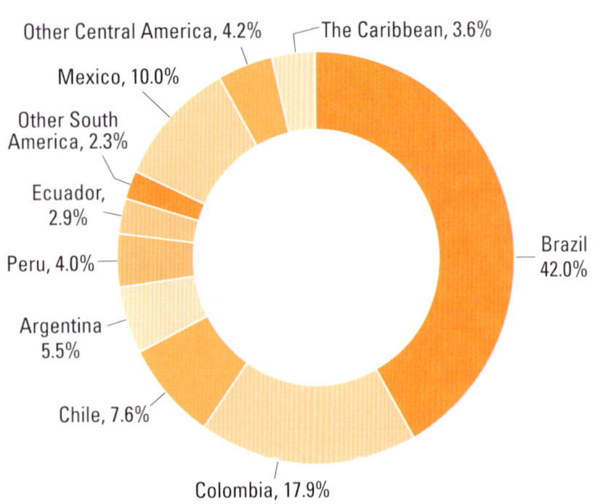

Note: analysis excludes Cuba, Suriname and Venezuela

© IISS

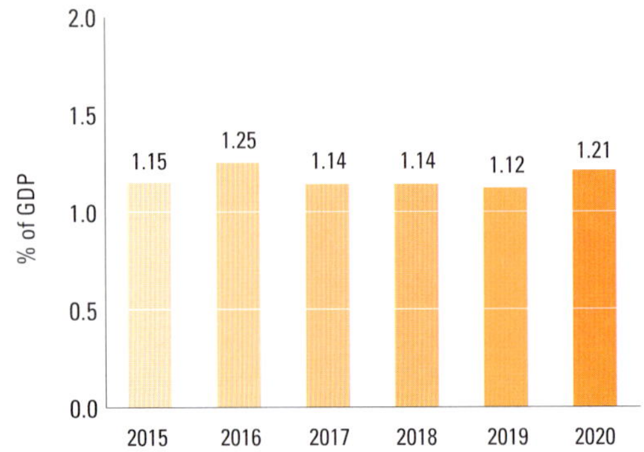

▲ Figure 24 **Latin America and the Caribbean regional defence spending** as % of GDP (average)

Defence spending

The region's defence spending continues to languish at low levels, due to lacklustre economic performance, social challenges and the absence of major external security threats, having varied between US$57 billion and US$67bn in real terms over the past decade. Between 2010 and 2019, the average annual increase amounted to 0.3% in real terms. Sizeable increases between 2011 and 2014 were balanced by cuts in 2015, 2016 and 2019. The reduction of just 1% in 2019 was modest, with increases in Brazil and Colombia insufficient to counter cuts in Argentina and Mexico.

Regional trends are heavily influenced by fluctuations in the Brazilian defence budget. Since 2010, Brazil has accounted for an average of 41% of regional spending. Over the past decade, Brazil's defence budget has been stable, hovering at around US$24–US$26bn in real terms, accounting for between 1.3% and 1.5% of GDP. With most other countries in the region allocating a lower proportion of economic output to defence, the regional average over this period is much lower, closer to 1.1% of GDP, well below the global average of 2.1% of GDP.

The government in Brazil was planning a 34% or R39bn (US$7.5bn) real-terms cut to the Ministry of

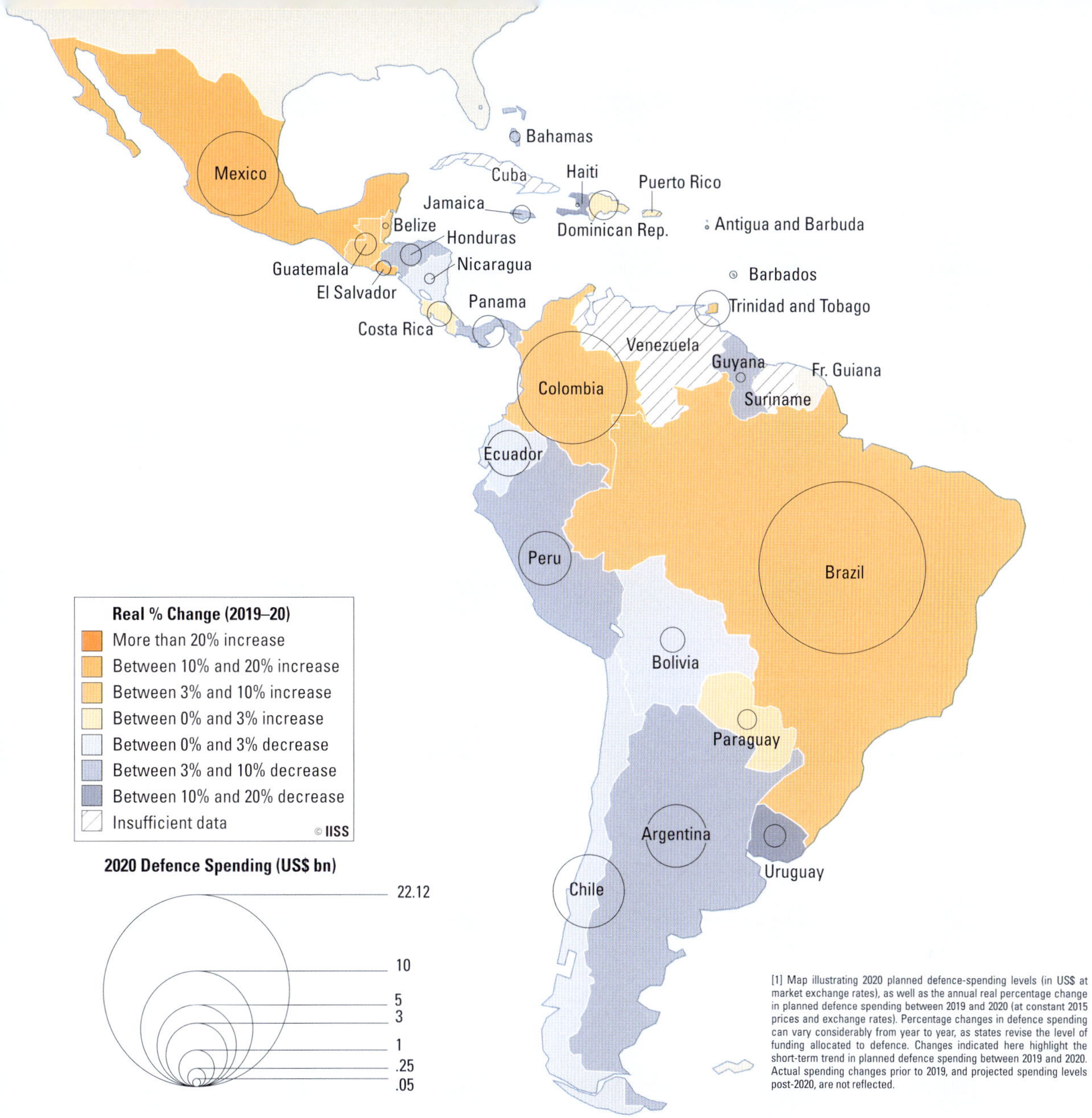

[1] Map illustrating 2020 planned defence-spending levels (in US$ at market exchange rates), as well as the annual real percentage change in planned defence spending between 2019 and 2020 (at constant 2015 prices and exchange rates). Percentage changes in defence spending can vary considerably from year to year, as states revise the level of funding allocated to defence. Changes indicated here highlight the short-term trend in planned defence spending between 2019 and 2020. Actual spending changes prior to 2019, and projected spending levels post-2020, are not reflected.

▲ Map 8 **Latin America and the Caribbean regional defence spending**[1]

Defence (MoD) budget in its original 2020 draft budget law (PLOA 2020), which was released on 31 August 2019. It would have reduced the MoD's final allocation to R68.6bn (US$13.2bn) from R112bn (US$28.4bn) in 2019, and this would have led total regional defence spending to contract by 15% in real terms as a result. The final budget law for 2020 (LOA 2020), agreed on 17 January 2020, allocated a slightly higher figure of R73.1bn (US$14.1bn) to the MoD. Meanwhile the projection for 2021 given in the draft budget law (PLOA 2021), released on 31 August 2020, proposed an MoD budget of R65.2bn (US$12.6bn). However, parliamentary amendments to government financing proposed over the course of 2020 due to the impact of the coronavirus pandemic appear to reverse the cuts in the MoD budget. In figures outlined in a Ministry of the Economy presentation on the PLOA 2021, also published on 31 August 2020, the amendments result in a final allocation for total defence of R114bn (US$22.1bn) and R116bn (US$22.4bn) in 2020 and 2021 respectively. It is possible that a fraction of this funding may lie outside core MoD funding, but this difference is likely to be fairly marginal, as total defence funding in 2019 was only marginally above the MoD budget (R112bn or US$28.4bn to R107bn or US$27.1bn respectively). As such, the regional outlook is more stable than it first appeared.

Without the increase in August 2020, Brazil's defence budget in 2020 would have been smaller, in real terms, than it had been in 2008. Indeed, when the 2020 budget was announced in August 2019, defence chiefs expressed concern over the source of the savings; a shorter working week was considered as a partial solution. The increase likely reflects the enhanced role that Brazil's armed forces have played in domestic security in light of the pandemic. Furthermore, with personnel and pay costs accounting for two thirds of Brazil's defence budget, the planned reduction in 2020 would have primarily affected salaries or force numbers in the short-term. Economic planners may have felt that it was desirable to avoid potential cuts like these when the government is trying to stimulate consumption and avoid further unemployment. Investment accounts for just 10% of the defence budget and would also have been severely constrained had the 2020 budget been enacted as planned.

The region's second-largest defence market, Colombia, also managed to sustain growth in 2020 with the defence budget increasing by 3% in real terms to US$9.4bn (US$10.5bn in constant US$2015). Despite exposure to commodity-price fluctuations and having been significantly affected by the coronavirus pandemic, the World Bank suggested in June 2020 that Colombia's economic performance before the pandemic gave it more monetary- and fiscal-policy headroom in order to deal with the effects of the pandemic. Colombia's economy was therefore projected by the IMF, again in June 2020, to be only moderately affected by the coronavirus pandemic, with a comparatively mild 2.4% contraction envisaged for 2020 followed by a 3.7% increase in 2021. However, the outlook for Colombia was severely downgraded in the IMF's October 2020 World Economic Outlook to an 8.2% contraction for 2020, suggesting that the resilience of Colombia's finances may have been overestimated earlier in the crisis.

As recently as 2017, Argentina was the third-largest budget in the region with a defence budget of US$6.2bn in nominal terms, significantly higher than the US$4.6bn defence budget of the next-largest market, Mexico. However, since 2018 the country has faced several economic challenges that have seen the defence budget contract significantly, falling to US$2.9bn (US$4.1bn in constant US$2015) by 2020. Although when the government announced an increase to the state budget for 2020 this did include additional defence funding devoted to salaries. With Argentina struggling to increase its defence spending, Mexico has since 2018 had the region's third-largest defence budget. This comes despite the volatility in its defence-budget allocations, reflected by the near 8% cut to the 2019 total and then the sizeable 14% real increase in the defence budget for 2020, with funding reaching US$5.3 billion (US$5.8bn in constant US$2015). However, the economy – and consequently public finances – remain vulnerable in the short term, including to the economic effects of the pandemic. The historic volatility of defence-budget allocations suggests that the military is not protected from wider movements in public spending, which may indicate an increased likelihood of future cuts as the full economic impact of the pandemic becomes clear.

Defence industry

The lack of any significant and sustained growth in defence spending and the low levels of funding allocated to investment (procurement and research and development (R&D)) among the major regional players constrains capability development in Latin America. The five largest spenders (Brazil, Colombia, Mexico, Argentina, Chile) allocate between 2% and 10% of annual defence expenditure to investment, which is low by international standards. For instance, NATO recommends that its members direct at least 20% of the defence budget toward new equipment investment and R&D – a metric 11 members achieved in 2019.

Colombia's offset policy has yet to result in a wide-ranging or advanced domestic defence capability, with the country still heavily reliant on foreign imports for major programmes. Mexico, meanwhile, has established a small defence industry capable of producing small arms and light weapons, 4x4 tactical patrol vehicles, unmanned aerial vehicles, fast interceptor craft, and coastal and ocean patrol vessels.

Brazil has been keen to develop its defence-manufacturing capabilities, with successive defence strategies released since 2008 outlining investment cycles that would progress from off-the-shelf purchases toward the execution of programmes that would enhance domestic capabilities and enable import substitutions. This is dependent not just on specific investment allocations but also on defence-budget growth more generally. To that end a July 2020 update to the National Defence Strategy, sent

Brazil: defence-industrial ambitions

The 2008 National Defence Strategy established the importance of the nuclear, space and cyber sectors and looked to support the development of research and development (R&D) projects, the local defence, trade and industrial base (DTIB) and Brazil's military presence in the Amazonian region. Brazil requires offsets for any acquisition exceeding US$50 million in order to support the local DTIB and state R&D and production organisations.

It is estimated by analysts that Brazil's DTIB comprises around 1,120 companies, including around 110 Strategic Defence Companies. Successive governments have established a range of measures to support the DTIB, including the FINTECH Defesa digital banking platform (created in June 2020) and the Special Tax Regime for the Defense Industry (RETID, agreed in 2012), which exempted from select taxes defence-related items from companies accredited as Strategic Defence Companies. The delivery of some equipment as part of Brazil's modernisation effort has been implemented through local development and production, with examples including the KC-390 airlifter produced by Embraer, the *ASTROS* II MK6 missile and rocket-launching system produced by Avibras, and Iveco's VBTP-MSR *Guarani* armoured vehicle. Nonetheless, a significant amount of equipment has been purchased abroad, including: *Igla*-S, RBS 70/NG and *Gepard* 1A2 air defence systems; Mi-35M and S-70B *Sea Hawk* helicopters; C295, *Turbo Trader* and P-3AM *Orion* aircraft, and *ScanEagle* and *Hermes* 450/900 uninhabited aerial vehicles; a range of naval vessels and LMV, AAV7A1 RAM/RS and *Piranha* IIIC armoured vehicles. In July 2019 Brazil was designated by the US as a major non-NATO ally, potentially easing access to US weapons and training. In March 2020 the two nations also signed a bilateral agreement on Research Development, Test and Evaluation (RDT&E), paving the way for potential future technological cooperation.

Brazil has also engaged with foreign defence suppliers as a means of securing technology transfers in order to further develop the local DTIB. Key projects include the PROSUB submarine development programme, the PFCT programme to locally build four MEKO A-100MB frigates, and the F-X2 programme to introduce *Gripen* multi-role combat aircraft into Brazilian service and at the same time develop the aerospace manufacturing and support base. The first *Gripen* E arrived in Brazil in September 2020 under the October 2014 contract for 28 F-39E *Gripens* and eight F-39Fs, and local production of aerostructures is under way at Saab's purpose-built plant in Brazil, Saab Aeronáutica Montagens; final assembly of future aircraft will also be undertaken by Embraer.

The PROSUB submarine development programme includes the construction by Odebrecht of the Island of Madeira submarine base (Base de Submarinos da Ilha da Madeira, BSIM) at the Itaguaí Naval Complex. This was inaugurated in July 2020 and is scheduled to be complete in 2022. One nuclear- and four conventionally powered submarines, procured under the PROSUB programme in 2009, will be based at the facility. (It is understood that Brazil's existing *Tupi*- and *Tikuna*-class U209/1400 series boats will, at least for the present, remain at their current base in Niterói.) The S-BR conventionally powered submarines are based on the French *Scorpène* design and are being built by Itaguaí Construções Navais, a joint venture between the navy, Odebrecht and France's Naval Group. The first boat, *Riachuelo*, is due to be commissioned in December 2020. As part of the SN-BR nuclear-powered submarine project, Brazil is also developing the technologies necessary to produce the boat's nuclear power plant under the PNM programme. The nuclear-powered submarine will be called *Álvaro Alberto* and is planned to be complete by 2029. Meanwhile, the Navy Strategic Plan 2040, unveiled on 10 September, outlined the acquisition of *Tamandaré*-class frigates, the NPa500-BR coastal patrol vessel and NPaOc-BR offshore patrol vessel projects (both part of the PRONAPA patrol vessel programme), Antarctic (NApAnt) and multi-purpose logistic (NApLMP) support ships, training ships, mine hunters, hydrographic survey ships (the NHo project), armoured vehicles and helicopters. The plan also mentioned an aircraft-carrier capability, previously discussed under the PRONAE programme. The main contract for four *Tamandaré*-class frigates under the PFCT programme (Programa Fragatas Classe Tamandaré) was signed on 5 March 2020 between EMGEPRON, on behalf of the navy, and the Águas Azuis Joint Venture, formed by Germany's ThyssenKrupp Marine Systems and Brazil's Embraer and Atech. The programme is valued at R9.1 billion (US$2.25bn) and deliveries are scheduled to take place from 2025–28. Brazil selected the MEKO A-100MB as the base design for the platforms and they will be equipped with weapons including locally developed MANSUP anti-ship missiles and MBDA's *Sea Ceptor* surface-to-air missiles. An additional two vessels could be acquired at a later date.

to the National Congress for approval, included an aspiration for Brazil to invest 2% of GDP in defence. However, economic challenges between 2014 and 2017 reduced the speed and scale of Brazil's military modernisation and while important developments took place in 2020, it remains unclear whether Brazil will be able to meet the 2% aspiration, not least because of the still unfolding economic impact of the coronavirus pandemic.

ARGENTINA

Argentina's armed forces are attempting to balance the requirement to maintain a credible military capability with the effect on the defence budget of persistent national economic challenges, and the additional economic impact of the coronavirus pandemic. The armed forces have experienced a significant decline in prestige since the end of the military dictatorship in 1983, exacerbated by defeat in the Falklands War in 1982 and the human-rights excesses committed during the military government. These factors led Argentina to reduce defence spending from 4.7% of GDP in 1979 to just 0.6% of GDP in 2020, which was also a means by which the country's new leaders sought to limit the political power previously held by the armed forces. Moreover, the state was also then focused on other national priorities.

A complicating factor is that for at least the last 30 years it has proven difficult to discern a consistent approach in Argentina's strategic policy orientation, and this uncertainty has complicated the task of defence planners. The strategic interest of governments since the end of the military government has varied between a focus on the Non-Aligned Movement, closer relations with the United States and Western states, and also with the socialist countries of Latin America. Most recently, under the Partido Justicialista administration of Alberto Fernández, there seems to be a blend between the latter two approaches, which is perhaps a reflection of internal dynamics within the ruling party. There has also been a return to heightened rhetoric concerning Argentina's claim to the Falkland Islands, and Fernández was reported as referring, in early 2020, to the 'usurpation of the Malvinas, South Georgia and South Sandwich Islands'. Argentina's claim had been less of a priority under the previous Macri government, which focused on improving the economic and political relationship with the United Kingdom.

Relations with China appear to be on an upward trajectory. In August 2020, Beijing renewed the currency swap with Argentina for US$18.5 billion and in the same week Argentina ratified the 2012 agreement with China concerning the installation of a satellite-tracking base in Bajada del Agrio in the province of Neuquén. Questions have been raised in Argentina over the purpose of the base, including over its possible use for military purposes and the degree of practical control that the Argentine state holds over its activities.

Plans to rebuild military capability

The deterioration in Argentina's military capabilities became more pronounced in the last decade. There was a reduction in inventory numbers and in the number of modernisation programmes as well as high-profile defence crises, most notably around the loss of the submarine *San Juan* with all hands in the South Atlantic in November 2017, and the sinking – while in port in 2013 – of the destroyer *Santísima Trinidad* (out of service at the time). A decline in salaries and incentives generated an increasing flow of personnel to the civil sector, while the recruitment to all ranks fell. As a result, analysts report that some units are understaffed, with this adding to equipment challenges in causing problems for effective capability generation. However, with unemployment that is increasing in light of the economic effects of the coronavirus pandemic, a greater number of young people are applying for voluntary military service, perhaps with an eye to employment stability as well as a military role.

The 2015–19 government of President Mauricio Macri was the first to emphasise the need to rebuild military capabilities, but this was not translated into policy and there was no significant increase in defence spending. Continued economic problems, as well as the relative de-prioritisation of military affairs, meant that investments declined further. Before coming to power (he was sworn in on 10 December 2019), Alberto Fernández raised the same concerns as his predecessor. In 2019, when still a deputy in Congress, current Defence Minister Agustín Rossi presented in Congress a bill to create the National Defence Fund (FONDEF), which outlined progressive funding increases to allow for the modernisation of military materiel and investment in the defence-industrial base. The bill proposed that 0.35% of total government income for 2020 should be allocated to FONDEF, with progressive increases to 0.8% of state income by 2023. The fund would be maintained at this level after

▲ Map 9 **Argentina: key military units and defence-industrial locations**

2023. This translates to approximately P14bn (US$199 million) in 2020 and potentially P30bn (US$426m) by 2023, although this depends on the health of the national economy and of state finances. The Chamber of Deputies and then the Senate approved the bill, the latter in September 2020. In October 2020 it was formally launched by the president at the Tandanor shipyard in Buenos Aires. Earlier, in May 2020, the government created a commission – chaired by former defence ministers Nilda Garré and Horacio Jaunarena – to outline military-reform plans, although neither its scope nor expected policy outcomes were specified.

The coronavirus pandemic generated an additional challenge for the armed forces as they were deployed to assist in supporting public-health measures and controlling compliance with social-distancing measures. This highlighted gaps in force dispositions across the country as well as in logistics support capacities, and also drew attention to reduced equipment-capability levels. The air force's airlift capacity is one example, as this was heavily employed not just in repatriating Argentine citizens from Brazil, Ecuador and Peru but also in a sustained logistics effort within the country; helicopters from the army and the air force were similarly employed. However, these deployments did have the effect of raising the profile of the armed forces within society, and some analysts assert that this increased support for military modernisation. The loss of an A-4AR *Skyhawk* fighter in early August again highlighted not just the relative obsolescence of this platform but also the difficulties that the armed forces have had in maintaining their ageing equipment.

Despite constrained resources, President Fernández announced at a 'camaraderie meeting' with the armed forces on 22 July 2020 that from October, salary payments would include some supplements that had hitherto been paid in a non-remunerative manner, which meant that they had not been counted when calculating military pensions. Although this is understood to have been a long-standing request of military personnel, it is not yet known how the state will pay for the increase in pensions for current retirees, which analysts estimate would amount to an increase of around 30%.

As well as budgetary issues, the government has also started to outline policy changes. The first came on 26 June 2020 with the publication of Decree 571/2020. This was intended to repeal Decrees 683 of 23 July 2018 and 703 of 30 July 2018. Passed under the Macri administration, these had enabled the armed forces to deploy against non-state threats. Instead, Decree 572/2020 was intended to revalidate Decrees 727 of 12 June 2006 and 1691 of 22 November 2006, which had been passed under the Kirchner administration. Decree 727, in particular, amended the April 1988 Law No. 23,554 on National Defence and imposed limits on the activities of the armed forces in that they could act only 'in the face of external aggressions perpetrated by armed forces belonging to other states'. The 2018 decrees had stated that it was possible to deploy against any external actor, not only foreign armed forces. It is now unclear how the legislative framework can enable the deployment of the armed forces against terrorist groups or criminals from outside the country.

Organisation and modernisation

While the president remains the commander in chief, and the defence ministry has executive power over the armed forces, the Joint Chiefs of Staff has over the past two decades transformed into the body providing operational management of the armed forces. While it is understood that no significant organisational change is planned, the possible redeployment of units has been raised several times, with the coronavirus pandemic reportedly raising concerns over the level of military presence in remote areas. That said, the only development foreseen is the creation of a base in the city of Ushuaia to support Antarctic activities, though this project has been ongoing for several years.

Force-modernisation plans have been stymied by economic problems, compounded by challenges related to the maintenance requirements of increasingly ageing platforms. The army has for some time been looking to modernise its main battle tank, the Tanque Argentino Mediano (TAM), and while this project reached the prototype phase in 2015, and 74 were envisaged, progress since then has been glacial. Moreover, the balance of the armoured fleet is under discussion, with potential projects also for the procurement of 6x6 or 8x8 wheeled armoured vehicles. In July 2020, the US approved the potential sale to Argentina of *Stryker* armoured vehicles, though there has been little progress since. Other modernisation or replacement programmes are planned for trucks, rifles and the SA-315B *Lama* and UH-1H-II *Huey* II helicopters.

The navy's main modernisation programmes include four *Bouchard*-class offshore patrol vessels. The first of these, the former French *L'Adroit*, arrived in Argentina in early 2020 and three further OPVs are being built to the same design by Kership. Naval-modernisation plans also include the purchase of an ice-capable ship to support the *Almirante Irízar* icebreaker, and a dock landing ship. The plan to revive the *Super Etendard* fleet with five former French *Super Etendard Modernisé* aircraft is progressing slowly, and they are expected to be operational in 2021. Meanwhile, the modernisation of the MEKO 360 destroyers and the MEKO 140 corvettes has been stopped, as has the plan to revive Argentina's submarine capability. The Macri administration began negotiations to replace Argentina's P-3B *Orion* aircraft with P-3Cs and, while this has been approved for export by the US, the degree of progress in 2020 remains uncertain.

The air force's top priority, other than fleet maintenance and rebuilding operational capability, relates to inventory modernisation. The *Mirage* fleet was retired in 2015 and while there remains an aspiration to replace this, as well as the A4-AR *Skyhawk* (*Fighting Hawk*), funds have yet to be allocated. More progress is being made in relation to the modernisation of the C-130H *Hercules* fleet and the delivery of *Pampa* III trainers, and the defence minister was reported as saying that *Pampa* deliveries could speed up if funds are released as part of the FONDEF programme. It is understood that the air force also aspires to purchase a Boeing 737-700 as part of a plan to replace the Fokker F28 fleet.

Defence economics

Argentina's defence-modernisation aspirations depend on improvements in the broader economy, with the country continuing to face an economic crisis. Output has stalled since 2011 and the country has been in recession since 2017. The effects of the coronavirus pandemic, including the consequent drop in domestic and international activity, will exacerbate Argentina's already poor economic environment and the IMF has projected a potential 10% contraction in real GDP in 2020. The contraction in government revenues, combined with the sharp rise in social-assistance spending, will lead to a primary fiscal deficit that is expected to reach more than 5% of GDP, the highest in the last 30 years. The government is partly compensating for this public-spending increase by reducing capital expenditures by more than 25%, which directly affects purchases of defence materiel. The government has achieved some success in renegotiating foreign debt, but in the short term financing requirements are being addressed through monetary issuance. The risk is that this increases inflation in 2021, after a decline in 2020 to levels of around 40% (against 54% in 2019).

As recently as 2017, Argentina's defence budget was the third-largest budget in the region after Brazil and Mexico, reaching US$6.2bn in real terms that year. However, since 2018 the country has faced several economic challenges, including high debt levels, currency depreciation, recession and high inflation. Consequently, the defence budget saw significant annual cuts in real terms in 2018, 2019 and again in 2020, averaging 15%. The defence budget was actually increased to P205bn (US$2.9bn in nominal terms) in 2020, up from P158bn (US$2.2bn in nominal terms) in 2019, but inflation in excess of 50% over the last year served to wipe out any nominal increases in Argentinian public expenditure. When looked at in real terms, the Argentinian defence budget was cut by US$650m or 15% to US$4.1bn in 2020, a budget that is 33% lower than 2017 spending levels in real terms. The severe devaluation of the peso against the dollar that has gathered pace since 2018 also erodes the defence budget in dollar terms. In August, the president announced that the state budget for 2020 would increase with a further P24bn (US$340m) provided to defence, of which 85% will be allocated to pay and salaries.

Defence industry

Defence minister Agustín Rossi has indicated that recapitalisation of the armed forces should as far as possible be fulfilled by procurement from national industry, though this leads to limitations in relation to the provision of complex military systems. The defence ministry owns three principal companies in the land, sea and air domains. Aviation firm Fábrica Argentina de Aviones (FAdeA) specialises in the production, modernisation and maintenance of aircraft, and is currently manufacturing the *Pampa* III at a low rate, modernising the *Hercules* fleet for the air force (it is working on a fifth airframe) and working with the army to modernise Bell-206 helicopters. Land-sector firm Fabricaciones Militares combines a series of factories for light weapons, explosives, rockets and other products. Shipbuilding firm Tandanor has in recent years been focused on the maintenance and modernisation of ships and submarines. The Rio Santiago shipyard, meanwhile, has long experience in shipbuilding and is producing four small training vessels for the navy.

State-owned technology firm Invap makes some products for military use, mainly radars, and has also prototyped UAVs. The *Arsat* SG-1 satellite, which is under development by Invap and the National Commission for Space Activities, is also intended to have a military role. In addition, there are some small private companies that focus on defence, such as FixView, which produces gyro-stabilised platforms and has developed pods for surveillance and reconnaissance for the air force. There have been few defence exports, though in 2019 FAdeA was engaged in (ultimately unsuccessful) negotiations with Guatemala for the export of *Pampa* III aircraft.

Antigua and Barbuda ATG

East Caribbean Dollar EC$		2019	2020	2021
GDP	EC$	4.49bn	3.75bn	
	US$	1.66bn	1.39bn	
per capita	US$	17,195	14,159	
Growth	%	3.4	-17.3	
Inflation	%	1.5	1.0	
Def bdgt [a]	EC$	19.5m	21.3m	
	US$	7.2m	7.9m	
US$1=EC$		2.70	2.70	

[a] Budget for the Ministry of Legal Affairs, Public Safety, Immigration & Labour

Real-terms defence budget trend (US$m, constant 2015)

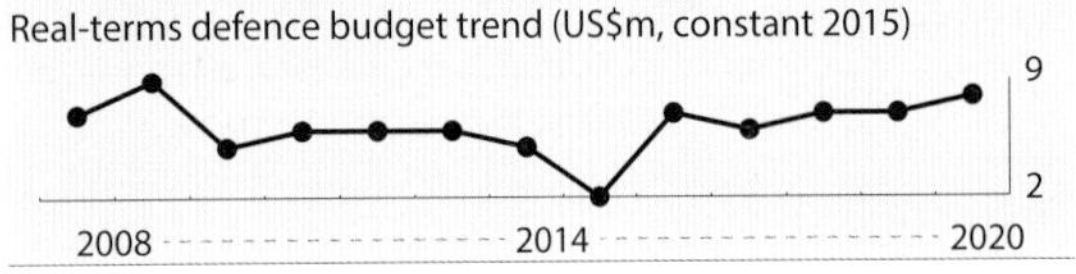

Population 98,179

Age	0–14	15–19	20–24	25–29	30–64	65 plus
Male	11.5%	3.9%	4.1%	3.6%	20.2%	3.8%
Female	11.1%	3.9%	4.2%	3.9%	24.7%	5.1%

Capabilities

The Antigua and Barbuda Defence Force (ABDF) focuses on internal security and disaster relief, and also contributes to regional counter-narcotics efforts. It comprises a light-infantry element, which carries out internal-security duties, and a coastguard, which is tasked with fishery protection and counter-narcotics. Antigua and Barbuda is a member of the Caribbean Community and the Caribbean Regional Security System. There are defence ties with the UK and the US. The ABDF has participated in US SOUTHCOM's *Tradewinds* disaster-response exercise, though it has no independent capacity to deploy forces other than in its immediate neighbourhood. There is no heavy land-forces equipment, while the coastguard maintains ex-US patrol vessels and a number of smaller boats. Aside from limited maintenance facilities, there is no indigenous defence industry.

ACTIVE 180 (Army 130 Coast Guard 50)

(all services form combined Antigua and Barbuda Defence Force)

RESERVE 80 (Joint 80)

ORGANISATIONS BY SERVICE

Army 130

FORCES BY ROLE

MANOEUVRE

Light

1 inf bn HQ

1 inf coy

COMBAT SERVICE SUPPORT

1 spt gp (1 engr unit, 1 med unit)

Coast Guard 50

EQUIPMENT BY TYPE

PATROL AND COASTAL COMBATANTS • PB 2: 1 *Dauntless*; 1 *Swift*

Argentina ARG

Argentine Peso P		2019	2020	2021
GDP	P	21.5tr	27.0tr	
	US$	444bn	383bn	
per capita	US$	9,890	8,433	
Growth	%	-2.1	-11.8	
Inflation	%	53.5	n.k	
Def bdgt	P	158bn	205bn	246bn
	US$	3.28bn	2.91bn	
US$1=P		48.25	70.46	

Real-terms defence budget trend (US$bn, constant 2015)

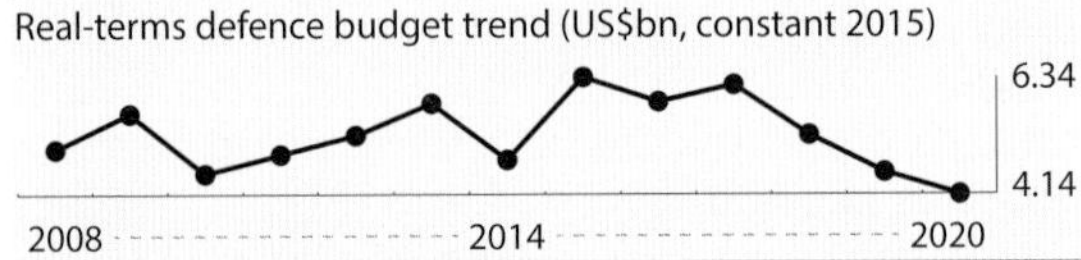

Population 45,479,118

Age	0–14	15–19	20–24	25–29	30–64	65 plus
Male	12.4%	4.0%	3.8%	3.7%	20.5%	5.1%
Female	11.6%	3.8%	3.6%	3.6%	20.8%	7.0%

Capabilities

Argentina's armed forces have sufficient training and equipment to fulfil internal-security tasks, although any power-projection ambition is limited by funding shortfalls. The armed forces principally focus on border security, surveillance and counter-narcotics operations, and there is some cooperation with Bolivia and Paraguay. In 2020, the government repealed 2018 legislation, passed under the previous administration, that had allowed greater latitude in deploying the armed forces to tackle external challenges including from organised-crime groups. There is military cooperation with the US and the armed forces participate in multinational exercises and bilateral peacekeeping exercises with neighbour Chile. There are limited deployment capabilities, and the equipment inventory is increasingly obsolete, with modernisation hampered by funding problems. The air force faces significant equipment-availability challenges and the navy has seen its capability decline in areas such as anti-submarine warfare, mine warfare and airborne early warning. Argentina possesses an indigenous defence-manufacturing and maintenance capacity covering land, sea and air systems, although industry fortunes have dipped in recent years because of a lack of investment.

ACTIVE 72,100 (Army 42,800 Navy 16,400 Air 12,900) Paramilitary 31,250

ORGANISATIONS BY SERVICE

Army 42,800

Regt and gp are usually bn-sized

FORCES BY ROLE

MANOEUVRE

Mechanised

1 (1st) div (1 armd bde (1 armd recce regt, 3 tk regt, 1 mech inf regt, 1 SP arty gp, 1 cbt engr bn, 1 int coy,

1 sigs sqn, 1 log coy), 1 (3rd) jungle bde (2 jungle inf regt, 2 jungle inf coy, 1 arty gp, 1 engr coy, 1 int coy, 1 sigs coy, 1 log coy, 1 med coy); 1 (12th) jungle bde (2 jungle inf regt, 1 jungle inf coy, 1 arty gp, 1 engr bn, 1 int coy, 1 sigs coy, 1 log coy, 1 med coy), 2 engr bn, 1 sigs bn, 1 log coy)

1 (3rd) div (1 armd bde (1 armd recce sqn, 3 tk regt, 1 mech inf regt, 1 SP arty gp, 1 cbt engr sqn, 1 int coy, 1 sigs sqn, 1 log coy); 1 mech bde (1 armd recce regt, 1 tk regt, 2 mech inf regt, 1 SP arty gp, 1 cbt engr bn, 1 int coy, 1 sigs coy, 1 log coy); 1 mech bde (1 armd recce regt, 1 tk regt, 2 mech inf regt, 1 SP arty gp, 1 cbt engr bn, 1 int coy, 1 sigs coy, 1 log coy); 1 int bn, 1 sigs bn, 1 log coy, 1 AD gp (2 AD bn))

1 (Rapid Deployment Force) div (1 SF gp; 1 mech bde (1 armd recce regt, 3 mech inf regt, 1 arty gp, 1 MRL gp, 1 cbt engr coy, 1 sigs coy,1 log coy); 1 AB bde (1 recce sqn, 2 para regt, 1 air aslt regt, 1 arty gp, 1 cbt engr coy, 1 sigs coy, 1 log coy))

Light

1 (2nd) mtn inf div (1 mtn inf bde (1 recce regt, 3 mtn inf regt, 1 mtn inf coy, 2 arty gp, 1 cbt engr bn, 1 sigs coy, 1 log coy); 1 mtn inf bde (1 recce regt, 3 mtn inf regt, 1 mtn inf coy, 1 arty gp, 1 cbt engr bn, 1 sigs coy, 1 log coy); 1 mtn inf bde (1 recce regt, 2 mtn inf regt, 2 arty gp, 1 cbt engr bn, 1 sigs coy, 1 construction coy, 1 log coy), 1 arty gp, 1 AD bn, 1 sigs bn)

1 mot cav regt (presidential escort)

COMBAT SUPPORT

1 engr bn
1 CBRN coy
1 sigs gp (1 EW bn, 1 sigs bn, 1 maint bn)
1 sigs bn
1 sigs coy

COMBAT SERVICE SUPPORT

3 maint bn

HELICOPTER

1 avn gp (bde) (1 avn bn, 1 tpt hel bn, 1 atk/ISR hel sqn)

EQUIPMENT BY TYPE

ARMOURED FIGHTING VEHICLES

MBT 231: 225 TAM, 6 TAM S21
LT TK 117: 107 SK-105A1 *Kuerassier;* 6 SK-105A2 *Kuerassier;* 4 *Patagón*
RECCE 47 AML-90
IFV 232: 118 VCTP (incl variants); 114 M113A2 (20mm cannon)
APC 278
APC (T) 274: 70 M113A1-ACAV; 204 M113A2
APC (W) 4 WZ-551B1

ENGINEERING & MAINTENANCE VEHICLES

ARV *Greif*

ANTI-TANK/ANTI-INFRASTRUCTURE

MSL • SP 3 M1025 HMMWV with TOW-2A
RCL 105mm 150 M-1968

ARTILLERY 1,108

SP 155mm 42: 23 AMX F3; 19 VCA 155 *Palmaria*
TOWED 172: **105mm** 64 Model 56 pack howitzer; **155mm** 108: 28 CITEFA M-77/CITEFA M-81; 80 SOFMA L-33
MRL 8: **105mm** 4 SLAM *Pampero;* **127mm** 4 CP-30
MOR 886: **81mm** 492; **SP 107mm** 25 M106A2; **120mm** 330 Brandt; **SP 120mm** 39 TAM-VCTM

AIRCRAFT

TPT • Light 14: 1 Beech 80 *Queen Air;* 3 C-212-200 *Aviocar;* 2 Cessna 208EX *Grand Caravan;* 1 Cessna 500 *Citation* (survey); 1 Cessna 550 *Citation Bravo;* 3 DA42 (to be converted to ISR role); 2 DHC-6 *Twin Otter;* 1 *Sabreliner* 75A (*Gaviao* 75A)
TRG 5 T-41 *Mescalero*

HELICOPTERS

MRH 5: 4 SA315B *Lama;* 1 Z-11
TPT 62: **Medium** 3 AS332B *Super Puma;* **Light** 59: 1 Bell 212; 25 Bell 205 (UH-1H *Iroquois* – 6 armed); 5 Bell 206B3; 13 UH-1H-II *Huey* II; 15 AB206B1

AIR DEFENCE

SAM • Point-defence RBS-70
GUNS • TOWED 229: **20mm** 200 GAI-B01; **30mm** 21 HS L81; **35mm** 8 GDF Oerlikon (*Skyguard* fire control)

Navy 16,400

Commands: Surface Fleet, Submarines, Naval Avn, Marines

FORCES BY ROLE

SPECIAL FORCES

1 (diver) SF gp

EQUIPMENT BY TYPE

SUBMARINES • SSK 1:

1 *Santa Cruz* (GER TR-1700) (undergoing MLU) with 6 single 533mm TT with SST-4 HWT
(1 *Salta* (GER T-209/1100) (non-operational since 2013) with 8 single 533mm TT with Mk 37/SST-4 HWT)

PRINCIPAL SURFACE COMBATANTS 5

DESTROYERS • DDH 1 *Hercules* (UK Type-42) (utilised as a fast troop-transport ship), with 1 114mm gun (capacity 2 SH-3H *Sea King* hel)
FRIGATES • FFGHM 4 *Almirante Brown* (GER MEKO 360) with 2 quad lnchr with MM40 *Exocet* AShM, 1 octuple *Albatros* lnchr with *Aspide* SAM, 2 triple ILAS-3 (B-515) 324mm TT with A244/S LWT, 1 127mm gun (capacity 1 AS555 *Fennec* hel)

PATROL AND COASTAL COMBATANTS 20

CORVETTES 7:
FSGH 6 *Espora* (GER MEKO 140) with 2 twin lnchr with MM38 *Exocet* AShM, 2 triple 324mm ILAS-3 (B-515) ASTT with A244/S LWT, 1 76mm gun (capacity 1 AS555 *Fennec* hel)
FSG 1 *Drummond* (FRA A-69) (2 laid up in 2019) with 2 twin lnchr with MM38 *Exocet* AShM, 2 triple 324mm ILAS-3 (B-515) ASTT with A244/S LWT, 1 100mm gun
PSOH 1 *Bouchard* (ex-FRA *L'Adroit*)
PSO 2: 1 *Irigoyen* (ex-US *Cherokee*); 1 *Teniente Olivieri* (ex-US oilfield tug)
PCFGT 1 *Intrepida* (GER Lurssen 45m) with 2 single lnchr with MM38 *Exocet* AShM, 2 single 533mm TT with SST-4 HWT, 1 76mm gun
PCF 1 *Intrepida* (GER Lurssen 45m) with 1 76mm gun
PCO 1 *Murature* (ex-US *King* – trg/river-patrol role) with 2 105mm gun
PB 7: 4 *Baradero* (ISR *Dabur*); 2 *Punta Mogotes* (ex-US *Point*); 1 *Zurubi*

Latin America and the Caribbean

AMPHIBIOUS 6 **LCVP**
LOGISTICS AND SUPPORT 16
ABU 3 *Red*
AFS 4 *Puerto Argentina* (ex-RUS *Neftegaz*)
AGB 1 *Almirante Irizar* (damaged by fire in 2007; returned to service in mid-2017)
AGHS 3: 1 *Austral* (ex-GER *Sonne*); 1 *Cormoran*; 1 *Puerto Deseado* (ice-breaking capability, used for polar research)
AGOR 1 *Commodoro Rivadavia*
AK 2 *Costa Sur* (capacity 4 LCVP)
AOR 1 *Patagonia* (ex-FRA *Durance)* with 1 hel platform
AXS 1 *Libertad*

Naval Aviation 2,000

EQUIPMENT BY TYPE
AIRCRAFT 25 combat capable
FGA 7 *Super Etendard* (9 more in store)
ATK 1 AU-23 *Turbo Porter*
ASW 7: 3 S-2T *Tracker*†; 4 P-3B *Orion*†
TPT • Light 7 Beech 200F/M *King Air*
TRG 10 T-34C *Turbo Mentor**
HELICOPTERS
ASW 2 SH-3H (ASH-3H) *Sea King*
MRH 4 AS555 *Fennec*
TPT • Medium 4 UH-3H *Sea King*
AIR-LAUNCHED MISSILES
AAM • IR R-550 *Magic*
AShM AM39 *Exocet*

Marines 2,500

FORCES BY ROLE
MANOEUVRE
Amphibious
1 (fleet) force (1 cdo gp, 1 (AAV) amph bn, 1 mne bn, 1 arty bn, 1 ADA bn)
1 (fleet) force (2 mne bn, 2 navy det)
1 force (1 mne bn)

EQUIPMENT BY TYPE
ARMOURED FIGHTING VEHICLES
RECCE 12 ERC-90F *Sagaie*
APC • APC (W) 31 VCR
AAV 11 LVTP-7
ENGINEERING & MAINTENANCE VEHICLES
ARV AAVR 7
ANTI-TANK/ANTI-INFRASTRUCTURE
RCL 105mm 30 M-1974 FMK-1
ARTILLERY 89
TOWED 19: **105mm** 13 Model 56 pack howitzer; **155mm** 6 M114
MOR 70: **81mm** 58; **120mm** 12
AIR DEFENCE
SAM • Point-defence RBS-70
GUNS 40mm 4 Bofors 40L

Air Force 12,900

4 Major Comds – Air Operations, Personnel, Air Regions, Logistics, 8 air bde

Air Operations Command

FORCES BY ROLE
GROUND ATTACK
2 sqn with A-4/OA-4 (A-4AR/OA-4AR) *Skyhawk*
2 (tac air) sqn with EMB-312 *Tucano* (on loan for border surv/interdiction)
ISR
1 sqn with Learjet 35A
SEARCH & RESCUE/TRANSPORT HELICOPTER
2 sqn with Bell 212; Bell 412; Mi-171, SA-315B *Lama*
TANKER/TRANSPORT
1 sqn with C-130H *Hercules*; KC-130H *Hercules*; L-100-30
TRANSPORT
1 sqn with DHC-6 *Twin Otter*; Saab 340
1 sqn with F-28 *Friendship*
1 sqn with Learjet 35A; Learjet 60
1 (Pres) flt with B-757-23ER; S-70A *Black Hawk*, S-76B
TRAINING
1 sqn with AT-63 *Pampa*
1 sqn with EMB-312 *Tucano*
1 sqn with Grob 120TP
1 sqn with IA-63 *Pampa* III*
1 sqn with T-6C *Texan* II
1 hel sqn with Hughes 369; SA-315B *Lama*
TRANSPORT HELICOPTER
1 sqn with Hughes 369; MD-500; MD-500D

EQUIPMENT BY TYPE
AIRCRAFT 47 combat capable
ATK 21: 19 A-4 (A-4AR) *Skyhawk*†; 2 OA-4 (OA-4AR) *Skyhawk*†
ELINT 1 Learjet 35A
TKR 2 KC-130H *Hercules*
TPT 20: **Medium** 4: 3 C-130H *Hercules*; 1 L-100-30; **Light** 13: 4 Cessna 182 *Skylane*; 2 DHC-6 *Twin Otter*; 3 Learjet 35A (of which 2 test and calibration and 1 medevac); 1 Learjet 60 (VIP); 1 PA-28-236 *Dakota*; 2 Saab 340 (jointly operated with LADE); **PAX** 3: 1 B-737; 1 B-757-23ER; 1 F-28 *Fellowship*
TRG 61: 20 AT-63 *Pampa** (LIFT); 11 EMB-312 *Tucano*; 8 Grob 120TP; 6 IA-63 *Pampa* III*; 6 P2002JF *Sierra*; 10 T-6C *Texan* II (8 EMB-312 *Tucano* in store)
HELICOPTERS
MRH 27: 4 Bell 412EP; 11 Hughes 369; 3 MD-500; 4 MD-500D; 5 SA315B *Lama*
TPT 12: **Medium** 3: 2 Mi-171E; 1 S-70A *Black Hawk* (VIP); **Light** 9: 7 Bell 212; 2 S-76B (VIP)
AIR DEFENCE
GUNS 88: **20mm**: 86 Oerlikon/Rh-202 with 9 Elta EL/M-2106 radar; **35mm**: 2 Oerlikon GDF-001 with *Skyguard* radar
AIR-LAUNCHED MISSILES
AAM • IR AIM-9L *Sidewinder*; R-550 *Magic*; *Shafrir* 2‡

Paramilitary 31,250

Gendarmerie 18,000

Ministry of Security
FORCES BY ROLE
COMMAND
7 regional comd
SPECIAL FORCES
1 SF unit

MANOEUVRE
Other
17 paramilitary bn
Aviation
1 (mixed) avn bn

EQUIPMENT BY TYPE

ARMOURED FIGHTING VEHICLES
RECCE S52 *Shorland*
APC (W) 87: 47 *Grenadier*; 40 UR-416
ARTILLERY • MOR 81mm
AIRCRAFT
TPT 13: **Light** 12: 3 Cessna 152; 3 Cessna 206; 1 Cessna 336; 1 PA-28 *Cherokee*; 2 PC-6B *Turbo Porter*; 2 PC-12; **PAX** 1 Learjet 35
HELICOPTERS
MRH 2 MD-500C
TPT • Light 18: 3 AW119 *Koala*; 3 Bell 206 *Jet Ranger* (AB-206); 7 AS350 *Ecureuil*; 1 H135; 1 H155; 3 R-44 *Raven* II
TRG 1 S-300C

Prefectura Naval (Coast Guard) 13,250

Ministry of Security

EQUIPMENT BY TYPE

PATROL AND COASTAL COMBATANTS 71
PCO 7: 1 *Correa Falcon*; 1 *Delfin*; 5 *Mantilla* (F30 *Halcón* – undergoing modernisation)
PCC 1 *Mariano Moreno*
PB 58: 1 *Dorado*; 25 *Estrellemar*; 2 *Lynch* (US *Cape*); 18 *Mar del Plata* (Z-28); 1 *Surel*; 8 Damen Stan 2200; 3 Stan Tender 1750
PBF 4 *Shaldag* II
PBR 1 *Tonina*
LOGISTICS & SUPPORT 11
AAR 1 *Tango*
AFS 1 *Prefecto Garcia*
AG 2
ARS 1 *Prefecto Mansilla*
AX 5: 1 *Mandubi*; 4 other
AXS 1 *Dr Bernardo Houssay*
AIRCRAFT
MP 1 Beech 350ER *King Air*
TPT • Light 6: 5 C-212 *Aviocar*; 1 Beech 350ER *King Air*
TRG 2 Piper PA-28 *Archer* III
HELICOPTERS
SAR 3 AS565MA *Panther*
MRH 1 AS365 *Dauphin 2*
TPT 5: **Medium** 3: 1 H225 *Puma*; 2 SA330L (AS330L) *Puma*; **Light** 2 AS355 *Ecureuil* II
TRG 4 S-300C

DEPLOYMENT

CENTRAL AFRICAN REPUBLIC: UN • MINUSCA 1

CYPRUS: UN • UNFICYP 282; 2 inf coy; 1 hel flt with 2 Bell 212

MIDDLE EAST: UN • UNTSO 3

WESTERN SAHARA: UN • MINURSO 2

Bahamas BHS

Bahamian Dollar B$		2019	2020	2021
GDP	B$	13.6bn	11.6bn	
	US$	13.6bn	11.6bn	
per capita	US$	35,664	30,027	
Growth	%	1.2	-14.8	
Inflation	%	1.3	1.8	
Def bdgt	B$	92.3m	85.8m	
	US$	92.3m	85.8m	
US$1=B$		1.00	1.00	

Real-terms defence budget trend (US$m, constant 2015)

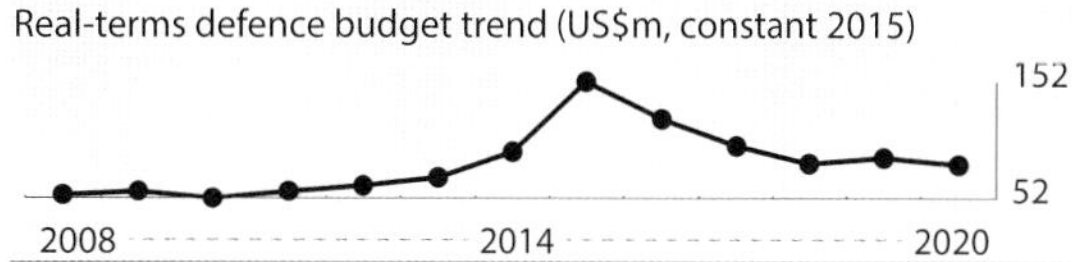

Population 337,721

Age	0–14	15–19	20–24	25–29	30–64	65 plus
Male	11.2%	3.7%	4.1%	4.2%	22.5%	3.4%
Female	10.9%	3.6%	4.0%	4.1%	23.1%	5.3%

Capabilities

The Royal Bahamas Defence Force (RBDF) is a primarily naval force primarily tasked with disaster relief, maritime security and countering narcotics trafficking. Its single commando squadron is tasked with base and internal security. The Bahamas is a member of the Caribbean Community, and the RBDF maintains training relationships with the UK and US. The RBDF has participated in US SOUTHCOM's *Tradewinds* disaster-response exercise. There is little independent capacity to deploy abroad, aside from recent regional disaster-relief efforts. The RBDF's Sandy Bottom Project, the largest-ever capital investment in the service, includes the acquisition of patrol craft and the development of bases and port facilities. A new permanent naval base on Grand Bahama is under discussion to bolster counter-narcotics tasks. The maritime wing is focused around patrol vessels and smaller patrol boats, while the air wing has a small inventory of light aircraft. Apart from limited maintenance facilities, the Bahamas has no indigenous defence industry.

ACTIVE 1,500

ORGANISATIONS BY SERVICE

Royal Bahamas Defence Force 1,500

FORCES BY ROLE

MANOEUVRE
Amphibious
1 mne coy (incl marines with internal- and base-security duties)

EQUIPMENT BY TYPE

PATROL AND COASTAL COMBATANTS 16
PCC 2 *Bahamas*
PB 14: 4 *Arthur Dion Hanna* (Damen Stan Patrol 4207); 2 *Dauntless*; 4 *Lignum Vitae* (Damen 3007); 2 Sea Ark 12m; 2 Sea Ark 15m
LOGISTICS & SUPPORT • **AKR** 1 *Lawrence Major* (Damen 5612)

AIRCRAFT • TPT • Light 3: 1 Beech A350 *King Air*; 1 Cessna 208 *Caravan*; 1 P-68 *Observer*

FOREIGN FORCES

Guyana Navy: Base located at New Providence Island

Barbados BRB

Barbados Dollar B$		2019	2020	2021
GDP	B$	10.4bn	9.26bn	
	US$	5.21bn	4.63bn	
per capita	US$	18,139	16,082	
Growth	%	-0.1	-11.6	
Inflation	%	4.1	2.9	
Def bdgt [a]	B$	76.7m	81.0m	
	US$	38.4m	40.5m	
US$1=B$		2.00	2.00	

[a] Defence & security expenditure

Real-terms defence budget trend (US$m, constant 2015)

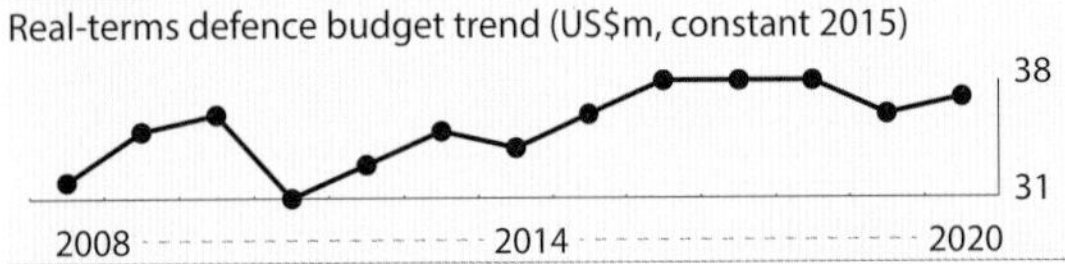

Population 294,560

Age	0–14	15–19	20–24	25–29	30–64	65 plus
Male	8.7%	3.1%	3.1%	3.4%	24.5%	5.6%
Female	8.7%	3.1%	3.1%	3.5%	25.2%	8.0%

Capabilities

Maritime security and resource protection are the main tasks of the Barbados Defence Force (BDF), but it has a secondary public-safety role in support of the police force. The BDF has undertaken counter-narcotics work, while troops have also been tasked with support to law enforcement. There are plans to improve disaster-relief capabilities. The Caribbean Regional Security System is headquartered in Barbados; it is also a member of the Caribbean Community. The BDF has participated in US SOUTHCOM's *Tradewinds* disaster-response exercise. There is limited capacity to independently deploy within the region, such as on hurricane-relief duties. The inventory consists principally of a small number of patrol vessels. Apart from limited maintenance facilities, Barbados has no indigenous defence industry.

ACTIVE 610 (Army 500 Coast Guard 110)

RESERVE 430 (Joint 430)

ORGANISATIONS BY SERVICE

Army 500

FORCES BY ROLE

MANOEUVRE

Light

1 inf bn (cadre)

Coast Guard 110

HQ located at HMBS Pelican, Spring Garden

EQUIPMENT BY TYPE

PATROL AND COASTAL COMBATANTS 6

PB 6: 1 *Dauntless*; 2 *Enterprise* (Damen Stan 1204); 3 *Trident* (Damen Stan Patrol 4207)

Belize BLZ

Belize Dollar BZ$		2019	2020	2021
GDP	BZ$	3.68bn	3.11bn	
	US$	1.84bn	1.56bn	
per capita	US$	4,498	3,734	
Growth	%	-2.0	-16.0	
Inflation	%	0.2	0.8	
Def bdgt [a]	BZ$	46.7m	49.1m	
	US$	23.3m	24.5m	
US$1=BZ$		2.00	2.00	

[a] Excludes funds allocated to Coast Guard and Police Service

Real-terms defence budget trend (US$m, constant 2015)

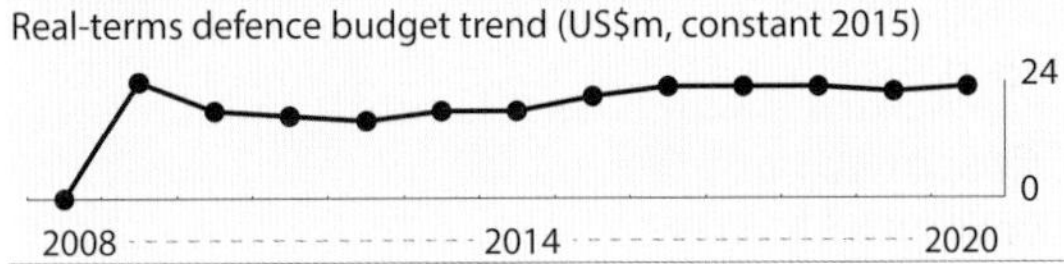

Population 399,598

Age	0–14	15–19	20–24	25–29	30–64	65 plus
Male	16.6%	5.4%	4.4%	4.5%	16.9%	2.2%
Female	15.9%	5.1%	4.1%	4.3%	18.1%	2.3%

Capabilities

Belize maintains a small Defence Force (BDF) and coastguard to meet national-security needs, particularly control of the borders with Guatemala and Mexico. Territorial defence and combating transnational crime are key objectives of the 2018–21 National Security and Defence Strategy. The UK has a long-standing security relationship with Belize and maintains a small training unit there, and the BDF also trains with US SOUTHCOM. Overall training levels are limited but generally sufficient for the BDF's tasks. Belize is a member of the Caribbean Community. The BDF does not deploy internationally and logistics support is adequate for border-security missions. The conventional-equipment inventory is limited and there is no significant domestic defence industry.

ACTIVE 1,500 (Army 1,500) **Paramilitary 550**

RESERVE 700 (Joint 700)

ORGANISATIONS BY SERVICE

Army ε1,500

FORCES BY ROLE

MANOEUVRE

Light

2 inf bn (3 inf coy)

COMBAT SERVICE SUPPORT

1 spt gp

EQUIPMENT BY TYPE

ANTI-TANK/ANTI-INFRASTRUCTURE • RCL 84mm *Carl Gustaf*

ARTILLERY • MOR 81mm 6

Air Wing

EQUIPMENT BY TYPE

AIRCRAFT

TPT • Light 2: 1 BN-2B *Defender*†; 1 Cessna 182 *Skylane*†

TRG 1 T-67M-200 *Firefly*

HELICOPTERS

TPT • Light 3: 2 Bell 205 (UH-1H *Iroquois*); 1 Bell 407

Reserve

FORCES BY ROLE

MANOEUVRE

Light

1 inf bn (3 inf coy)

Paramilitary 550

Coast Guard 550

EQUIPMENT BY TYPE

All operational patrol vessels under 10t FLD

FOREIGN FORCES

United Kingdom BATSUB 12

Bolivia BOL

Bolivian Boliviano B		2019	2020	2021
GDP	B	283bn	267bn	
	US$	41.2bn	38.9bn	
per capita	US$	3,566	3,322	
Growth	%	2.2	-7.9	
Inflation	%	1.8	1.7	
Def bdgt	B	3.29bn	3.28bn	
	US$	479.4m	478.7m	
US$1=B		6.86	6.86	

Real-terms defence budget trend (US$m, constant 2015)

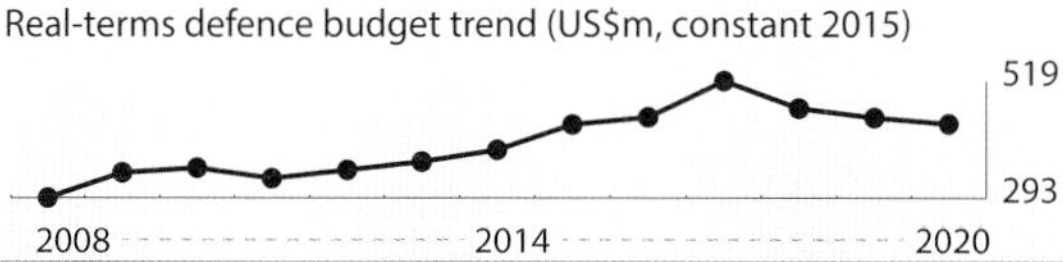

Population 11,639,909

Age	0–14	15–19	20–24	25–29	30–64	65 plus
Male	15.5%	5.1%	4.7%	4.4%	17.4%	2.5%
Female	14.9%	4.9%	4.6%	4.3%	18.6%	3.2%

Capabilities

The armed forces are constitutionally tasked with maintaining sovereignty and territorial defence, though principal tasks are counter-narcotics and internal and border security. Joint task forces have been formed and dispatched to border regions to combat smuggling activities, and a series of border posts are being established. Airspace control is an emerging strategic priority. There is defence-technology cooperation with Russia, but China remains a significant supplier of military materiel and delivered logistics vehicles in late 2018. Bolivia cooperates with Peru on countering illicit trafficking and disaster relief. The armed forces have stressed the need to improve conditions for personnel, amid greater internal deployments to border areas on counter-trafficking tasks. An Aerospace Research and Development Centre was created in 2018 in the military-engineering school with the objective of developing munitions and ISR UAVs. There is some local maintenance, repair and overhaul capacity.

ACTIVE 34,100 (Army 22,800 Navy 4,800 Air 6,500) **Paramilitary 37,100**

Conscript liability 12 months voluntary conscription for both males and females

ORGANISATIONS BY SERVICE

Army 9,800; 13,000 conscript (total 22,800)

FORCES BY ROLE

COMMAND

6 mil region HQ

10 div HQ

SPECIAL FORCES

3 SF regt

MANOEUVRE

Reconnaissance

1 mot cav gp

Armoured

1 armd bn

Mechanised

1 mech cav regt

2 mech inf regt

Light

1 (aslt) cav gp

5 (horsed) cav gp

3 mot inf regt

21 inf regt

Air Manoeuvre

2 AB regt (bn)

Other

1 (Presidential Guard) inf regt

COMBAT SUPPORT

6 arty regt (bn)

6 engr bn

1 int coy

1 MP bn

1 sigs bn

COMBAT SERVICE SUPPORT

2 log bn

AVIATION

2 avn coy

AIR DEFENCE

1 ADA regt

EQUIPMENT BY TYPE

ARMOURED FIGHTING VEHICLES

LT TK 54: 36 SK-105A1 *Kuerassier*; 18 SK-105A2 *Kuerassier*

RECCE 24 EE-9 *Cascavel*

APC 148+

APC (T) 87+: 50+ M113, 37 M9 half-track

APC (W) 61: 24 EE-11 *Urutu*; 22 MOWAG *Roland*; 15 V-100 *Commando*
AUV 19 *Tiger* 4×4
ENGINEERING & MAINTENANCE VEHICLES
ARV 4 *Greif*; M578 LARV
ANTI-TANK/ANTI-INFRASTRUCTURE
MSL
SP 2 *Koyak* with HJ-8
MANPATS HJ-8
RCL 90mm M67; **106mm** M40A1
ARTILLERY 311+
TOWED 61: **105mm** 25 M101A1; **122mm** 36 M-30 (M-1938)
MOR 250+: **81mm** 250 M29; Type-W87; **107mm** M30; **120mm** M120
AIRCRAFT
TPT • Light 3: 1 Fokker F-27-200; 1 Beech 90 *King Air*; 1 C-212 *Aviocar*
HELICOPTERS
MRH 6 H425
TRG 1 Robinson R44
AIR DEFENCE • GUNS • TOWED 37mm 18 Type-65

Navy 4,800

Organised into six naval districts with HQ located at Puerto Guayaramerín

EQUIPMENT BY TYPE
PATROL AND COASTAL COMBATANTS 9
PBR 9: 1 *Santa Cruz*; 6 Type 928 YC; 2 others
LOGISTICS AND SUPPORT 3
AG 1 *Mojo Huayna*
AH 2

Marines 1,700 (incl 1,000 Naval Military Police)

FORCES BY ROLE
MANOEUVRE
Mechanised
1 mech inf bn
Amphibious
6 mne bn (1 in each Naval District)
COMBAT SUPPORT
4 (naval) MP bn

Air Force 6,500 (incl conscripts)

FORCES BY ROLE
GROUND ATTACK
1 sqn with K-8WB *Karakorum*
ISR
1 sqn with Cessna 206; Cessna 402; Learjet 25B/25D (secondary VIP role)
SEARCH & RESCUE
1 sqn with AS332B *Super Puma*; H125 *Ecureuil*; H145
TRANSPORT
1 (TAM) sqn with B-727; B-737; MA60
1 (TAB) sqn with C-130A *Hercules*; MD-10-30F
1 sqn with C-130B/H *Hercules*
1 sqn with F-27-400M *Troopship*
1 (VIP) sqn with Beech 90 *King Air*; Beech 200 *King Air* Beech 1900; *Falcon* 900EX; *Sabreliner* 60; *Falcon* 50EX
6 sqn with Cessna 152/206; IAI-201 *Arava*; PA-32 *Saratoga*; PA-34 *Seneca*
TRAINING
1 sqn with DA40; T-25; Z-242L
1 sqn with Cessna 152/172
1 sqn with PC-7 *Turbo Trainer*
1 hel sqn with R-44 *Raven* II
TRANSPORT HELICOPTER
1 (anti-drug) sqn with Bell 205 (UH-1H *Iroquois*)
AIR DEFENCE
1 regt with Oerlikon; Type-65

EQUIPMENT BY TYPE
AIRCRAFT 22 combat capable
TPT 78: **Heavy** 1 MD-10-30F; **Medium** 4: 1 C-130A *Hercules*; 2 C-130B *Hercules*; 1 C-130H *Hercules*; **Light** 65: 1 *Aero Commander* 690; 3 Beech 90 *King Air*; 2 Beech 200 *King Air*; 1 Beech 250 *King Air*; 1 Beech 350 *King Air*; 3 C-212-100; 6 Cessna 152; 2 Cessna 172; 19 Cessna 206; 3 Cessna 210 *Centurion*; 1 Cessna 402; 8 DA40; 1 F-27-400M *Troopship*; 4 IAI-201 *Arava*; 2 Learjet 25B/D; 2 MA60†; 1 PA-32 *Saratoga*; 4 PA-34 *Seneca*; 1 *Sabreliner* 60; **PAX** 8: 1 B-727; 3 B-737-200; 1 *Falcon* 50EX; 1 *Falcon* 900EX (VIP); 2 RJ70
TRG 36: 6 K-8WB *Karakorum**; 6 T-25; 16 PC-7 *Turbo Trainer**; 8 Z-242L
HELICOPTERS
MRH 1 SA316 *Alouette* III
TPT 35: **Medium** 6 H215 *Super Puma*†; **Light** 29: 2 H125 *Ecureuil*; 19 Bell 205 (UH-1H *Iroquois*); 2 H145; 6 R-44 *Raven* II
AIR DEFENCE • GUNS 18+: **20mm** Oerlikon; **37mm** 18 Type-65

Paramilitary 37,100+

National Police 31,100+

FORCES BY ROLE
MANOEUVRE
Other
27 frontier sy unit
9 paramilitary bde
2 (rapid action) paramilitary regt

Narcotics Police 6,000+

FOE (700) – Special Operations Forces

DEPLOYMENT

CENTRAL AFRICAN REPUBLIC: UN • MINUSCA 5

DEMOCRATIC REPUBLIC OF THE CONGO: UN • MONUSCO 4

SOUTH SUDAN: UN • UNMISS 4

SUDAN: UN • UNISFA 3

Brazil BRZ

Brazilian Real R		2019	2020	2021
GDP	R	7.26tr	7.07tr	
	US$	1.84tr	1.36tr	
per capita	US$	8,751	6,450	
Growth	%	1.1	-5.8	
Inflation	%	3.7	2.7	
Def bdgt [a]	R	107bn	115bn	116bn
	US$	27.1bn	22.1bn	
US$1=R		3.95	5.18	

[a] Includes military pensions

Real-terms defence budget trend (US$bn, constant 2015)

Population 211,715,973

Age	0–14	15–19	20–24	25–29	30–64	65 plus
Male	10.8%	4.1%	4.1%	3.9%	22.5%	3.9%
Female	10.3%	3.9%	4.0%	3.8%	23.4%	5.3%

Capabilities

The armed forces are among the most capable in the region. Brazil seeks to enhance its power-projection capabilities, boost surveillance of the Amazon region and coastal waters, and further develop its defence industry. Security challenges from organised crime have seen the armed forces deploy on internal-security operations. Brazil maintains military ties with most of its neighbours including personnel exchanges and joint military training with Chile and Colombia. There is also defence cooperation with France, Sweden and the US, centred on procurement, technical advice and personnel training. Brazil's air-transport fleet enables it to independently deploy forces. It contributes small contingents to several UN missions. Brazil is attempting to modernise its equipment across all domains. Major platform developments include PROSUB (one nuclear-powered and four diesel-electric submarines) and the acquisition in 2018 of a former UK helicopter carrier. Projects to boost aerospace capabilities are also underway including the FX-2 project to procure the Saab *Gripen* combat aircraft, as well as the plan to introduce the Embraer KC-390 transport aircraft. Brazil has a well-developed defence-industrial base, across all domains, with a capability to design and manufacture equipment. Aerospace firms Avibras and Embraer also export some products. Local companies are also involved in the SISFRON border-security programme. There are industrial partnerships, including technology transfers and R&D support, with France's Naval Group (PROSUB) and Sweden's Saab (FX-2 fighter).

ACTIVE 366,500 (Army 214,000 Navy 85,000 Air 67,500) **Paramilitary 395,000**

Conscript liability 12 months (can go to 18; often waived)

RESERVE 1,340,000

ORGANISATIONS BY SERVICE

Space

EQUIPMENT BY TYPE

SATELLITES • COMMUNICATIONS 1 SGDC-1 (civil–military use)

Army 102,000; 112,000 conscript (total 214,000)

FORCES BY ROLE

COMMAND

8 mil comd HQ
12 mil region HQ
7 div HQ (2 with regional HQ)

SPECIAL FORCES

1 SF bde (1 SF bn, 1 cdo bn)
1 SF coy

MANOEUVRE

Reconnaissance

3 mech cav regt

Armoured

1 (5th) armd bde (1 mech cav sqn, 2 tk regt, 2 mech inf bn, 1 SP arty bn, 1 engr bn, 1 sigs coy, 1 log bn)
1 (6th) armd bde (1 mech cav sqn, 2 tk regt, 2 mech inf bn, 1 SP arty bn, 1 AD bty, 1 engr bn, 1 sigs coy, 1 log bn)

Mechanised

4 (1st, 3rd & 4th) mech cav bde (1 armd cav regt, 3 mech cav regt, 1 arty bn, 1 engr coy, 1 sigs coy, 1 log bn)
1 (2nd) mech cav bde (1 armd cav regt, 2 mech cav regt, 1 SP arty bn, 1 engr coy, 1 sigs coy, 1 log bn)
1 (15th) mech inf bde (3 mech inf bn, 1 arty bn, 1 engr coy, 1 log bn)

Light

1 (3rd) mot inf bde (1 mech cav sqn, 1 mech inf bn, 1 mot inf bn, 1 inf bn, 1 arty bn, 1 engr coy, 1 sigs coy, 1 log bn)
1 (4th) mot inf bde (1 mech cav sqn, 1 mot inf bn, 1 inf bn, 1 mtn inf bn, 1 arty bn, 1 sigs coy, 1 log bn)
1 (7th) mot inf bde (3 mot inf bn, 1 arty bn)
1 (8th) mot inf bde (1 mech cav sqn, 3 mot inf bn, 1 arty bn, 1 log bn)
1 (10th) mot inf bde (1 mech cav sqn, 4 mot inf bn, 1 inf coy, 1 arty bn, 1 engr coy, 1 sigs coy)
1 (13th) mot inf bde (1 mot inf bn, 2 inf bn, 1 inf coy, 1 arty bn)
1 (14th) mot inf bde (1 mech cav sqn, 3 inf bn, 1 arty bn)
1 (11th) lt inf bde (1 mech cav regt, 3 inf bn, 1 arty bn, 1 engr coy, 1 sigs coy, 1 MP coy, 1 log bn)
10 inf bn
1 (1st) jungle inf bde (1 mech cav sqn, 2 jungle inf bn, 1 arty bn)
4 (2nd, 16th, 17th & 22nd) jungle inf bde (3 jungle inf bn)
1 (23rd) jungle inf bde (1 cav sqn, 4 jungle inf bn, 1 arty bn, 1 sigs coy, 1 log bn)

Air Manoeuvre

1 AB bde (1 cav sqn, 3 AB bn, 1 arty bn, 1 engr coy, 1 sigs coy, 1 log bn)

1 (12th) air mob bde (1 cav sqn, 3 air mob bn, 1 arty bn, 1 engr coy, 1 sigs coy, 1 log bn)
Other
1 (9th) mot trg bde (3 mot inf bn, 1 arty bn, 1 log bn)
1 (18th) sy bde (2 sy bn, 2 sy coy)
1 sy bn
7 sy coy
3 gd cav regt
1 gd inf bn
COMBAT SUPPORT
3 SP arty bn
6 fd arty bn
1 MRL bn
1 STA bty
6 engr bn
1 engr gp (1 engr bn, 4 construction bn)
1 engr gp (4 construction bn, 1 construction coy)
2 construction bn
1 CBRN bn
1 EW coy
1 int coy
8 MP bn
2 MP coy
4 sigs bn
2 sigs coy
COMBAT SERVICE SUPPORT
5 log bn
1 tpt bn
4 spt bn
HELICOPTER
1 avn bde (3 hel bn, 1 maint bn)
1 hel bn
AIR DEFENCE
1 ADA bde (5 ADA bn)
EQUIPMENT BY TYPE
ARMOURED FIGHTING VEHICLES
MBT 296: 41 *Leopard* 1A1BE; 220 *Leopard* 1A5BR; 35 M60A3/TTS
LT TK 50 M41C
RECCE 408 EE-9 *Cascavel*
IFV 6 VBTP-MR *Guarani* 30mm
APC 1,333
APC (T) 660: 198 M113A1; 386 M113BR; 12 M113A2; 64 M577A2
APC (W) 673: 223 EE-11 *Urutu*; 450 VBTP-MR *Guarani* 6×6
ENGINEERING & MAINTENANCE VEHICLES
AEV 6+: *Greif*; 2 Sabiex HART; 4+ Pionierpanzer 2 *Dachs*
ARV 4+: BPz-2; 4 M88A1; M578 LARV
VLB 4+: XLP-10; 4 *Leopard* 1 with *Biber*
ANTI-TANK/ANTI-INFRASTRUCTURE
MSL • MANPATS *Eryx*; *Milan*; MSS-1.2 AC
RCL 194+: **84mm** *Carl Gustaf*; **106mm** 194 M40A1
ARTILLERY 1,881
SP 169: **155mm** 169: 37 M109A3; 100 M109A5; 32 M109A5+
TOWED 431: **105mm** 336: 233 M101/M102; 40 L118 Light Gun; 63 Model 56 pack howitzer; **155mm** 95 M114
MRL 127mm 36: 18 ASTROS II Mk3M; 18 ASTROS II Mk6
MOR 1,245: **81mm** 1,168: 453 L16, 715 M936 AGR; **120mm** 77 M2
HELICOPTERS
MRH 51: 22 AS565 *Panther* (HM-1); 12 AS565 K2 *Panther* (HM-1); 17 AS550A2 *Fennec* (HA-1 – armed)
TPT 38: **Heavy** 11 H225M *Caracal* (HM-4); **Medium** 12: 8 AS532 *Cougar* (HM-3); 4 S-70A-36 *Black Hawk* (HM-2); **Light** 15 AS350L1 *Ecureuil* (HA-1)
AIR DEFENCE
SAM • Point-defence RBS-70; 9K38 *Igla* (SA-18 *Grouse*); 9K338 *Igla*-S (SA-24 *Grinch*)
GUNS 100:
SP 35mm 34 *Gepard* 1A2
TOWED 66: **35mm** 39 GDF-001 towed (some with *Super Fledermaus* radar); **40mm** 27 L/70 (some with BOFI)

Navy 85,000

Organised into 9 districts with HQ I Rio de Janeiro, HQ II Salvador, HQ III Natal, HQ IV Belém, HQ V Rio Grande, HQ VI Ladario, HQ VII Brasilia, HQ VIII Sao Paulo, HQ IX Manaus

FORCES BY ROLE
SPECIAL FORCES
1 (diver) SF gp
EQUIPMENT BY TYPE
SUBMARINES • SSK 5:
2 *Tupi* (GER T-209/1400) (of which 1 in refit) with 8 single 533mm TT with Mk 24 *Tigerfish* HWT
2 *Tupi* (GER T-209/1400) with 8 single 533mm TT with Mk 48 HWT
1 *Tikuna* (GER T-209/1450) (in refit) with 8 single 533mm TT with Mk 24 *Tigerfish* HWT
PRINCIPAL SURFACE COMBATANTS 8
FRIGATES 8
FFGHM 7:
1 *Greenhalgh* (ex-UK *Broadsword*) with 4 single lnchr with MM38 *Exocet* AShM, 2 sextuple lnchr with *Sea Wolf* SAM, 6 single 324mm STWS Mk.2 ASTT with Mk 46 LWT (capacity 2 *Super Lynx* Mk21A hel)
1 *Greenhalgh* (ex-UK *Broadsword*) with 4 single lnchr with MM40 *Exocet* Block 2 AShM, 2 sextuple lnchr with *Sea Wolf* SAM, 6 single 324mm STWS Mk.2 ASTT with Mk 46 LWT (capacity 2 *Super Lynx* Mk21A hel)
5 *Niterói* with 2 twin lnchr with MM40 *Exocet* Block 2 AShM, 1 octuple *Albatros* lnchr with *Aspide* SAM, 2 triple 324mm SVTT Mk 32 ASTT with Mk 46 LWT, 1 twin 375mm Bofors ASW Rocket Launcher System A/S mor, 1 115mm gun (capacity 1 *Super Lynx* Mk21A hel)
FFGH 1 *Barroso* with 2 twin lnchr with MM40 *Exocet* Block 2 AShM, 2 triple 324mm SVTT Mk 32 ASTT with Mk 46 LWT, 1 115mm gun (capacity 1 *Super Lynx* Mk21A hel)
PATROL AND COASTAL COMBATANTS 44
CORVETTES • FSGH 1 *Inhaúma* with 2 twin lnchr with MM40 *Exocet* Block 2 AShM, 2 triple 324mm SVTT Mk

32 ASTT with Mk 46 LWT, 1 115mm gun (1 *Super Lynx* Mk21A hel)
PSO 3 *Amazonas* with 1 hel landing platform
PCO 6: 4 *Bracui* (ex-UK *River*); 1 *Imperial Marinheiro* with 1 76mm gun; 1 *Parnaiba* with 1 hel landing platform
PCC 2 *Macaé* (FRA Vigilante)
PCR 5: 2 *Pedro Teixeira* with 1 hel landing platform; 3 *Roraima*
PB 23: 12 *Grajau*; 6 *Marlim* (ITA *Meatini* derivative); 5 *Piratini* (US PGM)
PBR 4 LPR-40

MINE WARFARE • MINE COUNTERMEASURES 4
MSC 4 *Aratu* (GER *Schutze*)

AMPHIBIOUS
PRINCIPAL AMPHIBIOUS SHIPS 2
LPD 1 *Bahia* (ex-FRA *Foudre*) (capacity 4 hels; 8 LCM, 450 troops)
LPH 1 *Atlantico* (ex-UK *Ocean*) (capacity 18 hels; 4 LCVP; 40 vehs; 800 troops)
LANDING SHIPS 2
LST 1 *Mattoso Maia* (ex-US *Newport*) with 1 Mk 15 *Phalanx* CIWS (capacity 3 LCVP; 1 LCPL; 400 troops)
LSLH 1 *Almirante Saboia* (ex-UK *Sir Bedivere*) (capacity 1 med hel; 18 MBT; 340 troops)
LANDING CRAFT 16:
LCM 12: 10 EDVM-25; 2 *Icarai* (ex-FRA CTM)
LCT 1 *Marambaia* (ex-FRA CDIC)
LCU 3 *Guarapari* (LCU 1610)

LOGISTICS AND SUPPORT 41
ABU 5: 4 *Comandante Varella*; 1 *Faroleiro Mario Seixas*
ABUH 1 *Almirante Graca Aranah* (lighthouse tender)
AFS 1 *Potengi*
AGHS 5: 1 *Caravelas* (riverine); 4 *Rio Tocantin*
AGOS 2: 1 *Ary Rongel* with 1 hel landing platform; 1 *Almirante Maximiano* (capacity 2 AS350/AS355 *Ecureuil* hel)
AGS 8: 1 *Aspirante Moura*; 1 *Cruzeiro do Sul*; 1 *Antares*; 3 *Amorim do Valle* (ex-UK *River* (MCM)); 1 *Rio Branco*; 1 *Vital de Oliveira*
AGSH 1 *Sirius*
AH 5: 2 *Oswaldo Cruz* with 1 hel landing platform; 1 *Dr Montenegro*; 1 *Tenente Maximianol* with 1 hel landing platform; 1 *Soares de Meirelles*
AOR 1 *Almirante Gastão Motta*
AP 3: 1 *Almirante Leverger*; 1 *Paraguassu*; 1 *Pará* (all river transports)
ASR 2: 1 *Felinto Perry* (NOR *Wildrake*) with 1 hel landing platform; 1 *Guillobel*
ATF 2 *Tritao*
AX 1 *Brasil* (*Niterói* mod) with 1 hel landing platform
AXL 3 *Nascimento*
AXS 1 *Cisne Branco*

Naval Aviation 2,100

FORCES BY ROLE
GROUND ATTACK
1 sqn with A-4M (AF-1B) *Skyhawk*; TA-4M (AF-1C) *Skyhawk*
ANTI SURFACE WARFARE
1 sqn with *Super Lynx* Mk21A/B
ANTI SUBMARINE WARFARE
1 sqn with S-70B *Seahawk* (MH-16)
TRAINING
1 sqn with Bell 206B3 *Jet Ranger* III
TRANSPORT HELICOPTER
3 sqn with AS350 *Ecureuil* (armed)
1 sqn with AS350 *Ecureuil* (armed); AS355 *Ecureuil* II (armed); H135 (UH-17)
1 sqn with AS332 *Super Puma*; AS532 *Cougar* (UH-14)
1 sqn with H225M *Caracal* (UH-15)

EQUIPMENT BY TYPE
AIRCRAFT 3 combat capable
ATK 3: 2 A-4M (AF-1B) *Skyhawk*; 1 TA-4M (AF-1C) *Skyhawk* (15 A-4 (AF-1) *Skyhawk*; 2 TA-4 (AF-1A) *Skyhawk* in store)
HELICOPTERS
ASW 18: 9 *Super Lynx* Mk21A; 3 *Super Lynx* Mk21B; 6 S-70B *Seahawk* (MH-16)
CSAR 3 H225M *Caracal* (UH-15A)
TPT 54: **Heavy** 7 H225M *Caracal* (UH-15); **Medium** 7: 5 AS332 *Super Puma*; 2 AS532 *Cougar* (UH-14); **Light** 40: 15 AS350 *Ecureuil* (armed); 8 AS355 *Ecureuil* II (armed); 15 Bell 206B3 *Jet Ranger* III (IH-6B); 2 H135 (UH-17)
AIR-LAUNCHED MISSILES • AShM: AM39 *Exocet*; *Sea Skua*; AGM-119 *Penguin*

Marines 16,000

FORCES BY ROLE
SPECIAL FORCES
1 SF bn
MANOEUVRE
Amphibious
1 amph div (1 lt armd bn, 3 mne bn, 1 arty bn)
1 amph aslt bn
7 (regional) mne gp
1 rvn bn
COMBAT SUPPORT
1 engr bn
COMBAT SERVICE SUPPORT
1 log bn

EQUIPMENT BY TYPE
ARMOURED FIGHTING VEHICLES
LT TK 18 SK-105 *Kuerassier*
APC 60
APC (T) 30 M113A1 (incl variants)
APC (W) 30 *Piranha* IIIC
AAV 47: 13 AAV-7A1; 20 AAVP-7A1 RAM/RS; 2 AAVC-7A1 RAM/RS (CP); 12 LVTP-7
ENGINEERING VEHICLES • ARV 2: 1 AAVR-7; 1 AAVR-7A1 RAM/RS
ANTI-TANK/ANTI-INFRASTRUCTURE
MSL• MANPATS RB-56 *Bill*; MSS-1.2 AC
ARTILLERY 65
TOWED 41: **105mm** 33: 18 L118 Light Gun; 15 M101; **155mm** 8 M114
MRL 127mm 6 ASTROS II Mk6
MOR 81mm 18 M29
AIR DEFENCE • GUNS 40mm 6 L/70 (with BOFI)

Air Force 67,500

Brazilian airspace is divided into 7 air regions, each of which is responsible for its designated air bases. Air assets are divided among 4 designated air forces (I, II, III & V) for operations (IV Air Force temporarily deactivated)

FORCES BY ROLE

FIGHTER
4 sqn with F-5EM/FM *Tiger* II

FIGHTER/GROUND ATTACK
2 sqn with AMX (A-1A/B)

GROUND ATTACK/ISR
4 sqn with EMB-314 *Super Tucano* (A-29A/B)*

MARITIME PATROL
1 sqn with P-3AM *Orion*
2 sqn with EMB-111 (P-95A/B/M)

ISR
1 sqn with AMX-R (RA-1)*
1 sqn with Learjet 35 (R-35A); EMB-110B (R-95)

AIRBORNE EARLY WARNING & CONTROL
1 sqn with EMB-145RS (R-99); EMB-145SA (E-99)

SEARCH & RESCUE
1 sqn with C295M *Amazonas* (SC-105); UH-60L *Black Hawk* (H-60L)

TANKER/TRANSPORT
1 sqn with C-130H/KC-130H *Hercules*

TRANSPORT
1 VIP sqn with A319 (VC-1A); EMB-190 (VC-2); AS355 *Ecureuil* II (VH-55); H135M (VH-35); H225M *Caracal* (VH-36)
1 VIP sqn with EMB-135BJ (VC-99B); ERJ-135LR (VC-99C); ERJ-145LR (VC-99A); Learjet 35A (VU-35); Learjet 55C (VU-55C)
2 sqn with C-130E/H *Hercules*
2 sqn with C295M (C-105A)
7 (regional) sqn with Cessna 208/208B (C-98); Cessna 208-G1000 (C-98A); EMB-110 (C-95); EMB-120 (C-97)
1 sqn with ERJ-145 (C-99A)
1 sqn with EMB-120RT (VC-97), EMB-121 (VU-9)

TRAINING
1 sqn with EMB-110 (C-95)
2 sqn with EMB-312 *Tucano* (T-27) (incl 1 air show sqn)
1 sqn with T-25A/C

ATTACK HELICOPTER
1 sqn with Mi-35M *Hind* (AH-2)

TRANSPORT HELICOPTER
1 sqn with H225M *Caracal* (H-36)
1 sqn with AS350B *Ecureuil* (H-50); AS355 *Ecureuil* II (H-55)
1 sqn with Bell 205 (H-1H); H225M *Caracal* (H-36)
2 sqn with UH-60L *Black Hawk* (H-60L)

ISR UAV
1 sqn with *Hermes* 450/900

EQUIPMENT BY TYPE

AIRCRAFT 211 combat capable
FTR 49: 46 F-5EM *Tiger* II; 3 F-5FM *Tiger* II
FGA 48: 31 AMX (A-1); 8 AMX A-1M; 8 AMX-T (A-1B); 1 AMX A-1BM
ASW 9 P-3AM *Orion*
MP 18: 10 EMB-111 (P-95A *Bandeirulha*)*; 8 EMB-111 (P-95BM *Bandeirulha*)*
ISR: 8: 4 AMX-R (RA-1)*; 4 EMB-110B (R-95)
ELINT 6: 3 EMB-145RS (R-99); 3 Learjet 35A (R-35A)
AEW&C 5 EMB-145SA (E-99)
SAR 6: 2 C295M *Amazonas* (SC-105); 4 EMB-110 (SC-95B)
TKR/TPT 5: 2 KC-130H; 3 KC-390
TPT 181: **Medium** 20: 4 C-130E *Hercules*; 16 C-130H *Hercules*; **Light** 153: 11 C295M (C-105A); 7 Cessna 208 (C-98); 9 Cessna 208B (C-98); 13 Cessna 208-G1000 (C-98A); 52 EMB-110 (C-95A/B/C/M); 16 EMB-120 (C-97); 4 EMB-120RT (VC-97); 5 EMB-121 (VU-9); 7 EMB-135BJ (VC-99B); 3 EMB-201R *Ipanema* (G-19); 2 EMB-202A *Ipanema* (G-19A); 4 EMB-550 Legacy 500 (IU-50); 2 ERJ-135LR (VC-99C); 7 ERJ-145 (C-99A); 1 ERJ-145LR (VC-99A); 9 Learjet 35A (VU-35); 1 Learjet 55C (VU-55); **PAX** 8: 1 A319 (VC-1A); 3 EMB-190 (VC-2); 4 Hawker 800XP (EU-93A – calibration)
TRG 214: up to 50 EMB-312 *Tucano* (T-27); 39 EMB-314 *Super Tucano* (A-29A)*; 44 EMB-314 *Super Tucano* (A-29B)*; 81 T-25A/C

HELICOPTERS
ATK 12 Mi-35M *Hind* (AH-2)
MRH 2 H135M (VH-35)
TPT 57: **Heavy** 13 H225M *Caracal* (11 H-36 & 2 VH-36); **Medium** 16 UH-60L *Black Hawk* (H-60L); **Light** 28: 24 AS350B *Ecureuil* (H-50); 4 AS355 *Ecureuil* II (H-55/VH-55)

UNINHABITED AERIAL VEHICLES
ISR 5+: **Heavy** *Heron* 1; **Medium** 5: 4 *Hermes* 450; 1 *Hermes* 900

AIR-LAUNCHED MISSILES
AAM • **IR** MAA-1 *Piranha*; R-550 *Magic* 2; *Python* 3; **IIR** *Python* 4; **SARH** Super 530F; **ARH** *Derby*
AShM AM39 *Exocet*
ARM MAR-1 (in development)

Paramilitary 395,000 opcon Army

Public Security Forces 395,000

State police organisation technically under army control. However, military control is reducing, with authority reverting to individual states

DEPLOYMENT

CENTRAL AFRICAN REPUBLIC: EU • EUTM RCA 3; **UN** • MINUSCA 8

CYPRUS: UN • UNFICYP 3

DEMOCRATIC REPUBLIC OF THE CONGO: UN • MONUSCO 26

LEBANON: UN • UNIFIL 221; 1 FFGHM

SOUTH SUDAN: UN • UNMISS 11

SUDAN: UN • UNAMID 1; **UN** • UNISFA 3

WESTERN SAHARA: UN • MINURSO 4

Chile CHL

Chilean Peso pCh		2019	2020	2021
GDP	pCh	198tr	197tr	
	US$	282bn	245bn	
per capita	US$	14,772	12,612	
Growth	%	1.1	-6.0	
Inflation	%	2.3	2.9	
Def bdgt [a]	pCh	3.11tr	3.21tr	
	US$	4.42bn	4.00bn	
US$1=pCh		703.25	801.49	

[a] Includes military pensions

Real-terms defence budget trend (US$bn, constant 2015)

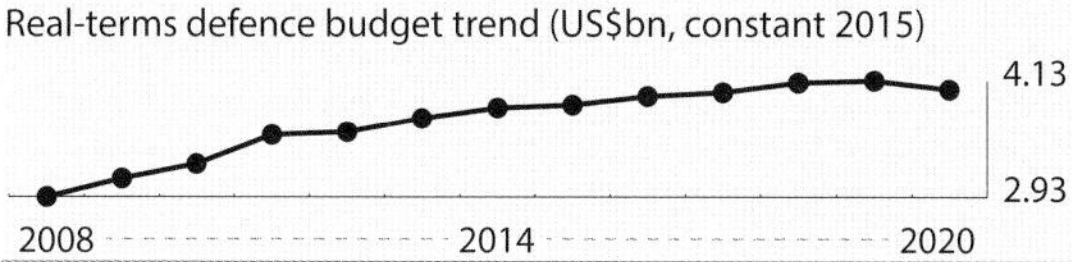

Population 18,186,770

Age	0–14	15–19	20–24	25–29	30–64	65 plus
Male	10.1%	3.3%	3.7%	4.0%	23.0%	5.0%
Female	9.7%	3.2%	3.6%	3.9%	23.6%	6.9%

Capabilities

Chile's 2017 defence white paper noted core roles of assuring sovereignty and territorial integrity, but also indicated an increasing shift towards non-traditional military roles such as disaster relief, humanitarian assistance and peacekeeping. Personnel-exchange programmes and R&D cooperation are under way with Brazil and Colombia. Defence cooperation with the US is centred on procurement, technical advisory and personnel training. There is routine national training, and the armed forces routinely participate in international exercises. Chile has a limited capacity to deploy independently beyond its borders. There are plans to upgrade Chile's F-16s to prolong service life, though these are understood to be dependent on available funding. Capability priorities reflect a focus on littoral and blue-water surveillance. Two frigates arrived in 2020, after purchase from Australia. Chile has a developed defence-industrial base, with ENAER conducting aircraft maintenance. ASMAR and FAMAE are key maritime and land firms respectively, with the former building a new icebreaker that will enhance Chile's ability to support operations in Antarctica.

ACTIVE 77,200 (Army 46,350 Navy 19,800 Air 11,050) Paramilitary 44,700

Conscript liability Army 12 months; Navy 18 months; Air Force 12 months. Legally, conscription can last for 2 years

RESERVE 40,000 (Army 40,000)

ORGANISATIONS BY SERVICE

Space

EQUIPMENT BY TYPE

SATELLITES

ISR 1 SSOT (Sistema Satelital de Observación de la Tierra)

Army 46,350

6 military administrative regions

FORCES BY ROLE

Currently being reorganised into 1 SF bde, 4 armd bde, 1 armd det, 4 mot bde, 2 mot det, 4 mtn det and 1 avn bde

COMMAND

6 div HQ

SPECIAL FORCES

1 SF bde (1 SF bn, 1 (mtn) SF gp, 1 para bn, 3 cdo coy, 1 log coy)

MANOEUVRE

Reconnaissance

4 cav sqn
2 recce sqn
2 recce pl

Armoured

1 (1st) armd bde (1 armd recce pl, 1 armd cav gp, 1 mech inf bn, 2 arty gp, 1 AT coy, 1 engr coy, 1 sigs coy)
2 (2nd & 3rd) armd bde (1 armd recce pl, 1 armd cav gp, 1 mech inf bn, 1 arty gp, 1 AT coy, 1 engr coy, 1 sigs coy)
1 (4th) armd bde (1 armd recce pl, 1 armd cav gp, 1 mech inf bn, 1 arty gp, 1 engr coy)
1 (5th) armd det (1 armd cav gp, 1 mech inf coy, 1 arty gp)

Mechanised

1 (1st) mech inf regt

Light

1 (1st) mot inf bde (1 recce coy, 1 mot inf bn, 1 arty gp, 3 AT coy, 1 engr bn)
1 (4th) mot inf bde (1 mot inf bn, 1 MRL gp, 2 AT coy, 1 engr bn)
1 (24th) mot inf bde (1 mot inf bn, 1 arty gp, 1 AT coy)
1 (Maipo) mot inf bde (3 mot inf regt, 1 arty regt)
1 (6th) reinforced regt (1 mot inf bn, 1 arty gp, 1 sigs coy)
1 (10th) reinforced regt (1 mot inf bn, 2 AT coy, 1 engr bn)
1 (11th) mot inf det (1 inf bn, 1 arty gp)
1 (14th) mot inf det (1 mot inf bn, 1 arty gp, 1 sigs coy, 1 AT coy)
4 mot inf regt
1 (3rd) mtn det (1 mtn inf bn, 1 arty gp, 1 engr coy)
1 (9th) mtn det (1 mtn inf bn, 1 engr coy, 1 construction bn)
2 (8th & 17th) mtn det (1 mtn inf bn, 1 arty coy)

COMBAT SUPPORT

1 engr regt
4 sigs bn
1 sigs coy
2 int regt
1 MP regt

COMBAT SERVICE SUPPORT

1 log div (2 log regt)
4 log regt
6 log coy
1 maint div (1 maint regt)

AVIATION

1 avn bde (1 tpt avn bn, 1 hel bn, 1 spt bn)

EQUIPMENT BY TYPE

ARMOURED FIGHTING VEHICLES

MBT 246: 115 *Leopard* 1; 131 *Leopard* 2A4

IFV 191: 173 *Marder* 1A3; 18 YPR-765 PRI
APC 548
APC (T) 369 M113A1/A2
APC (W) 179 *Piranha*
ENGINEERING & MAINTENANCE VEHICLES
AEV 9 Pionierpanzer 2 *Dachs*
ARV 35 BPz-2
VLB 16 *Biber*
MW 3+: *Bozena* 5; 3 *Leopard* 1
ANTI-TANK/ANTI-INFRASTRUCTURE
MSL • MANPATS *Spike*-LR; *Spike*-ER
RCL 84mm *Carl Gustaf*; **106mm** 213 M40A1
ARTILLERY 1,407
SP 155mm 48: 24 M109A3; 24 M109A5+
TOWED 240: **105mm** 192: 88 M101; 104 Model 56 pack howitzer; **155mm** 48 M-68
MRL 160mm 12 LAR-160
MOR 1,107: **81mm** 743: 303 ECIA L65/81; 175 FAMAE; 265 Soltam; **120mm** 293: 173 ECIA L65/120; 17 FAMAE; 93 M-65; **SP 120mm** 71: 35 FAMAE (on *Piranha* 6x6); 36 Soltam (on M113A2)
AIRCRAFT
TPT • Light 8: 2 C-212-300 *Aviocar*; 3 Cessna 208 *Caravan*; 3 CN235
HELICOPTERS
ISR 9 MD-530F *Lifter* (armed)
TPT 17: **Medium** 12: 8 AS532AL *Cougar*; 2 AS532ALe *Cougar*; 2 SA330 *Puma*; **Light** 5: 4 H125 *Ecureuil*; 1 AS355F *Ecureuil* II
AIR DEFENCE
SAM • Point-defence *Mistral*
GUNS 41:
SP 20mm 17 *Piranha*/TCM-20
TOWED 20mm 24 TCM-20

Navy 19,800

5 Naval Zones; 1st Naval Zone and main HQ at Valparaiso; 2nd Naval Zone at Talcahuano; 3rd Naval Zone at Punta Arenas; 4th Naval Zone at Iquique; 5th Naval Zone at Puerto Montt

FORCES BY ROLE
SPECIAL FORCES
1 (diver) SF comd

EQUIPMENT BY TYPE
SUBMARINES • SSK 4:
2 *O'Higgins* (*Scorpène*) with 6 single 533mm TT with SM39 *Exocet* Block 2 AShM/*Black Shark* HWT
2 *Thomson* (GER T-209/1400) (of which 1 in refit) with 8 single 533mm TT with SM39 *Exocet* Block 2 AShM/*Black Shark* HWT/SUT HWT
PRINCIPAL SURFACE COMBATANTS 8
FRIGATES • FFGHM 8:
2 *Almirante Cochrane* (ex-UK *Norfolk* Type-23) with 2 quad lnchr with RGM-84C *Harpoon* Block 1B AShM, 1 32-cell VLS with *Sea Wolf* SAM, 2 twin 324mm ASTT with Mk 46 mod 2 LWT, 1 114mm gun (capacity 1 AS-532SC *Cougar*) (MLU begun 2018)
1 *Almirante Cochrane* (ex-UK *Norfolk* Type-23) with 2 quad lnchr with RGM-84C *Harpoon* Block 1B AShM, 1 32-cell VLS with *Sea Ceptor* SAM, 2 twin 324mm ASTT with Mk 46 mod 2 LWT, 1 114mm gun (capacity 1 AS-532SC *Cougar*)
2 *Almirante Latorre* (ex-AUS *Adelaide*) with 1 Mk 13 GMLS with RGM-84L *Harpoon* Block II AShM/SM-2 Block IIIA SAM, 1 8-cell Mk 41 VLS with RIM-162B ESSM SAM, 2 triple 324mm SVTT Mk 32 ASTT with MU90 LWT, 1 76mm gun (capacity 2 AS532SC *Cougars*)
2 *Almirante Riveros* (ex-NLD *Karel Doorman*) with 2 quad lnchr with MM40 *Exocet* Block 3 AShM, 1 8-cell Mk 48 VLS with RIM-7P *Sea Sparrow* SAM, 4 single 324mm SVTT Mk 32 mod 9 ASTT with Mk 46 mod 5 HWT, 1 76mm gun (capacity 1 AS532SC *Cougar*)
1 *Almirante Williams* (ex-UK *Broadsword* Type-22) with 2 quad lnchr with RGM-84 *Harpoon* AShM, 2 8-cell VLS with *Barak*-1 SAM; 2 triple 324mm ASTT with Mk 46 LWT, 1 76mm gun (capacity 1 AS532SC *Cougar*)
PATROL AND COASTAL COMBATANTS 12
PSOH 4: 2 *Piloto Pardo*; 2 *Piloto Pardo* with 1 76mm gun (ice-strengthened hull)
PCG 3:
2 *Casma* (ISR *Sa'ar* 4) with 6 single lnchr with *Gabriel* I AShM, 2 76mm guns
1 *Casma* (ISR *Sa'ar* 4) with 4 single lnchr with *Gabriel* I AShM, 2 twin lnchr with MM40 *Exocet* AShM, 2 76mm guns
PCO 5 *Micalvi*
AMPHIBIOUS
PRINCIPAL AMPHIBIOUS SHIPS • LPD 1 *Sargento Aldea* (ex-FRA *Foudre*) with 3 twin *Simbad* lnchr with *Mistral* SAM (capacity 4 med hel; 1 LCT; 2 LCM; 22 tanks; 470 troops)
LANDING SHIPS 3
LSM 1 *Elicura*
LST 2 *Maipo* (FRA *Batral*) with 1 hel landing platform (capacity 7 tanks; 140 troops)
LANDING CRAFT 3
LCT 1 CDIC (for use in *Sargento Aldea*)
LCM 2 (for use in *Sargento Aldea*)
LOGISTICS AND SUPPORT 12
ABU 1 *George Slight Marshall* with 1 hel landing platform
AFD 3
AGOR 1 *Cabo de Hornos*
AGHS 1 *Micalvi*
AOR 2: 1 *Almirante Montt* (ex-US *Henry J Kaiser*) with 1 hel landing platform; 1 *Araucano*
AP 1 *Aguiles* (1 hel landing platform)
ATF 2 *Veritas*
AXS 1 *Esmeralda*

Naval Aviation 600

EQUIPMENT BY TYPE
AIRCRAFT 14 combat capable
ASW 4: 2 C295ASW *Persuader*; 2 P-3ACH *Orion*
MP 4: 1 C295MPA *Persuader*; 3 EMB-111 *Bandeirante**
ISR 7 P-68
TRG 7 PC-7 *Turbo Trainer**
HELICOPTERS
ASW 5 AS532SC *Cougar*
MRH 8 AS365 *Dauphin*

TPT 8: **Medium** 2 H215 (AS332L1) *Super Puma*; **Light** 6: 4 Bo-105S; 2 Bell 206 *Jet Ranger*
AIR-LAUNCHED MISSILES • AShM AM39 *Exocet*

Marines 3,600

FORCES BY ROLE
MANOEUVRE
Amphibious
1 amph bde (2 mne bn, 1 cbt spt bn, 1 log bn)
2 coastal def unit

EQUIPMENT BY TYPE
ARMOURED FIGHTING VEHICLES
LT TK 15 FV101 *Scorpion*
APC • APC (W) 25 MOWAG *Roland*
AAV 12 AAV-7
ARTILLERY 39
TOWED 23: **105mm** 7 KH-178; **155mm** 16 M-71
MOR 81mm 16
COASTAL DEFENCE • AShM MM38 *Exocet*
AIR DEFENCE • SAM • Point-defence 14: 4 M998 *Avenger*; 10 M1097 *Avenger*

Coast Guard

Integral part of the Navy

EQUIPMENT BY TYPE
PATROL AND COASTAL COMBATANTS 55
PBF 26 *Archangel*
PB 29: 18 *Alacalufe* (*Protector*); 4 *Grumete Diaz* (*Dabor*); 6 *Pelluhue*; 1 *Ona*

Air Force 11,050

FORCES BY ROLE
FIGHTER
1 sqn with F-5E/F *Tiger* III+
2 sqn with F-16AM/BM *Fighting Falcon*
FIGHTER/GROUND ATTACK
1 sqn with F-16C/D Block 50 *Fighting Falcon* (*Puma*)
ISR
1 (photo) flt with; DHC-6-300 *Twin Otter*; Learjet 35A
AIRBORNE EARLY WARNING
1 flt with B-707 *Phalcon*
TANKER/TRANSPORT
1 sqn with B-737-300; C-130B/H *Hercules*; KC-130R *Hercules*; KC-135 *Stratotanker*
TRANSPORT
3 sqn with Bell 205 (UH-1H *Iroquois*); C-212-200/300 *Aviocar*; Cessna O-2A; Cessna 525 *Citation* CJ1; DHC-6-100/300 *Twin Otter*; PA-28-236 *Dakota*; Bell 205 (UH-1H *Iroquois*)
1 VIP flt with B-767-300ER; B-737-500; Gulfstream IV
TRAINING
1 sqn with EMB-314 *Super Tucano**
1 sqn with Cirrus SR-22T; T-35A/B *Pillan*
TRANSPORT HELICOPTER
1 sqn with Bell 205 (UH-1H *Iroquois*); Bell 206B (trg); Bell 412 *Twin Huey*; S-70A *Black Hawk*
AIR DEFENCE
1 AD regt (5 AD sqn) with *Crotale*; NASAMS; *Mistral*; M163/M167 *Vulcan*; Oerlikon GDF-005

EQUIPMENT BY TYPE
AIRCRAFT 92 combat capable
FTR 48: 10 F-5E *Tigre* III+; 2 F-5F *Tigre* III+; 29 F-16AM *Fighting Falcon*; 7 F-16BM *Fighting Falcon*
FGA 10: 6 F-16C Block 50 *Fighting Falcon*; 4 F-16D Block 50 *Fighting Falcon*
ATK 16 C-101CC *Aviojet* (A-36 *Halcón*)
ISR 3 Cessna O-2A
AEW&C 1 B-707 *Phalcon*
TKR 5: 2 KC-130R *Hercules*: 3 KC-135 *Stratotanker*
TPT 33: **Medium** 2: 1 C-130B *Hercules*; 1 C-130H *Hercules*; **Light** 26: 2 C-212-200 *Aviocar*; 1 C-212-300 *Aviocar*; 4 Cessna 525 *Citation* CJ1; 3 DHC-6-100 *Twin Otter*; 7 DHC-6-300 *Twin Otter*; 2 Learjet 35A; 7 PA-28-236 *Dakota*; **PAX** 5: 1 B-737-300; 1 B-737-500 (VIP); 1 B-767-300ER (VIP); 2 Gulfstream IV (VIP)
TRG 53: 8 Cirrus SR-22T; 18 EMB-314 *Super Tucano**; 27 T-35A/B *Pillan*
HELICOPTERS
MRH 12 Bell 412EP *Twin Huey*
TPT 25: **Medium** 7: 1 S-70A *Black Hawk*; 6 S-70i (MH-60M) *Black Hawk*; **Light** 18: 13 Bell 205 (UH-1H *Iroquois*); 5 Bell 206B (trg)
UNINHABITED AERIAL VEHICLES
ISR • Medium 3 *Hermes* 900
AIR DEFENCE
SAM
Short-range 17: 5 *Crotale*; 12 NASAMS
Point-defence *Mistral* (including some *Mygale*/*Aspic*)
GUNS • TOWED 20mm M163/M167 *Vulcan*; **35mm** Oerlikon GDF-005
AIR-LAUNCHED MISSILES
AAM • IR AIM-9J/M *Sidewinder*; *Python* 3; *Shafrir*‡; **IIR** *Python* 4; **ARH** AIM-120C AMRAAM; *Derby*
ASM AGM-65G *Maverick*
BOMBS
Laser-guided *Paveway* II
INS/GPS guided JDAM

Paramilitary 44,700

Carabineros 44,700

Ministry of Interior; 15 zones, 36 districts, 179 *comisaria*

EQUIPMENT BY TYPE
ARMOURED FIGHTING VEHICLES
APC • APC (W) 20 MOWAG *Roland*
ARTILLERY • MOR 81mm
AIRCRAFT
TPT • Light 4: 1 Beech 200 *King Air*; 1 Cessna 208; 1 Cessna 550 *Citation* V; 1 PA-31T *Cheyenne* II
HELICOPTERS • TPT • Light 16: 5 AW109E *Power*; 1 AW139; 1 Bell 206 *Jet Ranger*; 2 BK-117; 5 Bo-105; 2 H135

DEPLOYMENT

BOSNIA-HERZEGOVINA: EU • EUFOR • *Operation Althea* 16

CYPRUS: UN • UNFICYP 12

INDIA/PAKISTAN: UN • UNMOGIP 2

MIDDLE EAST: UN • UNTSO 3

Colombia COL

Colombian Peso pC		2019	2020	2021
GDP	pC	1062tr	1008tr	
	US$	324bn	265bn	
per capita	US$	6,423	5,207	
Growth	%	3.3	-8.2	
Inflation	%	3.5	2.4	
Def bdgt [a]	pC	33.5tr	35.8tr	39.3tr
	US$	10.2bn	9.40bn	
FMA (US)	US$	38.5m	20.0m	20.0m
US$1=pC		3281.39	3806.40	

[a] Includes Defence and Security

Real-terms defence budget trend (US$bn, constant 2015)

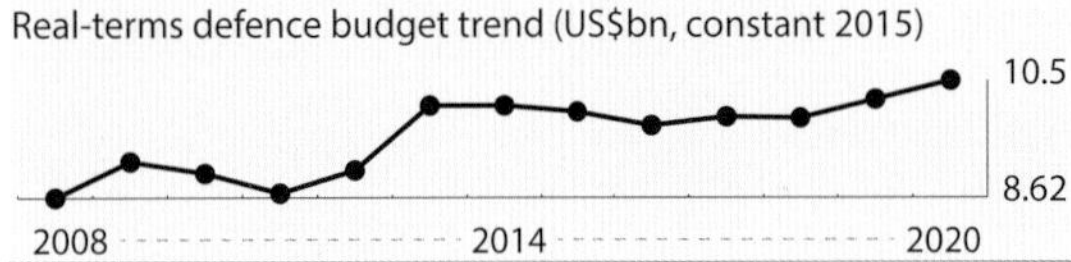

Population 49,084,841

Age	0–14	15–19	20–24	25–29	30–64	65 plus
Male	11.9%	4.1%	4.3%	4.3%	21.3%	3.5%
Female	11.3%	3.9%	4.1%	4.2%	22.2%	4.9%

Capabilities

Colombia's armed forces have improved their training and overall capabilities in recent decades. Internal security remains a priority, and the armed forces are focused on conducting counter-insurgency and counter-narcotics operations. While the armed forces are looking towards new security roles and organisations for the post-FARC era, the emergence of a FARC splinter group in mid-2019 means that internal security remains a key concern. In response to the humanitarian and security challenge from Venezuela, Colombia is strengthening cooperation with Brazil on border controls. There are military ties with Argentina, Chile and Peru. The US is Colombia's closest international military partner, with cooperation in equipment procurement, technical advice and personnel training. In May 2018, Colombia joined NATO as a global partner. The forces train regularly, including multilateral exercises. Although the equipment inventory mainly comprises legacy systems, Colombia has the capability to independently deploy its forces beyond national borders. The army is planning to modernise its armoured vehicles, while the navy has improved its offshore-patrol capacities in recent years. The air force's ground-attack capabilities remain limited, though a number of multi-role and transport helicopters have been procured. Colombia's defence industry is active in all domains. CIAC is developing its first indigenous UAVs, while CODALTEC is developing an air-defence system for regional export. COTECMAR has supplied patrol boats and amphibious ships for national and export markets.

ACTIVE 293,200 (Army 223,150, Navy 56,400 Air 13,650) Paramilitary 187,900

Conscript liability 18-months duration with upper age limit of 24, males only

RESERVE 34,950 (Army 25,050 Navy 6,500 Air 3,400)

ORGANISATIONS BY SERVICE

Army 223,150

FORCES BY ROLE

SPECIAL FORCES

1 SF div (3 SF regt)
1 (anti-terrorist) SF bn

MANOEUVRE

Mechanised

1 (1st) mech div (1 (2nd) mech bde (2 mech inf bn, 1 mtn inf bn, 1 engr bn, 1 MP bn, 1 spt bn, 2 Gaula anti-kidnap gp); 1 (10th) mech bde (1 (urban) spec ops bn, 1 armd recce bn, 1 mech cav bn, 1 mech inf bn, 1 mtn inf bn, 3 sy bn, 2 arty bn, 1 engr bn, 1 spt bn, 2 Gaula anti-kidnap gp))

Light

1 (2nd) inf div (1 (1st) inf bde (1 mech cav bn, 2 inf bn, 1 mtn inf bn, 1 sy bn, 1 arty bn, 1 spt bn, 1 Gaula anti-kidnap gp); 1 (5th) inf bde (3 inf bn, 1 jungle inf bn, 1 sy bn, 1 arty bn, 1 engr bn, 1 spt bn, 1 Gaula anti-kidnap gp); 1 (30th) inf bde (1 mech cav bn, 2 inf bn, 1 sy bn, 1 arty bn, 1 engr bn, 1 spt bn); 1 AD bn; 1 sy gp (1 (urban) spec ops bn, 4 COIN bn, 3 sy bn); 1 (rapid reaction) sy bde)

1 (3rd) inf div (1 (3rd) inf bde (2 inf bn, 1 mtn inf bn, 1 COIN bn, 1 arty bn, 1 engr bn, 1 cbt spt bn, 1 MP bn, 1 log bn, 1 Gaula anti-kidnap gp); 1 (23rd) inf bde (1 cav gp, 1 lt inf bn, 1 jungle inf bn, 1 spt bn, 1 log bn); 1 (29th) mtn bde (1 mtn inf bn, 1 inf bn, 2 COIN bn, 1 spt bn, 1 log bn); 1 lt cav bde (2 lt cav gp); 1 mtn inf bn; 2 (rapid reaction) sy bde)

1 (4th) inf div (1 (7th) air mob bde (1 (urban) spec ops bn, 2 air mob inf bn, 1 lt inf bn, 1 COIN bn, 1 engr bn, 1 spt bn, 1 log bn, 1 Gaula anti-kidnap gp); 1 (22nd) jungle bde (1 air mob inf bn, 1 lt inf bn, 1 jungle inf bn, 1 COIN bn, 1 spt bn, 1 log bn); 1 (31st) jungle bde (1 lt inf bn, 1 jungle inf bn))

1 (5th) inf div (1 (6th) lt inf bde (2 lt inf bn, 1 mtn inf bn, 3 COIN bn, 1 EOD bn, 2 spt bn, 1 Gaula anti-kidnap gp); 1 (8th) inf bde (1 inf bn, 1 mtn inf bn, 1 arty bn, 1 engr bn, 1 spt bn, 1 Gaula anti-kidnap gp); 1 (9th) inf bde (1 (urban) spec ops bn, 2 inf bn, 1 arty bn, 1 COIN bn, 1 sy bn, 1 spt bn, 1 Gaula anti-kidnap gp); 1 (13th) inf bde (1 recce bn, 3 inf bn, 1 mtn inf bn, 1 air mob bn, 1 COIN bn, 1 arty bn, 1 engr bn, 2 MP bn, 1 spt bn, 1 Gaula anti-kidnap gp))

1 (6th) inf div (1 (12th) inf bde (1 (urban) spec ops bn, 1 inf bn, 1 jungle inf bn, 1 mtn inf bn, 1 COIN bn, 1 engr bn, 1 spt bn, 1 Gaula anti-kidnap gp); 1 (26th) jungle bde (1 jungle inf bn, 1 spt bn); 1 (27th) jungle inf bde (1 inf bn, 1 jungle inf bn, 1 sy bn, 1 arty bn, 1 engr bn, 1 spt bn); 1 (13th) mobile sy bde; 2 COIN bn)

1 (7th) inf div (1 (4th) inf bde (1 (urban) spec ops bn; 1 mech cav gp, 3 inf bn, 1 sy bn, 1 arty bn, 1 engr bn, 1 MP bn, 1 spt bn, 2 Gaula anti-kidnap gp); 1 (11th) inf bde (1 inf bn, 1 air mob bn, 1 sy bn, 1 spt bn, 2 Gaula anti-kidnap gp); 1 (14th) inf bde (2 inf bn, 1 sy bn, 1 engr bn, 1 spt bn); 1 (15th) jungle bde (1 inf bn, 2 COIN bn, 1 engr bn); 1 (17th) inf bde (2 inf bn, 1 engr bn, 1 spt bn))

1 (8th) inf div (1 (16th) lt inf bde (1 recce bn, 1 inf bn, 1 spt bn, 1 Gaula anti-kidnap gp); 1 (18th) inf bde (1 air mob gp, 2 sy bn, 1 arty bn, 1 engr bn, 1 spt bn); 1 (28th) jungle bde (2 inf, 2 COIN, 1 spt bn); , 1 (rapid reaction) sy bde, 3 COIN bn)
3 COIN mobile bde (each: 4 COIN bn, 1 spt bn)

COMBAT SUPPORT
1 cbt engr bde (1 SF engr bn, 1 (emergency response) engr bn, 1 EOD bn, 1 construction bn, 1 demining bn, 1 maint bn)
1 int bde (2 SIGINT bn, 1 log bn, 1 maint bn)

COMBAT SERVICE SUPPORT
2 spt/log bde (each: 1 spt bn, 1 maint bn, 1 supply bn, 1 tpt bn, 1 medical bn, 1 log bn)

AVIATION
1 air aslt div (1 counter-narcotics bde (4 counter-narcotics bn, 1 spt bn); 1 (25th) avn bde (4 hel bn; 5 avn bn; 1 avn log bn); 1 (32nd) avn bde (1 avn bn, 2 maint bn, 1 trg bn, 1 spt bn); 1 SF avn bn)

EQUIPMENT BY TYPE

ARMOURED FIGHTING VEHICLES
RECCE 121 EE-9 *Cascavel*
IFV 60: 28 *Commando Advanced*; 32 LAV III
APC 114
APC (T) 54: 28 M113A1 (TPM-113A1); 26 M113A2 (TPM-113A2)
APC (W) 56 EE-11 *Urutu*
PPV 4+: some *Hunter* XL; 4 RG-31 *Nyala*
AUV 38 M1117 *Guardian*

ANTI-TANK/ANTI-INFRASTRUCTURE
MSL
SP 77 *Nimrod*
MANPATS TOW; *Spike*-ER
RCL 106mm 73 M40A1

ARTILLERY 1,796
TOWED 120: **105mm** 107: 22 LG1 MkIII; 85 M101; **155mm** 13 155/52 APU SBT-1
MOR 1,676: **81mm** 1,507; **120mm** 169

AIRCRAFT
ELINT 3: 2 Beech B200 *King Air*; 1 Beech 350 *King Air*
TPT • Light 23: 2 An-32B; 2 Beech B200 *King Air*; 3 Beech 350 *King Air*; 1 Beech C90 *King Air*; 2 C-212 *Aviocar* (Medevac); 8 Cessna 208B *Grand Caravan*; 1 Cessna 208B-EX *Grand Caravan*; 4 *Turbo Commander* 695A

HELICOPTERS
MRH 19: 8 Mi-17-1V *Hip*; 6 Mi-17MD; 5 Mi-17V-5 *Hip*
TPT 92: **Medium** 53: 46 UH-60L *Black Hawk*; 7 S-70i *Black Hawk*; **Light** 39: 24 Bell 205 (UH-1H *Iroquois*); 15 Bell 212 (UH-1N *Twin Huey*)

AIR DEFENCE • GUNS • TOWED 40mm 4 M1A1

Navy 56,400 (incl 12,100 conscript)

HQ located at Puerto Carreño

EQUIPMENT BY TYPE

SUBMARINES 4
SSK 2 *Pijao* (GER T-209/1200) each with 8 single 533mm TT each with *SeaHake* (DM2A3) HWT
SSC 2 *Intrépido* (GER T-206A) (of which 1 in refit) each with 8 single 533mm TT each with *SeaHake* (DM2A3) HWT

PATROL AND COASTAL COMBATANTS 65
CORVETTES 6
FSGHM 4 *Almirante Padilla* with 2 quad lnchr with *Hae Sung* I AShM, 2 twin *Simbad* lnchr with *Mistral* SAM, 2 triple 324mm ILAS-3 (B-515) ASTT each with A244/S LWT, 1 76mm gun (capacity 1 Bo-105/AS555SN *Fennec* hel)
FS 2:
1 *Almirante Tono* (ex-ROK *Po Hang* (Flight IV)) with 2 76mm
1 *Narino* (ex-ROK *Dong Hae*) with 2 triple 324mm SVTT Mk 32 ASTT with Mk 46 LWT
PSOH 3 *20 de Julio* (CHL *Piloto Pardo*)
PCO 2: 1 *San Andres* (ex-US *Balsam*); 1 *Valle del Cauca Durable* (ex-US *Reliance*) with 1 hel landing platform
PCC 3 *Punta Espada* (CPV-46)
PCR 10: 2 *Arauca* with 1 76mm guns; 8 *Nodriza* (PAF I-IV) with hel landing platform
PB 11: 1 *11 de Noviembre* (CPV-40) (GER Fassmer) with 1 *Typhoon* CIWS; 2 *Castillo y Rada* (Swiftships 105); 2 *Jaime Gomez* (ex-US Peterson Mk 3); 1 *José Maria Palas* (Swiftships 110); 3 *Point*; 2 *Toledo* (US Bender Marine 35m)
PBR 30: 5 *Diligente*; 7 LPR-40; 3 Swiftships; 9 *Tenerife* (US Bender Marine 12m); 2 PAF-L; 4 others

AMPHIBIOUS 15
LCT 5 *Golfo de Tribuga*
LCU 2 *Morrosquillo* (LCU 1466)
UCAC 8 *Griffon* 2000TD

LOGISTICS AND SUPPORT 9
ABU 1 *Quindio*
AG 2: 1 *Inirida*; 1 *Luneburg* (ex-GER, depot ship for patrol vessels)
AGHS 2: 1 *Caribe*; 1 *Roncador*
AGOR 2 *Providencia*
AGS 1 *Gorgona*
AXS 1 *Gloria*

Naval Aviation 150

EQUIPMENT BY TYPE

AIRCRAFT
MP 3 CN235 MPA *Persuader*
ISR 1 PA-31 *Navajo* (upgraded for ISR)
TPT • Light 11: 1 C-212 (Medevac); 4 Cessna 206; 3 Cessna 208 *Caravan*; 1 PA-31 *Navajo*; 1 PA-34 *Seneca*; 1 Beech 350 *King Air*

HELICOPTERS
SAR 2 AS365 *Dauphin*
MRH 8: 1 AS555SN *Fennec*; 3 Bell 412 *Twin Huey*; 4 Bell 412EP *Twin Huey*
TPT • Light 9: 1 Bell 212; 5 Bell 212 (UH-1N); 1 BK-117; 2 Bo-105

Marines 22,250

FORCES BY ROLE

SPECIAL FORCES
1 SF bde (4 SF bn)

MANOEUVRE
Amphibious
1 mne bde (1 SF (Gaula) bn, 5 mne bn, 2 rvn bn, 1 spt bn)

1 mne bde (1 SF bn, 2 mne bn, 2 rvn bn, 1 spt bn)
1 rvn bde (1 SF bn, 1 mne bn, 2 rvn bn, 1 spt bn)
1 rvn bde (4 rvn bn)
1 rvn bde (3 rvn bn)
COMBAT SERVICE SUPPORT
1 log bde (6 spt bn)
1 trg bde (7 trg bn, 1 spt bn)
EQUIPMENT BY TYPE
ARTILLERY • MOR 82: **81mm** 74; **120mm** 8
AIR DEFENCE • SAM Point-defence *Mistral*

Air Force 13,650

FORCES BY ROLE
FIGHTER/GROUND ATTACK
2 sqn with *Kfir* C-10/C-12/TC-12
GROUND ATTACK/ISR
1 sqn with A-37B/OA-37B *Dragonfly*
1 sqn with AC-47T
1 sqn with EMB-312 *Tucano**
2 sqn with EMB-314 *Super Tucano** (A-29)
EW/ELINT
2 sqn with Beech 350 *King Air;* Cessna 208; Cessna 560; C-26B *Metroliner*; SA 2-37; 1 *Turbo Commander* 695
TRANSPORT
1 (Presidential) sqn with B-737BBJ; EMB-600 *Legacy*; Bell 412EP; F-28 *Fellowship*; UH-60L *Black Hawk*
1 sqn with B-727; B-737-400; C-130B/H *Hercules*; C-212; C295M; CN235M; ; IAI *Arava*; KC-767
1 sqn with Beech C90 *King Air*; Beech 350C *King Air*; Cessna 208B; Cessna 550; EMB-110P1 (C-95)
TRAINING
1 sqn with Lancair *Synergy* (T-90 *Calima*)
1 sqn with T-37B
1 hel sqn with Bell 206B3
1 hel sqn with TH-67
HELICOPTER
1 sqn with AH-60L *Arpia* III
1 sqn with UH-60L *Black Hawk* (CSAR)
1 sqn with Hughes 500M
1 sqn with Bell 205 (UH-1H)
1 sqn with Bell 206B3 *Jet Ranger* III
1 sqn with Bell 212
EQUIPMENT BY TYPE
AIRCRAFT 72 combat capable
FGA 22: 10 *Kfir* C-10; 9 *Kfir* C-12; 3 *Kfir* TC-12
ATK 12: 6 A-37B/OA-37B *Dragonfly*; 6 AC-47T *Spooky (Fantasma)*
ISR 13: 1 Beech C90 *King Air*; 1 C-26B *Metroliner*; 5 Cessna 560 *Citation* II; 6 SA 2-37
ELINT 13: 4 Beech 350 *King Air*; 6 Cessna 208 *Grand Caravan*; 2 Cessna 337G; 1 *Turbo Commander* 695
TKR/TPT 1 KC-767
TPT 65: **Medium** 8: 3 C-130B *Hercules*; 4 C-130H *Hercules*; 1 B-737F; **Light** 49: 10 ATR-42; 2 Beech 300 *King Air*; 2 Beech 350C *King Air*; 1 Beech 350i *King Air* (VIP); 4 Beech C90 *King Air*; 4 C-212; 6 C295M; 1 Cessna 182R; 12 Cessna 208B (medevac); 1 Cessna 550; 2 CN235M; 2 EMB-110P1 (C-95); 1 EMB-170-100LR; 1 IAI-201 *Arava*; **PAX** 8: 2 B-727; 1 B-737-400; 1 B-737BBJ; 1 EMB-600 *Legacy*; 1 F-28-1000 *Fellowship*; 1 F-28-3000 *Fellowship*; 1 Learjet 60
TRG 78: 14 EMB-312 *Tucano**; 24 EMB-314 *Super Tucano* (A-29)*; 23 Lancair *Synergy* (T-90 *Calima*); 17 T-37B
HELICOPTERS
MRH 17: 6 AH-60L *Arpia* III; 8 AH-60L *Arpia* IV; 1 Bell 412EP *Twin Huey* (VIP); 2 Hughes 500M
TPT 48: **Medium** 13 UH-60L *Black Hawk* (incl 1 VIP hel); **Light** 35: 12 Bell 205 (UH-1H *Iroquois*); 12 Bell 206B3 *Jet Ranger* III; 11 Bell 212
TRG 30 TH-67
UNAMMED AERIAL VEHICLES • ISR • Medium 8: 6 *Hermes* 450; 2 *Hermes* 900
AIR-LAUNCHED MISSILES
AAM • IR *Python* 3; **IIR** *Python* 4; *Python* 5; **ARH** *Derby*; I-*Derby* ER (reported)
ASM *Spike*-ER; *Spike*-NLOS
BOMBS
Laser-guided *Paveway* II
INS/GPS guided *Spice*

Paramilitary 187,900

National Police Force 187,900

EQUIPMENT BY TYPE
AIRCRAFT
ELINT 5 C-26B *Metroliner*
TPT • Light 42: 5 ATR-42; 3 Beech 200 *King Air*; 2 Beech 300 *King Air*; 2 Beech 1900; 1 Beech C99; 4 BT-67; 2 C-26 *Metroliner*; 3 Cessna 152; 3 Cessna 172; 9 Cessna 206; 2 Cessna 208 *Caravan*; 2 DHC-6 *Twin Otter*; 1 DHC-8; 3 PA-31 *Navajo*
HELICOPTERS
MRH 4: 1 Bell 407GXP; 1 Bell 412EP; 2 MD-500D
TPT 80: **Medium** 22: 10 UH-60A *Black Hawk*; 9 UH-60L *Black Hawk*; 3 S-70i *Black Hawk*; **Light** 58: 34 Bell 205 (UH-1H-II *Huey II*); 6 Bell 206B; 5 Bell 206L/L3/L4 *Long Ranger*; 8 Bell 212; 5 Bell 407

DEPLOYMENT

CENTRAL AFRICAN REPUBLIC: UN • MINUSCA 2

EGYPT: MFO 275; 1 inf bn

LEBANON: UN • UNIFIL 1

FOREIGN FORCES

United States US Southern Command: 50

Costa Rica CRI

Costa Rican Colon C		2019	2020	2021
GDP	C	36.3tr	34.6tr	
	US$	62.1bn	59.6bn	
per capita	US$	12,244	11,629	
Growth	%	2.1	-5.5	
Inflation	%	2.1	0.8	
Sy Bdgt [a]	C	259bn	267bn	267bn
	US$	443m	461m	
FMA (US)	US$	7.5m	0m	0m
US$1=C		584.08	580.07	

[a] Paramilitary budget

Real-terms defence budget trend (US$m, constant 2015)

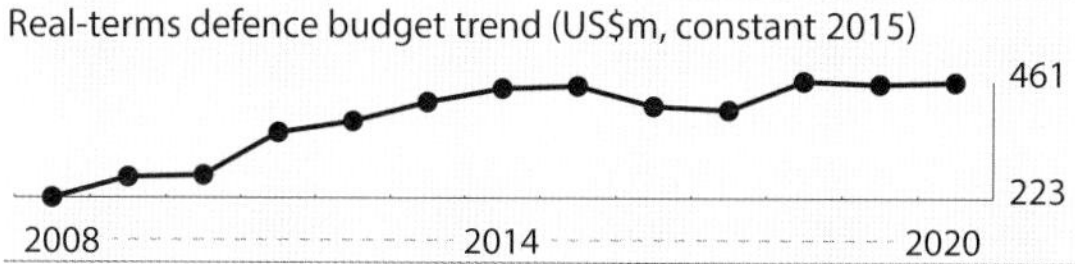

Population 5,097,988

Age	0–14	15–19	20–24	25–29	30–64	65 plus
Male	11.3%	3.7%	4.0%	4.2%	22.8%	4.0%
Female	10.8%	3.6%	3.9%	4.1%	22.9%	4.7%

Capabilities

Costa Rica's armed forces were constitutionally abolished in 1949, and the country relies on police and coast-guard organisations for internal-security, maritime and air domain-awareness and counter-narcotics tasks. A new National Security Strategy was adopted in 2018 in order to help tackle rising crime. Colombia and the US have provided assistance and training, focused on policing and internal-security tasks rather than conventional military operations. The Special Intervention Unit (UEI) has received specialist training from non-regional states, including the US. The Public Force, Coast Guard and Air Surveillance units have little heavy equipment, and recent modernisation has depended on donations from countries such as China and the US. Apart from limited maintenance facilities, Costa Rica has no domestic defence industry.

Paramilitary 9,950

ORGANISATIONS BY SERVICE

Paramilitary 9,950

Special Intervention Unit

FORCES BY ROLE

SPECIAL FORCES

1 spec ops unit

Public Force 9,000

11 regional directorates

Coast Guard Unit 550

EQUIPMENT BY TYPE

PATROL AND COASTAL COMBATANTS 10

PB 10: 2 *Cabo Blanco* (US *Swift* 65); 1 *Isla del Coco* (US *Swift* 105); 3 *Libertador Juan Rafael Mora* (ex-US *Island*); 2 *Point*; 1 *Primera Dama* (US *Swift* 42); 1 *Puerto Quepos* (US *Swift* 36)

Air Surveillance Unit 400

EQUIPMENT BY TYPE

AIRCRAFT • TPT • Light 14: 2 Cessna T210 *Centurion*; 4 Cessna U206G *Stationair*; 2 PA-31 *Navajo*; 2 PA-34 *Seneca*; 1 Piper PA-23 *Aztec*; 1 Cessna 182RG; 2 Y-12E

HELICOPTERS

MRH 3: 1 MD-500E; 2 MD-600N

TPT • Light 4 Bell 212 (UH-1N)

Cuba CUB

Cuban Peso P		2019	2020	2021
GDP	US$			
per capita	US$			
Growth				
Inflation				
Def exp	P			
	US$			
US$1=P				

Population 11,059,062

*definitive data not available

Age	0–14	15–19	20–24	25–29	30–64	65 plus
Male	8.4%	3.0%	3.2%	3.2%	24.8%	7.2%
Female	7.9%	2.8%	2.9%	3.0%	25.1%	8.6%

Capabilities

Cuba's armed forces are principally focused on protecting territorial integrity, and rely on a mass-mobilisation system. Military capability is limited by equipment obsolescence and a largely conscript-based force. Cuba maintains military ties with China and Russia, and the latter has stepped in to supply oil and fuel following Venezuela's economic collapse. Defence cooperation with Russia is largely centred around technical and maintenance support. Cooperation with China appears to be on a smaller scale and involves training agreements and personnel exchanges. Training levels are uncertain and flying hours are likely to be low due to the limited availability of serviceable aircraft. The armed forces are no longer designed for expeditionary operations and have little logistical capability to support operational deployments abroad. The inventory is almost entirely composed of legacy Soviet-era systems with varying degrees of obsolescence. Serviceability appears a problem, with much equipment at a low level of availability and maintenance demands growing as fleets age. Much of the aviation fleet is reported to be in storage. It is unlikely that Havana will be in a position to finance significant equipment recapitalisation in the near term. Cuba has little in the way of domestic defence industry, bar some upgrade and maintenance capacity. In the past, Cuba sent maintenance staff to South Africa, highlighting not just revenue-raising requirements for the forces but also the potential effect such activities might have on remaining maintenance capacity in Cuba.

ACTIVE 49,000 (Army 38,000 Navy 3,000 Air 8,000) Paramilitary 26,500

Conscript liability 2 years

RESERVE 39,000 (Army 39,000) Paramilitary 1,120,000

Ready Reserves (serve 45 days per year) to fill out Active and Reserve units; see also Paramilitary

ORGANISATIONS BY SERVICE

Army ε38,000

FORCES BY ROLE

COMMAND
3 regional comd HQ
3 army comd HQ

COMMAND
3 SF regt

MANOEUVRE

Armoured
1 tk div (3 tk bde)

Mechanised
2 (mixed) mech bde

Light
2 (frontier) bde

Air Manoeuvre
1 AB bde

AIR DEFENCE
1 ADA regt
1 SAM bde

Reserves 39,000

FORCES BY ROLE

MANOEUVRE

Light
14 inf bde

EQUIPMENT BY TYPE†

ARMOURED FIGHTING VEHICLES
MBT ε900 T-34/T-54/T-55/T-62
LT TK PT-76
ASLT BTR-60 100mm
RECCE BRDM-2;
AIFV ε50 BMP-1/1P
APC ε500 BTR-152/BTR-50/BTR-60

ANTI-TANK/ANTI-INFRASTRUCTURE
MSL
SP 2K16 *Shmel* (AT-1 *Snapper*)
MANPATS 9K11 *Malyutka* (AT-3 *Sagger*)
GUNS 600+: **57mm** 600 ZIS-2 (M-1943); **85mm** D-44

ARTILLERY 1,715+
SP 40+: **100mm** AAPMP-100; CATAP-100; **122mm** 2S1 *Gvozdika*; AAP-T-122; AAP-BMP-122; *Jupiter* III; *Jupiter* IV; **130mm** AAP-T-130; *Jupiter* V; **152mm** 2S3 *Akatsiya*
TOWED 500: **122mm** D-30; M-30 (M-1938); **130mm** M-46; **152mm** D-1; M-1937 (ML-20)
MRL • SP 175: **122mm** BM-21 *Grad*; **140mm** BM-14
MOR 1,000: **82mm** M-41; **82mm** M-43; **120mm** M-43; M-38

AIR DEFENCE
SAM
Short-range 2K12 *Kub* (SA-6 *Gainful*)
Pont-defence 200+: 200 9K35 *Strela*-10 (SA-13 *Gopher*); 9K33 *Osa* (SA-8 *Gecko*); 9K31 *Strela*-1 (SA-9 *Gaskin*); 9K36 *Strela*-3 (SA-14 *Gremlin*); 9K310 *Igla*-1 (SA-16 *Gimlet*); 9K32 *Strela*-2 (SA-7 *Grail*)‡
GUNS 400
SP 23mm ZSU-23-4; **30mm** BTR-60P SP; **57mm** ZSU-57-2
TOWED 100mm KS-19/M-1939/**85mm** KS-12/**57mm** S-60/**37mm** M-1939/**30mm** M-53/**23mm** ZU-23

Navy ε3,000

Western Comd HQ at Cabanas; Eastern Comd HQ at Holquin

EQUIPMENT BY TYPE

SUBMARINES • **SSW** 1 *Delfin*

PATROL AND COASTAL COMBATANTS 9
PCG 2 *Rio Damuji* with two single P-22 (SS-N-2C *Styx*) AShM, 2 57mm guns, 1 hel landing platform
PCM 1 Project 1241PE (FSU *Pauk* II) with 1 quad lnchr (manual aiming) with 9K32 *Strela*-2 (SA-N-5 *Grail*) SAM, 2 RBU 1200 A/S mor, 1 76mm gun
PBF 6 Project 205 (FSU *Osa* II)† each with 4 single lnchr (for P-20U (SS-N-2B *Styx*) AShM – missiles removed to coastal-defence units)

MINE WARFARE AND MINE COUNTERMEASURES 5
MHI 3 *Korund* (Project 1258 (*Yevgenya*))†
MSC 2 *Yakhont* (FSU Project 1265 (*Sonya*))†

LOGISTICS AND SUPPORT 2
ABU 1
AX 1

Coastal Defence

EQUIPMENT BY TYPE

ARTILLERY • **TOWED 122mm** M-1931/37; **130mm** M-46; **152mm** M-1937
COASTAL DEFENCE • **AShM** 4+: *Bandera* IV (reported); 4 4K51 *Rubezh* (SSC-3 *Styx*)

Naval Infantry 550+

FORCES BY ROLE

MANOEUVRE

Amphibious
2 amph aslt bn

Anti-aircraft Defence and Revolutionary Air Force ε8,000 (incl conscripts)

Air assets divided between Western Air Zone and Eastern Air Zone

FORCES BY ROLE

FIGHTER/GROUND ATTACK
3 sqn with MiG-21bis/UM *Fishbed*; MiG-23ML/UB/UM *Flogger*; MiG-29/MiG-29UB *Fulcrum*

TRANSPORT
1 (VIP) tpt sqn with An-24 *Coke*; Mi-8P *Hip*

ATTACK HELICOPTER
2 sqn with Mi-17 *Hip* H; Mi-35 *Hind*

TRAINING
2 (tac trg) sqn with L-39C *Albatros* (basic); Z-142 (primary)

EQUIPMENT BY TYPE

AIRCRAFT 31 combat capable
FTR 26: 16 MiG-23ML *Flogger*; 1+ MiG-23UB *Flogger*; 4 MiG-23UM *Flogger*; 2 MiG-29 *Fulcrum*†; 3 MiG-29UB *Fulcrum*†; (6 MiG-23ML *Flogger*; 2 MiG-23UM *Flogger* in store)
FGA 5: 3 MiG-21bis *Fishbed*; 2 MiG-21UM *Fishbed*
ISR 1 An-30 *Clank*†
TPT 25: **Heavy** 2 Il-76 *Candid*; **Light** 11: 1 An-24 *Coke* (Aerogaviota); 3 An-26 *Curl* (Aerogaviota); 5 ATR-42-500 (Cubana & Aergaviota); 2 ATR-72-500 (Cubana); **PAX** 12: 6 An-158 (Cubana); 3 Il-96-300 (Cubana); 3 Tu-204E-100 (Cubana)
TRG 25+: up to 25 L-39 *Albatros*; some Z-142C

HELICOPTERS
ATK 4 Mi-35 *Hind*† (8 more in store)
MRH 8 Mi-17 *Hip* H (12 more in store)
TPT • **Medium** 2 Mi-8P *Hip*

AIR DEFENCE • SAM
Medium-range S-75 *Dvina* (SA-2 *Guideline*); S-75 *Dvina* mod (SA-2 *Guideline* – on T-55 chassis)
Short-range S-125 *Pechora* (SA-3 *Goa*); S-125 *Pechora* mod (SA-3 *Goa* – on T-55 chassis)

AIR-LAUNCHED MISSILES
AAM • **IR** R-3‡ (AA-2 *Atoll*); R-60 (AA-8 *Aphid*); R-73 (AA-11A *Archer*); **IR/SARH** R-23/24‡ (AA-7 *Apex*); R-27 (AA-10 *Alamo*)
ASM Kh-23‡ (AS-7 *Kerry*)

Paramilitary 26,500 active

State Security 20,000

Ministry of Interior

Border Guards 6,500

Ministry of Interior

PATROL AND COASTAL COMBATANTS 20
PCC 2 *Stenka*
PB 18 *Zhuk*

Youth Labour Army 70,000 reservists

Civil Defence Force 50,000 reservists

Territorial Militia ε1,000,000 reservists

FOREIGN FORCES

United States US Southern Command: 800 (JTF-GTMO) at Guantanamo Bay

Dominican Republic DOM

Dominican Peso pRD		2019	2020	2021
GDP	pRD	4.56tr	4.43tr	
	US$	89.0bn	77.9bn	
per capita	US$	8,596	7,445	
Growth	%	5.1	-6.0	
Inflation	%	1.8	3.3	
Def bdgt	pRD	31.6bn	33.3bn	33.2bn
	US$	616m	585m	
US$1=pRD		51.24	56.90	

Real-terms defence budget trend (USm, constant 2015)

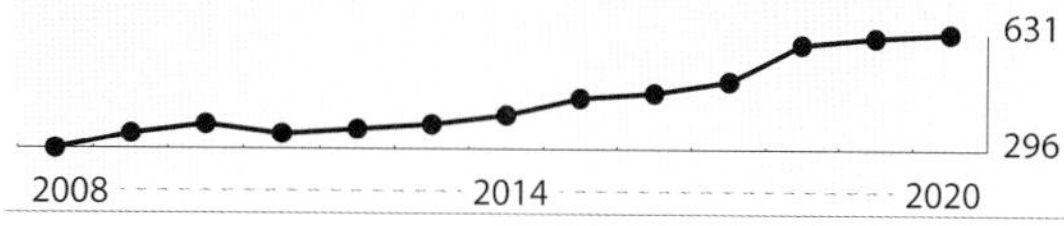

Population 10,499,707

Age	0–14	15–19	20–24	25–29	30–64	65 plus
Male	13.6%	4.7%	4.6%	4.3%	20.4%	3.0%
Female	13.2%	4.5%	4.4%	4.2%	19.8%	3.3%

Capabilities

The principal tasks for the Dominican armed forces include internal- and border-security missions, as well as disaster relief. Training and operations increasingly focus on counter-narcotics and include collaboration with the police in an inter-agency task force. The US sends training teams to the country under the terms of a 2015 military-partnership agreement, and the navy has trained with French forces. The Dominican Republic has participated in US SOUTHCOM's *Tradewinds* disaster-response exercise. The army has strengthened its presence along the border with Haiti, establishing new surveillance posts. There is little capacity to deploy and sustain forces abroad. The army's limited number of armoured vehicles are obsolete and likely difficult to maintain. The air force operates a modest number of light fixed- and rotary-wing assets, and the navy a small fleet of mainly ex-US patrol craft of varying size. Aside from maintenance facilities, the country does not have a domestic defence industry.

ACTIVE 56,050 (Army 28,750 Navy 11,200 Air 16,100) Paramilitary 15,000

ORGANISATIONS BY SERVICE

Army 28,750

5 Defence Zones

FORCES BY ROLE

SPECIAL FORCES
3 SF bn

MANOEUVRE
Light
4 (1st, 2nd, 3rd & 4th) inf bde (3 inf bn)
2 (5th & 6th) inf bde (2 inf bn)
Air Manoeuvre
1 air cav bde (1 cdo bn, 1 (6th) mtn bn, 1 hel sqn with Bell 205 (op by Air Force); OH-58 *Kiowa*; R-22; R-44 *Raven* II)

Other
1 (Presidential Guard) gd regt
1 (MoD) sy bn
COMBAT SUPPORT
1 cbt spt bde (1 lt armd bn; 1 arty bn; 1 engr bn; 1 sigs bn)
EQUIPMENT BY TYPE
ARMOURED FIGHTING VEHICLES
LT TK 12 M41B (76mm)
APC • APC (W) 8 LAV-150 *Commando*
ANTI-TANK/ANTI-INFRASTRUCTURE
RCL 106mm 20 M40A1
GUNS 37mm 20 M3
ARTILLERY 104
TOWED 105mm 16: 4 M101; 12 *Reinosa* 105/26
MOR 88: **81mm** 60 M1; **107mm** 4 M30; **120mm** 24 Expal Model L
HELICOPTERS
ISR 8: 4 OH-58A *Kiowa*; 4 OH-58C *Kiowa*
TPT • Light 6: 4 R-22; 2 R-44 *Raven* II

Navy 11,200
HQ located at Santo Domingo
FORCES BY ROLE
SPECIAL FORCES
1 (SEAL) SF unit
MANOEUVRE
Amphibious
1 mne sy unit
EQUIPMENT BY TYPE
PATROL AND COASTAL COMBATANTS 18
PCO 1 *Almirante Didiez Burgos* (ex-US *Balsam*)
PCC 2 *Tortuguero* (ex-US *White Sumac*)
PB 15: 2 *Altair* (Swiftships 35m); 4 *Bellatrix* (US Sewart Seacraft); 1 *Betelgeuse* (Damen Stan Patrol 2606); 2 *Canopus* (Swiftships 110); 3 *Hamal* (Damen Stan Patrol 1505); 3 *Point*
AMPHIBIOUS • LANDING CRAFT
LCU 1 *Neyba* (ex-US LCU 1675)
LOGISTICS AND SUPPORT 1
AX 1 *Almirante Juan Bautista Cambiaso*

Air Force 16,100
FORCES BY ROLE
GROUND ATTACK
1 sqn with EMB-314 *Super Tucano**
SEARCH & RESCUE
1 sqn with Bell 205 (UH-1H *Huey II*); Bell 205 (UH-1H *Iroquois*); Bell 430 (VIP); OH-58 *Kiowa* (CH-136); S-333
TRANSPORT
1 sqn with C-212-400 *Aviocar*; PA-31 *Navajo*
TRAINING
1 sqn with T-35B *Pillan*
AIR DEFENCE
1 ADA bn with 20mm guns
EQUIPMENT BY TYPE
AIRCRAFT 8 combat capable
ISR 1 AMT-200 *Super Ximango*
TPT • Light 13: 3 C-212-400 *Aviocar*; 1 Cessna 172; 1 Cessna 182; 1 Cessna 206; 1 Cessna 207; 1 *Commander* 690; 3 EA-100; 1 PA-31 *Navajo*; 1 P2006T
TRG 12: 8 EMB-314 *Super Tucano**; 4 T-35B *Pillan*
HELICOPTERS
ISR 9 OH-58 *Kiowa* (CH-136)
TPT • Light 16: 8 Bell 205 (UH-1H *Huey* II); 5 Bell 205 (UH-1H *Iroquois*); 1 H155 (VIP); 2 S-333
AIR DEFENCE • GUNS 20mm 4

Paramilitary 15,000

National Police 15,000

Ecuador ECU

United States Dollar $		2019	2020	2021
GDP	US$	107bn	93.1bn	
per capita	US$	6,222	5,316	
Growth	%	0.1	-11.0	
Inflation	%	0.3	n.k.	
Def bdgt	US$	1.59bn	1.55bn	

Real-terms defence budget trend (US$bn, constant 2015)

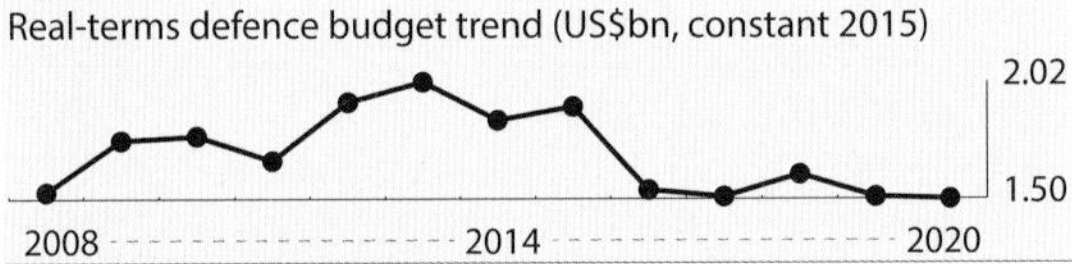

Population	16,904,867

Age	0–14	15–19	20–24	25–29	30–64	65 plus
Male	13.2%	4.5%	4.5%	4.2%	19.3%	3.8%
Female	12.6%	4.3%	4.4%	4.2%	20.5%	4.3%

Capabilities

Ecuador's armed forces are able to fulfil internal-security tasks, although the crisis in Venezuela and resulting refugee flows have added to existing security challenges in the northern border area. Due to the security conditions along the northern border area, the armed forces created a joint task force for counter-insurgency and counter-narcotics operations. Defence cooperation with Peru includes demining efforts on the border. Military ties with Washington have been revived, and Ecuador's navy hosted the 2020 iteration of the *Unitas* exercise. The armed forces train regularly and have participated in large regional and international military exercises. There is limited capability to independently deploy beyond its borders. The equipment inventory is derived from a variety of sources, though obsolescence and low availability is a challenge. Modernisation plans are modest in scope and currently focused on aviation as well as maritime-patrol capabilities. Ecuador's defence industries are centred on the army's Office of Industries (DINE), which produces military equipment through army-run enterprises. The state-owned shipyard ASTINAVE has some construction, maintenance and repair capabilities, although the navy's submarines are being modernised in Chile.

ACTIVE 41,250 (Army 25,650 Navy 9,400 Air 6,200)
Paramilitary 500
Conscript liability Voluntary conscription

RESERVE 118,000 (Joint 118,000)
Ages 18–55

ORGANISATIONS BY SERVICE

Army 25,650

FORCES BY ROLE
gp are bn sized
COMMAND
4 div HQ
SPECIAL FORCES
1 (9th) SF bde (3 SF gp, 1 SF sqn, 1 para bn, 1 sigs sqn, 1 log comd)
MANOEUVRE
Mechanised
1 (11th) armd cav bde (3 armd cav gp, 1 mech inf bn, 1 SP arty gp, 1 engr gp)
1 (5th) inf bde (1 SF sqn, 2 mech cav gp, 2 inf bn, 1 cbt engr coy, 1 sigs coy, 1 log coy)
Light
1 (1st) inf bde (1 SF sqn, 1 armd cav gp, 1 armd recce sqn, 3 inf bn, 1 med coy)
1 (3rd) inf bde (1 SF gp, 1 mech cav gp, 1 inf bn, 1 arty gp, 1 hvy mor coy, 1 cbt engr coy, 1 sigs coy, 1 log coy)
1 (7th) inf bde (1 SF sqn, 1 armd recce sqn, 1 mech cav gp, 3 inf bn, 1 jungle bn, 1 arty gp, 1 cbt engr coy, 1 sigs coy, 1 log coy, 1 med coy)
1 (13th) inf bde (1 SF sqn, 1 armd recce sqn, 1 mot cav gp, 3 inf bn, 1 arty gp, 1 hvy mor coy, 1 cbt engr coy, 1sigs coy, 1 log coy)
2 (17th & 21st) jungle bde (3 jungle bn, 1 cbt engr coy, 1 sigs coy, 1 log coy)
1 (19th) jungle bde (3 jungle bn, 1 jungle trg bn, 1 cbt engr coy, 1 sigs coy, 1 log coy)
COMBAT SUPPORT
1 (27th) arty bde (1 SP arty gp, 1 MRL gp, 1 ADA gp, 1 cbt engr coy, 1 sigs coy, 1 log coy)
1 (23rd) engr bde (3 engr bn)
2 indep MP coy
1 indep sigs coy
COMBAT SERVICE SUPPORT
1 (25th) log bde (1 log bn, 1 tpt bn, 1 maint bn, 1 med bn)
9 indep med coy
AVIATION
1 (15th) avn bde (2 tpt avn gp, 2 hel gp, 1 mixed avn gp)
AIR DEFENCE
1 ADA gp

EQUIPMENT BY TYPE
ARMOURED FIGHTING VEHICLES
LT TK 25 AMX-13
RECCE 42: 10 EE-3 *Jararaca*; 32 EE-9 *Cascavel*
APC 151
APC (T) 102: 82 AMX-VCI; 20 M113
APC (W) 49: 17 EE-11 *Urutu*; 32 UR-416
ARTILLERY 486
SP 155mm 5 Mk F3
TOWED 106: **105mm** 84: 36 M101; 24 M2A2; 24 Model 56 pack howitzer; **155mm** 22: 12 M114; 10 M198
MRL 122mm 18 BM-21 *Grad*
MOR 81mm 357 M29
AIRCRAFT
TPT • Light 11: 1 Beech 200 *King Air*; 2 C-212; 1 CN235; 2 Cessna 172; 2 Cessna 206; 1 Cessna 500 *Citation* I; 1 IAI-201 *Arava*; 1 M-28 *Skytruck*
TRG 4: 2 MX-7-235 *Star Rocket*; 2 T-41D *Mescalero*
HELICOPTERS
MRH 33: 7 H125M (AS550C3) *Fennec*; 6 Mi-17-1V *Hip*; 2 SA315B *Lama*; 18 SA342L *Gazelle* (13 with HOT for anti-armour role)
TPT 13: **Medium** 9: 5 AS332B *Super Puma*; 2 Mi-171E; 2 SA330 *Puma*; **Light** 4: 2 H125 (AS350B2) *Ecureuil*; 2 H125 (AS350B3) *Ecureuil*
AIR DEFENCE
SAM • Point-defence *Blowpipe*; 9K32 *Strela*-2 (SA-7 *Grail*)‡; 9K38 *Igla* (SA-18 *Grouse*)
GUNS 240
SP 20mm 44 M163 *Vulcan*
TOWED 196: **14.5mm** 128 ZPU-1/-2; **20mm** 38: 28 M-1935, 10 M167 *Vulcan*; **40mm** 30 L/70/M1A1
AIR-LAUNCHED MISSILES • ASM HOT

Navy 9,400 (incl Naval Aviation, Marines and Coast Guard)

EQUIPMENT BY TYPE
SUBMARINES 2
SSK 2 *Shyri* (GER T-209/1300) with 8 single 533mm TT each with A184 mod 3 HWT
PRINCIPAL SURFACE COMBATANTS • FRIGATES 2
FFGH 2 *Moran Valverde* (ex-UK *Leander* batch II) with 1 quad lnchr with MM40 *Exocet* AShM, 2 triple 324mm ILAS-3 (B-515) ASTT with A244 LWT, 1 Mk 15 *Phalanx* CIWS, 1 twin 114mm gun (capacity 1 Bell 230 hel)
PATROL AND COASTAL COMBATANTS 9
CORVETTES • FSGM 6
5 *Esmeraldas* (ITA Tipo 550) with 2 triple lnchr with MM40 *Exocet* AShM, 1 quad *Albatros* lnchr with *Aspide* SAM, 2 triple 324mm ILAS-3 (B-515) ASTT with A244 LWT, 1 76mm gun, 1 hel landing platform
1 *Esmeraldas* (ITA Tipo 550) with 2 triple lnchr with MM40 *Exocet* AShM, 1 quad *Albatros* lnchr with *Aspide* SAM, 1 76mm gun, 1 hel landing platform
PCFG 3 *Quito* (GER Lurssen TNC-45 45m) with 4 single lnchr with MM38 *Exocet* AShM, 1 76mm gun (upgrade programme ongoing)
LOGISTICS AND SUPPORT 7
AE 1 *Calicuchima*
AG 1 *Hualcopo* (ex-PRC *Fu Yuan Yu Leng* 999) (utilised in transport ship role)
AGOS 1 *Orion* with 1 hel landing platform
AGS 1 *Sirius*
ATF 1 *Chimborazo*
AWT 1 *Atahualpa*
AXS 1 *Guayas*

Naval Aviation 380

EQUIPMENT BY TYPE

AIRCRAFT

MP 1 CN235-300M

ISR 3: 2 Beech 200T *King Air*; 1 Beech 300 *Catpass King Air*

TPT • Light 3: 1 Beech 200 *King Air*; 1 Beech 300 *King Air*; 1 CN235-100

TRG 3 T-35B *Pillan*

HELICOPTERS

TPT • Light 9: 3 Bell 206A; 3 Bell 206B; 1 Bell 230; 2 Bell 430

UNINHABITED AERIAL VEHICLES

ISR 4: **Heavy** 2 *Heron*; **Medium** 2 *Searcher* Mk.II

Marines 1,950

FORCES BY ROLE

SPECIAL FORCES

1 cdo bn

MANOEUVRE

Amphibious

5 mne bn

EQUIPMENT BY TYPE

ARTILLERY • MOR 32+ **81mm/120mm**

AIR DEFENCE • SAM • Point-defence 9K38 *Igla* (SA-18 *Grouse*)

Air Force 6,200

Operational Command

FORCES BY ROLE

FIGHTER

1 sqn with *Cheetah* C/D

GROUND ATTACK

1 sqn with EMB-314 *Super Tucano**

Military Air Transport Group

FORCES BY ROLE

ISR

1 sqn with Beech 350i *King Air*; Gulfstream G-1159; *Sabreliner* 40

SEARCH & RESCUE/TRANSPORT HELICOPTER

1 sqn with AW119 *Koala*; Bell 206B *Jet Ranger* II

1 sqn with Cessna 206; PA-34 *Seneca*; H145

TRANSPORT

1 sqn with C295M

1 sqn with DHC-6-300 *Twin Otter*

1 sqn with B-727; B-737-200; EMB-135BJ *Legacy* 600; L-100-30

TRAINING

1 sqn with DA20-C1

1 sqn with G-120TP

EQUIPMENT BY TYPE

AIRCRAFT 29 combat capable

FGA 12: 10 *Cheetah* C; 2 *Cheetah* D

TPT 20: **Medium** 1 L-100-30; (2 C-130B *Hercules*; 1 C-130H *Hercules* in store); **Light** 12: 1 Beech E90 *King Air*; 1 Beech 350i *King Air*; 3 C295M; 1 Cessna 206; 3 DHC-6 *Twin Otter*; 1 EMB-135BJ *Legacy* 600; 1 PA-34 *Seneca*; 1 *Sabreliner* 40; **PAX** 7: 2 A320 (operated by TAME); 2 B-727; 1 B-737-200; 1 *Falcon* 7X; 1 Gulfstream G-1159

TRG 36: 11 DA20-C1; 17 EMB-314 *Super Tucano**; 8 G-120TP

HELICOPTERS • TPT • Light 12: 4 AW119 *Koala*; 6 Bell 206B *Jet Ranger* II; 2 H145

AIR-LAUNCHED MISSILES • AAM • IR *Python* 3; R-550 *Magic*; **IIR** *Python* 4

AIR DEFENCE

SAM • Point-defence 10+: 10 9K33 *Osa* (SA-8 *Gecko*); 9K310 *Igla*-1 (SA-16 *Gimlet*)

GUNS • TOWED 52: **23mm** 34 ZU-23; **35mm** 18 GDF-002 (twin)

Paramilitary 500

Coast Guard 500

EQUIPMENT BY TYPE

PATROL AND COASTAL COMBATANTS 22

PCC 5: 3 *Isla Fernandina* (*Vigilante*); 2 *Isla San Cristóbal* (Damen Stan Patrol 5009)

PB 14: 1 *10 de Agosto*; 2 *Espada*; 2 *Manta* (GER Lurssen 36m); 1 *Point*; 4 *Rio Coca*; 4 *Isla Santa Cruz* (Damen Stan 2606)

PBR 3: 2 *Río Esmeraldas*; 1 *Rio Puyango*

DEPLOYMENT

SOUTH SUDAN: UN • UNMISS 3

SUDAN: UN • UNAMID 1; **UN** • UNISFA 2

WESTERN SAHARA: UN • MINURSO 4

El Salvador SLV

United States Dollar $		2019	2020	2021
GDP	US$	27.0bn	24.8bn	
per capita	US$	4,187	3,821	
Growth	%	2.4	-9.0	
Inflation	%	0.1	0.2	
Def bdgt	US$	145m	172m	248m

Real-terms defence budget trend (US$m, constant 2015)

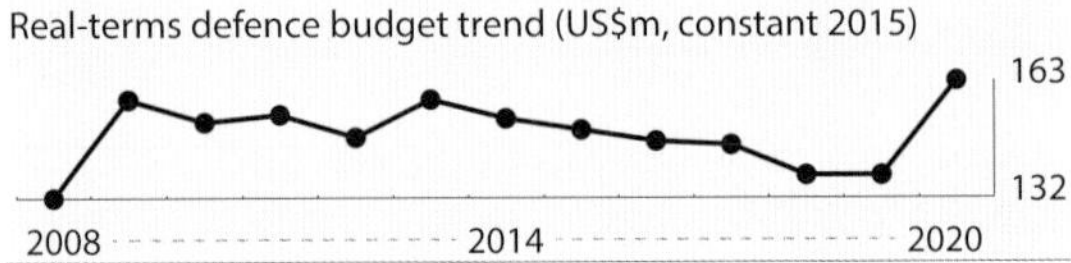

Population 6,481,102

Age	0–14	15–19	20–24	25–29	30–64	65 plus
Male	13.2%	4.6%	5.0%	4.7%	17.2%	3.3%
Female	12.6%	4.4%	4.9%	4.8%	21.1%	4.3%

Capabilities

The armed forces' primary challenge is tackling organised crime and narcotics trafficking in support of the police. A Territorial Control Plan, implemented in 2019, has seen mixed military and police patrols deployed to areas with high crime rates. El Salvador participates in a tri-national border task force with Guatemala and Honduras. There are training programmes with regional states and with the US focused on internal security, disaster relief and support to civilian authorities. El Salvador has deployed on UN peacekeeping missions up to company strength but lacks the logistical support to sustain independent international deployments. The armed forces are dependent on a Cold War-era inventory, and the operational status of many platforms indicates adequate support and maintenance. El Salvador lacks a significant defence industry but has successfully produced light armoured vehicles based upon commercial vehicles.

ACTIVE 24,500 (Army 20,500 Navy 2,000 Air 2,000)
Paramilitary 17,000

Conscript liability 12 months (selective); 11 months for officers and NCOs

RESERVE 9,900 (Joint 9,900)

ORGANISATIONS BY SERVICE

Army 20,500

FORCES BY ROLE

SPECIAL FORCES
1 spec ops gp (1 SF coy, 1 para bn, 1 (naval inf) coy)

MANOEUVRE

Reconnaissance
1 armd cav regt (2 armd cav bn)

Light
6 inf bde (3 inf bn)

Other
1 (special) sy bde (2 border gd bn, 2 MP bn)

COMBAT SUPPORT
1 arty bde (2 fd arty bn, 1 AD bn)
1 engr comd (2 engr bn)

EQUIPMENT BY TYPE

ARMOURED FIGHTING VEHICLES
RECCE 5 AML-90 (4 more in store)
APC • APC (W) 38: 30 VAL *Cashuat* (mod); 8 UR-416

ANTI-TANK/ANTI-INFRASTRUCTURE
RCL 399: **106mm** 20 M40A1 (incl 16 SP); **90mm** 379 M67

ARTILLERY 229+
TOWED 66: **105mm** 54: 36 M102; 18 M-56 (FRY); **155mm** 12 M198
MOR 163+: **81mm** 151 M29; **120mm** 12+: 12 UBM 52; (some M-74 in store)

AIR DEFENCE • **GUNS** 35: **20mm** 31 M 55; 4 TCM-20

Navy 2,000

EQUIPMENT BY TYPE

PATROL AND COASTAL COMBATANTS 10
PB 10: 3 Camcraft (30m); 1 *Point;* 1 Swiftships 77; 1 Swiftships 65; 4 Type-44 (ex-US)

AMPHIBIOUS • LANDING CRAFT • **LCM** 4

Naval Inf (SF Commandos) 90

FORCES BY ROLE

SPECIAL FORCES
1 SF coy

Air Force 2,000

FORCES BY ROLE

FIGHTER/GROUND ATTACK/ISR
1 sqn with A-37B/OA-37B *Dragonfly*; O-2A/B *Skymaster**

TRANSPORT
1 sqn with BT-67; Cessna 210 *Centurion*; Cessna 337G; *Commander* 114; IAI-202 *Arava*; SA-226T *Merlin* IIIB

TRAINING
1 sqn with R-235GT *Guerrier*; T-35 *Pillan*; T-41D *Mescalero*; TH-300

TRANSPORT HELICOPTER
1 sqn with Bell 205 (UH-1H *Iroquois*); Bell 407; Bell 412EP *Twin Huey*; MD-500E; UH-1M *Iroquois*

EQUIPMENT BY TYPE

AIRCRAFT 25 combat capable
ATK 14 A-37B *Dragonfly*
ISR 11: 6 O-2A/B *Skymaster**; 5 OA-37B *Dragonfly**
TPT • Light 10: 2 BT-67; 2 Cessna 210 *Centurion*; 1 Cessna 337G *Skymaster*; 1 *Commander* 114; 3 IAI-201 *Arava*; 1 SA-226T *Merlin* IIIB
TRG 9: 5 R-235GT *Guerrier*; 3 T-35 *Pillan*; 1 T-41D *Mescalero*

HELICOPTERS
MRH 14: 4 Bell 412EP *Twin Huey*; 8 MD-500E; 2 UH-1M *Iroquois*
TPT• Light 9: 8 Bell 205 (UH-1H *Iroquois*); 1 Bell 407 (VIP tpt, govt owned)
TRG 5 TH-300

AIR-LAUNCHED MISSILES • **AAM • IR** *Shafrir*‡

Paramilitary 17,000

National Civilian Police 17,000

Ministry of Public Security

AIRCRAFT
ISR 1 O-2A *Skymaster*
TPT • Light 1 Cessna 310

HELICOPTERS
MRH 2 MD-520N
TPT • Light 3: 1 Bell 205 (UH-1H *Iroquois*); 2 R-44 *Raven* II

DEPLOYMENT

LEBANON: UN • UNIFIL 52; 1 inf pl

MALI: UN • MINUSMA 205; 1 hel sqn with 3 MD-500E

SOUTH SUDAN: UN • UNMISS 4

SUDAN: UN • UNISFA 1

WESTERN SAHARA: UN • MINURSO 2

FOREIGN FORCES

United States US Southern Command: 1 Forward Operating Location (Military, DEA, USCG and Customs personnel)

Guatemala GUA

Guatemalan Quetzal q		2019	2020	2021
GDP	q	590bn	590bn	
	US$	76.7bn	76.2bn	
per capita	US$	4,354	4,240	
Growth	%	3.8	-2.0	
Inflation	%	3.7	2.1	
Def bdgt	q	2.63bn	2.83bn	2.63bn
	US$	341m	365m	
US$1=q		7.70	7.74	

Real-terms defence budget trend (US$m, constant 2015)

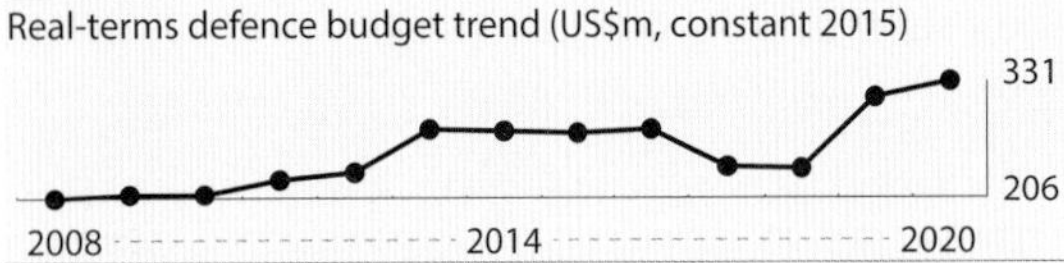

Population 17,153,288

Age	0–14	15–19	20–24	25–29	30–64	65 plus
Male	17.2%	5.1%	4.9%	4.4%	15.9%	2.1%
Female	16.5%	5.0%	4.8%	4.5%	17.0%	2.6%

Capabilities

The armed forces are refocusing on border security, having drawn down their decade-long direct support for the National Civil Police in 2018 as part of the inter-agency Plan Fortaleza. Guatemala maintains an inter-agency task force with neighbouring El Salvador and Honduras. The army has trained with US SOUTHCOM, as well as with regional partners such as Brazil and Colombia. Training for conventional military operations is limited by budget constraints and the long focus on providing internal security. Guatemala has participated in UN peacekeeping missions to company level. The equipment inventory is small and ageing. The US has provided several soft-skinned vehicles to the army, while there has been modest recapitalisation of the air force's fixed-wing transport and surveillance capacity. Funding is being sought for additional maritime- and air-patrol capabilities. Aside from limited maintenance facilities, the country has no domestic defence industry.

ACTIVE 18,050 (Army 15,550 Navy 1,500 Air 1,000) Paramilitary 25,000

RESERVE 63,850 (Navy 650 Air 900 Armed Forces 62,300)

(National Armed Forces are combined; the army provides log spt for navy and air force)

ORGANISATIONS BY SERVICE

Army 15,550

15 Military Zones

FORCES BY ROLE

SPECIAL FORCES
- 1 SF bde (1 SF bn, 1 trg bn)
- 1 SF bde (1 SF coy, 1 ranger bn)
- 1 SF mtn bde

MANOEUVRE

Light
- 1 (strategic reserve) mech bde (1 inf bn, 1 cav regt, 1 log coy)
- 6 inf bde (1 inf bn)

Air Manoeuvre
- 1 AB bde with (2 AB bn)

Amphibious
- 1 mne bde

Other
- 1 (Presidential) gd bde (1 gd bn, 1 MP bn, 1 CSS coy)

COMBAT SUPPORT
- 1 engr comd (1 engr bn, 1 construction bn)
- 2 MP bde with (1 MP bn)

Reserves

FORCES BY ROLE

MANOEUVRE

Light
- ε19 inf bn

EQUIPMENT BY TYPE

ARMOURED FIGHTING VEHICLES
- RECCE (7 M8 in store)
- APC 47
  - APC (T) 10 M113 (5 more in store)
  - APC (W) 37: 30 *Armadillo*; 7 V-100 *Commando*

ANTI-TANK/ANTI-INFRASTRUCTURE
- RCL 120+: **75mm** M20; **105mm** 64 M-1974 FMK-1 (ARG); **106mm** 56 M40A1

ARTILLERY 149
- TOWED **105mm** 76: 12 M101; 8 M102; 56 M-56
- MOR 73: **81mm** 55 M1; **107mm** (12 M30 in store); **120mm** 18 ECIA

AIR DEFENCE • GUNS • TOWED 32: **20mm** 16 GAI-D01; 16 M-55

Navy 1,500

EQUIPMENT BY TYPE

PATROL AND COASTAL COMBATANTS 10
- PB 10: 6 *Cutlass*; 1 *Dauntless*; 1 *Kukulkan* (US *Broadsword* 32m); 2 *Utatlan* (US *Sewart*)

AMPHIBIOUS • LANDING CRAFT 3
- LCT 1 *Quetzal* (COL *Golfo de Tribuga*)
- LCP 2 *Machete*

LOGISTICS AND SUPPORT • AXS 3

Marines 650 reservists

FORCES BY ROLE

MANOEUVRE

Amphibious
- 2 mne bn(-)

Air Force 1,000

2 air comd

FORCES BY ROLE

TRANSPORT
- 1 sqn with Beech 90/200/300 *King Air*
- 1 (tactical support) sqn with Cessna 206

TRAINING
- 1 sqn with T-35B *Pillan*

TRANSPORT HELICOPTER
1 sqn with Bell 212 (armed); Bell 407GX; Bell 412 *Twin Huey* (armed)

EQUIPMENT BY TYPE

Serviceability of ac is less than 50%

AIRCRAFT

TPT • Light 16: 1 Beech 90 *King Air*; 2 Beech 200 *King Air*; 2 Beech 300 *King Air* (VIP); 2 Cessna 206; 3 Cessna 208B *Grand Caravan*; 1 DHC-6 *Twin Otter*; 2 PA-28 *Archer* III; 1 PA-31 *Navajo*; 2 PA-34 *Seneca*; (5 Cessna R172K *Hawk* XP in store)

TRG 1 SR22; (4 T-35B *Pillan* in store)

HELICOPTERS

MRH 2 Bell 412 *Twin Huey* (armed)

TPT • Light 6: 2 Bell 206B *Jet Ranger*; 2 Bell 212 (armed); 2 Bell 407GX

Tactical Security Group

Air Military Police

Paramilitary 25,000

National Civil Police 25,000

FORCES BY ROLE

SPECIAL FORCES
1 SF bn

MANOEUVRE

Other
1 (integrated task force) paramilitary unit (incl mil and treasury police)

DEPLOYMENT

CENTRAL AFRICAN REPUBLIC: UN • MINUSCA 4

DEMOCRATIC REPUBLIC OF THE CONGO: UN • MONUSCO 155; 1 SF coy

LEBANON: UN • UNIFIL 1

MALI: UN • MINUSMA 2

SOUTH SUDAN: UN • UNMISS 5

SUDAN: UN • UNISFA 3

Guyana GUY

Guyanese Dollar G$		2019	2020	2021
GDP	G$	1.08tr	1.42tr	
	US$	5.17bn	6.81bn	
per capita	US$	6,594	8,649	
Growth	%	5.4	26.2	
Inflation	%	2.1	1.0	
Def bdgt	G$	14.0bn	13.7bn	
	US$	67.2m	65.8m	
US$1=G$		208.49	208.74	

Real-terms defence budget trend (US$m, constant 2015)

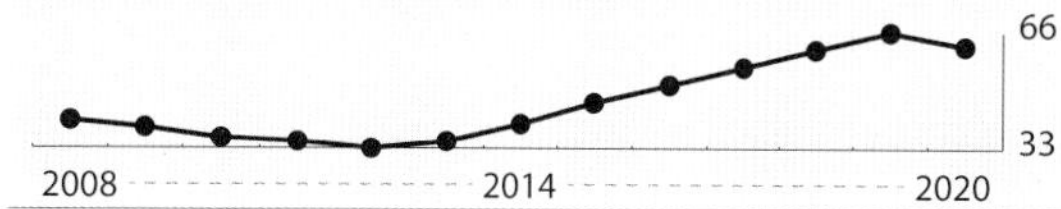

Population 750,204

Age	0–14	15–19	20–24	25–29	30–64	65 plus
Male	12.2%	5.4%	5.5%	4.7%	19.8%	2.8%
Female	11.7%	5.2%	5.2%	4.4%	18.9%	4.2%

Capabilities

The Guyana Defence Force (GDF) is focused on border control, support for law-enforcement operations and assistance to the civil power. The government is planning to restructure the GDF to improve its flexibility. Guyana is part of the Caribbean Basin Security Initiative. It has close military ties with Brazil, with whom it cooperates on border security via annual military regional exchange meetings. The country also has bilateral agreements with France, China and the US. The GDF trains regularly and takes part in bilateral and multinational exercises. A training initiative with China helped two Guyanese pilots to acquire air-combat certification although Guyana has no combat aircraft in its inventory. There is no expeditionary or associated logistics capability. Equipment is mostly composed of second-hand platforms, mainly of Brazilian and North American manufacture. The air force has expanded its modest air-transport capabilities with some second-hand utility aircraft. Apart from maintenance facilities, there is no defence-industrial sector.

ACTIVE 3,400 (Army 3,000 Navy 200 Air 200)

Active numbers combined Guyana Defence Force

RESERVE 670 (Army 500 Navy 170)

ORGANISATIONS BY SERVICE

Army 3,000

FORCES BY ROLE

SPECIAL FORCES
1 SF coy

MANOEUVRE

Light
1 inf bn

Other
1 (Presidential) gd bn

COMBAT SUPPORT
1 arty coy

1 (spt wpn) cbt spt coy
1 engr coy

EQUIPMENT BY TYPE

ARMOURED FIGHTING VEHICLES

RECCE 9: 6 EE-9 *Cascavel* (reported); 3 S52 *Shorland*

ARTILLERY 54

TOWED **130mm** 6 M-46†

MOR 48: **81mm** 12 L16A1; **82mm** 18 M-43; **120mm** 18 M-43

Navy 200

EQUIPMENT BY TYPE

PATROL AND COASTAL COMBATANTS 5

PCO 1 *Essequibo* (ex-UK *River* (MCM))

PB 4 *Barracuda* (ex-US Type-44) (of which 1 under repair)

Air Force 200

FORCES BY ROLE

TRANSPORT

1 unit with Bell 206; Cessna 206; Y-12 (II)

EQUIPMENT BY TYPE

AIRCRAFT • TPT • **Light** 6: 2 BN-2 *Islander*; 1 Cessna 206; 2 SC.7 3M *Skyvan*; 1 Y-12 (II)

HELICOPTERS

MRH 1 Bell 412 *Twin Huey*†

TPT • **Light** 2 Bell 206

Haiti HTI

Haitian Gourde G		2019	2020	2021
GDP	G	732bn	860bn	
	US$	8.71bn	8.35bn	
per capita	US$	773	732	
Growth	%	-1.2	-4.0	
Inflation	%	17.3	22.4	
Def bdgt	G	1.09bn	998m	
	US$	13.0m	9.68m	
US$1=G		84.13	103.09	

Real-terms defence budget trend (US$m, constant 2015)

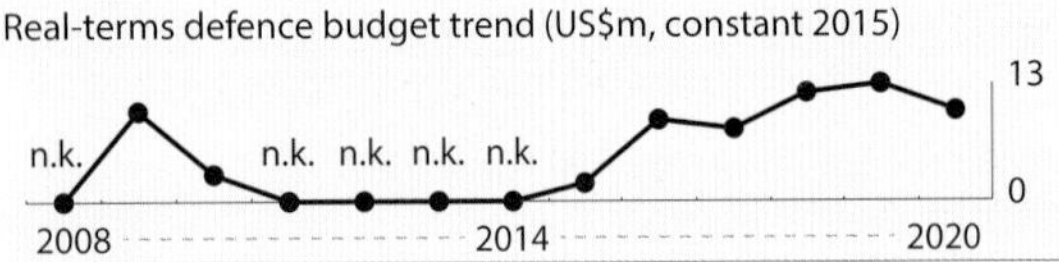

Population 11,067,777

Age	0–14	15–19	20–24	25–29	30–64	65 plus
Male	15.5%	5.2%	5.1%	4.7%	16.9%	1.9%
Female	15.7%	5.2%	5.1%	4.7%	17.5%	2.4%

Capabilities

Haiti possesses almost no military capability. A small coastguard is tasked with maritime security and law enforcement and the country's army is still in the very early stages of development, though it is hoped this will eventually number around 5,000 personnel. Plans for military expansion were outlined in the 2015 White Paper on Security and Defence. A road map for the re-establishment of the Haitian armed forces was distributed to ministers in early 2017 and in March 2018 an army high command was established. Ecuador and Brazil have both pledged to assist with training the new army. The army's primary missions will reportedly be disaster relief and border security. The initial 500 troops are focused on engineering and medical capability for disaster-relief tasks. However, it is unclear whether budgetary allocations are sufficient to fund the capability levels required. Haiti is a member of the Caribbean Community and has participated in US SOUTHCOM's *Tradewinds* disaster-response exercise. There is no heavy military equipment, and no defence industry.

ACTIVE 500 (Army 500) **Paramilitary 50**

ORGANISATIONS BY SERVICE

Army 500

FORCES BY ROLE

MANOEUVRE

1 inf bn (forming)

Paramilitary 50

Coast Guard ε50

EQUIPMENT BY TYPE

PATROL AND COASTAL COMBATANTS

PB 8: 5 *Dauntless*; 3 3812-VCF

Honduras HND

Honduran Lempira L		2019	2020	2021
GDP	L	615bn	599bn	
	US$	24.9bn	24.0bn	
per capita	US$	2,551	2,412	
Growth	%	2.7	-6.6	
Inflation	%	4.4	3.3	
Def bdgt [a]	L	8.53bn	8.48bn	8.46bn
	US$	346m	340m	
US$1=L		24.68	24.96	

[a] Defence & national-security budget

Real-terms defence budget trend (US$m, constant 2015)

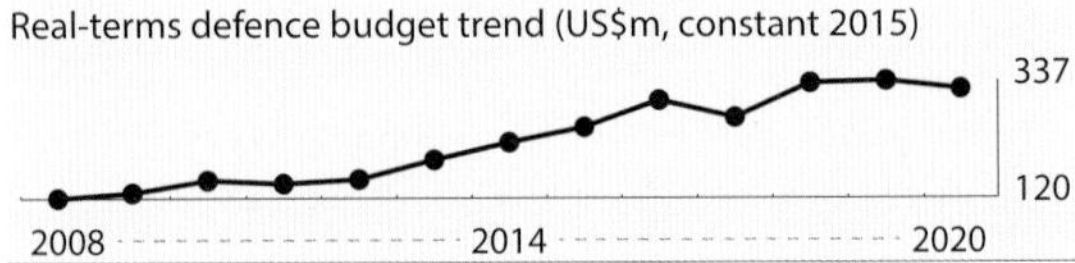

Population 9,235,340

Age	0–14	15–19	20–24	25–29	30–64	65 plus
Male	15.3%	5.5%	5.0%	4.3%	16.2%	2.4%
Female	14.9%	5.4%	5.1%	4.5%	18.4%	3.0%

Capabilities

The armed forces have been deployed in support of the police to combat organised crime and narcotics trafficking since 2011. Honduras maintains diplomatic relations with Taiwan, which has supplied surplus military equipment, and also receives US security assistance. Honduras hosts a US base at Soto Cano airfield and is also part of a tri-national border-security task force with neighbouring El Salvador and Guatemala. Training remains focused on internal- and border-security requirements, and training for conventional military action is limited. Honduras does not have the capability to maintain substantial overseas deployments. Most equipment is ageing, with serviceability in doubt. There have been reports of security assistance from Israel. Apart from limited maintenance facilities, the country has no domestic defence industry.

ACTIVE 14,950 **(Army 7,300 Navy 1,350 Air 2,300 Military Police 4,000) Paramilitary 8,000**

RESERVE 60,000 **(Joint 60,000; Ex-servicemen registered)**

ORGANISATIONS BY SERVICE

Army 7,300

FORCES BY ROLE

SPECIAL FORCES

1 (special tac) spec ops gp (2 spec ops bn, 1 inf bn; 1 AB bn; 1 arty bn)

MANOUEVRE

Mechanised

1 inf bde (1 mech cav regt, 1 inf bn, 1 arty bn)

Light

1 inf bde (3 inf bn, 1 arty bn)

3 inf bde (2 inf bn)

1 indep inf bn

Other

1 (Presidential) gd coy

COMBAT SUPPORT

1 engr bn

1 sigs bn

EQUIPMENT BY TYPE

ARMOURED FIGHTING VEHICLES

LT TK 12 FV101 *Scorpion*

RECCE 57: 1 FV105 *Sultan* (CP); 3 FV107 *Scimitar*; 40 FV601 *Saladin*; 13 RBY-1

ANTI-TANK/ANTI-INFRASTRUCTURE

RCL 50+: **84mm** *Carl Gustaf*; **106mm** 50 M40A1

ARTILLERY 118+

TOWED 28: **105mm:** 24 M102; **155mm:** 4 M198

MOR 90+: **81mm**; **120mm** 60 FMK-2; **160mm** 30 M-66

Navy 1,350

EQUIPMENT BY TYPE

PATROL AND COASTAL COMBATANTS 18

PCO 1 *General Cabañas* (ISR OPV 62 *Sa'ar*)

PB 17: 2 *Lempira* (Damen Stan Patrol 4207 – leased); 1 *Chamelecon* (Swiftships 85); 1 *Tegucilgalpa* (US *Guardian* 32m); 4 *Guanaja* (ex-US Type-44); 3 *Guaymuras* (Swiftships 105); 5 *Nacaome* (Swiftships 65); 1 *Rio Coco* (US PB Mk III)

AMPHIBIOUS • LANDING CRAFT 3

LCT 1 *Gracias a Dios* (COL *Golfo de Tribuga*)

LCM 2 LCM 8

Marines 1,000

FORCES BY ROLE

MANOEUVRE

Amphibious

2 mne bn

Air Force 2,300

FORCES BY ROLE

FIGHTER/GROUND ATTACK

1 sqn with A-37B *Dragonfly*

1 sqn with F-5E/F *Tiger* II

GROUND ATTACK/ISR/TRAINING

1 unit with Cessna 182 *Skylane*; EMB-312 *Tucano*; MXT-7-180 *Star Rocket*

TRANSPORT

1 sqn with Beech 200 *King Air*; C-130A *Hercules*; Cessna 185/210; IAI-201 *Arava*; PA-42 *Cheyenne*; *Turbo Commander* 690

1 VIP flt with PA-31 *Navajo*; Bell 412EP/SP *Twin Huey*

TRANSPORT HELICOPTER

1 sqn with Bell 205 (UH-1H *Iroquois*); Bell 412SP *Twin Huey*

AIR DEFENCE

1 ADA bn

EQUIPMENT BY TYPE

AIRCRAFT 17 combat capable

FTR 11: 9 F-5E *Tiger* II†; 2 F-5F *Tiger* II†

ATK 6 A-37B *Dragonfly*

TPT 17: **Medium** 1 C-130A *Hercules*; **Light** 16: 1 Beech 200 *King Air*; 2 Cessna 172 *Skyhawk*; 2 Cessna 182 *Skylane*; 1 Cessna 185; 3 Cessna 208B *Grand Caravan*; 1 Cessna 210; 1 EMB-135 *Legacy* 600; 1 IAI-201 *Arava*; 1 L-410 (leased); 1 PA-31 *Navajo*; 1 PA-42 *Cheyenne*; 1 *Turbo Commander* 690

TRG 15: 9 EMB-312 *Tucano*; 6 MXT-7-180 *Star Rocket*

HELICOPTERS

MRH 7: 1 Bell 412EP *Twin Huey* (VIP); 4 Bell 412SP *Twin Huey*; 2 Hughes 500

TPT • Light 7: 6 Bell 205 (UH-1H *Iroquois*); 1 H125 *Ecureuil*

AIR DEFENCE • GUNS **20mm** 48: 24 M-55A2; 24 TCM-20

AIR-LAUNCHED MISSILES • **AAM • IR** *Shafrir*‡

Military Police 4,000

FORCES BY ROLE

MANOUEVRE

Other

8 sy bn

Paramilitary 8,000

Public Security Forces 8,000

Ministry of Public Security and Defence; 11 regional comd

DEPLOYMENT

WESTERN SAHARA: UN • MINURSO 9

FOREIGN FORCES

United States US Southern Command: 370; 1 avn bn with CH-47F *Chinook*; UH-60 *Black Hawk*

Jamaica JAM

Jamaican Dollar J$		2019	2020	2021
GDP	J$	2.12tr	2.02tr	
	US$	15.9bn	14.2bn	
per capita	US$	5,826	5,221	
Growth	%	0.9	-8.6	
Inflation	%	3.9	5.1	
Def bdgt	J$	34.9bn	33.8bn	
	US$	262m	238m	
US$1=J$		133.50	141.87	

Real-terms defence budget trend (US$m, constant 2015)

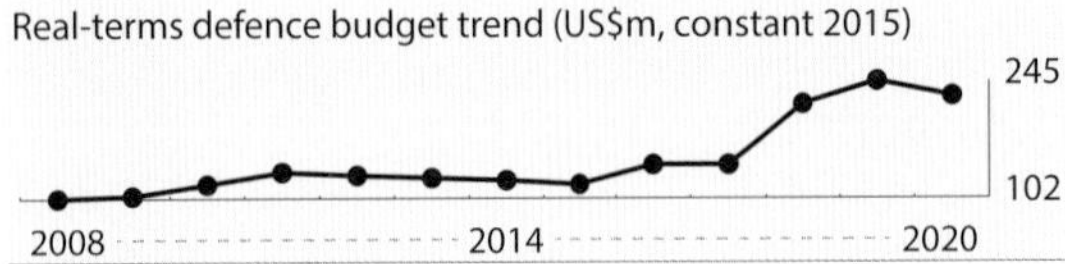

Population	2,808,570

Age	0–14	15–19	20–24	25–29	30–64	65 plus
Male	12.8%	4.6%	4.5%	3.9%	19.4%	4.3%
Female	12.4%	4.5%	4.4%	3.9%	20.5%	4.8%

Capabilities

The Jamaica Defence Force (JDF) is focused principally on maritime and internal security, including support to police operations. Jamaica maintains military ties, including for training purposes, with Canada, the UK and the US and is a member of the Caribbean Community. The defence force has participated in US SOUTHCOM's *Tradewinds* disaster-response exercise. Jamaica is host to the Caribbean Special Tactics Centre, which trains special-forces units from Jamaica and other Caribbean nations. The JDF does not have any capacity to support independent deployment abroad. Funds have been allocated to procure new vehicles and helicopters. Other than limited maintenance facilities, Jamaica has no domestic defence industry.

ACTIVE 5,950 (Army 5,400 Coast Guard 300 Air 250)
(combined Jamaican Defence Force)

RESERVE 2,580 (Army 2,500 Coast Guard 60 Air 20)

ORGANISATIONS BY SERVICE

Army 5,400

FORCES BY ROLE
MANOEUVRE
Mechanised
1 (PMV) lt mech inf coy
Light
4 inf bn
COMBAT SUPPORT
1 engr regt (4 engr sqn)
1 MP bn
1 cbt spt bn
COMBAT SERVICE SUPPORT
1 spt bn (1 med coy, 1 log coy, 1 tpt coy)
EQUIPMENT BY TYPE
ARMOURED FIGHTING VEHICLES
AUV 12 *Bushmaster*
ARTILLERY • **MOR 81mm** 12 L16A1

Reserves

FORCES BY ROLE
MANOEUVRE
Light
3 inf bn
COMBAT SERVICE SUPPORT
1 spt bn

Coast Guard 300

EQUIPMENT BY TYPE
PATROL AND COASTAL COMBATANTS 11
PCC 1 *Nanny of the Maroons* (Damen Fast Crew Supplier 5009)
PBF 3
PB 7: 2 *County* (Damen Stan Patrol 4207); 4 *Dauntless*; 1 *Paul Bogle* (US 31m)

Air Wing 250

Plus National Reserve

FORCES BY ROLE
MARITIME PATROL/TRANSPORT
1 flt with Beech 350ER *King Air*; BN 2A *Defender*; Cessna 210M *Centurion*
SEARCH & RESCUE/TRANSPORT HELICOPTER
1 flt with Bell 407
1 flt with Bell 412EP
TRAINING
1 unit with Bell 206B3 *Jet Ranger*; DA40-180FP *Diamond Star*
EQUIPMENT BY TYPE
AIRCRAFT
MP 1 Beech 350ER *King Air*
TPT • **Light** 4: 1 BN-2A *Defender*; 1 Cessna 210M *Centurion*; 2 DA40-180FP *Diamond Star*
HELICOPTERS
MRH 2 Bell 412EP
TPT • **Light** 6: 1 Bell 206B3 *Jet Ranger*; 3 Bell 407; 2 Bell 429

Mexico MEX

Mexican Peso NP		2019	2020	2021
GDP	NP	24.2tr	22.7tr	
	US$	1.26tr	1.04tr	
per capita	US$	9,862	8,069	
Growth	%	-0.3	-9.0	
Inflation	%	3.6	3.4	
Def bdgt [a]	NP	97.6bn	115bn	136bn
	US$	5.07bn	5.27bn	
FMA (US)	US$	5m	0m	0m
US$1=NP		19.26	21.84	

[a] National-security expenditure

Real-terms defence budget trend (US$bn, constant 2015)

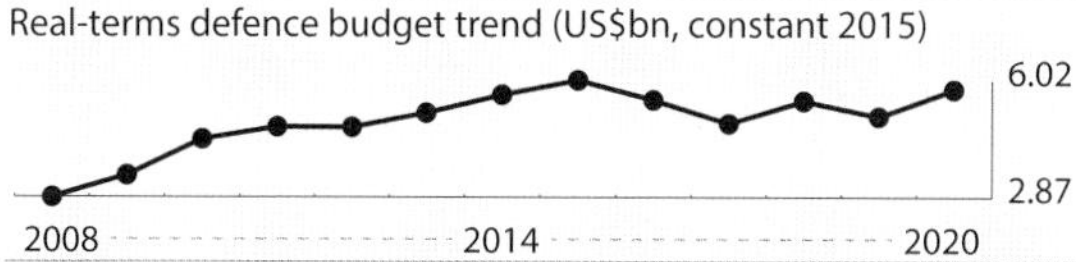

Population 128,649,565

Age	0–14	15–19	20–24	25–29	30–64	65 plus
Male	13.3%	4.4%	4.3%	4.1%	19.6%	3.4%
Female	12.7%	4.2%	4.2%	4.1%	21.6%	4.3%

Capabilities

Mexico has the most capable armed forces in Central America. They have been committed to providing internal-security support within Mexico for nearly a decade. Under the National Plan for Peace and Security 2018–24, the armed forces are now to hand over lead responsibility for tackling drug cartels and other organised crime to the National Guard gendarmerie. Mexico has a close defence relationship with the US, which has provided equipment and training to Mexican forces under the Mérida Initiative, as well as through bilateral programmes via the Pentagon. The armed forces have a moderate capability to deploy independently, but do not do so in significant numbers. There are plans to recapitalise diverse and ageing conventional combat platforms across all three services. In 2020 Mexico brought back to service some of its ageing F-5 combat aircraft. State-owned shipyards have produced patrol craft for the navy. Army factories have produced light armoured utility vehicles for domestic use. Airbus Helicopters operates a manufacturing plant in Querétaro.

ACTIVE 216,000 (Army 157,500 Navy 50,500 Air 8,000) Paramilitary 111,900

Conscript liability 12 months (partial, selection by ballot) from age 18, serving on Saturdays; voluntary for women; conscripts allocated to reserves.

RESERVE 81,500 (National Military Service)

ORGANISATIONS BY SERVICE

Space

EQUIPMENT BY TYPE

SATELLITES • COMMUNICATIONS 2 *Mexsat*

Army 157,500

12 regions (total: 46 army zones)

FORCES BY ROLE

SPECIAL FORCES
1 (1st) SF bde (5 SF bn)
1 (2nd) SF bde (7 SF bn)
1 (3rd) SF bde (4 SF bn)

MANOEUVRE

Reconnaissance
3 (2nd, 3rd & 4th Armd) mech bde (2 armd recce bn, 2 lt mech bn, 1 arty bn, 1 (Canon) AT gp)
25 mot recce regt

Light
1 (1st) inf corps (1 (1st Armd) mech bde (2 armd recce bn, 2 lt mech bn, 1 arty bn, 1 (Canon) AT gp), 3 (2nd, 3rd & 6th) inf bde (each: 3 inf bn, 1 arty regt, 1 (Canon) AT gp), 1 cbt engr bde (3 engr bn))
3 (1st, 4th & 5th) indep lt inf bde (2 lt inf bn, 1 (Canon) AT gp)
92 indep inf bn
25 indep inf coy

Air Manoeuvre
1 para bde with (1 (GAFE) SF gp, 3 bn, 1 (Canon) AT gp)

COMBAT SUPPORT
1 indep arty regt

EQUIPMENT BY TYPE

ARMOURED FIGHTING VEHICLES
RECCE 255: 19 DN-5 *Toro*; 127 ERC-90F1 *Lynx* (7 trg); 40 M8; 37 MAC-1; 32 VBL
IFV 390 DNC-1 (mod AMX-VCI)
APC 309
APC (T) 73: 40 HWK-11; 33 M5A1 half-track
APC (W) 236: 95 BDX; 16 DN-4; 2 DN-6; 28 LAV-100 (*Pantera*); 26 LAV-150 ST; 25 MOWAG *Roland*; 44 VCR (3 amb; 5 cmd post)
AUV 347: 100 DN-XI; 247 *SandCat*

ENGINEERING & MAINTENANCE VEHICLES
ARV 7: 3 M32 *Recovery Sherman*; 4 VCR ARV

ANTI-TANK/ANTI-INFRASTRUCTURE
MSL • SP 8 VBL with *Milan*
RCL • 106mm 1,187+ M40A1 (incl some SP)
GUNS 37mm 30 M3

ARTILLERY 1,390
TOWED 123: **105mm** 123: 40 M101; 40 M-56; 16 M2A1, 14 M3; 13 NORINCO M90
MOR 1,267: **81mm** 1,100: 400 M1; 400 Brandt; 300 SB
120mm 167: 75 Brandt; 60 M-65; 32 RT-61

AIR DEFENCE • GUNS • TOWED 80: **12.7mm** 40 M55; **20mm** 40 GAI-B01

Navy 50,500

Two Fleet Commands: Gulf (6 zones), Pacific (11 zones)

EQUIPMENT BY TYPE

PRINCIPAL SURFACE COMBATANTS • FRIGATES 1
FFGHM 1 *Benito Juárez* (Damen SIGMA 10514) with 2 quad lnchr with RGM-84L *Harpoon* Block II AShM, 1 8-cell Mk 56 VLS with RIM-162 ESSM SAM, 1 21-cell Mk 49 lnchr with RIM-116C RAM Block 2 SAM, 2 triple

324mm SVTT Mk 32 ASTT with Mk 54 LWT, 1 57mm gun (capacity 1 med hel)

PATROL AND COASTAL COMBATANTS 128

PSOH 8:

4 *Oaxaca* with 1 76mm gun (capacity 1 AS565MB *Panther* hel)

4 *Oaxaca* (mod) with 1 57mm gun (capacity 1 AS565MB *Panther* hel)

PCOH 16:

4 *Durango* with 1 57mm gun (capacity 1 Bo-105 hel)

4 *Holzinger* (capacity 1 MD-902 *Explorer*)

3 *Sierra* with 1 57mm gun (capacity 1 MD-902 *Explorer*)

5 *Uribe* (ESP *Halcon*) (capacity 1 Bo-105 hel)

PCO 9: 6 *Valle* (US *Auk* MSF) with 1 76mm gun; 3 *Valle* (US *Auk* MSF) with 1 76mm gun, 1 hel landing platform

PCGH 1 *Huracan* (ex-ISR *Aliya*) with 4 single lnchr with *Gabriel* II AShM, 1 Mk 15 *Phalanx* CIWS

PCC 2 *Democrata*

PBF 72: 6 *Acuario*; 2 *Acuario B*; 48 *Polaris* (SWE CB90); 16 *Polaris* II (SWE IC 16M)

PB 20: 3 *Azteca*; 3 *Cabo* (ex-US *Cape Higgon*); 2 *Lago*; 2 *Punta* (US *Point*); 10 *Tenochtitlan* (Damen Stan Patrol 4207)

AMPHIBIOUS • LANDING SHIPS

LST 4: 2 *Monte Azules* with 1 hel landing platform; 1 *Papaloapan* (ex-US *Newport*) with 2 twin 76mm guns, 1 hel landing platform; 1 *Papaloapan* (ex-US *Newport*) with 1 hel landing platform

LOGISTICS AND SUPPORT 26

AGOR 2 *Altair* (ex-US *Robert D. Conrad*)

AGS 8: 5 *Arrecife*; 1 *Onjuku*; 1 *Rio Hondo*; 1 *Rio Tuxpan*

AK 1 *Rio Suchiate*

AOTL 2 *Aguascalientes*

AP 2: 1 *Isla Maria Madre* (Damen Fast Crew Supplier 5009); 1 *Nautla*

ATF 4 *Otomi* with 1 76mm gun

ATS 4 *Kukulkan*

AX 2 *Huasteco* (also serve as troop transport, supply and hospital ships)

AXS 1 *Cuauhtemoc*

Naval Aviation 1,250

FORCES BY ROLE

MARITIME PATROL

5 sqn with MX-7-180 *Star Rocket*; T-6C+ *Texan* II

1 sqn with Beech 350ER *King Air*; CN235-300 MPA *Persuader*

TRANSPORT

1 (VIP) sqn with DHC-8 *Dash* 8

TRANSPORT HELICOPTER

2 sqn with AS555 *Fennec*; AS565MB/AS565MBe *Panther*; MD-902

5 sqn with Mi-17-1V/V-5 *Hip*

TRAINING

1 sqn with Z-242L; Z-143Lsi

EQUIPMENT BY TYPE

AIRCRAFT 3 combat capable

MP 6 CN235-300 MPA *Persuader*

ISR 2 Z-143Lsi

TPT 20: **Light** 18: 5 Beech 350ER *King Air* (4 used for ISR); 3 Beech 350i *King Air*; 4 C295M; 2 C295W; 1 DHC-8 *Dash 8*; 2 Learjet 31A; 1 Learjet 60; **PAX** 2: 1 CL-605 *Challenger*; 1 Gulfstream 550

TRG 47: 7 MX-7-180 *Star Rocket*; 13 T-6C+ *Texan* II; 27 Z-242L

HELICOPTERS

MRH 23: 2 AS555 *Fennec*; 17 Mi-17-1V *Hip*; 4 Mi-17V-5 *Hip*

SAR 14: 4 AS565MB *Panther*; 10 AS565MBe *Panther*

TPT 26: **Heavy** 3 H225M *Caracal*; **Medium** 9 UH-60M *Black Hawk*; **Light** 14: 1 AW109SP; 5 MD-902 (SAR role); 8 S-333

TRG 4 Schweizer 300C

Marines 21,500

FORCES BY ROLE

SPECIAL FORCES

3 SF unit

MANOEUVRE

Light

32 inf bn(-)

Air Manoeuvre

1 AB bn

Amphibious

1 amph bde (4 inf bn, 1 amph bn, 1 arty gp)

Other

1 (Presidential) gd bn (included in army above)

COMBAT SERVICE SUPPORT

2 spt bn

EQUIPMENT BY TYPE

ARMOURED FIGHTING VEHICLES

APC • APC (W) 29: 3 BTR-60 (APC-60); 26 BTR-70 (APC-70)

ANTI-TANK/ANTI-INFRASTRUCTURE

RCL 106mm M40A1

ARTILLERY 22+

TOWED 105mm 16 M-56

MRL 122mm 6 *Firos*-25

MOR 81mm some

AIR DEFENCE • SAM • Point-defence 9K38 *Igla* (SA-18 *Grouse*)

Air Force 8,000

FORCES BY ROLE

FIGHTER

1 sqn with F-5E/F *Tiger* II

GROUND ATTACK/ISR

4 sqn with T-6C+ *Texan* II*

1 sqn with PC-7*

ISR/AEW

1 sqn with Beech 350ER *King Air*; EMB-145AEW *Erieye*; EMB-145RS

TRANSPORT

1 sqn with C295M; PC-6B

1 sqn with B-737; Beech 90 *King Air*

1 sqn with C-27J *Spartan*; C-130K-30 *Hercules*; L-100-30

5 (liaison) sqn with Cessna 182

1 (anti-narcotic spraying) sqn with Bell 206

1 (Presidential) gp with AS332L *Super Puma*; AW109SP; B-737; B-757; B-787; Gulfstream 150/450/550; H225; Learjet 35A; Learjet 36; *Turbo Commander* 680

1 (VIP) gp with B-737; Beech 200 *King Air*; Beech 350i *King Air*; Cessna 501 *Citation*; CL-605 *Challenger*; Gulfstream 550; Learjet 35A; Learjet 45; S-70A-24 *Black Hawk*

TRAINING

1 sqn with Cessna 182
1 sqn with PC-7; T-6C+ *Texan* II
1 sqn with Grob G120TP

TRANSPORT HELICOPTER

4 sqn with Bell 206B; Bell 407GX
1 (anti-narcotic spraying) sqn with Bell 206
1 sqn with MD-530MF/MG
1 sqn with Mi-17 *Hip*
1 sqn with H225M *Caracal*; Bell 412EP *Twin Huey*; S-70A-24 *Black Hawk*
1 sqn with UH-60M *Black Hawk*

ISR UAV

1 unit with *Hermes* 450; *Hermes* 900; S4 *Ehécatl*

EQUIPMENT BY TYPE

AIRCRAFT 80 combat capable

FTR 5: 4 F-5E *Tiger* II; 1 F-5F *Tiger* II
ISR 2 Cessna 501 *Citation*
ELINT 8: 6 Beech 350ER *King Air*; 2 EMB-145RS
AEW&C 1 EMB-145AEW *Erieye*
TPT 100: **Medium** 7: 4 C-27J *Spartan*; 2 C-130K-30 *Hercules*; 1 L-100-30; **Light** 80: 2 Beech 90 *King Air*; 1 Beech 200 *King Air*; 1 Beech 350i *King Air*; 6 C295M; 2 C295W; 59 Cessna 182; 1 Cessna 501 *Citation*; 2 Learjet 35A; 1 Learjet 36; 1 Learjet 45XP; 3 PC-6B; 1 *Turbo Commander* 680; **PAX** 13: 6 B-737; 1 B-757; 1 B-787; 1 CL-605 *Challenger*; 2 Gulfstream 150; 1 Gulfstream 450; 1 Gulfstream 550
TRG 100: 25 Grob G120TP; 20 PC-7* (30 more possibly in store); 55 T-6C+ *Texan* II*

HELICOPTERS

MRH 41: 14 Bell 407GXP; 11 Bell 412EP *Twin Huey*; 16 Mi-17 *Hip* H
ISR 12: 4 MD-530MF; 8 MD-530MG
TPT 109: **Heavy** 12 H225M *Caracal*; **Medium** 28: 3 AS332L *Super Puma* (VIP); 2 H225 (VIP); 6 S-70A-24 *Black Hawk*; 17 UH-60M *Black Hawk* **Light** 69: 5 AW109SP; 45 Bell 206; 13 Bell 206B *Jet Ranger* II; 6 Bell 206L

UNINHABITED AERIAL VEHICLES • ISR 9: **Medium** 4: 3 *Hermes* 450; 1 *Hermes* 900; **Light** 5 S4 *Ehécatl*

Paramilitary 111,900

Federal Ministerial Police 4,500

EQUIPMENT BY TYPE

HELICOPTERS

TPT • Light 25: 18 Bell 205 (UH-1H); 7 Bell 212

UNINHABITED AERIAL VEHICLES

ISR • Heavy 2 *Dominator* XP

National Guard 90,000

Public Security Secretariat. Gendarmerie created in 2019 from elements of the Army, Navy, Air Force and Federal Police

FORCES BY ROLE

MANOEUVRE

Other

12 sy bde (3 sy bn)

EQUIPMENT BY TYPE

HELICOPTERS

MRH 5: 1+ Bell 407GX; 4 Mi-17 *Hip* H
TPT • Medium 7 UH-60M *Black Hawk*

Rural Defense Militia 17,400

FORCES BY ROLE

MANOEUVRE

Light

13 inf unit
13 (horsed) cav unit

DEPLOYMENT

CENTRAL AFRICAN REPUBLIC: UN • MINUSCA 2

INDIA/PAKISTAN: UN • UNMOGIP 1

MALI: UN • MINUSMA 2

WESTERN SAHARA: UN • MINURSO 3

Nicaragua NIC

Nicaraguan Gold Cordoba Co		2019	2020	2021
GDP	Co	415bn	409bn	
	US$	12.5bn	11.9bn	
per capita	US$	1,920	1,832	
Growth	%	-3.9	-5.5	
Inflation	%	5.4	4.4	
Def bdgt	Co	2.60bn	2.71bn	
	US$	78.5m	78.9m	
US$1=Co		33.08	34.34	

Real-terms defence budget trend (US$m, constant 2015)

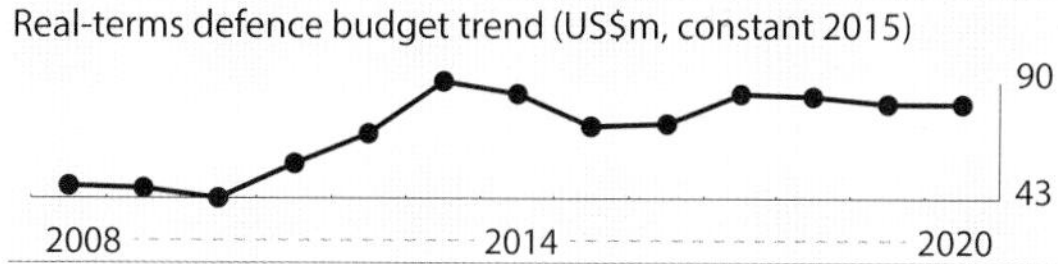

Population 6,203,441

Age	0–14	15–19	20–24	25–29	30–64	65 plus
Male	13.1%	4.8%	5.0%	5.2%	18.1%	2.6%
Female	12.5%	4.7%	5.0%	5.2%	20.5%	3.3%

Capabilities

Nicaragua's armed forces are primarily a territorial light-infantry force, with limited coastal-patrol capability. They are tasked with border and internal security, as well as with support for disaster-relief efforts and ecological protection. Nicaragua has training relationships with Russia and the US, as well as with neighbouring and regional states, including Cuba and Venezuela. Training is largely focused on key internal- and border-security tasks, although the mechanised brigade has received Russian training. The armed forces do not undertake significant international deployments and lack the logistical support for large-scale military operations, although the mechanised brigade can deploy internally. Equipment primarily consists of ageing Cold War-era platforms. Russia has supplied some second-hand tanks and armoured vehicles to help re-equip the mechanised brigade and has supported the establishment of a repair workshop to maintain the vehicles in-country. Barring maintenance facilities there is no domestic defence industry.

ACTIVE 12,000 (Army 10,000 Navy 800 Air 1,200)

ORGANISATIONS BY SERVICE

Army ε10,000

FORCES BY ROLE

SPECIAL FORCES
1 SF bde (2 SF bn)

MANOEUVRE
Mechanised
1 mech inf bde (1 armd recce bn, 1 tk bn, 1 mech inf bn, 1 arty bn, 1 MRL bn, 1 AT coy)
Light
1 regional comd (3 lt inf bn)
1 regional comd (2 lt inf bn; 1 arty bn)
3 regional comd (2 lt inf bn)
2 indep lt inf bn
Other
1 comd regt (1 inf bn, 1 sy bn, 1 int unit, 1 sigs bn)
1 (ecological) sy bn

COMBAT SUPPORT
1 engr bn

COMBAT SERVICE SUPPORT
1 med bn
1 tpt regt

EQUIPMENT BY TYPE

ARMOURED FIGHTING VEHICLES
MBT 82: 62 T-55 (65 more in store); 20 T-72B1
LT TK (10 PT-76 in store)
RECCE 20 BRDM-2
IFV 17+ BMP-1
APC • APC (W) 90+: 41 BTR-152 (61 more in store); 45 BTR-60 (15 more in store); 4+ BTR-70M

ENGINEERING & MAINTENANCE VEHICLES
AEV T-54/T-55 AEV
VLB TMM-3

ANTI-TANK/ANTI-INFRASTRUCTURE
MSL
SP 12 9P133 *Malyutka* (AT-3 *Sagger*)
MANPATS 9K11 *Malyutka* (AT-3 *Sagger*)
RCL 82mm B-10
GUNS 281: **57mm** 174 ZIS-2; (90 more in store); **76mm** 83 ZIS-3; **100mm** 24 M-1944

ARTILLERY 766
TOWED 12: **122mm** 12 D-30; (**152mm** 30 D-20 in store)
MRL 151: **107mm** 33 Type-63: **122mm** 118: 18 BM-21 *Grad*; 100 *Grad* 1P (BM-21P) (single-tube rocket launcher, man portable)
MOR 603: **82mm** 579; **120mm** 24 M-43; (**160mm** 4 M-160 in store)

AIR DEFENCE • SAM • **Point-defence** 9K36 *Strela*-3 (SA-14 *Gremlin*); 9K310 *Igla*-1 (SA-16 *Gimlet*); 9K32 *Strela*-2 (SA-7 *Grail*)‡

Navy ε800

EQUIPMENT BY TYPE

PATROL AND COASTAL COMBATANTS
PB 12: 3 *Dabur*; 1 *Rio Segovia* (*Zhuk* 1400ME); 4 Rodman 101; 2 *Soberania* (ex-JAM Damen Stan Patrol 4207); 2 others

Marines

FORCES BY ROLE

MANOEUVRE
Amphibious
1 mne bn

Air Force 1,200

FORCES BY ROLE

TRANSPORT
1 sqn with An-26 *Curl*; Beech 90 *King Air*; Cessna U206; Cessna 404 *Titan* (VIP)

TRAINING
1 unit with Cessna 172; PA-18 *Super Cub*; PA-28 *Cherokee*

TRANSPORT HELICOPTER
1 sqn with Mi-17 *Hip* H (armed)

AIR DEFENCE
1 gp with ZU-23

EQUIPMENT BY TYPE

AIRCRAFT
TPT • Light 9: 3 An-26 *Curl*; 1 Beech 90 *King Air*; 1 Cessna 172; 1 Cessna U206; 1 Cessna 404 *Titan* (VIP); 2 PA-28 *Cherokee*
TRG 2 PA-18 *Super Cub*

HELICOPTERS
MRH 7 Mi-17 *Hip* H (armed)†
TPT • Medium 2 Mi-171E

AIR DEFENCE • **GUNS 23mm** 18 ZU-23

AIR-LAUNCHED MISSILES • **ASM** 9M17 *Skorpion* (AT-2 *Swatter*)

Panama PAN

Panamanian Balboa B		2019	2020	2021
GDP	B	66.8bn	60.3bn	
	US$	66.8bn	60.3bn	
per capita	US$	15,834	14,090	
Growth	%	3.0	-9.0	
Inflation	%	-0.4	-0.8	
Def bdgt [a]	B	805m	753m	
	US$	805m	753m	
US$1=B		1.00	1.00	

[a] Public-security expenditure

Real-terms defence budget trend (US$m, constant 2015)

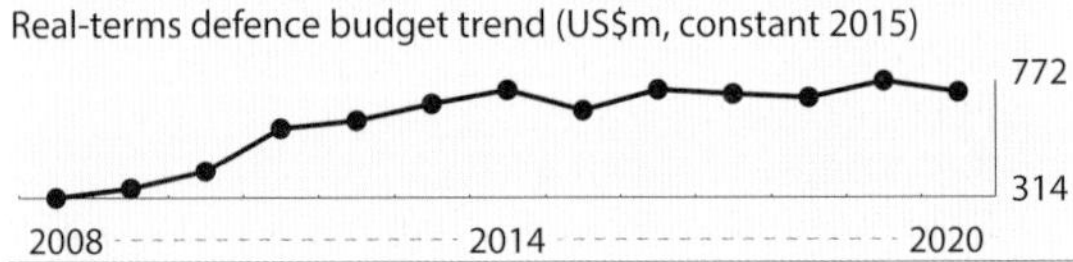

Population 3,894,082

Age	0–14	15–19	20–24	25–29	30–64	65 plus
Male	13.0%	4.2%	4.2%	3.9%	20.7%	4.1%
Female	12.5%	4.1%	4.1%	3.8%	20.4%	4.9%

Capabilities

Panama abolished its armed forces in 1990, but retains a border service, a police force and an air/maritime service for low-level security tasks. The primary security focus is on the southern border with Colombia, and the majority of the border service is deployed there. Both Colombia and the US have provided training and support. Training is focused on internal and border security rather than conventional military operations and there is no capability to mount significant external deployments. None of Panama's security services maintain heavy military equipment, focusing instead on light transport, patrol and surveillance capabilities. Aside from limited maintenance facilities, the country has no domestic defence industry.

Paramilitary 27,700

ORGANISATIONS BY SERVICE

Paramilitary 27,700

National Border Service 4,000

FORCES BY ROLE
SPECIAL FORCES
1 SF gp
MANOEUVRE
Other
1 sy bde (5 sy bn(-))
1 indep sy bn

National Police Force 20,000

No hvy mil eqpt, small arms only

FORCES BY ROLE
SPECIAL FORCES
1 SF unit
MANOEUVRE
Other
1 (presidential) gd bn(-)

National Aeronaval Service 3,700

FORCES BY ROLE
TRANSPORT
1 sqn with C-212M *Aviocar*; Cessna 210; PA-31 *Navajo*; PA-34 *Seneca*
1 (Presidential) flt with ERJ-135BJ; S-76C
TRAINING
1 unit with Cessna 152; Cessna 172; T-35D *Pillan*
TRANSPORT HELICOPTER
1 sqn with AW139; Bell 205; Bell 205 (UH-1H *Iroquois*); Bell 212; Bell 407; Bell 412EP; H145; MD-500E

EQUIPMENT BY TYPE
PATROL AND COASTAL COMBATANTS 15
PCO 1 *Independencia* (ex-US *Balsam*)
PCC 1 *Saettia*
PB 13: 1 *Cocle* (ex-US Swift); 1 *Chiriqui* (ex-US PB MkIV); 2 *Panquiaco* (UK Vosper 31.5m); 5 3 *De Noviembre* (ex-US *Point*), 4 Type-200
AMPHIBIOUS • LANDING CRAFT 1
LCU 1 *General Estaban Huertas*
LOGISTICS AND SUPPORT 2
AG 1 *Lina María*
AKR 1 *Manuel Amador Guerror* (Damen Stan Lander 5612)
AIRCRAFT
TPT • Light 16: 1 Beech 100 *King Air*; 1 Beech 350 *King Air*; 2 DHC-6-400 *Twin Otter*; 3 C-212M *Aviocar*; 1 Cessna 152, 1 Cessna 172; 2 Cessna 208B; 1 Cessna 210; 1 ERJ-135BJ; 1 PA-31 *Navajo*; 2 PA-34 *Seneca*
TRG (2 T-35D *Pillan* in store)
HELICOPTERS
MRH 10: 8 AW139; 1 Bell 412EP; 1 MD-500E
TPT • Light 5: 1 AW109; 2 Bell 212; 2 Bell 407

Paraguay PRY

Paraguayan Guarani Pg		2019	2020	2021
GDP	Pg	234tr	232tr	
	US$	37.4bn	35.6bn	
per capita	US$	5,233	4,909	
Growth	%	0.0	-4.0	
Inflation	%	2.8	2.9	
Def bdgt	Pg	1.77tr	1.88tr	1.89tr
	US$	284m	288m	
US$1=Pg		6240.70	6519.86	

Real-terms defence budget trend (US$m, constant 2015)

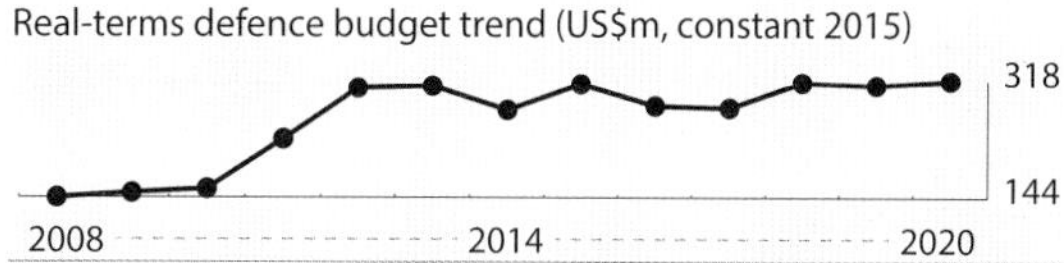

Population	7,191,685

Age	0–14	15–19	20–24	25–29	30–64	65 plus
Male	11.9%	4.3%	4.7%	4.7%	20.8%	3.7%
Female	11.5%	4.2%	4.6%	4.7%	20.7%	4.2%

Capabilities

The armed forces are small by regional standards and the equipment inventory for all services is ageing and largely obsolete. The country faces internal challenges from insurgency and transnational organised crime, chiefly drug trafficking. Conscript numbers have reduced in recent years, and there are a significant number of higher ranks in the force structure. Key formations have long been under-strength. Paraguay has had a consistent if limited tradition of contributing to UN peacekeeping operations since 2001. There is only limited ability to self-sustain forces abroad, and no effective power-projection capacity. There is a small force of river-patrol craft, and some of the oldest vessels have finally been retired. Armoured capability is very limited. Recent acquisitions of heavy equipment have been confined to small quantities of engineering and transport capabilities. There is some local maintenance capacity but the effectiveness of systems is limited by age. While there is some R&D and manufacturing cooperation with local research institutes, there is no traditional defence-industrial base.

ACTIVE 13,950 (Army 7,400 Navy 3,800 Air 2,750)
Paramilitary 14,800
Conscript liability 12 months

RESERVE 164,500 (Joint 164,500)

ORGANISATIONS BY SERVICE

Army 7,400

Much of the Paraguayan army is maintained in a cadre state during peacetime; the nominal inf and cav divs are effectively only at coy strength. Active gp/regt are usually coy sized

FORCES BY ROLE

MANOEUVRE

Light

3 inf corps (total: 6 inf div(-), 3 cav div(-), 6 arty bty)

Other

1 (Presidential) gd regt (1 SF bn, 1 inf bn, 1 sy bn, 1 log gp)

COMBAT SUPPORT

1 arty bde with (2 arty gp, 1 ADA gp)

1 engr bde with (1 engr regt, 3 construction regt)

1 sigs bn

Reserves

FORCES BY ROLE

MANOEUVRE

Light

14 inf regt (cadre)

4 cav regt (cadre)

EQUIPMENT BY TYPE

ARMOURED FIGHTING VEHICLES

RECCE 28 EE-9 *Cascavel*

APC • APC (W) 12 EE-11 *Urutu*

ARTILLERY 99

TOWED 105mm 19 M101

MOR 81mm 80

AIR DEFENCE • GUNS 22:

SP 20mm 3 M9 half track

TOWED 19: **40mm** 13 M1A1, 6 L/60

Navy 3,800

EQUIPMENT BY TYPE

PATROL AND COASTAL COMBATANTS 18

PCR 1 *Itaipú* (BRZ *Roraima*) with 1 hel landing platform

PBR 17: 1 *Capitan Cabral*; 2 *Capitan Ortiz* (ROC *Hai Ou*); 2 *Novatec;* 4 Type-701 (US Sewart); 3 *Croq* 15 (AUS Armacraft); 5 others

AMPHIBIOUS • LANDING CRAFT • LCVP 3

Naval Aviation 100

FORCES BY ROLE

TRANSPORT

1 (liaison) sqn with Cessna 150; Cessna 210 *Centurion;* Cessna 310

TRANSPORT HELICOPTER

1 sqn with AS350 *Ecureuil* (HB350 *Esquilo*)

EQUIPMENT BY TYPE

AIRCRAFT • TPT • Light 6: 2 Cessna 150; 1 Cessna 210 *Centurion;* 2 Cessna 310

HELICOPTERS • TPT • Light 2 AS350 *Ecureuil* (HB350 *Esquilo*)

Marines 700; 200 conscript (total 900)

FORCES BY ROLE

MANOEUVRE

Amphibious

3 mne bn(-)

ARTILLERY • TOWED 105mm 2 M101

Air Force 2,750

FORCES BY ROLE

GROUND ATTACK/ISR

1 sqn with EMB-312 *Tucano**

TRANSPORT

1 gp with C-212-200/400 *Aviocar*; DHC-6 *Twin Otter*

1 VIP gp with Beech 58 *Baron*; Bell 427; Cessna U206 *Stationair*; Cessna 208B *Grand Caravan*; Cessna 402B; PA-32R *Saratoga* (EMB-721C *Sertanejo*)

TRAINING

1 sqn with T-25 *Universal*; T-35A/B *Pillan*

TRANSPORT HELICOPTER

1 gp with AS350 *Ecureuil* (HB350 *Esquilo*); Bell 205 (UH-1H *Iroquois*)

MANOEUVRE

Air Manoeuvre

1 AB bde

EQUIPMENT BY TYPE

AIRCRAFT 6 combat capable

TPT 16: **Light** 15: 1 Beech 58 *Baron*; 4 C-212-200 *Aviocar*; 1 C-212-400 *Aviocar*; 2 Cessna 208B *Grand Caravan*; 1 Cessna 310, 2 Cessna 402B, 2 Cessna U206 *Stationair*; 1 DHC-6 *Twin Otter*; 1 PA-32R *Saratoga* (EMB-721C *Sertanejo*); **PAX** 1 Cessna 680 *Sovereign*

TRG 21: 6 EMB-312 *Tucano**; 6 T-25 *Universal*; 6 T-35A *Pillan*; 3 T-35B *Pillan*

HELICOPTERS • TPT • Light 11: 3 AS350 *Ecureuil* (HB350 *Esquilo*); 6 Bell 205 (UH-1H *Iroquois*); 1 Bell 407; 1 Bell 427 (VIP)

Paramilitary 14,800

Special Police Service 10,800; 4,000 conscript (total 14,800)

DEPLOYMENT

CENTRAL AFRICAN REPUBLIC: UN • MINUSCA 3

CYPRUS: UN • UNFICYP 12

DEMOCRATIC REPUBLIC OF THE CONGO: UN • MONUSCO 7

SOUTH SUDAN: UN • UNMISS 3

Peru PER

Peruvian Nuevo Sol NS		2019	2020	2021
GDP	NS	770bn	686bn	
	US$	231bn	196bn	
per capita	US$	6,958	5,845	
Growth	%	2.2	-13.9	
Inflation	%	2.1	1.8	
Def bdgt	NS	7.47bn	7.45bn	7.05bn
	US$	2.24bn	2.13bn	
FMA (US)	US$	1.8m	0m	0m
US$1=NS		3.34	3.50	

Real-terms defence budget trend (US$bn, constant 2015)

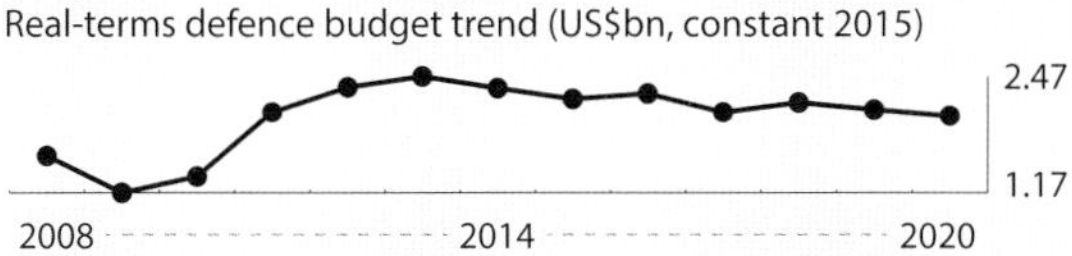

Population	31,914,989

Age	0–14	15–19	20–24	25–29	30–64	65 plus
Male	12.9%	4.3%	4.4%	4.4%	19.3%	3.8%
Female	12.5%	4.2%	4.4%	4.5%	21.1%	4.3%

Capabilities

Peru's armed forces are primarily orientated towards preserving territorial integrity and security, focusing on counter-insurgency and counter-narcotics operations, while strengthening their disaster-relief capabilities. Training and capabilities are sufficient for domestic-security tasks, although the armed forces are limited by an ageing equipment inventory and economic constraints. There is a cooperation agreement with Colombia on air control, humanitarian assistance and counter-narcotics. The armed forces train regularly and take part in national and multilateral exercises. The armed forces are capable of independently deploying externally and contribute to UN missions abroad. There has been some aviation modernisation, though not across the whole fixed- and rotary-wing fleets. The navy is modernising its ageing submarines. The state-owned shipyard SIMA and aviation firm SEMAN are key players in Peru's defence industry, both in terms of manufacturing and maintenance. SEMAN completed in 2017 final assembly for the last South Korean-designed KT-1 trainer, and the navy in 2018 commissioned the first locally built and South Korean-designed multipurpose vessel.

ACTIVE 81,000 (Army 47,500 Navy 24,000 Air 9,500)
Paramilitary 77,000

Conscript liability 12 months voluntary conscription for both males and females

RESERVE 188,000 (Army 188,000)

ORGANISATIONS BY SERVICE

Space

EQUIPMENT BY TYPE

SATELLITES • ISR PERÚSAT-1

Army 47,500

4 mil region

FORCES BY ROLE

SPECIAL FORCES
1 (1st) SF bde (2 spec ops bn, 2 cdo bn, 1 cdo coy, 1 CT coy, 1 airmob arty gp, 1 MP coy, 1 cbt spt bn)
1 (3rd) SF bde (1 spec ops bn, 2 cdo bn, 1 airmob arty gp, 1 MP coy)
1 (6th) SF bde (2 spec ops bn, 2 cdo bn, 1 cdo coy, 1 MP coy)

MANOEUVRE

Armoured
1 (3rd) armd bde (2 tk bn, 1 armd inf bn, 1 arty gp, 1 AT coy, 1 AD gp, 1 engr bn, 1 cbt spt bn)
1 (9th) armd bde (2 tk bn, 1 armd inf bn, 1 SP arty gp, 1 ADA gp)

Mechanised
1 (3rd) armd cav bde (3 mech cav bn, 1 mot inf bn, 1 arty gp, 1 AD gp, 1 engr bn, 1 cbt spt bn)
1 (1st) cav bde (4 mech cav bn, 1 MP coy, 1 cbt spt bn)

Light
2 (2nd & 31st) mot inf bde (4 mot inf bn, 1 arty gp, 1 MP coy, 1 log bn)
3 (1st, 7th & 32nd) inf bde (3 inf bn, 1 MP coy, 1 cbt spt bn)
1 (33rd) inf bde (4 inf bn)
1 (4th) mtn bde (1 armd regt, 3 mot inf bn, 1 arty gp, 1 MP coy, 1 cbt spt bn)
1 (5th) mtn bde (1 armd regt, 2 mot inf bn, 3 jungle coy, 1 arty gp, 1 MP coy, 1 cbt spt bn)
1 (6th) jungle inf bde (4 jungle bn, 1 engr bn, 1 MP coy, 1 cbt spt bn)
1 (35th) jungle inf bde (1 SF gp, 3 jungle bn, 3 jungle coy, 1 jungle arty gp, 1 AT coy, 1 AD gp, 1 jungle engr bn)

COMBAT SUPPORT
1 arty gp (bde) (4 arty gp, 2 AD gp, 1 sigs gp)
1 (3rd) arty bde (4 arty gp, 1 AD gp, 1 sigs gp)
1 (22nd) engr bde (3 engr bn, 1 demining coy)

COMBAT SERVICE SUPPORT
1 (1st Multipurpose) spt bde

AVIATION
1 (1st) avn bde (1 atk hel/recce hel bn, 1 avn bn, 2 aslt hel/tpt hel bn)

AIR DEFENCE
1 AD gp (regional troops)

EQUIPMENT BY TYPE

ARMOURED FIGHTING VEHICLES
MBT 165 T-55; (75† in store)
LT TK 96 AMX-13
RECCE 95: 30 BRDM-2; 15 Fiat 6616; 50 M9A1
APC 295
APC (T) 120 M113A1
APC (W) 175: 150 UR-416; 25 Fiat 6614

ENGINEERING & MAINTENANCE VEHICLES
ARV M578
VLB GQL-111

ANTI-TANK-ANTI-INFRASTRUCTURE
MSL
SP 22 M1165A2 HMMWV with 9K135 *Kornet* E (AT-14 *Spriggan*)
MANPATS 9K11 *Malyutka* (AT-3 *Sagger*); HJ-73C; 9K135 *Kornet* E (AT-14 *Spriggan*); *Spike*-ER
RCL 106mm M40A1

ARTILLERY 1,011
SP 155mm 12 M109A2
TOWED 290: **105mm** 152: 44 M101; 24 M2A1; 60 M-56; 24 Model 56 pack howitzer; **122mm**; 36 D-30; **130mm** 36 M-46; **155mm** 66: 36 M114, 30 Model 50
MRL 122mm 35: 22 BM-21 *Grad*; 13 Type-90B
MOR 674+: **81mm/107mm** 350; **SP 107mm** 24 M106A1; **120mm** 300+ Brandt/Expal Model L
AIRCRAFT
TPT • Light 17: 2 An-28 *Cash*; 3 An-32B *Cline;* 1 Beech 350 *King Air*; 1 Beech 1900D; 4 Cessna 152; 1 Cessna 208 *Caravan*; 1 Cessna 560 *Citation*; 2 Cessna U206 *Stationair;* 1 PA-31T *Cheyenne* II; 1 PA-34 *Seneca*
TRG 4 IL-103
HELICOPTERS
MRH 7 Mi-17 *Hip* H
TPT 35: **Heavy** (3 Mi-26T *Halo* in store); **Medium** 22 Mi-171Sh; **Light** 13: 2 AW109K2; 9 PZL Mi-2 *Hoplite*; 2 R-44
TRG 4 F-28F
AIR DEFENCE
SAM • Point-defence 9K36 *Strela*-3 (SA-14 *Gremlin*); 9K310 *Igla*-1 (SA-16 *Gimlet*); 9K32 *Strela*-2 (SA-7 *Grail*)‡
GUNS 165
SP 23mm 35 ZSU-23-4
TOWED 23mm 130: 80 ZU-23-2; 50 ZU-23

Navy 24,000 (incl 1,000 Coast Guard)

Commands: Pacific, Lake Titicaca, Amazon River

EQUIPMENT BY TYPE
SUBMARINES • SSK 6:
4 *Angamos* (GER T-209/1200) with 8 single 533mm TT with SST-4 HWT (of which 1 in refit)
2 *Islay* (GER T-209/1100) with 8 single 533mm TT with SUT 264 HWT
PRINCIPAL SURFACE COMBATANTS • FRIGATES 7
FFGHM 7:
2 *Aguirre* (ex-ITA *Lupo*) with 8 single lnchr with *Otomat* Mk2 AShM, 1 octuple Mk 29 lnchr with RIM-7P *Sea Sparrow* SAM, 2 triple 324mm ASTT with A244 LWT, 1 127mm gun (capacity 1 Bell 212 (AB-212)/SH-3D *Sea King*)
2 *Aguirre* (ex-ITA *Lupo*) with 2 twin lnchr with MM40 *Exocet* Block 3 AShM, 1 octuple Mk 29 lnchr with RIM-7P *Sea Sparrow* SAM, 2 triple 324mm ASTT with A244 LWT, 1 127mm gun (capacity 1 Bell 212 (AB-212)/SH-3D *Sea King*)
1 *Carvajal* (ITA *Lupo* mod) with 8 single lnchr with *Otomat* Mk2 AShM, 1 octuple *Albatros* lnchr with *Aspide* SAM, 2 triple 324mm ASTT with A244 LWT, 1 127mm gun (capacity 1 Bell 212 (AB-212)/SH-3D *Sea King*)
2 *Carvajal* (ITA *Lupo* mod) with 8 single lnchr with MM40 *Exocet* Block 3 AShM, 1 octuple *Albatros* lnchr with *Aspide* SAM, 2 triple 324mm ASTT with A244 LWT, 1 127mm gun (capacity 1 Bell 212 (AB-212)/SH-3D *Sea King*)
PATROL AND COASTAL COMBATANTS 13
CORVETTES 7
FSG 6 *Velarde* (FRA PR-72 64m) with 4 single lnchr with MM38 *Exocet* AShM, 1 76mm gun
FS 1 *Ferré* (ex-ROK *Po Hang*) with 1 76mm gun
PCR 6: 2 *Amazonas* with 1 76mm gun; 2 *Manuel Clavero*; 2 *Marañon* with 2 76mm guns
AMPHIBIOUS
PRINCIPAL AMPHIBIOUS SHIPS • LPD 1 *Pisco* (IDN *Makassar*) (capacity 2 LCM; 3 hels; 24 IFV; 450 troops)
LANDING SHIPS • LST 2 *Paita* (capacity 395 troops) (ex-US *Terrebonne Parish*)
LANDING CRAFT • UCAC 7 *Griffon* 2000TD (capacity 22 troops)
LOGISTICS AND SUPPORT 26
AG 5 *Rio Napo*
AGOR 1 *Humboldt*
AGORH 1 *Carrasco*
AGS 5: 1 *Zimic* (ex-NLD *Dokkum*); 2 *Van Straelen*; 1 *La Macha*, 1 *Stiglich* (river survey vessel for the upper Amazon)
AH 4 (river hospital craft)
AO 2 *Noguera*
AOR 1 *Mollendo*
AORH 1 *Tacna* (ex-NLD *Amsterdam*)
AOT 2 *Bayovar*
ATF 1 *Morales*
AWT 1 *Caloyeras*
AXS 2: 1 *Marte*; 1 *Union*

Naval Aviation ε800

FORCES BY ROLE
MARITIME PATROL
1 sqn with Beech 200T; Bell 212 ASW (AB-212 ASW); F-27 *Friendship*; Fokker 60; SH-2G *Super Seasprite*; SH-3D *Sea King*
TRANSPORT
1 flt with An-32B *Cline;* Cessna 206; Fokker 50
TRAINING
1 sqn with F-28F; T-34C *Turbo Mentor*
TRANSPORT HELICOPTER
1 (liaison) sqn with Bell 206B *Jet Ranger II*; Mi-8 *Hip*
EQUIPMENT BY TYPE
AIRCRAFT
MP 8: 4 Beech 200T; 4 Fokker 60
ELINT 1 F-27 *Friendship*
TPT • Light 5: 2 An-32B *Cline*; 1 Cessna 206; 2 Fokker 50
TRG 5 T-34C *Turbo Mentor*
HELICOPTERS
ASW 6: 2 Bell 212 ASW (AB-212 ASW); 1 SH-2G *Super Seasprite*; 3 SH-3D *Sea King*
MRH 3 Bell 412SP
TPT 11: **Medium** 8: 2 Mi-8 *Hip*; 6 UH-3H *Sea King;* **Light** 3 Bell 206B *Jet Ranger* II
TRG 5 F-28F
MSL • AShM AM39 *Exocet*

Marines 4,000

FORCES BY ROLE
SPECIAL FORCES
3 cdo gp
MANOEUVRE
Light
2 inf bn
1 inf gp

Amphibious
1 mne bde (1 SF gp, 1 recce bn, 2 inf bn, 1 amph bn, 1 arty gp)
Jungle
1 jungle inf bn

EQUIPMENT BY TYPE
ARMOURED FIGHTING VEHICLES
RECCE 7 RAM Mk3
APC • APC (W) 47+: 32 LAV II; V-100 *Commando*; 15 V-200 *Chaimite*
ANTI-TANK/ANTI-INFRASTRUCTURE
RCL 84mm *Carl Gustaf*; **106mm** M40A1
ARTILLERY 18+
TOWED 122mm D-30
MOR 18+: **81mm** some; **120mm** ε18
AIR DEFENCE • GUNS 20mm SP (twin)

Air Force 9,500

Divided into five regions – North, Lima, South, Central and Amazon

FORCES BY ROLE
FIGHTER
1 sqn with MiG-29S/SE *Fulcrum* C; MiG-29UBM *Fulcrum* B
FIGHTER/GROUND ATTACK
1 sqn with *Mirage* 2000E/ED (2000P/DP)
2 sqn with A-37B *Dragonfly*
1 sqn with Su-25A *Frogfoot* A; Su-25UBK *Frogfoot* B
ISR
1 (photo-survey) sqn with Learjet 36A; SA-227-BC *Metro* III (C-26B)
TRANSPORT
1 sqn with B-737; An-32 *Cline*
1 sqn with DHC-6 *Twin Otter*; DHC-6-400 *Twin Otter*; PC-6 *Turbo Porter*
1 sqn with L-100-20
TRAINING
2 (drug interdiction) sqn with EMB-312 *Tucano*
1 sqn with MB-339A*
1 sqn with Z-242
1 hel sqn with Schweizer 300C
ATTACK HELICOPTER
1 sqn with Mi-25/Mi-35P *Hind*
TRANSPORT HELICOPTER
1 sqn with Mi-17 *Hip* H
1 sqn with Bell 206 *Jet Ranger*; Bell 212 (AB-212); Bell 412 *Twin Huey*
1 sqn with Bo-105LS
AIR DEFENCE
6 bn with S-125 *Pechora* (SA-3 *Goa*)

EQUIPMENT BY TYPE
AIRCRAFT 60 combat capable
FTR 19: 9 MiG-29S *Fulcrum* C; 3 MiG-29SE *Fulcrum* C; 5 MiG-29SMP *Fulcrum*; 2 MiG-29UBM *Fulcrum* B
FGA 12: 2 *Mirage* 2000ED (2000DP); 10 *Mirage* 2000E (2000P) (some†)
ATK 19: 15 A-37B *Dragonfly*; 2 Su-25A *Frogfoot* A; 2 Su-25UBK *Frogfoot* B; (8 Su-25A *Frogfoot* A; 6 Su-25UBK *Frogfoot* B in store)
ISR 5: 2 Learjet 36A; 3 SA-227-BC *Metro* III (C-26B)
TPT 37: **Medium** 6: 4 C-27J *Spartan*; 2 L-100-20; **Light** 29: 4 An-32 *Cline*; 7 Cessna 172 *Skyhawk*; 3 DHC-6 *Twin Otter*; 12 DHC-6-400 *Twin Otter*; 1 Learjet 45 (VIP); 1 PA-44; 1 PC-6 *Turbo-Porter*; **PAX** 2 B-737
TRG 73: 6 CH-2000; 19 EMB-312 *Tucano*†; 20 KT-1P; 10 MB-339A*; 6 T-41A/D *Mescalero*; 12 Z-242
HELICOPTERS
ATK 18: 16 Mi-25 *Hind* D; 2 Mi-35P *Hind* E
MRH 19: 2 Bell 412 *Twin Huey*; 17 Mi-17 *Hip* H
TPT 23: **Medium** 3 Mi-171Sh; **Light** 20: 8 Bell 206 *Jet Ranger*; 6 Bell 212 (AB-212); 6 Bo-105LS
TRG 4 Schweizer 300C
AIR DEFENCE • SAM
Short-range S-125 *Pechora* (SA-3 *Goa*)
Point-defence *Javelin*
AIR-LAUNCHED MISSILES
AAM • IR R-3 (AA-2 *Atoll*)‡; R-60 (AA-8 *Aphid*)‡; R-73 (AA-11A *Archer*); R-550 *Magic*; **IR/SARH** R-27 (AA-10 *Alamo*); **ARH** R-77 (AA-12 *Adder*)
ASM AS-30; Kh-29L (AS-14A *Kedge*)
ARM Kh-58 (AS-11 *Kilter*)

Paramilitary 77,000

National Police 77,000 (100,000 reported)

EQUIPMENT BY TYPE
ARMOURED FIGHTING VEHICLES
APC (W) 120: 20 BMR-600; 100 MOWAG *Roland*
AIRCRAFT
TPT • Light 1 An-32B *Cline*
HELICOPTERS
MRH 1 Mi-17 *Hip* H

General Police 43,000

Security Police 21,000

Technical Police 13,000

Coast Guard 1,000

Personnel included as part of Navy

EQUIPMENT BY TYPE
PATROL AND COASTAL COMBATANTS 38
PSOH 1 *Carvajal* (ITA *Lupo* mod) with 1 127mm gun (capacity 1 Bell 212 (AB-212)/SH-3D *Sea King*)
PCC 8: 4 *Río Pativilca* (ROK *Tae Geuk*); 4 *Rio Nepena*
PB 10: 6 *Chicama* (US *Dauntless*); 1 *Río Chira*; 3 *Río Santa*
PBR 19: 1 *Río Viru*; 8 *Parachique*; 10 *Zorritos*
LOGISTICS AND SUPPORT • AH 1 *Puno*
AIRCRAFT
TPT • Light 3: 1 DHC-6 *Twin Otter*; 2 F-27 *Friendship*

Rondas Campesinas

Peasant self-defence force. Perhaps 7,000 rondas 'gp', up to pl strength, some with small arms. Deployed mainly in emergency zone

DEPLOYMENT

CENTRAL AFRICAN REPUBLIC: UN • MINUSCA 216; 1 engr coy

DEMOCRATIC REPUBLIC OF THE CONGO: UN • MONUSCO 7

LEBANON: UN • UNIFIL 1

SOUTH SUDAN: UN • UNMISS 5

SUDAN: UN • UNISFA 1; UN • UNISFA 2

Suriname SUR

Suriname Dollar srd		2019	2020	2021
GDP	srd	27.6bn	39.3bn	
	US$	3.70bn	2.54bn	
per capita	US$	6,191	4,199	
Growth	%	0.3	-13.1	
Inflation	%	4.4	49.8	
Def bdgt	srd	n.k.	n.k.	
	US$	n.k.	n.k.	
US$1=srd		7.46	15.50	

Population 609,569

Age	0–14	15–19	20–24	25–29	30–64	65 plus
Male	11.9%	4.3%	4.5%	4.0%	22.8%	2.9%
Female	11.5%	4.1%	4.3%	3.8%	22.2%	3.7%

Capabilities

The armed forces are principally tasked with preserving territorial integrity. They also assist the national police in internal- and border-security missions, as well as tackling transnational criminal activity and drug trafficking and have also been involved in disaster-relief and humanitarian-assistance operations. Suriname is a member of the Caribbean Disaster Emergency Management Agency and the Caribbean Basin Security Initiative. Ties with Brazil, China, India and the US have been crucial for the supply of equipment, including a limited number of armoured vehicles and helicopters, as well as training activity. The armed forces take part in US SOUTHCOM's *Tradewinds* disaster-response exercise. The armed forces are not sized or equipped for power projection. Resource challenges and limited equipment serviceability means the armed forces are constrained in providing sufficient border and coastal control and surveillance. There is no capability to design and manufacture modern military equipment and Suriname has looked to its foreign-military cooperation to improve not just trade training but also maintenance capacity.

ACTIVE 1,840 (Army 1,400 Navy 240 Air 200)
(All services form part of the army)

ORGANISATIONS BY SERVICE

Army 1,400

FORCES BY ROLE
MANOEUVRE
Mechanised
1 mech cav sqn
Light
1 inf bn (4 coy)
COMBAT SUPPORT
1 MP bn (coy)

EQUIPMENT BY TYPE
ARMOURED FIGHTING VEHICLES
RECCE 6 EE-9 *Cascavel*
APC • APC (W) 15 EE-11 *Urutu*
ANTI-TANK/ANTI-INFRASTRUCTURE
RCL 106mm M40A1
ARTILLERY • **MOR 81mm** 6

Navy ε240

EQUIPMENT BY TYPE
PATROL AND COASTAL COMBATANTS 13
PB 8: 2 FPB 72; 1 FPB 98 Mk I; 3 Rodman 101†; 2 others
PBR 5 Rodman 55

Air Force ε200

EQUIPMENT BY TYPE
AIRCRAFT 2 combat capable
TPT • Light 2: 1 BN-2 *Defender**; 1 Cessna 182
TRG 1 PC-7 *Turbo Trainer**
HELICOPTERS • **MRH** 3 SA316B *Alouette* III (*Chetak*)

Trinidad and Tobago TTO

Trinidad and Tobago Dollar TT$		2019	2020	2021
GDP	TT$	163bn	153bn	
	US$	24.1bn	22.7bn	
per capita	US$	17,276	16,197	
Growth	%	0.0	-5.6	
Inflation	%	1.0	0.0	
Def bdgt	TT$	6.12bn	6.44bn	5.23bn
	US$	906m	953m	
US$1=TT$		6.75	6.75	

Real-terms defence budget trend (US$m, constant 2015)

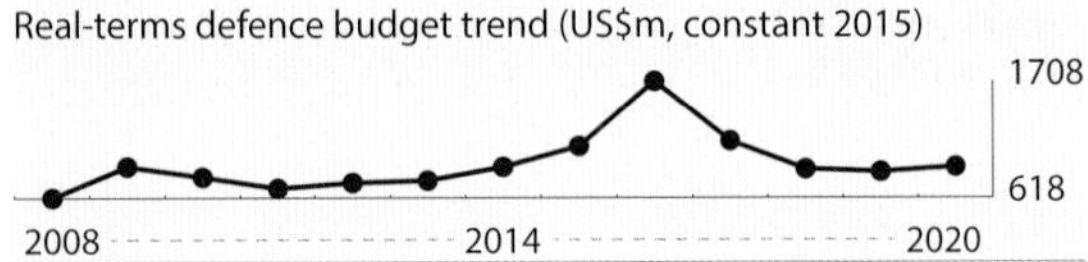

Population 1,208,789

Age	0–14	15–19	20–24	25–29	30–64	65 plus
Male	9.7%	3.0%	2.9%	3.2%	26.6%	5.3%
Female	9.3%	2.8%	2.6%	2.9%	24.8%	6.8%

Capabilities

The Trinidad and Tobago Defence Force (TTDF) focuses on border protection and maritime security, as well as counter-narcotics tasks. A larger role in law-enforcement support is planned for the army. Trinidad and Tobago is a member of the Caribbean Community, and cooperates with other countries in the region in disaster-relief efforts. It has taken part in US SOUTHCOM's *Tradewinds* disaster-response exercise and has sent personnel to the US and UK for training. Trinidad and Tobago has no capacity to deploy and maintain troops abroad, and bar limited maintenance facilities has no domestic defence industry.

ACTIVE 4,600 (Army 3,000 Coast Guard 1,600)
(All services form the Trinidad and Tobago Defence Force)

ORGANISATIONS BY SERVICE

Army ε3,000

FORCES BY ROLE

SPECIAL FORCES

1 SF unit

MANOEUVRE

Light

2 inf bn

COMBAT SUPPORT

1 engr bn

COMBAT SERVICE SUPPORT

1 log bn

EQUIPMENT BY TYPE

ANTI-TANK/ANTI-INFRASTRUCTURE

RCL 84mm *Carl Gustaf*

ARTILLERY • **MOR 81mm** 6 L16A1

Coast Guard 1,600

FORCES BY ROLE

COMMAND

1 mne HQ

EQUIPMENT BY TYPE

PATROL AND COASTAL COMBATANTS 15

PCO 1 *Nelson* II (ex-PRC *Shuke* III)

PCC 6: 2 *Point Lisas* (Damen Fast Crew Supplier 5009); 4 *Speyside* (Damen Stan Patrol 5009)

PB 8: 2 *Gaspar Grande*; 6 *Scarlet Ibis* (Austal 30m) (1 *Cascadura* (SWE *Karlskrona* 40m) non-operational)

Air Wing 50

EQUIPMENT BY TYPE

AIRCRAFT

TPT • Light 2 SA-227 *Metro* III (C-26)

HELICOPTERS

MRH 2 AW139

TPT • Light 1 S-76

Uruguay URY

Uruguayan Peso pU		2019	2020	2021
GDP	pU	1.98tr	2.10tr	
	US$	56.7bn	54.1bn	
per capita	US$	16,111	15,332	
Growth	%	0.2	-4.5	
Inflation	%	7.9	10.0	
Def bdgt	pU	14.8bn	14.7bn	
	US$	423m	379m	
US$1=pU		34.86	38.86	

Real-terms defence budget trend (US$m, constant 2015)

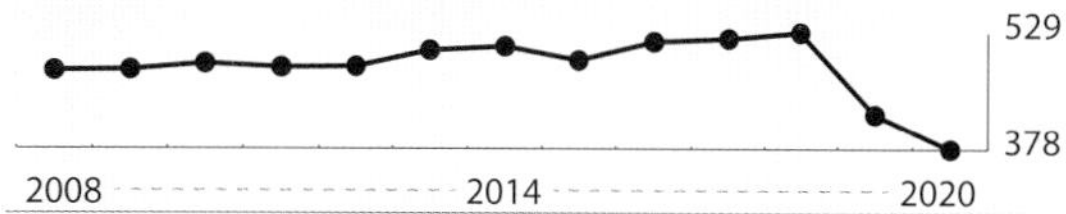

Population 3,387,605

Age	0–14	15–19	20–24	25–29	30–64	65 plus
Male	9.9%	3.7%	4.0%	3.8%	21.0%	5.9%
Female	9.6%	3.6%	3.9%	3.7%	22.0%	8.8%

Capabilities

Principal tasks for the armed forces are assuring sovereignty and territorial integrity. In 2019, parliament approved a new Military Law, which aims, among other measures, to reduce the number of senior officers and address promotion issues across all services. Uruguay and Argentina have a joint peacekeeping unit and conduct joint exercises. In 2018 a defence-cooperation agreement was signed with Russia, including training exchanges. The armed forces participate regularly in multinational exercises and deployments, notably on UN missions. The air force is focused on the counter-insurgency role, but ambitions to purchase a light fighter aircraft remain hampered by funding problems. The acquisition of air-defence radars may have improved the armed forces' ability to monitor domestic airspace, but the lack of interdiction capability will continue to limit the capacity to respond to contingencies. Much equipment is second-hand, and there is little capacity for independent power projection. Maintenance work is sometimes outsourced to foreign companies, such as Chile's ENAER.

ACTIVE 21,100 (Army 13,500 Navy 5,000 Air 2,600)
Paramilitary 1,400

ORGANISATIONS BY SERVICE

Army 13,500

Uruguayan units are substandard size, mostly around 30%. Div are at most bde size, while bn are of reinforced coy strength. Regts are also coy size, some bn size, with the largest formation being the 2nd armd cav regt

FORCES BY ROLE

COMMAND

4 mil region/div HQ

MANOEUVRE

Mechanised

2 (1st & 2nd Cav) mech bde (1 armd cav regt, 2 mech cav regt)

1 (3rd Cav) mech bde (2 mech cav regt, 1 mech inf bn)
3 (2nd, 3rd & 4th Inf) mech bde (2 mech inf bn; 1 inf bn)
1 (5th Inf) mech bde (1 armd cav regt; 1 armd inf bn; 1 mech inf bn)
Light
1 (1st Inf) inf bde (2 inf bn)
Air Manoeuvre
1 para bn
COMBAT SUPPORT
1 (strategic reserve) arty regt
5 fd arty gp
1 (1st) engr bde (2 engr bn)
4 cbt engr bn
AIR DEFENCE
1 AD gp
EQUIPMENT BY TYPE
ARMOURED FIGHTING VEHICLES
MBT 15 *Tiran-5*
LT TK 47: 22 M41A1UR; 25 M41C
RECCE 15 EE-9 *Cascavel*
IFV 18 BMP-1
APC 376
APC (T) 27: 24 M113A1UR; 3 MT-LB
APC (W) 349: 54 *Condor*; 48 GAZ-39371 *Vodnik*; 53 OT-64; 47 OT-93; 147 *Piranha*
ENGINEERING & MAINTENANCE VEHICLES
AEV MT-LB
ANTI-TANK/ANTI-INFRASTRUCTURE
MSL • MANPATS *Milan*
RCL 106mm 69 M40A1
ARTILLERY 185
SP 122mm 6 2S1 *Gvozdika*
TOWED 44: **105mm** 36: 28 M101A1; 8 M102; **155mm** 8 M114A1
MOR 135: **81mm** 91: 35 M1, 56 Expal Model LN; **120mm** 44 Model SL
UNINHABITED AERIAL VEHICLES • ISR • Light 1 *Charrua*
AIR DEFENCE • GUNS • TOWED 14: **20mm** 14: 6 M167 *Vulcan*; 8 TCM-20 (w/Elta M-2106 radar)

Navy 5,000
HQ at Montevideo
EQUIPMENT BY TYPE
PATROL AND COASTAL COMBATANTS 16
CORVETTES • FS 1 *Uruguay* (ex-PRT *Joao Belo*) (FRA *Commandant Riviere*) with 2 triple 324mm SVTT Mk 32 ASTT with Mk 46 LWT, 2 100mm gun
PB 15: 2 *Colonia* (ex-US *Cape*); 1 *Paysandu*; 9 Type-44; 3 PS
MINE WARFARE • MINE COUNTERMEASURES 2
MSO 2 *Temerario* (*Kondor* II)
AMPHIBIOUS 3: 2 **LCVP**; 1 **LCM**
LOGISTICS AND SUPPORT 10
AAR 2 *Islas de Flores* (ex-GER *Hermann Helms*)
ABU 1 *Sirius*
AG 2: 1 *Artigas* (GER *Freiburg*, general spt ship with replenishment capabilities); 1 *Maldonado* (also used as patrol craft)
AGS 2: 1 *Helgoland*; 1 *Trieste* (in refit)
ARS 1 *Vanguardia*
AXS 2: 1 *Capitan Miranda*; 1 *Bonanza*

Naval Aviation 210
FORCES BY ROLE
MARITIME PATROL
1 flt with Beech 200T*; Cessna O-2A *Skymaster*
SEARCH & RESCUE/TRANSPORT HELICOPTER
1 sqn with AS350B2 *Ecureuil* (*Esquilo*); Bell 412SP *Twin Huey*
TRANSPORT/TRAINING
1 flt with T-34C *Turbo Mentor*
TRAINING
1 hel sqn with Bell 412SP *Twin Huey*; OH-58 *Kiowa*
EQUIPMENT BY TYPE
AIRCRAFT 2 combat capable
ISR 4: 2 Beech 200T*; 2 Cessna O-2A *Skymaster*
TRG 2 T-34C *Turbo Mentor*
HELICOPTERS
ISR 1 OH-58 *Kiowa*
MRH 4: 2 Bell 412 (AB-412); 2 Bell 412SP *Twin Huey*
TPT • Light 1 AS350B2 *Ecureuil* (*Esquilo*)

Naval Infantry 700
FORCES BY ROLE
MANOEUVRE
Amphibious
1 mne bn(-)

Air Force 2,600
FORCES BY ROLE
FIGHTER/GROUND ATTACK
1 sqn with A-37B *Dragonfly*
ISR
1 flt with EMB-110 *Bandeirante*
TRANSPORT
1 sqn with C-130B *Hercules*; C-212 *Aviocar*; EMB–110C *Bandeirante*; EMB-120 *Brasilia*
1 (liaison) sqn with Cessna 206H; T-41D
1 (liaison) flt with Cessna 206H
TRAINING
1 sqn with PC-7U *Turbo Trainer*
1 sqn with Beech 58 *Baron* (UB-58); SF-260EU
TRANSPORT HELICOPTER
1 sqn with AS365 *Dauphin*; Bell 205 (UH–1H *Iroquois*); Bell 212
EQUIPMENT BY TYPE
AIRCRAFT 13 combat capable
ATK 12 A-37B *Dragonfly*
ISR 1 EMB-110 *Bandeirante**
TPT 24: **Medium** 2 C-130B *Hercules*; **Light** 22: 1 BAe-125-700A; 2 Beech 58 *Baron* (UB-58); 6 C-212 *Aviocar*; 9 Cessna 206H; 1 Cessna 210; 2 EMB-110C *Bandeirante*; 1 EMB-120 *Brasilia*; **PAX** 1 C-29 *Hawker*
TRG 17: 5 PC-7U *Turbo Trainer*; 12 SF-260EU
HELICOPTERS
MRH 2 AS365N2 *Dauphin* II
TPT • Light 9: 5 Bell 205 (UH–1H *Iroquois*); 4 Bell 212

Paramilitary 1,400

Guardia Nacional Republicana 1,400

DEPLOYMENT

CENTRAL AFRICAN REPUBLIC: UN • MINUSCA 2

DEMOCRATIC REPUBLIC OF THE CONGO: UN • MONUSCO 926; 1 inf bn; 1 log coy; 1 hel sqn

EGYPT: MFO 41; 1 engr/tpt unit

INDIA/PAKISTAN: UN • UNMOGIP 3

LEBANON: UN • UNIFIL 2

SYRIA/ISRAEL: UN • UNDOF 191; 1 mech inf coy

Venezuela VEN

Venezuelan Bolivar Fuerte Bs		2019	2020	2021
GDP	Bs	n.k	n.k	
	US$	64.0bn	48.6bn	
per capita	US$	2,299	1,739	
Growth	%	-35.0	-25.0	
Inflation	%	19906.0	6500.0	
Def bdgt	Bs	n.k	n.k	
	US$ [a]	n.k	n.k	
US$1=Bs		n.k	n.k	

[a] US dollar figures should be treated with caution due to high levels of currency volatility as well as wide differentials between official and parallel exchange rates

Real-terms defence budget trend (US$bn, constant 2015)

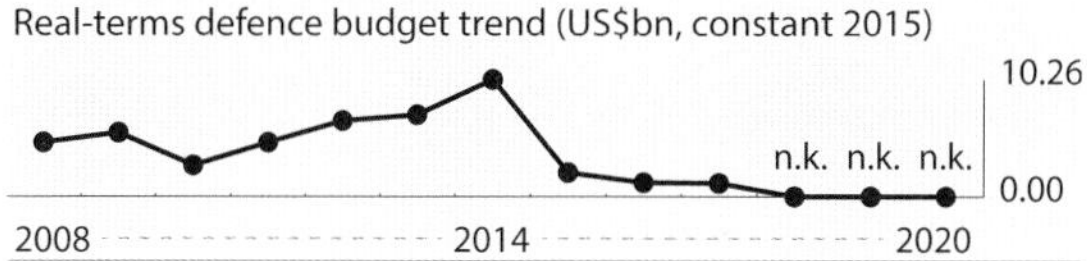

Population 28,644,603

Age	0–14	15–19	20–24	25–29	30–64	65 plus
Male	13.1%	4.2%	4.0%	4.1%	20.6%	3.7%
Female	12.5%	4.0%	4.0%	4.1%	21.3%	4.4%

Capabilities

The armed forces and national guard are tasked with protecting sovereignty, assuring territorial integrity and assisting with internal-security and counter-narcotics operations. They have sufficient capabilities and funding to fulfil internal-security tasks and fulfil their regime-protection role, but economic challenges will affect future equipment availability and training levels. Incidents such as the apparent assassination attempt against President Maduro in 2018 and the attack on the supreme court in 2017 point to some internal stresses in the armed forces. Venezuela is almost completely isolated regionally, with frictions relating to the humanitarian crisis leading to troop deployments near the Colombia–Venezuela border. There are close ties with China and Russia. Caracas relies on both countries for procurements and technical support. The armed forces train regularly and civil–military cooperation has increased. Venezuela has also taken part in joint combined exercises with China, Cuba and Russia. There is little logistics capability to support deployment abroad. Equipment is relatively modern and much is of Chinese and Russian manufacture. However, the economic crisis has affected the government's ability to sustain military expenditure. Maintenance and further procurement may suffer as a consequence. Venezuela's defence industry is based on a series of small, state-owned companies, mainly focused on the production of small arms and munitions. Local platform production has been limited to small coastal-patrol boats.

ACTIVE 123,000 (Army 63,000 Navy 25,500 Air 11,500 National Guard 23,000) **Paramilitary 220,000**
Conscript liability 30 months selective, varies by region for all services

RESERVE 8,000 (Army 8,000)

ORGANISATIONS BY SERVICE

Space

EQUIPMENT BY TYPE
SATELLITES • COMMUNICATIONS 1 *Venesat*-1

Army ε63,000

FORCES BY ROLE
MANOEUVRE
Armoured
1 (4th) armd div (1 armd bde, 1 lt armd bde, 1 AB bde, 1 arty bde)
Mechanised
1 (9th) mot cav div (1 mot cav bde, 1 ranger bde, 1 sy bde)
Light
1 (1st) inf div (1 SF bn, 1 armd bde, 1 mech inf bde, 1 ranger bde, 1 inf bde, 1 arty unit, 1 spt unit)
1 (2nd) inf div (1 mech inf bde, 1 inf bde, 1 mtn inf bde)
1 (3rd) inf div (1 inf bde, 1 ranger bde, 1 sigs bde, 1 MP bde)
1 (5th) inf div (1 SF bn, 1 cav sqn, 2 jungle inf bde, 1 engr bn)
COMBAT SUPPORT
1 cbt engr corps (3 engr regt)
COMBAT SERVICE SUPPORT
1 log comd (2 log regt)
AVIATION
1 avn comd (1 tpt avn bn, 1 atk hel bn, 1 ISR avn bn)

Reserve Organisations 8,000

FORCES BY ROLE
MANOEUVRE
Armoured
1 armd bn
Light
4 inf bn
1 ranger bn
COMBAT SUPPORT
1 arty bn
2 engr regt

EQUIPMENT BY TYPE
ARMOURED FIGHTING VEHICLES
MBT 173: 81 AMX-30V; 92 T-72B1
LT TK 109: 31 AMX-13; 78 *Scorpion*-90
RECCE 121: 42 *Dragoon* 300 LFV2; 79 V-100/V-150

IFV 237: 123 BMP-3 (incl variants); 114 BTR-80A (incl variants)
APC 81
APC (T) 45: 25 AMX-VCI; 12 AMX-PC (CP); 8 AMX-VCTB (Amb)
APC (W) 36 *Dragoon* 300
ENGINEERING & MAINTENANCE VEHICLES
ARV 5: 3 AMX-30D; BREM-1; 2 *Dragoon* 300RV; *Samson*
VLB *Leguan*
NBC VEHICLES 10 TPz-1 *Fuchs* NBC
ANTI-TANK/ANTI-INFRASTRUCTURE
MSL • MANPATS IMI MAPATS
RCL 106mm 175 M40A1
GUNS • SP 76mm 75 M18 *Hellcat*
ARTILLERY 515+
SP 60: **152mm** 48 2S19 *Msta*-S (replacing Mk F3s); **155mm** 12 Mk F3
TOWED 92: **105mm** 80: 40 M101A1; 40 Model 56 pack howitzer; **155mm** 12 M114A1
MRL 56: **122mm** 24 BM-21 *Grad*; **160mm** 20 LAR SP (LAR-160); **300mm** 12 9A52 *Smerch*
GUN/MOR 120mm 13 2S23 NONA-SVK
MOR 294+: **81mm** 165; **SP 81mm** 21 *Dragoon* 300PM; AMX-VTT; **120mm** 108: 60 Brandt; 48 2S12
AIRCRAFT
TPT • Light 28: 1 Beech 90 *King Air*; 1 Beech 200 *King Air*; 1 Beech 300 *King Air*; 1 Cessna 172; 6 Cessna 182 *Skylane*; 2 Cessna 206; 2 Cessna 207 *Stationair*; 1 IAI-201 *Arava*; 2 IAI-202 *Arava*; 11 M-28 *Skytruck*
HELICOPTERS
ATK 9 Mi-35M2 *Hind*
MRH 32: 10 Bell 412EP; 2 Bell 412SP; 20 Mi-17V-5 *Hip* H
TPT 9: **Heavy** 3 Mi-26T2 *Halo*; **Medium** 2 AS-61D; **Light** 4: 3 Bell 206B *Jet Ranger*, 1 Bell 206L3 *Long Ranger* II

Navy ε22,300; ε3,200 conscript (total ε25,500)

EQUIPMENT BY TYPE
SUBMARINES 1
SSK 1 *Sábalo* (1 in refit, 1 more non-operational) (GER T-209/1300) with 8 single 533mm TT with SST-4 HWT
PRINCIPAL SURFACE COMBATANTS • FRIGATES 2
FFGHM 2 *Mariscal Sucre* (ITA *Lupo* mod)† (1 more non-operational) with 8 single lnchr with *Otomat* Mk2 AShM, 1 octuple *Albatros* lnchr with *Aspide* SAM, 2 triple 324mm ASTT with A244 LWT, 1 127mm gun (capacity 1 Bell 212 (AB-212) hel)
PATROL AND COASTAL COMBATANTS 10
PSOH 3 *Guaiqueri* with 1 *Millennium* CIWS, 1 76mm gun
PBG 3 *Federación* (UK Vosper 37m) with 2 single lnchr with *Otomat* Mk2 AShM
PB 4: 3 *Constitucion* (UK Vosper 37m) with 1 76mm gun; 1 *Fernando Gomez de Saa* (Damen Stan Patrol 4207)
AMPHIBIOUS
LANDING SHIPS • LST 4 *Capana* (ROK *Alligator*) capacity 12 tanks; 200 troops)
LANDING CRAFT 3:
LCU 2 *Margarita* (river comd)
UCAC 1 *Griffon* 2000TD
LOGISTICS AND SUPPORT 10
AGOR 1 *Punta Brava*
AGS 2 *Gabriela*
AKR 4 *Los Frailes*
AORH 1 *Ciudad Bolivar*
ATF 1 *Almirante Francisco de Miranda* (Damen Salvage Tug 6014)
AXS 1 *Simon Bolivar*

Naval Aviation 500

FORCES BY ROLE
ANTI-SUBMARINE WARFARE
1 sqn with Bell 212 ASW (AB-212 ASW)
MARITIME PATROL
1 flt with C-212-200 MPA
TRANSPORT
1 sqn with Beech 200 *King Air*; C-212 *Aviocar*; *Turbo Commander* 980C
TRAINING
1 hel sqn with Bell 206B *Jet Ranger* II; TH-57A *Sea Ranger*
TRANSPORT HELICOPTER
1 sqn with Bell 412EP *Twin Huey*; Mi-17V-5 *Hip* H
EQUIPMENT BY TYPE
AIRCRAFT 2 combat capable
MP 2 C-212-200 MPA*
TPT • Light 7: 1 Beech C90 *King Air*; 1 Beech 200 *King Air*; 4 C-212 *Aviocar*; 1 *Turbo Commander* 980C
HELICOPTERS
ASW 4 Bell 212 ASW (AB-212 ASW)
MRH 12: 6 Bell 412EP *Twin Huey*; 6 Mi 17V 5 *Hip* H
TPT • Light 1 Bell 206B *Jet Ranger* II (trg)
TRG 1 TH-57A *Sea Ranger*

Marines ε15,000

FORCES BY ROLE
COMMAND
1 div HQ
SPECIAL FORCES
1 spec ops bde
MANOEUVRE
Amphibious
1 amph aslt bde
3 mne bde
3 (rvn) mne bde
COMBAT SUPPORT
1 cbt engr bn
1 MP bde
1 sigs bn
COMBAT SERVICE SUPPORT
1 log bn
EQUIPMENT BY TYPE
ARMOURED FIGHTING VEHICLES
LT TK 10 VN-16
IFV 21: 11 VN-1; 10 VN-18
APC • APC (W) 37 EE-11 *Urutu*
AAV 11 LVTP-7
ENGINEERING & MAINTENANCE VEHICLES
ARV 1 VN-16 ARV
AEV 1 AAVR7
ANTI-TANK/ANTI-INFRASTRUCTURE
RCL 84mm *Carl Gustaf*; **106mm** M40A1

ARTILLERY 30
TOWED 105mm 18 M-56
MOR 120mm 12 Brandt
PATROL AND COASTAL COMBATANTS
PBR 23: 18 *Constancia*; 2 *Manaure*; 3 *Terepaima* (*Cougar*)
AMPHIBIOUS • LANDING CRAFT • 1 **LCU**; 1 **LCM**; 12 **LCVP**

Coast Guard 1,000

EQUIPMENT BY TYPE
PATROL AND COASTAL COMBATANTS 22
PSO 2 *Guaicamacuto* (of which 1 in refit) with 1 *Millennium* CIWS, 1 76mm gun (capacity 1 Bell 212 (AB-212) hel)
PB 20: 12 *Gavion*; 2 *Pagalo* (Damen Stan Patrol 2606); 4 *Petrel* (US *Point*); 2 *Protector*
LOGISTICS AND SUPPORT 4
AG 1 *Los Taques* (salvage ship)
AKSL 1
AP 2

Air Force 11,500

FORCES BY ROLE
FIGHTER/GROUND ATTACK
1 sqn with F-5 *Freedom Fighter* (VF-5)
2 sqn with F-16A/B *Fighting Falcon*
4 sqn with Su-30MKV *Flanker*
2 sqn with K-8W *Karakorum**
GROUND ATTACK/ISR
1 sqn with EMB-312 *Tucano**
ELECTRONIC WARFARE
1 sqn with *Falcon* 20DC; SA-227 *Metro* III (C-26B)
TRANSPORT
1 sqn with Y-8; C-130H *Hercules*; KC-137
1 sqn with A319CJ; B-737
4 sqn with Cessna T206H; Cessna 750
1 sqn with Cessna 500/550/551; *Falcon* 20F; *Falcon* 900
1 sqn with G-222; Short 360 *Sherpa*
TRAINING
1 sqn with Cessna 182N; SF-260E
2 sqn with DA40NG; DA42VI
1 sqn with EMB-312 *Tucano**
TRANSPORT HELICOPTER
1 VIP sqn with AS532UL *Cougar*; Mi-172
3 sqn with AS332B *Super Puma*; AS532 *Cougar*
2 sqn with Mi-17 *Hip* H

EQUIPMENT BY TYPE
AIRCRAFT 82 combat capable
FTR 18: 15 F-16A *Fighting Falcon*†; 3 F-16B *Fighting Falcon*†
FGA 22 Su-30MKV *Flanker*
EW 4: 2 *Falcon* 20DC; 2 SA-227 *Metro* III (C-26B)
TKR 1 KC-137
TPT 75: **Medium** 14: 5 C-130H *Hercules* (some in store); 1 G-222; 8 Y-8; **Light** 56: 6 Beech 200 *King Air*; 2 Beech 350 *King Air*; 10 Cessna 182N *Skylane*; 12 Cessna 206 *Stationair*; 4 Cessna 208B *Caravan*; 1 Cessna 500 *Citation* I; 3 Cessna 550 *Citation* II; 1 Cessna 551; 1 Cessna 750 *Citation* X; 2 Do-228-212; 1 Do-228-212NG; 11 Quad City *Challenger* II; 2 Short 360 *Sherpa*; **PAX** 5: 1 A319CJ; 1 B-737; 1 *Falcon* 20F; 2 *Falcon* 900
TRG 84: 24 DA40NG; 6 DA42VI; 18 EMB-312 *Tucano**; 24 K-8W *Karakorum**; 12 SF-260E
HELICOPTERS
MRH 8 Mi-17 (Mi-17VS) *Hip* H
TPT 22: **Medium** 14: 3 AS332B *Super Puma*; 7 AS532 *Cougar*; 2 AS532UL *Cougar*; 2 Mi-172 (VIP); **Light** 8+ Enstrom 480B
AIR-LAUNCHED MISSILES
AAM • IR AIM-9L/P *Sidewinder*; R-73 (AA-11A *Archer*); PL-5E; R-27T/ET (AA-10B/D *Alamo*); **IR** *Python* 4; **SARH** R-27R/ER (AA-10A/C *Alamo*); **ARH** R-77 (AA-12 *Adder*)
ASM Kh-29L/T (AS-14A/B *Kedge*); Kh-59M (AS-18 *Kazoo*)
AShM Kh-31A (AS-17B *Krypton*); AM39 *Exocet*
ARM Kh-31P (AS-17A *Krypton*)

Air Defence Command (CODAI)

Joint service command with personnel drawn from other services

FORCES BY ROLE
AIR DEFENCE
5 AD bde
COMBAT SERVICE SUPPORT
1 log bde (5 log gp)

EQUIPMENT BY TYPE
AIR DEFENCE
SAM
Long-range S-300VM
Medium-range 9K317M2 *Buk*-M2E (SA-17 *Grizzly*); S-125 *Pechora*-2M (SA-26)
Point-defence 9K338 *Igla*-S (SA-24 *Grinch*); ADAMS; *Mistral*; RBS-70
GUNS 440+
SP 40mm 12+: 6+ AMX-13 *Rafaga*; 6 M42
TOWED 428+: **20mm**: 114 TCM-20; **23mm** ε200 ZU-23-2; **35mm**; **40mm** 114+: 114+ L/70; Some M1

National Guard (Fuerzas Armadas de Cooperacion) 23,000

(Internal sy, customs) 9 regional comd

EQUIPMENT BY TYPE
ARMOURED FIGHTING VEHICLES
APC • APC (W) 44: 24 Fiat 6614; 20 UR-416
ARTILLERY • MOR 50 **81mm**
PATROL AND COASTAL COMBATANTS
PB 34: 12 *Protector*; 12 *Punta*; 10 *Rio Orinoco* II
AIRCRAFT
TPT • Light 34: 1 Beech 55 *Baron*; 1 Beech 80 *Queen Air*; 1 Beech 90 *King Air*; 1 Beech 200C *King Air*; 3 Cessna 152 *Aerobat*; 2 Cessna 172; 2 Cessna 402C; 4 Cessna U206 *Stationair*; 6 DA42 MPP; 1 IAI-201 *Arava*; 12 M-28 *Skytruck*
TRG 3: 1 PZL 106 *Kruk*; 2 PLZ M2-6 *Isquierka*
HELICOPTERS
MRH 13: 8 Bell 412EP; 5 Mi-17V-5 *Hip* H
TPT • Light 19: 9 AS355F *Ecureuil* II; 4 AW109; 5 Bell 206B/L *Jet Ranger*/*Long Ranger*; 1 Bell 212 (AB 212);
TRG 5 F-280C

Paramilitary ε220,000

Bolivarian National Militia ε220,000

Arms procurements and deliveries – Latin America and the Caribbean

Significant events in 2020

FEBRUARY — JAMAICA PROCUREMENT PLANS

The Government of Jamaica outlined plans to procure, for the Jamaica Defence Force, aircraft worth J$3.03bn (US$22.59m), armoured vehicles worth J$617.78m (US$4.60m) and equipment for Jamaica's patrol boats worth J$2.88bn (US$21.47m) in the fiscal year ending 31 March 2021. Jamaica received two *County*-class (Damen Stan Patrol 4207) patrol boats in 2016, six Bell helicopters and one Beech 350ER *King Air* maritime-patrol aircraft in 2017, and six *Bushmaster* armoured utility vehicles in June 2020. Since 2016, Jamaica's defence spending has more than doubled and in 2019 it reached 1.67% of GDP. The JDF plays an important role supporting the police in countering organised crime and associated violence.

FEBRUARY — COLOMBIA *KFIR* REPLACEMENT OFFERS

Colombia received additional offers for its five-year-long effort to replace *Kfir* fighter aircraft, with the Spanish government reportedly offering 17 second-hand Tranche 1 Eurofighter *Typhoon* aircraft (that would be upgraded), while Airbus offered 15 new-build Tranche 3 aircraft *Typhoons*. New Saab *Gripens* and F-16s are also under consideration as well as an offer from the original equipment manufacturer IAI to upgrade the *Kfirs*. Colombia currently operates 22, and the aircraft will soon be in urgent need of replacement or upgrade. A decision is planned to be made in 2021, though the impact of the coronavirus pandemic on government finances may delay a decision.

APRIL — EMBRAER–BOEING JV TERMINATED

Boeing Brasil–Commercial, a planned joint venture between Boeing (49%) and Embraer (51%), was cancelled by Boeing. The firm stated that although the planned partnership received unconditional approval from the regulatory authorities, apart from the European Commission, there were elements of the planned agreement that could not be resolved, something which Embraer denied. Although the statement issued by Boeing claimed that the partnership to jointly market the KC-390 will continue, Embraer stated in its Q1 earnings presentation that this has also been cancelled. Like many aerospace companies, the coronavirus pandemic has significantly affected Embraer, which was already facing challenges before the pandemic.

MAY — TKMS ACQUIRES OCEANA ESTALEIRO

Brazil's navy awarded a R9.1bn (US$2.25bn) contract to a consortium made up of ThyssenKrupp Marine Systems (TKMS), Atech, and Embraer Defense and Security in March for the construction of an initial batch of four *Tamandaré* frigates. As part of this project, TKMS purchased the Oceana Estaleiro shipyard in Brazil which was selected to build the vessels. Deliveries are planned between 2025 and 2028. Brazil also has a decade-long requirement for 6,000-tonne frigates and OPVs under the PROSUPER programme. While Brazil has not been able to fund that programme, TKMS's investment in local shipbuilding facilities could prove of benefit to the company in future navy surface-combatant programmes.

JULY — *GRIPEN* SUBCOMPONENT PRODUCTION BEGINS IN BRAZIL

Saab began production of *Gripen* E/F aerostructures at the Saab Aeronáutica Montagens sub-assembly plant in São Bernardo do Campo. The facility, which opened in May 2018, will manufacture the front section of the single-seat and much of the fuselage of the twin-seat *Gripens*. These parts will then be delivered to Embraer and Saab for final assembly. This is a part of the F-X2 programme that the government signed in October 2014 with Saab in collaboration with Embraer for 36 *Gripens* worth R$13bn (US$5.52bn) including support packages, local production and technology-transfer agreements. The first *Gripen* E for the Brazilian Air Force was built in Sweden and its first flight in Brazil took place in September 2020. The first 15 Brazilian-assembled *Gripens* are planned to be delivered in 2021 and deliveries of all 36 are planned to be completed by the end of 2026.

Table 9: Brazil: military aerospace procurement since 2005

Contract date	Programme	Equipment	Type	Quantity	Value (US$)	Prime contractor(s)	Deliveries	Service
Apr 2005	CL-X	C295M (C-105A)	Light transport aircraft	12	$296.46m	M Airbus Defence and Space	2006–09	Air Force
Apr 2005	P-3BR	P-3AM *Orion*	Anti-submarine-warfare aircraft upgrade	9	$505.42m	M Airbus Defence and Space	2011–14	Air Force
c. 2007	H-60	UH-60L *Black Hawk*	Medium transport helicopter	6	n.k.	Sikorsky	2008	Air Force
c. 2008	F-5M	F-5E/F	Fighter aircraft	11	n.k.	Jordanian government surplus	2009	Air Force
Oct 2008	AH-X	Mi-35M *Hind*	Attack helicopter	12	≥$250m	Rostvertol	2010–14	Air Force
Dec 2008	H-XBR	H225M	Heavy transport helicopter	50	$2.09bn	Helibras M Airbus Helicopters	2010–ongoing	All
Apr 2009	AF-1M	AF-1B/C	Attack-aircraft upgrade	8	n.k.	Embraer	2015–ongoing	Navy
Jun 2009	H-60	UH-60L *Black Hawk*	Medium transport helicopter	10	n.k.	Sikorsky	2011–13	Air Force
Dec 2009	*Panther* K2	AS565 K2 *Panther* (HM-1)	Multi-role-helicopter upgrade	34	$173.78m	Helibras	2014–ongoing	Army
Dec 2009	SH-16	S-70B *Seahawk*	Anti-submarine-warfare helicopter	6	$264.34m	Sikorsky	2012–15	Navy
Dec 2010	F-5M (Lot 2)	F-5FM	Fighter-aircraft upgrade	3	$107.41m	Embraer	2014–ongoing	Air Force
Dec 2010	*Esquilo/Fennec* AvEx	H125M (AS550U2) *Fennec* (HA-1)	Multi-role-helicopter upgrade	36	$94.97m	Helibras	2014–ongoing	Army
		H125 (AS350L1) *Ecureuil* (HA-1)	Light-transport-helicopter upgrade					
Jan 2011	RQ-450	*Hermes* 450	Medium ISR uninhabited aerial vehicle	2	n.k.	Elbit Systems	2011	Air Force
c. 2012	RQ-450	*Hermes* 450	Medium ISR uninhabited aerial vehicle	2	$22.33m	Elbit Systems	2013	Air Force
Dec 2012	E-99M	EMB-145SA (E-99)	Airborne early-warning and control aircraft upgrade	5	$215m	Embraer	2014–22*	Air Force
Mar 2014	RQ-900	*Hermes* 900	Medium ISR uninhabited aerial vehicle	1	n.k.	Elbit Systems	2014	Air Force
May 2014	KC-390	KC-390	Tanker/transport aircraft	28	$1.17bn	Embraer	2019–26*	Air Force
Jul 2014	AH-11B	*Super Lynx* Mk21B (AH-11B)	Anti-submarine-warfare helicopter upgrade	8	$160m	Leonardo MW	2019–ongoing	Navy
Jul 2014	CL-X2	C295M (SC-105 *Amazonas*)	Search-and-rescue aircraft	3	$245.38m	M Airbus Defence and Space	2017–20*	Air Force
Oct 2014	F-X2	*Gripen* E/F (F-39)	Fighter/ground attack aircraft	36	$4.62bn	Saab	2019–24*	Air Force
Feb 2019	UH-17	H135 (UH-17)	Light transport helicopter	3	n.k.	Helibras	2020–ongoing	Navy

*Planned; M – multinational

Chapter Nine

Sub-Saharan Africa

- Fighting in Ethiopia, in late 2020, posed a risk not just for Ethiopian security, but also for regional security. This was because of the risk that the conflict would overspill, because of the population displacement it generates, and also because it raised questions over Ethiopia's hitherto large contribution to regional peacekeeping missions.
- The conflict in Ethiopia also highlighted the growth in certain military and security capabilities sourced from China. China launched a remote-sensing satellite for Ethiopia in 2019 and has supplied military equipment to Addis Ababa. Indeed, the seizure of heavy equipment by Tigray People's Liberation Front rebels in late 2020 included what was believed to be at least one Chinese-manufactured PHL-03 multiple rocket launcher formerly in Ethiopian service.
- China's military assistance to African states includes financial support (including for continental-level security institutions), training support and exchange agreements as well as material assistance, including in medical supplies and conventional-equipment deliveries. While the trade in Chinese arms in the region traditionally focused on Soviet-era designs, weapons of Chinese design are increasingly being operated by regional armed forces. These range from armed uninhabited aerial vehicles to armoured vehicles and artillery pieces, including multiple rocket launchers.
- More spending by Nigeria, along with smaller increases in neighbouring states, meant that West African spending accounted for 36.8% of total regional spending, for the first time exceeding that of Southern Africa.
- Maritime insecurity in West Africa and the Gulf of Guinea has seen states modestly boost naval capabilities in recent years, but piracy and terrorist threats elsewhere – such as in the Mozambique channel – could expose the relative lack of maritime capacity in regional states. Although South Africa's navy has patrolled the Mozambique channel since 2011, patrols have been restricted because of concerns over funding as well as operational availability.

Sub-Saharan Africa defence spending, 2020 – top 5

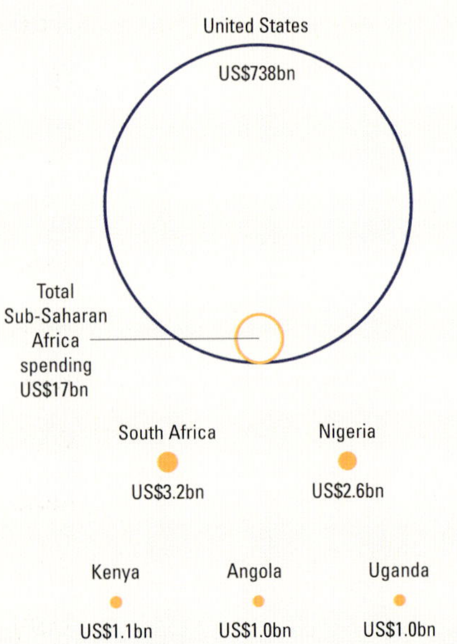

Active military personnel – top 10
(10,000 per unit)

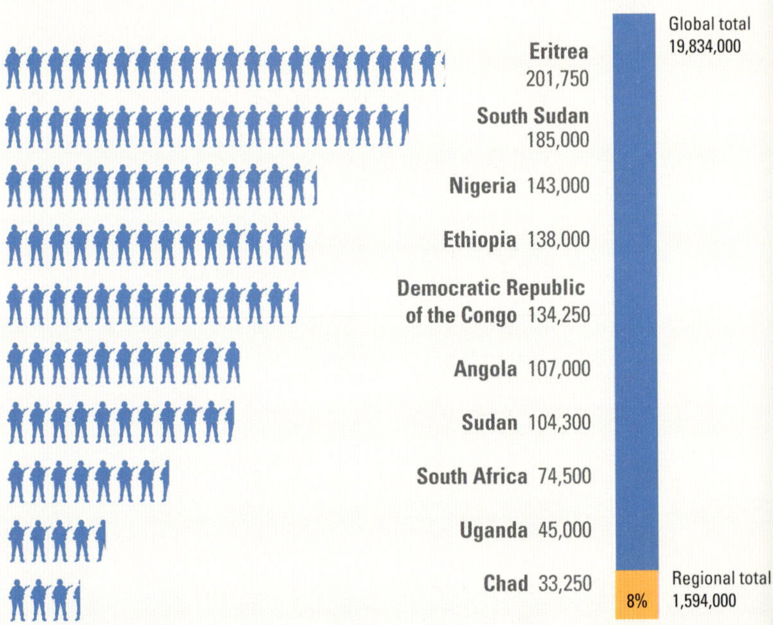

Sub-Saharan Africa: medium and heavy airlift, 2020 (1 per unit)

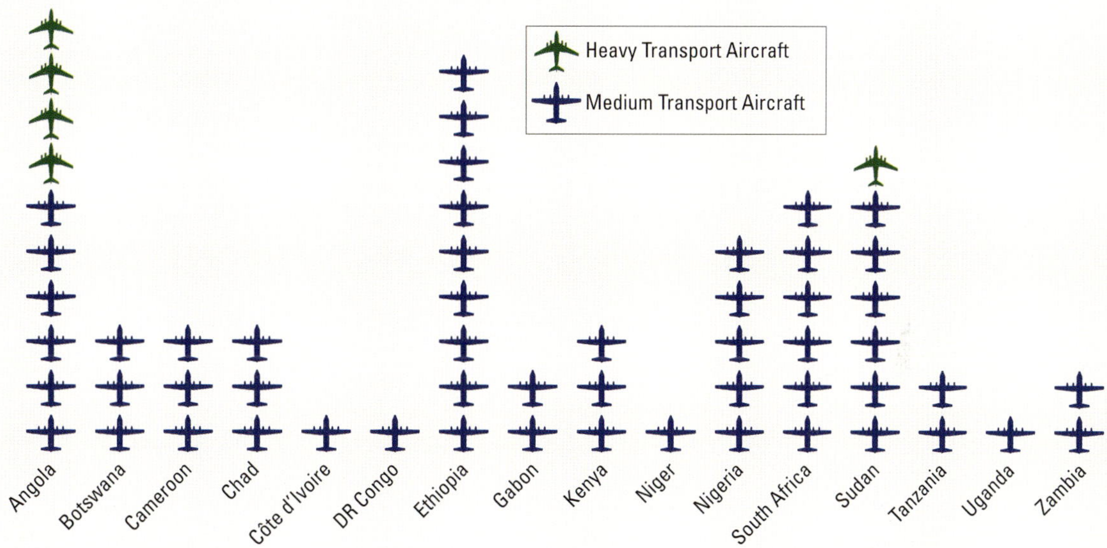

Sub-Saharan Africa: Selected derivatives of China's ZSL-92 6x6 wheeled APC (5 per unit)

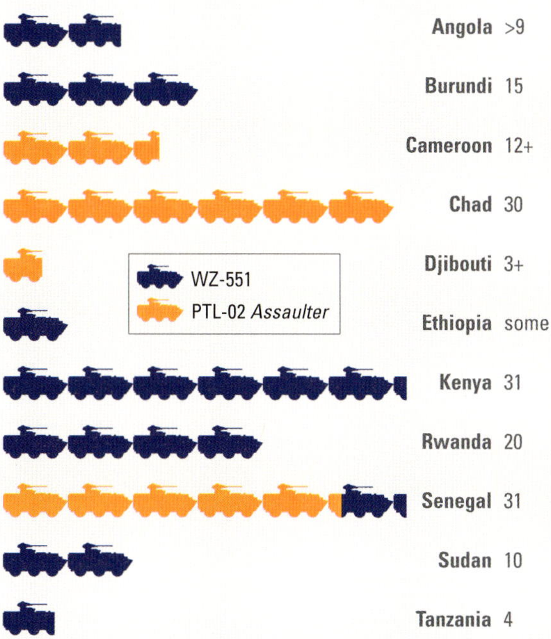

Sub-Saharan Africa: selected amphibious capabilities

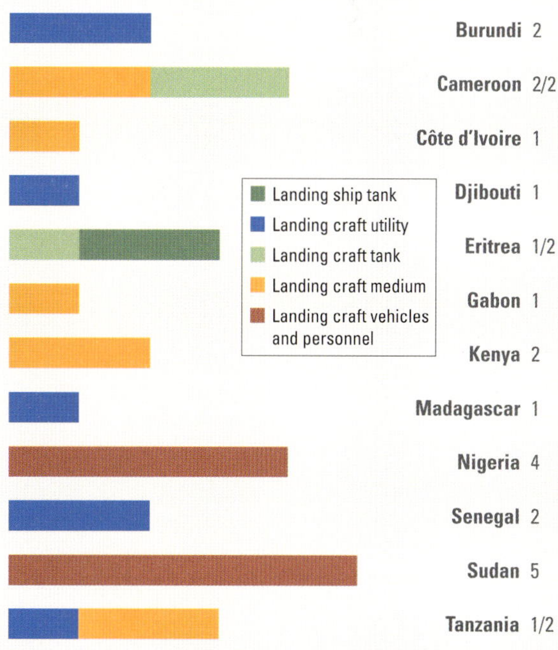

Sub-Saharan Africa

Military and security forces across the region continue to face problems arising from the activity of non-state actors; weak defence organisations; capability challenges caused by equipment fleets that vary in number, age and maintenance levels; the changing priorities of external economic, political and security partners; and, in 2020, the economic and political effects of the coronavirus pandemic. However, despite this shared context, areas within sub-Saharan Africa display distinct characteristics relating to conflict dynamics, national defence priorities, equipment inventories and military structures.

West Africa

Fragile state structures and persistent Islamist insurgency, features of the enduring conflicts in northeastern Nigeria and the broader Sahel, continue to prompt deployments and operations by regional armed forces. Military deployments by international forces also continue. The French-led *Operation Barkhane* in Mali and the broader Sahel is particularly important to counter-insurgency efforts, as persistent capability challenges experienced by regional security forces – such as those organised in the G5 Sahel Joint Force – make it difficult for them to confront the threat effectively without outside assistance. Nigeria's armed forces, meanwhile, continue to experience difficulties in effectively countering Boko Haram.

In West Africa, the character of insurgent operations has posed particular problems for regional military structures. Regional armed forces face constraints in effectively generating capabilities and defence organisations suitable for the type of mobile warfare demanded by the region's topography and the nature of the insurgent challenge. Adversaries have favoured mobile tactics, allowing them to both rapidly disperse and concentrate for strike operations, even if these tactics present them with logistical challenges.

While armed and paramilitary forces in the region have traditionally fielded modest inventories of armoured vehicles (particularly mine- and ambush-protected vehicles), the focus on mobile-combat tasks has led to the introduction of more lightly armoured vehicles, including 4x4s. They still comprise a significant portion of regional states' inventories, operating in light-cavalry and logistics roles. These vehicles are attractive because of the relative ease with which they can be repaired, their cost, fuel economy and – for these reasons combined – the degree of autonomy they provide to the government forces and insurgent formations that utilise them. They are also relatively easy to adapt with pintle-mounted light weapons and, as seen in Libya and Syria, even anti-aircraft guns.

However, a series of recent attacks have highlighted the difficulties that can arise when using lightly armoured vehicles in environments that may require high mobility but also a level of protection against small-arms fire and attacks by explosive devices. This arguably includes the Tongo Tongo attack by terrorists in Niger in October 2017 that killed four United States soldiers and a number of Nigerien troops. Although a range of factors were held responsible by the subsequent US investigation, it was directed that armoured vehicles should be provided to troops as an option. To the south, Nigeria's armed forces are now procuring (albeit limited numbers of) protected patrol vehicles and armoured vehicles. Mali is another example, with the government having received since 2017 both *Casspir* and Stark Motors' *Storm Light* armoured vehicles (the former from Germany and the latter from Qatar).

Such concerns and developments pose a continuing problem for the region's armed forces as they look to balance troop-protection requirements with the need to conduct mobile warfare at a significant geographic scale, and do so in an environment of constrained defence budgets. In this context, it is all the more important that forces build their logistics-support capacities and develop – or secure from partners – the support of a broader range of capabilities, including fixed- and rotary-wing air-transport capabilities.

East Africa

Until November, the security situation in East Africa and the Horn of Africa was dominated by the enduring challenges generated by al-Shabaab's presence in Somalia. The outbreak of conflict in Ethiopia between government forces and the Tigray People's Liberation Front threatened stability in the sub-region. This is not only because the conflict affected Eritrea and led to refugee flows within and from Ethiopia's north, but also because Ethiopia was a major contributor to the AMISOM mission in Somalia as well as to the UNMISS operation in South Sudan, and withdrawing forces to return home risked affecting these operations.

Over the past decade, the sub-region has become a focus of both global and regional political competition. External states including China, France, Japan, Saudi Arabia, Spain, Turkey, the United Arab Emirates and the US, as well as the European Union, have displayed increased interest in providing military assistance to the region and establishing military bases there. Some of these arrangements are long-standing. For decades, France has maintained military forces in Djibouti; the US developed a base after 9/11 and China opened its first overseas military base there in 2017. Djibouti has also seen deployments by EU member states as part of EU counter-piracy initiatives.

There is also competition for influence and support among several regional players, with Iran, Qatar, Saudi Arabia, Turkey and the UAE reportedly engaged in providing economic support and/or security assistance, as well as expressing interest in military basing. For instance, Saudi and UAE naval forces stepped up activity in the Red Sea in the wake of maritime attacks emanating from Iranian-backed Houthi forces based in Yemen. As well as the security factors that drive this external concern, rifts in the region caused by lack of consensus on such issues as the sharing of Nile waters, natural-resource exploitation, approaches to Somalia and Eritrea, and Islamist extremism mean that regional states remain susceptible to external influences. This susceptibility may be increased by the economic impact of the coronavirus pandemic and the resulting need for financial resources to bolster local economies.

In response to the range of external actors vying for influence around the Red Sea and Gulf of Aden, the Intergovernmental Authority on Development (IGAD) established the Task Force on the Red Sea and the Gulf of Aden in April 2019. The task force is intended to build a common IGAD position and develop a regional plan of action for the Red Sea and Gulf of Aden. Together with the African Union's ownership of the negotiations over the Grand Ethiopian Renaissance Dam, this may indicate that regional states view multilateral consensus and action as important, possibly in helping to insulate the area from wider regional and global political dynamics. While for regional states it may reinforce the value of the African multilateral and institutional framework – such as the African Union, Regional Economic Communities and the African Standby Force – neighbouring states have established new formations. For instance, in early 2020 the Council of Arab and African States Bordering the Red Sea and Gulf of Aden met for the first time in Riyadh. This saw ministers from Djibouti, Egypt, Eritrea, Jordan, Saudi Arabia, Somalia, Sudan and Yemen discuss primarily security matters (when first proposed the forum was to focus on topics including economic and environmental issues).

Even before fighting erupted in Ethiopia in November, the Ethiopian National Defense Force (ENDF) had been strained by domestic deployments arising from political unrest and increasing militarisation in ethnic police forces. Analysts understood that this had reduced the number of recruits joining the federal forces. The fighting with the TPLF in the north – which was ongoing at the time of writing – highlighted not just the TPLF's residual military capability, but also the equipment capabilities that the DNDF had held in the north, including Chinese-origin MRL systems

In response to the coronavirus pandemic, security forces across the region were employed to enforce emergency legislation. Only Ethiopia declared a national state of emergency; Kenya and Uganda declared public-health emergencies. Other countries adopted a variety of restrictions. However, the pandemic is likely to have a lasting political, economic and security impact in the region, and there were reports of concern over the nature of some police and military actions to enforce coronavirus containment measures.

Central and Southern Africa

The Southern African Development Community (SADC) is faced with two ongoing conflicts and the risk that other disputes might escalate. SADC members' armed forces face obstacles in dealing with these challenges; they lack both personnel and key capabilities, notably the ability to deploy forces promptly to a crisis, and then sustain such a deployment.

Conflict has persisted in the northeast of the Democratic Republic of the Congo (DRC) since

1996. Indeed, a key challenge for international forces deployed in the country is that local militias and guerrillas seeking to target Rwanda and Uganda are now so well established that these countries' forces are effectively engaged more in counter-insurgency operations than in peacekeeping. Substantial criminal activity complicates the situation. The 13,500-strong UN mission in the Democratic Republic of the Congo (MONUSCO) has had some success – principally due to its Force Intervention Brigade (comprising infantry from Malawi, South Africa and Tanzania, with South African helicopters) – but it is too weak to stabilise the northeast of the country, while DRC forces remain relatively ineffective. Elsewhere in the sub-region, instability in the Central African Republic, the intermittent civil war in South Sudan and instability in Burundi present continued risk.

The insurgency in Mozambique's northern Cabo Delgado province has seen guerrillas take and hold towns, and the situation is complicated by drug trafficking and other criminal activity. Government forces lack the strength, mobility and air support to have significant effect, while private military companies have not produced decisive results. Analysts understand that Mozambique has informally asked South Africa for air and naval support, the latter to block guerrilla resupply via the Ruvuma River. Both conflicts have the further complication of guerrillas having affiliated themselves to the Islamic State, which risks seeing ISIS tactics and strategies applied.

Piracy and maritime terrorism in the Mozambique Channel would represent a risk for Mozambique and Madagascar as both lack significant naval capability. It would also affect South Africa, which imports oil through the channel and has trade flowing through it. The South African Navy has patrolled the channel since 2011 but tight funding and reduced operational availability have restricted these patrols.

The key weakness of the SADC's armed forces is the lack of airlift, a critical capability given the geographic scale of the region. Only Angola and South Africa have an appreciable fixed- and rotary-wing airlift inventory, but both face serviceability issues. These countries also possess combat-aircraft and attack-helicopter fleets, but their regional mobility is hampered by the lack of aerial-refuelling aircraft. A further disadvantage is their inability to patrol SADC waters effectively. South Africa and Namibia may have blue-water capable ships, but both fleets face serviceability challenges.

DEFENCE ECONOMICS

Macroeconomics

The International Monetary Fund (IMF) projects that economic output in sub-Saharan African will contract on average by 2.8% in 2020 before expanding by 3.8% in 2021. As such, real GDP is projected to recover to pre-coronavirus-pandemic levels by the end of 2021.

Although this is the region's most severe economic contraction since the 1970s, it is more muted than in other commodity-dependent regions like Latin America and the Middle East and North Africa. This reflects the relatively lower infection rates seen in the region and, as a result, more muted lockdown measures. Even so, government efforts to support the economy were constrained by falling revenues and limited fiscal space.

The regional economic outlook worsened over the course of 2020 as the IMF and World Bank downgraded projections for tourism- and commodity-dependent economies. Severe contractions in leading regional economies, such as Angola, Nigeria and South Africa (–4.0%, –4.3% and –8.0%) served as a drag on overall output. Lower growth rates are also expected in other key economies like Ethiopia and Kenya. Ethiopia's economy grew by 9% in 2019 but will only see a modest 2% increase in output in 2020, while Kenya's 5.4% growth in 2019 will be followed by a 1% increase in 2020.

The region was also affected by disruption to trading partners and global supply chains, as well as the collapse of global oil prices and the fall in tourism revenues. Before the pandemic, the region's economic outlook was variable, with more diversified economies expected to continue posting annual real GDP growth rates of 5%, while those more dependent on commodities would be constrained by low oil prices and weak agricultural output.

High inflation is a concern for several states. In Zimbabwe, the inflation rate reached 255% in 2019 and is expected to increase to 319% in 2020. In South Sudan and Sudan, inflation rates averaged above 50% in 2019, while Angola, Ethiopia, Liberia, Nigeria and Sierra Leone all posted double-digit inflation rates.

The World Bank determined that sharp currency appreciations and supply-chain disruptions would likely cause regional inflation rates to increase further in 2020. The expansionary measures used to counter the pandemic's economic impact will also increase inflation, again pointing to the limited headroom that countries in the region have when determining a fiscal-policy response.

According to the IMF, fiscal measures – including additional spending, loans and guarantees – aimed at countering the pandemic account for an average of 20% of GDP for advanced economies and 6% of GDP among emerging markets. The fiscal burden on low-income economies is much lower, closer to 1.8% of GDP, though the risk remains that the pandemic may worsen and costs increase. Moreover, as advanced economies try to pay for the unprecedented mobilisation of fiscal support measures, this could affect foreign-aid expenditure. The IMF projects that the pandemic will reduce per-capita GDP levels to 2010 levels, wiping out almost a decade of progress in development and poverty reduction.

Defence economics

Economic constraints have continued to subdue defence expenditure, with the region in 2020 again accounting for just 0.9% of the global total. Between 2015 and 2019, regional defence spending had been on a downward trajectory, with annual real-terms reductions of 4.1% on average across the region. Revisions in Ugandan and Tanzanian expenditure wiped out the small 0.6% increase in 2019, with final expenditure falling by 0.8%. Somewhat surprisingly, in the face of the economic headwinds caused by the pandemic, regional expenditure is on track to increase by 8.4% in real terms in 2020 to reach US$17.7 billion (US$17.0bn in current terms). The increase brings regional spending back to 2017 levels in real terms, though spending remains below the 2014 peak (driven by spending increases in Angola at the time).

▼ Map 10 **Sub-Saharan Africa regional defence spending**[1]

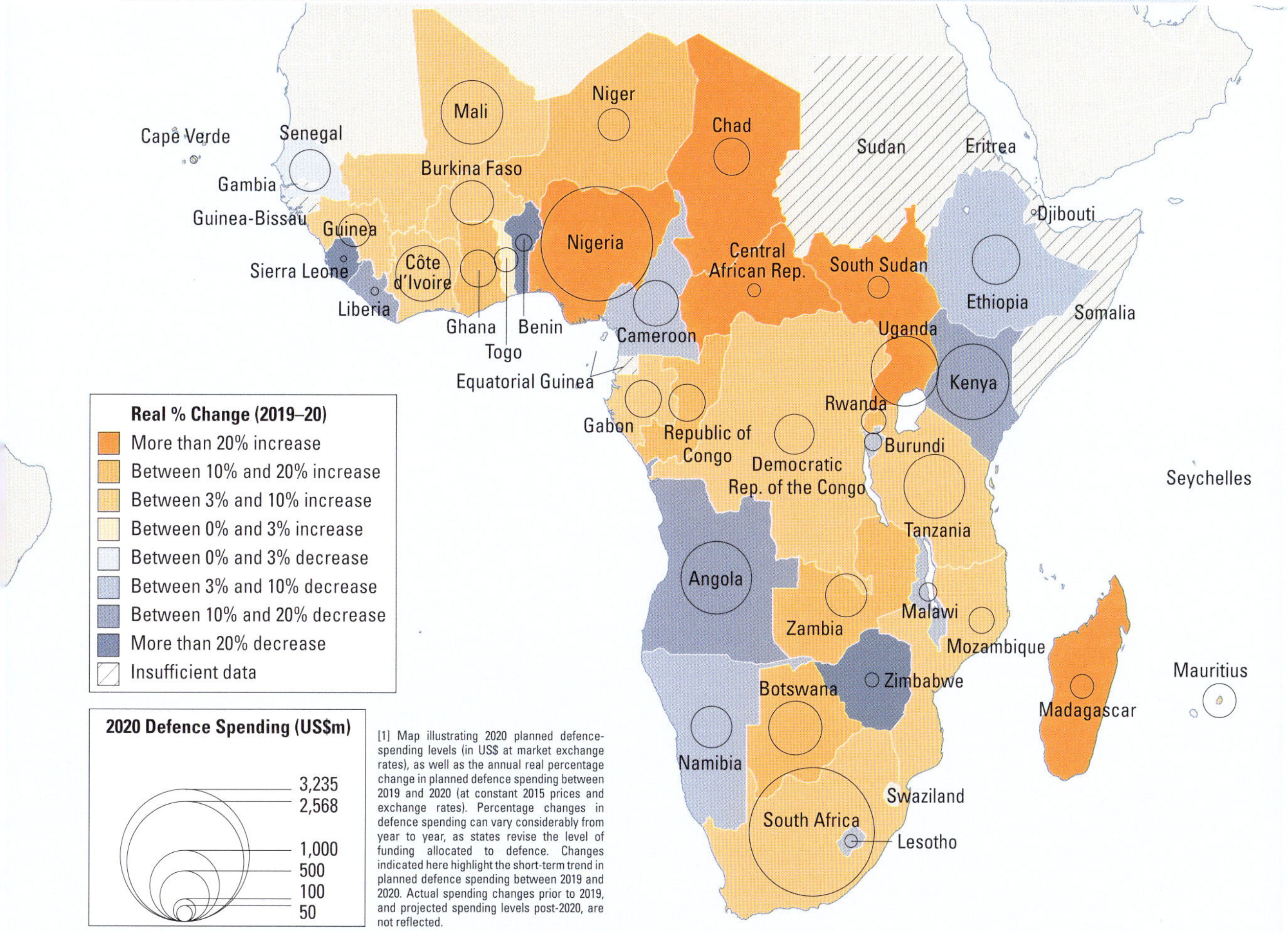

[1] Map illustrating 2020 planned defence-spending levels (in US$ at market exchange rates), as well as the annual real percentage change in planned defence spending between 2019 and 2020 (at constant 2015 prices and exchange rates). Percentage changes in defence spending can vary considerably from year to year, as states revise the level of funding allocated to defence. Changes indicated here highlight the short-term trend in planned defence spending between 2019 and 2020. Actual spending changes prior to 2019, and projected spending levels post-2020, are not reflected.

Fuelling the 2020 increase was a dramatic 36% surge in Nigeria's defence budget. The initial 2020 budget appropriation bill allocated 878bn naira (US$2.5bn) to defence, which would have been a 33% increase over 2019 levels. In response to the pandemic, a revised appropriation bill was introduced in June 2020 that enacted several spending measures to support healthcare and regional social-care programmes. The projected budget deficit widened as the oil price benchmark was lowered to US$20/barrel from US$57/barrel in the initial budget, while total public spending increased from NGN10.3 trillion (US$29.4bn) to NGN10.8tr (US$30.8bn). The defence ministry received a further uplift to NGN900bn (US$2.6bn).

The 2020 boost to Nigerian defence spending mostly benefited the army, particularly personnel costs, which make up 90% of the service's total budget. The 2020 increase brings the Nigerian defence budget up to 0.58% of GDP – the highest since at least 2008 but still very low by international standards, where the global average is 2.1% of GDP.

The 2020 surge in Nigerian spending, combined with smaller increases in Chad, Côte d'Ivoire, Mali and Niger, means that West African spending now accounts for 36.8% of total spending in sub-Saharan Africa, for the first time exceeding the share allocated by Southern African states. As spending in South Africa stagnated, and as Angolan spending declined after 2015, Southern Africa's share of the overall regional total has fallen from 60.9% in 2009 to 36.3% in 2020.

South Africa's defence budget was also adjusted over the course of the year. The original budget, tabled in March 2020, allocated 52.4bn rand (US$3.1bn) to the Department of Defence (DoD). It was adjusted and 're-tabled' on 20 July, allocating R55.3bn (US$3.2bn) to defence. This resulted in a budget that is 4% higher in real terms than 2019 spending but is the same in real terms as the 2008 budget, highlighting a decade of stagnation in military spending. Even before the pandemic, South African public spending had been restricted by a poorly performing economy and rising unemployment, among other factors.

The Medium Term Expenditure Framework (MTEF) to 2022/23 projects that spending will fall to R50.9bn (US$3.0bn) in 2021/22 before recovering to R53bn (US$3.1bn) in 2022/23, indicating that there is no significant change planned to the spending pattern of previous decades. The 2020 increase is primarily evident in the 'Force Employment' budget, which covers operational capability and border security. Investment spending continued to fall as a proportion of defence spending, from an estimated 11.4% in 2019 to 9.7% in the 2020 adjusted budget. According to the MTEF, this figure will fall to 2.7% by 2022.

Angola's defence budget declined for the fifth consecutive year in 2020, falling by 12% in real terms to just US$1.8bn; the budget was US$5.4bn at its peak in 2014. Angola's budget was also revised during 2020, with defence allocations increasing from an initial limit of 540 billion new Angolan kwanza (Kz) (US$954m) to Kz586bn (US$1.0bn), which means the 2020 budget is in nominal terms the same as in

▼ Figure 25 **Sub-Saharan Africa regional defence spending** as % of GDP (average)

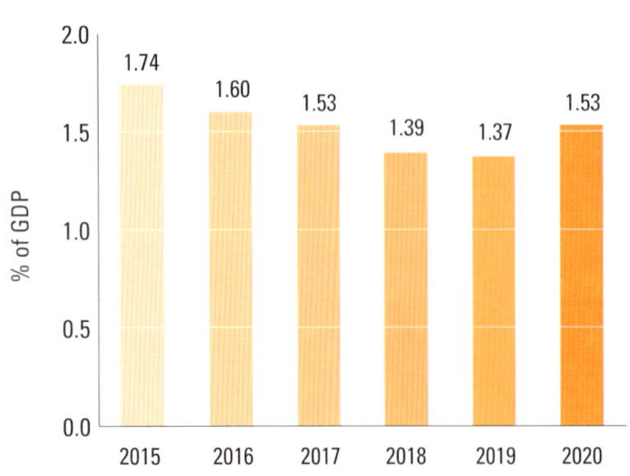

Note: Analysis excludes Djibouti, Equatorial Guinea, Eritrea, Gambia, Guinea-Bissau, Seychelles, Somalia, Sudan

▼ Figure 26 **Sub-Saharan Africa defence spending by country and sub-region, 2020**

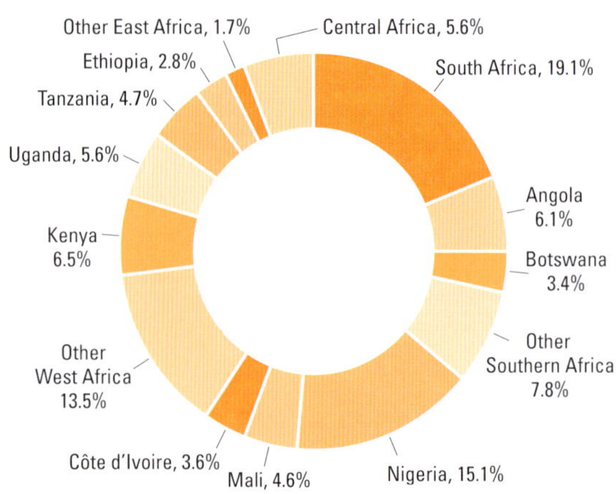

Note: Analysis excludes Djibouti, Equatorial Guinea, Eritrea, Gambia, Guinea-Bissau, Seychelles, Somalia, Sudan

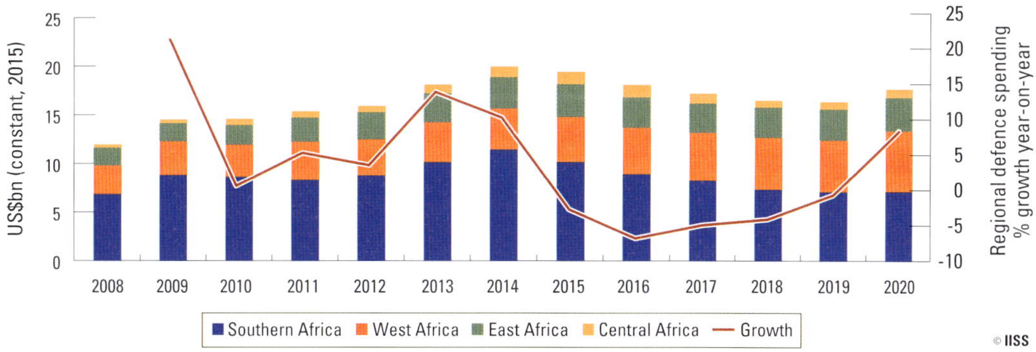

▲ Figure 27 **Sub-Saharan Africa: total defence spending by sub-region, 2008–20**

2019. The 12% real cut is a result of inflation, which is projected to average 20% in 2020. The defence budget increased from 2017–19. However, inflation rates in those years – of 30%, 20% and 17% respectively – negated these increases and led to cuts averaging 14% in real terms.

The picture is different in East Africa. Consistent spending in Tanzania and Ethiopia, combined with a 2020 increase in Uganda's defence budget, means that the sub-region now accounts for 21.3% of total defence spending in sub-Saharan Africa; in 2009 the figure was 11.8%. However, spending in Central Africa continues to decline as a proportion of total regional spending. While the Republic of the Congo posted a 19% increase in its 2020 defence spending, cuts in both the Democratic Republic of the Congo and Republic of the Congo between 2017 and 2019 mean that the sub-region now accounts for 5.6% of the overall regional total, down from a high of 7.4% in 2016.

Defence industry

The lack of sustained or meaningful growth in defence spending and low levels of investment in defence research and development restrict the potential for the development of local defence industries in sub-Saharan Africa. In South Africa, which has perhaps the most advanced sector in the region, funding challenges and the decreasing sums allocated to investment in the defence budget have affected capability development.

The South African DoD Strategic Plan 2020–2025, published in July 2020, acknowledged that the economic outlook and persistently low defence budget allocations had constrained the implementation of the National Policy on Defence (elaborated in the 2015 Defence Review), 'with a profound adverse impact on the availability and modernisation of required defence capabilities'. The 2015 review outlined force-recapitalisation goals and wider targets for the national defence industry, including the support of exports and the promotion of self-sufficiency. However, progress has been slow and the defence industry faces considerable challenges. State-owned Denel was already facing a severe liquidity crisis before the pandemic and has reported 2019/20 losses of R1.7 billion (US$99m).

Meanwhile, Nigeria's efforts to bolster indigenous capabilities have also been restricted by economic conditions, such as the persistently low oil price since 2015, which has hampered increased investment. Conflicting budget priorities, including the drive to professionalise the armed forces and the need to counter Boko Haram, create a higher personnel and operational burden within the budget, further constricting funds for research and development. Allocations to the state-owned Defence Industries Corporation of Nigeria (DICON) have been maintained over the last five years. However, as a proportion of the defence budget, allocations have fallen from 1.1% in 2016 to just 0.4% in 2020.

SENEGAL

Senegal maintains small but relatively capable armed forces. It has experienced peaceful transitions of political power since gaining independence from France in 1960 and the armed forces remain removed from domestic politics. Despite limited public documentation on Senegalese defence plans and priorities, insights can be gleaned from deployments, operations and procurements, as well as remarks by senior leaders. President Macky Sall elaborated key defence and security priorities in his opening address to the 2019 Dakar International Forum on Peace and Security in Africa. He noted 'classic threats', such as conflict and instability stemming from political disputes, but focused on the challenges of violent extremism and terrorism. These were, he said, 'more difficult to prevent and stamp out'. To better tackle these challenges, the capacity of security forces needed to be improved with better education, equipment and training. The president recognised that enhancing cooperation between the African Union and United Nations would help, as would efforts to operationalise the African Standby Force concept. However, Sall continued that there was also a need to boost border surveillance and improve regional cooperation, with outside assistance required to tackle asymmetric threats. That said, though relatively small, the armed and security forces are well balanced in regional terms. They have a mix of equipment (and capability-development plans to grow equipment inventories in specific areas) that reflects the requirement for the forces to act in regional and wider contingencies; tackle the insurgency in the Casamance region; address growing concerns over maritime insecurity; and maintain a gendarmerie as an internal security force. Senegal's armed forces are today important regional security actors and have taken part in regional and continental deployments.

Operations

Senegal routinely takes part in multinational exercises and contributes forces to peacekeeping and counter-insurgency missions, such as the United Nations Multidimensional Integrated Stabilization Mission in Mali (MINUSMA). The armed forces deployed operationally on 19 January 2017 to enter the Gambia. This came as part of an Economic Community of West African States (ECOWAS) mission to restore order and ensure a peaceful transfer of power after former president Yahya Jammeh lost the 1 December 2016 presidential election to Adama Barrow. As well as dispatching its own troops, Senegal hosted Nigerian light-attack aircraft that were deployed in case of contingencies around the operation. A mooted plan to deploy some 2,000 troops to Yemen to join the Saudi-led coalition was not implemented because of political and popular opposition.

As of end-July 2020, Senegal had some 1,000 troops and just over 300 police personnel deployed to MINUSMA. It is understood by analysts that around 700 troops operate in the force-protection unit. With the G5 Sahel Joint Force still experiencing challenges in its mission to tackle insurgent groups, analysts understand that Senegal could increase its deployments further, perhaps even as part of a mooted deployment of ECOWAS forces to the area flanked by the borders of Burkina Faso, Mali and Niger. At the 2019 Dakar International Forum, Senegal was asked to join the Partnership for Security and Stability in the Sahel (P3S). This initiative was launched by France and Germany in 2019 and is intended to improve the domestic defence and internal security of Sahel countries by 'improving international coordination, supporting security sector reform and strengthening security forces while enhancing their accountability'. Senegal's response to this initiative remains unclear, though there have been calls for more coordination within and around the G5 Sahel, while the president has called for a more robust mandate for MINUSMA.

Security issues closer to home also focus the attention of defence planners. Relations with Mauritania have improved since 2018. Following clashes over border tensions in 1989 and 1991, bilateral relations had remained under-developed. This owes much to the need for cooperation to effectively exploit energy reserves, including the cross-border offshore gas field known as the Greater Tortue Complex. A proposal for joint patrols of the Senegal River was aired in late 2018. However, the long-running insurgency in the Casamance region continues, despite a ceasefire in 2014, and according to analysts it is sustained in part by funds from trafficking and racketeering. Continued conflict has stalled any move to implement demining initiatives in the region. Armed clashes are reported to still occur infrequently.

Internal security receives significant attention, with the gendarmerie benefiting from investments in infrastructure, training and equipment. The gendarmerie has territorial units located across the country, as well as mobile formations. The Gendarmerie Intervention Legion's armoured unit fields Panhard PVP and *Bastion* and *Ejder Yalcin* armoured vehicles. Other units operate a range of smaller lightly armed vehicles. Under its Emergency Trust Fund for Africa, the European Union sponsors the GAR-SI (Rapid Action Groups – Surveillance and Intervention in the Sahel) programme. Units have been formed in several countries in the Sahel region. In Senegal, the project is organised by the French Gendarmerie. Senegal's GAR-SI unit consists of some 150 gendarmes with vehicles, including light-armoured URO ST-5s, and is relatively well trained and well equipped (reportedly with night-vision goggles and mini-uninhabited aerial vehicles). The GAR-SI formations are intended to tackle a variety of challenges, including organised crime and armed groups. Its units have comparatively significant weaponry, are mobile and can conduct relatively autonomous operations in coordination with other Senegalese forces and those of neighbouring countries. Senegal's GAR-SI units operate in the east and experienced their first significant incident on 22–23 January 2020, when a GAR-SI patrol came under fire on the border with Mali. It remains unclear whether this attack can be attributed to jihadists or was linked to banditry.

Foreign defence relations

Senegal's armed forces benefit from military-cooperation agreements with several states. Defence relations with France and the United States are particularly important. French forces are actively involved in providing training and logistics support for the armed forces and the internal security forces. France maintains forces and equipment in Senegal (the Éléments français au Sénégal, or EFS) that play a central role in this relationship. Its 'regional cooperation and operational hub', mainly located in Dakar, includes some 350 personnel engaged in organising training for Senegal and other regional states and supporting France's broader regional policies and missions. Military cooperation with the US is not as far-reaching, at least in terms of its physical footprint, but is also important for training and equipment support plans, particularly relating to counter-terrorist imperatives.

Senegal established a regional counter-terrorism training centre in 2018 with funding from the US Department of State's Anti-Terrorism Assistance (ATA) programme. It is located in Thiès in an existing Senegalese training centre. According to the US, the centre will provide training in crisis response, explosives countermeasures, post-blast investigations and rural border-patrol operations. Thiès is also the location of Senegal's commando battalion and other general reserve units. In early 2020, US Africa Command (AFRICOM) identified Senegal as one of four African countries that will act as a pillar for its Africa Enlisted Development Strategy (AEDS). This is intended to improve professional military education in African armed forces, streamline training for personnel, improve 'train the trainer' capability and cultivate regional training centres of excellence. The Vermont National Guard works with the Senegalese armed forces as part of the US Department of Defense's State Partnership Program. Train-the-trainer initiatives are also a focus of defence assistance by the United Kingdom, and in early 2020 UK Royal Marines were involved in training activities alongside US personnel during AFRICOM's *Flintlock* special-operations exercise. Meanwhile, since 2013 Spain has maintained an aviation detachment (Marfil) in Dakar to support regional counter-insurgency operations. Senegal also has a defence relationship with Indonesia, and negotiations for an additional CN235 aircraft from PT Dirgantara Indonesia were proceeding in 2020 but reportedly delayed by travel restrictions due to the coronavirus pandemic.

The armed forces

The 'Horizon 2025' programme outlines the planned modernisation of the Senegalese armed forces and their infrastructure. Personnel numbers are due to increase. Efforts to improve operational capability are focused on developing greater mobility and firepower – including by acquiring new equipment – and broader efforts designed to improve force health, such as the distribution of improved personal equipment. Meanwhile, it is understood that lessons from the intervention in the Gambia have led to a focus not only on army but also navy and air-force requirements. Equipment designed to tackle internal tasks and regional counter-insurgency is under development, as are capabilities that can deliver broader surveillance coverage. As well as maritime-patrol capabilities, such as the CN235 project, there are also plans to expand marine surveillance by increasing ship numbers, both perhaps in anticipation of the planned start of offshore gas development as well as in response to broader security challenges in the Gulf of Guinea.

Defence planners have paid particular attention to the development of doctrine and personnel training. Senegal routinely takes part in multinational exercises and there is close cooperation with French forces in the country. It also runs its own training activities. While the armed forces are in general structured around regional military zones, the commando battalion is a mobile force. As well as a reconnaissance section, one company specialises in airborne operations, another in amphibious operations and another is dedicated to urban or counter-terrorist tasks. There are also several units badged as special-forces formations.

Equipment developments

Land

The army's inventory of armoured fighting vehicles has been strengthened by the acquisition from China of the wheeled PTL-02 *Assaulter*. There is an apparent focus on vehicles that deliver mobility but still offer a degree of ballistic and blast protection, such as the IAI Ramta RAM Mk3, the RG-31, the *Casspir* and the *Oncilla* light-armoured vehicle (based on the Ukrainian *Dozor* and used by the artillery battalion). However, while maintaining such a range of vehicles in service might deliver protection and mobility to troops, the fleet's diversity raises challenges in terms of crew training, operational procedures and maintenance. (Several vehicles repatriated from operations in Darfur were overhauled between 2018 and 2020.) As well as the capability offered by the main gun on the PTL-02s, the army's artillery inventory comprises mainly HM-2/M-101 105mm towed howitzers and 155mm TR-F1s, while BM-21 *Grad* Multiple Rocket Launchers were observed on parade in 2017. There remain unconfirmed reports by analysts that *Spike* or *Lahat* anti-tank missiles may be mounted on RAM vehicles.

Air

In April 2018, the air force signed a contract for four L-39NG light-attack aircraft. The contract reportedly includes training for aircrew and ground personnel as well as logistics support. Deliveries were scheduled for 2020–21 but were postponed until 2022 for financial reasons. There remain uncertainties over the project's full implementation. Air-force training capacity increased in early 2020 with the arrival of two KT-1 trainers (designated KA-1S) built by Korea Aerospace Industries. Two more were expected to enter service before the end of 2020, but the delivery timeline was unclear; delivery of the initial two aircraft was subject to change because of the coronavirus pandemic. It has been reported that the aircraft will be capable of reconnaissance tasks, advanced training and possibly light attack. They would add to the number of TB-30 light trainers in service (boosted by the French donation of three in 2019). They are reportedly intended to train not just Senegalese aircrew but also those of other regional states.

Senegal is looking to boost its maritime-patrol capabilities, notably through the procurement of two NC-212s and a CN235 Maritime Patrol Aircraft variant. Negotiations with PT Dirgantara Indonesia over the CN235 purchase have, according to analysts, been postponed because of the pandemic. In addition, Senegal is reported to have acquired small tactical UAVs in order to improve its aerial-surveillance capability, though it remains unclear whether these are operated by the air force or army. For airspace surveillance, in 2018 Senegal procured the Thales *Ground Master* 400 long-range radar as well as a *SkyView* command-and-control centre, both to be located at the Dakar-Ouakam air base.

Maritime

The prospect of offshore gas exploitation and the resultant security requirements, as well as the increase in piracy to the south in the Gulf of Guinea, has spurred a greater focus on naval capabilities. The navy cooperates with other agencies, such as the Customs Administration, and some Customs boats are crewed by naval personnel, who also provide training. The largest platform is currently the *Fouladou* (OPV-190 Mk II). However, three OPV 58 vessels were contracted from the Kership group in November 2019. Deliveries are expected from mid-2023 and there have been reports that these may be equipped with *Mistral* surface-to-air missiles and *Marte* Mk2 anti-ship missiles. This would mark a significant capability development in the region. Apart from a few local maintenance capabilities, such as the maintenance workshop at the St Louis naval base inaugurated in 2019, the country has no significant defence-industrial base. However, as the navy expands in size and capability there will need to be similar investment in maintenance and logistics support.

Defence economics

While there is limited transparency in Senegal's defence strategy and priorities, the allocation breakdown provided in annual budget documentation is among the more transparent in the region. In 2020, the US$346 million budget for the Ministry of the Armed Forces was divided between three programmes: 'Defence of the national territory', 'Steering, coordination and administrative management' and 'Public security and maintenance of order'. The respective split between these three programmes in 2020 was 62.3%, 4.4% and 33.3%. The functional breakdown within each strand gives greater detail than programme-level spending. Figures for goods and services, personnel, investment executed by the state, and current and capital transfers are provided annually. Therefore it is possible to determine that in 2020, 58.5% of the US$322m Armed Forces budget was allocated to military personnel, 21.2% to operations and maintenance (O&M), or 18.8% to investment and the remaining 1.5% to other expenses.

Between 2014 and 2019, the budget for the Ministry of the Armed Forces grew by 10% annually in real terms, increasing from US$203m to US$330m. As a proportion of GDP, the defence budget grew from 1.2% in 2014 to 1.4% in 2019. However, the rate of growth has shown signs of slowing since 2018 and the 2020 defence budget was cut by 0.9% in nominal terms (1.9% real) to US$320m in constant (2015) US dollars, or 1.32% of GDP. Examining the functional breakdown provided in previous years indicates steady growth in the personnel and O&M budget since 2018 but dramatic nominal cuts to investment of 10.9% in 2019 and 13% in 2020. In real terms, these cuts translate to 11.8% and 13.9% respectively. In 2018, investment spending was higher than O&M within the budget (25.1% compared to 18.2%) but the share has since shifted, with O&M spending exceeding investment this year.

Angola ANG

New Angolan Kwanza Kz		2019	2020	2021
GDP	Kz	32.6tr	35.5tr	
	US$	89.4bn	62.7bn	
per capita	US$	2,968	2,021	
Growth	%	-0.9	-4.0	
Inflation	%	17.1	21.0	
Def bdgt	Kz	586bn	586bn	
	US$	1.61bn	1.04bn	
USD1=Kz		364.83	566.26	

Real-terms defence budget trend (US$bn, constant 2015)

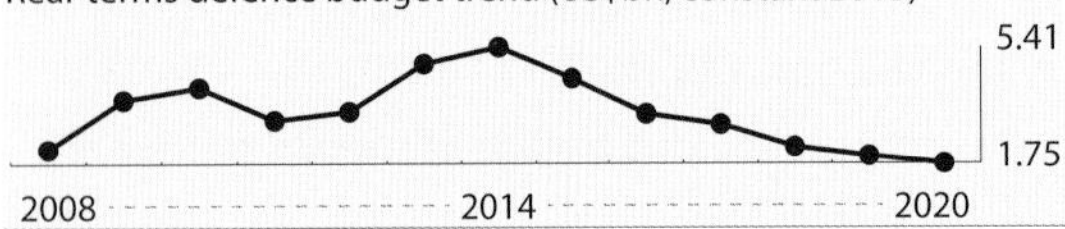

Population 32,522,339

Age	0–14	15–19	20–24	25–29	30–64	65 plus
Male	23.9%	5.1%	4.0%	3.3%	11.6%	1.0%
Female	24.0%	5.3%	4.2%	3.6%	12.8%	1.3%

Capabilities

Though numerically one of the region's largest and best-equipped armed forces, Angola's available inventory faces maintenance and readiness challenges. The armed forces are constitutionally tasked with ensuring sovereignty and territorial integrity, and there is growing attention on maritime security and the protection of offshore resources. Defence ties with Russia continue, mainly in the form of equipment deliveries, though there are plans to boost defence-industrial cooperation. There are increasing military ties with China, and Luanda is looking to Beijing for help in military modernisation and defence-industrial development. Angola retains conscription and, in recent years, force health and education have been investment priorities. The armed forces train regularly and have participated in multinational exercises. Angola is the only regional state with a strategic-airlift capacity, though availability remains an issue. Improving the military-logistics system has been identified as a key requirement, but progress is unclear. Modernisation plans have been curtailed by the fall in oil prices though there have nonetheless been some orders and acquisitions. Defence industry is limited to in-service maintenance facilities, but Angola has ambitions to develop greater capacity by partnering with countries such as Brazil, China, Portugal and Russia.

ACTIVE 107,000 (Army 100,000 Navy 1,000 Air 6,000) **Paramilitary 10,000**

Conscript liability 2 years

ORGANISATIONS BY SERVICE

Army 100,000

FORCES BY ROLE

MANOEUVRE

Armoured

1 tk bde

Light

1 SF bde

1 (1st) div (1 mot inf bde, 2 inf bde)

1 (2nd) div (3 mot inf bde, 3 inf bde, 1 arty regt)

1 (3rd) div (2 mot inf bde, 3 inf bde)

1 (4th) div (1 tk regt, 5 mot inf bde, 2 inf bde, 1 engr bde)

1 (5th) div (2 inf bde)

1 (6th) div (1 mot inf bde, 2 inf bde, 1 engr bde)

COMBAT SUPPORT

Some engr units

COMBAT SERVICE SUPPORT

Some log units

EQUIPMENT BY TYPE†

ARMOURED FIGHTING VEHICLES

MBT 300: ε200 T-55AM2; 50 T-62; 50 T-72

LT TK 10 PT-76

ASLT 9+ PTL-02 *Assaulter*

RECCE 600 BRDM-2

IFV 250 BMP-1/BMP-2

APC 246

APC (T) 31 MT-LB

APC (W) 170+: ε170 BTR-152/BTR-60/BTR-80; WZ-551 (CP)

PPV 45 *Casspir* NG2000

ENGINEERING & MAINTENANCE VEHICLES

ARV T-54/T-55

MW *Bozena*

ARTILLERY 1,439+

SP 16+: **122mm** 2S1 *Gvozdika*; **152mm** 4 2S3 *Akatsiya*; **203mm** 12 2S7 *Pion*

TOWED 575: **122mm** 523 D-30; **130mm** 48 M-46; **152mm** 4 D-20

MRL 98+: **122mm** 98: 58 BM-21 *Grad*; 40 RM-70; **240mm** BM-24

MOR 750: **82mm** 250; **120mm** 500

ANTI-TANK/ANTI-INFRASTRUCTURE

MSL • MANPATS 9K11 (AT-3 *Sagger*)

RCL 500: 400 **82mm** B-10/107mm B-11†; **106mm** 100 M40†

GUNS • SP 100mm SU-100†

AIR DEFENCE

SAM • Point-defence 9K32 *Strela*-2 (SA-7 *Grail*)‡; 9K36 *Strela-3* (SA-14 *Gremlin*); 9K310 *Igla*-1 (SA-16 *Gimlet*)

GUNS

SP 23mm ZSU-23-4

TOWED 450+: **14.5mm** ZPU-4; **23mm** ZU-23-2; **37mm** M-1939; **57mm** S-60

Navy ε1,000

EQUIPMENT BY TYPE

PATROL AND COASTAL COMBATANTS 24

PCO 2 *Ngola Kiluange* with 1 hel landing platform (Ministry of Fisheries)

PCC 5 *Rei Bula Matadi* (Ministry of Fisheries)

PBF 8: 3 HSI 32; 5 PVC-170

PB 9: 4 *Mandume*; 5 *Comandante Imperial Santana* (Ministry of Fisheries)

Coastal Defence

EQUIPMENT BY TYPE

COASTAL DEFENCE • AShM 4K44 *Utyos* (SS-C-1B *Sepal* – at Luanda)

Air Force/Air Defence 6,000

FORCES BY ROLE

FIGHTER
1 sqn with MiG-21bis/MF *Fishbed*
1 sqn with Su-27/Su-27UB/Su-30K *Flanker*
FIGHTER/GROUND ATTACK
1 sqn with MiG-23BN/ML/UB *Flogger*
1 sqn with Su-22 *Fitter* D
GROUND ATTACK
1 sqn with Su-25 *Frogfoot*
MARITIME PATROL
1 sqn with Cessna 500 *Citation* 1; C-212 *Aviocar*
TRANSPORT
3 sqn with An-12 *Cub*; An-26 *Curl*; An-32 *Cline*; An-72 *Coaler*; BN-2A *Islander*; C-212 *Aviocar*; Do-28D *Skyservant*; EMB-135BJ *Legacy* 600 (VIP); Il-76TD *Candid* MA60
TRAINING
1 sqn with Cessna 172K/R
1 sqn with EMB-312 *Tucano*
1 sqn with L-29 *Delfin*; L-39 *Albatros*
1 sqn with PC-7 *Turbo Trainer*; PC-9*
1 sqn with Z-142
ATTACK HELICOPTER
2 sqn with Mi-24/Mi-35 *Hind*; SA342M *Gazelle* (with HOT)
TRANSPORT HELICOPTER
2 sqn with AS565; SA316 *Alouette* III (IAR-316) (trg)
1 sqn with Bell 212
1 sqn with Mi-8 *Hip*; Mi-17 *Hip* H
1 sqn with Mi-171Sh
AIR DEFENCE
5 bn/10 bty with S-125 *Pechora* (SA-3 *Goa*); 9K35 *Strela*-10 (SA-13 *Gopher*)†; 2K12 *Kub* (SA-6 *Gainful*); 9K33 *Osa* (SA-8 *Gecko*); 9K31 *Strela*-1 (SA-9 *Gaskin*); S-75M *Volkhov* (SA-2 *Guideline*)

EQUIPMENT BY TYPE†

AIRCRAFT 97 combat capable
FTR 36: 6 Su-27/Su-27UB *Flanker*; 12 Su-30K *Flanker*; 18 MiG-23ML *Flogger*
FGA 41: 20 MiG-21bis/MF *Fishbed*; 8 MiG-23BN/UB *Flogger*; 13 Su-22 *Fitter* D
ATK 10: 8 Su-25 *Frogfoot*; 2 Su-25UB *Frogfoot*
MP 1 Cessna 500 *Citation* I
TPT 58: **Heavy** 4 Il-76TD *Candid*; **Medium** 6 An-12 *Cub*; **Light** 48: 12 An-26 *Curl*; 2 An-32 *Cline*; 8 An-72 *Coaler*; 8 BN-2A *Islander*; 2 C-212 *Aviocar*; 5 Cessna 172K; 6 Cessna 172R; 1 Do-28D *Skyservant*; 1 EMB-135BJ *Legacy* 600 (VIP); 2 MA60
TRG 48: 13 EMB-312 *Tucano*; 6 EMB-314 *Super Tucano**; 6 K-8W *Karakorum*; 6 L-29 *Delfin*; 2 L-39C *Albatros*; 5 PC-7 *Turbo Trainer*; 4 PC-9*; 6 Z-142
HELICOPTERS
ATK 56: 34 Mi-24 *Hind*; 22 Mi-35 *Hind*
MRH 64: 8 AS565 *Panther*; 4 AW139; 9 SA316 *Alouette* III (IAR-316) (incl trg); 8 SA342M *Gazelle*; 27 Mi-8 *Hip*/Mi-17 *Hip* H; 8 Mi-171Sh *Terminator*
TPT • Light 10: 2+ AW109E; 8 Bell 212
AIR DEFENCE • SAM 113
Medium-range 40 S-75M *Volkhov* (SA-2 *Guideline*)‡
Short-range 28: 16 2K12 *Kub* (SA-6 *Gainful*) (upgraded to 2K12-ML standard); 12 S-125 *Pechora* (SA-3 *Goa*)
Point-defence 45: 10 9K35 *Strela*-10 (SA-13 *Gopher*)†; 15 9K33 *Osa* (SA-8 *Gecko*); 20 9K31 *Strela*-1 (SA-9 *Gaskin*)
AIR-LAUNCHED MISSILES
AAM
IR R-3 (AA-2 *Atoll*)‡; R-60 (AA-8 *Aphid*); R-73 (AA-11A *Archer*)
IR/SARH R-23/24 (AA-7 *Apex*)‡; R-27 (AA-10 *Alamo*)
ASM AT-2 *Swatter*; HOT
ARM Kh-28 (AS-9 *Kyle*)

Paramilitary 10,000

Rapid-Reaction Police 10,000

Benin BEN

CFA Franc BCEAO fr		2019	2020	2021
GDP	fr	8.43tr	8.77tr	
	US$	14.4bn	15.3bn	
per capita	US$	1,218	1,259	
Growth	%	6.9	2.0	
Inflation	%	-0.9	2.5	
Def bdgt	fr	39.9bn	32.0bn	
	US$	68.1m	55.8m	
US$1=fr		585.90	573.80	

Real-terms defence budget trend (US$m, constant 2015)

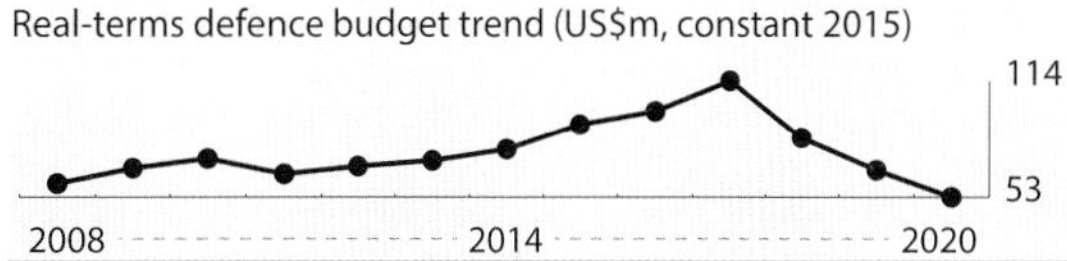

Population	12,864,634

Age	0–14	15–19	20–24	25–29	30–64	65 plus
Male	23.0%	5.5%	4.6%	3.6%	11.4%	1.1%
Female	22.6%	5.5%	4.8%	4.0%	12.7%	1.3%

Capabilities

The armed forces focus on border- and internal-security issues, as well as combating illicit trafficking. Border patrols and security have increased following concern over the regional threat from Islamist groups. Maritime security is a priority in light of continuing piracy in the Gulf of Guinea. A National Guard was reportedly created in 2020 but its precise composition and responsibilities have yet to be defined. There is a military-cooperation agreement with France, whose Senegal-based forces have delivered training to boost Benin's border-surveillance capacity. The US has provided similar training to the army and national police. US forces have also delivered pre-deployment training to the armed forces and training in professional ethics, anti-corruption and accountability to the Republican Police. Benin contributes personnel to the Multi-National Joint Task Force fighting Boko Haram. There is a limited capacity to deploy beyond neighbouring states without external support. There is some maintenance capability but no defence manufacturing sector.

ACTIVE 7,250 (Army 6,500 Navy 500 Air 250)
Paramilitary 4,800
Conscript liability 18 months (selective)

ORGANISATIONS BY SERVICE

Army 6,500

FORCES BY ROLE

MANOEUVRE

Armoured

2 armd sqn

Light

1 (rapid reaction) mot inf bn

8 inf bn

Air Manoeuvre

1 AB bn

COMBAT SUPPORT

2 arty bn

1 engr bn

1 sigs bn

COMBAT SERVICE SUPPORT

1 log bn

1 spt bn

EQUIPMENT BY TYPE

ARMOURED FIGHTING VEHICLES

LT TK 18 PT-76†

RECCE 34: 3 AML-90; 14 BRDM-2; 7 M8; 10 VBL

APC 34 • **APC (T)** 22 M113; **APC (W)** 2 *Bastion* APC; **PPV** 10 *Casspir* NG

ARTILLERY 16+

TOWED 105mm 16: 12 L118 Light Gun; 4 M101

MOR 81mm some; **120mm** some

Navy ε500

EQUIPMENT BY TYPE

PATROL AND COASTAL COMBATANTS • **PB** 6: 2 *Matelot Brice Kpomasse* (ex-PRC); 3 FPB 98; 1 27m (PRC)

Air Force 250

EQUIPMENT BY TYPE

AIRCRAFT

TPT 3: **Light** 1 DHC-6 *Twin Otter*†; **PAX** 2: 1 B-727; 1 HS-748†

TRG 2 LH-10 *Ellipse*

HELICOPTERS

TPT • **Light** 5: 4 AW109BA; 1 AS350B *Ecureuil*†

Paramilitary 4,800

Police Republicaine 4,800

EQUIPMENT BY TYPE

ARMOURED FIGHTING VEHICLES

APC • **PPV** *Casspir* NG

DEPLOYMENT

CENTRAL AFRICAN REPUBLIC: UN • MINUSCA 7

CHAD: Lake Chad Basin Commission • MNJTF 150

DEMOCRATIC REPUBLIC OF THE CONGO: UN • MONUSCO 2

MALI: UN • MINUSMA 257; 1 mech inf coy(+)

SOUTH SUDAN: UN • UNMISS 4

SUDAN: UN • UNISFA 2

Botswana BWA

Botswana Pula P		2019	2020	2021
GDP	P	197bn	179bn	
	US$	18.5bn	15.9bn	
per capita	US$	7,773	6,558	
Growth	%	3.0	-9.6	
Inflation	%	2.8	1.6	
Def bdgt	P	5.75bn	6.42bn	
	US$	539m	571m	
US$1=P		10.68	11.25	

Real-terms defence budget trend (US$m, constant 2015)

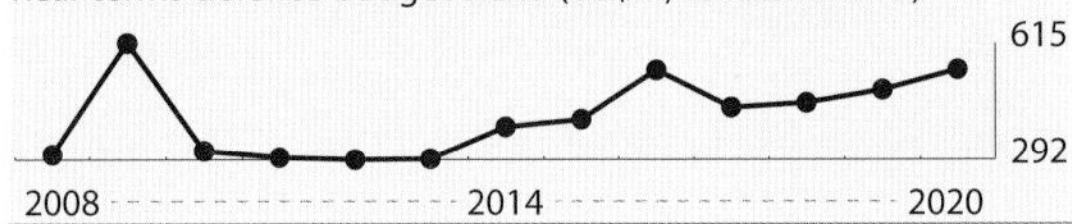

Population 2,317,233

Age	0–14	15–19	20–24	25–29	30–64	65 plus
Male	15.4%	4.6%	4.4%	4.3%	17.0%	2.3%
Female	15.1%	4.7%	4.6%	4.4%	19.8%	3.2%

Capabilities

The Botswana Defence Force (BDF) comprises ground forces and a small but comparatively well-equipped air wing. The BDF's major task is to ensure territorial integrity, as well as tasks including tackling poaching. There is a history of involvement in peacekeeping operations. The BDF has reportedly been working on a defence doctrine that is believed to be influenced by US concepts and practices. Botswana has a good relationship with the US and regularly sends its officers to train there. The armed forces also train with other African nations, including Namibia, with whom it holds biannual exercises. The operations centre for the SADC Standby Force is located in Gaborone. Recent personnel priorities include improving conditions of service and overhauling retirement ages. Relations with China are improving, and some military personnel have travelled there for training. The air force has a modest airlift capacity and the BDF is able to deploy a small force by air if required. There is interest in replacing the ageing fleet of F-5 combat aircraft, though a successor type has yet to be ordered. While reports suggest a limited maintenance capacity, there is no defence-manufacturing sector.

ACTIVE 9,000 (Army 8,500 Air 500)

ORGANISATIONS BY SERVICE

Army 8,500

FORCES BY ROLE

MANOEUVRE

Armoured

1 armd bde(-)

Light

2 inf bde (1 armd recce regt, 4 inf bn, 1 cdo unit, 1 engr regt, 1 log bn, 2 ADA regt)

COMBAT SUPPORT

1 arty bde

1 engr coy

1 sigs coy

COMBAT SERVICE SUPPORT
1 log gp
AIR DEFENCE
1 AD bde(-)

EQUIPMENT BY TYPE

ARMOURED FIGHTING VEHICLES
LT TK 45: ε20 SK-105 *Kurassier*; 25 FV101 *Scorpion*
RECCE 72+: RAM-V-1; ε8 RAM-V-2; 64 VBL
APC 157:
APC (W) 145: 50 BTR-60; 50 LAV-150 *Commando* (some with 90mm gun); 45 *Piranha* III
PPV 12 *Casspir*
AUV 6 FV103 *Spartan*
ENGINEERING & MAINTENANCE VEHICLES
ARV *Greif*; M578
MW *Aardvark* Mk2
ANTI-TANK/ANTI-INFRASTRUCTURE
MSL
SP V-150 TOW
MANPATS TOW
RCL 84mm *Carl Gustaf*
ARTILLERY 78
TOWED 30: **105mm** 18: 12 L118 Light Gun; 6 Model 56 pack howitzer; **155mm** 12 Soltam
MRL 122mm 20 APRA-40
MOR 28: **81mm** 22; **120mm** 6 M-43
AIR DEFENCE
SAM
Short-range VL MICA
Point-defence *Javelin*; 9K310 *Igla*-1 (SA-16 *Gimlet*); 9K32 *Strela*-2 (SA-7 *Grail*)‡; *Mistral*
GUNS • TOWED 20mm 7 M167 *Vulcan*

Air Wing 500

FORCES BY ROLE

FIGHTER/GROUND ATTACK
1 sqn with F-5A *Freedom Fighter*; F-5D *Tiger* II
ISR
1 sqn with O-2 *Skymaster*
TRANSPORT
2 sqn with BD-700 *Global Express*; BN-2/-2B *Defender**; Beech 200 *King Air* (VIP); C-130B *Hercules*; C-212-300/400 *Aviocar*; CN-235M-100; Do-328-110 (VIP)
TRAINING
1 sqn with PC-7 MkII *Turbo Trainer**
TRANSPORT HELICOPTER
1 sqn with AS350B *Ecureuil*; Bell 412EP/SP *Twin Huey*; EC225LP *Super Puma*

EQUIPMENT BY TYPE

AIRCRAFT 28 combat capable
FTR 13: 8 F-5A *Freedom Fighter*; 5 F-5D *Tiger* II
ISR 5 O-2 *Skymaster*
TPT 20: **Medium** 3 C-130B *Hercules*; **Light** 16: 4 BN-2 *Defender**; 6 BN-2B *Defender**; 1 Beech 200 *King Air* (VIP); 1 C-212-300 *Aviocar*; 1 C-212-400 *Aviocar*; 2 CN-235M-100; 1 Do-328-110 (VIP); **PAX** 1 BD700 *Global Express*
TRG 5 PC-7 MkII *Turbo Trainer**
HELICOPTERS
MRH 7: 2 Bell 412EP *Twin Huey*; 5 Bell 412SP *Twin Huey*
TPT 9: **Medium** 1 EC225LP *Super Puma*; **Light** 8 AS350B *Ecureuil*

Burkina Faso BFA

CFA Franc BCEAO fr		2019	2020	2021
GDP	fr	9.23tr	9.23tr	
	US$	15.7bn	16.1bn	
per capita	US$	775	769	
Growth	%	5.7	-2.0	
Inflation	%	-3.2	2.0	
Def bdgt	fr	210bn	223bn	
	US$	358m	388m	
US$1=fr		585.90	573.80	

Real-terms defence budget trend (US$m, constant 2015)

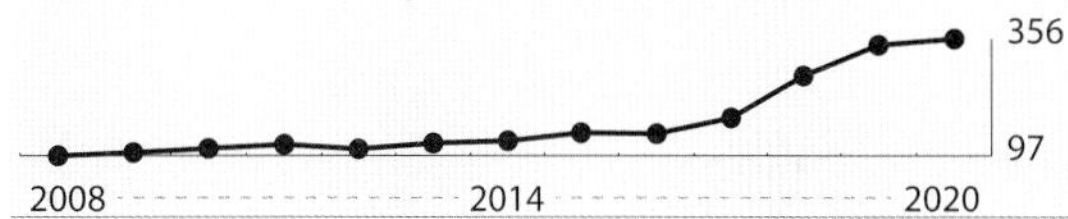

Population	20,835,401					
Age	**0–14**	**15–19**	**20–24**	**25–29**	**30–64**	**65 plus**
Male	22.1%	5.8%	4.4%	3.7%	11.6%	1.4%
Female	21.5%	5.6%	4.6%	4.1%	13.6%	1.8%

Capabilities

Burkina Faso's security forces have in recent years been challenged by an increased terrorist threat. The country is part of the G5 Sahel grouping and, as part of its support for this, France has supplied armed 4x4 vehicles. The terrorist threat has led Ouagadougou to refocus its military efforts to the north. There is cooperation with France and Mali, particularly on border security. Aviation capacities are slowly improving with the arrival of additional helicopters, though financial challenges hinder broader capability developments. Deployment capacity is limited to neighbouring countries without external support. While there are maintenance facilities there is no defence-manufacturing sector.

ACTIVE 11,200 (Army 6,400 Air 600 Gendarmerie 4,200) Paramilitary 250

ORGANISATIONS BY SERVICE

Army 6,400

Three military regions. In 2011, several regiments were disbanded and merged into other formations, including the new 24th and 34th *régiments interarmes*

FORCES BY ROLE

MANOEUVRE
Mechanised
1 cbd arms regt
Light
1 cbd arms regt
6 inf regt
Air Manoeuvre
1 AB regt (1 CT coy)

COMBAT SUPPORT
1 arty bn (2 arty tp)
1 engr bn

EQUIPMENT BY TYPE

ARMOURED FIGHTING VEHICLES
RECCE 91+: 19 AML-60/AML-90; 8+ *Bastion Patsas*; 24 EE-9 *Cascavel*; 30 *Ferret*; 2 M20; 8 M8
APC 86
APC (W) 25: 13 Panhard M3; 12 *Bastion* APC
PPV 61: 6 *Gila*; 31 *Puma* M26-15; 24 Stark Motors *Storm*
AUV 4 *Cobra*

ENGINEERING & MAINTENANCE VEHICLES
MW 3 *Shrek*-M

ANTI-TANK/ANTI-INFRASTRUCTURE
RCL 75mm Type-52 (M20); **84mm** *Carl Gustaf*

ARTILLERY 50+
TOWED 14: **105mm** 8 M101; **122mm** 6
MRL 9: **107mm** ε4 Type-63; **122mm** 5 APR-40
MOR 27+: **81mm** Brandt; **82mm** 15; **120mm** 12

AIR DEFENCE
SAM • Point-defence 9K32 *Strela*-2 (SA-7 *Grail*)‡
GUNS • TOWED 42: **14.5mm** 30 ZPU; **20mm** 12 TCM-20

Air Force 600

FORCES BY ROLE

GROUND ATTACK/TRAINING
1 sqn with SF-260WL *Warrior**; Embraer EMB-314 *Super Tucano**

TRANSPORT
1 sqn with AT-802 *Air Tractor*; B-727 (VIP); Beech 200 *King Air*; CN235-220; PA-34 *Seneca*; *Tetras*

ATTACK/TRANSPORT HELICOPTER
1 sqn with AS350 *Ecureuil*; Mi-8 *Hip*; Mi-17 *Hip* H; Mi-35 *Hind* AW 139; UH-1Y *Huey*

EQUIPMENT BY TYPE

AIRCRAFT 5 combat capable
ISR 1 DA42M (reported)
TPT 9: **Light** 8: 1 AT-802 *Air Tractor*; 2 Beech 200 *King Air*; 1 CN235-220; 1 PA-34 *Seneca*; 3 *Tetras*; **PAX** 1 B-727 (VIP)
TRG 5: 3 EMB-314 *Super Tucano**; 2 SF-260WL *Warrior**

HELICOPTERS
ATK 2 Mi-35 *Hind*
MRH 3: 2 Mi-17 *Hip* H; 1 AW139
TPT 3: **Medium** 1 Mi-8 *Hip*; **Light** 2: 1 AS350 *Ecureuil*; 1 UH-1H *Huey*

Gendarmerie 4,200

FORCES BY ROLE

SPECIAL FORCES
1 spec ops gp (USIGN)

EQUIPMENT BY TYPE

ARMOURED FIGHTING VEHICLES
APC • APC (W) some *Bastion* APC

Paramilitary 250

People's Militia (R) 45,000 reservists (trained)

Security Company 250

DEPLOYMENT

CENTRAL AFRICAN REPUBLIC: UN • MINUSCA 4

DEMOCRATIC REPUBLIC OF THE CONGO: UN • MONUSCO 5

MALI: UN • MINUSMA 651; 1 inf bn

SUDAN: UN • UNISFA 1

FOREIGN FORCES

France *Operation Barkhane* 4000; 1 SF gp; 2 *Tiger*; 2 AS532UL; 2 H225M; 3 *Gazelle*

Burundi BDI

Burundi Franc fr		2019	2020	2021
GDP	fr	5.74tr	6.00tr	
	US$	3.11bn	3.13bn	
per capita	US$	270	264	
Growth	%	1.8	-3.2	
Inflation	%	-0.7	7.6	
Def bdgt	fr	117bn	119bn	129bn
	US$	63.2m	62.3m	
US$1=fr		1845.43	1914.77	

Real-terms defence budget trend (US$m, constant 2015)

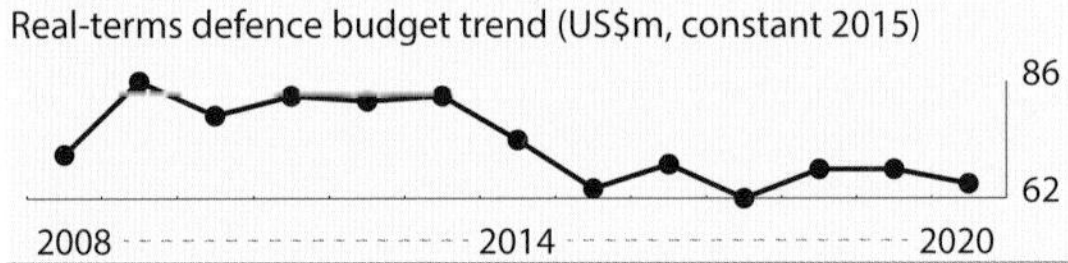

Population 11,865,821

Age	0–14	15–19	20–24	25–29	30–64	65 plus
Male	22.1%	5.4%	4.4%	3.6%	12.7%	1.3%
Female	21.8%	5.4%	4.5%	3.7%	13.3%	1.8%

Capabilities

The political crisis in 2015 tested the cohesion of the armed forces and largely halted military-training activity with international partners. Cross-border security challenges continue, though there are signs of improving relations with Rwanda. Burundi signed a cooperation agreement with Russia in 2018 on counter-terrorism and joint training. The experience accumulated during UN operations, where troops have gained valuable military experience, boosts training levels. The armed forces have a limited capability to deploy externally and maintain a deployment to the AMISOM mission in Somalia. Peacekeeping missions help to fund the armed forces, though financial and equipment deficiencies otherwise restrict military effectiveness. Apart from limited maintenance facilities, the country has no domestic defence-industrial capability.

ACTIVE 30,050 (Army 30,000 Navy 50) **Paramilitary 21,000**

ORGANISATIONS BY SERVICE

Army 30,000

FORCES BY ROLE

MANOEUVRE

Mechanised

2 lt armd bn (sqn)

Light

7 inf bn

Some indep inf coy

COMBAT SUPPORT

1 arty bn

1 engr bn

AIR DEFENCE

1 AD bn

EQUIPMENT BY TYPE

ARMOURED FIGHTING VEHICLES

RECCE 55: 6 AML-60; 12 AML-90; 30 BRDM-2; 7 S52 *Shorland*

APC 94

APC (W) 60: 20 BTR-40; 10 BTR-80; 9 Panhard M3; 15 Type-92; 6 *Walid*

PPV 34: 12 *Casspir*; 12 RG-31 *Nyala*; 10 RG-33L

AUV 15 *Cougar* 4×4

ARTILLERY 120

TOWED 122mm 18 D-30

MRL 122mm 12 BM-21 *Grad*

MOR 90: **82mm** 15 M-43; **120mm** ε75

ANTI-TANK/ANTI-INFRASTRUCTURE

MSL • MANPATS *Milan* (reported)

RCL 75mm Type-52 (M20)

AIR DEFENCE

SAM • Point-defence 9K32 *Strela-2* (SA-7 *Grail*)‡

GUNS • TOWED 150+: **14.5mm** 15 ZPU-4; 135+ **23mm** ZU-23/**37mm** Type-55 (M-1939)

Air Wing 200

EQUIPMENT BY TYPE

AIRCRAFT 1 combat capable

TPT • Light 2 Cessna 150L†

TRG 1 SF-260W *Warrior**

HELICOPTERS

ATK 2 Mi-24 *Hind*

MRH 2 SA342L *Gazelle*

TPT • Medium (2 Mi-8 *Hip* non-op)

Reserves

FORCES BY ROLE

MANOEUVRE

Light

10 inf bn (reported)

Navy 50

EQUIPMENT BY TYPE

PATROL AND COASTAL COMBATANTS • **PB** 4

AMPHIBIOUS • **LCT** 2

Paramilitary ε1,000

General Administration of State Security ε1,000

DEPLOYMENT

CENTRAL AFRICAN REPUBLIC: UN • MINUSCA 753; 1 inf bn

MALI: UN • MINUSMA 1

SOMALIA: AU • AMISOM 4,000; 5 inf bn

SUDAN: UN • UNISFA 4

Cameroon CMR

CFA Franc BEAC fr		2019	2020	2021
GDP	fr	22.8tr	22.4tr	
	US$	38.9bn	39.0bn	
per capita	US$	1,524	1,493	
Growth	%	3.9	-2.8	
Inflation	%	2.5	2.8	
Def bdgt	fr	247bn	234bn	
	US$	422m	408m	
US$1=fr		585.89	573.81	

Real-terms defence budget trend (US$m, constant 2015)

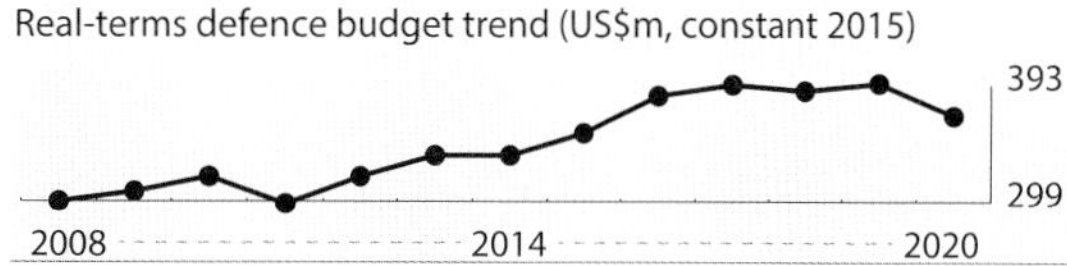

Population 27,744,989

Age	0–14	15–19	20–24	25–29	30–64	65 plus
Male	21.4%	5.4%	4.6%	4.0%	13.0%	1.5%
Female	21.0%	5.4%	4.6%	4.0%	13.5%	1.7%

Capabilities

Internal security is a key concern for Cameroon's armed forces, as is the cross-border challenge from Boko Haram. In 2018, a fifth military region was created in response to security challenges in the west arising from separatist activity. Cameroon is part of the Multi-National Joint Task Force engaged on operations against Boko Haram. There are long-standing military ties with France, including for support and training. There is a military-assistance agreement with China and the two countries have cooperated over a floating dock at Kribi, which it is hoped will improve operational readiness. The AU maintains its continental logistics base at Douala. The armed forces are considered well organised, though allegations of abuses led the US in February 2019 to halt some military assistance. Although deployments continue to UN peacekeeping operations, there is only limited organic power-projection capability without external support. Much of the equipment inventory is ageing, but infantry fighting vehicles and protected patrol vehicles have been acquired from China and South Africa and gifted by the US. The armed forces are improving their ISR capability with fixed-wing aircraft and small UAVs. Additional patrol vessels have in recent years improved maritime capability. Cameroon has no defence-industrial capacity, bar maintenance facilities.

ACTIVE 25,400 (Army 23,500 Navy 1,500 Air 400) **Paramilitary 9,000**

ORGANISATIONS BY SERVICE

Army 23,500

5 Mil Regions

FORCES BY ROLE

MANOEUVRE

Light

1 rapid reaction bde (1 armd recce bn, 1 AB bn, 1 amph bn)
1 mot inf bde (4 mot inf bn, 1 spt bn)
5 mot inf bde (3 mot inf bn, 1 spt bn)
6 rapid reaction bn
4 inf bn

Air Manoeuvre

1 cdo/AB bn

Other

1 (Presidential Guard) gd bn

COMBAT SUPPORT

1 arty regt (5 arty bty)
5 engr regt

AIR DEFENCE

1 AD regt (6 AD bty)

EQUIPMENT BY TYPE

ARMOURED FIGHTING VEHICLES

ASLT 18: 6 AMX-10RC; ε12 PTL-02 mod (*Cara* 105)
RECCE 64: 31 AML-90; 15 *Ferret*; 8 M8; 5 RAM Mk3; 5 VBL
IFV 42: 8 LAV-150 *Commando* with 20mm gun; 14 LAV-150 *Commando* with 90mm gun; 12 *Ratel*-20 (Engr); ε8 Type-07P
APC 64
APC (T) 12 M3 half-track
APC (W) 36: 15 *Bastion* APC (reported); 21 LAV-150 *Commando*
PPV 16 Gaia *Thunder*
AUV 6+: 6 *Cougar* 4×4; *Panthera* T6

ENGINEERING & MAINTENANCE VEHICLES

ARV WZ-551 ARV

ANTI-TANK/ANTI-INFRASTRUCTURE

MSL
SP 24 TOW (on Jeeps)
MANPATS *Milan*
RCL 53: **75mm** 13 Type-52 (M20); **106mm** 40 M40A2

ARTILLERY 106+

SP 155mm 18 ATMOS 2000
TOWED 52: **105mm** 20 M101; **130mm** 24: 12 M-1982 (reported); 12 Type-59 (M-46); **155mm** 8 M-71
MRL 122mm 20 BM-21 *Grad*
MOR 16+: **81mm** (some SP); **120mm** 16 Brandt

AIR DEFENCE • GUNS

SP 20mm RBY-1 with TCM-20
TOWED 54: **14.5mm** 18 Type-58 (ZPU-2); **35mm** 18 GDF-002; **37mm** 18 Type-63

Navy ε1,500

HQ located at Douala

EQUIPMENT BY TYPE

PATROL AND COASTAL COMBATANTS 16

PCC 3: 1 *Dipikar* (ex-FRA *Flamant*); 2 *Le Ntem* (PRC *Limam El Hidrami*)
PB 11: 2 Aresa 2400; 2 Aresa 3200; 2 Rodman 101; 4 Rodman 46; 1 *Quartier Maître Alfred Motto*†
PBR 2 *Swift*-38

AMPHIBIOUS • LANDING CRAFT 4

LCU 2 Type-067 (*Yunnan*)
LCM 2: 1 Aresa 2300; 1 *Le Moungo*

Fusiliers Marin

FORCES BY ROLE

MANOEUVRE

Amphibious

3 mne bn

Air Force 300–400

FORCES BY ROLE

FIGHTER/GROUND ATTACK

1 sqn with *Alpha Jet**†

TRANSPORT

1 sqn with C-130H/H-30 *Hercules*; PA-23 *Aztec*
1 VIP unit with AS332 *Super Puma*; AS365 *Dauphin* 2; Bell 206B *Jet Ranger*; Gulfstream III

TRAINING

1 unit with *Tetras*

ATTACK HELICOPTER

1 sqn with SA342 *Gazelle* (with HOT); Mi-24 *Hind*

TRANSPORT HELICOPTER

1 sqn with Bell 206L-3; Bell 412; SA319 *Alouette* III

EQUIPMENT BY TYPE

AIRCRAFT 6 combat capable

ISR 2 Cessna 208B *Grand Caravan*
TPT 17: **Medium** 3: 2 C-130H *Hercules*; 1 C-130H-30 *Hercules*; **Light** 13: 1 CN235 (1 IAI-201 *Arava* in store); 2 J.300 *Joker*; 1 MA60; 2 PA-23 *Aztec*; 7 *Tetras*; **PAX** 1 Gulfstream III
TRG 6 *Alpha Jet**†

HELICOPTERS

ATK 2 Mi-24 *Hind*
MRH 15: 1 AS365 *Dauphin* 2; 4 Bell 412 *Twin Huey*; 2 Mi-17 *Hip* H; 2 SA319 *Alouette* III; 4 SA342 *Gazelle* (with HOT); 2 Z-9
TPT 6: **Medium** 4: 2 AS332 *Super Puma*; 2 SA330J *Puma*; **Light** 2: 1 Bell 206B *Jet Ranger*; 1 Bell 206L3 *Long Ranger*

AIR-LAUNCHED MISSILES

ASM HOT

Fusiliers de l'Air

FORCES BY ROLE

MANOEUVRE

Other

1 sy bn

Paramilitary 9,000

Gendarmerie 9,000

FORCES BY ROLE

MANOEUVRE

Reconnaissance

3 (regional spt) paramilitary gp

DEPLOYMENT

CENTRAL AFRICAN REPUBLIC: UN • MINUSCA 761; 1 inf bn

DEMOCRATIC REPUBLIC OF THE CONGO: UN • MONUSCO 4

MALI: UN • MINUSMA 2

FOREIGN FORCES

United States 300; MQ-1C *Gray Eagle*

Cape Verde CPV

Cape Verde Escudo E		2019	2020	2021
GDP	E	195bn	184bn	
	US$	1.98bn	1.87bn	
per capita	US$	3,602	3,358	
Growth	%	5.7	-6.8	
Inflation	%	1.1	1.0	
Def bdgt	E	1.05bn	1.12bn	
	US$	10.7m	11.4m	
US$1=E		98.49	98.35	

Real-terms defence budget trend (US$m, constant 2015)

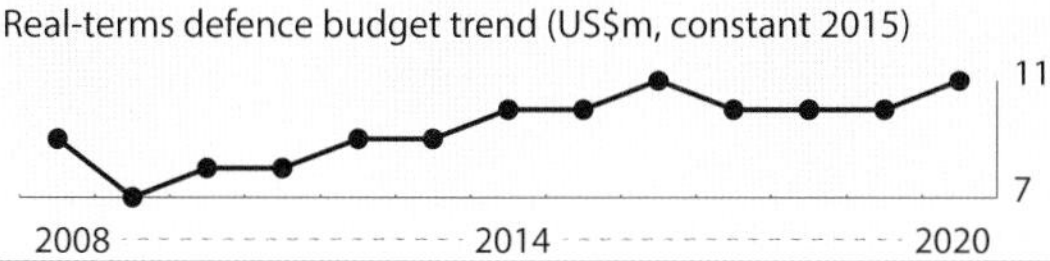

Population 583,255

Age	0–14	15–19	20–24	25–29	30–64	65 plus
Male	14.1%	4.7%	4.7%	4.7%	18.4%	2.1%
Female	13.9%	4.7%	4.7%	4.7%	20.1%	3.4%

Capabilities

In its legislative programme for 2016–21, the government outlined defence priorities including territorial defence, maritime security, and EEZ and airspace protection. Although the armed forces are small and presently have limited capability, the government has suggested reorganising around marines, engineering and paramilitary national-guard units. The government is interested in greater regional and international defence engagement; international partners provide some maritime-security training support. The armed forces take part in multinational regional exercises and cooperative activities. Equipment capabilities remain limited and there is no defence industry, beyond limited maintenance facilities.

ACTIVE 1,200 (Army 1,000 Coast Guard 100 Air 100)

Conscript liability Selective conscription (14 months)

ORGANISATIONS BY SERVICE

Army 1,000

FORCES BY ROLE

MANOEUVRE

Light

2 inf bn (gp)

COMBAT SUPPORT

1 engr bn

EQUIPMENT BY TYPE

ARMOURED FIGHTING VEHICLES

RECCE 10 BRDM-2

ARTILLERY • MOR 18: **82mm** 12; **120mm** 6 M-1943

AIR DEFENCE

SAM • Point-defence 9K32 *Strela* (SA-7 *Grail*)‡

GUNS • TOWED 30: **14.5mm** 18 ZPU-1; **23mm** 12 ZU-23

Coast Guard ε100

EQUIPMENT BY TYPE

PATROL AND COASTAL COMBATANTS 5

PCC 2: 1 *Guardião*; 1 *Kondor I*

PB 2: 1 *Espadarte*; 1 *Tainha* (PRC 27m)

PBF 1 *Archangel*

AIRCRAFT • TPT • Light 1 Do-228

Air Force up to 100

FORCES BY ROLE

MARITIME PATROL

1 sqn with An-26 *Curl*

EQUIPMENT BY TYPE

AIRCRAFT • TPT • Light 3 An-26 *Curl*†

Central African Republic CAR

CFA Franc BEAC fr		2019	2020	2021
GDP	fr	1.33tr	1.36tr	
	US$	2.28bn	2.32bn	
per capita	US$	480	480	
Growth	%	3.0	-1.0	
Inflation	%	2.7	2.9	
Def exp	fr	19.0bn	23.8bn	
	US$	32.4m	40.6m	
US$1=fr		585.87	585.01	

Real-terms defence budget trend (US$m, constant 2015)

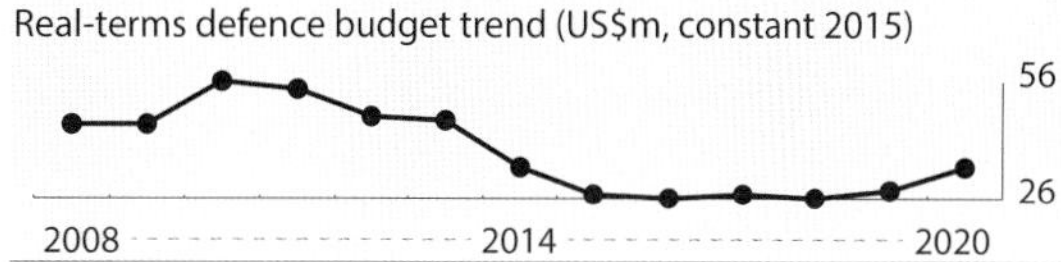

Population 5,990,855

Age	0–14	15–19	20–24	25–29	30–64	65 plus
Male	19.8%	5.3%	4.7%	4.2%	14.4%	1.3%
Female	19.6%	5.3%	4.6%	4.1%	14.6%	2.0%

Capabilities

Effective military and security organisations remain largely absent in the wake of violence in 2013 and the armed forces remain unable to fully address the country's internal-security challenges. Under the National Recovery and Peacebuilding Plan 2017–21, attempts to improve security focus on DDR and SSR. The UN's MINUSCA mission remains the principal security provider in the country. The CAR benefits from defence partnerships with France and Russia. Moscow has been deepening its military ties in the country and has donated small arms and armoured vehicles and has dispatched teams of military instructors to the CAR. Apart from some equipment deliveries, the country remains under a UN arms embargo, though the terms of this were eased in late 2019. The armed forces receive training from UN forces and an EU training mission. Poor infrastructure and logistics capacity limit the armed forces' ability to provide security across the country. There is no independent capability to deploy troops externally, while the lack of financial resources and defence-industrial capacity makes equipment maintenance problematic.

ACTIVE 9,150 (Army 9,000 Air 150) Paramilitary 1,000

Conscript liability Selective conscription 2 years; reserve obligation thereafter, term n.k.

ORGANISATIONS BY SERVICE

Army ε9,000

FORCES BY ROLE

MANOEUVRE
Light
1 inf bn
Amphibious
1 amph coy

EQUIPMENT BY TYPE

ARMOURED FIGHTING VEHICLES
MBT 3 T-55†
RECCE 29: 8 *Ferret*†; 21 BRDM-2
IFV 18 *Ratel*
APC • APC (W) 14+: 4 BTR-152†; 10+ VAB†
ARTILLERY • **MOR** 12+: **81mm**†; **120mm** 12 M-1943†
ANTI-TANK/ANTI-INFRASTRUCTURE
RCL 106mm 14 M40†
PATROL AND COASTAL COMBATANTS • **PBR** 9†

Air Force 150

EQUIPMENT BY TYPE

AIRCRAFT • **TPT** 6: **Medium** (1 C-130A *Hercules* in store); **Light** 6: 3 BN-2 *Islander*; 1 Cessna 172RJ *Skyhawk*; 2 J.300 *Joker*
HELICOPTERS • **TPT** • **Light** 1 AS350 *Ecureuil*

FOREIGN FORCES

MINUSCA unless stated
Argentina 1
Bangladesh 1,299; 1 cdo coy; 1 inf bn; 1 med coy; 1 hel coy
Benin 7
Bhutan 4
Bolivia 5
Bosnia-Herzegovina EUTM RCA 2
Brazil 8 • EUTM RCA 3
Burkina Faso 4
Burundi 753; 1 inf bn
Cambodia 214; 1 engr coy
Cameroon 761; 1 inf bn
Colombia 2
Congo 12
Côte d'Ivoire 6
Czech Republic 3
Egypt 1,000; 1 inf bn; 1 tpt coy
France 9 • EUTM RCA 40 • Army 160
Gabon 447; 1 inf bn(-)
Gambia 8
Georgia EUTM RCA 35
Ghana 14
Guatemala 4
Indonesia 212; 1 engr coy
Italy EUTM RCA 3
Jordan 9
Kenya 15
Lithuania EUTM RCA 2
Mauritania 452; 1 inf bn(-)
Mexico 2
Moldova 3
Morocco 763; 1 inf bn
Nepal 737; 1 inf bn; 1 MP pl
Niger 4
Nigeria 3
Pakistan 1,252; 1 inf bn; 2 engr coy; 1 hel sqn
Paraguay 3
Peru 216; 1 engr coy
Philippines 3
Poland EUTM RCA 1
Portugal 186; 1 AB coy • EUTM RCA 45
Romania EUTM RCA 13
Russia 13
Rwanda 1,357; 2 inf bn; 1 fd hospital
Senegal 5; 1 atk hel sqn
Serbia 76; 1 med coy • EUTM RCA 7
Sierra Leone 5
Slovakia EUTM RCA 4
Spain EUTM RCA 8
Sri Lanka 110; 1 avn unit
Sweden EUTM RCA 9
Tanzania 455; 1 inf bn(-)
Togo 10
Tunisia 2
United States 7
Uruguay 2
Vietnam 6
Zambia 923; 1 inf bn

Chad CHA

CFA Franc BEAC fr		2019	2020	2021
GDP	fr	6.41tr	6.15tr	
	US$	10.9bn	10.5bn	
per capita	US$	686	640	
Growth	%	3.0	-0.7	
Inflation	%	-1.0	2.8	
Def bdgt	fr	120bn	157bn	
	US$	205m	269m	
US$1=fr		585.91	585.07	

Real-terms defence budget trend (US$m, constant 2015)

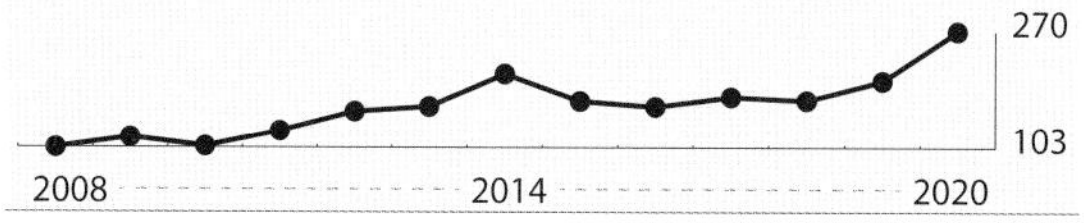

Population 16,877,357

Age	0–14	15–19	20–24	25–29	30–64	65 plus
Male	24.0%	5.6%	4.3%	3.5%	11.0%	1.0%
Female	23.4%	5.5%	4.3%	3.6%	12.3%	1.4%

Capabilities

Chad's principal security concerns relate to instability in West Africa and the Sahel and counter-insurgency operations against Boko Haram in the Lake Chad Basin area. Although the armed forces are combat experienced, some observers judge a requirement for improved strategy and doctrine, and command and control. The country is a key contributor to the G5 Sahel and is an important component of the Multi-National Joint Task Force fighting Boko Haram. There is close defence cooperation with France and *Operation Barkhane* is headquartered in N'Djamena. Chadian military skills are widely recognised by partners, though training levels are not uniform across the force. A lack of logistical capacity has hindered routine rotations for deployed forces. The country's ISR capability improved with the arrival of aircraft from the US, following improvements in ground-attack and medium-airlift capability. Barring maintenance facilities, there is no domestic defence-industrial capacity.

ACTIVE 33,250 (Army 27,500 Air 350 State Security Service 5,400) Paramilitary 11,900

Conscript liability Conscription authorised

ORGANISATIONS BY SERVICE

Army ε27,500

7 Mil Regions

FORCES BY ROLE

MANOEUVRE

Armoured

1 armd bn

Light

7 inf bn

COMBAT SUPPORT

1 arty bn

1 engr bn

1 sigs bn

COMBAT SERVICE SUPPORT

1 log gp

EQUIPMENT BY TYPE

ARMOURED FIGHTING VEHICLES

MBT 60 T-55

ASLT 30 PTL-02 *Assaulter*

RECCE 309+: 132 AML-60/AML-90; 22 *Bastion Patsas*; ε100 BRDM-2; 20 EE-9 *Cascavel*; 4 ERC-90F *Sagaie*; 31+ RAM Mk3

IFV 131: 80 BMP-1; 42 BMP-1U; 9 LAV-150 *Commando* with 90mm gun

APC • APC (W) 103: 4 *Bastion* APC; 24 BTR-80; 12 BTR-3E; ε20 BTR-60; ε10 *Black Scorpion*; 25 VAB-VTT; 8 WZ-523

ARTILLERY 26+

SP 122mm 10 2S1 *Gvozdika*

TOWED 105mm 5 M2

MRL 11+: **107mm** some Type-63; **122mm** 11: 6 BM-21 *Grad*; 5 Type-81

MOR 81mm some; **120mm** AM-50

ANTI-TANK/ANTI-INFRASTRUCTURE

MSL • MANPATS *Eryx*; *Milan*

RCL 106mm M40A1

AIR DEFENCE

SAM

Short-range 2K12 *Kub* (SA-6 *Gainful*)

Point-defence 9K310 *Igla*-1 (SA-16 *Gimlet*)

GUNS • TOWED 14.5mm ZPU-1/ZPU-2/ZPU-4; **23mm** ZU-23

Air Force 350

FORCES BY ROLE

GROUND ATTACK

1 unit with PC-7; PC-9*; SF-260WL *Warrior**; Su-25 *Frogfoot*

TRANSPORT

1 sqn with An-26 *Curl*; C-130H-30 *Hercules*; Mi-17 *Hip* H; Mi-171

1 (Presidential) Flt with B-737BBJ; Beech 1900; DC-9-87; Gulfstream II

ATTACK HELICOPTER

1 sqn with AS550C *Fennec*; Mi-24V *Hind*; SA316 *Alouette* III

MANOEUVRE

Other

1 sy bn

EQUIPMENT BY TYPE

AIRCRAFT 10 combat capable

FTR (1 MiG-29S *Fulcrum* C in store)

ATK 7: 6 Su-25 *Frogfoot* (2 more in store); 1 Su-25UB *Frogfoot* B (1 more in store)

ISR 2 Cessna 208B *Grand Caravan*

TPT 10: **Medium** 3: 2 C-27J *Spartan*; 1 C-130H-30 *Hercules*; **Light** 4: 3 An-26 *Curl*; 1 Beech 1900; **PAX** 3: 1 B-737BBJ; 1 DC-9-87; 1 Gulfstream II

TRG 4: 2 PC-7 (only 1*); 1 PC-9 *Turbo Trainer**; 1 SF-260WL *Warrior**

HELICOPTERS

ATK 5 Mi-24V *Hind*

MRH 8: 3 AS550C *Fennec*; 3 Mi-17 *Hip* H; 2 SA316

TPT • Medium 2 Mi-171

State Security Service General Direction (DGSSIE) 5,400

Paramilitary 11,900 active

Gendarmerie 4,500

National and Nomadic Guard (GNNT) 7,400

Police Mobile Intervention Group (GMIP)

DEPLOYMENT

MALI: UN • MINUSMA 1,456; 1 SF coy; 2 inf bn

FOREIGN FORCES

Benin MNJTF 150
France *Operation Barkhane* 1,500; 1 mech inf BG; 1 FGA det with 4 *Mirage* 2000D; 1 tpt det with 1 C-130H; 2 CN235M; 1 UAV det with 1 MQ-9A *Reaper*

Congo, Republic of COG

CFA Franc BEAC fr		2019	2020	2021
GDP	fr	7.35tr	5.83tr	
	US$	12.5bn	10.0bn	
per capita	US$	2,746	2,128	
Growth	%	-0.6	-7.0	
Inflation	%	2.2	2.5	
Def bdgt	fr	176bn	179bn	174bn
	US$	300m	306m	
US$1=fr		585.87	585.06	

Real-terms defence budget trend (US$m, constant 2015)

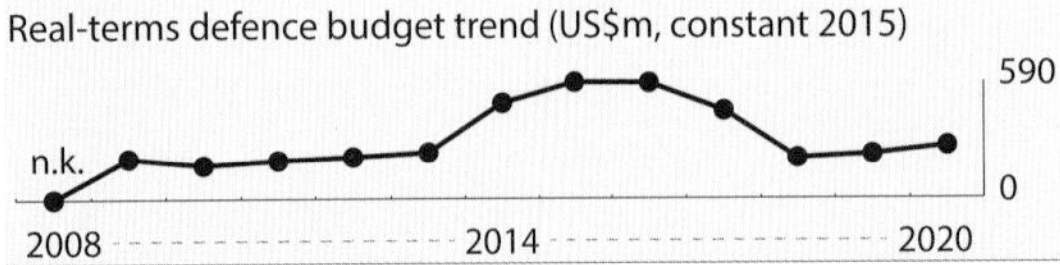

Population 5,293,070

Age	0–14	15–19	20–24	25–29	30–64	65 plus
Male	21.0%	4.7%	3.9%	3.6%	15.5%	1.4%
Female	20.6%	4.6%	3.9%	3.6%	15.3%	1.8%

Capabilities

Congo's small armed forces have low levels of training and limited overall capability, and utilise ageing equipment. They have struggled to recover from the brief but devastating civil war in the late 1990s. France provides advisory assistance and capacity-building support in military administration and military and police capability. A military-cooperation agreement was signed with Russia in 2019. The troop contingent deployed to the CAR was withdrawn by the government in mid-2017, amid allegations of indiscipline. Deployment capability is limited to neighbouring countries without external support. The air force is effectively grounded for lack of spares and serviceable equipment. The navy is largely a riverine force, despite the need for maritime security on the country's small coastline. A modernisation effort is under way and several MRAPs have been bought. Maintenance facilities are limited and the country has no domestic defence-industrial capability.

ACTIVE 10,000 (Army 8,000 Navy 800 Air 1,200)
Paramilitary 2,000

ORGANISATIONS BY SERVICE

Army 8,000

FORCES BY ROLE
MANOEUVRE
Armoured
2 armd bn
Light
2 inf bn (gp) each with (1 lt tk tp, 1 arty bty)
1 inf bn
Air Manoeuvre
1 cdo/AB bn
COMBAT SUPPORT
1 arty gp (with MRL)
1 engr bn

EQUIPMENT BY TYPE†
ARMOURED FIGHTING VEHICLES
MBT 40: 25 T-54/T-55; 15 Type-59; (some T-34 in store)
LT TK 13: 3 PT-76; 10 Type-62
RECCE 25 BRDM-1/BRDM-2
APC 133+
APC (W) 78+: 28 AT-105 *Saxon*; 20 BTR-152; 30 BTR-60; Panhard M3
PPV 55: 18 *Mamba*; 37 *Marauder*
ARTILLERY 56+
SP 122mm 3 2S1 *Gvozdika*
TOWED 15+: **122mm** 10 D-30; **130mm** 5 M-46; **152mm** D-20
MRL 10+: **122mm** 10 BM-21 *Grad*; **140mm** BM-14; **140mm** BM-16
MOR 28+: **82mm**; **120mm** 28 M-43
ANTI-TANK/ANTI-INFRASTRUCTURE
RCL 57mm M18
GUNS 15: **57mm** 5 ZIS-2 (M-1943); **100mm** 10 M-1944
AIR DEFENCE • GUNS
SP 23mm ZSU-23-4 *Shilka*
TOWED 14.5mm ZPU-2/ZPU-4; **37mm** 28 M-1939; **57mm** S-60; **100mm** KS-19

Navy ε800

EQUIPMENT BY TYPE
PATROL AND COASTAL COMBATANTS 8
PCC 4 *5 Février 1979*
PBR 4

Air Force 1,200

FORCES BY ROLE
FIGHTER/GROUND ATTACK
1 sqn with *Mirage* F-1AZ
TRANSPORT
1 sqn with An-24 *Coke*; An-32 *Cline*; CN235M-100

EQUIPMENT BY TYPE†
3 combat capable
AIRCRAFT
FGA up to 3 *Mirage* F-1AZ
TPT • Light 4: 1 An-24 *Coke*; 2 An-32 *Cline*; 1 CN235M-100

HELICOPTERS
ATK (2 Mi-35P *Hind* in store)
TPT • Medium (3 Mi-8 *Hip* in store)
AIR-LAUNCHED MISSILES • AAM • IR R-3 (AA-2 *Atoll*)‡

Paramilitary 2,000 active

Gendarmerie 2,000

FORCES BY ROLE
MANOEUVRE
Other
20 paramilitary coy

Presidential Guard some

FORCES BY ROLE
MANOEUVRE
Other
1 paramilitary bn

DEPLOYMENT

CENTRAL AFRICAN REPUBLIC: UN • MINUSCA 12

Côte d'Ivoire CIV

CFA Franc BCEAO fr		2019	2020	2021
GDP	fr	34.3tr	35.3tr	
	US$	58.6bn	61.5bn	
per capita	US$	2,230	2,281	
Growth	%	6.5	1.8	
Inflation	%	0.8	1.2	
Def bdgt [a]	fr	315bn	349bn	
	US$	538m	609m	
US$1=fr		585.91	573.81	

[a] Defence budget only – order and security expenses excluded

Real-terms defence budget trend (US$m, constant 2015)

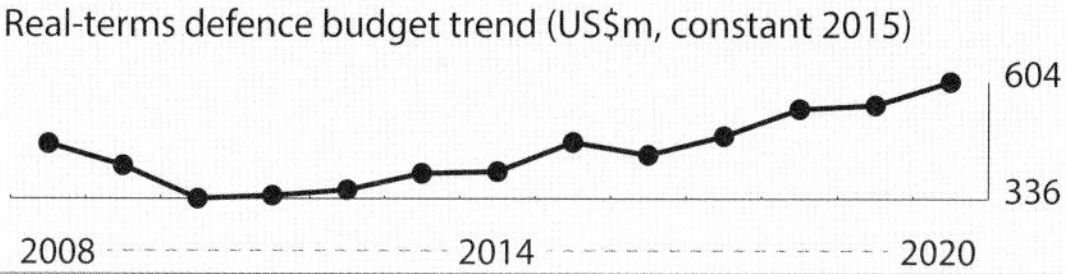

Population 27,481,086

Age	0–14	15–19	20–24	25–29	30–64	65 plus
Male	19.3%	5.4%	4.7%	4.2%	15.4%	1.3%
Female	19.2%	5.4%	4.7%	4.1%	14.8%	1.6%

Capabilities

The armed forces are still regenerating and SSR initiatives continue. A 2015 law detailed defence zones and military regions as well as the creation of a general staff and general inspectorate for the armed forces. It stressed the armed forces' role in assisting society. In 2016 a Military Programme Law for 2016–20 was adopted. The plan was to reduce military strength in order to enable an increase in the gendarmerie, though the precise outcome remains unclear. An aviation academy was established in Abidjan as part of the SSR process, with some training capacity for rotary-wing aircrew and maintenance personnel. The authorities have moved to standardise promotion and salary structures to boost professionalisation, and are also looking to improve military infrastructure. There is close defence cooperation with France, which has a significant training mission in the country. Except for limited maintenance facilities, there is no domestic defence-industrial capability.

ACTIVE 27,400 (Army 23,000 Navy 1,000 Air 1,400 Special Forces 2,000) Paramilitary n.k.

Moves to restructure and reform the armed forces continue

ORGANISATIONS BY SERVICE

Army ε23,000

FORCES BY ROLE
MANOEUVRE
Armoured
1 armd bn
Light
7 inf bn
Air Manoeuvre
1 cdo/AB bn
COMBAT SUPPORT
1 arty bn
1 engr bn
COMBAT SERVICE SUPPORT
1 log bn
AIR DEFENCE
1 AD bn

EQUIPMENT BY TYPE
ARMOURED FIGHTING VEHICLES
MBT 10 T-55†
RECCE 18: 13 BRDM-2; 5 *Cayman* BRDM
IFV 10 BMP-1/BMP-2†
APC 41
APC (W) 40: 9 *Bastion* APC; 6 BTR-80; 12 Panhard M3; 13 VAB
PPV 1 *Snake*
ENGINEERING & MAINTENANCE VEHICLES
VLB MTU
ANTI-TANK/ANTI-INFRASTRUCTURE
MSL • MANPATS 9K111-1 *Konkurs* (AT-5 *Spandrel*) (reported); 9K135 *Kornet* (AT-14 *Spriggan*) (reported)
RCL 106mm ε12 M40A1
ARTILLERY 36+
TOWED 4+: **105mm** 4 M-1950; **122mm** (reported)
MRL 122mm 6 BM-21
MOR 26+: **81mm**; **82mm** 10 M-37; **120mm** 16 AM-50
AIRCRAFT • TPT • Medium 1 An-12 *Cub*†
AIR DEFENCE
SAM • Point-defence 9K32 *Strela*-2 (SA-7 *Grail*)‡ (reported)
GUNS 21+
SP 20mm 6 M3 VDAA
TOWED 15+: **20mm** 10; **23mm** ZU-23-2; **40mm** 5 L/60

Navy ε1,000

EQUIPMENT BY TYPE
PATROL AND COASTAL COMBATANTS 4
PB 4: 3 *L'Emergence*; 1 *Atchan* 2

AMPHIBIOUS • LANDING CRAFT 1
LCM 1 *Aby*

Air Force ε1,400

EQUIPMENT BY TYPE†
AIRCRAFT
TPT 3: **Light** 2: 1 An-26 *Curl*; 1 C295W; **PAX** 1 B-727
HELICOPTERS
ATK 2 Mi-24 *Hind*
MRH 2 Mi-8P *Hip*
TPT • Medium 2 SA330L *Puma* (IAR-330L)

Special Forces ε2,000

FORCES BY ROLE
SPECIAL FORCES
1 spec ops bde

Paramilitary n.k.

Republican Guard n.k.

Gendarmerie n.k.

EQUIPMENT BY TYPE†
ARMOURED FIGHTING VEHICLES
RECCE 3 *Cayman* BRDM
APC • APC (W) some VAB; **PPV** some Streit *Spartan*
PATROL AND COASTAL COMBATANTS • PB 1 *Bian*

DEPLOYMENT

CENTRAL AFRICAN REPUBLIC: UN • MINUSCA 6
MALI: UN • MINUSMA 564; 1 inf bn; 1 sy coy
MALI: UN • MINURSO 2

FOREIGN FORCES

France 950; 1 (Marine) inf bn; 2 SA330 *Puma*; 1 *Gazelle*

Democratic Republic of the Congo DRC

Congolese Franc fr		2019	2020	2021
GDP	fr	82.2tr	89.4tr	
	US$	49.8bn	46.1bn	
per capita	US$	509	457	
Growth	%	4.4	-2.2	
Inflation	%	4.7	11.5	
Def bdgt	fr	555bn	641bn	
	US$	336m	330m	
US$1=fr		1650.00	1941.77	

Real-terms defence budget trend (US$m, constant 2015)

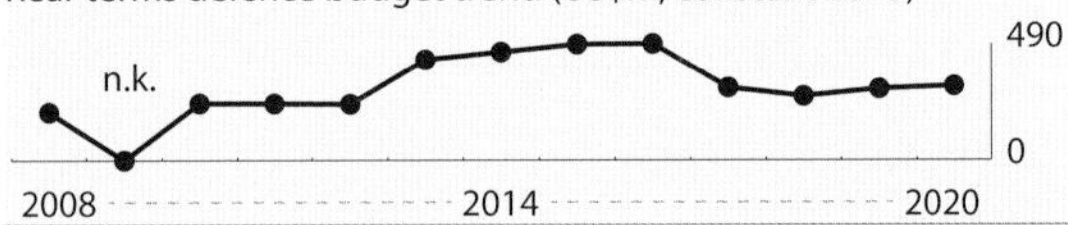

Population 101,780,263

Age	0–14	15–19	20–24	25–29	30–64	65 plus
Male	23.3%	5.2%	4.5%	3.8%	12.0%	1.1%
Female	23.0%	5.2%	4.5%	3.8%	12.1%	1.4%

Capabilities

On paper, the DRC has the largest armed forces in Central Africa. However, given the country's size and the poor levels of training, morale and equipment, they are unable to provide security throughout the country. Kinshasa has pursued several military-modernisation programmes, though plans to re-examine doctrine and organisation have seen little positive impact. When conflict finally abates in the east, significant attention to wide-ranging DDR and SSR will be required, to continue the work intermittently undertaken over the past decade. The mandate of the UN's MONUSCO mission was renewed in December 2019 and the UN's Force Intervention Brigade (FIB) remains active in the east. Training will have improved for units operating with the FIB, and there is also foreign assistance in training and capacity-building. The armed forces have incorporated a number of non-state armed groups. There remains significant scope to improve training, recruitment and retention. Deployment capability is limited and the lack of logistics vehicles significantly reduces transport capacity. The lack of sufficient tactical airlift and helicopters is a brake on military effectiveness and there is some reliance on MONUSCO capabilities, which are also insufficient given the geographical scale of the country. Much of the inventory is in poor repair and, while new equipment has been acquired, the absence of any defence sector apart from limited maintenance capability will also hinder military efficiency.

ACTIVE ε134,250 (Central Staffs ε14,000, Army 103,000 Republican Guard 8,000 Navy 6,700 Air 2,550)

ORGANISATIONS BY SERVICE

Army (Force Terrestre) ε103,000

The DRC has 11 Military Regions. In 2011, all brigades in North and South Kivu provinces were consolidated into 27 new regiments, the latest in a sequence of reorganisations designed to integrate non-state armed groups. The actual combat effectiveness of many formations is doubtful

FORCES BY ROLE
MANOEUVRE
Light
6 (integrated) inf bde
ε3 inf bde (non-integrated)
27+ inf regt
COMBAT SUPPORT
1 arty regt
1 MP bn

EQUIPMENT BY TYPE†
(includes Republican Guard eqpt)
ARMOURED FIGHTING VEHICLES
MBT 174: 12–17 Type-59†; 32 T-55; 25 T-64BV-1; 100 T-72AV
LT TK 40: 10 PT-76; 30 Type-62†
RECCE up to 52: up to 17 AML-60; 14 AML-90; 19 EE-9 *Cascavel*; 2 RAM-V-2
IFV 20 BMP-1
APC 104+:
APC (T) 9: 3 BTR-50; 6 MT-LB
APC (W) 95+: 30–70 BTR-60PB; 58 Panhard M3†; 7 TH 390 *Fahd*
ANTI-TANK/ANTI-INFRASTRUCTURE
RCL 57mm M18; **73mm** SPG-9; **75mm** M20; **106mm** M40A1
GUNS 85mm 10 Type-56 (D-44)
ARTILLERY 726+
SP 16: **122mm** 6 2S1 *Gvozdika*; **152mm** 10 2S3 *Akatsiya*
TOWED 125: **122mm** 77 M-30 (M-1938)/D-30/Type-60; **130mm** 42 Type-59 (M-46)/Type-59-I; **152mm** 6 D-20 (reported)
MRL 57+: **107mm** 12 Type-63; **122mm** 24+: 24 BM-21 *Grad*; some RM-70; **128mm** 6 M-51; **130mm** 3 Type-82; **132mm** 12
MOR 528+: **81mm** 100; **82mm** 400; **107mm** M30; **120mm** 28: 10 Brandt; 18 other
AIR DEFENCE
SAM • Point-defence 9K32 *Strela*-2 (SA-7 *Grail*)‡
GUNS • TOWED 64: **14.5mm** 12 ZPU-4; **37mm** 52 M-1939

Republican Guard 8,000

FORCES BY ROLE
MANOEUVRE
Armoured
1 armd regt
Light
3 gd bde
COMBAT SUPPORT
1 arty regt

Navy 6,700 (incl infantry and marines)

EQUIPMENT BY TYPE
PATROL AND COASTAL COMBATANTS 1
PB 1 Type-062 (*Shanghai* II)

Air Force 2,550

EQUIPMENT BY TYPE
AIRCRAFT 4 combat capable
ATK 4 Su-25 *Frogfoot*
TPT 4: **Medium** 1 C-130H *Hercules*; **Light** 1 An-26 *Curl*; **PAX** 2 B-727
HELICOPTERS
ATK 7: 4 Mi-24 *Hind*; 3 Mi-24V *Hind*
TPT • Medium 3: 1 AS332L *Super Puma*; 2 Mi-8 *Hip*

Paramilitary

National Police Force

Incl Rapid Intervention Police (National and Provincial)

People's Defence Force

FOREIGN FORCES

All part of MONUSCO unless otherwise specified
Algeria 2
Bangladesh 1,703; 1 inf bn; 1 engr coy; 1 avn coy; 1 hel coy
Belgium 1
Benin 2
Bhutan 1
Bolivia 4
Bosnia-Herzegovina 3
Brazil 26
Burkina Faso 5
Cameroon 4
Canada (*Operation Crocodile*) 7
China, People's Republic of 232; 1 engr coy; 1 fd hospital
Czech Republic 2
Egypt 10
France 4
Gambia 2
Ghana 195; 1 inf bn(-)
Guatemala 155; 1 SF coy
India 1,867; 2 inf bn; 1 med coy
Indonesia 1,033; 1 inf bn; 1 engr coy
Ireland 3
Jordan 10
Kenya 9
Malawi 864; 1 inf bn
Malaysia 2
Mali 3
Mongolia 2
Morocco 1,372; 1 inf bn; 1 fd hospital
Nepal 898; 1 inf bn; 1 engr coy
Niger 4
Nigeria 12
Pakistan 1,970; 2 inf bn; 1 hel sqn
Paraguay 7
Peru 7
Poland 2
Romania 6
Russia 8
Senegal 5
Serbia 1
Sierra Leone 2

South Africa (*Operation Mistral*) 1,133; 1 inf bn; 1 hel sqn
Switzerland 1
Tanzania 959; 1 SF coy; 1 inf bn
Tunisia 7
Ukraine 259; 1 atk hel sqn
United States 4
Uruguay 926; 1 inf bn; 1 log coy; 1 hel sqn
Zambia 6

Djibouti DJB

Djiboutian Franc fr		2019	2020	2021
GDP	fr	595bn	606bn	
	US$	3.35bn	3.41bn	
per capita	US$	3,103	3,074	
Growth	%	7.5	-1.0	
Inflation	%	3.3	2.9	
Def exp	fr	n.k	n.k	
	US$	n.k	n.k	
FMA (US)	US$	5m	5m	5m
US$1=fr		177.73	177.73	

Population 921,804

Age	0–14	15–19	20–24	25–29	30–64	65 plus
Male	15.0%	4.8%	4.8%	4.5%	14.6%	1.8%
Female	14.9%	5.1%	5.7%	5.9%	20.8%	2.2%

Capabilities

Djibouti's strategic location and relative stability have led a number of foreign states to station forces in the country. The armed forces' main responsibility is internal and border security, as well as counter-insurgency operations. The 2017 defence white paper highlighted a requirement to modernise key capabilities, but funds remain limited. Djibouti maintains close defence cooperation with France as it hosts its largest foreign military base. The US also operates its Combined Joint Task Force–Horn of Africa from Djibouti. Japan has based forces there for regional counter-piracy missions and the EU and NATO have at various times maintained a presence to support their operations. Djibouti also hosts an Italian base with a focus on anti-piracy activities. China's first overseas military base, including dock facilities, was officially opened in Djibouti in 2017. France and the US provide training assistance. EU NAVFOR Somalia has delivered maritime-security training to the navy and coastguard. Djibouti participates in a number of regional multinational exercises and contributes personnel to the AMISOM mission in Somalia but has limited capacity to independently deploy beyond its territory. Army equipment consists predominantly of older French and Soviet-era equipment. There are some maintenance facilities, but no defence manufacturing sector.

ACTIVE 10,450 (Army 8,000 Navy 200 Air 250 Gendarmerie 2,000) Paramilitary 2,650

ORGANISATIONS BY SERVICE

Army ε8,000

FORCES BY ROLE

4 military districts (Tadjourah, Dikhil, Ali-Sabieh and Obock)

MANOEUVRE

Mechanised

1 armd regt (1 recce sqn, 3 armd sqn, 1 (anti-smuggling) sy coy)

Light

4 inf regt (3-4 inf coy, 1 spt coy)
1 rapid reaction regt (4 inf coy, 1 spt coy)

Other

1 (Republican Guard) gd regt (1 sy sqn, 1 (close protection) sy sqn, 1 cbt spt sqn (1 recce pl, 1 armd pl, 1 arty pl), 1 spt sqn)

COMBAT SUPPORT

1 arty regt
1 demining coy
1 sigs regt
1 CIS sect

COMBAT SERVICE SUPPORT

1 log regt
1 maint coy

EQUIPMENT BY TYPE

ARMOURED FIGHTING VEHICLES

ASLT 3+ PTL-02 *Assaulter*
RECCE 38: 4 AML-60†; 17 AML-90; 2 BRDM-2; 15 VBL
IFV 28: 8 BTR-80A; 16-20 *Ratel*
APC 43
APC (W) 30+: 12 BTR-60†; 4+ AT-105 *Saxon*; 14 *Puma*
PPV 13: 3 *Casspir*; 10 RG-33L
AUV 22: 10 *Cougar* 4×4 (one with 90mm gun); 2 CS/VN3B; 10 PKSV

ANTI-TANK/ANTI-INFRASTRUCTURE

RCL 106mm 16 M40A1

ARTILLERY 76

SP 155mm 10 M109L
TOWED 122mm 9 D-30
MRL 12: **107mm** 2 PKSV AUV with PH-63; **122mm** 10: 6 (6-tube Toyota Land Cruiser 70 series); 2 (30-tube Iveco 110-16); 2 (30-tube)
MOR 45: **81mm** 25; **120mm** 20 Brandt

AIR DEFENCE • GUNS 15+

SP 20mm 5 M693
TOWED 10: **23mm** 5 ZU-23-2; **40mm** 5 L/70

Navy ε200

EQUIPMENT BY TYPE

PATROL AND COASTAL COMBATANTS 12

PBF 2 Battalion-17
PB 10: 1 *Plascoa*†; 2 Sea Ark 1739; 1 *Swari*†; 6 others

AMPHIBIOUS • LANDING CRAFT 1

LCT 1 EDIC 700

Air Force 250

EQUIPMENT BY TYPE

AIRCRAFT

TPT • Light 6: 1 Cessna U206G *Stationair*; 1 Cessna 208 *Caravan*; 2 Y-12E; 1 L-410UVP *Turbolet*; 1 MA60

HELICOPTERS

ATK (2 Mi-35 *Hind* in store)
MRH 6: 4 AS365 *Dauphin*; 1 Mi-17 *Hip* H; 1 Z-9WE
TPT 3: **Medium** 1 Mi-8T *Hip*; **Light** 2 AS355F *Ecureuil* II

Gendarmerie 2,000+
Ministry of Defence

FORCES BY ROLE

MANOEUVRE

Other

1 paramilitary bn

EQUIPMENT BY TYPE

AFV • **AUV** 2 CS/VN3B

PATROL AND COASTAL COMBATANTS • 1 **PB**

Paramilitary ε2,650

National Police Force ε2,500
Ministry of Interior

Coast Guard 150
EQUIPMENT BY TYPE

PATROL AND COASTAL COMBATANTS • **PB** 11: 2 *Khor Angar*; 9 other

DEPLOYMENT

SOMALIA: AU • AMISOM 1,800; 2 inf bn

WESTERN SAHARA: UN • MINURSO 1

FOREIGN FORCES

China 240: 1 mne coy(-); 1 med unit; 2 ZTL-11; 8 ZBL-08; 1 LPD; 1 ESD

France 1,450: 1 SF unit; 1 (Marine) combined arms regt (2 recce sqn, 2 inf coy, 1 arty bty, 1 engr coy); 1 hel det with 2 SA330 *Puma*; 2 SA342 *Gazelle;* 1 LCM; 1 air sqn with 4 *Mirage* 2000-5; 1 CN235M; 2 SA330 *Puma*

Germany *Operation Atalanta* 1 AP-3C *Orion*

Italy BMIS 117

Japan 170; 2 P-3C *Orion*

Spain *Operation Atalanta* 1 P-3M *Orion*

United States US Africa Command: 4,000; 1 tpt sqn with C-130H/J-30 *Hercules;* 1 spec ops sqn with MC-130H; PC-12 (U-28A); 1 CSAR sqn with HH-60G *Pave Hawk;* 1 CISR sqn with MQ-9A *Reaper;* 1 naval air base

Equatorial Guinea EQG

CFA Franc BEAC fr		2019	2020	2021
GDP	fr	6.93tr	5.75tr	
	US$	11.8bn	10.0bn	
per capita	US$	8,690	7,131	
Growth	%	-6.1	-6.0	
Inflation	%	1.2	3.0	
Def exp	fr	n.k	n.k	
	US$	n.k	n.k	
US$1=fr		585.92	573.83	

Population 836,178

Age	0–14	15–19	20–24	25–29	30–64	65 plus
Male	19.7%	5.4%	4.7%	4.0%	14.5%	1.6%
Female	19.1%	5.3%	4.5%	3.9%	15.0%	2.3%

Capabilities

The army dominates the armed forces, with internal security the principal task. Equatorial Guinea has been trying for several years to modernise its armed forces. France maintains a military detachment in Malabo, advising on defence-institutional development and delivering capacity-building support at the naval-focused regional school at Tica. These personnel also engage in some training activities with French forces based in Gabon and more generally in the region as part of France's *Corymbe* mission. There is only limited capability for power projection and deployments are limited to neighbouring countries without external support. Recent naval investments include both equipment and onshore-infrastructure improvements at Bata and Malabo, although naval capabilities overall remain limited. Maritime-security concerns in the Gulf of Guinea have resulted in an increased emphasis on boosting maritime-patrol capacity. There is limited maintenance capacity and no defence industrial sector.

ACTIVE 1,450 (Army 1,100 Navy 250 Air 100)

ORGANISATIONS BY SERVICE

Army 1,100
FORCES BY ROLE

MANOEUVRE

Light

3 inf bn(-)

EQUIPMENT BY TYPE

ARMOURED FIGHTING VEHICLES

MBT 3 T-55

RECCE 6 BRDM-2

IFV 20 BMP-1

APC 35

APC (W) 10 BTR-152

PPV 25 *Reva* (reported)

Navy ε250
EQUIPMENT BY TYPE

PRINCIPAL SURFACE COMBATANTS • **FRIGATES** 1

FF 1 *Wele Nzas* with 2 MS-227 *Ogon'* 122mm MRL, 2 AK630 CIWS, 2 76mm guns

PATROL AND COASTAL COMBATANTS 10

CORVETTES • **FSG** 1 *Bata* with 2 *Katran*-M RWS with *Barrier* SSM, 2 AK630 CIWS, 1 76mm gun

PCC 2 OPV 62

PBF 2 *Shaldag* II

PB 5: 1 *Daphne*; 2 *Estuario de Muni*; 2 *Zhuk*

LOGISTICS AND SUPPORT

AKRH 1 *Capitán David Eyama Angue Osa* with 1 76mm gun

Air Force 100
EQUIPMENT BY TYPE

AIRCRAFT 4 combat capable

ATK 4: 2 Su-25 *Frogfoot*; 2 Su-25UB *Frogfoot* B

TPT 4: **Light** 3: 1 An-32B *Cline*; 2 An-72 *Coaler*; **PAX** 1 *Falcon* 900 (VIP)

TRG 2 L-39C *Albatros*

HELICOPTERS

ATK 5 Mi-24P/V *Hind*

MRH 1 Mi-17 *Hip* H

TPT 4: **Heavy** 1 Mi-26 *Halo*; **Medium** 1 Ka-29 *Helix*; **Light** 2 Enstrom 480

Paramilitary

Guardia Civil

FORCES BY ROLE
MANOEUVRE
Other
2 paramilitary coy

Coast Guard n.k.

Eritrea ERI

Eritrean Nakfa ERN		2019	2020	2021
GDP	ERN	29.9bn	31.3bn	
	US$	1.98bn	2.08bn	
per capita	US$	567	585	
Growth	%	3.8	-0.6	
Inflation	%	-16.4	4.7	
Def exp	ERN	n.k	n.k	
	US$	n.k	n.k	
USD1=ERN		15.07	15.07	

Population 6,081,196

Age	0–14	15–19	20–24	25–29	30–64	65 plus
Male	19.2%	5.6%	4.6%	3.7%	14.5%	1.6%
Female	19.0%	5.7%	4.7%	3.8%	15.3%	2.4%

Capabilities

Eritrea has maintained large armed forces mainly because of historical military tensions and conflict with Ethiopia, though tensions have reduced after a September 2018 peace agreement. A UN arms embargo was lifted in November 2018. While the peace agreement may have afforded them the opportunity to consider restructuring and recapitalisation, fighting in northern Ethiopia in late 2020 will have refocused attention on border security. Maritime insecurity, including piracy, remains a challenge. The UAE has established a military presence in Eritrea, and port and airfield facilities at Assab have been used to support Gulf states participating in the Yemen campaign. Eritrea maintains a large army due to mandatory conscription. For some the term of service is reportedly indefinite, and significant numbers of conscripts have chosen to leave the country or otherwise evade service. These factors likely affect overall military cohesion and effectiveness. Eritrea has not demonstrated any capacity to deploy beyond its borders. The armed forces' inventory primarily comprises outdated Soviet-era systems and modernisation was restricted by the UN arms embargo, although Russia has since agreed to supply Eritrea with lightweight multipurpose helicopters. The arms embargo will have resulted in serviceability issues, notwithstanding allegations of external support. The navy remains capable of only limited coastal-patrol and interception operations. There is some maintenance capability, but no defence manufacturing sector.

ACTIVE 201,750 (Army 200,000 Navy 1,400 Air 350)

Conscript liability 18 months (4 months mil trg) between ages 18 and 40

RESERVE 120,000 (Army ε120,000)

ORGANISATIONS BY SERVICE

Army ε200,000

Heavily cadreised

FORCES BY ROLE
COMMAND
4 corps HQ
MANOEUVRE
Mechanised
1 mech bde
Light
19 inf div
1 cdo div

Reserve ε120,000

FORCES BY ROLE
MANOEUVRE
Light
1 inf div

EQUIPMENT BY TYPE
ARMOURED FIGHTING VEHICLES
MBT 270 T-54/T-55
RECCE 40 BRDM-1/BRDM-2
IFV 15 BMP-1
APC 35
APC (T) 10 MT-LB†
APC (W) 25 BTR-152/BTR-60
ENGINEERING & MAINTENANCE VEHICLES
ARV T-54/T-55 reported
VLB MTU reported
ANTI-TANK/ANTI-INFRASTRUCTURE
MSL • MANPATS 9K11 *Malyutka* (AT-3 *Sagger*); 9K111-1 *Konkurs* (AT-5 *Spandrel*) **GUNS 85mm** D-44
ARTILLERY 258
SP 45: **122mm** 32 2S1 *Gvozdika*; **152mm** 13 2S5 *Giatsint-S*
TOWED 19+: **122mm** D-30; **130mm** 19 M-46
MRL 44: **122mm** 35 BM-21 *Grad*; **220mm** 9 9P140 *Uragan*
MOR 150+: **82mm** 50+; **120mm/160mm** 100+
AIR DEFENCE
SAM • Point-defence 9K32 *Strela-2* (SA-7 *Grail*)‡
GUNS 70+
SP 23mm ZSU-23-4 *Shilka*
TOWED 23mm ZU-23

Navy 1,400

EQUIPMENT BY TYPE
PATROL AND COASTAL COMBATANTS 12
PBF 9: 5 Battalion-17; 4 *Super Dvora*
PB 3 Swiftships
AMPHIBIOUS 3
LANDING SHIP 2
LST 2: 1 *Chamo*† (Ministry of Transport); 1 *Ashdod*†
LANDING CRAFT 1
LCU 1 T-4† (in harbour service)

Air Force ε350

FORCES BY ROLE
FIGHTER/GROUND ATTACK
1 sqn with MiG-29/MiG-29SE/MiG-29UB *Fulcrum*
1 sqn with Su-27/Su-27UBK *Flanker*

TRANSPORT
1 sqn with Y-12(II)
TRAINING
1 sqn with L-90 *Redigo*
1 sqn with MB-339CE*
TRANSPORT HELICOPTER
1 sqn with Bell 412EP *Twin Huey*
1 sqn with Mi-17 *Hip H*

EQUIPMENT BY TYPE

AIRCRAFT 14 combat capable
FTR 8: 4 MiG-29 *Fulcrum*; 2 MiG-29UB *Fulcrum*; 1 Su-27 *Flanker*; 1 Su-27UBK *Flanker*
FGA 2 MiG-29SE *Fulcrum*
TPT • Light 5: 1 Beech 200 *King Air*; 4 Y-12(II)
TRG 16+: 8 L-90 *Redigo*; 4 MB-339CE*; 4+ Z-143/Z-242
HELICOPTERS
MRH 8: 4 Bell 412EP *Twin Huey* (AB-412EP); 4 Mi-17 *Hip* H
AIR-LAUNCHED MISSILES
AAM • IR R-60 (AA-8 *Aphid*); R-73 (AA-11A *Archer*); **IR/SARH** R-27 (AA-10 *Alamo*)

FOREIGN FORCES

United Arab Emirates *Operation Restoring Hope* 500; 2 FSGHM; 4 Bell 407MRH; 2 CH-47F *Chinook*; 4 UH-60M *Black Hawk*; 1 C-130H; 1 DHC-6-400; 1 DHC-8

Ethiopia ETH

Ethiopian Birr EB		2019	2020	2021
GDP	EB	2.70tr	3.34tr	
	US$	92.8bn	95.6bn	
per capita	US$	961	974	
Growth	%	9.0	1.9	
Inflation	%	15.8	20.2	
Def bdgt	EB	15.0bn	16.5bn	
	US$	516m	472m	
US$1=EB		29.06	34.99	

Real-terms defence budget trend (US$m, constant 2015)

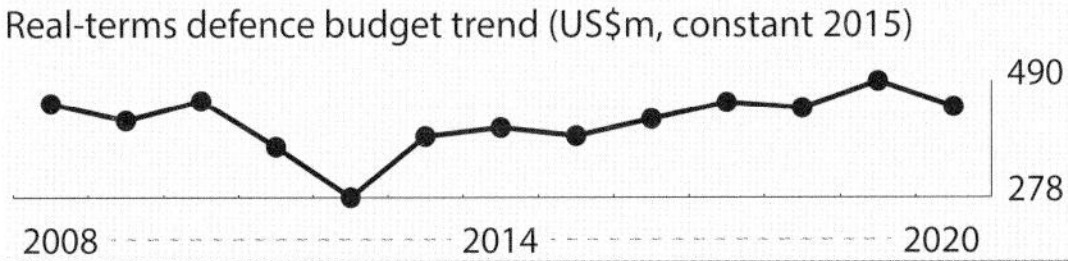

Population 108,113,150

Age	0–14	15–19	20–24	25–29	30–64	65 plus
Male	20.0%	5.2%	4.5%	3.9%	14.6%	1.6%
Female	19.8%	5.2%	4.5%	4.0%	14.8%	1.8%

Capabilities

Ethiopia maintains one of the region's largest and most capable armed forces. It is a significant contributor to regional peacekeeping. The September 2018 peace agreement with Eritrea may affect future military dispositions. Military clashes in late 2020 between central authorities and forces in Tigray province highlighted internal political challenges as well as the persistence of non-state armed capacity. Countering al-Shabaab remains an ongoing military commitment and Addis Ababa continues to provide military support to the Federal Government of Somalia. The armed forces are experienced by regional standards, following a history of combat operations. Training and experience are also gained through international peacekeeping deployments. Ethiopia has demonstrated the capability to make significant contributions to the UN missions in Darfur and South Sudan and the AMISOM mission in Somalia. The country's inventory comprises mostly Soviet-era equipment, though surplus stocks have been acquired from Hungary, Ukraine and the US, and modern air-defence systems have been purchased from Russia. There is a modest local defence-industrial base, primarily centred on small arms, with some licensed production of light armoured vehicles. There is adequate maintenance capability but only a limited capacity to support advanced platforms.

ACTIVE 138,000 (Army 135,000 Air 3,000)

ORGANISATIONS BY SERVICE

Army 135,000

4 Mil Regional Commands (Northern, Western, Central and Eastern) each acting as corps HQ

FORCES BY ROLE

MANOEUVRE
Light
1 (Agazi Cdo) SF comd
1 (Northern) corps (1 mech div, 4 inf div)
1 (Western) corps (1 mech div, 3 inf div)
1 (Central) corps (1 mech div, 5 inf div)
1 (Eastern) corps (1 mech div, 5 inf div)

EQUIPMENT BY TYPE

ARMOURED FIGHTING VEHICLES
MBT 461+: 246+ T-54/T-55/T-62; 215 T-72B/T-72UA1
RECCE ε100 BRDM-1/BRDM-2
IFV ε20 BMP-1
APC 300+
APC (T) some Type-89
APC (W) 300+: ε300 BTR-60/BTR-152; some Type-92
PPV Gaia *Thunder*
AUV some *Ze'ev*
ENGINEERING & MAINTENANCE VEHICLES
ARV T-54/T-55 reported; 4 BTS-5B
VLB MTU reported
MW *Bozena*
ANTI-TANK/ANTI-INFRASTRUCTURE
MSL • MANPATS 9K11 *Malyutka* (AT-3 *Sagger*); 9K111 *Fagot* (AT-4 *Spigot*); 9K135 *Kornet*-E (AT-14 *Spriggan*)
RCL 82mm B-10; **107mm** B-11
GUNS 85mm D-44
ARTILLERY 524+
SP 10+: **122mm** 2S1 *Gvozdika*; **152mm** 10 2S19 *Msta*-S
TOWED 464+: **122mm** 464 D-30/M-30 (M-1938); **130mm** M-46; **155mm** AH2
MRL 300mm PHL-03; **122mm** ε50 BM-21 *Grad*
MOR 81mm M1/M29; **82mm** M-1937; **120mm** M-1944
AIR DEFENCE
SAM
Medium-range S-75 *Dvina* (SA-2 *Guideline*)
Short-range S-125 *Pechora* (SA-3 *Goa*)
Point-defence 9K32 *Strela*-2 (SA-7 *Grail*)‡; 96K6 *Pantsir*-S1 (SA-22 *Greyhound*)

GUNS
SP 23mm ZSU-23-4 *Shilka*
TOWED 23mm ZU-23; 37mm M-1939; 57mm S-60

Air Force 3,000

FORCES BY ROLE
FIGHTER/GROUND ATTACK
1 sqn with MiG-23ML *Flogger* G/MiG-23UB *Flogger* C
1 sqn with Su-27/Su-27UB *Flanker*
TRANSPORT
1 sqn with An-12 *Cub*; An-26 *Curl*; An-32 *Cline*; C-130B *Hercules*; DHC-6 *Twin Otter*; L-100-30; Yak-40 *Codling* (VIP)
TRAINING
1 sqn with L-39 *Albatros*
1 sqn with G 120TP
ATTACK/TRANSPORT HELICOPTER
2 sqn with Mi-24/Mi-35 *Hind*; Mi-8 *Hip*; Mi-17 *Hip* H; SA316 *Alouette* III

EQUIPMENT BY TYPE
AIRCRAFT 22 combat capable
FTR 11: 8 Su-27 *Flanker*; 3 Su-27UB *Flanker*
FGA 8 MiG-23ML/UB *Flogger* G/C
ATK 3: 1 Su-25T *Frogfoot*; 2 Su-25UB *Frogfoot*
TPT 15: **Medium** 9: 3 An-12 *Cub*; 2 C-130B *Hercules*; 2 C-130E *Hercules*; 2 L-100-30; **Light** 6: 1 An-26 *Curl*; 1 An-32 *Cline*; 3 DHC-6 *Twin Otter*; 1 Yak-40 *Codling* (VIP)
TRG 18: 6 G 120TP; 12 L-39 *Albatros*
HELICOPTERS
ATK 18: 15 Mi-24 *Hind*; 3 Mi-35 *Hind*
MRH 19: 1 AW139; 6 SA316 *Alouette* III; 12 Mi-8 *Hip*/Mi-17 *Hip* H
AIR-LAUNCHED MISSILES
AAM • IR R-3 (AA-2 *Atoll*)‡; R-60 (AA-8 *Aphid*); R-73 (AA-11A *Archer*); **IR/SARH** R-23/R-24 (AA-7 *Apex*); R-27 (AA-10 *Alamo*)

DEPLOYMENT

MALI: UN • MINUSMA 1

SOMALIA: AU • AMISOM 4,000; 5 inf bn

SOUTH SUDAN: UN • UNMISS 2,131; 3 inf bn

SUDAN: UN • UNAMID 808; 1 inf bn; **UN** • UNISFA 3,663; 3 inf bn; 2 arty coy; 1 engr coy; 1 sigs coy; 8 fd hospital; 1 hel sqn

Gabon GAB

CFA Franc BEAC fr		2019	2020	2021
GDP	fr	9.89tr	8.86tr	
	US$	16.9bn	15.1bn	
per capita	US$	8,111	7,185	
Growth	%	3.8	-2.7	
Inflation	%	2.0	3.0	
Def bdgt [a]	fr	156bn	156bn	
	US$	267m	267m	
US$1=fr		585.89	585.07	

[a] Includes funds allocated to Republican Guard

Real-terms defence budget trend (US$m, constant 2015)

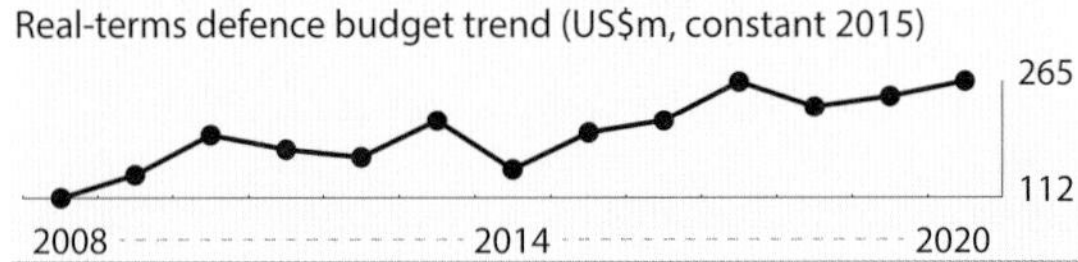

Population 2,230,908

Age	0–14	15–19	20–24	25–29	30–64	65 plus
Male	18.6%	5.9%	5.5%	5.0%	14.9%	2.0%
Female	17.9%	5.5%	5.0%	4.2%	13.5%	2.0%

Capabilities

Oil revenues have allowed the government to support small but regionally capable armed forces, while the country has benefited from the long-term presence of French troops acting as a security guarantor. There is regular training with French forces and with regionally deployed French naval forces, as well as with the US and other international partners. Gabonese forces have taken part in the US Navy-led *Obangame Express* exercise. A new maritime-operations centre was built by the US in 2019. Military medicine is well regarded. The armed forces retain sufficient airlift to ensure mobility within the country and even a modest capability to project power by sea and air. Apart from limited maintenance facilities, there is no domestic defence-industrial capacity.

ACTIVE 4,700 (Army 3,200 Navy 500 Air 1,000) Paramilitary 2,000

ORGANISATIONS BY SERVICE

Army 3,200

Republican Guard under direct presidential control

FORCES BY ROLE
MANOEUVRE
Light
1 (Republican Guard) gd gp (bn)
(1 armd/recce coy, 3 inf coy, 1 arty bty, 1 ADA bty)
8 inf coy
Air Manoeuvre
1 cdo/AB coy
COMBAT SUPPORT
1 engr coy

EQUIPMENT BY TYPE
ARMOURED FIGHTING VEHICLES
RECCE 77: 24 AML-60/AML-90; 12 EE-3 *Jararaca*; 14 EE-9 *Cascavel*; 6 ERC-90F4 *Sagaie*; 7 RAM V-2; 14 VBL

IFV 17: 12 EE-11 *Urutu* (with 20mm gun); 5 VN-1†
APC 77
APC (W) 35: 9 LAV-150 *Commando*; 5 *Bastion* APC; 3 WZ-523; 5 VAB; 12 VXB-170; 1 *Pandur*
PPV 42: 8 *Aravis*; 34 Ashok Leyland MPV
ANTI-TANK/ANTI-INFRASTRUCTURE
MSL • MANPATS *Milan*
RCL 106mm M40A1
ARTILLERY 67
TOWED 105mm 4 M101
MRL 24: **107mm** 16 PH-63; **140mm** 8 *Teruel*
MOR 39: **81mm** 35; **120mm** 4 Brandt
AIR DEFENCE • GUNS 41
SP 20mm 4 ERC-20
TOWED 37+: **14.5mm** ZPU-4; **23mm** 24 ZU-23-2; **37mm** 10 M-1939; **40mm** 3 L/70

Navy ε500

HQ located at Port Gentil

EQUIPMENT BY TYPE
PATROL AND COASTAL COMBATANTS 9
PB 9: 4 *Port Gentil* (FRA VCSM); 4 Rodman 66; 1 *Patra*†
AMPHIBIOUS • LANDING CRAFT 1
LCM 1 Mk 9 (ex-UK)

Air Force 1,000

FORCES BY ROLE
FIGHTER/GROUND ATTACK
1 sqn with *Mirage* F-1AZ
TRANSPORT
1 (Republican Guard) sqn with AS332 *Super Puma*; ATR-42F; *Falcon* 900; Gulfstream IV-SP/G650ER
1 sqn with C-130H *Hercules*; CN-235M-100
ATTACK/TRANSPORT HELICOPTER
1 sqn with Bell 412 *Twin Huey* (AB-412); SA330C/H *Puma*; SA342M *Gazelle*

EQUIPMENT BY TYPE
AIRCRAFT 8 combat capable
FGA 6 *Mirage* F-1AZ
MP (1 EMB-111* in store)
TPT 6: **Medium** 1 C-130H *Hercules*; (1 L-100-30 in store); **Light** 2: 1 ATR-42F; 1 CN-235M-100; **PAX** 3: 1 *Falcon* 900; 1 Gulfstream IV-SP; 1 Gulfstream G650ER
TRG 2 MB-326 *Impala* I* (4 CM-170 *Magister* in store)
HELICOPTERS
MRH 2: 1 Bell 412 *Twin Huey* (AB-412); 1 SA342M *Gazelle*; (2 SA342L *Gazelle* in store)
TPT 7: **Medium** 4: 1 AS332 *Super Puma*; 3 SA330C/H *Puma*; **Light** 3: 2 H120 *Colibri*; 1 H135
AIR-LAUNCHED MISSILES • AAM • IR U-*Darter* (reported)

Paramilitary 2,000

Gendarmerie 2,000

FORCES BY ROLE
MANOEUVRE
Armoured
2 armd sqn
Other
3 paramilitary bde
11 paramilitary coy
Aviation
1 unit with AS350 *Ecureuil*; AS355 *Ecureuil* II

EQUIPMENT BY TYPE
HELICOPTERS • TPT • Light 4: 2 AS350 *Ecureuil*; 2 AS355 *Ecureuil* II

DEPLOYMENT

CENTRAL AFRICAN REPUBLIC: UN • MINUSCA 447; 1 inf bn(-)

FOREIGN FORCES

France 450; 1 inf bn

Gambia GAM

Gambian Dalasi D		2019	2020	2021
GDP	D	91.4bn	94.1bn	
	US$	1.8bn	1.8bn	
per capita	US$	774	746	
Growth	%	6.1	-1.8	
Inflation	%	7.1	6.1	
Def bdgt	D	n.k	n.k	
	US$	n.k	n.k	
US$1=D		50.28	52.13	

Population 2,173,999

Age	0–14	15–19	20–24	25–29	30–64	65 plus
Male	18.1%	5.2%	4.7%	4.4%	15.3%	1.7%
Female	17.9%	5.3%	4.9%	4.6%	16.0%	2.0%

Capabilities

Reform of Gambia's security structure, and the armed forces, has been a key objective of the SSR process that was implemented following political instability in 2016–17. A National Security Policy was launched in mid-2019. The SSR process is supported by the AU, ECOWAS, EU, France and the US. Gambia's small forces have traditionally focused on maritime security and countering human trafficking. France and the US provide some military training, and Turkey and the UK also delivered training in 2019. There is also cooperation with neighbouring states and with the AU, which maintains a technical-support mission to assist in the SSR process, including on defence reform, military reorganisation and the rule of law. The ECOMIG deployment remains in place, with its mandate extended until early 2021, and will transform into a police mission. The armed forces participate in some multinational exercises and have deployed in support of UN missions in Africa. The equipment inventory is limited, with serviceability in doubt for some types. Gambia has no significant defence-industrial capabilities.

ACTIVE 4,100 (Army 3,500 Navy 300 National Guard 300)

ORGANISATIONS BY SERVICE

Gambian National Army 3,500

FORCES BY ROLE
MANOEUVRE
Light
4 inf bn
COMBAT SUPPORT
1 engr sqn

Air Wing

EQUIPMENT BY TYPE
AIRCRAFT
TPT 5: **Light** 2 AT-802A *Air Tractor*; **PAX** 3: 1 B-727; 1 CL-601; 1 Il-62M *Classic* (VIP)

Gambia Navy 300

EQUIPMENT BY TYPE
PATROL AND COASTAL COMBATANTS 8
PBF 4: 2 Rodman 55; 2 *Fatimah* I
PB 4: 1 *Bolong Kanta*†; 3 *Taipei* (ROC *Hai Ou*) (one additional damaged and in reserve)

Republican National Guard 300

FORCES BY ROLE
MANOEUVRE
Other
1 gd bn (forming)

DEPLOYMENT

CENTRAL AFRICAN REPUBLIC: UN • MINUSCA 8

DEMOCRATIC REPUBLIC OF THE CONGO: UN • MONUSCO 2

MALI: UN • MINUSMA 5

MIDDLE EAST: UN • UNTSO 1

SOUTH SUDAN: UN • UNMISS 4

SUDAN: UN • UNAMID 135; 1 inf coy

FOREIGN FORCES

Ghana ECOMIG 50
Nigeria ECOMIG 197
Senegal ECOMIG 250

Ghana GHA

Ghanaian New Cedi C		2019	2020	2021
GDP	C	349bn	388bn	
	US$	67.0bn	67.3bn	
per capita	US$	2,221	2,188	
Growth	%	6.5	0.9	
Inflation	%	7.2	10.6	
Def bdgt	C	1.3bn	1.5bn	
	US$	244m	268m	
US$1=C		5.22	5.77	

Real-terms defence budget trend (US$m, constant 2015)

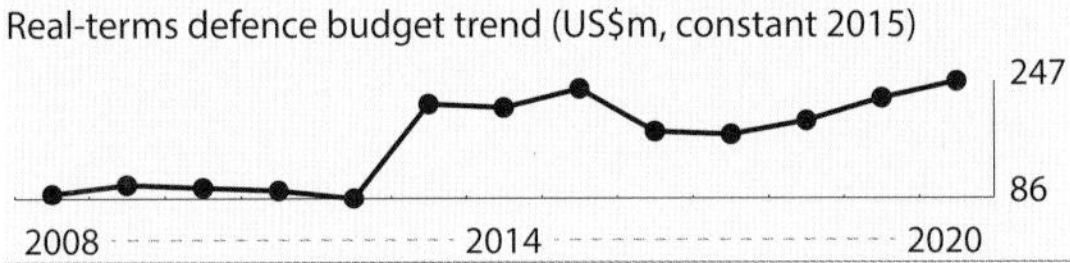

Population 29,340,248

Age	0–14	15–19	20–24	25–29	30–64	65 plus
Male	18.8%	4.9%	4.3%	3.8%	15.4%	2.0%
Female	18.6%	5.0%	4.4%	3.9%	16.4%	2.4%

Capabilities

Ghana's armed forces are among the most capable in the region, with a long-term development plan covering both the current and next decade. The ability to control its EEZ is of increasing importance, and this underpins the navy's expansion plans, including the opening of a new forward-operating base. Internal and maritime security are central military tasks, along with peacekeeping missions. The US delivers training and support, and there is also significant and long-standing defence engagement with the UK. Air-force training, close-air support and airlift capabilities have developed in recent years. There are plans to boost training and exercises, as well as to improve military infrastructure. The army is a regular contributor to UN peacekeeping operations. Ghana has started to develop forward-operating bases, principally with the objective of protecting oil resources. There is a limited defence-industrial base, including maintenance facilities and ammunition manufacturing as well as a more recent armoured-vehicle production capability.

ACTIVE 15,500 (Army 11,500 Navy 2,000 Air 2,000)

ORGANISATIONS BY SERVICE

Army 11,500

FORCES BY ROLE
COMMAND
2 comd HQ
MANOEUVRE
Reconnaissance
1 armd recce regt (2 recce sqn)
1 armd recce regt (forming)
Light
1 (rapid reaction) mot inf bn
6 inf bn
Air Manoeuvre
2 AB coy

COMBAT SUPPORT
1 arty regt (1 arty bty, 2 mor bty)
1 fd engr regt (bn)
1 sigs regt
1 sigs sqn
COMBAT SERVICE SUPPORT
1 log gp
1 tpt coy
2 maint coy
1 med coy
1 trg bn
EQUIPMENT BY TYPE
ARMOURED FIGHTING VEHICLES
RECCE 3 EE-9 *Cascavel*
IFV 48: 24 *Ratel*-90; 15 *Ratel*-20; 4 *Piranha* 25mm; 5+ Type-05P 25mm
APC 105
APC (W) 55+: 46 *Piranha*; 9+ Type-05P
PPV 50 Streit *Typhoon*
AUV 33 *Cobra*/*Cobra* II
ARTILLERY 87+
TOWED 122mm 6 D-30
MRL 3+: **107mm** Type-63; **122mm** 3 Type-81
MOR 78: **81mm** 50; **120mm** 28 Tampella
ENGINEERING & MAINTENANCE VEHICLES
AEV 1 Type-05P AEV
ARV *Piranha* reported
ANTI-TANK/ANTI-INFRASTRUCTURE
RCL 84mm *Carl Gustaf*
AIR DEFENCE
SAM • Point-defence 9K32 *Strela*-2 (SA-7 *Grail*)‡
GUNS • TOWED 8+: **14.5mm** 4+: 4 ZPU-2; ZPU-4; **23mm** 4 ZU-23-2

Navy 2,000

Naval HQ located at Accra; Western HQ located at Sekondi; Eastern HQ located at Tema

EQUIPMENT BY TYPE
PATROL AND COASTAL COMBATANTS 14
PCO 2 *Anzone* (US)
PCC 10: 2 *Achimota* (GER Lurssen 57m) with 1 76mm gun; 2 *Dzata* (GER Lurssen 45m); 4 *Snake* (PRC 47m); 2 *Yaa Asantewa (ex-GER Albatros)*
PBF 1 *Stephen Otu* (ROK *Sea Dolphin*)
PB 1 *David Hansen* (US)

Special Boat Squadron

FORCES BY ROLE
SPECIAL FORCES
1 SF unit

Air Force 2,000

FORCES BY ROLE
GROUND ATTACK
1 sqn with K-8 *Karakorum**; L-39ZO*; MB-339A*
ISR
1 unit with DA42
TRANSPORT
1 sqn with BN-2 *Defender*; C295; Cessna 172
TRANSPORT HELICOPTER
1 sqn with AW109A; Bell 412SP *Twin Huey*; Mi-17V-5 *Hip* H; SA319 *Alouette* III; Z-9EH
EQUIPMENT BY TYPE†
AIRCRAFT 8 combat capable
ATK (3 MB-326K in store)
TPT 10: **Light** 10: 1 BN-2 *Defender*; 3 C295; 3 Cessna 172; 3 DA42; (**PAX** 1 F-28 *Fellowship* (VIP) in store)
TRG 8: 4 K-8 *Karakorum**; 2 L-39ZO*; 2 MB-339A*
HELICOPTERS
MRH 10: 1 Bell 412SP *Twin Huey*; 3 Mi-17V-5 *Hip* H; 2 SA319 *Alouette* III; 4 Z-9EH
TPT 6: **Medium** 4 Mi-171Sh; **Light** 2 AW109A

DEPLOYMENT

CENTRAL AFRICAN REPUBLIC: UN • MINUSCA 14
CYPRUS: UN • UNFICYP 1
DEMOCRATIC REPUBLIC OF THE CONGO: UN • MONUSCO 195; 1 inf bn(-)
GAMBIA: ECOWAS • ECOMIG 50
LEBANON: UN • UNIFIL 864; 1 recce coy; 1 mech inf bn
MALI: UN • MINUSMA 152; 1 engr coy
SOUTH SUDAN: UN • UNMISS 871; 1 inf bn
SUDAN: UN • UNAMID 5; **UN** • UNISFA 8
SYRIA/ISRAEL: UN • UNDOF 2
WESTERN SAHARA: UN • MINURSO 14

Guinea GUI

Guinean Franc fr		2019	2020	2021
GDP	fr	127tr	140tr	
	US$	13.8bn	14.2bn	
per capita	US$	1,012	1,019	
Growth	%	5.6	1.4	
Inflation	%	9.5	9.1	
Def bdgt	fr	1.79tr	2.02tr	
	US$	195m	206m	
US$1=fr		9183.68	9793.94	

Real-terms defence budget trend (US$m, constant 2015)

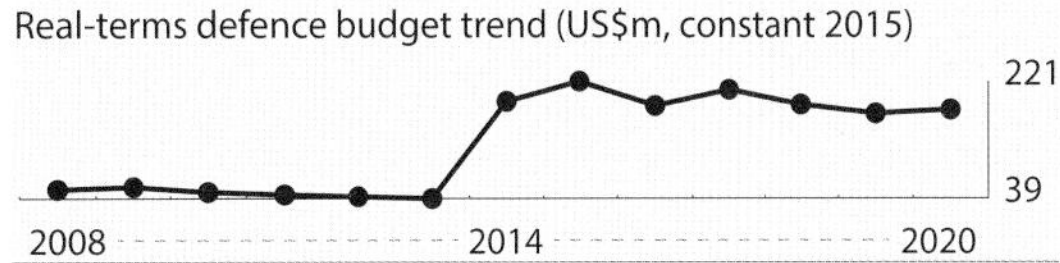

Population 12,527,440

Age	0–14	15–19	20–24	25–29	30–64	65 plus
Male	20.8%	5.3%	4.4%	3.7%	14.0%	1.7%
Female	20.4%	5.2%	4.4%	3.7%	14.1%	2.2%

Capabilities

Guinea's armed forces remain limited in size and capacity. Reforms since 2010 have been intended to bring these forces under political control and begin professionalisation. Violence surrounding the presidential election in late 2020 reportedly led to the army being deployed internally. Piracy in the Gulf of Guinea is a key concern, as is illegal trafficking and fishing. A military-programme law for the period 2015–20 has not been fully implemented due to funding issues. Defence cooperation with France and the US has led to financial and training assistance, including for personnel earmarked for deployment to Mali. Much of the country's military equipment is ageing and of Soviet-era vintage; serviceability will be questionable for some types. There is limited organic airlift and France is supporting the development of a light aviation observation capability. Guinea is also attempting to improve its logistics and military-health capacities. There are no significant defence-industrial capabilities.

ACTIVE 9,700 (Army 8,500 Navy 400 Air 800)
Paramilitary 2,600
Conscript liability 9–12 months (students, before graduation)

ORGANISATIONS BY SERVICE

Army 8,500

FORCES BY ROLE
MANOEUVRE
Armoured
1 armd bn
Light
1 SF bn
5 inf bn
1 ranger bn
1 cdo bn
Air Manoeuvre
1 air mob bn
Other
1 (Presidential Guard) gd bn
COMBAT SUPPORT
1 arty bn
1 AD bn
1 engr bn

EQUIPMENT BY TYPE
ARMOURED FIGHTING VEHICLES
MBT 38: 30 T-34; 8 T-54
LT TK 15 PT-76
RECCE 27: 2 AML-90; 25 BRDM-1/BRDM-2
IFV 2 BMP-1
APC 59
APC (T) 10 BTR-50
APC (W) 30: 16 BTR-40; 8 BTR-60; 6 BTR-152
PPV 19: 10 *Mamba*†; some *Puma* M26; 9 *Puma* M36
ENGINEERING & MAINTENANCE VEHICLES
ARV T-54/T-55 reported
ANTI-TANK/ANTI-INFRASTRUCTURE
MSL • MANPATS 9K11 *Malyutka* (AT-3 *Sagger*); 9K111-1 *Konkurs* (AT-5 *Spandrel*)
RCL 82mm B-10
GUNS 6+: **57mm** ZIS-2 (M-1943); **85mm** 6 D-44
ARTILLERY 47+
TOWED 24: **122mm** 12 M-1931/37; **130mm** 12 M-46
MRL 220mm 3 9P140 *Uragan*
MOR 20+: **82mm** M-43; **120mm** 20 M-1938/M-1943
AIR DEFENCE
SAM • Point-defence 9K32 *Strela*-2 (SA-7 *Grail*)‡
GUNS • TOWED 24+: **30mm** M-53 (twin); **37mm** 8 M-1939; **57mm** 12 Type-59 (S-60); **100mm** 4 KS-19

Navy ε400

EQUIPMENT BY TYPE
PATROL AND COASTAL COMBATANTS 4
PB 4: 1 Swiftships†; 3 RPB 20

Air Force 800

EQUIPMENT BY TYPE†
AIRCRAFT
FGA (3 MiG-21 *Fishbed* non-op)
TPT • Light 4: 2 An-2 *Colt*; 2 *Tetras*
HELICOPTERS
ATK 4 Mi-24 *Hind*
MRH 5: 2 MD-500MD; 2 Mi-17-1V *Hip* H; 1 SA342K *Gazelle*
TPT 2: **Medium** 1 SA330 *Puma*; **Light** 1 AS350B *Ecureuil*
AIR-LAUNCHED MISSILES
AAM • IR R-3 (AA-2 *Atoll*)‡

Paramilitary 2,600 active

Gendarmerie 1,000

Republican Guard 1,600

People's Militia 7,000 reservists

DEPLOYMENT

MALI: UN • MINUSMA 863; 1 inf bn

SOUTH SUDAN: UN • UNMISS 2

SUDAN: UN • UNISFA 1

WESTERN SAHARA: UN • MINURSO 4

Guinea-Bissau GNB

CFA Franc BCEAO fr		2019	2020	2021
GDP	fr	843bn	814bn	
	US$	1.44bn	1.39bn	
per capita	US$	811	767	
Growth	%	4.5	-2.9	
Inflation	%	0.2	2.0	
Def bdgt	fr	n.k	n.k	
	US$	n.k	n.k	
US$1=fr		585.75	584.99	

Real-terms defence budget trend (US$m, constant 2015)

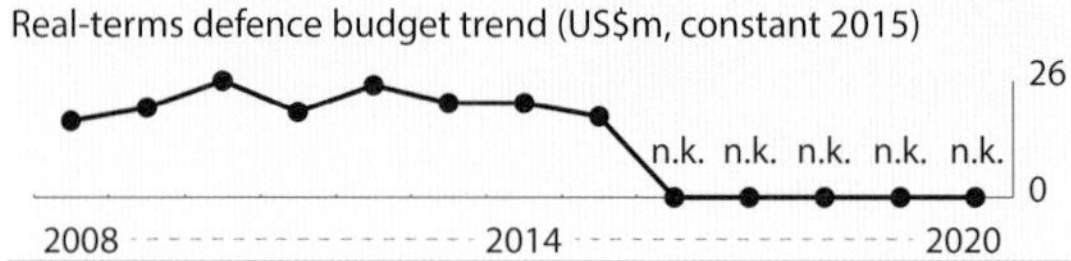

Population 1,927,104

Age	0–14	15–19	20–24	25–29	30–64	65 plus
Male	21.7%	5.5%	4.5%	3.7%	12.1%	1.3%
Female	21.5%	5.6%	4.8%	4.1%	13.4%	1.8%

Capabilities

Guinea-Bissau's armed forces have limited capabilities and are in the midst of DDR and SSR programmes. The UN expressed concern about the armed forces' role in politics, following political disputes after the late-2019 election. Defence policy is focused mainly on tackling internal-security challenges, in particular drug trafficking. It was announced in September 2020 that an ECOWAS mission, which had been present since a coup in 2012, would be withdrawn. Training remains limited and there are problems with recruitment and retention, as well as in developing adequate non-commissioned-officer structures. Much of the country's military equipment is ageing and maintenance likely limits military effectiveness. There is no defence manufacturing sector.

ACTIVE 4,450 (Army 4,000 Navy 350 Air 100)

Conscript liability Selective conscription

Personnel and eqpt totals should be treated with caution. A number of draft laws to restructure the armed services and police have been produced

ORGANISATIONS BY SERVICE

Army ε4,000

FORCES BY ROLE

MANOEUVRE

Reconnaissance

1 recce coy

Armoured

1 armd bn (sqn)

Light

5 inf bn

COMBAT SUPPORT

1 arty bn

1 engr coy

EQUIPMENT BY TYPE

ARMOURED FIGHTING VEHICLES

MBT 10 T-34

LT TK 15 PT-76

RECCE 10 BRDM-2

APC • **APC (W)** 55: 35 BTR-40/BTR-60; 20 Type-56 (BTR-152)

ANTI-TANK/ANTI-INFRASTRUCTURE

RCL 75mm Type-52 (M20); **82mm** B-10

GUNS 85mm 8 D-44

ARTILLERY 26+

TOWED 122mm 18 D-30/M-30 (M-1938)

MOR 8+: **82mm** M-43; **120mm** 8 M-1943

AIR DEFENCE

SAM • **Point-defence** 9K32 *Strela*-2 (SA-7 *Grail*)‡

GUNS • **TOWED** 34: **23mm** 18 ZU-23; **37mm** 6 M-1939; **57mm** 10 S-60

Navy ε350

EQUIPMENT BY TYPE

PATROL AND COASTAL COMBATANTS 4

PB 4: 2 *Alfeite*†; 2 Rodman 55m

Air Force 100

EQUIPMENT BY TYPE

AIRCRAFT • **TPT** • **Light** 1 Cessna 208B

Kenya KEN

Kenyan Shilling sh		2019	2020	2021
GDP	sh	9.74tr	10.6tr	
	US$	95.4bn	101.0bn	
per capita	US$	2,004	2,075	
Growth	%	5.4	1.0	
Inflation	%	5.2	5.3	
Def bdgt [a]	sh	126bn	117bn	
	US$	1.23bn	1.11bn	
US$1=sh		102.09	105.36	

[a] Includes national-intelligence funding

Real-terms defence budget trend (US$m, constant 2015)

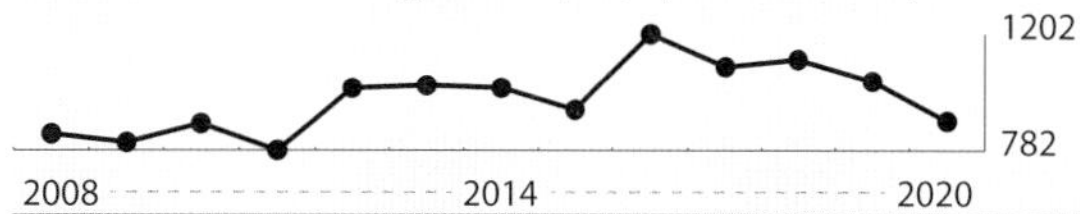

Population 53,527,936

Age	0–14	15–19	20–24	25–29	30–64	65 plus
Male	19.5%	5.7%	4.6%	3.7%	15.1%	1.4%
Female	19.3%	5.6%	4.6%	3.8%	15.1%	1.7%

Capabilities

The armed forces are concerned with threats to regional stability and tackling security challenges, particularly from neighbouring Somalia. A separate coastguard service, established in late 2018, is intended to boost maritime security. A long-standing defence and security agreement with the UK includes a permanent UK training unit and dedicated support for maritime security and a counter-IED training centre. There are also significant defence ties with the US and evidence of developing relationships with the Chinese and Jordanian armed forces. Involvement in a number of regional security missions and multinational exercises may also foster improved levels of cooperation and interoperability. Training has received attention, given the need to prepare for AU deployments. Regular operational deployments have increased military experience and confidence. Kenya's armed forces regularly participate in multinational exercises. Kenya remains a key contributor to AMISOM in Somalia, demonstrating limited capacity to project power immediately beyond its own territory. The armed forces also provide smaller contributions to other UN missions and are a leading element of the East African Standby Force. Recent equipment investments have focused on improving counter-insurgency capabilities, including the delivery of light-attack helicopters, armoured vehicles and ISR systems. The air force has also commissioned into service several fixed-wing transport aircraft to support regional deployments. There is a limited defence industry focused on equipment maintenance and the manufacture of small-arms ammunition.

ACTIVE 24,100 (Army 20,000 Navy 1,600 Air 2,500) **Paramilitary 5,000**

ORGANISATIONS BY SERVICE

Army 20,000

FORCES BY ROLE

MANOEUVRE

Armoured

1 armd bde (1 armd recce bn, 2 armd bn)

Light
1 spec ops bn
1 ranger bn 1 inf bde (3 inf bn)
1 inf bde (2 inf bn)
1 indep inf bn
Air Manoeuvre
1 air cav bn
1 AB bn
COMBAT SUPPORT
1 arty bde (2 arty bn, 1 mor bty)
1 ADA bn
1 engr bde (2 engr bn)
EQUIPMENT BY TYPE
ARMOURED FIGHTING VEHICLES
MBT 78 Vickers Mk 3
RECCE 92: 72 AML-60/AML-90; 12 *Ferret*; 8 S52 *Shorland*
APC 200
APC (W) 95: 52 UR-416; 31 Type-92; 12 *Bastion* APC; (10 M3 Panhard in store)
PPV 105 *Puma* M26-15
ENGINEERING & MAINTENANCE VEHICLES
ARV 7 Vickers ARV
MW *Bozena*
ARTILLERY 111
SP 155mm 2+ *Nora* B-52
TOWED 105mm 47: 40 L118 Light Gun; 7 Model 56 pack howitzer
MOR 62: **81mm** 50; **120mm** 12 Brandt
ANTI-TANK/ANTI-INFRASTRUCTURE
MSL • MANPATS *Milan*
RCL 84mm *Carl Gustaf*
HELICOPTERS
MRH 42: 2 Hughes 500D†; 12 Hughes 500M†; 10 Hughes 500MD *Scout Defender*† (with TOW); 9 Hughes 500ME†; 6 MD-530F; 3 Z-9W
AIR DEFENCE • GUNS • TOWED 94: **20mm** 81: 11 Oerlikon; ε70 TCM-20; **40mm** 13 L/70
AIR-LAUNCHED MISSILES • ASM TOW

Navy 1,600 (incl 120 marines)

EQUIPMENT BY TYPE
PATROL AND COASTAL COMBATANTS 7
PCO 1 *Jasiri* with 1 AK630 CIWS, 1 76mm gun
PCF 2 *Nyayo*
PCC 3: 1 *Harambee* II (ex-FRA P400); 1 *Shujaa* with 1 76mm gun; 1 *Shujaa*
PBF 1 *Archangel*
AMPHIBIOUS • LANDING CRAFT 2
LCM 2 *Galana*
LOGISTICS AND SUPPORT • AP 2

Air Force 2,500

FORCES BY ROLE
FIGHTER/GROUND ATTACK
2 sqn with F-5E/F *Tiger* II
TRANSPORT
Some sqn with DHC-5D *Buffalo*†; DHC-8†; F-70† (VIP); Y-12(II)†; C-27J *Spartan*
TRAINING
Some sqn with *Bulldog* 103/*Bulldog* 127†; EMB-312 *Tucano*†*; *Hawk* Mk52†*
TRANSPORT HELICOPTER
1 sqn with SA330 *Puma*†
EQUIPMENT BY TYPE†
AIRCRAFT 37 combat capable
FTR 21: 17 F-5E *Tiger* II; 4 F-5F *Tiger* II
TPT 19: **Medium** 3 C-27J *Spartan*; **Light** 15: 4 DHC-5D *Buffalo*†; 3 DHC-8†; 8 Y-12(II)†; (6 Do-28D-2 in store);
PAX 1 F-70 (VIP)
TRG 29: 8 *Bulldog* 103/127†; 11 EMB-312 *Tucano*†*; 5 Grob 120A; 5 *Hawk* Mk52†*
HELICOPTERS
ATK 3 AH-1F *Cobra*
MRH 9 H125M (AS550) *Fennec*
TPT 20: **Medium** 12: 2 Mi-171; 10 SA330 *Puma*†; **Light** 8 Bell 205 (UH-1H *Huey* II)
AIR-LAUNCHED MISSILES
AAM • IR AIM-9 *Sidewinder*
ASM AGM-65 *Maverick*

Paramilitary 5,000

Police General Service Unit 5,000

EQUIPMENT BY TYPE
ARMOURED FIGHTING VEHICLES
APC • PPV 25 CS/VP3
AUV some Streit *Cyclone*
PATROL AND COASTAL COMBATANTS 5
PB 5 (2 on Lake Victoria)

Air Wing

EQUIPMENT BY TYPE
AIRCRAFT • TPT • Light 6: 2 Cessna 208B *Grand Caravan*; 3 Cessna 310; 1 Cessna 402
HELICOPTERS
MRH 3 Mi-17 *Hip* H
TPT 5: **Medium** 1 Mi-17V-5; **Light** 4: 2 AW139; 1 Bell 206L *Long Ranger*; 1 Bo-105
TRG 1 Bell 47G

Coast Guard

Ministry of Interior
EQUIPMENT BY TYPE
PATROL AND COASTAL COMBATANTS 1
PCC 1 *Doria* with 1 hel landing platform

DEPLOYMENT

CENTRAL AFRICAN REPUBLIC: UN • MINUSCA 15

DEMOCRATIC REPUBLIC OF THE CONGO: UN • MONUSCO 9

LEBANON: UN • UNIFIL 3

MALI: UN • MINUSMA 12

SOMALIA: AU • AMISOM 4,000: 3 inf bn; **UN** • UNSOS 1

SOUTH SUDAN: UN • UNMISS 16

SUDAN: UN • UNAMID 87; 1 MP coy

FOREIGN FORCES

United Kingdom BATUK 350; 1 trg unit

Lesotho LSO

Lesotho Loti M		2019	2020	2021
GDP	M	35.9bn	35.3bn	
	US$	2.43bn	1.91bn	
per capita	US$	1,185	924	
Growth	%	1.0	-4.8	
Inflation	%	5.2	4.3	
Def bdgt	M	643m	621m	
	US$	43.5m	33.5m	
US$1=M		14.79	18.53	

Real-terms defence budget trend (US$m, constant 2015)

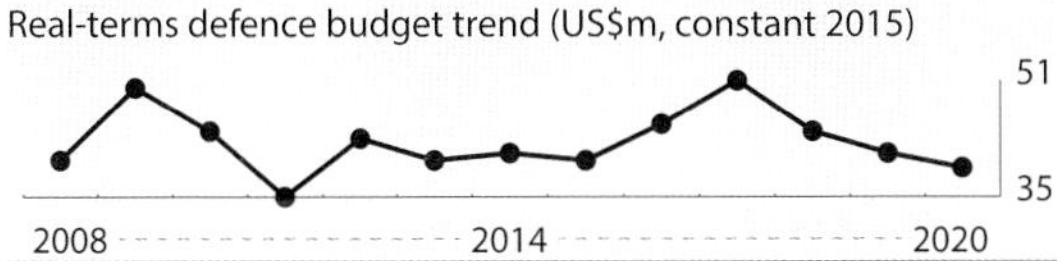

Population 1,969,334

Age	0–14	15–19	20–24	25–29	30–64	65 plus
Male	15.7%	4.9%	4.4%	4.2%	17.4%	2.9%
Female	15.6%	5.1%	4.9%	4.9%	17.3%	2.7%

Capabilities

Lesotho has a small ground force and an air wing for light transport and liaison. Lesotho is a SADC member state, and a force from that organisation deployed to Lesotho in late 2017 to support the government following the assassination of the army chief. The mission concluded at the end of November 2018. Lesotho's armed forces are charged with protecting territorial integrity and sovereignty and ensuring internal security. The armed forces are a voluntary service. In April 2020, the army was briefly deployed internally by the prime minister, though they were later recalled, and the prime minister resigned. The armed forces, and military facilities, were also utilised in the country's coronavirus response. There is no independent capacity to deploy and sustain missions beyond national borders. Lesotho's limited inventory is obsolescent by modern standards though, while there is little possibility of significant recapitalisation, there is an aspiration to acquire light helicopters. Barring limited maintenance capacity, there is no defence-industrial base.

ACTIVE 2,000 (Army 2,000)

ORGANISATIONS BY SERVICE

Army ε2,000

FORCES BY ROLE

MANOEUVRE

Reconnaissance

1 recce coy

Light

7 inf coy

Aviation

1 sqn

COMBAT SUPPORT

1 arty bty(-)

1 spt coy (with mor)

EQUIPMENT BY TYPE

ARMOURED FIGHTING VEHICLES

MBT 1 T-55

RECCE 30: 4 AML-90; 2 BRDM-2†; 6 RAM Mk3; 10 RBY-1; 8 S52 *Shorland*

ANTI-TANK/ANTI-INFRASTRUCTURE

RCL 106mm 6 M40

ARTILLERY 12

TOWED 105mm 2

MOR 81mm 10

Air Wing 110

AIRCRAFT

TPT • Light 3: 2 C-212-300 *Aviocar*; 1 GA-8 *Airvan*

HELICOPTERS

MRH 3: 1 Bell 412 *Twin Huey*; 2 Bell 412EP *Twin Huey*

TPT • Light 4: 1 Bell 206 *Jet Ranger*; 3 H125 (AS350) *Ecureuil*

Liberia LBR

Liberian Dollar L$		2019	2020	2021
GDP	L$	3.18bn	3.07bn	
	US$	3.18bn	3.07bn	
per capita	US$	694	654	
Growth	%	-2.5	-3.0	
Inflation	%	27.0	11.9	
Def bdgt	L$	13.9m	12.2m	12.9m
	US$	13.9m	12.2m	
US$1=L$		1.00	1.00	

Real-terms defence budget trend (US$m, constant 2015)

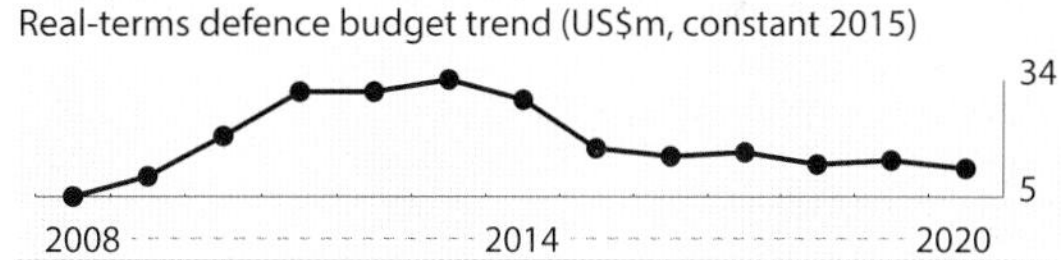

Population 5,073,296

Age	0–14	15–19	20–24	25–29	30–64	65 plus
Male	21.9%	5.4%	4.7%	3.3%	13.2%	1.4%
Female	21.4%	5.4%	4.8%	3.4%	13.6%	1.4%

Capabilities

A revised National Security Strategy was produced in 2017, reportedly clarifying the roles of Liberia's security institutions. The government is emphasising national security as part of its development agenda. Military personnel have indicated that priorities include improving training, operational readiness and personnel welfare. There are plans to establish an air wing to boost the country's search-and-rescue, movement and logistics, medevac and maritime-patrol capacities. Plans to increase establishment strength to 5,000 remain aspirational. US military assistance has in recent years focused on areas such as force health, including schemes to improve recruitment and retention, as well as maritime security and military medicine. The armed forces are able to deploy and sustain small units, such as to the MINUSMA mission in Mali. Equipment recapitalisation will be dependent on finances, as well as the development of a supporting force structure. Liberia has no domestic defence industry, bar limited maintenance-support capacities.

ACTIVE 2,010 (Army 1,950, Coast Guard 60)

ORGANISATIONS BY SERVICE

Army 1,950

FORCES BY ROLE

MANOEUVRE

Light

1 (23rd) inf bde with (2 inf bn, 1 engr coy, 1 MP coy)

COMBAT SERVICE SUPPORT

1 trg unit (forming)

ARMOURED FIGHTING VEHICLES

APC • PPV 3+ Streit *Cougar*

Coast Guard 60

All operational patrol vessels under 10t FLD

DEPLOYMENT

MALI: UN • MINUSMA 117; 1 inf coy

SOUTH SUDAN: UN • UNMISS 2

SUDAN: UN • UNISFA 2

Madagascar MDG

Malagsy Ariary fr		2019	2020	2021
GDP	fr	51.1tr	51.6tr	
	US$	14.1bn	14.2bn	
per capita	US$	525	515	
Growth	%	4.8	-3.2	
Inflation	%	5.6	4.3	
Def bdgt	fr	278bn	407bn	
	US$	76.9m	112.0m	
US$1=fr		3618.43	3635.33	

Real-terms defence budget trend (US$m, constant 2015)

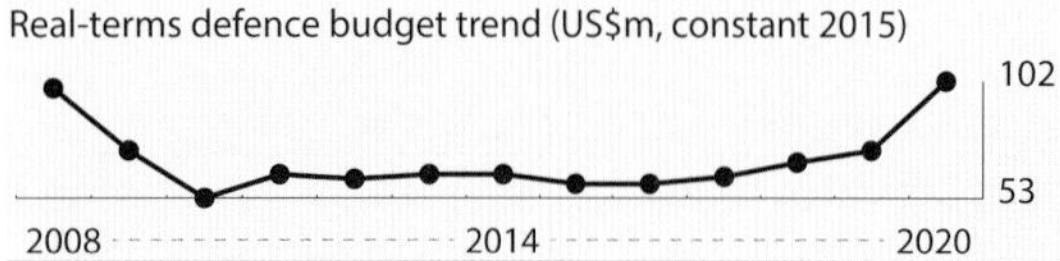

Population 26,955,737

Age	0–14	15–19	20–24	25–29	30–64	65 plus
Male	19.6%	5.3%	4.8%	4.2%	14.6%	1.6%
Female	19.3%	5.2%	4.8%	4.2%	14.7%	1.9%

Capabilities

Principal defence aspirations include ensuring sovereignty and territorial integrity, while maritime security is also an area of focus. The army is the largest armed service. The armed forces intervened in domestic politics in 2009. Madagascar is a member of SADC and its regional Standby Force. In 2018, the country signed an 'umbrella defence agreement' with India to explore closer defence ties and an intergovernmental agreement with Russia on military cooperation. There is no independent capacity to deploy and support an operation beyond national borders. The equipment inventory is obsolescent and with economic development a key government target, equipment recapitalisation is unlikely to be a key priority. A small number of second-hand transport aircraft and helicopters were acquired in 2019, modestly boosting military mobility. In 2020, UAE-manufactured protected patrol vehicles were observed at a parade.

ACTIVE 13,500 (Army 12,500 Navy 500 Air 500)
Paramilitary 8,100

Conscript liability 18 months (incl for civil purposes)

ORGANISATIONS BY SERVICE

Army 12,500+

FORCES BY ROLE

MANOEUVRE

Light

2 (intervention) inf regt

10 (regional) inf regt

COMBAT SUPPORT

1 arty regt

3 engr regt

1 sigs regt

COMBAT SERVICE SUPPORT

1 log regt

AIR DEFENCE

1 ADA regt

EQUIPMENT BY TYPE

ARMOURED FIGHTING VEHICLES

LT TK 12 PT-76

RECCE 73: ε35 BRDM-2; 10 FV701 *Ferret*; ε20 M3A1; 8 M8

APC • APC (T) ε30 M3A1 half-track

AUV 6 *Panthera* T4

ANTI-TANK/ANTI-INFRASTRUCTURE

RCL 106mm M40A1

ARTILLERY 25+

TOWED 17: **105mm** 5 M101; **122mm** 12 D-30

MOR 8+: **82mm** M-37; **120mm** 8 M-43

AIR DEFENCE • GUNS • TOWED 70: **14.5mm** 50 ZPU-4; **37mm** 20 PG-55 (M-1939)

Navy 500 (incl some 100 Marines)

EQUIPMENT BY TYPE

PATROL AND COASTAL COMBATANTS 8

PCC 1 *Trozona*

PB 7 (ex-US CG MLB)

AMPHIBIOUS • **LCT** 1 (ex-FRA EDIC)

Air Force 500

FORCES BY ROLE

TRANSPORT

1 sqn with An-26 *Curl*; Yak-40 *Codling* (VIP)

1 (liaison) sqn with Cessna 310; Cessna 337 *Skymaster*; PA-23 *Aztec*

TRAINING

1 sqn with Cessna 172; J.300 *Joker*; *Tetras*

TRANSPORT HELICOPTER
1 sqn with SA318C *Alouette* II

EQUIPMENT BY TYPE

AIRCRAFT • TPT 17: **Light** 15: 1 An-26 *Curl*; 4 Cessna 172; 1 Cessna 310; 2 Cessna 337 *Skymaster*; 1 CN235M; 2 J.300 *Joker*; 1 PA-23 *Aztec*; 1 *Tetras*; 2 Yak-40 *Codling* (VIP); **PAX** 2 B-737

HELICOPTERS
MRH 3 SA318C *Alouette* II
TPT • Light 4: 3 AS350 *Ecureuil*; 1 BK117

Paramilitary 8,100

Gendarmerie 8,100

Malawi MWI

Malawian Kwacha K		2019	2020	2021
GDP	K	5.71tr	6.26tr	
	US$	7.66bn	8.33bn	
per capita	US$	378	399	
Growth	%	4.5	0.6	
Inflation	%	9.4	9.3	
Def bdgt	K	50.8bn	51.1bn	
	US$	68.2m	68.0m	
US$1=K		745.23	751.53	

Real-terms defence budget trend (US$m, constant 2015)

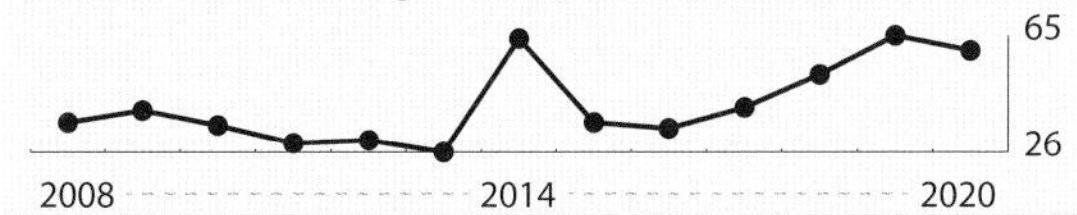

Population	21,196,629

Age	0–14	15–19	20–24	25–29	30–64	65 plus
Male	22.8%	5.5%	4.7%	3.8%	11.5%	1.2%
Female	23.0%	5.6%	4.8%	3.8%	11.8%	1.5%

Capabilities

The Malawi Defence Forces (MDF) are constitutionally tasked with ensuring sovereignty and territorial integrity. Additional tasks include providing military assistance to civil authorities and support to the police, and in recent years the army has been used to help with infrastructure development and attempts to control illegal deforestation. The army is the largest element of the armed forces. Counter-trafficking is a role for the MDF's small air force, previously an air wing, and its naval unit. Development priorities include improving combat readiness and military medicine and engineering. Malawi is a member of the SADC and its Standby Force. The armed forces have contributed to AU and UN peace-keeping operations, including in Côte d'Ivoire and the DRC. There is no independent capacity to deploy and support an operation beyond national borders. The UK provided training and support for the armed forces' deployment to the DRC, where troops contribute to the Force Intervention Brigade and also provide support to the MDF's counter-poaching tasks.

ACTIVE 10,700 (Army 10,500 Air Force 200)
Paramilitary 4,200

ORGANISATIONS BY SERVICE

Army 10,500

FORCES BY ROLE

MANOEUVRE
Mechanised
1 mech bn
Light
1 inf bde (4 inf bn)
1 inf bde (1 inf bn)
Air Manoeuvre
1 para bn

COMBAT SUPPORT
3 lt arty bty
1 engr bn

COMBAT SERVICE SUPPORT
12 log coy

EQUIPMENT BY TYPE

ARMOURED FIGHTING VEHICLES
RECCE 66: 30 *Eland-90*; 8 FV701 *Ferret*; 20 FV721 *Fox*; 8 RAM Mk3
APC • PPV 31: 14 *Casspir*; 9 *Marauder*; 8 *Puma* M26-15

ARTILLERY 107
TOWED 105mm 9 L118 Light Gun
MOR 81mm 98: 82 L16A1; 16 M3

AIR DEFENCE • GUNS • TOWED 72: **12.7mm** 32; **14.5mm** 40 ZPU-4

Navy 220

EQUIPMENT BY TYPE

PATROL AND COASTAL COMBATANTS • **PB** 3: 1 *Kasungu* (ex-FRA *Antares*); 2 *Mutharika* (PRC)

Air Force 200

EQUIPMENT BY TYPE

AIRCRAFT • TPT • Light 1 Do-228

HELICOPTERS • TPT 8: **Medium** 3: 1 AS532UL *Cougar*; 1 SA330H *Puma*; 1 H215 *Super Puma* **Light** 5: 1 AS350L *Ecureuil*; 4 SA341B *Gazelle*

Paramilitary 4,200

Police Mobile Service 4,200

EQUIPMENT BY TYPE

ARMOURED FIGHTING VEHICLES
RECCE 8 S52 *Shorland*

AIRCRAFT
TPT • Light 4: 3 BN-2T *Defender* (border patrol); 1 SC.7 3M *Skyvan*

HELICOPTERS • MRH 2 AS365 *Dauphin 2*

DEPLOYMENT

DEMOCRATIC REPUBLIC OF THE CONGO: UN • MONUSCO 864; 1 inf bn

SUDAN: UN • UNAMID 1; **UN •** UNISFA 2

WESTERN SAHARA: UN • MINURSO 2

Mali MLI

CFA Franc BCEAO fr		2019	2020	2021
GDP	fr	10.2tr	10.1tr	
	US$	17.3bn	17.7bn	
per capita	US$	907	899	
Growth	%	5.1	-2.0	
Inflation	%	-2.9	0.5	
Def bdgt [a]	fr	423bn	452bn	
	US$	722m	788m	
US$1=fr		585.90	573.80	

[a] Defence and security budget

Real-terms defence budget trend (US$m, constant 2015)

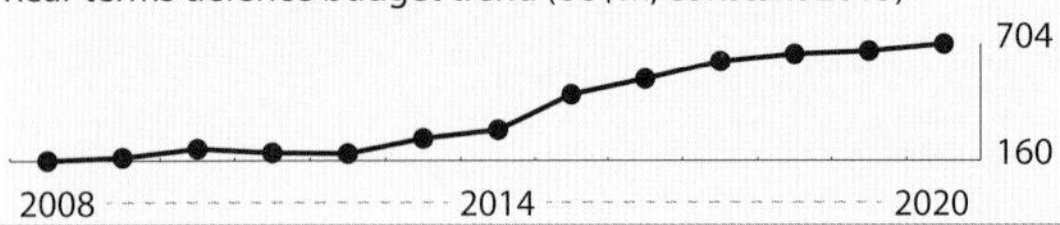

Population 19,553,397

Age	0–14	15–19	20–24	25–29	30–64	65 plus
Male	24.0%	5.2%	3.9%	3.0%	11.1%	1.5%
Female	23.7%	5.5%	4.5%	3.7%	12.4%	1.5%

Capabilities

The armed forces remain focused on countering rebel and Islamist groups. A defence-reform process is ongoing, but the objectives of the 2015–19 military-programming law were only partly achieved. Improvements are still required including in recruitment and training, as well as in improving basic administrative support. Mali is supported by neighbouring states in the G5 Sahel partnership, and has received training assistance from the EU, France and the US. However, a military coup in August cast doubt on international support. The EU suspended its training mission after the coup, while the US also suspended support. Military leaders have reportedly since indicated an 18-month timeline for new elections. France maintains bases, personnel and equipment in Mali as part of *Operation Barkhane*, while some other states supporting *Barkhane* also deploy personnel and equipment in-country. The armed forces also participate in multinational exercises, particularly those focused on counter-terrorism capabilities, but the forces still suffer from operational deficiencies as well as broader institutional weakness. Despite vehicle deliveries by external partners, and the acquisition of several aircraft, the armed forces remain underequipped. Mali does not possess a defence manufacturing industry and, with equipment and maintenance capabilities limited, equipment serviceability will likely be variable.

ACTIVE 13,000 (Army 13,000) **Paramilitary 7,800**

ORGANISATIONS BY SERVICE

Army ε13,000

FORCES BY ROLE

The remnants of the pre-conflict Malian army are being reformed into new combined-arms battlegroups, each of which comprise one lt mech coy, three mot inf coy, one arty bty and additional recce, cdo and cbt spt elms

MANOEUVRE

Light

9 mot inf bn

1 inf coy (Special Joint Unit)

Air Manoeuvre

1 para bn

COMBAT SUPPORT

1 engr bn

COMBAT SERVICE SUPPORT

1 med unit

EQUIPMENT BY TYPE

ARMOURED FIGHTING VEHICLES

LT TK 2+ PT-76

RECCE BRDM-2†

APC 172:

APC (W) 46+: 27 *Bastion* APC; 10+ BTR-60PB; 9 BTR-70

PPV 126: 29 *Casspir*; 13 *Marauder*; 24 Stark Motors *Storm Light*; 30 Streit *Cougar*; 4 Streit *Gladiator*; 5+ Streit *Python*; 21 Streit *Typhoon*†

ARTILLERY 30+

TOWED 122mm D-30

MRL 122mm 30+ BM-21 *Grad*

Air Force

FORCES BY ROLE

TRANSPORT

1 sqn with BT-67; C295W; Y-12E

TRAINING

1 sqn with *Tetras*

TRANSPORT/ATTACK HELICOPTER

1 sqn with H215; Mi-24D *Hind*; Mi-35M *Hind*

EQUIPMENT BY TYPE

AIRCRAFT 3 combat capable

ISR 1 Cessna 208 *Caravan*

TPT • Light 11: 1 BT-67; 1 C295W; 7 *Tetras*; 2 Y-12E (1 An-24 *Coke*; 2 An-26 *Curl*; 2 BN-2 *Islander* all in store)

TRG 3 A-29 *Super Tucano** (6 L-29 *Delfin*; 2 SF-260WL *Warrior** all in store)

HELICOPTERS

ATK 4: 2 Mi-24D *Hind*; 2 Mi-35M *Hind*

TPT • Medium 2 H215 (AS332L1) *Super Puma*; (1 Mi-8 *Hip* in store); **Light** (1 AS350 *Ecureuil* in store)

Paramilitary 7,800 active

Gendarmerie 1,800

FORCES BY ROLE

MANOEUVRE

Other

8 paramilitary coy

1 air tpt gp (2 sy coy; 1 tpt coy)

EQUIPMENT BY TYPE

ARMOURED FIGHTING VEHICLES

APC • PPV 1+ RG-31 *Nyala*

National Guard 2,000

FORCES BY ROLE

MANOEUVRE

Reconnaissance

6 (camel) cav coy

Light
1 inf coy (Anti-terrorist special force)

EQUIPMENT BY TYPE

ARMOURED FIGHTING VEHICLES

APC • PPV 1+ RG-31 *Nyala*

National Police 1,000

Militia 3,000

DEPLOYMENT

DEMOCRATIC REPUBLIC OF THE CONGO: UN • MONUSCO 3

FOREIGN FORCES

All under MINUSMA comd unless otherwise specified
Albania EUTM Mali 4
Austria 1 • EUTM Mali 13
Bangladesh 1,310; 1 inf bn; 1 engr coy; 2 sigs coy; 1 tpt coy
Belgium 38 • EUTM Mali 15
Benin 257; 1 mech inf coy
Bhutan 5
Bosnia-Herzegovina 2
Bulgaria EUTM Mali 5
Burkina Faso 651; 1 inf bn
Burundi 1
Cambodia 297; 2 engr coy; 1 EOD coy
Cameroon 2
Canada 4
Chad 1,456; 1 SF coy; 2 inf bn
China 426; 1 sy coy; 1 engr coy; 1 fd hospital
Côte d'Ivoire 564; 1 inf bn; 1 sy coy
Czech Republic 3 • EUTM Mali 120
Denmark 11
Egypt 1,149; 1 SF coy; 1 sy bn; 1 MP coy
El Salvador 205; 1 hel sqn with 3 MD-500E
Estonia 3 • *Operation Barkhane* 95 • EUTM Mali 10
Ethiopia 1
Finland 4 • EUTM Mali 12
France 24 • *Operation Barkhane* 2,500; 1 mech inf BG; 1 log bn; 1 tpt unit with 1 CN235M; 1 PC-6B; 1 hel unit with 5 *Tiger*; 5 AS532UL *Cougar*; 5 NH90 TTH; 4 SA342 *Gazelle* • EUTM Mali 13
Gambia 5
Georgia EUTM Mali 1
Germany 361; 1 obs; 1 sy coy; 1 int coy; 1 UAV sqn • EUTM Mali 50
Ghana 152; 1 engr coy
Greece EUTM Mali 2
Guatemala 2
Guinea 863; 1 inf bn
Hungary EUTM Mali 20
Indonesia 6
Ireland EUTM Mali 20
Italy 2 • EUTM Mali 12
Jordan 64
Kenya 12
Latvia 1 • EUTM Mali 3
Liberia 117; 1 sy coy
Lithuania 37 • EUTM Mali 2
Luxembourg EUTM Mali 2
Mauritania 7
Mexico 2
Moldova EUTM Mali 3
Nepal 198; 1 EOD coy
Netherlands 5 • EUTM Mali 4
Niger 870; 1 inf bn
Nigeria 82; 1 fd hospital
Norway 13; 1 tpt flt with 1 C-130J
Pakistan 142; 1 hel sqn
Portugal 63; 1 tpt flt with 1 C295M • EUTM Mali 11
Romania 124; 1 hel sqn • EUTM Mali 1
Senegal 1,001; 1 inf bn; 1 engr coy
Serbia EUTM Mali 3
Sierra Leone 25
Slovakia EUTM Mali 15
Slovenia EUTM Mali 8
Spain EUTM Mali 235
Sri Lanka 249; 1 sy coy
Sweden 105; 1 int coy • EUTM Mali 6
Switzerland 5
Togo 936; 1 inf bn; 1 fd hospital
Tunisia 85; 1 tpt flt with 1 C-130J-30
Ukraine 9
United Kingdom 3 • *Operation Barkhane* 90; 1 hel flt with 3 CH-47SD *Chinook* HC5; • EUTM Mali 8
United States 9

Mauritius MUS

Mauritian Rupee R		2019	2020	2021
GDP	R	498bn	444bn	
	US$	14.0bn	11.3bn	
per capita	US$	11,090	8,951	
Growth	%	3.0	-14.2	
Inflation	%	0.5	2.6	
Def bdgt [a]	R	7.76bn	8.86bn	8.42bn
	US$	219m	226m	
US$1=R		35.47	39.16	

[a] Police service budget

Real-terms defence budget trend (US$m, constant 2015)

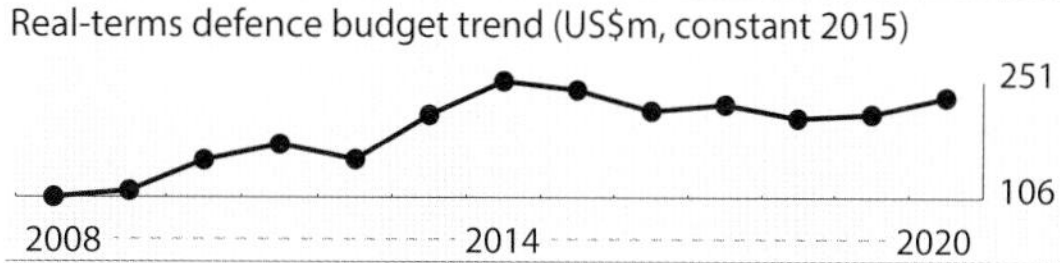

Population 1,379,365

Age	0–14	15–19	20–24	25–29	30–64	65 plus
Male	9.9%	3.6%	3.6%	3.9%	23.5%	4.6%
Female	9.5%	3.4%	3.5%	3.8%	24.1%	6.5%

Capabilities

The country has no standing armed forces; instead, security tasks are met by the police force's Special Mobile Force (SMF), formed as a motorised infantry battalion. The SMF is tasked with ensuring internal and external territorial and maritime security. India provides support to the Mauritian National Coast Guard, which is a branch of the police force. The SMF trains along traditional military lines but has no ability to deploy beyond national territory. There is no defence industry, beyond very limited maintenance facilities.

ACTIVE NIL Paramilitary 2,550

ORGANISATIONS BY SERVICE

Paramilitary 2,550

Special Mobile Force ε1,750

FORCES BY ROLE
MANOEUVRE
Reconnaissance
2 recce coy
Light
5 (rifle) mot inf coy
COMBAT SUPPORT
1 engr sqn
COMBAT SERVICE SUPPORT
1 spt pl

EQUIPMENT BY TYPE
ARMOURED FIGHTING VEHICLES
IFV 2 VAB with 20mm gun
APC • **APC (W)** 12: 3 *Tactica*; 9 VAB
ARTILLERY • **MOR 81mm** 2

Coast Guard ε800

EQUIPMENT BY TYPE
PATROL AND COASTAL COMBATANTS 17
PCO 1 *Barracuda* with 1 hel landing platform
PCC 2 *Victory* (IND *Sarojini Naidu*)
PB 14: 10 (IND Fast Interceptor Boat); 1 P-2000; 1 SDB-Mk3; 2 *Rescuer* (FSU *Zhuk*)
AIRCRAFT • **TPT** • **Light** 4: 1 BN-2T *Defender*; 3 Do-228-101

Police Air Wing

EQUIPMENT BY TYPE
HELICOPTERS
MRH 9: 1 H125 (AS555) *Fennec*; 2 *Dhruv*; 1 SA315B *Lama* (*Cheetah*); 5 SA316 *Alouette* III (*Chetak*)

Mozambique MOZ

Mozambique New Metical M		2019	2020	2021
GDP	M	957bn	999bn	
	US$	15.2bn	14.6bn	
per capita	US$	488	455	
Growth	%	2.3	-0.5	
Inflation	%	2.8	3.6	
Def bdgt	M	7.94bn	9.11bn	
	US$	126m	133m	
US$1=M		62.97	68.60	

Real-terms defence budget trend (US$m, constant 2015)

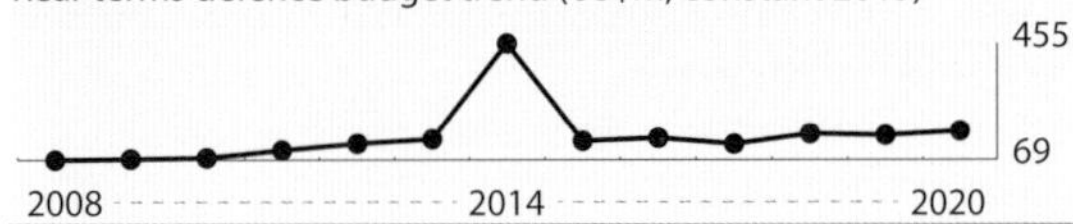

Population 30,098,197

Age	0–14	15–19	20–24	25–29	30–64	65 plus
Male	23.1%	5.4%	4.5%	3.5%	11.2%	1.4%
Female	22.5%	5.4%	4.6%	3.8%	13.0%	1.5%

Capabilities

Mozambique faces a continuing internal threat from Islamist groups, with attacks in the north of the country. The armed forces are tasked with ensuring territorial integrity and internal security, as well as tackling piracy and human trafficking. Private military companies, however, have reportedly been drafted since the fourth quarter of 2019 to help counter the insurrection. Integrating RENAMO personnel into the armed forces is a long-standing objective, though tackling Islamist militancy is now the likely priority. Mozambique has defence relationships with China, Portugal and Russia. In 2017, the UN raised concerns that Mozambique was receiving defence support from North Korea, a claim the government denied. The defence ministry is reportedly to implement a military HIV policy, including more screening, to try to reduce HIV incidence. The armed forces have no capacity to deploy externally without assistance. Soviet-era equipment comprises the majority of the inventory and maintaining this will be problematic, not least in the absence of any local defence industry and continued economic challenges.

ACTIVE 11,200 (Army 10,000 Navy 200 Air 1,000)
Conscript liability 2 years

ORGANISATIONS BY SERVICE

Army ε9,000–10,000

FORCES BY ROLE
SPECIAL FORCES
3 SF bn
MANOEUVRE
Light
7 inf bn
COMBAT SUPPORT
2-3 arty bn
2 engr bn

COMBAT SERVICE SUPPORT
1 log bn

EQUIPMENT BY TYPE†
Equipment estimated at 10% or less serviceability
ARMOURED FIGHTING VEHICLES
MBT 60+ T-54
RECCE 30 BRDM-1/BRDM-2
IFV 40 BMP-1
APC 326
APC (T) 30 FV430
APC (W) 285: 160 BTR-60; 100 BTR-152; 25 AT-105 *Saxon*
PPV 11+: 11 *Casspir*; some Tata Motors MRAP
AUV 9+ *Tiger* 4×4
ANTI-TANK/ANTI-INFRASTRUCTURE
MSL • MANPATS 9K11 *Malyutka* (AT-3 *Sagger*); 9K111 *Fagot* (AT-4 *Spigot*)
RCL 75mm; **82mm** B-10; **107mm** 24 B-12
GUNS 85mm 18: 6 D-48; 12 PT-56 (D-44)
ARTILLERY 126
TOWED 62: **100mm** 20 M-1944; **105mm** 12 M101; **122mm** 12 D-30; **130mm** 6 M-46; **152mm** 12 D-1
MRL 122mm 12 BM-21 *Grad*
MOR 52: **82mm** 40 M-43; **120mm** 12 M-43
AIR DEFENCE • GUNS 290+
SP 57mm 20 ZSU-57-2
TOWED 270+: **20mm** M-55; **23mm** 120 ZU-23-2; **37mm** 90 M-1939; (10 M-1939 in store); **57mm** 60 S-60; (30 S-60 in store)

Navy ε200

EQUIPMENT BY TYPE
PATROL AND COASTAL COMBATANTS 14
PBF 10: 2 DV 15; 6 HSI 32; 2 *Namilti* (ex-IND C-401)
PB 4: 3 *Ocean Eagle* 43 (capacity 1 *Camcopter* S-100 UAV); 1 *Pebane* (ex-ESP *Conejera*)
UNINHABITED AERIAL VEHICLES
ISR • Light 1 S-100 *Camcopter*

Air Force 1,000

FORCES BY ROLE
FIGHTER/GROUND ATTACK
1 sqn with MiG-21bis *Fishbed*; MiG-21UM *Mongol* B
TRANSPORT
1 sqn with An-26 *Curl*; FTB-337G *Milirole*; Cessna 150B; Cessna 172; PA-34 *Seneca*
ATTACK/TRANSPORT HELICOPTER
1 sqn with Mi-24 *Hind*†

EQUIPMENT BY TYPE
AIRCRAFT 8 combat capable
FGA 8: 6 MiG-21bis *Fishbed*; 2 MiG-21UM *Mongol* B
ISR 2 FTB-337G *Milirole*
TPT 6: **Light** 5: 1 An-26 *Curl*; 2 Cessna 150B; 1 Cessna 172; 1 PA-34 *Seneca*; (4 PA-32 *Cherokee* non-op); **PAX** 1 Hawker 850XP
TRG 2 L-39 *Albatros*
HELICOPTERS
ATK 2 Mi-24 *Hind*†
TPT • Medium 2 Mi-8 *Hip*
AD • SAM • TOWED: (S-75 *Dvina* (SA-2 *Guideline*) non-op‡; S-125 *Pechora* SA-3 *Goa* non-op‡)

Namibia NAM

Namibian Dollar N$		2019	2020	2021
GDP	N$	180bn	175bn	
	US$	12.5bn	10.3bn	
per capita	US$	5,072	4,052	
Growth	%	-1.0	-5.9	
Inflation	%	3.7	2.3	
Def bdgt	N$	5.88bn	5.88bn	
	US$	407m	344m	
US$1=N$		14.45	17.10	

Real-terms defence budget trend (US$m, constant 2015)

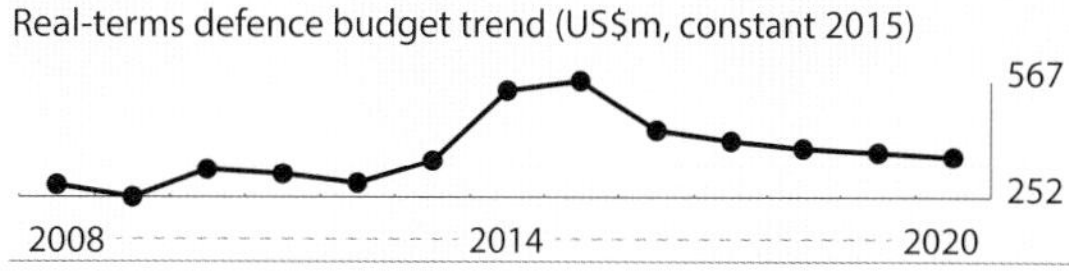

Population 2,630,073

Age	0–14	15–19	20–24	25–29	30–64	65 plus
Male	18.0%	5.4%	4.8%	4.3%	14.9%	1.7%
Female	17.7%	5.3%	4.8%	4.4%	16.6%	2.2%

Capabilities

The defence authorities aim to develop a small, mobile professional force. According to the constitution, the Namibian Defence Force's (NDF's) primary mission is territorial defence. Secondary roles include assistance to the civil power, and supporting the AU, SADC and UN. The NDF Development Strategy 2012–22 states that the NDF design should be based on a conventional force with a force-projection capability. The navy exercises with the SADC as part of the SADC's Standing Maritime Committee. There are annual meetings of a permanent commission on defence and security with Zambia and an MoU on training and cooperation has been signed with Botswana. While the NDF receives a comparatively large proportion of the state budget, there have been recent problems in adequately funding training. Namibia has deployed on AU and UN missions, but there is only limited capacity for independent power projection. The NDF is equipped for the most part with ageing or obsolescent systems, but economic difficulties make recapitalisation unlikely in the near term. There is a limited defence manufacturing sector mainly focused on armoured vehicles, tactical communications and ammunition.

ACTIVE 9,900 (Army 9,000 Navy 900) Paramilitary 6,000

ORGANISATIONS BY SERVICE

Army 9,000

FORCES BY ROLE
MANOEUVRE
Reconnaissance
1 recce regt
Light
3 inf bde (total: 6 inf bn)
Other
1 (Presidential Guard) gd bn

COMBAT SUPPORT
1 arty bde with (1 arty regt)
1 AT regt
1 engr regt
1 sigs regt
COMBAT SERVICE SUPPORT
1 log bn
AIR DEFENCE
1 AD regt
EQUIPMENT BY TYPE
ARMOURED FIGHTING VEHICLES
MBT T-54/T-55†; T-34†
RECCE 12 BRDM-2
IFV 7: 5 Type-05P mod (with BMP-1 turret); 2 *Wolf Turbo* 2 mod (with BMP-1 turret)
APC 61
APC (W) 13: 10 BTR-60; 3 Type-05P
PPV 48: 20 *Casspir*; 28 *Wolf Turbo* 2
ENGINEERING & MAINTENANCE VEHICLES
ARV T-54/T-55 reported
ANTI-TANK/ANTI-INFRASTRUCTURE
RCL 82mm B-10
GUNS 12+: **57mm** ZIS-2; **76mm** 12 ZIS-3
ARTILLERY 72
TOWED 140mm 24 G-2
MRL 122mm 8: 5 BM-21 *Grad*; 3 PHL-81
MOR 40: **81mm**; **82mm**
AIR DEFENCE
SAM • Point-defence FN-6 (CH-SA-10)
GUNS 65
SP 23mm 15 *Zumlac*
TOWED 50+: **14.5mm** 50 ZPU-4; **57mm** S-60

Navy ε900

EQUIPMENT BY TYPE
PATROL AND COASTAL COMBATANTS 7
PSO 1 *Elephant* with 1 hel landing platform
PCC 3: 2 *Daures* (ex-PRC *Haiqing* (Type-037-IS)) with 2 FQF-3200 A/S mor; 1 *Oryx*
PB 3: 1 *Brendan Simbwaye* (BRZ *Grajaú*); 2 *Terrace Bay* (BRZ *Marlim*)
AIRCRAFT • TPT • Light 1 F406 *Caravan II*
HELICOPTERS • TPT • Medium 1 S-61L

Marines ε700

Air Force

FORCES BY ROLE
FIGHTER/GROUND ATTACK
1 sqn with F-7 (F-7NM); FT-7 (FT-7NG)
ISR
1 sqn with O-2A *Skymaster*
TRANSPORT
Some sqn with An-26 *Curl*; *Falcon* 900; Learjet 36; Y-12
TRAINING
1 sqn with K-8 *Karakorum**
ATTACK/TRANSPORT HELICOPTER
1 sqn with H425; Mi-8 *Hip*; Mi-25 *Hind* D; SA315 *Lama* (*Cheetah*); SA316B *Alouette* III (*Chetak*)

EQUIPMENT BY TYPE
AIRCRAFT 12+ combat capable
FTR 8: 6 F-7 (F-7NM); 2 FT-7 (FT-7NG)
ISR 5 Cessna O-2A *Skymaster*
TPT 6: **Light** 5: 2 An-26 *Curl*; 1 Learjet 36; 2 Y-12; **PAX** 1 *Falcon* 900
TRG 4+ K-8 *Karakorum**
HELICOPTERS
ATK 2 Mi-25 *Hind* D
MRH 5: 1 H425; 1 SA315 *Lama* (*Cheetah*); 3 SA316B *Alouette* III (*Chetak*)
TPT • Medium 1 Mi-8 *Hip*

Paramilitary 6,000

Police Force • Special Field Force 6,000 (incl Border Guard and Special Reserve Force)

DEPLOYMENT

SOUTH SUDAN: UN • UNMISS 2
SUDAN: UN • UNAMID 1; **UN** • UNISFA 6

Niger NER

CFA Franc BCEAO fr		2019	2020	2021
GDP	fr	7.57tr	7.79tr	
	US$	12.9bn	13.0bn	
per capita	US$	554	536	
Growth	%	5.9	0.5	
Inflation	%	-2.5	4.4	
Def bdgt	fr	101bn	121bn	
	US$	172m	202m	
US$1=fr		585.90	600.22	

Real-terms defence budget trend (US$m, constant 2015)

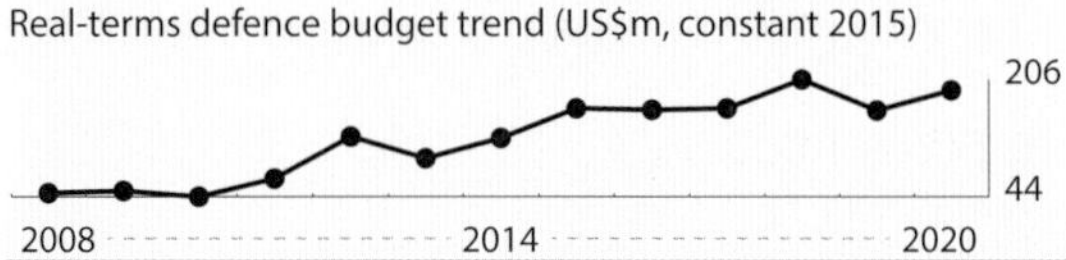

Population	22,772,361

Age	0–14	15–19	20–24	25–29	30–64	65 plus
Male	25.5%	5.6%	4.2%	3.1%	9.8%	1.3%
Female	25.1%	5.7%	4.4%	3.3%	10.6%	1.4%

Capabilities

Principal military roles include maintaining internal and border security, in light of the regional threat from Islamist groups. The country is a member of the G5 Sahel group and part of the Multi-National Joint Task Force fighting Boko Haram in the Lake Chad Basin. France has conducted joint counter-terrorism operations with Niger's armed forces. Niamey hosts air contingents from France, Germany (an air-transport base to supply its troops in neighbouring Mali) and the US, which maintains a detachment of UAVs. Niger's armed forces are combat experienced and relatively well trained, and there is training support from France, Italy and the US. Combat operations have also been conducted with US forces. There is limited capacity to deploy beyond neighbouring countries without external support. Operations in austere environments have demonstrated adequate sustainment and manoeuvre capacity. While there have been moves to integrate better-protected armoured vehicles, the armed forces are generally under-equipped and under-resourced. Apart from limited maintenance facilities, the country has no domestic defence-industrial capability.

ACTIVE 5,300 (Army 5,200 Air 100) **Paramilitary 5,400**

Conscript liability Selective conscription, 2 years

ORGANISATIONS BY SERVICE

Army 5,200

3 Mil Districts

FORCES BY ROLE

MANOEUVRE

Reconnaissance

4 armd recce sqn

Light

7 inf coy

Air Manoeuvre

2 AB coy

COMBAT SUPPORT

1 engr coy

COMBAT SERVICE SUPPORT

1 log gp

AIR DEFENCE

1 AD coy

EQUIPMENT BY TYPE

ARMOURED FIGHTING VEHICLES

RECCE 132: 35 AML-20/AML-60; 90 AML-90; *Bastion Patsas*; 7 VBL

APC 104

APC (W) 42: 13 *Bastion* APC; 22 Panhard M3; 20 WZ-551

PPV 62+: 41 *Mamba* Mk7; 21 *Puma* M26-15; some *Puma* M36

ANTI-TANK/ANTI-INFRASTRUCTURE

RCL 14: **75mm** 6 M20; **106mm** 8 M40

ARTILLERY • **MOR** 40: **81mm** 19 Brandt; **82mm** 17; **120mm** 4 Brandt

AIR DEFENCE • **GUNS** 39

SP 20mm 10 Panhard M3 VDAA

TOWED 20mm 29

Air Force 100

EQUIPMENT BY TYPE

AIRCRAFT 2 combat capable

ATK 2 Su-25 *Frogfoot*

ISR 6: 4 Cessna 208 *Caravan*; 2 DA42 MPP *Twin Star*

TPT 7: **Medium** 1 C-130H *Hercules*; **Light** 5: 1 An-26 *Curl*; 2 Cessna 208 *Caravan*; 1 Do-28 *Skyservant*; 1 Do-228-201; **PAX** 1 B-737-700 (VIP)

HELICOPTERS

ATK 2 Mi-35P *Hind*

MRH 5: 2 Mi-17 *Hip*; 3 SA342 *Gazelle*

Paramilitary 5,400

Gendarmerie 1,400

Republican Guard 2,500

National Police 1,500

DEPLOYMENT

CENTRAL AFRICAN REPUBLIC: UN • MINUSCA 4

DEMOCRATIC REPUBLIC OF THE CONGO: UN • MONUSCO 4

MALI: UN • MINUSMA 870; 1 inf bn

FOREIGN FORCES

Belgium METT *Maradi* 100

France *Operation Barkhane* 600; 1 FGA det with 4 *Mirage* 2000D; 1 tkr/tpt det with 1 *Atlantique*-2; 1 C-135FR; 1 C-130J-30; 1 UAV det with 3 MQ-9A *Reaper*

Germany *Operation Barkhane* 2 C-160

Italy MISIN 295; 1 inf coy; 1 engr unit; 1 CBRN unit; 1 med unit; 1 trg unit; 1 ISR unit

United States 800; 1 ISR UAV sqn with MQ-9A *Reaper*

Nigeria NGA

Nigerian Naira N		2019	2020	2021
GDP	N	146tr	155tr	
	US$	448bn	443bn	
per capita	US$	2,230	2,149	
Growth	%	2.2	-4.3	
Inflation	%	11.4	12.9	
Def bdgt	N	594bn	900bn	841bn
	US$	1.83bn	2.57bn	
US$1=N		325.00	350.50	

Real-terms defence budget trend (US$bn, constant 2015)

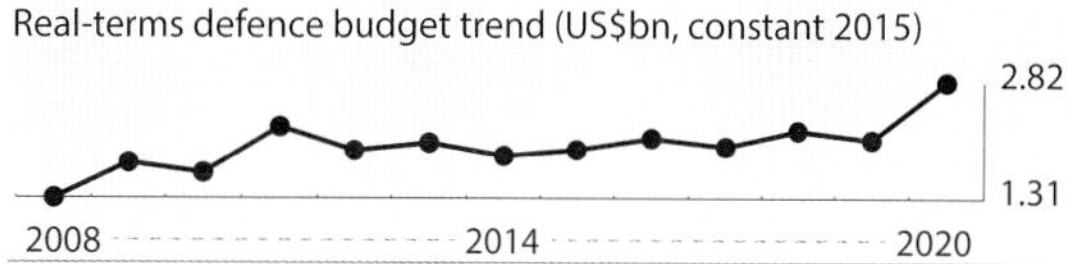

Population 214,028,302

Age	0–14	15–19	20–24	25–29	30–64	65 plus
Male	21.3%	5.7%	4.6%	3.7%	13.7%	1.6%
Female	20.4%	5.5%	4.5%	3.6%	13.8%	1.7%

Capabilities

Nigeria is the region's principal military power and faces numerous security challenges, including from Boko Haram and militants in the Delta. Reform initiatives have developed, after the challenge from Boko Haram exposed relative military weaknesses. There have been operational changes, including attempts to implement counter-insurgency tactics, forward-operating bases and quick-reaction groups. Nigeria is part of the Multi-National Joint Task Force and is a key member of the ECOWAS Standby Force. Military and security assistance is either discussed or under way with

Germany, the UK and the US. The UK bases its British Defence Staff West Africa in Nigeria. Efforts have been made to improve training, notably in the air force, with the establishment of Air Training Command and Ground Training Command. Contractors have also been used to improve training levels. Nigeria is able to mount regional operations, though its deployment capacities remain limited. Efforts are under way to refurbish stored aircraft and develop support capacities in advance of the arrival of *Super Tucano* aircraft. A number of small coastal-patrol boats have been acquired in light of security requirements in the Delta region. Nigeria is developing its defence-industrial capacity, including local production facilities for small arms and protected patrol vehicles.

ACTIVE 143,000 (Army 100,000 Navy 25,000 Air 18,000) Paramilitary 80,000

Reserves planned, none org

ORGANISATIONS BY SERVICE

Army 100,000

FORCES BY ROLE

SPECIAL FORCES

1 spec ops bn

3 (mobile strike team) spec ops units

1 ranger bn

MANOEUVRE

Armoured

1 (3rd) armd div (1 armd bde, 1 arty bde)

Mechanised

1 (1st) mech div (1 recce bn, 1 mech bde, 1 mot inf bde, 1 arty bde, 1 engr regt)

1 (2nd) mech div (1 recce bn, 1 armd bde, 1 arty bde, 1 engr regt)

1 (81st) composite div (1 recce bn, 1 mech bde, 1 arty bde, 1 engr regt)

Light

1 (6th) inf div (1 amph bde, 2 inf bde)

1 (7th) inf div (1 spec ops bn, 1 recce bn(-), 1 armd bde, 7 (task force) inf bde, 1 arty bde, 1 engr regt)

1 (8th Task Force) inf div (2 inf bde)

1 (82nd) composite div (1 recce bn, 3 mot inf bde, 1 arty bde, 1 engr regt)

1 (Multi-National Joint Task Force) bde (2 inf bn(-))

Other

1 (Presidential Guard) gd bde (4 gd bn)

AIR DEFENCE

1 AD regt

EQUIPMENT BY TYPE

ARMOURED FIGHTING VEHICLES

MBT 316+: 100 T-55†; 10 T-72AV; 31 T-72M1; 173 Vickers Mk 3; 2 + VT-4

LT TK 154 FV101 *Scorpion*

ASLT 3+ ST-1

RECCE 385: 88 AML-60; 40 AML-90; 70 EE-9 *Cascavel*; 45 ERC-90F1 *Lynx*; 50 FV721 *Fox*; 20 FV601 *Saladin* Mk2; 72 VBL

IFV 32: 10 BTR-4EN; 22 BVP-1

APC 601+

APC (T) 314: 249 4K-7FA *Steyr*; 65 MT-LB

APC (W) 178+: 10 FV603 *Saracen*; 110 AVGP *Grizzly* mod/*Piranha* I 6x6; 47 BTR-3UN; 5 BTR-80; some EE-11 *Urutu* (reported); 6 *Ezugwu*

PPV 109+: 14 *Caiman*; some *Conqueror*; some CS/VP3; some *Marauder*; 7+ *Maxxpro*; 8 Proforce *Ara*-1; 13 Proforce *Ara*-2; 23 REVA III 4×4; 10 Streit *Spartan*; 9 Streit *Cougar* (*Igirigi*); 25 Streit *Typhoon*

AUV 107 *Cobra*

ENGINEERING & MAINTENANCE VEHICLES

ARV 17+: AVGP *Husky*; 2 *Greif*; 15 Vickers ARV

VLB MTU-20; VAB

ANTI-TANK/ANTI-INFRASTRUCTURE

MSL • MANPATS *Shershen*

RCL 84mm *Carl Gustaf*; **106mm** M40A1

ARTILLERY 518+

SP 43+: **105mm** 4+ SH-5; **122mm** some SH-2; **155mm** 39 *Palmaria*

TOWED 104: **105mm** 49 M-56; **122mm** 48 D-30/D-74; **130mm** 7 M-46; (**155mm** 24 FH-77B in store)

MRL 122mm 41: 9 BM-21 *Grad*; 25 APR-21; 7 RM-70

MOR 330+: **81mm** 200; **82mm** 100; **120mm** 30+

AIR DEFENCE

SAM • Point-defence 16+: 16 *Roland*; *Blowpipe*; 9K32 *Strela*-2 (SA-7 *Grail*)‡

GUNS 89+

SP 23mm 29 ZSU-23-4 *Shilka*

TOWED 60+: **20mm** 60+; **23mm** ZU-23; **40mm** L/70

Navy 25,000 (incl Coast Guard)

Western Comd HQ located at Apapa; Eastern Comd HQ located at Calabar; Central Comd HQ located at Brass

EQUIPMENT BY TYPE

PRINCIPAL SURFACE COMBATANTS • FRIGATES

FFGHM (1 *Aradu* (GER MEKO 360) (non-operational) with 8 single lnchr with *Otomat* Mk1 AShM, 1 octuple *Albatros* lnchr with *Aspide* SAM, 2 triple 324mm ASTT with A244/S LWT, 1 127mm gun (capacity 1 med hel))

PATROL AND COASTAL COMBATANTS 117

CORVETTES • FSM (1 *Erinomi* (UK Vosper Mk 9) (non-operational) with 1 triple lnchr with *Seacat*† SAM, 1 twin 375mm Bofors ASW Rocket Launcher System A/S mor, 1 76mm gun)

PSOH 4: 2 *Centenary* with 1 76mm gun (capacity 1 Z-9 hel); 2 *Thunder* (ex-US *Hamilton*) with 1 76mm gun

PCFG 1 *Siri* (FRA *Combattante* IIIB)† with 2 twin lnchr with MM38 *Exocet* AShM, 1 76mm gun

PCF 2 *Siri* (FRA *Combattante* IIIB) with 1 76mm gun

PCO 4 *Kyanwa* (ex-US CG *Balsam*)

PCC 2 *Ekpe* (GER Lurssen 57m)† with 1 76mm gun

PBF 25: 1 C-*Falcon*; 12 *Manta* MkIII (Suncraft 17m); 4 *Manta* MkII; 3 *Shaldag* II; 2 *Torie* (Nautic Sentinel 17m); 3 *Wave Rider*

PB 79: 1 *Andoni*; 1 *Dorina* (FPB 98); 2 FPB 110 MkII; 8 *Okpoku* (FPB 72); 1 *Karaduwa*; 1 *Sagbama*; 2 *Sea Eagle* (Suncraft 38m); 15 *Stingray* (Suncraft 16m); 40 Suncraft 12m; 4 Swiftships; 2 *Town* (of which one laid up); 2 *Yola*†

MINE WARFARE • MINE COUNTERMEASURES 2

MCC 2 *Ohue* (ITA *Lerici* mod)†

AMPHIBIOUS • LANDING CRAFT 4

LCVP 4 *Stingray 20*

LOGISTICS AND SUPPORT • **AX** 1 *Prosperity*

Naval Aviation

EQUIPMENT BY TYPE

HELICOPTERS

MRH 2 AW139 (AB-139)

TPT • Light 3 AW109E *Power*†

Special Boat Service 200

EQUIPMENT BY TYPE

FORCES BY ROLE

SPECIAL FORCES

1 SF unit

Air Force 18,000

FORCES BY ROLE

Very limited op capability

FIGHTER/GROUND ATTACK

1 sqn with F-7 (F-7NI); FT-7 (FT-7NI)

MARITIME PATROL

1 sqn with ATR-42-500 MP; Do-128D-6 *Turbo SkyServant*; Do-228-100/200

TRANSPORT

2 sqn with C-130H *Hercules*; C-130H-30 *Hercules*; G-222

1 (Presidential) gp with B-727; B-737BBJ; BAe-125-800; Beech 350 *King Air*; Do-228-200; *Falcon* 7X; *Falcon* 900; Gulfstream IV/V

TRAINING

1 unit with *Air Beetle*†

1 unit with *Alpha Jet**

1 unit with L-39 *Albatros*†*; MB-339AN*

1 unit with *Super Mushshak*; DA40NG

1 hel unit with Mi-34 *Hermit* (trg)

ATTACK HELICOPTER

1 sqn with Mi-24/Mi-35 *Hind*†

TRANSPORT HELICOPTER

1 sqn with H215 (AS332) *Super Puma*; (AS365N) *Dauphin*; AW109LUH; H135

EQUIPMENT BY TYPE†

AIRCRAFT 60 combat capable

FTR 12: 10 F-7 (F-7NI); 2 FT-7 (FT-7NI)

ELINT 2 ATR-42-500 MP

ISR 1 Beech 350 *King Air*

TPT 34: **Medium** 5: 1 C-130H *Hercules* (4 more in store†); 1 C-130H-30 *Hercules* (2 more in store); 3 G.222† (2 more in store†); **Light** 20: 3 Beech 350 *King Air*; 1 Cessna 550 *Citation*; 8 Do-128D-6 *Turbo SkyServant*; 1 Do-228-100; 2 Do-228-101; 5 Do-228-200 (incl 2 VIP); **PAX** 9: 1 B-727; 1 B-737BBJ; 1 BAe 125-800; 2 *Falcon* 7X; 2 *Falcon* 900; 1 Gulfstream IV; 1 Gulfstream V

TRG 118: 58 *Air Beetle*† (up to 20 awaiting repair); 3 *Alpha Jet* A*; 10 *Alpha Jet* E*; 2 DA40NG; 23 L-39ZA *Albatros**†; 12 MB-339AN* (all being upgraded); 10 *Super Mushshak*

HELICOPTERS

ATK 16: 2 Mi-24P *Hind*; 4 Mi-24V *Hind*; 3 Mi-35 *Hind*; 2 Mi-35P *Hind*; 5 Mi-35M *Hind*

MRH 11+: 6 AW109LUH; 2 Bell 412EP; 3+ SA341 *Gazelle*

TPT 23: **Medium** 12: 2 AW101; 5 H215 (AS332) *Super Puma* (4 more in store); 3 AS365N *Dauphin*; 1 Mi-171Sh; 1 Mi-171E; **Light** 11: 4 H125 (AS350B) *Ecureuil*; 1 AW109; 2 AW109M; 1 Bell 205; 3 H135

UNINHABITED AERIAL VEHICLES 2+

CISR • Heavy 1+ CH-3

ISR 1: **Medium** (9 *Aerostar* non-operational); **Light** 1+ *Tsaigami*

AIR-LAUNCHED MISSILES

AAM • IR R-3 (AA-2 *Atoll*)‡; PL-9C

ASM AR-1

BOMBS • INS/GPS guided FT-9

Paramilitary ε80,000

Security and Civil Defence Corps 80,000

EQUIPMENT BY TYPE

ARMOURED FIGHTING VEHICLES

APC 80+

APC (W) 74+: 70+ AT105 *Saxon*†; 4 BTR-3U; UR-416

PPV 6 *Springbuck* 4x4

AIRCRAFT • TPT • Light 4: 1 Cessna 500 *Citation* I; 2 PA-31 *Navajo*; 1 PA-31-350 *Navajo Chieftain*

HELICOPTERS • TPT • Light 5: 2 Bell 212 (AB-212); 2 Bell 222 (AB-222); 1 Bell 429

DEPLOYMENT

CENTRAL AFRICAN REPUBLIC: UN • MINUSCA 3

DEMOCRATIC REPUBLIC OF THE CONGO: UN • MONUSCO 12

GAMBIA: ECOWAS • ECOMIG 197

GUINEA-BISSAU: ECOWAS • ECOMIB 100

LEBANON: UN • UNIFIL 1

MALI: UN • MINUSMA 82; 1 fd hospital

SOUTH SUDAN: UN • UNMISS 10

SUDAN: UN • UNAMID 6; UN • UNISFA 3

WESTERN SAHARA: UN • MINURSO 3

FOREIGN FORCES

United Kingdom 80 (trg teams)

Rwanda RWA

Rwandan Franc fr		2019	2020	2021
GDP	fr	9.11tr	9.89tr	
	US$	10.1bn	10.4bn	
per capita	US$	816	823	
Growth	%	9.4	2.0	
Inflation	%	2.4	6.9	
Def bdgt	fr	101bn	121bn	151bn
	US$	112m	127m	
US$1=fr		899.44	948.78	

Real-terms defence budget trend (US$m, constant 2015)

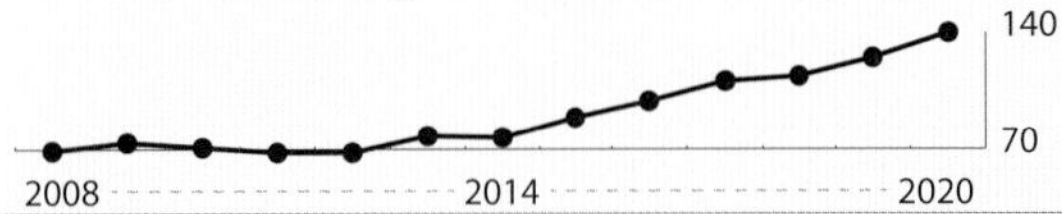

Population 12,712,431

Age	0–14	15–19	20–24	25–29	30–64	65 plus
Male	20.2%	5.4%	4.7%	3.8%	13.8%	1.1%
Female	19.8%	5.3%	4.7%	3.9%	15.8%	1.6%

Capabilities

Rwanda is one of the principal security actors in East Africa, with disciplined and well-trained armed forces. Their key missions are to defend territorial integrity and national sovereignty. The country fields a relatively large army, but units are lightly equipped, with little mechanisation. Rwanda signed a Mutual Defence Treaty with Kenya and Uganda in 2014 and participates in the East African Standby Force. A law on downsizing and demobilising elements of the armed forces was published in October 2015 and there have in recent years been official retirement ceremonies for those reaching rank-related retirement ages. The lack of fixed-wing aircraft limits the armed forces' ability to independently deploy much overseas beyond personnel. There have been some acquisitions of modern artillery and armoured vehicles. There is limited maintenance capacity but no defence manufacturing sector.

ACTIVE 33,000 (Army 32,000 Air 1,000) Paramilitary 2,000

ORGANISATIONS BY SERVICE

Army 32,000

FORCES BY ROLE

MANOEUVRE

Light

2 cdo bn

4 inf div (3 inf bde)

COMBAT SUPPORT

1 arty bde

EQUIPMENT BY TYPE

ARMOURED FIGHTING VEHICLES

MBT 34: 24 T-54/T-55; 10 *Tiran*-5

RECCE 106: ε90 AML-60/AML-90; 16 VBL

IFV 35+: BMP; 15 *Ratel*-90; 20 *Ratel*-60

APC 60+

APC (W) 20+: BTR; *Buffalo* (Panhard M3); 20 WZ-551 (reported)

PPV 40 RG-31 *Nyala*

AUV 30 *Cobra*

ENGINEERING & MAINTENANCE VEHICLES

ARV T-54/T-55 reported

ANTI-TANK/ANTI-INFRASTRUCTURE

MSL • SP HJ-9A (on *Cobra*)

ARTILLERY 177+

SP 17: **122mm** 12: 6 CS/SH-1; 6 SH-3; **155mm** 5 ATMOS 2000

TOWED 35+: **105mm** some; **122mm** 6 D-30; **152mm** 29 Type-54 (D-1)†

MRL 10: **122mm** 5 RM-70; **160mm** 5 LAR-160

MOR 115: **81mm**; **82mm**; **120mm**

AIR DEFENCE SAM • Point-defence 9K32 *Strela*-2 (SA-7 *Grail*)‡

GUNS ε150: **14.5mm**; **23mm**; **37mm**

Air Force ε1,000

FORCES BY ROLE

ATTACK/TRANSPORT HELICOPTER

1 sqn with Mi-17/Mi-17MD/Mi-17V-5/Mi-17-1V *Hip* H; Mi-24P/V *Hind*

EQUIPMENT BY TYPE

HELICOPTERS

ATK 5: 2 Mi-24V *Hind* E; 3 Mi-24P *Hind*

MRH 12: 1 AW139; 4 Mi-17 *Hip* H; 1 Mi-17MD *Hip* H; 1 Mi-17V-5 *Hip* H; 5 Mi-17-1V *Hip* H

TPT • Light 1 AW109S

Paramilitary

District Administration Security Support Organ ε2,000

DEPLOYMENT

CENTRAL AFRICAN REPUBLIC: UN • MINUSCA 1,357; 2 inf bn; 1 fd hospital

SOUTH SUDAN: UN • UNMISS 1,790; 3 inf bn; 2 hel sqn

SUDAN: UN • UNAMID 1,131; 2 inf bn; **UN** • UNISFA 7

Senegal SEN

CFA Franc BCEAO fr		2019	2020	2021
GDP	fr	13.8tr	14.0tr	
	US$	23.6bn	24.4bn	
per capita	US$	1,446	1,455	
Growth	%	5.3	-0.7	
Inflation	%	1.0	2.0	
Def bdgt	fr	201bn	199bn	
	US$	342m	346m	
US$1=fr		585.89	573.82	

Real-terms defence budget trend (US$m, constant 2015)

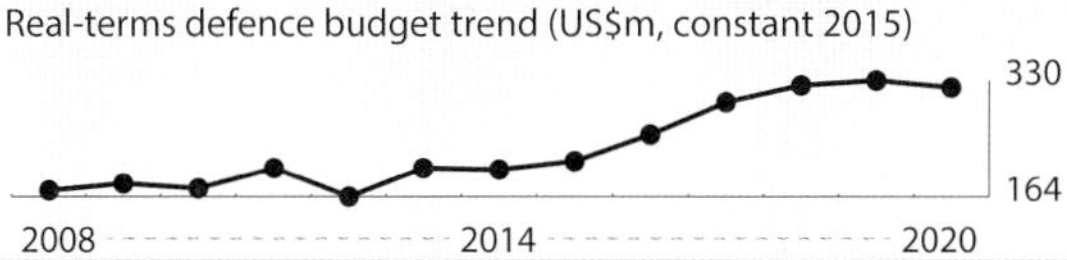

Population 15,736,368

Age	0–14	15–19	20–24	25–29	30–64	65 plus
Male	20.3%	5.5%	4.7%	4.0%	12.6%	1.3%
Female	20.1%	5.5%	4.7%	4.2%	15.4%	1.8%

Capabilities

Senegal's armed forces have strong international relationships and are experienced in foreign deployments. Their focus is internal and border security, including counter-insurgency in the country's south and Islamist activity in neighbouring states, as well as combating narcotics trafficking. Under the 'Horizon 2025' programme, the defence authorities intend to reorganise and re-equip key defence organisations and renew infrastructure. Areas for improvement include mobility and firepower. Despite limited resources, there are plans to improve operational capabilities and training. France remains Senegal's principal defence partner and retains a military presence in the country. French military forces deliver training assistance to the armed forces and gendarmerie, including in search and rescue. The US also provides security assistance, including to the national police and gendarmerie. A US-funded counter-terrorism training centre was opened in September 2018. The UK trains personnel in tasks relating to international peacekeeping operations. The armed forces are able to deploy personnel using organic airlift, but short-notice movements of heavy equipment would be problematic without external assistance. Air-force and naval modernisation is a priority. A contract was signed in 2019 for three OPVs for the navy, but deliveries of four L-39NG light attack aircraft have reportedly been delayed. Barring limited maintenance facilities, the country has no domestic defence-industrial capability.

ACTIVE 13,600 (Army 11,900 Navy 950 Air 750)
Paramilitary 5,000

Conscript liability Selective conscription, 24 months

ORGANISATIONS BY SERVICE

Army 11,900 (incl conscripts)

7 Mil Zone HQ

FORCES BY ROLE

MANOEUVRE

Reconnaissance

5 armd recce bn

Light

1 cdo bn

6 inf bn

Air Manoeuvre

1 AB bn

Other

1 (Presidential Guard) horse cav bn

COMBAT SUPPORT

1 arty bn

1 engr bn

3 construction coy

1 sigs bn

COMBAT SERVICE SUPPORT

1 log bn

1 med bn

1 trg bn

EQUIPMENT BY TYPE

ARMOURED FIGHTING VEHICLES

ASLT 27 PTL-02 *Assaulter*

RECCE 165: 30 AML-60; 74 AML-90; 20 BRDM-2; 10 M8; 4 M20; 27 RAM Mk3

IFV 26 *Ratel*-20

APC 91

APC (T) 12 M3 half-track

APC (W) 22: 2 *Oncilla;* 16 Panhard M3; 4 WZ-551 (CP)

PPV 57: 8 *Casspir;* 39 *Puma* M26-15; 10 *Puma* M36

ENGINEERING & MAINTENANCE VEHICLES

ARV 2 *Puma* M36 ARV

ANTI-TANK/ANTI-INFRASTRUCTURE

MSL • MANPATS *Milan*

ARTILLERY 82

TOWED 20: **105mm** 6 HM-2/M101; **155mm** 14: ε6 Model-50; 8 TR-F1

MRL 122mm 6 BM-21 *Grad* (UKR *Bastion*-1 mod)

MOR 56: **81mm** 24; **120mm** 32

AIR DEFENCE • GUNS • TOWED 39: **14.5mm** 6 ZPU-4 (tch); **20mm** 21 M693; **40mm** 12 L/60

Navy (incl Coast Guard) 950

FORCES BY ROLE

SPECIAL FORCES

1 cdo coy

EQUIPMENT BY TYPE

PATROL AND COASTAL COMBATANTS 7

PCO 1 *Fouladou* (OPV 190 Mk II)

PCC 1 *Njambour* (FRA SFCN 59m) with 2 76mm gun

PBF 3: 1 *Ferlo* (RPB 33); 2 *Shaldag* II

PB 2: 1 *Conejera;* 1 *Kedougou*

AMPHIBIOUS • LANDING CRAFT 2

LCT 2 EDIC 700

LOGISTICS AND SUPPORT• AG 1

Air Force 750

FORCES BY ROLE

MARITIME PATROL/SEARCH & RESCUE

1 sqn with C-212 *Aviocar;* CN235; Bell 205 (UH-1H *Iroquois*)

ISR

1 unit with BN-2T *Islander* (anti-smuggling patrols)

TRANSPORT

1 sqn with B-727-200 (VIP); F-27-400M *Troopship*

TRAINING
1 sqn with R-235 *Guerrier**; TB-30 *Epsilon*; KA-1S*
ATTACK/TRANSPORT HELICOPTER
1 sqn with AS355F *Ecureuil* II; Bell 206; Mi-35P *Hind*; Mi-171Sh

EQUIPMENT BY TYPE
AIRCRAFT 3 combat capable
TPT 10: **Light** 8: 1 BN-2T *Islander* (govt owned, mil op); 1 C-212-100 *Aviocar*; 2 CN235; 2 Beech B200 *King Air*; 2 F-27-400M *Troopship* (3 more in store); **PAX** 2: 1 A319; 1 B-727-200 (VIP)
TRG 9: 2+ KA-1S*; 1 R-235 *Guerrier**; 6 TB-30 *Epsilon*; KA-1S*
HELICOPTERS
ATK 4: 2 Mi-24V *Hind* D; 2 Mi-35P *Hind*
MRH 1 AW139
TPT 8: **Medium** 2 Mi-171Sh; **Light** 6: 1 AS355F *Ecureuil* II; 1 Bell 205 (UH-1H *Iroquois*); 2 Bell 206; 2 PZL Mi-2 *Hoplite*

Paramilitary 5,000

Gendarmerie 5,000

EQUIPMENT BY TYPE
ARMOURED FIGHTING VEHICLES
RECCE 13: 2 *Bastion Patsas*; 11 RAM Mk3
APC 56
APC (W) 24: 7 *Bastion* APC; 5 EE-11 *Urutu*; 12 VXB-170†
PPV 32: 24 *Ejder Yalcin*; 8 *Gila*

DEPLOYMENT

CENTRAL AFRICAN REPUBLIC: UN • MINUSCA 5

DEMOCRATIC REPUBLIC OF THE CONGO: UN • MONUSCO 5

GAMBIA: ECOWAS • ECOMIG 250

MALI: UN • MINUSMA 1,001; 1 inf bn; 1 engr coy

SOUTH SUDAN: UN • UNMISS 2

SUDAN: UN • UNAMID 1

FOREIGN FORCES

France 350; 1 *Falcon* 50MI
Spain *Operation Barkhane* 60; 2 C295M

Seychelles SYC

Seychelles Rupee SR		2019	2020	2021
GDP	SR	23.2bn	20.7bn	
	US$	1.65bn	1.20bn	
per capita	US$	17,127	12,323	
Growth	%	3.9	-13.8	
Inflation	%	1.8	3.9	
Def exp	SR	n.k	n.k	
	US$	n.k	n.k	
US$1=SR		14.05	17.30	

Population 95,981

Age	0–14	15–19	20–24	25–29	30–64	65 plus
Male	9.7%	3.2%	3.3%	4.0%	28.1%	3.4%
Female	9.2%	2.9%	2.9%	3.5%	25.0%	4.9%

Capabilities

The Seychelles maintains one of the smallest standing armed forces in the world. Its proximity to key international shipping lanes increases its strategic significance. The Seychelles People's Defence Force (PDF) primarily focuses on maritime security and counter-piracy operations. The country hosts US military forces conducting maritime-patrol activities on a rotational basis, including the operation of unarmed UAVs. India maintains strong defence ties with the Seychelles, donating equipment, providing maintenance and supporting efforts to enhance its maritime-patrol and surveillance capability. There are plans to improve defence cooperation with China, which has already led to some equipment deliveries. The Seychelles continues to participate in and host a number of multi-national maritime-security exercises. The PDF does not deploy overseas and has a limited capacity to deploy and support troops operating in the archipelago. Modern platforms in the air force and coastguard comprise donations from China, India and the UAE. There are limited maintenance facilities but no domestic defence manufacturing sector.

ACTIVE 420 (Land Forces 200; Coast Guard 200; Air Force 20)

ORGANISATIONS BY SERVICE

People's Defence Force

Land Forces 200

FORCES BY ROLE
SPECIAL FORCES
1 SF unit
MANOEUVRE
Light
1 inf coy
Other
1 sy unit
COMBAT SUPPORT
1 MP unit

EQUIPMENT BY TYPE
ARMOURED FIGHTING VEHICLES
RECCE 6 BRDM-2†
ARTILLERY • **MOR 82mm** 6 M-43†

AIR DEFENCE • GUNS • TOWED 14.5mm ZPU-2†; ZPU-4†; **37mm** M-1939†

Coast Guard 200 (incl 80 Marines)

EQUIPMENT BY TYPE

PATROL AND COASTAL COMBATANTS 10

PCO 3: 1 *Andromache* (ITA *Pichiotti* 42m); 2 *Topaz* (ex-IND *Trinkat*)

PBF 3: 1 *Hermes* (ex-IND *Coastal Interceptor Craft*); 2 *Wave Rider*

PB 4: 1 *Etoile* (*Shanghai* II mod); 2 *Le Vigilant* (ex-UAE Rodman 101); 1 *Fortune* (UK *Tyne*)

Air Force 20

EQUIPMENT BY TYPE

AIRCRAFT

TPT • Light 5: 1 DHC-6-320 *Twin Otter*; 2 Do-228; 2 Y-12

Sierra Leone SLE

Sierra Leonean Leone L		2019	2020	2021
GDP	L	38.0tr	41.5tr	
	US$	4.21bn	4.14bn	
per capita	US$	539	518	
Growth	%	5.4	-3.1	
Inflation	%	14.8	15.7	
Def bdgt	L	96.0bn	70.8bn	120bn
	US$	10.6m	7.1m	
US$1=L		9017.35	10022.94	

Real-terms defence budget trend (US$m, constant 2015)

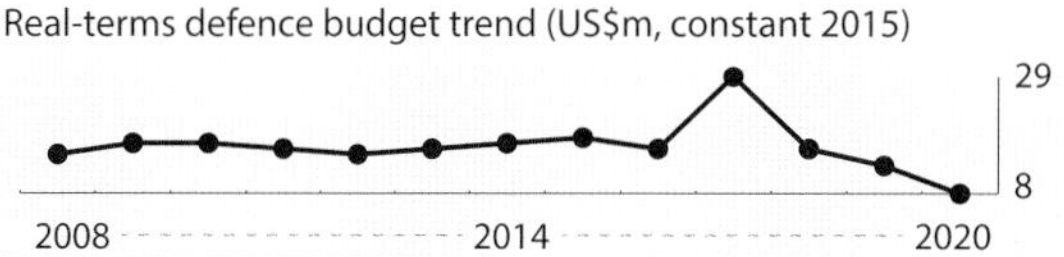

Population	6,624,933

Age	0–14	15–19	20–24	25–29	30–64	65 plus
Male	20.7%	5.1%	4.1%	3.6%	13.6%	1.5%
Female	20.7%	5.2%	4.4%	3.9%	14.9%	2.2%

Capabilities

The armed forces' primary task is to ensure internal and border security and provide forces for continental peacekeeping missions. With international support, there remains a focus on building defence institutions, generating formal defence documentation and improving planning functions. The UK is heavily involved in supporting the development of Sierra Leone's security institutions and improving training. UK training is also intended to boost the capacity of the police force. Freetown's Horton Academy trains local and regional personnel in SSR issues, as well as general military training tasks for local staff. Defence ties with China include personnel exchanges and support for the development of military infrastructure. The armed forces' ability to deploy anything other than small units is constrained by force size and logistics-support capacity. Limited capability in areas including air and maritime surveillance was modestly boosted by the arrival of additional inshore-patrol craft in 2020. There is limited maintenance capacity and no defence-manufacturing capability.

ACTIVE 8,500 (Joint 8,500)

ORGANISATIONS BY SERVICE

Armed Forces 8,500

FORCES BY ROLE

MANOEUVRE

Reconnaissance

1 recce unit

Light

3 inf bde (total: 12 inf bn)

COMBAT SUPPORT

1 engr regt

1 int unit

1 MP unit

1 sigs unit

COMBAT SUPPORT

1 log unit

1 fd hospital

EQUIPMENT BY TYPE

ARMOURED FIGHTING VEHICLES

APC • PPV 4: 3 *Casspir*; 1 *Mamba* Mk5

ANTI-TANK/ANTI-INFRASTRUCTURE

RCL 84mm *Carl Gustaf*

ARTILLERY 37

TOWED 122mm 6 Type-96 (D30)

MOR 31: **81mm** ε27; **82mm** 2; **120mm** 2

HELICOPTERS • MRH 2 Mi-17 *Hip* H/Mi-8 *Hip*†

AIR DEFENCE • GUNS 14.5mm 3

Maritime Wing ε200

EQUIPMENT BY TYPE

PATROL AND COASTAL COMBATANTS 2

PB 2: 1 Type-062/I (ex-PRC *Shanghai* III)†; 1 *Isle of Man*

DEPLOYMENT

CENTRAL AFRICAN REPUBLIC: UN • MINUSCA 5

DEMOCRATIC REPUBLIC OF THE CONGO: UN • MONUSCO 2

LEBANON: UN • UNIFIL 3

MALI: UN • MINUSMA 25

SOMALIA: UN • UNSOM 1; **UN** • UNSOS 1

SOUTH SUDAN: UN • UNMISS 1

SUDAN: UN • UNAMID 2; **UN** • UNISFA 6

Somalia SOM

Somali Shilling sh		2019	2020	2021
GDP	sh	4.94bn	4.92bn	
	US$	4.94bn	4.92bn	
per capita	US$	n.k.	n.k.	
Growth	%	2.9	-1.5	
Inflation	%	n.k.	n.k.	
Def bdgt	sh	n.k.	n.k.	
	US$	n.k.	n.k.	
US$1=sh		1.00	1.00	

*Definitive economic data unavailable

Population 11,757,124

Age	0–14	15–19	20–24	25–29	30–64	65 plus
Male	21.2%	5.4%	4.6%	3.7%	14.7%	0.9%
Female	21.2%	5.4%	4.5%	3.5%	13.7%	1.4%

Capabilities

Internal stability remains fragile following decades of conflict and insurgency, with al-Shabaab and other extremist groups active in the country. Deployed international forces look to provide security, stabilisation and capacity-building assistance, with a transition plan in place for the country to assume security responsibility in 2021. The Somali National Army (SNA) remains weak in terms of both organisation and military capability although a multi-clan US-mentored force, the Danab Brigade, has displayed greater capability and is slated for modest expansion. US forces are deployed independently to Somalia and target militant groups. Plans to professionalise, legitimatise and unite the loose collections of clan-based militia groups that form the SNA have yet to be fully realised. Although training programmes for enlisted and, more recently, senior commissioned personnel have been delivered by a number of countries, organisations and private-security companies, there are no common training standards throughout the army. There is no capacity to deploy beyond national borders, while there is minimal national infrastructure available to support domestic operations. The equipment inventory is limited and eclectic, and government plans to re-establish and equip Somalia's air and maritime forces remain unfulfilled. There is no domestic defence-industrial capability.

ACTIVE 19,800 (Army 19,800)

ORGANISATIONS BY SERVICE

Army 19,800 (plus further militias (to be integrated))

FORCES BY ROLE

COMMAND

4 div HQ

MANOEUVRE

Light

Some cdo bn(+)

12 inf bde (3 inf bn)

2 indep inf bn

Other

1 gd bn

EQUIPMENT BY TYPE

ARMOURED FIGHTING VEHICLES

APC 56+

APC (W) 38+: 25+ AT-105 *Saxon*; 13 *Bastion* APC; Fiat 6614

PPV 18+: *Casspir*; MAV-5; 9+ *Kirpi*; 9+ *Mamba* Mk5; RG-31 *Nyala*

AUV 12 *Tiger* 4×4

Paramilitary

Coast Guard

All operational patrol vessels under 10t FLD

FOREIGN FORCES

Under UNSOM command unless stated

Burundi AMISOM 4,000; 5 inf bn

Djibouti AMISOM 1,800; 2 inf bn

Ethiopia AMISOM 4,000; 6 inf bn

Finland EUTM Somalia 10

India 1

Italy EUTM Somalia 148

Kenya AMISOM 4,000; 3 inf bn • UNSOS 1

Mauritania UNSOS 1

Pakistan UNSOS 1

Portugal EUTM Somalia 4

Romania EUTM Somalia 1

Serbia EUTM Somalia 6

Sierra Leone 1 • UNSOS 1

Spain EUTM Somalia 20

Sweden EUTM Somalia 9

Turkey 1 • Army: 200 (trg base)

Uganda 624; 1 sy bn • AMISOM 5,800; 7 inf bn • UNSOS 1

United Kingdom 1 • UNSOS 13 • EUTM Somalia 3; Army: 65 (trg team)

United States Africa Command 700

TERRITORY WHERE THE GOVERNMENT DOES NOT EXERCISE EFFECTIVE CONTROL

Data presented here represents the de facto situation. This does not imply international recognition as a sovereign state. Much of this equipment is in poor repair or inoperable

Somaliland

Army ε12,500

FORCES BY ROLE

MANOUEVRE

Armoured

2 armd bde

Mechanised

1 mech inf bde

Light

14 inf bde

COMBAT SUPPORT

2 arty bde

COMBAT SERVICE SUPPORT
1 spt bn

EQUIPMENT BY TYPE†

ARMOURED FIGHTING VEHICLES
MBT T-54/55
RECCE Fiat 6616
APC • APC(W) Fiat 6614

ARTILLERY • **MRL** various incl BM-21 *Grad*

AIR DEFENCE • **GUNS • 23mm** ZU-23-2

Ministry of the Interior

Coast Guard 600

All operational patrol vessels under 10t FLD

Puntland

Army ε3,000 (to be integrated into Somali National Army)

Maritime Police Force ε1,000

EQUIPMENT BY TYPE

AIRCRAFT • **TPT** 4: **Light** 3 Ayres S2R; **PAX** 1 DC-3

HELICOPTERS • **MRH** SA316 *Alouette* III

PATROL AND COASTAL COMBATANTS
All operational patrol vessels under 10t FLD

South Africa RSA

South African Rand R		2019	2020	2021
GDP	R	5.08tr	4.83tr	
	US$	351bn	283bn	
per capita	US$	5,978	4,736	
Growth	%	0.2	-8.0	
Inflation	%	4.1	3.3	
Def bdgt	R	50.5bn	55.3bn	
	US$	3.50bn	3.24bn	
US$1=R		14.45	17.10	

Real-terms defence budget trend (US$bn, constant 2015)

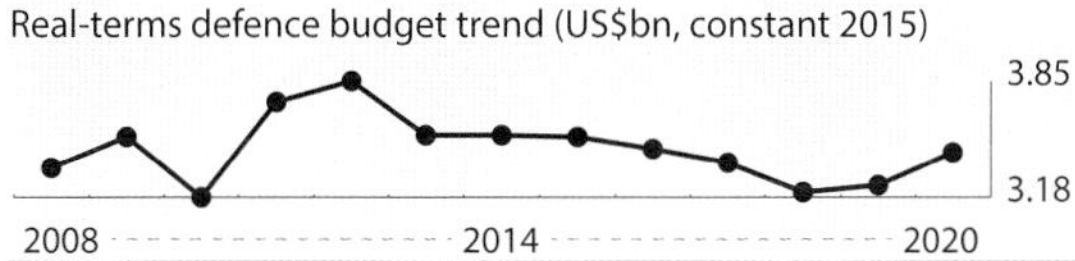

Population 56,463,617

Age	0–14	15–19	20–24	25–29	30–64	65 plus
Male	14.0%	4.2%	4.1%	4.4%	20.1%	2.6%
Female	14.0%	4.2%	4.3%	4.6%	20.0%	3.5%

Capabilities

South Africa's armed forces are on paper the most capable in the region, but economic and structural problems are eroding capability. Principal roles include maintaining territorial integrity and supporting the police service. The Department of Defence Strategic Plan 2015–2020 is the South African National Defence Force's (SANDF) primary policy instrument. The first of five strategic-planning milestones is to arrest the decline of critical military capabilities. However, the 2018/2019 Defence Annual Report said that funding constraints meant that the armed forces' level of ambition was not 'sustainable' and that the goals outlined in the 2015 Defence Review were not being met. South Africa contributes to UN operations and has been a key component of the Force Intervention Brigade in the DRC since its inception. It is a member of the SADC Standby Force. Historically, South African forces have also played a key role in training and supporting other regional forces. The SANDF can independently deploy its forces and participates in peacekeeping missions and national and multinational exercises. The SANDF has a credible modernisation plan on paper, but funding problems undermine these ambitions and some programmes are behind schedule. Budget cuts are likely to have an adverse effect on training. There is concern in the army over the obsolescence of principal equipment. South Africa has the continent's most capable defence industry, including the state-owned Armaments Corporation of South Africa and weapons manufacturer Denel. As of late 2020, Denel was facing serious challenges and there were reports that the Treasury was extending additional funding to the firm.

ACTIVE 74,500 (Army 38,200 Navy 6,800 Air 9,800 South African Military Health Service 7,400 Other 12,300)

RESERVE 15,050 (Army 12,250 Navy 850 Air 850 South African Military Health Service Reserve 1,100)

ORGANISATIONS BY SERVICE

Space

EQUIPMENT BY TYPE

SATELLITES • **ISR** 1 *Kondor*-E

Army 38,200

FORCES BY ROLE

Regt are bn sized. A new army structure is planned with 3 mixed regular/reserve divisions (1 mechanised, 1 motorised and 1 contingency) comprising 12 brigades (1 armoured, 1 mechanised, 7 motorised, 1 airborne, 1 air-landed and 1 sea landed)

COMMAND
2 bde HQ

SPECIAL FORCES
2 SF regt(-)

MANOEUVRE
Reconnaissance
1 armd recce regt
Armoured
1 tk regt(-)
Mechanised
2 mech inf bn
Light
8 mot inf bn
1 lt inf bn
Air Manoeuvre
1 AB bn
1 air mob bn
Amphibious
1 amph bn

COMBAT SUPPORT
1 arty regt
1 engr regt
1 construction regt
3 sigs regt

COMBAT SERVICE SUPPORT
1 engr spt regt
AIR DEFENCE
1 ADA regt

Reserve 12,250 reservists (under-strength)

FORCES BY ROLE
MANOEUVRE
Reconnaissance
3 armd recce regt
Armoured
4 tk regt
Mechanised
6 mech inf bn
Light
14 mot inf bn
3 lt inf bn (converting to mot inf)
Air Manoeuvre
1 AB bn
2 air mob bn
Amphibious
1 amph bn
COMBAT SUPPORT
7 arty regt
2 engr regt
AIR DEFENCE
5 AD regt

EQUIPMENT BY TYPE
ARMOURED FIGHTING VEHICLES
MBT 24 *Olifant* 2 (133 *Olifant* 1B in store)
ASLT 50 *Rooikat*-76 (126 in store)
IFV 534 *Ratel*-20/*Ratel*-60/*Ratel*-90
APC • PPV 810: 370 *Casspir*; 440 *Mamba*
ENGINEERING & MAINTENANCE VEHICLES
ARV *Gemsbok*
VLB *Leguan*
MW *Husky*
ANTI-TANK/ANTI-INFRASTRUCTURE
MSL
SP ZT-3 *Swift*
MANPATS *Milan* ADT/ER
RCL 106mm M40A1 (some SP)
ARTILLERY 1,240
SP 155mm 2 G-6 (41 in store)
TOWED 155mm 6 G-5 (66 in store)
MRL 127mm 6 *Valkiri* Mk II MARS *Bataleur*; (26 *Valkiri* Mk I and 19 *Valkiri* Mk II in store)
MOR 1,226: **81mm** 1,190 (incl some SP on *Casspir* & *Ratel*); **120mm** 36
UNINHABITED AERIAL VEHICLES
ISR • Light up to 4 *Vulture*
AIR DEFENCE
SAM • Point-defence *Starstreak*
GUNS 40
SP 23mm (36 *Zumlac* in store)
TOWED 35mm 40: 22 GDF-002; 18 GDF-005A/007

Navy 6,800

Fleet HQ and Naval base located at Simon's Town; Naval stations located at Durban and Port Elizabeth

EQUIPMENT BY TYPE
SUBMARINES 2
SSK 2 *Heroine* (Type-209/1400 mod) (1 additional boat in refit since 2014, awaiting funds to complete) with 8 533mm TT with SUT 264 HWT
PRINCIPAL SURFACE COMBATANTS • FRIGATES 4
FFGHM 4 *Valour* (MEKO A200) with 2 quad lnchr with MM40 *Exocet* Block 2 AShM (upgrade to Block 3 planned); 2 16-cell VLS with *Umkhonto*-IR SAM, 1 Denel Dual Purpose Gun (DPG) CIWS, 1 76mm gun (capacity 1 *Super Lynx* 300 hel)
PATROL AND COASTAL COMBATANTS 4
PCC 3 *Warrior* (ISR *Reshef*) with 1 76mm gun
PB 1 *Tobie* (2 additional in reserve)
MINE WARFARE • MINE COUNTERMEASURES 3
MHC 3 *River* (GER *Navors*) (limited operational roles; training and dive support) (1 additional in reserve)
LOGISTICS AND SUPPORT 2
AGHS 1 *Protea* (UK *Hecla*) with 1 hel landing platform
AORH 1 *Drakensberg* (capacity 2 *Oryx* hels; 100 troops)

Maritime Reaction Squadron

FORCES BY ROLE
MANOEUVRE
Amphibious
1 mne patrol gp
1 diving gp
1 mne boarding gp
COMBAT SERVICE SUPPORT
1 spt gp

Air Force 9,800

Air Force HQ, Pretoria, and 4 op gps
Command & Control: 2 Airspace Control Sectors, 1 Mobile Deployment Wg, 1 Air Force Command Post

FORCES BY ROLE
FIGHTER/GROUND ATTACK
1 sqn with *Gripen* C/D (JAS-39C/D)
GROUND ATTACK/TRAINING
1 sqn with *Hawk* Mk120*
TRANSPORT
1 (VIP) sqn with B-737 BBJ; Cessna 550 *Citation* II; *Falcon* 50; *Falcon* 900
1 sqn with C-47TP
2 sqn with Beech 200/300 *King Air*; C-130B/BZ; C-212
ATTACK HELICOPTER
1 (cbt spt) sqn with AH-2 *Rooivalk*
TRANSPORT HELICOPTER
4 (mixed) sqn with AW109; BK-117; *Oryx*

EQUIPMENT BY TYPE
AIRCRAFT 50 combat capable
FGA 26: 17 *Gripen* C (JAS-39C); 9 *Gripen* D (JAS-39D)
TPT 23: **Medium** 6: 2 C-130B *Hercules*; 4 C-130BZ *Hercules*; **Light** 13: 3 Beech 200C *King Air*; 1 Beech 300 *King Air*; 3 C-47TP (maritime); 2 C-212-200 *Aviocar*†; 1 C-212-300 *Aviocar*†; 2 Cessna 550 *Citation* II; 1 PC-12; (9

Cessna 208 *Caravan* in store) **PAX** 4: 1 B-737BBJ; 2 *Falcon* 50; 1 *Falcon* 900
TRG 59: 24 *Hawk* Mk120*; 35 PC-7 Mk II *Astra*
HELICOPTERS
ATK 11 AH-2 *Rooivalk*
MRH 4 *Super Lynx* 300
TPT 69: **Medium** 36 *Oryx*; **Light** 33: 25 AW109; 8 BK-117
AIR-LAUNCHED MISSILES • **AAM** • **IIR** IRIS-T
BOMBS • **Laser-guided** GBU-12 *Paveway* II

Ground Defence

FORCES BY ROLE
MANOEUVRE
Other
12 sy sqn (SAAF regt)

South African Military Health Service 7,400

DEPLOYMENT

DEMOCRATIC REPUBLIC OF THE CONGO: UN • MONUSCO • *Operation Mistral* 1,133; 1 inf bn; 1 hel sqn

MOZAMBIQUE CHANNEL: Navy • 1 FFGHM

SUDAN: UN • UNAMID (*Operation Cordite*) 4

South Sudan SSD

South Sudanese Pound ssp		2019	2020	2021
GDP	ssp	780bn	728bn	
	US$	4.93bn	4.18bn	
per capita	US$	369	303	
Growth	%	0.9	4.1	
Inflation	%	51.2	27.1	
Def bdgt [a]	ssp	10.7bn	16.4bn	
	US$	67.9m	93.9m	
US$1=ssp		158.06	174.34	

[a] Defence and Veterans Affairs budget

Real-terms defence budget trend (US$m, constant 2015)

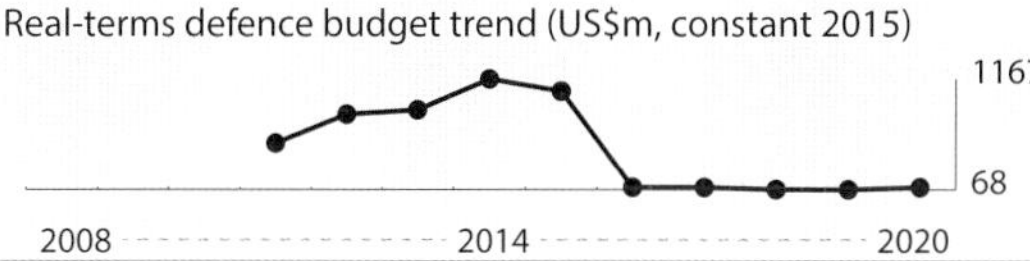

Population 10,561,244

Age	0–14	15–19	20–24	25–29	30–64	65 plus
Male	21.2%	5.9%	5.1%	4.1%	13.8%	1.5%
Female	20.4%	5.9%	4.5%	3.5%	13.2%	1.1%

Capabilities

South Sudan's civil war formally ended in 2020. The deal built on a 2018 accord that laid out a peace framework for opposition and government forces, resulting in changes to the number and demarcation of federal states, which had been a stumbling block in peace negotiations. Although major fighting between government and opposition forces has halted, violence has continued between various armed groups and militias. Progress on professionalising and restructuring the rebranded South Sudan People's Defence Force (SSPDF) remains unclear, and allegations of child-soldier recruitment and the use of sexual and gender-based violence continue. In 2018, South Sudan reportedly signed a defence agreement with South Africa, including scope to cooperate on procurement and training. There is no capacity to deploy and sustain military units beyond national borders. Equipment is primarily of Soviet origin with some light arms of Chinese origin, and there have been efforts to expand the small air force. Sanctions remain in place, with both the EU and UN arms embargoes widened in 2018 to include all types of military equipment. South Sudan has no established domestic defence industry but has reportedly sought to develop an ammunition-manufacturing capacity in recent years.

ACTIVE 185,000 (Army 185,000)

ORGANISATIONS BY SERVICE

Army ε185,000

FORCES BY ROLE
3 military comd
MANOEUVRE
Light
8 inf div
COMBAT SUPPORT
1 engr corps

EQUIPMENT BY TYPE
ARMOURED FIGHTING VEHICLES
MBT 80+: some T-55†; 80 T-72AV†
APC • **PPV** Streit *Typhoon*; Streit *Cougar*; *Mamba*
ANTI-TANK/ANTI-INFRASTRUCTURE
MSL • **MANPATS** HJ-73; 9K115 *Metis* (AT-7 *Saxhorn*)
RCL 73mm SPG-9 (with SSLA)
ARTILLERY
SP 122mm 2S1 *Gvozdika*; **152mm** 2S3 *Akatsiya*
TOWED 130mm Some M-46
MRL 122mm BM-21 *Grad*; **107mm** PH-63
MOR 82mm; **120mm** Type-55 look-alike
AIR DEFENCE
SAM
Short-range 16 S-125 *Pechora* (SA-3 *Goa*) (reported)
Point-defence 9K32 *Strela*-2 (SA-7 *Grail*)‡; QW-2
GUNS 14.5mm ZPU-4; **23mm** ZU-23-2; **37mm** Type-65/74

Air Force

EQUIPMENT BY TYPE
AIRCRAFT 2 combat capable
TPT • **Light** 1 Beech 1900
TRG ε2 L-39 *Albatros**
HELICOPTERS
ATK 5: 2 Mi-24V *Hind*; 3 Mi-24V-SMB *Hind*
MRH 9 Mi-17 *Hip* H
TPT 3: **Medium** 1 Mi-172 (VIP); **Light** 2 AW109 (civ livery)

FOREIGN FORCES

All UNMISS, unless otherwise indicated
Australia 10
Azerbaijan 2
Bangladesh 1,618; 1 inf coy; 2 rvn coy; 2 engr coy
Benin 4
Bhutan 2
Bolivia 4

Brazil 11
Cambodia 79; 1 MP unit
Canada 9
China, People's Republic of 1,050; 1 inf bn; 1 engr coy; 1 fd hospital
Ecuador 3
Egypt 4
El Salvador 4
Ethiopia 2,131; 3 inf bn
Fiji 4
Gambia 4
Germany 7
Ghana 871; 1 inf bn
Guatemala 5
Guinea 2
India 2,378; 2 inf bn; 1 engr coy; 1 sigs coy; 1 fd hospital
Indonesia 4
Japan 6
Jordan 5
Kenya 16
Korea, Republic of 272; 1 engr coy
Kyrgyzstan 1
Liberia 2
Moldova 5
Mongolia 865; 1 inf bn
Morocco 1
Namibia 5
Nepal 1,737; 2 inf bn
New Zealand 3
Nigeria 10
Norway 15
Pakistan 11
Paraguay 3
Peru 5
Philippines 2
Poland 1
Romania 6
Russia 6
Rwanda 1,790; 3 inf bn; 2 hel sqn
Senegal 2
Sierra Leone 1
Sri Lanka 164; 1 fd hospital; 1 hel sqn
Switzerland 1
Tanzania 10
Thailand 266; 1 engr coy
Togo 2
Tunisia 2
Uganda 2
Ukraine 1
United Kingdom 7
United States 7
Vietnam 68
Zambia 11
Zimbabwe 5

Sudan SDN

Sudanese Pound sdg		2019	2020	2021
GDP	sdg	2.02tr	4.48tr	
	US$	33.4bn	32.6bn	
per capita	US$	772	735	
Growth	%	-2.5	-8.4	
Inflation	%	51.0	141.6	
Def exp	sdg	n.k	n.k	
	US$	n.k	n.k	
US$1=sdg		60.47	137.56	

Population 45,561,556

Age	0–14	15–19	20–24	25–29	30–64	65 plus
Male	21.3%	5.9%	4.7%	3.7%	12.9%	1.6%
Female	20.7%	5.8%	4.5%	3.5%	13.8%	1.4%

Capabilities

In April 2019, President Omar al-Bashir was ousted after 30 years in power and months of protests. By August, opposition forces and the armed forces comprised a Sovereign Council that is due to lead a transition to civilian rule. The armed forces remain focused on internal security as well as border issues, though a peace deal was signed with five rebel groups in August 2020. This could lead to some former rebels being integrated into the armed forces. The hybrid AU–UN mission, UNAMID, remains in Darfur. Sudan is part of the Saudi-led coalition intervention in Yemen. A defence agreement with Iran in 2008 reportedly included assistance in developing the domestic arms industry. The armed forces are conscript-based and will have gained operational experience from internal-security deployments and the Saudi-led coalition. By regional standards, Sudan's armed forces are relatively well equipped, with significant holdings of both ageing and modern systems. While there is a UN arms embargo in place, it is limited to equipment used within the Darfur region. Recent acquisitions have included Russian and Ukrainian government surplus, and also new Chinese jet trainers. The state-run Military Industry Corporation manufactures a range of ammunition, small arms and armoured vehicles for the domestic and export market. The majority of the corporation's products are based on older Chinese and Russian systems.

ACTIVE 104,300 (Army 100,000 Navy 1,300 Air 3,000) **Paramilitary 20,000**

Conscript liability 2 years for males aged 18–30

RESERVE NIL Paramilitary 85,000

ORGANISATIONS BY SERVICE

Space

EQUIPMENT BY TYPE

SATELLITES • ISR 1 SRSS-1

Army 100,000+

FORCES BY ROLE

SPECIAL FORCES

5 SF coy

MANOEUVRE
Reconnaissance
1 indep recce bde
Armoured
1 armd div
Mechanised
1 mech inf div
1 indep mech inf bde
Light
15+ inf div
6 indep inf bde
Air Manoeuvre
1 air aslt bde
Amphbious
1 mne div
Other
1 (Border Guard) sy bde
COMBAT SUPPORT
3 indep arty bde
1 engr div (9 engr bn)

EQUIPMENT BY TYPE
ARMOURED FIGHTING VEHICLES
MBT 465: 20 M60A3; 60 Type-59/Type-59D; 305 T-54/T-55; 70 T-72AV; 10 *Al-Bashier* (Type-85-IIM)
LT TK 115: 70 Type-62; 45 Type-63
RECCE 206: 6 AML-90; 70 BRDM-1/2; 50–80 FV701 *Ferret*; 30–50 FV601 *Saladin*
IFV 152+: 135 BMP-1/2; 10 BTR-3; 7 BTR-80A; WZ-523 IFV
APC 415+
APC (T) 66: 20-30 BTR-50; 36 M113
APC (W) 349+: 10 BTR-70M *Kobra* 2; 50–80 BTR-152; 20 OT-62; 50 OT-64; 3+ *Rakhsh*; 10 WZ-551; WZ-523; 55-80 V-150 *Commando*; 96 *Walid*
PPV some *Sarsar*-2
AUV 4+ Nimr *Ajban* 440A
ANTI-TANK/ANTI-INFRASTRUCTURE
MSL • MANPATS 9K11 *Malyutka* (AT-3 *Sagger*); HJ-8; 9K135 *Kornet* (AT-14 *Spriggan*)
RCL 106mm 40 M40A1
GUNS 76mm ZIS-3; **100mm** M-1944; **85mm** D-44
ARTILLERY 860+
SP 66: **122mm** 56 2S1 *Gvozdika*; **155mm** 10 Mk F3
TOWED 128+: **105mm** 20 M101; **122mm** 21+: 21 D-30; D-74; M-30; **130mm** 75 M-46/Type-59-I; **155mm** 12 M114A1
MRL 666+: **107mm** 477 Type-63; **122mm** 188: 120 BM-21 *Grad*; 50 *Saqr*; 18 Type-81; **302mm** 1+ WS-1
MOR 81mm; **82mm**; **120mm** AM-49; M-43; W86
AIR DEFENCE
SAM • Point-defence 4+: 9K32 *Strela*-2 (SA-7 *Grail*)‡; FN-6; 4+ 9K33 *Osa* (SA-8 *Gecko*)
GUNS 966+
SP 20: **20mm** 8 M163 *Vulcan*; 12 M3 VDAA
TOWED 946+: 740+ **14.5mm** ZPU-2/**14.5mm** ZPU-4/**37mm** Type-63/**57mm** S-60/**85mm** M-1944; **20mm** 16 M167 *Vulcan*; **23mm** 50 ZU-23-2; **37mm** 80 M-1939; (30 M-1939 unserviceable); **40mm** 60

Navy 1,300

EQUIPMENT BY TYPE
PATROL AND COASTAL COMBATANTS 11
PBR 4 *Kurmuk*
PB 7: 1 13.5m; 1 14m; 2 19m; 3 41m (PRC)
AMPHIBIOUS • LANDING CRAFT • LCVP 5
LOGISTICS AND SUPPORT 4
AG 3
AXL 1 *Petrushka* (ex-RUS)

Air Force 3,000

FORCES BY ROLE
FIGHTER
2 sqn with MiG-29SE/UB *Fulcrum*
FIGHTER/GROUND ATTACK
1 sqn with FTC-2000*
GROUND ATTACK
1 sqn with Su-24M/MR *Fencer*
1 sqn with Su-25K/Su-25UB *Frogfoot*
TRANSPORT
Some sqn with An-30 *Clank*; An-32 *Cline*; An-72 *Coaler*; An-74TK-200/300; C-130H *Hercules*; Il-76 *Candid*; Y-8
1 VIP unit with *Falcon* 20F; *Falcon* 50; *Falcon* 900; F-27; Il-62M *Classic*
TRAINING
1 sqn with K-8 *Karakorum**
ATTACK HELICOPTER
2 sqn with Mi-24/Mi-24P/Mi-24V/Mi-35P *Hind*
TRANSPORT HELICOPTER
2 sqn with Mi-8 *Hip*; Mi-17 *Hip* H; Mi-171
AIR DEFENCE
5 bty with S-75 *Dvina* (SA-2 *Guideline*)‡

EQUIPMENT BY TYPE
AIRCRAFT 59 combat capable
FTR 22: 20 MiG-29SE *Fulcrum* C; 2 MiG-29UB *Fulcrum* B
ATK 20: 6 Su-24M/MR *Fencer*; 14 Su-25K/UB *Frogfoot*; (15 A-5 *Fantan* in store)
ISR 2 An-30 *Clank*
TPT 24: **Heavy** 1 Il-76 *Candid*; **Medium** 6: 4 C-130H *Hercules*; 2 Y-8; **Light** 13: ε3 An-26 *Curl*; 2 An-32 *Cline*; 2 An-72 *Coaler*; 4 An-74TK-200; 2 An-74TK-300; **PAX** 4: 1 *Falcon* 20F (VIP); 1 *Falcon* 50 (VIP); 1 *Falcon* 900; 1 Il-62M *Classic*
TRG 21+: 6 FTC-2000*; 11 K-8 *Karakorum**; some SAFAT-03; 3 Utva-75
HELICOPTERS
ATK 40: 25 Mi-24 *Hind*; 2 Mi-24P *Hind*; 7 Mi-24V *Hind* E; 6 Mi-35P *Hind*
MRH ε3 Mi-17 *Hip* H
TPT 27: **Medium** 23: 21 Mi-8 *Hip*; 2 Mi-171; **Light** 4: 1 Bell 205; 3 Bo-105
UNINHABITED AERIAL VEHICLES
CISR • Heavy CH-3; CH-4
AIR DEFENCE • SAM • Medium-range: 90 S-75 *Dvina* (SA-2 *Guideline*)‡
AIR-LAUNCHED MISSILES • AAM • IR R-3 (AA-2 *Atoll*)‡; R-60 (AA-8 *Aphid*); R-73 (AA-11A *Archer*); **ARH** R-77 (AA-12A *Adder*)

Paramilitary 20,000

Popular Defence Force 20,000 (org in bn 1,000); 85,000 reservists (total 105,000)

mil wing of National Islamic Front

DEPLOYMENT

SAUDI ARABIA: *Operation Restoring Hope* 3 Su-24M/MR *Fencer*

YEMEN: *Operation Restoring Hope* 650; 1 mech BG; T-72AV, BTR-70M *Kobra* 2

FOREIGN FORCES

All UNAMID, unless otherwise indicated
Bangladesh 4 • UNISFA 3
Benin UNISFA 2
Bhutan 1 • UNISFA 2
Bolivia UNISFA 3
Brazil 1 • UNISFA 3
Burkina Faso UNISFA 1
Burundi UNISFA 4
Cambodia 1 • UNISFA 1
China, People's Republic of 367; 1 engr coy
Ecuador 1 • UNISFA 2
Egypt 161; 1 inf coy
El Salvador UNISFA 1
Ethiopia 808; 1 inf bn • UNISFA 3,663; 3 inf bn; 2 arty coy; 1 engr coy; 1 sigs coy; 8 fd hospital; 1 hel sqn
Gambia 135; 1 inf coy
Germany 1
Ghana 5 • UNISFA 8
Guatemala UNISFA 3
Guinea UNISFA 1
India UNISFA 3
Indonesia 3 • UNISFA 4
Jordan 4
Kenya 87; 1 MP coy
Kyrgyzstan 2 • UNISFA 1
Liberia UNISFA 2
Malawi 1 • UNISFA 2
Malaysia UNISFA 2
Mongolia 1 • UNISFA 5
Namibia 1 • UNISFA 6
Nepal 5 • UNISFA 8
Nigeria 6 • UNISFA 3
Pakistan 907; 1 inf bn; 2 fd hospital • UNISFA 2
Peru 1 • UNISFA 2
Russia UNISFA 2
Rwanda 1,131; 2 inf bn • UNISFA 7
Senegal 1
Sierra Leone 2 • UNISFA 6
South Africa 4
Tanzania 342; 1 inf bn(-) • UNISFA 3
Thailand 3
Togo 2
Uganda UNISFA 2
Ukraine UNISFA 3
Zambia 5 • UNISFA 3
Zimbabwe 4 • UNISFA 10

Tanzania TZA

Tanzanian Shilling sh		2019	2020	2021
GDP	sh	140tr	148tr	
	US$	60.8bn	64.1bn	
per capita	US$	1,080	1,106	
Growth	%	7.0	1.9	
Inflation	%	3.4	3.6	
Def bdgt	sh	1.67tr	1.85tr	
	US$	728m	801m	
US$1=sh		2300.51	2313.33	

Real-terms defence budget trend (US$m, constant 2015)

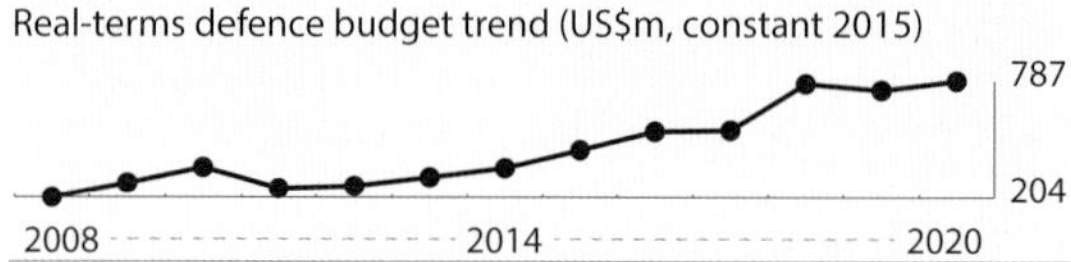

Population 58,552,845

Age	0–14	15–19	20–24	25–29	30–64	65 plus
Male	21.6%	5.6%	4.6%	3.8%	13.1%	1.3%
Female	21.1%	5.6%	4.6%	3.8%	13.2%	1.8%

Capabilities

Non-state actors pose the principal threat to Tanzania's security, with terrorism, poaching and piracy of concern. A developing relationship with China has led to a series of procurements and training contacts. There are also defence-related ties with Israel, Pakistan and Russia. The armed forces take part in multinational exercises in Africa and have provided some training assistance to other African forces. Training relationships also exist with extra-regional armed forces, including the US. Tanzania's contribution to the UN's Force Intervention Brigade in the eastern DRC, notably its special forces, will have provided lessons for force development. However, there is only a limited capacity to project power independently beyond the country's borders. Budget constraints have limited recapitalisation ambitions and, although heavy equipment is ageing, airlift capacity has improved with the delivery of new helicopters. There are local ammunition facilities, but otherwise Tanzania relies on imports for its military equipment.

ACTIVE 27,000 (Army 23,000 Navy 1,000 Air 3,000)
Paramilitary 1,400

Conscript liability Three months basic military training combined with social service, ages 18–23

RESERVE 80,000 (Joint 80,000)

ORGANISATIONS BY SERVICE

Army ε23,000

FORCES BY ROLE
SPECIAL FORCES
1 SF unit

MANOEUVRE
Armoured
1 tk bde
Light
5 inf bde
COMBAT SUPPORT
4 arty bn
1 mor bn
2 AT bn
1 engr regt (bn)
COMBAT SERVICE SUPPORT
1 log gp
AIR DEFENCE
2 ADA bn

EQUIPMENT BY TYPE†
ARMOURED FIGHTING VEHICLES
MBT 45: 30 T-54/T-55; 15 Type-59G
LT TK 57+: 30 FV101 *Scorpion*; 25 Type-62; 2+ Type-63A
RECCE 10 BRDM-2
APC • APC (W) 14: ε10 BTR-40/BTR-152; 4 Type-92
ANTI-TANK/ANTI-INFRASTRUCTURE
RCL 75mm Type-52 (M20)
GUNS 85mm 75 Type-56 (D-44)
ARTILLERY 344+
TOWED 130: **122mm** 100: 20 D-30; 80 Type-54-1 (M-30); **130mm** 30 Type-59-I
GUN/MOR 120mm 3+ Type-07PA
MRL 61+: **122mm** 58 BM-21 *Grad*; **300mm** 3+ A100
MOR 150: **82mm** 100 M-43; **120mm** 50 M-43

Navy ε1,000

EQUIPMENT BY TYPE
PATROL AND COASTAL COMBATANTS 14
PCC 2 *Mwitongo* (ex-PRC *Haiqing*)
PHT 2 Type-025 (*Huchuan*) each with 2 single 533mm ASTT
PB 10: 2 *Ngunguri*; 2 Type-062 (PRC *Shanghai* II); 2 VT 23m; 4 *Mambwe* (Damen Fast Crew Supplier 3307)
AMPHIBIOUS 3
LCT 1 *Kasa*
LCM 2 *Mbono* (ex-PRC *Yunnan*)

Air Defence Command ε3,000

FORCES BY ROLE
FIGHTER
3 sqn with F-7/FT-7; FT-5; K-8 *Karakorum**
TRANSPORT
1 sqn with Cessna 404 *Titan*; DHC-5D *Buffalo*; F-28 *Fellowship*; F-50; Gulfstream G550; Y-12 (II)
TRANSPORT HELICOPTER
1 sqn with Bell 205 (AB-205); Bell 412EP *Twin Huey*

EQUIPMENT BY TYPE†
AIRCRAFT 17 combat capable
FTR 11: 9 F-7 (F-7TN); 2 FT-7 (FT-7N)
ISR 1 SB7L-360 *Seeker*
TPT 12: **Medium** 2 Y-8; **Light** 7: 2 Cessna 404 *Titan*; 3 DHC-5D *Buffalo*; 2 Y-12(II); **PAX** 3: 1 F-28 *Fellowship*; 1 F-50; 1 Gulfstream G550
TRG 9: 3 FT-5 (JJ-5); 6 K-8 *Karakorum**
HELICOPTERS
MRH 1 Bell 412EP *Twin Huey*
TPT 2: **Medium** 1+ H225M; **Light** 1 Bell 205 (AB-205)
AIR DEFENCE
SAM
Short-range 2K12 *Kub* (SA-6 *Gainful*)†; S-125 *Pechora* (SA-3 *Goa*)†
Point-defence 9K32 *Strela*-2 (SA-7 *Grail*)‡
GUNS 200
TOWED 14.5mm 40 ZPU-2/ZPU-4†; **23mm** 40 ZU-23-2; **37mm** 120 M-1939

Paramilitary 1,400 active

Police Field Force 1,400

18 sub-units incl Police Marine Unit

Air Wing

EQUIPMENT BY TYPE
AIRCRAFT • TPT • Light 1 Cessna U206 *Stationair*
HELICOPTERS
TPT • Light 4: 2 Bell 206A *Jet Ranger* (AB-206A); 2 Bell 206L *Long Ranger*
TRG 2 Bell 47G (AB-47G)/Bell 47G2

Marine Unit 100

EQUIPMENT BY TYPE
PATROL AND COASTAL COMBATANTS
All operational patrol vessels under 10t FLD

DEPLOYMENT

CENTRAL AFRICAN REPUBLIC: UN • MINUSCA 455; 1 inf bn(-)

DEMOCRATIC REPUBLIC OF THE CONGO: UN • MONUSCO 959; 1 SF coy; 1 inf bn

LEBANON: UN • UNIFIL 120; 1 MP coy

SOUTH SUDAN: UN • UNMISS 10

SUDAN: UN • UNAMID 342; 1 inf bn(-); **UN** • UNISFA 3

Togo TGO

CFA Franc BCEAO fr		2019	2020	2021
GDP	fr	3.20tr	3.35tr	
	US$	5.46bn	5.72bn	
per capita	US$	675	690	
Growth	%	5.3	0.0	
Inflation	%	0.7	1.4	
Def bdgt	fr	61.9bn	66.6bn	
	US$	106m	114m	
US$1=fr		585.93	585.03	

Real-terms defence budget trend (US$m, constant 2015)

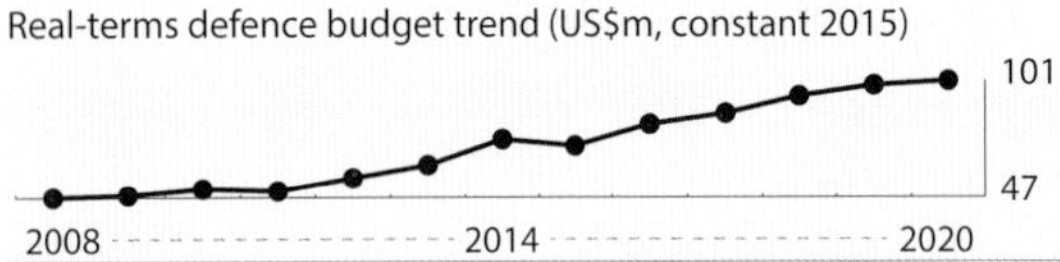

Population 8,608,444

Age	0–14	15–19	20–24	25–29	30–64	65 plus
Male	19.9%	5.1%	4.4%	4.1%	14.6%	1.5%
Female	19.8%	5.1%	4.4%	4.1%	15.0%	2.0%

Capabilities

Defence authorities are increasingly concerned by the challenge from piracy and other illicit maritime activities in the Gulf of Guinea. Regional cooperation is being strengthened as a result. France continues to deliver military training, including for Togolese peacekeeping contingents. There is also a peacekeeping training centre in Lomé. The US Africa Contingency Operations Training and Assistance programme has also provided training assistance. Togo's deployment capabilities are limited without external support, while financial challenges limit capability development more broadly. There are limited maintenance facilities, but no defence manufacturing sector.

ACTIVE 8,550 **(Army 8,100 Navy 200 Air 250)**
Paramilitary 750

Conscript liability Selective conscription, 2 years

ORGANISATIONS BY SERVICE

Army 8,100+

FORCES BY ROLE

MANOEUVRE

Reconnaissance

1 armd recce regt

Light

2 cbd arms regt

2 inf regt

1 rapid reaction force

Air Manoeuvre

1 cdo/para regt (3 cdo/para coy)

Other

1 (Presidential Guard) gd regt (1 gd bn, 1 cdo bn, 2 indep gd coy)

COMBAT SUPPORT

1 cbt spt regt (1 fd arty bty, 2 ADA bty, 1 engr/log/tpt bn)

EQUIPMENT BY TYPE

ARMOURED FIGHTING VEHICLES

MBT 2 T-54/T-55

LT TK 9 FV101 *Scorpion*

RECCE 86: 3 AML-60; 7 AML-90; 29 *Bastion Patsas*; 36 EE-9 *Cascavel*; 6 M8; 3 M20; 2 VBL

IFV 20 BMP-2

APC 34

APC (T) 4 M3A1 half-track

APC (W) 30 UR-416

ANTI-TANK/ANTI-INFRASTRUCTURE

RCL 75mm Type-52 (M20)/Type-56; **82mm** Type-65 (B-10)

GUNS 57mm 5 ZIS-2

ARTILLERY 30+

SP 122mm 6

TOWED 105mm 4 HM-2

MRL 122mm Type-81 mod (SC6 chassis)

MOR 82mm 20 M-43

AIR DEFENCE • GUNS • TOWED 43 **14.5mm** 38 ZPU-4; **37mm** 5 M-1939

Navy ε200 (incl Marine Infantry unit)

EQUIPMENT BY TYPE

PATROL AND COASTAL COMBATANTS 3

PBF 1 *Agou* (RPB 33)

PB 2 *Kara* (FRA *Esterel*)

Air Force 250

FORCES BY ROLE

TRANSPORT

1 sqn with Beech 200 *King Air*

1 VIP unit with DC-8; F-28-1000

TRAINING

1 sqn with TB-30 *Epsilon**

TRANSPORT HELICOPTER

1 sqn with SA315 *Lama*; SA316 *Alouette* III; SA319 *Alouette* III; SA342L1 *Gazelle*

EQUIPMENT BY TYPE†

AIRCRAFT 3 combat capable

TPT 5: **Light** 2 Beech 200 *King Air*; **PAX** 3: 1 DC-8; 2 F-28-1000 (VIP)

TRG 3 TB-30 *Epsilon** (3 *Alpha Jet**; 4 EMB-326G* in store)

HELICOPTERS

MRH 6: 2 SA315 *Lama*; 1 SA316 *Alouette* III; 1 SA319 *Alouette* III; 2 SA342L1 *Gazelle*

TPT • Medium 2 Mi-8T *Hip* C (1 SA330 *Puma* in store)

Paramilitary 750

Gendarmerie 750

Ministry of Interior

FORCES BY ROLE

2 reg sections

MANOEUVRE

Other

1 (mobile) paramilitary sqn

DEPLOYMENT

CENTRAL AFRICAN REPUBLIC: UN • MINUSCA 10

MALI: UN • MINUSMA 936; 1 inf bn; 1 fd hospital

SOUTH SUDAN: UN • UNMISS 2

SUDAN: UN • UNAMID 2

WESTERN SAHARA: UN • MINURSO 2

Uganda UGA

Ugandan Shilling Ush		2019	2020	2021
GDP	Ush	138tr	143tr	
	US$	36.5bn	37.7bn	
per capita	US$	916	915	
Growth	%	6.7	-0.3	
Inflation	%	2.9	4.2	
Def bdgt	Ush	2.07tr	3.62tr	4.35tr
	US$	547m	953m	
US$1=Ush		3781.14	3799.87	

Real-terms defence budget trend (US$m, constant 2015)

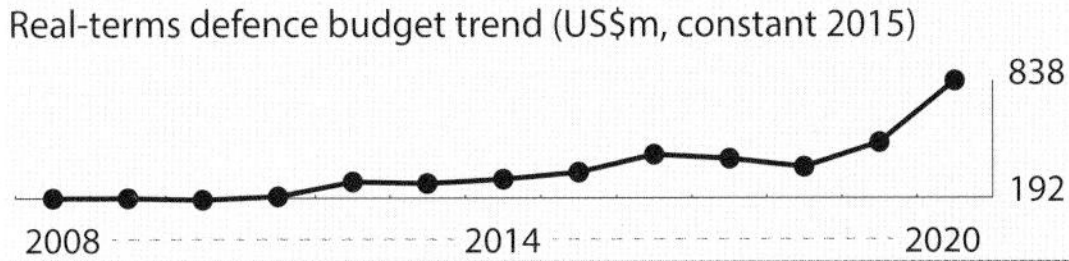

Population 43,252,966

Age	0–14	15–19	20–24	25–29	30–64	65 plus
Male	24.4%	5.6%	4.2%	3.3%	10.1%	1.0%
Female	23.8%	5.7%	4.7%	4.0%	11.8%	1.4%

Capabilities

Uganda's armed forces are well equipped and are important contributors to East African security. Operational experience and training have led to improvements in administration and planning, as well as in military skills including counter-IED and urban patrolling. A number of years spent targeting the Lord's Resistance Army has also developed counter-insurgency experience. There are plans to establish a National Defence College. Uganda is one of the largest contributors to the East Africa Standby Force and in 2014 signed a Mutual Defence Treaty with Kenya and Rwanda. Training levels are adequate, particularly for the special forces, and are improving after recent experiences. There is regular training with international partners, including at Uganda's own facilities. Airlift is limited though rotary-wing aviation has improved in recent years, partly due to US assistance. Motorised infantry formations still lack sufficient transport and logistics capacity. Mechanised forces are relatively well equipped, though heavy equipment is disparate and ageing. Improvements include the arrival of MRAP and other protected vehicles. There is limited defence-industrial capacity, though there is some manufacturing of light armoured vehicles. Uganda's 2015–19 Security Sector Development plan included the establishment of an engineering centre and a defence-research facility.

ACTIVE 45,000 (Ugandan People's Defence Force 45,000) Paramilitary 1,400

RESERVE 10,000

ORGANISATIONS BY SERVICE

Ugandan People's Defence Force ε40,000–45,000

FORCES BY ROLE

MANOEUVRE

Armoured

1 armd bde

Light

1 cdo bn

5 inf div (total: 16 inf bde)

Other

1 (Special Forces Command) mot bde

COMBAT SUPPORT

1 arty bde

AIR DEFENCE

2 AD bn

EQUIPMENT BY TYPE†

ARMOURED FIGHTING VEHICLES

MBT 239+: 140 T-54/T-55; 45 T-55AM2; 10 T-72B1; 44 T-90S; ZTZ-85-IIM

LT TK ε20 PT-76

RECCE 46: 40 *Eland*-20; 6 FV701 *Ferret*

IFV 31 BMP-2

APC 165

APC (W) 58: 15 BTR-60; 20 *Buffel*; 4 OT-64; 19 *Bastion* APC

PPV 107: 42 *Casspir*; 15 *Hizir*; 40 *Mamba*; 10 RG-33L

AUV 15 *Cougar*

ENGINEERING & MAINTENANCE VEHICLES

ARV T-54/T-55 reported

VLB MTU reported

MW *Husky*

ARTILLERY 333+

SP 155mm 6 ATMOS 2000

TOWED 243+: **122mm** M-30; **130mm** 221; **155mm** 22: 4 G-5; 18 M-839

MRL 6+: **107mm** (12-tube); **122mm** 6+: BM-21 *Grad*; 6 RM-70

MOR 78+: **81mm** L16; **82mm** M-43; **120mm** 78 *Soltam*

AIR DEFENCE

SAM

Short-range 4 S-125 *Pechora* (SA-3 *Goa*)

Point-defence 9K32 *Strela*-2 (SA-7 *Grail*)‡; 9K310 *Igla*-1 (SA-16 *Gimlet*)

GUNS • TOWED 20+: **14.5mm** ZPU-1/ZPU-2/ZPU-4; **37mm** 20 M-1939

Marines ε400

All operational patrol vessels under 10t FLD

FORCES BY ROLE

MANOEUVRE

Amphibious

1 mne bn

Air Wing

FORCES BY ROLE

FIGHTER/GROUND ATTACK

1 sqn with Su-30MK2 *Flanker*

TRANSPORT
1 unit with Y-12
1 VIP unit with Gulfstream 550; L-100-30
TRAINING
1 unit with L-39ZA/ZO *Albatros**
ATTACK/TRANSPORT HELICOPTER
1 sqn with Bell 206 *Jet Ranger*; Bell 412 *Twin Huey*; Mi-Mi-17/171E *Hip*; Mi-24V/P *Hind* E/F; Mi-17A1 (VIP)

EQUIPMENT BY TYPE
AIRCRAFT 13 combat capable
FGA 6 Su-30MK2 *Flanker* (3+ MiG-21bis *Fishbed*; 1 MiG-21UM *Mongol* B in store)
TPT 10: **Medium** 1 L-100-30; **Light** 8: 4 Cessna 172; 2 Cessna 208B; 2 Y-12; **PAX** 1 Gulfstream 550
TRG 7 L-39ZA/ZO *Albatros**
HELICOPTERS
ATK ε7 Mi-24V/P *Hind* E/F
MRH 7: 2 Bell 412 *Twin Huey*; 5 Mi-17/171E *Hip*
TPT 3: **Medium** 1 Mi-171A1 (VIP); **Light** 1 Bell 206A *Jet Ranger*
AIR-LAUNCHED MISSILES
AAM • IR R-73 (AA-11A *Archer*); **SARH** R-27 (AA-10 *Alamo*); **ARH** R-77 (AA-12 *Adder*) (reported)
ARM Kh-31P (AS-17A *Krypton*) (reported)

Paramilitary ε600 active

Border Defence Unit ε600
Equipped with small arms only

DEPLOYMENT

SOMALIA: AU • AMISOM 5,800; 7 inf bn; **UN** • UNSOM 624; 1 sy bn; **UN** • UNSOS 1

SOUTH SUDAN: UN • UNMISS 2

SUDAN: UN • UNISFA 2

Zambia ZMB

Zambian Kwacha K		2019	2020	2021
GDP	K	311bn	338bn	
	US$	24.2bn	18.9bn	
per capita	US$	1,318	1,001	
Growth	%	1.4	-4.8	
Inflation	%	9.8	14.5	
Def bdgt	K	5.07bn	6.57bn	5.64bn
	US$	394m	367m	
US$1=K		12.89	17.88	

Real-terms defence budget trend (US$m, constant 2015)

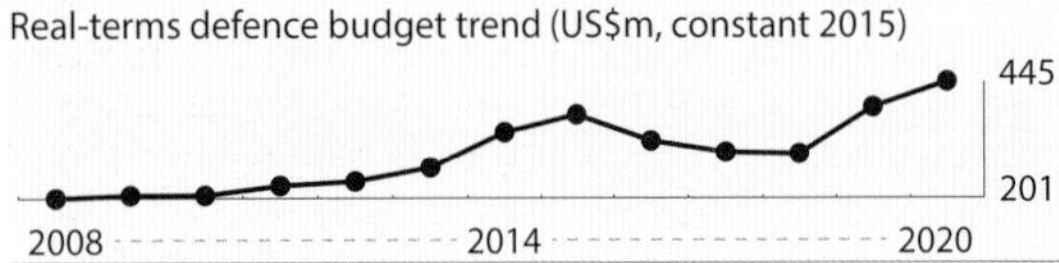

Population 17,426,623

Age	0–14	15–19	20–24	25–29	30–64	65 plus
Male	23.0%	5.4%	4.6%	3.8%	12.1%	1.0%
Female	22.8%	5.4%	4.6%	3.8%	12.2%	1.3%

Capabilities

Zambia faces no immediate external threat, though its border with the Democratic Republic of the Congo presents a security challenge. Ties have developed with China over the past decade, including in military training and weapons sales. The armed forces' principal tasks are ensuring territorial integrity and border security, and there is also a commitment to international peacekeeping operations, though these roles could be challenged by equipment obsolescence and a comparatively small establishment strength. Zambia is a member of the AU and SADC and the services have participated in exercises with international and regional partners including for the SADC Standby Force. Zambia's largest peacekeeping contribution is to the MINUSCA operation in the Central African Republic (CAR). In April 2017, Zambia signed a defence deal with Russia for spare-parts support. The armed forces are all-volunteer. The US has provided funding and material support for army and air-force pre-deployment training for the CAR peacekeeping mission. The armed forces have limited capacity to independently deploy and sustain forces beyond national borders. While there is a need to modernise the inventory, funds remain limited. The country has no defence-industrial base, apart from limited ammunition production.

ACTIVE 15,100 (Army 13,500 Air 1,600) **Paramilitary 1,400**

RESERVE 3,000 (Army 3,000)

ORGANISATIONS BY SERVICE

Army 13,500

FORCES BY ROLE
COMMAND
3 bde HQ
SPECIAL FORCES
1 cdo bn
MANOEUVRE
Armoured
1 armd regt (1 tk bn, 1 armd recce regt)
Light
6 inf bn
COMBAT SUPPORT
1 arty regt (2 fd arty bn, 1 MRL bn)
1 engr regt

EQUIPMENT BY TYPE
Some equipment†
ARMOURED FIGHTING VEHICLES
MBT 30: 20 Type-59; 10 T-55
LT TK 30 PT-76
RECCE 70 BRDM-1/BRDM-2 (ε30 serviceable)
IFV 23 *Ratel*-20
APC • APC (W) 33: 13 BTR-60; 20 BTR-70
ENGINEERING & MAINTENANCE VEHICLES
ARV T-54/T-55 reported
ANTI-TANK/ANTI-INFRASTRUCTURE
MSL • MANPATS 9K11 *Malyutka* (AT-3 *Sagger*)
RCL 12+: **57mm** 12 M18; **75mm** M20; **84mm** *Carl Gustaf*
ARTILLERY 182
TOWED 61: **105mm** 18 Model 56 pack howitzer; **122mm** 25 D-30; **130mm** 18 M-46
MRL 122mm 30 BM-21 *Grad* (ε12 serviceable)
MOR 91: **81mm** 55; **82mm** 24; **120mm** 12

AIR DEFENCE
SAM • MANPAD 9K32 *Strela-2* (SA-7 *Grail*)‡
GUNS • TOWED 136: **20mm** 50 M-55 (triple); **37mm** 40 M-1939; **57mm** ε30 S-60; **85mm** 16 M-1939 KS-12

Reserve 3,000

FORCES BY ROLE
MANOEUVRE
Light
3 inf bn

Air Force 1,600

FORCES BY ROLE
FIGHTER/GROUND ATTACK
1 sqn with K-8 *Karakorum**
1 sqn with L-15*
TRANSPORT
1 sqn with MA60; Y-12(II); Y-12(IV); Y-12E
1 (VIP) unit with AW139; HS-748
1 (liaison) sqn with Do-28
TRAINING
2 sqn with MB-326GB; MFI-15 *Safari*
TRANSPORT HELICOPTER
1 sqn with Mi-17 *Hip* H
1 (liaison) sqn with Bell 47G; Bell 205 (UH-1H *Iroquois*/AB-205)
AIR DEFENCE
3 bty with S-125 *Pechora* (SA-3 *Goa*)

EQUIPMENT BY TYPE†
Very low serviceability
AIRCRAFT 21 combat capable
TPT 25: **Medium** 2 C-27J *Spartan*; **Light** 21: 5 Do-28; 2 MA60; 4 Y-12(II); 5 Y-12(IV); 5 Y-12E; **PAX** 2: 1 Gulfstream G650ER; 1 HS-748
TRG 51: 15 K-8 *Karakorum**; 6 L-15*; 10 MB-326GB; 8 MFI-15 *Safari*; 12 SF-260TW
HELICOPTERS
MRH 5: 1 AW139; 4 Mi-17 *Hip* H
TPT • Light 12: 9 Bell 205 (UH-1H *Iroquois*/AB-205); 3 Bell 212
TRG 5 Bell 47G
UNINHABITED AERIAL VEHICLES 3+
ISR • Medium 3+ *Hermes* 450
AIR DEFENCE
SAM • Short-range S-125 *Pechora* (SA-3 *Goa*)
AIR-LAUNCHED MISSILES
AAM • IR PL-5E-II
ASM 9K11 *Malyutka* (AT-3 *Sagger*)

Paramilitary 1,400

Police Mobile Unit 700

FORCES BY ROLE
MANOEUVRE
Other
1 police bn (4 police coy)

Police Paramilitary Unit 700

FORCES BY ROLE
MANOEUVRE
Other
1 paramilitary bn (3 paramilitary coy)

EQUIPMENT BY TYPE
ARMOURED FIGHTING VEHICLES
APC • PPV 8+: 3+ *Marauder*; 5+ CS/VP3

DEPLOYMENT

CENTRAL AFRICAN REPUBLIC: UN • MINUSCA 923; 1 inf bn

DEMOCRATIC REPUBLIC OF THE CONGO: UN • MONUSCO 6

SOUTH SUDAN: UN • UNMISS 11

SUDAN: UN • UNAMID 5; **UN** • UNISFA 3

Zimbabwe ZWE

Zimbabwe Dollar Z$		2019	2020	2021
GDP	Z$	159bn	785bn	
	US$	18.7bn	14.0bn	
per capita	US$	1,254	922	
Growth	%	-6.5	-10.4	
Inflation	%	255.3	622.8	
Def bdgt	Z$	547m	3.1bn	
	US$	64.2m	38.9m	
US$1=Z$		8.52	80.00	

Real-terms defence budget trend (US$m, constant 2015)

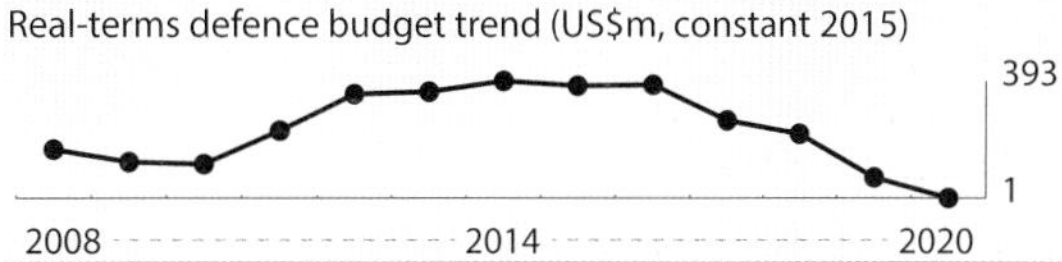

Population 14,546,314

Age	0–14	15–19	20–24	25–29	30–64	65 plus
Male	19.0%	5.3%	4.6%	4.4%	14.1%	1.8%
Female	19.3%	5.5%	4.8%	4.4%	14.2%	2.7%

Capabilities

Political instability and a weak economy remain key challenges. Sporadic domestic troop deployments continue, some two years after the presidential election resulted in victory for Emmerson Mnangagwa. Principal military tasks include ensuring sovereignty, territorial integrity and border security, as well as providing internal-security support to the police. The armed forces take an active political role. Zimbabwe is a member of the AU and the SADC and takes part in SADC Standby Force exercises. There are defence ties with China, and an emergent defence relationship with Belarus, while Russia in mid-2019 reportedly said it would consider alternative payment means for military cooperation. Military leaders have identified training as a development priority. Small numbers of personnel have deployed on peacekeeping operations, but there is no capacity to sustain a force far beyond national borders. Equipment recapitalisation is also a priority, though much will depend on domestic economic health and perhaps the extent to which China and Russia may provide support. There are plans to revive state-owned small-arms and munitions manufacturer Zimbabwe Defence Industries.

ACTIVE 29,000 (Army 25,000 Air 4,000) **Paramilitary 21,800**

ORGANISATIONS BY SERVICE

Army ε25,000

FORCES BY ROLE

COMMAND
1 SF bde HQ
1 mech bde HQ
5 inf bde HQ

SPECIAL FORCES
1 SF regt

MANOEUVRE
Armoured
1 armd sqn
Mechanised
1 mech inf bn
Light
15 inf bn
1 cdo bn
Air Manoeuvre
1 para bn
Other
3 gd bn
1 (Presidential Guard) gd gp

COMBAT SUPPORT
1 arty bde
1 fd arty regt
2 engr regt

AIR DEFENCE
1 AD regt

EQUIPMENT BY TYPE

ARMOURED FIGHTING VEHICLES
MBT 40: 30 Type-59†; 10 Type-69†
RECCE 115: 20 *Eland*-60/90; 15 FV701 *Ferret*†; 80 EE-9 *Cascavel* (90mm)
IFV 2+ YW307
APC • APC (T) 30: 8 ZSD-85 (incl CP); 22 VTT-323

ENGINEERING & MAINTENANCE VEHICLES
ARV T-54/T-55 reported; ZJX-93 ARV
VLB MTU reported

ARTILLERY 254
SP 122mm 12 2S1 *Gvozdika*
TOWED 122mm 20: 4 D-30; 16 Type-60 (D-74)
MRL 76: **107mm** 16 Type-63; **122mm** 60 RM-70
MOR 146: **81mm/82mm** ε140; **120mm** 6 M-43

AIR DEFENCE
SAM • Point-defence 9K32 *Strela*-2 (SA-7 *Grail*)‡
GUNS • TOWED 116: **14.5mm** 36 ZPU-1/ZPU-2/ZPU-4; **23mm** 45 ZU-23-2; **37mm** 35 M-1939

Air Force 4,000

FORCES BY ROLE

FIGHTER
1 sqn with F-7 II†; FT-7†

FIGHTER/GROUND ATTACK
1 sqn with K-8 *Karakorum**

GROUND ATTACK/ISR
1 sqn with Cessna 337/O-2A *Skymaster**

ISR/TRAINING
1 sqn with SF-260F/M; SF-260TP*; SF-260W *Warrior**

TRANSPORT
1 sqn with BN-2 *Islander*; CASA 212-200 *Aviocar* (VIP)

ATTACK/TRANSPORT HELICOPTER
1 sqn with Mi-35 *Hind*; Mi-35P *Hind* (liaison); SA316 *Alouette* III; AS532UL *Cougar* (VIP)
1 trg sqn with Bell 412 *Twin Huey*, SA316 *Alouette* III

AIR DEFENCE
1 sqn

EQUIPMENT BY TYPE

AIRCRAFT 45 combat capable
FTR 9: 7 F-7 II†; 2 FT-7†
ISR 2 O-2A *Skymaster*
TPT • Light 25: 5 BN-2 *Islander*; 7 C-212-200 *Aviocar*; 13 Cessna 337 *Skymaster**; (10 C-47 *Skytrain* in store)
TRG 33: 10 K-8 *Karakorum**; 5 SF-260M; 8 SF-260TP*; 5 SF-260W *Warrior**; 5 SF-260F

HELICOPTERS
ATK 6: 4 Mi-35 *Hind*; 2 Mi-35P *Hind*
MRH 9: 8 Bell 412 *Twin Huey*; 1 SA316 *Alouette* III
TPT • Medium 2 AS532UL *Cougar* (VIP)

AIR-LAUNCHED MISSILES • AAM • IR PL-2; PL-5 (reported)

AD • GUNS 100mm (not deployed); **37mm** (not deployed); **57mm** (not deployed)

Paramilitary 21,800

Zimbabwe Republic Police Force 19,500
incl air wg

Police Support Unit 2,300
PATROL AND COASTAL COMBATANTS
All operational patrol vessels under 10t FLD

DEPLOYMENT

SOUTH SUDAN: UN • UNMISS 5

SUDAN: UN • UNAMID 4; **UN** • UNISFA 10

Arms procurements and deliveries – Sub-Saharan Africa

Significant events in 2020

DEC. 2019 — NIGERIA–AUTOKRAZ JOINT PRODUCTION AGREEMENT

Nigeria's Proforce signed a partnership agreement with Ukraine's AutoKrAZ for technology transfer and joint vehicle production of KrAZ military and civilian vehicles in Nigeria. Nigeria procured KrAZ 4×4s for the police force in 2008 and 6×6 vehicles for the armed forces in 2015. Proforce has a growing presence across sub-Saharan Africa with its own *Ara/Thunder* mine-resistant ambush-protected vehicles (MRAPs) and the PF1/PF2 wheeled armoured personnel carrier, and the company is one of several growing defence companies in Nigeria.

FEBRUARY — MILKOR EXPANDS GLOBAL FOOTPRINT

South African company Milkor opened its first international office in India in February. The 39-year-old firm is set to open a second in the United Arab Emirates (UAE) during IDEX Abu Dhabi 2021 in February 2021. The company is best known for its grenade launcher which has been sold to the Indian Army. More recently, Milkor has displayed models and prototypes for 4×4 armoured vehicles, UAVs and fast patrol boats. India has recently approved changes to procurement policy to support its 'Make in India' initiative and the UAE has ambitious plans for its own local industry. Securing contracts in both those countries may increasingly depend on having a strong local presence.

JUNE — HENSOLDT SOUTH AFRICA ACQUISITIONS

Hensoldt South Africa has signed an agreement to acquire two business units of Tellumat, including one located in Pretoria that sells radar-equipped air-traffic-management systems and a business unit based in Cape Town dealing in UAVs, IFF and tactical-communications systems. Tellumat recently generated sales of more than R500m (US$28.76m) and the units of the firm being acquired made up 40% of that. Hensoldt South Africa itself is only a year old, having been formed by the consolidation of two Hensoldt businesses: GEW Technology and Hensoldt Optronics South Africa. The acquisition will see Hensoldt further expand its portfolio as well as its presence in the country.

JULY — OTT TECHNOLOGIES ACQUIRES LMT

OTT Technologies acquired fellow South African firm Land Mobility Technologies (LMT) Products. The majority shareholder, Denel, lost control of LMT Products in 2019 after the company entered voluntary business rescue. Despite generating revenues of more than R103m (US$7.22m) in the 2018–19 tax year, LMT – like Denel – has had significant cash-flow problems. Both LMT Products and OTT Technologies operate in the light and armoured vehicles sector and have sold widely, with particular success in the Middle East. As Denel's financial woes continue, the sale of different subsidiaries and parts of the business could be explored further.

OCTOBER — DENEL TROUBLES CONTINUE

Financial problems continued for South African state-owned defence firm Denel. The company requested R3.8bn (US$258.18m) worth of extra support from the government over the coming years as it struggles with liquidity problems. Although Denel received a cash injection of R1.8bn (US$126.13m) in 2019, it had requested R2.8bn (US$196.21m). Similarly, Denel was offered R72m (US$4.89m) for the 2020–21 fiscal year although it had requested R576m (US$39.13m). Despite the company's effort to sell off non-core businesses and underperforming operations such as LMT Products, Denel made a R1.75bn (US$122.63m) loss in FY2019/20. Denel is responsible for several key South African defence programmes, such as the Project *Hoefyster* armoured-vehicle programme, and if the company were to go bankrupt it would have significant consequences for the South African armed forces.

Figure 28: **G5 Sahel nations: acquisition since 2012**

Conflict and instability have spread across the Sahel region since violence in Mali erupted in 2012, and have prompted military intervention by several outside countries, most notably France. In addition, five regional states have over the past eight years acquired a variety of equipment to improve their counter-insurgency efforts. These states – Burkina Faso, Chad, Mali, Mauritania and Niger – have since 2004 as the 'G5 Sahel' coordinated certain activities to tackle common challenges. Much of the equipment they have acquired, typically 4x4 armoured vehicles, has been donated by the European Union, the US and Germany and has often been accompanied by training support. Other countries with security interests in the region, such as the UAE and Qatar, have also donated equipment of this type, although it is less clear if those donations have come with the post-delivery support given by Western countries and organisations. Three G5 Sahel members have acquired the *Super Tucano* light attack/training aircraft from Brazilian company Embraer. Likewise, Burkina Faso, Chad and Mali have bolstered their rotary-wing fleets with small batches of helicopters and Mali, Mauritania and Niger have made small improvements to their tactical-airlift capability. While the sources of this equipment are varied, the batches are relatively small and low-tech. It is likely that the five Sahel nations will continue to rely on outside assistance to provide a number of key capabilities such as airborne and ground-based intelligence, surveillance and reconnaissance, intra-theatre lift, and air-to-surface attack-capable platforms.

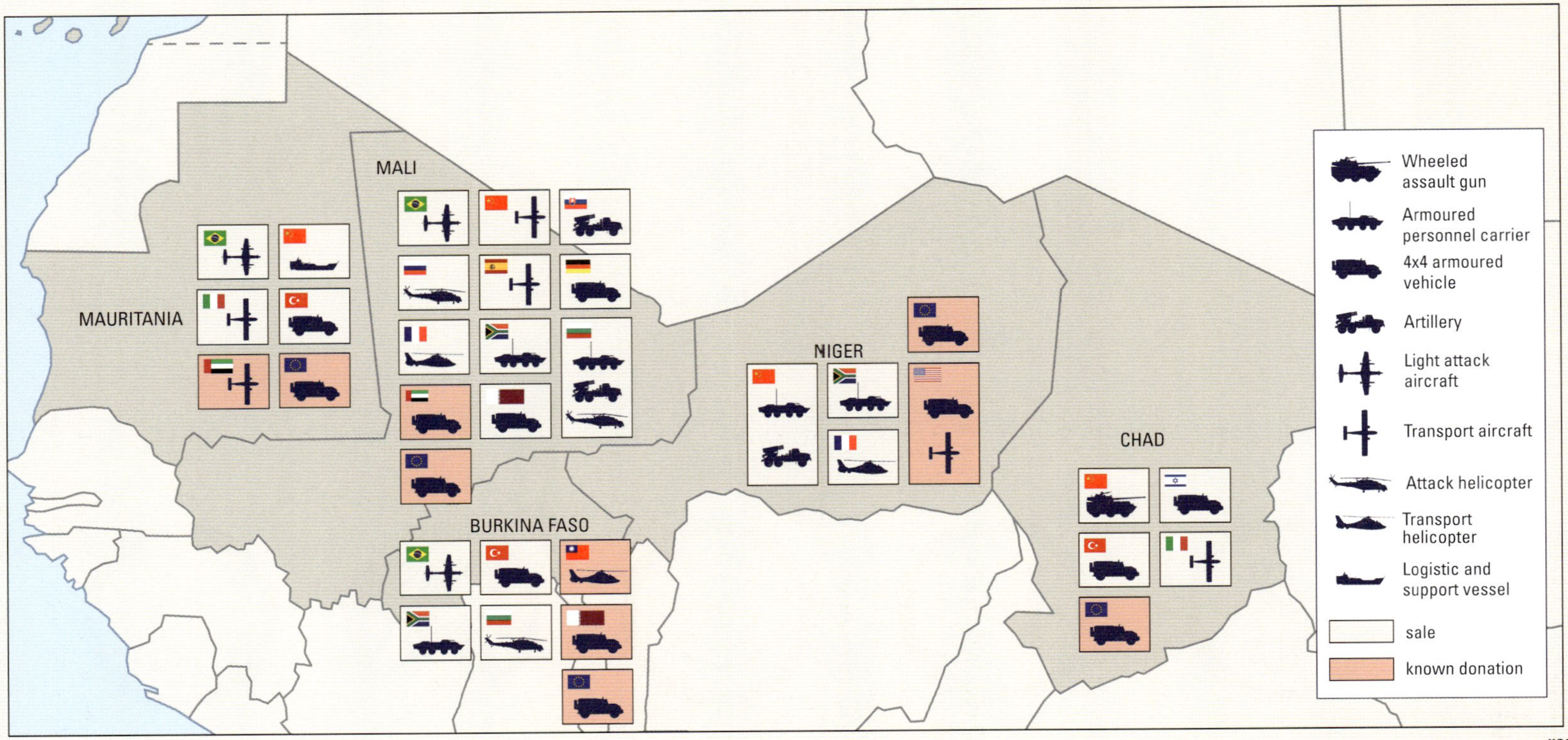

© IISS

Chapter Ten
Military cyber capabilities

The Military Balance+ database is integrating information on military-owned cyber capacities, principally relating to indicators of military-owned cyber capability, framed by a prose assessment. However, military-owned cyber capabilities only constitute one element of national cyber power. As a result, care must be taken when looking at this information for judgements of national cyber power. In a study of national power and cyber capabilities, the IISS has found that few countries have comprehensively transformed in order to integrate cyberspace operations into their operational concepts, doctrines and force structures. This consideration, along with low availability of key data, has led the IISS to gather indicative data in four areas. *Strategy and doctrine* includes military cyber strategies and doctrines as well as broader national military or security strategies containing specific references to military uses of cyberspace. *Principal defence cyber units* identifies commands and units formally acknowledged as primarily for military cyber purposes, subdivided into command-level and subordinate formations. *Defence satellites* details selected satellites important because of their use of or potential vulnerabilities to cyber capabilities. *Defence cyber exercises* provide evidence of the dissemination of operational concepts and capabilities for cyberspace operations through national military cyber exercises.

United Kingdom

The United Kingdom's cyber capability is principally centred on the National Cyber Security Centre, and related cyber-intelligence capability in the Government Communications Headquarters (GCHQ). The UK does not have a unified military cyber command. Strategic Command, established in 2020 from the previous Joint Forces Command, commands UK military cyber formations. UK armed forces have their own cyber-focused formations. There is well-developed doctrine and strategy, within the framework of overall civil-sector strategies. Offensive cyber is covered in detail in published UK military doctrine, including its uses to create freedom of manoeuvre, to project power, for military effect and for deterrence. The UK's development of an offensive cyber capability has been a joint venture between GCHQ and the Ministry of Defence. In 2014, GCHQ and the Ministry of Defence organised a National Offensive Cyber Programme, which was replaced in 2020 by a new model (still short of a dedicated military cyber command) in the form of a National Cyber Force run jointly by the two.

Indicator 1: Strategy and doctrine
National/defence: National Security Strategy 2010, 2015; National Cyber Security Strategy 2009, 2011, 2016

Defence/services: Integrated Operating Concept 2025, 2020; UK Cyber Doctrine, JDP 0-50; Joint Doctrine for Cyber and Electromagnetic Activities 2018

Indicator 2: Principal defence cyber units
National Cyber Force (Ministry of Defence, Government Communications Headquarters (GCHQ), Secret Intelligence Service (SIS))

Strategic Command *with* 1 (6 UK) cbt spt div, 1 ISR bde, 1 (77) info ops bde, Army Cyber Information Security Operations Centre (13 Signal Regiment), RN and RAF formations

Indicator 3: Defence satellites
Comms: 3 *Skynet-4*; 4 *Skynet-5*

Indicator 4: Defence cyber exercises (national): *Mercury Cypher; Cyber Spartan; Information Warrior*

United States

The United States is widely perceived as the most capable military cyber state. A National Cyber Strategy and a Department of Defense (DoD) Cyber Strategy in September 2018 were intended to overcome diverse approaches and doctrine that had arisen due to the extended development of US cyber capabilities. Earlier in 2018, Cyber Command became a unified combatant command (CYBERCOM). The director of the National Security Agency (NSA) serves as the Commander of US Cyber Command. Its capabilities are regulated by US governmental authorities, and there is careful signalling of their use under CYBERCOM's 'defend-forward' strategy and its retaliatory premise. Cyber Command draws on five service-specific components: Army Cyber Command, Fleet Cyber Command, Air Forces Cyber, Marine Corps Forces Cyber Command, Coast Guard, as well as National Guard units (army and air force). Some of the 133 Cyber Mission Teams are aligned to individual services while others are aligned to support combatant commands and synchronise cyberspace operations with operations in the land, sea, air and space domains. The US aspiration is to provide cyber-attack options in all phases of operations and at every level of command. There is growing focus on integrating cyber and other capabilities into multi-domain task forces.

Indicator 1: Strategy and doctrine
National/defence: National Security Strategy 2018; DoD Cyber Strategy 2018
Defence/services: Commander's Vision statement, Cyber Command 2018; JP 3-13 Information Operations 2018; JP 3-12 Cyberspace Operations 2018

Indicator 2: Principal Military Cyber Units
US Cyber Command:
Army Cyber Comd, with 1 (1st) info ops comd (2 IO bn), 1 (780th) MI bde (cyber) (2 MI bn, 1 (915th) cyber warfare bn; Cyber Protection Bde; 1 ICEWS bn (MDTF (developing)); 41 cyber mission teams

Fleet Cyber Comd (10th Fleet) , *with* Naval Network Warfare Comd, Navy Cyber Defense Ops Comd, Cryptologic Warfare Group 6, Navy Cyber Warfare Dev Gp; 40 cyber mission teams

Air Forces Cyber (16th Air Force), *with* 1 (67th) cyberspace wg (1st cyberspace ops gp), 1 (70th) ISR wg (6 ISR gp), 1 (363rd) ISR wg (3 ISR gp), 1 (480th) ISR wg (6 ISR gp); 1 (688th) cyberspace wg (1 cbt comms gp, 1 cyberspace ops gp, 1 cyberspace engr installation gp, 1 cyberspace ops gp); 39 cyber mission teams

Marine Forces Cyber Comd *with* 1 USMC cyber ops gp, 1 USMC cyber warfare gp; 13 cyber mission teams

Coast Guard Cyber Comd

Nat Gd Cyber Bde

Indicator 3: Defence satellite capability
1 Counterspace; 17 ISR; 46 Comms; 30 PNT; 27 ELINT/SIGINT; 6 Space Surveillance; 8 Early Warning

Indicator 4: National military cyber exercises
Cyber Blitz (to become *Multi-Domain Operations Live*, in PACOM in 2021); *Cyber Flag*; *Cyber Guard*; *Cyber Shield*

France

France's Cyber Defence Command (COMCYBER) was established in 2017, under the defence ministry. French offensive military cyber operations are subordinate to its Cyber Command but distributed down to the tactical level. There is a separation between offensive and defensive cyber operations. France's leading agency for cyber security (ANSSI) is exclusively dedicated to defensive operations. Each branch of the armed forces is responsible for their own defensive cyber operations and operates its own Security Operations Centre (SOC). The Centre for the Analysis of Cyber Defence (CALID) is the Ministry's SOC. It assesses global cyber risk so that Cyber Command can act and advise political authorities. The Centre for the Review of Information Systems Security (CASSI) conducts penetration testing and security audits on military systems. The 807th Signals Company, based in Rennes, is COMCYBER's deployable branch, sent to theatres in order to secure communications and weapons systems.

Indicator 1: Strategy and doctrine
National/defence: National Cyber Security Strategy 2011; Defence and National Security Strategic Review 2017

Defence/services: Military Doctrine for Offensive Cyber Operations, 2019; Ministerial Policy for Cyber Defence, 2019; Cyber Strategy for the Armed Forces, 2019; Strategic Review of Cyber Defence, 2018

Indicator 2: Principal Military Cyber Units
Cyber Defence Command (Centre for Analysis of Cyber Defence, Centre for the Review of Information Systems Security, Cyber Defence Reserve and Operational Preparation Centre); Information and Communication Systems Command (5 sigs regt, 1 (807th) sigs coy)

Indicator 3: Defence satellite capability
Comms: 2 *Syracuse-2*, 1 *Athena-Fidus* ISR: 2 *Pleiades*, 2 *Helios* 2A/2B

Indicator 4: National military cyber exercises
DEFNET

China

In 2015, the Strategic Support Force (SSF) was set up as part of People's Liberation Army (PLA) organisational reforms, and combined space-, cyber-, electronic- and psychological-warfare capacities. The SSF reports to the Central Military Commission and has two equal branches – the Network Systems Department, which has control of the cyber force responsible for information operations, and the Space Systems Department for space operations. The SSF is responsible for both strategic information support – the management of technical-intelligence collection, strategic-intelligence support to the theatre commands, enabling joint operations – and strategic-information operations, which involves coordinating its three warfare branches to 'paralyze [and] sabotage' enemy operational and war-command 'systems-of-systems'. Consolidating these functions into the SSF reflects the PLA's new conception of space, cyber and the electromagnetic spectrum as a singular military domain rather than adjunct functions serving other forms of combat. The implications of the SSF for China's military cyber capability are twofold. First, a more unified force will be able to prosecute the type of complex, multi-dimensional information operations that the PLA foresees in future conflicts. Second, the SSF will improve China's military readiness and help the PLA shift more smoothly from a peacetime to a wartime posture. In combining espionage and attack functions across electronic-, cyber- and space-warfare units – and by bringing them under a single command – the PLA aims to survey the battlefield, prepare cross-discipline operations and develop specific capabilities that can be continuously adapted to match the requirements of fast-moving situations.

Indicator 1: Strategy and doctrine
National/defence: Science of Military Strategy 2013; 2015 Military Strategy

Indicator 2: Principal Military Cyber Units
Strategic Support Force, Network Systems Department (Unit 61398, Unit 61419, Unit 61786, Unit 61486, 56th, 57th, 58th Research Institute); Network-electronic countermeasures teams in Theatre Commands; Central Military Commission Joint Staff Department (Information Assurance Base; Network-electronic Bureau)

Indicator 3: Defence satellite capability
1 Counterspace (reported); 29 ISR; 9 Comms; 45 PNT; 8 Meteorology and oceanography; 41 ELINT/SIGINT

Russia

Russian strategy and doctrine have historically considered cyber-security and cyber operations to be a component of wider information operations; this can blur the distinction between military and civilian capabilities. Russia's leading cyber agencies are represented at higher levels within the Security Council: its members include the Defence Minister, the head of the Federal Security Service (FSB) and the Chief of the General Staff. The 2015 Military Doctrine said that Russian territory included cyberspace, which the armed forces were tasked with protecting. The 2011 document 'Conceptual Views on the Activity of the Russian Federation Armed Forces in the Information Space' offered some information on how the armed forces see their role in cyberspace but it focused on situational and threat awareness and force protection. In 2017 'Information-operations troops' were formed but while these were thought by some to have cyber responsibilities, most indications are that they are being used mainly for traditional information activities and psychological operations. The Main Directorate of the General Staff – still known by its older GRU acronym – and its subordinate formations including the 85th Main Special Service Center (Unit 26165) and the 72nd Special Service Center (Unit

54777) are, according to US authorities, principal actors in offensive cyber and influence operations.

Indicator 1: Strategy and doctrine
Information Security Doctrine 2016; National Security Strategy 2015; Military Doctrine 2014; Conceptual Views on the Activity of the Russian Federation Armed Forces in the Information Space 2011

Indicator 2: Principal Military Cyber Units
Main Directorate of the General Staff (GRU), Information Operations Troops

Indicator 3: Defence satellite capability
1 Counterspace; 11 ISR; 58 Comms; 28 PNT; 5 ELINT/SIGINT; 4 Early Warning

Indicator 4: National military cyber exercises
Included in strategic-level exercises (*Kavkaz, Tsentr, Vostok, Zapad*)

Explanatory notes

The Military Balance provides an assessment of the armed forces and defence economics of 171 countries and territories. Each edition contributes to the provision of a unique compilation of data and information, enabling the reader to discern trends by studying editions as far back as 1959. The data in the current edition is accurate according to IISS assessments as of November 2020, unless specified. Inclusion of a territory, country or state in *The Military Balance*, or terminology or boundaries in mapping, does not imply legal recognition or indicate support for any government.

General arrangement and contents

The introduction is an assessment of global defence developments and key themes in the 2021 edition. There are three analytical essays, followed by a graphical section analysing comparative defence statistics by domain, as well as key trends in defence economics.

Regional chapters begin with analysis of the military and security issues that drive national-defence policy developments, and key trends in regional defence economics. These are followed by focused analysis, for certain countries, of defence policy and capability issues, and defence economics. Next, detailed data on regional states' military forces and equipment, and defence economics, is presented in alphabetical order. Graphics assessing important regional arms procurements and deliveries complete each region.

The book closes with a reference section containing comparisons of defence economics and personnel statistics.

The Military Balance wall chart

The Military Balance 2021 wall chart is an assessment of submarine holdings and trends in anti-submarine warfare. Using data and graphics, the chart assesses the number and type of submarines in service, under construction or planned, in 43 states, with detail on service status, propulsion and significant weapons systems. There is a map detailing the principal submarine-manufacturing states, and major export contracts, and also an assessment of likely trends in sub-surface warfare.

Using *The Military Balance*

The country entries assess personnel strengths, organisation and equipment holdings of the world's armed forces. Force-strength and equipment-inventory data is based on the most accurate data available, or on the best estimate that can be made. In estimating a country's total capabilities, old equipment may be counted where it is considered that it may still be deployable.

The data presented reflects judgements based on information available to the IISS at the time the book is compiled. Where information differs from previous editions, this is mainly because of changes in national forces, but it is sometimes because the IISS has reassessed the evidence supporting past entries. For instance, defence budget calculations for several countries in the Middle East and North Africa will have shifted this year as security line items were removed. Given this, care must be taken in constructing time-series comparisons from information given in successive editions.

Abbreviations and definitions

Qualifier	
'At least'	Total is no less than the number given
'Up to'	Total is at most the number given, but could be lower
'About'	Total could be higher than given
'Some'	Precise inventory is unavailable at time of press
'In store'	Equipment held away from front-line units; readiness and maintenance varies
Billion (bn)	1,000 million (m)
Trillion (tr)	1,000 billion
$	US dollars unless otherwise stated
ε	Estimated
*	Aircraft counted by the IISS as combat capable
-	Part of a unit is detached/less than
+	Unit reinforced/more than
†	IISS assesses that the serviceability of equipment is in doubt[a]
‡	Equipment judged obsolete (weapons whose basic design is more than four decades old and which have not been significantly upgraded within the past decade)[a]

[a] Not to be taken to imply that such equipment cannot be used

Country entries

Information on each country is shown in a standard format, although the differing availability of information and differences in nomenclature result in some variations. Country entries include economic, demographic and military data. Population figures are based on demographic statistics taken from the US Census Bureau. Military data includes personnel numbers, conscript liability where relevant, outline organisation, number of formations and units, and an inventory of the major equipment of each service. Details of national forces stationed abroad and of foreign forces stationed within the given country are also provided.

Arms procurements and deliveries

A series of thematic tables, graphics and text follow the regional data. These are designed to illustrate key trends, principal programmes and significant events in regional defence procurements. More detailed information on defence procurements, organised by country, equipment type and manufacturing company, can be found on the IISS Military Balance+ database (*https://www.iiss.org/militarybalanceplus*). The information in this section meets the threshold for a *Military Balance* country entry and as such does not feature information on sales of small arms and light weapons.

Defence economics

Country entries include annual defence budgets (and expenditure where applicable), selected economic-performance indicators and demographic aggregates. All country entries are subject to revision each year as new information, particularly regarding actual defence expenditure, becomes available. On pp. 517–22, there are also international comparisons of defence expenditure and military personnel, giving expenditure figures for the past three years in per capita terms and as a % of gross domestic product (GDP). The aim is to provide a measure of military expenditure and the allocation of economic resources to defence.

Individual country entries show economic performance over the past two years and current demographic data. Where this data is unavailable, information from the last available year is provided. All financial data in the country entries is shown in both national currency and US dollars at current – not constant – prices. US-dollar conversions are calculated from the exchange rates listed in the entry.

Definitions of terms

Despite efforts by NATO and the UN to develop a standardised definition of military expenditure, many countries prefer to use their own definitions (which are often not made public). In order to present a comprehensive picture, *The Military Balance* lists three different measures of military-related spending data.

- For most countries, an official defence-budget figure is provided.
- For those countries where other military-related outlays, over and above the defence budget, are known or can be reasonably estimated, an additional measurement referred to as defence expenditure is also provided. Defence-expenditure figures will naturally be higher than official budget figures, depending on the range of additional factors included.
- For NATO countries, a defence-budget figure, as well as defence expenditure reported by NATO in local currency terms and converted using IMF exchange rates, is quoted.

NATO's military-expenditure definition (the most comprehensive) is cash outlays of central or federal governments to meet the costs of national armed forces. The term 'armed forces' includes strategic, land, naval, air, command, administration and support forces. It also includes other forces if they are trained, structured and equipped to support defence forces and are realistically deployable. Defence expenditures are reported in four categories: Operating Costs, Procurement and Construction, Research and Development (R&D) and Other Expenditure. Operating Costs include salaries and pensions for military and civilian personnel; the cost of maintaining and training units, service organisations, headquarters and support elements; and the cost of servicing and repairing military equipment and infrastructure. Procurement and Construction expenditure covers national equipment and infrastructure spending, as well as common infrastructure programmes. R&D is defence expenditure up to the point at which new equipment can be put in service, regardless of whether new equipment is actually procured. Foreign Military Assistance (FMA) contributions are also noted – primarily the IISS tracks Foreign Military Financing (FMF) allocations from the US.

For many non-NATO countries the issue of transparency in reporting military budgets is fundamental. Not every UN member state reports defence-budget data (even fewer report real defence expenditures) to their electorates, the UN, the IMF or other multinational organisations. In the case of governments with a proven record of transparency, official figures generally conform to the standardised definition of defence budgeting, as adopted by the UN, and consistency problems are not usually a major issue. The IISS cites official defence budgets as reported by either national governments, the UN, the OSCE or the IMF.

For those countries where the official defence-budget figure is considered to be an incomplete measure of total military-related spending, and appropriate additional data is available, the IISS will use data from a variety of sources to arrive at a more accurate estimate of true defence expenditure. The most frequent instances of budgetary manipulation or falsification typically involve equipment procurement, R&D, defence-industrial investment, covert weapons programmes, pensions for retired military and civilian personnel, paramilitary forces and non-budgetary sources of revenue for the military arising from ownership of industrial, property and land assets. There will be several countries listed in *The Military Balance* for which only an official defence-budget figure is provided but where, in reality, true defence-related expenditure is almost certainly higher.

Percentage changes in defence spending are referred to in either nominal or real terms. Nominal terms relate to the percentage change in numerical spending figures, and do not account for the impact of price changes (i.e. inflation) on defence spending. By contrast, real terms account for inflationary effects, and may therefore be considered a more accurate representation of change over time.

The principal sources for national economic statistics cited in the country entries are the IMF, the OECD, the World Bank and three regional banks (the Inter-American, Asian and African Development banks). For some countries, basic economic data is difficult to obtain. GDP figures are nominal (current) values at market prices. GDP growth is real, not nominal growth, and inflation is the year-on-year change in consumer prices. When real-terms defence-spending figures are mentioned, these are measured in constant 2015 US dollars.

General defence data

Personnel

The 'Active' total comprises all servicemen and women on full-time duty (including conscripts and long-term assignments from the Reserves). When a gendarmerie or equivalent is under control of the defence ministry, they may be included in the active total. Only the length of conscript liability is shown; where service is voluntary there is no entry. 'Reserve' describes formations and units not fully manned or operational in peacetime, but which can be mobilised by recalling reservists in an emergency. Some countries have more than one category of reserves, often kept at varying degrees of readiness. Where possible, these differences are denoted using the national descriptive title, but always under the heading of 'Reserves' to distinguish them from full-time active forces. All personnel figures are rounded to the nearest 50, except for organisations with under 500 personnel, where figures are rounded to the nearest ten.

Other forces

Many countries maintain forces whose training, organisation, equipment and control suggest that they may be used to support or replace regular military forces, or be used more broadly by states to deliver militarily relevant effect; these are called 'paramilitary'. They include some forces that may have a constabulary role. These are detailed after the military forces of each country, but their personnel numbers are not normally included in the totals at the start of each entry.

Units and formation strength

Company	100–200
Battalion	500–1,000
Brigade	3,000–5,000
Division	15,000–20,000
Corps or Army	50,000–100,000

Forces by role and equipment by type

Quantities are shown by function (according to each nation's employment) and type, and represent what are believed to be total holdings, including active and reserve operational and training units. Inventory totals for missile systems relate to launchers and not to missiles. Equipment held 'in store' is not counted in the main inventory totals.

Deployments

The Military Balance mainly lists permanent bases and operational deployments, including peacekeeping operations, which are often discussed in the regional text. Information in the country-data sections details, first, deployments of troops and, second, military observers and, where available, the role and equipment of deployed units. Personnel figures are not generally included for embassy staff, standing multinational headquarters, or deployments of purely maritime and aerospace assets, such as Iceland Air Policing or anti-piracy operations.

Land forces

To make international comparison easier and more consistent, *The Military Balance* categorises forces by role and translates national military terminology for unit and formation sizes. Typical personnel strength, equipment holdings and organisation of formations such as brigades and divisions vary from country to country. In addition, some unit terms, such as 'regiment', 'squadron', 'battery' and 'troop', can refer to significantly different unit sizes in different countries. Unless otherwise stated, these terms should be assumed to reflect standard British usage where they occur.

Naval forces

Classifying naval vessels according to role is complex. A post-war consensus on primary surface combatants revolved around a distinction between independently operating cruisers, air-defence escorts (destroyers) and anti-submarine-warfare escorts (frigates). However, ships are increasingly performing a range of roles. Also, modern ship design has meant that the full-load displacement (FLD) of different warship types has evolved and in some cases overlaps. For these reasons, *The Military Balance* now classifies vessels by an assessed combination of role, equipment fit and displacement.

Air forces

Aircraft listed as combat capable are assessed as being equipped to deliver air-to-air or air-to-surface ordnance. The definition includes aircraft designated by type as bomber, fighter, fighter/ground attack, ground attack and anti-submarine warfare. Other aircraft considered to be combat capable are marked with an asterisk (*). Operational groupings of air forces are shown where known. Typical squadron aircraft strengths can vary both between aircraft types and from country to country. When assessing missile ranges, *The Military Balance* uses the following range indicators:

- Short-range ballistic missile (SRBM): less than 1,000 km;
- Medium-range ballistic missile (MRBM): 1,000–3,000 km;
- Intermediate-range ballistic missile (IRBM): 3,000–5,000 km;
- Intercontinental ballistic missile (ICBM): over 5,000 km.

Cyber

The Military Balance+ database is integrating information on military-owned cyber capacities. The research taxonomy focuses on enablers, including indicators of capability from the armed forces. In last year's *Military Balance*, a potential collection plan was identified utilising six categories and these have been refined further to: military or militarily relevant cyber strategy and doctrine; command-level organisations and principal cyber units; relevant satellite capability, principally communications and ISR; and military cyber exercises; with these categories framed by a brief prose assessment. However, military-owned cyber capabilities only constitute one part of national cyber capabilities and, because of this, care must be taken when looking at this information for judgements of national cyber 'power'. An IISS study on *Cyber Capabilities and National Power*, to be released in 2021, contains a broader assessment of the methodological considerations and whole-of-nation cyber capabilities that are of most relevance to national power.

Attribution and acknowledgements

The International Institute for Strategic Studies owes no allegiance to any government, group of governments, or any political or other organisation. Its assessments are its own, based on the material available to it from a wide variety of sources. The cooperation of governments of all listed countries has been sought and, in many cases, received. However, some data in *The Military Balance* is estimated. Care is taken to ensure that this data is as accurate and free from bias as possible. The Institute owes a considerable debt to a number of its own members, consultants and all those who help compile and check material. The Director-General and Chief Executive and staff of the Institute assume full responsibility for the data and judgements in this book. Comments and suggestions on the data and textual material contained within the book, as well as on the style and presentation of data, are welcomed and should be communicated to the Editor of *The Military Balance* at: IISS, Arundel House, 6 Temple Place, London, WC2R 2PG, UK, email: *milbal@iiss.org*.

Principal land definitions

FORCES BY ROLE

Command:	free-standing, deployable formation headquarters (HQs).
Special Forces (SF):	elite units specially trained and equipped for unconventional warfare and operations in enemy-controlled territory. Many are employed in counter-terrorist roles.
Manoeuvre:	combat units and formations capable of manoeuvring. These are subdivided as follows:
Reconnaissance:	combat units and formations whose primary purpose is to gain information.
Armoured:	units and formations principally equipped with main battle tanks (MBTs) and infantry fighting vehicles (IFVs) to provide heavy mounted close-combat capability. Units and formations intended to provide mounted close-combat capability with lighter armoured vehicles, such as light tanks or wheeled assault guns, are classified as light armoured.
Mechanised:	units and formations primarily equipped with lighter armoured vehicles such as armoured personnel carriers (APCs). They have less mounted firepower and protection than their armoured equivalents, but can usually deploy more infantry.
Light:	units and formations whose principal combat capability is dismounted infantry, with few, if any, organic armoured vehicles. Some may be motorised and equipped with soft-skinned vehicles.
Air Manoeuvre:	units and formations trained and equipped for delivery by transport aircraft and/or helicopters.
Amphibious:	amphibious forces are trained and equipped to project force from the sea.
Other Forces:	includes security units such as Presidential Guards, paramilitary units such as border guards and combat formations permanently employed in training or demonstration tasks.
Combat Support:	combat support units and formations not integral to manoeuvre formations. Includes artillery, engineers, military intelligence, nuclear, biological and chemical defence, signals and information operations.
Combat Service Support (CSS):	includes logistics, maintenance, medical, supply and transport units and formations.

EQUIPMENT BY TYPE

Light Weapons:	small arms, machine guns, grenades and grenade launchers and unguided man-portable anti-armour and support weapons have proliferated so much and are sufficiently easy to manufacture or copy that listing them would be impractical.
Crew-Served Weapons:	crew-served recoilless rifles, man-portable ATGW, MANPADs and mortars of greater than 80 mm calibre are listed, but the high degree of proliferation and local manufacture of many of these weapons means that estimates of numbers held may not be reliable.
Armoured Fighting Vehicles (AFVs):	armoured combat vehicles with a combat weight of at least six metric tonnes, further subdivided as below:
Main Battle Tank (MBT):	armoured, tracked combat vehicles, armed with a turret-mounted gun of at least 75 mm calibre and with a combat weight of at least 25 metric tonnes.
Light Tank (LT TK):	armoured, tracked combat vehicles, armed with a turret-mounted gun of at least 75 mm calibre and with a combat weight of less than 25 metric tonnes.
Wheeled Assault Gun (ASLT):	armoured, wheeled combat vehicles, armed with a turret-mounted gun of at least 75 mm calibre and with a combat weight of at least 15 metric tonnes.
Armoured Reconnaissance (RECCE):	armoured vehicles primarily designed for reconnaissance tasks with no significant transport capability and either a main gun of less than 75 mm calibre or a combat weight of less than 15 metric tonnes, or both.
Infantry Fighting Vehicle (IFV):	armoured combat vehicles designed and equipped to transport an infantry squad and armed with a cannon of at least 20 mm calibre.

Armoured Personnel Carrier (APC):	lightly armoured combat vehicles designed and equipped to transport an infantry squad but either unarmed or armed with a cannon of less than 20mm calibre.
Airborne Combat Vehicle (ABCV):	armoured vehicles designed to be deployable by parachute alongside airborne forces.
Amphibious Assault Vehicle (AAV):	armoured vehicles designed to have an amphibious ship-to-shore capability.
Armoured Utility Vehicle (AUV):	armoured vehicles not designed to transport an infantry squad, but capable of undertaking a variety of other utility battlefield tasks, including light reconnaissance and light transport.
Specialist Variants:	variants of armoured vehicles listed above that are designed to fill a specialised role, such as command posts (CP), artillery observation posts (OP), signals (sigs) and ambulances (amb), are categorised with their parent vehicles.
Engineering and Maintenance Vehicles:	includes armoured engineer vehicles (AEV), armoured repair and recovery vehicles (ARV), assault bridging (VLB) and mine warfare vehicles (MW).
Nuclear, Biological and Chemical Defence Vehicles (NBC):	armoured vehicles principally designed to operate in potentially contaminated terrain.
Anti-Tank/Anti-Infrastructure (AT):	guns, guided weapons and recoilless rifles designed to engage armoured vehicles and battlefield hardened targets.
Surface-to-Surface Missile Launchers (SSM):	launch vehicles for transporting and firing surface-to-surface ballistic and cruise missiles.
Artillery:	weapons (including guns, howitzers, gun/howitzers, multiple-rocket launchers, mortars and gun/mortars) with a calibre greater than 100mm for artillery pieces and 80mm and above for mortars, capable of engaging ground targets with indirect fire.
Coastal Defence:	land-based coastal artillery pieces and anti-ship-missile launchers.
Air Defence (AD):	guns, directed-energy (DE) weapons and surface-to-air missile (SAM) launchers designed to engage fixed-wing, rotary-wing and uninhabited aircraft. Missiles are further classified by maximum notional engagement range: point-defence (up to 10 km); short-range (10–30 km); medium-range (30–75 km); and long-range (75 km+). Systems primarily intended to intercept missiles rather than aircraft are categorised separately as Missile Defence.

Principal naval definitions

To aid comparison between fleets, the following definitions, which do not always conform to national definitions, are used as guidance:

Submarines:	all vessels designed to operate primarily under water. Submarines with a dived displacement below 250 tonnes are classified as midget submarines (SSW); those below 500 tonnes are coastal submarines (SSC).
Principal Surface Combatants:	all surface ships designed for combat operations on the high seas, with an FLD above 2,200 tonnes. Aircraft carriers (CV), including smaller support carriers (CVS) embarking STOVL aircraft and helicopter carriers (CVH), are vessels with a flat deck primarily designed to carry fixed- and/or rotary-wing aircraft, without specialised amphibious capability. Other principal surface combatants include cruisers (C) (FLD above 9,750 tonnes), destroyers (DD) (FLD 4,500–9,749 tonnes with a primary area air-defence weapons fit and role) and frigates (FF) (FLD 2,200–9,000 tonnes and a primary anti-submarine/general-purpose weapons fit and role).

Patrol and Coastal Combatants:	surface vessels designed for coastal or inshore operations. These include corvettes (FS), which usually have an FLD between 500 and 2,199 tonnes and are distinguished from other patrol vessels by their heavier armaments. Also included in this category are offshore-patrol ships (PSO), with an FLD greater than 1,500 tonnes; patrol craft (PC), which have an FLD between 250 and 1,499 tonnes; and patrol boats (PB) with an FLD between ten and 250 tonnes. Vessels with a top speed greater than 35 knots are designated as 'fast'.
Mine warfare vessels:	all surface vessels configured primarily for mine laying (ML) or countermeasures. Countermeasures vessels are either: sweepers (MS), which are designed to locate and destroy mines in an area; hunters (MH), which are designed to locate and destroy individual mines; or countermeasures vessels (MC), which combine both roles.
Amphibious vessels:	vessels designed to transport combat personnel and/or equipment onto shore. These include aviation-capable amphibious assault ships (LHA), which can embark rotary-wing or STOVL air assets and may have a well deck for LCACs and landing craft; aviation-capable amphibious assault ships with a well dock for LCACs and landing craft (LHD), which can embark rotary-wing or STOVL assets; landing platform helicopters (LPH), which have a primary role of launch and recovery platform for rotary-wing or STOVL assets; landing platform docks (LPD), which do not have a through deck but do have a well dock and carry both combat personnel and equipment; and land ships docks (LSD) with a well dock but focused more on equipment transport. Landing ships (LS) are amphibious vessels capable of ocean passage and landing craft (LC) are smaller vessels designed to transport personnel and equipment from a larger vessel to land or across small stretches of water. Landing ships have a hold; landing craft are open vessels. Landing craft air cushioned (LCAC) are differentiated from utility craft air cushioned (UCAC) in that the former have a bow ramp for the disembarkation of vehicles and personnel.
Auxiliary vessels:	ocean-going surface vessels performing an auxiliary military role, supporting combat ships or operations. These generally fulfil five roles: replenishment (such as oilers (AO) and solid stores (AKS)); logistics (such as cargo ships (AK) and logistics ships (AFS)); maintenance (such as cable-repair ships (ARC) or buoy tenders (ABU)); research (such as survey ships (AFS)); and special purpose (such as intelligence-collection ships (AGI) and ocean-going tugs (ATF)).
Weapons systems:	weapons are listed in the following order: land-attack cruise missiles (LACM), anti-ship missiles (AShM), surface-to-air missiles (SAM), heavy (HWT) and lightweight (LWT) torpedoes, anti-submarine weapons (A/S), CIWS, guns and aircraft. Missiles with a range less than 5 km and guns with a calibre less than 57mm are generally not included.
Organisations:	naval groupings such as fleets and squadrons frequently change and are shown only where doing so would aid qualitative judgements.
Legacy platforms:	legacy-generation platforms, unless specifically modified for a new role, may be listed with their original designations although they may not conform fully with current guidance criteria.

Principal aviation definitions

Bomber (Bbr):	comparatively large platforms intended for the delivery of air-to-surface ordnance. Bbr units are units equipped with bomber aircraft for the air-to-surface role.
Fighter (Ftr):	aircraft designed primarily for air-to-air combat, which may also have a limited air-to-surface capability. Ftr units are equipped with aircraft intended to provide air superiority, which may have a secondary and limited air-to-surface capability.
Fighter/Ground Attack (FGA):	multi-role fighter-size platforms with significant air-to-surface capability, potentially including maritime attack, and at least some air-to-air capacity. FGA units are multi-role units equipped with aircraft capable of air-to-air and air-to-surface attack.
Ground Attack (Atk):	aircraft designed solely for the air-to-surface task, with limited or no air-to-air capability. Atk units are equipped with fixed-wing aircraft.

Attack Helicopter (Atk hel):	rotary-wing platforms designed for delivery of air-to-surface weapons, and fitted with an integrated fire-control system.
Anti-Submarine Warfare (ASW):	fixed- and rotary-wing platforms designed to locate and engage submarines, many with a secondary anti-surface-warfare capability. ASW units are equipped with fixed- or rotary-wing aircraft.
Anti-Surface Warfare (ASuW):	ASuW units are equipped with fixed- or rotary-wing aircraft intended for anti-surface-warfare missions.
Maritime Patrol (MP):	fixed-wing aircraft and unmanned aerial vehicles (UAVs) intended for maritime surface surveillance, which may possess an anti-surface-warfare capability. MP units are equipped with fixed-wing aircraft or UAVs.
Electronic Warfare (EW):	fixed- and rotary-wing aircraft and UAVs intended for electronic warfare. EW units are equipped with fixed- or rotary-wing aircraft or UAVs.
Intelligence/ Surveillance/ Reconnaissance (ISR):	fixed- and rotary-wing aircraft and UAVs intended to provide radar, visible-light or infrared imagery, or a mix thereof. ISR units are equipped with fixed- or rotary-wing aircraft or UAVs.
Combat/Intelligence/ Surveillance/ Reconnaissance (CISR):	aircraft and UAVs that have the capability to deliver air-to-surface weapons, as well as undertake ISR tasks. CISR units are equipped with armed aircraft and/or UAVs for ISR and air-to-surface missions.
COMINT/ELINT/ SIGINT:	fixed- and rotary-wing platforms and UAVs capable of gathering electronic (ELINT), communications (COMINT) or signals intelligence (SIGINT). COMINT units are equipped with fixed- or rotary-wing aircraft or UAVs intended for the communications-intelligence task. ELINT units are equipped with fixed- or rotary-wing aircraft or UAVs used for gathering electronic intelligence. SIGINT units are equipped with fixed- or rotary-wing aircraft or UAVs used to collect signals intelligence.
Airborne Early Warning (& Control) (AEW (&C)):	fixed- and rotary-wing platforms capable of providing airborne early warning, with a varying degree of onboard command and control depending on the platform. AEW(&C) units are equipped with fixed- or rotary-wing aircraft.
Search and Rescue (SAR):	units are equipped with fixed- or rotary-wing aircraft used to recover military personnel or civilians.
Combat Search and Rescue (CSAR):	units are equipped with armed fixed- or rotary-wing aircraft for recovery of personnel from hostile territory.
Tanker (Tkr):	fixed- and rotary-wing aircraft designed for air-to-air refuelling. Tkr units are equipped with fixed- or rotary-wing aircraft used for air-to-air refuelling.
Tanker Transport (Tkr/Tpt):	platforms capable of both air-to-air refuelling and military airlift.
Transport (Tpt):	fixed- and rotary-wing aircraft intended for military airlift. Light transport aircraft are categorised as having a maximum payload of up to 11,340 kg; medium up to 27,215 kg; and heavy above 27,215 kg. Light transport helicopters have an internal payload of up to 2,000 kg; medium transport helicopters up to 4,535 kg; heavy transport helicopters greater than 4,535 kg. PAX aircraft are platforms generally unsuited for transporting cargo on the main deck. Tpt units are equipped with fixed- or rotary-wing platforms to transport personnel or cargo.
Trainer (Trg):	fixed- and rotary-wing aircraft designed primarily for the training role; some also have the capacity to carry light to medium ordnance. Trg units are equipped with fixed- or rotary-wing training aircraft intended for pilot or other aircrew training.
Multi-Role Helicopter (MRH):	rotary-wing platforms designed to carry out a variety of military tasks including light transport, armed reconnaissance and battlefield support.
Uninhabited Aerial Vehicles (UAVs):	remotely piloted or controlled unmanned fixed- or rotary-wing systems. Light UAVs are those weighing 20–150 kg; medium: 150–600 kg; and large: more than 600 kg.
Loitering Munition:	an air vehicle with an integral warhead designed to be able to fly in a search or holding pattern and to be used to identify and attack a target

Reference

Table 10 **List of abbreviations for data sections**

AAA	anti-aircraft artillery
AAM	air-to-air missile
AAR	search-and-rescue vessel
AAV	amphibious assault vehicle
AB	airborne
ABM	anti-ballistic missile
ABU/H	sea-going buoy tender/with hangar
ABCV	airborne combat vehicle
ac	aircraft
ACV	armoured combat vehicle
ACS	crane ship
AD	air defence
ADA	air-defence artillery
adj	adjusted
AE	auxiliary, ammunition carrier
AEM	missile support ship
AEV	armoured engineer vehicle
AEW	airborne early warning
AFD/L	auxiliary floating dry dock/small
AFS/H	logistics ship/with hangar
AFSB	afloat forward staging base
AFV	armoured fighting vehicle
AG	misc auxiliary
AGB/H	icebreaker/with hangar
AGE/H	experimental auxiliary ship/with hangar
AGF/H	command ship/with hangar
AGHS	hydrographic survey vessel
AGI	intelligence collection vessel
AGM	space tracking vessel
AGOR	oceanographic research vessel
AGOS	oceanographic surveillance vessel
AGS/H	survey ship/with hangar
AH	hospital ship
AIP	air-independent propulsion
AK/L	cargo ship/light
aka	also known as
AKEH	dry cargo/ammunition ship
AKR/H	roll-on/roll-off cargo ship/with hangar
AKS/L	stores ship/light
ALCM	air-launched cruise missile
amb	ambulance
amph	amphibious/amphibian
AO/S	oiler/small
AOE	fast combat support ship
AOR/L/H	fleet replenishment oiler with RAS capability/light/with hangar
AOT/L	oiler transport/light
AP	armour-piercing/anti-personnel/transport ship
APB	barracks ship
APC	armoured personnel carrier
AR/C/D/L	repair ship/cable/dry dock/light
ARG	amphibious ready group
ARH	active radar homing
ARL	airborne reconnaissance low
ARM	anti-radiation missile
armd	armoured
ARS/H	rescue and salvage ship/with hangar
arty	artillery
ARV	armoured recovery vehicle
AS	anti-submarine/submarine tender
ASBM	anti-ship ballistic missile
ASCM	anti-ship cruise missile
AShM	anti-ship missile
aslt	assault
ASM	air-to-surface missile
ASR	submarine rescue craft
ASTT	anti-submarine torpedo tube
ASW	anti-submarine warfare
ASuW	anti-surface warfare
AT	anti-tank
ATBM	anti-tactical ballistic missile
ATF	ocean going tug
ATGW	anti-tank guided weapon
Atk	attack/ground attack
ATS	tug, salvage and rescue ship
AUV	armoured utility vehicle
avn	aviation
AWT	water tanker
AX/L/S	training craft/light/sail
BA	Budget Authority (US)
Bbr	bomber
BCT	brigade combat team
bde	brigade
bdgt	budget
BG	battlegroup
BMD	ballistic-missile defence
BMEWS	ballistic missile early warning system
bn	battalion/billion
bty	battery
C2	command and control
C4	command, control, communications, and computers
casevac	casualty evacuation
cav	cavalry
cbt	combat
CBRN	chemical, biological, radiological, nuclear, explosive
cdo	commando
C/G/H/M/N	cruiser/with surface-to-surface missile/with hangar/with SAM/nuclear-powered
CISR	combat ISR
CIMIC	civil–military cooperation
CIWS	close-in weapons system
COIN	counter-insurgency
comd	command
COMINT	communications intelligence
comms	communications
coy	company
CP	command post
CPX	command post exercise
CS	combat support
CSAR	combat search and rescue
CSS	combat service support
CT	counter-terrorism
CV/H/L/N/S	aircraft carrier/helicopter/light/nuclear powered/STOVL
CW	chemical warfare/weapons
DD/G/H/M	destroyer/with surface-to-surface missile/with hangar/with SAM
DDR	disarmament, demobilisation and reintegration
DE	directed energy
def	defence
det	detachment
div	division
ECM	electronic countermeasures
ELINT	electronic intelligence
elm	element/s
engr	engineer
EOD	explosive ordnance disposal
EPF	expeditionary fast transport vessel
eqpt	equipment
ESB	expeditionary sea base
ESD	expeditionary transport dock
EW	electronic warfare
excl	excludes/excluding
exp	expenditure/expeditionary
FAC	forward air control
fd	field
FF/G/H/M	frigate/with surface-to-surface missile/with hangar/with SAM
FGA	fighter/ground attack
FLD	full-load displacement
flt	flight
FMA	Foreign Military Assistance
FRS	fleet replacement squadron
FS/G/H/M	corvette/with surface-to-surface missile/with hangar/with SAM
Ftr	fighter
FTX	field training exercise
FY	fiscal year
GBU	guided bomb unit
gd	guard
GDP	gross domestic product
GLCM	ground-launched cruise missile
GMLS	Guided Missile Launching System
gp	group
HA/DR	humanitarian assistance/disaster relief
hel	helicopter
how	howitzer
HQ	headquarters
HUMINT	human intelligence
HWT	heavyweight torpedo
hy	heavy
IBU	inshore boat unit
ICBM	intercontinental ballistic missile
IFV	infantry fighting vehicle
IIR	imaging infrared
IMINT	imagery intelligence
imp	improved
indep	independent

inf	infantry
info ops	information operations
INS	inertial navigation system
int	intelligence
IOC	Initial operating capability
IR	infrared
IRBM	intermediate-range ballistic missile
ISD	in-service date
ISR	intelligence, surveillance and reconnaissance
ISTAR	intelligence, surveillance, target acquisition and reconnaissance
LACM	land-attack cruise missile
LC/A/AC/H/M/P/T/U/VP	landing craft/assault/air cushion/heavy/medium/personnel/tank/utility/vehicles and personnel
LCC	amphibious command ship
LGB	laser-guided bomb
LHA	aviation-capable amphibious assault ship
LHD	aviation-capable amphibious assault ship with well dock
LIFT	lead-in ftr trainer
LKA	amphibious cargo ship
LLI	long-lead items
lnchr	launcher
LoA	letter of offer and acceptance
log	logistic
LoI	letter of intent
LP/D/H	landing platform/dock/helicopter
LRIP	low-rate initial production
LS/D/L/H/M/T	landing ship/dock/logistic/with hangar/medium/tank
lt	light
LWT	lightweight torpedo
maint	maintenance
MANPAD	man-portable air-defence system
MANPATS	man-portable anti-tank system
MBT	main battle tank
MC/C/CS/D/I/O	mine countermeasure coastal/command and support/diving support/inshore/ocean
MCM	mine countermeasures
MCMV	mine countermeasures vessel
MD	military district
mech	mechanised
med	medium/medical
medevac	medical evacuation
MH/C/D/I/O	mine hunter/coastal/drone/inshore/ocean
mil	military
MIRV	multiple independently targetable re-entry vehicle
mk	mark (model number)
ML	minelayer
MLU	mid-life update
mne	marine
mnv enh	manoeuvre enhancement
mod	modified/modification
mor	mortar
mot	motorised/motor
MoU	memorandum of understanding
MP	maritime patrol/military police
MR	maritime reconnaissance/motor rifle
MRBM	medium-range ballistic missile
MRH	multi-role helicopter
MRL	multiple rocket launcher
MS/C/D/I/O/R	mine sweeper/coastal/drone/inshore/ocean/river
msl	missile
mtn	mountain
MW	mine warfare
n.a.	not applicable
n.k.	not known
NBC	nuclear, biological, chemical
NCO	non-commissioned officer
nm	nautical mile
nuc	nuclear
O & M	operations and maintenance
obs	observation/observer
OCU	operational conversion unit
OP	observation post
op/ops	operational/operations
OPFOR	opposition training force
org	organised/organisation
OPV	offshore patrol vessel
para	paratroop/parachute
PAX	passenger/passenger transport aircraft
PB/F/G/I/M/R/T	patrol boat/fast/with surface-to-surface missile/inshore/with SAM/riverine/with torpedo
PC/C/F/G/H/I/M/O/R/T	patrol craft/coastal/fast/with surface-to-surface missile/with hangar/inshore/with CIWS missile or SAM/offshore/riverine/with torpedo
pdr	pounder
pers	personnel
PGM	precision-guided munitions
PH/G/M/T	patrol hydrofoil/with surface-to-surface missile/with SAM/with torpedo
pl	platoon
PKO	peacekeeping operations
PoR	programme of record
PPP	purchasing-power parity
PPV	protected patrol vehicle
PRH	passive radar-homing
prepo	pre-positioned
PSO/H	peace support operations or offshore patrol ship/with hangar
ptn	pontoon bridging
quad	quadruple
R&D	research and development
RCL	recoilless launcher
recce	reconnaissance
regt	regiment
RFI	request for information
RFP	request for proposals
RL	rocket launcher
ro-ro	roll-on, roll-off
RRC/F/U	rapid-reaction corps/force/unit
RV	re-entry vehicle
rvn	riverine
SAM	surface-to-air missile
SAR	search and rescue
SARH	semi-active radar homing
sat	satellite
SDV	swimmer delivery vehicles
SEAD	suppression of enemy air defence
SF	special forces
SHORAD	short-range air defence
SIGINT	signals intelligence
sigs	signals
SLBM	submarine-launched ballistic missile
SLCM	submarine-launched cruise missile
SLEP	service-life-extension programme
SP	self-propelled
Spec Ops	special operations
SPAAGM	self-propelled anti-aircraft gun and missile system
spt	support
sqn	squadron
SRBM	short-range ballistic missile
SS	submarine
SSA/N	auxiliary support submarine/nuclear-powered
SSB/N	ballistic missile submarine/nuclear-powered
SSC	coastal submarine
SSG	conventionally-powered attack submarine with dedicated launch tubes for guided missiles
SSGN	nuclear-powered submarine with dedicated launch tubes for guided missiles
SSK	conventionally-powered attack submarine
SSM	surface-to-surface missile
SSN	nuclear-powered attack submarine
SSR	security-sector reform
SSW	midget submarine
str	strength
STOVL	short take-off and vertical landing
surv	surveillance
sy	security
t	tonnes
tac	tactical
tch	technical
temp	temporary
tk	tank
tkr	tanker
TMD	theatre missile defence
torp	torpedo
tpt	transport
tr	trillion
trg	training
TRV	torpedo recovery vehicle
TT	torpedo tube
UAV	unmanned/uninhabited aerial vehicle
UCAC	utility craft air cushioned
UCAV	unmanned combat air vehicle
utl	utility
UUV	unmanned/uninhabited underwater vehicle
veh	vehicle
VLB	vehicle launched bridge
VLS	vertical launch system
VSHORAD	very short-range air defence
WFU	withdrawn from use
wg	wing

Table 11 **International comparisons of defence expenditure and military personnel**

	Defence spending (current US$m)			Defence spending per capita (current US$)			Defence spending % of GDP			Active armed forces (000)	Estimated reservists (000)	Active paramilitary (000)
	2018	2019	2020	2018	2019	2020	2018	2019	2020	2021	2021	2021
North America												
Canada	19,125	18,731	20,049	533	518	532	1.11	1.08	1.25	67	36	5
United States	631,161	684,568	738,000	1,917	2,063	2,219	3.06	3.19	3.55	1,388	845	0
Total	**650,286**	**703,299**	**758,049**	**1,225**	**1,291**	**1,375**	**2.09**	**2.14**	**2.40**	**1,456**	**881**	**5**
Europe												
Albania	133	143	187	43	47	61	0.88	0.93	1.33	8	0	0
Austria	3,391	3,238	3,471	386	367	392	0.74	0.73	0.80	22	126	0
Belgium	4,845	4,817	5,453	419	414	465	0.89	0.91	1.08	26	5	0
Bosnia-Herzegovina	171	165	169	45	43	44	0.85	0.83	0.89	11	0	0
Bulgaria	959	2,067	1,158	136	295	166	1.45	3.05	1.70	37	3	0
Croatia	966	1,048	1,121	226	247	265	1.58	1.73	1.98	15	18	3
Cyprus	431	402	406	348	321	320	1.73	1.64	1.75	15	50	1
Czech Republic	2,710	2,910	3,278	254	272	306	1.09	1.16	1.35	25	0	0
Denmark	4,559	4,557	4,909	785	780	836	1.28	1.31	1.45	15	44	0
Estonia	645	688	717	518	557	583	2.13	2.21	2.37	7	18	0
Finland	3,757	3,920	4,105	679	706	737	1.36	1.46	1.53	24	216	3
France	53,160	52,106	55,034	789	771	811	1.91	1.92	2.16	203	41	101
Germany	45,510	48,398	51,347	566	603	641	1.15	1.25	1.36	184	30	0
Greece	4,858	4,806	4,984	451	447	470	2.23	2.29	2.56	143	221	4
Hungary	1,665	1,824	2,041	169	186	209	1.05	1.13	1.36	28	20	12
Iceland	39	55	52	115	159	148	0.15	0.23	0.25	0	0	0
Ireland	1,118	1,113	1,189	221	217	230	0.29	0.28	0.30	9	4	0
Italy	28,701	27,049	29,344	461	434	470	1.38	1.35	1.59	166	18	176
Latvia	724	710	758	377	373	403	2.13	2.10	2.31	6	11	0
Lithuania	1,057	1,061	1,163	378	384	426	1.99	1.97	2.12	22	7	15
Luxembourg	403	341	390	665	552	621	0.57	0.48	0.57	0	0	1
Macedonia, North	125	151	188	59	71	89	0.99	1.23	1.51	8	5	8
Malta	70	84	81	156	184	177	0.47	0.56	0.57	2	0	0
Montenegro	79	74	74	129	120	121	1.44	1.43	1.50	2	0	10
Netherlands	11,254	12,060	12,615	656	700	730	1.23	1.33	1.42	34	6	7
Norway	6,756	6,689	6,493	1,258	1,234	1,188	1.56	1.66	1.77	23	40	0
Poland	11,893	11,824	12,875	310	308	336	2.03	2.00	2.22	114	0	75

Table 11 **International comparisons of defence expenditure and military personnel**

	Defence spending (current US$m)			Defence spending per capita (current US$)			Defence spending % of GDP			Active armed forces (000)	Estimated reservists (000)	Active paramilitary (000)
	2018	2019	2020	2018	2019	2020	2018	2019	2020	2021	2021	2021
Portugal	2,573	2,674	2,857	248	259	277	1.07	1.12	1.29	27	212	25
Romania	4,359	4,939	5,207	203	231	244	1.80	1.98	2.09	69	53	57
Serbia	704	903	878	99	128	125	1.39	1.76	1.69	28	50	4
Slovakia	1,300	1,862	2,123	239	342	390	1.23	1.77	2.08	16	0	0
Slovenia	594	628	606	283	299	288	1.10	1.17	1.17	7	1	0
Spain	15,059	12,881	12,985	305	259	260	1.06	0.92	1.04	123	15	76
Sweden	6,193	6,373	7,050	617	630	691	1.11	1.20	1.33	15	10	0
Switzerland	4,811	5,343	5,741	580	640	683	0.68	0.76	0.81	20	123	0
Turkey	12,858	12,082	11,038	158	148	135	1.65	1.59	1.70	355	379	157
United Kingdom*	58,931	59,226	61,526	905	905	936	2.06	2.09	2.33	149	79	0
Total	**297,361**	**299,209**	**313,611**	**385**	**395**	**413**	**1.29**	**1.39**	**1.50**	**1,955**	**1,805**	**732**
Russia and Eurasia												
Armenia	513	644	621	169	212	206	4.12	4.71	4.85	45	210	4
Azerbaijan	1,709	1,787	2,267	170	176	222	3.63	3.72	5.44	67	300	15
Belarus	599	651	616	63	69	65	1.00	1.03	1.07	45	290	110
Georgia	321	310	288	65	63	58	2.03	1.95	1.89	21	0	5
Kazakhstan	1,500	1,595	1,439	80	84	75	0.84	0.88	0.87	39	0	32
Kyrgyzstan	n.k.	n.k.	n.k.	n.k.	n.k.	n.k.	n.k.	n.k.	n.k.	11	0	10
Moldova	38	43	42	11	12	13	0.45	0.36	0.38	5	58	1
Russia [a]	45,041	46,384	43,184	317	327	305	2.70	2.72	2.95	900	2,000	554
Tajikistan	86	89	89	10	10	10	1.14	1.09	1.12	9	0	8
Turkmenistan*	n.k.	n.k.	n.k.	n.k.	n.k.	n.k.	n.k.	n.k.	n.k.	37	0	5
Ukraine	3,184	3,989	4,320	72	91	98	2.50	2.65	3.05	209	900	102
Uzbekistan	n.k.	n.k.	n.k.	n.k.	n.k.	n.k.	n.k.	n.k.	n.k.	48	0	20
Total**	**52,991**	**55,491**	**52,865**	**106**	**116**	**117**	**2.05**	**2.12**	**2.40**	**1,435**	**3,758**	**865**
Asia												
Afghanistan	2,038	1,963	2,006	58	55	55	11.08	10.40	10.56	179	0	99
Australia	31,973	29,912	31,325	1,362	1,262	1,230	2.25	2.16	2.35	59	30	0
Bangladesh	2,575	3,653	3,676	16	23	23	0.94	1.21	1.16	163	0	64
Brunei	365	432	437	811	945	942	2.69	3.21	4.11	7	1	1
Cambodia*	964	1,043	1,032	59	62	61	3.95	3.90	3.92	124	0	67
China	170,504	181,135	193,295	122	130	139	1.23	1.23	1.27	2,035	510	500
Fiji	51	56	49	56	60	52	0.93	1.03	1.23	4	6	0

Table 11 International comparisons of defence expenditure and military personnel

	Defence spending (current US$m)			Defence spending per capita (current US$)			Defence spending % of GDP			Active armed forces (000)	Estimated reservists (000)	Active paramilitary (000)
	2018	2019	2020	2018	2019	2020	2018	2019	2020	2021	2021	2021
India	57,830	60,794	64,138	45	46	48	2.13	2.12	2.47	1,459	1,155	1,586
Indonesia	7,507	7,600	8,375	29	29	31	0.72	0.68	0.77	396	400	280
Japan	47,011	48,229	49,653	373	383	396	0.95	0.95	1.01	247	56	14
Korea, DPR of	n.k.	n.k.	n.k.	n.k.	n.k.	n.k.	n.k.	n.k.	n.k.	1,280	600	189
Korea, Republic of	39,217	40,071	40,404	763	776	779	2.27	2.43	2.55	599	3,100	14
Laos	n.k.	n.k.	n.k.	n.k.	n.k.	n.k.	n.k.	n.k.	n.k.	29	0	100
Malaysia	3,890	3,360	3,720	122	104	114	1.08	0.92	1.11	113	52	23
Mongolia	104	98	103	34	31	32	0.81	0.73	0.77	10	137	8
Myanmar	1,915	2,132	2,107	34	38	37	2.87	3.11	2.97	406	0	107
Nepal	431	398	437	14	13	14	1.49	1.30	1.36	97	0	15
New Zealand	2,353	2,673	3,255	518	584	661	1.14	1.30	1.68	9	3	0
Pakistan	11,357	10,175	10,774	55	48	46	3.63	3.71	3.55	652	0	291
Papua New Guinea	63	79	94	9	11	13	0.26	0.32	0.40	4	0	0
Philippines	2,592	3,541	3,651	24	33	33	0.76	0.95	1.01	143	131	12
Singapore	10,944	11,339	10,887	1,825	1,858	1,753	2.93	3.05	3.23	51	253	8
Sri Lanka	1,795	1,666	1,589	80	73	69	2.03	1.98	1.96	255	6	62
Taiwan	10,864	11,010	13,867	461	467	587	1.79	1.80	2.18	163	1,657	12
Thailand	6,724	7,316	6,992	98	106	101	1.33	1.35	1.37	361	200	94
Timor-Leste	21	31	42	16	23	30	1.31	1.93	2.17	2	0	0
Vietnam*	4,784	5,221	5,681	49	53	58	1.58	1.59	1.68	482	5,000	40
Total **	**417,873**	**433,925**	**457,588**	**281**	**289**	**292**	**2.09**	**2.13**	**2.27**	**9,327**	**13,295**	**3,585**
Middle East and North Africa												
Algeria	9,591	10,306	9,922	230	243	231	5.52	6.09	6.73	130	150	187
Bahrain	1,409	1,411	1,405	977	957	934	3.74	3.66	4.06	8	0	11
Egypt	2,908	3,353	4,106	29	33	39	1.68	1.54	1.49	439	479	397
Iran	18,490	17,189	14,059	223	205	166	4.24	2.94	2.30	610	350	40
Iraq	7,956	9,855	10,267	198	239	264	3.64	4.39	5.76	193	0	148
Israel	16,245	16,804	16,644	1,928	1,965	1,918	5.22	5.09	5.20	170	465	8
Jordan	1,635	1,691	1,723	156	158	159	4.79	4.75	4.87	101	65	15
Kuwait	7,462	7,864	7,763	2,558	2,660	2,593	5.30	5.81	7.14	18	24	7
Lebanon	2,122	1,928	1,921	348	330	351	4.05	3.87	10.52	60	0	20
Libya	n.k.	n.k.	n.k.	n.k.	n.k.	n.k.	n.k.	n.k.	n.k.	n.k.	n.k.	n.k.

Table 11 **International comparisons of defence expenditure and military personnel**

	Defence spending (current US$m)			Defence spending per capita (current US$)			Defence spending % of GDP			Active armed forces (000)	Estimated reservists (000)	Active paramilitary (000)
	2018	2019	2020	2018	2019	2020	2018	2019	2020	2021	2021	2021
Mauritania	159	161	196	41	41	49	2.26	2.12	2.64	16	0	5
Morocco	4,842	4,817	5,961	141	139	168	4.11	4.07	5.31	196	150	50
Oman	7,437	7,530	7,484	2,128	2,113	2,059	9.32	9.86	12.01	43	0	4
Palestinian Territories	n.k.	n.k.	n.k.	n.k.	n.k.	n.k.	n.k.	n.k.	n.k.	0	0	n.k.
Qatar	6,384	6,399	6,466	2,701	2,659	2,645	3.48	3.64	4.37	17	0	5
Saudi Arabia	58,133	50,933	48,533	1,757	1,514	1,420	7.39	6.42	7.13	227	0	25
Syria	n.k.	n.k.	n.k.	n.k.	n.k.	n.k.	n.k.	n.k.	n.k.	169	0	100
Tunisia	881	999	1,145	76	86	98	2.38	2.79	3.02	36	0	12
United Arab Emirates	19,444	19,329	19,826	2,004	1,964	1,984	4.61	4.59	5.60	63	0	0
Yemen	n.k.	n.k.	n.k.	n.k.	n.k.	n.k.	n.k.	n.k.	n.k.	40	0	0
Total**	**165,097**	**160,568**	**157,420**	**969**	**957**	**942**	**4.48**	**4.48**	**5.51**	**2,533**	**1,683**	**1,034**
Latin America and the Caribbean												
Antigua and Barbuda	7	7	8	74	74	80	0.44	0.43	0.57	0	0	0
Argentina	4,317	3,281	2,911	97	73	64	0.83	0.74	0.76	72	0	31
Bahamas	85	92	86	255	275	254	0.65	0.68	0.74	2	0	0
Barbados	40	38	41	138	131	138	0.79	0.74	0.88	1	0	0
Belize	23	23	25	60	59	61	1.29	1.27	1.58	2	1	0
Bolivia	503	479	479	44	42	41	1.24	1.16	1.23	34	0	37
Brazil	27,551	27,123	22,120	132	129	104	1.46	1.47	1.62	367	1,340	395
Chile	4,701	4,416	4,001	262	245	220	1.58	1.56	1.63	77	40	45
Colombia	10,570	10,216	9,404	219	210	192	3.18	3.17	3.56	293	35	188
Costa Rica	451	443	461	90	88	90	0.75	0.72	0.77	0	0	10
Cuba	n.k.	n.k.	n.k.	n.k.	n.k.	n.k.	n.k.	n.k.	n.k.	49	39	27
Dominican Republic	603	616	584	59	59	56	0.70	0.69	0.75	56	0	15
Ecuador	1,698	1,590	1,545	103	95	91	1.58	1.48	1.66	41	118	1
El Salvador	141	145	172	23	23	27	0.55	0.54	0.69	25	10	17
Guatemala	255	341	365	15	20	21	0.35	0.45	0.48	18	64	25
Guyana	60	67	66	81	90	88	1.26	1.30	0.97	3	1	0
Haiti	8	13	10	1	1	1	0.13	0.15	0.12	1	0	0
Honduras	331	346	340	36	37	37	1.41	1.39	1.42	15	60	8
Jamaica	230	262	238	82	93	85	1.47	1.65	1.67	6	3	0
Mexico	5,314	5,067	5,266	42	40	41	0.44	0.40	0.51	216	82	112
Nicaragua	82	79	79	13	13	13	0.63	0.63	0.66	12	0	0

Table 11 **International comparisons of defence expenditure and military personnel**

	Defence spending (current US$m)			Defence spending per capita (current US$)			Defence spending % of GDP			Active armed forces (000)	Estimated reservists (000)	Active paramilitary (000)
	2018	2019	2020	2018	2019	2020	2018	2019	2020	2021	2021	2021
Panama	738	805	753	194	209	193	1.14	1.21	1.25	0	0	28
Paraguay	311	284	288	44	40	40	0.77	0.76	0.81	14	165	15
Peru	2,286	2,238	2,127	73	71	67	1.03	0.97	1.09	81	188	77
Suriname	n.k.	n.k.	n.k.	n.k.	n.k.	n.k.	n.k.	n.k.	n.k.	2	0	0
Trinidad and Tobago	921	906	953	758	747	789	3.87	3.76	4.20	5	0	0
Uruguay	561	423	379	166	125	112	0.94	0.75	0.70	21	0	1
Venezuela	n.k.	n.k.	n.k.	n.k.	n.k.	n.k.	n.k.	n.k.	n.k.	123	8	220
Total**	**61,787**	**59,302**	**52,702**	**123**	**120**	**116**	**1.14**	**1.12**	**1.21**	**1,534**	**2,152**	**1,251**
Sub-Saharan Africa												
Angola	2,158	1,607	1,035	71	51	32	2.04	1.80	1.65	107	0	10
Benin	90	68	56	8	6	4	0.63	0.47	0.36	7	0	5
Botswana	518	539	571	230	236	246	2.78	2.92	3.59	9	0	0
Burkina Faso	313	358	388	16	18	19	1.93	2.27	2.42	11	0	0
Burundi	65	63	62	6	5	5	2.05	2.03	1.99	30	0	21
Cameroon	430	422	408	17	16	15	1.11	1.09	1.05	25	0	9
Cape Verde	11	11	11	20	19	19	0.57	0.54	0.61	1	0	0
Central African Rep	31	32	41	5	6	7	1.36	1.42	1.75	9	0	1
Chad	182	205	269	12	13	16	1.65	1.88	2.56	33	0	12
Congo	293	300	306	58	58	58	2.18	2.39	3.07	10	0	2
Côte d'Ivoire	555	538	609	21	20	22	0.96	0.92	0.99	27	0	n.k.
Dem Republic of the Congo	297	336	330	3	4	3	0.63	0.67	0.72	134	0	0
Djibouti	n.k.	n.k.	n.k.	n.k.	n.k.	n.k.	n.k.	n.k.	n.k.	10	0	3
Equatorial Guinea	n.k.	n.k.	n.k.	n.k.	n.k.	n.k.	n.k.	n.k.	n.k.	1	0	0
Eritrea	n.k.	n.k.	n.k.	n.k.	n.k.	n.k.	n.k.	n.k.	n.k.	202	120	0
Ethiopia	437	516	472	4	5	4	0.55	0.56	0.49	138	0	0
Gabon	261	267	267	123	123	120	1.55	1.58	1.76	5	0	2
Gambia	n.k.	n.k.	n.k.	n.k.	n.k.	n.k.	n.k.	n.k.	n.k.	4	0	0
Ghana	219	244	268	8	9	9	0.33	0.36	0.40	16	0	0
Guinea	197	195	206	17	16	16	1.62	1.42	1.45	10	0	3
Guinea-Bissau	n.k.	n.k.	n.k.	n.k.	n.k.	n.k.	n.k.	n.k.	n.k.	4	0	0
Kenya	1,286	1,232	1,111	27	25	21	1.46	1.29	1.10	24	0	5
Lesotho	48	44	34	24	22	17	1.94	1.79	1.76	2	0	0

Table 11 **International comparisons of defence expenditure and military personnel**

	Defence spending (current US$m)			Defence spending per capita (current US$)			Defence spending % of GDP			Active armed forces (000)	Estimated reservists (000)	Active paramilitary (000)
	2018	2019	2020	2018	2019	2020	2018	2019	2020	2021	2021	2021
Liberia	13	14	12	3	3	2	0.40	0.44	0.40	2	0	0
Madagascar	73	77	112	3	3	4	0.53	0.54	0.79	14	0	8
Malawi	51	68	68	3	3	3	0.74	0.89	0.82	11	0	4
Mali	726	722	788	39	38	40	4.27	4.17	4.46	13	0	8
Mauritius	222	219	226	163	159	164	1.57	1.56	2.00	0	0	3
Mozambique	129	126	133	5	5	4	0.88	0.83	0.91	11	0	0
Namibia	450	407	344	178	158	131	3.32	3.26	3.36	10	0	6
Niger	230	172	202	12	8	9	1.79	1.33	1.56	5	0	5
Nigeria	1,750	1,827	2,568	9	9	12	0.44	0.41	0.58	143	0	80
Rwanda	107	112	127	9	9	10	1.11	1.11	1.22	33	0	2
Senegal	347	342	346	23	22	22	1.49	1.45	1.42	14	0	5
Seychelles	n.k.	n.k.	n.k.	n.k.	n.k.	n.k.	n.k.	n.k.	n.k.	0	0	0
Sierra Leone	13	11	7	2	2	1	0.33	0.25	0.17	9	0	0
Somalia	n.k.	n.k.	n.k.	n.k.	n.k.	n.k.	n.k.	n.k.	n.k.	20	0	0
South Africa	3,622	3,495	3,235	65	63	57	0.98	0.99	1.14	75	0	15
South Sudan	70	68	94	7	7	9	1.51	1.38	2.25	185	0	0
Sudan	n.k.	n.k.	n.k.	n.k.	n.k.	n.k.	n.k.	n.k.	n.k.	104	0	20
Tanzania	758	728	801	14	13	14	1.34	1.20	1.25	27	80	1
Togo	104	106	114	13	13	13	1.95	1.94	1.99	9	0	1
Uganda	393	547	953	10	13	22	1.17	1.50	2.53	45	10	1
Zambia	334	394	367	20	23	21	1.24	1.63	1.94	15	3	1
Zimbabwe	206	64	39	15	4	3	0.99	0.34	0.28	29	0	22
Total**	**16,994**	**16,476**	**16,978**	**34**	**33**	**32**	**1.39**	**1.37**	**1.53**	**1,594**	**213**	**255**
Summary												
North America	**650,286**	**703,299**	**758,049**	**1,225**	**1,291**	**1,375**	**2.09**	**2.14**	**2.40**	**1,456**	**881**	**5**
Europe	**297,361**	**299,209**	**313,611**	**385**	**395**	**413**	**1.29**	**1.39**	**1.50**	**1,955**	**1,805**	**732**
Russia and Eurasia	**52,991**	**55,491**	**52,865**	**106**	**116**	**117**	**2.05**	**2.12**	**2.40**	**1,435**	**3,758**	**865**
Asia	**417,873**	**433,925**	**457,588**	**281**	**289**	**292**	**2.09**	**2.13**	**2.27**	**9,327**	**13,295**	**3,585**
Middle East and North Africa	**165,097**	**160,568**	**157,420**	**969**	**957**	**942**	**4.48**	**4.48**	**5.51**	**2,533**	**1,683**	**1,034**
Latin America and the Caribbean	**61,787**	**59,302**	**52,702**	**123**	**120**	**116**	**1.14**	**1.12**	**1.21**	**1,534**	**2,152**	**1,251**
Sub-Saharan Africa	**16,994**	**16,476**	**16,978**	**34**	**33**	**32**	**1.39**	**1.37**	**1.53**	**1,594**	**213**	**255**
Global totals	**1,662,390**	**1,728,270**	**1,809,213**	**262**	**265**	**268**	**1.81**	**1.85**	**2.08**	**19,834**	**23,787**	**7,727**

Totals may not sum precisely due to rounding. * Estimates. **Totals exclude defence-spending estimates for states where insufficient official information is available in order to enable approximate comparisons of regional defence spending between years. Defence spending per capita (current US$) and defence spending % of GDP totals are regional averages. [a] 'National Defence' budget chapter. Excludes other defence-related expenditures included under other budget lines (e.g. pensions) – see Table 5, p.174. Defence Spending as % of GDP includes US foreign military financing programmes – other figures do not.

Table 12 Index of country/territory abbreviations

AFG	Afghanistan
ALB	Albania
ALG	Algeria
ANG	Angola
ARG	Argentina
ARM	Armenia
ATG	Antigua and Barbuda
AUS	Australia
AUT	Austria
AZE	Azerbaijan
BDI	Burundi
BEL	Belgium
BEN	Benin
BFA	Burkina Faso
BGD	Bangladesh
BHR	Bahrain
BHS	Bahamas
BIH	Bosnia-Herzegovina
BIOT	British Indian Ocean Territory
BLG	Bulgaria
BLR	Belarus
BLZ	Belize
BOL	Bolivia
BRB	Barbados
BRN	Brunei
BRZ	Brazil
BWA	Botswana
CAM	Cambodia
CAN	Canada
CAR	Central African Republic
CHA	Chad
CHE	Switzerland
CHL	Chile
CIV	Côte d'Ivoire
CMR	Cameroon
COG	Republic of Congo
COL	Colombia
CPV	Cape Verde
CRI	Costa Rica
CRO	Croatia
CUB	Cuba
CYP	Cyprus
CZE	Czech Republic
DJB	Djibouti
DNK	Denmark
DOM	Dominican Republic
DPRK	Korea, Democratic People's Republic of
DRC	Democratic Republic of the Congo
ECU	Ecuador
EGY	Egypt
EQG	Equitorial Guinea
ERI	Eritrea
ESP	Spain
EST	Estonia
ETH	Ethiopia
FIN	Finland
FJI	Fiji
FLK	Falkland Islands
FRA	France
GAB	Gabon
GAM	Gambia
GEO	Georgia
GER	Germany
GF	French Guiana
GHA	Ghana
GIB	Gibraltar
GNB	Guinea-Bissau
GRC	Greece
GRL	Greenland
GUA	Guatemala
GUI	Guinea
GUY	Guyana
HND	Honduras
HTI	Haiti
HUN	Hungary
IDN	Indonesia
IND	India
IRL	Ireland
IRN	Iran
IRQ	Iraq
ISL	Iceland
ISR	Israel
ITA	Italy
JAM	Jamaica
JOR	Jordan
JPN	Japan
KAZ	Kazakhstan
KEN	Kenya
KGZ	Kyrgyzstan
KWT	Kuwait
LAO	Laos
LBN	Lebanon
LBR	Liberia
LBY	Libya
LKA	Sri Lanka
LSO	Lesotho
LTU	Lithuania
LUX	Luxembourg
LVA	Latvia
MDA	Moldova
MDG	Madagascar
MEX	Mexico
MHL	Marshall Islands
MKD	Macedonia, North
MLI	Mali
MLT	Malta
MMR	Myanmar
MNE	Montenegro
MNG	Mongolia
MOR	Morocco
MOZ	Mozambique
MRT	Mauritania
MUS	Mauritius
MWI	Malawi
MYS	Malaysia
NAM	Namibia
NCL	New Caledonia
NER	Niger
NGA	Nigeria
NIC	Nicaragua
NLD	Netherlands
NOR	Norway
NPL	Nepal
NZL	New Zealand
OMN	Oman
PT	Palestinian Territories
PAN	Panama
PAK	Pakistan
PER	Peru
PHL	Philippines
POL	Poland
PNG	Papua New Guinea
PRC	China, People's Republic of
PRT	Portugal
PRY	Paraguay
PYF	French Polynesia
QTR	Qatar
ROC	Taiwan (Republic of China)
ROK	Korea, Republic of
ROM	Romania
RSA	South Africa
RUS	Russia
RWA	Rwanda
SAU	Saudi Arabia
SDN	Sudan
SEN	Senegal
SER	Serbia
SGP	Singapore
SLB	Solomon Islands
SLE	Sierra Leone
SLV	El Salvador
SOM	Somalia
SSD	South Sudan
STP	São Tomé and Príncipe
SUR	Suriname
SVK	Slovakia
SVN	Slovenia
SWE	Sweden
SYC	Seychelles
SYR	Syria
TGO	Togo
THA	Thailand
TJK	Tajikistan
TKM	Turkmenistan
TLS	Timor-Leste
TTO	Trinidad and Tobago
TUN	Tunisia
TUR	Turkey
TZA	Tanzania
UAE	United Arab Emirates
UGA	Uganda
UK	United Kingdom
UKR	Ukraine
URY	Uruguay
US	United States
UZB	Uzbekistan
VEN	Venezuela
VNM	Vietnam
YEM	Yemen, Republic of
ZMB	Zambia
ZWE	Zimbabwe

Table 13 Index of countries and territories